Forest Ecology

Forest Ecology

FIFTH EDITION

Daniel M. Kashian
Wayne State University,
Detroit, USA

Donald R. Zak
University of Michigan,
Ann Arbor, USA

Burton V. Barnes (deceased)
University of Michigan,
Ann Arbor, USA

Stephen H. Spurr (deceased)
University of Texas at Austin,
Austin, USA

This fifth edition first published 2023

Edition History
John Wiley & Sons Ltd. (4e, 1998); Ronald Press (3e, 1980; 2e, 1973; 1e, 1964)

Registered Offices
John Wiley & Sons, Inc., 111 River Street, Hoboken, NJ 07030, USA
John Wiley & Sons Ltd., The Atrium, Southern Gate, Chichester, West Sussex, PO19 8SQ, UK

Editorial Office
9600 Garsington Road, Oxford, OX4 2DQ, UK

For details of our global editorial offices, customer services, and more information about Wiley products, visit us at www.wiley.com.

Wiley also publishes its books in a variety of electronic formats and by print-on-demand. Some content that appears in standard print versions of this book may not be available in other formats.

Library of Congress Cataloging-in-Publication Data
Names: Kashian, Daniel M., author. | Zak, Donald R., author. | Barnes, Burton Verne, 1930-2014, author. | Spurr, Stephen H., 1918-1990, author.
Title: Forest ecology / Daniel M. Kashian, Wayne State University, Detroit, USA, Donald R. Zak, University of Michigan, Ann Arbor, USA, Burton V. Barnes (deceased), University of Michigan, Ann Arbor, USA, Stephen H. Spurr (deceased), University of Texas at Austin, Austin, USA.
Description: Fifth edition. | Hoboken, NJ : Wiley, 2022. | Precedey by: Forest ecology / Burton V. Barnes ... [et al.]. 4th ed. c1998. | Includes bibliographical references and index.
Identifiers: LCCN 2022037810 (print) | LCCN 2022037811 (ebook) | ISBN 9781119476085 (paperback) | ISBN 9781119476054 (adobe pdf) | ISBN 9781119476146 (epub)
Subjects: LCSH: Forest ecology.
Classification: LCC QK938.F6 F635 2022 (print) | LCC QK938.F6 (ebook) | DDC 577.3–dc23/eng/20220831
LC record available at https://lccn.loc.gov/2022037810
LC ebook record available at https://lccn.loc.gov/2022037811

Cover Design: Wiley
Cover Image: © U.S. Forest Service Photo

Set in 9.5/12pt STIXTwoText by Straive, Pondicherry, India
SKY10084631_091324

Dedication

We dedicate the 5th edition of Forest Ecology to the co-author of the 2nd and 3rd editions and the lead author of the 4th edition, *Burton V. Barnes* (1930–2014). It would be no easy task to find a more accomplished and humble leader in his field. He excelled at his science, was a truly beloved teacher, and helped to shape the world view of thousands of colleagues, friends, students, managers, and scientists alike, all with an unmatched humor and a love of the natural world.

Burt Barnes was world-renowned as an expert in the ecology of North American aspens and the ecological classification of forest ecosystems. His professional training was in forest ecology, botany, and genetics, but he dabbled heavily in glacial geomorphology, soil science, phytogeography, and woody plant physiology. Perhaps his greatest love, however, was teaching, especially in the field, which drove his motivation for this textbook. Generations of students have been touched by his love for the art and science of teaching field ecology, which they will forever pass on to future generations. We, as authors of this edition, have been personally and professionally shaped by him as a mentor, colleague, and friend. His legacy is therefore unending. To him we say, in his own words, "Thanks for everything you have done—and will do."

Contents

PART 2 The Forest Tree

3 Forest Tree Variation 35

4 Regeneration Ecology 69

10 Fire 219

11 Site Quality and Ecosystem Evaluation and Classification 237

PART 4 Forest Communities

21 Invasive Species in Forest Ecosystems 587

22 Forest Landscape Ecology 611

Preface

Forest Ecology deals with forest ecosystems—spatial and volumetric segments of the Earth—and their climate, landforms, soils, and biota. It is designed as a textbook for people interested in forest ecosystems—either in the context of courses in forest ecology and environmental science or as an ecological reference for those in professional practice. This book is meant to provide basic ecological concepts and principles for field ecologists, foresters, naturalists, botanists, and others interested in the conservation and restoration of forest ecosystems.

Ecology, in general, has undergone several sea changes since the appearance of the first edition in 1964, with enormous increases in public interest and scientific development of theory and research. Ecology and the issues associated with it have become part of our modern lexicon. The great number of advances in our ecological knowledge, as well as increased public interest, presents forest ecologists with both opportunities and challenges to sustainably manage ecosystems using our best understanding of ecosystem properties and processes. This book will hopefully be useful in that process by providing an understanding of the ecological relationships of individual trees and forest ecosystems.

The book has six major subdivisions. "Forest Ecology and Landscape Ecosystems" introduces forests as whole ecosystems rather than tree communities, and at multiple scales. "The Forest Tree" considers the genetic variations among individual trees, the causes of diversity within and between species, regeneration ecology, and selected aspects of tree structure and function. "The Physical Environment" treats the physical factors of forest ecosystems that form the forest site—the influences of light, temperature, physiography, soil, and fire on the individual forest plant and on plant communities. The concluding chapter in this part considers methods of evaluation and classification of the forest site and ecosystems. In Part 4, "Forest Communities," we consider the forest community of trees and associated plants and animals that form a key structural component of forest ecosystems—one part of the whole. We also consider the importance and measurement of diversity of species and ecosystems. In "Forest Ecosystem Dynamics," we examine the functional relationships of the physical environment and the biota. We first examine changes in communities and ecosystems over tens of thousands of years. We then consider the extent to which disturbance, an ecosystem process, initiates change (termed succession) over shorter time scales of centuries. Chapters on carbon balance and nutrient cycling present a detailed consideration of the pattern in which carbon (i.e., energy) and plant nutrients flow within forest ecosystems and how natural and human-induced disturbances alter these patterns. Finally, "Forests of the Future" explores the role of humans in the sustainability of forests. Here we emphasize two of the most pressing issues in forest ecology today, climate change and invasive species, and present a review of landscape ecology which has humans at its center. We end the book with a treatment of sustainability itself.

In this edition, we have made great attempts to maintain the core organization and readability of previous editions while adding those areas most relevant to forest ecology that have developed over the last quarter-century. We have retained the important focus on landscape ecosystems (rather than organisms and communities) that was developed in the fourth edition and have added critical new ecological concepts and research that have developed in genetics, diversity, climate change, invasive species, and sustainability. The ecological literature has only become more voluminous over the past 25 years, and as in previous editions our use of the literature was selective, rather than exhaustive. New references were most often chosen based on their accessibility to students and practitioners as understandable examples of important ecological concepts. At the same time, we have retained many older references that still provide excellent examples of fundamental ecological concepts, many of which would otherwise be lost in obscurity.

As before, we have integrated woody plant physiology into multiple chapters, rather than developing it in a single chapter, and in this edition have done the same with climate because of its overriding influence on so many ecosystem processes. We have also further limited our treatment of forest ecology with examples from temperate and boreal forests, with special emphasis on North America and Europe. With a primary focus on the ecological principles of forests that form the basis for management, we have largely avoided specific treatments of forest management techniques and strategies throughout the book, although by necessity we have provided some examples of forest management in the chapters on diversity and invasive species.

The revised edition would not have come to pass without the patience and dedication of many colleagues who helped in its preparation. We are especially indebted to the following for their reviews of one or more chapters: Dennis Albert, Brian Buma, Mark Dixon, Tom Dowling, Jennifer Fraterrigo, Stephen Handler, Donna Kashian, Doug Pearsall, David Rothstein, Madelyn Tucker, and Chris Webster. Other important contributions, either through provision of material, enlightening discussions, or enthusiastic encouragement, were made by Jonathon Adams, Virginia Laetz, and Dan Binkley. Victoria Meller was extremely supportive of D. M. Kashian and his time spent away from campus in the completion of this revision, as were the graduate students in the Kashian lab. Perhaps without knowing it, the graduate and undergraduate students in the 2018 and 2020 Terrestrial Ecology classes at Wayne State had an immense role in the thought processes and revisions made in the 5th edition. D. R. Zak was partially supported by the National Science Foundation and the Department of Energy during the preparation of this text.

DANIEL M. KASHIAN
Detroit, Michigan, USA

DONALD R. ZAK
Ann Arbor, Michigan, USA

Forest Ecology and Landscape Ecosystems

PART 1

Forest ecologists work to understand the dynamics, structure, and function of forest ecosystems. The first step in doing so is to approach ecosystems as geographic units or landscape ecosystems—real, three-dimensional, defined locations at the Earth's surface. Landscape ecosystems include organisms, species, and communities, but also the air above and the soil below these living things such that organisms are but one important part of a larger landscape unit. Our perception of forest ecosystems in this book is that ecosystems are not simply extensions of forest communities, whereby ecosystems are conceived as organisms and their environment. Instead, units of whole ecosystems that integrate factors of climate, landforms, soils, co-occurring biota, and all the interactions between them are the most appropriate units of study rather than their individual parts that too often claim our immediate attention.

In understanding forest ecosystems, particularly in the context of an ecology course or in solving ecological problems, ecosystems need to be conceived at several spatial and temporal scales. It is very daunting to attempt to gain such understanding, and certain well-known ecologists have suggested that the environmental complex is unknowable and inexpressible. However, conceiving landscapes as ecosystems and proceeding "from above," that is, from the biggest to the smallest units to understand their spatial relationships, makes synthesis possible and manageable. Although local sites and stands with their familiar species appear a convenient starting point, we know that these are, in turn, affected by geological and climatic factors of higher levels within which they are embedded. Landscape ecosystems are nested within one another in a hierarchy of spatial sizes such that it is best to consider the big picture first—the comprehensive view from the outside—rather than the minutia of details as seen from inside of the ecosystem. Our perspective in understanding ecosystems involves establishing a framework for studying the components—a framework of the spatial and temporal, hierarchical pattern of ecosystems of all sizes making up the ecosphere.

Therefore, in the first part of this book, before immersing in the detail of organisms, sites, and their interrelationships, we wish to first examine landscape ecosystems, and then place them within a perspective of spatial levels and their processes at different

Forest Ecology, Fifth Edition. Daniel M. Kashian, Donald R. Zak, Burton V. Barnes, and Stephen H. Spurr.

temporal scales. In Chapter 1, we present the basic concepts of forest ecology within a framework of landscape ecosystems. In Chapter 2, we present the concept of scale and consider the hierarchy of these landscape ecosystems. These fundamental concepts provide the basis for studying the variation, life history, and ecology of individual organisms that follow in Part 2.

Concepts of Forest Ecology

CHAPTER 1

A **forest** is an ecological system dominated by trees and other woody vegetation. More than simply a stand of trees or a community of woody and herbaceous plants, a forest is a complex ecological system, or **ecosystem**, characterized by a layered structure of functional parts. **Ecology** is the study of ecological systems and their interacting abiotic and biotic components. **Forest ecology**, therefore, addresses the structure, composition, and function of forests. In forest ecology, we study forest organisms and their responses to physical factors of the environment across forested landscapes. Forests are widespread on land surfaces in humid climates outside of the polar regions. It is with forests in general, and with the temperate North American forest in particular, that this book is concerned.

There are many ways to study forest ecosystems. Most simply, a forest may be considered in terms of the trees that give the forest its characteristic aboveground appearance or **physiognomy**. Thus, we think of a beech–sugar maple forest, a ponderosa pine forest, or of other **forest types**, for which the naming of the predominant trees alone serves to characterize the forest ecosystem. Forest types are often considered to be composed of **forest stands**, which are trees in a local setting possessing sufficient uniformity of species composition, age, spatial arrangement, or condition to be distinguishable from adjacent stands (Ford-Robertson 1983).

A broader concept of a forest may take into account the interrelationships that exist between forest trees and other organisms. Certain herbs and shrubs are commonly found in beech–sugar maple forests, and these may differ from those found in ponderosa pine or loblolly pine forests. Similar interrelationships may be demonstrated, for example, for birds, mammals, arthropods, mosses, fungi, and bacteria. Thus, part of the forest ecosystem is the assemblage of plants and animals living together in a **biotic community**. The **forest community**, then, is an aggregation of plants and animals living together and occupying a common area. It is thus a more organismally complex unit than the forest type.

A third approach is to focus on geographic or **landscape ecosystems**. This approach is centered conceptually and in practice on whole ecosystems and not just their parts. When our primary focus is real live chunks of Earth space, that is, landscapes and waterscapes (oceans, lakes, rivers; hereafter included as parts of a landscape), we can effectively study their parts (e.g., organisms, soils, and landforms) while recognizing that each is but one part of a functioning whole. We emphasize this focus on ecosystems rather than on the individual organisms and species that are parts of them.

In the past, the forest stand or the species has been the focus in natural resource fields such as forestry and wildlife. However, we are really managing whole forest ecosystems, despite their incredible complexity, because the diverse biota is inseparable from the physical environment that supports it. A consideration of the field of ecology from this viewpoint provides an overall perspective.

Forest Ecology, Fifth Edition. Daniel M. Kashian, Donald R. Zak, Burton V. Barnes, and Stephen H. Spurr.

ECOLOGY

Broader fields of scientific inquiry are difficult to limit and define, and ecology is one of the most indistinct. In 1866, Ernst Haeckel proposed the term **oecology**, from the Greek *oikos* meaning home or place to live, as the fourth field of biology dealing with environmental relationships of organisms. Thus, ecology literally means "the knowledge of home," or "home wisdom." Since its introduction, the term has been applied at one time or another to almost every aspect of scientific investigation involving the relationship of organisms to one another or to the environment (Rowe 1989). Haeckel's organismal focus of ecology has since been redefined and expanded to include the physical aspects of the environment that provide life for those organisms (Hagen 1992; Golley 1993). Thus, Rowe (1989, p. 230) suggests:

> *Ecology is, or should be, the study of ecological systems that are home to organisms at the surface of the earth. From this larger-than-life perspective, ecology's concerns are with volumes of earth space, each consisting of an atmospheric layer lying on an earth/water layer with organisms sandwiched at the solar-energized interfaces. These three-dimensional air/organisms/earth systems are real ecosystems—the true subjects of ecology.*

This approach to ecology emphasizes whole ecosystems as well as organisms, both volumetric and having structure and function.

LANDSCAPE ECOSYSTEMS

The British botanist–ecologist Arthur Tansley (1935) introduced the term ecosystem, writing with an emphasis on "the whole 'system,' including not only the organism complex but also the whole complex of physical factors." He also noted that from the point of view of the ecologist, ecosystems "are the basic units of nature on the face of the Earth." Tansley was a biologist and vegetation ecologist, and so his idea of ecosystem was centered on organisms (species or communities) rather than geographic or landscape entities. With this **bioecosystem** approach, "ecosystem" derives its meaning from particular plant or animal organisms of interest, and an "abiotic" environment defined by the organisms as relevant or not is considered with lesser emphasis. In this approach, every organism defines its own ecosystem, nearly infinite in number and difficult to study and use as a basis for management and conservation.

On the other hand, others (e.g., Rowe 1961a and Troll 1968, 1971) view ecosystems centered on geographic or landscape units (i.e., **geoecosystems)** of which organisms are but one important structural component (Rowe and Barnes 1994). We term these units **landscape ecosystems** in part to differentiate them from geology-based units of study (e.g., Huggett 1995). Landscape ecosystems are geographic objects, with a defined place on the Earth. Landscape ecosystems have three dimensions (volume) just as organisms do, including landforms and biota at the Earth's surface as well as the air above them and the soils below them (Figure 1.1). Other terms have been introduced to express the same idea, but are less commonly used, such as the ecotope (Troll 1963a, 1968) and the ecoterresa (Jenny 1980). This geographic/volumetric concept has been discussed and adapted by professional and academic ecological societies (Christensen et al. 1996), and is useful to field ecologists, naturalists, foresters and other land managers, and natural resource professionals. The concept is described in detail in Chapters 2 and 11.

In addition to being geographic and volumetric, landscape ecosystems are hierarchical, extending downward from the largest ecosystem we know, the **ecosphere** (Cole 1958), through multiple levels of ecological organization (Figure 1.2). These levels include macrolevel units of continents and seas, each of which contains mesolevel units of regional ecosystems (major physiographic units and their included organisms), which in turn contain local ecosystems (Hills 1952), the smallest level of homogeneous environment with organisms enveloped in it. We therefore conceive the ecosphere and its landscapes as ecosystems, large and small, nested

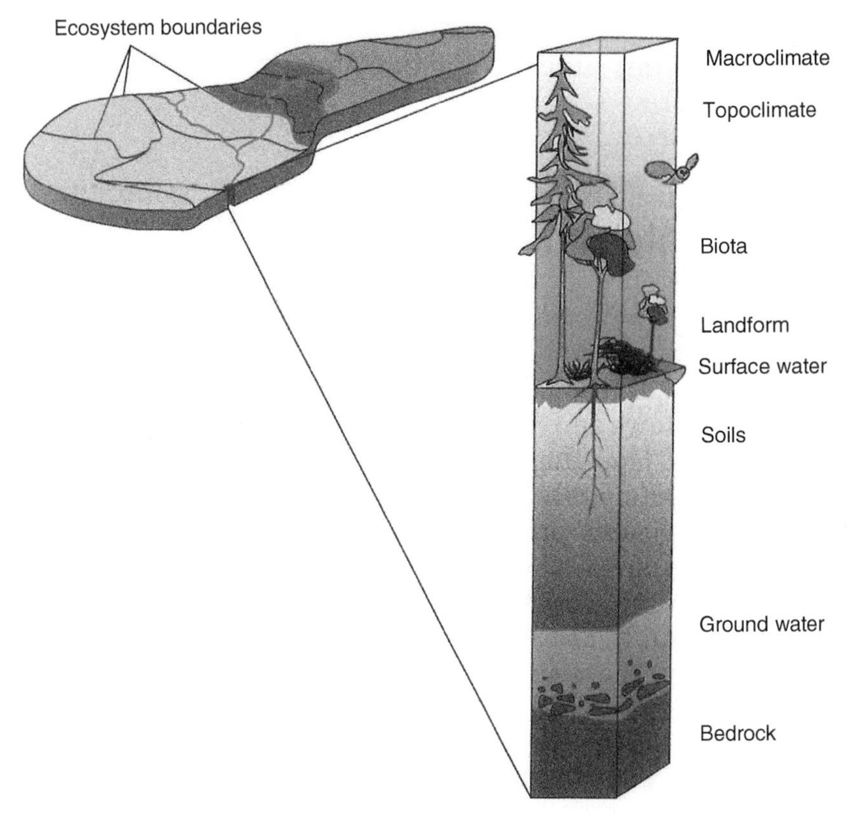

FIGURE 1.1 The three-dimensional, volumetric nature of a landscape ecosystem. Ecosystems comprise the atmosphere (macroclimate as well as the climate affected by surface relief), landforms and soils (underlain by ground water and bedrock), and the biota that provide a physical connection between the air and the Earth. *Source:* Bailey (2009) / Springer Nature.

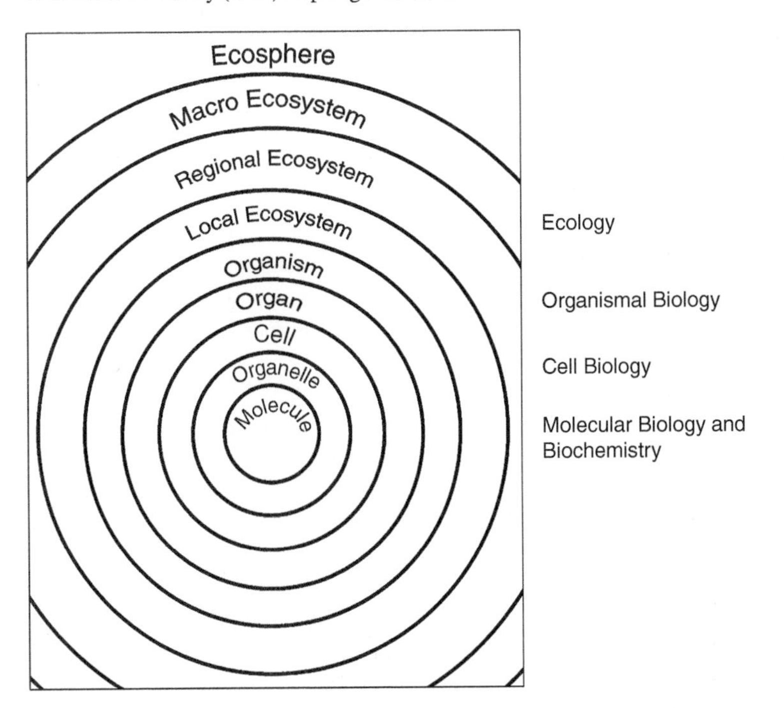

FIGURE 1.2 Objects of study from the most inclusive (ecosphere) to the least inclusive levels of organization (cell and organelle and molecule below it). Note that each higher level envelops the lower ones as parts of its whole. Some corresponding fields of study are also shown. Aggregates of organisms, such as populations and communities, are components of ecosystems at all scales. Like other components such as atmosphere, landform, and soil, they do not appear in this diagram of first-order objects of study, but are shown in Figures 1.3 and 1.4. *Source:* Modified from Rowe (1961a).

within one another in an ecological hierarchy, having processes at each level with their own spatial and temporal scales (see Chapter 2).

Two landscape ecosystems in Figure 1.3 (Rowe 1984b) illustrate characteristic ecosystem differences in hilly or mountainous terrain. The two ecosystems are distinguished by different geomorphologies (convex upper slope versus concave lower slope, and high versus low topographic slope position) that mediate microclimate, soil water, and nutrient availability. The vertical dashed line is placed at an ecologically significant boundary that spatially separates the two ecosystems. Organisms in these ecosystems are sandwiched between the air and the Earth. Also shown in Figure 1.3 are the traditional fields of study in which individuals seek to understand each of the ecosystem components, although forest ecologists aim to understand the integrated effects of all of these components.

In many parts of this book, we focus on organisms, species, and communities, but always remembering that they are parts of volumetric, hierarchical ecosystems. For studies of organisms in their immediate surroundings, a biological approach is often useful. Nevertheless, for management of ecosystems, studies of forest productivity, and the conservation and restoration of forest ecosystems, a landscape ecosystem approach is eminently practical and theoretically sound. Forest ecologists not only study (i) organisms of these systems and their aggregates as communities

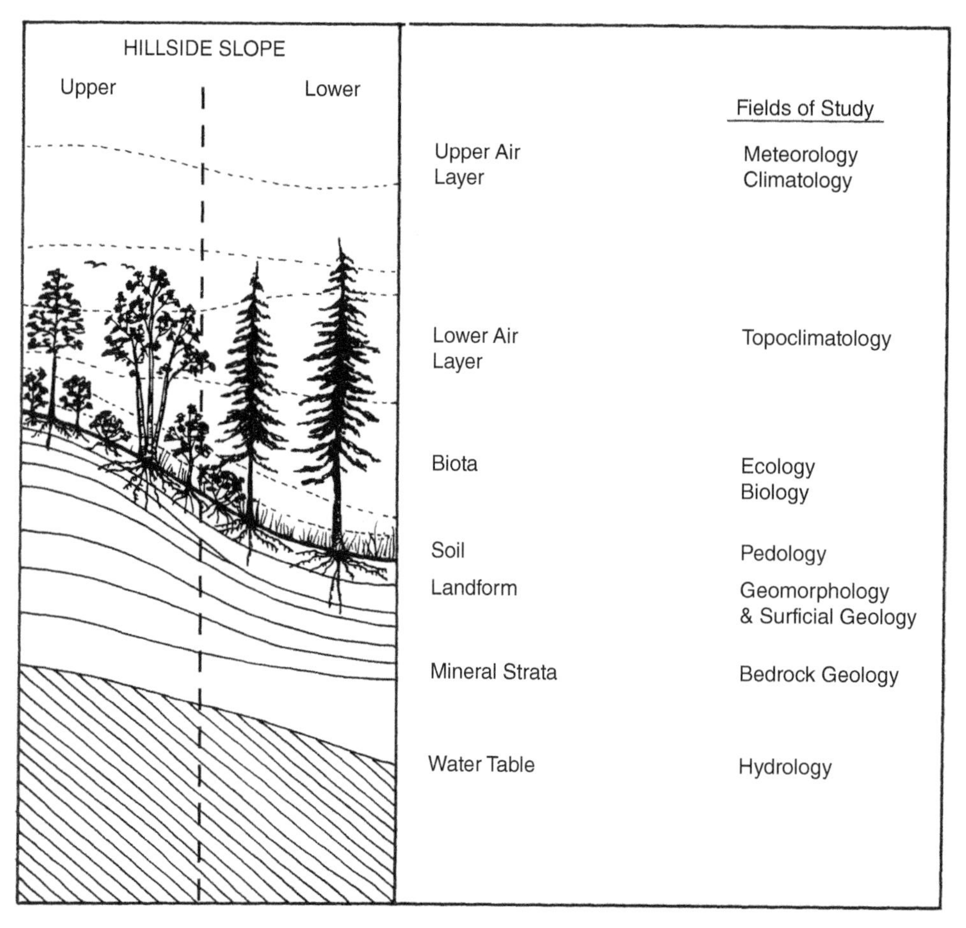

FIGURE 1.3 Structural profile illustrating landscape ecosystems of upper and lower slopes. Air–earth layers surround the organisms at the Earth's energized surface. The vertical line is set at an ecologically significant topographic break, dividing the upper and lower slope ecosystems. *Source:* Rowe (1984b) / United States Department of Agriculture / Public Domain.

and populations (see Chapters 3–5, 12, and 13), but also (ii) the functioning of local ecosystems that involves complex interactions among organisms and their supporting environment (Chapters 6–11, 13–14, and 16–19), and (iii) the spatial patterns of occurrence and interrelationships of entire forest ecosystems (Chapters 2, 18, 19, and 22).

LANDSCAPE ECOSYSTEM AND COMMUNITY

The term ecosystem was introduced in 1935 by Tansley, but terms in various languages, such as taiga, heath, bald, Auenwald (river floodplain forest), pampas, prairie, chaparral, maquis, hammock, muskeg, and bog, have long been used to depict interactions among air, organisms, and soil. Very often the terms emphasize a distinctive plant community and may therefore imply that an ecosystem is simply an extension of a community. Tansley's definition of ecosystem as "organism + environment" leads to this view, and general definitions such as "ecosystem = biotic community + environment" reinforce this view. Although indispensable for forest ecologists and managers, vegetation may not always coincide with climate–landform–soil-based ecosystems because of disturbance and/or unknown historical factors. Thus, we emphasize the importance of geography and physiography within a regional climatic setting as the basis of understanding not only vegetation but whole ecosystem structure and function.

This conceptual approach is not to say that communities are unimportant; they form the key ecosystem component whose response is essential in affecting ecosystem change and indicating the integrated effects of many site factors (Chapter 11). Communities and populations, however, fundamentally differ from entities such as ecosystems, organisms, and cells because they are aggregates of individuals but *not functional systems*. Entities such as ecosystem, organism, cell, and molecule (Figure 1.2) are "volumetric" levels of organization because they have structurally joined parts that form a functioning unit (Rowe 1961a, 1992c). By contrast, communities and populations are assemblages or aggregates of spatially separated trees and understory plants that have no necessary, physical, structural connections.

ECOSYSTEM STRUCTURE AND FUNCTION

An ecosystem's structure describes the spatial arrangement of its parts. In local forest ecosystems, multiple strata of atmosphere and soil (Figure 1.3), as well as surface relief, influence species composition and its patterns of occurrence. Plants provide a physical connection between the soil strata and the aboveground layers of air. The layered structure and related physical and chemical properties of soil are well studied and understood, but similar properties of the atmosphere are not (Rowe 1961a; Woodward 1987), particularly as they affect forest organisms at the Earth's surface. Vegetation itself is also structured vertically, and the vertical structure and species composition of communities as well as species interactions provide insights critical in understanding properties and processes of ecosystems. In addition to species composition, vegetation structure is shaped by tree form and stem density, canopy characteristics, shrub and herbaceous plant abundance, and the amount and distribution of dead vegetation, among other characteristics. Thus, the physiognomy of vegetation varies markedly in different regional and local ecosystem types (Chapters 5 and 6).

The horizontal spatial patterning of these structural components occurs at multiple spatial scales, but is most often described across broad scales. Landscapes are structured horizontally into mosaics of ecosystems that reflect differences in climate, geology, and physiography and their relative effects on vegetation. The natural communities of these systems reflect limiting factors of climate, soil water, nutrients, and disturbances, and in turn modify the physical factors. Different ecosystem mosaics characterize mountains, plains, and river valleys due to fundamental differences in their physical factors and the vegetation adapted thereto (Chapter 8). The diversity of landscape ecosystems can be assessed by understanding such mosaics (Lapin and Barnes 1995). Such an understanding provides a spatial ecosystem framework for programs in biodiversity that seek to conserve and manage the diversity of organisms and maintain and increase populations of rare and endangered species (Chapters 14 and 22).

Local landscape ecosystems, besides having a structure of interconnected parts, are functional units characterized by many processes that define their properties. Ecosystem-level processes are part of the entire system and are not restricted only to physical or biotic parts. These processes drive or mediate the flow of energy or matter and/or the cycling of materials in the system. Organic matter decomposition; cycling of water, nutrients, and carbon; and biomass accumulation are considered ecosystem-level processes. Other processes are often more associated with plant species, populations, or communities, such as photosynthesis and respiration, reproduction, regeneration, mortality, and succession. Despite their basis in organisms, these are also ecosystem processes because they are mediated and regulated by characteristic ecosystem factors of temperature, water, nutrients, and disturbances (such as fire, windstorm, and flooding). Ecosystem function is often described with box-and-arrow diagrams, flow charts, and simulation modeling as ways of disentangling very complex systems (Figure 1.4).

Landscape ecosystem structure and function are tightly coupled by the physical environment, the frequency and severity of disturbances that reset succession, and the life histories of the plants and animals that comprise the biotic community. There is increasing evidence that many aspects of an ecosystem's function are linked to the diversity of its biota (Chapter 14). In turn, an

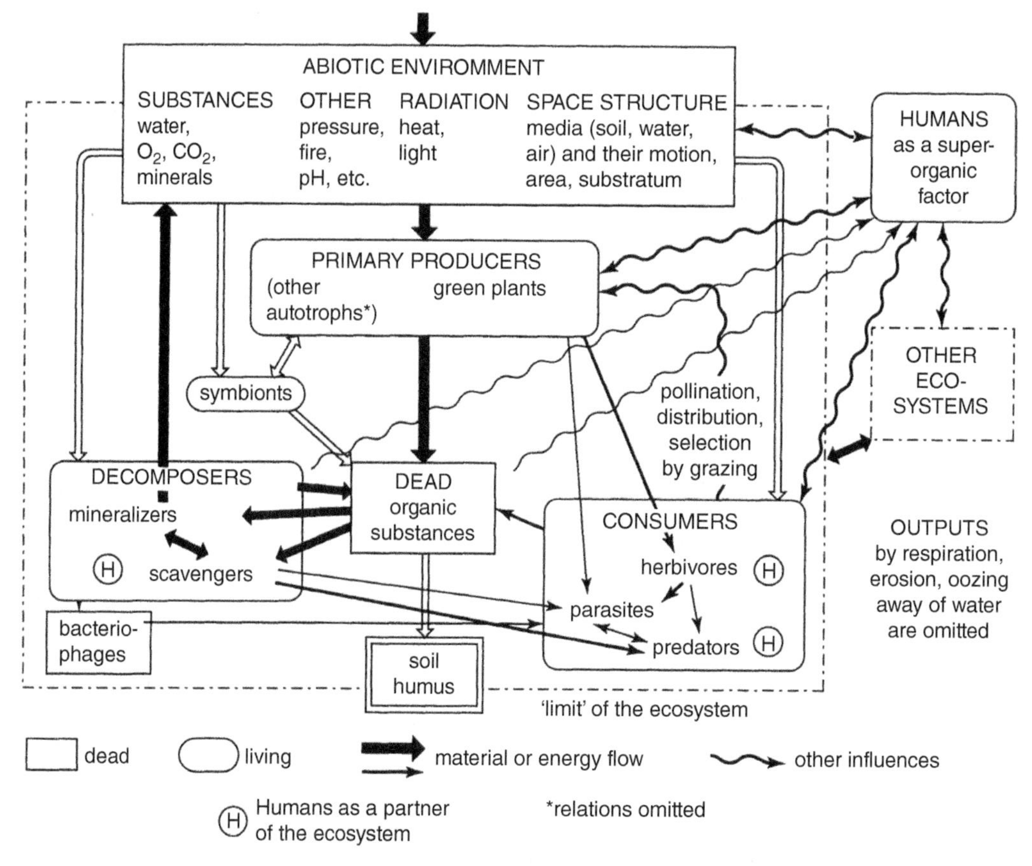

FIGURE 1.4 A model of a landscape ecosystem detailing the flow of energy and matter among biota. The model also describes the interactions between the physical environment and the biota. *Source:* Reprinted from Ellenberg (1988) / Cambridge University Press.

ecosystem's biodiversity is strongly shaped by its physical factors such as climate, physiography, and soil, which provide the context within which organisms survive, adapt, and evolve. In addition, the functional aspects of an ecosystem are strongly affected by its geographical context and by its spatial position relative to its surrounding neighbors on a landscape. Adjacent ecosystems, especially in mountainous or hilly terrain, affect one another by the lateral exchanges of materials and energy. Water and snow, soil, organic matter, nutrients, and seeds are transported downhill; the effects on other ecosystems depend on the size, shape, and composition of the systems. Therefore, an understanding of the spatial pattern and configuration of landscape ecosystems can play an important role in the management of ecosystems and their biota. An excellent overview of function in terrestrial ecosystems is given by Chapin et al. (2012), its variation across landscapes considered in Lovett et al. (2005), and the history of the ecosystem concept itself is reviewed by Golley (1993).

EXAMPLES OF LANDSCAPE ECOSYSTEMS

There are many examples of landscape ecosystems, some of which are easily discerned in the field, whereas others must be carved out of geographic continua. For example, bogs in the glaciated terrain of northern and boreal regions are easily discernible ecosystem types. They exhibit distinctive physiography, soil, and vegetation, and they recur in the landscape. The distinct terraces of river floodplains distinguish local ecosystem types that differ in microclimate, soil, drainage, vegetation, and their dynamics. Mountains are characterized by gradients of elevation, aspect, and slope steepness along which strikingly different ecosystems can be purposefully delimited and mapped, although their boundaries may not be sharply demarcated. On old lake-bed terrain, dry, sandy beach ridges are easily distinguished from adjacent depressions of swampy land. Differences in water drainage and oxygen availability in these adjacent systems result in major differences in their native tree- and ground-flora vegetation. Unseen but critical are the very different processes that also distinguish these ecosystems.

Ecosystem components will change over time, giving rise to a sequence of different landscape or waterscape ecosystems on a given place on the Earth's surface. Striking changes in the fossil record illustrate long-term changes in physiography, soil, and biota (Chapter 15), but short-term changes also occur. Shorter-term changes are most obvious when natural or human-caused disturbances affect vegetation. A site-specific location may exhibit a range of different forest communities over 100 to 300 years, young and old, as disturbances or lack thereof affect the biota. Ecosystem structure and function at this site change over time. The landforms and associated pattern of parent material, soil, and climate also change over time, but more slowly than the suite of species available to recolonize the disturbed site. Thus, what we term the **ecosystem site type** or simply **ecosystem type** (land area supporting potentially equivalent ecosystems) can be distinguished and mapped regardless of the forest community currently present.

One such example is illustrated in Figure 1.5 where deciduous forests occur on glaciated terrain in southern Michigan. In this setting, three local *ecosystem types* are distinguished by differences in physiography (outwash and moraine landforms); soil; drainage; and overstory, understory, and ground-cover vegetation. The relatively fine-textured, silty soil on the outwash plain (type 1) supports forest dominated by white oak, whereas the drier, coarse-textured, sandy outwash soil (type 2) supports a black oak community. The fine-textured, moist, clayey soil on the rolling moraine landform (type 3) supports a community dominated by northern red oak. The moraine formerly supported a beech–sugar maple forest. However, recurrent fires through the drier white and black oak ecosystems killed the fire-sensitive mesic species of the adjacent moraine ecosystem and led to dominance of northern red oak. An occasional beech is still found, and red maple has invaded the shaded understory.

The western portion of ecosystem type 1 was clear-cut about 90 years ago, and a *cover type* markedly different from the adjacent old-growth white oak forest has formed (Figure 1.5, type

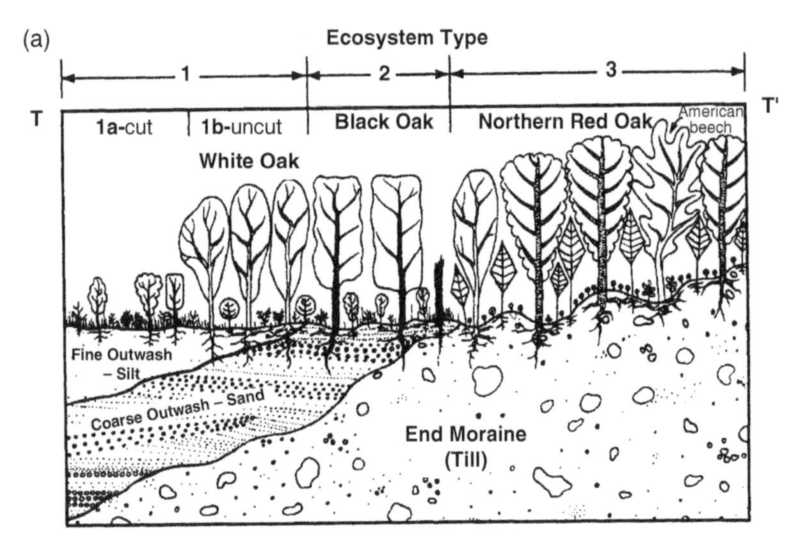

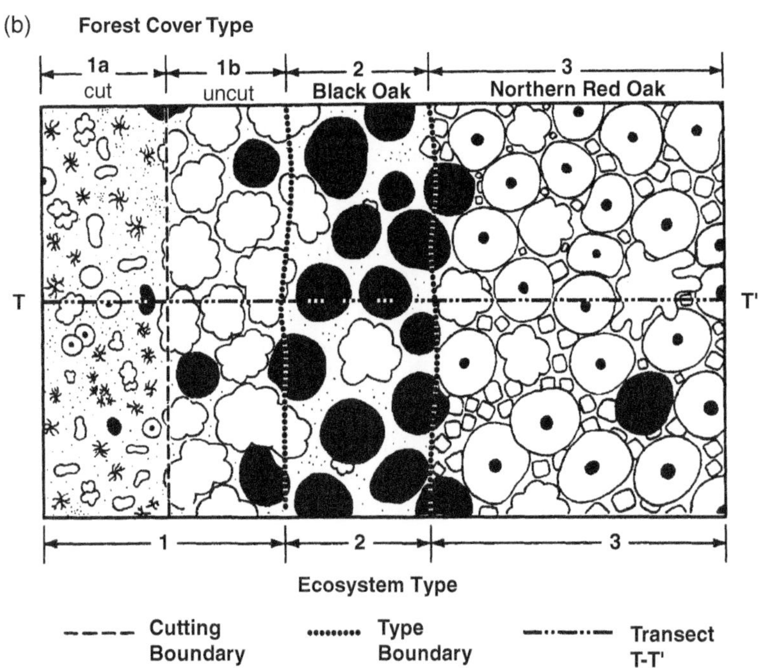

FIGURE 1.5 Three local *landscape ecosystem types* in glacial terrain that are differentiated by landform, parent material, soil, and vegetation: (a) Lateral transect from T to T′ showing vegetation and underlying geological parent material. (b) Top view showing distribution of forest trees, cover types, and ecosystem types. Following clear-cutting of part of type 1, two forest cover types (1a—early successional oak forest; 1b—old-growth white-oak forest) are distinguished. The diagram illustrates that different forest cover types (1a and 1b) are not necessarily different *ecosystem types*. They represent two of many possible ecosystem derivatives (disturbed in 1a versus relatively undisturbed in 1b) of a given ecosystem type. See text for discussion.

la). The overstory tree layer of the disturbed area is now dominated by white, black, and red oaks, white ash, black maple, American elm, black cherry, and sassafras. These species either sprouted from the base of cut trees, were already present in the white oak forest understory, or seeded in from adjacent communities following cutting. Today, we can recognize two (types 1a and 1b) of the many compositionally different forest communities that might occur at the site of ecosystem type 1; these easily could be mapped as two different forest cover types. Because the cut-over area (type 1a) has physiography and soil like type 1b, its vegetation may gradually become similar in structure to those of ecosystem type 1b, providing that no major changes in climate, soil, or ecosystem processes (including changed browsing pressure by herbivores) were caused by clear-cutting. Thus, a given local ecosystem type, defined by relatively stable features of physiography and soil, may have a suite of disturbance-induced cover types (Simpson et al. 1990). Therefore, recognizing forest cover types alone is not necessarily likely to provide a useful estimate of site potential for management or conservation. However, cover types are extremely useful in management planning for wildlife, timber, water, and recreational use of existing forest communities.

It is often the conspicuous forest cover type that receives our immediate attention. However, the enormous complexity of geographic space and changes in its component ecosystems through time require that major attention be directed to atmospheric, geologic, physiographic, and soil properties of forest landscapes. In summary, for every landscape, a combination of factors should be used to distinguish the pattern of local and regional landscape and waterscape ecosystems that have similar ecological potential in the long run. Understanding ecological units at multiple spatial scales (Chapters 2, 11, and 22) is needed to provide the basis for monitoring ecosystem change over time and for ecosystem management.

AN APPROACH TO THE STUDY OF FOREST ECOLOGY

This book presents the scope of forest ecology as having at least two major parts. The first is *analysis*: studying the details of ecosystem components (atmosphere, organisms, and Earth). The second part is *synthesis*: understanding the broad ecological framework within which ecosystem components are integrated. We therefore emphasize a "big picture" approach even as we discuss much smaller details of the organic and inorganic parts of ecosystems, their interactions, and the structure and function of entire ecosystems. The explicit separation and analysis of the biotic and physical parts of ecosystems are not realistic or desirable from an ecological perspective, but may serve as reasonable subdivisions to provide both analysis and synthesis. The sequence of presentation is (i) understanding forest ecosystems at multiple spatial scales, (ii) the variation, life history, and structure of the forest tree, (iii) the physical environment of forests, (iv) forest communities, (v) forest ecosystem dynamics, and (vi) the outlook for and consideration of future forests.

In Part 1 of this book, landscape ecosystems are the primary objects of interest, but we observe the importance of understanding spatial scale in examining them. Spatial scale influences the observations we make about ecosystems because ecological processes vary as spatial scale changes. Advances in hierarchy theory in ecology have helped us to understand ecosystems as volumetric segments of the ecosphere within which smaller volumetric units are nested.

In Part 2, we consider that the forest tree largely owes its appearance, rate of growth, and size to the environment in which it has grown throughout its life. In other words, the **phenotype**, the individual as it appears in the forest, is the product of the environment acting on its **genotype**, its individual genetic constitution. An understanding of trees as parts of ecosystems depends, in part, upon the genetics of the trees themselves and the way their genetic heritage affects their response to the environment they live in. In addition, life history features of regeneration as well as anatomical and physiological aspects of forest trees in relation to environment are considered.

We examine the physical factors of forest ecosystems in Part 3. The **site** is the sum total of environmental factors surrounding and available to the plant at a specific geographic place. These factors are primarily the atmospheric, physiographic, and soil components of the physical environment, but also include important influences of growing vegetation which itself affects the microclimate and soil. **Climate** includes various atmospheric factors that vary across hours, days, months, and years, such as solar radiation, air temperature, precipitation, humidity, wind, and carbon dioxide content. We provide special consideration to **sunlight** (solar radiation) and **temperature**, and we examine the relations and dynamics of **water** throughout the text in appropriate contexts. Physiological processes related to photosynthesis, respiration, and growth are included in Part 3 in chapters on light and temperature, but also in later chapters on carbon balance and nutrient cycling.

Physiography—comprising landforms and soil parent material——plays a key role in affecting climate as well as the amount and rates at which radiation and moisture are received and distributed in forest ecosystems. Below the ground surface, the supply of soil moisture and nutrients, the physical structure of the soil, microbial communities in the soil, and the nature and decomposition pattern of organic matter, that is, factors pertaining to **edaphic** characteristics of soil, all affect the growth and development of plants. We also treat fire as a site factor. The study of the environmental factors and their effects on individual plants constitutes the field of **autecology**, and we summarize those aspects of forest autecology most pertinent to an understanding of forest ecosystems. Finally, we examine site quality and its evaluation, focusing on the degree to which individual site factors and combinations of them are used to estimate the productivity of forest ecosystems and determine management prescriptions for them.

In Part 4, we consider forest communities and their plants and animals as integral parts of ecosystems. First, we examine the important roles of animals in affecting all phases of plant development, as well as their effects on forest communities and ecosystem processes. Forest communities are then treated, emphasizing their composition, occurrence, and the interactions (with emphasis on competition and mutualism) of their constituent individuals. Finally, we examine concepts of biological and ecological diversity, including their importance, measurement, and conservation.

Forest ecosystems are ever changing, and we examine their dynamics in Part 5. We first consider change in ecosystems over hundreds to millions of years. Primary bases for this change come from geologic and physiographic studies of the land itself and determination of plant species and their communities as deduced from the paleoecological record. Disturbance and succession as ecosystem processes are treated next, followed by chapters on whole ecosystem functioning, including the carbon balance of trees and ecosystems and the dynamics of nutrients.

Part 6 examines future forest ecosystems in considering their sustainability. We first examine the effects of climate change on forest ecosystems, including effects on individuals, populations, communities, and whole ecosystems. We also explore the potential for mitigation of climate change using forest management. We next consider the importance of invasive plants and animals for forest ecosystem structure and function. Finally, we place humans in an appropriate context of forest ecosystems in considering forest landscape ecology and the emerging concept of forest sustainability.

APPLICABILITY TO FOREST MANAGEMENT

The management of forests is, as it should be, a continuously evolving effort as new science, technology, and consideration of the role of humans in forest ecosystems develop. Management of forests may include decisions made at broad or local scales, but consideration is increasingly given to whole systems, their complexity and heterogeneity, as well as their long-term sustainability and integrity (Franklin et al. 2018) rather than single-focus use of their parts for recreation, wildlife, or timber. To ecologists, understanding ecosystems for their own sake has significant value in itself, but the principles of forest ecology are essential for their guidance in conservation and management practices. Spurr (1945) defined **silviculture**, or applied forest ecology, as the theory and practice of controlling forest establishment, composition, and growth. Modern texts suggest that current silviculture is most concerned with managing forests for future products and services in a manner that has social stability (Ashton and Kelty 2018). Nevertheless, we explicitly emphasize that silviculture is the theory and practice of controlling forest ecosystem composition, structure, function, and heterogeneity, often at multiple spatial and temporal scales. Even local management operations may have significant and often long-term effects on surrounding landscapes far removed from the specific site of human activity.

The management, conservation, and restoration of forested landscape ecosystems rely heavily on understanding the structure and function of forest ecosystems and the autecology of their organisms (i.e., forest ecology). More than ever, understanding how ecological systems work is critical for making sound decisions regarding human intervention in forests. Understanding the structure and function of whole ecosystems will also equip us to best deal with the ecological consequences of past human activities, such as deforestation and its wide range of impacts, habitat loss, and fragmentation. Human activities associated with industrialization, such as air pollution or the introduction of destructive new species, have resulted in local tree mortality and, in certain ecosystems, widespread death and decline of forest trees and other organisms. Increasing the carbon dioxide concentration in the atmosphere due to the burning of fossil fuels and deforestation are creating climatic changes and concomitant changes in forest ecosystems that are quickly increasing in their severity. Invasive plants and animals introduced into forests are well known for disrupting ecosystem structure and function on relatively short time scales. In all cases, one must understand the system in order to understand the extent to which it is disrupted.

At the time of the last edition of this textbook over 20 years ago, a new paradigm in forestry had occurred that shifted the view of the forest from one of single commodities (timber, wildlife, water, and recreation) to forestry understood as sustaining ecosystems and their structure, diversity, heterogeneity, and function (Rowe 1994; Kohm and Franklin 1997). That paradigm emphasized maintaining the integrity of ecosystems across landscapes by sustaining their natural patterns and processes even when heavily manipulated by humans. This paradigm has been further refined over the past two decades to include an emphasis on considering or even incorporating natural disturbances, increasing ecosystem resistance and resilience, embracing structural complexity, and maintaining a range of ecosystem conditions across broad scales (Franklin et al. 2018). Whether intentionally or otherwise, this paradigm has the landscape ecosystem approach at its core because it focuses on the maintenance of landscapes *as complete ecosystems*. The only way to assure the sustained yields of forests, wildlife, and water, now and in the future, is to maintain the integrity of their processes and keep their biota in a healthy state. Such a practice demands an intimate familiarity with forest ecology as we have described it.

SUGGESTED READINGS

Bailey, R.G. (2009). *Ecosystem Geography*, 2e. (Chapters 1 and 11). New York: Springer-Verlag 251 pp. + 1 map.

Chapin, F.S. III, Matson, P.A., Vitousek, P., and Chapin, M.C. (2012). *Principles of Terrestrial Ecosystem Ecology*, 2e. New York: Springer 520 pp.

Golley, F.B. (1993). *A History of the Ecosystem Concept in Ecology*. New Haven, CT: Yale Univ. Press 254 pp.

Franklin, J.F., Johnson, K.N., and Johnson, D.L. (2018). *Ecological Forest Management*. (Chapter 1). Long Grove, IL: Waveland Press 646 pp.

Kohm, K.A. and Franklin, J.F. (ed.) (1997). *Creating a Forestry for the 21st Century*. Washington, D.C: Island Press 475 pp.

Rowe, J.S. (1961). The level-of-integration concept and ecology. *Ecology* 42: 420–427.

Rowe, J.S. (1992). The ecosystem approach to forestland management. *For. Chron.* 68: 222–224.

Rowe, J.S. (1994). A new paradigm for forestry. *For. Chron.* 70: 565–568.

Tansley, A.G. (1935). The use and abuse of vegetational concepts and terms. *Ecology* 16: 284–307.

Landscape Ecosystems at Multiple Scales

CHAPTER 2

Forest ecologists often by default are also landscape ecologists—people whose interests and practices necessarily encompass ecosystems of many different spatial scales. A **landscape** is a heterogeneous land area composed of a group of intermeshed ecosystems, each with interacting atmosphere, earth, and biota. Landscape ecology is the study of the spatial patterns of ecosystems, the causes of the pattern, and how the pattern affects ecological processes. Many of the core concepts of landscape ecosystems overlap with those of landscape ecology, but the domain of interest of the two fields is not identical. In this chapter, we examine the spatial occurrence and complexity of landscape ecosystems, whereas the discipline of landscape ecology is considered in Chapter 22.

OVERVIEW OF SPATIAL AND TEMPORAL SCALES

Forest ecologists ask questions, conduct field experiments, and resolve problems at different scales of space and time, and these scales directly influence the problems posed, the questions asked, and the techniques employed. **Scale** describes the dimensions of an object or process in space or time (Turner and Gardner 2015). Scale is often confused with biological or ecological levels of organization, such as cells, tissues, organs, organisms, and ecosystems (see Figure 1.2, Chapter 1), but the concepts are quite distinct. For example, organisms are obviously an important level of organization, but they may occur over very different spatial scales (e.g., 1–10 µm for bacteria to 84 m tall giant sequoia) with very different life spans (e.g., 24 hours for mayflies to nearly 5000 years for bristlecone pine). A given scale is defined by its **grain** (its finest spatial or temporal resolution) and its **extent** (the size of a study area or the length of time considered), and there is some trade-off between the two characteristics. For example, forest ecologists might increase grain size (e.g., sample plot size) in a larger study area (e.g., the size of a county) because of their interest in capturing variation in species composition at a larger spatial scale. However, they might reduce grain size (using smaller plots or quadrats) if they are interested in describing species variation across a 1 ha woodlot. As a result, no single set of spatial and temporal scales is applicable to all the questions addressed by forest ecologists (Turner and Gardner 2015).

Processes that occur at a given scale do so within a hierarchy. The process in question is provided context (influenced and constrained) by the processes at the hierarchical level above it, while the level below it provides its mechanisms. For example, it is vital in understanding the growth of trees in forested wetlands to examine those processes that occur at broader and finer spatial scales than those of an individual tree. Wetland tree growth is constrained by broader-scale processes such as landscape position and soils, but the mechanisms that influence tree growth, such as tree physiology, occur at finer scales. A classical hierarchical model shown in Figures 2.1 and 2.2 suggests that many forest managers, ecologists, and naturalists would have an interest at a microscale (up to 100 ha and 500 years), which would encompass the size and life span of a typical forest in North America (Delcourt and Delcourt 1988). Disturbance events such as clear-cuts, some wildfires, and wind events, as well as biotic responses such as gap-phase replacement, competition,

Forest Ecology, Fifth Edition. Daniel M. Kashian, Donald R. Zak, Burton V. Barnes, and Stephen H. Spurr.

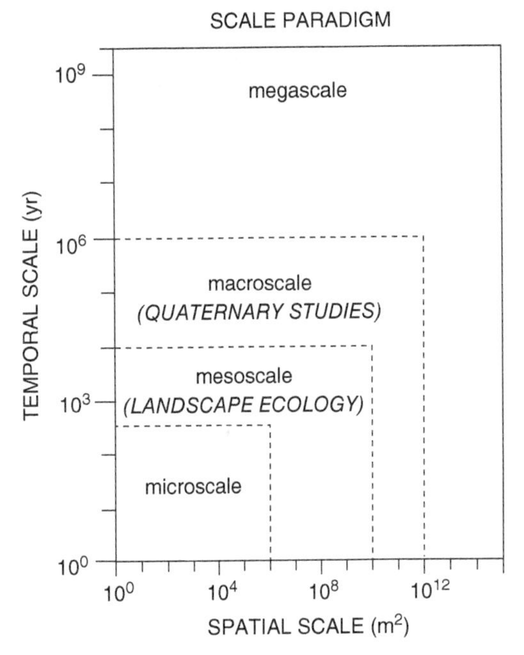

FIGURE 2.1 General spatial–temporal scales for the hierarchical characterization of disturbance regimes, ecosystem and biotic responses, and vegetation units. Quaternary studies at the macroscale refer to investigations of species migrations and extinctions in the time frame of 10 000–1 000 000 years in the Quaternary Period (see Chapter 15) on a continental scale. *Source:* After Delcourt and Delcourt (1988) / Springer Nature.

and productivity might occur at this spatiotemporal scale (Figure 2.2). Many landscape ecologists are concerned with the mesoscale (100–10 000 ha and 500–100 000 years), which encompasses longer-term processes such as fire regimes, insect or pathogen outbreaks, and soil development, as well as biotic processes such as secondary succession. Seasonal patterns of precipitation and temperature and longer-term (decades to centuries) weather trends and climatic fluctuations affect the distribution, growth, and interactions of plants and animals at microscale and mesoscale. Macroscale and megascale (Figures 2.1 and 2.2) extend in space and time to include not only events of ecosystem change and plant migration occurring at the level of physiographic regions (macroscale), but also plate tectonics and evolution of biota on the spatial scale of continents and the ecosphere (megascale).

The disturbance regimes, ecological responses, and vegetation units that are of concern to forest ecologists may occur over a range of spatiotemporal scales rather than cleanly within one shown in the hierarchical framework in Figure 2.2. Disturbance events such as wildfire or insect and pathogen outbreaks may occur at a local site or over thousands of square kilometers. Wind may uproot a single tree in an old-growth forest or may affect huge land areas for much longer intervals, such as the 150 000 ha (370 000 acres) of forest blown down in the Boundary Waters Canoe Area in northern Minnesota in 1999 (Mattson and Shriner 2001). Likewise, geomorphic processes such as soil creep, debris avalanches, stream transport, and deposition of sediments affect vegetation at levels from individual plants to large forest stands and occur on areas extending from the size of sample plots (m^2) and stream watersheds (ha) to mountain ranges (km^2). Vegetation change or succession is another good example of an ecosystem process that may be examined at different spatial and temporal scales. For example, vegetation change may occur in just a few years in small tree-fall gaps, as widespread secondary succession over decades or centuries, or as evolution over millennia on a continental spatial scale. Even the same disturbance type may occur at different scales among ecosystems. For example, subalpine lodgepole pine forests of Yellowstone National Park were historically burned by large fires (>5000 ha) every 100–300 years (Romme and Despain 1989), which spans both the microscale and mesoscale in Figure 2.1. By contrast, hemlock-northern hardwood forests of the Lake States were likely burned by very small fires

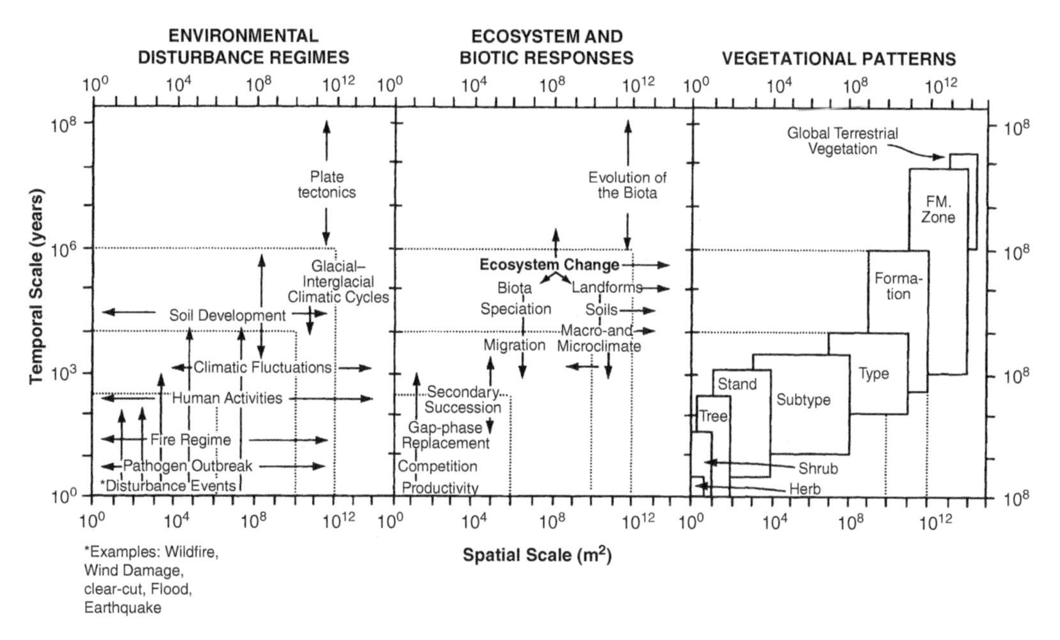

FIGURE 2.2 Disturbance regimes, ecosystem and biotic responses, and vegetational units viewed in the context of four spatial–temporal scales shown in Figure 2.1. *Source:* Modified from Delcourt and Delcourt, 1988 / with permission of Elsevier.

(<5 ha) over very long intervals (approximately 1000 years; Frelich and Lorimer 1991), which also spans both the microscale and mesoscale.

SPATIAL SCALES OF HIERARCHICAL LANDSCAPE ECOSYSTEMS

The ecosphere is the largest and most all-inclusive ecosystem that includes within it volumetric segments (i.e., landscape ecosystems). Three general levels of landscape ecosystems are typically identified (Rowe and Sheard 1981; Rowe 1992b; Bailey 2009). Macroscale landscape ecosystems (scale in this context refers to spatial scale) are seas and continents and their major units, essentially encompassing the macro- and mega- scales shown in Figure 2.2. Continents contain mesoscale landscape ecosystems, usually major physiographic features such as broad plains, rolling hills, and mountain ranges. These major physiographic features contain microscale landscape ecosystems that are local landforms supporting upland forests, swamps, and lakes. This hierarchical framework proceeds from the more complex and heterogeneous ecosystem units at broad spatial scales to less complex and relatively homogeneous at fine spatial scales. Note the three levels of the hierarchy cited in the earlier text are arbitrary, as the complexity of a landscape and the ecological questions at hand will determine the number of levels necessarily distinguished.

Ecosystems at each level, whether at the broad regional scale (macroscale and mesoscale landscape ecosystems) or at a local scale (microscale), all include climate, physiography, soil, and organisms. It would be ideal to delineate and map landscape ecosystems by integrating all these components at all levels, but this formidable task has been accomplished only at local scales (e.g., Barnes et al. 1982; Pregitzer and Barnes 1984; Spies and Barnes 1985a; Simpson et al. 1990). Despite failing to integrate multiple factors, single-component classifications of ecosystems by climate, vegetation, or physiography are useful to illustrate the spatial pattern of the focal component. In the sections provided in the following text, we provide examples of these approaches as an avenue to emphasize the complexity of landscape ecosystems at multiple spatial scales.

CLIMATIC CLASSIFICATION

Climate is the source of energy and water for all ecosystems, and thus it is the most important ecological factor at the macroscale. Together with physiography, climate strongly controls genetic differentiation of organisms and hence their distribution in space and time. Climate may be classified at the broadest spatial scale, such as the "ecoclimatic zones" of Bailey (1988; Figure 2.3), to illustrate the pattern of climate across the ecosphere. In this way climate may be used to distinguish macroscale and mesoscale ecosystem units to which other ecosystem components can be linked. Each of the 13 ecoclimatic zones in Figure 2.3, adapted from the climatic classification of Köppen (1931) (modified by Trewartha 1968), is characterized by a particular climatic regime (the characteristic seasons and daily ranges in temperature and moisture). Each climatic zone essentially corresponds to a general zonal soil type (usually at the level of soil order) and a broad, late-successional vegetation type (Table 2.1). These climatic zones span the continents and in many cases include high mountains. Mountains provide vertical stacks of ecoclimatic zones or sequences of altitudinal belts that are related to elevation, slope aspect, and associated climatic and soil characteristics (Figure 2.4) within the context of the horizontal ecoclimatic zones. The sequence of altitudinal belts may differ according to the geographic position of the zone in which it is located.

In this example, climate is the primary factor for classification of mountainous landscapes into horizontal and vertical zones at the broadest spatial scales. At the scale of continents, gross physiographic features and vegetative cover also play a major role in affecting climate both horizontally and vertically. At the mesoscale, Bailey (2009, 1988) subdivides climatic zones using physiography (specific landforms and their surficial form and parent material) as the key factor. Physiographic features affect the distribution and composition of vegetation by influencing both macroclimate and microclimate (Chapter 8).

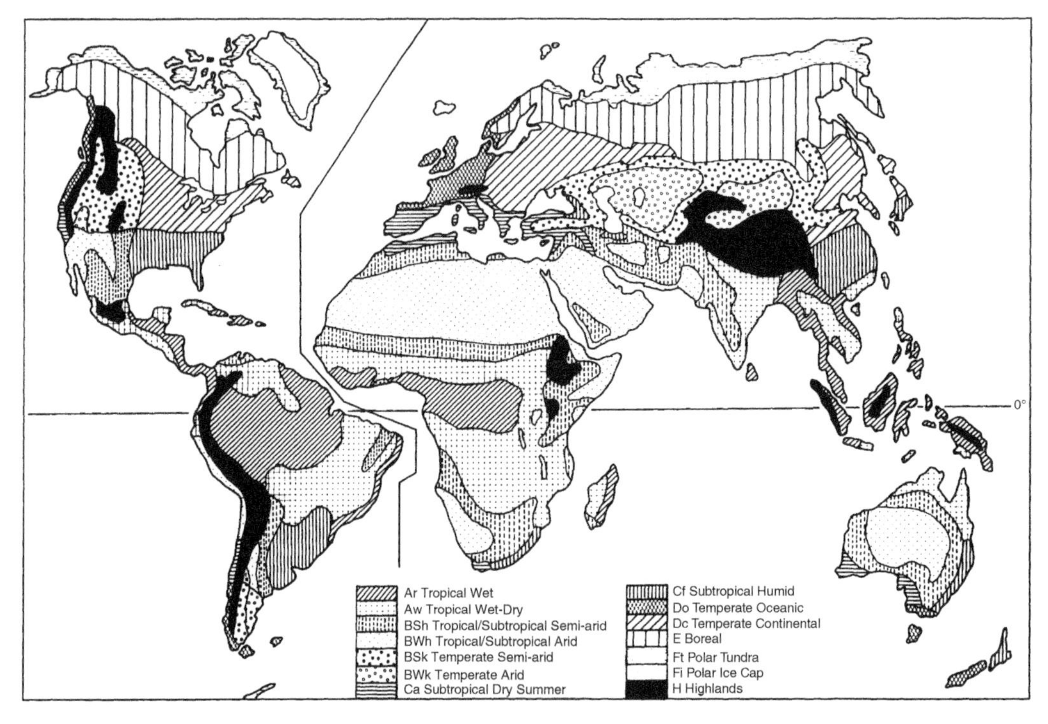

FIGURE 2.3 Ecoclimatic zones of the world. *Source:* Bailey (2009) / Springer Nature.

Table 2.1 Zonal relationships between climate, soil orders, and vegetation.[a]

Ecoclimatic zone[b]	Soil order[c]	Vegetation
Ar	Oxisols (latosols)	Evergreen tropical rainforest (selva)
Aw	Oxlsols (latosols)	Tropical deciduous forest or savanna
BS	Mollisols, aridisols (chestnut, brown soils, and sierozems)	Shortgrass
BW	Aridisols (desert)	Shrubs or sparse grasses
Cs	Entisols, inceptisols (Mediterranean brown earths)	Sclerophyllous woodlands
Cf	Ultisols (red and yellow podzolic)	Coniferous and mixed coniferous–deciduous forest
Do	Alfisols (brown forest and gray-brown podzolic)	Coniferous forest
Dc	Alfisols (gray-brown podzolic)	Deciduous and mixed coniferous–deciduous forest
E	Spodosols and associated histosols (podzolic)	Boreal coniferous forest (taiga)
Ft	Entisols, inceptisols, and associated histosols (tundra humus soils with solifluction)	Tundra vegetation (treeless)

[a] After Walter (1984, p. 3).
[b] Abbreviations of the zones are shown in Figure 2.3.
[c] Names in parentheses are soil orders (USDA Natural Resources Soil Conservation Service, 1999).
Source: Bailey (1988) / United States Department of Agriculture / Public Domain.

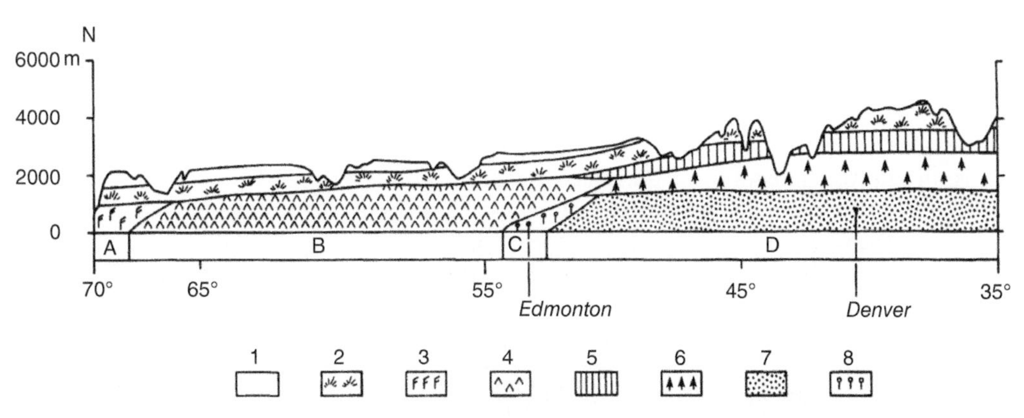

FIGURE 2.4 Vertical zonation (cover types 1–8) within boreal (B) and temperate semi-arid (D) ecoclimatic zones (Figure 2.2) from 70° to 35° N latitude along the eastern slopes of the Rocky Mountains from approximately 102° to 115° W longitude. Climatic zones: A, transition; B, boreal; C, transition; D, temperate semi-arid. Vertical zones: 1, ice region; 2, mountain vegetation above tree line; 3, boreal and subpolar open coniferous woodland; 4, boreal evergreen coniferous forest; 5, boreal evergreen mountain coniferous forest; 6, coniferous dry forest; 7, shortgrass dry steppe; 8, boreal evergreen coniferous forest with cold-deciduous broad-leaved trees. *Source:* Rowe (1988) / United States Department of Agriculture / Public Domain.

PHYSIOGRAPHY

Physiography is the Earth's surface relief and its associated geologic substance or parent material and is perhaps the single most useful component in distinguishing landscape ecosystems. The Earth's structures and forms mediate the fluxes of energy and materials at the Earth's surface where atmospheric and earth layers meet, and thus physiography often has a strong influence on climate in a given place. Vegetation often corresponds to landforms and to the soil that develops from their parent material. Therefore, the physiographic regions of North America (Figure 2.5) provide a useful basis for distinguishing broadscale landscape ecosystems at the macroscale and mesoscale.

Fenneman (1938) created a hierarchical physiographic classification in eastern North America that provided the basis for Braun's vegetation classification (1950) and Omernick's ecoregion classification (1987). Fenneman's physiographic classification provided the landscape framework for Braun's classification of deciduous forest regions and sections of the eastern United States (Figure 2.6). The boundaries of these broad regional geologic units also matched well with major climatic regions (Bryson and Hare 1974), such as those identified by the Holderidge (1947, 1967) model of world bioclimatic formations or life zones. Holderidge's classification identifies a series of broad "life zones," such as rain forest, wet forest, dry forest, tundra, and steppe, based on "biotemperature" (the sum of average monthly temperature above 0 °C divided by 12), precipitation, and the ratio of potential evapotranspiration/precipitation (E/P). Lindsey and Sawyer (1970) and Lindsey and Escobar (1976) used Holderidge's classification to compare the macroclimatic conditions among eight of Braun's forest regions using data from 1100 National Weather Service stations. The physiographically based forest regions are well separated, with exceptions where forest regions had similar canopy species composition among the regions at low and moderate elevations. This example illustrates that physiography is an effective way to delineate landscape ecosystems at regional scales, particularly when coupled with the distribution of pre-European colonization or near-natural vegetation.

VEGETATION TYPES AND BIOMES

The physiognomy (form) and species composition of vegetation may also serve to delineate the extent and boundaries of ecosystems at broad scales (macroscale and mesoscale). In fact, vegetation has been the traditional criterion for distinguishing broad landscape units, at least in part because it is easier to describe and map than physiography or climate. Moreover, vegetation is an effective integrator of other ecosystem factors. We emphasize, however, that any single ecosystem component—vegetation, climate, soil, or landform—is an incomplete indicator of the whole.

At very broad spatial scales, macroscale ecosystems roughly coincide with the formations, plant community types, life zones, biotic provinces, or biomes that biogeographers traditionally recognize to reflect macroclimatic regimes and major physiographic regions. The term **formation** is applied to macroscale plant communities, but the term **biome** (biotic region) is more commonly used because plants and animals are interrelated parts of these units. Broadscale climate patterns such as air masses and movement patterns correlate well with the distribution of some biomes (Tallis 1991; Figure 2.7). This relationship is particularly true for the boreal forest, whose southern limit corresponds closely with the typical southern extent of Arctic air masses during the winter and whose northern limit corresponds closely to the southern extent of Arctic air masses during the summer (Bryson 1966; Barry 1967).

Thus, the location of major biotic regions (Figure 2.8) indicates the location and extent of the Earth's macroscale ecosystems, if we conceive these ecosystems as extensions of the biota

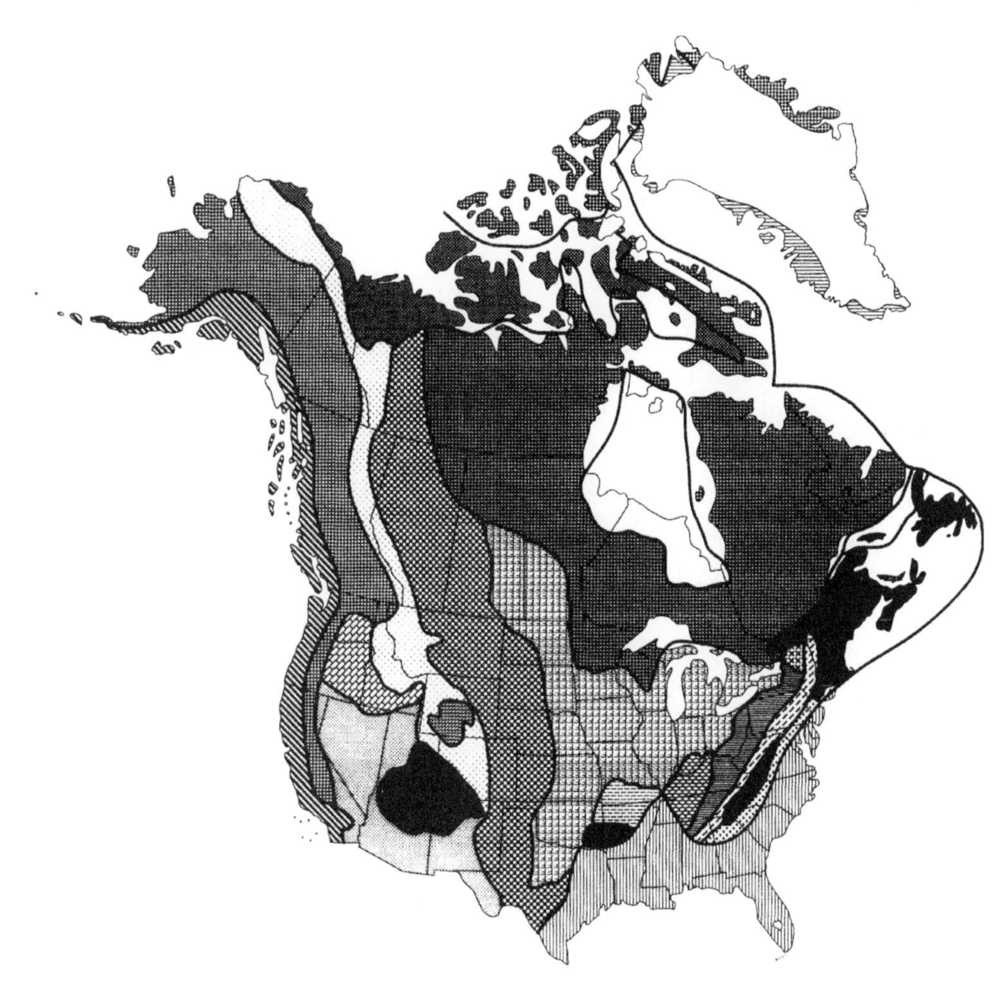

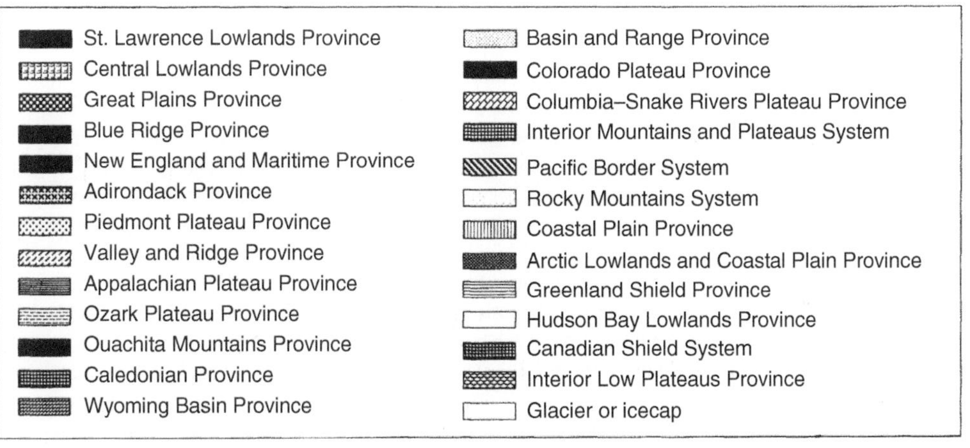

FIGURE 2.5 Physiographic regions of North America and their regional landforms. *Source:* Brouillet and Whetstone (1993) / Oxford University Press.

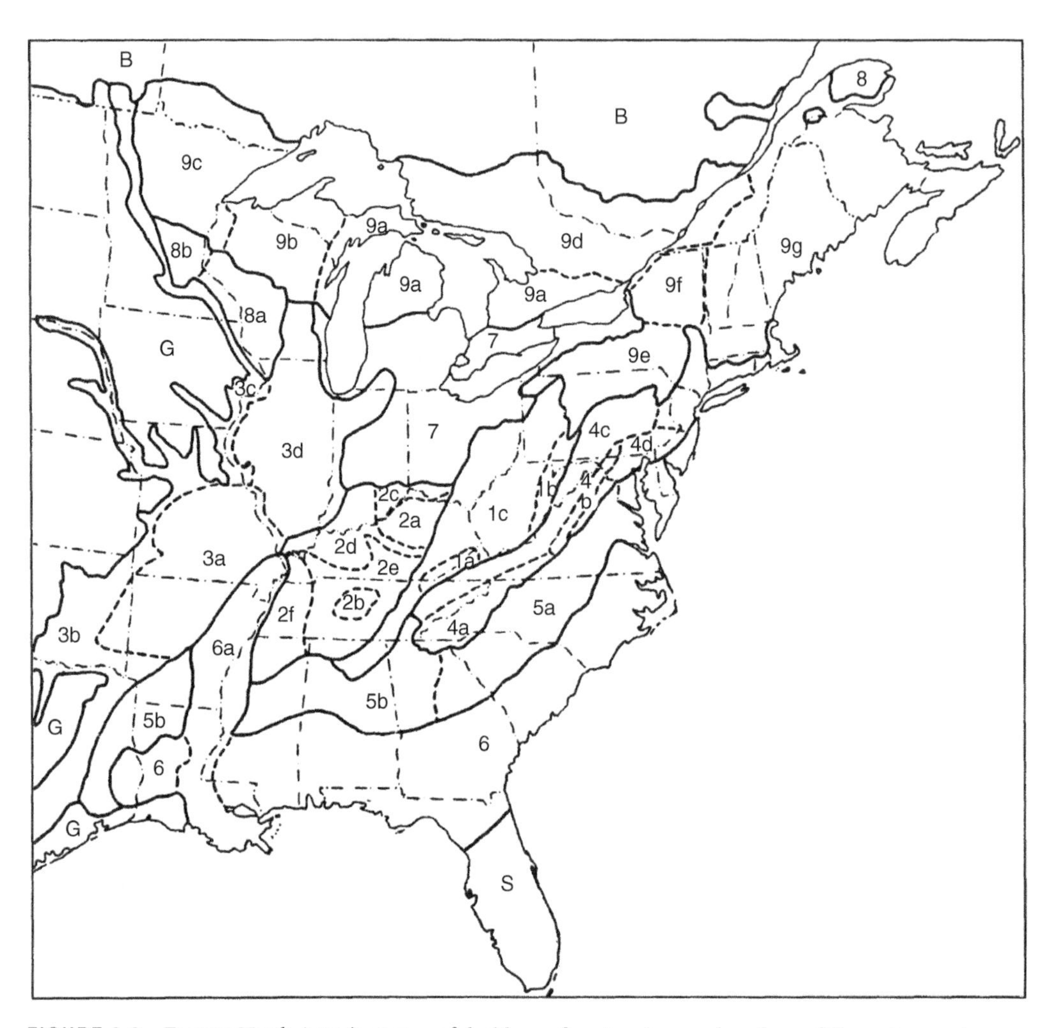

FIGURE 2.6 Eastern North American map of deciduous forest regions and sections of the eastern and midwestern United States. Formation, region, and section names: B = boreal or spruce–fir forest formation; G = grassland or prairie formation; S = subtropical broad-leaved evergreen forest formation. Regions and sections of the deciduous forest formation: 1 = mixed mesophytic forest region: a, Cumberland Mountains; b, Allegheny Mountains; c, Cumberland and Allegheny Plateaus. 2 = western mesophytic forest region, sections: a, bluegrass; b, Nashville Basin; c, area of Illinoian glaciation; d, hill; e, Mississippian Plateau; f, Mississippi Embayment. 3 = oak–hickory forest region: southern division, sections: a, interior highlands; b, forest–prairie transition area; northern division, sections: c, Mississippi Valley; d, Prairie Peninsula. 4 = oak–chestnut forest region, sections: a, Southern Appalachians; b, Northern Blue Ridge; c, ridge and valley; d, Piedmont; e, glaciated. 5 = oak–pine forest region, sections: a, Atlantic slope; b, Gulf slope. 6 = southeastern evergreen forest region, sections: a, Mississippi Alluvial Plain. 7 = beech–maple forest region. 8 = maple–basswood forest region, sections: a, driftless; b, big woods. 9 = hemlock–white pine–northern hardwoods region: Great Lakes-St. Lawrence division, sections: a, Great Lake; b, Superior Upland; c, Minnesota; d, Laurentian; Northern Appalachian highland division, sections: e, Allegheny; f, Adirondack; g, New England. *Source:* Based on a map of physiographic provinces and sections of Fenneman (1938). Braun (1950) / McGraw-Hill.

(bioecosystems). Many textbooks with this biotic emphasis include similar diagrams and descriptions of these major units. Detailed descriptions of the major vegetation types shown in Figure 2.8, and subdivisions of these types are provided by several authors (Rowe 1972; Vankat 1979; Barbour and Billings 1988; Barbour and Christensen 1993). In addition, books in the series "Ecosystems of the World" (for example, *Tropical Rain Forest Ecosystems* (Golley 1983), *Temperate Broad-Leaved*

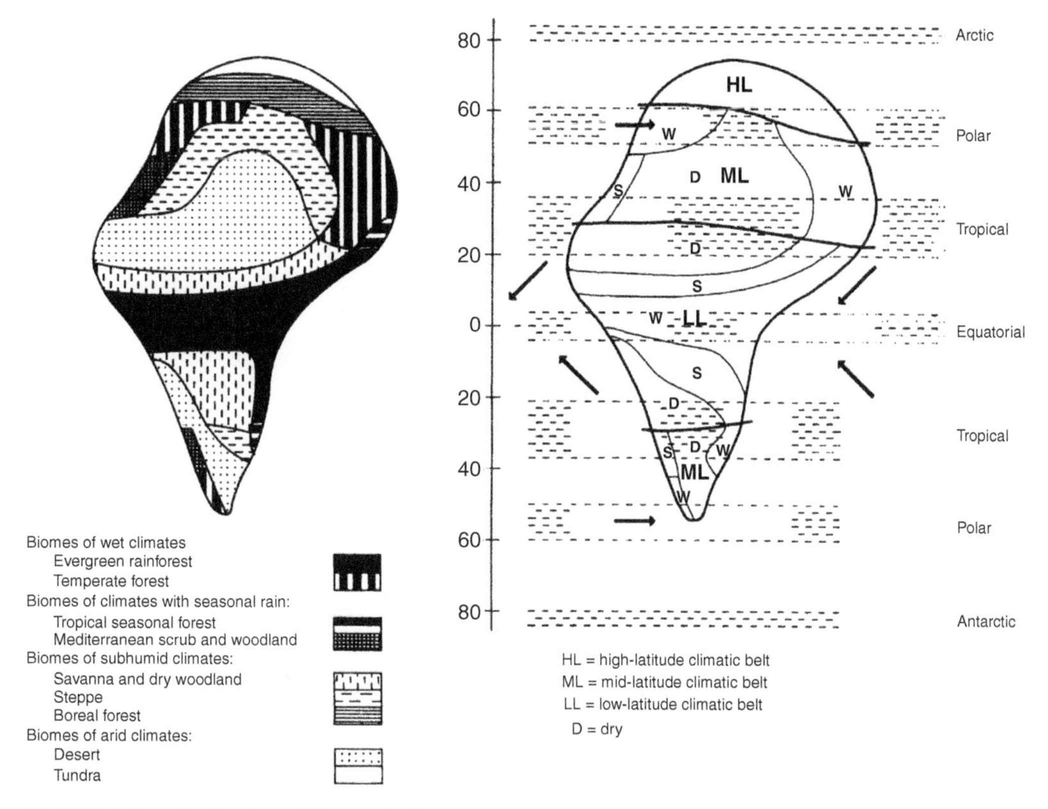

FIGURE 2.7 Distribution of biomes (left) and source regions and air masses (right) in relation to a hypothetical land mass. Source regions are shown shaded; arrows show direction of prevailing winds. HL = high-latitude climatic belt; ML = mid-latitude climatic belt; LL = low-latitude climatic belt; D = dry; W = wet; S = seasonal precipitation. *Source:* Tallis (1991) /Springer Nature. Reproduced with permission from *Plant Community History* by Chapman and Hall.

Evergreen Forests (Ovington 1983), and *Temperate Deciduous Forests* (Röhrig and Ulrich 1991)) provide comprehensive accounts of many forest communities and ecosystems.

Formations or biomes are broadscale groupings of ecosystems based on similar plant community form and structure, or plant **physiognomy**. Plant physiognomy is a direct response to physical factors, such as the temperate and tropical rain forests of warm and wet climates in contrast to tundra and sclerophyllous scrub communities of cold and/or dry climates. Macroclimate (general climate over a large geographical area) is often perceived as the primary environmental factor controlling the form, vertical layering, and composition of these broad units. However, physiographic features and vegetation also influence climate, especially where abrupt discontinuities affect the exchanges of energy and materials at the surface (Hare and Ritchie 1972). Figure 2.8 shows differences in the distribution of vegetation such as tundra, coniferous forests, deciduous forests, grassland, and tropical rain forest that reflect, to a certain degree, the "ecoclimatic zones" of Figure 2.3. There are, however, notable differences between the two figures even at the broad scales of these major units.

One well-known classification system based on vegetation is the United States National Vegetation Classification (USNVC 2019). The USNVC is a hierarchical system with eight levels that emphasizes physiognomy (in this case dominant growth forms) at its highest three levels (Formation Class, Formation Subclass, and Formation) as it reflects environment at broad (global to continental) spatial scales. For deciduous forests in the northern Lake States, for example, these three highest levels would be forest and woodland (Formation Class), Temperate and Boreal Forest

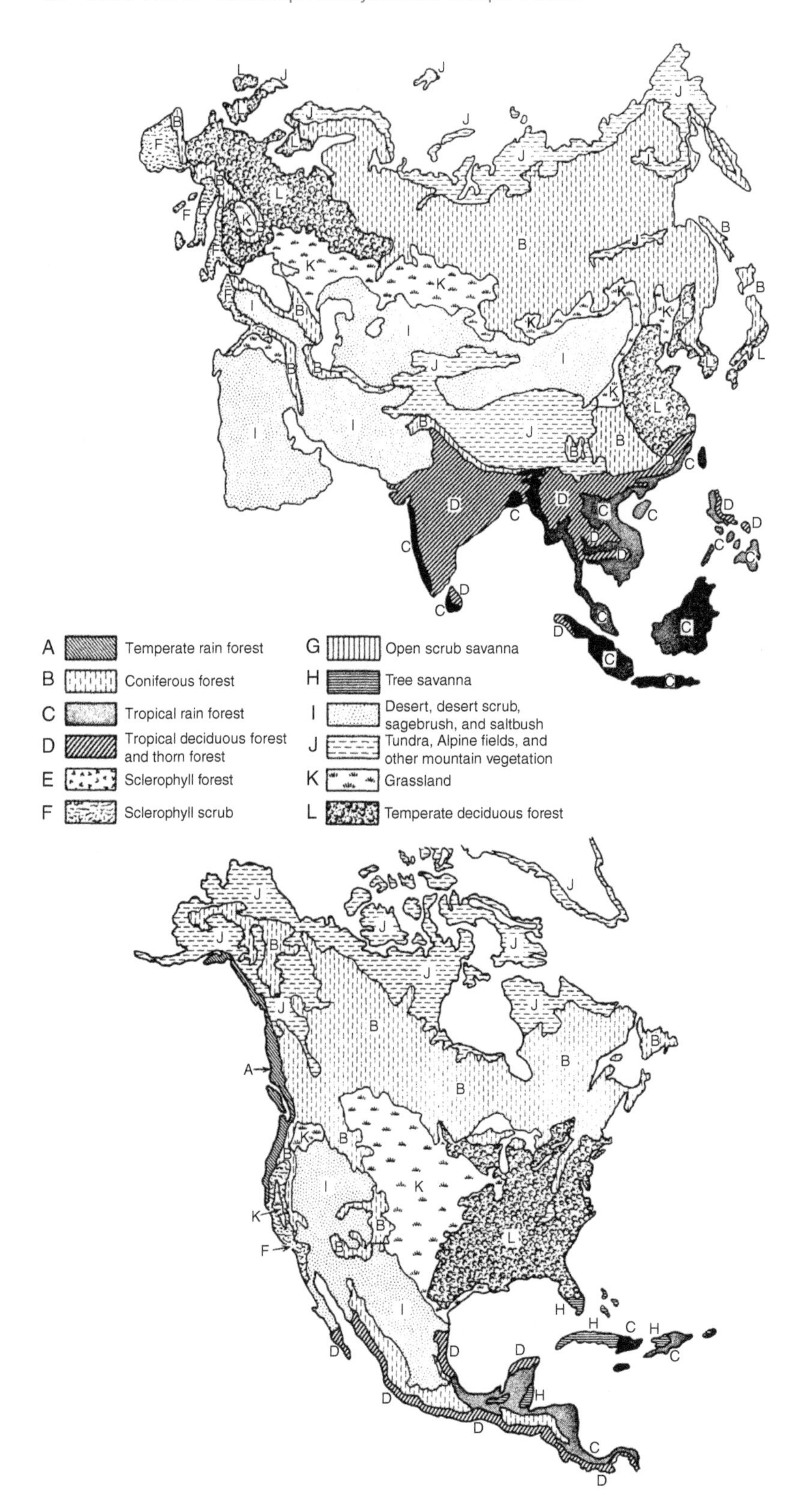

FIGURE 2.8 The major biotic regions or biomes of Eurasia and North America. *Source:* Dasman (1984) / John Wiley & Sons.

and Woodland (Formation Subclass), and Cool Temperate Forest and Woodland (Formation). At increasingly lower levels of the hierarchy (Division, Macrogroup, and Group), biogeography and floristics (similarities in species composition) are emphasized as well as physiognomy reflecting environmental factors at spatial scales of continents to regions. The northern Lake States deciduous forest would fall within the sugar maple–American beech–northern red oak Forest and Woodland Division, perhaps the sugar maple–yellow birch–eastern Hemlock Forest Macrogroup, and the American beech–sugar maple–yellow birch Forest Group. The lowest levels of the hierarchy (Alliance and Association) are based mainly on physiognomy and compositional similarity reflecting local to regional environmental factors. The northern Lake States deciduous forest might fall within the sugar maple–American basswood–white ash Forest Alliance and the sugar maple–yellow birch–American basswood Forest (USNVC 2019). Although the classification interprets physical factors (primarily climate but also geology, soils, and hydrology at lower levels) as the ecological context for vegetation at each level, it does so as it is *reflected by the vegetation*, and thus heavily emphasizes the biotic component of ecosystems rather than the ecosystems as a whole. Vegetation classification is certainly useful for applications focusing on vegetation alone, but such a classification should not be confused with one that classifies landscape ecosystems.

Classifications and maps of the individual ecosystem components of physiography, climate, and vegetation are useful, but their boundaries do not typically correspond well. Landscape ecosystems are volumetric units of many interrelated components, and thus mapping them demands the simultaneous integration of physiographic features with those of climate, soil, and vegetation at multiple spatial scales. Mapping necessarily depends on visible features that can be observed on satellite and aerial imagery as well as on the ground, that is, landform and vegetation. However, these individual components, though easy to observe and conceive, are not ecosystems themselves. In particular, biomes are not ecosystems, but they are often treated as such when in fact they are only reflections of the physical features of a given place. Likewise, physiographic features are not ecosystems because they lack the biotic component that would make them ecological. Fortunately, the ecological relationships of these components provide useful insights to ecosystem components that are much harder to observe: the mobile animal community, soils, the hydrologic regime, and the transparent and continuously shifting atmospheric conditions that comprise climate.

DISTINGUISHING AND MAPPING LANDSCAPE ECOSYSTEMS AT MULTIPLE SPATIAL SCALES

A useful model to describe the progressive dividing of the ecosphere into a hierarchical series of ecosystems would be Chinese boxes or Russian dolls, with a series of nested units fitting one inside the other. The *division* of landscapes into smaller hierarchical units from the top-down is termed **regionalization**. Regionalization provides the basis for a logical *grouping* or **classification** of ecosystems that have practical significance for management or conservation. No one standardized scheme encompasses the enormous diversity of ecosystem patterns around the world, as the number and name of ecosystem levels will vary depending on the size and ecological configuration of the landscape, the purposes for which the classification is developed, and personal preferences for names of the levels. One example of such a hierarchical system of units was proposed by Klijn and Udo de Haes (1994), using the prefix "eco" to indicate entire ecosystems in a holistic sense (Table 2.2). Each level in the hierarchy can be matched with general levels of macro-ecosystem, meso-ecosystem, and micro-ecosystem that correspond to the spatial scales (macro, meso, and micro) shown in Figures 2.1 and 2.2. The corresponding ecological units of the system in wide use today by the US Forest Service (Cleland et al. 1997) also correspond well to the hierarchical levels of this example (Table 2.2).

A key feature of ecosystem geography is that ecosystems at different spatial scales are nested within one another. Climate and gross physiography of a broadscale ecosystem (macroscale) affect

Table 2.2 Overview of hierarchical ecosystem classification at various spatial scales within the general framework of macroecosystem, mesoecosystem, and microecosystem levels.

General ecosystem level[a]	Ecosystem unit name[b]	Mapping scale[b]	Size of basic mapping unit[b]	US Forest Service ecological unit name[c]
MACRO-ECOSYSTEM	Ecozone	1: >50 000 000	>62 500 km^2	Domain Division
	Ecoprovince	1: 10 000 000–50 000 000	2500–62 500 km^2	Province
MESO-ECOSYSTEM	Ecoregion	1: 2 000 000–10 000,000	100–2500 km^2	Section
	Ecodistrict	1: 500 000–2 000 000	625–10 000 ha	Subsection
	Ecosection	1: 100 000–500 000	25–625 ha	Landtype association
MICRO-ECOSYSTEM	Ecoseries	1: 25 000–100 000	1.5–25 ha	Landtype and landtype phase
	Ecotype	1: 5000–25 000	0.25–1.5 ha	Landtype phase
	Eco-element	1: <5000	<0.25 ha	Landtype phase

[a] Correspondence with spatial scales shown in Figures 2.1 and 2.2 (Delcourt and Delcourt, 1988): Macroecosystem level approximates megascale and macroscale combined; mesoecosystem level approximates mesoscale; microecosystem level approximates microscale.
[b] After Klijn and Udo de Haes (1994).
[c] US Forest Service National Hierarchical Framework of Ecological Units (Cleland et al., 1997).
Source: Modified from Klijn and Udo de Haes (1994).

ecological properties and processes of those at finer scales (mesoscale) and the fine-scale (microscale) ecosystems they in turn contain. Climate and physiography of one broadscale ecosystem will therefore affect its constituent ecosystems differently than the climate and physiography of an adjacent broadscale ecosystem would affect its own finer-scale ecosystems. It therefore follows that a given management practice in one broadscale ecosystem will not necessarily have the same results in a different broadscale ecosystem. Recognizing ecosystem hierarchies therefore avoids the mistake of extrapolating management policies and practices too broadly across the landscape. Mapping ecosystem units at multiple scales discourages managing isolated bits and pieces of ecosystems (species, stands, soils, etc.) and encourages a more integrated perspective of the linkages among ecosystems and their components. Robert Bailey's (2009) book, *Ecosystem Geography*, provides a detailed treatment of the theory and principles of ecosystem classification and mapping at multiple scales.

REGIONAL LANDSCAPE ECOSYSTEMS

The process of delineating broadscale regional ecosystems has been described in detail (Bailey 1983, 1988, 2009; Bailey et al. 1985; Bailey and Hogg 1986; Omernik 2004, 1995; Omernik and Griffith 2014). Broadscale maps of ecoregions in the United States have been developed by Omernik (1987) and Bailey (1995, 1994), the former used extensively by the US Environmental Protection Agency and the latter by the US Forest Service. A detailed map and description of regional landscape ecosystems of Michigan, Minnesota, and Wisconsin are also available (Albert 1995). Ecological regionalization has a longer history in Canada, beginning with Halliday's (1937) macroscale forest classification, which was revised by Rowe (1972) as the *Forest Regions of Canada*. Mesoscale classification then followed by Hills (1960) and Wickware and Rebec (1989) in Ontario, Loucks (1962) in the Maritimes, and MacKinnon et al. (1992) in British Columbia. A comprehensive and multipurpose classification of seven hierarchical levels from broad "ecozones" to site-specific "ecoelements" provided a Canada-wide system (Wicken 1986).

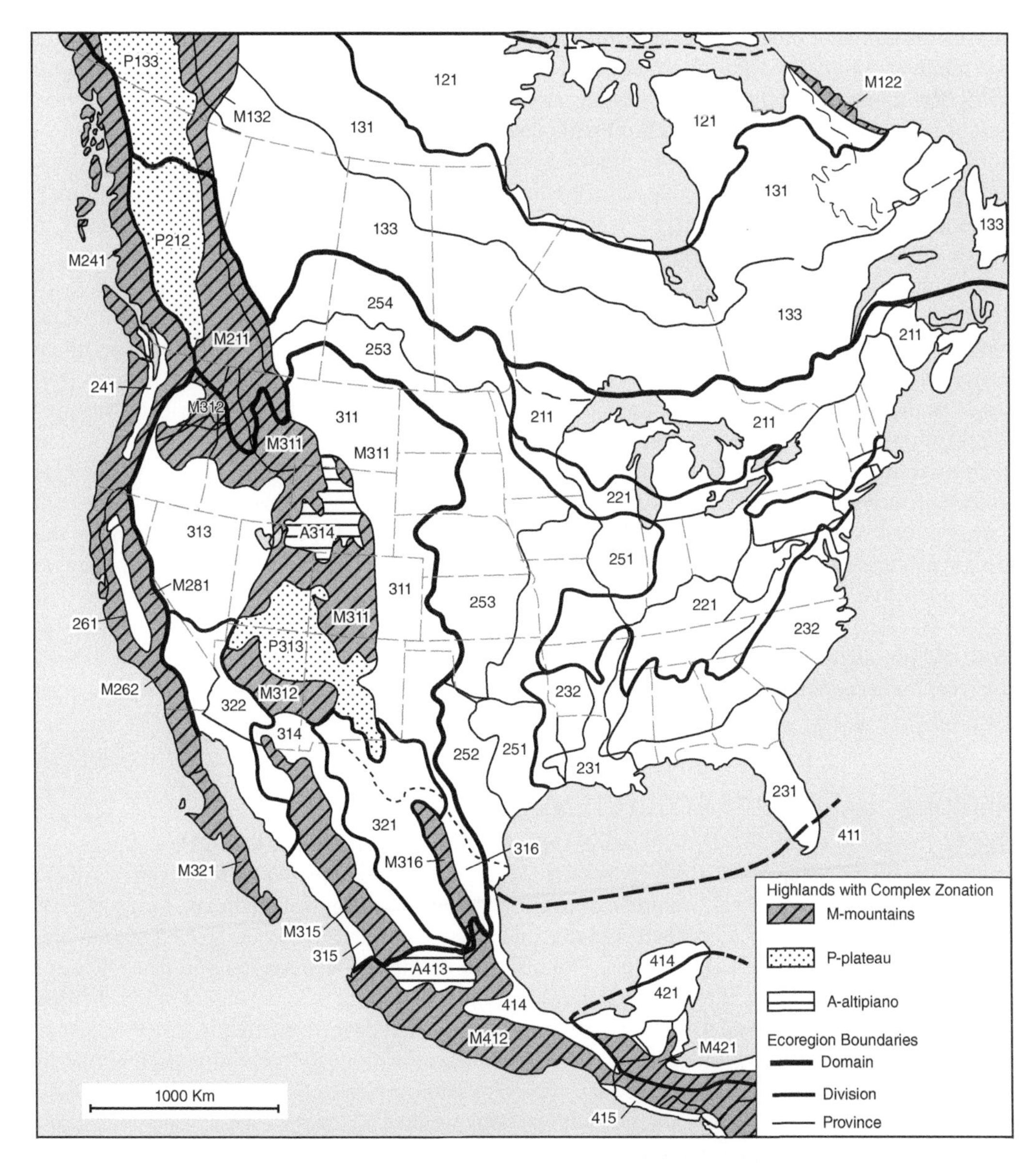

FIGURE 2.9 Ecoregion map for North America. Ecoregion code: 100-series, polar domain; 200-series, humid temperate domain; 300-series, dry domain; 400-series, humid tropical domain. Key to letter symbols: M = mountains; P = plateau; A = altiplano. *Source:* Bailey and Hogg (1986) / Environmental Conservation.

The highest hierarchical levels of the regionalization of North America (Bailey and Hogg 1986) are termed domain, division, and province, and are reasonably equivalent to ecosystems at the macroscale (Figure 2.9). Domains are distinguished by climate (polar, humid temperate, dry, and humid tropical) and correspond broadly with "ecoclimatic zones" (Figure 2.3), physiographic regions (Figure 2.6), and the biotic regions of North America (Figure 2.8). The four domains differ among each other, but also themselves include great spatial heterogeneity in climate, physiography, soil, and vegetation. For example, the humid temperate domain (Figure 2.9) includes temperate rain forests of the Pacific Northwest, sclerophyllous scrub of southern California, grasslands of the Great Plains, deciduous forests of the eastern United States, and coniferous forests of the northeastern United States and adjacent Canada (Figure 2.8). As domains are subdivided into divisions, provinces, and finer subdivisions below them, the heterogeneity of ecosystem components decreases. Divisions are delineated based on finer distinctions in climate, and provinces are distinguished according to potential natural vegetation (Küchler 1964). Vertical zonation of landscape ecosystems in mountainous terrain based on elevational belts and associated mountain physiography of ridges, valleys, and slope–aspect orientation of hillsides is superimposed on the geographic framework we described in the earlier text (cross-hatched areas in Figure 2.9; see also Figure 2.4 and related discussion).

Some argument has been made that watersheds or similar hydrologic units are suitable for ecological classification (Lotspeich 1980; Montgomery et al. 1995), and in some cases are preferable to ecoregions delineated with climate and physiography because their definitions are much clearer and more consistent than ecoregions (Omernik and Bailey 1997). However, using watersheds as the basis for ecosystem delineation includes three major pitfalls as described by Bailey (2009). First, watersheds are not always clearly defined in many parts of the United States, and in such cases, it is extremely difficult to elucidate the boundaries between them. Second, hydrological units such as watersheds may be difficult to delineate on a map where surface-water flow does not coincide well with the movement and flow of groundwater. Finally, it is quite common for the stream system of a watershed to flow though very different climates and physiographic systems, such that some watersheds may include a great deal of heterogeneity in site factors that cannot be easily reconciled even at very broad scales. Whereas broadscale areas with similar climate and physiography include ecologically relatable units with similar environments, watersheds often contain ecosystems that are ecologically dissimilar and therefore are too heterogeneous to be useful (Bailey 2009).

REGIONAL LANDSCAPE ECOSYSTEMS OF MICHIGAN

One of the early examples of delineating **regional landscape ecosystems** at the mesoscale (within the province level of Bailey's ecoregions; Figure 2.9) was completed for Michigan by Albert et al. (1986). Using a multiscale, multifactor integrated approach, regional landscape ecosystems of Michigan were determined, described, and mapped at three hierarchical levels termed region, district, and subdistrict (Albert et al. 1986). The first two of these levels (region and district) were determined by integrating climatic and physiographic/soil classifications (Figure 2.10). At this spatial scale, detailed analytical methods were necessary to first develop a climatic classification of Michigan using multivariate procedures and separate classifications for growing season temperature, winter temperature, and growing season precipitation variables (Denton and Barnes 1988). These classifications were then combined with physiographic data to create a single classification. The unique geographic location of Michigan consists of two large peninsulas oriented in a perpendicular direction that project into the Great Lakes, which dramatically influences the climatic differences of both peninsulas. The distinctiveness of warm, vegetationally diverse southern Lower Michigan (Region I, Figure 2.10) and cold Upper Michigan (Region IV, Figure 2.10) is in marked contrast to lake-moderated Regions II and III where district-level ecosystems parallel lakes Huron, Michigan, and Superior. Within districts, major physiographic features and their associated soil and vegetation were used to identify subdistricts. Macroclimate is increasingly homogeneous within a given district and even more so within a subdistrict.

District 1 in southeastern Michigan (approximately 56 000 km^2) provides a good example of the spatial pattern of local ecosystems distinguished at the subdistrict level. Subdistrict 1.1 (shaded area in Figure 2.10) is the heat island of the Detroit metropolitan area, which is superimposed on an extensive glacial lake plain of Subdistrict 1.2. The lake plain extends south into Ohio and north toward Lake Huron where it forms major parts of Districts 5, 6, and 7 (Figure 2.10). This relatively flat plain of lacustrine clay and sandy soil is interspersed with sand beach ridges and adjacent wet

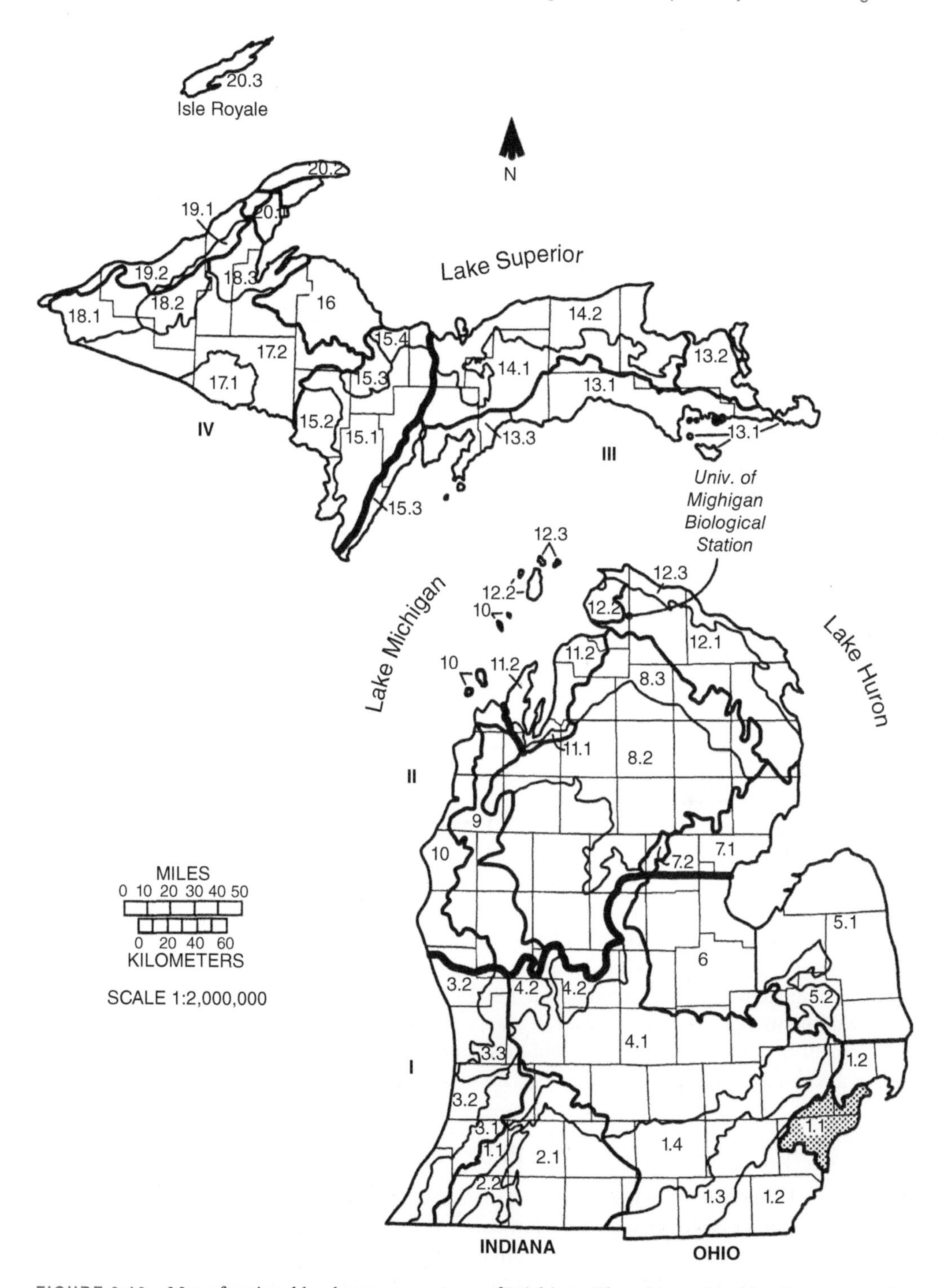

FIGURE 2.10 Map of regional landscape ecosystems of Michigan. Three hierarchical levels are mapped: Regions I–IV; districts 1 to 20, and subdistricts within the districts. Region I = Southern Lower Michigan, districts: 1, Washtenaw; 2, Kalamazoo; 3, Allegan; 4, Ionia; 5, Huron; 6, Saginaw. Region II = Northern Lower Michigan, districts: 7, Arenac; 8, High Plains; 9, Newaygo; 10, Manistee; 11, Leelanau; 12, Presque Isle. Region III = Eastern Upper Michigan, districts: 13, Mackinac; 14, Luce. Region IV = Western Upper Michigan, districts: 15, Dickinson; 16, Michigamme; 17, Iron; 18, Bergland; 19, Ontonagon; 20, Keweenaw. *Source:* Albert et al. (1986) / United States Department of Agriculture / Public Domain.

depressions. In contrast to the flat lake-plain landform, Subdistrict 1.3 (just to the west) is characterized by large, end-moraine ridges of rolling terrain and fine- to medium-textured soils. Adjacent to it and further west is Subdistrict 1.4, characterized by hilly ice-contact terrain of kettle (wet or dry depressions) and kame (steep hills) topography with distinctive microclimatic patterns very different from the other subdistricts. Within each of these subdistricts of uniform macroclimate, local landscape ecosystems and their diverse forest communities occur in distinctive patterns reflecting those of the diverse landforms.

LOCAL LANDSCAPE ECOSYSTEMS

Thus far we have described the delineation of regional ecosystem units at multiple spatial scales—from domains to districts and subdistricts over six hierarchical levels. This regionalization process was based upon progressive division of the landscape using gross climatic and physiographic features and associated soil and natural vegetation. At some lower level in the hierarchy, we eventually reach a land area within which the *regional macroclimate is relatively homogeneous*. At this level in the hierarchy, gross physiography and recurrent patterns of vegetation become most important in distinguishing **local landscape ecosystems** (microscale). At this scale, major landforms and local landscape ecosystem types occur in a mosaic or complex network of interrelated systems. Thus, delineation of local ecosystems cannot proceed exactly as it did at regional levels, where boundaries are drawn around large areas distinguished by major discontinuities of macroclimate, gross physiography, and vegetation (as in Figures 2.9 and 2.10). At a local level, **physiographic systems** or major landforms (e.g., outwash plain, mountain slope or plateau, and moraine) are recognized that recur within a climatically homogeneous area (subdistrict level). Physiographic systems have a profound influence on topoclimate, soil development, the biotic community, and successional trends (Chapter 8). Within these physiographic systems, local ecosystem types can then be grouped appropriately by these major physiographic or landform complexes, the local landforms within them (such as floodplains on outwash physiographic systems or ground moraine on moraine physiographic systems), and with even more specific site factors of microclimate, soil texture, and nutrient availability (Pregitzer and Barnes 1984; Spies and Barnes 1985a; Archambault et al. 1990; Bailey 1995).

In any regionalization of a large land area, identifying the largest unit of relatively homogeneous macroclimate (i.e., the district level in Figure 2.10) is of the utmost importance. Homogeneous macroclimate within such units has great ecological significance because of the intense genetic adaptations of plants to climatic factors along latitudinal and elevational gradients (Chapter 3). In an area of homogeneous macroclimate, individual species and groups of plant species (Chapter 11) indicate different site-factor complexes of water, nutrients, and light (Spies and Barnes 1985b; Archambault et al. 1989). The caveat is that these same species may not indicate the same relationships in areas of different macroclimate (e.g., different districts in Michigan).

LOCAL LANDSCAPE ECOSYSTEMS IN UPPER MICHIGAN

Even within an area of homogeneous macroclimate and physiography, local ecosystem types occur and recur in a complex spatial mosaic, particularly in glacial terrain (Figure 2.11). In the ice-contact terrain of the Sylvania Wilderness Area in Upper Michigan, daily, monthly, and yearly microclimate of local ecosystems, as well as their soil and vegetation, is markedly different. The presence of many lakes, bogs, and coniferous swamps (types 16, 17, 18, and 19) and adjacent sandy uplands (hemlock-northern hardwood types 5, 9, 10, and 11) illustrates the local complexity of ice-block disintegration features that were formed as the last glacial ice sheet retreated. Many tiny (<0.1 ha) depressions forming black-ash swamps (type 22) dot the entire area. A lake- and swamp-edge ecosystem (type 14), characterized by sandy soil and fire-regenerated pines and white birches,

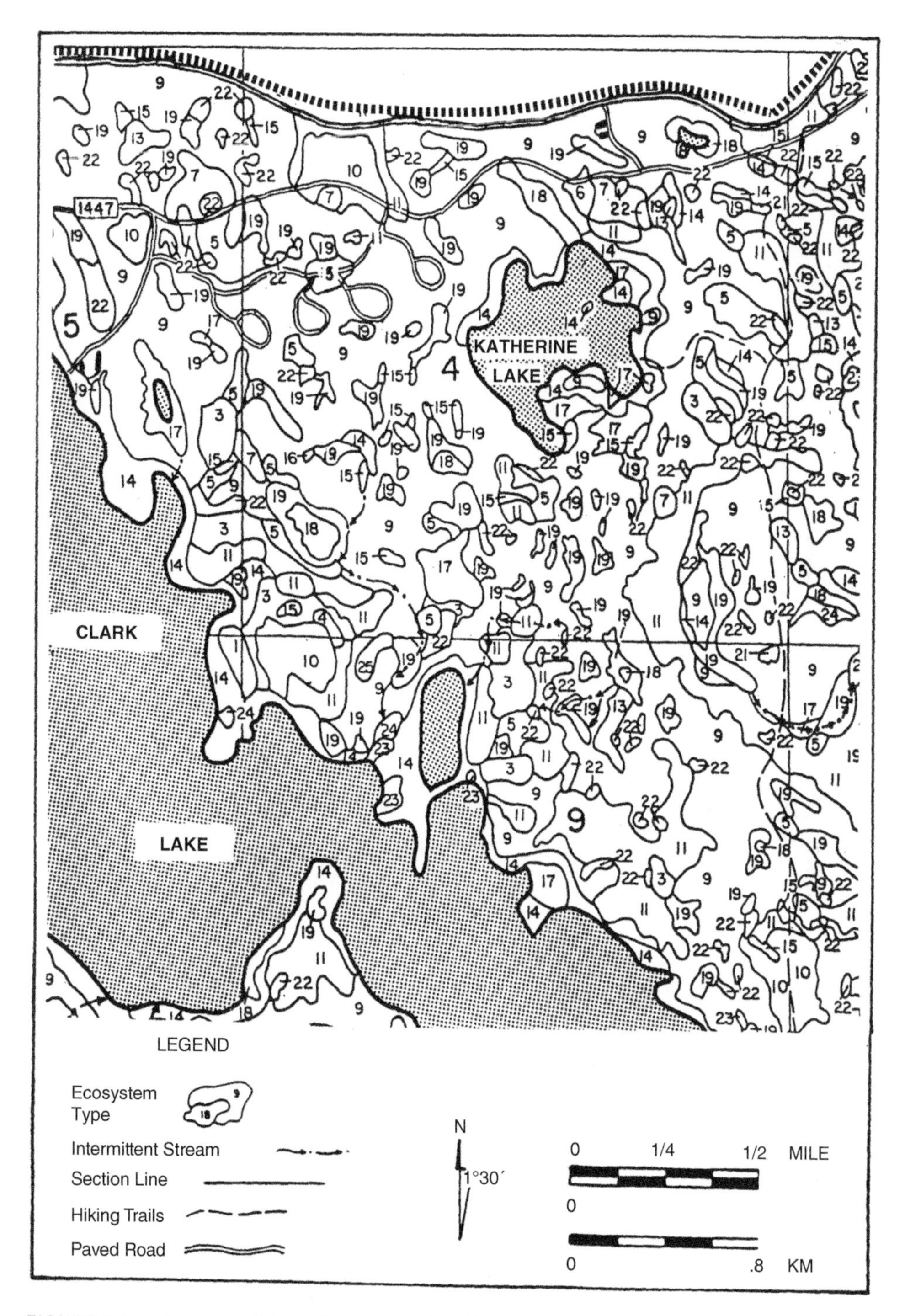

FIGURE 2.11 Example of the patterns of local landscape ecosystems that occur within the local level. The map illustrates local landscape ecosystem types surrounding part of Clark Lake in the north central part of the Sylvania Wilderness Area, Ottawa National Forest, Upper Michigan. Area shown mapped: 13.9 km^2 (5.4 mi^2); 1390 ha (3435 acres).

repeatedly occurs along the fire-prone margins of lakes and large swamps. In the unique shoreline landscape position, offshore winds continuously dry the site and coniferous needles of the forest floor, which ignite readily when lightning strikes the tall red and white pines. Geological processes that formed the Sylvania Wilderness have therefore created a remarkable diversity of ecosystem sizes, shapes, and spatial patterns of wetland, upland, and lake-shore types. Managing areas with local ecosystems as small as 0.1 ha, especially when they occur in an enormously complex pattern over thousands of square kilometers, is clearly a major challenge for wildlife ecologists, foresters, and land managers. Insights on how to deal with such complexity are considered in Chapters 11 and 22.

Multiscale, multifactor landscape ecosystem classification has proven to be incredibly useful as a framework for many different aspects of ecological research, as reflected by its adoption across the United States, Canada, and in many parts of the world. Since the last revision of this textbook, for example, landscape ecosystem classifications have been completed for portions of North Carolina (McNab et al. 1999), Alabama (Carter et al. 1999), the Gulf Coastal Plain in Georgia (Goebel et al. 2001), northern Michigan (Kashian et al. 2003), South Carolina (Abella et al. 2003), Indiana (Dolan and Parker 2005), Vermont (Ferree and Thompson 2008), the southern Appalachian Mountains (Stottlemyer et al. 2009), and Nova Scotia (Cameron and Williams 2011), among many other areas. In the late 1990s, The Nature Conservancy, a global environmental conservation organization, developed a continental-scale ecoregional system based on a regional landscape ecosystem framework and has used it to prioritize areas for conservation in the Western Hemisphere (The Nature Conservancy 1996). Landscape ecosystem classifications resembling the system we described in the earlier text have also been completed in Italy (Blasi et al. 2000), the Czech Republic (Kusbach and Mikeska 2003), and Iran (Azizi Jalilian et al. 2020).

SUGGESTED READINGS

Albert, D. A. (1995). Regional landscape ecosystems of Michigan, Minnesota, and Wisconsin: a working map and classification. *USDA For. Serv. Gen. Tech. Report NC-178*. North Central For. Exp. Sta., St. Paul, MN. 250 pp. + map.

Bailey, R.G. (2009). *Ecosystem Geography*, 2e. New York: Springer-Verlag 251 pp. + 2 maps.

Bailey, R.G. and Hogg, H.C. (1986). A world ecoregions map for resource reporting. *Environ. Conserv.* 13: 195–202.

Denton, S.R. and Barnes, B.V. (1988). An ecological climatic classification of Michigan: a quantitative approach. *For. Sci.* 34: 119–138.

Omernik, J.M. and Griffith, G.E. (2014). Ecoregions of the conterminous United States: evolution of a hierarchical spatial framework. *Environ. Manag.* 54: 1249–1266.

Rowe, J.S. (1992). The ecosystem approach to forestland management. *For. Chron.* 68: 222–224.

Rowe, J.S. and Sheard, J.W. (1981). Ecological land classification: a survey approach. *Environ. Manag.* 5: 451–464.

Spies, T.A. and Barnes, B.V. (1985). A multi-factor ecological classification of the northern hardwood and conifer ecosystems of Sylvania Recreation Area, Upper Peninsula of Michigan. *Can. J. For. Res.* 15: 949–960.

The Forest Tree

PART 2

In the three chapters of this part, our focus is on forest organisms and their populations in relation to the forest ecosystems of which they are a part. In Chapter 3, the genetic differentiation of populations is seen to be closely related to the specific climatic and physiographic conditions in which plants grow. In addition, we examine the considerable within-plant variation that is important in the survival and persistence of woody plant populations. The processes and factors affecting variation within and among populations and at the species level, such as hybridization and polyploidy, are also considered.

In Chapter 4, the life history of forest organisms is considered with emphasis on the regeneration of woody plants. The important processes of reproduction, dispersal, germination, and establishment are discussed. In Chapter 5, the focus is on how trees grow in relation to site conditions. We examine the structural parts of woody plants, the crown and its architecture, and the structure and growth of stems and roots. We also emphasize the effects of water stress on the growth of plants.

Forest Ecology, Fifth Edition. Daniel M. Kashian, Donald R. Zak, Burton V. Barnes, and Stephen H. Spurr.

Forest Tree Variation

CHAPTER 3

Genetic adaptations of woody plants inseparably link these organisms to the environment. Genetics are the governing biochemical control mechanism that can be passed on from generation to generation; without such a mechanism, a forest tree or any other organism could not perpetuate itself. Both genetic and environmental factors together control the form and development of an organism, and it is therefore a futile exercise to argue which controls the phenotype. Environmental factors are the most obvious because they are generally visible and readily accessible, and it is natural that ecologists were long preoccupied with their study. Genetic factors have since become more fully appreciated and assessed in forest trees.

Organisms are able to live and reproduce in a given range of environments because of their **adaptedness** (Dobzhansky 1968), an ability critical to forest ecology. In the present chapter, we emphasize variation in organisms: sources of variation; the kinds and extent of variation within and among individuals, populations, and species; and how physical and biotic factors elicit adaptive changes in tree populations. Detailed treatments are available for the fields of population genetics (Gillespie 2004; Hartl and Clark 2006; Hamilton 2009; Charlesworth and Charlesworth 2010; Hedrick 2011; Allendorf et al. 2012; Hartl 2020), geographic variation in trees (Morgenstern 1996), and forest genetics and tree improvement (White et al. 2007).

COMPONENTS OF PHENOTYPIC VARIATION

Ecologists work with organisms both as individuals and as aggregations of individuals, such as populations and communities. The genetic constitution of an individual is termed its **genotype**. We do not physically witness a plant's genotype because from the moment of fertilization the genotype is influenced by the environment—the plant's internal environment of cells and biochemical reactions and the external environment of light, temperature, and moisture. We can only see the result—the **phenotype**—the observable properties of an organism produced by the genotype together with the environment. This relationship can be expressed for the entire organism, or for individual characters, by the simple formula $\mathbf{P} = \mathbf{G} + \mathbf{E} + \mathbf{GE}$. The phenotype (**P**) is the sum total of the effects of these components: the genetic information coded in the chromosomes (**G**), the environment (**E**), and the interaction of the genotype and the environment (**GE**). The genotype–environment interaction is often disregarded, but in certain cases it is of major importance. The genotype itself may be "observed" using biological assays, whereby the data obtained can be compared against either a second individual's gene sequence or a database of sequences. Genes control physiological functions through complex pathways, and many of these influence the morphology of the plant. The basic interrelationship of the genotype, environment, and plant processes of the phenotype is illustrated in Figure 3.1.

Ecologists often examine the degree to which a phenotypic character is controlled by the genotype or by the environment. For example, consider two adjacent trees of the same species, where one is a dominant tree exhibiting excellent growth in a forest plantation and the second is dying. The dominant clearly has a superior phenotype, but to what degree is its genotype responsible, and to

Forest Ecology, Fifth Edition. Daniel M. Kashian, Donald R. Zak, Burton V. Barnes, and Stephen H. Spurr.

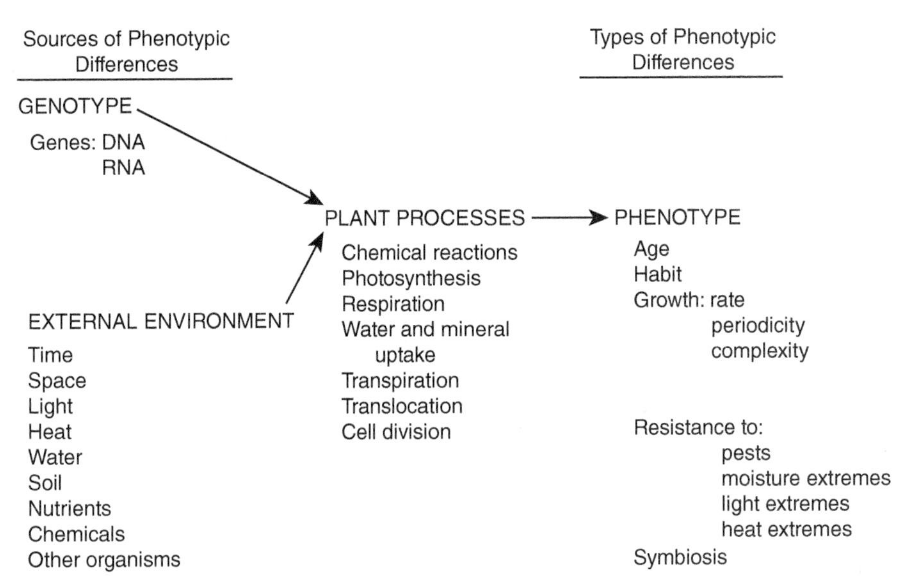

FIGURE 3.1 Relationship of an individual's phenotype to its genotype and environment. Differences in plant phenotypes have their origin in the genetic constitution (genotype) of an individual and the effects of environmental factors on the genotype. Phenotypic differences of the individual become apparent as physiological processes occur in the internal environment of plant cells, tissues, and organs.
Source: Courtesy of J. W. Hanover.

what degree does a favorable microenvironment or the random factor of better handling in the nursery account for the difference? Do environment and genotype contribute equally, or is one more important than the other, and by how much? Despite its poor phenotype, the genotype of the dying tree may or may not be superior to that of the dominant. Both the genetic and environmental components are always involved, and no two phenotypes are exactly alike, but the relative degrees of genetic and nongenetic influences on a given character are important factors.

To determine the relative effects of genotype and environment, forest geneticists compute genetic ($\mathbf{V_g}$) and environmental ($\mathbf{V_e}$) variance in the process of quantifying the strength of genetic control for a given character. The total phenotypic variance ($\mathbf{V_p}$) is equal to the sum of these two values ($V_p = V_g + V_e$). The strength of genetic control of a character, termed **heritability** (i.e., ability of a character to be passed on to successive generations), is then determined using the ratio of genetic variance to the total phenotypic variance (heritability $= V_g/V_p \times 100$). A high ratio of the genetic to total variance (for example, 80) indicates a strong genetic control for the trait. In trees, strong genetic control (high heritability) has been shown for stem straightness, the specific gravity of stemwood, susceptibility to leaf rusts, and date of spring bud burst. Traits highly influenced by the environment, such as height (strongly influenced by soil water and fertility) and diameter (strongly influenced by stand density), are predictably under weak genetic control, and, accordingly, have low heritabilities.

PLASTICITY OF THE PHENOTYPE

A given genotype may exhibit one phenotype in one environment and exhibit a different phenotype in another environment. This variability, termed **phenotypic plasticity** (Bradshaw 1965; Schlichting 1986), is the degree to which a character of a given genotype can be modified (a nongenetic change) by environmental conditions. Bradshaw (1965) cites the size of vegetative parts, number of shoots, leaves, and flowers, and elongation rate of stems to be plastic traits; nonplastic characters include leaf shape, leaf margin serration, and floral characteristics. Though determined for herbaceous species, there is good reason to believe these findings are consistent for woody plants. In general, characters such as stem elongation are formed over long time periods of meristematic activity, are more subject to environmental influences, and are therefore more plastic. By contrast, characters such as reproductive structures are formed rapidly, or traits such as leaf shape are impressed at an early stage of shoot development and are therefore less plastic.

Because trees are sessile and long-lived, plasticity has substantial adaptive value. An example of plasticity with adaptive value is the rooting habit of many tree species, particularly Norway spruce, white spruce, and balsam fir, that may develop either shallow or deep roots depending on the soil environment (e.g., shallow in a poorly drained swamp versus deep in a sandy loam upland soil). High plasticity for certain establishment and growth characters allows individuals of a species to regenerate and maintain themselves on a variety of sites, as well as to endure decades or centuries of fluctuating climate as adults.

The differences in phenotypes can be summarized using three hypothetical situations involving individuals of a given species:

Situation A	Situation B	Situation C
$P1 = G1 + E1$	$P1 = G1 + E1$	$P1 = G1 + E1$
$P2 = G2 + E2$	$P2 = G1 + E2$	$P2 = G2 + E1$
$P3 = G3 + E3$	$P3 = G1 + E3$	$P3 = G3 + E1$

The phenotypes in A illustrate the typical situation in the field, where all phenotypes have different genotypes, but the environments are different enough to produce differences in the phenotypes. Situations B and C illustrate experimental situations where we can either hold constant the genotype (in B) or test different genotypes in a given environment (in C).

Situation B illustrates plasticity, because different phenotypes of a single genotype (G1) result from environmental differences; the environment has modified growth and development. In nature, the degree of plasticity of a character is difficult or impossible to measure precisely because each individual typically has a different genotype (as in A), and the extent of environmental modification can therefore only be *inferred*. For example, individual genotypes of an even-aged forest in rolling terrain may occur from a dry ridgetop to a moist, fertile valley, and an increase in tree height is observable as we progress from ridge top into the valley. If there lack major differences among genotypes along the gradient, environment is likely to be the major factor controlling the observed phenotypic differences in height. However, precise determination of plasticity for representative genotypes would require experiments based on situation C.

In situation C, where the environment is constant for all individuals, phenotypic differences result from differences among genotypes, and the amount of genetic variation can be estimated directly from the phenotypes. In practice, the environment cannot be held constant, but the ideal may be approached using growth chambers or relatively uniform field test plots and a replicated experimental design. This "common garden" approach is widely applied in determining genetic differences among selected individuals or populations (e.g., Pregitzer et al. 2013; Potts and Hunter 2021).

SOURCES OF VARIATION

As we have seen, phenotypes vary partly due to the genotype and partly due to the environment. The major sources of genetic variation are **mutation** and gene **recombination**. Gene mutations increase the number of **alleles** (the different forms of a gene) available for recombination at each locus (position on a chromosome where a gene is located), thereby adding to the pool of genetic variability.

Continuous variation is typical for most characters of plants. This is because most characters are affected by many genes (termed **polygenic**) that simultaneously segregate and interact. Continuous variation also arises from nongenetic causes. Relatively few traits are controlled by a single gene with major effects, such as chlorophyll deficiency (albinism) in seedlings of species of the pine family (Franklin 1970). Single-gene mutations may also result in differences in leaf morphology of closely related taxa. For example, the rare Virginia round-leaf birch of the southern Appalachian Mountains is similar to sweet birch except for its round leaves (Sharik et al. 1990). Similarly, a single-gene mutation in a white ash population may have produced the single-leaf ash, an ash with simple rather than compound adult leaves (Wagner et al. 1988). In addition to mutation, additional sources of variation include:

Recombination, which spreads mutations and extracts maximum variability from them. Recombination is the major source of genetic variation of individuals that reproduce sexually. Natural selection then acts upon the raw material of variation distributed by recombination.

Gene flow, which is the exchange of genes between different populations. Migrants, in the form of pollen and seed, bring new genetic material to a population from other populations. The process is called **hybridization** when the two populations involved are substantially different (such as species).

Nongenetic variation, whose major sources are (i) the internal environment of the plant, and (ii) the external or physical environment (climate, landform, soil, biotic factors). Factors of the physical environment that modify plants and elicit genetic adaptations are considered in Chapters 6–11.

Variation within an individual that is not directly related to factors of the external environment is far less appreciated. All cells of a tree have the same genetic information, but the plant's internal environment may affect its expression of genes and hence the traits we observe and measure. Striking physiological changes occur as a seedling develops into an adult tree. A series of developmental phases is recognized, particularly the differences between the juvenile and adult phases, apparently driven by changes that take place in apical meristems as they age. The most universal feature is the inability of trees in the juvenile stage to flower (Chapter 4). Classic studies by Schaffalitzky de Muckadell (1959, 1962) examined many other characters that differed between juvenile and adult phases. In European beech and oak, as well as American beech and many oaks in North America, the brown and withered leaves are retained over winter by trees in the juvenile phase, but the leaves are deciduous in the adult phase. Entire portions of the lower trunk and branches, even of very old trees, remain in the juvenile phase and exhibit the juvenile trait of leaf retention. Reciprocal grafting experiments of juvenile and adult branches in European beech have shown that the juvenile phase leafs out later than the adult phase. This juvenile trait may be of adaptive value given that late spring frosts can pose a serious problem for regenerating beech.

THE EVOLUTIONARY SEQUENCE

The basic unit of evolutionary biology in sexual populations is the local interbreeding population,[1] a group of potentially interbreeding genotypes that constitute the gene pool for that population. The heritable variation of these genotypes allows natural selection to bring about evolutionary changes. **Natural selection** is the differential and nonrandom reproduction of genotypes. Ultimately it is this differential reproduction of genotypes, rather than just differential mortality of

[1] A population is any group of individuals considered together because of a particular spatial, temporal, or other relationship (Heslop-Harrison 1967).

individuals, that brings about evolutionary change. Natural selection is therefore the mechanism for **evolution**, the cumulative change in genetic makeup of populations over successive generations—in simplest form, a change in gene frequencies.

Of the thousands or millions of zygotes of a species on a given site that might develop to maturity and contribute offspring to the next generation, only a few survive. Sugar maple provides an excellent example of the raw materials available to selection and its severity. Curtis (1959) reported that 55.7% of 6 678 400 potentially viable sugar maple seeds germinated, leaving 3 673 100 seedlings per hectare. Only 198 740 seedlings remained by late summer of the same year, and only 35 380 seedlings remained alive 2 years later—less than 3% of those that germinated. He estimated that the canopy gap resulting from the death of a mature sugar maple tree (about 6–7 m in diameter) would initially support about 15 000 seedlings, which would be reduced to about 150 over the first 3 years, and eventually only 1 or 2 trees would occupy the opening. Selection acts in every phase of the plant's life, but it is most effective on young seedlings eliminating in some species all that are not well adapted to their immediate environment.

SEXUAL AND ASEXUAL SYSTEMS

The sexual breeding system characterizes most woody species and is dominated by cross-pollination. Sexual systems allow evolution to proceed much faster because new genotypes are constantly generated. Most woody species also have some form of asexual reproduction or **apomixis**—any means of reproduction, including vegetative propagation, which does not involve fertilization. In angiosperms, vegetative reproduction includes sprouting from the basal part of the stem (oaks, hickories, ashes) and roots (aspens, sweetgum) following death or removal of the aboveground stem. Conifers rarely form sprouts (except for redwood and juvenile individuals of loblolly pine), but the production of adventitious roots from branches (termed **layering)** in larches, spruces, and firs is an important mechanism of asexual reproduction in swamps or mesic uplands. Amazingly, some species such as hawthorns may form seeds asexually. Whatever the mechanism, asexual reproduction promotes uniformity and gives the plant immediate fitness in the prevailing environment.

Aspens of North America and Eurasia sucker from roots to form natural colonies termed clones—an excellent example of asexual reproduction. A **clone** is the aggregate of stems produced asexually from one sexually produced seedling. The clone is the most common growth habit of aspens across the world (Barnes 1967). The tendency for aspens to reproduce asexually may be a major factor in their ability to compete successfully in conifer-dominated landscapes, where fire is common and stimulates sprouting. Aspens occur in large clones in parts of western North America, some over 40 ha in size (Kemperman and Barnes 1976), and may live almost indefinitely by recurrent suckering so long as fire is not excluded from the landscape.

GENETIC DIVERSITY OF WOODY SPECIES

Genetic diversity refers to the variation or differences between organisms at the molecular level. It reflects the presence of different alleles in a gene pool, and hence different genotypes within populations. As differing forms of genes are alleles, a gene with identical alleles across all individuals in a population is considered **monomorphic**. If a gene has more than one allele in a population, it is considered to be **polymorphic**. Increasing polymorphisms within a population increases genetic diversity, which allows populations to better adapt to environmental variability. In a heterogeneous or unpredictable environment, small genetic differences among individuals increase the likelihood that some individuals will exhibit genetically determined traits that allow them to survive and reproduce better than others, and thus undergo natural selection.

As described in the earlier text, the interactions of many genes with each other and the environment typically make it difficult to infer amounts of genetic diversity using phenotype alone.

Studies of genetic variation are typically conducted at the molecular level, and have become more predictive as modern advances in molecular biotechnology techniques provide an assortment of very powerful methodologies to detect differences in DNA (Hartl 2020). Genetic variation was first widely detected at the molecular level, using variations in allelic variants (**allozymes**) in proteins at a single-gene locus. Determining gene frequencies for the allelic variants for many loci and for many populations and individuals allowed the genetic diversity of species and their populations to be estimated. Later, researchers were able to directly examine variation in DNA sequences with a wide array of polymerase chain reaction (PCR)-based marker systems, allowing for more precise detection of genetic variation and estimation of levels of diversity. A detailed treatment of the measurement of genetic diversity is presented by Hartl (2020) and Cutter (2019).

Trees generally have more genetic diversity than herbaceous plants (Hamrick and Godt 1989; Hamrick 2004; Nybom 2004). A series of early studies using allozymes (Hamrick and Godt 1989; Hamrick et al. 1992) found that the genetic diversity of long-lived woody plants is 15% greater than that of annual plants, 42% higher than herbaceous perennials, and 53% greater than short-lived woody species. Geographic range is the best predictor of allozyme variation in long-lived woody species. Genetic diversity is highest in widespread and regionally distributed species (Scots pine, white spruce) and lowest in endemic woody species (red pine, Torrey pine, and balsam poplar).

Despite their high levels of genetic diversity within populations, trees have relatively little diversity among populations; about 91% of the total genetic diversity of trees is within their populations (Hamrick 2004). This contrast is likely related to trees' tall stature and low population densities (creating the potential for longer dispersal distances and greater potential for gene flow; Hamrick and Godt 1996), predominance of outcrossing (which reduces the probability of allele loss), more chromosomes (which affects chromosomal recombination), more diversifying selection, and delayed maturity (which allows colonization of available space by a diverse group of offspring; Austerlitz et al. 2000; Petit and Hampe 2006). These and other factors affecting gene flow in trees have great implications for their genetic diversity. For example, a study of global wind patterns and tree genetic diversity found that tree populations are more genetically similar when linked by stronger winds, and downwind populations have higher genetic diversity (Kling and Ackerly 2021).

There is evidence that trees are able to quickly adapt to new environmental conditions at local scales (Petit et al. 2004) while simultaneously having low nucleotide substitution rates and low speciation rates (Petit and Hampe 2006). The large effective population sizes of temperate trees, which reduces the likelihood that rare alleles are lost to genetic drift, coupled with the occurrence of trees in heterogeneous microhabitats, are one explanation for their high levels of adaptive genetic variation (Sork et al. 2013). Understanding the life history and ecological traits of plant species and populations with high genetic diversity could be a powerful tool for prioritizing conservation efforts (Chung et al. 2020).

In short, the genetic diversity present in trees is rather amazing relative to most other higher plants. In the sections that follow, we again emphasize the inseparability of organism and environment in describing some of the patterns of this amazing diversity and the processes that account for it.

GENECOLOGY

Genecology is the study of variation in plant species from an ecological viewpoint. Specifically, genecology is concerned with the adaptive properties of any sexual population in relation to the environment—species, subspecies, race, or local interbreeding population (Langlet 1971). Early genecological studies (Turreson 1923) demonstrated that phenotypic variation among populations that was correlated to ecological conditions often had a genetic basis and was not simply the result of environmentally induced modification of individuals. Genecology has major practical implications for tree introduction attempts around the world, which have typically employed a trial-and-error method too problematical and costly as a general practice. Understanding which environmental and biotic factors elicit a genetic response, how finely populations are adapted to these factors, and the patterns of adaptation along major environmental gradients would allow predictions of how a given species population would perform in a new environment.

Comparative cultivation of seedling populations of forest trees, originating from environmentally different sites, was first explored as early as the mid-1700s and refined throughout the nineteenth century. Cieslar in Austria (1895, 1899) and Engler in Switzerland (1905, 1908) were instrumental in showing that the genetic makeup of forest trees in the Alps left them adapted to the climatic conditions of their respective environments (Langlet 1971). Cieslar published evidence in 1895 of a continuous gradient of juvenile height growth for Norway spruce. This work demonstrated genetic adaptation to growing-season conditions grading from low to high altitude and confirmed observations of the previous century that seedlings of lowland areas proved worthless on mountain sites.

Turesson transplanted whole individuals from markedly different habitats into standard conditions within a common garden (rather than cultivating seedlings from seed as Cieslar and Engler did). The phenotypes Turesson observed in nature usually differed in growth habit (i.e., procumbent or erect) and in various morphological characters. If maintained in a common garden, these differences indicated genetic differences among the populations studied. Natural populations were presumably exposed to the factors of their respective environments for generations. As such, forces of natural selection guided the genetic differentiation of each population to adjust it to the daily, seasonal, yearly, and even longer-term climatic and soil water fluctuations of its respective environment.

Because whole plants are preconditioned to the environment in which they grow (Rowe 1964) and are difficult to transplant, forest scientists classically collected seeds from the desired populations, termed **provenances**, raised the seedlings in a common garden, and studied the differences among the provenances. Such experiments are termed **provenance** or **seed-source** tests (e.g., George et al. 2017; Kashian and Barnes 2021). This type of testing reveals (i) significant genetic differences among populations for specific characters, and (ii) the degree of genetic differentiation among provenances under the environmental conditions of the common garden. Provenance testing does not directly indicate which mechanisms cause the differences, although these may be inferred.

PATTERNS OF GENECOLOGICAL DIFFERENTIATION

Genetically based ecological differentiation or divergence of populations is a recurrent feature of plants, but whether it is discontinuous or whether it is continuous in nature has been met with some controversy. A **cline** is a gradation in measurable characters which might be continuous or discontinuous, stepped or smooth, or sloping in various ways (Huxley 1938, 1939). The term cline does not mean or necessarily imply a genetically based gradation in a character and could refer to any gradation of phenotypic characters observed along a natural gradient.

Genecological differentiation occurs as a multidimensional response of individuals of a population to their environment and is unique for each population and species. In the following text, we present the following generalizations of differentiation in forest species.

1. The pattern of differentiation is determined largely by the total natural range of a species, the distribution pattern (continuous, discontinuous, mosaic) of a species within this range, and the way in which the conditioning environmental factors vary. If a species that exhibits continuous distribution over a wide range, particularly in latitude or elevation, is subjected to more or less continuously varying climatic factors, genetic variation tends to be

continuous. By contrast, a discontinuous pattern may result if the species' distribution or the conditioning factors are discontinuous. The pattern of variation is often represented as a series of contour lines whose spacing reflects the rate of change in the conditioning factors.

2. The spatial scale at which a given genecological study is conceived and conducted affects its results (Heslop-Harrison 1964). Thus, broadscale studies are less likely to detect local discontinuities (Chapter 22). For example, a wide-ranging investigation of a species along a north–south gradient may expose a clinal variation pattern that may mask other clines associated with elevation at a given latitude, or local discontinuities that may arise from marked changes in microclimate or soil drainage.
3. Incomplete sampling and certain methods of data analysis may suggest a discontinuity even in an extremely continuous cline (Langlet 1959).
4. A discontinuous pattern of genetically based variation is typically the exception rather than the rule. This is because the major factors of climate that elicit genetic variation are continuous. Clinal patterns have a fundamental basis in the genetic system of many forest trees favoring high rates of recombination and outbreeding (cross-pollination) and associated features of (i) long-lived individuals, (ii) high site stability, (iii) high and selective seedling mortality, and (iv) physiological mechanisms in the adult to tolerate fluctuating environmental conditions. By contrast, discontinuous variation is favored by inbreeding, low rates of gene recombination, short life span, and fluctuating sites and communities, and is therefore more typical in herbaceous than in tree species. These and other evolutionary aspects of genecological differentiation are discussed by Petit and Hampe (2006), Hamrick (2004, 1989), Hamrick et al. (1992), and Linhart (1989).

Genecological Categories The continuum of genetic differentiation occurring among species populations along ecological or geographic gradients may be subdivided into classes. Species populations that differ significantly in one or more morphological and physiological characters in common garden tests are generically called **races**. Races differ in the frequencies of one or more gene alleles (Dobzhansky 1951) and may be further described as geographic or local races, and climatic, edaphic (soil), or photoperiodic races. Races are basic elements of evolution, but if genetic differences are especially great, the populations may be recognized as formal taxonomic categories or taxa such as varieties, subspecies, or species. The term **ecotype**, introduced by Tureson (1922a,b), is also sometimes used to designate genetically different population units. He defined ecotype as arising from the genotypic response of a population to a "definite habitat" or "particular habitat" (i.e., a local ecological race). However, the term has been used in many different contexts, each having a different genecological significance, and fails to provide a useful concept for a specialized local race. We therefore use "race" to distinguish units of genetic differentiation below the taxonomic rank of variety.

FACTORS ELICITING GENECOLOGICAL DIFFERENTIATION

Genetic differences in growth and other characters are often obvious when populations are grown at latitudes or elevations substantially different from those of their native habitats. Factors affecting the length and nature of the growing season in the native habitat, such as mean and extreme temperatures, occurrence of early and late frosts, thermoperiod, photoperiod, and amount and periodicity of rainfall, are major selective forces affecting survival, growth, and reproduction that elicit genetic differences in plant populations.

Plant associates and selective pressures of insects, mammals, and birds also influence various characters (see Chapter 12). Coevolutionary systems are evident for animals and reproductive traits of various woody species. For example, Janzen (1969) listed 31 traits of woody legumes in Central America that eliminate or lower the destruction of seeds by bruchid beetles. Most defense mechanisms against these predators are deterrents, such as biochemical repellents (alkaloids and free amino acids) or increasing the number of seeds to the point of predator satiation, likely accompanied by a decrease in seed size.

Growth Cessation A plant's efficient use of its growing season is closely related to the nature and patterns of genetic differentiation. Temperate plants will be damaged or killed by early autumn frosts if they grow too late in the season, but they may be eventually overtopped or suppressed by competitors if they annually cease growth too soon. Species are able to anticipate seasonal fluctuations by responding genetically to the more annually consistent factors of their native environment (such as day length and heat sum) than to variable factors such as frost occurrence. Response to a photoperiodic signal of shortening days late in the growing season in temperate regions sets in motion a gradual and complex process of acclimation to dormancy (Chapter 7). This response of plants to the timing of light and darkness (usually expressed as day length) is called **photoperiodism**; it is a biological clock enabling plants to adjust their metabolism to certain seasonal fluctuations. Unlike other environmental factors, day length changes everywhere (except at the equator) in a consistent annual cycle, and plants are remarkably precise in monitoring these changes.

Photoperiod strongly influences when woody plants enter into dormancy, particularly in species with northern ranges. These species are genetically adapted to a photoperiod that enables them to become dormant before factors of their prevailing environment, such as freezing temperatures, become limiting. Early autumn frosts and cold winters are factors that significantly affect survival and thus are highly selective for plants in northern climates or at high elevations. Hence, photoperiod may be highly developed as a reliable mechanism in triggering the dormancy sequence. For example, there exists a very strong relationship between growth cessation and latitude for balsam poplar that reflects a close adaptation to day length (Figure 3.2). Balsam poplar occurs in a broad range within the boreal forest of North America and especially Canada, extending over 26° of latitude from southern Ontario to northern Alaska. The relationship is very strong because latitude, and its associated day-length regime, characterizes very well the differences in climate that occur across this broad range. The relationship is remarkable because species that range over mountainous terrain and encounter widely varying temperature and moisture conditions within any given latitude are far less likely to exhibit a close relationship between growth cessation and latitude.

Most genecological and provenance tests feature populations grown in day length regimes different from their native habitats. For example, high-latitude provenances of black cottonwood (native to western North America) ceased height growth in June when planted at a low-latitude site near Boston, Massachusetts (Figure 3.3; Pauley and Perry 1954; Pauley 1958). Southern provenances moved north to the test site continued height growth until September and October; some individuals continued to grow until their terminal shoots were killed by the first severe frost. Generally, movement of a southern population northward into longer days prolongs the active growth period and results in larger plant size of the southern populations compared to the northern populations native to that site. However, if such a move is too far north, plants become susceptible to early frosts, which may lead to injury, decreased growth, or death. Conversely, movement of northern populations southward into shorter summer days reduces the active growth period in comparison to native plants at or south of the test site. In the black cottonwood example, individuals of high-latitude provenances grew only about 15–20 cm, whereas those from southern localities grew about 2 m (Pauley 1958). When trees from the high latitude at the test site were exposed to longer days simulated by artificial light, they grew over 1.3 m (Pauley and Perry 1954), demonstrating that growth is strongly regulated by day length.

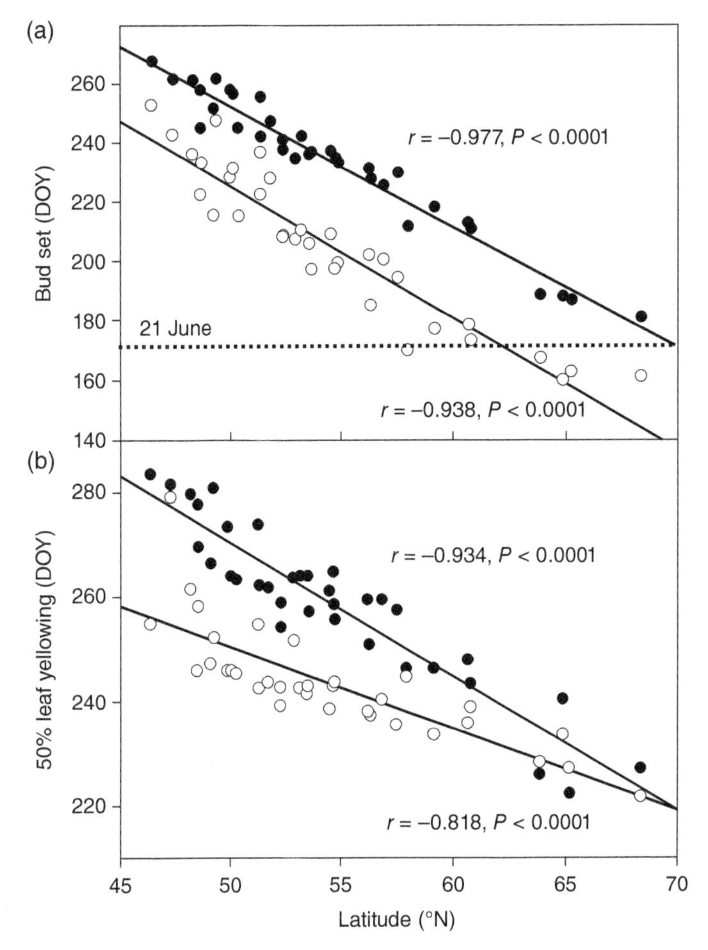

FIGURE 3.2 Relationship between latitude of origin and date of growth cessation in place of origin for 35 balsam poplar provenances from Canada. Indicators of growth cessation include: (a) date of final bud set, and (b) date of 50% canopy color change. Dark circles indicate experiment performed in Vancouver, British Columbia, and open circles in Indian Head, Saskatchewan. *Source:* After Soolanayakanahally et al. (2013) / John Wiley & Sons.

Although a genetically based clinal response was shown in relation to latitude for black cottonwood, the response is not simple and direct (compare with the balsam poplar example shown in Figure 3.2) as evidenced by substantial variation among provenances from 44° to 48° (Figure 3.3). These provenances included a variety of sources sampled from the Pacific Coast to western Montana and over an elevational range from sea level to 1525 m. The difference in latitude is far less than the variation in the length of growing season among these sources due to elevation, aspect, and microsite conditions. Clinal genetic adaptation to growing season length was found within the narrow latitudinal range of 45–47° (Figure 3.4), and elevation of the source probably explains much of the variation not accounted for by latitude. Populations at low and high elevations at a given latitude have growing seasons of different lengths much like populations among latitudes. Likewise, they become accordingly adapted to different photoperiods, and in particular to a critical day length in autumn that regulates their entrance into dormancy. High-elevation populations necessarily cease growth earlier than low-elevation populations due to earlier occurrence of killing frosts, and they adapt to longer day lengths (occurring earlier in the year) than those at low elevations. Further, a high-elevation

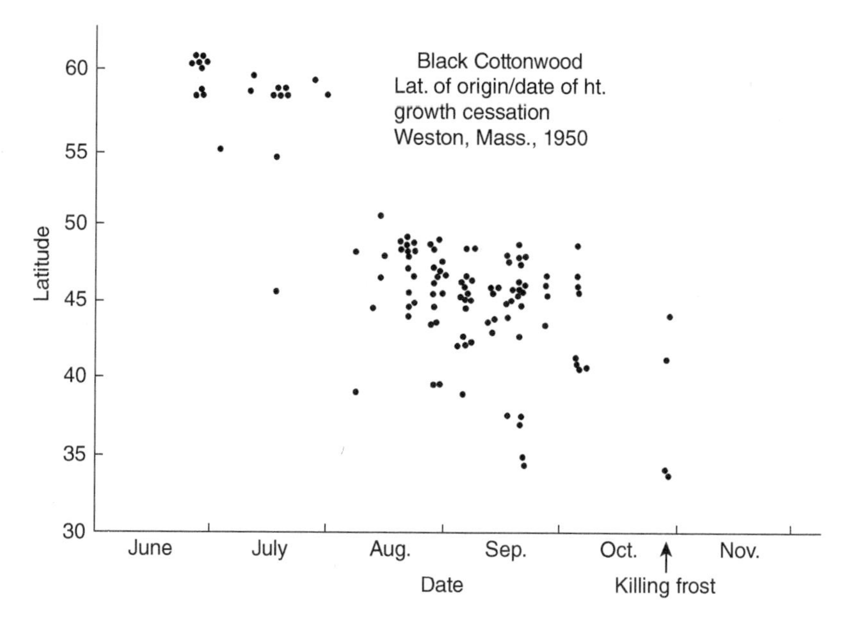

FIGURE 3.3 Relation between latitude and time of height growth cessation in black cottonwood. *Source:* After Pauley and Perry (1954) / Arnold Arboretum of Harvard University.

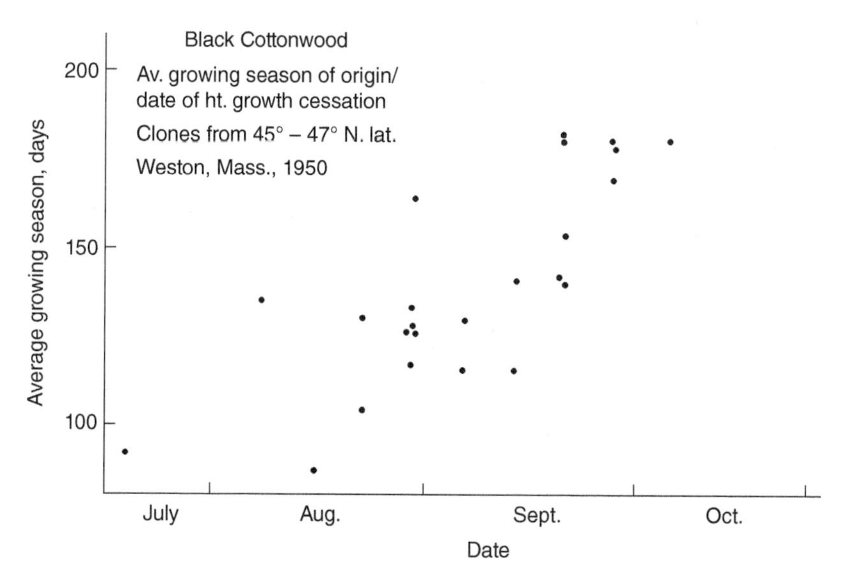

FIGURE 3.4 Relation between length of growing season and time of height-growth cessation in black cottonwood. *Source:* After Pauley and Perry (1954) / Arnold Arboretum of Harvard University.

population may have the same length of growing season as a low-elevation population that occurs several degrees of latitude farther north, because of the compensation of latitude for elevation. Such populations, if interchanged, would have a similar photoperiodic adaptation mechanism through this equivalence in growing season length, and may show only negligible differences in growth rate.

The interrelationship between elevation and latitude often goes unrecognized in genecological studies. Correlations of growth cessation or plant size and latitude of source are often confounded by elevational differences. Wiersma (1962) modified a formula developed for Swedish conditions to avoid this problem, using growing season (number of days ≥6 °C) to relate latitude to elevation. He reported that a displacement of 1° of latitude north is equivalent to a displacement of 100 m upward in altitude. Wiersma (1963) recomputed correlations of latitude of source and various characters from published papers using this adjustment and found greatly improved relationships. Similarly, Sharik and Barnes (1976) found that adjusting latitude for elevation substantially improved the correlation between latitude of origin and height growth cessation for yellow birch and black birch populations, as did Barnes (1977) for a provenance test of larch of European origin.

The functioning of this photoperiodic mechanism over gradients of latitude and elevation has been demonstrated experimentally for Norway spruce, which is wide ranging in northern and central Europe. The critical night length that stimulates bud set is about 6–7 hours in southern populations, but only 2 hours in northern ones (Figure 3.5). A similar clinal pattern of variation is also evident for elevation. Norway spruce populations at 700 m in Austria exhibited a critical night length of 7 hours, and populations at 1400 m had a critical night length of 5 hours. A comparison of populations on the west coast of Norway and those in northern Finland shows a strong growing season length at a given latitude and elevation (four versus two dark hours) (Figure 3.5). In addition to controlling growth cessation, photoperiod also affects the optimum growth of plants during the growing season itself. Southern populations of both Scots and lodgepole pines grew vigorously in a 16-hour day, whereas northern populations grew optimally in an 18-hour day (Ekberg et al. 1979).

Growth Resumption While growth cessation for temperate trees is based largely on photoperiod, the resumption of vegetative growth in the spring (**flushing** or **bud burst**) and flowering occurs following a winter chilling requirement, and thus is controlled largely by temperature (Sarvas 1969; Hunter and Lechowicz 1992; Chapter 7). Northern trees are genetically adapted to initiate flushing or flowering once a certain number of "heat units" have accumulated above a threshold temperature (Owens et al. 1977). This relationship has been reported for many conifers (Campbell and Sugano 1979) and for 26 hardwood species of eastern North America (Hunter and Lechowicz 1992).

The mean temperature or heat sum[2] of spring flowering for both northern and southern populations of a given species is the *same fraction* of the whole year's average heat sum for their respective locality. That is, northern (or high-elevation) individuals of a species are genetically adapted to flower and flush at a lower absolute heat sum than individuals of the same species at a more southerly location where the total heat sum is greater. Therefore, northern plants in nature may flower and flush relatively late in the spring if warming is delayed and slow. When their progeny are grown at warmer, lower latitudes or lower elevations, northern plants tend to flush *earlier* than the native populations and vice versa when southern populations are moved north. Notably, frost must be a significant selective force for this north–south relationship to hold; species whose northern sources do not flush earlier (Nienstaedt 1974) are species with more southerly ranges (black walnut, sweetgum, tulip tree, American sycamore) or with a coastal, ocean-moderated distribution such as Sitka spruce (Burley 1966).

EXAMPLES OF GENECOLOGICAL DIFFERENTIATION

Much of the information regarding genetic differences in physiological and morphological characters among populations comes from provenance or common garden tests established to find the most suitable provenances for planting in one or more localities. Many of the early tests were therefore not designed to answer genecological questions of how and why populations are adapted to their environments. Today, many forest scientists instead sample tissues from natural populations, genotype them with high-throughput sequencing methods (providing enormous amounts of information), and

[2] A **temperature** or **heat sum** is the product of temperature above a certain base or threshold level (such as 0° or 5 °C) and the time duration of that temperature. It may be expressed in degree-hours or degree-days.

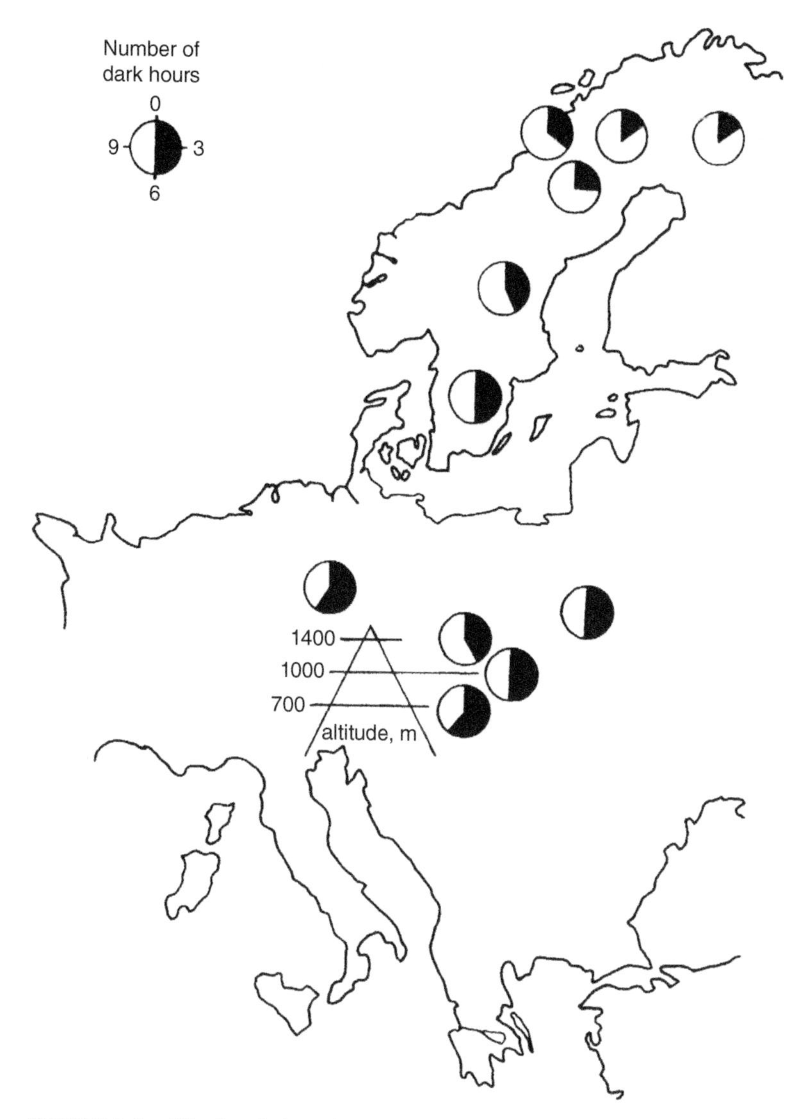

FIGURE 3.5 Clinal variation of the critical night length for bud set in Norway spruce populations of different geographic origin in Fennoscandia and central Europe. Critical night length is based on the number of dark hours bringing about bud set in 50% of the plants after pretreatment with continuous light. *Source:* Ekberg et al. (1979) / John Wiley & Sons.

then perform analyses using a genome (an organism's complete set of genetic instructions) sequence for reference (Eckert et al. 2010). Compelling arguments have been made to retain common garden tests in combination with genotyping in understanding local adaptations (de Villemereuil et al. 2016). Provenance studies, particularly those monitored for several decades (Morgenstern and Mullin 1990), have uncovered major adaptive responses, primarily along latitudinal and altitudinal gradients.

Eastern North American Species Common garden tests have revealed an enormous amount of genetic variation in deciduous angiosperms. Photoperiodic races associated with latitude have been reported for many species (Barnes 1991; Morgenstern 1996) and for a variety of characters, including survival, time of growth cessation and flushing, height and diameter growth, frost resistance, winterkill and crown dieback, tree form, and foliage color. Similarly, there is much evidence that clinal genecological differentiation is widespread in many eastern conifers, including jack pine (Mátyás and Yeatman 1992; Morgenstern 1996), loblolly pine (Wells and Wakeley 1966; Wells et al. 1991), eastern white pine (Wright 1970; Morgenstern 1996), and black and white spruce (Park and Fowler 1988; Morgenstern 1996). The rate of change in clinal variation is well illustrated by a study of the genetic variation of slash pine seedlings (Figure 3.6; Squillace 1966), which also revealed weakly defined or highly fluctuating gradients. Many of the 25 characters studied revealed reversals in the general cline, as illustrated in the variation pattern of needle length (Figure 3.6). From a minimum of 16 cm in southernmost Florida, needle length of the progenies increased to its maximum values, 19–20 cm, in south-central Florida and then progressively decreased northward.

There has been tremendous interest in the variation patterns of the resistance of loblolly pine, a wide-ranging and commercially important species of the South, to fusiform rust (Wells and Wakeley 1966; Wells et al. 1991). In general, sources west of the Mississippi River and in southeastern Louisiana are resistant to the rust, but sources east of this area, except those from the northern

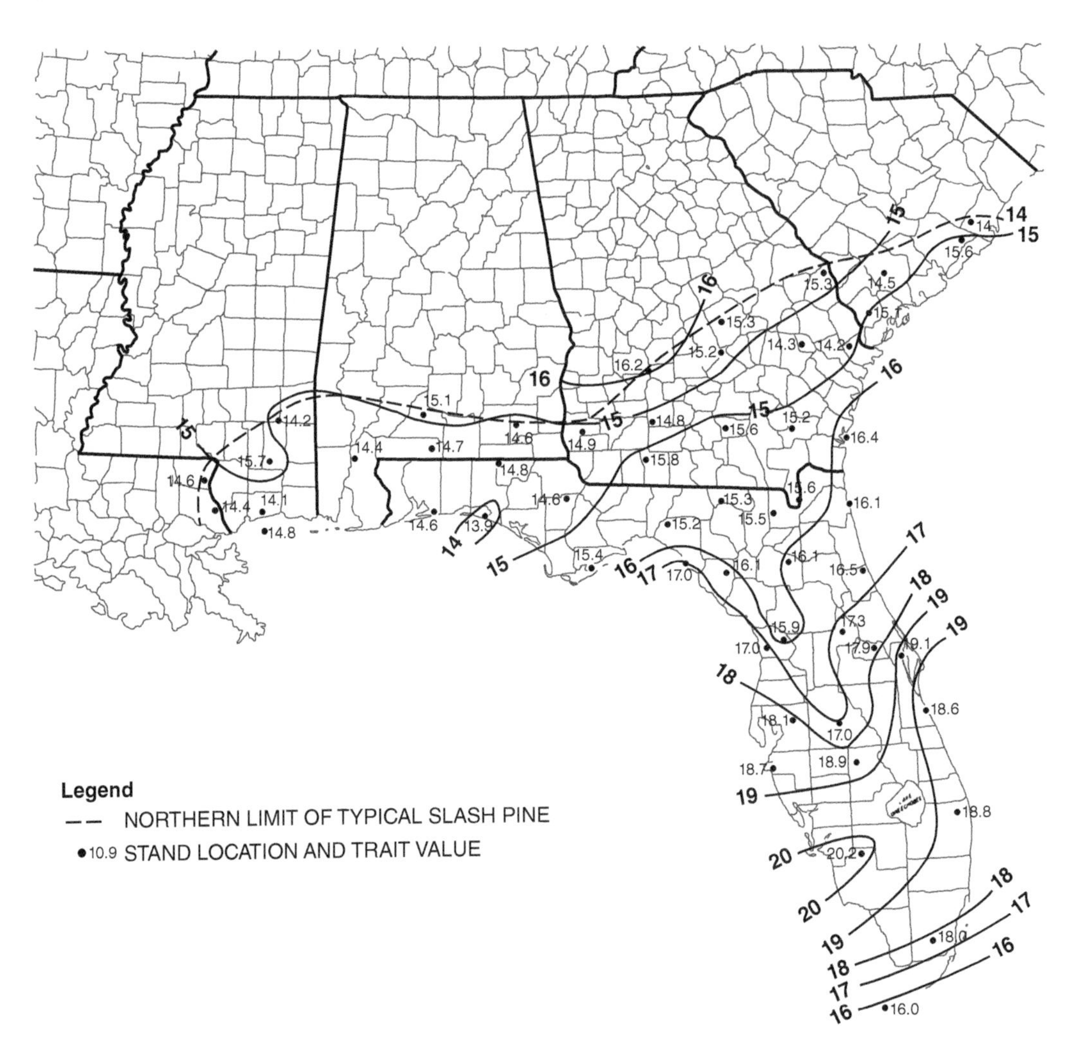

FIGURE 3.6 The pattern of variation of needle length (cm) in seedling progenies of slash pine. *Source:* Squillace (1966) / with permission of Oxford University Press.

Atlantic Coast, are susceptible. This pattern is probably explained by past climatic conditions. The wetter conditions in the West and along the northern Atlantic Coast during the last glacial maximum (18 000 years ago; Figure 3.7) were optimal for selection for fusiform rust resistance, whereas drier conditions in Florida gave rise to susceptible populations that migrated into present day Alabama, Georgia, and South Carolina. It is probable that the DeSoto Canyon (Figure 3.7), east of the Mississippi River, acted to isolate eastern and western populations as selection for resistance occurred.

Scots Pine Scots pine is the Earth's most wide-ranging pine and has been investigated in many studies at European (dating from 1745) and North American test sites. Genetic differences in height growth, foliage color, stem form, rooting habit, resistance to insects, fruitfulness, and time of bud set have been demonstrated (Langlet 1959; Troeger 1960; Wright et al. 1966, 1967). Many of these characters exhibit a clinal pattern. Parts of a cline such as races, ecotypes, or varieties are used as the basis for selecting seed-collection zones, particularly where major differences in tree

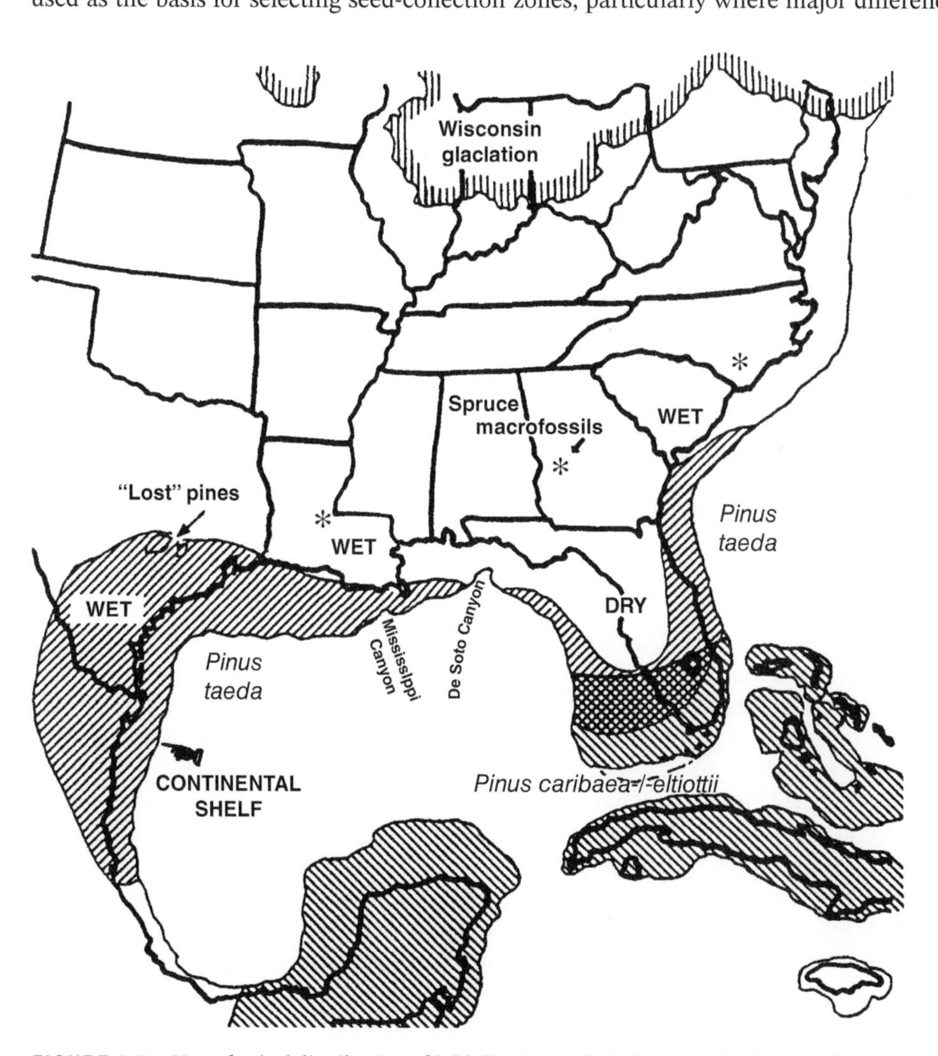

FIGURE 3.7 Hypothetical distribution of loblolly pine and slash pine at the Wisconsin glacial maximum, 18 000 years ago. Caribbean pine (*Pinus caribaea*) and slash pine (*P. elliottii*) may have been a single species at that time and are shown as such. *Source:* After Wells et al. (1991). Reprinted by permission of the Bundesforschungsanstalt für Forst- und Holzwirtschaft, Grosshansdorf, Germany.

characteristics are important. American Christmas tree growers, for example, prefer Scots pine varieties from Spain and southern Europe because they remain green in winter rather than those of northern Europe that turn yellow (Wright et al. 1966).

Wide-Ranging Western North American Conifers Western North America has provided a wealth of research examining the genetic response of species to heterogeneous environments. The heterogeneity of physiography and climate across a wide range of latitude, longitude, and elevations from Mexico to northern British Columbia and Alberta, Canada have elicited both very localized differentiation and broad patterns of variation. Douglas-fir and lodgepole pine exhibit the most variation because of the diversity of sites they occupy (Critchfield 1957, 1985; Wheeler and Critchfield 1985; Rehfeldt 1988). Both species have coastal populations yet range eastward to high elevations in the Rocky Mountains and arid interior lands. Ponderosa pine also exhibits broad, subcontinental variability, substantial variation related to elevation in various parts of its range, and significant local differentiation. We describe in the following text the different kinds of genetic responses of ponderosa pine and Douglas-fir to their heterogeneous environments.

Ponderosa pine Ponderosa pine ranges in North America from central Mexico to southern British Columbia, Canada. It is classified in the subsection Ponderosae of the genus *Pinus* and is subdivided into two varieties: var. *scopulorum* in the Rocky Mountain portion of its range (with two major races; Figure 3.8) and var. *ponderosa* in the western part of its range (with three races; Figure 3.9; Conkle and Critchfield 1988).

The major races of ponderosa pine exhibit differences in morphology and biochemistry as evidenced by range-wide provenance testing, likely related to water use and drought. The Rocky Mountain variety was named for the compact, brushlike, bushy-tuft (scopulate) appearance of its foliage (Figure 3.8) in contrast to the open, plumelike foliage of far-western populations (Figure 3.9). The western variety is morphologically distinct because of its general lack of two-needle fascicles compared to the Rocky Mountain variety (Figure 3.10). The number of needles per fascicle is influenced by climate and site conditions, and there are fewer per fascicle in younger trees (Haller 1965). Two-needle fascicles reduce water loss by having less surface area and fewer stomates, and they require less energy to produce than three-needle fascicles. These features are of survival value in dry, harsh Rocky Mountain conditions.

Studies of the monoterpene components of xylem resin by Smith (1977) also illustrate differences among geographic races (Figure 3.11). Beginning in the southern California race, a clockwise pattern of decrease is seen for α-pinene and limonene through the Pacific, North Plateau, and Rocky Mountain races to the southwestern race. Conversely, 3-carene is negligible in southern California, but present in significant amounts in the other races. Conkle and Critchfield (1988) emphasize the correspondence of physiographic barriers and the distinct monoterpene races and their sharp transition zones.

Despite clear morphological and biochemical differences among races of ponderosa pine, genetic marker studies are yet unable to detect such differences in races at the molecular level and will likely rely on future genome sequencing to do so. Nevertheless, physiological traits related to drought tolerance have been demonstrated to have a direct genetic basis (though not yet specific to ponderosa pine). Epigenetic studies of several pine species have demonstrated shifts in gene expression of several physiological traits that control drought tolerance when seedlings of ponderosa pine are drought-stressed (see summary in Moran et al. 2017). Genes upregulated (activated) during drought stress include those related to physiological traits such as water uptake or the production of protective molecules and heat shock proteins; those downregulated are those involved in growth, including cell division. Most research in this area thus far has been limited to pine seedlings rather than adults. Given that such shifts in gene expression may vary widely among trees of different ages, and most gene expression studies do not examine difference among populations

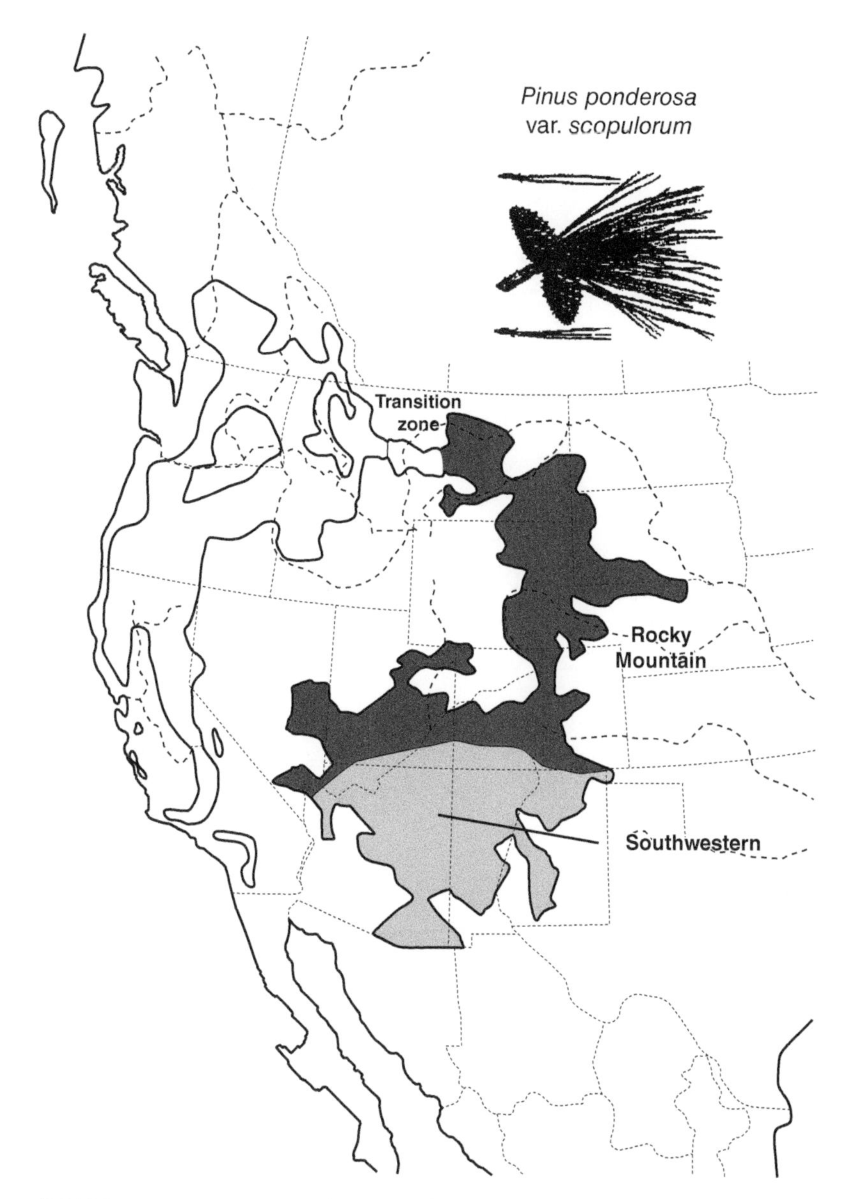

FIGURE 3.8 Eastern geographic races of ponderosa pine, excluding poorly understood populations in central Mexico. *Source:* After Conkle and Critchfield (1988) / United States Department of Agriculture / Public Domain.

(Hamanishi and Campbell 2011), several researchers have suggested that a combination of gene expression/epigenetic studies and provenance testing is the new frontier of understanding differences in drought-tolerance mechanisms among races of coniferous species (Moran et al. 2017). The basics of epigenetics are discussed near the end of this chapter.

Each of these lines of evidence illustrates the enormous variation that can be expected over a diverse range of regional and local ecosystems that differ in macroclimate and microclimate, elevation, and evolutionary history. Great genetic variation also occurs within these varieties and races, and it is strongly related to length of the frost-free period along elevational gradients (Conkle 1973; Read 1980; Rehfeldt 1986a,b, 1990).

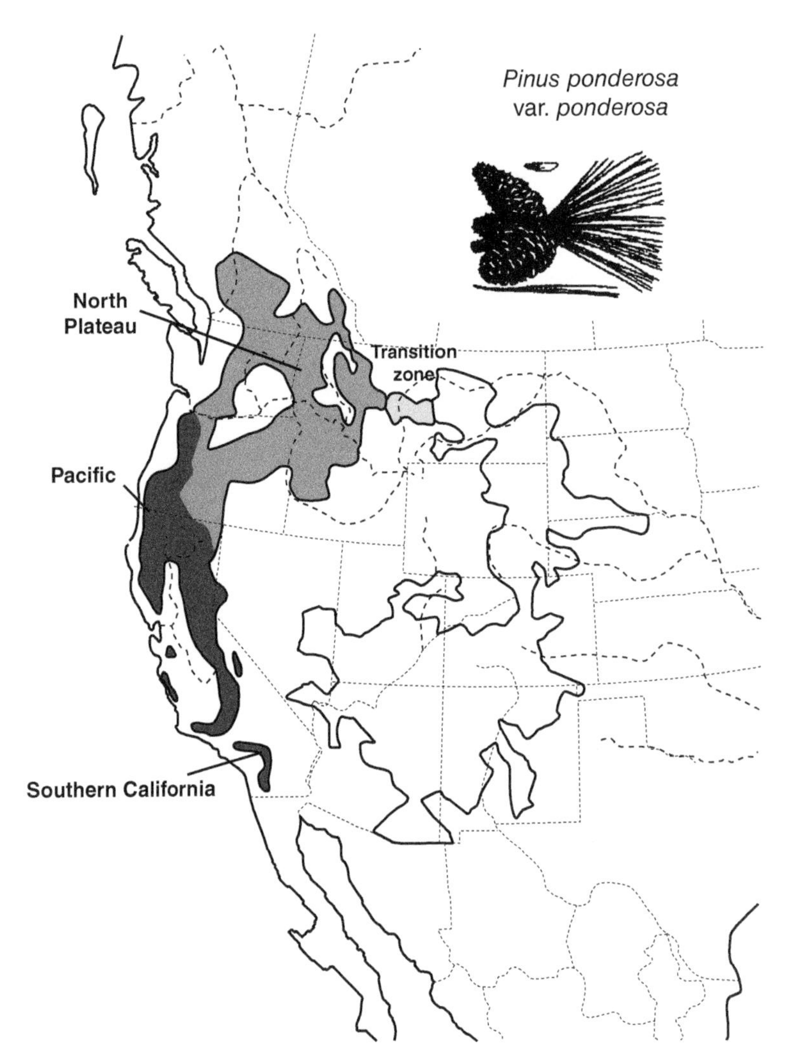

FIGURE 3.9 Western geographic races of ponderosa pine. *Source:* After Conkle and Critchfield (1988) / United States Department of Agriculture / Public Domain.

Douglas-fir The genetic differentiation of Douglas-fir has been investigated more than any other North American tree species. Two major geographic varieties—(i) the coastal or green type, *Pseudotsuga menziesii var. menziesii* and (ii) the interior, Rocky Mountain, or blue type, *Pseudotsuga menziesii var. glauca*—span an enormous geographic range in western North America. The two types are found on extremely variable sites in coastal and mountain areas of the Pacific Northwest and in the Rocky Mountains from Mexico to Alberta and British Columbia, Canada. Broadscale provenance testing of Douglas-fir has been conducted extensively in species-poor Europe; the most extensive test included 182 native provenances with plantings across 30 countries. Overall, there was a relatively clear separation between coastal and interior varieties in many traits, including frost sensitivity, phenology, morphology, and growth traits associated with latitude and elevation (Kleinschmit and Bastien 1992). Many populations of the coastal variety that perform best in Europe and the Pacific Northwest come from areas of high genetic diversity (Silen 1978; Kleinschmit and Bastien 1992).

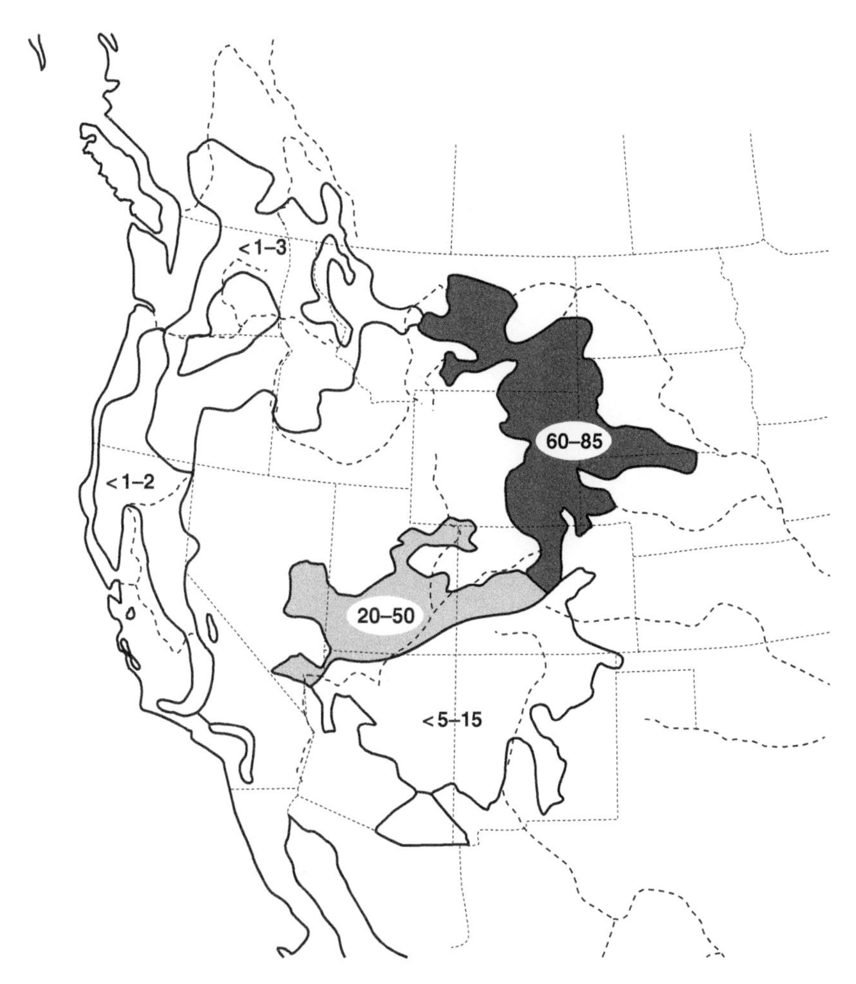

FIGURE 3.10 Average frequencies of two-needle fascicles on (mature/young) native ponderosa pines (Haller 1965). Similar results were reported by Wiedman (1939) and Read (1980) for native and plantation grown pines. *Source:* After Conkle and Critchfield (1988) / United States Department of Agriculture / Public Domain.

A recent study of coastal Douglas-fir in Washington and Oregon examined growth and phenological traits and their relationships to environmental variation in seedlings of 1338 parents from 1048 locations (St. Clair et al. 2005). The objective of the study was to identify genetic variation and races *within* the coastal variety of Douglas-fir, which appeared to form two races located on the east and west sides of the Cascade Mountains in Washington as a result of variation in minimum winter temperatures and late spring and early fall frost dates. The dates of bud set, emergence, and growth were strongly related to elevation and cool-season temperatures, and variation in bud burst and growth partitioning to stem diameter versus height was related to latitude and summer drought. Seedlings from the east side of the Washington Cascades were smaller, set their buds later, and burst their buds earlier than populations from the west side when planted in a common garden (St. Clair et al. 2005). Douglas-fir not only exhibits high genetic variation throughout its range but displays remarkable variation in morphological and physiological characters within physiographic regions of both coastal (Campbell and Sorensen 1978; Campbell 1986, 1991) and Rocky Mountain races (Rehfeldt 1989).

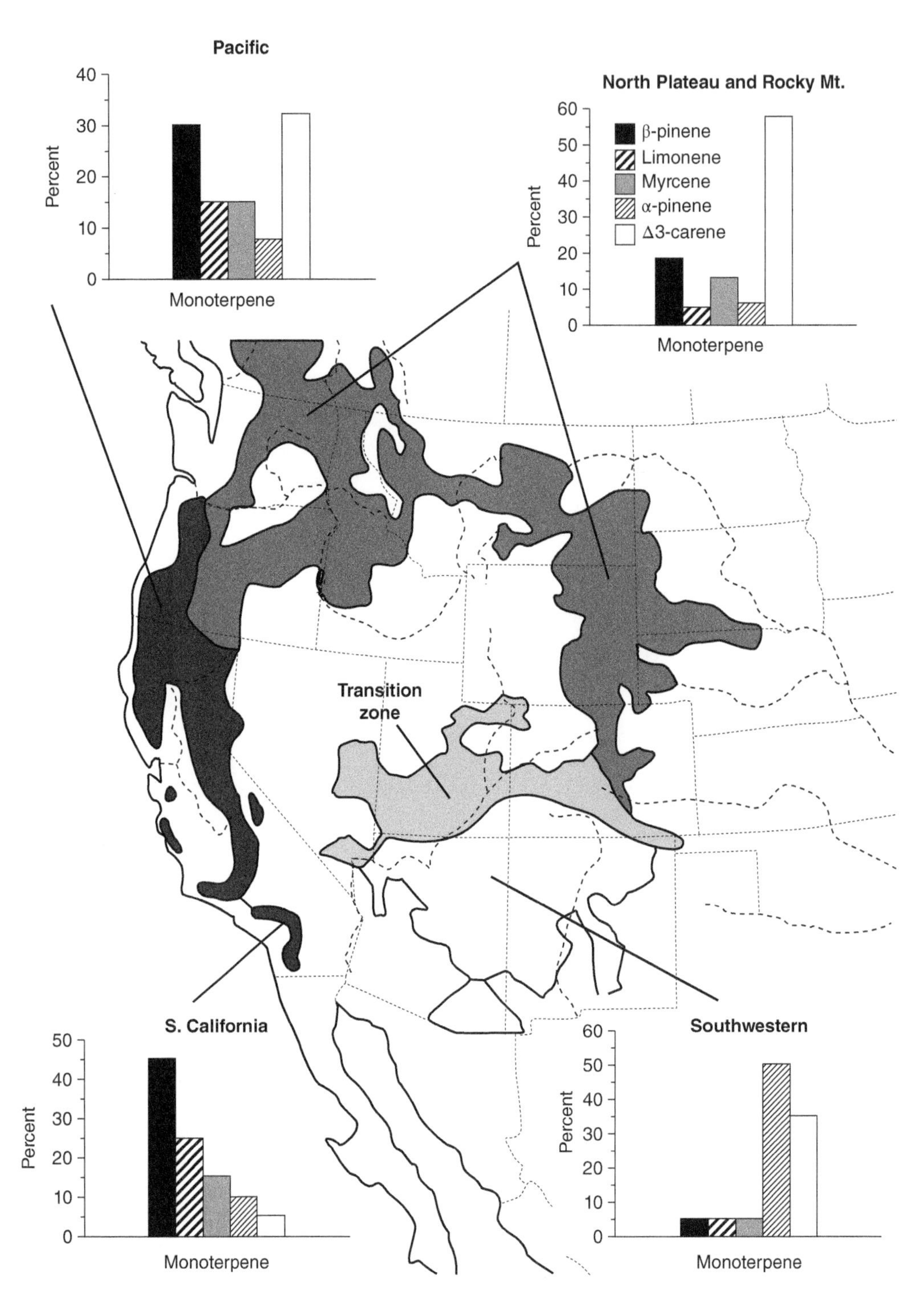

FIGURE 3.11 Average amounts of the major monoterpene components in xylem resin of native ponderosa pines (Smith 1977). *Source:* After Conkle and Critchfield (1988) / United States Department of Agriculture / Public Domain.

Local Genecological Differentiation As described in the earlier text, ecologically based genetic differentiation occurs at the macro or range-wide scale for virtually all tree species and mirrors the variation of the environment. The question of **local adaptation** thus becomes relevant—the degree to which populations are adapted to local segments of geographic and ecological gradients and fine-scale differences in soil water, elevation, aspect, and soil type. Early evidence for herbaceous species demonstrated extensive local adaptations to soil and microclimatic conditions (Antonovics 1971; Bradshaw 1971; Snaydon and Davies 1972, 1976). More recent evidence suggests that trees are also able to adapt to local environmental conditions relatively quickly (Petit et al. 2004). Chemical and physical properties of soils elicit sharp discontinuities in plant distribution, and strong edaphic preferences are quite common for woody species, but examples of intraspecific genetic adaptation to local soil type are not well documented.

Local differentiation has been shown for populations on adjacent sites representing the extremes of soil water conditions. For example, Wright et al. (1992) demonstrated genetic differences between populations of jack pine growing on adjacent wet and dry sites. Those originating from the wet site had significantly more tertiary roots than those from dry sites, which probably enhances nutrient uptake in the spring when the soil is likely to be saturated and anaerobic. Likewise, genetic differences were reported between a few populations of northern white-cedar from wet and dry sites (Habeck 1958; Musselman et al. 1975), but failed to be demonstrated for upland and lowland populations of black spruce (Fowler and Mullin 1977; Boyle et al. 1990).

Local races are often formed by the contrasting ecological conditions created by slope aspect. Differentiation caused by aspect has been demonstrated for Douglas-fir seedlings originating from north and south slopes in southern and central Oregon (Hermann and Lavender 1968; Campbell 1979), and for Sitka spruce in southeastern Alaska (Campbell et al. 1989). Elevation and aspect are often associated with local differentiation, presumably due to the heterogeneity of microclimate created by these physiographic factors. For example, microevolution of the coastal race of Douglas-fir was demonstrated by Campbell (1979) in a single 6100 ha (61 km^2) watershed in central Oregon with elevations ranging from 500 to 1600 m. Seedlings of 193 trees located throughout the watershed were grown in a common garden. Substantial genetic differentiation was noted in 14 of 16 traits, including seed germination rate, height, bud burst, bud set, and dry weight. Variation was mainly associated with elevation for traits related to vegetative growth. Estimates of maladaptation showed that none of the seedlings of a subpopulation would be adapted to locations 670 m higher or lower on the same slope.

It was once assumed that vast stands of wind-pollinated trees are genetically homogeneous, but significant genetic heterogeneity may be identified among local populations. Studies of local populations of lodgepole pine (Knowles and Grant 1985), trembling aspen (Mitton and Grant 1980), ponderosa pine (Linhart et al. 1981), and Engelmann spruce and subalpine fir (Shea 1990) demonstrate significant genetic heterogeneity over short distances, for example, within 2 ha for ponderosa pine (Linhart 1989). All trees occupying an area of approximately 2 ha in a Colorado ponderosa pine stand were mapped and their genetic constitution determined for seven polymorphic loci. Most trees fell into one of six clusters that differed from one another at one or more loci (Linhart 1989), a population structure promoted by the patchy nature of disturbance and regeneration in ponderosa pine forests. However, this fine-grained genetic heterogeneity is not necessarily related to adaptive differences among sites, but localized ecological races are possible wherever selection pressures are strong enough and gene exchange somewhat limited. For example, Shea (1990) found significant differences in allele frequencies between wet and dry sites dominated by Engelmann spruce and subalpine fir.

FACTORS AFFECTING DIFFERENTIATION: GENE FLOW AND SELECTION PRESSURE

Gene flow in plants among populations via pollen, seeds, and other propagules is a cohesive force acting to keep populations from diverging, and thus influences the extent of differentiation. By contrast, gene flow is impeded by isolating factors such as spatial distance between populations, or ecological isolation (south versus north slopes, wetlands versus uplands). Selection is also important, because populations will remain similar if subjected to similar selection forces, but will differentiate if they are not. The tall stature of trees, among other characteristics of the life form, creates the potential for high gene flow (Hamrick and Godt 1996), particularly over short distances, and local differentiation is therefore unlikely. Gene flow over short distances occurs mainly by high selection pressure.

Gene flow is restricted by several factors besides limited dispersal of pollen and seeds, including: (i) a limited number of sexually mature trees, (ii) individual variation in flowering times, and (iii) biological and ecological factors that control zygote viability and seedling establishment, such as genetic incompatibility, frost, drought, shade, and herbivory. The age of seed-bearing and periodicity of seed crops vary widely among species, and trees also vary in their reproductive capacity; some are highly fruitful, some moderately so, and others are completely barren year after year. Only a few breeding trees in a population contribute substantially to subsequent generations of the many trees that could potentially exchange genes. For example, Schmidt (1970) found in an 89-year-old Scots pine stand that only 30% of the trees were responsible for 71% of the female strobili and 64% of the male strobili.

The phenology of flowering (time of pollen release and female receptivity) is vital in pollination and is closely related to air temperature and relative humidity. The phenology of trees within a given population is more likely to be synchronized with that of trees close together than with those progressively farther away. Although the timing of flowering generally favors local gene exchange, gene flow from distant sources is a definite possibility. Despite its potential to be transported many miles, viable pollen may reach the distant population too early or too late to compete effectively with local pollen. This effect becomes even more important in Scots pine and probably other pines and conifers, given that the capacity of the pollen chamber is limited (Sarvas 1962), and pollen grains have an unequal chance to fertilize the eggs. Only two of the many pollen grains reaching the micropyle of the ovule have an opportunity to fertilize the eggs of each ovule, one of which eventually develops into the embryo. Therefore, pollen grains of neighboring trees have a higher probability of achieving fertilization than those of distant trees because of the greater probability of their being first to reach the micropyle.

The range of seed dispersal in forest trees is limited, except perhaps for very small-seeded species such as birches, hemlocks, poplars, and willows. Bird dispersal in oaks and beeches (Johnson and Webb 1989) and pines (Vander Wall and Balda 1977) may extend dispersal to 20 km or more (Chapter 12). By contrast, pollen may be carried great distances by wind (Lanner 1966). Andersson (1963) reported that pollen was blown from Germany to southern Sweden, a distance of 72 km, and that in a high pollen year in Sweden, clouds of pollen were heavy enough to be mistaken for forest fires. Genetic markers have provided more precise estimates of effective dispersal distance of pollen in trees. For example, a study of Scots pine in Spain by Robledo-Arnuncio and Gil (2005) found that over 4% of matings resulted from pollen dispersed greater than 30 km away. A study of European ash in Scotland documented more than half of the individuals successfully pollinated occurred from populations greater than 10 km away (Bacles et al. 2006). And finally, about 27% of individuals of black cottonwood in the Pacific Northwest were pollinated by populations 3–10 km away (DiFazio et al. 2004).

Widespread pollen and seed dispersal might lead to the conclusion that populations over large areas are prevented from diverging because of widespread gene exchange. Extensive gene flow of trees suggests that adaptation would occur at broad scales and that local adaptation is unlikely in the absence of very strong selective forces (Boshier et al. 2015). Several recent studies have shown explicitly with genetic markers that adaptation in trees for various traits may occur at fine spatial scales, including those of maritime pine (Budde et al. 2014; Vizcaíno-Palomar et al. 2014), European beech (Csilléry et al. 2014), sugar pine (Eckert et al. 2015), whitebark pine (Lind et al. 2017), Pacific silver fir (Roschanski et al. 2016), and the model organism black cottonwood (Holliday et al. 2016). The combination of high local differentiation for adaptive traits in the face of extensive gene flow is rather paradoxical (Petit and Hampe 2006).

The mechanisms of local adaptation in trees are long studied. Most pollen and seeds of trees are dispersed close to the source, and their frequency declines rapidly with distance (Ellstrand 1992). Individuals of most wind-pollinated species in the north temperate zone in natural stands are pollinated and fertilized by their neighbors less than about 100 m away (Koski 1970; Shea 1990; Adams 1992). Monitoring flow and destination of pollen from individual trees and from different stands remains extremely difficult, and the extent of gene exchange is difficult to estimate but is probably often underestimated (Savolainen et al. 2007). Gene flow may be sometimes restricted to nearest neighbors, favoring inbreeding, whereas at other times gene exchange may occur over considerable distances, thus significantly affecting the gene pool of receptor populations.

Natural selection guides the genetic makeup of populations in a given ecosystem against the prevailing level of gene flow. Environmental selection pressures play a significant role over distances of only 2–4 m in herbaceous species (Aston and Bradshaw 1966). A classic study by Barber and Jackson (1957) demonstrated the effect of intense selection in glaucous and nonglaucous (green) phenotypes of urn tree (eucalyptus) in Tasmania. Green phenotypes are found at low elevations and more sheltered sites, while glaucous individuals (having leaves covered with a whitish wax effective in freeze avoidance) are more frequent in exposed and colder environments along a gradient from low to high elevation. Clines of glaucousness are correlated with frost occurrence, with more glaucous populations occurring in the frostier localities (Barber 1955). The change from green to glaucous types occurred completely over a vertical distance of 122–152 m (0.8–1.6 km ground distance) in the adult populations. Gene flow is likely via insect and bird pollinators, as indicated by the production of glaucous seedlings from nonglaucous mother trees and vice versa. Nevertheless, intense selection eliminates glaucous seedlings as they mature in low elevation forests, and at the higher elevations green seedlings are eliminated, building high genetic diversity over a short distance even in the face of considerable gene flow. More generally, the annual variation in selective forces experienced by woody plants, such as temperature, precipitation, or frosts, may have a stabilizing rather than a selective effect that may contribute to local adaptation (Savolainen et al. 2007; Boshier et al. 2015).

ECOLOGICAL CONSIDERATIONS AT THE SPECIES LEVEL

Populations diversify in response to genetic differentiation as they change through time and radiate into new areas. Mechanisms that act to separate species are those that favor reproductive isolation (Table 3.1). Many mechanisms, called **pre-zygotic barriers**, act singly or in concert to prevent gene flow among plant populations, and this process is an important route of speciation. *Geographic separation* is the most obvious cause of reproductive isolation, resulting when populations are separated by physical barriers such as mountain ranges, waterways, or wide canyons, and the gene pools thereby occur in different regions. Such isolation may result in **allopatric speciation**, whereby each population undergoes unique development based on the demands of the environment of their given region or differences in their specific gene pool. Plant populations may also experience mechanisms that favor reproductive isolation even when the populations co-occur in the same region. **Sympatric speciation** occurs in such instances, whereby resulting species are able to maintain their distinctness even when individuals of each are within effective pollinating

Table 3.1 Pre- and post-zygotic barriers that contribute to reproductive isolation in plants. Except for geographic separation, barriers are listed in the order they occur.

Timing	Type of barrier	Effect
Pre-zygotic barriers (fertilization and/or zygote formation prevented)		
Pre-pollination	Geographic	Populations are physically separated into different regions
Pre-pollination	Ecological/niche differentiation	Seeds, seedlings, and/or adults of different populations in the same region have the highest fitness in differing habitats
Pre-pollination	Phenological	Populations in the same region differ in phenology (flowering times)
Pre-pollination	Ethological	Adults of populations in the same region have specialized pollinators specific to one population and not the other
Post-pollination	Pollen interactions	Populations in the same region may pollinate each other, but incompatible gametes reduce or eliminate fertilization
Post-zygotic barriers (fertilization occurs, but zygote is inviable, weak, or sterile)		
Post-pollination	Hybrid seed formation	Hybrid zygotes are less likely to mature into seeds
Post-pollination	Hybrid inviability or weakness	Hybrid seeds are less likely to germinate and establish than parental seeds or establish as weak hybrids
Post-pollination	Hybrid sterility	Hybrid plants are more likely to be sterile than parental plants
Post-pollination	Hybrid offspring inviability and sterility	First-generation hybrids are normal, fertile, and vigorous, but hybrid offspring are inviable or sterile

Source: Adapted from Baack et al. (2015).

range of one another and hybridization would be possible. Sympatric species may hybridize and form hybrid individuals, such as in loblolly pine and shortleaf pine or yellow birch and paper birch, but the parent species remain distinct.

Many pre-zygotic barriers that occur prior to pollination prevent gene flow among populations or species occurring in the same region (Baack et al. 2015). *Ecological separation* or *niche differentiation* (see niche discussion in the following text) promotes reproductive isolation when populations co-occur in the same region but occupy different habitats and become reproductively isolated. Ecological separation is effective when trees experience reduced fitness in some habitats compared to others at some point in their lifestyle. For example, black ash and white ash co-occur in the eastern United States, but black ash is restricted to poorly drained soils where white ash is unable to persist, and white ash to uplands where black ash is unable to compete. *Phenological separation* reproductively isolates populations and species when phenology—especially flowering periods—is not synchronized. *Ethological separation* occurs when species or populations have specialized pollinators, a barrier especially common in the tropics. Finally, *gametic incompatibility* is a pre-zygotic barrier that occurs *after* pollination, when different populations or species may pollinate each other, but incompatible gametes prevent or reduce fertilization (Table 3.1; Stebbins 1971; Baack et al. 2015).

Following various degrees of population differentiation, populations exhibiting strong genetic differences in morphology and physiology are classified as formal taxa (species, subspecies, or varieties). Stebbins (1971) has described this process as occurring in four stages. First, a single population that exists in a homogeneous environment may eventually differentiate into races and subspecies as individuals of the population migrate into new environments. Next, some of these races and subspecies may become geographically isolated as differentiation and migration continues. Third, the geographically isolated races or subspecies further differentiate at a genetic level and may eventually become reproductively isolated. Finally, changes in the environment may allow populations once geographically isolated to co-occur in the same area (sympatric), remaining distinct because they have differentiated to the point of reproductive isolation (Stebbins 1971). Not all species have the same degree of divergence, as noted in their degree of morphological difference and in their reproductive isolation. "Good" species, from the standpoint of reproductive isolation, are generally sympatric.

There exist several additional isolating mechanisms, called **post-zygotic barriers**, that operate even when pollination and fertilization are able to occur between co-occurring species and hybrids are formed (Stebbins 1971). First and foremost, outcrossed zygotes that form when populations or species pollinated each other are often less likely to mature into seeds than parental zygotes. Similarly, outcrossed zygotes may be unable to germinate or become established (*hybrid inviability*) or if established may be unable to form reproductive organs (*hybrid weakness*). An example of a weak hybrid is the hybrid that occurs between bigtooth and trembling aspen (*Populus* × *smithii*) in the northeastern United States and southeastern Canada. Alternatively, *hybrid sterility* may occur because reproductive organs develop abnormally, or chromosome incompatibility leads to meiosis failure. Notably, the hybrids created from the outcrossing of sympatric species may be vigorous and fertile, but their offspring (F_2 generation) are largely inviable or sterile (Stebbins 1971; Baack et al. 2015).

In some cases, hybridization may dissolve or "swamp out" species distinctions in a flood of intermediates. In this situation, complete reproductive isolation has not yet occurred, and the populations may be more appropriately regarded as subspecies. This situation has been documented in Canada, where intermingling occurs between the morphologically similar white spruce and Engelmann spruce, and balsam poplar and black cottonwood. Subspecies have been recognized in both cases (Taylor 1959; Brayshaw 1965; Viereck and Foote 1970).

Biologists have long recognized that no single definition of a species is entirely applicable to classify the enormous diversity of organisms nor to serve the various purposes desired by different scientists. This problem of defining speciation has been considered concisely by Stebbins (1971), Heslop-Harrison (1967), and Solbrig (1970), and in detail by Stebbins (1950 and 1970) and Grant (1963, 1971, 1977). Ecological and genetic studies have identified the ability for populations to diverge even when there is considerable gene exchange (Arnold 2015), and the emerging ability to study entire plant genomes (genomics) has largely reshaped our understanding of species and speciation. Genome sequencing has provided many examples of gene exchange in nature between good species, and additional examples should be expected as additional species are sequenced. One modern definition of a species is a group of genotypes that are more similar to each other than to other groups across the genome as a whole (Mallet 2006). Such a definition focuses on distinct genetic backgrounds among species in light of the fact that some genes or genomic regions may be similar among species (Hartl 2020).

As the field of genomics continues to develop, its methodologies will undoubtedly be critical to understanding how speciation occurs at the molecular level. Genomics may be used to (i) identify the specific loci that control species differences or cause reproductive isolation; (ii) quantify the number of loci subject to selection that are needed for the evolution of new species and how they are distributed through the genome; and (iii) identify and quantify gene flow between populations (Jiggins 2019). In general, genomic approaches utilize many types of genetic markers to compare species, populations, and individuals, and the genetic processes that occur among them. For example, Zheng et al. (2021) sequenced 55 trees from three hornbeam species (genus *Carpinus*) in China to examine the gene flow and selection that occurred during the speciation process. One of the species, *Carpinus tibetana*, was only recently identified and named, and exhibited very low genetic diversity attributed to its extremely small population size. The other two species exhibited evidence of gene flow between them; divergent and selected genomic regions included genes associated with temperature, stress response, and plant development that were likely adaptations to local environments (Zheng et al. 2021). The use of genomics to study speciation is a rapidly developing field, but the basics have been reviewed by Jiggins (2019), Trense and Tietze (2018), and Campbell et al. (2018), among others.

NICHE

The genetic differentiation among species that has resulted from ecological interactions has served to place each species into a different **niche**. The term niche expresses where, when, and how a species is genetically adapted to persist with other species it interacts with on a site. The niche of a species is the result of the multidimensional specialization of that species in space and time. By definition, no two species can occupy the same niche without one being outcompeted by the other (Chapter 13). Three components are recognized in examining how woody species differentiate their niches: a spatial component (the physical site conditions to which the species is adapted), a temporal component (whether a species dominates early or late in succession), and a functional component—physiologically based genetic adaptations, sometimes termed natural history or life history traits, such as number of seeds produced; dispersal time and mechanism; growth rate; and tolerance of shade, drought, fire, and flooding. A species' functional component is the set of genetic adaptations that enable it to (i) occupy a geographic range and the characteristic local sites it includes, and (ii) dominate at a characteristic time during succession on a given site. These three niche components identify where (spatial), when (temporal), and how (functional) a species competes and persists in regional and local landscape ecosystems. We use the term niche as the most concise formulation of this genetic specialization.

The **spatial component** of the niche is illustrated by silver maple, which occupies river floodplains, whereas the related sugar maple thrives on upland sites. Their niche differentiation results from the different physiological adaptations of these species, which make them relatively more competitive on these respective sites. The **temporal component** is illustrated by paper birch and eastern hemlock on an upland site in the hemlock–northern hardwood forest. Paper birch is a pioneer species that dominates a site following fire, early in the course of succession. Hemlock seedlings establish simultaneously on the same site, but grow slowly under the birch canopy, and then dominate the site a century or more later as the birches decline and die. In this case, the species are niche-differentiated by the physiological adaptations that enable them to dominate the site at different times rather than by site conditions. The **functional component** is illustrated by the physiological adaptations of paper birch and hemlock. Birches colonize a burned site quickly and their seedlings grow rapidly with leaves of high photosynthetic efficiency on sites with high light levels. By contrast, hemlock seedlings are photosynthetically efficient at low light levels and, using other physiological adaptations to obtain soil water and nutrients, they are able to develop under the birches and replace them in 100–200 years.

HYBRIDIZATION

Hybridization, the crossing between individuals of populations having different adaptive gene complexes (races, subspecies, species), is frequent in natural populations of many woody plant groups. Many tree and shrub hybrids were discovered and reported during the twentieth century. For example, widespread disturbances by humans have enabled the European white poplar to initiate naturally occurring hybrids with a native poplar on three continents: the gray poplar (*Populus* × *canescens*) in Europe, Rouleau's poplar (*Populus* × *rouleauiana*) in eastern North America (where the white poplar was introduced as an ornamental tree during European colonization), and the hairy poplar (*Paulownia × tomentosa*) in China. Similarly, the introduction of eastern cottonwood into gardens of France and England enabled it to hybridize with the native European cottonwood to produce the highly successful black poplar hybrid (FAO 1980). Although most natural hybrids demonstrate hybrid weakness, two of the best examples of so-called hybrid vigor in trees are clones of the European black poplar hybrid and Rouleau's poplar (Little et al. 1957; Spies and Barnes 1981).

Many hybrids were formerly treated as normal divergent species and given binomial names, for example, *Populus × acuminata* (hybrid between narrowleaf cottonwood and Fremont cottonwood (Crawford 1974; Eckenwalder 1977, 1996)). Hybrids are now considered **nothospecies** (hybrid species) and may be designated by a taxonomic formula, as in the case of the hybrid between shingle oak and northern red oak (*Quercus imbricaria × Quercus rubra*), or as a binomial with the multiplication sign (×, signifying hybrid) placed directly before the epithet, for example, *Quercus × runcinata* (Wagner 1983). These and most hybrids were originally identified as such using morphological characteristics, such as leaf shape, intermediate between the parent species, but molecular markers are more typically used in recent decades.

Hybridization is of major evolutionary significance because it acts as an evolutionary catalyst (Stebbins 1969, 1970; Arnold 2015). An estimated 70% of all flowering plants originated from past natural hybridization between different species or genera (Whitham et al. 1991; Arnold 1994). Many hybrids of woody species are of major ecological and practical significance, with importance to management and horticulture due to their rapid growth, good form, disease resistance, or frost hardiness (Duffield and Snyder 1958; Nikles 1970; Wright 1976; Zobel and Talbert 1984). Importantly, hybridization may also act as a mechanism for reversing the divergence of species (White et al. 2007). A recent study in China used genetic markers to identify loss of genetic diversity and an associated higher extinction risk of an endangered oak species due to introgression (An et al. 2017).

Natural hybridization is probably more common today than it was in pre-European colonization forests due to massive disturbances associated with humans that affected habitats and plant populations in the late nineteenth century and most of the twentieth century. Such disturbances have greatly increased the likelihood of hybridization and gene flow by creating open, disturbed sites and thereby reducing competition and favoring establishment and survival of hybrid plants. Natural hybrids often occur on disturbed sites and in zones of contact between species (Brayton and Mooney 1966; Remington 1968). Disturbed areas often act as intermediate or hybrid habitats (Anderson 1948) where neither parent is well adapted. Hybrids between the narrowleaf cottonwood and Fremont cottonwood in northern Utah (Whitham 1989) illustrate the occurrence of hybrids in a contact zone between parent species. The hybrids occupy a zone of overlap between the morphologically distinct parents that occur in upper and lower elevations, respectively, along the Weber River. Leaf-gall-producing aphids are concentrated on leaves of hybrids in a 13 km zone where the host's parents interbreed, yet pure Fremont cottonwood is totally resistant to the aphid, and aphids rarely colonize leaves of pure narrowleaf cottonwood. Susceptibility to this parasite illustrates one kind of weakness that is often found in hybrids. Despite the boom of new hybrid reports in the twentieth century, relatively few detailed ecological studies have compared the establishment of hybrids to their parents or the presumed differences between the so-called hybrid habitat and that of the parents.

Hybridization may enrich the gene pool of species by the process of introgressive hybridization or **introgression**: the infiltration of genes (or small portions of a genome) from one species into another due to hybridization and repeated backcrossing. Introgression is simply gene flow between species and occurs in three phases: (i) initial formation of F_1 hybrids, (ii) their backcrossing to one or other of the parental species, and (iii) natural selection of certain favorable recombinant types (Davis and Heywood 1963; Figure 3.12). When the third of these phases is achieved—that is, the recipient species benefits by natural selection from the transfer of genetic

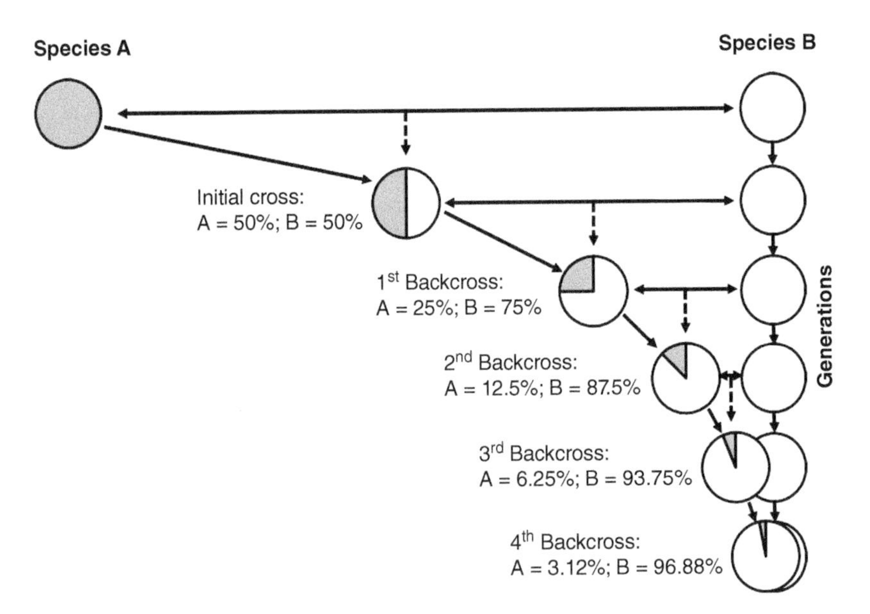

FIGURE 3.12 Hypothetical example of introgression—the interbreeding of species followed by backcrossing of some genes from one parent into at least some members of the population of the other.

material—the process is often called **adaptive introgression** (Suarez-Gonzalez et al. 2018). Hybridization between two closely related species is likely to result in higher gene flow than when the species have well-integrated but different gene pools. Genes from one population will be incorporated into the gene pool of the other (regardless of taxonomic level—species, subspecies, etc.) if they improve the well-integrated harmony of the foreign gene pool (Whitney et al. 2006), but their frequency reduced if they tend to disrupt the harmony (Barton 2001; Shaw and Mullen 2011). Roughly 25% of plant species exchange genes with relatives, twice the rate estimated for animals (Mallet et al. 2016).

Oaks are notorious for hybridization (Little 1979; Miller and Lamb 1985; White et al. 2007). The *Quercus* genus currently includes more than 450 species, many of which hybridize (Rushton 1993). In California alone, more than 20 hybrids have been recognized, 11 of which are named (Pavlik et al. 1991), and introgression is therefore likely among closely related oak species. Hybridization and possible backcrossing among oaks were graphically illustrated by Wagner and Schoen (1976) between shingle oak, having entire, nonlobed leaf blades, and associated northern red and black oaks that have multiple leaf lobes (Figure 3.13). Some plants are possible backcrosses to northern red oak (Figure 3.13b) or shingle oak (Figure 3.13e), and introgression is suspected.

Woody plant populations exhibit enormous variation in morphological characters, such that most modern studies of parent and hybrid variation employ molecular genetic techniques to estimate the extent of hybridization and gene flow. Prior to the availability of genomic and genetic markers, divergence caused by intense selection was often misinterpreted as introgression (Barber and Jackson 1957). Thus, in determining the amount of gene flow between species, it is important to establish through more than observations of morphological characters of phenotypes that (i) hybridization and backcrossing have actually taken place, and (ii) increased variation of the parent species occurs outside the area of hybridization and is due to hybridization rather than intense selection along an environment gradient.

The use of chloroplast DNA (cpDNA) markers has brought about a better understanding of the contributions of hybridization to adaptation and speciation. Chloroplasts contain a separate

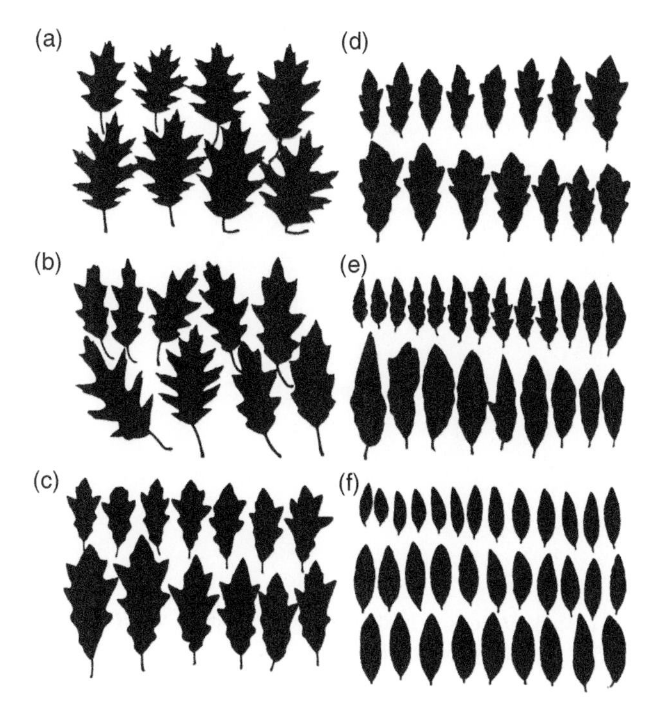

FIGURE 3.13 Silhouettes of leaves from (a) northern red oak, (b)–(e) hybrid of northern red oak and shingle oak, and (f) shingle oak. *Source:* Wagner and Schoen (1976) / FAO.

genome from that of the nucleus, and the cpDNA in most cases is inherited from only the maternal parent. Detecting similar cpDNA haplotypes—sets of polymorphisms in the chloroplasts that are inherited together from a single parent—is useful in inferring evolutionary relationships among species, as well as evidence of hybridization and/or introgression. For example, Saeki et al. (2011) identified 38 cpDNA haplotypes in red maple and 7 in silver maple in eastern North America. The greater haplotype diversity of red maple probably resulted from its greater ecological range and its ability to be a habitat generalist compared to silver maple, which is restricted to river floodplains. Moreover, five of the seven silver maple cpDNA haplotypes were shared with red maple in areas of geographic overlap (the Lower Mississippi Valley and a northern area centered on Vermont and northern Michigan), suggesting that introgression was likely (Figure 3.14). The shared haplotypes in the Lower Mississippi Valley were likely derived from silver maple, whose preferred habitat is centered on this region, and the shared haplotype in the post-glacial northern landscapes likely originated from red maple (Saeki et al. 2011). A similar study of birches in eastern North America also documented high haplotype diversity and evidence of introgression (Thomson et al. 2015). Genetic markers have uncovered evidence of introgression in several additional tree genera, including poplars (Lexer et al. 2005) and willows (Hardig et al. 2000), as well as oaks (Belahbib et al. 2001).

POLYPLOIDY

Variation at the species level is not only influenced by the combination of two different chromosomal sets of genes, as in hybridization, but also by the *number of similar sets* of chromosomes that individuals of a species possess. **Polyploids** are organisms with three or more sets of chromosomes (compared to haploid organisms with one set, or diploid organisms with two). The ploidy level of a species (triploid = 3 sets, tetraploid = 4 sets, etc.) is measured in relation to the

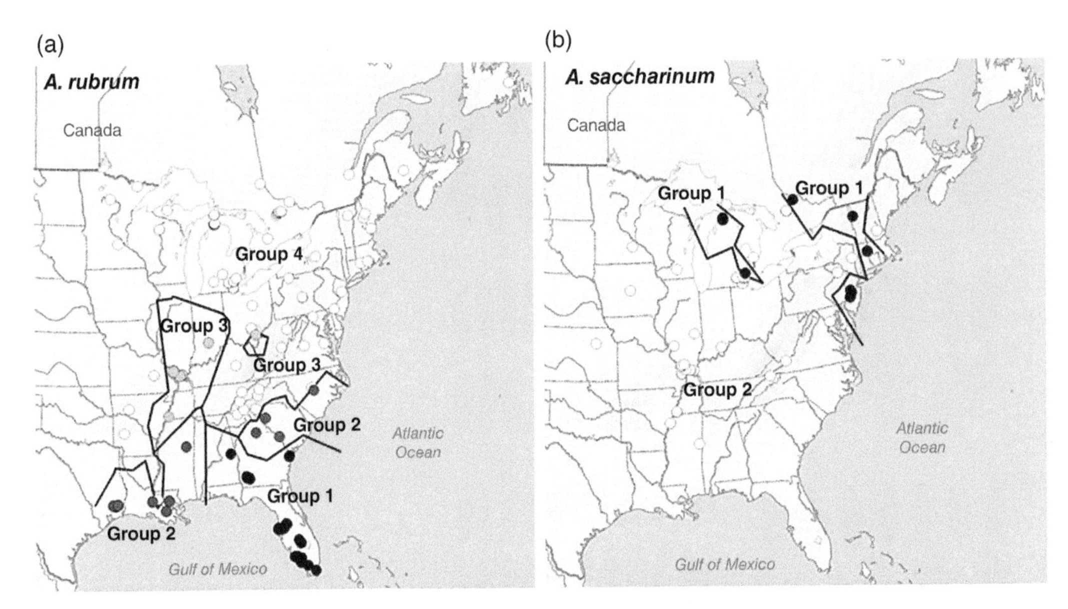

FIGURE 3.14 Maps of genetic boundaries for (a) red maple and (b) silver maple in eastern North America denote two regions of geographic overlap and likely areas of introgression. *Source:* From Saeki et al. (2011) / John Wiley & Sons.

base or x chromosome number established for the genus or family, usually the lowest haploid (gametic) number for the group. For example, the base number for the genus *Pinus* is 12 (x = 12), and all pine species have 24 chromosomes (2x = 24, because they are diploid). By contrast, the birches have a base number of 14 chromosomes and exhibit many polyploid levels among species. These include diploids (black birch and river birch; 2x = 28 chromosomes), tetraploids (bog birch, paper birch, and European white birch; 4x = 56), and a hexaploid (yellow birch; 6x = 84).

Polyploids arise when related species hybridize and the chromosomes of the hybrids then double, a process that produces new species. Recent research has also denoted this process **whole genome duplication**. Polyploids are of considerable evolutionary and ecological significance in part because they exhibit a wide geographical distribution, including alpine, arctic, and tropical environments. In addition, they occur on a wide range of sites from mesic to more extreme sites characterized by harsh conditions (Stebbins 1985).

Polyploids are successful largely because of their greater ability to colonize new or disturbed habitats relative to that of their diploid ancestors (Stebbins 1985). These sites may be associated with zones of contact and hybridization between genetically differentiated diploid populations. The resultant hybrid polyploids often contain gene combinations for aggressive colonization that are buffered and maintained by the polyploid condition. Many polyploids are notable for their ability to colonize and persist on cold, harsh sites, particularly those that were formed during and following Pleistocene glaciation.

The success of some polyploids under severe site conditions relative to related diploids is explained in part by the uniformity-promoting mechanism of polyploidy. Polyploidy is analogous to a sponge, absorbing mutations but rarely expressing them. By contrast, diploids more easily express mutations or new recombinations because of the low chromosome number and fewer of each kind of chromosome. Given that polyploids may have four, six, or more chromosomes of each kind, new gene combinations are less likely to produce a major change in the phenotype. Thus, polyploids represent an efficient buffering system, resisting the effects of natural selection on particular genes and promoting and preserving phenotypic uniformity (Mosquin 1966). The narrow adaptational limits of high latitude and weedy polyploids are an adaptive feature corresponding to the narrow and relatively uniform environments of boreal, arctic, and disturbed or weedy habitats. The abundance of many polyploids, such as high-latitude birches and willows, results from the widespread availability of their preferred habitats.

Polyploidy may affect how an individual interacts with the environment (Maherali et al. 2009; Segraves and Anneberg 2016), and not all polyploids are found in the harshest environments. A study of polyploids in paper birch (Li et al. 1996) found several traits that reduce water loss, such as reduced photosynthesis and stomatal conductance during water stress, as well as smaller stomata, higher stomatal density, and higher leaf pubescence. Notably, polyploid paper birches tend to be found in warmer and drier ecosystems compared to diploids (Li et al. 1996). Trembling aspen, which is either diploid or triploid, also provides an excellent example of the potential adaptive importance of polyploidy (Kemperman and Barnes 1976; Mock et al. 2012). Triploid aspens have been found to have higher stomatal conductance and water use efficiency, higher carbon uptake, bigger leaves, and faster growth rates (Greer et al. 2018), and are most frequently found at lower latitudes in warmer and drier climates (Greer et al. 2016). By contrast, aspen in the cooler and wetter portion of its range is nearly completely diploid (Mock et al. 2012; Blonder et al. 2020). Triploids have also been characterized as having less resilience to climate-induced stress (Greer et al. 2018).

Certain tropical floras also exhibit high levels of polyploidy, some of which are of very ancient origin. As in temperate zones during recent centuries, newly opened habitats were available during the period when angiosperms diversified, and increasing polyploidy probably facilitated the establishment and spread of new groups of angiosperms during the early stages of their history (Stebbins 1970). Thus, the ability of polyploids to colonize newly opened environments is the common denominator of their success in diverse regions of the world, whether the time of their origin was ancient or modern. Temperate species such as American basswood and tulip tree are examples of ancient polyploids that have outlived their ancestors. By contrast, a newly evolved polyploid birch, an octoploid, was discovered on a disturbed lake margin in southern Michigan (Barnes and Dancik 1985). Overall, polyploidy illustrates a significant and intimate ongoing process of site–plant relationships throughout evolutionary time.

Polyploidy is rare in gymnosperms; only about 5% of gymnosperms exhibit polyploidy (Khoshoo 1959; Ahuja 2005). For many decades, very few conifers were known to be polyploids (Khoshoo 1959), the most notable being coastal redwood, a hexaploid with 66 chromosomes. All known coniferous polyploids reside in the cypress family (Cupressaceae) and none in the pine family (Pinaceae). A recent screen of the ploidy level of over 96% of the genus *Juniperus* found a much higher frequency of polyploidy than previously thought, with 15 *Juniperus* taxa being tetraploid and one being hexaploid (Farhat et al. 2019). It is likely that polyploids are more common in *Juniperus* because of the overlap in their geographical ranges, providing ample opportunity for hybridization between species. By contrast, about 50–80% of all angiosperm species are known to be polyploids. Amazingly, about 70% of all angiosperms are estimated to have experienced one or more episodes of polyploidy in their ancestry (Masterson 1994), and 2–4% of all speciation events in angiosperms involved polyploids (Otto and Whitton 2000). Early estimates reported that 70% of monocots were polyploids (Goldblatt 1980), and between 80 and 90% were reported for grasses (Stebbins 1985). Some woody angiosperm genera have no polyploid species (*Juglans*, *Aesculus*, *Cercis*), whereas many others each have species of various ploidy levels (*Prunus*, *Salix*, *Betula*, *Alnus*, *Magnolia*, *Acer*, and many others) (Wood et al. 2009). Modern genomic techniques have shifted the focus from the proportion of polyploid angiosperms to the number of polyploid episodes per lineage. Recent discussions of the evolutionary aspects of polyploidy are provided by Jiao et al. (2011), Soltis (2005), Soltis et al. (2009, 2014, 2015), De Bodt et al. (2005), Van de Peer et al. (2017), and Rice et al. (2019).

THE FITNESS–FLEXIBILITY COMPROMISE

A species must compromise between having flexibility—the potential to change—and having adaptedness to its present environment if it is to survive and persist over time. *Flexibility* that will permit further adaptive change is favored by variability-promoting mechanisms such as cross-pollination and a high rate of recombination (Mosquin 1966). Inbreeding, apomixis, polyploidy, and a low rate of recombination are uniformity-promoting devices that favor *fitness* in a given environment. Most woody species exhibit mixtures of uniformity- and variability-promoting devices, shaped by selection, that serve such a compromise well. The proportion of each type of device exhibited by a given species is closely related to the limiting ecological factors of the environment that supports it. For example, trembling aspen is widely distributed in northern, glaciated, and disturbed sites, and eastern cottonwood is also wide-ranging but primarily found at lower latitudes in river floodplains. Both species are primarily dioecious (male and female flowers borne on different individuals) and thus are effective cross-pollinators. A great amount of variation is generated in these species to be widely circulated through abundant seed production and widespread wind dispersal. Aspen is highly proficient at clonal growth via vegetative propagation by root suckers, which assures genetic uniformity; this adaptation has been widely favored by fire. By contrast, cottonwood is found in a virtually fire-free environment and rarely produces root suckers in nature, but its branches and young shoots root easily in soil. This trait is selectively advantageous and of immediate fitness in river bottoms subject to flooding, whereby branches are broken off by ice and debris and deposited in new soil. Both species have strongly developed mechanisms, closely linked to their respective environments, that provide both fitness and flexibility.

Pines are typically cross-pollinating species, but maintain a certain amount of self-pollinating ability in their breeding systems. Fire plays a major role in establishment of pines around the world such that the colonizing ability of one or a few survivors following fire, if necessary, by self-fertilization, is of major significance. Post-fire colonization is likely to be accompanied at first by an increase in the degree of inbreeding, but outbreeding is restored as stand density increases (Bannister 1965). Fire-imposed selection pressure has been instrumental in promoting a very high degree of self-fertility in red pine (Fowler 1965a), and unlike most other pines, seedlings resulting from self-pollination are as vigorous as those from cross-pollination (Fowler 1965b). A high degree of self-fertility has been achieved at the cost of variability, however, because red pine is one of the most uniform of all woody-plant species and has a relatively restricted range compared to its associates, jack pine and eastern white pine.

EPIGENETICS

A developing area of research in understanding phenotypic plasticity is **epigenetics**, which is the study of heritable changes in gene expression and function unexplainable by changes in DNA sequences (Richards 2006; Bird 2007). These changes occur due to a set of molecular processes that alter gene expression by activating, deactivating, or reducing the activity of particular genes. In turn, these changes in gene expression can produce phenotypic variation in plants that is either perpetuated during development and/or is heritable across generations. Because epigenetic changes may affect plant phenotype and ultimately fitness, be inherited across multiple generations, and vary across populations and individuals, they are likely to contribute to the ability of plants to colonize, adapt, or evolve in variable environments (Amaral et al. 2020). For example, epigenetic phenomena have been shown to be responsible for a "memory" analog in plants to stressors such as pathogens, herbivory, or drought. Upon experiencing a stressful environmental event, epigenetic changes allow plants to store information about the stress such that the individual may respond differently if the stress is repeated (Lämke and Bäurle 2017). This epigenetically controlled memory may last from several days to years, and may even be passed on to offspring, helping trees to quickly respond to recurring biotic or abiotic stressors (Hilker and Schmülling 2019). Examples of studies that document epigenetic responses in trees to environmental stressors include those examining salt stress (Liu et al. 2019), heavy metal exposure (Cicatelli et al. 2013), drought (Sow et al. 2021), temperature extremes (Dewan et al. 2018), and pathogenic infection (Sollars and Buggs 2018).

Perhaps the most well-documented example of epigenetic memory is found in Norway spruce, a species native to areas of mild summers and cold winters in Europe. The temperatures that occur during embryo development and seed maturation have been shown to create important and long-lasting changes in the phenotype of seedlings that develop from those seeds, such as changes in the timing of bud set or bud break (Johnsen et al. 2005; Besnard et al. 2008; Yakovlev et al. 2010; Carneros et al. 2017). The ability of a seedling to "remember" the environmental conditions present when seeds were produced has important implications for their ability to rapidly adapt to the environment at hand, which has evolutionary significance if the environment is novel or changing. A study by Kvaalen and Johnsen (2008) showed that this epigenetic memory system in Norway spruce is able to produce differences in bud set or bud break resembling those seen among populations separated by as much as six degrees of latitude. In addition to evolutionary consequences, this epigenetic phenomenon will likely be important for tree breeding, conservation of genetic resources, and forest management (Amaral et al. 2020).

Much of this chapter has focused on the importance of genetic variation in species, for which there are decades of research with regard to forest ecosystems. As it is a rapidly emerging field, our focus on species' epigenetic diversity and its potential for phenotypic variation and adaptation is purposely reserved. However, knowledge of epigenetics will likely revolutionize our understanding of individual tree variation within the next decade. Detailed reviews of epigenetics in forest trees are provided by Amaral et al. (2020), Carbó et al. (2019), Sow et al. (2018), Pascual et al. (2014), and Yakovlev et al. (2012). Implications of epigenetics for ecology in general are described by Herrel et al. (2020), Rey et al. (2020), Kilvitis et al. (2014), Richards et al. (2010), and Bossdorf et al. (2008).

SUGGESTED READINGS

Adams, W.T., Strauss, S.H., Copes, D.L., and Griffin, A.R. (ed.) (1992). *Population Genetics of Forest Trees*. Boston, MA: Kluwer 420 pp.

Hamrick, J.L. and Godt, M.J.W. (1996). Effects of life history traits on genetic diversity in plant species. *Philos. Trans. R. Soc. B: Biol. Sci.* 351: 1291–1298.

Morgenstsern, E.K. (1996). *Geographic Variation in Forest Trees*. Vancouver, BC: Univ. British Columbia Press 209 pp.

Petit, R. and Hampe, A. (2006). Some evolutionary consequences of being a tree. *Annu. Rev. Ecol. Evol. Syst.* 37: 187–214.

Saeki, I., Dick, C.W., Barnes, B.V., and Murakami, N. (2011). Comparative phylogeography of red maple (*Acer rubrum* L.) and silver maple (*Acer saccharinum* L.): impacts of habitat specialization, hybridization and glacial history. *J. Biogeogr.* 38: 992–1005.

Stebbins, G.L. (1985). Polyploidy, hybridization, and invasion of new habitats. *Ann. Mo. Bot. Gard.* 72: 824–832.

White, T.L., Adams, W.T., and Neale, D.B. (2007). *Forest Genetics*. Cambridge, MA: CABI Publishing 708 pp.

Wood, T.E., Takebayashi, N., Barker, M.S. et al. (2009). The frequency of polyploid speciation in vascular plants. *Proc. Natl. Acad. Sci.* 106: 13875–13879.

Regeneration Ecology

CHAPTER 4

In Chapters 4 and 5, we consider the life history of forest trees within the context of landscape ecosystems. In this chapter, we examine woody plant regeneration as a critical part of the forest tree's life cycle, emphasizing the environmental factors influencing this process from reproduction to establishment. In Chapter 5, we discuss aspects of the structure and function of shoots, crowns, and roots as related to site conditions and growth of forest trees. As elsewhere, we do not attempt a full treatment of tree physiology; excellent treatments of the physiology of woody plants are available in the books of Kozlowski (1971a,b), Zimmermann and Brown (1971), Fitter and Hay (1987), Raghavendra (1991), Kozlowski and Pallardy (1997a,b), and Pallardy (2008). The life history of plants has been examined from the perspective of plant population biology by Grime (1979, 1988) and Silvertown and Lovett-Doust (1993).

The growth and form of a plant and its associated populations are affected by the physiological processes that occur in cells and organs. In turn, physical factors of the ecosystems in which plants reproduce, establish, and grow direct these physiological processes. Individuals of a population are recruited and established from a "bank" of seeds stored on or in the forest floor (Phases I and II; Figure 4.1). Once established, plants grow in height and mass, which requires space, light, nutrients, and moisture that may be insufficient to allow vigorous growth of all individuals (Phase III). During Phase III, some plants die (unbranched stems, T), others thrive (shown by branched systems), and others may persist for many years as part of the ground cover ("stored" as seedlings or sprouts) until their recruitment proceeds into the understory and overstory. The individuals are eventually restrained by limited abiotic resources and biotic limits (such as herbivory) of the ecosystem, indicated by the vertical bars on either side of the population (Figure 4.1). The environment of the understory and forest floor changes as the plant canopy develops (shown by the dashed arrow) that, in turn, affects the recruitment of new individuals. Seeds are dispersed to the forest floor in Phase IV, where reproduction occurs. In this chapter, we emphasize regeneration as a complex of ecosystem processes involving sexual and vegetative reproduction, dispersal, and establishment in relation to environmental factors.

REGENERATION

The process of regeneration, including seed production and maturation prior to dispersal, allows plants to maintain and/or expand their populations over time. In sexual reproduction, **seed production** is followed by **dispersal** of fruits and seeds, **germination,** and finally the **establishment** of seedlings on the forest floor, as follows:

Reproduction → Fruit and seed dispersal → Active or dormant seed bank → Seed germination → Plant establishment

Regeneration may also occur via vegetative reproduction, whereby stems of existing plants develop to maintain and expand the forest community.

Forest Ecology, Fifth Edition. Daniel M. Kashian, Donald R. Zak, Burton V. Barnes, and Stephen H. Spurr.

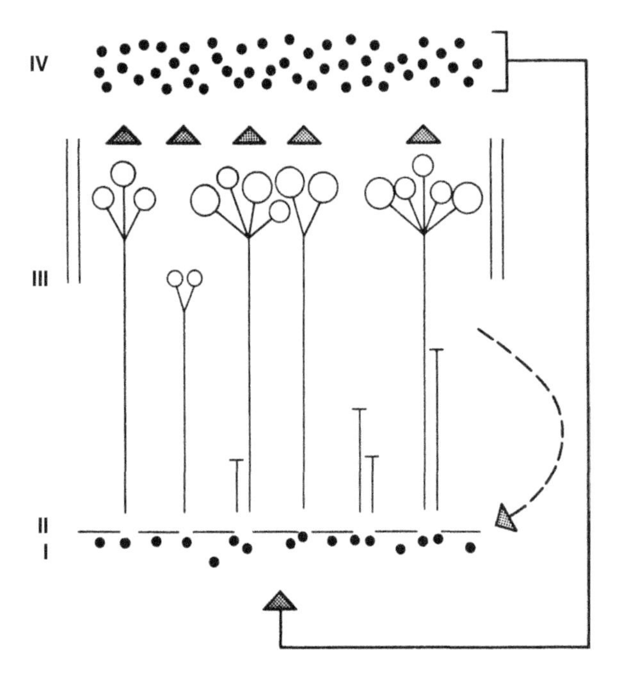

FIGURE 4.1 The plant life cycle with trees shown as a series of repeating modular units of shoots.
Phases shown:
I The bank of seeds on and in the forest floor.
II The establishment and recruitment of seedlings.
III Growth in height, mass, and number of modular units; vertical bars represent environmental constraints on growth; dashed line indicates the influence of the overstory on establishment and recruitment.
IV Seed production and dispersal.
Source: After Harper (1977) / Elsevier.

Whether by sexual or asexual reproduction, regeneration is an ecological process ensuring that successive generations of plants develop as part of a given landscape ecosystem. The established plants that do not die grow upward into successive forest strata (e.g., from the ground cover layer to the understory) via the process of **recruitment**. Finally, even fewer individuals in the understory are recruited into the overstory or canopy layer of the forest. At each step in the regeneration process, physical site factors such as light, temperature, moisture, nutrients, wind, and disturbance regimes are critical in directing the outcome, as are biotic factors such as herbivory, disease, competition, and mutualisms with plants and animals.

The functional characteristics and site factors of many widespread models of regeneration in forest ecosystems illustrate the interconnectedness of environment and biota and under which conditions each "strategy" is successful (Table 4.1). Episodic regeneration and recruitment of a new generation as a single cohort are common with the strategies of vegetative reproduction, fire-induced cone opening, or wind-dispersed seeds, but continuous or sporadic regeneration and recruitment over a period of years are more common with persistent juveniles (Table 4.1). Seeds in the active seed bank have no dormancy requirement (such as aspens, cottonwoods, and many pines) and germinate readily once seedbed conditions are favorable. In temperate forests, seeds of many trees and shrubs lie dormant over winter and germinate the following spring.

Sexual reproduction is the basic mode by which plants maintain their populations and persist in space and time. Sexual reproduction allows plants to adapt to changing environments because the process forms a zygote that is genetically different from either parent and from other offspring. By contrast, vegetative or asexual reproduction (i.e., **clone** formation) in plants creates

Table 4.1 Regenerative strategies of widespread occurrence in woody plants.

Regeneration method		Functional characteristics	Conditions under which strategy appears to enjoy a selective advantage
Vegetative: clone maintenance or expansion	V	New vegetative shoots remaining attached to parent plant until well established	Productive or unproductive sites subject to low or high intensities of disturbance
Active seed bank	B_a	Viable seeds that have no dormancy requirement; reside in seed bank less than a year	Sites with favorable weather or seasonally predictable disturbance by climatic or biotic factors
Dormant and persistent seed bank	B_d	Viable but dormant seeds present throughout the year; some persisting more than a year	Sites subjected to temporally unpredictable disturbance
Fire-induced opening of cones	F	Heat of fire opens cones where seeds are stored; seeds germinate immediately following favorable moisture and site conditions	Sites prone to relatively frequent, intense fires
Widely wind-dispersed seeds	W	Propagules numerous and exceedingly buoyant in air; widely dispersed and often of limited viability	Sites subjected to spatially unpredictable disturbance
Locally dispersed seeds	L	Propagules few and heavy; dispersed by gravity and animals; seed buried	Sites predictable in vicinity of parent plant
Persistent juveniles	B_j	Offspring derived from an independent propagule; seedling capable of long-term persistence in a juvenile state	Sites subjected to low intensities of disturbance

Source: Modified from Grime (1988).

genetically identical stems or **ramets** from a sexually produced parent plant (the **genet)** to form a spontaneous clone. Almost all woody plants are capable of some form of cloning. The multi-stemmed clone best illustrates the concept of an individual tree as a population of *modular units*, the shoots, where each shoot becomes a ramet of the original plant (Figure 4.1). By fragmenting its genotype, the plant gains growing space, water, and nutrients and eventually increases its capacity for sexual reproduction. Vigorous asexual reproduction by a plant is not known to diminish its sexual ability to produce flowers and seeds.

SEXUAL REPRODUCTION

The sexual reproduction and regeneration process is rather complex and involves several key factors (Figure 4.2). This process includes not only the production of fruits and seeds from flowers but also seed dispersal, formation or addition to a seed bank, germination and establishment, recruitment, and finally sexual maturation and flower production to begin the cycle anew. We begin the following text with maturation as the initiating factor.

MATURATION AND THE ABILITY TO FLOWER

Forest trees progress from *a juvenile phase*, characterized by no flowering,[1] to the *adult phase* when flowering occurs. **Maturation** refers to the gradual set of complex physiological changes in woody plant meristems that occur with increasing chronological age (Wareing 1987). Many characteristics

[1]Angiosperms produce flowers, whereas conifers do not bear flowers but produce cones (strobili) that bear naked ovules. For simplicity, we will use the term flowering to denote the reproductive process of both groups.

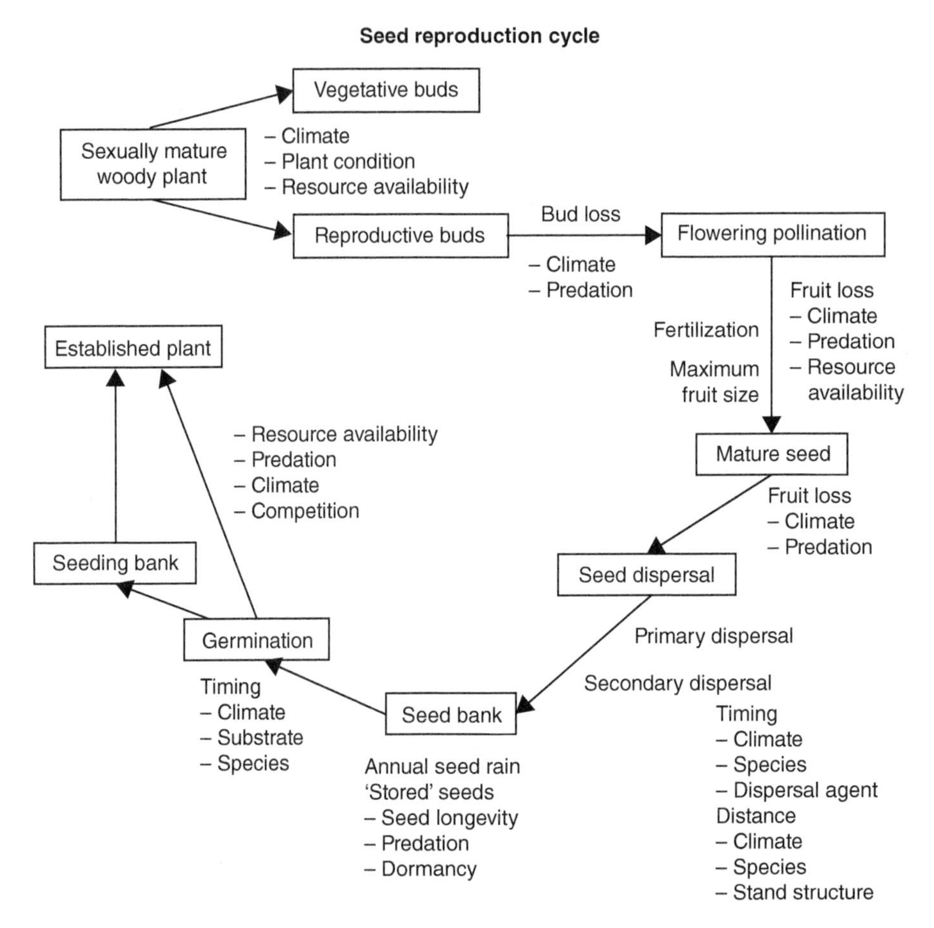

FIGURE 4.2 The sexual regeneration process of woody plants. *Source:* Zasada et al. (1992) / Cambridge University Press.

differ between the juvenile and the adult phases, including ability to flower, leaf morphology and retention, disease resistance, rooting ability, and growth rate (Bonga and Aderkas 1993; Greenwood and Hutchison 1993). We emphasize here that maturation is not the same as tree age; seedlings and young trees only have meristems in the juvenile phase, but older trees do not necessarily contain all meristems in the adult phase. In older trees, meristems in the juvenile phase occur at branch positions close to the trunk and near the base of the crown, apparently where meristems remain dormant for long periods (Bonga and Aderkas 1993). The adult phase appears first at the top of the crown as a tree ages and increases in size, apparently where the highest number of divisions have occurred in the meristems. Overwinter retention of dead leaves in oaks and beeches is indicative of the juvenile phase in a tree's crown.

The duration of the juvenile period varies markedly among species, from 1 to over 40 years (Owens 1991). Some conifer species may remain in the juvenile stage for 45 years or even their entire lives, to the point that species having very different juvenile versus adult morphologies have been misclassified into the incorrect genus (Pallardy 2008). Flowering appears to occur sooner in fast-growing, shade-intolerant species such as paper birch and Virginia pine compared to slow-growing, shade-tolerant trees such as American beech and eastern hemlock. For example, the juvenile period is estimated to last 5–10 years in Scots pine, whereas in European beech it is 30–40 years (Wareing 1959). The length of the juvenile phase may also be greatly influenced by the site conditions at different geographic locations (Kozlowski and Pallardy 1997b; Ross et al. 1983). Overall, plants with a reduced juvenile phase are short-lived, and those with a long juvenile phase are long-lived with a long reproductive life (Harper and White 1974; Bender et al. 2000).

There is inconsistent evidence that trade-offs exist between reproduction and growth (see discussion in the following text; Obeso 2002; Knops et al. 2007), but onset of the adult phase is more closely related to tree size than to age for many species (although height and diameter of young plants are often correlated). Small, suppressed trees in the forest understory may never flower, even at ages of 50–100 or more years. For many species, the attainment of a certain minimum size, rather than age, is the critical factor in attaining the adult or flowering phase. For example, European larch normally remains in the juvenile phase for 10–15 years, but was "forced" to flower in just 4 years when seedlings were quickly grown to the size threshold in a greenhouse (Wareing and Robinson 1963).

We emphasize here that flowering itself is not directly triggered by the attainment of a critical size, but through the expression of flowering-time genes that accumulate with growth. For example, a critical level of FT2 expression is required to initiate flowering (ending the juvenile phase) in eastern cottonwood (Hsu et al. 2006). Moreover, trees in natural stands vary widely in their genetic disposition to flower, particularly in number of reproductive buds produced and the ratio of female to male buds. Even some adult trees situated favorably in the overstory canopy do not flower or rarely flower, whereas adjacent trees are highly fruitful. Although trees attaining the adult phase are able to flower annually or seasonally thereafter, they may not do so every year due to a variety of environmental and genetic conditions (Pallardy 2008).

Variation in flowering among individuals of the same species may also be explained by sexual differentiation in some species. Species of several genera (*Acer*, *Ailanthus*, *Diospyros*, *Fraxinus*, *Gymnocladus*, *Maclura*, *Populus*, *Sassafras*, etc.) bear unisexual male and female flowers on different individuals (**dioecious** condition). The dioecious trait ensures outcrossing between genetically different individuals and precludes self-pollination (selfing). Selfing and inbreeding generally lead to growth depression in trees (red pine is a notable exception; see Chapter 3). Angiosperms rarely produce viable self-pollinated seeds. Conifers (especially pines) are more likely to produce viable self-pollinated seeds, although the resulting seedlings are often outcompeted and eliminated at an early stage in natural populations. Some tree species, including most conifers, bear both male and female unisexual flowers on the same tree (**monoecious** condition). Notably, some individual monoecious trees may be predominantly male, whereas others are predominantly female.

Increasing Seed Production Assuming there is a trade-off between reproduction and growth, a long juvenile phase is desirable for growth, but undesirable for fruit or seed production. Reducing the length of the juvenile phase is most effective when seedlings are grown so that they attain a large size as rapidly as possible. This increased growth rate may be accomplished with appropriate light conditions, by applying fertilizers, or using other cultural methods. Gibberellic acid application and other cultural treatments cause early flowering in most conifer species (Ross et al. 1983), but the use of asexually reproduced clones rather than sexually reproduced seedlings is often used for tree crops in high-yield plantations (Ahuja and Libby 1993). Trees of many species in the adult phase may be stimulated to increase flower and seed production by a variety of cultural methods (Matthews 1963; Kozlowski 1971a).

REPRODUCTIVE CYCLES

The reproductive cycle of trees and shrubs is closely adapted to and influenced by the physical site factors where they grow. For example, many woody species that grow on river floodplains and other wetland sites (most willows, cottonwoods, elms, silver maple, river birch, red maple, etc.) flower in the early spring and disseminate seeds four to six weeks thereafter. The seeds are dispersed onto moist seedbeds (e.g., recently flooded river sites and wetlands) where they germinate readily (active seed bank, denoted by B_a, and widely dispersed seeds, denoted by W, in Table 4.1). For these species, only a brief period of a few weeks is needed to proceed from pollination to seed dispersal, and the small seeds develop quickly. Millions of seeds are produced, some of which inevitably find a favorable seedbed for establishment so that a large food reserve in the seed is unnecessary. By contrast, most other trees in the north temperate zone have fruits and seeds that require two to four months to develop throughout the growing season and are disseminated in the fall or winter. In these cases, the trees produce medium-size and large seeds that lie dormant over the winter and germinate the following spring (dormant and persistent seed bank, denoted by B_d, in Table 4.1). The seedlings of such species (such as upland oaks and hickories) are soon subjected to drying soils and soil-water stress, requiring that a root system develops to cope with these droughty conditions. The large amount of stored food in the seeds of oaks and hickories provides the resources for this early root development (see discussion later in this chapter). A description of the reproductive cycle of many North American forest trees is available from the United States Department of Agriculture (USDA, 1990).

Virtually all tree species initiate reproductive buds or primordia (earliest bud stage) during the growing season of the year before **anthesis**, or the opening of the flowers. Flower buds are often visible along the current year's shoots of conifers and in the axils of leaves of hardwoods in the fall of the year prior to flowering. The reproductive cycle of Douglas-fir includes the initiation of lateral bud primordia in April along the vegetative shoot that develops inside the terminal vegetative bud (denoted by letter A in Figure 4.3). Some or many of these primordia may become pollen or seed cone buds depending on the favorability of the internal environment. When the vegetative bud bursts, needles flush (denoted by letter B in Figure 4.3), the shoot elongates (denoted by letter C in Figure 4.3), and the lateral buds become visible (by late July or August) along the young shoot (denoted by letters D and E in Figure 4.3). The new lateral buds will enlarge and flush the following spring to produce new vegetative shoots or male or female cones. Lateral buds at the base of the young shoot tend to become pollen cones, whereas buds toward the tip of the shoot become either seed cones or vegetative shoots; lateral bud fate is obvious by October.

Lateral bud primordia may follow any one of multiple development pathways. Some bud primordia abort early, degenerate, and leave no trace. Others, called **latent buds**, form bud scales and then cease to develop. If the terminal bud is removed, for example, by herbivory, the latent buds may develop into vegetative buds. Latent buds play an important role in oaks, for example, whose foliage is killed by spring frost or defoliated by insects, because they are able to develop an entirely new set of leaves. Lateral bud primordia not aborted or developed into latent buds develop into pollen cone buds, seed cone buds, or vegetative buds. The fate of lateral bud primordia is largely determined by the internal nutrition and hormonal relations in the shoot and tree. The same number of primordia may be initiated in two consecutive years, but the ratio of vegetative to reproductive buds may differ significantly. Thus, the inter-annual variation in seed cone production in Douglas-fir, and probably many other forest trees, results from the proportion of primordia that develop into cones, not the number of primordia initiated.

In Douglas-fir and most conifers, pollination and fertilization take place in the same year; cones mature during the summer, and seeds are released in the fall. This cycle is similar in many hardwoods. By contrast, fertilization occurs 12 months after pollination and the reproductive cycle is a full year longer in the pines and the red oak group.

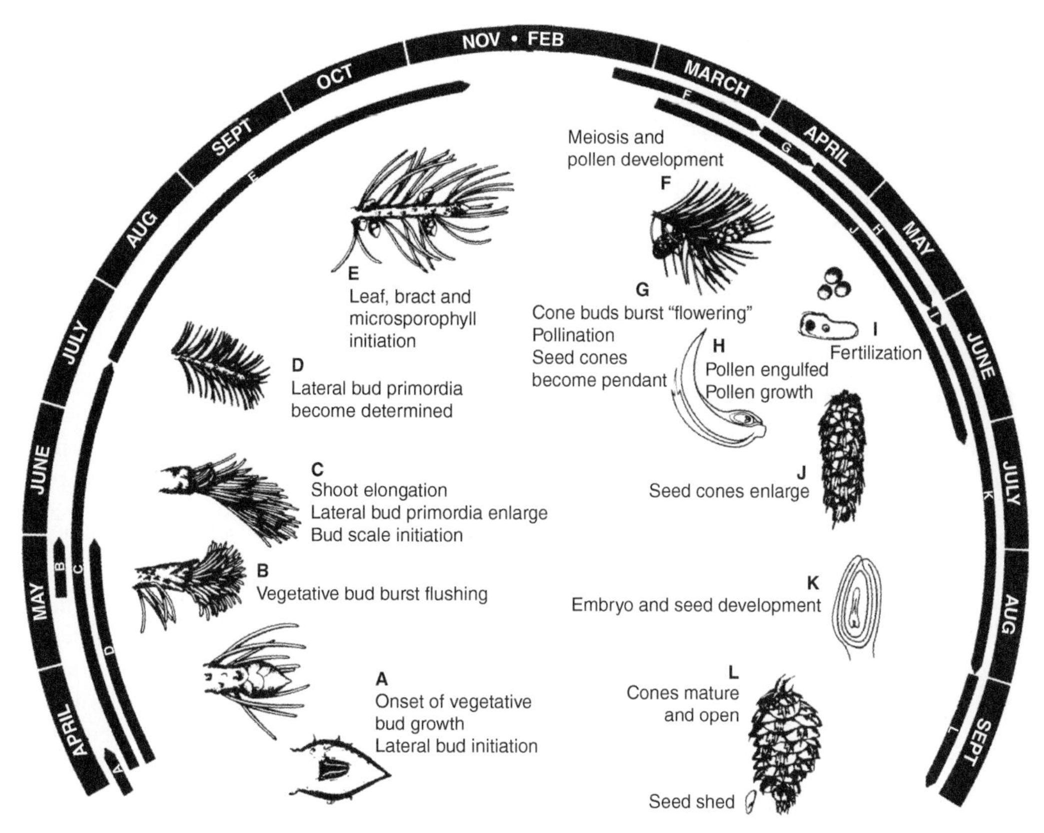

FIGURE 4.3 Reproductive cycle of Douglas-fir. The entire cycle extends over 17 months. Lateral buds are initiated in April and differentiate into vegetative, pollen, or seed cone buds during the ensuing 10 weeks. Pollination of the seed cones occurs the following April and the mature seeds are shed in September of the second year. The various stages are identified by letters A–L and are briefly described. The approximate length of each stage is shown by the arrows. *Source:* Allen and Owens (1972) / Canadian Forest Service.

POLLINATION

The timing of pollen release and female receptivity in deciduous species is closely related to whether pollination occurs by wind or animals. Wind-pollinated species such as aspens, birches, elms, and red maple flower in the early spring before leaf flush. Insect-pollinated species, such as basswood, tulip tree, and black cherry, typically flower during leaf flush. Conifers are exclusively wind-pollinated, which is promoted by the overproduction of pollen and the positioning of the female cones at the ends of shoots in the upper third of the crown. Wind-pollinated trees produce far more pollen than insect-pollinated species; a single anther of wind-pollinated birch may produce as many as 10 000 pollen grains compared to an insect-pollinated maple, which produces an order of magnitude fewer pollen grains (Faegri and Iverson 1989). Wind pollination is much more common in temperate and boreal forests than in tropical forests and in early-successional forests compared to late-successional forests (Pallardy 2008).

Weather conditions, especially temperature, affect the shedding of pollen and the receptivity of female flowers. Although anthesis of male and female is synchronized, it is not precisely synchronous on the same tree, thus reducing the likelihood of self-pollination. Flowering is rapid in wind-pollinated species. Most of the pollen for individual stands of Scots pine in Finland is shed within three to seven days, and over an even shorter period for some individual trees (Sarvas 1962).

Pollen discharge was highly correlated with high temperature and low humidity, and the day of maximum shed usually coincided with the warmest day of the flowering season. Moreover, pollen discharge was delayed in unfavorable springs until the occurrence of one or several favorable days adequate for pollen shed and spreading. In 14 years of monitoring pollination of Scots pine and European white birch, not once was a major part of the pollen crop destroyed by unfavorable weather. These data highlight the distinct advantage of rapidity for wind pollination (Sarvas 1962). By contrast, severe weather conditions may cause failure of fertilization (as opposed to pollination) and the process of ovule development. For example, regeneration failure of European linden in northwestern England is reported due to temperatures too low to permit fertilization (Pigott and Huntley 1981).

PERIODICITY OF SEED CROPS

Abundant flowering and seed production occur irregularly and erratically from year to year in natural populations of forest trees. This cyclic nature of seed production, highlighted by a year of exceptionally high production known as a **mast**, greatly promotes the establishment of tree seedlings and may be one of the most important life history traits in woody plants. Periodicity of fruiting and seed production is a pattern of repeated cycles of mast years followed by one or more years of lesser production. This pattern is particularly pronounced in beeches, oaks, and chestnuts, and is thought to be an antipredator adaptation. Many early-successional species—typically fast-growing and shade-intolerant species such as aspens, birches, cherries, cottonwoods, junipers, and many pines—exhibit mast years more frequently with fewer years between them than late-successional, slower-growing, more shade-tolerant species such as beeches, firs, sugar maple, and northern red and white oaks. In species prone to seed predation, masting was initially hypothesized to satiate seed predators because more seeds are produced than can be consumed (Silvertown 1980; Nilsson and Wästljung 1987; Kelly and Sullivan 1997). Mast flowering may have also evolved because of wind pollination, because many wind-pollinated species are pollen-limited due to an economy of scale (i.e., wind pollination is far more efficient when pollen density is high; Norton and Kelly 1988; Smith 1990). Both predator satiation and wind pollination may lead to masting, as in a study of yellow birch in New Hampshire (Kelly et al. 2001). Finally, some authors have suggested that masting is advantageous because it enhances seed dispersal during mast years (Crone and Rapp 2014; Pearse et al. 2016).

Trees tend to exhibit a regular cycle of seed production when studied over a long period of time, although some baseline number of seeds may be produced each year by the dominant trees. For example, Scots pine in central Finland produced 100 seeds m^{-2} every 5 years, and 150 seeds m^{-2} at 8-year intervals over a 50-year period (Koski and Tallqvist 1978). Similarly, abundant cone crops of western white pine occur about every three to four years, each followed by a marked drop in cone production the following year (Figure 4.4; Rehfeldt et al. 1971; Eis 1976). Notably, individual trees tend to exhibit similar periodicity even when there is considerable variation in inherent productive ability among trees (Figure 4.4). Good cone and seed crops typically occur at three- to seven-year intervals in pines, but they may occur nearly every year in some species (such as Virginia, Monterey, jack, and lodgepole pines). A three-year cycle of good cone crops has been reported for several northwestern fir species (Franklin 1968), and a five-year cycle for Douglas-fir. The periodicity of flowering and the minimum seed-bearing age for most North American trees and shrubs are summarized by the US Department of Agriculture (1974, 1990).

Factors that influence the initiation of flower primordia in turn affect seed production. Very often primordia initiation has been related to climate variables—particularly temperature and moisture—but their influence varies depending when they occur during the growing season. Wet and cool growing seasons prior to the initiation of flower primordia have been correlated with high seed production in temperate forests for spruces (Buechling et al. 2016), oaks (Fearer et al. 2008),

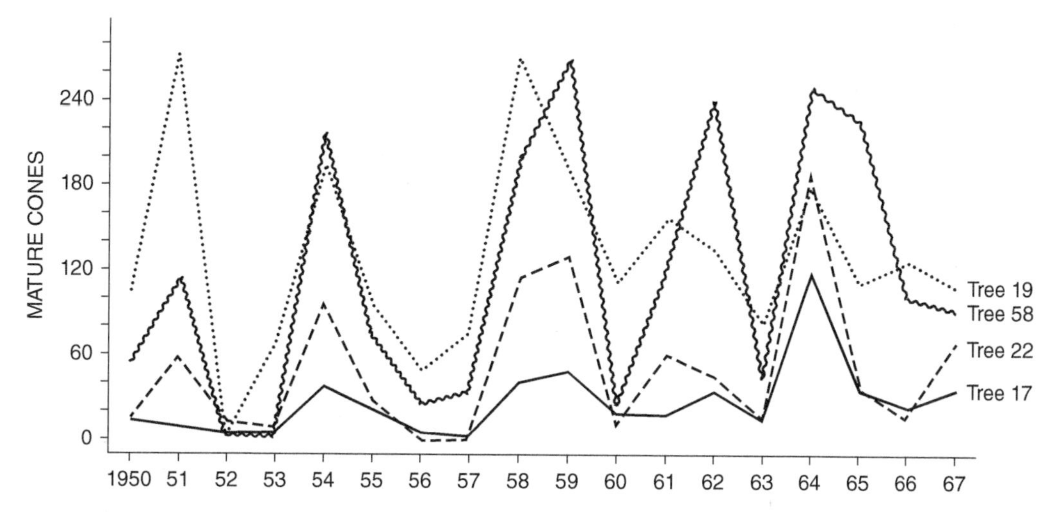

FIGURE 4.4 Periodicity of cone production of four western white pines in northern Idaho over 18 years. *Source:* After Rehfeldt et al. (1971) / with permission of Oxford University Press.

beeches (Piovesan and Adams 2001), and Douglas-fir (Lowry 1966). At the same time, warm summer temperatures during the early phase of floral primordia development have also been related to high seed production in these same species. The relationship is complex because two, and sometimes three, reproductive cycles are proceeding simultaneously (in pines and the red oak group), making it difficult to determine which portion of the cycle is being affected by temperature and moisture conditions. In addition to climate, temporal changes in resource uptake and utilization (particularly nitrogen) have been linked to mast events (Miyazaki 2013; Han and Kabeya 2017), but this relationship remains relatively unexplored. The maturing seed crop is a large sink for photosynthate and associated plant nutrients, and thus strongly influences the number of new flower buds initiated and the development of those already formed.

In addition to flower primordia initiation, seed production is also governed by factors that may preclude fertilization and ovule development following the initiation of flower primordia or cause the loss of maturing flower buds, fruits, or cones (Figure 4.2; Sork et al. 1993). Newly formed reproductive buds and immature fruits and cones may be killed or injured by spring frosts. Wind pollination and dispersal are negatively affected by high precipitation at anthesis (Allen and Platt 1990; Pearse et al. 2016), and low temperatures limit fertilization and ovule development. High temperatures or severe drought during fruit or seed maturation may cause immature fruits and cones to abort or be greatly reduced in size, although several studies have shown that moderate increases in temperature during seed maturation increase seed production (e.g., Lowry 1966; Sork et al. 1993; Fearer et al. 2008; Buechling et al. 2016). The seed crop is also subject to mechanical damage by strong winds, hail, and insect damage. In general, no single factor appears to drive masting in all species; one hypothesis is that mast seeding may be the result of an ancient ecological relationship in ancestral populations, probably related to both climate and resource availability, which was amplified though selection in the tree species we see today (Allen et al. 2017).

The abundance and periodicity of flowering are strongly influenced by physical site factors of light, temperature, and moisture as well as the genetic and physiological mechanisms of the plant. In a closed-canopy forest, most seed production results from the large dominant trees that have crowns well exposed to sunlight; smaller trees with narrow and suppressed crowns yield few if any seeds. Large, open-grown trees, free from aboveground competition for light and belowground competition for water, typically flower more frequently and abundantly than equally large individuals of the same species in a closed-canopy forest. Such open-grown trees may produce large quantities of seed each year if they are well pollinated. Isolated, open-grown conifers may produce many seeds by self-pollination, decreasing their viability and reducing the growth of seedlings they produce. In addition, more favorable site conditions are likely to result in a greater flower and seed crops. A classic study by Sarvas (1962) found that on poor sites in southern Finland (height of dominant trees, 16 m at 100 years) fewer than 30 seeds were produced per square meter. On sites of medium fertility (height of dominants, 23 m) and high fertility (height of dominants 27 m), 60 seeds m^{-2} and 90 seeds m^{-2} were produced, respectively. Similarly, pollen yields were much higher on fertile sites (35 kg ha^{-1}) than on infertile sites (9 kg ha^{-1}).

In a study of mast fruiting in black, northern red, and white oaks in east-central Missouri, Sork et al. (1993) considered both weather variables and the effects of prior seed production on the current year's acorn crop. They found that the size of an acorn crop was determined by both abundance of flowers produced and their survival to mature fruit production. Furthermore, they found that the length of reproductive cycles differed among the species (two, three, and four years for black, white, and northern red oak, respectively), although evidence exists for widespread synchronous flowering of oak species in mast years (Downs and McQuilken 1944; Beck 1977). They concluded that there was an inherent mast-year periodicity for each species that additionally may be affected by weather at critical times. Favorable spring temperatures during the season of fruit maturation appeared to be critical for good acorn crops of all species, whereas summer drought had a negative effect.

EFFECTS OF REPRODUCTION ON VEGETATIVE GROWTH

Flowering and the production of fruit and seed crops reduce vegetative growth. The common hypothesis to explain this relationship is that photosynthate is a limiting resource that must be allocated among reproduction and growth. Reproductive structures (flower, fruits, and cones) are major sinks for photosynthate that will be unavailable to support growth, representing an important trade-off (Harper 1977; Obeso 2002). Reproduction also affects terminal and lateral buds, which receive less photosynthate and nutrients so that immature shoots telescoped within them (next year's shoots) are also reduced in size. These patterns have been long documented for fruit trees (Pallardy 2008). Many correlational studies have also inferred this trade-off in forest trees, noting that heavy seed crops decrease both height and radial growth. For example, Holmsgaard and Bang (1989) reported that the stem volume increment of Norway spruce was reduced 12–25% the year of an unusually large or "bumper" crop and from 4 to 9% in the two following years. For the evergreen Mediterranean tree holm oak, acorn production was negatively correlated with annual and late summer–autumn stem growth during mast years, but these trade-offs were greater in smaller trees on poor sites (Martín et al. 2015). The negative relationship between cone production and radial growth has been documented over a large geographic area for firs, larches, spruces, pines, Douglas-fir, cedars, and hemlocks in North America (Koenig and Knops 1998). The strength of this trade-off appears to be species-specific (Berdanier and Clark 2016) and increases with tree age (Genet et al. 2010).

The evidence for a trade-off between reproduction and stem growth in trees is not clear, however (Obeso 2002). A study of five oak species in California found that radial growth and acorn production were indeed negatively correlated, but only in species that produce acorns in the same year that pollination occurs; species requiring two years for seed maturation showed no such relationship (Knops et al. 2007). Importantly, increased rainfall increases radial growth, but decreases seed production because it interferes with effective wind pollination. As a result, growth and reproduction are independent of each other, but vary in opposite directions due to the same environmental conditions, and therefore their negative correlation does not imply a causal mechanism (Figure 4.5). Likewise, a study of European beech found that radial growth was affected by reproductive effort and summer drought when they occurred simultaneously, but unaffected when one factor was

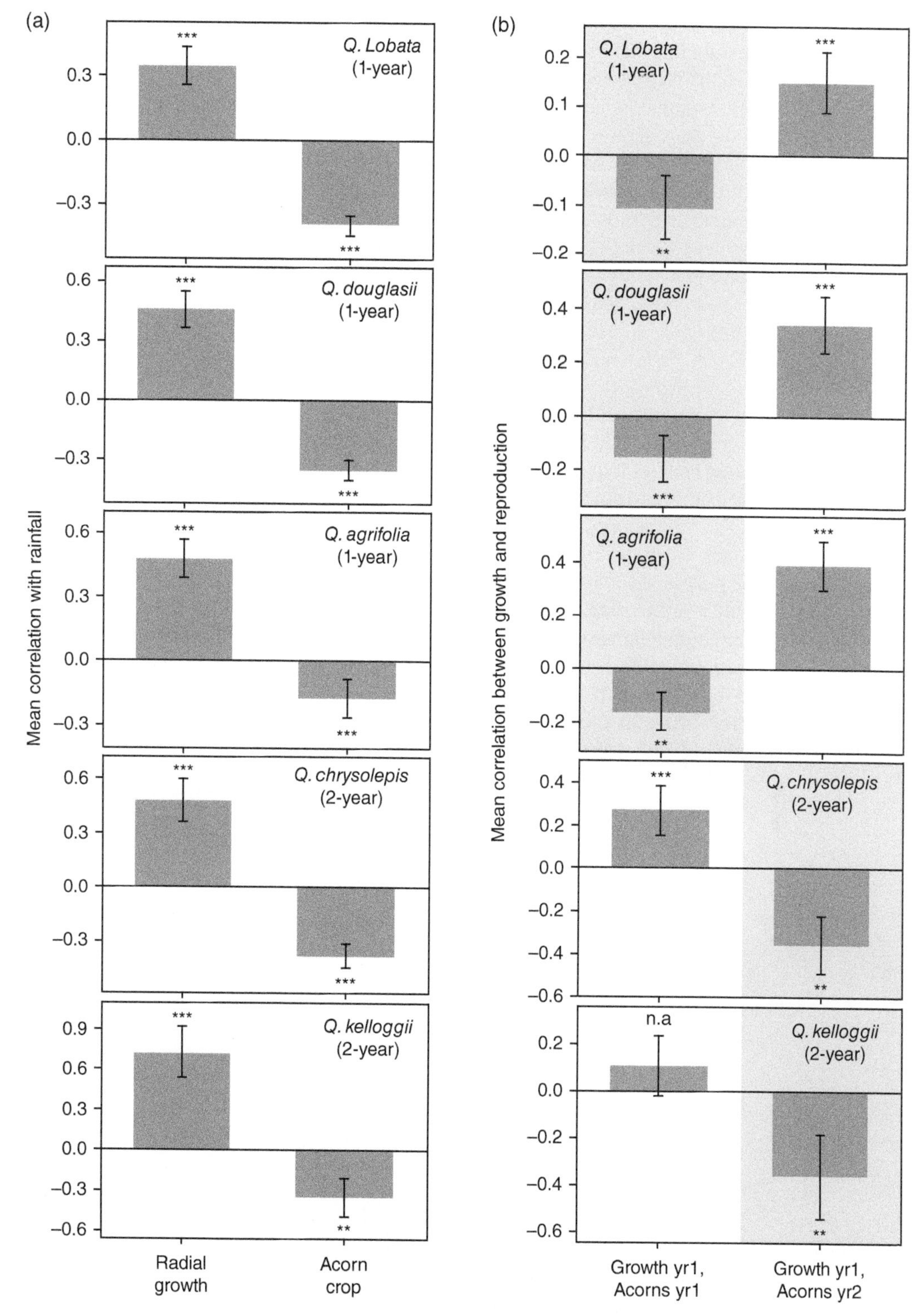

FIGURE 4.5 Correlation of (a) rainfall with growth and reproduction, and (b) growth with reproduction for five California oak species. Stem growth of all species increases with increasing rainfall, but reproduction decreases. Species whose seeds mature in one year exhibit a negative correlation between growth and reproduction in the mast year; this relationship occurs in the second year for species requiring two years for seed maturation. *Source:* Knops et al. (2007). Copyright (2007) National Academy of Sciences, U.S.A.

absent (Hacket-Pain et al. 2017). These studies suggest that if a trade-off exists between growth and reproduction, environmental factors play an important role in the strength of the trade-off.

Crown dieback as well as reduced radial growth has been documented in the event of a heavy seed crop. For example, an enormous yellow birch seed crop of 1967 in Ontario was eight times greater than the good seed year that occurred the previous year (Gross and Harnden 1968; Gross 1972). Resource allocation to flowering and fruiting suppressed early leaf expansion, and leaves in heavily fruiting crown areas were only 75% of their normal size. These small leaves led to poor shoot growth and small buds or lack of buds on the shoots, marked crown dieback occurred in 1968, and new crowns developed below the dead branches. Radial increment in 1967 was only 47% of the annual average increment, and diameter growth of current shoots was less than 5% of that in the previous year. Negligible fruit crops were produced the two following years.

DISPERSAL

Seeds are dispersed by gravity, wind, water, animals, and by combinations of these agents to microsites where they may germinate and become established seedlings. The method by which seeds of a species are dispersed enables it to compete in a particular spatial and temporal niche. In general, most tree species disperse seeds locally, with a high proportion falling within perhaps 40–50 m of the parent. Seeds of certain species are widely distributed by water and wind, particularly early-successional trees such as willows, sycamores, cottonwoods, birches, and aspens (widely dispersed seeds in Table 4.1). Most seeds of even these light-seeded, shade-intolerant, short-lived species are dispersed locally, but some are blown by wind or carried by water far from the parent to recently disturbed sites where establishment is more probable. Widespread seed dispersal was a successful means of sexually reproducing early-successional species in pre-European colonization forests, where disturbances were common.

Most trees and shrubs of northern latitudes disperse their seeds locally, with a seed rain decreasing exponentially with distance from the tree (Figure 4.6). In a study of Engelmann spruce

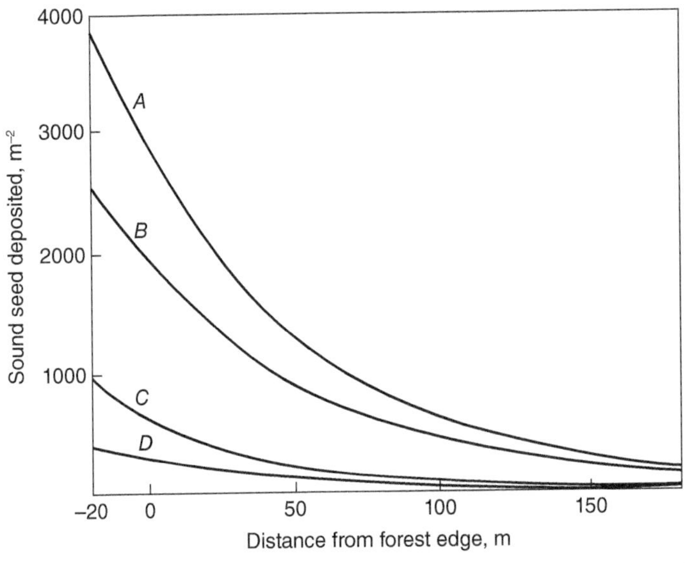

FIGURE 4.6 The distribution of seeds of Engelmann spruce in good seed years from the edges of forest stands into adjacent openings. (A) Togwotee Pass (1964), Teton National Forest, Wyoming; (B) Falls Creek (1952), Flathead National Forest, Montana; (C) Griffin Top (1964), Dixie National Forest, Utah; (D) Fisher Creek (1964), Payette National Forest, Idaho. *Source:* After Harper (1977) / Elsevier.

seed dispersal from four stands into adjacent openings, Dobbs (1976) found that most seeds fall near the parent, and approximately 70% of the seeds fell within 50 m of the edge of the standing timber. For white spruce in British Columbia, at 50, 100, and 200 m from the stand, seed rains of 19%, 12%, and 4%, respectively, were reported (Dobbs 1976). Only 9%, 5%, and 1% were reported for the same species and distances in Alaska (Zasada 1985). Seed rain in the interior of a forest stand is considerably shorter than into an adjacent opening because many seeds are intercepted by tree crowns of the overstory and understory.

Heavy-seeded species such as walnuts and oaks have seeds dispersed close to the parents, primarily by animals (locally dispersed seeds in Table 4.1). Dispersal relatively close (10–30 m) to the parent usually occurs onto soil-site conditions similar to those where the parent tree became established and is considered to be a selective advantage. Some species in this category have evolved additional adaptations to ensure that they are able to establish and compete with the plants and animals of the parent sites. Black walnut, for example, is highly sensitive to site conditions, and is not competitive where the soil is too wet, too dry, or too shallow. Its heavy seeds are an adaptation for this site specificity, ensuring that they reach the ground under the trees. Squirrels and other animals disperse the seeds to microsites away from the parent tree, but similar in soil-site conditions.

Most pines have moderately heavy, winged seeds that are wind-disseminated mostly within 50 m of the parent tree. In a study of longleaf pine, about 80% of the seeds caught in traps placed at increasing distances from the parent fell within 40 m of the tree. Similarly, for shortleaf pine in Arkansas, Yocom (1968) reported that one-half of the seeds trapped in a forest opening fell 10 m from the edge of the stand, and 85% fell within 50 m of the stand edge. Because pines are typically regenerated by fire and fire is likely within the lifetime of the parent tree, regeneration is likely to occur in the vicinity of the parent. Conifers having wingless and/or heavy seeds such as junipers and the stone pines (limber, whitebark, and Swiss stone, among others) and pinyon pines (wings remain attached to the cone scales as they open) are dispersed primarily by birds. Bird-and-mammal dispersal is extremely important for these and many other forest species and is discussed in detail in Chapter 12.

If seeds are stored in a seed bank on the tree, dispersal may occur immediately after seed maturation or periodically, depending on the establishment adaptations of the seedlings to conditions of different ecosystems (Zasada et al. 1992). Seeds of elms, aspens, and silver maple, all spring germinators, are dispersed immediately after ripening over a relatively short period in the spring, germinating and establishing on moist or wet sites. Some pine species (jack, lodgepole, Virginia, Monterey, etc.) have seeds that mature within cones that may remain closed (termed **serotinous** cones) for short or long periods of time (fire-induced opening of cones, denoted by F in Table 4.1; Chapter 10). These species are adapted to dry, fire-prone sites. Cones of some serotinous species may open periodically when threshold temperatures are achieved. Other cones remain closed until opened by fire-generated heat, dispersing the seeds onto the fire-prepared seedbed for germination following summer rains. Most other pines have a four- to eight-week seed dispersal period from cones that open shortly after ripening. This duration may be affected by weather conditions (Figure 4.7). For eastern white pine in Maine in 1965, 75% of the filled seeds were shed over a three-week period under warm and dry conditions. By contrast, cool, moist conditions of 1968 delayed and lengthened the dispersal period, requiring seven weeks to shed the same amount of seeds. Seeds of hemlocks, spruces, and firs are typically dispersed over many months in moist, cool climates. Dispersal begins in September or October for firs and continues over the winter as the cones disintegrate and release the seeds. In black spruce, cones are retained on the tree for several years in a "semi-serotinous" state, opening in warm and dry conditions (or when a fire occurs) and closing in cool and moist conditions. Semi-serotiny is advantageous because it ensures a continuous seed supply and that the seeds are not all susceptible to wildfire on the forest floor.

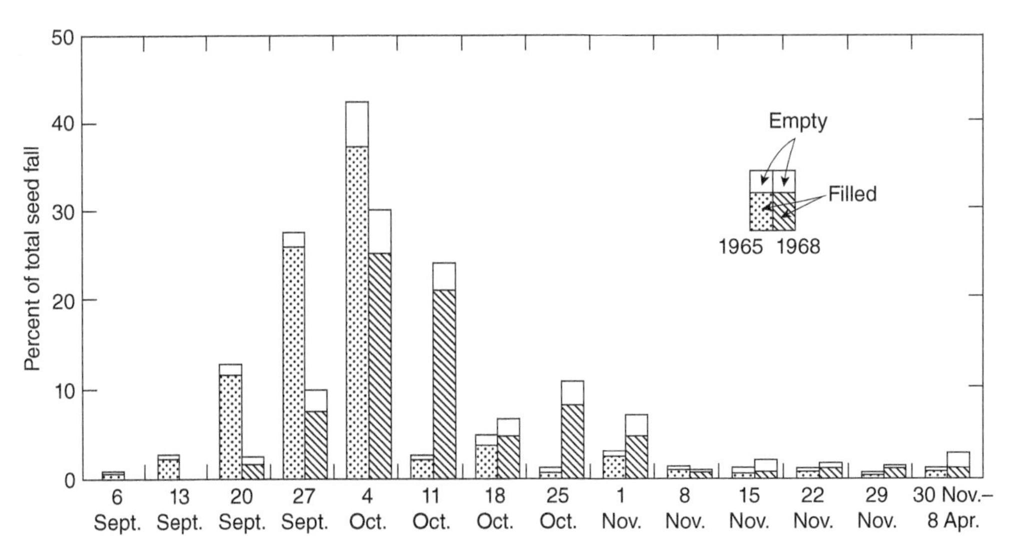

FIGURE 4.7 The time of seed fall of eastern white pine in 1965 and 1968, southwestern Maine. *Source:* After Graber (1970) / United States Department of Agriculture / Public Domain.

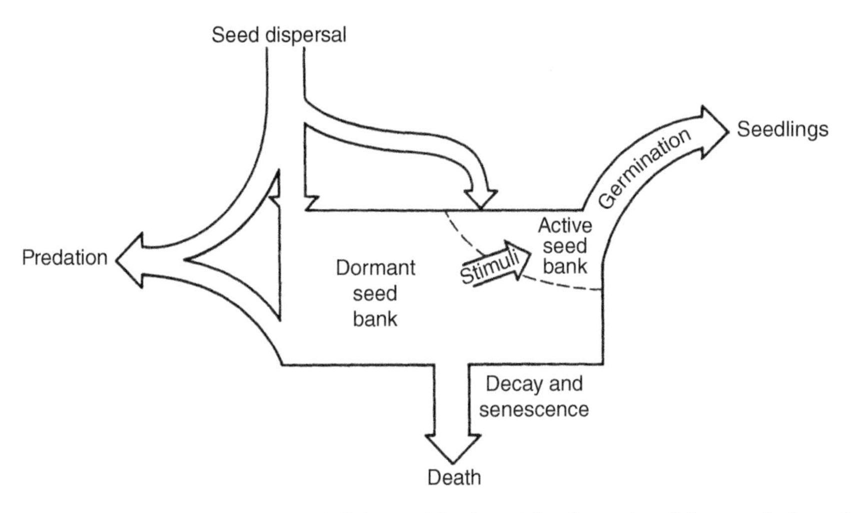

FIGURE 4.8 Conceptual model of the seed bank and the dynamics of the population of seeds. *Source:* Modified from Harper (1977), © 1977 by J. L. Harper. Reproduced by permission of Academic Press, Inc.

SEED BANK, DORMANCY, AND GERMINATION

Seeds disseminated onto the forest floor are stored in the seed bank until they germinate (Figure 4.8). Seeds of many forest trees arrive at the forest floor in a dormant state and remain in the dormant seed bank until internal and external conditions are favorable for germination. The residence period of viable tree seeds in the dormant seed bank (Table 4.1) is relatively short; being relatively large, they are eaten by animals, and they decompose easily. Seeds of other species, particularly those disseminated in the spring (e.g., eastern cottonwood and silver maple), are capable of germinating just a few days or weeks following dispersal, entering directly into the active seed bank (Figure 4.8 and Table 4.1). Non-dormant seeds are typically adapted to certain favorable germination conditions present immediately following dispersal. This adaptation applies to seeds of spring-disseminated, river floodplain species. Also, seeds of many species of the hard-pine group (including jack, lodgepole, longleaf, pitch, ponderosa, and Virginia) lack a dormancy requirement. Being strongly adapted to fire, their seeds are ready to germinate following rains on the fire-prepared seedbed. Other pines, including all pines of the soft-pine group (white and stone pines, among others), exhibit a pronounced period of seed dormancy, although they also depend on fire for establishment.

To germinate, viable seeds must (i) imbibe water, (ii) activate metabolic processes, and (iii) initiate growth of the embryo. Seeds are considered to be in a state of dormancy if they are unable to satisfy any of these requirements (i.e., blockage of any of these processes). Morphological and physiological changes must occur before dormancy is broken and the seed is capable of germinating. These changes take place slowly for seeds that overwinter on the forest floor, and the seeds typically germinate the following spring (dormant seed bank in Table 4.1). The moist, cool conditions of the forest floor over a period of weeks or months decrease germination inhibitors, increase germination-promoting hormones, and create favorable internal conditions for germination. Seeds of other species, including basswood, white ash, and black cherry, may require much longer periods (one to three years after dispersal) before their dormancy requirements are satisfied. Likewise, seeds of many shrub species (*Ribes*, *Rhus*, *Ceanothus*, and many chaparral species) may be stored for several years in the soil and germinate vigorously after fire.

ESTABLISHMENT FOLLOWING SEXUAL REPRODUCTION

Sexually produced seeds of woody plants produce roots, stems, and leaves in forming a seedling upon germination that must cope with the environment. Many seedlings perish soon after germination, but those that survive and grow vigorously are regarded as established. Establishment may take one to five years or more depending on the species and site conditions and is the most critical stage in the life history of a tree. As discussed in the earlier text, species have evolved many adaptations to produce and disperse seeds in such a way that maximizes the chance that many seedlings will encounter favorable conditions and survive. The regeneration ecology of each species is closely tied to the physical and biotic factors of the sites where it typically establishes. Notably, seedlings must survive on their seedbed, but they must also compete both with other seedlings of their own species as well as plants already established on the site. Assuming natural processes and disturbance regimes remain, these interactions bring about an apparent order and some degree of organization that enable us to associate particular forest communities with specific sites.

Establishment strategies may be categorized into **epigeous** (above the soil surface) and **hypogeous** (below the soil surface) germination and early seedling development. In the epigeous condition (nearly all conifers and most angiosperms), the cotyledons are elevated above the surface by the elongating hypocotyl, often still with the pericarp (fruit wall) attached (Figure 4.9a,b). In hypogeous species (mostly nut-producing species such as oaks, hickories, walnuts, and buckeyes), the cotyledons remain below the surface, attached to the seedling for weeks or months while the epicotyl grows upward and develops true leaves (Figure 4.9c,d). Four stages in cotyledon development are recognized (Marshall and Kozlowski 1976):

1. **Storage:** Cotyledon cells are packed with stored foods such as fats, carbohydrates, proteins, and mineral nutrients, which are utilized in the first days of seedling growth.
2. **Transition:** Following exposure to light, chloroplasts develop, and chlorophyll synthesis begins, stomates develop, epidermal cells expand, and large intercellular spaces form in the mesophyll.
3. **Photosynthesis:** The process of photosynthesis begins to further the development of the shoot and tap and lateral roots. Appreciable photosynthesis begins four to six days after radicle emergence in black locust, red maple, and American elm (Marshall and Kozlowski 1976). Peak photosynthesis occurs 8–15 days after radicles emerge and continues for about four weeks.
4. **Senescence:** Dry weight decreases and some mineral nutrients are translocated back into the seedling as the cotyledon function declines.

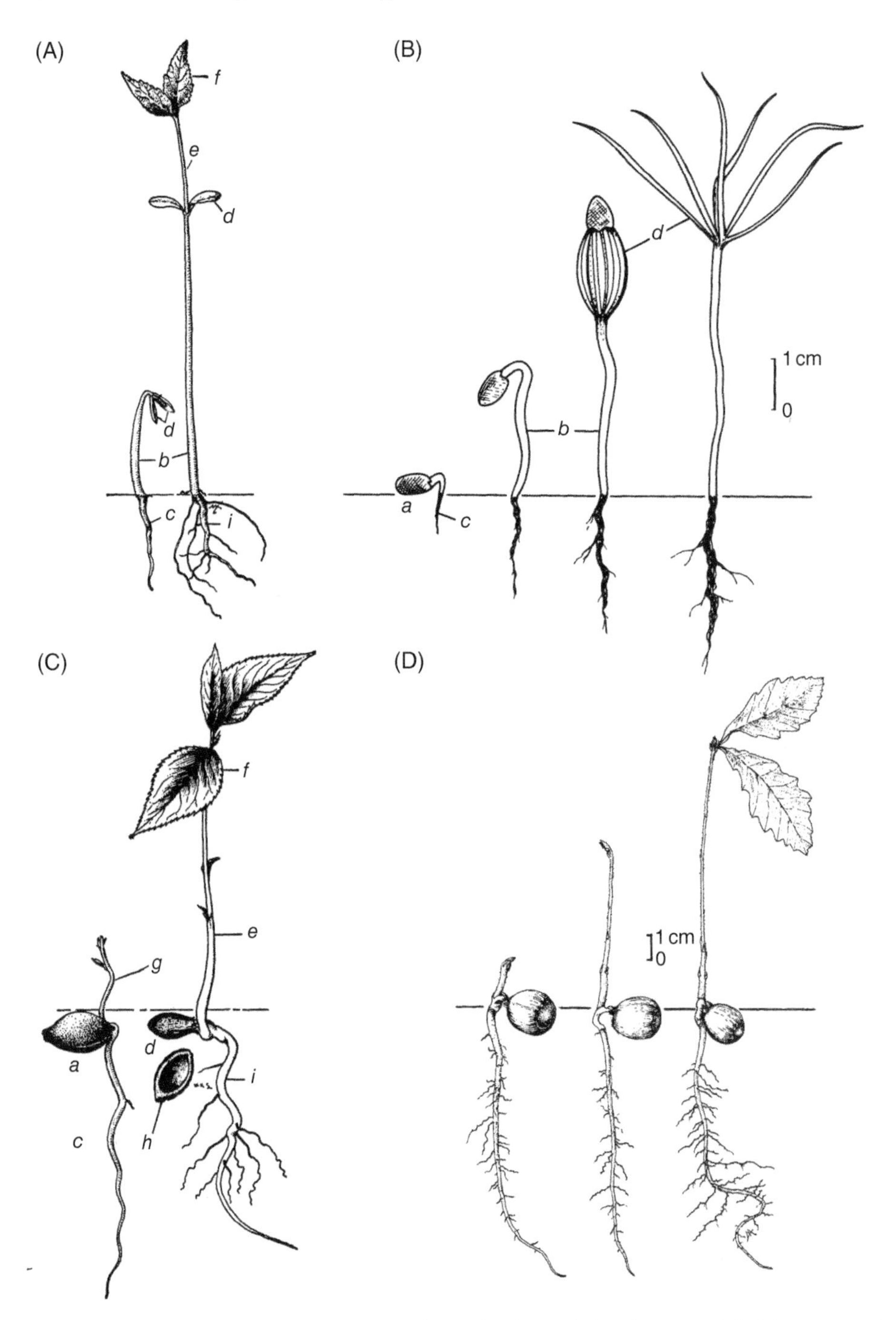

FIGURE 4.9 Epigeous and hypogeous seed germination and development. (A) Epigeous germination of pin cherry seedlings at 1 and 10 days; (B) epigeous germination of red pine at 1, 2, 6, and 10 days; (C) hypogeous germination of Allegheny plum seedlings at 1 and 9 days; (D) hypogeous germination of bur oak seedlings at 1, 5, and 12 days. (a) Seed; (b) hypocotyl; (c) radicle; (d) cotyledons; (e) epicotyl; (f) leaves; (g) plumule; (h) pericarp; (i) primary root. The plumule consists of the epicotyl and the emerging leaves. *Source:* A, C, and D after USDA (1974) / United States Department of Agriculture / Public Domain; B by W. H. Wagner, Jr.

Epigeous species store relatively little food in the endosperm or cotyledons and rely strongly on the cotyledons for photosynthesis to stimulate early root development. Thus, the cotyledons are extremely important to the development of seedlings during the first few weeks, and any damage such that might be caused by animals or frost inhibits seedling growth. By contrast, hypogeous seedlings have large, fleshy cotyledons that remain below ground during seedling development and are enclosed in the pericarp (Figure 4.9c,d). The large amount of stored food favors extensive root development prior to the development of a transpiring shoot and leaf system. The cotyledons also store considerable water and enough food to reestablish the epicotyl should it become damaged. Furthermore, the cotyledons are protected from browsing animals because they are underground and inside the pericarp.

The hypogeous system is typical of large-seeded woody species whose seeds are typically buried by small animals and thereby dispersed relatively close to the parent tree (locally dispersed seeds, denoted by L in Table 4.1). Many oaks and hickories grow on dry sites characterized by summer drought, and their seedlings typically establish themselves in the shade of trees whose roots compete for soil water and nutrients. In this context, production of fewer well-provisioned seeds appears to be an adaptive advantage. Mesic sites with available soil water throughout the growing season often support mixed hardwood forests with both hypogeous (bitternut hickory and northern red oak) and epigeous (sugar maple, white ash, and beech) systems.

Young seedlings pass through a succulent stage during the first several weeks, when tissues are soft and highly susceptible to fungal infections, damage by insect larvae and other animals, smothering, and desiccation. Species whose seedlings have shallow young roots are especially subject to mortality compared to those with deeply penetrating "tap" roots. There follows a juvenile period when the seedling becomes increasingly hardy but still subject to mortality once tissues begin to harden. A classic study of seedling survival of Douglas-fir (Lawrence and Rediske 1962) documented low cumulative germination and survival resulting from high seed destruction by fungi and animals prior to germination (46%) and high first-year germination failure (27%), as well as other factors (Figure 4.10).

The kinds of sites and seedbeds where seedlings of different species become established are important for understanding site–species relationships and for predicting seedling survival. Most tree species depend on disturbances (e.g., fire, uprooting of trees by wind, or flooding) to provide suitable seedbeds for their establishment. Fast-growing, light-demanding species (pines, willows, cottonwoods, aspens, birches, etc.) are favored by catastrophic disturbances that provide open and extreme sites. The adaptations of fire-dependent species are presented in Chapter 10.

The establishment of many tree species is favored or not severely limited by partial shade of an overstory, but heavy shade typically prevents establishment and/or persistence of most species. Similarly, most seedlings can establish on continuously moist forest soils, but only few are able to establish and persist as soils dry out and soil-water stress becomes severe. Each species therefore has a unique set of adaptations that facilitates establishment under certain physical and biotic conditions. In the following text, we divide (and simplify) the broad continuum of regeneration systems into three groups of species that differ in establishment pattern in relation to site conditions and the type and severity of disturbance.

1. **Early-successional species** (often called **pioneer** species) seed into open areas following major disturbances such as fires, flooding, windstorms, and landslides. Relatively little tree competition is present, and the environment is often harsh (hot, cold, dry, wet, and/or windy). Germination and growth on these sites are rapid, and seedling roots penetrate deeply enough into the soil (as in the pines) to make the seedlings that eventually become established resistant to yearly summer drought. Seedlings of floodplain species tolerate variable water levels and are able to generate adventitious roots from their stems if covered with water or silt. Fire-dependent pioneer species, such as the southeastern pines of North

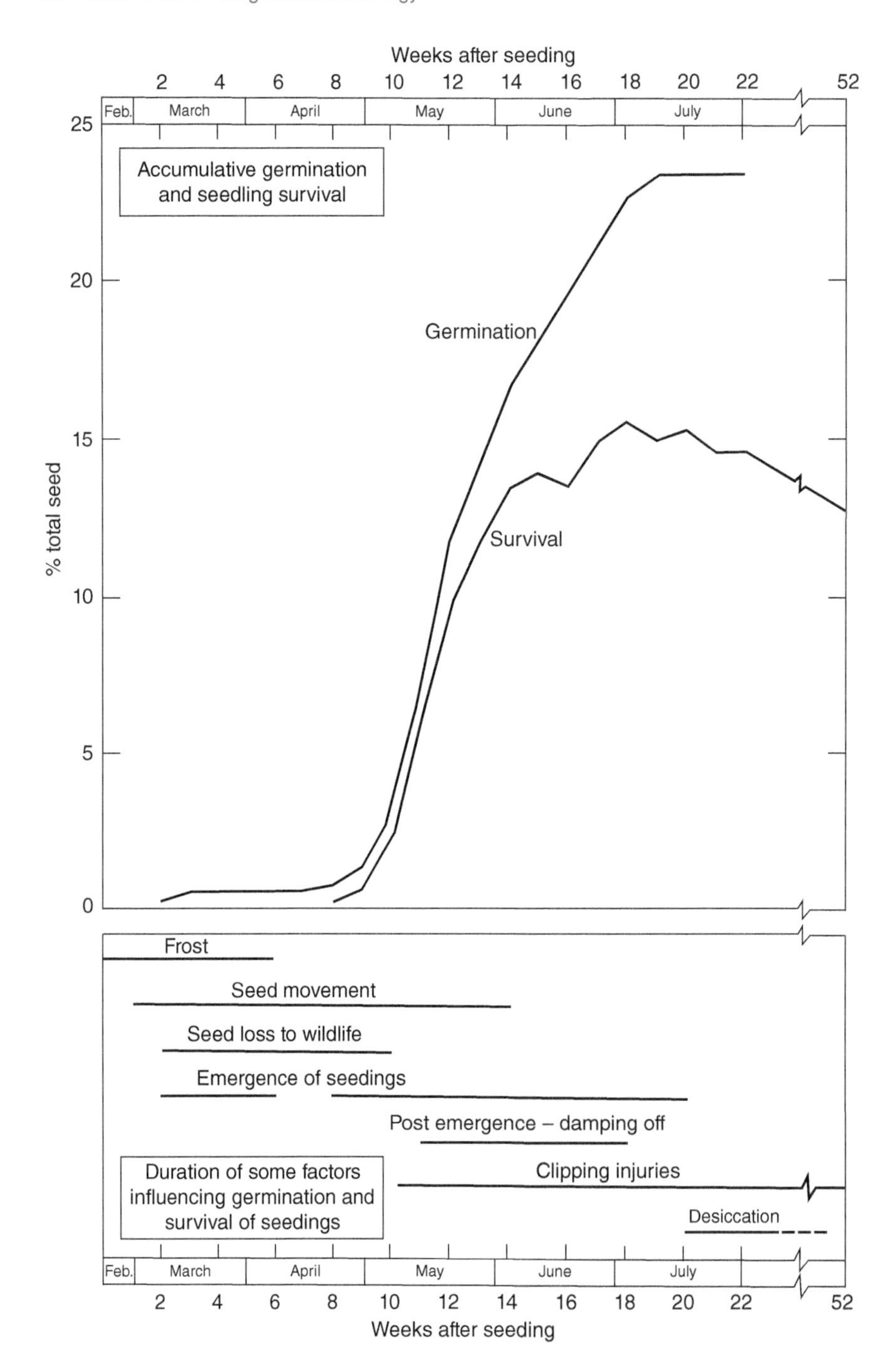

FIGURE 4.10 The cumulative germination and survival of Douglas-fir seeds during the first growing season. 440 Scandium[46]-tagged seeds were seeded by hand in February and their fate carefully followed. *Source:* Lawrence and Rediske (1962) / with permission of Oxford University Press.

America (longleaf, shortleaf, loblolly, and slash), typically establish in dense, even-aged stands on recently burned areas, and the seedlings develop rapidly during the first year (Figure 4.11; Wakeley 1954). Tap root development is rapid because considerable soil water is needed to meet transpiration demands in the needles and shoots which develop rapidly in the ensuing juvenile period.

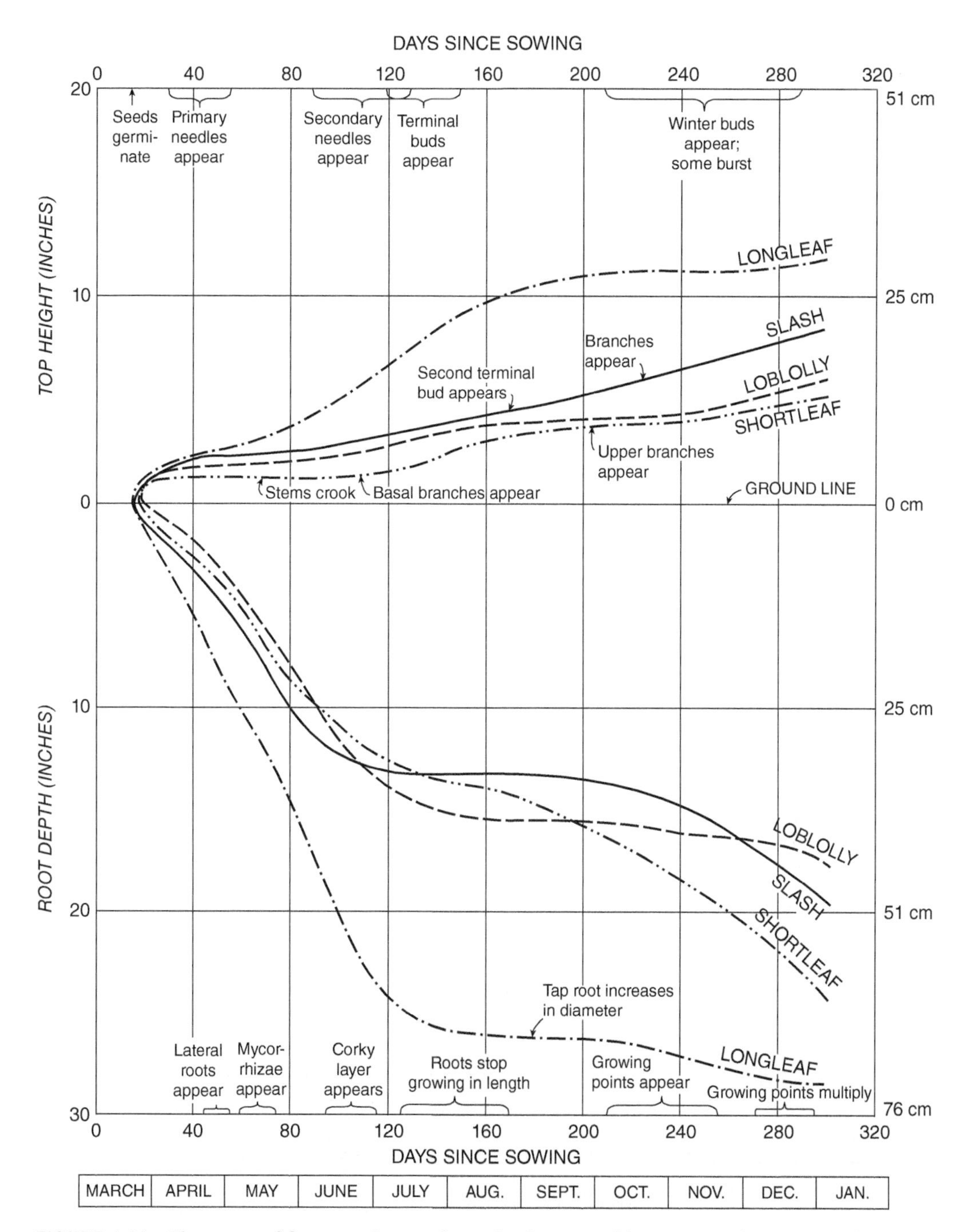

FIGURE 4.11 The course of first-year shoot and root development of four species of pines under favorable conditions for establishment (in a forest nursery) in Louisiana. *Source:* Wakeley (1954) / United States Department of Agriculture / Public Domain.

2. **Gap-phase species** establish under the existing forest canopy and are shade-tolerant enough to persist until a small disturbance enables them to penetrate a gap in the canopy (Chapter 16). Common gap species include white ash, black cherry, white and northern red oak, red maple, yellow birch, basswood, black walnut, slippery elm, white spruce, and to a lesser extent Douglas-fir and white pines. Notably, a seedling of any species may reach the

overstory canopy via a gap, if positioned in the right place at the right time. Gap-phase species are able to tolerate the deeply shaded forest understory, but can also grow quickly enough to colonize the gap before the gap is filled by nearby crowns or by competing species. Gap size is important for this process, as light-requiring pioneer species are better able to regenerate in large gaps than gap-phase species or the slow-growing, very shade-tolerant species that can reach the canopy in small gaps. The phenomenon of gap or patch dynamics is discussed in detail in Chapter 16.

3. **Shade/understory-tolerant species** have seedlings able to establish in the shaded understory on moist or **mesic** sites (i.e., where soil water is available throughout the growing season) and persist for years or decades (persistent juveniles, denoted by B_j in Table 4.1). Seedlings of these species (e.g., sugar maple, American beech, hemlocks, true firs, and western redcedar) slowly penetrate the canopy as overstory trees die or windthrow provides openings for them.

Seedlings of both gap-phase and shade-tolerant species (groups 2 and 3) establish under a forest canopy and must persist in the understory until conditions are favorable for their growth into the overstory. Species in these groups have evolved many different kinds of adaptations that allow their seedlings to persist in the understory and respond to release depending on (i) the physical and biotic conditions of the understory, (ii) the species' intrinsic growth rate, and (iii) the type of disturbance that provides canopy openings. Both groups (but especially group 3) accumulate a few to hundreds or millions of seedlings per hectare in the understory. Both groups also have different functional adaptations that enable them to tolerate and survive the physical rigor (low light and soil-water stress) and biotic hazards (herbivores and diseases) of the understory environment. Gap-phase species better tolerate soil-water stress than do shade-tolerant species, but the reverse is true for low-light conditions. In addition, deciduous gap-phase species are able to sprout after fire or animal damage and continue to persist in the understory. Most shade-tolerant species have slow-growing seedlings which require more soil water, typically dominate on mesic sites, and may accumulate large seedling populations that experience high mortality. Seedlings of shade-tolerant species may endure for 20–30 years or more to shade out any seedlings that might try to establish under them; understory hemlocks may even be 100 or more years old. Both gap-phase and shade-tolerant trees that eventually reach the overstory were recruited from a bank of established and persisting seedlings.

Two associated species of the northern hardwood forest, yellow birch (gap-phase) and sugar maple (shade-tolerant), have contrasting establishment strategies. These species grow side by side in the northern forest, but with very different physiological adaptations and realized niches. Seed dispersal for sugar maple occurs with leaf drop (before snowfall) in the early fall. Yellow birch seeds (more than 50 times smaller than sugar maple seeds) also mature in the early fall, but they are dispersed gradually throughout the winter and may be blown for great distances on top of the snow (Figure 4.12a; Tubbs 1965). Sugar maple seeds germinate under the snow in the early spring within a layer of leaves where temperatures are only about 1 °C (Figure 4.12b,c), while yellow birch seeds remain on top of the snow (Figure 4.12a). Radicles of maple seedlings penetrate the wet leaf mat and establish by the millions in a good seed year, often forming a carpet of first-year seedlings on the forest floor. By contrast, birch seeds come to rest on the forest floor after snowmelt and germinate in late spring at higher temperatures (about 10 °C). Birch seed radicles are tiny and unable to penetrate the thick leaf mat, which tends to dry out rapidly. Most yellow birch seedlings desiccate and die, but some birch seeds come to rest by chance on rotting logs, moss-covered rocks, or mineral soil of mounds formed by uprooted trees where they may establish if sufficient light is available and the substrate does not dry out.

Sugar maple seedlings are highly shade-tolerant and are able to survive the forest floor environment better than most other tree species in its geographic range. Yellow birch seedlings cannot endure heavy shade, but grow much faster than sugar maple at higher light levels. Thus, an

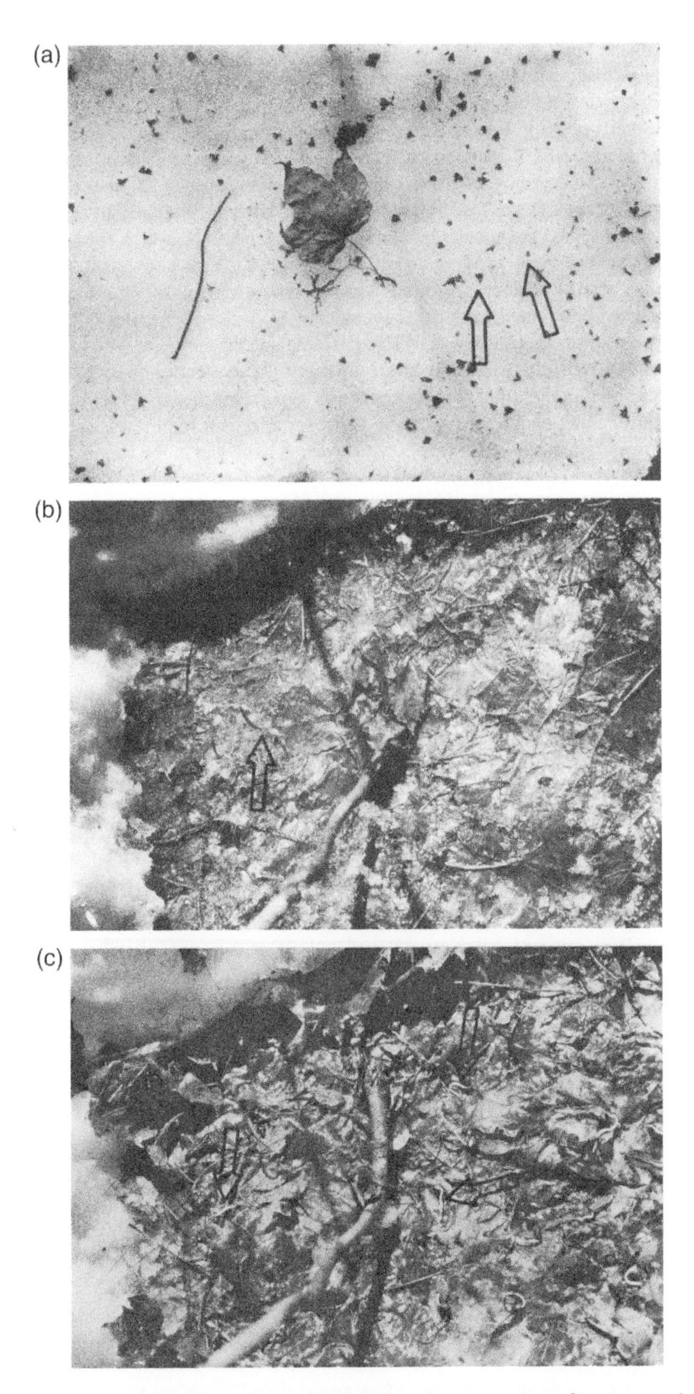

FIGURE 4.12 Sites of seed dispersal and germination of sugar maple and yellow birch in the northern forest. (a) Yellow birch bracts and seeds litter the surface of melting snow on April 30 in a northern hardwood stand on the Upper Peninsula Experimental Forest, Michigan. The overstory is composed primarily of sugar maple of seeding age, but seldom are maple seeds observed on top of or within the snow cover even after a bumper crop. Right arrow indicates a yellow birch seed; left arrow points to a bract. (b) Removing the snow from the exact spot shown in (a) to the top of the previous fall's leaf layer reveals no seeds of any species. The leaves are compressed into a soggy mat that is often partially frozen. Spring ephemerals have pushed through the mat (arrow). (c) When the top leaf layer shown in (b) is removed, the ability of sugar maple to germinate underneath a snow cover in early spring is revealed. Arrows point to germinated seeds. In those areas sampled on the Experimental Forest, the bulk of sugar maple seed was found under a layer of leaves, whereas yellow birch seeds were distributed on the top of the snow as illustrated in (a). *Source:* After Tubbs (1965); U. S. Forest Service photos.

appropriately positioned yellow birch is able to outcompete the slower-growing maples for a place in the overstory canopy upon the formation of a large gap. Sugar maple is able to dominate both understory and overstory in many ecosystems of the upland hemlock–northern hardwood forest because of its effective seed production, dispersal, germination, and establishment adaptations.

POST-ESTABLISHMENT DEVELOPMENT

Tree species exhibit rapid growth in the juvenile phase when environmental conditions are favorable. Growth in trees slows greatly with increasing age and begins to decline in old age. The high rate of stem mortality that occurs during the establishment period decreases with time. Trees continue to compete for light, water, and nutrients as they increase in size, and their density (number of stems per unit area) decreases (Chapter 13). Relationships of shoot, crown, and root structure to regeneration ecology and tree growth are considered in detail in Chapter 5.

VEGETATIVE REPRODUCTION

Reproduction by vegetative means rather than sexually with flowers and fruits may be critical for the survival of woody plant populations in some situations, such as following disturbances. Vegetative reproduction is achievable (and is a major fitness trait) for all woody angiosperms once established (Chapter 3), but is less common in conifers. This difference between conifers and angiosperms may be one of the major reasons for angiosperm dominance since the Cretaceous Period. Vegetative reproduction is advantageous because it allows the plant to survive and reestablish itself in place and sometimes to expand its spatial coverage of an area. It also allows plants to quickly colonize disturbed areas, as in pioneer species such as aspens, willows, sumacs, dogwoods, and many shrub species. Vegetative reproduction produces ramets genetically identical to the parent plant which grow rapidly from an already-established and functioning root system—quite unlike seedling establishment. Ramets may persist for long periods of time once established. In fire-prone ecosystems, woody species that form clones resprout vigorously after fire, preventing erosion and rapidly recoupling nutrient cycles.

Vegetative regeneration contributes meristems to the bud bank from the crown, basal stem and root collar, and underground roots and rhizomes (Figure 4.13). Vegetative buds are released by temperature, fire, wind breakage, herbivores, or other stimuli, forming new shoots and initiating clonal regeneration and expansion. Most woody plants regenerate vegetatively through sprouting (development of new shoots from adventitious buds) and rooting of stems. Sprouting occurs in different parts of the plant:

Basal stem: New shoots may develop from dormant or adventitious buds at the basal part of the stem (i.e., at the root collar where stem joins root) of an established plant. Sprouts from basal stems may form multiple-stemmed shrubs or trees such as oaks, hickories, basswoods, ashes, walnuts, birches, alders, prickly-ash (Reinartz and Popp 1987), and species of many other genera.

Root: New shoots may arise from adventitious buds on roots. Root sprouting is common in clone-forming trees and shrubs such as aspens, sassafras, sumacs, and sweetgum, as well as in American beech.

Rhizome: New shoots may develop from underground horizontal stems. The distal end of a rhizome becomes erect and bears leaves and flowers. Sprouting from rhizomes is typical of many shrubs such as salmonberry (Tappeiner et al. 1991) and other *Rubus* species, and trees such as flowering dogwood, striped maple, Gambel oak (Tiedemann et al. 1987), and dwarf chestnut oak.

Lignotuber: New shoots may arise from a buried mass of stem tissue termed the lignotuber, as in Gambel oak and eucalypts.

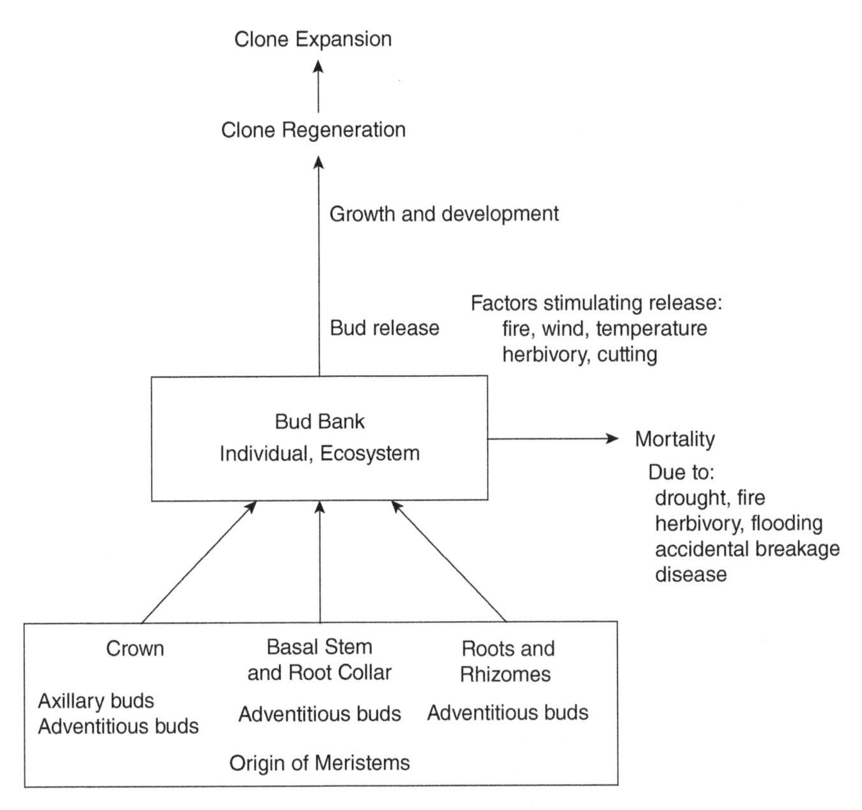

FIGURE 4.13 Conceptual model of vegetative regeneration and clonal expansion of woody plants. *Source:* Zasada et al. (1992). Reprinted with permission of Cambridge University Press.

Vegetative reproduction that occurs by rooting of aerial stems and subsequent shoot development includes:

Stolons or runners: Stolons are the arching branches of shrubs, such as red-osier dogwood and *Rubus* species, which take root when they come in contact with the soil surface and form new plants. Runners (procumbent stems) of plants such as creeping strawberry-bush and Virginia creeper take root at various points along their length as they encounter the soil surface.

Fragmentation: Branches that are broken off by ice, flood debris, or wind or senesced from willow, cottonwood, and other stream-side species may take root and become established after being buried in soil of the stream bank or the active floodplain.

Layering: Lower branches of boreal and northern conifers such as spruces, firs, and larches are often pressed into the soil by the weight of snow or by woody debris. These branches take root and form clones around the perimeter of the parent tree.

Tipping: Northern white-cedars in swamps are often uprooted just enough by wind so that they very slowly tip over and eventually lie procumbent on the peaty surface of the swamp (Curtis 1959). The lateral tree branches turn upright and develop into independent trees as moss and organic matter cover the old tree, and roots form at the base of each branch.

In addition to maintaining individual genotypes following disturbance, vegetative reproduction expands the area occupied by many shrubs and trees such as blueberries, dogwoods, sumacs, sassafras, beeches, aspens, sweetgum, and sycamores. Trembling aspen clones, consisting

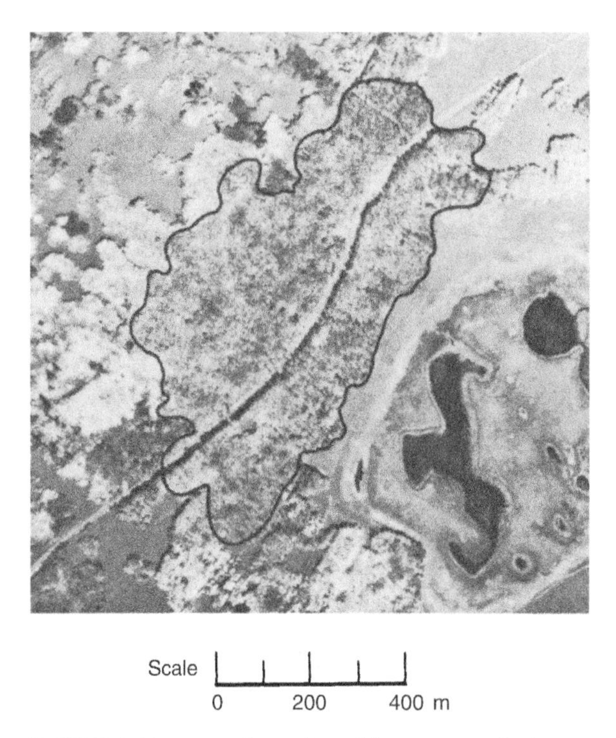

FIGURE 4.14 Large clone of trembling aspen, 43 ha in extent, with the boundary outlined; south-central Utah. Note the smaller clones around the large clone; the differences in tone indicate differences in fall coloration of clone foliage. *Source:* After Kemperman and Barnes (1976). Reprinted with permission of National Research Council Canada.

of many genetically identical ramets, are very common following wildfires. In the Intermountain West of North America, trembling aspen clones may encompass very large areas; one such clone in south-central Utah commonly known as the "Pando Clone" is 43 ha in area (Figure 4.14; Barnes 1975; Kemperman and Barnes 1976; DeWoody et al. 2008; Mock et al. 2008) and is reported to be one of the world's largest organisms (Grant et al. 1992; Grant 1993). Vegetative reproduction is much more common in maintaining and expanding some clonal species than is the production of new seedlings, including trembling aspen (Romme et al. 2005) and the northern subshrub wintergreen (Moola and Vasseur 2008).

Rather than spreading expansively by sprouting from roots, many forest shrubs spread by the repeated rooting of aerial stems that root and produce new ramets when pressed into the forest floor. For example, clones of the shade-tolerant vine maple develop in the understory of Douglas-fir forests in the Pacific Northwest when stems are pinned to the forest floor by falling trees and branches and then take root (O'Dea et al. 1995; Figure 4.15). Vine maple establishes by seed following fire or windstorms and then develops as multiple-stemmed groups. The maple branches are vinelike and flexible and droop to the forest floor over time. Almost 99% of the stems that reproduce vegetatively by layering are pinned to the forest floor by fallen trees or branches, and repeated layering events by the genet and its ramets extends the clone over a large area (Figure 4.15). This clonal process allows vine maple to persist in the understories of Douglas-fir forests, perpetuated in part by the mutualistic relationship with the dominant trees themselves.

Most vegetative reproduction by conifers is by layering in the northern, boreal, and alpine areas, but even in these examples, conifers reproduce primarily by sexual means. Only a few conifer species are able to sprout vigorously and regenerate themselves after fire, grazing, or cutting. Examples include the redwood, which sprouts vigorously from root collar, stump, and stem, and

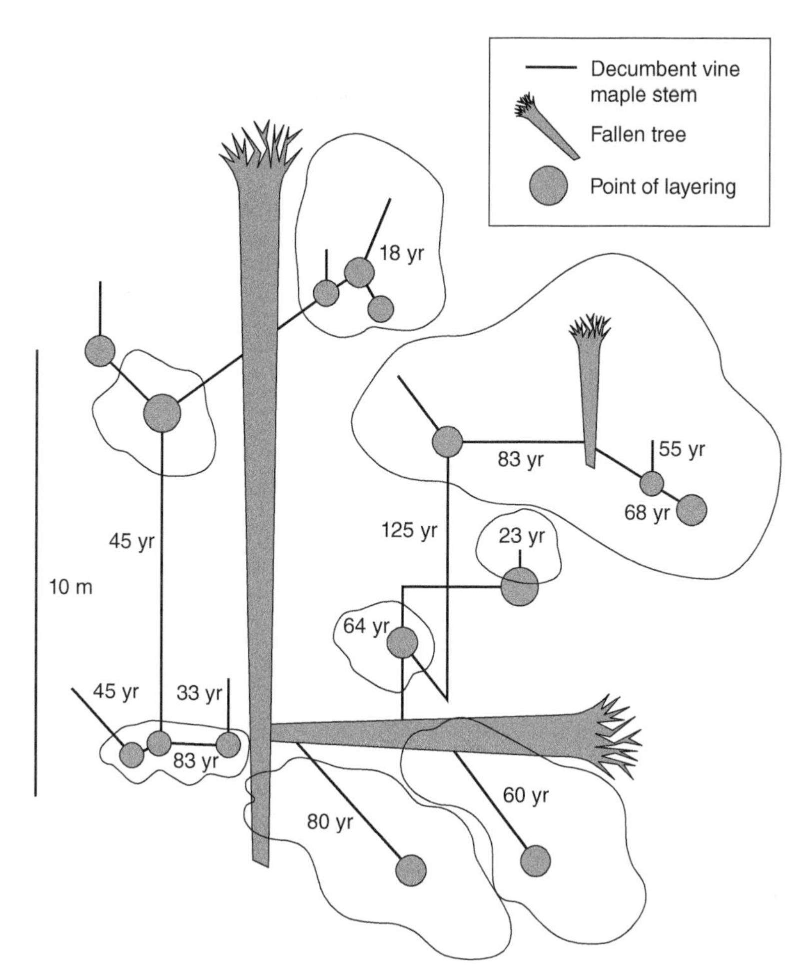

FIGURE 4.15 Top view of a vine maple clone in a 130-year-old Douglas-fir forest showing decumbent stems and points of layering. Aerial stems are not shown. Irregular enclosed areas indicate vine maple crown coverage. Note the large range in age among ramets of the clone. *Source:* O'Dea et al. (1995) / John Wiley & Sons.

various hard pines that sprout from dormant buds along the stem after fire or cutting, including pitch pine, Virginia pine, and pond pine (Kozlowski 1971a). Near tree line in the southern Rocky Mountains, Marr (1977) reported "tree islands" of subalpine fir and Engelmann spruce that moved along the ground by repeated layering and growth to leeward. Movement of 5 m in 11 years was common, and some clonal islands apparently moved at least 15 m. In this example, seedlings apparently became established in sheltered microsites, but the clones expanded and colonized adjacent microsites inhospitable to establishment.

In addition to the adaptive benefits described in the earlier text, vegetative reproduction ensures the survival of a single genet either by resprouting from its roots or protected stem following destruction of the aboveground parts or by spreading mortality risk of the genet among existing ramets that are capable of independent survival (Cook 1979). For example, persistence of green ash in the presence of emerald ash borer in eastern North America is likely to be maintained by basal sprouting from top-killed trees after attack by the invasive beetle (Kashian 2016). In a five-year study of pure green ash stands in southeastern Michigan, basal sprouting was the dominant mode of regeneration where 58% of the trees had been killed by the beetle. Sprouts grew far faster than either established ash seedlings or surviving overstory ash trees, and 27% of the sprouts produced significant seed during a mast year that occurred during the study. Thus, the presence of vegetative reproduction from top-killed ash facilitated continued sexual reproduction that will likely lead to the species' persistence, although at much lower abundance with far smaller and shorter-lived trees (Kashian 2016). Sprouting in this example, which notably does not occur from every killed tree or in every population of green ash, is therefore critical to ash persistence. Vegetative reproduction probably contributes more to short-term persistence and spatial dominance of woody species, but it is clearly sexual reproduction that provides the genetic variation for persistence in the face of longer-term environmental change.

SUGGESTED READINGS

Grubb, P.J. (1977). The maintenance of species-richness in plant communities: the importance of the regeneration niche. *Biol. Rev.* 52: 107–145.

Knops, J.M., Koenig, W.D., and Carmen, W.J. (2007). Negative correlation does not imply a tradeoff between growth and reproduction in California oaks. *Proc. Natl. Acad. Sci.* 104: 16982–16985.

Kozlowski, T.T. (1971). *Growth and Development of Trees, Vol. I. Seed Germination, Ontogeny, and Shoot Growth*. New York: Academic Press 443 pp.

Obeso, J.R. (2002). The costs of reproduction in plants. *New Phytol.* 155: 321–348.

Owens, J.N. (1991). Flowering and seed set. In: *Physiology of Trees* (ed. A.S. Raghavendra), 247–271. New York: Wiley.

Pallardy, S.G. (2009). *Physiology of Woody Plants*, 3e. San Diego, CA: Academic Press 454 pp.

Silvertown, J.W. (1980). The evolutionary ecology of mast seeding in trees. *Biol. J. Linn. Soc.* 14: 235–250.

Zasada, J.C., Shank, T.L., and Nygren, M. (1992). The reproductive process in boreal forest trees. In: *A Systems Analysis of the Global Boreal Forest* (ed. H.H. Shugart, R. Leemans and G.B. Bonan), 211–233. Cambridge: Cambridge Univ. Press.

Tree Structure and Growth

CHAPTER 5

Two kinds of growth occur when a seedling develops into a large tree: **primary growth**, or the extension of each growing point forming the shoots of the crown and the roots, and **secondary growth**, involving the expansion of stem and root diameter. Each species has a characteristic aboveground and belowground structure and form that represent a combination of these growth processes, following an inherent architectural model and influenced by environmental factors. In this chapter, we emphasize the tree crown, stem/shoots, and selected aspects of root structure and growth in relation to environmental factors, especially soil water. We consider growth and development of shoots and roots in detail, but a treatment of other important aspects of tree and shrub physiology is beyond our scope. For detailed treatments of tree physiology, the reader is directed to books by Kramer and Boyer (1995), Kozlowski and Pallardy (1997a), Larcher (2003), Pallardy (2008), and Lambers and Oliveira (2019). For considerations of tree structure and function, classic works by Büsgen and Münch (1929) and Zimmermann and Brown (1971) are notable as well as by Oldeman (1990).

TREE FORM

Tree form, habit, or architecture is as incredible as it is fascinating. Diverse regional and local ecosystems favor many species, vertical vegetation strata, and long periods of rhythmic or continuing growth, especially in tropical forests. Tree architecture is determined by many factors such as apical control of growth, shoot types, branching orientation, timing of meristematic activity, herbivory, and shedding of shoots. Tree architecture is also shaped by physical factors of light, temperature, soil water, and nutrient availability. Readers are urged to examine more detailed treatments of tree architecture by Wilson (1984), Barthelemy et al. (1991), Bell (1991), Turnbull (2005), and Hollender and Dardick (2015).

Generally, tree form refers to the shape, size, and/or habit of the crown of a tree. The form of a woody plant is determined by its genetically pre-determined architectural plan, and by the environmental influences that affect its development. Each species has obtained a precisely determined growth plan or architectural model selected by and adapted to the physical and biotic factors of the ecosystems where it lives. In northern and alpine environments, where snow, ice, and wind are strong selective factors, trees (primarily conifers) are adapted to grow in a conical form (Figure 5.1a). It is clear that this trait is under strong genetic control when species from high altitudes (spruces, larches, and firs) maintain their conical form in parks far from their native sites. The conical form results when the terminal shoot or leader grows systematically faster than the lateral branches below it, creating a prominent, central stem. This tree form is called **excurrent** and is typical of northern conifers and a few deciduous trees, such as sweetgum and tulip tree.

The excurrent form results from strong **apical control**, whereby the terminal leader inhibits the growth of the lateral branches below it, and thus grows faster than them (Cline 1997; Sterck 2005). Apical control is a process that shapes the growth and development of the entire tree, including the length and angle of branches and stem diameter (Wilson 2000; Sterck 2005).

Forest Ecology, Fifth Edition. Daniel M. Kashian, Donald R. Zak, Burton V. Barnes, and Stephen H. Spurr.

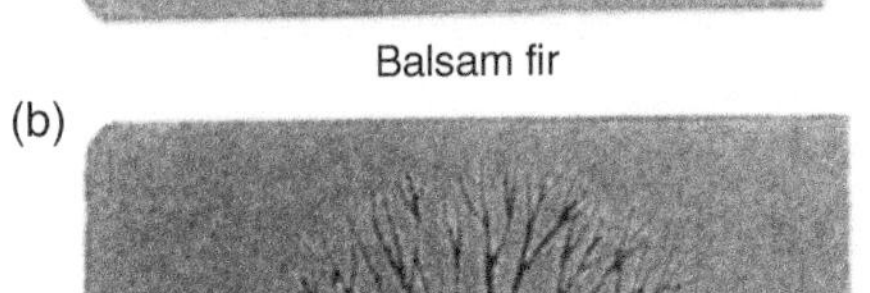

FIGURE 5.1 Examples of (a) excurrent and (b) decurrent or deliquescent forms of trees. *Source:* After Hosie (1969). Reproduced by permission from *Native Trees of Canada*, 7th ed. by R. C. Hosie; published by the Canadian Forest Service and Fitzhenry and Whiteside Ltd., 1969.

Apical dominance is related to apical control, but specifically refers only to the effect of the terminal shoot on the current year of growth rather than on the tree as a whole (Brown et al. 1967). Together, apical dominance and control are strong determinants of tree architecture. The subtle difference between these processes is important because apical control is not simply a summation of annual apical dominance, and the molecular mechanisms driving apical dominance are far better understood compared to those driving apical control (Hollender and Dardick 2015).

Climate has a strong influence on tree form such that not even all conifers are excurrent. For example, pines of the warm and humid southeastern United States, such as loblolly and longleaf, exhibit a strong central stem, but a more rounded crown than spruces and firs of the far north. In the arid Southwest, pinyon pines and junipers tend to have short, compact, rounded, and bushy crowns. This form is related to the high temperature and soil-water stress of their environment as well as the sparse nature of the forests in the region. These conditions favor extensive root development and compact crowns rather than tall excurrent growth that would severely expose the crown to prolonged, drying winds.

In many deciduous species, the lateral branches grow nearly as fast or faster than the terminal leader. Weak apical dominance and control may allow the main stem to fork repeatedly, giving rise to a broad, spreading form (Figure 5.1b). This **decurrent** or **deliquescent** form is typical of elms, oaks, maples, and many other species that grow in less harsh environments compared to conifers. In these species, a multiple-terminal leader system is often developed, most pronounced in American elm and many oaks. The characteristic "vase shape" of many elms is related to the branching and re-branching of lateral shoots that results in an effective disappearance of a main stem (Pallardy 2008). Such broad, spreading crowns may have an adaptive function in spacing out the relatively large deciduous leaves to allow access to light compared to the densely packed needles of many excurrent conifers.

Within the group of decurrent angiosperms, some species that grow along rivers and streams or along a forest edge, besides having spreading crowns, may exhibit markedly leaning trunks compared to upland or forest-interior species. For example, Loehle (1986) reported that black willow, sycamore, and river birch had a lean from the vertical of 21°–22°, whereas the average lean of three upland oaks was only 7°. The growth response toward light, **phototropism**, markedly affects tree form of certain species and enables these shade-intolerant species to obtain light from directly over the river. However, other bottomland species such as American elm, eastern cottonwood, and red or green ash rarely display such a pronounced lean, but instead exhibit different branching architectures to optimize light capture by the crown for photosynthesis.

ARCHITECTURAL MODELS

The total diversity of tree forms were generalized into 23 architectural models (which also apply to shrubs and herbs) by Hallé and colleagues (1978; Figure 5.2). Most of these models describe tropical species, but several are directly applicable to temperate conifers and deciduous trees. A given species may share features of more than one model, as for eastern hemlock (Hibbs 1981). The models were developed based on whether shoot growth is rhythmic (seasonal) or continuous, whether the tree is unbranched (as in some palms) or branched, whether or not branching is **orthotropic** (erect) or **plagiotropic** (horizontal), and whether flowers are located on terminal or lateral shoots. Hallé et al. (1978) identified several of these models appropriate for temperate plants and named them after a well-known botanist (Figure 5.2): Leeuwenberg (sumacs, red-osier dogwood, and rosebay rhododendron), Rauh (most species of pines, red and live oak, and probably most oaks), Massart (redwood and true fir species such as balsam fir and European silver fir), Tomlinson (most multiple-stem palms and many sedges and grasses), Troll (American beech and American elm), and McClure (many bamboo species). Note that excurrent forms of pines are grouped with spreading forms of oaks, indicating the broad nature of these architectural plans. A set of non-standard tree forms, as summarized by Hollender and Dardick (2015), have also been described as applicable to crop trees and those used for landscape design. The forms include columnar/pillar, weeping, dwarf/semi-dwarf, and compact/highly branched, and most have been at least partially described at the molecular level.

Architectural models for trees are most easily observed when trees are in the seedling or sapling stage because their form has been least subject to the environmental stressors or damage that occur as they mature. Therefore, examining tree architecture clarifies how woody species respond to stresses and crown damage over time. When new shoots arise from a damaged tree, they arise according to the species' original architectural growth plan. This process, termed **reiteration** (Barthelemy et al. 1991), also occurs in old crowns and in spontaneous vegetative reproduction from roots or stems. This replication of architecture without an obvious stimulus has been proposed as a way for some tree species to maintain their crown and thus extend their longevity (Ishii et al. 2007). Thus, the crown shape of woody plants results from a genetically determined architectural plan, modified by the environment, and reiterated as the plant matures.

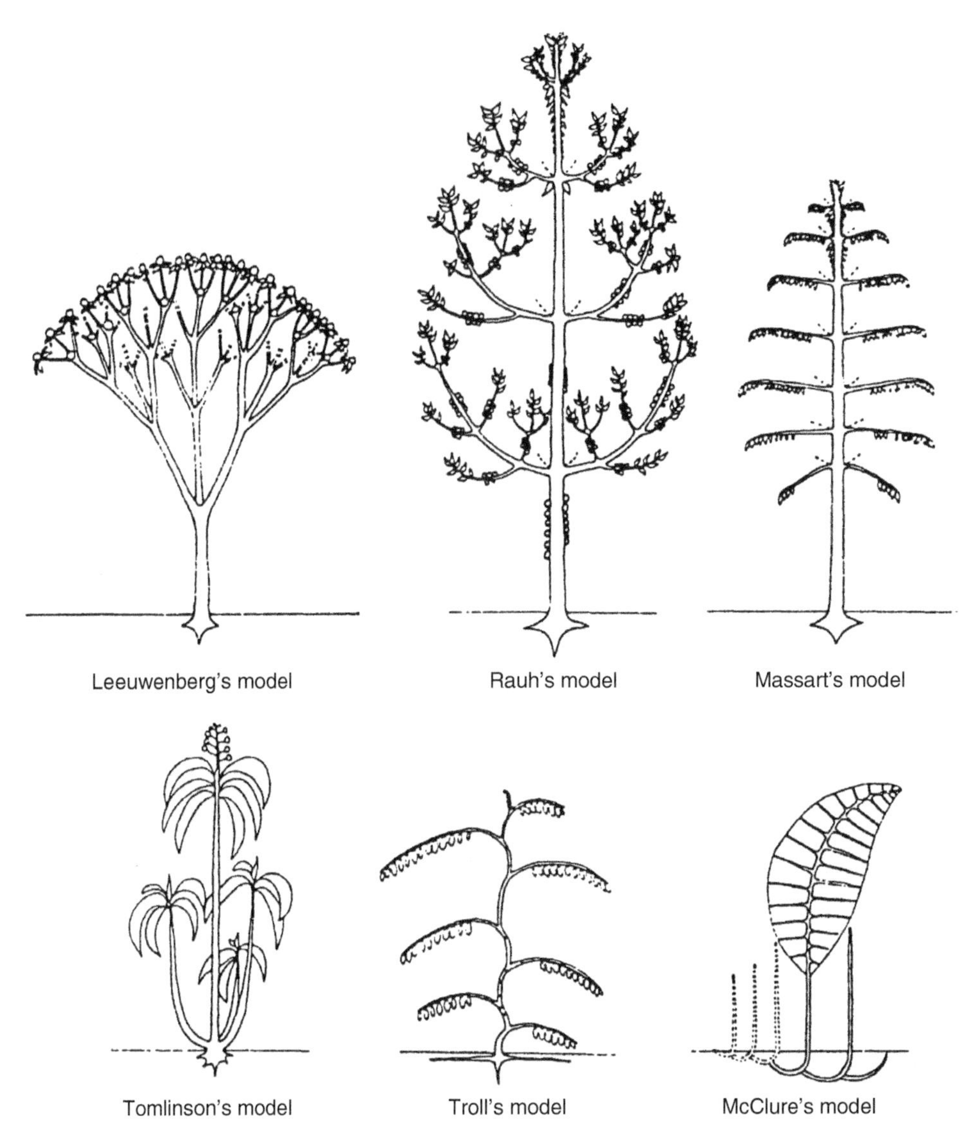

FIGURE 5.2 Generalized architectural models of trees, selected from the 23 models of Hallé et al. (1978). Besides tropical species, the following temperate groups or species are characterized: Leeuwenberg's model—sumacs, rosebay rhododendron; Rauh's model—most pines, red and live oaks; Massart's model—redwood and true firs; Tomlinson's model—multiple-stem palms; Troll's model—American beech, American elm; McClure's model—many bamboo species. *Source:* Tomlinson (1983) / with permission of Sigma Xi, The Scientific Research Honor Society.

The canopy architecture of individual species is combined in a given temperate or tropical ecosystem to produce a particular canopy structure, usually of several layers. Multiple tree and shrub layers characterize moist tropical and temperate forests (for example, see Figure 6.11 in Chapter 6), whereas dry tropical or temperate forests may exhibit a totally different vertical canopy structure. Canopy structure directly influences the amount of leaf surface area present to capture light for photosynthesis. Trees with narrow columnar forms are associated with dry or high-elevation sites, whereas broader crowns are associated with mesic or humid sites (Landsberg 1995). Tropical studies have suggested that the tallest trees have the widest and flattest crowns with upwardly angled branching, the second layer trees are as deep as they are wide, and in the third layer, trees crowns are conical with horizontal branching (Pallardy 2008). The interrelated factors of tree form, vertical canopy structure, and leaf area, as they influence interception of light, photosynthesis, and overall tree productivity, are discussed in Chapter 6.

Short and Long Shoots The kinds of shoots and the patterns of shoot growth influence tree form and competitive ability in different ecosystems, as well as reproductive and regeneration potential. Plants have an *open system of growth* whereby the shoot (the stem with its collection of leaves and buds) is the *modular unit* of construction. In trees, the modular unit implies that buds open and expand, the shoot elongates, and then develops a new bud for the process to repeat. Multiple kinds of shoots make up the modular units and also the patterns of shoot growth that affect tree structure and function.

All woody plants bear long shoots, and some species have both long shoots and short shoots. Long shoots contain **nodes**, the areas where leaves and buds are found on the stem, and **internodes**, the areas between successive nodes. Long shoots lengthen mostly from elongation of the internodes. The annual growth of long shoots of yellow birch may range from a few centimeters to several meters. Annual long-shoot growth of over 3 m is not uncommon for young eastern cottonwood trees on alluvial sites.

Short shoots, also called **dwarf shoots** in conifers or **spur shoots** in deciduous trees, exhibit very little or no internode elongation (a few millimeters or less) and typically mark annual shoot growth only by groups of adjacent ringlike scars left by the bud scales. For example, we collected a 28-year-old short shoot of yellow birch that measured only 7 cm long (compare to the length of long shoots above). Short shoots are characteristic of beeches, birches, maples, and fruit trees such as cherries and apples. In the gymnosperms, short shoots are found in the deciduous genera including larches and ginkgo. In the pines, needle-bearing **fascicles** are modified short shoots. They are determinate and persist more than one season, but do not extend into long shoots. Temperate trees tend to feature lateral short shoots, but short shoots are terminal in most tropical trees. Long shoots may be either terminal or lateral in both temperate and tropical forests and often occur in the upper crown where the main canopy growth occurs. In European white birch, for example, short shoots make up over 90% of all shoots on a tree, but 80% of the shoots at the top of the tree are long shoots. Leaves on short shoots are produced at the same place every year and their relatively fixed position allows them to serve as primary photosynthetic units (Bell 1991, p. 254). For example, the broad-paired, short-shoot leaves of hobblebush, a shrub that grows in shaded understories of conifer forests of the Appalachian Mountains, are positioned to maximize light capture.

Short shoots bear only "early" leaves whose primordia are pre-formed in the dormant or resting bud; they also may bear flowers, cones, or thorns. Short shoots are characterized by **determinate** (fixed) **growth**, meaning that shoots are pre-formed in the dormant bud, expanding in the spring, and then ceasing growth before setting buds late in the growing season. In contrast to short shoots, long shoots may be either determinate or indeterminate. **Indeterminate** (free) **growth** features shoots where the basal portion is pre-formed in the bud and elongates in the spring, but the shoot continues to grow throughout the growing season if light and other site conditions are favorable. Indeterminate long shoots are advantageous in that they extend the framework of the plant into new space. Indeterminate shoots are especially well developed in well-lighted crown positions of fast growing, pioneer species of pines, birches, cherries, cottonwoods, balsam poplars, aspens, sassafras, and sycamores (Marks 1975). Determinate long shoots bear only early leaves, whereas indeterminate long shoots bear early leaves along their basal portion and "late" leaves, those formed during the current growing season (rather than pre-formed in the bud), along the upper portion. Late leaves produced after the early leaves develop may bear little resemblance to early leaves (Figure 5.3), but represent the internal environment at the time and place of primordia formation and development rather than the external environment.

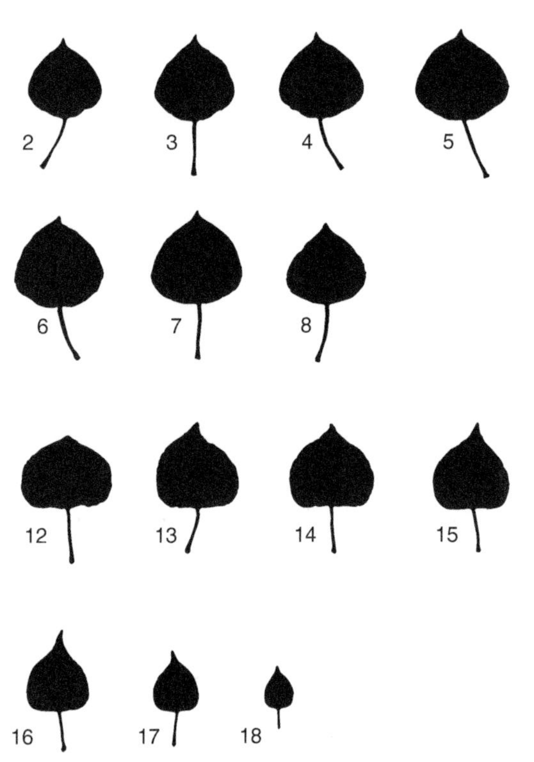

FIGURE 5.3 Silhouettes of leaves from a leader shoot of trembling aspen. Leaves 2–7 are typical early leaves; 8 is a transition form to late leaves; and leaves 12–18 are typical late leaves. *Source:* Barnes (1969) / with permission of JD Sauerlander's Verlag.

Patterns of Intermittent Growth Most woody plants exhibit intermittent (periodic, episodic, and rhythmic) rather than continuous growth. Some trees exhibit continuous growth by producing a pair of leaves every few months, such as in palms and mangroves, but even these do not constantly extend their shoots (Borchert 1991). Meristems or buds typically are induced into a resting or dormant state by cold temperature or soil-water stress; buds in a resting state resume growth within one to two weeks once the favorable conditions resume, but dormant buds do not. Intermittent growth in temperate areas (evidenced by leaf flushing and shoot extension) is controlled by temperature; in the tropics it is controlled by soil water. Intermittent growth in some trees that is unrelated to environmental changes is called **rhythmic growth** (Borchert 1991).

Long shoots are responsible for height growth and crown extension and therefore are an advantage to use the growing season effectively. Zimmermann and Brown (1971) illustrated various ways temperate trees make use of favorable growing conditions by designating four types of long-shoot extension:

1. A single flush of terminal growth followed by formation of a resting bud (seasonally determinate long shoots).
2. Recurrent flushes of terminal growth with terminal bud formation at the end of each flush.
3. A flush of growth followed by shoot-tip abortion.
4. A sustained production of leaves including early- and late-leaves up to the time of terminal bud formation (seasonally indeterminate long shoots).

Type 1: As described in the earlier text for determinate growth, many northern species make a burst of growth in spring or early summer and then form terminal or end buds (Figures 5.4 and 5.5) before mid-summer soil-water stress intensifies. Ecological conditions largely determine when the burst of growth begins. For example, red and eastern white pines exhibit different timing of shoot growth in North Carolina as compared to that in New Hampshire (Figure 5.4). Once

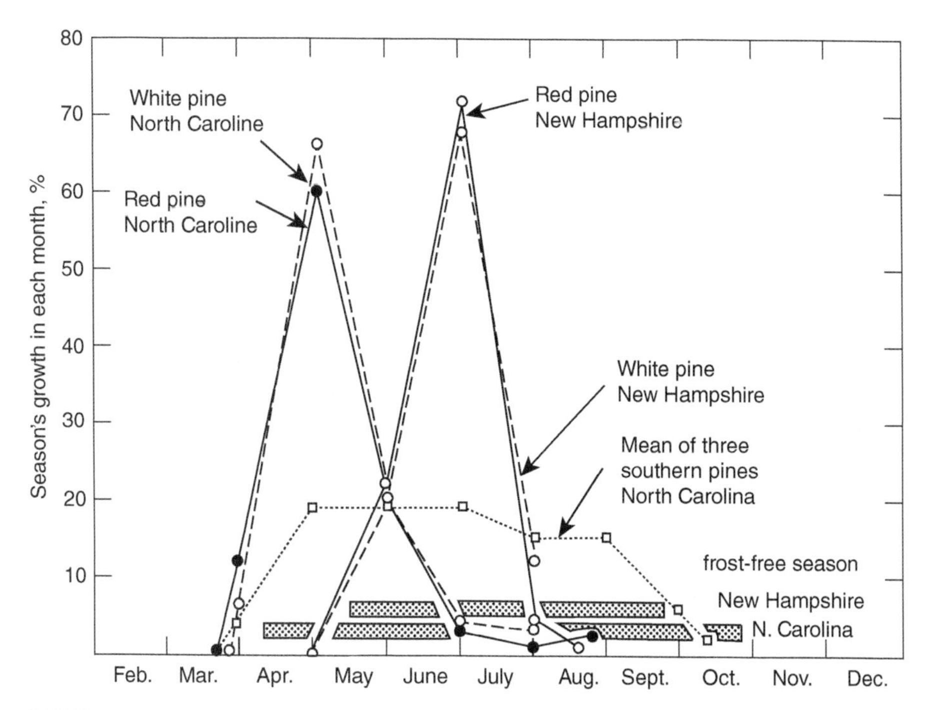

FIGURE 5.4 Variations in seasonal height-growth patterns of red pine and eastern white pine in North Carolina and in New Hampshire, and of three southern pines (loblolly, shortleaf, and slash) in North Carolina. The northern pines have pre-formed shoots and usually have one annual growth flush, whereas the southern pines grow in recurrent flushes. *Source:* Kramer (1943) / with permission of Oxford University Press.

the terminal bud forms, shoot and leaf primordia develop within the bud scales and expand the following year. The current year's shoot growth is therefore primarily dependent on environmental conditions of the previous year when leaf primordia were formed. Other example species of this type of fixed growth include some oaks, hickories, and buckeyes.

Type 2: Trees in regions with periodically favorable growth conditions, such as the pines of the southeastern United States and Central America and Monterey pine in California, typically exhibit recurrent flushes of growth during the growing season. This type of growth is also the most typical one among conifer and deciduous trees in the subtropical and tropical regions. The environmental conditions of the site, especially soil water, determine the number of flushes of growth.

Type 3: In 16 tree genera (Zimmermann and Brown 1971), the shoot tip aborts either following the extension of the pre-formed leaves (beeches) or after many late leaves have been produced later in the growing season (black willow, birches, and mimosa). In many cases, shoot-tip abortion is associated with the shortening days of summer. Shoot extension occurs from the last fully formed lateral bud the following spring. The ecological significance of shoot-tip abortion is unclear, but it may occur as a way to continue shoot extension as long as favorable conditions exist.

Type 4: In a few southern species such as tulip tree, sweetgum, and eastern cottonwood, some shoots grow continuously, including both early and late-leaf development, before setting a terminal bud. The potential of severe frost damage is low for these species within their range, and selection pressures for shoot-tip abortion may be weak. These species also occupy moist sites so that early cessation of growth for drought avoidance is unimportant. Species exhibiting this type of shoot extension and true terminal bud formation tend to have symmetrical crowns and a more pronounced main stem than in trees exhibiting shoot-tip abortion.

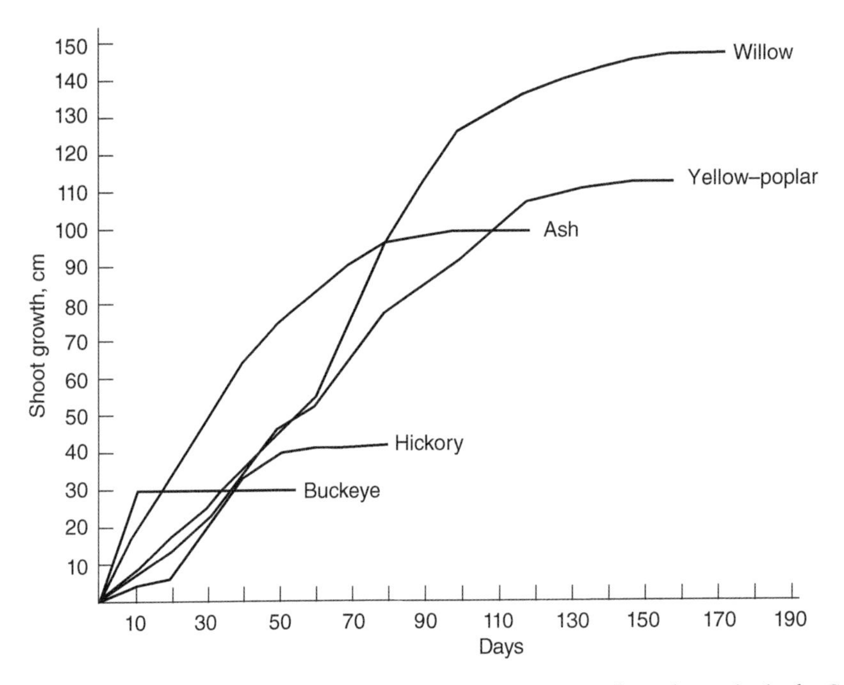

FIGURE 5.5 Rate and duration of shoot growth among several woody species in the Georgia Piedmont. Measurements of shoot elongation were made biweekly on 9 trees ranging from 8 to 15 years in age: painted buckeye, mockernut hickory, red ash, yellow-poplar (tulip tree), and black willow. *Source:* Zimmermann and Brown (1971) / Springer Nature.

Sylleptic and Proleptic Shoots It is common that recently formed buds open and extend late during the same growing season in which they are formed. This growth occurs laterally and produces long shoots. Development of a lateral shoot from a lateral bud before the bud is fully formed is referred to as **sylleptic** growth, which forms a sylleptic shoot. This pattern is far less common in temperate regions than in the tropics, although sassafras, alternate-leaf dogwood, sweetgum, and alders may regularly form sylleptic shoots (Wilson 1984). In **prolepsis**, lateral buds at the base of a terminal bud develop lateral shoots after a rest period. Proleptic growth typically takes place following the winter-dormant period in temperate areas, but lateral meristems of some temperate species may resume growth briefly after setting buds if late season weather is warm and moist. These shoots may not adequately harden before winter and may be damaged by freezing temperatures. Proleptic shoots are common in fast-growing pines (Rudolph 1964). **Lammas** growth occurs when the *terminal* shoot resumes growth following bud set, usually in temperate areas. When lateral meristems develop late in the season (prolepsis) but the terminal shoot does not, several laterals may dominate the top of the tree in the next several growing seasons and eventually replace the terminal shoot. A heavily forked form is the result as often seen in young, open-grown white pines (Wilson 1984). The term for lammas growth was coined when late-season shoots apparently occurred about the time of European holidays, Saint's Day (June 24), and in England, Lammas Day (August 1).

ROOTS

Two of the many important functions of tree roots are the firm anchoring of the tree in the soil and the absorption of water and nutrients. The entire root system serves to anchor the plant, whereas absorption is accomplished mainly by fine (<0.5 mm), short-lived, non-woody roots. Larger roots also store carbohydrates and other materials, synthesize organic compounds, trans-

Table 5.1 Comparative root development of seedlings of deciduous and coniferous species.

Species	Age (months)	Number of roots	Total root length (meters)	Growing conditions
Black locust	4	7124	325.5	Greenhouse
Loblolly pine	4	419	1.6	Greenhouse
Flowering dogwood	6	2657	51.4	Greenhouse
Loblolly pine	6	767	3.9	Greenhouse
White oak	12	196	2.3	Forest
Loblolly pine	12	148	1.0	Forest

Source: Kozlowski and Scholtes (1948) / with permission of Oxford University Press.

port water and nutrients to the crown, secrete chemicals, and, for some species, generate vegetative shoots. The growth and functioning of roots have long received attention from many authors, and recent major reviews include Baluška et al. (1995), Pregitzer and Friend (1996), and Gašpariková et al. (2001).

Root structure and function differ between angiosperms and conifers. A classic study of root development demonstrated that deciduous seedlings developed more extensive root systems than loblolly pine in greenhouse and forest environments (Table 5.1; Kozlowski and Scholtes 1948). Decomposition of leaves from deciduous trees is rapid (Ellenberg 1988, p. 59), such that the extensive and finely divided root systems of these species may have evolved to efficiently intercept and absorb the nutrient ions released from decomposing leaves (Kozlowski et al. 1991). Fine roots of deciduous species are generally smaller in diameter than those of conifers (Voigt 1968) to maximize surface area and associated nutrient absorption (Pregitzer et al. 2002). By contrast, fine roots of the more primitive conifers probably have a greater ability to extract nutrient ions from the soil, especially important where soil nutrient status is poor (Chapter 19).

KINDS, FORMS, AND OCCURRENCE

Tree root systems consist of a framework of relatively large, woody, long-lived roots supporting a mass of small, short-lived, non-woody absorbing roots, many of which are associated with fungi (these root–fungal relationships are discussed in Chapter 19). Seedlings of many species develop tap roots that provide early stability and survival on dry and fire-prone sites. Tap roots are common in pines and in deciduous species that have large seeds with heavy food reserves (oaks, hickories, chestnuts, walnuts, etc.). Roots of the main framework grow quickly (up to 1 m per year) for distances up to 20 m or more (Wilson 1984). Vertical **sinker** roots may develop from the lateral system as it develops, reaching into lower soil horizons and helping to anchor the root system.

The form and structure of the root system are largely genetically controlled, but site conditions also influence root depth and extent. Root systems vary widely among species; trees able to produce deep roots or highly branching roots are able to extract water and nutrients from a larger soil area than trees with more restricted root systems (Kozlowski 1971b). Thus, species adapted to droughty sites such as pines and upland oaks and hickories often have deep-penetrating tap roots. By contrast, root systems are more often shallow where the water table is near the surface as in swamps or peat bogs, or where deep root growth is restricted by bedrock or dense or poorly aerated soil. Shallow rooting on such sites often leads to the common uprooting of trees by wind (i.e., **windthrow**). Lateral root extent also may vary with soil, extending further with drier, better aerated, sandy soils compared to clay soils (Rogers and Booth 1960). A sample of the great variety of rooting forms that develop in relation to different soil and water table conditions is presented in Figure 5.6.

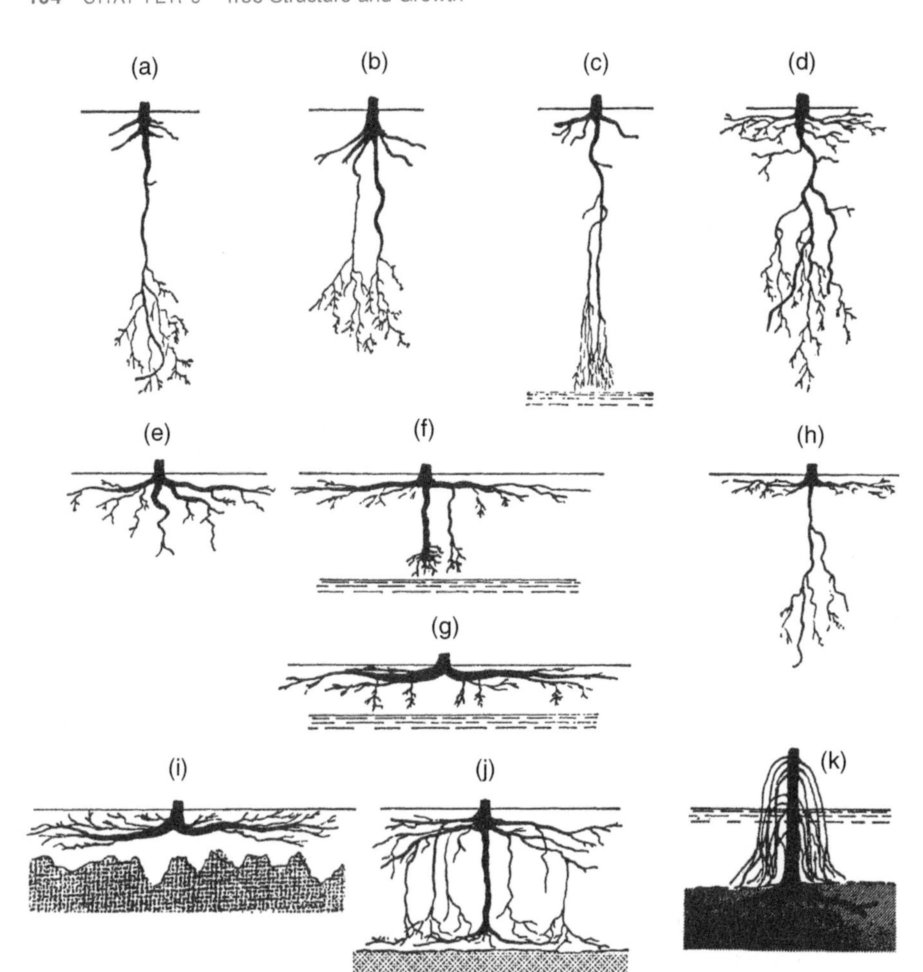

FIGURE 5.6 Modification of root systems of forest trees by site. (a) and (b) Taproots with reduced upper laterals: patterns found in coarse sandy soils underlain by fine-textured substrata. (c) Taproot with long tassels, a structure induced by extended capillary fringe. (d) Superficial laterals and deep network of fibrous roots outlining an interlayer of porous materials. (e) Flattened heart-root formed in lacustrine clay over a sand bed. (f) Plate-shaped root developed in a soil with a reasonably deep groundwater table. (g) Plate-shaped root formed in organic soils with shallow groundwater table. (h) Bimorphic system of platelike crown and taproot, found in leached soils with a surface rich in organic matter. (i) Flat root of angiosperms in strongly leached soils with raw humus at surface and hardpan below. (j) Two parallel plate roots connected by vertical sinkers in a hardpan spodosol. (k) Pneumatophores of mangrove trees in tidal lands. *Source:* Wilde (1958) / John Wiley & Sons.

Aerial roots such as strangling roots and stilt and prop roots are formed by some tropical trees (Bell 1991). Yellow birch and eastern hemlock approximate this root form in eastern North America by germinating on decaying logs and stumps and producing roots that grow downward and into the soil. When the woody debris decays away, the roots are exposed and stilted (Figure 5.7). Older yellow birch trees are able to generate aerial roots from the stem cambium in humid environments such as the southern Appalachian Mountains. These roots develop downward, either on the outside of the trunk or through the decaying heartwood of the standing tree, and into the soil (Spurr and Barnes 1980, p. 91).

Photo by Dan Kashian

FIGURE 5.7 Stilted roots of an eastern hemlock in northern Lower Michigan. A seed of this eastern hemlock germinated on a decaying log. The roots of the growing seedling grew through and around the log and into the ground. The log has since decayed away, leaving space beneath some of the roots.

FINE ROOT RELATIONS

Lateral roots are formed along the main roots that may become long roots or remain short, such that the root system consists of several orders of woody, perennial long roots and one or more orders of small or fine non-woody roots (Figure 5.8). Fine roots (<1 mm) make up very little of the biomass (weight) of a tree's root system but most of its length and surface area (Jackson et al. 1997; Pallardy 2008). In a study of northern red oak, for example, Lyford (1980) found that fine roots branch and re-branch to form many root orders down to a minimum size of about 0.07 mm (Figure 5.8). A mature northern red oak (40–70 years old) in Massachusetts would therefore have 90 million fine-root tips in its system, and an average of 1000 fine roots per cubic centimeter of forest floor! These fine roots were estimated to occupy only 3% of the cube, but have the surface area of more than 6 cm^2, not counting fungal hyphae or root hairs.

In most soils, fine roots are concentrated in the organic and upper mineral soil horizons (Kozlowski 1971b) where it is well aerated, soil water is most readily available following precipitation events, and nutrients are at their highest concentration. A classic study of two pine and two oak forests of the Piedmont of North Carolina found that 94–97% of the roots <2.5 mm in diameter were located in the upper 13 cm of soil (Coile 1937). In northern red oak, many roots arising from the lateral root system grow upward and into the organic layers of the forest floor, where they proliferate into many small-diameter, non-woody roots (Figure 5.8; Lyford 1980). This upward growth has also been described for some temperate conifers (Coutts and Nicoll 1991).

The number of fine roots per unit volume varies greatly during the growing season and by soil depth. Variation during the growing season occurs in part because fine-root turnover is extremely rapid. Over 60 years ago, Kalela (1957) reported a continual flux of death and replacement

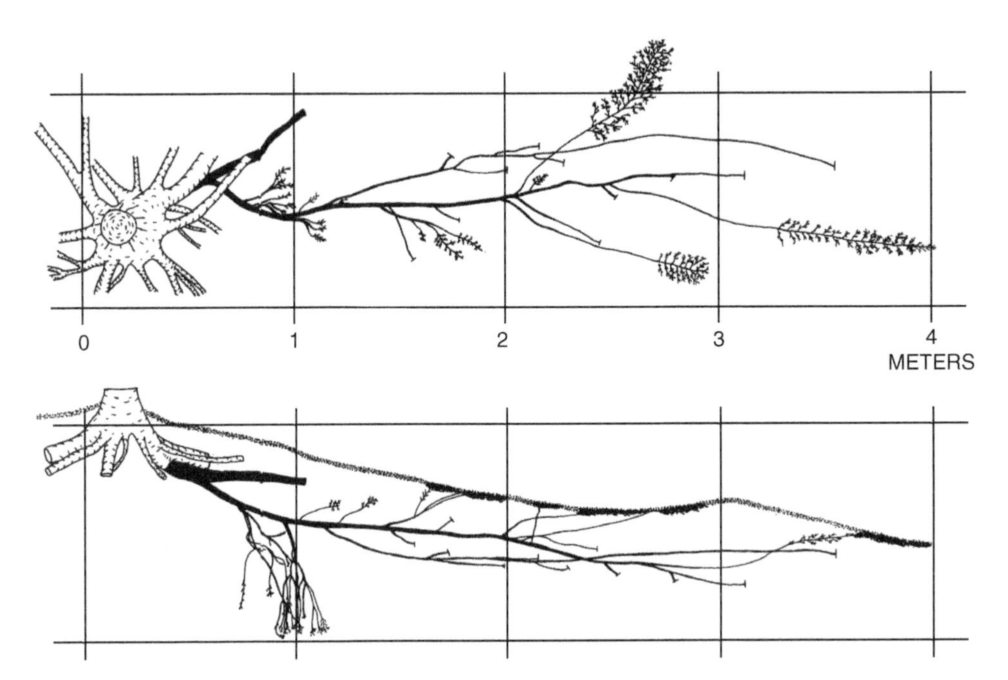

FIGURE 5.8 Diagrams of a horizontal woody third-order lateral root of red oak emphasizing roots that ascend to the surface and elaborate into many small-diameter, non-woody roots in the forest floor. Top view of roots (top panel) illustrates proliferation of small roots in forest floor. Side view (bottom panel) illustrates sinker roots as well as those conforming to the surface topography and ascending into forest floor. Squares are 1 m on a side. *Source:* Wilde (1958) / FAO.

of fine roots for Scots pine stands in sandy soils of Finland. Forty-five years later, Hobbie et al. (2002) found that about 70% of fine-root carbon was new each year in a study of Douglas-fir in the Pacific Northwest. As much as 33% of annual photosynthate may be allocated to fine roots (Jackson et al. 1997). Kalela had concluded that fine roots were sensitive to local ecological conditions, with roots dying *en masse* when conditions were unfavorable (drought, cold, lack of photosynthate, etc.), but were rapidly formed again once conditions became favorable. Using this reasoning, one might predict that fine-root mortality may peak during the winter due to low temperatures and lack of photosynthate. The degree of fine-root turnover described by Kalela has been confirmed in studies of carbon balance and nutrient cycling (Chapters 18 and 19), but the timing of the turnover has varied in studies over the last half-century. For example, fine roots of ponderosa-pine seedlings lived the longest when produced in the fall, but longevity was shorter when they were produced at higher temperatures in the spring and summer (Johnson et al. 2000). Hendrick and Pregitzer (1992) observed that fine-root mortality in a northern hardwood forest was continuous throughout the growing season, with a peak in autumn. In a beech forest in northern Germany, peak concentration occurred in the spring and in organic matter layers where soil water, aeration, and mineralization of nutrients were most favorable.

A final point about fine roots is that their branching patterns vary widely among species. A study of nine North American tree species (Pregitzer et al. 2002) revealed that the majority of fine roots for most species were <0.5 mm in diameter. Sweetgum roots were thick and unbranched compared to those of sugar maple or white oak, which were much thinner and more intricately branched (Figure 5.9). In general, fine roots of deciduous species tend to be smaller and more finely branched than coniferous species, although all have fine roots dominated by short, thin individual roots. Besides representing an analog to differences in canopy form and architecture,

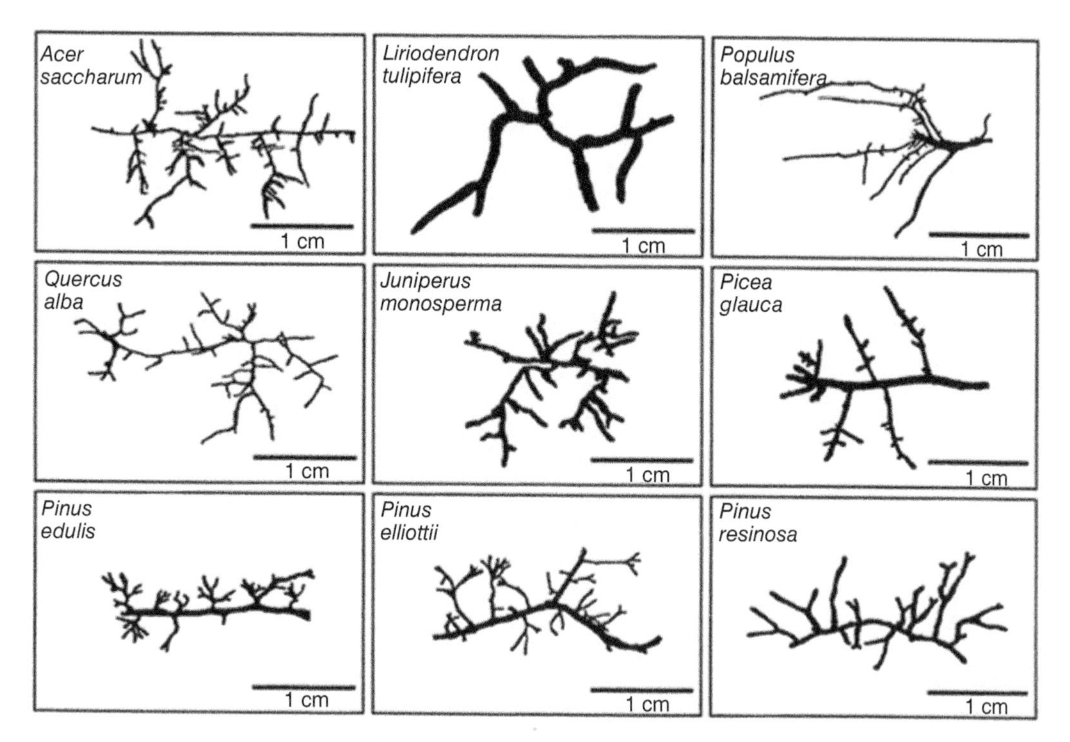

FIGURE 5.9 Digitized representative samples of fine-root-branching patterns of nine North American tree species. All species include short, thin branches, but deciduous species (except for sweetgum, *Liriodendron tulipifera*) tend to have thinner and more finely branched roots compared to conifers. *Source:* Pregitzer et al. (2002) / John Wiley & Sons.

these differences in fine-root architectures among species suggest that assuming roots as large as 2 mm function as fine roots across all species may be inaccurate. In many species, roots as small as 1 mm may still primarily serve transport or storage rather than absorption functions; the position of a root on an individual branch is probably more important for its function than its diameter (Pregitzer et al. 2002).

HORIZONTAL AND VERTICAL ROOT DEVELOPMENT

Perennial long roots have the potential to horizontally develop a significant distance from the main stem of the tree. Lateral roots of pines, birches, and black locust may extend 10 or 20 m as they continually occupy new volumes of soil (Lyr and Hoffmann 1967). In a meta-analysis of rooting extent, Stone and Kalisz (1991) found that the maximum distance of lateral rooting for most tree species was less than 30 m; elms, walnuts, oaks, and aspens were able to exceed 30 m, and willows achieved 40 m. Lateral spread is related to the nature of the rooting medium and is more extreme in sandy soils than in clay. For example, in a sandy Wisconsin soil, roots of 20-year-old red pine trees extended about seven times the average height of the trees (Kozlowski 1971b). In a classic study of the rooting of fruit trees, lateral roots were found to spread about three times as far as the crown on sandy soils, but only twice as far on loamy soils, and one-and-a-half times as far on clay soils (Rogers and Booth 1960). The importance of distant roots for soil water and nutrient absorption appears to be relatively low (Stone and Kalisz 1991).

Rooting depth is particularly important for accessing soil water in deeper soil layers, and may have survival value for some species in times of drought. Deep-rooting trees found on prairies, such as certain species of oaks, locusts, and walnuts, are well-documented examples of the ability of trees to survive drought by accessing deep soil water. Reviewing the literature for 113 temperate tree species, Stone and Kalisz (1991) reported that 70% had rooting depths shallower than 5 m and 33% shallower than 3 m. Although rooting depths of 2–3 m are not uncommon for many forest trees, deep-rooting species (rooting depths of 3–6 m or more) include eastern redcedar, black and honey locust, bur oak, and osage-orange (Kozlowski 1971b). Extraordinary rooting depths of 61 m have been reported for one-seed juniper by Cannon (1960), who located them in uranium mines on the Colorado Plateau. Roots of the desert shrub mesquite were found 53 m below the surface in southern Arizona (Phillips 1963). Stone and Kalisz (1991) cited various studies reporting rooting as deep as 8 m for locust, 10 m for apple, and 24 m for oak and ponderosa pine.

PERIODICITY OF PRIMARY ROOT GROWTH

Much like shoots, extension growth of roots is intermittent. For many temperate tree species, root growth initiates earlier in the spring and proceeds later into the fall than growth of shoots. Specifically, root growth tends to peak when soil temperature and moisture conditions are most favorable in the spring and fall. Root growth often decreases during summer drought periods and slows or stops with the onset of winter and low soil temperatures. In a study of root and shoot growth of red and sugar maple in Virginia, root growth began in early March, about one month before budbreak (Harris and Fanelli 1999). Root growth rate decreased sharply at budbreak, but resumed and continued as shoots grew. Root growth continued even after leaf senescence occurred, until soil temperatures deceased below 7 °C (45 °F) in the fall (Figure 5.10). Many exceptions to these trends have been observed, however. For example, short-shoot needles expand before root growth begins in larch species (Lyr and Hoffmann 1967), and a series of horticultural trees in New York State began root growth only after shoot growth was initiated (Harris et al. 1995).

Roots tend to cease growth during the winter, but do not enter a dormant phase as shoots do within winter buds. Rather, roots of many species may grow during the winter if temperature, soil water, and aeration are favorable (Lyford and Wilson 1966; Lyr and Hoffmann 1967). Winter root growth has been reported from regions where winters are mild and soils are frost-free, such as in the southern United States, coastal British Columbia, Crimea, and in parts of Europe. Woods (1957) reported considerable growth of longleaf pine roots in the surface layer of sandy soils of the southeastern coastal plain of the United States during the winter, and Kramer (1969) reported root growth during every month of the year for other pine species in the same region.

ROOT GRAFTING

Natural root grafting among individuals of a species is relatively common in woody plants (Graham and Bormann 1966), especially in pure stands (commonly pines, spruces, and other conifers) that exhibit a high concentration of roots near the soil surface. Grafting occurs when the cambium layers of two closely associated roots fuse together, thus connecting their phloem and xylem tissues, allowing the transport of carbohydrates, growth substances, and pathogens between the two trees. For example, root grafts among shade-intolerant lodgepole pines have been shown to support shaded trees that would have otherwise been outcompeted via the transmission of carbohydrate (Fraser et al. 2006). Water transport between two trees is less common and requires a more complete fusion of root tissue. However, a study in red pine plantations observed the survival of girdled trees for 18 years due to continued water transport supported by root grafting between girdled and ungirdled stems (Stone 1974). Grafting has proposed to have

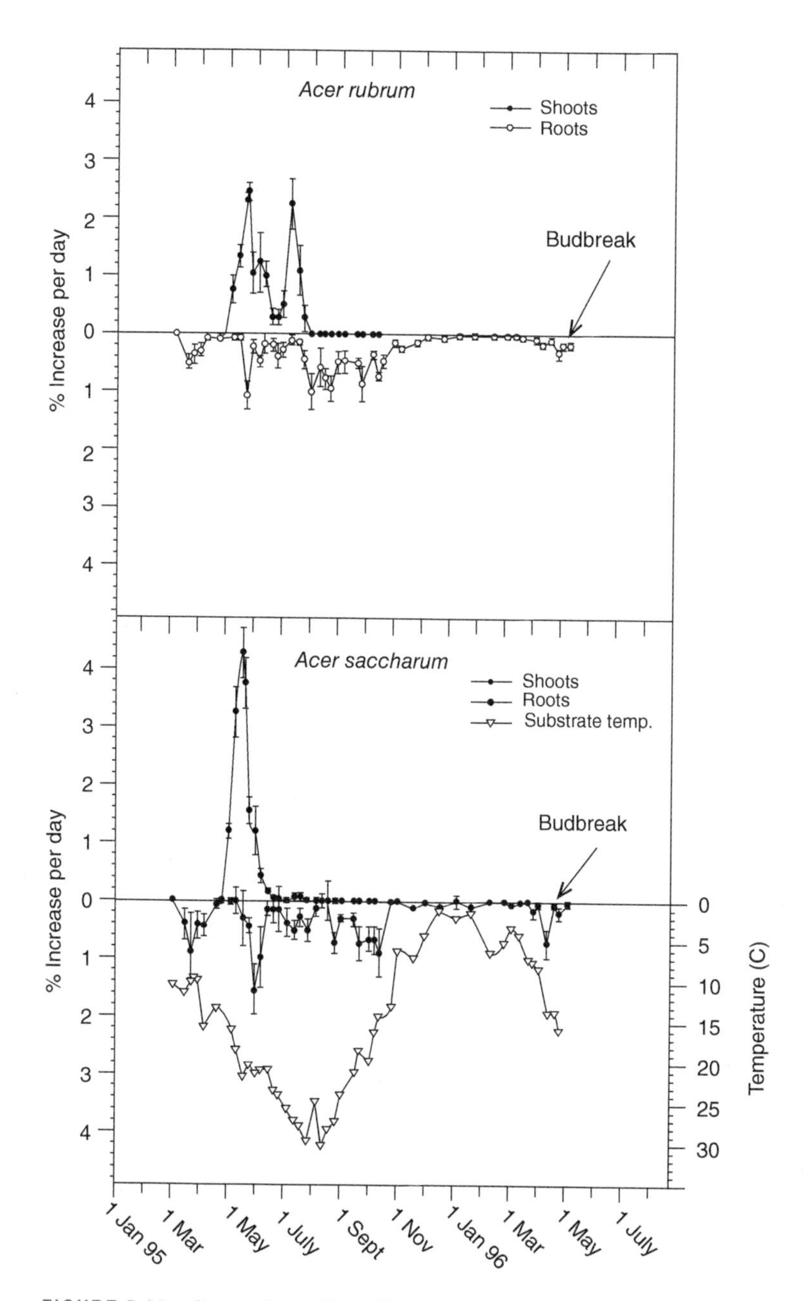

FIGURE 5.10 Comparison of weekly extension for shoots and growing roots of red and sugar maple over a 14-month period. Root and shoot growth follow distinct but different seasonal patterns. The growth of roots for both species initiates before shoots and continues longer into the growing season. *Source:* Harris and Fanelli (1999) / with permission of Horticultural Research Institute.

some adaptive significance, such as increased mechanical support or the formation of beneficial communal root systems, but relatively little empirical data exist to suggest that it occurs in response to more than incidental conditions (Loehle and Jones 1990; Lev-Yadun and Sprugel 2011).

Root grafting plays a critical role in the transmission of systemic diseases such as Dutch elm disease and oak wilt (caused by *Ophiostoma ulmi* or *Ophiostoma novo-ulmi* and *Bretziella fagacearum*, respectively) and in root rots caused by *Armillaria mellea*, *Heterobasidion annosum*, and *Phellinus weirii*. In American elm, the roots of nearly all trees within 2 m of one another and about 30% of those within 8 m are grafted (Verrall and Graham 1935; Himelick and Neely 1962). This interconnection permits rapid transmission of the fungus from infected to uninfected trees through the vessels of the roots. In the root rots, the fungus spreads in part by growing and decaying its way between adjacent trees via root grafts (as in *H. annosum*).

SPECIALIZED ROOTS AND BUTTRESSES

Specialized roots may develop in relation to site conditions. Perhaps the best example of such structures are the pneumatophores (roots that function as a respiratory organ; Figure 5.6k) produced by various species in periodically flooded river valleys or subtropical and tropical swamps (Kozlowski 1971b). In bald cypress, "knee roots" are formed at the air–water interface in flooded sites, but do not form when bald cypress trees are grown in upland, non-flooded sites unless they are injured. The aerial portion of the root exhibits high cambial activity compared to the submerged portion of the root. These knee roots are not important in root aeration in bald cypress, but may strengthen the roots for firmer anchorage (Gill 1970). Pneumatophores in other species may contain lenticels and facilitate the exchange of gasses with the roots.

Buttresses—flattened, vertical plates that develop at the base of the stem—serve to hold the tree upright when site conditions limit the development of suitable vertical root growth. They are most common in tropical species, but occur in temperate trees such as American elm and Lombardy poplar. In bald cypress and water tupelo of the southern United States, buttresses develop as a response to the level of flooding (Varnell 1998). On flooded sites, buttresses form because (i) ethylene accumulates at the waterline, and (ii) growth regulators are unable to translocate beyond the waterline. As a result, the number or size of individual cells increases dramatically at this location on the stem (a process called **hypertrophy**) where the trunk is neither completely aerated nor completed saturated. Stem diameter at any height therefore represents the frequency of occurrence of flooding at that height.

Buttressing may also develop from mechanical stresses. Based on studies of 191 trees of African whitewood in Ghana, Johnson (1972) reported that small buttresses were found on well-drained upland soils, whereas large buttresses occurred on poorly drained mid- or lower-slope sites. When soil conditions permit deep rooting, the vertical root system is able to absorb stresses that would otherwise concentrate at the lateral root–stem juncture. By contrast, limited vertical root development due to impermeable layers or a high water table concentrate stresses at the stem base, thus inducing marked buttress growth. Wind and associated swaying may also result in buttressing. For example, a study of Sitka spruce by Nicoll and Ray (1996) documented considerable buttressing on the leeward side of spruce stems compared to the windward sides to support the stem against prevailing winds. Notably, entire spruce root systems also had more structural root mass on the leeward side than the windward side of the tree.

STEMS

Trees and shrubs are called "woody" because they form consecutive layers of structural tissues to the primary stem. These layers are termed **secondary growth**, and they strengthen the stem, support the crown, and increase the transport of food, water, nutrients, and hormones between the shoots and the roots. A meristematic sheath of cells called the **vascular cambium** surrounds the stem, shoots, and roots (Figure 5.11). It creates the successive layers of secondary growth and consists of a zone of actively dividing cells. The vascular cambium builds tissues in two directions: toward the center of the tree and toward the outside of the tree. Toward the center of the tree, the cambium gives rise to **xylem**, which conducts water and the nutrients dissolved in it, typically

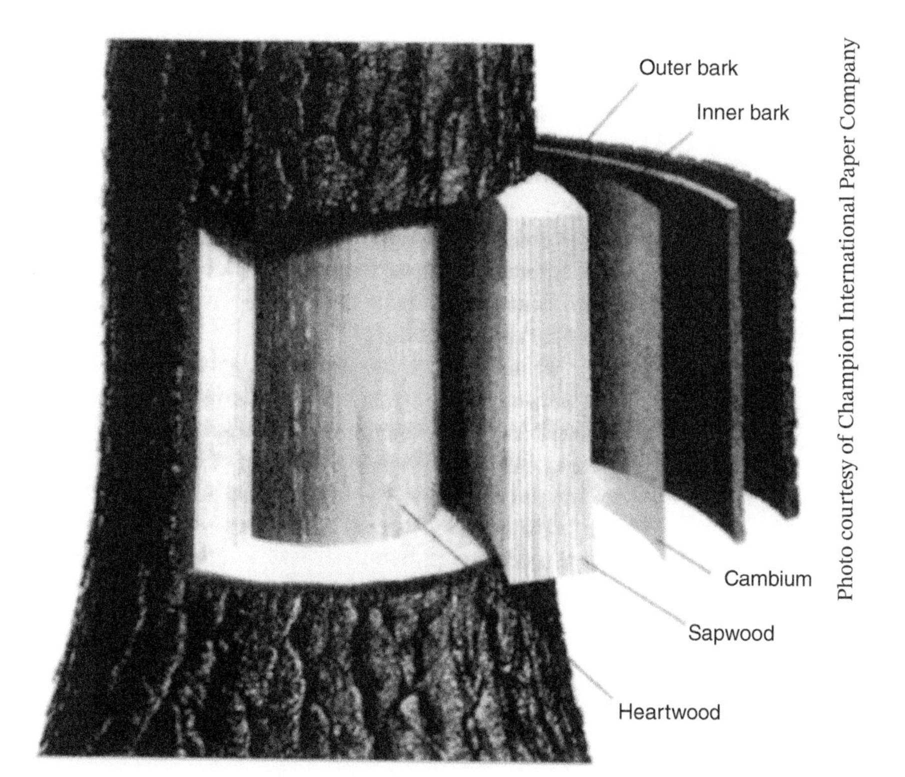

FIGURE 5.11 Generalized structure of a tree trunk (i.e., a stem) showing the position of major tissues. Tissues include the outer bark (dead and compressed phloem cells), the inner bark (living phloem-food-conducting cells), cambium (sheath of living cells that give rise to phloem and xylem cells), sapwood (outer band of xylem-water-conducting tissue), and heartwood (inner core of xylem-water-conducting cells).

from the roots toward the crown. The xylem cells become lignified and form the dead, woody axis of the tree. It is remarkable that the indispensable function of conducting water rapidly over long distances necessarily requires dead cells (Zimmermann 1971). Dead xylem also provides the structure for height growth that distinguishes forests from prairie and tundra. There are two types of xylem—the central core, or **heartwood**, and the outer portion, or **sapwood** (Figure 5.11). Most water is conducted through sapwood, but heartwood is purely structural and readily decayed in old trees. Heartwood is often much denser than sapwood (Woodcock and Shier 2002). Huge, completely hollow trees (for example, basswood, American beech, sycamore, etc.) may live for decades supported only by a thin shell of phloem, cambium, and sapwood.

Toward the outside of the tree, the cambium generates the **phloem** or inner bark, which conducts photosynthate created in the leaves to the rest of the plant. Phloem cells are not lignified and function for only a few years before collapsing and becoming part of the outer bark (Figure 5.11). Phloem cells are therefore as regularly constructed as xylem cells, though they do not accumulate. The anatomies of the xylem, phloem, and bark are summarized by Esau (1977) and Bell (1991), and the physiology by Zimmermann and Brown (1971) and Pallardy (2008).

XYLEM CELLS AND GROWTH RINGS

Diameter growth of trees results from the annual formation of layers of xylem cells by the cambium. In conifers, the primary water-conducting cells are the vertically oriented (perpendicular to the

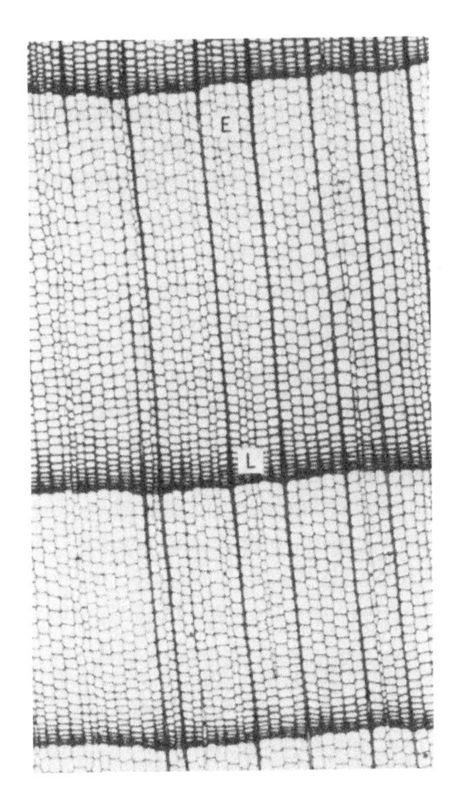

FIGURE 5.12 Transverse section of wood of ponderosa pine showing a transition in size and shape of earlywood (E) and latewood (L) in successive growth rings. *Source:* Zimmermann and Brown, 1971; reproduced with permission from Springer Nature.; Photo by Claud L. Brown.

ground in the stem), overlapping **tracheids** (shown in cross section in Figure 5.12). These cells are thick-walled, tapered, 3–5 mm long and up to 100 times longer than wide (Kozlowski 1971b), and arranged in uniform radial rows in conifers. Parenchyma cells and ray tracheids are oriented horizontally (parallel to the ground in the stem) and transport water, food, and other substances laterally. Perforations in the walls of the tracheids allow water and other substances to transfer among adjacent cells. Conifer growth rings are distinguished by differences in cell diameter and the cell-wall thickness of the tracheids. Thus, an annual growth ring consists of both a light and a dark portion, with the dark portion produced most recently (Figure 5.12).

In angiosperms, xylem cells are both more specialized and more diverse, including vessels, tracheids, fibers, and parenchyma. The majority of the xylem cells in angiosperms are fibers, but cells called **vessels** are those that conduct most of the water through the stem. Vessels are single cells whose end walls have many perforations, such that many cells lined up end to end act as tubes that may be up to several meters long. This system of vessels allows rapid transport of water and nutrients to the shoots. In **ring-porous** angiosperms (including ashes, hickories, oaks, elms, and black locust), vessels formed early in the growing season have much larger diameters (up to 100 times as great) than those formed late in the growing season (Figure 5.13). Growth rings of ring-porous species are easily distinguished because of the large vessels in the earlywood. **Diffuse-porous** species (such as aspens, willows, birches, basswood, and maples) have smaller vessels that are more uniformly distributed throughout the growth ring (Figure 5.14). Growth rings of diffuse-porous trees are not as clearly distinguished as those of ring-porous trees.

PERIODICITY AND CONTROL OF SECONDARY GROWTH

In north temperate species, annual diameter growth typically results from the growth of a single annual ring. The widths of annual rings vary among species, tree ages, and site conditions. Where tree growth is affected more by weather than site conditions, secondary growth is sharply reflective

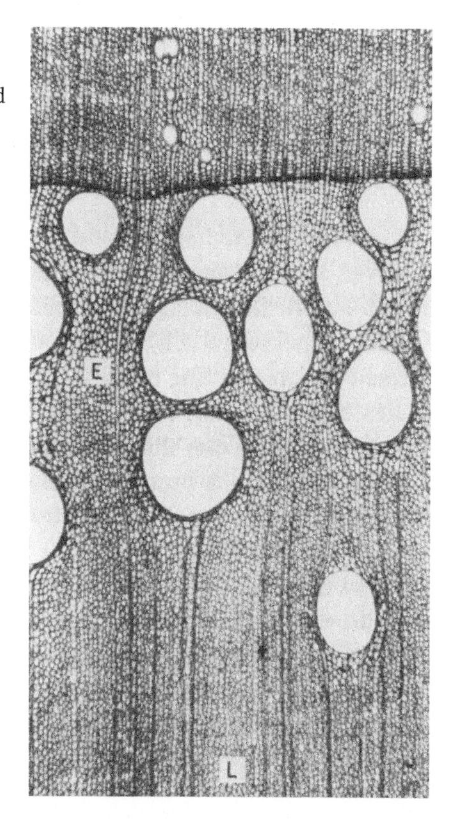

FIGURE 5.13 Transverse section of wood of a ring-porous black oak showing very large vessels in the first formed earlywood (E) and the preponderance of fibers in the latewood (L). *Source:* with permission from Springer Nature.; Photo by Claud L. Brown.

of temperature and/or precipitation during the growing season when the ring was generated. Where the growing season is very short, such as in subarctic regions and high elevations, temperature is often more limiting than precipitation (Kozlowski et al. 1991). By contrast, growth may be nearly continuous throughout the year in tropical and subtropical areas, and many tropical trees lack annual growth rings.

Most secondary growth occurs in temperate areas during the spring and early summer when soil conditions are favorable, but the duration of the growth period varies widely among species. For 21 tree species on one site in the Georgia Piedmont, the length of the most rapid period of growth varied from 70 to 209 days (Jackson 1952). In a study of European species, Norway spruce, larch, and ash formed growth rings rapidly, with little wood laid down after July (Ladefoged 1952). Species of birch, beech, alder, maple, and oak laid down wood from May until early September, with up to one-third of the ring being formed in the late summer. For seven boreal tree species in northern Quebec, radial growth was similar among species but did not appreciably begin until early June, explained mainly by increasing soil and air temperature (Tardif et al. 2001).

The initiation of cambial growth in the spring occurs when growth-regulating hormones, auxin, and other substances (promoters and inhibitors) from expanding buds and leaves trigger activity. In conifers, cambial activity begins in the crown at the base of actively developing buds and leafy shoots prior to extension and moves rapidly downward throughout the shoots, branches, and the main stem. The process is similar but far slower in diffuse-porous hardwoods. Cambial activation is almost simultaneous throughout the entire stem in ring-porous species, which is an important adaptation for forming new vessels to replace those becoming nonfunctional from the previous year. The newly formed vessels are vital in ring-porous species to supply water to developing foliage.

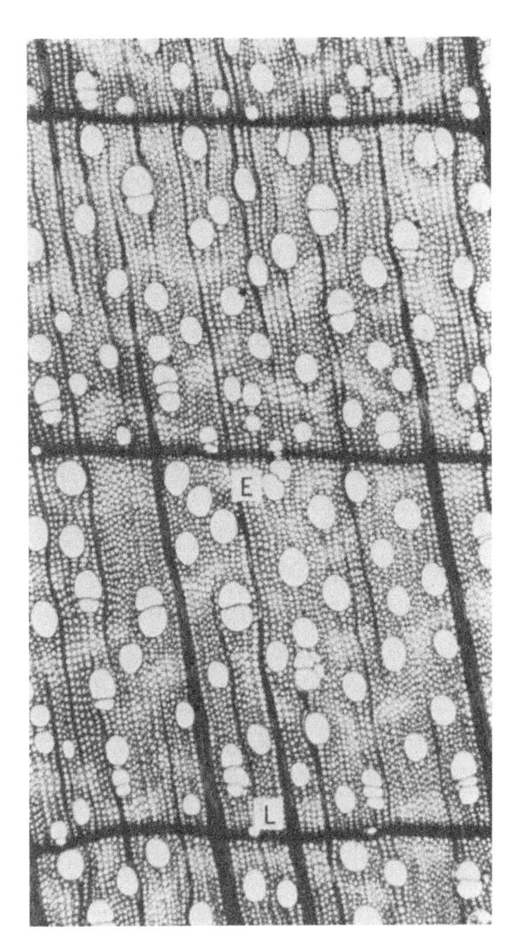

FIGURE 5.14 Transverse section of a diffuse-porous hardwood, tulip tree, showing fairly uniform distribution of vessels throughout the growth ring and radial flattening of the last formed latewood cells. *Source:* Zimmermann and Brown, 1971; reproduced with permission from Springer Nature.; Photo by Claud L. Brown.

Control of Earlywood and Latewood Formation Much research has focused upon the formation of earlywood and latewood and their properties and physiology, particularly because of their importance to pulpwood and timber quality. Earlywood is formed by rapid cell division early in the growing season; cell enlargement is prolonged, and the duration of cell wall thickening is short, creating larger tracheids with walls only thick enough to avoid cell implosion (Hacke et al. 2001; Pratt et al. 2007). These earlywood cells supply most of the demand of the crown for water. Latewood is formed when the rate of cell division decreases later in the growing season, the cell enlargement duration period is shortened, and cell walls thicken for a longer period (Cuny et al. 2014). The narrow diameter of these cells limits their utility for water transport (Sperry et al. 2006). As a result, latewood tends to be denser than earlywood because it contains thicker cell walls per unit volume. The specific gravity of latewood in conifers is about two to three times that of earlywood (Kozlowski 1971b), but the density of both earlywood and latewood varies among years. The density of earlywood in a given year is related to temperature in the previous growing season, specifically as it relates to photosynthetic activity and its impact on cell size. By contrast, latewood density in a given year is related to early spring temperatures in the same year, specifically as it relates to the activity of the cambium and thickness of the cell walls (Björklund et al. 2017). The type of cells produced (earlywood or latewood) is also likely to be influenced by the balance of growth hormones, especially as they interact with environmental factors, but their production is unlikely to be controlled by hormonal signaling alone (Buttò et al. 2020).

Dendrochronology, or the study of tree rings, has long been used to assess global climate change because tree rings are able to record the natural variation of the past climatic record. For example, both ring width and maximum wood density of the rings have been used to reconstruct changes in average summer temperatures over the last 6000 years in northern Sweden, Finland, and Siberia (Pearce 1996). At these high latitudes (inside the Arctic Circle and near the northern limit of Scots pine), small changes in average summer temperature greatly affect the number of days warmer than the threshold for radial growth (5 °C/41 °F). The width and density of tree rings, and particularly of latewood cells within the rings, dramatically reflect these changes in temperature. Thus, understanding the structure of wood is at the forefront of distinguishing long- and short-term patterns of climate change. Accurately understanding and interpreting patterns of tree-ring width and density also permits *crossdating* ring samples from a tree, whereby each individual tree ring is assigned its exact year of formation by matching patterns of wide and narrow rings between samples from the same tree and between trees from different locations. The body of research utilizing dendrochronology to study forests in North America and elsewhere is enormous and well beyond our scope. The science of dendrochronology is treated extensively by Trouet (2020), Amoroso et al. (2017), Speer (2010), Cook and Kairiukstis (1990), Schweingruber (1989), and Fritts (1976).

WINTER FREEZING AND WATER TRANSPORT

Winter freezing of water in xylem cells is problematic for water conduction in trees of northern latitudes, particularly for ring-porous species with very large earlywood vessels. The presence of ice is less of an issue than dissolved gasses in liquid water, which form bubbles when frozen. When the water thaws, the bubbles create gaps or embolisms in the water columns, breaking their continuity and forming an air lock against transpiration. Zimmermann and Brown (1971) proposed three coping mechanisms of trees in this situation:

1. **Small-diameter xylem cells:** Very small conducting cells in some species may form bubbles small enough to redissolve when the water thaws, rejoining the vertical water column before transpiration begins. Conifers (firs, spruces, etc.) and certain diffuse-porous trees with very small vessels (alders, willows, trembling aspen, paper birch, etc.) are thus easily able to survive the extreme freezing and thawing regime of the boreal or alpine forest. Notably, these species are common in boreal and alpine regions. No ring-porous species are boreal, and most (such as oaks, hickories, and walnuts) are most abundant in the central or southern portions of North America.
2. **Root pressure:** Some diffuse-porous species such as birches may develop enough root pressure to fill the air gaps formed in small vessels.
3. **Spring formation of large vessels:** Ring-porous species that have large earlywood vessels conduct much of their water within the most recent growth ring. These same species are able to very quickly activate their cambium and produce new vessels prior to leaf flush in the spring.

Spring vessel formation is advantageous because water transport through large vessels is very rapid upon leaf flush as it is required by the developing and transpiring leaves. However, many species within this group, such as oaks, hickories, walnuts, and ashes, must delay their flushing in the spring until the xylem system is ready to supply the water required by the foliage. Delayed flushing has the disadvantage that the leaves must develop very rapidly in the warm temperatures of late spring when transpiration is high, and that a delay in flushing ultimately results in loss of growth relative to those species that flush earlier. Spring vessel formation is also advantageous because late-flushing species have a greater probability of escaping late frost than early flushing species. An additional disadvantage, however, is the dependence on the most recent ring for water conduction, because injury or disease to the ring would cause significant damage to the tree. American chestnut and American elm are examples of species that have been dramatically affected by diseases that block their water vessels (see Chapter 21).

WATER DEFICITS AND TREE GROWTH

Tree growth responds more to water stress than any other perennial factor of the forest site, and thus forest-site productivity depends heavily on soil-water availability in many parts of the world. Environmental soil-water stress affects height and diameter growth, flowering and fruiting, seedling germination and establishment (Chapter 4), and virtually every aspect of plant morphology and physiology (see also Chapters 9, 11, 18, and 19). A definitive review of soil-water deficit is by Zahner (1968), with more recent reviews by Kramer and Boyer (1995) and Pallardy (2008).

In temperate climates, soil-water deficits tend to occur during the middle of the growing season when air temperatures are high, soil water replenished by winter snowmelt and spring rains is no longer present, and fully flushed and growing forests have removed soil water via transpiration. The effect of soil-water stress on height growth is seen clearly in the low heights that trees attain in dry climates or on dry sites compared with those growing on moist sites, as well as reduced shoot growth of seedlings and saplings under soil-water stress.

Effects of soil-water stress vary among species, and they depend partly on the species' seasonal pattern of shoot elongation. Species that maintain shoot elongation throughout the growing season such as birches, tulip tree, and loblolly pine are affected by late-season droughts. Species that complete height growth and set buds by midsummer, including red pine, eastern and western white pine, white ash, sugar maple, American beech, and northern red and white oaks, are unaffected by late-summer droughts of the current year. However, a late-summer water deficit in the current year will affect height growth during the following growing season. This pattern occurs because photosynthate accumulates following bud set and is stored for use in the next growing season, but the amount of photosynthate produced and stored will be reduced under drought conditions. For example, 72% of the variation in annual shoot growth in young red pine was explained by water deficits of both the previous and current growing seasons together (Zahner and Stage 1966). However, height growth of tree species that exhibit shoot elongation throughout the growing season is unaffected by soil-water conditions in the previous year (Zahner 1968).

Diameter growth is also affected by soil-water stress. Periodic changes (daily and annual) in the radial growth of conifers and angiosperms are directly related to soil-water availability (Figure 5.15). For example, the period of highest radial growth for seven boreal species in northern Quebec was explained by precipitation and subsequent soil-water availability (Tardif et al. 2001). In a study of deciduous forests in Indiana, 13 of 20 species exhibited decreased radial growth in response to soil-water stress; the 7 remaining species unaffected by drought were classified as xerophytic rather than mesophytic (Brzostek et al. 2014). Water absorption by roots may lag behind transpiration later in the growing season in temperate regions, dehydrating crown tissues, slowing photosynthesis, and limiting growth below the potential for that time of year (Zahner 1968). Normal radial growth usually resumes with late-season rains if the mid-season water deficit has not been severe.

Plants resist drought by avoiding and tolerating drought stress (Levitt 1980b). Woody plants mainly avoid drought stress using morphological and anatomical adaptations that maintain relatively high tissue-water content in environments where water is limited by minimizing water loss and maximizing water uptake. Major avoidance mechanisms include: (i) the ability of roots to extract large amounts of water from the soil; (ii) high root-to-shoot ratio; (iii) the ability to reduce leaf surface area (including rolling, folding, and shedding of leaves, and maintaining a dense

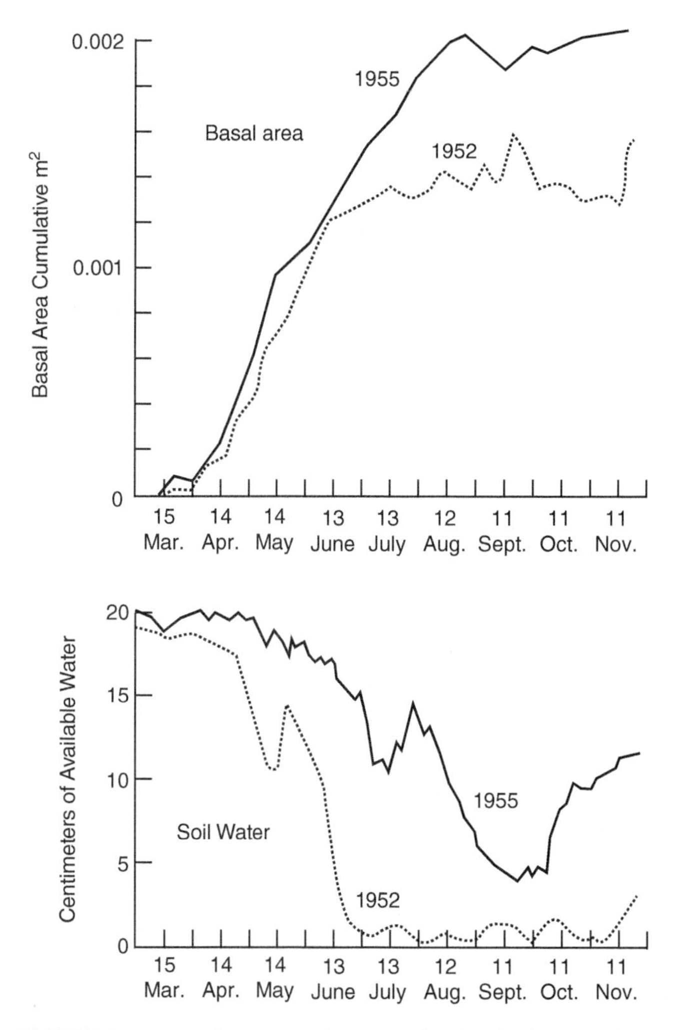

FIGURE 5.15 Basal area growth per tree for shortleaf pine and trends of available moisture for relatively wet (1955) and dry (1952) growing seasons. Note that the growth rate slowed in mid-June of 1952 but did not slow until mid-August of 1955, at the time in each year when available soil water had been depleted to about 5 cm or below. *Source:* Zahner (1968) / Elsevier.

pyramidal crown); (iv) the ability to reduce transpiration via stomatal control—closing stomata most of the day and/or the ability to close stomata very rapidly in response to stress; (v) thick cuticle to reduce cuticular transpiration, and/or (vi) a high proportion of water-conducting tissue to nonconducting tissue. The ability of trees to absorb large quantities of water may be more important for drought avoidance than reducing water loss through transpiration. For example, drought avoidance by stomatal closure has the disadvantage of reduced photosynthetic rates, but species efficient at absorbing water can maintain high tissue-water content even when stomata open for photosynthesis and transpiration occurs. Drought tolerance is the ability of plants to endure low tissue-water content through adaptive traits that maintain cell turgor using osmotic adjustment and cellular elasticity. Drought tolerance is less common than drought avoidance in woody plants, but many species withstand considerable dehydration before stomatal closure. Woody plants are heavily influenced by many other site factors in addition to drought, and these are discussed in Part 3.

SUGGESTED READINGS

Barthelemy, D., Edelin, C., and Halle, F. (1991). Canopy architecture. In: *Physiology of Trees* (ed. A.S. Raghavendra), 1–20. New York: John Wiley.

Hallé, F., Oldeman, R.R.A.A., and Tomlinson, P.B. (1978). *Tropical Trees and Forests*. Berlin: Springer-Verlag 441 pp.

Kozlowski, T.T. (1982). Water supply and tree growth. Part I. Water deficits. *For. Abstr.* 43: 57–95.

Kramer, P.J. and Boyer, J.S. (1995). *Water Relations of Plants and Soils*. New York: Academic Press 495 pp.

Lyford, W. H. (1980). Development of the root system of northern red oak *(Quercus rubra* L.) *Harvard Forest Paper* No. 21:1–30.

Pallardy, S.G. (2008). *Physiology of Woody Plants*, 3e. San Diego, CA: Academic Press 454 pp.

Pregitzer, K.S., DeForest, J.L., Burton, A.J. et al. (2002). Fine root architecture of nine North American trees. *Ecol. Monogr.* 72: 293–309.

Tomlinson, P.B. (1983). Tree architecture. *Am. Scientist.* 71: 141–149.

Zimmermann, M.H. and Brown, C.L. (1971). *Trees: Structure and Function*. New York: Springer-Verlag 336 pp.

The Physical Environment

PART 3

FOREST ENVIRONMENT

We emphasized in Part 1 that ecosystems integrate factors of climate, landforms, soils, biota, and all their interrelationships. This integration defines ecosystems as more than simply organisms (biotic) and their environments (abiotic). Although integrating these ecosystem components is our goal, it is useful to examine the physical factors of ecosystems as they affect plant distribution and processes. The sum total of the physical factors of an ecosystem determines the **forest site**. Importantly, plants and animals that occupy a site modify the local physical environment. For example, decomposing organic matter forms soil horizons as it interacts with mineral soil particles and water. The fact that abiotic and biotic factors are so heavily integrated emphasizes the difficulty of treating factors separately, despite the usefulness and simplicity of such an approach. The chapters in this section fall within the subdivision of ecology known as **autecology**, the study of organismal relationships to environment—in this case those of forest trees or other organisms.

We examine in this section the physical factors that influence plants where they live. The forest site is the place where plants live—defined by a specific geographic location and by the suite of physiographic, edaphic, hydrologic, and climatic factors that sustain the plants. To varying degrees, these abiotic factors are modified by the plants and animals of the locality. A given physical site factor may directly affect plants and other forest organisms, but more often its action is affected, intensified, or diminished by interactions with biotic factors and other physical factors.

The term *habitat* is widely used in the literature, conventionally when the major focus is on organisms. For example, "the white-tailed deer's habitat is the northern white-cedar swamp" (as if the swamp belonged to the deer), where typically the vegetation that provides shelter and food defines the habitat. The white-cedar's habitat, however, is defined by physical site factors including air, soil, and drainage. The term habitat may therefore take on several meanings. Because our focus is on ecosystems rather than organisms, we use the preferred term *site* (or habitat supporting plants) whenever possible to refer to the physical environment (as modified by biotic effects) of landscape ecosystems.

Forest Ecology, Fifth Edition. Daniel M. Kashian, Donald R. Zak, Burton V. Barnes, and Stephen H. Spurr.

SITE FACTORS

It may be fairly easy to enumerate the physical factors that support the existence and growth of forest organisms. However, it is exceedingly difficult to understand and evaluate the sum total of the *interactions* among the physical factors and their interactions with biotic factors that make up the complex we term *site.*

The site supplies a set of very basic factors. The tree grows with its crown in the air, gathering light, warmth, carbon dioxide, and oxygen. The tree also grows with its roots in the soil, gathering the mineral nutrients and water necessary for photosynthesis and other processes. The availability of each of these factors to the tree, however, depends upon an endless system of changing climate, day length, and soil development—changes that are in part related to the developing vegetation and associated animals.

ORGANIZATION OF SITE FACTORS

Biotic and physical site factors may be organized into broad groups that may be considered separately. The following diagram shows that site factors interact to yield the light, heat, water, etc. that are directly available and used by the plant. Physical and biotic factors have multiple effects on the plant.

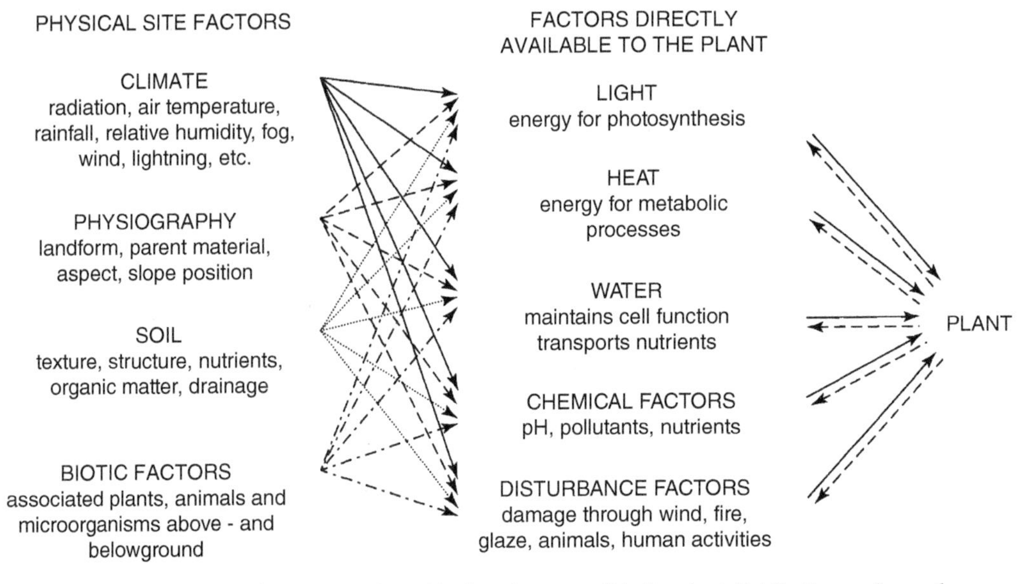

Relationships between site factors directly and indirectly responsible for plant distribution and growth. Source: Modified from Ellenberg 1968. Reprinted from *Wege der Geobotanik zum Verstdändnis der Pflanzendecke* © by Springer-Verlag Berlin, Heidelberg 1968. Reprinted with permission of Springer-Verlag New York, Inc.

Climate is an overriding driver of ecosystems on the Earth. The physical factors of climate related to the atmosphere affect the aboveground growth and persistence of plants, such as solar radiation, air temperature, air humidity, wind, lightning, and the

CO_2 content of the air. Climate also determines belowground temperature, moisture (via precipitation), CO_2, and the weathering of nutrients from rock substrates. Climate factors are repeatedly discussed and integrated into nearly every chapter of this book. The reception of light and the effects of light on plant growth are given special consideration in Chapter 6, as is the influence of temperature in Chapter 7.

Physiographic and soil factors (Chapters 8 and 9) include the topographic and structural features of landscapes together with all the physical, chemical, and biological properties of the soil. The relief and form of the land, the nature of the parent material, physical and chemical properties of soil, and soil biota are particularly important in plant growth and in determining site quality.

Biotic activities, both visible and microscopic, change climate, physiography, and soil—and therefore affect site quality. Large organisms (such as trees, grazing animals, and humans) create the most obvious changes in the microclimate and the soil. Small organisms occurring in great numbers (including fungi, bacteria, earthworms, rodents, and many others) can also substantially change the site. Because of the influence of biota on site, a discussion on various biotic factors is integrated into most of the chapters in this part. In addition, the roles of animals on site quality are discussed later in Chapter 12 (Part 4).

Fire is also a critical factor affecting forest species and site quality, and it is considered as a physical force in Chapter 10. For example, the burning of soil organic matter and the heating of the surface horizons of mineral soil result in changes in the physical and chemical properties of the soil and its soil biota.

Once the individual factors affecting forest site have been considered, it is necessary to integrate them into a whole if we are truly interested in accurately characterizing the forest site. Approaches to the determination of forest site quality and the evaluation of ecosystems are considered in Chapter 11.

INTERRELATIONSHIPS AMONG SITE FACTORS

It is both conventional and convenient to discuss the various factors of the forest site one at a time, but we approach such a discussion with caution. The consideration of factors individually does not imply that any is an independent force affecting forest biota. It is easy to study ecology in terms of simple direct relationships, making such statements as “at a temperature of 55 °C, the plant suffers direct heat injury” or “precipitation limits the distribution of this species.” Such statements ignore the reality that plants live in the total complex of the environment, where a change in any one factor may well bring about a changed requirement or tolerance of the plant for other factors. The individual environmental factors are not isolated, independent forces operating on the plant but are interdependent and interrelated influences that in the end must be considered together.

The *law of the minimum* holds that the rate of growth of an organism is controlled by that factor available in the smallest amount—often referred to as the "limiting factor." Ecologists have restated the law of the minimum to account for the interactions existing between the various single factors that operate at different scales. For instance, upon observing that seedlings die under dense forest canopies, an ecologist may conclude that low light intensity is responsible for the mortality. In fact, many environmental factors are affected by dense forest cover. Light intensity is certainly low, but so also is the supply of water in a soil permeated by the roots of the many trees present. The temperature regime and wind speed under a dense canopy are greatly changed from those in the open or even under an open forest. The organic matter and the soil microorganisms, too, will be greatly affected. In short, the whole environmental complex is related to forest canopy cover, and it is misleading, if not downright incorrect, to attribute changes in plant response to any single factor (Chapin et al. 1987).

IMPORTANCE OF SITE IN FOREST ECOLOGY

In this book, we consider site before biota for good reason. In a forest ecosystem, the site is more concrete, more stable, and more easily defined than the biota that occupy it. Therefore, perhaps counterintuitively, the site constitutes a better basis for the description of the forest ecosystem than do the trees, other plants, and animals. The emphasis on earth science as the basis for ecology is seen in the books on **geoecology** by Matthews (1992) and Huggett (1995).

The forest site occupies a fixed geographical area and is capable of fairly precise definition, especially using physiographic features or landforms. In contrast, various types of forest communities may develop on a particular site type, depending upon chance, past history, or changing environmental conditions (Chapter 1). Moreover, several quite different forest communities may be characterized by a given species or even a given group of species. Ideally, it is the combination of site and biota, the landscape ecosystem itself, that is necessary in describing forest landscapes and in understanding how forest ecosystems function.

Light

CHAPTER 6

Earth's ecological systems are driven by the energy provided by light. Light energy is used in **photosynthesis**, and its signals are used in the photoregulation of plant growth and development. Photosynthesis is carried out primarily in the chloroplasts of higher plants as indicated in this simplified equation:

$$\underset{\text{Carbon}}{CO_2} + \underset{\text{Water}}{H_2O} \xrightarrow{\text{light energy}} \underset{\text{Glucose}}{(CH_2O)_n} + \underset{\text{Oxygen}}{O_2}$$

This process features photons, packets of light energy, that excite electrons and move them from a lower to higher energy state. Green plant chloroplasts have a reaction center containing two molecules of chlorophyll, the green pigment that captures light energy. When the chloroplast absorbs a photon of light, an electron is transferred out of the reaction center, leaving the system with enough energy to split water in the cell and release its oxygen:

$$2H_2O \rightarrow O_2 + H^+ + 4e^-$$

The hydrogen ions and the electrons released in the cell bind to carbon dioxide to form sugar molecules, the source of energy used in **respiration**. Respiration is the process by which tissue biosynthesis and maintenance occur (Chapter 18). In the simplest terms, this process is how light energy is converted to and stored as chemical energy in photosynthetic products. Photosynthesis is the initial, basic step in the transfer of energy from light to the food chains of the ecosphere.

In addition to using light energy, plants respond and adapt to changes in environmental factors using the signals provided by light. Photosensory systems of plants as well as animals, fungi, and microorganisms acquire and process information about light direction, duration, intensity, and spectral quality. In higher plants, photoregulation occurs at all phases of the plant life cycle, from seed formation and germination to the genetic adaptation of populations to their site conditions (Chapters 3 and 4). Dormancy is initiated by seasonal changes in light quality, and daily changes in light conditions elicit leaf orientation and chloroplast distribution within leaf cells.

In this chapter, we examine the distribution of light at the Earth's surface, the interception of light by plant canopies and its use in photosynthesis, and the effects of light on tree growth and leaf morphology. In Chapter 18, we focus on the ecological aspects of carbon fixation and the allocation of photosynthate to growth, maintenance, storage, and defense. These processes have important implications for the cycling and storage of carbon in forest ecosystems.

Forest Ecology, Fifth Edition. Daniel M. Kashian, Donald R. Zak, Burton V. Barnes, and Stephen H. Spurr.

DISTRIBUTION OF LIGHT REACHING THE ECOSPHERE

Solar radiation may be reflected or absorbed by nearly anything in the atmosphere, such as clouds, smoke, ozone, smog, or greenhouse gases. Radiation reaching the Earth's surface on overcast days may be one-quarter to one-half of that for cloudless days under the same conditions. Most radiation under overcast conditions is scattered by the atmosphere and other interceptors, and then reaches the ground indirectly. The total amount of radiation reaching the surface is even further diminished by absorption and scattering when the sun is lower in the sky.

Total solar radiation per unit surface area is not the same thing as radiation in the visible range, or what we perceive as **light**. Energy is measured in Joules (J), or when integrated over time, **watts** ($1\,W = 1\,J\,s^{-1}$). Solar radiation is divided into the ultraviolet, the visible, and the infrared spectra. Wavelengths at the ultraviolet end of the spectrum (100–400 nm) are shorter and nearly completely absorbed by the atmosphere. Long wavelengths fall into the infrared spectrum (700–1000 nm), which contains about 50% of total radiation. Most of the radiation arriving at the Earth's surface is shortwave radiation with wavelengths shorter than 200 nm. Radiation emitted from the Earth into space—including the reflection of about 30% of incoming radiation known as **albedo**—is mostly longwave radiation, or radiation in the infrared region.

Radiation with wavelengths between 400 and 700 nm—a very narrow portion of total solar radiation—is visible to the human eye. Light in many historical studies in ecology was based on measurements that represent the amount of light detectable by the human eye—called **illuminance**. Illuminance is commonly expressed in the English system as foot-candles and in the metric system as luxes. **Photosynthetically active radiation** (PAR) of visible light is a measure of the total number of photons within the visible spectrum striking a surface and describes the effects of light on plants. Plants selectively absorb radiation for photosynthesis, and leaves appear green to the human eye because plants reflect most green light (wavelengths 500–580 nm).

The amount of solar radiation, or **insolation**, reaching a horizontal surface on the Earth ranges from 800 to $1200\,W\,m^{-2}$. These values change based not only on atmospheric content, as described in the earlier text, but also because of the slope and aspect of the ground. When radiation strikes the ground at angles greater or less than 90°, the radiation is distributed over a larger surface area and the amount of energy received (the photon flux, in moles or $\mu E\,m^{-2}\,s^{-1}$) is less (Figure 6.1). In the Northern Hemisphere, south-facing slopes receive more radiation per unit area than north-facing slopes because they face the sun at its highest angle during the day. Moreover, eastern slopes receive their highest irradiance in the morning, whereas western slopes receive their highest irradiance in the afternoon (Figure 6.2).

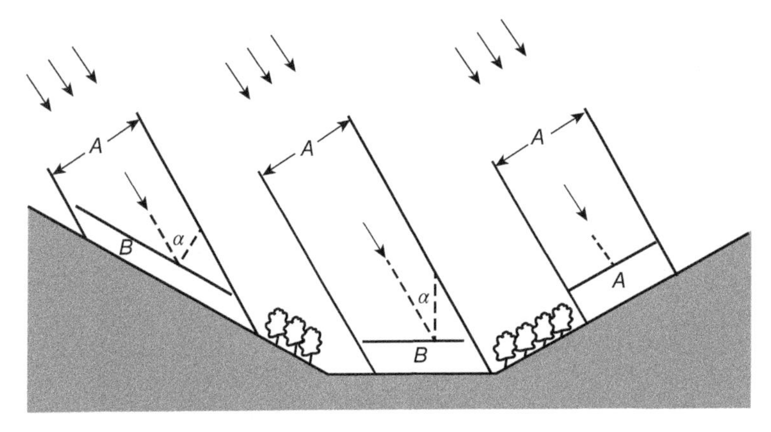

FIGURE 6.1 Distribution of insolation at different aspects and slopes. Photon flux changes according to the cosine of the angle of incidence (∝). Angle of incidence varies with the orientation of the receiving surface as well as with the position of the light source. Energy received at B = photon flux (moles $m^{-2}\,s^{-1}$) of Acos∝. *Source:* Hart (1988) / Unwin Hyman.

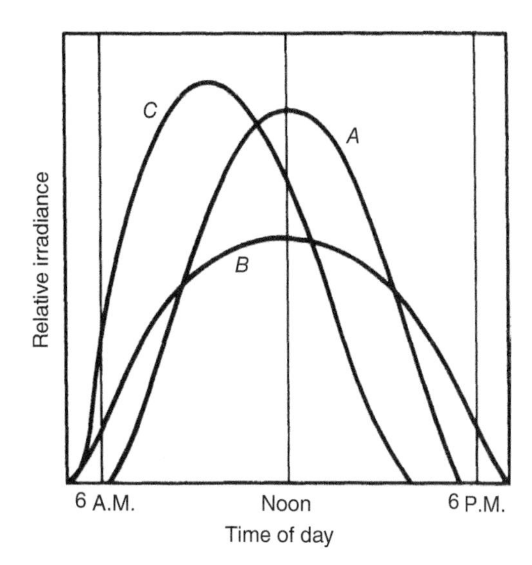

FIGURE 6.2 Relative irradiance received throughout the day on a 100% (45°) slope in the southern Appalachians (north latitude 35° 30′) on June 21. (A) South; (B) north; (C) east. The west slope curve is a mirror image of that for the east slope; the maximum irradiance occurs in the afternoon. *Source:* Bryam and Jemison (1943) / United States Department of Agriculture / Public Domain.

PLANT INTERCEPTION OF RADIATION

Plants utilize solar radiation efficiently by the way their leaves absorb, transmit, and reflect it. Leaf **absorption** is the fraction of incident light absorbed by the leaf, and **transmittance** is the amount of light that passes through the leaf and comes out the other side. Absorption is logarithmically related to transmittance, with absorption = 0 corresponding to 100% transmittance, and absorption = 1 corresponding to 10% transmittance. **Reflectance** is the amount of incident light neither absorbed nor transmitted by leaves. Only about 4–5% of total solar energy is absorbed and converted to chemical energy, but only 1–2% of total solar energy is used in photosynthesis, a remarkably small amount considering the amount of plant biomass produced globally on an annual basis (see Chapter 18). Plant leaves are most efficient at absorbing ultraviolet and visible light, but inefficient at absorbing near-infrared light (the 0.7–1.5-μm range, high in energy—see Figure 6.3). Leaves have high absorptivity of far-infrared light where solar irradiance is very low (Gates 1968). By contrast, reflectance and transmittance are highest for near-infrared light and lowest for visible and ultraviolet light. These properties are similar among many species, even when comparing plants from many different environments and with varying leaf anatomy. Thicker leaves seem to decrease transmittance and increase reflectance in many species (Knapp and Carter 1998). The widest difference among species for absorption and transmittance occurs in the portion of visible light that appears green (wavelengths 500–600 nm; Knapp and Carter 1998; Hagemeier and Leuschner 2019).

CANOPY STRUCTURE AND LEAF AREA

Interception of solar radiation depends very strongly on the structure of the forest canopy, determined largely by the surface area, shape, density, height, and depth of many adjacent tree crowns. Our understanding of canopy structure effects on light is based heavily on wet tropical forests. These forests have high tree and shrub diversity that incorporates wide ranges of shade tolerance and crown architecture, so they are able to produce a multi-layered canopy structure (Figure 6.4a). As a result, virtually all vertical space may be occupied in many forests of the wet tropics, extending from the tallest overstory trees down to herbaceous plants occupying the forest floor, and light intensity at the forest floor is extremely low. By contrast, canopy structure is often much less complex in temperate forests. In many trembling aspen or paper birch forests, for instance, overstory trees form a single-layered canopy with relatively sparse crowns that allow ample light to reach the

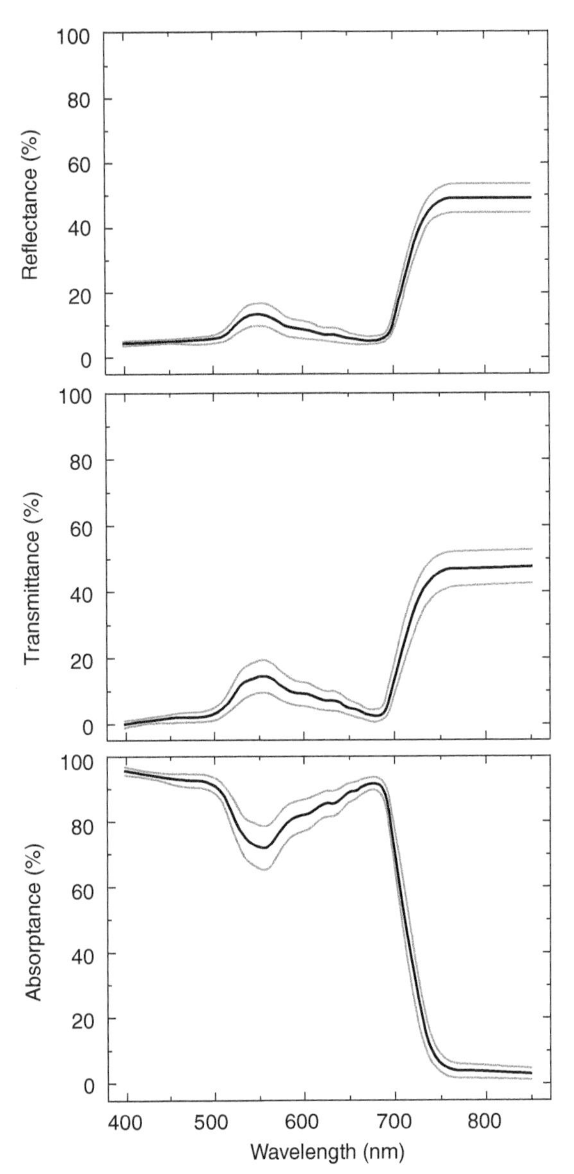

FIGURE 6.3 Mean reflectance, transmittance, and absorptance (thick lines) of the leaves of 26 species. The thin lines around the mean represent +1 standard deviation of the mean. *Source:* Knapp and Carter (1998) / with permission of John Wiley & Sons.

well-developed shrub and ground flora (Figure 6.4c). Coniferous forests of the Pacific Northwest may also have a single canopy layer, but their vertical canopy depths extend deeply toward the forest floor (Figure 6.4b), utilizing available sunlight to a much greater extent than aspen or birch forests. Canopy depth varies substantially among shade-intolerant, mid-tolerant, and tolerant species, with higher shade tolerance permitting leaf maintenance lower in the canopy. In the Pacific Northwest, for example, the canopy of grand fir, a very shade-tolerant conifer, is much deeper than that of mid-tolerant Douglas-fir and intolerant ponderosa pine (Figure 6.5).

Structural differences in the canopy determine the amount of leaf surface area present to capture light energy, quantified by forest ecologists using **leaf area index** (LAI). For deciduous canopies, LAI is defined as the one-sided green area of leaves (m^2) per unit ground surface area (m^2), also called the projected leaf area. Because their leaves have no anatomical top or bottom surface, LAI for conifers may be defined as the projected LAI, half the total needle surface per unit

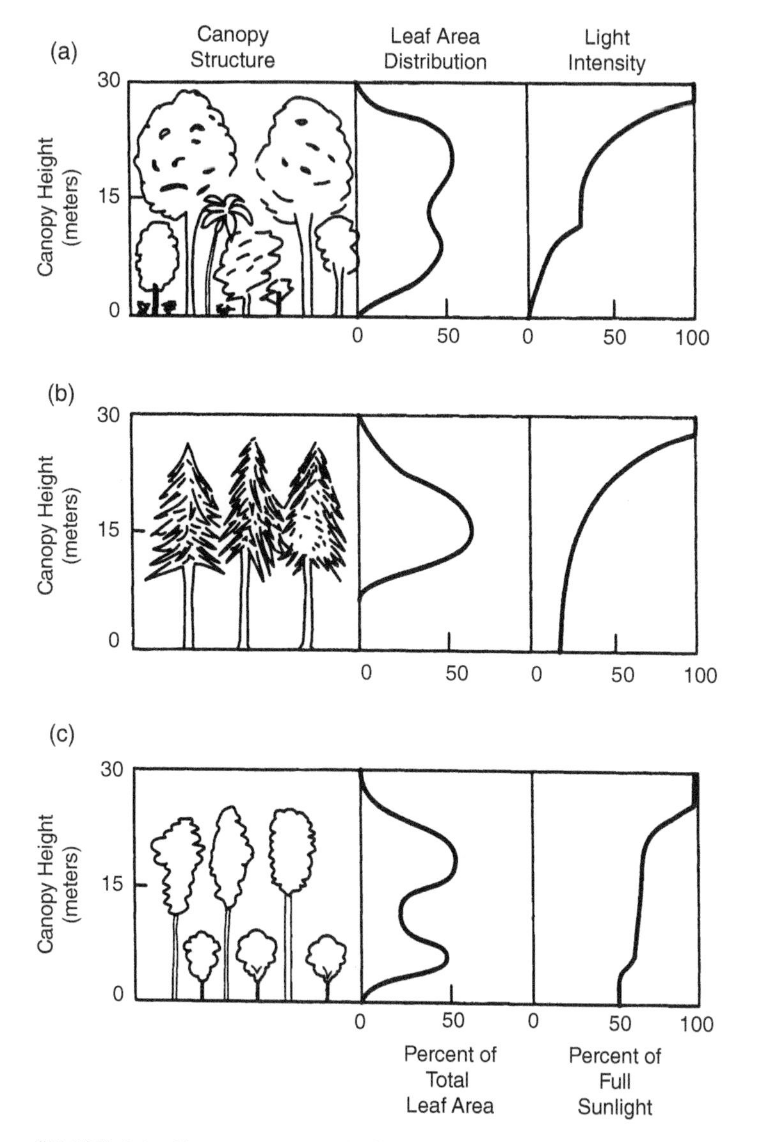

FIGURE 6.4 Canopy structure, leaf area distribution, and light intensity in (a) a wet tropical, (b) western conifer, and (c) a trembling aspen ecosystem. Notice the relationship between canopy structure and the distribution of leaf area within each forest. Differences in the vertical distribution of leaf area result in dramatically different light profiles within these ecosystems.

ground area, or total needle surface per ground area, all with units $m^2 m^{-2}$. LAI varies widely among ecosystems, ranging from $1 m^2 m^{-2}$ in sparse juniper woodlands in southwest Wisconsin (Schuler and Smith 1988) to $8.4 m^2 m^{-2}$ for a sugar maple-dominated forest in northern Wisconsin (Fassnacht and Gower 1997) to $13 m^2 m^{-2}$ in coastal old-growth Douglas-fir–western hemlock ecosystems of the Pacific Northwest (Thomas and Winner 2000). Moreover, LAI may vary significantly with forest age and tree density even within the same forest type (Kashian et al. 2005a). At least some inaccurately high estimates of LAI in old forests (as high as $22 m^2 m^{-2}$) result from using methodology for estimating LAI developed in younger forests.

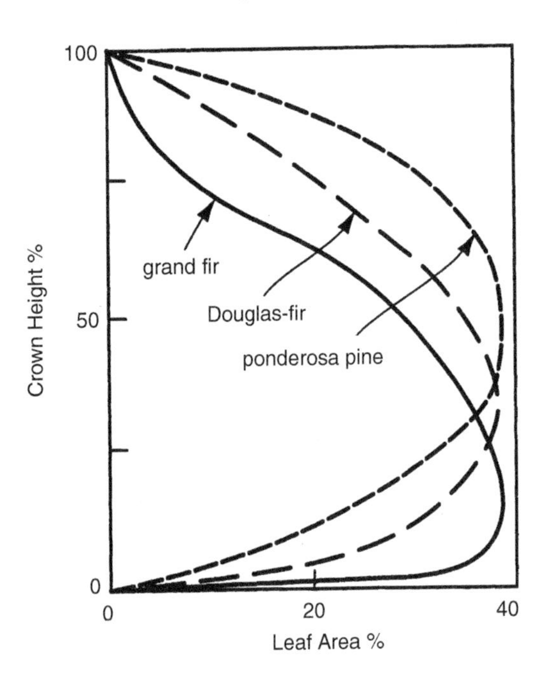

FIGURE 6.5 Differences in the vertical distribution of leaf area in grand fir, Douglas-fir, and ponderosa pine. The crown of grand fir, a very shade-tolerant species, has a deeper vertical distribution of leaves, compared to the shade-intolerant ponderosa pine.

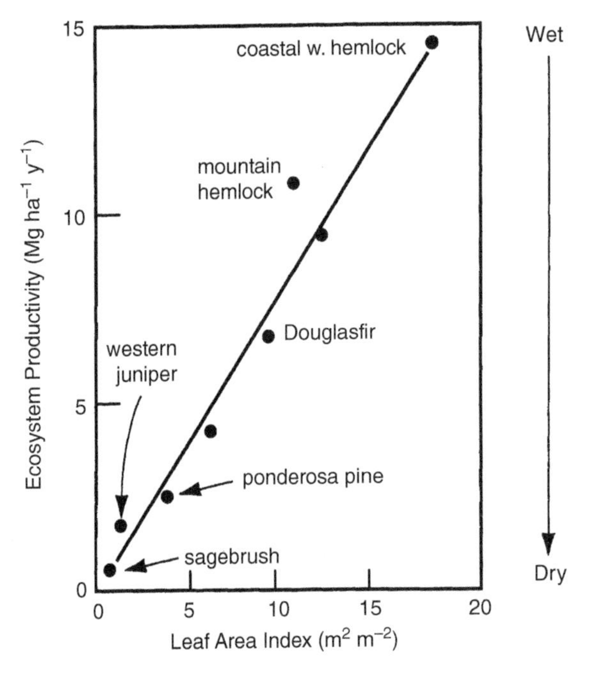

FIGURE 6.6 The relationship between leaf area index and ecosystem productivity among selected forest ecosystems of the Pacific Northwest. Productivity is expressed in megagrams of carbon per hectare per year.

LAI and productivity are closely related in many forest ecosystems because leaf area determines the amount of photosynthetically active surface area available to convert light energy into plant biomass. The amount of leaf area also determines the amount of surface area from which water can be lost via transpiration, however, such that climate and site, leaf area, and productivity are all interrelated. In Rocky Mountain juniper woodlands found east of the Cascade Mountains, for example, low amounts of precipitation in the semi-arid climate constrain leaf area, which in turn reduces ecosystem productivity (Figure 6.6). By contrast, high amounts of precipitation on the coast of the Pacific Northwest support western hemlock, which attains a much greater leaf area, is less affected by water loss in a humid climate, and ultimately results in greater rates of ecosystem productivity. The relationships between water availability, leaf area, and productivity are discussed in detail in Chapter 18.

Tree species vary widely in their ability to intercept solar radiation, and this variation is obvious even for broad vegetation types. Relative illumination beneath a deciduous forest in the winter may approach 50–80% of full sunlight, 10–15% under open, even-aged pine stands, <1–5% under temperate hardwoods in summer, and as low as 0.1–2% beneath tropical rain forests (Huber 1978). As we suggested in the earlier text, we observe differences in light interception among canopy tree species because they differ in shade tolerance and crown architecture. Shade-tolerant species are able to produce dense or deep (extending closer to the ground) crowns whose inner or lower foliage is able to survive in lower light conditions. In a study of light interception by canopy trees in southern New England, canopies dominated by eastern hemlock or American beech transmitted <1% of PAR to the understory (Canham et al. 1994). Both species are very shade-tolerant and exhibit dense canopies; eastern hemlock, notably, exhibits a canopy along 95% of its total tree height. By contrast, canopies dominated by mid-tolerant species with sparser crowns, such as northern red oak or white ash, transmitted >6% PAR to the understory. Compared to eastern hemlock, canopy depth of white ash and northern red oak represented only 39 and 47% of total tree height, respectively. Because transmission of PAR through canopies of northern red oak is nearly 6 times that through eastern hemlock canopies, the presence of a single hemlock tree could potentially create a zone of deep shade in a northern red oak forest (Canham et al. 1994).

Because tree species differ in the amount and quality of light they transmit, it follows that a mixture of species within a forest canopy is likely to be more efficient at intercepting incoming radiation. Forest canopies of mixed species have been shown to accomplish **canopy packing**, whereby species that have complementary canopy architecture and physiology are able to more efficiently use available growing space (Pretzsch 2014; Jucker et al. 2015), thereby increasing light interception and absorbance and reducing transmittance to the understory. Often the effect of a species-diverse tree canopy has a greater impact on light interception than the identity of the species themselves (Sercu et al. 2017). For example, a study conducted across Europe showed that mixed canopies of European beech and Scots pine had a 14% higher absorption of PAR than pure canopies of either species (Forrester et al. 2018). Although sometimes the presence of different canopy shapes from different species alone increases the interception of light by the forest canopy, other times the presence of multiple species changes the crown shape of the species present to increase canopy packing even further (Figure 6.7).

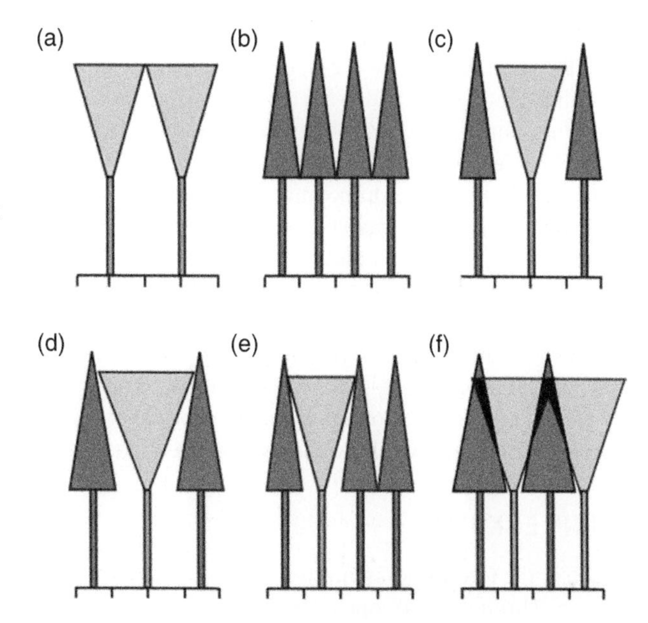

FIGURE 6.7 Potential effects of mixed species on canopy packing and thus light interception and absorption. When two species have (a) and (b) complementary shapes, their presence has (c) an additive effect of space filling. Canopy packing may increase even further (and thus light interception increases) if mixing allows (d) wider crown expansion, (e) higher stand density, or (f) both. *Source:* Pretzsch (2014) / with permission of Elsevier.

Technological advances in forest canopy measurement and characterization over the past two decades have allowed a greater appreciation for the complexity and importance of canopy structure. **Canopy structural complexity** (CSC) is a concept that describes the vertical and horizontal arrangement and complexity of both photosynthetic and non-photosynthetic tissue within forest canopies. The concept of CSC encompasses LAI, which estimates photosynthetic surface area in a single dimension, but is multidimensional, also including variables relating to canopy height, arrangement, openness, and variability (Atkins et al. 2018a). In essence, CSC is able to quantify the spatial pattern of leaf area, both vertically and horizontally, in a forest canopy. Increases in some CSC parameters such as canopy variability and height are correlated with higher levels of light absorption, whereas increases in other parameters such as canopy openness and clumping are correlated with lower levels of light absorption (Atkins et al. 2018b). Measures of CSC are novel and are still developing, but they are likely to better predict how forest canopies utilize incoming solar radiation than LAI alone (Atkins et al. 2018b). Likewise, CSC is emerging as an improved predictor of ecosystem productivity compared to LAI (Fotis et al. 2018; Gough et al. 2019). It appears that among all measures of CSC, maximum canopy height may be a primary driver of CSC because it allows the development of more complex arrangements of vegetation that improve the efficiency of resource use (including light) and thus increases productivity (Gough et al. 2020).

LIGHT QUALITY BENEATH THE FOREST CANOPY

As we suggested in the earlier text, differences in the amount of available light during the growing season are the main drivers of plant response in the understory of forests. Here, we describe the quality of radiation reaching the understory. Beneath forest canopies, PAR consists of (i) small amounts of light transmitted through leaves, (ii) unscattered radiation reflected downward by foliage, known as beam enrichment, and (iii) light that passes unabated through openings in the canopy. The primary source of understory PAR is through canopy openings, but transmitted light and beam enrichment are particularly important beneath closed or very dense canopies (Hutchinson and Matt 1976). Diffuse transmission represented only 0.5% and 0.7% of total understory PAR in a Douglas-fir–western hemlock forest in the Pacific Northwest and a northern hardwood forest in southern Ohio, respectively, but >6% understory PAR in red spruce–balsam fir forests of western Tennessee (Canham et al. 1990).

Much of the light transmitted or reflected by the forest canopy to the understory is in the far-red spectrum (700–750 nm), and light quality is very often reported as a red (600–700 nm) to far-red (R/FR) ratio. Most of the UV-B, blue, and red wavelengths are absorbed by the canopy, and the remaining light transmitted to the understory is enriched in green and far-red light spectra. In general, therefore, shady understory light conditions are typically characterized by low PAR and a low R/FR ratio (Messier and Bellefleur 1988; Fiorucci and Fankhauser 2017). However, a classic study by Vézina and Boulter (1966) documented differences in light quality among forest ecosystem types. Light beneath a sugar maple canopy was dominated by light in the far-red spectrum with lesser amounts of green, blue, and red light and ultraviolet radiation (Figure 6.8). Note that red light increases sharply in Figure 6.8 in the morning and early afternoon, when direct sunlight reached the understory through small canopy openings (sunflecks; see the following text). By contrast, no significant differences were found among various wavelengths in the visible spectrum beneath the red pine canopy, given its lack of selectivity in transmission. The proportion of radiation transmitted by forest canopies is also higher on cloudy or overcast days compared to clear days, and the sugar maple canopy in particular was less selective than on clear days (Vézina and Boulter 1966).

SUNFLECKS

Much of the forest floor in many forest ecosystems is in heavy shade for the majority of the growing season. Through small openings in the canopy, direct-beam radiation reaches the understory to

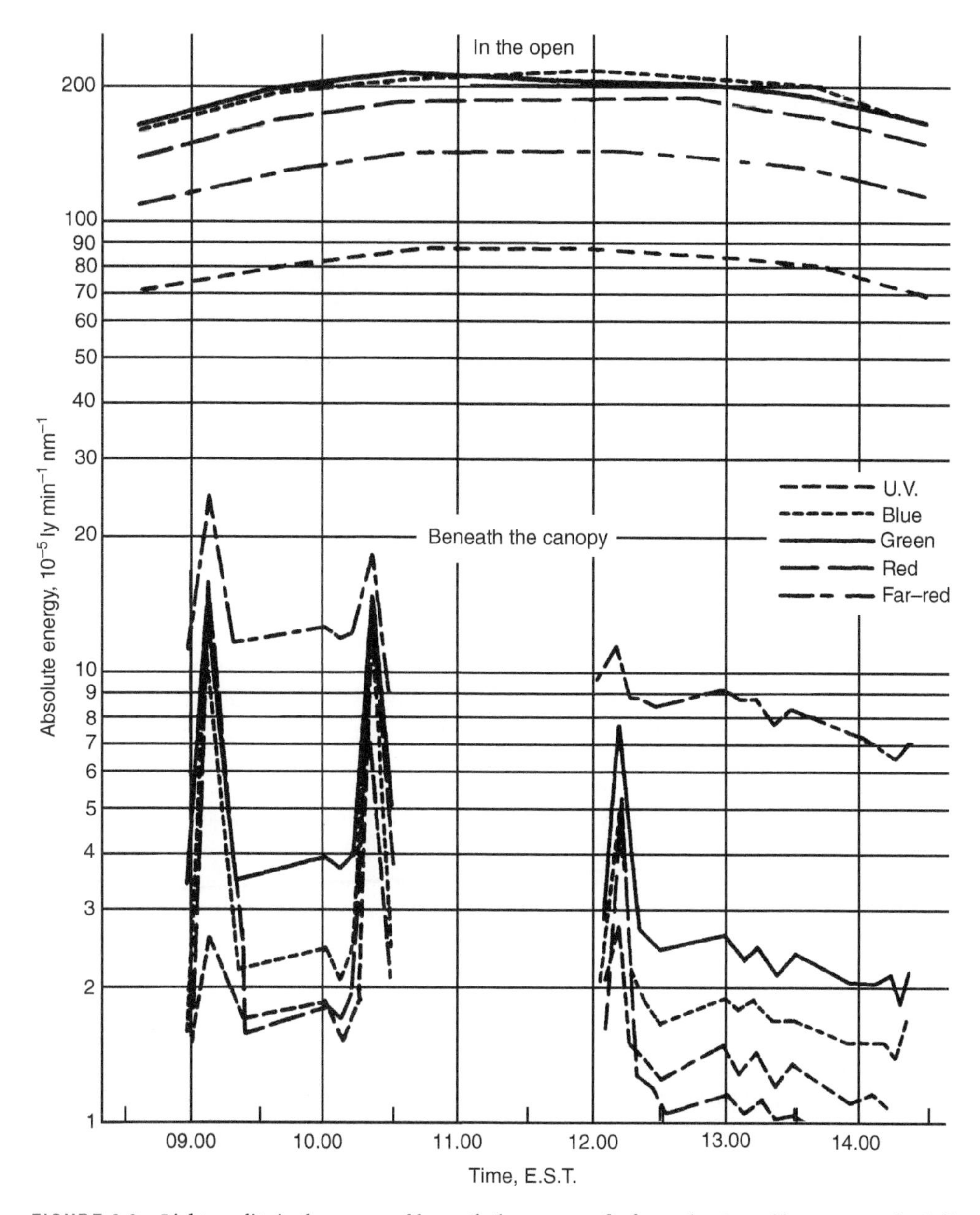

FIGURE 6.8 Light quality in the open and beneath the canopy of a forest dominated by sugar maple. Spikes in the understory light regime represent the presence of sunflecks. *Source:* Vézina and Boulter (1966) / Canadian Science Publishing.

create **sunflecks**. These short-lived patches of sunlight sweep over the ground over the course of the day, momentarily bathing small areas with unfiltered radiation (Figure 6.9), though with reduced PAR compared to full sunlight. A key feature of sunflecks is that they are not present at any given location of the understory for more than 10% of the time, yet they disproportionately contribute to the amount of PAR available for understory plants (Chazdon 1988). Sunflecks are often clustered in time, separated by long periods of few or no sunflecks (Vierling and Wessman 2000). Canham et al. (1990) reported that the proportion of total growing season PAR received as sunflecks beneath closed forest canopies at several temperate sites and one tropical rain forest

Photo by Dan Kashian

FIGURE 6.9 Sunflecks are short-lived patches of direct solar radiation reaching the understory through small holes in the canopy. These patches sweep light over the ground during the course of a day, creating an important source of radiation for understory vegetation. Sunflecks in this photo provide enough light irradiance to support a thick seedling layer of red ash beneath a closed canopy.

site ranged from 47% to 68% of the total amount of light penetrating through canopy openings. Notably, sunfleck duration averaged only 5.7–7.1 minutes. Most sunflecks are even shorter in duration, however, with 95% lasting less than 2 minutes (Pearcy et al. 1994).

The ability of an understory plant to utilize sunflecks depends on a complex balance of rapid changes in chloroplasts, electron transport carriers, and the potential accumulation of pools of photosynthetic metabolites. Plants may require several minutes to adjust their photosynthetic mechanism to adapt to this new high radiant energy flux (Fitter and Hay 1987). In addition, stomatal opening on the illuminated leaves must occur quickly enough to allow carbon dioxide uptake to occur for photosynthesis while the sunfleck is still in place. Some tree seedlings, such as sugar maple and white ash that open stomata rapidly and exhibit slow stomatal closure with increasing light, are probably able to make use of sunflecks (Davies and Kozlowski 1974). In general, however, understory plants benefit little from individual short-duration sunflecks because stomata take at least 5–10 minutes to open (Pallardy 2008). However, the induction time of photosynthesis of a leaf is reduced each time it is exposed to relatively high light levels. This means that sunflecks, even if short in duration, will prepare the leaf for better utilization of the next subsequent sunfleck. Thus, the sunflecks most beneficial to understory plants are those of long duration or which occur in rapid succession. A full description of the physiological responses of plants to sunflecks is provided by Way and Pearcy (2012).

Notably, the effects of sunflecks are not always beneficial. Leaves of understory plants are typically well adapted to shade, and brief exposure to the bright light of sunflecks has the potential to damage leaves due to photoinhibition, overheating, or water stress. A study by Ustin et al. (1984) compared red-fir seedlings on north and south slopes in upper montane forests of the central Sierra Nevada Mountains of California (Ustin et al. 1984). On the south slopes, plots with few seedlings had mean daily irradiance twice as high as plots with many seedlings, primarily due to sunflecks that occurred 3.5 times as often at irradiance levels above $1025\,\mu E\,m^{-2}s^{-1}$. Photon flux densities $>500\,\mu E\,m^{-2}s^{-1}$ exceeded the light saturation point for net photosynthesis in red fir, and thus substantially increased the energy load (heat) without corresponding increases in carbon gain. Sunfleck tolerance strategies are many and may include physical restructuring of chloroplasts, changes in leaf angles in some species, utilization of leaf pigments, and the emission of isoprene, the latter for the protection of photosynthesis against heat stress (Way and Pearcy 2012).

LIGHT AND GROWTH OF TREES

Growth and development of forest trees are tightly linked to photosynthesis, which of course is dependent on light. **Absorption spectra** are described in the earlier text as the wavelengths of visible light absorbed by plants; here we discuss **action spectra**, which are the specific wavelengths most effective for photosynthesis. We described in the earlier text that green foliage reflects visible light in the green wavelengths (Figure 6.3), and thus absorption is highest in the blue-violet and yellow-red wavelengths. We might predict that plants absorb only the wavelengths that they will use to drive photosynthesis, and thus the action spectra for most plants also include blue-violet and yellow-red wavelengths. Several studies have shown that not all green wavelengths are reflected and often >70% are absorbed (Figure 6.3). Action spectra for most green plants are highest in red wavelengths (600–700 nm), followed by blue wavelengths (400–500 nm), and then by green wavelengths (500–600 nm; McCree 1972). Several classic studies of woody plants identified action spectra even before the concept was articulated. For example, Burns (1942) showed that orange-red wavelengths were most effective in stimulating photosynthesis in Norway spruce, eastern white pine, and red pine, whereas the shorter violet-blue waves were relatively unimportant. Similarly, light absorption in Scots pine seedlings was highest in the red and the blue regions, although the photosynthetic efficiency of blue light was lower (Linder 1971). An important point is that absorption in both the blue and red regions is needed for maximum photosynthetic efficiency.

Many studies have examined how variation in the amount of light affects the rate of photosynthesis. Respiration is the opposite reaction to photosynthesis—the process that draws energy from the carbon that is fixed into biomass via photosynthesis and releases CO_2 to the atmosphere (Chapter 18). **Light respiration** occurs mainly in leaves, results in a relatively high loss of fixed carbon, and is genetically controlled. **Dark respiration** occurs in all living cells, acts to build and maintain living tissues, and is more closely coupled to environmental factors. Net photosynthesis (photosynthesis minus dark respiration) must be positive—meaning more CO_2 is taken up than given off—for a leaf to fix carbon into its tissues. When light levels are low, the rate of photosynthesis is also low, and the leaf (and ultimately the plant) loses more CO_2 than it gains. The **light compensation point** is the light intensity at which carbon gain from photosynthesis and carbon loss from respiration are equal, and net CO_2 is neither lost nor gained. In dense forests with very dark understories, light intensity in the understory is generally too low for photosynthesis to exceed respiration without the presence of sunflecks and adaptions by understory plants to trap diffuse light (Huber 1978).

Light compensation points vary widely among species, and particularly among species with differing shade tolerances (Figure 6.10). A study of 115 species of woody plant seedlings found that compensation points were lower for shade-tolerant species than for shade-intolerant ("sun-loving") and mid-tolerant species because of lower dark respiration rates in shade-tolerant species (Craine and Reich 2005). For example, the light compensation points of ponderosa pine and Scots pine (shade intolerant) are nearly three times as high as for eastern white pine (mid-tolerant), but those for eastern hemlock, American beech, and sugar maple (shade tolerant) are extremely low (Burns 1923). Shade plants generally require 0.5% of full sunlight at the compensation point, and sun plants require 1.5% (Perry et al. 1969). Given that understory light intensity may be very near the compensation point for as much as 90% of the growing season (Ellsworth and Reich 1993), low compensation points and dark respiration rates for shade-tolerant species are critical for them to achieve positive net photosynthesis (Figure 6.10).

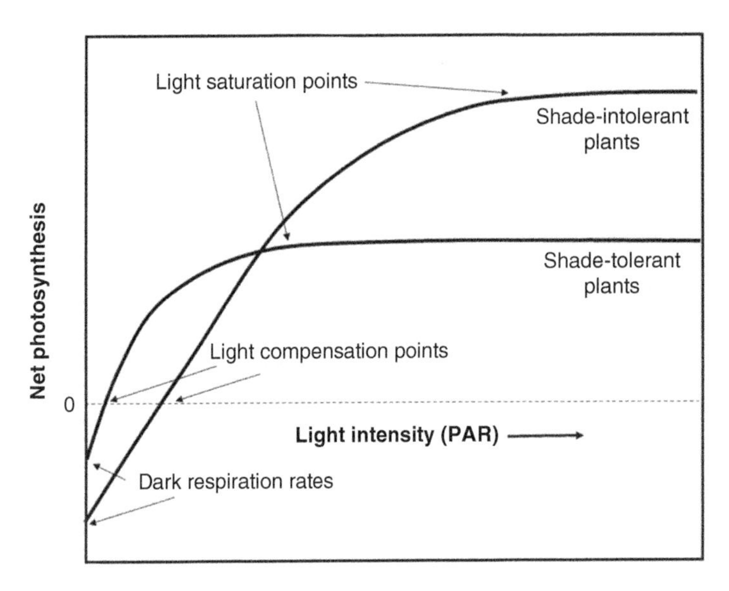

FIGURE 6.10 Hypothetical photosynthetic light saturation curves for plants of differing shade tolerance. Carbon gained through photosynthesis is greater than that lost through dark respiration when net photosynthesis is positive. Net photosynthesis increases asymptotically with increasing light intensity to a saturation point for all plants, light is limiting at intensities less than the saturation point, and CO_2 is limiting once the saturation point is reached. Light saturation is lower, dark respiration rates are lower (less negative), and light compensation points are less for shade-tolerant plants.

Woody plants continue to increase their photosynthetic rate as light intensity increases above the compensation point, eventually reaching an asymptote at 25–50% full sunlight. The point at which photosynthesis no longer increases with increasing light intensity (and carbon fixation no longer increases) is called the **light saturation point**. At this point, the reaction is not light-limited, but is instead affected by the supply of CO_2, PO_4^{3-}, and enzymatic processes. The light saturation point is higher for shade-intolerant woody plants, such that they are able to attain higher levels of net photosynthesis (Figure 6.10). Classic studies by Kramer and Decker (1944) and Kozlowski (1949) identified a much higher light saturation point for loblolly pine seedlings compared to associated hardwood species, which reached their maximum photosynthetic rate at 30% or less full sunlight. Such a finding partly explains better relative competitive ability of hardwoods under open pine stands where the relative illumination is typically near 30% full sunlight and is common for many hardwood species that succeed pine (Figure 6.11). Higher potential rates of photosynthesis generally trade off with the higher physiological cost of synthesizing and maintaining higher proportions of photosynthetic machinery per unit of biomass.

The relationships between net photosynthesis and light intensity are similar for sun leaves and shade leaves in a single canopy as they are for shade-intolerant and shade-tolerant plants. These relationships affect the branching patterns of species depending on their shade tolerance and growth rate. Fast-growing shade-intolerant species tend to have branches more widely spaced than those of shade-tolerant species whose shade leaves acclimate to the light conditions of the inner crown by anatomical features (see discussion in the following text). Many floodplain species (willows, cottonwoods, silver maple, American elm), for example, have broad and often wide-spreading crowns that facilitate light capture by foliage of the external and internal crown.

Many other factors affect the competitive ability of different species under given light conditions, including the color, shape, presence of waxes and hair, and the arrangement of

FIGURE 6.11 A conspicuous hardwood understory in an open 80-year-old loblolly pine stand, North Carolina. *Source:* Roth (1990). Reprinted with permission of Gebrüder Borntraeger Verlagsbuchhandlung, Stuttgart, Germany.

leaves. These factors regulate the amount of light that reaches the chlorophyll and drives photosynthesis. For example, pines have rounded needles arranged in dense clusters, often inclined upward rather in a plane perpendicular to incident sunlight, which results in scattered light and shading of one needle by one or more others (Kramer and Clark 1947; Carter and Smith 1985). Many hardwoods associated with pines in the eastern United States, such as dogwood, have broad, thin leaves, arranged to minimize leaf shading and oriented perpendicular to the direction of incident radiation. In this comparison, chlorophyll in a pine needle receives a much smaller fraction of the light incident to it than does chlorophyll in a dogwood leaf. As such, pine is able to maintain a high net photosynthetic rate in high light conditions of the overstory, and dogwood is able to efficiently photosynthesize in low light conditions of the understory.

Conifers and evergreens in general have a relatively low photosynthetic capacity per unit area of foliage (Waring 1991; Reich et al. 1995). From the tropics to the boreal forest, the maximum photosynthetic capacities of evergreen species are usually half to less than a quarter of that of associated deciduous species. Moreover, many deciduous species have equal photosynthetic efficiencies at low and high light intensities, whereas many evergreen species are more efficient when light intensity is high (Pallardy 2008). Such a pattern does not always imply that conifers and evergreen species are less competitive than deciduous species, however, in part because conifers are able to tolerate lower resources on poorer sites. Moreover, in environments where deciduous and coniferous trees occur together, conifers often develop a canopy dense enough to capture most of the incident radiation in the visible range (Jarvis and Laverenz 1983). Once a dense, closed-canopy stand of conifers dominates an area, as after fire, all but the most shade-tolerant co-occurring deciduous species are at a disadvantage. Finally, the ability of evergreen species to achieve some carbon fixation via photosynthesis while associated deciduous plants are dormant also improves their competitive ability over the long term.

Both the angle and azimuth of the leaf blade affect its interception of solar radiation. In many deciduous species, leaf petioles are able to orient the leaves to either increase absorption of radiation or avoid it. Positioning the leaf perpendicular to incoming radiation maximizes the amount of light absorbed and heat received. Normally this orientation is achieved over several to many days during the normal growth process, but many species are able to re-orient leaves at the temporal scale of minutes or hours. For example, woody leguminous species such as redbud, locust, or Kentucky coffeetree contain a small group of cells at the base of the leaf petiole that facilitates the movement of individual leaves into acute angles in response to radiation and, most often, its associated temperature. The horizontally flattened petioles of aspen species around the world (trembling aspen, big tooth aspen, European and Asian aspens) enable the leaf blade to move even in slight breezes, which permits light to bathe leaves more effectively than if they were stationary. Aspen leaf movement also reduces excess heat on the leaf surface when irradiance is high.

LIGHT AND SEEDLING SURVIVAL AND GROWTH

In general, at least 20% of full sunlight is required for survival of woody plants over a period of years, the actual amount varying with the species and with growing conditions. Measuring the height or weight (dry matter) growth of plants over a period of years under different degrees of irradiance has shown that several pine species grow best in full sunlight, although diameter growth is more affected than height growth at 50% sunlight (Pearson 1936, 1940; Shirley 1945b; Logan 1966a). Similarly, height growth of other coniferous species is virtually unaffected by partial light, such as white spruce, white pine, and Douglas-fir (Gustafson 1943; Shirley 1945b; Logan 1966a; Lassoie 1982). Several studies suggest that height growth of many hardwood species actually improves in partial sunlight. Height growth data for six hardwood species in Missouri (McDermott 1954) showed that five of the six species grew tallest in either 33% or 50% relative illumination (Table 6.1). The maximum seedling height of white birch, yellow birch, silver maple, and American elm seedlings in Ontario was at 45% relative illumination, but it was at 25% for highly shade-tolerant basswood and sugar maple (Logan 1965, 1966b). At least some recent studies support the idea that shade-tolerant tree seedlings have higher growth and survival in low light than shade-intolerant species, but shade-intolerant species grow faster at higher light intensity (e.g., Walters and Reich 1996, 2000a). Notably, fewer species achieve greater dry weight under partial light compared to full light because of increased allocation to roots (Chapter 18).

Root development of seedlings is also limited by low light conditions, such that seedlings of most forest trees have relatively shallow and poorly developed root systems in shaded conditions. In Table 6.1, American elm may achieve its greatest height growth at 33% sunlight, but aboveground biomass is 3.2 times that of belowground biomass under these conditions. Elms developing in full sunlight are shorter, but the root system is far better developed under these conditions (top/root ratio = 1.7). The reduction of root growth in low light is critical for the competitive ability and survival of woody plant seedlings in a forest understory, where root competition substantially reduces the amount of available soil water and nutrients. Although there are many causes of

Table 6.1 Seedling height (HT) and top/root weight ratio (T/R) of newly germinated seedlings as influenced by amount of sunlight (relative illumination, RI).

Species	20% RI		33% RI		50% RI		100% RI	
	HT (cm)	T/R	HT (cm)	T/R	HT (cm)	T/R	HT (cm)	T/R
American elm (*Ulmus americana*) 13 weeks	73	3.1	81	3.2	69	3.0	37	1.7
Winged elm (*Ulmus alata*) 13 weeks	59	5.2	71	3.9	76	4.4	28	1.9
Sycamore (*Platanus occidentalis*) 15 weeks	43	4.6	41	3.5	40	4.1	33	2.0
River birch (*Betula nigra*) 10 weeks	26	13.4	33	9.3	52	8.3	41	3.6
Red maple (*Acer rubrum*) 13 weeks	26	4.1	27	3.7	29	3.5	26	2.0
Alder (*Alnus rugosa*) 14 weeks	21	3.4	20	3.5	38	4.7	30	2.8

Source: Based on McDermott (1954).

seedling mortality, growth of seedlings in low light often occurs too slowly to develop a root system capable of taking up soil water and nutrients necessary for long-term survival of the plant (Coomes and Grubb 2000).

These patterns of biomass allocation to aboveground versus belowground structures are not likely to be a direct result of light availability alone. Instead, they have been proposed as a trade-off between maximizing photosynthetic efficiency in low light by allocating more biomass to aboveground structures for light capture at the expense of roots and maximizing the efficiency of water and nutrient capture in high light conditions where transpiration rates are higher and belowground resources may be less available (Smith and Huston 1989; Pallardy 2008). As a result, some species may tolerate deeper shade on soils with higher water and nutrient availability, and less shade where soils are poorer (Kobe et al. 1995; Walters and Reich 2000a). A detailed discussion of aboveground and belowground carbon allocation is presented in Chapter 18.

LIGHT AND TREE MORPHOLOGY AND ANATOMY

Morphological differences between sun-grown and shade-grown plants of the same species may result from their respective exposure to varying light conditions. These changes are of ecological importance in understanding a species' plasticity—its capacity to adjust to shaded conditions—or its response to sudden release following removal of the overstory by windstorm or by cutting.

Perhaps most studied for their responses to changes in radiation are the leaves of trees and shrubs, especially because the leaf is the plants' principal photosynthetic organ. In most cases, a species is able to adjust the physiology and structure of its leaves developed in the shade compared to those developed in full or near-full sunlight. Similar to how the rates of photosynthesis vary among shade and sun leaves, the altered structure and physiology of leaves on a plant grown entirely in the shade resemble those of leaves found at the base or interior of a tree canopy. Typically, shade leaves have a larger surface area per unit weight, and are thinner and less deeply lobed, with a thinner epidermis, less conducting tissue, fewer stomata, more intercellular space and spongy mesophyll, more chlorophyll per unit dry weight, and lower respiration (Pallardy 2008;

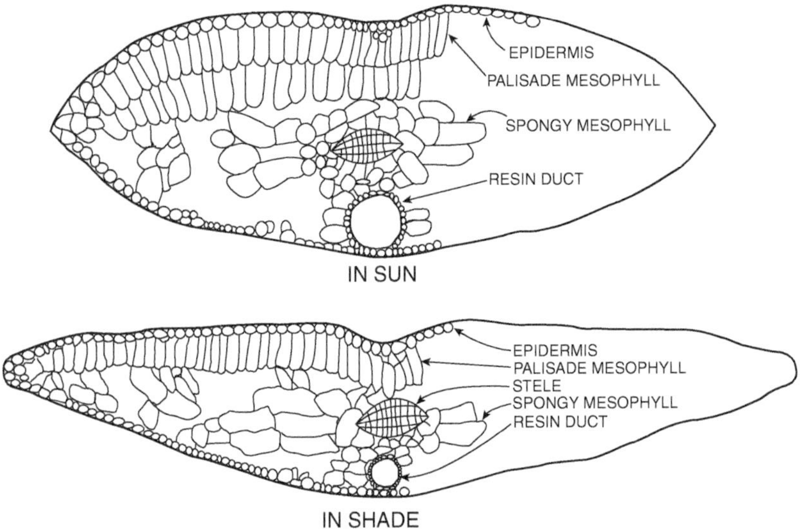

FIGURE 6.12 Structural variability in western hemlock needles grown in full sun (top) and in the understory beneath a dense Douglas-fir canopy. Note the thinner shape, larger surface area, thinner epidermis, and reduced palisade in the shade-grown needle. *Source:* Tucker and Emmington (1977) / with permission of Oxford University Press.

Larcher 2003) than comparable sun leaves of the same tree (Figure 6.12). The higher surface area of shade leaves increases their ability to gather light per unit of respiring surface area, and the thinner epidermis transmits more light energy compared to sun leaves (Daubenmire 1974). A study by Eschrich et al. (1989) also found that shade leaves of European beech had one layer of palisade mesophyll, but two layers when grown in the sun.

Although shade leaves develop in response to reduced light, temperature is also an important factor resulting in differing structure and anatomy. Temperature generally increases with irradiance, and water stress generally increases with temperature. Sun leaves therefore tend to experience higher water stress compared to shade leaves, but leaves within the crown and in its lower portions on clear days are under markedly less water stress compared to the sun leaves at the top of a tree. Therefore, many characteristics of sun leaves are as likely to be important adaptations to reduce water stress with increasing temperature as they are for differences in light. For example, the deep lobing of sun leaves apparently aids in the convective cooling of surfaces. Likewise, thicker leaves with a thicker epidermis act to reduce water loss in sun leaves, and their thicker cuticles reflect incoming radiation to reduce temperature.

The plasticity in leaf anatomy described in the earlier text varies among species, even among those in the same genus. The leaves of black oak, a xeric, shade-intolerant species, exhibit far more anatomical plasticity compared to those of northern red oak, a mesic, mid-tolerant species (Ashton and Berlyn 1994). Some species exhibit little plasticity in leaf anatomy and thus little difference between their sun leaves and their shade leaves. The anatomy of such plants is suited for a limited set of environmental conditions, and thus they are usually found only in the overstory or only in the understory. A study by Jackson (1967) found that leaves of shade-tolerant trees such as red maple, American beech, and flowering dogwood had one layer of palisade mesophyll regardless of light intensity, but those of shade-intolerant species such as tulip tree or black cherry were more plastic in the layers of palisade they exhibited across light regimes.

LIGHT AND EPICORMIC SPROUTING

Light may have important effects upon bark thickness and the stimulation of adventitious buds. Adventitious buds in the bark may be activated to suddenly form **epicormic branches** when the bark is exposed to direct solar radiation, such as by the thinning of forest stands via management or disturbance. Under sudden changes from shade to light, epicormic branches may develop in most forest trees, both conifers and hardwoods, although they are much more common in deciduous trees (Pallardy 2008). The occurrence of epicormic branches after thinning is particularly serious with many of the oaks, maples, birches, and other hardwoods where stem quality is of paramount economic importance. Although light has often been assumed to be the factor that triggers hormonal responses within a tree that lead to epicormic sprouting, the direct influence of light on sprouting is not clear (Meier et al. 2012). Some recent research suggests that shading the stem limits the initiation of epicormic branching (e.g., Gordon et al. 2006), but just as many studies report weaker correlative results (e.g., Wignall et al. 1987). A study by Nicolini et al. (2001) found heavy epicormic sprouting in natural stands of European beech without significant canopy gaps or sudden changes in light.

Many other factors in addition to light have been attributed to the initiation of epicormic sprouting, such as various stressors, low tree vigor, weakened crowns, or tree pruning (Meier et al. 2012). Many field observations have shown that epicormic branches form on the exposed boles of many trees with weak, poorly developed crowns. European foresters have been able to reduce the amount of epicormic branching on oaks by maintaining vigorous crowns and keeping the bole clothed with understory trees such as beech and hornbeam. It is clear that carefully controlled experimentation is required to determine which factor or combination of factors actually initiates the change of physiological circumstances that lead to the development of epicormic branches.

PHOTOCONTROL OF PLANT RESPONSE

Most critical plant processes are subject to the same photocontrol system, including stem elongation, root development, dormancy, germination, flowering, and fruit development. This system, whereby light is absorbed by a reversible pigment system in the plant, is the physiological basis of **photoperiodism** and affects seasonal rhythm, timing, and amount of tree growth. Photoperiodism was discussed in Chapter 3 as the response of the plant to the relative length of the day and night and the changes in this relationship throughout the year.

Garner and Allard (1920) and Garner (1923) proposed the term photoperiodism, and showed that both the initiation of reproductive processes in plants as well as vegetative development could be greatly affected by varying the relative length of day and night. Specifically, it is the length of an interrupted dark period that is the main factor in photoperiodism that drives plant responses, though we typically speak of "day length" rather than "night length." Many plants were not found to be especially sensitive to day length in these early studies. In other plants, known as short-day plants, flowering could be induced when the plant was exposed to days shorter than a certain critical length. Likewise, some plants could be termed long-day plants because they responded to days longer than a given critical length or even to continuous illumination (Galston 1994). The photoperiodic effect on the flowering of woody plants is not well studied, in part because it has been difficult to clearly assign long-day and short-day flowering types to trees and shrubs as has been done in herbaceous species. Most woody plants are therefore considered to be day-neutral, meaning that day length is unimportant for flowering.

Photocontrol is important not just for how it affects the growth and development of plants, but also that plant responses to differing day lengths are clearly different for different species and even individuals within a species. As described in Chapter 3, photoperiodic requirements vary with the latitude and altitude of the source population with a species. Many tree species are therefore able to differentiate on a genecological basis because they have been naturally selected for a particular photoperiod associated with the limiting factors of the growing season.

Plant response to light occurs because light is absorbed by the phytochrome system of the plant (Taiz and Zeiger 1991; Galston 1994). **Phytochrome** is a protein that acts as one of the more important photoreceptors in plants. Phytochrome is expressed across many tissues and developmental stages in plants (Li et al. 2011) and directs plant physiological changes in response to changes in red and far-red light conditions. Phytochrome exists in two reversible forms:

$$\text{Pr} \underset{\text{far-red light } 0.73\,\mu m}{\overset{\text{red light } 0.66\,\mu m}{\rightleftarrows}} \text{Pfr} \xrightarrow{\text{darkness}} \text{Pr}$$

One form, **Pr**, has an absorption peak in the red region of the spectrum (650–670 nm) and is the physiologically inactive form, appearing turquoise in color to the human eye. The other form, **Pfr**, is physiologically active, has an absorption peak in the far-red region (705–740 nm), and appears more green in color. Absorption of light at the appropriate wavelength converts one phytochrome form to the other; when the red-absorbing form receives red light, it is changed into the far-red-absorbing form (Pfr). When the far-red-absorbing form receives far-red light, it reverses form back to Pr. Besides reversibility, the system is characterized by a slow drift in darkness from the far-red absorbing form to the red-absorbing form. The detection of phytochrome, its properties, and effects in morphogenesis of plants are reviewed by Vince-Prue (1975), Larcher (1995), Hart (1988), and Pallardy (2008).

The phytochrome system drives growth cessation in short-day conditions and continuous growth over long days. Red light dominates the spectrum near the end of the daylight period, with a red light to far-red light ratio (R/FR) of around 1.15 (Smith 1982) and more than 70% of the pigment is in the Pfr form at this time. The pigment reverts to the Pr form once the sun sets and far-red light is dominant (R/FR 0.05–0.7; Smith 1982), and less than 10% remains in the Pfr form depending on the length of the dark period (Downs 1962). As days shorten, less red light is available and thus less of the pigment will revert to the physiologically active Pfr form; this process is exacerbated by lengthening nights. Woody plants eventually cease growth and enter dormancy as the physiologically inactive Pr form accumulates. Conversely, under long days or continuous light, growth is promoted because the pigment is in the Pfr form. This physiological system is responsible for the development of a continuous sequence of photoperiodic races of trees in the north temperate zone across species' ranges. Unlike growth cessation, spring leaf flushing in woody plants is controlled mainly by temperature rather than photoperiod once the plant's chilling requirement has been met (Chapter 3).

In addition to sensing seasonal changes in light, phytochrome allows plants to sense the presence of neighboring plants. Light reflected from tree leaves or transmitted through tree canopies or neighboring vegetation is enriched in far-red light (lower R/FR), reducing Pfr and increasing Pr. The accumulation of Pr beyond a specific threshold will initiate a series of signal transduction pathways that will in turn induce a set of shade-avoidance responses in the plant (Casal 2012). Shade-avoidance responses include rapid elongation of stems and petioles, thinner leaves, reduced branching, and reduced chlorophyll. These responses theoretically represent an effort to quickly elevate the plant toward higher-quality light in crowded populations. Shade avoidance is much more common in herbaceous plants compared to tree seedlings and is typically measured in trees as the ratio of vertical to horizontal shoot growth. A study of 12 deciduous trees in Ontario revealed that shade-tolerant understory tree species such as musclewood and hop-hornbeam had low shade avoidance in low light compared to canopy tree species such as sugar maple (Henry and Aarssen 2001). Thus, shade avoidance in shade-tolerant trees may be most important for those species that rarely leave the understory.

The signaling action of phytochrome is sensitive to temperature and acts as a light-independent thermal sensor similar to the way it is able to sense light. Recent work on herbaceous plants has shown that Pfr reverts to Pr through a process called thermal reversion, which depends strongly on temperature and is hastened as temperature increases. Thermal reversion is important in controlling the amount of time Pfr is able to signal physiological changes in the early hours of darkness (Jung et al. 2016). Higher temperatures are also effective at reducing Pfr in low light such as that present in a shaded understory (Legris et al. 2016). Understanding that Pfr is high at the end of the daylight period, stem elongation such as that occurring as shade avoidance will be suppressed in the early hours of darkness (Van Buskirk et al. 2014). The fact that phytochrome responds to temperature as well as light therefore suggests that it controls plant elongation at night as well as during the day. The enhancement effect of cooler temperatures on Pfr (Jung et al. 2016) may have implications for a plant's cold tolerance, although this relationship has yet to be explored.

LIGHT AND ECOSYSTEM CHANGE

We cannot close a chapter about light without briefly discussing its importance for vegetation change in ecosystems. Vegetation is one of the most visible and measurable expressions of the change experienced by landscape ecosystems over time. Vegetation change, often termed succession (Chapter 17), occurs within the context of site factors (landform, soil properties, temperature, drainage, etc.), often following disturbances. Light is a critical part of the process of succession, but it is fundamentally different from the other site factors. We rarely can find adjacent sites that are permanently dark and light, but we can easily locate adjacent sites with vastly different soil moisture, soil fertility, and temperature regimes.

Though strongly constrained by site factors, forest vegetation itself modifies the site factors and disturbance regimes over time and thus plays a key role in ecosystem change. Vegetation does little to affect landforms, at least from a geological sense, but it has a stronger effect on soil development at the temporal scale of centuries to millenia. The dominant trees in an ecosystem, however, may affect an enormous change in light availability over mere years or decades—independent of the other site factors. The occurrence and species composition of the forest below the dominant overstory (subdominant overstory, understory, shrub, and herbaceous layers) are therefore heavily dependent upon light levels mediated by the density and crown architecture of species in the overstory canopy (Figure 6.4). These light levels in turn affect or determine many of the concepts discussed in this chapter, such as the complexity of the canopy, shade tolerance, shade avoidance, root development of the understory, and the frequency and duration of sunflecks at the forest floor.

Once the overstory fully develops in an ecosystem, trees having greater shade tolerance than those than compose the overstory tend to establish and grow up in the shaded understory. Many site factors in addition to light strongly influence species composition and density in any ecosystem. However, the plants themselves probably have the biggest impact on light availability compared to all site factors, and light is a key factor in the dynamics of vegetation (see Chapter 17). In the next chapter we focus on temperature, which is closely associated with light availability in forest ecosystems.

SUGGESTED READINGS

Craine, J.M. and Reich, P.B. (2005). Leaf-level light compensation points in shade-tolerant woody seedlings. *New Phytol.* 166: 710–713.

Gough, C.M., Atkins, J.W., Fahey, R.T. et al. (2020). Community and structural constraints on the complexity of eastern North American forests. *Glob. Ecol. Biogeogr.* 29: 2107–2118.

Hart, J.W. (1988). *Light and Plant Growth*, 204. Boston, MA: Unwin Hyman.

Kozlowski, T.T. and Pallardy, S.G. (1997). *Growth Control in Woody Plants*. San Diego, CA: Academic Press 641 pp.

Larcher, W. (2003). *Physiological Plant Ecology*, 4e, 534. New York: Springer-Verlag.

Pallardy, S.G. (2008). *Physiology of Woody Plants*, 3e. San Diego, CA: Academic Press 454 pp.

Pretzsch, H. (2014). Canopy space filling and tree crown morphology in mixed-species stands compared with monocultures. *For. Ecol. Manage.* 327: 251–264.

Way, D.A. and Pearcy, R.W. (2012). Sunflecks in trees and forests: from photosynthetic physiology to global change biology. *Tree Physiol.* 32: 1066–1081.

Temperature

CHAPTER 7

Similar to light, the temperature regime at the Earth's surface is determined by the energy provided by incoming solar radiation. Incoming radiation interacts with secondary heat transfers resulting from terrestrial radiation and air movement to produce the mean annual temperature at a given location. The surface layers of air near the Earth are heated during the day to an extent that depends on the amount of infrared radiation received. Thus, the greatest amount of heating occurs at tropical latitudes, at high elevations, and where the air is free from water vapor, clouds, and particulates. By contrast, nighttime temperatures are most determined by the rate and amount of heat radiation from terrestrial objects and the atmosphere that were heated during the day.
In this chapter, we first consider broadscale patterns of temperature as determined by climate without considering the interrelated effects of precipitation or water vapor. We then examine temperatures at the soil surface, where the critical phases of plant germination and establishment take place, within the forest, and in relation to topography and surficial features. Our main focus in this chapter is temperature and plant growth, particularly how plants resist freezing, the process of dormancy, and the distribution of plants in relation to cold hardiness. Detailed discussion of the physiology of plants and plant growth as they relate to temperature is presented by Fitter and Hay (1987), Sakai and Larcher (1987), and Pallardy (2008).

GEOGRAPHICAL PATTERNS OF TEMPERATURE

Broad patterns of temperature at the surface of the Earth are controlled largely by the rotation of the Earth around its axis (which controls daily variation in temperature of a specific place) and around the sun (which, combined with the Earth's tilted axis, controls seasonal temperature). These patterns make latitude an important determinant of broadscale temperature; generally, temperatures are warmest near the equator because incoming solar radiation occurs most directly at the center of the globe. Tropical areas (between about 23.5°N and 23.5°S latitude) have an average temperature of about 79 °F (26 °C). Geometrically, the center of the globe is also the area least affected by the Earth's tilted axis, such that the variation between winter and summer temperature is small. In fact, the temperature difference between day and night in tropical areas is likely larger than that between summer and winter. Average temperatures increase with latitude either north or south of the equator because the angle of incidence of solar radiation is less direct, and seasonal variability increases because the Earth's tilt places these areas markedly closer to the sun in summer and further in winter. For comparison, the average temperature at International Falls, Minnesota (a continental region at about 49°N latitude) is 38 °F (3 °C), with an average of 56 °F (13 °C) in the summer and 19 °F (−7 °C) in the winter.

Temperatures are not strictly determined by latitude, as the proximity of an area to large bodies of water such as oceans may alter temperature at a specific latitude. The most common example of the influence of water bodies is in comparing maritime to continental locations. The

Forest Ecology, Fifth Edition. Daniel M. Kashian, Donald R. Zak, Burton V. Barnes, and Stephen H. Spurr.

German-Russian climatologist Wladimir Köppen first documented the difference between these two types of geographical regions in the late 1880s. Maritime climates tend to be cool with relatively little daily or seasonal variation in temperature. This is because the oceans adjacent to maritime regions moderate the temperature as the warmer water heats the air in the winter and land heats the air in the summer via conduction, and these heat transfers create convective currents within the atmosphere. Land has a lower heat capacity compared to water, and thus continental climates lose and gain heat much quicker than maritime climates. As a result, continental climates are cold, lack the moderating influence of water bodies, and tend to have wider variation in seasonal temperatures. For example, Vancouver, British Columbia lies at the same approximate latitude as International Falls but in a maritime climate, and its average temperature is higher (52 °F/11 °C). Moreover, Vancouver's average temperature varies much less, from 68 °F (20 °C) in the summer to 43 °F (6 °C) in the winter. There are many other meteorological phenomena related to atmospheric and oceanic currents and precipitation that contribute to the difference between continental and maritime climates in addition to temperature.

Freshwater lakes, if large enough, can have a similar moderating influence on otherwise continental climates, though their influence is less than that of oceans. Perhaps the best example of the moderating influence of large lakes occurs in the Great Lakes region. Climate on the eastern shore of Lake Michigan has a higher average annual temperature, higher average temperature during the summer, and less extreme minimum temperatures in the winter compared to only 10 miles (16 km) further inland. This region is renowned for its production of cherry crops that are otherwise produced in the maritime climate of the Pacific Northwest. Even a small inland lake (<8 mi^2/21 km^2) in the center of Michigan's Lower Peninsula has been shown to slightly moderate the climate on its eastern shore (Albert et al. 1986).

In addition to latitude and proximity to large water bodies, altitude has a strong effect on temperature at broad scales. Temperature generally decreases with increasing elevation in much the same way that it decreases with increasing latitude; a long-used conversion factor is that 100 m of elevation is equivalent to 1° of latitude in terms of temperature (Wiersma 1963). As such, the form of woody plants near the tree line in mountainous terrain begins to resemble those in northern boreal and subarctic regions. Latitude and altitude do not act independently on vegetation, and the variation in tree line across the globe is a good indicator of how altitude and latitude work in concert. In general, tree line occurs at lower elevations at high latitudes and at higher elevations near the equator, and this trend is less extreme for mountains in maritime climates compared to continental climates (Figure 7.1). Although Grace (1989) hypothesized that trees cannot grow where the combination of latitude and altitude results in an average temperature of the warmest month <50 °F (10 °C), mean growing season temperature worldwide at tree line occurs between 42 and 45.5 °F (5.5 and 7.5 °C; Körner and Paulsen 2004).

One important factor determined by temperature at a particular area is **growing season length**. The growing season length for vegetation in a particular area is the number of days when aboveground plant growth is able to take place during the year. Quantitatively, growing season length is the number of days when soil temperature at a depth of 50 cm (20 in.) is greater than 5 °C (41 °F). However, it is often more simply estimated as the number of days between the average date of the last killing frost recorded in the spring and the first killing frost recorded in the fall, where a killing frost is considered to be −2 °C (28 °F) or lower. As one might expect, growing season length is longest where freezing temperatures are least likely (the tropics, maritime areas, or at lower elevations) and shortest where they are most likely (closest to the poles, continental areas, or at higher elevations). Notably, some tree processes occur well outside of their growing season, such as respiration and even photosynthesis in conifers.

Another important aspect of temperature for vegetation is the **heat sum**, which is a measure of the amount of heat to which plants are exposed. A heat sum for a given species in a given place is typically measured in **degree days**. Degree days are calculated by summing over

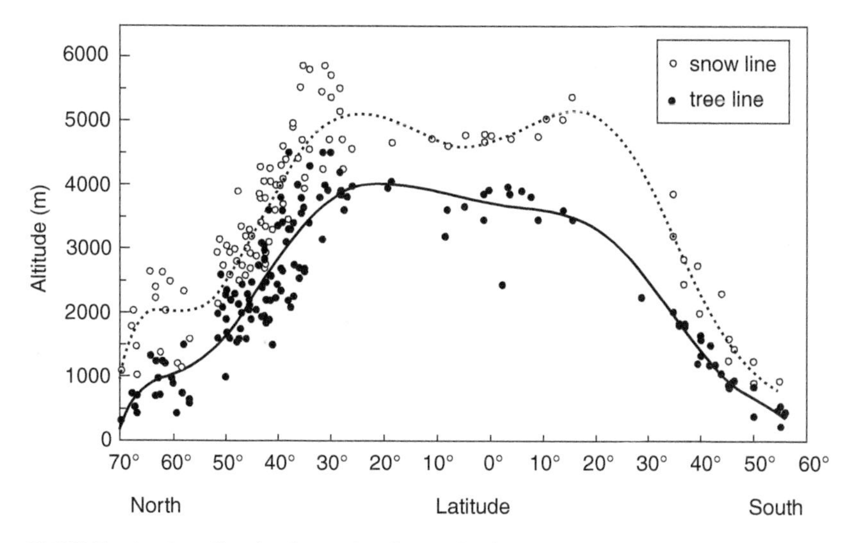

FIGURE 7.1 Tree-line (and associated snow line) position around the globe by latitude. With some variation, tree line occurs at lower elevation nearest the poles and higher elevation nearest the equator. *Source:* Körner (1998) / with permission of Springer Nature.

the course of a growing season the product of (1) the number of degrees over a base temperature in degrees Celsius and (2) the number of daylight hours of that temperature. The most common base temperatures used for calculating a heat sum are 5 °C (41 °F) and 10 °C (50 °F), with the assumption that significant plant growth will not occur below the base temperature. The importance of the heat sum for woody plants is that an important event will occur once a certain heat sum has been achieved, such as flowering, bud burst in the spring (Worrall 1983; Hari and Häkkinen 1991), or other important phenological events. Therefore, populations of a species in a location are able to adapt to and are able to monitor the temperature regime of that location. For example, red maple is one of the earliest flowering trees in the spring and has been documented to flower at 1–27 degree days, but white ash in the same location flowers at 30–50 degree days (Burns and Honkala 1990). Importantly, however, northern and southern populations of widely distributed species of woody plants are able to adapt their respective heat sums to the temperature regime at hand, which eventually plays an important role in genecological differentiation (Chapter 3).

TEMPERATURES AT THE SOIL SURFACE

When we speak of variation in soil temperature, we are speaking chiefly about warming of the ground exposed to the sun during daylight and its subsequent cooling at night. Solar radiation is absorbed during the day both by the soil surface and the air immediately above it, and these processes largely determine daytime temperatures. At night, however, temperatures depend on the amount of heat absorbed by terrestrial objects including soil and how quickly the heat is lost to the atmosphere (typically as infrared radiation) once solar radiation is no longer present. These factors are in turn determined by the amount of vegetation and litter cover. Soil temperature is therefore influenced both by the physical properties of the soil itself as well as the climate in which the soil occurs. From a vegetation perspective, soil temperature is important because it affects the biological activities of plant roots and the soil organisms that are mutualistic with them. Most chemical and biological processes increase with temperature, such that soil temperature has a strong effect on the ability of tree roots to grow or uptake water and nutrients, and on microbial and other activities that affect important belowground processes such as mineralization and decomposition (Davidson et al. 1998; Heinze et al. 2017). Soil temperature would also be expected to have an important influence on the establishment and survival of tree seedlings.

The amount of water contained in a soil strongly controls its temperature regime. The **heat capacity** of soils, or the amount of heat necessary to raise its temperature, has many implications for the effects of soil moisture on soil temperature. As discussed in the earlier text, water has a high heat capacity relative to air, meaning that almost five times as much heat is required to raise the temperature of a volume of water compared to the same volume of air, and thus heating and cooling occur more slowly in wet soil compared to dry soil. In addition, heat moves from warm areas to cold areas, and the rate of this movement is known as **thermal conductivity**. Organic soils have lower thermal conductivity compared to mineral soils, and soils with smaller particles (clays) have lower thermal conductivity compared to sandy soils (Weil and Brady 2017). As a result, tree roots found on drier, sandier soils are more likely to suffer freezing injuries compared to wetter or more clayey soils because the latter retain their heat longer into the fall. Finally, soils lose an enormous amount of heat when water evaporates from the soil surface during the day. A study by Maguire (1955) showed that watering bare soil in California, for example, resulted in its cooling by 9–23 °C (48–73 °F) over the next three days.

Soil temperature changes with depth, with surface soil much more prone to variability in temperature compared to subsurface soil. Subsequently, deeper soils take much longer to reach maximum and minimum soil temperatures compared to the surface, if they ever achieve those extremes at all. Temperatures measured deep in the soil frequently remain constant throughout the year and approximate the mean annual temperature of the area (Poulson and White 1969).

Soil color also affects soil temperature. Dark soils absorb more radiant energy than light soils, which have a higher albedo (Oke 1987). The importance of soil color for temperature was classically demonstrated by Isaac (1938) in southern Washington. He found that soils blackened by recent fire reached a temperature of 73 °C (163 °F) when the air temperature was 38 °C (100 °F), but comparable gray mineral soil heated up only to 64 °C (147 °F) and yellow mineral soil to 62 °C (144 °F). These trends logically extend to wetter soils, with a slightly dampened effect (so to speak). A higher moisture content of soil darkens their color, such that wetter soils absorb more solar radiation than drier soils, but the high heat capacity of the soil water results in a slowed warming of the wetter soils.

Of course, the soil surface is often not barren and is in fact covered with layers of organic matter. Organic layers on the forest floor tend to heat very rapidly during the day without transmitting much of this heat to the soil below it. At night, heat loss from the mineral soil is reduced by an organic layer, which itself cools quickly without incoming radiant energy. This tendency for high variability in temperature is one reason that organic matter is not an ideal place for seedling establishment, as either its high daytime temperatures in the summer or the freezing nighttime temperatures in the fall are potentially lethal to young seedlings. In eastern Connecticut, a surface of eastern white pine needles litter in a small clearing was recorded to be 68 °C (154 °F) on a day when the air temperature reached only 24 °C (75 °F), but the surface temperature of bare mineral soil was 46 °C (115 °F) and that of a moss surface layer reached only 39 °C (102 °F) (Smith 1951). Similarly, soil at 10 cm depth beneath bare mineral soil in high-elevation spruce–fir forests of British Columbia was found to be 8 °C (46 °F) warmer than sites covered by forest floor and vegetation (Balisky and Burton 1995). Snow is also an excellent insulator and reduces heat loss from the soil during the winter months (Hardy et al. 2001).

A striking example of the importance of soil temperature is found for forest ecosystems of the Chena River floodplain in interior Alaska (Viereck 1970). The annual soil temperature regime varies markedly for a sequence of ecosystems dominated by willow, balsam poplar, white spruce, and black spruce at successive distances from the river. Near the river's edge the gravelly soil is cold in winter and warm in summer. However, this wide annual fluctuation is reduced inland until in soil under black spruce, soils are cold throughout the summer and waterlogged when not frozen.

Many other examples could be given of the importance of soil temperature, illustrating not only differences among landscape ecosystems but also its influence in bringing about vegetation changes over time at a specific site.

TEMPERATURE WITHIN THE FOREST

Broad-level patterns in temperature are but one aspect of climate that sets the template for smaller, local variations in temperature experienced by the woody plants in a forest. When considering the effects of water as well as temperature, these local variations in climate are termed **microclimate**. Microclimate influences the distribution of local ecosystems as well as the plants that occur within and among them. The most obvious differences in microclimate are those that occur with slope and aspect in topographically diverse terrain, which we discuss as part of physiography in Chapter 8. Here and in the next section, we discuss the importance of the forest itself in affecting the temperature component of microclimate. In the following section, we discuss the influence of local topography on microclimate.

Within a forest, the canopy and tree stems shelter the region near the ground from solar radiation and reduce the mixing of air, such that the temperature and its variation are significantly different from areas in the open (Geiger et al. 2009). Shading by the forest canopy results in reduced warming of air temperature during the day, but infrared heat re-radiated from plants and soil at night tends to be captured beneath the canopy. As a result, air temperature warms less during the day and cools off less during the night compared to open areas, and the forest canopy therefore has a moderating effect on the air and soil temperature beneath it. Relative humidity is closely related to air temperature and also increases at night beneath a forest canopy. Data collected within and adjacent to an eastern white pine plantation by Spurr (1957) gave a summer air temperature range of 15.9 °C (60.2 °F) within the forest compared to 21.6 °C (70.9 °F) in the open (Table 7.1). Throughout the year, at a height of 1 m from the ground, the maxima and means were lower and the minima higher within the forest as contrasted to the open station. The mobility of air, even beneath a forest canopy, allows some degree of mixing with nearby areas such that the shading provided by the canopy has a larger effect on soil temperature than air temperature (Morecroft et al. 1998; Porte et al. 2004). Many important forest processes respond to the moderating effect of forests on air temperature, soil temperature, and relative humidity, including the survival and growth of tree seedlings.

Table 7.1 Mean weekly maxima, minima, and mean temperatures (°C) in the open and under a dense 20-year-old eastern white pine plantation.[a]

	Winter	Spring	Summer	Fall
Open				
Maximum	5.1	22.8	29.7	14.2
Minimum	-18.4	-2.2	8.1	-7.3
Range	23.5	25.0	21.6	21.5
Mean	-6.7	10.4	18.9	3.4
Under forest				
Maximum	2.7	19.9	25.6	11.0
Minimum	-16.7	-2.8	9.7	-5.7
Range	19.4	20.2	15.9	16.7
Mean	-7.1	9.8	17.7	2.7

[a] *Petersham, MA; by 13-week quarters, 1943–1944.*

Forest structure and species composition have important effects on the strength of the moderating effect on air temperature, in large part because their differences in leaf area influence the degree of absorption of incoming solar radiation. In a study of deciduous, spruce–fir, and pine forests, von Arx et al. (2012) found that forest cover in general decreased daily maximum air temperature as much as 5.1 °C (41.2 °F). Deciduous and non-pine conifer forests moderated daytime maximum and average air temperature about twice as much as pine forests, and pine forests cooled significantly less at night. The moderating effect of the canopy varies with its density regardless of species composition, with higher-leaf area canopies having a greater moderating effect (von Arx et al. 2013). Forests having old-growth characteristics (tall canopies, high biomass, complex vertical structure) had reduced maximum temperature in the warmest month of the growing season, reduced mean monthly maximum temperature, and reduced variability in mean weekly temperature (Frey et al. 2016).

Temperature also varies vertically within a forest. Many researchers have documented changes in temperature with increasing height above the ground beneath both deciduous and coniferous forest canopies (Harley et al. 1996; Zweifel et al. 2002), but it is important to recognize the effects of the canopy itself on temperature variation. The classic study of variation in temperature along a vertical gradient was performed in an American beech-dominated forest in Ohio by Christy (1952). Summer and fall temperatures were similar at the soil surface and beneath the canopy, but the widest range of temperatures occurred in the canopy itself. In the winter and spring, temperature variation was similar at all vertical strata because of the absence of the canopy that allowed for greater mixing of the air within the forest. The effects of both changing temperature and varying temperature at increasing distance from the forest floor are significant for their potential effects on photosynthesis and carbon capture within the canopy (Ellsworth and Reich 1993; Bauerle et al. 2009).

TEMPERATURE VARIATION WITH LOCAL TOPOGRAPHY

The effects of local topography upon local temperature have long been studied (most recently Kelsey et al. 2019; Arduini et al. 2020; and Rupp et al. 2020). Common among these studies is the pattern that low, often concave landforms accumulate cold air on still, cold nights. Cold air is denser and flows downward from adjacent high land. Radiant cooling on high ground forms a shallow layer of cold air near the ground, and minimum temperatures at night remain fairly high. When cold air pools in narrow valleys or small, closed depressions, these processes may result in air temperatures near the feature's bottom as much as 8 °C (46 °F) lower than that of the "thermal belt" further up the slopes. The resulting condition is known as an **inversion** in mountainous terrain, in that the temperature increases with elevation in contrast to the usual decrease in temperature with elevation. Smaller concave areas are known as **frost pockets** because the temperatures commonly occurring near the ground result in late spring frosts, early fall frosts, and consequent short growing seasons. These areas of cold air pooling need not be deep. Spurr (1957) found that minimum temperatures in a small depression only 1 m deep were comparable to those in a nearby deep valley 60 m below the general land level. Interestingly, the same low landforms tend to accumulate radiant energy during the day and reach high maximum temperatures near the ground, while mounds, ridges, and other higher convex surfaces tend to remain cooler.

An excellent example of the effects of cold-air drainage on forests can be found in the jack pine forests of northern Lower Michigan. Jack pine is tolerant of low-nutrient, dry, cold sites. Northern pin oak is a secondary tree species that also tolerates dry, infertile sites, but is far more sensitive to frost. Ice-block depressions on this landscape are sites where cold air frequently pools at night. Even shallow depressions exhibit distinctive effects of this phenomenon, where regeneration of pine is sparse near the bottom of the depressions and oak is limited by freezing temperatures (Kashian et al. 2003). Where the depressions are deep, the slopes of the depressions are often steep enough to exhibit the effects of aspect. South-facing slopes of these deep depressions are often very dry and devoid of trees, and oaks are limited to the higher terrain (Figure 7.2).

Despite the importance of local topographic influence on temperature variation for the distribution of plants and tree growth in frost pockets, soil-water drainage is likely to be an important interacting factor. The same characteristics of a landform that ensure cold-air drainage, such as the existence of convex surfaces, will make a soil very well drained. Thus, a very well-drained site is apt

Photos by Dan Kashian

FIGURE 7.2 Effects of microclimate on vegetation as mediated by topography. The top image shows a shallow ice-block depression (approximately 6 m/20 ft deep) after a recent wildfire in a cold, continental region of Lower Michigan. The center of the depression lacks pine regeneration due to frequent frosts during the growing season. Note the bushy, resprouting oaks that are frost-sensitive and limited to the upper rim of the depression. The lower image is a large ice-block depression (approximately 26 m/85 ft deep); the sides of the depression are steep enough to lack significant tree growth on its south-facing slope (foreground). Again, note the oaks (seen at the horizon) are limited to the very top of the depression on the north-facing slope.

to have a warmer microclimate suitable for plants of a more southern distribution. Likewise, a very poorly drained soil is apt to result from a concave land surface where both cold air and water will pool. Very poorly drained sites are apt to be characterized by temperature extremes and a short growing season and might prove suitable for plants of a more northern distribution. Such examples are common in northern Ohio and southern Michigan where boreal and northern forest species (black spruce, tamarack, northern white-cedar, black ash, yellow birch, leatherleaf, among others), relics from their late-glacial northward retreat, persist in wet depressions that also act as frost pockets.

As a final note, cold-air drainage may be influenced by the vegetation itself as well as by local topography. Given that a forest moderates the area beneath its canopy, as discussed in the earlier text, the canopy tends to warm more during the day than the soil surface, and cold-air drainage occurs in the subcanopy and understory while the warmer air above the canopy moves upslope (Froelich and Schmid 2006). As the canopy cools faster than the ground at night, this cold-air drainage is likely to weaken (Mahrt et al. 2000). Similarly, a phenomenon resembling topographic frost pockets may result from the creation of small clearings in a forest by windthrow, insects or diseases, or cutting. The tree crowns of the surrounding forest act to channel cold air into the clearings on still, clear nights; larger clearings are more prone to winds that will help to drain the cold air away and are thus less likely to pool the cold air.

TEMPERATURE AND PLANT GROWTH

We often consider plants as entire organisms and how their growth might be affected by external factors such as temperature. In reality, different parts of the organism experience temperature differently, especially when the plant is large (such as an overstory tree). In that the canopy of a forest moderates the temperature beneath it, we would expect that the parts of a tree that occur below the canopy would experience temperature differently than the canopy itself. The temperature of tree stems, for example, approximates the air temperature in the shade, but may be quickly heated by sunlight. High tree-stem temperature of sunlit trees may be modified by upward water movement through the xylem, and bark—depending on its thickness and color—may keep stems warmer than the ambient air temperature in the winter and cooler in the summer (Nicolai 1986).

Tree roots experience temperature of the soil environment; as discussed in the earlier text, soil temperature is typically cooler than the air in the summer and warmer in the winter. Roots are more tolerant to a wider range of temperatures than are shoots, although optimum temperatures for root growth vary by species (Gray and Brady 2016), even for species growing on the same site. Thus, roots are able to grow well into the fall after shoot growth has ceased, and can begin growth in the spring when soil temperatures remain cold but air temperatures have stimulated shoot growth. In a study of six European tree species, Alvarez-Uria and Körner (2007) found that most root growth occurred at temperatures warmer than 9 °C (48 °F) and was restricted below 6 °C (43 °F). Root growth of ponderosa pine is more dependent on soil temperature and shoot growth more dependent upon air temperature (Larson 1967). Optimum root growth occurred in 15 °C (59 °F) air and 23 °C (73 °F) soil. Optimum root growth for northern red oak, basswood, and white ash also occurs at relatively high soil temperatures (Larson 1970). Temperature of the crown also affects root growth, because increases in respiration and transpiration that occur with increasing temperature utilize carbohydrates and water, making them less available for use by the roots. For example, root growth of silver maple seedlings decreased markedly as temperature of the crown increased from 5 to 30 °C (41–86 °F; Richardson 1956).

In contrast to stems and roots, the leaves of plants are able to regulate their temperature by dissipating some of the energy they absorb as a way of avoiding injurious or lethal high temperatures. Re-radiation alone dissipates approximately 50% of absorbed heat energy (Gates 1980).

Additional heat loss is achieved through transpiration, which cools the leaf as a large amount of heat energy is consumed in changing water to water vapor. A third process, convection, transfers heat from the leaf to the surrounding cooler air as it moves across the leaf's **boundary layer** (the thin layer of still air found on all surfaces). Cooler nighttime air temperatures relative to the leaf surface reverse this process, and heat is transferred from the air to the leaf by both convection and conduction (Nobel 2009). We will discuss a series of plant adaptations that interact with these processes to stabilize leaf temperature in a given environment, maintaining heat in cooler air and facilitating heat loss in warmer air.

As temperature increases, plant activities increase up to an optimum temperature and then decrease until lethal high temperatures are reached. Temperate tree species generally increase their rate of photosynthesis from freezing to an optimum at 15–25 °C (59–77 °F; Pallardy 2008). Notably, respiration often peaks at temperatures higher than photosynthesis, such that net photosynthesis is inhibited. Plant processes generally function as long as living cells and their proteins are stable and their enzymes active, usually 0–40 °C (32–104 °F) depending on species, age, season, and population source. Physiologically, temperature most strongly influences (i) the activity of enzymes that catalyze photosynthesis and respiration, (ii) the solubility of carbon dioxide and oxygen in plant cells, (iii) transpiration, (iv) the ability of roots to absorb water and minerals from the soil, and (v) membrane permeability. A detailed discussion of effects of temperature on photosynthesis is presented in Chapter 18.

Plant growth and metabolism depend on water in liquid form, and these processes slow or stop when water is frozen into ice. Photosynthesis can occur when the ambient air temperature is below freezing, down to about −8 °C (18 °F), because solar and terrestrial radiation often warms tissues to near or above freezing. Low temperatures inhibit the rate of photosynthesis, but appreciable photosynthesis may occur in winter in conifers. For example, although the optimum temperature range for photosynthesis in Douglas-fir is between 10 and 25 °C (50 and 77 ºF), net photosynthesis at 0 °C is still 70% of the amount occurring at 10 °C (Lassoie 1982). Winter photosynthesis in many conifers, however, is limited because low soil temperatures tend to reduce metabolic activity and reduce membrane permeability so that uptake of water and nutrients needed for photosynthesis is low.

Photosynthesis decreases when temperature increases beyond the optimum, respiration eventually exceeds photosynthesis, and cell death occurs when temperatures approach 55 °C (131 °F). These temperatures are those of the plant tissues themselves, not air temperature, and thus air temperatures must be even higher to warm up the plant tissues to the lethal point. In this manner, prolonged exposure to high but sublethal air temperatures may eventually kill plants that might survive short exposure to very high air temperatures. Exposure time is critical for survival of seedlings because extremely high temperatures may be reduced by even the lightest shade, as is the case when cotyledons of the seedlings shade the basal portion of the seedling stem. A good treatment of high temperature stress and heat injury of plants is presented by Levitt (1980a) and Pallardy (2008).

As discussed in Chapter 6, leaf arrangement and orientation change in response to light irradiance and in the process may reduce the amount of solar energy absorbed and hence prevent overheating of the leaf. Red maple seedlings are able to deflect their leaf blades almost vertically downward in high light and return to the horizontal position when shaded (Grime 1966). This process may partially explain the versatility of red maple in colonizing both dry, exposed sites and shaded habitats. Leaf morphology, coloration, leaf pubescence, and maintenance of protein integrity also enable woody plants to function effectively at high temperatures (Levitt 1980a).

Near-lethal temperatures for most woody plants are largely confined to the exposed ground-air boundary in forests. As a result, direct heat injury in forest trees occurs most significantly in small seedlings, which have relatively unprotected live tissues in this critical zone. Heat damage in seedlings occurs when the soil surface reaches about 52 °C (126 °F; Helgerson 1989). High temperatures may indirectly cause damage to trees via their effects on water loss, particularly from leaves. Many mature hardwoods and conifers suffer leaf damage due to water deficiency of cells along the leaf margin and the tips of conifer needles. For example, many woody species suffered leaf damage in northern California when temperatures suddenly rose above 38 °C (100 °F) in areas that had had an exceptionally cool spring (Treshow 1970).

Various phases of the temperature regime, such as day temperature, night temperature, heat sums, and the difference between day and night temperature, all affect growth, each in a species-specific or even a population-specific way. A series of experiments in the 1950s and 1960s demonstrated that the response of tree seedlings to the temperature regime varied widely by species (Kramer 1957; Hellmers 1962, 1966a,b; Hellmers et al. 1970). Engelmann spruce seedling growth responded most to night temperature (Figure 7.3), but redwood responded to day temperature, reaching its maximum growth in the moderate range of 15–19 °C (59–66 °F). Both species grew best at the same day temperature (19 °C/59 °F), but spruce grew better when nights were warm and also tolerated warmer days better than redwood. Redwood grew best when day and night temperatures differed only slightly, consistent with the lower diurnal variation in its native maritime climate. By contrast, Engelmann spruce was apparently adapted to the greater diurnal changes of continental, high-elevation environments.

Several species (Jeffrey pine, erectcone pine, eastern hemlock) show a marked growth response to the heat sum during a day, regardless of the time of application. For example, Jeffrey pine exhibit the most growth when 300–400 degree-hours are accumulated, regardless of the specific different day and night temperatures. Still other species show a primary growth response when nights are considerably cooler than days. This difference between day and night temperatures is termed **thermoperiod**. Low night temperatures coupled with moderate day temperatures are important in the flowering and fruit set, flavor, and quality of various crop plants and fruit trees (Treshow 1970).

Strong responses to thermoperiod are common in seedlings of many forest trees. Maximum shoot growth of loblolly pine seedlings occur with thermoperiods of 12 °C (54 °F), with night

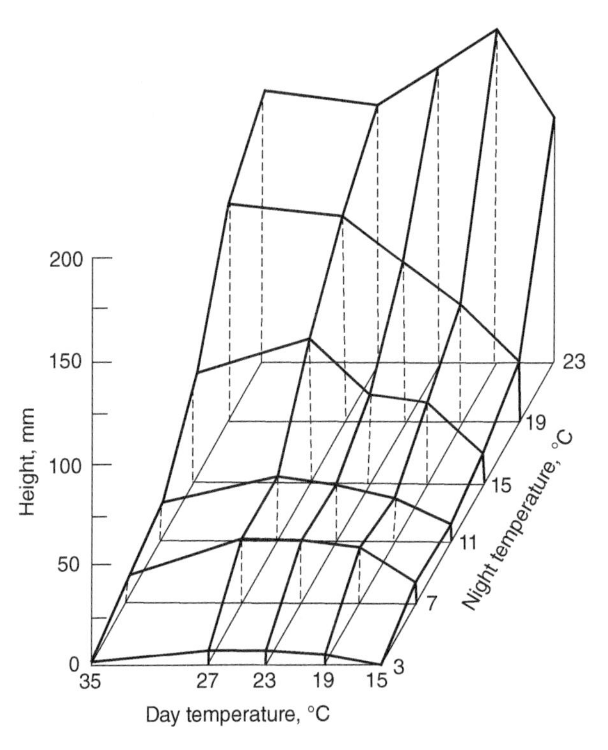

FIGURE 7.3 Average height of 36 Engelmann spruce seedlings from each of 30 combinations of day and night temperatures where the plants were grown for 24 weeks. *Source:* Hellmers et al. (1970) / with permission of Oxford University Press.

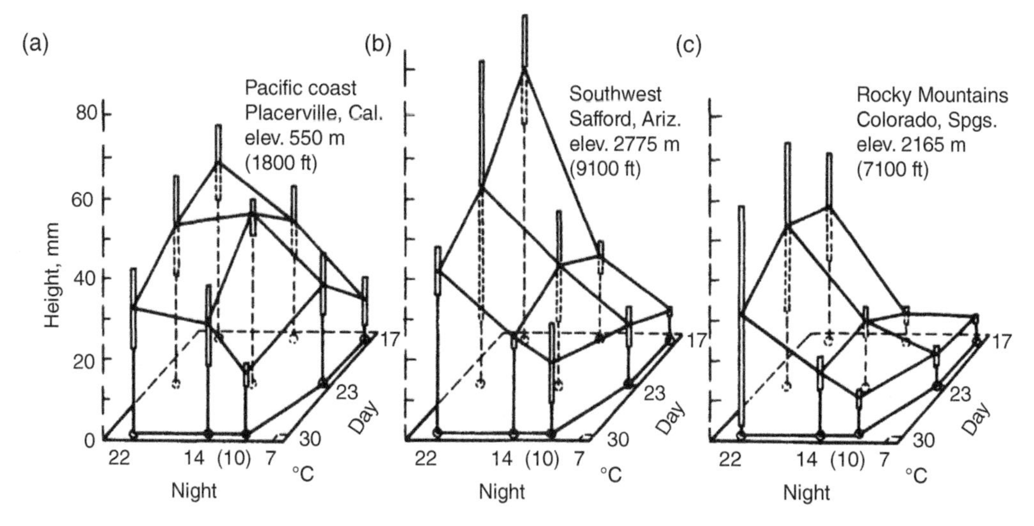

FIGURE 7.4 (a)–(c) Mean height growth of ponderosa pine progenies from three provenances in three regions. Treatment included 16-h days and nine combinations of three day and three night temperatures. The vertical bars show the range of the mean ± standard errors of the mean. *Source:* Callaham (1962) / John Wiley & Sons.

temperature colder than day temperature (Kramer 1957). A similar response has been consistently reported for Douglas-fir, although the optimal temperature differential has varied among the studies conducted and the provenances tested (Hellmers and Sundahl 1959; Lavender and Overton 1972). Atlantic white-cedar and Fraser fir both maximize total seedling growth at a thermoperiod of 8 °C (46 °F) with warm days and cool nights (Hinesley 1981; Jull et al. 1999). Red fir exhibit two effects of thermoperiod (Hellmers 1966a). First, for maximum height growth to occur, a warm day must be followed by a cool night, whereas a cool day must have a cold night. Second, maximum height growth is obtained when the thermoperiod is 13 °C (55 °F). Although maximum growth under a 17 °C (63 °F) day and a 23 °C (73 °F) day is nearly equal, this growth occurs only when the cooler day was followed by a 4 °C (39 °F) night and the warmer day with a 10 °C (50 °F) night (a 13 °C thermoperiod in both cases).

In contrast to the three species previously described in the earlier text requiring cold nights and warm days, seedlings of ponderosa pine grow best when nights are warmer than days (Callaham 1962; Larson 1967). Furthermore, pines from diverse parts of the range show significantly different responses to the temperature regime (Figure 7.4). Seedlings from east of the Rocky Mountains (Figure 7.4c) require a high night temperature for optimum growth, but seedlings from the Southwest grow remarkably fast under cold days and hot nights. Meanwhile, seedlings from the west slope of the Sierra Nevada Mountains in California maximize growth with lower night temperature. Similarly, red maple height growth varies in response to day and night temperatures based on where it was located across its range (Perry 1962).

COLD INJURY TO PLANTS

Lethal cold temperatures occur periodically throughout the entire zone of tree growth in the temperate and boreal regions of the Earth. Thus, they affect to a greater or lesser degree the distribution and growth of trees in these zones. In addition to growing season temperature and water availability, cold temperatures may be important for the distribution of temperate trees (Chapter 2). For example, the northern and western extents of loblolly pine are strongly correlated with a combination of cold temperatures occurring simultaneously with low soil water (Hocker 1956).

Plant tissues, particularly those sensitive or actively growing, may be damaged or killed when freezing occurs quickly and ice crystals form within the protoplasm. Even slow freezing may kill plants at temperatures of −15 °C (5 °F) to −45 °C (−49 °F) when cooling occurs at rates commonly found in nature (Weiser 1970). During this process, intercellular water freezes, dehydrating cell contents until a point is reached when only "bound" (unfreezable) water remains in the protoplasm (Wolfe et al. 2002). Further decreases in temperature pull the bound water away from the protoplasm, initiating protein denaturation and ultimately killing the cell. In tropical plants, death may occur at above-freezing temperatures ranging from 0 to 10 °C (32–50 °F). Recent work has suggested that the primary sites of freezing damage in plants are cell membranes. Frost injury is indicated first by an increase in the leakage of electrolytes across the plasma membrane and out of cells. Lowering temperatures also affect the phase and configuration of the lipid bilayer of the cell membrane (Uemura et al. 2006; Ambroise et al. 2020).

Freezing temperatures have strong effects on whole plant organs as well as cells and tissues, and their extent of damage depends strongly on the timing of their occurrence. If plants are flowering, freezing temperatures may kill flowers and thus limit fruit production. For example, a study of Gambel oak in Utah found that low-elevation oaks suffered freezing of their catkins and a subsequent failure of the acorn crop, while oaks at high elevations did not (Neilson and Wullstein 1980). Freezing can also kill or damage portions of the cambium in tree stems. Outer portions of the stem freeze and thus contract more quickly than the inner portions, which can create cracking of the stem. If freezing occurs during the growing season before xylem cells in the stem have thickened and lignified, ice formation between cells may crush the outermost zone of weaker cells and leave a scar within the stem that is evident in a stem cross section. These "frost rings" have been used to identify significant past cooling events during the life of a tree, such as the eruptions of major volcanoes (LaMarche and Hirschboeck 1984).

Aboveground portions of the plant are most commonly the focus of studies of cold injury to plants, but roots are similarly or even more affected. Root damage can occur for a variety of physical reasons related to soil moisture in cold temperatures, but like shoots, cellular damage is more likely the reason for cold injury than mechanical damage. Fine roots are most susceptible to freezing damage, which are responsible for water and mineral uptake and whose regrowth uses resources at the expense of the crown (Gaul et al. 2008). An important difference between roots and shoots is that shoots tend to increase their frost hardiness as the tree ages, whereas the roots do so much less. For example, some oak species exhibit an increase of 9–10 °C (48–50 °F) of frost hardiness of the shoot cambium over a seedling's first three years, but roots increase only 1–2 °C (34–36 °F; Sakai and Larcher 1987). This difference appears to be related to a higher water content in roots compared to shoots and the added insulation provided by the soil. A detailed discussion of the frost hardiness and cold tolerance of roots is provided by Ambroise et al. (2020).

Most trees in the temperate and boreal zones become increasingly inactive as the day length shortens and temperatures decrease near the end of the growing season. Protoplasm water content is reduced as the concentration of other cytoplasmic materials increases during the onset of dormancy. Water moves out of cells and freezes in extracellular spaces, leaving the cells undamaged, and thus many species are able to survive and even avoid injury by subfreezing temperatures. This ability of individuals to initiate and continue the reduction of water in their cells to correspond with the progressively decreasing temperatures of their native site in autumn is critical to the cold-acclimation process. Cold hardiness occurs slowly, and modern plants have been selected naturally to survive the gradual lowering of temperatures in autumn and winter and their gradual rise in spring.

DORMANCY

We describe plants that have transitioned to a hardy or dormant condition from a tender or sensitive condition as acclimated or *hardened*. Three stages of dormancy are normally recognized (Vegis 1964; Perry 1971; Sakai and Larcher 1987): pre-dormancy (early rest), true dormancy (winter rest), and post-dormancy (after rest) (Figures 7.5 and 7.6). Short days trigger growth cessation and initiate metabolic changes characteristic of pre-dormancy at temperatures of 10–20 °C (50–68 °F; Figures 7.5 and 7.6). During this stage, growth is reduced, and starch accumulates in the roots (Dumont et al. 2011). Pre-dormancy may be reversed by warmer temperatures and increased photoperiod. Otherwise, these changes facilitate further metabolic changes in the second stage of acclimation, sometimes called full dormancy, triggered by temperatures below about 5 °C (41 °F) and especially subfreezing temperatures including the first frost of autumn (Figure 7.6). Full dormancy involves the production of proteins, membrane lipids, and metabolites involved in frost hardiness (Beck et al. 2004). Plants have increased cold resistance in full dormancy, which can be reversed after a chilling requirement. The third stage of acclimation is the attainment of true dormancy, which is exclusive to woody plants. This stage is induced by low temperatures of −30 °C (−22 °F) to −50 °C (−58 °F). During this stage, water remaining within the cell vitrifies or turns to a glassy material lacking the crystalline structure of ice that is lethal to cells (Strimbeck et al. 2015). Truly dormant buds and seeds cannot be immediately induced to normal growth. Some species, such as birches, European

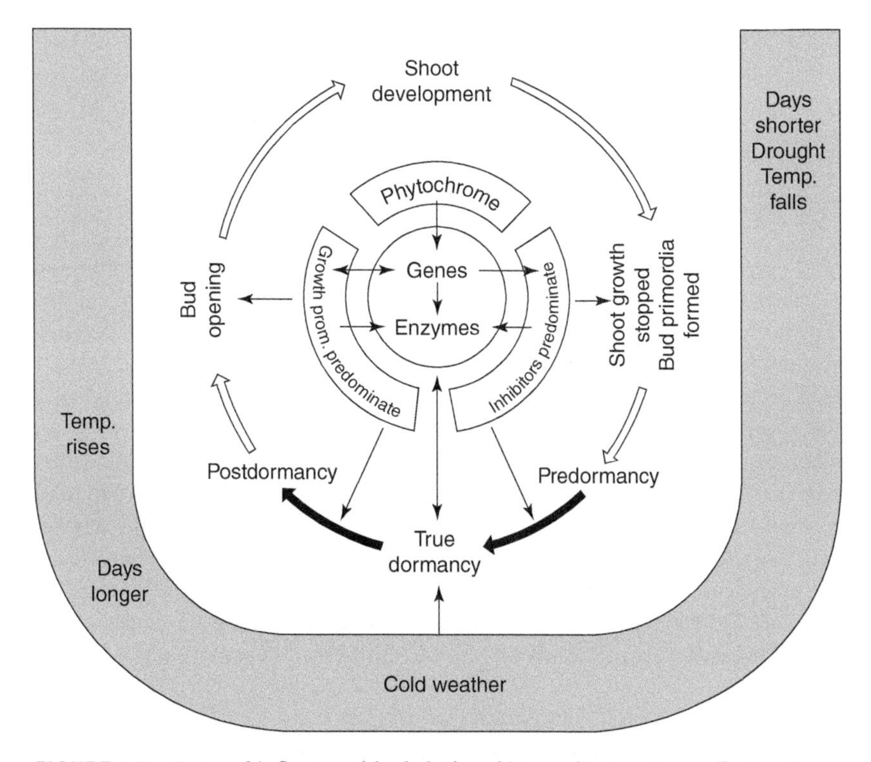

FIGURE 7.5 External influences (shaded U) and internal interactions affecting the seasonal alternation of vegetative and floral activity and dormancy in woody plants. Note that the growth inhibitors that influence bud formation in the fall and the growth promotors that initiate bud opening in the spring are activated by the phytochrome system. *Source:* Sakai and Larcher (1987) / Springer Nature.

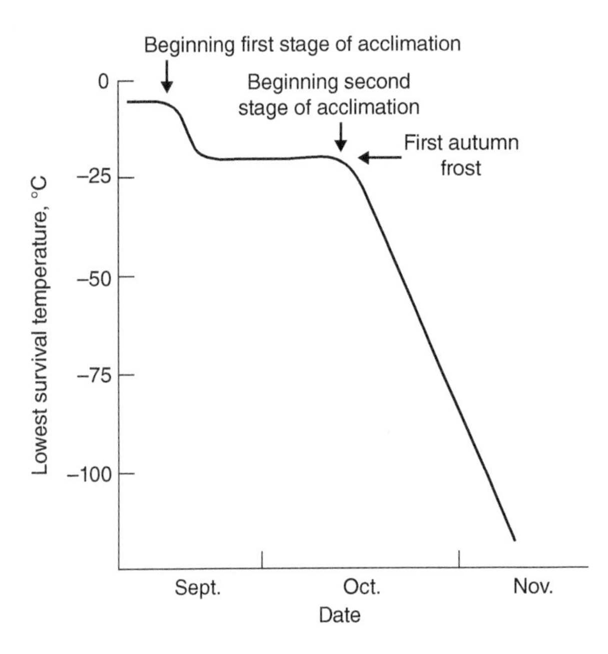

FIGURE 7.6 A typical seasonal pattern of cold resistance in the living bark of red-osier dogwood (*Cornus sericea*) stems in Minnesota. In nature, acclimation in this hardy shrub and in a number of other woody species proceeds in two distinct stages, as shown. The beginning of the second stage of acclimation characteristically coincides with the first autumn frost. *Source:* Weiser (1970) / with permission of American Association for the Advancement of Science.

beech, and some oaks, do not enter true dormancy and will pass readily from pre-dormancy to post-dormancy. Plants in their native ecosystems are not at risk from damage when truly dormant. However, plants are susceptible to severe injury or death from abrupt departures from the general temperature rise or decline. Alternate descriptions of the phases of dormancy are common in the literature (e.g., Lang et al. 1987).

FROST HARDINESS AND COLD RESISTANCE

The timing of the cold acclimation process is a major factor in the development of frost hardiness, and geographical races within a species may develop as a result. The more rapid changes in day length that occur in northern latitudes lead to earlier onset of the acclimation process in northern races, which are then more resistant to freezing compared to southern races. Examples of eastern tree species that have developed racial differences in forest hardiness include eastern cottonwood (Mohn and Pauley 1969), red maple (Perry and Wang 1960), sugar maple (Kriebel 1957), sweetgum (Williams and McMillan 1971), and white ash (Alexander et al. 1984), among others. Flint (1972) studied the hardening of northern red oak populations grown from seed and found a strong relationship between the temperatures at which stems are killed by frost and the latitude, annual minimum temperature, and estimated extreme minimum temperature of their original site. Oaks whose sources experienced colder average minimum temperatures required colder temperatures to kill the stems (Figure 7.7). When examining the progression of hardening and de-hardening of twigs from four populations at the extremes of the range, northern sources hardened earlier (Figure 7.8).

Geographic races are genetically adapted to minimum temperatures in midwinter as well as to the continuously changing gradient of temperature (Chapter 3; Figure 7.8). Overall, species are able to resist freezing injury well below the minimum temperatures they typically confront in their native sites. Woody plants of the boreal and northern forest (paper birch, spruces, pines, tamarack, trembling aspen, willows) survive prolonged subzero temperatures. Amazingly, if the freezing rate to −30 °C (−22 °F) is gradual and plants are fully acclimated, these and numerous hardy species are known to withstand experimental freezing to −196 °C (−321 °F), the temperature of liquid

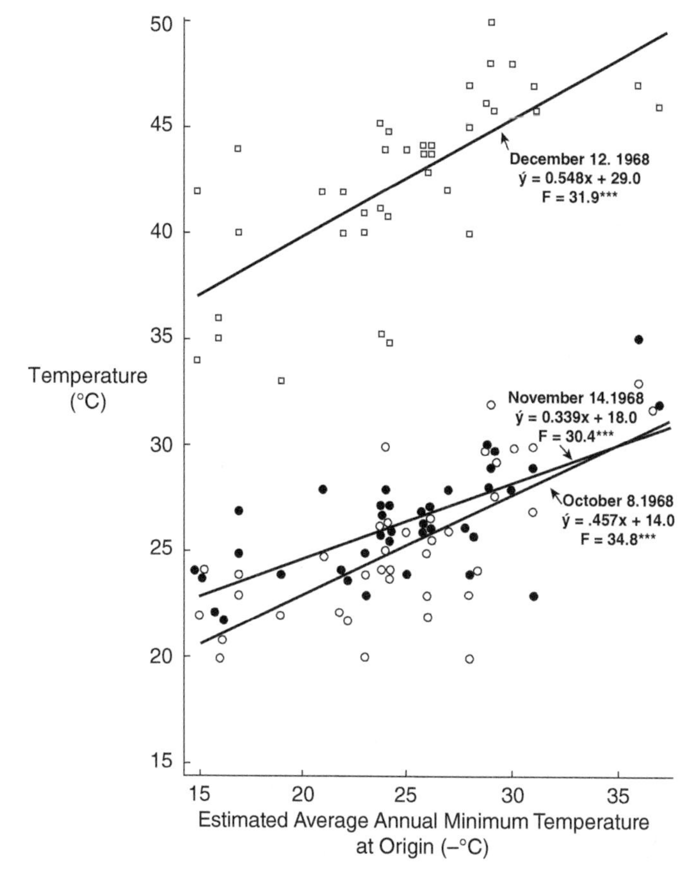

FIGURE 7.7 Regression of killing temperature on estimated average annual minimum temperature at seed-source origin of northern red oak at three sampling dates: 9 October (○), 14 November (●), and 12 December (□) 1969. *Source:* Flint (1972) / John Wiley & Sons.

nitrogen (Weiser 1970; Sakai and Weiser 1973). In contrast to boreal and alpine species, many trees from regions characterized by mild fall and winter temperatures such as those of the Pacific Coast region and the Lower Mississippi Valley and southern Coastal Plain are injured at temperatures colder than −20 °C (−4 °F) to −30 °C (−22 °F).

Plants cope with freezing temperatures either by tolerating them (tolerating the presence of ice crystals and related dehydration in extracellular spaces) or by avoiding them (preventing the formation of ice crystals in the first place). These two mechanisms are the basis for freezing resistance and may be found simultaneously occurring in the same plant (Levitt 1980a). In terms of tolerance, freezing within the protoplasm is the only source of direct injury by freezing to plants, and there is no tolerance to intracellular freezing at temperatures occurring in nature (i.e., it is always fatal). However, ice formation between cells—even the formation of masses several times larger than the cells themselves—is rather common. Thus, the main way plants achieve frost resistance ("hardiness") is by tolerating extracellular freezing, which involves both the tolerance and avoidance of dehydration caused by cold temperatures (Figure 7.9). These mechanisms are discussed in detail by Levitt (1980a) and Sakai and Larcher (1987).

Freeze avoidance by plants occurs by either avoiding freezing temperatures altogether or by avoiding ice formation (adaptations 2, 3, and 4 in Figure 7.9). Most obviously, plants may live where frost is uncommon, absent, or infrequent, but they may have also insulating mechanisms

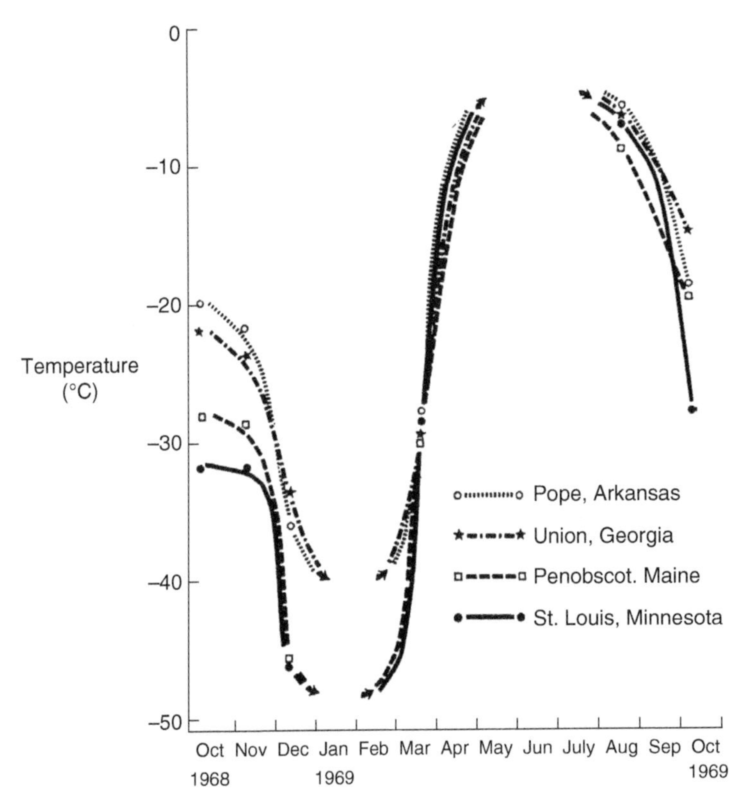

FIGURE 7.8 Progress of hardening and de-hardening of twigs from four geographic sources representing the corners of the natural range of northern red oak. Curves show the respective temperatures that cause freezing death to oak twigs. *Source:* Flint (1972) / John Wiley & Sons.

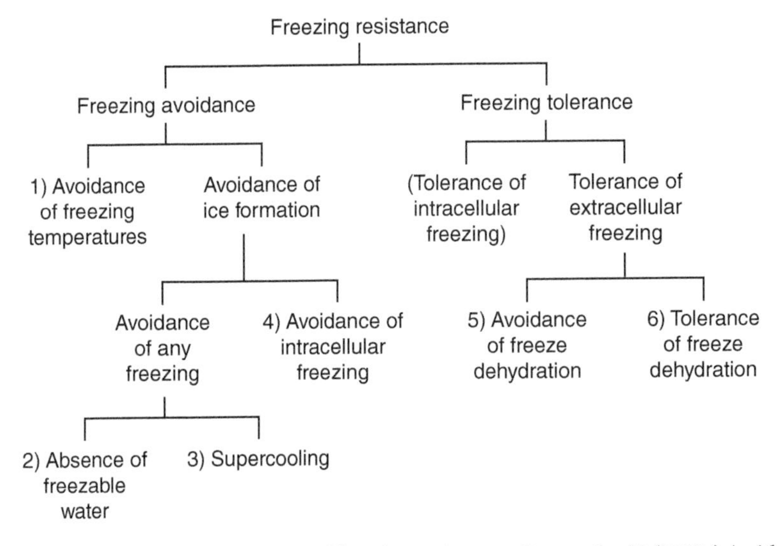

FIGURE 7.9 Six mechanisms of freezing resistance. *Source:* Levitt (1980a) / with permission of Elsevier.

such as bud scales and bark that provide at least initial protection. If the plant cannot avoid freezing, it may avoid ice formation and thus freezing injury via three adaptations: antifreeze, dehydration, and supercooling (undercooling) (Levitt 1980a). Reducing water in plant tissues, and thus decreasing the probability of ice formation, may be achieved in part by increasing water loss via transpiration and decreasing water uptake by roots via increased suberization. Increasing both the proportion of unfreezable (bound) water and the concentration of solutes (especially sugars and carbohydrates) also reduces the likelihood of ice formation. Sugar accumulation in the vacuoles can either prevent or reduce ice formation and thus also favor the avoidance of freeze dehydration (5 in Figure 7.9).

Supercooling is an important mechanism of freeze avoidance that reduces the freezing point of plant water below that of pure water (Levitt 1980a). The supercooling point is the lowest subfreezing temperature attained before ice will form. Cytoplasm freezes between −1 and −3 °C (27–30 °F); this level of supercooling is the mode of freezing resistance in the meristematic tissues of flower buds of many woody plants, such as azalea, blueberries, cherries, and plums (Weiser 1970; George and Burke 1977). Supercooling represents a temporary, unstable state that may provide protection against brief frosts (Sakai and Larcher 1987). In very small cells that are gradually hardened, more persistent, deep supercooling may occur to very low temperatures (George et al. 1974; Quamme 1985; Sakai and Larcher 1987). Deep supercooling of the ray parenchyma cells in the xylem of many woody species prevents freezing at temperatures as low as −37 to −47 °C (−35 to −53 °F). The distributions of 49 North American tree species were found to be related to the level of deep supercooling (George et al. 1974; George and Burke 1976). Plants protected by this resistance mechanism are limited to an area where minimum winter temperatures rarely are less than −45 °C (−49 °F).

Freeze avoidance can also be accomplished by plants through the synthesis of specific solutes within their fluids. Given that cytoplasm will freeze at lower temperatures as solute concentration increases, these solutes are effective at lowering the water freezing point (Zachariassen and Kristiansen 2000). Some extremely hardy boreal deciduous genera such as *Salix*, *Betula*, and *Populus* have no supercooling mechanism to protect them. Instead, plants of these genera permit freeze dehydration of their cells via very thin and elastic cell walls (Sakai and Larcher 1987). Most freezing tolerance depends on the capacity of living cytoplasm to tolerate freeze-induced dehydration even though supercooling is also an important resistance mechanism for some tissues and species.

De-acclimation and de-hardening need to occur to move plants out of dormancy. De-hardening occurs mainly due to increasing temperatures (Beck et al. 2004) and the length of the dormant period (Kalberer et al. 2006). Rapid thawing of tissues and cells is harmful to cell membranes, but de-hardening tends to occur more quickly than hardening, and the rate of de-hardening depends on the rate of increase in temperature. Moreover, it takes relatively little time to lose frost hardiness as the temperature rises. For example, a species of blueberry exposed to 5 °C (41 °F) during dormancy in midwinter took only seven days to lose more than 30 °C (86 °F) in frost hardiness (Taulavuori et al. 2002). Once the de-hardening process is complete, plants may be unable to re-harden again before an additional season of growth (Kalberer et al. 2006).

Thermotropic Movements in Rhododendrons Leaf movements and orientation are important adaptations in response to cold temperatures and freeze–thaw cycles in the genus *Rhododendron* (Nilsen 1992). This genus contains over 600 species of trees and shrubs distributed widely across the northern hemisphere; this distribution is related to such adaptations. Temperature-sensitive or **thermotropic** movements, leaf drooping, and curling into a pencil-shaped coil (Figure 7.10) are behaviors widely reported by scientists and rhododendron enthusiasts. Harshberger (1899) may have been the earliest to write about thermotropic

FIGURE 7.10 Drooping and curled leaves of rosebay rhododendron in winter. *Source:* Nicholas A. Tonelli/ Flickr/CC BY 2.0.

movments, observing that the petiole bends sharply downward through an angle of about 70°. Overall, rhododendron species that exhibit the most intense thermotropic movements are also the hardiest.

Leaf curling and drooping are distinct responses to climatic factors and have different ecological significances (Nilsen 1992). Leaf curling is a direct response to leaf temperature regardless of light or water availability. Leaf drooping is a response to leaf water potential, which in turn is influenced by leaf temperature, light intensity, and soil-water stress. Thus, lower temperature or higher light conditions result in increasing pendent behavior of the leaf. On winter days in the southern Appalachian Mountains when air temperature is constant and slightly below 0 °C, leaf angle of rhododendrons is vertical and leaf curling is 100% during the daylight hours. If light intensity increases and raises leaf temperature, there is some uncurling. Air temperature increasing from well below 0 °C to well above zero during the morning (diurnal changes of 24 °C/75 °F have been documented in the southern Appalachians) causes rapid leaf uncurling and a movement of leaves horizontally.

The ecological significance of these leaf movements is hypothesized to be the avoidance of cold injury due to freezing and thawing. The leaves of rosebay rhododendron in the southern Appalachians often freeze in winter at −8 °C (18 °F), but leaf temperature routinely decreases to −15 °C (5 °F; Nilsen 1985, 1987). At the same time, daytime winter temperatures may reach 15 °C (59 °F) and initiate rapid thawing. Therefore, rhododendrons in cold northern, alpine, or Arctic sites must have leaves adapted to tolerate both freezing and thawing, in repetition. Thermotropic movements may therefore be adaptations for reducing the rate of thaw. If frozen leaves were positioned horizontally, irradiance hitting a leaf may warm them too rapidly to avoid cell damage. It takes only seconds for leaf temperature to increase by 16 °C (61 °F) when a sunfleck hits a rhododendron leaf (Bao and Nilsen 1988). However, thawing in response to a sunfleck or by irradiance through a leafless canopy occurs much more slowly if the leaf is pendent and curled.

A second advantage of thermotropic movement relates to the interaction of cold temperature and high light on photosynthesis. Rhododendron leaves are evergreen, but do not photosynthesize in winter. High irradiance may cripple the plant's photosynthetic machinery when leaves are cold by damaging membranes of chloroplasts. This damage may impair photosynthesis by as much as 50% the following summer (Bao and Nilsen 1988). Notably, rhododendrons growing under a leafless tree canopy experience the highest radiation of the year when temperatures are the coldest (Nilsen 1985). The drooping of rhododendron leaves may therefore serve to protect cell membranes from damage in high light. Moreover, leaf curling reduces leaf surface area available to intercept light, and injury to the photosynthetic mechanism is reduced or avoided. Rosebay and many other rhododendron species grow in dim light of forest understories, and their evergreen leaves function for several years. Fully functional leaves are therefore critical for their survival, which would be threatened by winter injury to their photosynthetic mechanism and a subsequent reduction to their net carbon gain.

WINTER CHILLING AND GROWTH RESUMPTION

Buds and seeds of most temperate woody species require a period of **winter chilling** before growth is resumed in the spring, typically at temperatures near 3–4 °C (37–39 °F; Harrington et al. 2010). Once minimum temperature requirements are met, physiological changes take place that enable growth to resume in the post-dormancy period. Plants are most sensitive to freezing temperatures just after de-hardening and during leaf flushing (Kalberer et al. 2006), and thus a chilling requirement prevents freeze-related injury or death by premature resumption of growth triggered by a midwinter warm spell. The sensitivity of newly flushed leaves to cold temperatures was evident in the springs of 2007 and 2010 in the eastern United States. In those years, forests suffered major freeze damage due to uncharacteristically early-spring warm temperatures that were followed by late-spring frosts (Gu et al. 2008; Augspurger 2009; Hufkens et al. 2012).

The degree of dormancy attained by an individual tree and its chilling requirement vary with the climatic adaptation of a local race of species. For example, red maples from New York State attain true dormancy and have a chilling requirement of approximately 30 days or more (Perry and Wang 1960). By contrast, red maples in south Florida have no chilling requirement yet still drop their leaves in late autumn and resume growth in early spring. The southern maples are unable to withstand severe freezing temperature, but such events are rare in southern Florida.

When the chilling requirement is met and dormancy is broken, processes such as bud burst (leaf flushing), bud-scale initiation, and shoot axis elongation are regulated by temperature, a process known as **forcing**. In vegetative buds, dormancy breaks when mitotic activity begins, often well before the buds flush (several weeks earlier in conifers; Owens et al. 1977). After dormancy is broken, heat sums are very effective at predicting when a stage of development, such as flowering or flushing, will occur or how rapidly a shoot will elongate. Notably, winter chilling and forcing work in tandem in many woody plants: if chilling over winter is decreased, a higher heat sum is necessary for bud burst, and longer or colder chilling periods reduce the heat sum needed for bud burst (Cannell and Smith 1986). This pattern prevents very early bud burst when winters are warm and reduces the possibility of freeze damage. The pattern also suggests that warming winters may not be equally disruptive for all species. For example, in a study of chilling and forcing in tree species of the Pacific Northwest, western hemlock required less forcing for bud burst than Douglas-fir at low levels of chilling, but more forcing for bud burst than Douglas-fir at high levels of chilling (Harrington and Gould 2015). The processes involved in the seasonal growth–dormancy cycle for trees, including dormancy, chilling, and growth resumption (Figure 7.11), are also under strong genetic control, as discussed in Chapter 3.

Seeds of many temperate forest species also require a chilling period, typically accompanied by moist conditions, and thus overwinter on the forest floor before germinating the following spring. This period is termed **stratification**, and it alters the hormonal balance between growth

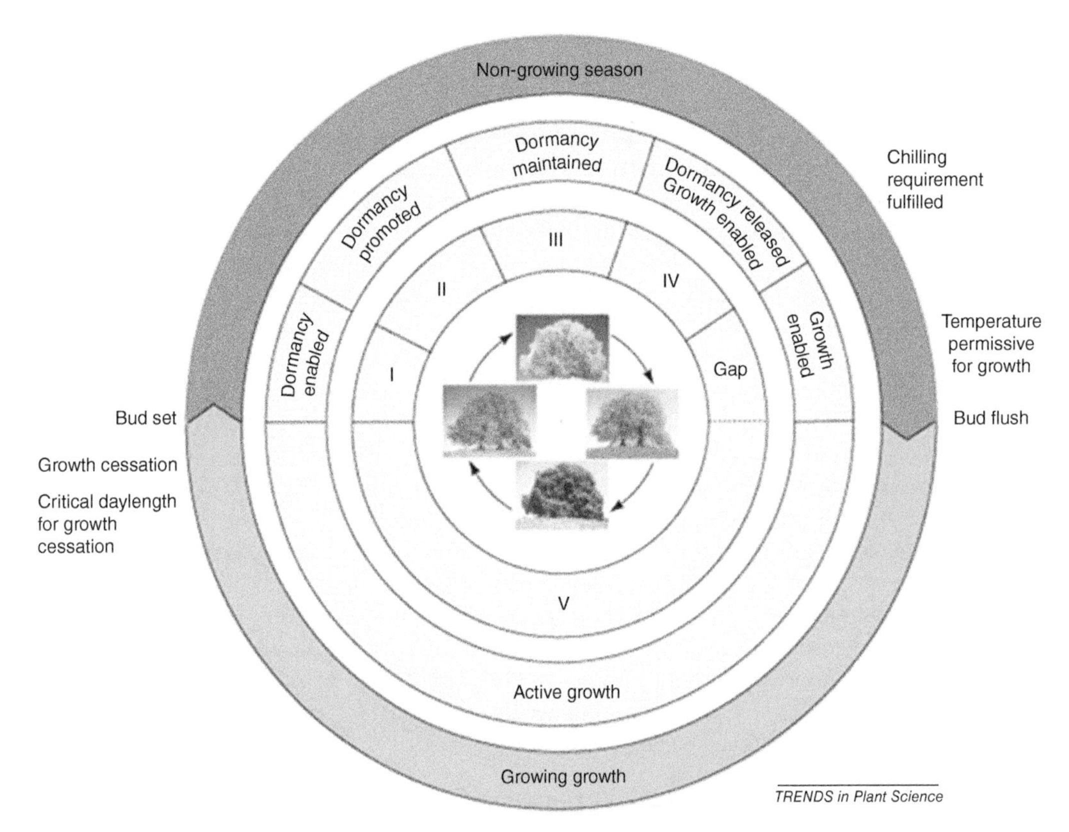

FIGURE 7.11 Transitions in seasonal growth–dormancy cycling in trees. The growing season is the length of time between bud flush and bud set, and the dormant season occurs opposite to the growing season. Growth cessation is achieved in response to photoperiod in the fall, and growth resumption by a chilling requirement in some species and the accumulation of a heat sum. Stage I approximates pre-dormancy; stage II, full dormancy; stage III, true dormancy; stage IV, post-dormancy; stage V, active growth. Growth is possible but does not occur during the "gap" between stages IV and V because of environmental constraints. *Source:* Rohde and Bhalerao (2007) / with permission of Elsevier.

inhibitors and promoters to favor germination when conditions are ideal for growth. Seeds of other species, including spring-fruiting trees (elms, cottonwoods, aspens, willows) and many fire-dependent species (including most hard pines), lack a cold requirement and may germinate after dispersal as soon as sufficient soil water is available.

NATURAL PLANT DISTRIBUTIONS AND COLD HARDINESS

Temperature and the ability of plants to resist freezing are critical factors limiting the natural distribution of plants. Winter cold hardiness appears to be related to the northern limit of species' distributions (Kreyling et al. 2014), but the frost hardiness and hardening capabilities of adult individuals as discussed in the earlier text suggest that minimum winter temperatures may not be a critical constraint. Instead, the timing of bud burst in relation to spring temperatures appears to be the main control of the northern limit of tree distributions (Kollas et al. 2014; Vitra et al. 2017). Freezing temperatures that occur in the spring just prior to and during bud burst are a main selective agent, because this is the period when freezing resistance is weakest and plants are unable to re-harden. Tree species are not limited by their relative abilities to withstand minimum winter temperatures at the northern extent of their range. Rather, they are limited by their relative abilities to withstand spring frosts that become more common further northward.

In a study of the frost hardening of four temperate deciduous genera, Vitra et al. (2017) found that hornbeams and cherries, which had a lower forcing requirement and thus flushed earlier in the spring, were able to withstand lower temperatures compared to beeches and oaks, which had a higher forcing requirement and flushed later in the spring. The greater freeze resistance of the early-flushing species thus seems to have developed from their vulnerability to late winter frosts, because their bud burst coincides with a period when both warm and freezing temperatures are possible. The relationship of the timing of bud burst to freeze resistance suggests that spring rather than winter temperatures are pivotal in shaping northern distribution of temperate trees. Notably, genera such as oaks that flush later do so at the cost of a reduced growing season, whereas early-flushing species with a shorter duration of dormancy benefit from a longer growing season at the cost of higher resource investment into frost resistance (Basler and Körner 2012; Muffler et al. 2016).

An additional factor limiting the distribution of tree species is the effect of cold temperatures on reproductive structures and flower buds, which tend to be more sensitive to frost than vegetative buds (Larcher and Bauer 1981). The failure to reproduce sexually can markedly affect migration and range extension, thus shaping species distribution. Perhaps more importantly, the ability of species to regenerate and thus establish at the limits of their range depends on the survival of seedlings in cold temperatures. Figure 7.12 illustrates frost resistance of several tissues of holm oak, a relatively hardy, evergreen Mediterranean oak, whose seeds germinate in autumn and winter when water availability is favorable (Larcher and Bauer 1981). Germinating seedlings and surface-feeding roots are especially prone to freezing compared to other anatomical parts and developmental stages. Adult trees are able to thrive where temperatures of −12 °C (10 °F) occur each winter, but natural regeneration of the species is completely suppressed if winter temperatures regularly drop below −8 °C (18 °F).

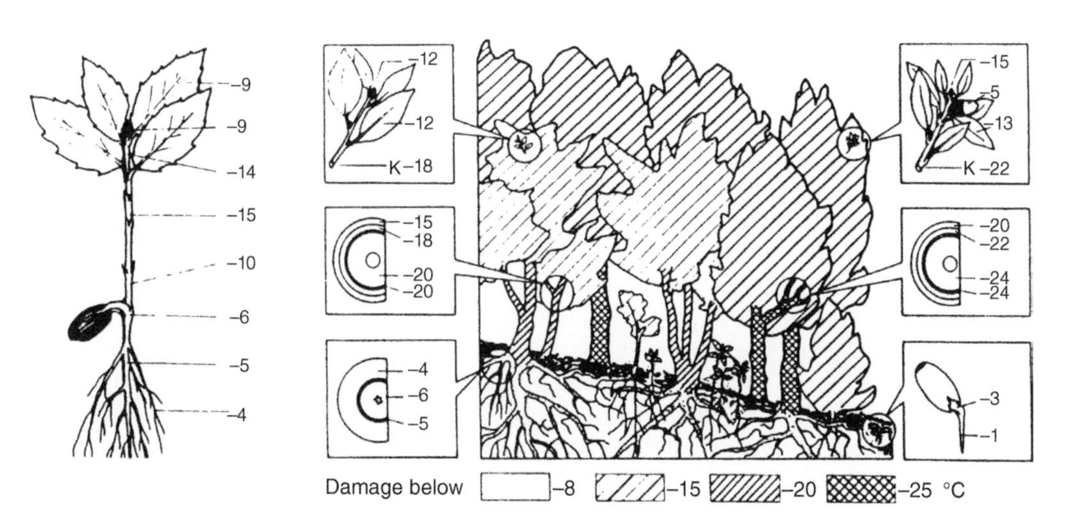

FIGURE 7.12 Freezing resistance of the various strata and age groups in a stand of holm oak (*Quercus ilex*). Numbers are degrees centigrade. *Source:* Sakai and Larcher (1987) / Springer Nature.

Similarly, in North America, many species with southerly distributions can survive, reach large size, and even produce fruit in climates far north of their natural range where their establishment has been provided by humans. For example, sweetgum is used for landscaping in many northern states far from the northernmost limits of its natural range. Northern catalpa has been planted up to 650 km north of the northern extent of its natural range from northeastern Missouri to southeastern Illinois. Osage-orange, whose northernmost natural range is in southern Oklahoma and Arkansas (34° N), has been widely planted and successful as far north as southern Michigan (42° 30′ N) or about 1000 km north of its northernmost natural range. At more northerly sites, these species fail to regenerate and compete in natural communities.

DECIDUOUSNESS AND TEMPERATURE

Compared to maintaining thick evergreen leaves throughout the year, trees conserve energy by shedding thin deciduous leaves in winter and protecting meristems from water loss (Breckle 2002). This is considered an important mechanism to avoid freezing during the winter and to reduce the loss of carbon via respiration when the potential to gain carbon through photosynthesis is low (Givnish 2002). Losing the ability to photosynthesize without leaves (excluding those thin-barked trees with stems that are able to photosynthesize) means that deciduous species must be able to accumulate enough photosynthate during the growing season to support a full year's maintenance, growth, and reproduction. In part, deciduous trees compensate for shorter growing seasons by producing thinner leaves with higher photosynthetic rates per unit mass compared to conifers, whose long-lived leaves are thicker and have lower rates of photosynthesis (Reich et al. 1999; Wright et al. 2004).

Deciduous woody species also address this trade-off by dominating forests in areas where growing seasons are long, warm, and humid with readily available soil water and nutrients (approximately from 35 to 45°N latitude and from 70 to 95°W longitude in North America). To the west, deciduous forests are replaced by grasslands (Figures 2.6 and 2.8, Chapter 2) and dry-site conifers (junipers, pinyon pines, and ponderosa pine) at low elevations where humidity is low, soils are dry, and fires are frequent. On the southern Coastal Plain, growing seasons are long, summers are warm, and winters are not extreme, but fire is frequent, and soils are nutrient-poor. These latter conditions also favor conifers and other evergreen species over deciduous species.

The relationship between deciduousness and temperature is most obvious at higher latitudes and elevations, where growing conditions become harsh and extreme, and conifers tend to replace deciduous forests. With increasing latitude and elevation, the growing season shortens, summers are cooler, soils are poor, and prolonged drying winds are common. Cold conditions inhibit biological activity including decomposition, reducing nutrient availability compared with comparable soils at lower latitudes and elevations. Moreover, cold soils result in unfavorable conditions for root growth and for absorption and conduction of water and nutrients. Therefore, deciduous and coniferous trees as well as forest shrubs and herbs that are better adapted to extreme conditions outcompete those adapted to more favorable conditions. If we consider leaf longevity as a proxy for the evergreen (high longevity) versus deciduous (low longevity) condition, then deciduousness (or lack thereof) is closely tied to the length of the growing season, which in turn is best predicted by temperature (van Ommen Kloeke et al. 2012). Temperature and other site factors such as soil water that significantly affect plant distribution are very strongly affected by physiographic features, which are considered in the next chapter.

SUGGESTED READINGS

Kozlowski, T.T., Kramer, P.J., and Pallardy, S.G. (1991). *The Physiological Ecology of Woody Plants*. New York: Academic Press 454 pp.

Larcher, W. and Bauer, H. (1981). Ecological significance of resistance to low temperature. In: *Physiological Plant Ecology I, Responses to the Physical Environment*, New Series, vol. 12A (ed. O.L. Lange, P.S. Nobel, C.B. Osmond and H. Ziegler), 403–437. New York: Springer-Verlag.

Levitt, J. (1980). *Responses of Plants to Environmental Stresses*, vol. 1. New York: Academic Press Chapter 5 (The Freezing Process), Chapter 6 (Freezing Injury), Chapter 7 (Freezing Resistance—types, measurement, and changes).

Sakai, A. and Larcher, W. (1987). *Frost Survival of Plants*. New York: Springer-Verlag 321 pp.

Weiser, C.J. (1970). Cold resistance and injury in woody plants. *Science* 169: 1269–1278.

Physiography

CHAPTER 8

Many of our most vivid and unforgettable mental images are of the landscape where we grew up or where we live. Moreover, the shape of the land and the silhouettes it creates provide one with a sense of place, and sometimes a sense of purpose. The author Edward Abbey was fond of relating the landscapes of the American Southwest to a sense of place and purpose, writing in his book *Desert Solitaire* (Abbey 1968):

> *May your mountains rise into and above the clouds. May your rivers flow without end, meandering through pastoral valleys tinkling with bells, past temples and castles and poets towers into a dark primeval forest where tigers belch and monkeys howl, through miasmal and mysterious swamps and down into a desert of red rock, blue mesas, domes and pinnacles and grottos of endless stone, and down again into a deep vast ancient unknown chasm where bars of sunlight blaze on profiled cliffs, where deer walk across the white sand beaches, where storms come and go as lightning clangs upon the high crags, where something strange and more beautiful and more full of wonder than your deepest dreams waits for you – beyond that next turning of the canyon walls.*

Indeed, mountains inspire a sense of natural grandeur and power with their distinctive shapes and striking seasonal changes. Prairie landscapes with their endless sweeps of flat or gently rolling land have an irresistible and memorable quality whose beauty and harshness are portrayed in many books and films. Lakes, rivers, and streams and their associated lake shores and stream banks provide unforgettable mental maps to which we can link diverse biotic and human patterns of occurrence and activities in space and time. Landscapes of rocky coastlines, sand hills, wetland depressions, tropical lowlands and highlands, tidal flats, arid plateaus, and mountain slopes all reveal close interrelationships between the form of the land or water with the climate above, the parent material and soil below, and the organisms sandwiched between. At different spatial scales, from subcontinental plains and mountains to pit-and-mound microsites in a hemlock–northern hardwood forest, physiographic features or landforms provide the best path for understanding the structure and dynamics of landscape ecosystems and the organisms they contain. We consider foundations of ecological geomorphology in the following sections by examining the concepts and ecological attributes of physiographic features and examples of their pervasive importance for the forest ecologist.

CONCEPTS AND TERMS

The term **physiography** is an abbreviation for physical geography—the surface features of an area. Surface features such as mountains or outwash plains have *form* at their surface that represents relief from the horizontal. These features also have *parent material* beneath their surface that represents the underlying geological material from which soil develops (Bailey 2009). Thus, in forest ecology, we define physiography as the form and substance of the surface features of a given

Forest Ecology, Fifth Edition. Daniel M. Kashian, Donald R. Zak, Burton V. Barnes, and Stephen H. Spurr.

regional or local area. Physiography differs from **geomorphology**, which is the science dealing with the nature and origin of topographic features of the Earth, in that it integrates both the shape of the surface as well as the geological surface materials (i.e., parent material) associated with the features. The specific physiographic features themselves are **landforms** (e.g., mountain slopes, plateaus, outwash plains, and river floodplains), created by erosion, sedimentation, or movement. We use the term landform in the context of specific physiographic features of the Earth's surface and the term physiography in the broad context of a major ecosystem component. Physiography is a major ecosystem component that not only provides spatial form and structure to a landscape and its ecosystems but significantly affects ecosystem function. For instance, land in a high position influences adjacent low-lying lands and their biota through its effect on climate and hydrology. Land adjacent to a river affects its course, rate of flow, and water quality; and the river affects the land as flood waters erode and deposit sediments, forming levees and floodplain bottoms.

We discussed in Chapter 2 that the major ecosystem components of climate, physiography, soil, and vegetation may be used to distinguish and map landscape ecosystems at regional and local levels, and also to understand ecosystem structure and function. Physiography is exceedingly important in this regard because it is the component least affected by short- and long-term natural and human disturbances, forming a relatively permanent framework of landscape ecosystems (Rowe 1984b). Physiography influences ecosystem function because it controls the climatic regime at and above the surface, and it controls soil development and acts to regulate soil processes below the surface. Landforms and their parent materials modify the fluxes of radiation, soil water, and nutrients thereby regulating plant establishment, distribution, growth, and productivity.

Let us consider a newly exposed glacial landform with different local climates on its slopes: one slope relatively warm and dry, and the other relatively cool and moist (Figure 8.1). Different plant species will initially colonize these different slopes based on their tolerances and requirements for temperature and soil water, and in turn animals will respond to the vegetation and microclimate. Across a landscape, the size, shape, and relation of one landform to another shape plant and animal communities, as well as genetic races that develop (Chapter 3). Soils develop from the parent material of the new landform as it interacts with local climate, biota, and time. Over time, similar ecosystems develop on similar landforms, and repetitive patterns of landforms give rise to repetitive patterns of vegetation. Thus, a landform map is a proxy for a soils map, as well as a map of local climate. In this regard, a landform map acts as a best approximation of a map of regional and local vegetation (Rowe 1969). In addition, a landform map reflects the history and patterns of disturbances such as wildfire and flooding, because the shape and parent material of specific landforms may influence the severity and frequency of these disturbances.

CHARACTERISTICS OF PHYSIOGRAPHY AND THEIR SIGNIFICANCE

We list and briefly describe the specific characteristics of physiography that are useful in understanding ecosystem structure and function in the hierarchical framework that follows. Many physiographic characteristics are significant at many different spatial scales, from continents to a local wetland depression less than 0.1 ha in area.

Physiographic Characteristics:

- Physiographic setting (e.g., glaciated or unglaciated terrain, coastal or interior landscapes)
- Specific landform (e.g., mountain, outwash plain, river terrace)
  - Elevation
  - Size
  - Form (shape or configuration)
    - Level landforms
    - Sloping landforms

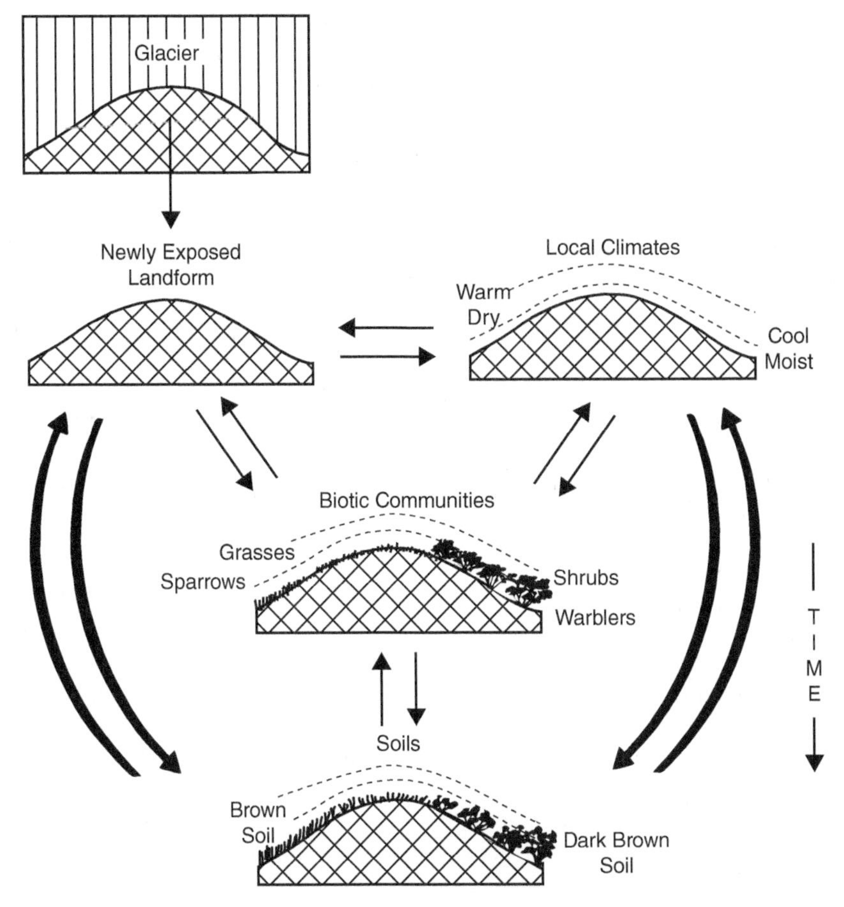

FIGURE 8.1 Landscape development since deglaciation. A hill of glacial till emerges from the melting ice and develops its local climates. These and the parent-material substrate develop the ecological basis for arriving plants that in turn provide habitat for adapted animals. Soils develop through interactions of landform, local climate, and biota. Arrows indicate interactions and feedback so that, for example, the properties of developing soils and communities influence the evolution (by erosion and deposition) of the landform. *Source:* Rowe (1984b) / United States Department of Agriculture / Public Domain.

Slope shape or configuration (straight or planar, concave or convex)

Slope position (upper slope, mid slope, lower slope)

Slope aspect (direction a slope faces)

Slope inclination (slope percent or degree of slope)

Parent material of the feature or landform (kind of substrate, including rock type, particle size, mineralogy)

Position of landform in the landscape (high-level, low-level)

Position of landform in relation to other landforms

PHYSIOGRAPHIC SETTING

Physiographic setting represents the broadest spatial scale of physiography that provides the context for specific landforms. An overriding consideration is whether the landscape has been glaciated by a continental ice sheet or not, because many landforms are specific to glaciated terrain.

The characteristics of features such as rivers, lake beds, mountains, and hills differ depending on whether the feature occurs in glaciated or unglaciated terrain. Similarly, coastal regions represent a very different physiographic setting than interior regions, and specific landforms that occur in both regions are quite different in form and substance between the two settings.

SPECIFIC LANDFORMS

Specific kinds of landforms and their sizes determine to a large degree the physical factors of ecosystems and the productivity and patterns of occurrence of the organisms with them. Specific landforms occur at spatial scales from continental to local (<1–100 ha). Mountain ranges, plains, and large river basins are examples of specific landforms that occur at the spatial scale of continents. Their properties and effects are appropriately considered in the classic treatments of physiography by Fenneman (1931, 1938) and have since been discussed in detail in most geography texts. We describe local landforms in the following sections.

ELEVATION

Elevation above sea level affects local climate, and for a given physiographic region or specific landform, elevation is especially important when the climate associated with it is limiting to plant establishment and growth. Two factors related to elevation that explain its effects on local climate are the impacts of elevation on temperature and precipitation. First, air temperature decreases as it moves upward through the Earth's atmosphere at a rate of change known as the **lapse rate**, such that temperatures experienced by plants are generally cooler at higher elevations than at lower elevations. Moist air that is lifted through the atmosphere cools as it rises, losing its ability to hold water and releasing the water as rain or snow called **orographic precipitation**. Much of this precipitation falls on the windward side of a mountain or mountain range, such that the leeward side tends to be drier (often called a **rain shadow**). In general, sites at high elevations are often cooler and wetter than sites at low elevations.

Elevation is therefore only a rough proxy for climate, because climatic factors also vary at a given elevation with latitude and other physiographic elements (aspect, slope percent, etc.). At a known latitude, however, elevation of a site may predict plant performance fairly well. For example, the vegetation of riparian ecosystems differs at low elevations compared to high elevations whether one is in the Cascade Mountains in Washington State or the Smoky Mountains in Tennessee. It follows that the vegetation of low-elevation ecosystems at high latitudes may resemble the vegetation of high-elevation ecosystems at lower latitudes (see Chapter 7). For example, communities dominated by red spruce at sea level in eastern Canada have many similar species to those found at elevations over 1500 m in the southern Appalachians.

FORM OF LANDFORMS

As we discussed in the earlier text, the two key elements of physiography are the shape and parent material of specific landforms. The shape of landforms, in particular, influences how radiation, soil water, and nutrients are received and distributed and in turn the vegetation patterns of an area.

Level Terrain Whether land is level or sloping is one of the initial distinctions made regarding landscapes. Landforms with characteristically level or nearly level terrain include plains, river terraces, old lake beds, prairies, and plateaus, all of which are either deposited or influenced by water. Some exceptions to this association exist, such as the exceedingly flat till plains of parts of Indiana and Ohio that were deposited by glacial ice as it rapidly retreated at a relatively uniform rate. The Tipton Till Plain in central Indiana is so flat and monotonous that one observer noted: "... the traveler may ride upon the railroad train for hours without seeing a greater elevation than a haystack or pile of sawdust" (Petty and Jackson 1966).

The lateral movement of water on level landforms is limited, and thus precipitation may more quickly move vertically through the soil depending on the nature of the parent material. Level terrain with coarse-textured parent materials has high vertical water movement and tends to be dry. In warm, dry seasons, water drainage is rapid and water nearest the soil surface is quickly absorbed by plant roots or is evaporated. If the soil has layers or bands of less-permeable material or bedrock at or near the surface, vertical water movement will be slowed or restricted. Even in very dry soils, such bands may subtly or markedly affect plant growth and community composition (Kashian et al. 2003). Level terrain is typically the location of ecosystems where water accumulates, such as lakes, marshes, swamps, and bogs. Most of the world's wetlands are found on relatively flat terrain, including coastal plains, old lake beds and areas surrounding lakes, and areas associated with river floodplains where groundwater is found at the surface.

Flat terrain also features strong effects of temperature on microclimate. Level terrain limits cold-air drainage, such that small differences in the terrain in the form of small depressions or flat land surrounded by even slightly higher land create natural sinks for accumulation of cold air at night (Chapter 7). Low nighttime temperatures in such frost pockets further affect the composition and growth of vegetation and animal populations.

Sloping Terrain Sloping terrain is ecologically different from flat terrain because water may move laterally, either internally or overland along a slope so that poorly drained soils may seldom occur. Sloping terrain generally features three shapes: straight or planar (surface with zero curvature), convex (curving outward, such as a hill), or concave (curving inward, such as a depression). These types are combined in a variety of ways in different landforms. On convex surfaces, water, organic matter, and dissolved nutrients tend to move out of the feature. Concave surfaces tend to accumulate soil water, nutrients, and organic matter, often making them more favorable for plant growth. Plants lower on convex surfaces often attain larger sizes than individuals of the same species upslope (Chapter 11).

Other important characteristics of sloping terrain are **aspect**, or the direction a slope faces, and **slope inclination**, or the steepness of the land from the horizontal. **Slope length** (in m or km), for example, such as a short, steep slope versus a long, gradual slope, may affect plant distribution in certain landscapes because a longer slope evokes steady rather than abrupt changes in environmental conditions. In a classic study of deciduous swamps in northern Ohio, Sampson (1930) reported that an abrupt elevation change of 18–25 cm above the wet swamp forest of American elm, black ash, and silver and red maple gave sufficient local aeration for American beech and sugar maple or even oak and hickory trees to occur. By contrast, areas with long, gentle slopes rising 3 or 4 ft (about 1 m) over a mile (6.4 km), although their elevation may be several feet higher than the abrupt rises, were still too wet to support American beech or sugar maple.

Glaciated landscapes often feature high and moderately-to-steeply sloping end and lateral moraines, as well as flat, gently sloping, or rolling ground moraines. **Ice-contact terrain** exhibits steep, hilly topography of kames and eskers with relatively abrupt and short slopes interspersed with ice-block depressions of variable depths termed kettles (often containing lakes and wetlands). Ocean and lake beach ridges, terraces, and dunes exhibit locally distributed low, rolling features. Similar to unglaciated terrain, the recurrence of landforms and their parent materials on the landscape provides predictable patterns of ecosystem types and plant communities.

Slope Characteristics Slope position, slope aspect, and slope inclination influence microclimate, soil depth, soil profile development, and the texture and structure of the surface soil. Because these microclimate and soil factors in turn influence vegetation, slope characteristics have a heavy bearing on the composition, development, and productivity of the ecosystem.

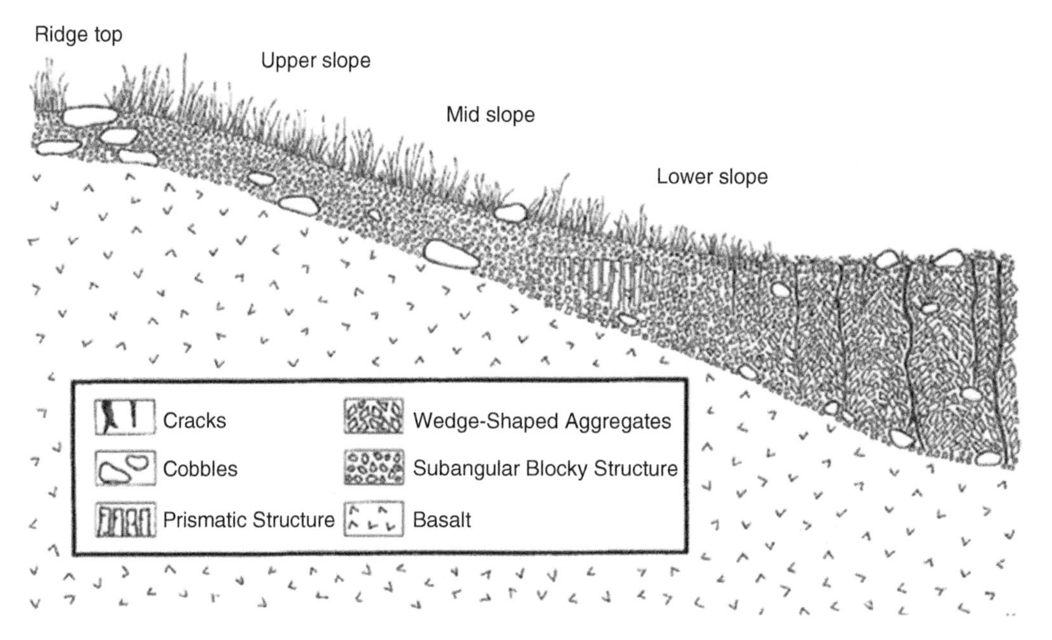

FIGURE 8.2 Cross-sectional diagram of changes in soil depth and morphology with slope position for a vernal pool in Riverside County, California. The diagram depicts only a 3-m rise in elevation over a 30-m horizontal distance. *Source:* Hawkins and Graham (2017).

Position on slope The tops of high features of the landscape, such as ridge tops or upper convex slope surfaces, tend to be drier than is the average for the region. These features are differentially exposed to intense solar radiation and high winds and are subject to higher rates of soil-water drainage and erosion. By contrast, lower slopes with concave surfaces tend to be moister than average for the region, are sheltered from high winds, accumulate water, organic matter, and soil, and are subject to cold-air drainage. Midslopes are generally intermediate in their characteristics (Figure 8.2). Soil development differs in response to these differences in moisture and temperature. Soil differences are most pronounced when a slope represents a large change in elevation, but even small elevational changes are subject to these processes (Hawkins and Graham 2017).

Slope position, used as a proxy for relative elevation (i.e., ridge top; upper, mid, lower slope), has a strong influence on site productivity (Chapter 11) and is one of the most useful criteria for classifying and mapping forest ecosystems. Many researchers have shown slope position to be the single most useful factor in evaluation of the growth potential of forest trees (Ralston 1964; Carmean 1975, 1977). Slope position is also convenient in that it is able to be remotely sensed, historically using aerial photography and more recently with digital elevation models (DEMs) and LiDAR (Light Detection and Ranging) technology.

Aspect The aspect of a slope, that is, its orientation with regard to the sun's position, has a direct influence on microclimate. In northern temperate regions, the sun moves across the southern horizon and is therefore to the south during the warmest part of the day. As a result, south-facing slopes receive the highest intensity of sunlight and thus its sites are generally hot and dry. In a detailed study of eight environmental factors of four southerly aspects in the desert foothills of Arizona, Haase (1970) found the sequence of warmest and driest aspects to be south, south-southwest, southwest, and south-southeast. On combining aspect with slope inclination (see the following text), the hottest and driest sites are those with slopes perpendicular to the sun's angle during the middle of the summer day, such that both steeper and more gentle slopes receive less radiation. The amount of sunlight received on a site governs air and soil temperature, precipitation, and soil water, all of which affect plant establishment and growth.

North-facing slopes receive less sunlight and are invariably cooler and moister in the Northern Hemisphere. East and west slopes show similar but less extreme variation. East-facing slopes are exposed to direct sunlight in the cool of the morning and are somewhat cooler and moister than west-facing slopes, which are exposed to the sun for an equal amount of time as east-facing slopes but receive the radiation in the hotter part of the day. As described in the earlier text, these differences in microclimate have important implications for site productivity. In mixed upland oak forests of the Appalachian Mountains, northeast aspects are the most productive, being approximately 15% more productive than south and west aspects, which are the least productive (Chapter 11). In a study of succession in subalpine landscapes of Colorado, succession proceeded more quickly on north-facing slopes than south-facing slopes, primarily driven by differences in soil moisture and air temperature (Donnegan and Rebertus 1999).

Slope inclination Slope gradient is measured in terms of percentage or in degrees, and steeper slopes provide greater surface per unit area measured horizontally. For this reason, good forest sites of moderate slope often contain more trees and produce greater timber yields per hectare measured horizontally than do comparable level sites. Movement of water, snow, and soil is more rapid on steep slopes compared to gentle slopes, and the danger of erosion, avalanche, and mass soil movement is much greater. Notably, uprooting is one of the most common effects of steep slopes on trees, and this process itself contributes to mass movement of soil (Norman et al. 1995). These processes favor species dependent on disturbances that expose mineral soil and cause canopy openings. For example, a thick forest canopy is rarely formed in southern Wisconsin on steep hillsides, so that light penetration occurs from the side and species such as black and white oak persist. Succession away from black and white oak is more rapid on level terrain where light available to plants of the forest floor is controlled by the tree canopy rather than disturbance events more common on steep slopes (Curtis 1959). On rocky landscapes, rockfalls on steep slopes may damage trees at a frequency that inhibits tree growth (Moos et al. 2021).

PARENT MATERIAL IN RELATION TO LANDFORM

The *substance* component of landform is important because it allows us to approximate broad classes of soils from what we understand about parent material. Parent materials are typically specific to landforms regardless of whether the materials developed in place or were deposited in association with glaciation (Chapter 9 and Figure 9.1). Soils that develop from residual limestone or sandstone rock support markedly different plant communities. Parent material develops by weathering of specific rock types, stream action, and geomorphic processes such as avalanche and debris flow. Wind initiates landforms, most commonly sand dunes (from sand parent material) and loess deposits (from silt), which reflect the particle size of the parent material. The landforms of river floodplains, including the front, levee, first bottom, and backswamp, reflect the nature of their parent material and flooding patterns. On glaciated landscapes, flat (old lake beds, glacial outwash plains) and hilly (moraines, ice-contact terrain) landforms have characteristic parent materials that distinguish them and, together with the shape of the feature, help determine forest composition, productivity, and successional pathways. Despite a perceived lack of major topographic differentiation on glaciated landscapes, landforms constrain the distribution of parent material in such a way that they may be the most important driver of ecosystem structure. For example, forest composition resulting from natural reforestation of a central Massachusetts landscape following clearing for agriculture 150 years ago was found to be more influenced by landform than agricultural history or natural disturbances including fire, hurricanes, or logging (Gerhardt and Foster 2002). The mineralogy and chemistry of parent material are discussed in more detail in Chapter 9.

POSITION OF LANDFORM IN THE LANDSCAPE

The position of a specific landform (e.g., outwash plain, ravine, ridge, etc.) on the landscape may influence plants and animals differently. Broadly, landforms in the rain shadow of a mountain range exhibit different climates, soils, and vegetation than similar landforms on the windward side where precipitation is markedly higher for a given elevation. For example, some landforms on the western (windward) side of the Cascade Range in Washington State may have up to 4 m of annual precipitation compared to less than 50 cm for similar landforms on the eastern side. A high, exposed ridge may support different plants and site potential compared to one with similar geology and parent material in a lower and protected position. In non-mountainous glacial terrain, an elevated, dry outwash terrace is more favorable for tree growth than an equally dry terrace below it in a cold, frost-prone basin. Thus, the position of a given landform on the landscape may provide different site conditions for plant establishment and growth.

A related phenomenon to landscape position is landform adjacency. The ecological effect of a forested wetland is much larger on land adjacent to it than land at a greater distance. At a broad scale, landforms adjacent to large water bodies (oceans or lakes) are more strongly affected than lands further inland or at further distance from a lake. At a local scale, a first-bottom ecosystem adjacent to a river is sharply different in soil and vegetation than an elevated, second-bottom ecosystem further from the river due to its greater likelihood and frequency of flooding.

In summary, many different physiographic characteristics act together to influence vegetation and structure and function of landscape ecosystems and have been used as the stable basis for the management of forested ecosystems. For example, Angus Hills (1952) pioneered the use of physiography on glacial landscapes (integrating slope position, aspect, slope gradient, and parent material) to determine forest composition and productivity in Ontario, Canada (Chapter 11). Hills demonstrated how topographic-mediated differences in microclimate (warm versus cold) and soil water can bring about marked differences in late-successional communities (Figure 8.3).

MULTIPLE ROLES OF PHYSIOGRAPHY

Physiography results from geologic and geomorphic processes, and its characteristics already described influence ecosystem structure and function. This interrelated suite of characters influences ecosystem processes at multiple spatial scales and influences soil development, climate, disturbance regimes, and the distribution of vegetation. The roles of landforms in affecting ecosystem processes have been summarized into four classes, all of which are interrelated (Swanson et al. 1988):

1. **Quantity of materials available** (Figure 8.4a): As described in the earlier text, the kind of landform and its elevation, slope aspect, inclination, and position influence the quantity of solar radiation, air, water, nutrients, and other materials received at the site. In addition to the effects described in the earlier text, these factors may be important determinants of other drivers of ecosystem structure and function such as soil moisture and disturbance (Méndez-Toribio et al. 2016).

2. **Flux and transport of materials** (Figure 8.4b): The same factors mediate the flow of air, water, materials (including organic and inorganic particulate matter, dissolved materials), organisms, propagules, and energy within and across ecosystem boundaries by: (i) affecting gravitational gradients, (ii) guiding flows of wind and water, and (iii) forming barriers to movement. River floodplains and corridors are perhaps the best example of landforms that broadly direct the flow of many materials.

3. **Disturbance** (Figure 8.4c): Landforms and their physiographic features influence the spatial pattern, frequency, and severity of disturbances caused by fire, wind, water,

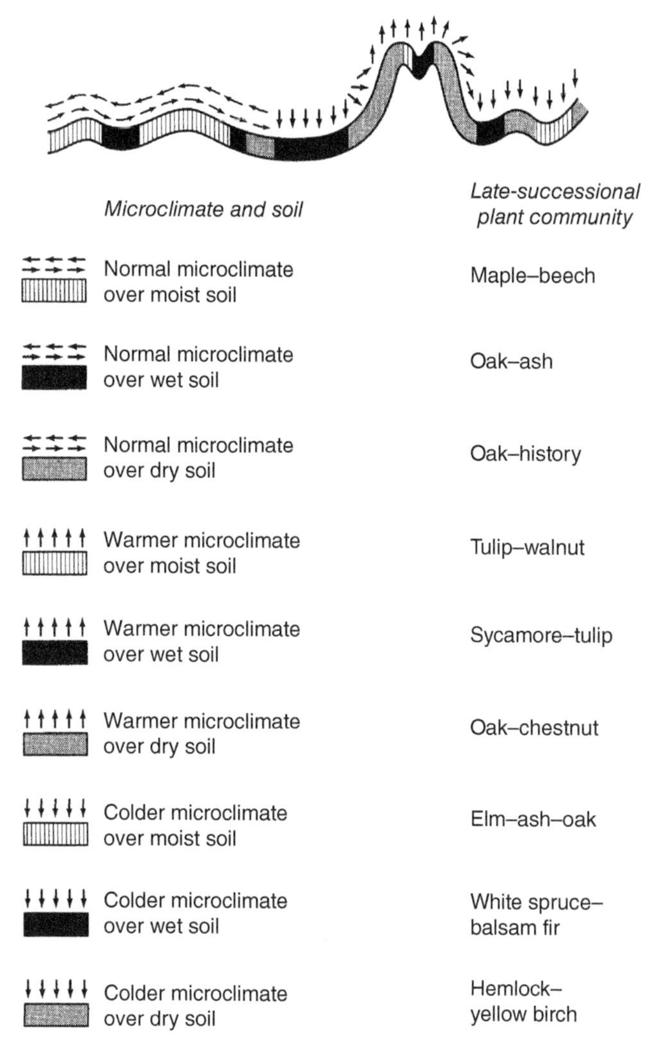

FIGURE 8.3 Physiographic diagram illustrating the effect of local topography (slope position and aspect) in determining late-successional species composition of landscape ecosystems in the temperate continental zone of southern Ontario, Canada. *Source:* Modified from Bailey (1988) and Hills (1952).

avalanche, and soil movement, as well as human-caused disturbances of erosion, pollutants, and barriers to natural fluxes of water and nutrients. Disturbances by wind are particularly affected by landforms, with wind speeds highest on exposed ridge tops if it moves perpendicular to a ridge, at mid-slope if the wind direction is at an acute angle, and at the valley bottom if the wind moves parallel to the ridge (Frelich 2002). Hurricanes tend to be most damaging to forests on aspects that face the wind direction (Boose et al. 1994).

4. **Spatial position of landforms** (Figure 8.4d): The juxtaposition of one landform in relation to another (landform adjacency) may affect regional and local climate and hydrology, movement of materials, and disturbance regimes. The arrangement of landforms across a landscape may be important in determining patterns of major disturbances such as large fires, windthrow areas, or landslide events, because all or portions of some landforms are more or less susceptible to a given disturbance.

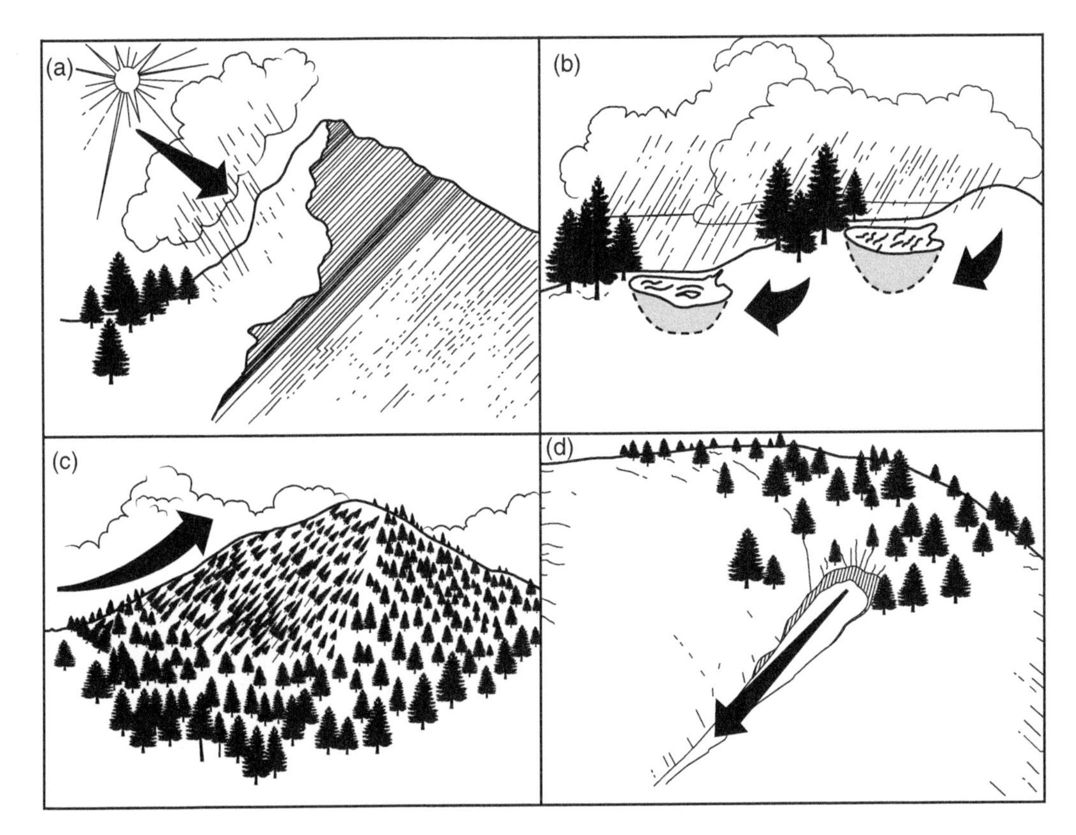

FIGURE 8.4 Summary of the roles of physiography as represented by four classes of landform effects: (a) influences on materials received (e.g., effects of aspect, slope position, and slope inclination on solar radiation and precipitation); (b) influences on fluxes and transport (e.g., effects of landscape position on water input to lakes); (c) influences on disturbances (e.g., effects of aspect and slope position on wind events); (d) spatial position of landforms (e.g., effects of adjacent landforms on the susceptibility to landslides). *Source:* Swanson et al. (1988) / with permission of Oxford University Press.

This simple four-class summary suggests that physiography directly or indirectly affects all aspects of plant life: regeneration, geographic distribution and abundance, growth and productivity, and death. In the following sections, we examine the influence of physiography at multiple spatial scales on landscape ecosystems and their biota, from the subcontinental to the microtopographic level.

PHYSIOGRAPHIC DIVERSITY, LANDSCAPE ECOSYSTEMS, AND VEGETATION

An interconnectedness of physiography and vegetation is clear throughout the world. Distinctive patterns of ecosystems and their vegetation are strongly related to geomorphology and geomorphic processes, and vegetation in turn influences physiographic processes.

MOUNTAINOUS PHYSIOGRAPHY

The influence of physiography on ecosystems is perhaps most obvious in mountainous terrain. Several examples serve to illustrate the intricacy and critical importance of physiography and mountain landforms.

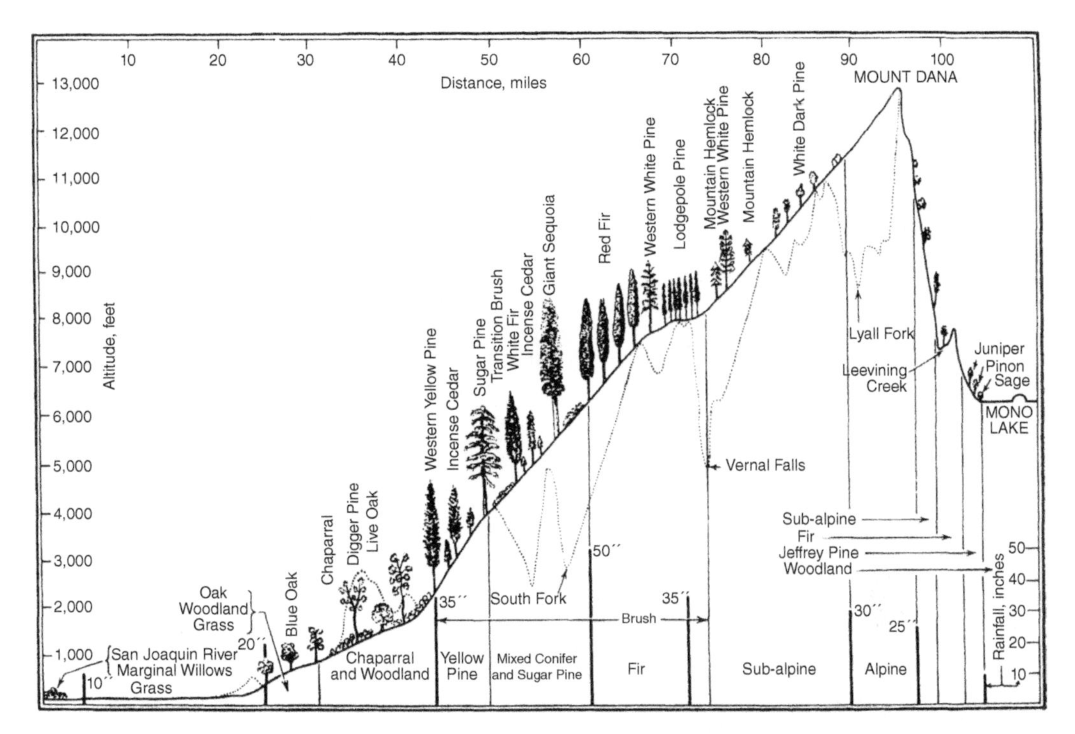

FIGURE 8.5 Physiographic diagram of the Sierra Nevada Mountains illustrating the distribution of dominant plants in relation to elevation and precipitation. Variations in topography along this generalized gradient are shown by the dotted lines. They define large stream valleys and ridges in this rugged terrain. The amount of precipitation along the transect is represented by vertical bars along the bottom of the diagram. *Source:* Harlow et al. (1996) / McGraw-Hill.

Mountainous Terrain of California and the Pacific Northwest A figurative transect across the eastern half of the San Joaquin Valley and over the Sierra Nevada Mountains in California illustrates how the physiographic position of valley and mountain features influence precipitation that, together with temperature, determines the distribution of vegetation from near sea level to over 3650 m (12 000 ft) (Figure 8.5). Conifer-dominated ecosystems appear on western slopes above about 610 m (2000 ft), just above the dry valley and oak woodland. Further inland and higher in elevation is a red fir-dominated forest, with annual precipitation at 125 cm compared to 50 cm in the oak woodland. Conifers maintain their dominance to nearly 3350 m (11 000 ft), as precipitation declines and where low temperatures and thin soils prevail, and the growing season is very short. Precipitation drops sharply on the opposite (eastern) slopes of the Sierra Nevada range within a significant rain shadow, where only the most drought-adapted species can survive.

The Pacific Northwest states of Washington and Oregon provide an excellent example of physiographic diversity, where highly varied landscapes result in an enormous diversity of regional and local ecosystems (Franklin and Dyrness 1980). Broad vegetation types are associated closely with similarly broad physiographic provinces, which in effect describe regional landscape ecosystems (Figure 8.6). Physiographic provinces delineate major landscapes characterized by differences in climate and geology, with substrates contrasting in physical and chemical properties. Variations in precipitation, temperature, and vegetation are associated with gradients in elevation, latitude, aspect, and distance from the ocean (Swanson et al. 1990). Coastal marine terraces and broad inland valleys (Puget Sound area and Willamette Valley) exhibit the lowest relief and support unique vegetation types (Figure 8.6b). At the other extreme are the mountains of the northern Cascade Range and Olympic Mountains of Washington (Figure 8.6a) where relief from ridge crest to adjacent valley floor may be 1000–1400 m, again with the vegetation reflecting these major differences in physiography and climate (Figure 8.6b). The arrangement of landforms and landform adjacency are important in provinces such as the Willamette Valley between the Coast and

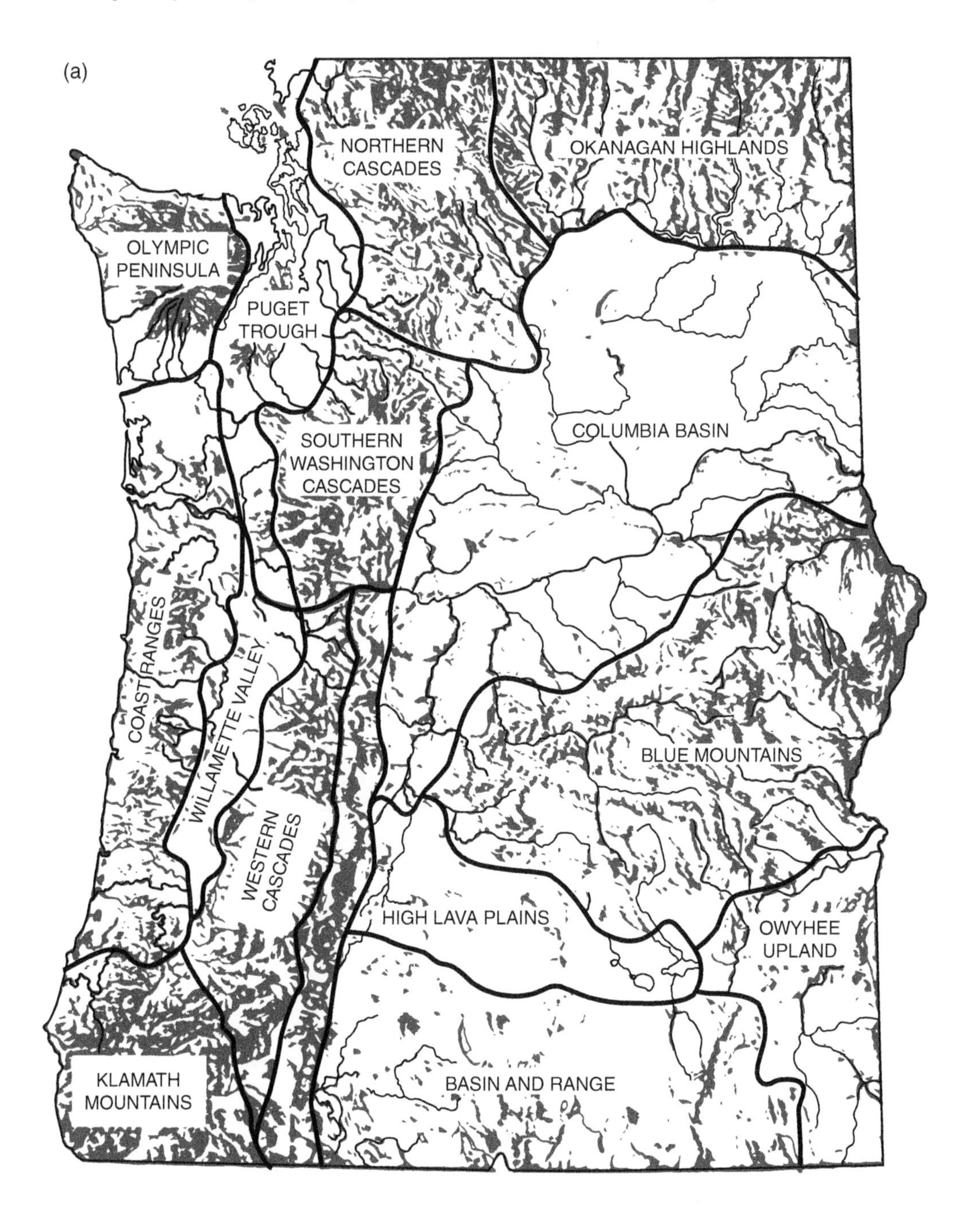

FIGURE 8.6 Comparison of physiographic provinces and generalized vegetation areas of Oregon and Washington. (a) Physiographic–geologic provinces. (b) Major vegetation areas. *Source:* Franklin and Dyrness (1980) / Oregon State University Press.

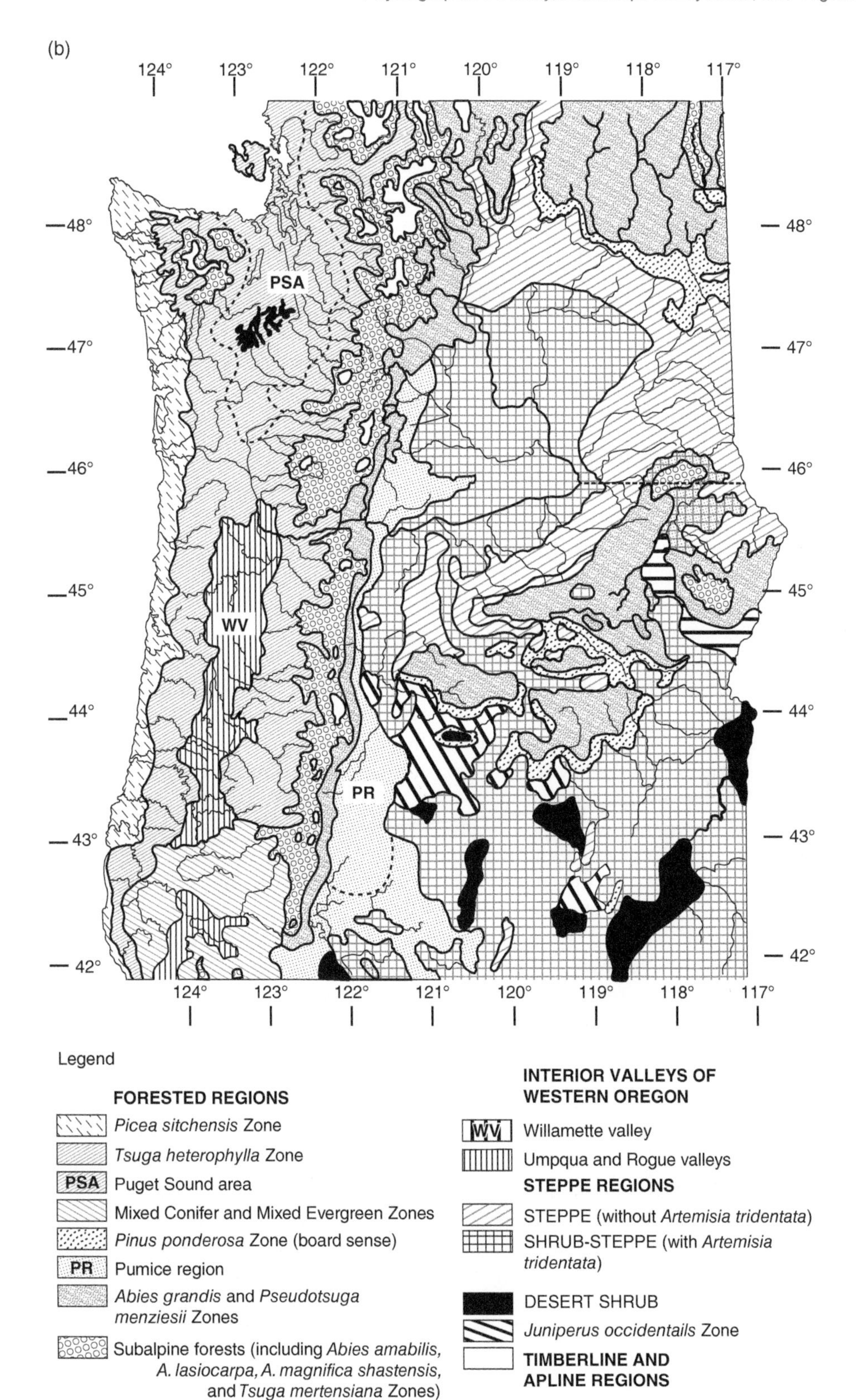

FIGURE 8.6 (Continued)

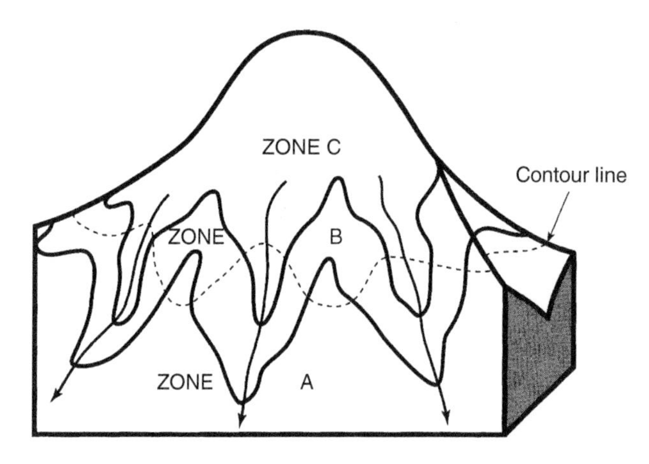

FIGURE 8.7 Interfingering of vegetation zones in mountainous topography. Streams are shown running downslope in valleys. A given vegetation zone is distributed higher in elevation on ridges than in valleys. *Source:* Franklin and Dyrness (1980) / Oregon State University Press.

Cascade ranges or the Columbia Basin in the rain shadow of the Cascade Mountains. In addition, the parent material of Pacific Northwest landforms is important in determining the great diversity of soils (Franklin and Dyrness 1980).

The broad zones of natural vegetation correspond closely to broad physiographic provinces in the Pacific Northwest, but the effect of landform on vegetation is also obvious at local scales. Vegetation zones may occur as sequential belts along an elevational gradient on mountain slopes, but more often their lower elevation limit occurs lower in valleys than on adjacent slopes (Figure 8.7). This "interfingering" occurs because the site conditions in valleys differ in temperature and soil water compared to adjacent slopes of similar elevation. Vegetation zones are relatively broad so that a given species, Douglas-fir for example, may occur in moist sites in the lower ponderosa pine zone and on relatively dry ridges higher up in the montane grand fir zone.

Franklin and Dyrness (1980) showed that forest vegetation zones in the Pacific Northwest may be differentiated using precipitation (drought stress) and temperature (temperature growth index), both of which are strongly influenced by physiography (Figure 8.8). Oak–juniper woodlands and ponderosa pine zones form a group of forests limited by moisture that occurs in the rain shadows of eastern Washington and Oregon and in southwestern Oregon. The mixed conifer and coastal temperate groups form extensive temperate forests. Within this group are the unique temperate conifer forests of the world (zones: 1, *Tsuga heterophylla* (western hemlock); 11, *Abies grandis* (grand fir); 12, mixed evergreen; including Sitka spruce, Douglas-fir, and others) (Franklin et al. 1981). These unique forests grow under highly favorable temperature and moisture conditions reaching extremely large size and old age. Finally, subalpine and boreal groups are severely constrained by temperatures, reflected in short growing seasons and harsh mountain environments.

Physiography and Forests of the Central Appalachians Forest species and communities are closely associated with local geologic and topographic conditions throughout the different sections of the Appalachian Mountains (southern, central, northern, Blue Ridge, Ridge and Valley, etc.; Hack and Goodlett 1960; Leak 1982). Hack and Goodlett (1960) studied a 142 km^2 (55 mi^2) densely forested area in the drainage of the Little River of northern Virginia, located in the Ridge and Valley section of the central Appalachians. Local distribution of species and forest communities between about 600 and 1300 m in elevation coincides with well-defined differences in landform and disposition of water. Dry pine forest, dry–mesic oak forest, and mesic "northern hardwood" forests are found in characteristic positions in first-order valleys (Figure 8.9). Pine and pine–oak forests, dominated by pitch pine, table mountain pine, and several oak species, are found on noses, ridges, and other slopes that are convex away from the mountain such as those shown in

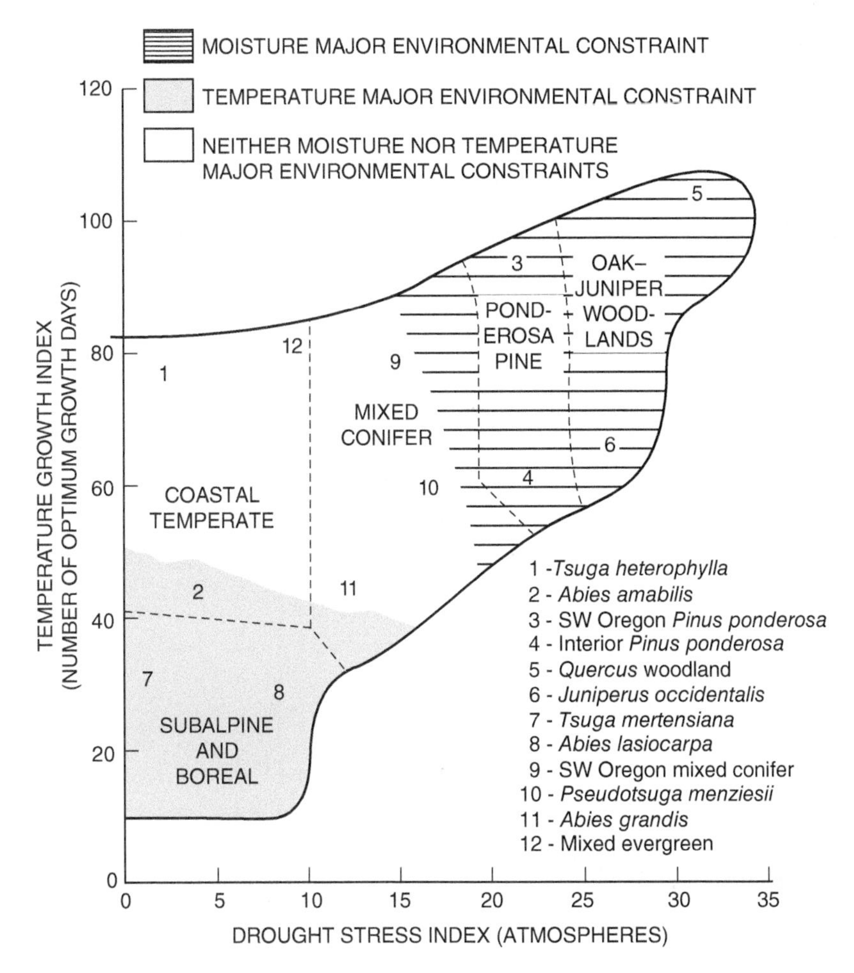

FIGURE 8.8 Distribution of some major forest zones of Oregon and Washington within an environmental field based on moisture (maximum plant moisture stress during the dry season) and temperature (optimum growth days). *Source:* Franklin and Dyrness (1980) / Oregon State University Press.

Figure 8.9a. Dry–mesic oak communities—variable in composition but dominated by oaks—are found on the straight slopes between nose and valley or in drier hollows. Mesic northern hardwood communities of sugar maple, basswood, yellow and black birches, and northern red oak are restricted to moist hollows and other concave slopes that concentrate water. These mesic communities occupy the floodplains of larger valleys, extend up the floors of smaller valleys, and in the first-order valleys extend into the hollows at the valley heads. In Valley 3 (Figure 8.9b), dry pine forests are largely found near the top of the valley on its west-facing slope, where temperatures are higher during the growing season and soil moisture is reduced. Mesic northern hardwood forest, by contrast, is found on the east-facing slope and in the hollow of the valley. Trees of the northern hardwood type in the hollow of north-facing Valley 3 are the tallest, largest, and have the greatest volume compared with those of the other two forest types.

The association between the local distribution of species and communities and their position in a valley is strongly determined by physiographic characteristics. Important characteristics include: (i) valley size, because a large area is essential for the northern hardwood community to persist in a hollow; (ii) orientation of the valleys and side slopes, shown by north and east slopes being commonly forested with northern hardwoods, whereas south and west slopes are typically

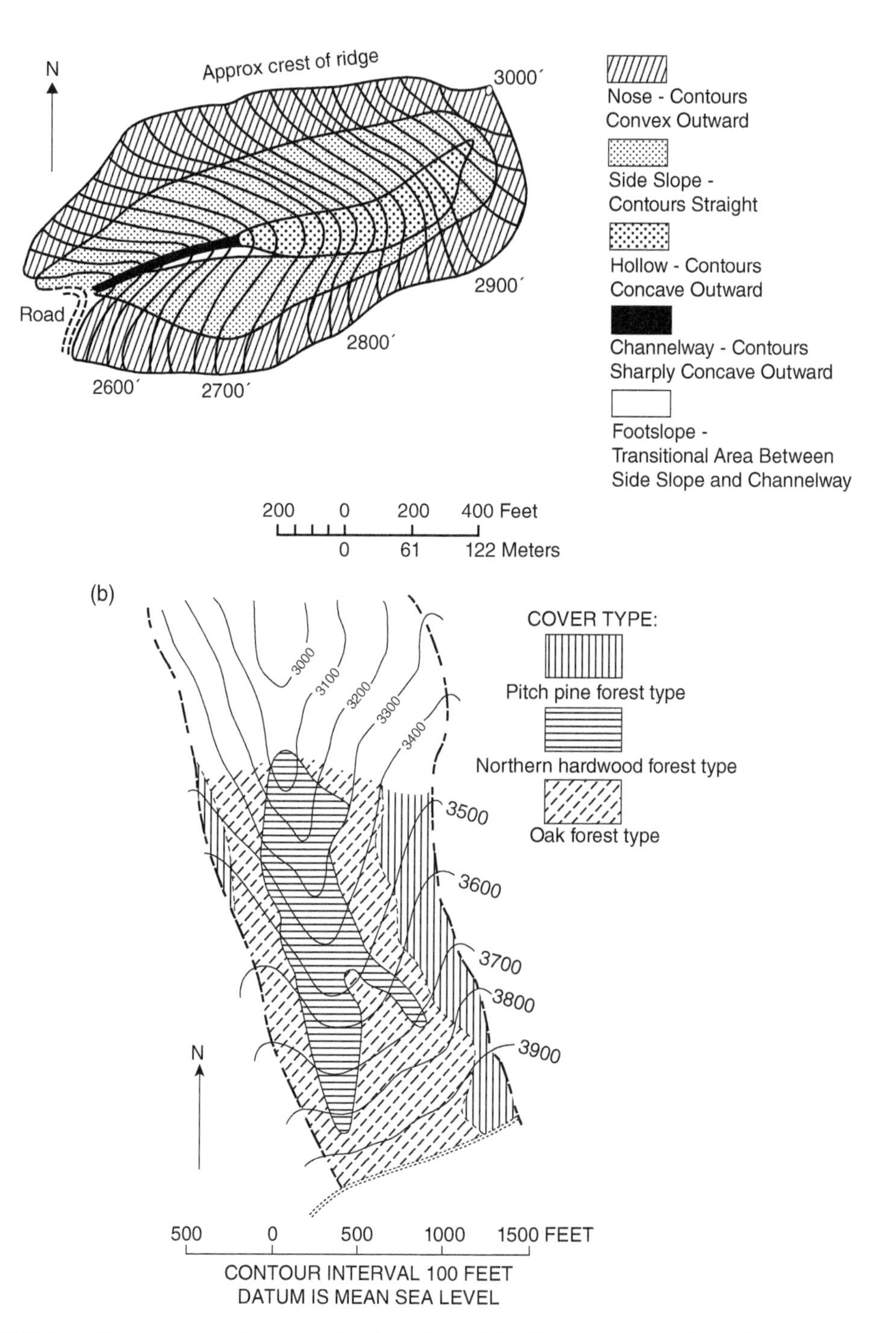

FIGURE 8.9 Contour maps of typical first-order valleys, Little River Basin, central Appalachian Mountains, northwestern Virginia. (a) Map of landforms of upper part of Valley 1, west side of Crawford Mountain. Legend gives landform types of first-order valleys; vegetation distribution is not shown. (b) Map of forest type distribution for Valley 3, northeast of Reddish Knob. *Source:* Hack and Goodlett (1960) / United States Government Printing Office/ Public Domian.

oak; and (iii) nature of the bedrock, because formations favoring soil-water accumulation and retention are more favorable for the more mesophytic species. Northern hardwoods occupy the hollow of north-facing Valley 3 (Figure 8.9b), but oak forest is found in the hollow as well as the side slopes in Valley 1 because it faces southwest (Figure 8.9a), leading to drier soils. Trees of the oak type in the hollow of Valley 1 are larger and taller than those of the same species on the side slope due to better soil-water and nutrient conditions, and ground-cover species differ between hollow and side slopes. These patterns suggest three different landscape ecosystem types (nose; side slope; and hollow, footslope, and channel) and two forest cover types (oak forest and northern hardwood forest) in Valley 1 (see Chapter 1).

Detailed studies of vegetation in five additional first-order valleys showed that physiography as controlled by geomorphic processes has predictable effects on vegetation, although species composition varied among valleys and among hollows (Hack and Goodlett 1960). Three valleys supported northern hardwood forest where topographic features provided a relatively moist environment. The three valleys faced favorable aspects (either northeast or northwest), were relatively large and deeply cut into the mountain, and had parent material favoring water retention. The topographic features of these valleys are controlled by geomorphic processes, specifically debris avalanches and floods, caused by runoff from cloudbursts, which produce slopes that are concave-upward, thus favoring mesic northern hardwood species. By contrast, the two valleys supporting oak forest were physiographically similar to Valley 1 (Figure 8.9a), providing a drier environment. The topography of the drier valleys resulted from soil creep, which produces convex-upward slopes that typically support drier species. Overall, the present landscape is the product of a long period of interaction between vegetative and geomorphic processes.

The central and southern Appalachians are some of the regions where the relationship of forest composition and physiography is best studied. Technological advances over the last two decades have allowed ecologists to quantitatively integrate the multiple components of physiography to interpret their ecological effects on forest composition and productivity. For example, Iverson et al. (1997) used DEMs and soil series maps to predict forest composition and productivity (site index; Chapter 11) in southeastern Ohio. Landscape features derived from DEMs and soil series maps included slope steepness and aspect, curvature of the landscape, soil-water-holding capacity, and downslope water flow. These factors were used to build a single **integrated moisture index** (Figure 8.10) that could be easily related to various ecological processes across the landscape. The resulting index was used to successfully predict, for example, that oak was most productive in the valley bottoms and least productive on upper south-facing slopes and ridgetops. This index derived from physiographic data has also been shown to be closely correlated to understory vegetation, species richness, litter depth, soil pH, and bird species distribution (Iverson et al. 1997). Similarly, Bolstad et al. (1998) predicted forest composition using elevation and terrain shape (cove, ridge, or side slope) extracted from DEMs for the southern Appalachians. Species characteristic of cove ecosystems were positively related to concave locations on the landscape, while dry oak and pine species were negatively related. Northern hardwood species were more closely related to elevation than to terrain shape. Methods such as these demonstrate the value of incorporating physiography into our understanding of forest vegetation composition and distribution and have since been embraced by many ecological modelers and geographers.

FLATLANDS

Flat and gently sloping terrain is clearly less topographically diverse compared to mountainous terrain. Nevertheless, subtle differences in topography and variable parent material result in distinctive effects on the distribution and composition of vegetation. Such subtle variations in physiography may affect the spread of fire, the rate of nutrient cycling, and forest composition, among other processes.

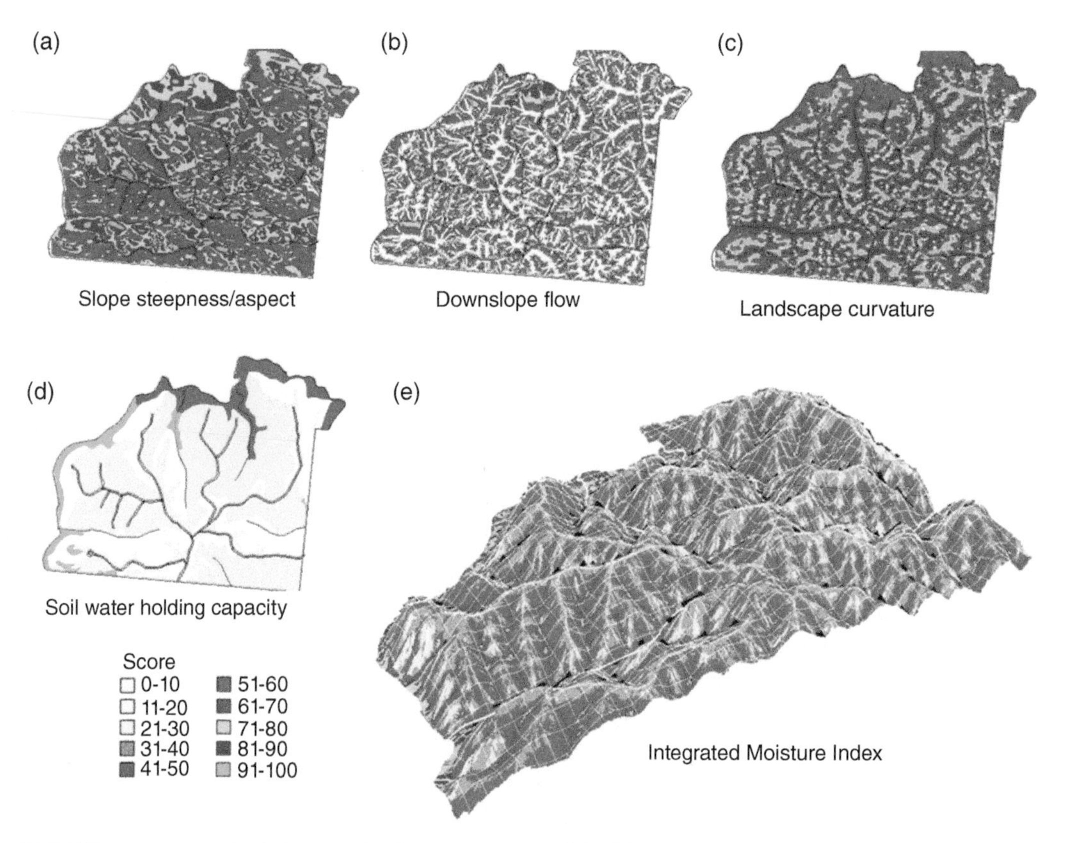

FIGURE 8.10 Maps depicting standardized scores (0–100) for (a) slope steepness and aspect, (b) downslope water flow, (c) landscape curvature, and (d) soil-water-holding capacity for the Vinton Furnace Experimental Forest in southeastern Ohio. These four physiographic factors were combined to create (e) an integrated soil moisture index for the landscape. *Source:* Maps (a)–(d) after Iverson et al. (1997); map (e) courtesy of Louis Iverson.

The Great Plains Grasslands dominated the flat or rolling plains and gentle slopes of the North American Great Plains at the time of European colonization (Wells 1970), and forests are not often considered as having broad ecological importance in the region. It is the relative *lack* of forests across this landscape, however, that is instructive about the dominant process driving the structure and function of the ecosystems of the Great Plains: fire. Vast, frequent, wind-driven prairie fires would readily destroy tree seedlings and saplings, particularly those of conifers that lack the ability to sprout, from areas where fires were too frequent. Forests of the plains therefore existed where they were sheltered from fires, such as along rivers, in rough, dissected, boulder-strewn escarpments of the uplands, and on relatively level sites on the leeward sides of lakes and rivers. As is true with more complex distributions of species on forested landscapes, physiography facilitates this treeless landscape through its influence on climate and fire (Wells 1970). Across the entire width of the Central Plains, rougher and more dissected topography favors the occurrence and spread of forest at the expense of grassland (Wells 1965).

Regional physiography drives the dry climate of the Great Plains (continental climate and their location east of a major mountain range), but other physiographic features also make fire common in this region. The landscape is vast and unbroken in its flat or rolling smoothness, and these features, as well as dry soils, support flammable, grassy vegetation. It is easy to imagine that grass fires swept by wind spreading across a flat or rolling plain, whether ignited by lightning or humans, would continue indefinitely until quenched by rain or stopped by an abrupt break in topography. Trees therefore persisted in escarpments with sparse grasses, perhaps temporarily encroaching on grassland during short fire-free periods. The mosaic of ecosystems of the Great Plains is therefore due to the regional flatness and continuity of the physiography as much as it is to climate. The interrelationships of climate, physiography, and the production of fuels from grassy vegetation create the "perfect storm" for frequent fire on this landscape.

Pine Savannas of the Western Great Lakes Region Fire is also favored by the flat and gently rolling physiography of the western Great Lakes region and adjacent Canada. In combination with hot spring–summer periods and droughty, sand soils, physiography favors frequent fires (though less frequent than in the Great Plains) and perpetuates fire-dependent species and communities. In parts of Minnesota, northern Wisconsin and Michigan, and southern Ontario, flat outwash or rolling plains, dry sand soils, and flammable vegetation led to frequent fires that maintained communities of nearly pure jack pine or jack pine and oak, often interspersed with large openings or areas of sparse and scattered trees (Curtis 1959; Vogl 1964, 1970; McAndrews 1966). Similar to the Great Plains without major topographic breaks, lightning or human-ignited fires are likely to have swept uninterrupted across the landscape until reaching a major water body or burning onto physiography supporting less flammable vegetation.

These "jack pine barrens" are unique to the upper Great Lakes region (Heinselman 1981) and occupied over 400 000 ha in Michigan (Voss 1972, p. 62) and over a million hectares in Wisconsin (Curtis 1959). The account of a reporter accompanying an expedition by botanists William J. Beal and Liberty Hyde Bailey across the northern Lower Peninsula of Michigan in June 1888 carried the title: A BARREN WASTE. MOSQUITOES THE LARGEST AND MOST PROMISING PRODUCT YET FOUND (Voss and Crow 1976). The landscape is vividly described:

> *These plains are clothed with a scant vegetation, the most conspicuous and common characteristic plant being the jack, or scrub pine . . .. Fires often sweep over the plains, destroying all vegetation. Young pines soon spring up in these burnt areas . . .. These groves of young trees are often miles in extent, and are so dense that the traveler is completely hidden from view at the distance of a few paces . . .. The sward-like glades are found to be clothed with stunted huckleberries or blueberries, miserable growths of sweet fern, and a few other pinched and starved plants which can endure the heat and dryness of the sands . . .. On the whole the plains are exceedingly uninviting in aspect.*

Physiography is particularly important in this landscape because subtle topographic differences are related to local soil differences, and both have large effects on vegetation composition and growth. For example, physiographic differences perceived to be minimal have strong effects on jack pine height growth, which in turn affect the duration of habitat suitability for the threatened (formerly endangered) Kirtland's warbler, which breeds nearly exclusively on this unique landscape (Kashian et al. 2003; Walker et al. 2003).

Till Plains of the Midwest The extensive till plains of Indiana and Ohio are relatively flat or gently rolling and crossed only by a number of low moraines, ridges, and shallow drainages. They form the central core of the classic beech–maple forest region described by Braun (1950; Figure 2.6, Chapter 2). The topography is extremely flat, but the parent material is heavily textured with soils of moderate to poor drainage. The soils favor the dominance of American beech and sugar maple in this region to the virtual exclusion of other tree species and the absence of fire. Leaf litter is moist and rapidly decomposing, leaving insufficient fuel to carry a fire. Thus, the till plains are an example of physiography where the substance (parent material) is more important than the form in its direct and indirect control of dominant species and exclusion of fire.

Southeastern and Southern Coastal Plain A relatively flat plain supporting a mosaic of landforms stretches along the southeastern and southern coast of the United States, from New Jersey to Texas. Landforms include flat plains and terraces, embayed rivers, and sandy ridges and swales near the Atlantic Coast giving way to rolling hills adjacent to the Piedmont Plateau. Over short distances and very low topographic relief on the Coastal Plain, vegetation may vary from grassland and savanna to shrubland, to needle- and broad-leaved sclerophyllous woodland, and to rich mesophytic forest (Christensen 1988a; Myers and Ewel 1990). The region also supports the most diverse assemblage of freshwater wetlands in North America (Ewel 1990). These stark vegetative differences result in part from the effects of physiography on water table, drainage patterns, and hydroperiod (length of time soils are saturated during a year); soil water and nutrient cycling; and fire frequency and severity.

Fire is an important disturbance throughout much of the Coastal Plain (Chapter 16), and the occurrence of rare local ecosystems there results from favorable physiographic and soil conditions together with their relative protection from fire based on their position in the landscape. For example, the positions of four distinctive ecosystem types in western Louisiana were reconstructed using the General Land Office Surveys of 1821 (Delcourt and Delcourt 1974; Figure 8.11). The mesic magnolia–holly–beech upland hardwood community occurred on thick loess deposits of the upland, the magnolia–beech–holly bottomland hardwoods in ravine and river lowlands, and the tupelo-gum–cypress community in swamp land adjacent to the Mississippi River. A mixed oak–pine–beech ecosystem was found where the loess cap thinned out toward the northeast (Figure 8.11). The dominance of southern magnolia, American beech, and their mesophytic associates in the Coastal Plain was largely due to their protection by firebreaks from natural and human-set fires.

Floodplains Large rivers all across the world exhibit similar fluvial (river) processes and patterns of physiography. Accordingly, specific fluvial landforms are identifiable, and complex patterns of ecosystems and microsites occur within them. Rivers and their landforms are linear

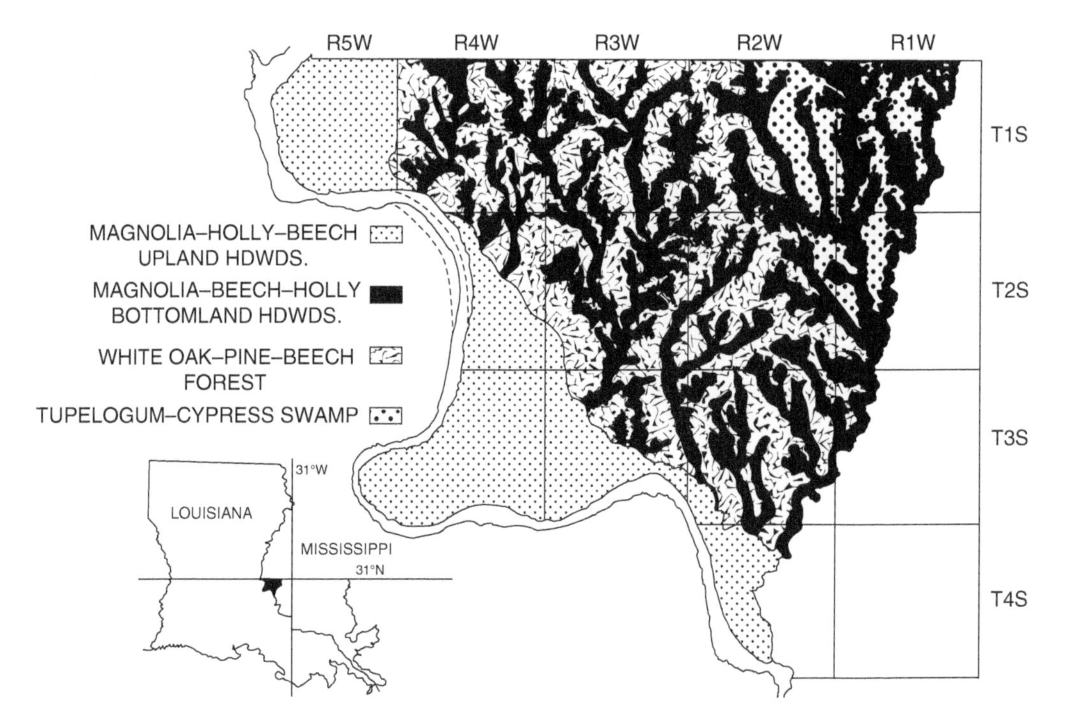

FIGURE 8.11 Pre-European colonization forest communities of West Feliciana Parish, Louisiana, as reconstructed from 1821 General Land Office survey records. The dashed line represents the position of the Mississippi River in 1821. *Source:* Delcourt and Delcourt (1974) / John Wiley & Sons.

features so that a principle of **similar effects** is the result, in that similar ecosystems with characteristic physiography, soils, and riverine vegetation may extend for tens or hundreds of kilometers along a major river (at least before river processes were altered by dams, channels, and human-made levees). Processes of flooding, sediment transport and deposition, and erosion and abrasion by ice and water movement result in a degree of physiographic uniformity on the landscape, whether a lowland river or a mountain stream. Nanson and Croke (1992) identified three basic types of floodplains, including high-energy floodplains with non-cohesive sediments (steep headwaters whose physiography is shaped episodically by extreme flow events); medium-energy floodplains with non-cohesive sediments (featuring regular flow events in broad valleys); and low-energy floodplains with fine-textured, cohesive sediments. Our emphasis in this section is on the third of these floodplain types. Notably, stream floodplains or **riparian** zones are zones of direct interaction (i.e., in flooding, bank cutting, and sedimentation) between terrestrial and aquatic ecosystems (Gregory et al. 1991). Riparian characteristics and processes have received detailed attention (e.g., Swanson et al. 1982b; Brinson 1990; Gregory et al. 1991; Malanson 1993; Ward et al. 2002).

The pattern of zonation of landforms, ecosystems, and their vegetation is similar for large, low-gradient, lowland rivers where landforms develop largely from lateral channel migration. In this floodplain type, constant river meandering in the floodplain cuts into outer banks while depositing sediments and forming point bars and new land downstream on inner banks (Figure 8.12). Sedimentation during floods gradually but constantly alters the topography of the floodplain. Coarser sediments are deposited adjacent to the river channel by floodwater, forming

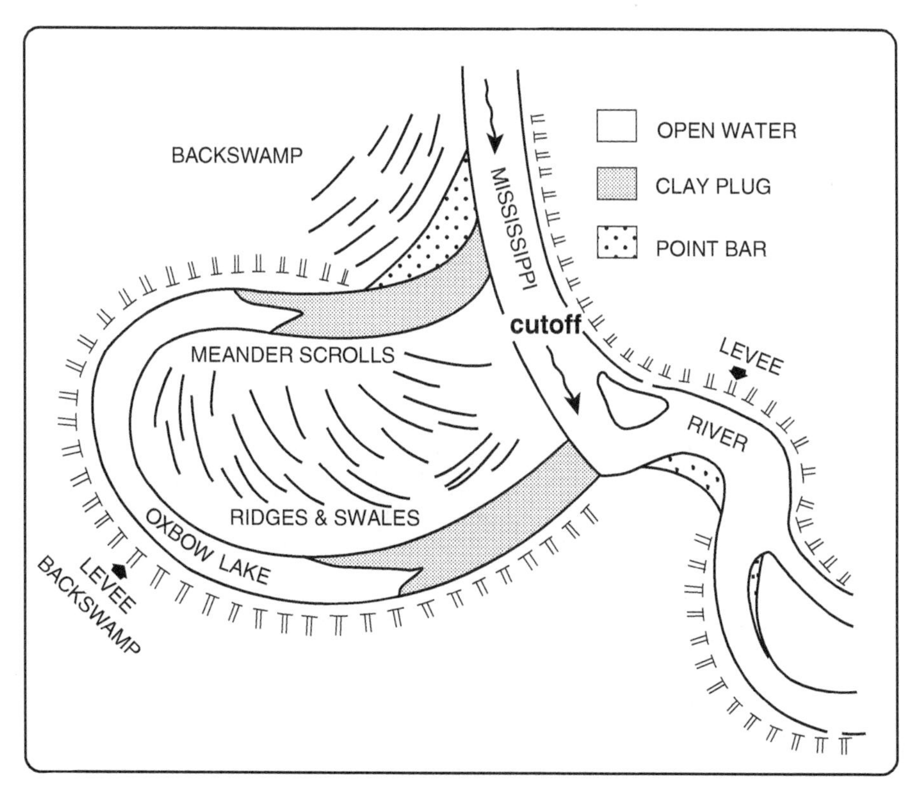

FIGURE 8.12 Typical section of Mississippi River floodplain near False River, Louisiana. Physiographic features include: (1) natural levees adjacent to the channel; (2) ridge-and-swale topography (meander scrolls) associated with lands on and between former point bar/levee deposits as the channel migrated laterally and downslope; (3) oxbow lake where former channel has been cut off; (4) backswamps of lower topography behind the levees, and (5) point bars on inside curve of channel where deposition is rapid. *Source:* Brinson (1990) / with permission of Elsevier.

well-drained ridges or natural levees. "Gallery" or riverside forests are typical of this landform where high light availability along the river enables trees to lean over the bank and develop large spreading crowns (Loehle 1986). However, such riverside species must tolerate periodic inundation during periods of flooding as well as drought when water levels recede, and the coarse sediments drain rapidly. Silts and clays are deposited beyond the levee in a poorly drained zone or backswamp of the first bottom. A series of bottoms or terraces typically lie beyond the first bottom and are progressively drier and less frequently flooded. Rivers further alter the physiography of their floodplains as entire channels may be abandoned, forming oxbow lakes that eventually fill in with sediment and vegetation. These depressions and lakes or sloughs support plant species best adapted to inundation and anaerobic soils. Many floodplains exhibit ridge-and-swale topography where channels, point bars, levees, and backswamps were cut off and abandoned by the meandering stream (Figure 8.12), creating an enormous number of landforms inhabited by plant species with different tolerances to hydroperiods and substrates.

The pattern of landform zonation and associated species occurrence is found across river bottomlands of the southeastern United States (Figure 8.13). Floodplains are variable in their number and kinds of landforms and the names associated with these features, but many landforms of riverine ecosystems are quite typical (Wharton et al. 1982): (i) the river channel margin where deposition builds a raised point bar that becomes new land and then the "front," (ii) the ridge or natural levee where coarse sediments are deposited, and (iii) a series of bottoms or terraces behind the levee. Local relief of ridges, swales (low areas between ridges), and oxbows or sloughs are found across the first bottom, although ridges and swales may differ in elevation by as little as 2 m. Finer sediments are deposited across the first bottom, forming low, broad areas of poorly drained,

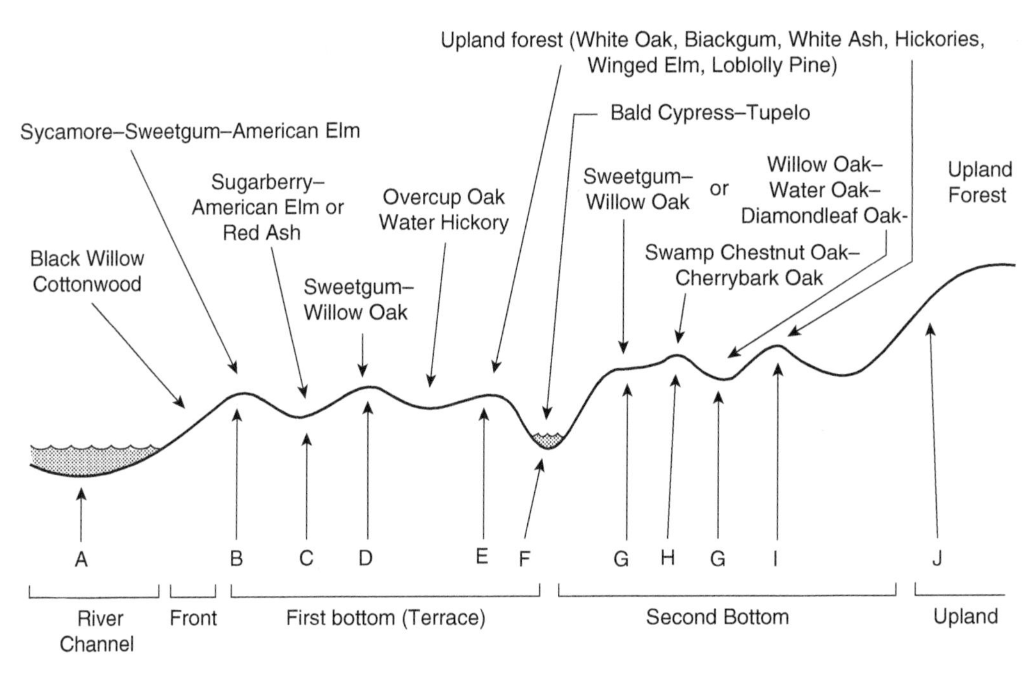

FIGURE 8.13 Correspondence between alluvial floodplain landforms and forest types for rivers of the southeastern US. A, river channel; B, natural levee, C, backswamp or first terrace flat; D, low first-terrace ridge; E, high first-terrace ridge; F, oxbow; G, second-terrace flats; H, low second-terrace ridge; I, high second-terrace ridge; J, upland. Vertical scale is exaggerated. *Source:* Wharton et al. (1982) / U.S. Department of the Interior / Public Domain.

slack-water clay or silty clay soils. Higher land of the second bottom is flooded less frequently and for a shorter time, and thus supports a different set of species (Wharton et al. 1982).

Variation in the types of landforms on floodplains may arise from geographic location, size, and velocity of the stream, kinds of materials transported, and the physiography and parent material of adjacent terrain (Brinson 1990). In glaciated landscapes, for example, different fluvial landforms may result when floodplains are located on different physiographic systems (outwash plain, moraine, etc.). In a study of landscape ecosystems in riverine systems of northern Lower Michigan (Baker and Barnes 1998), floodplains found on sandy outwash plains exhibited straighter and shallower river channel segments and broad, continuous first bottoms (Figure 8.14). Where outwash soils were deposited over moraines, floodplains had deeper and more sharply curving river segments, narrow and discontinuous first bottoms, larger second bottoms, and generally a more diverse set of fluvial landforms. These differences were attributed to differences in parent material related to the physiographic system: sandier material characteristic of outwash plains is more easily erodible and allows more lateral migration of the channel, but finer-textured glacial till

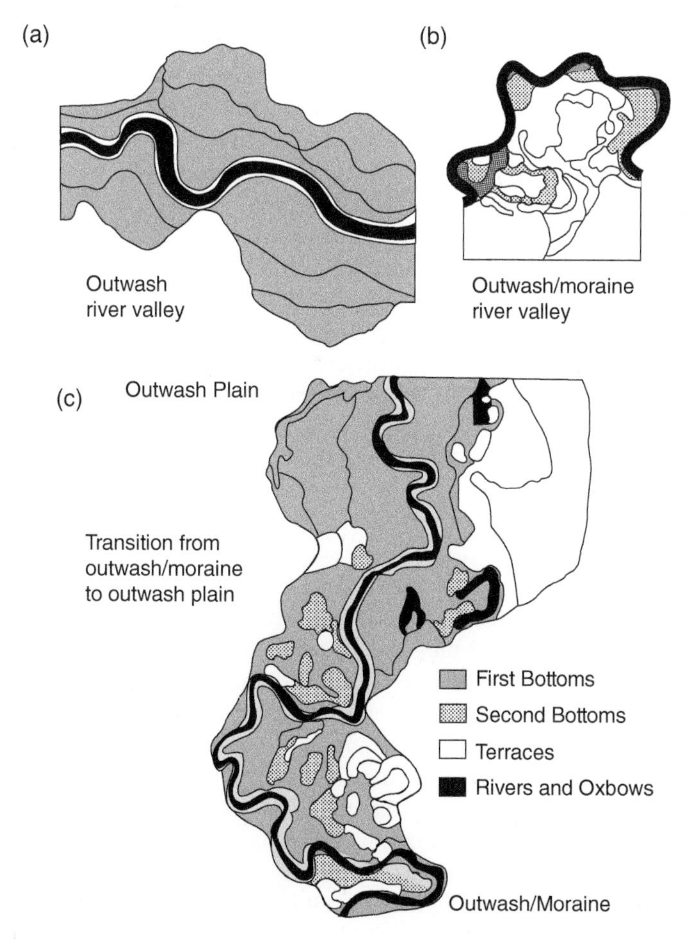

FIGURE 8.14 Typical sections of floodplain in outwash plains and outwash over moraine physiography along the Pere Marquette River, northern Lower Michigan. (a) River valleys in outwash plains have broad, continuous first-bottom floodplains with shallow, straighter river channels. (b) Valleys in outwash over moraine physiography have narrow and broken first bottoms and larger second bottoms with deeper, winding river channels. (c) These differences are evident at transitions across the landscape between the two physiographic systems. *Source:* Baker and Barnes (1998) / with permission of Canadian Science Publishing.

parent material of the outwash over moraines causes deepening rather than lateral movement of channels. Thus, broad physiography (outwash plain, moraine) affects the development of fluvial landforms, which in turn affects the distribution and diversity of landscape ecosystems, which in turn affects species occurrence and community composition (Baker and Barnes 1998).

The physiography of floodplains changes rapidly (temporal scales of years or decades) relative to uplands where the geomorphic processes of sediment erosion, transport, and deposition are far slower (temporal scales of centuries or millenia). The fact that fluvial landforms are vulnerable to changes over short time periods suggests that vegetation may have an important influence on the distribution and shape of fluvial landforms in floodplains. Indeed, a series of recent studies of European rivers suggests that vegetation interacts with water and sediment to affect the river channel and the landforms on its floodplain (Corenblit et al. 2007, 2009, 2016; Gurnell et al. 2012, 2016; Wohl 2013). In particular, aquatic vegetation, riparian trees, and dead wood in the river channel are effective trappers and stabilizers of sediment, leading to channel narrowing and bending (Gurnell and Grabowski 2016) and the formation and alteration of fluvial landforms nearest the channel, such as fronts, levees, and potentially first bottoms and backswamps. Thus, the dynamic nature of physiography within floodplains provides for a positive feedback between fluvial landforms and floodplain vegetation that likely varies among landscape ecosystems.

As described in the earlier text, species adapted to floodplains must tolerate a range of hydrological conditions depending on the landform where they grow. The physiological tolerances of species to flooding related to topographic positions and effects of flooding have been described in detail (Gill 1970; Hook 1984). Notably, diverse swamp ecosystems are associated with floodplains throughout eastern North America, especially on the Coastal Plain (Christensen 1988a; Ewel 1990). A great many studies emphasize floodplain landforms and vegetation for rivers from all regions of the United States including northern Alaska (Bliss and Cantlon 1957), Oregon (Hawk and Zobel 1974), North Dakota (Johnson et al. 1976), Texas (Chambless and Nixon 1975), New Jersey (Wistendahl 1958; Frye and Quinn 1979), Virginia (Osterkamp and Hupp 1984; Hupp and Osterkamp 1985), and Indiana (Lindsey et al. 1961). An excellent review of North American rivers is presented by Brinson (1990), and those of other continents are described in a book on forested wetlands of the world (Goodall et al. 1990). Several reviews have examined the association between vegetation and fluvial landforms (Ward et al. 2002; Lytle and Poff 2004).

PHYSIOGRAPHY AND FIREBREAKS

Firebreaks are places on a landscape that may interrupt the spread of a fire, such as rivers and streams, lakes, or rough topography. Whereas fire spreads rapidly over flat land, significant surface relief may disrupt the continuity of fuels, alter wind movements, retard air flow, and generally slow and stop advancing fires (Grimm 1984; Taylor and Skinner 2003). Topography serves as a natural firebreak most influentially in complex compared to gentle terrain (Heyerdahl et al. 2001; Povak et al. 2018) where physiographic variation is most complex. In the Rocky Mountains of North America, abrupt shifts between different aspects—such as valley bottoms and ridge tops or between north and south aspects—may serve as effective natural firebreaks, as do cooler and moister northern-facing slopes in general (Holsinger et al. 2016; Meigs et al. 2020). Several recent studies have shown that low places on the landscape, such as depressions and valley bottoms, also act as natural firebreaks because they generally have higher soil and fuel moisture than the surrounding landscape (Ouarmim et al. 2014; Krawchuk et al. 2016; Wilkin et al. 2016; Povak et al. 2018; Meigs et al. 2020).

There are many specific examples of how physiography affects fire spread and influences vegetation distribution and composition. As described in the earlier text, escarpments and rivers in the Great Plains acted as natural firebreaks and enabled trees to survive in an otherwise flat landscape prone to sweeping fires. Tall, dense, deciduous forests on prairie soils in southcentral

Minnesota, known as the Big Woods, existed more because of firebreaks than any other factor (Daubenmire 1936; Grimm 1984). Remarkably, these forested ecosystems were dominated by fire-sensitive species such as American elm, basswood, sugar maple, and hop-hornbeam. The location, size, shape, and orientation of physiographic features in the Boundary Waters Canoe Area of northern Minnesota (lakes, streams and wetlands, bedrock ridges, valleys and troughs) are related to historic fire patterns and distribution of vegetation because of how they either favored or inhibited the spread of fires (Heinselman 1973).

The highly shade-intolerant and fire-dependent longleaf pine dominated pre-European colonization forests throughout much of the Coastal Plain (Christensen 1988a; Myers 1990). However, in central Florida, late-successional communities lacking longleaf pine and dominated by evergreen and deciduous angiosperms occurred on fire-protected islands and peninsulas. Because of the pervasive presence of fire on uplands, mesophytic species such as American beech and southern magnolia were relegated to the sites most protected from fire, such as the high bluffs east of the Apalachicola River in the Florida Panhandle. A variety of mesophytic species occurred here because of its protected position and its rich, moist soils, including two rare species that are endemic to this bluff region—Florida yew and Florida torreya.

MICROLANDFORMS AND MICROTOPOGRAPHY

Microtopographic features, at scales less than 10 m^2 or even 1 m^2, also have important effects on ecosystems and specifically influence regeneration of plants, succession, and local soil development. Examples of microtopographic features include microsites created by tree-uprooting events, fallen trees on the forest floor, and the elevated bases of living and dead trees in wetlands. The form and substance of these biotic "landforms" create soil-site heterogeneity and provide favorable microsites for plant establishment and growth over very large, forested landscapes.

TREE UPROOTING AND PIT-AND-MOUND MICROTOPOGRAPHY

Probably the most widespread example of microphysiography and microsites on these landforms is the **pit-and-mound microtopography** created by the uprooting of trees. Trees are consistently felled by wind in forests, uprooted annually by strong storms, and over longer temporal scales by hurricanes or tornadoes. An uprooted tree falls with much of its large roots intact, lifting the soils among the roots and creating a depression or pit at the former position of the main mass of structural roots (Figure 8.15). The upthrown root and soil mass typically forms a mound adjacent to the pit as it decays, forming another kind of microtopographic feature on the forest floor that is itself colonized by woody plants. Pits and mounds usually occur in pairs, and their sizes vary because of soil-site conditions, tree size and rooting habit, and the quantity of soil that slumps or erodes back into the pit (Schaetzl et al. 1989a). Collectively over time, pairs of pits and mounds can form significant microtopography over large areas. For example, in a landscape of sandy, glacial drift in New Brunswick, 36% of the area was occupied by 1255 mounds per hectare and 12% by 1455 pits per hectare (Figure 8.16; Lyford and MacLean 1966). Approximately 10–50% of the forest floor in temperate forests may be covered by pit-and-mound microtopography (Schaetzl et al. 1989a). Microtopographic features usually persist for fewer than 500 years, although Schaetzl and Follmer (1990) documented a range of ages of these features from 10 to 2420 years using radiocarbon dating. A few mounds in hardwood forests in the Upper Peninsula of Michigan have been radiocarbon-dated to be as old as 4000–6000 years (Šamonil et al. 2013).

Tree uprooting and pit-and-mound topography do not occur randomly across a landscape; they are associated with ecosystems subject to severe windstorms, particularly those having parent material and soil that favor shallow rooting. Ecosystems with wet mineral or organic soils (high water tables), rocky soils, or soils developing root-restricting horizons (e.g., clay-heavy soils) more

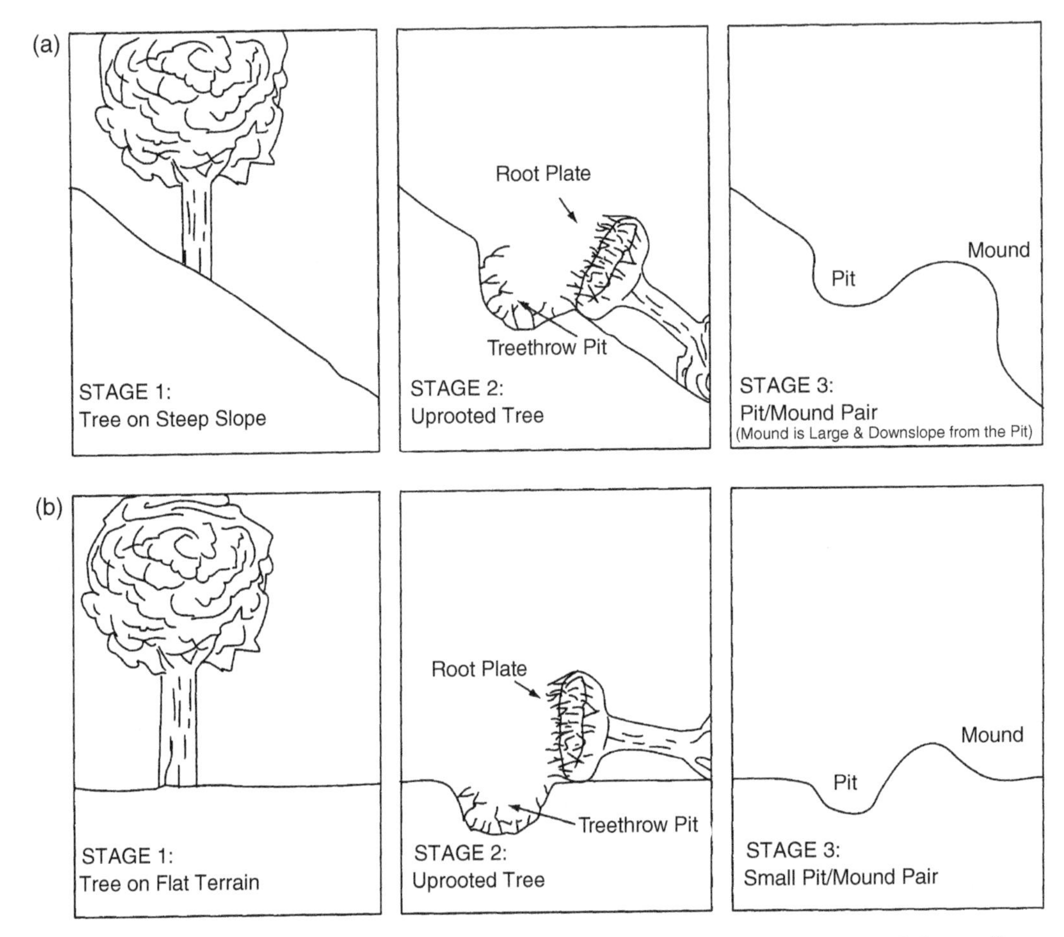

FIGURE 8.15 The uprooting process, showing the formation of a pit-and-mound pair by soil slump off the root plate, and the resulting mixed horizons within the mound. (a) Steep slope. (b) Flat terrain.
Source: Schaetzl and Follmer (1990) / with permission of Elsevier.

commonly feature tree uprooting. On glaciated upland landscapes, tree uprooting is much more common on clay-rich parent material (till) than loamy or sandy outwash parent material. In sloping or mountainous terrain, the abundance of pit-and-mound microlandforms is closely related to landforms positioned low in the landscape (i.e., low, wetland areas with high water tables) and those high in the landscape having steep slopes or exposed to strong winds. Pit-and-mound topography also tends to be more pronounced in old-growth (larger trees) compared to second-growth forests because the size and persistence of pits and mounds are closely related to the size of the uprooted trees (Sobhani et al. 2014; Plotkin et al. 2017).

The ecological importance of pit-and-mound topography is that it creates heterogeneous soil and microclimatic conditions where it occurs. Uprooting of mature trees may disturb an average of 12–16 m^2 of soil to a depth of 1 m or more (Lutz 1940; Peterson et al. 1990), bringing lower soil horizons to the surface and creating significant soil mixing. As sediments slump and erode from the root mass, mineral soil is exposed and often produces irregular or discontinuous horizons within the soil mound (Schaetzl et al. 1990). Soil development is often accelerated beneath pits more than on mounds (Stone 1975; Šamonil et al. 2015). Notably, the falling of trees downslope on mountain slopes results in net downslope transfer of sediment and can trigger mass movement and debris flow (Swanson et al. 1982a; Schaetzl et al. 1990).

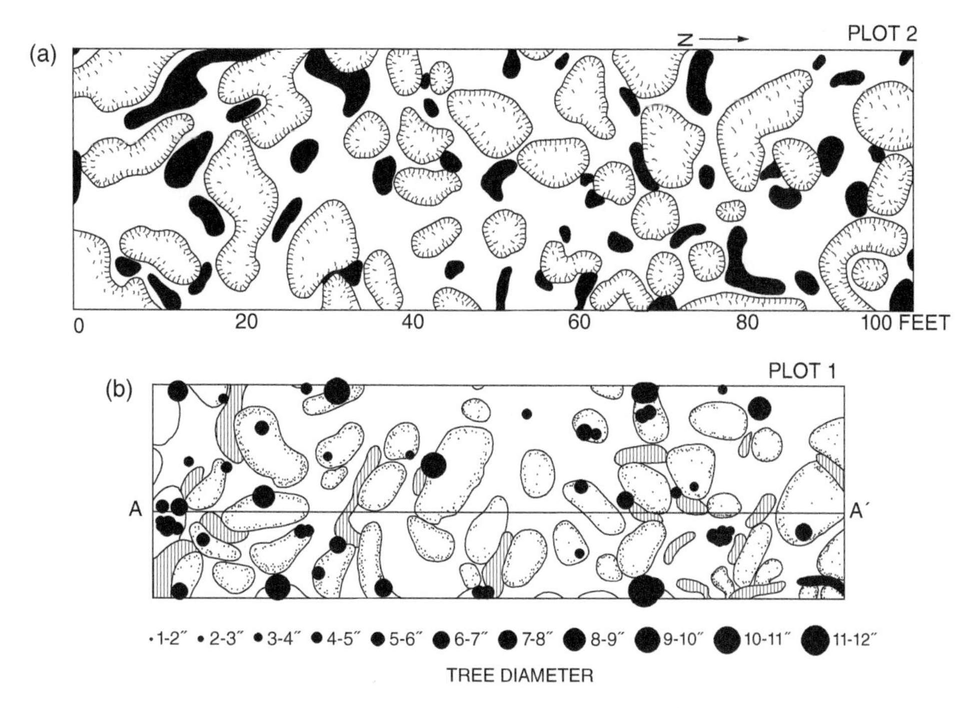

FIGURE 8.16 Pit-and-mound microtopography. (a) Map of the spatial distribution of pits and mounds in a 30 × 100-ft plot in red spruce–balsam fir forest near Fredericton, New Brunswick, Canada. Dark areas are pits; stippled areas are mounds. (b) Location and diameter size class of live trees (black circles) in relation to pits (areas with dark vertical lines) and mounds (lightly stippled areas). *Source:* Lyford and MacLean (1966).

Microsite heterogeneity in pits and on mounds provides a variety of environments for colonization by woody species, and establishment of trees is almost exclusively on the lower slopes of mounds where soil water and nutrients are most favorable (Figure 8.16). Pit centers may be colonized soon after tree fall (Peterson, et al. 1990), but they are a poor location for tree establishment and development because of wet conditions and deep accumulations of leaf litter (Schaetzl et al. 1989b). Mounds tend to be more exposed to wind and are cleared of litter, lack competing roots, and provide mineral-soil seedbeds ideal for species with tiny, wind-dispersed seeds (e.g., hemlock and birch). In addition to favorable soil conditions, the canopy gap created by the uprooting process provides sufficient light for establishment and growth of even shade-intolerant species, if only temporarily. Thus, the uprooting process and pit-and-mound microlandforms not only influence community composition of different ecosystems but also affect microscale ecosystem changes in light, topography, and soil conditions that, in turn, mediate vegetation change (Clinton and Baker 2000; see Chapter 17). Reviews of the extensive literature on uprooting and microtopography are provided by Schaetzl et al. (1989a,b, 1990) and Schaetzl and Follmer (1990).

MICROTOPOGRAPHY AND REGENERATION IN HARDWOOD SWAMPS

Microtopography and its effects in forest ecosystems are also associated with fallen and rotting tree trunks and other coarse woody debris on the forest floor. These organic sources, often called **nurse logs**, provide moist and competition-free establishment sites for seedlings of certain species. Nurse logs in swamps of the southeastern Coastal Plain often bear distinctive vegetation (Schlesinger 1978; Huenneke and Sharitz 1986). Swamps are waterlogged, periodically flooded systems that represent a harsh growing environment for plants, and many other kinds of microtopography provide favorable sites for regeneration. Titus (1990) identified 19 types of microsites in 4 different categories in a riverine swamp in Florida: (i) frequently submerged soil sites such as swamp bottoms, (ii) raised-soil sites (elevated soil, soil near a stump, cypress knee, or shrub base, etc.), (iii) deadwood sites (fallen log or branch), and (iv) live-wood sites (trunk, root, cypress knee, palm tree base). Distribution patterns of the seedlings of 25 woody species corresponded closely to flooding duration and microsite substrate type, with the highest seedling densities on raised-soil sites. The role of microtopography is enormous in wetland sites around the world where standing water and low oxygen adversely affect seed germination, seedling establishment, and growth of woody plants (Diamond et al. 2021).

SUGGESTED READINGS

Gerhardt, F. and Foster, D.R. (2002). Physiographical and historical effects on forest vegetation in Central New England, USA. *J. Biogeogr.* 29: 1421–1437.

Grimm, R.C. (1984). Fire and other factors controlling the Big Woods vegetation of Minnesota in the mid-nineteenth century. *Ecol. Monogr.* 54: 291–311.

Hack, J. T. and Goodlett, J. C. (1960). Geomorphology and forest ecology of a mountain region in the central Appalachians. *Geol. Survey Prof. Paper* 347. 66 pp + map.

Heinselman, M.L. (1981). Fire and succession in the conifer forests of northern North America. In: *Forest Succession, Concepts and Application* (ed. E.C. West, H.H. Shugart and D.B. Botkin), 374–405. New York: Springer-Verlag.

Huggett, R.J. and Cheesman, J. (2002). *Topography and the Environment.* Pearson Education 274 pp.

Nanson, G.C. and Croke, J.C. (1992). A genetic classification of floodplains. *Geomorphology* 4: 459–486.

Rowe, J.S. (1988). Landscape ecology: the study of terrain ecosystems. In: *Symp. Proc. Landscape Ecology and Management.* (ed. M.R. Moss). Montreal: Polysci. Publ. Inc.

Swanson, F.J., Kratz, T.K., Caine, N., and Woodmansee, R.G. (1988). Landform effects on ecological processes and features. *Bioscience* 38: 92–98.

Soil

CHAPTER 9

Forest trees, like all other terrestrial plants, require five primary resources for growth and development: solar radiation, carbon dioxide (CO_2), water (H_2O), nutrients, and a porous medium for physical support. Although plants obtain energy from the sun and CO_2 from the atmosphere, the remaining resources are provided by soil. Consequently, soil forms the "foundation" of forest ecosystems in more ways than one. As you will see, soil is critical to the cycling of nutrients (Chapter 19), a process that influences the growth of individual trees and the functioning of entire ecosystems. In this chapter, we provide an overview of the physical, chemical, and biological properties of soil that regulate the availability of soil resources to plants, particularly forest trees. We begin with a discussion of the soil-forming process and explain how climate, geology, and biota influence the distribution of forest soils.

There are many definitions of soil, as it pertains to the growth of terrestrial plants. For our purpose, we define soil as a porous medium consisting of minerals, organic matter, water, and gases. The combined influences of climate, topography, biota, and time differentiate geologic materials into soil. As such, soils are as diverse as the climates in which they occur, the landforms on which they develop, and the plants that grow upon and within them. One would expect soils supporting tropical rain forests to differ markedly in their physical, chemical, and biological properties from those beneath forests in temperate or boreal climates for the reasons we have briefly outlined in the text above. In the pages that follow, we review the processes that give rise to such differences and focus on how they influence plant growth and ecosystem function.

PARENT MATERIAL

The Earth's surface is blanketed by a wide array of geologic materials, differing in their chemical composition and degree of consolidation. The relatively unweathered geologic material from which a particular soil has developed is called **parent material**. It constitutes the basic substrate for soil formation and exerts a substantial influence on the physical, chemical, as well as biological characteristics of soil. Parent materials, as you will read, are typically associated with characteristic kinds of landforms. **Weathering** is an important component of soil formation, because physical abrasion and chemical dissolution differentiate freshly exposed geologic material (i.e., parent material) into soil. Living organisms play an integral role in this process too, wherein organic acids produced by plant roots and soil microorganisms solubilize minerals, allowing their elemental constituents to be leached and deposited at depth. Additionally, the hydrolysis of CO_2 resulting from root and microbial respiration produces acidity, which further contributes to the dissolution of minerals and the weathering process.

Parent materials are broadly classified as **consolidated** and **unconsolidated** (Figure 9.1). Consolidated parent materials include igneous, sedimentary, and metamorphic rocks. A description of them can be found in most introductory geology texts, and Fairbridge (1972) provides a particularly detailed discussion of their formation and chemical composition. Soil developing in

Forest Ecology, Fifth Edition. Daniel M. Kashian, Donald R. Zak, Burton V. Barnes, and Stephen H. Spurr.

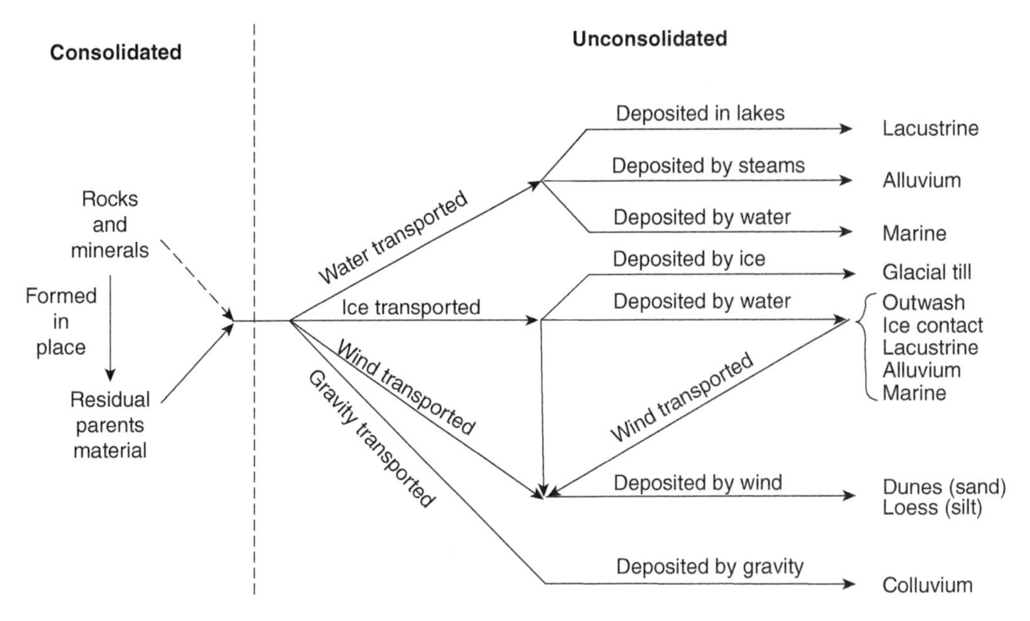

FIGURE 9.1 Diagram illustrating how various kinds of parent material are formed, transported, and deposited. *Source:* Modified from Brady (1990). Reprinted with permission of the Macmillan Publishing Co., Inc. as modified from *The Nature and Properties of Soils*, 11th ed., by Nyle C. Brady. Copyright ©1996 by Macmillan Publishing Co., Inc.

consolidated geologic substrate is said to be formed in residual parent material (Figure 9.1). Forest soils derived from residual parent materials occur throughout North America, primarily in areas that have not been influenced by glaciation, moving water, or oceanic uplift. Forests of the Piedmont Plateau and Appalachian Uplands in the eastern United States occur on these materials, as do forests of the Sierra Nevada, Cascade, and Rocky Mountains in the western United States.

Rates of soil formation on residual parent material composed of hard minerals can be quite slow, and, in some situations, deep soils may never develop because erosion rates exceed those of soil formation. This situation commonly occurs in mountainous regions in which steeply sloping topography and exposed rock combine to form relatively thin soils. Nevertheless, neither all consolidated parent materials weather at slow rates nor give rise to thin soils. The relative resistance of rock to physical and chemical weathering is as follows (Birkeland 1974):

$$\text{quartzite, chert} > \text{granite, basalt} > \text{sandstone, siltstone} > \text{dolomite, limestone.}$$

In general, rocks composed of insoluble minerals (i.e., quartzite, SiO_2) are relatively more resistant to weathering than those containing soluble minerals (i.e., calcite, $CaCO_3$ or dolomite, $CaCO_3{\cdot}MgCO_3$), which rapidly weather in warm, humid climates. Residual parent materials also yield very different chemical constituents as they are abraded and dissolved during the weathering process. For example, the average SiO_2 and Al_2O_3 contents of igneous and sedimentary rocks are similar; however, their calcium (Ca) and sulfur (S) contents can dramatically differ (Table 9.1).

There can be considerable differences in the chemical constituents among soils derived from residual parent material, with implications for tree growth. In the Ozark Mountains of Arkansas, for example, different forest ecosystems develop on two kinds of residual parent material: black and white oaks are overstory dominants on soil derived from chert (SiO_2), whereas eastern red cedar and northern red oak are dominant on limestone-derived ($CaCO_3$) soil (Read 1952). Although both of these parent materials are derived from sedimentary rock, differences in their chemical composition and weathering rate markedly influence the distribution of forest trees.

Table 9.1 Generalized chemical composition of igneous, metamorphic, and sedimentary rocks.

	SiO_2	Al_2O_3	Fe_2O_3	FeO	MgO	CaO	K_2O	P_2O_5	S
	%								
Igneous	69.9	15.2	2.0	2.0	2.2	4.0	3.3	0.20	0.04
Metamorphic	58.2	15.5	2.9	4.8	3.8	6.0	2.6	0.30	0.03
Sedimentary	49.9	13.0	3.0	2.8	3.1	11.7	2.0	0.16	0.18

Source: Modified from Ronov and Yaroshevsky (1972).

Unconsolidated parent materials are mineral particles that have been transported by water, ice, wind, or gravity (Figure 9.1). They are chemically similar to the rock from which they originate, but are distinguished from residual parent material by being moved from their point of geologic origin. The agent of transport has a substantial influence on the physical, chemical, and biological properties of soils formed in unconsolidated parent materials—differences that often influence forest composition and ecosystem productivity. In general, sediments deposited by water and wind have a narrow particle-size distribution, whereas those deposited by ice contain fragments that range in size from microscopic clay particles to boulders several meters in diameter. A complete discussion of transported parent materials and the soils that they give rise to can be found in Buol et al. (1980).

One example of how unconsolidated parent materials influence forest composition and ecosystem productivity comes from the glaciated portions of eastern North America. In this region, the Wisconsin Glaciation (14 000 years before present) left behind a landscape consisting almost exclusively of unconsolidated parent materials. Stratified materials were deposited by glacial meltwaters (outwash), semi-stratified materials were deposited near the margins of stagnant ice (ice contact), and unstratified materials were deposited directly by glacial ice (till). These parent materials differ markedly in their particle-size distribution and their ability to supply both water and nutrients for plant growth. In the northern portion of Michigan's Lower Peninsula, for example, dry-oak-dominated (northern pin, black, and white oaks) ecosystems consistently occur on sandy glacial outwash (72% coarse and medium sand). Mesic northern hardwood ecosystems occur on till-derived soils with lower sand contents (55% coarse and medium sand; Host et al. 1988). In addition to differences in species composition, aboveground productivity varies by a factor of three between the dry oak ecosystems ($1.3\,Mg\,ha^{-1}\,y^{-1}$) and the mesic northern hardwoods ($3.4\,Mg\,ha^{-1}\,y^{-1}$). Similar relationships also have been observed among geology, soil, forest composition, and ecosystem productivity in the glaciated portions of Wisconsin (Pastor et al. 1984). Clearly, the type of geologic material in which soils develop exerts a profound influence on the composition and function of forest trees and the ecosystem they compose.

SOIL FORMATION

As parent material weathers, and is colonized by plants and animals, it differentiates into more or less distinct horizontal zones, giving rise to a **soil profile.** The type of soil profile that develops depends upon the interaction of (i) climate, (ii) parent material, (iii) plants and animals occupying the soil, (iv) slope and aspect of the land, and (v) the amount of time that has elapsed. Soil formation is, in part, a chemical process resulting from the weathering of geologic material exposed to air and water, and in part a biological process resulting from the activities of organisms growing on and in soil, both roots and soil microorganisms.

SOIL PROFILE DEVELOPMENT

Forests comprise the natural vegetation in many of the moister parts of the Earth in which precipitation provides more water than can be evaporated or transpired by plants. Under such conditions two factors dominate the soil-forming process: (i) precipitation in excess of evaporation and transpiration moves downward through the soil dissolving and transporting soluble minerals, and(ii) tree roots remove both water and nutrients from the soil, transpiring most of the former and eventually returning most of the latter to the soil surface as leaves, twigs, fruits, cones, seeds, and fine roots. In temperate regions, the typical forest soil can be differentiated into five zones or **horizons**, so identified by the soil-forming process occurring within them (Figure 9.2). A soil horizon is differentiated from the overlying and underlying portions of soil by attributes that can be easily identified by field observation.

The accumulation of organic matter on the mineral soil surface, or **O horizon**, is an attribute unique to forest soils. This horizon consists of leaves, twigs, flowers, fruits, cones, and seeds that have been deposited on the soil surface; the fine roots (<0.5 mm in diameter) also permeate this soil horizon in some situations. The O horizon lies above the mineral soil and is distinguished from

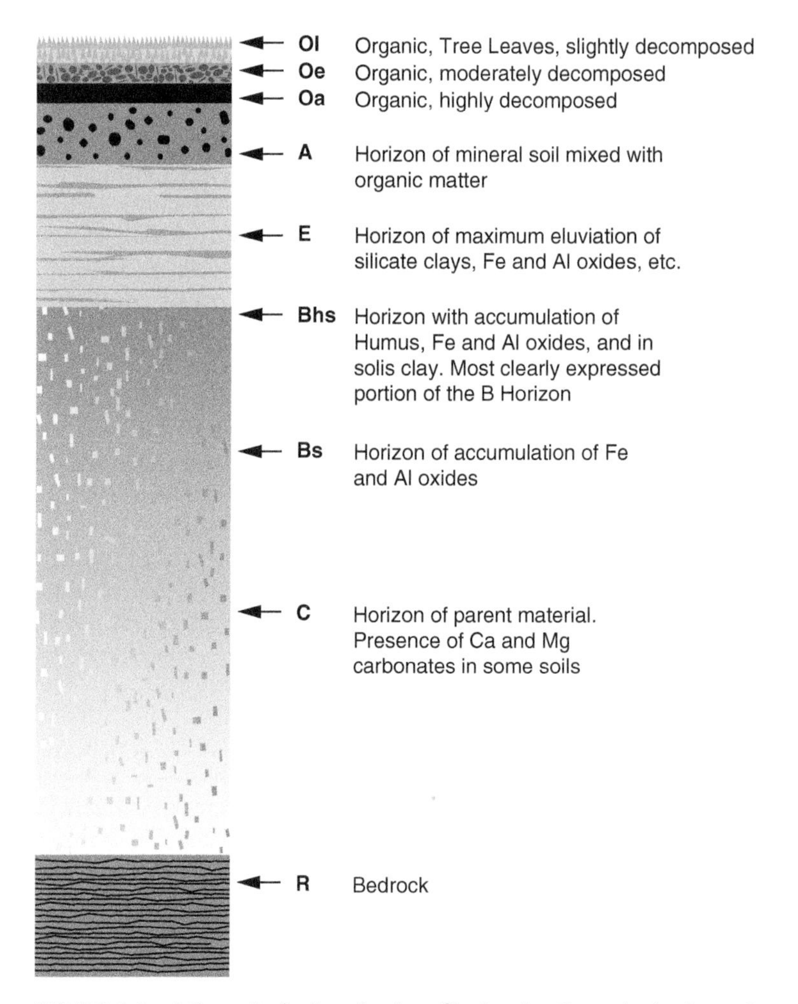

FIGURE 9.2 A theoretical mineral soil profile showing the major horizons that may be represented; realize that not all soils contain all of these horizons. Their development is a function of climate, parent material, topography, biota, and time. *Source:* Brady and Weil (1996) / Pearson. Reprinted with permission of the Macmillan Publishing Co., Inc. from *The Nature and Properties of Soils*, 11th ed., by Nyle C. Brady. Copyright ©1996 by Macmillan Publishing Co., Inc.

it by a high organic-matter content (>20% organic matter if soil has no clay; >30% if soil is more than 50% clay). It can be divided into three subordinate horizons, each reflecting different stages of decomposition: (i) the Oi horizon contains relatively "fresh" organic matter whose origin is easily recognized (e.g., a white pine needle, or a sugar maple leaf), (ii) the Oe horizon is composed of partially decomposed plant parts which are recognizable regarding their origin, and (iii) the Oa horizon reflects the latter stages of decomposition and consists of well-decomposed organic matter of unrecognizable origin (i.e., humus).

The **A horizon** marks the surface of the mineral soil and is characterized by: (i) the leaching or **eluviation** of soluble minerals that migrate in the downward flow of water, and (ii) the accumulation of organic matter originating from the overlying O horizon. In many forest soils, the A horizon is dark in color (e.g., black) and is well structured, owing to its relatively high organic-matter content (4–12%). The majority of aggregates contained within this horizon are of crumb (<1–5 mm diameter) or granular (<1–10 mm diameter) size (see soil structure that follows). The A horizon also is characterized by a large number of fine roots which actively forage for nutrients released during organic-matter decomposition. Because of the shallow distribution of fine roots and their decomposition products, most forest soils are characterized by a thin A horizon. By contrast, A horizons of grassland ecosystems support deeply rooted grasses that incorporate organic matter to depths of 50 cm.

In humid climates, the downward flux of water can move humus, silicate clays, Fe oxides, and Al oxides from the surface soil, leaving behind light-colored, weathering-resistant minerals, especially quartz (SiO_2). The layer so formed is an **E horizon**, and it is distinguished from the overlying A horizon by its light color (e.g., light gray or white). In cool, humid climates, E horizons often develop under coniferous forest growing on course-textured (sand to sandy loam) parent material. The acidity of this horizon is typically higher than the overlying or underlying horizons due to the loss of base cations and presence of inorganic acids. Root densities within the E horizon are low because few plant nutrients reside within this highly leached horizon.

Materials leached from either the A or E horizon migrate downward and are deposited at depth to form the B horizon. The B horizon results from the process of **illuviation** (i.e., accumulation) and is distinguished from other soil horizons by this important soil-forming process. Materials that accumulate in the B horizon include silicate clays, humus, and Fe and Al oxides, and can have a great impact on the physical, chemical, and biological properties of the B horizon. In arid and semi-arid regions, $CaCO_3$, $CaSO_4$, and other salts can accumulate in the B horizon.

The **C horizon** is the unconsolidated parent material underlying the A, E, and B horizons and is outside the influence of the processes giving rise to the horizons above it. The A, E, and B horizons can be derived from the same material contained within the C horizon. However, in areas where geologic activity has deposited a relatively thin layer of mineral material over a previously existing deposit, the A, E, or B horizon may be derived from a different parent material. This situation is common along streams and rivers, where flooding and the subsequent deposition of sediment can bury preexisting soil profiles. In regions where soil forms in residual parent material, the C horizon is replaced by an **R**, the horizon that denotes the presence of the underlying consolidated rock.

Note that the A, B, and C horizons refer to zones that have been leached, enriched, and unaffected by soil-forming processes, respectively. It does not necessarily follow that the upper mineral horizon is always the A horizon, or that all horizons are present in every soil. Following sheet erosion, the B or C horizon may be exposed on the surface. Likewise, very "young" or unweathered soils, like those forming on sand dunes or on any other recent geologic deposit, may lack a B horizon and consist only of an A horizon and a C horizon. In these landscape positions, clay, humus, or Al and Fe oxides often have not accumulated to any extent, thereby causing the absence of a B horizon.

PHYSICAL PROPERTIES OF SOIL

As plant roots grow within the soil, they anchor the aboveground portion of the plant and supply it with water and nutrients. Plant roots also require oxygen for respiration, the supply of which is controlled by the rate at which oxygen diffuses through water and other gases in soil. The physical properties of the solid, liquid, and gaseous phases of soil have a substantial influence on the supply of water, nutrients, and oxygen for plant and microbial metabolism, in addition to the availability of physical space to anchor aboveground plant structures. Providing physical support for aboveground tissues is of particular importance, because plants must properly orient leaves to capture solar energy for use in photosynthesis. In shallow and poorly drained soils, windthrow is common because belowground physical space is limited by the occurrence of shallow bedrock in the former and anoxic (without O_2) conditions in the latter. In the pages that follow, we discuss the physical properties of soil that are of particular importance for the growth of forest plants and the functioning of forest ecosystems.

SOIL TEXTURE

Soils are composed of mineral particles with a wide array of sizes and shapes. These particles, or soil separates, are grouped into three size classes: sand, silt, and clay. Sand particles range in size from 2.00 to 0.02 mm in diameter, silt ranges from 0.02 to 0.002 mm in diameter, and the clay fraction is less than 0.002 mm. By convention, soil is composed of mineral particles that are less than 2.0 mm in diameter; larger particles, like gravel and cobbles, are termed the coarse fraction, because they do not contribute to water and nutrient supply for plant growth. **Soil texture** refers to relative proportion of sand-, silt-, and clay-sized particles contained in a particular soil. This physical property plays an integral role in regulating the availability of water and nutrients for plant uptake, as well as the rate at which gases (O_2 and CO_2) are exchanged between soil and the overlying atmosphere.

In addition to grouping mineral particles by size, particles in different size classes can be distinguished by their physical and chemical properties. For example, sand- and silt-sized particles are chemically identical to the rock from which they originate and are called primary minerals. They are round or irregular in shape, composed primarily of quartz (SiO_2) or other silicate minerals such as orthoclase ($KAlSi_3O_8$) or plagioclase ([Ca, Na][Al, Si] $AlSi_2O_8$). As a consequence of their size and shape, sand particles have relatively low surface areas (1–2 $m^2\ g^{-1}$) with large pores between individual particles (1 g of soil would form a mound in the palm of your hand equal to the size of a United States quarter coin). These attributes provide sandy soils with good aeration, but limit the amount of water they can retain for plant use. Because silt-sized particles have higher surface areas (45 $m^2\ g^{-1}$), they hold relatively larger quantities of water compared to a soil consisting primarily of sand.

The clay fraction of soil found in the B horizon comprises **secondary minerals**, which result from the physical and chemical weathering of primary minerals. In temperate soils, clays consist primarily of aluminosilicate minerals that differ markedly from sand and silt in their shape and mineralogy. These minerals form plate-like structures or **micelles** that are referred to as phyllosilicate (leaf-like silicate) clays. Due to their shape, they have a high surface area (80–800 $m^2\ g^{-1}$), enabling them to hold relatively large quantities of water for plant use. Phyllosilicate minerals differ widely in their physical and chemical properties from Al and Fe oxides, which compose the clay fraction of soils in the humid tropics. Differences in clay minerals found in temperate and tropical soils will be discussed in more detail later in this chapter (see Section 9.4, Chemical Properties of Soil). The physical weathering of mineral material, such as through the grinding action of glacial ice moving across the landscape, also can produce clay-sized particles that can be found throughout a soil profile. These retain the chemical properties of the minerals

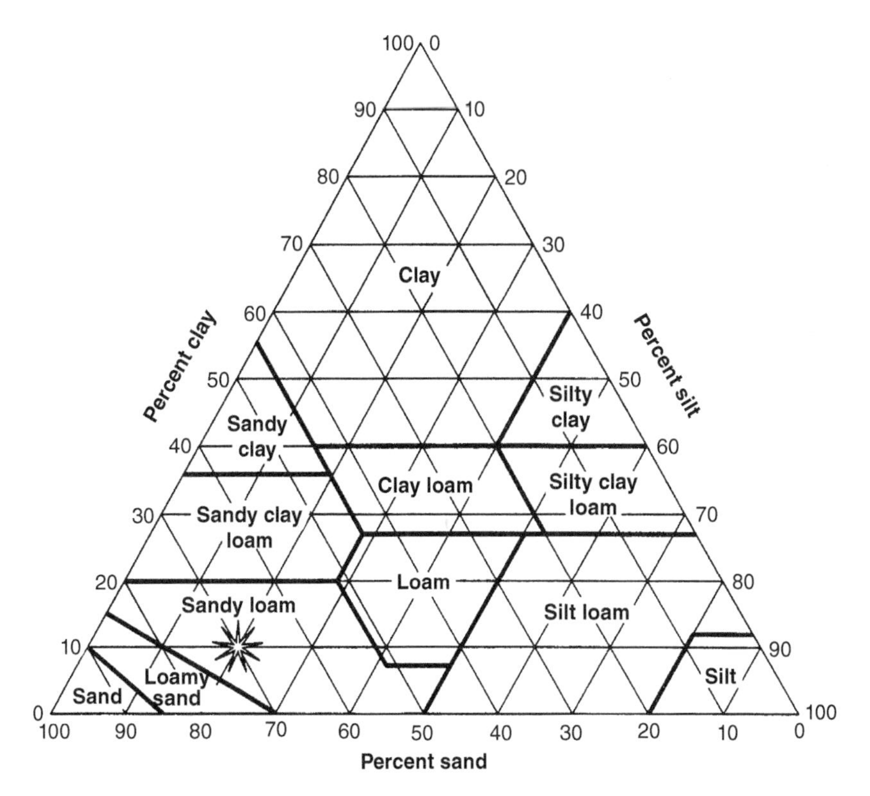

FIGURE 9.3 The soil textural triangle indicating the relative proportions of sand, silt, and clay composing each textural class. Note that some textural classes encompass a range of particle-size distributions (i.e., the clay class ranges from 60 to 100% clay), whereas other textural classes are narrowly defined (i.e., sand class ranges from 90 to 100% sand). The particular soil (70% sand, 20% silt, and 10% clay) represented by an asterisk in the sandy loam class can be used to illustrate the use of this diagram. Begin by finding the lines of equal value for 70% sand and 20% silt. The intersection of those lines with the 10% clay lines locates the position of the asterisk, and the textural class of the soil. Importantly, the heart center of this diagram (clay loam) is the particle-size distribution that provides the maximum supply of water and nutrients to plants.

from which they are derived, which markedly differ from phyllosilicates clays that are formed by weathering and deposition into the B horizon.

Soils are grouped into textural classes based on their sand, silt, and clay content. For example, a soil consisting of equal proportions of sand, silt, and clay is classified as a loam (Figure 9.3). For a soil to be classified as clay, 60% of its particles must be less than 0.002 mm in diameter. By contrast, sands are soils in which more than 90% of the mineral particles range from 0.02 to 2.00 mm in diameter. The textural classes illustrated in Figure 9.3 have been delineated with specific reference to plant growth. As you will read later in this chapter, water availability, nutrient supply, and aeration all are substantially influenced by this all-encompassing soil property.

SOIL STRUCTURE

Primary soil particles (i.e., sand, silt, and clay) are arranged into secondary structures called **aggregates** or **peds**, which result from the combined activities of plants and soil microorganisms. Plant roots enmesh and compress mineral particles and bind primary particles to one another. Further pressure can be exerted during wetting and drying cycles, because clays expand as they hydrate and contract as they dry. Organic compounds (polysaccharides) excreted by plant roots and produced by microbial activity during decay of plant litter function as cementing agents binding to the surface of one or more soil particles. In combination, these processes make soil aggregates highly stable structures that often remain intact even when immersed in water.

Aggregates range from single-grain structure, in which soil particles are totally unattached, to massive structure, in which all soil particles adhere to one another in large clods. Beach sand is an example of the former, whereas the latter often occurs in poorly drained soils with a high clay content. The degree of aggregation most conducive to plant growth lies somewhere between these extremes. Within this mid-range, aggregates vary in size from granules several millimeters in diameter to blocks, prisms, or columns several centimeters in size. Granular aggregates commonly occur in the A horizons of many forest soils, whereas blocky aggregates often occur in B horizons. Plant roots generally occupy the spaces between aggregates, rather than growing through or within them. These spaces, or macropores, have a substantial influence on the rate at which water and gases move into and through the rooting zone. As a result, a well-aggregated soil will typically hold more water and will have better aeration than an unstructured soil of the same texture.

SOIL COLOR

Color provides insight into many physical and chemical properties of soil, particularly organic-matter content and drainage (or aeration). The surface horizons of mineral soils are generally dark in color, reflecting organic-matter additions from leaf and root litter. A deep-colored surface soil, dark in color (e.g., black or dark brown), usually contains relatively greater amounts of organic matter than a thin, light-colored surface soil. The subsoil of sandy forest soils in northerly climates can contain accumulations of organic matter and iron that have been leached from the surface soil. This subsurface accumulation is easily identified by its dark chocolate-brown color (i.e., Bh or Bhs horizon).

In addition to providing insight into organic-matter content, soil color can provide qualitative information regarding soil drainage. Most soils contain large amounts of iron (Fe), which, in an oxidized state (Fe^{3+}), is bright orange or red. This condition occurs when soils are well aerated and O_2 rapidly diffuses into the soil profile. However, during prolonged periods of water saturation, O_2 in soil can be depleted if the demand by plant roots and soil microorganisms exceeds the diffusion rate of O_2 in water. In these situations, Fe is reduced (Fe^{2+}) producing compounds blue-gray to gray in color. Zones of mottling, a patchwork of yellow-orange and blue-gray colors, indicate the presence of both oxidized and reduced conditions. This situation occurs where the level of the water table fluctuates within the soil profile, and color can be used to determine if it lies within the rooting zone of plants. High water tables in poorly drained landscape positions greatly restrict the rooting depth of trees, making them prone to windthrow. Using soil color, one can easily identify soils in the field with poor drainage that often restrict the growth of many forest trees.

SOIL WATER

Water availability controls the global, regional, and local distribution of plants on Earth. For example, forests occur in regions in which the annual amount of water supplied by precipitation exceeds that which is lost through evaporation and transpiration. Although broadscale patterns of precipitation control the *total* amount of water entering ecosystems, it is the interaction of water molecules with soil particles that largely influences the amount of water that actually can be used by an individual plant for growth.

Water flows along a continuum extending from the atmosphere, through the plant, and into soil. The force driving water movement along this continuum is transpiration; water is actually "pulled" from soil and through plants into the atmosphere by this process. Transpiration at the leaf surface creates tensions that are translated down water columns extending through plants and into soil. Inasmuch as soils can be "dry" to plants, but can still contain substantial amounts of water. This relationship may occur if the force holding water in soil exceeds the force (i.e., tension) created by the transpiration of water at the leaf surface. Understanding the dynamics of water along the atmosphere–plant–soil continuum has clear relevance for the study of plant growth and ecosystem function, because these dynamics directly control the amount of water available for plant use.

Physical Properties of Water The physical properties of water greatly influence its availability to plants. Water (H—O—H) molecules have a net positive charge on one side of the molecule and a net negative charge to the other. Mineral particles in soil also have charged surfaces that attract water molecules, due to the dipole nature of the water molecule. The attraction of positively and negatively charged bodies is termed **adhesion**, a force in soil that greatly influences the amount of water available for plant use. Adsorped water molecules, those attracted to charged surfaces in soil, are linked to other water molecules through hydrogen bonding, a chemical bond linking the oxygen atom (−) of one water molecule to a hydrogen atom (+) of another. Hydrogen bonding gives rise to the cohesive force, or **cohesion**, joining water molecules into chains or polymers that extend away from the surface of mineral particles.

Due to the strong attraction of water molecules to charged surfaces in soil, adsorped water molecules are closely packed and exist in an energy state less than that of pure water. Although water in direct contact with mineral particles is strongly held to their surface by adhesion, that force diminishes as the distance from the solid surface increases, much like the attraction of a magnet for iron diminishes as the distance between them increases. When soils are saturated, some water molecules are only weakly attracted to the surface of mineral particles, because they lie at relatively large distances from any charged surface (i.e., weak force of attraction due to adhesion). Water draining from saturated soil does so because the Earth's gravitational pull exceeds the adhesive and cohesive forces holding a portion of soil water. As soil continues to dry, either through plant uptake or evaporation, the forces holding the remaining water molecules steadily increase as the layer of water surrounding particle surfaces diminishes. Adhesive and cohesive forces holding water in soil are tensions that must be overcome if plants are to extract water from soil. Because the forces holding water in soil can exceed those imposed by plants, only a proportion of the water in soil is available for plant use. That is why soils can contain sufficient quantities of water, but can "feel dry" to plants.

In addition to being attracted to charged surfaces, water molecules also are attracted to ions with net positive (cations) or negative (anions) charges. Salts, such as NaCl, dissolve in water because the attraction of water molecules for cations (Na^+) and anions (Cl^-) is much greater than the attraction between cations and ions. The strong attraction of water for positively or negatively charged ions causes water molecules to lose energy as they hydrate either type of ion. Because water molecules lose energy as they associate with cations and anions, water in soil has a lower energy status than pure water.

The semi-permeable membrane surrounding plant and microbial cells in soil (i.e., plasmalemma) functions as a "molecular sieve," allowing water to transverse while excluding larger, hydrated ions. The movement of water molecules across any semi-permeable membrane in response to differences in ion concentration (i.e., the energy status of water inside versus outside the cell) is **osmosis**. Water molecules moving across a semi-permeable membrane exert a force known as osmotic pressure. Osmosis, and the energy it produces, is of particular relevance to plant growth, because it influences the movement of water into and out of plant and microbial cells. For example, plants under salt stress suffer from a lack of available water, because dissolved ions in soil water lower its energy status to a point where it is less than that in the plant cell. Water flows from the plant into soil solution by osmosis in this particular situation. This can cause the water content of plant cells to decrease to such a low level that physiological processes are impaired, and the plant is no longer able to maintain turgor.

Soil Water Potential The movement of water in soil, its uptake by plant roots, and its loss to the atmosphere from the leaf surface are all energy-related phenomenon. In soil, adhesion, cohesion, the presence of dissolved ions, and the Earth's gravitational pull are the primary forces influencing the energy status of soil water, and hence the movement of soil water and the proportion of it are available for plant use. Forces in soil acting on water molecules can be pressures (gravity) or tensions (adhesion and cohesion), both of which are measured in megapascals (MPa; values less than 0 MPa are tensions and those greater than 0 MPa are pressures). In combination, these forces control how much of soil water is actually available for plant use.

Soil-water potential is the energy status of soil water; it also can be thought of as the effective concentration of water in soil. By definition, the potential of pure, liquid water at 20 °C and at standard atmospheric pressure is 0 MPa. Pure water is used as a standard reference point from which we measure the influence of adhesion, cohesion, dissolved ions, and gravity on the energy status of soil water.

Adhesion and cohesion, the forces holding water molecules to charged surfaces and to one another, give rise to the **matric potential** of soil water. Because adhesion and cohesion lower the energy status of soil water (i.e., relative to pure water), matric potentials are less than 0 MPa. As such, matric potentials are tensions holding water molecules to one another and to the surfaces of charged particles in soil. The presence of dissolved ions, which also lower the energy status of soil water relative to pure water, gives rise to the **osmotic potential** of soil water. **Gravitational potential** results from the downward force of gravity and its ability to extract water from soil. Gravitational potentials are greater than zero (i.e., pressures), because the downward pull of gravity extracts water from soil. In combination, matric, osmotic, and gravitational potentials give rise to the **total soil-water potential**, which represents the summed energy status of soil water. In most well-drained soils, matric potential is the most important factor regulating the supply of water to the root surface.

Because adsorption and dissolved ions lower the free energy status of soil water, plants must expend energy to remove water from soil. As such, the water potential (or energy status) of plants must be lower than that of soil, if water is to flow from soil into plant roots, up the stem, and out of leaves. Transpiration at the leaf surface drives the flow of water along the energy-related path from soil to the atmosphere. The **atmospheric water potential**, which greatly influences the transpiration rate of plants, is largely determined by relative humidity and air temperature. The concentration of water in the atmosphere is much less than the concentration of water in either plants or soils, and as a consequence, atmospheric water potentials are more negative (i.e., at a lower energy status) than those of plants or soil. It is not unusual for atmospheric water potentials to attain values of −100 MPa, values 10–100 times more negative than those in plants or soil (Bidwell 1974). Because water flows from a region of high potential (i.e., high-energy state) to one of low potential (i.e., low-energy state), water moves from soil into plant roots, through the vascular system of the plant, and into the atmosphere via stomates in leaves. Large negative atmospheric water potentials drive the process of transpiration and the flow of water from soil, through plants, and into the atmosphere.

Although large negative water potentials at the leaf surface are translated downward to the root surface, plants are generally unable to extract soil water held by potentials less than −1.5 MPa (i.e., a tension of 1.5 MPa; Figure 9.4). Under these conditions plants wilt and are unable to regain turgor even following the addition of water. At a potential of −1.5 MPa, soil water has attained the **permanent wilting point**, which defines the lower limit of plant available water (Figure 9.4). **Field capacity** represents the upper limit of plant available water and is the amount remaining in soil after it has freely drained due to the downward pull of gravity (a potential of +0.01 MPa). The quantity of water bounded by field capacity and the permanent wilting point represents the

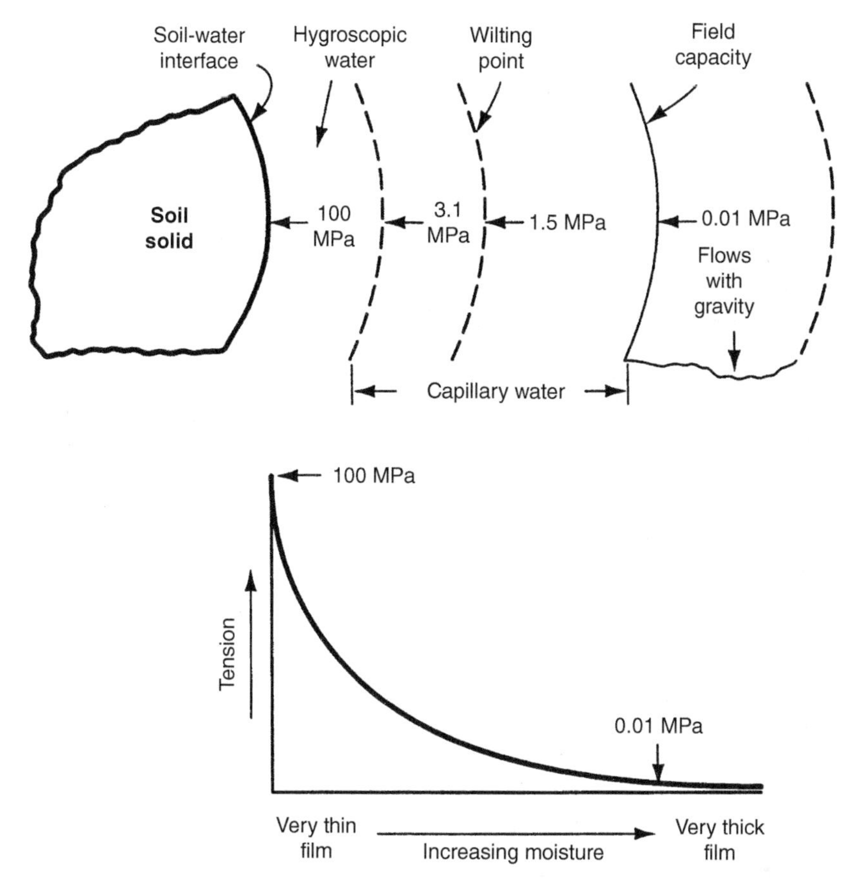

FIGURE 9.4 Diagrams showing the relationship between the thickness of water films and the tension with which water is held by soil particles. Tensions are presented in megapascals in the upper illustration. The thickness of water films in relationship to matric potential is presented in the lower figure. *Source:* Brady (1974) / Springer Nature. Reprinted with permission of the Macmillan Publishing Co., Inc. from *The Nature and Properties of Soils*, 8th ed., by Nyle C. Brady. Copyright ©1974 by Macmillan Publishing Co., Inc.

available water content (ml cm^{-3}) of soil. It differs substantially from the **saturation water content**, which is the total amount of water that can be stored in all soil pores (i.e., no air-filled pores; Figure 9.4). The reader is referred to Kramer (1983) for a complete treatment of plant–water relations and soil-water dynamics.

Soil texture substantially influences the available water content of soil. Figure 9.5 illustrates the relationship between water content (mL of water per cm^3 of soil) and soil-water potential for a clay, loam, and sand. At field capacity (i.e., water held by tension more negative than −0.01 MPa), the clay soil holds approximately 0.6 ml of water per cm^3 of soil, almost 2.5 times more water than the sand (calculated from Figure 9.5). The adhesive properties of water, in combination with the large surface area of clay-sized particles, allow the clay to hold more water at any given potential than the sand. Notice that silt loam in Figure 9.5 contains the greatest quantities of plant available water. It does so because of the favorable distribution of macropore and micropore spaces. The available water content of sand, calculated from Figure 9.5, is 0.15 ml cm^{-3}, approximately 50% of that held by the silt loam (0.36 ml cm^{-3}) and clay (0.33 ml cm^{-3}).

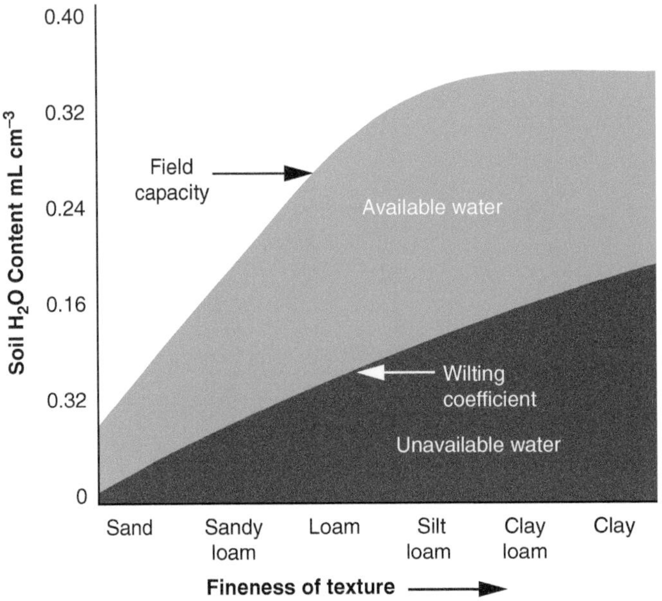

FIGURE 9.5 The relationship between soil texture and the available water content of soil. Note that field capacity increases from the sand to the silt loam and then levels off as the proportion of clay increases. Because the permanent wilting point increases linearly as a function of soil texture, the largest amount of plant available water occurs in soil with a silt loam texture. *Source:* Brady and Weil (1996) / Pearson. Reprinted with permission of the Macmillan Publishing Co., Inc. from *The Nature and Properties of Soils*, 11th ed., by Nyle C. Brady. Copyright ©1996 by Macmillan Publishing Co., Inc.

CHEMICAL PROPERTIES OF SOIL

Plant life is constructed from a surprisingly small suite of elements, whether we consider the majestic redwoods of northern California or the single-cell alga growing on the soil surface. In Table 9.2, we summarize the chemical-building blocks from which plant life is constructed and their source within terrestrial ecosystems. Each is required for plant growth and development, albeit in different quantities. **Macronutrients** are those elements required in relatively large

Table 9.2 Macronutrient elements required by plants and their source within terrestrial ecosystems.

Element	Symbol	Source
Carbon	C	Atmosphere
Hydrogen	H	Water
Oxygen	O	Atmosphere, water
Nitrogen	N	Organic matter, atmosphere
Phosphorus	P	Mineral soil, organic matter
Potassium	K	Mineral soil, organic matter
Sulfur	S	Mineral soil, organic matter, atmosphere
Magnesium	Mg	Mineral soil
Calcium	Ca	Mineral soil

Source: Brady (1990) / Springer Nature.

Table 9.3 The biochemical function of plant macronutrients, their form of uptake, and typical leaf concentrations in plants.

Element	Biochemical function(s)	Form assimilated	Leaf concentration
Carbon (C) Hydrogen (H) Oxygen (O)	Form the basic building blocks of all biologically active compounds	CO_2, H_2O	90–98%
Nitrogen (N)	Nucleic acids, amino acids, proteins, chlorophyll, anthocyanins, alkaloids	NH_4^+, NO_3^-	1–4%
Phosphorus (P)	Nucleic acids, nucleotides, sugar phosphates, phospholipids	$H_2PO_4^-$	0.1–0.4%
Potassium (K)	Enzyme co-factor, osmotic regulation, cell ion balance	K^+	1%
Calcium (Ca)	Pectin synthesis and cell-wall formation, metabolism/formation of nucleus and mitochondria, enzyme activator	Ca^{2+}	0.8%
Sulfur (S)	Amino acids, proteins, sulfolipids	SO_4^{2-}	0.2%
Magnesium (Mg)	Chlorophyll, enzyme co-factor	Mg^{2+}	0.2%

Source: Salisbury and Ross (1992) / Wadsworth Publishing.

amounts. They are commonly found as constituents of nucleic acids, proteins, carbohydrates, lipids, and chlorophyll (Table 9.3). Micronutrients (Fe, Mn, Bo, Mo, Cu, Zn, Cl, and Co), as their name implies, are required in relatively small amounts and occur as co-factors in enzymatic reactions. Although micronutrients are required in small amounts, they are nonetheless important in the biochemical functioning of plants and entire ecosystems; all are supplied to plants by chemical processes in soil. Further discussion regarding the biochemical and physiological functions of plant nutrients can be found in Salisbury and Ross (1992).

Although plants assimilate carbon dioxide and oxygen from the atmosphere, the majority of macronutrients and micronutrients are supplied by ion-exchange reactions, mineral weathering, or organic-matter decomposition—processes all occurring within soil. The supply of nutrients often limits the growth of individual plants and entire ecosystems. Nitrogen (N) availability, for example, is known to limit the growth of many boreal and temperate forests (Flanagan and Van Cleve 1983; Pastor et al. 1984), whereas phosphorus (P) has been observed to constrain forest growth in the humid tropics (Vitousek 1984; Vitousek and Sanford 1986). In the following discussion, we explore some of the chemical processes in soil that regulate the supply of nutrients for plant growth in both temperate and tropical forest soils. These processes, in combination with plant uptake and litter decomposition, control the cycling of nutrients within forest ecosystems (Chapter 19).

CLAY MINERALOGY

In studying Table 9.3, note that many macronutrients exist as cations, the ionic form assimilated by plant roots. The mineralogy of clay particles and the ion-exchange reactions mediated by their negatively charged surfaces substantially influence the supply of cations for plant growth. Ion-exchange reactions in soil also are an important mechanism influencing the retention and loss of nutrients from forest ecosystems, especially following disturbances such as harvesting, large-scale windthrow, or fire.

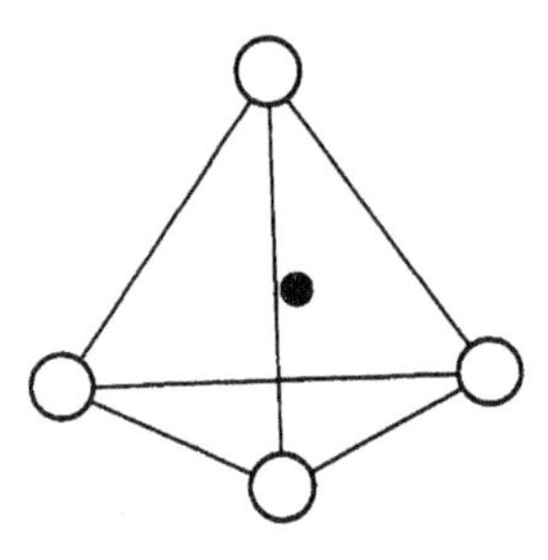

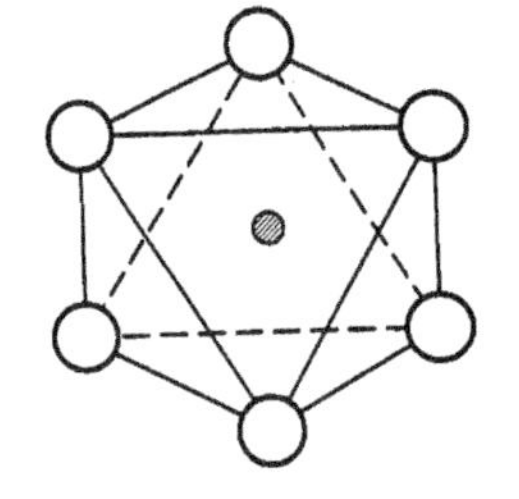

FIGURE 9.6 The structure of the silica tetrahedra and the alumina octrahedra that form the basic building blocks of phyllosilicate minerals.

Table 9.4 Physical and chemical properties of some common phyllosilicate minerals.

Property	Fine-grained mica	Montmorillonite	Kaolinite	Chlorite
Ratio of octahedral to tetrahedral sheets	2:1	2:1	1:1	2:1:1
Size (μm)	0.1–5.0	0.01–1.0	0.5–5.0	0.1–2.0
Shape	Plates Flakes	Flakes	Hexagonal Crystals	Variable
External surface ($m^2\ g^{-1}$)	50–100	70–120	10–30	70–100
Net negative charge ($cmol\ kg^{-1}$)[a]	100–180	80–120	2–5	15–40

[a] Centimoles of negative charge per kilogram of dry soil.
Source: Brady (1990) / Springer Nature.

Clay minerals form during the weathering process and, to some extent, chemically reflect the primary minerals from which they originate. Phyllosilicate minerals dominate the clay fraction of temperate soil and originate from a wide array of primary minerals including feldspar, orthoclase, and hornblende. Although phyllosilicate clays may differ in mineralogy, they are all formed from the same chemical-building blocks. The primary structures of these minerals are the silica tetrahedra (SiO_4) and the alumina-magnesia octahedra (AlO_6 or MgO_6; Figure 9.6). By sharing O atoms at their corners, these subunits can link to form tetrahedral ($[AlO_6]_n$) or octahedral sheets ($[MgO_6]_n$), secondary structures that confer many unique properties to phyllosilicate minerals.

Fine-grained mica, vermiculite, chlorite, montmorillonite, and kaolinite are common phyllosilicate minerals in temperate soil, and some of their physical and chemical properties are summarized in Table 9.4. The weathering process, which gives rise to clay minerals, substantially influences on the extent to which phyllosilicate clays attract and bind cation nutrients. As a consequence, unweathered soils and highly weathered soils differ markedly in clay mineralogy and consequently their ability to retain and release cation nutrients for plant growth.

Isomorphic substitution conveys a net negative charge on phyllosilicate minerals enabling them to attract cations and supply them for plant growth. This process occurs as one type of clay mineral (i.e., montmorillonite) weathers and gives rise to another (i.e., kaolinite). During the formation of phyllosilicate minerals, atoms of the same size, but of lower charge, replace either Si in tetrahedral sheets or Al in the octahedral sheets. The extent to which a particular clay is substituted is influenced by the chemical constituents of the parent material and the degree to which it has weathered. This property is a permanent attribute of phyllosilicate minerals and is relatively unaffected by changes in soil acidity. Because of a low degree of isomorphic substitution, kaolinite, a highly weathered clay mineral, has a low net negative charge that is balanced by a small number of cations adsorbed into its crystalline surface (Table 9.4). Montmorillonite, a clay mineral less weathered than kaolinite, is highly substituted in its octahedral sheets and has a relatively large net negative charge, therefore attracting a greater number of cations to its surface. The aforementioned examples illustrate how weathering directly controls the ability of phyllosilicate clays to adsorb and exchange cations with soil solution, a factor that has a profound effect on the supply of nutrients for plant growth and the ability of ecosystems to retain nutrients against leaching.

Following long periods of intense weathering (i.e., 100 000–1 000 000 years), Si is lost from the structure of phyllosilicate minerals leaving behind oxides of Fe and Al, which constitute the clay fraction of some highly weathered soils in the humid tropics. These oxides differ markedly in their chemical characteristics from the phyllosilicate minerals from which they originate. In contrast to the ordered crystalline structure of phyllosilicate minerals, oxides of Fe and Al are amorphous (without form) and exist in a less-ordered semi-crystalline state (Schwertmann and Taylor 1989; Sposito 1989). More importantly, these minerals exhibit a pH-dependent variable charge, in contrast to the stable, net negative charge of phyllosilicate minerals. This property results from the large number of hydroxyl groups (−OH) contained within these minerals; the formulae for goethite [$FeO_3(OH)_3$] and gibbsite [$Al(OH)_6$] illustrate this point.

Soil acidity in highly weathered tropical soils plays an important role in the functioning of clay minerals and their ability to supply plants with nutrients. In acidic soils, the hydroxyl groups of the Al and Fe oxides are protonated, conveying a net positive charge and the ability to adsorb and exchange anions. However, these minerals lose protons (H^+) in relatively alkaline soil (e.g., $AlO(OH)_5^- + H^+$), thereby producing a net negative charge and the ability to adsorb and exchange cations. As a consequence, soils dominated by Al and Fe oxides have a pH-dependent charge and also a point of zero charge at which neither cations nor anions are adsorbed; both cations and anions are susceptible to loss through leaching. Land management practices which alter soil pH clearly have the potential to alter the ability of tropical soils to adsorb and retain plant nutrients. For further elaboration on the dynamics of variable-charge soils and nutrient mobility within them, we refer readers to Sollins et al. (1988) and Uehara and Gillman (1981).

CATION EXCHANGE AND THE SUPPLY OF NUTRIENTS

The **cation-exchange capacity** of soil is a general measure of plant nutrient availability and represents the total amount of cations (centimoles of positive charge) that can be adsorped by a kilogram of soil. Cations in soil solution exist in an equilibrium with those adsorbed to cation-exchange sites on clay micelles. When cations in soil solution are assimilated by plants, those adsorbed to clay particles are released and a new equilibrium is established. As such, adsorbed cations can be thought of as a "reservoir" of plant nutrient in soil. Because cations vary in size and charge, they are adsorped and exchanged in a predictable manner reflecting their affinity for exchange sites on clay micelles. In general, cations with a small, hydrated radius and a large positive charge are most strongly held. We have listed the cations commonly found in soil solution in order of decreasing affinity:

$$Al^{3+} > H^+ > Ca^{2+} > Mg^{2+} > K^+ = NH_4^+ > Na^+.$$

Weakly adsorbed cations such as K^+ and NH_4^+ are more available for plant growth; they also are more susceptible to leaching than others higher in the order.

Table 9.5 The relationship among soil texture, clay mineralogy, and the cation-exchange capacity in surface forest soils of the eastern United States.

Vegetation/physiography	Texture	Clay (%)	Mineralogy	Cation-exchange capacity (cmol kg^{-1})	Organic matter (%)
I. Northeastern US					
Jack pine/outwash plain	Sand	2	Mixed	13	6.4
Northern pin oak–black oak–white pine/outwash plain	Sand	4	Mixed	11	6.0
Sugar maple–bass-wood/moraine	Loamy sand	6	Mixed	18	12.0
II. Southeastern US					
Loblolly pine/coastal plain	Silt loam	7	Montmorillonitic	11	3.9
Loblolly pine/coastal plain	Silt loam	20	Montmorillonitic	20	3.7
Loblolly pine/coastal plain	Silty clay	37	Montmorillonitic	28	2.3
Water oak–willow–oak/ central prairie region	Silty clay	39	Montmorillonitic	30	7.0
Loblolly pine/upper coastal plain	Loamy sand	5	Kaolinitic	4	0.9
Loblolly pine/upper coastal plain	Sandy loam	11	Kaolinitic	7	1.3
Loblolly pine/upper coastal plain	Clay loam	28	Kaolinitic	26	2.3

We summarize the cation-exchange capacity for forest soils of different texture and mineralogy to illustrate the combined influence of weathering and parent material on supply of cations for plant growth (Table 9.5). Note that cation-exchange capacity increases as the percentage of clay also increases. Also note that soils dominated by kaolinite, a weakly substituted and highly weathered phyllosilicate mineral, generally have lower cation-exchange capacities than those soils containing montmorillonite. The soils of mixed mineralogy contain a mixture of phyllosilicate clays and have relatively high cation-exchange capacities, even though clay contents are relatively low. As you will read later in this section, the high organic-matter contents of these soils greatly contribute to their ability to supply plants with cation nutrients.

The proportion of cation-exchange sites occupied by Ca^{2+}, Mg^{2+}, K^+, and Na^+ is the **percent base saturation** of soil. These cations are not technically bases, because they do not directly neutralize H^+ in soil solution. Rather, they reduce soil acidity when they are adsorbed in place of H^+, a topic that we will further elaborate in the following section. Percent base saturation is a chemical property of particular relevance to plant growth, because it is both a general measure of cation nutrient availability and soil-buffering capacity; that is, the ability of soil to resist a change in pH following the addition of H^+. In general, soils that have a high proportion of exchangeable bases have a high capacity to supply plants both with base cations and buffer acidic inputs.

SOIL ACIDITY

Soil pH is commonly used to quantify acidity and, by definition, is the negative log of the hydrogen ion concentration in soil solution ($pH = -\log[H^+]$). In combination with adsorption and exchange reactions, soil acidity substantially influences the supply of nutrients for plant growth. It does so

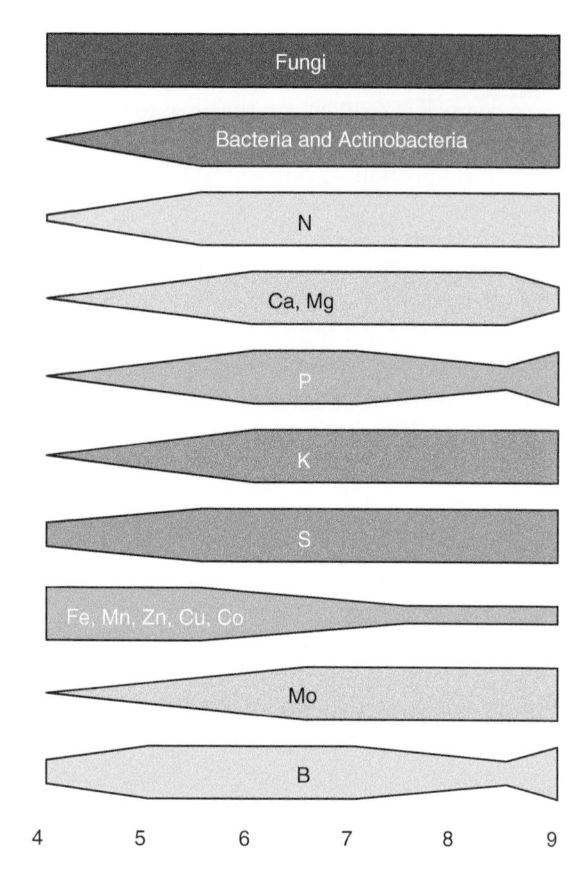

FIGURE 9.7 The relationship between soil pH and the availability of plant macronutrients and micronutrients is shown. The influence of soil pH on microbial activity within soil also is illustrated.

by controlling the solubility of minerals composing soil. Figure 9.7 illustrates the availability of plant nutrients along a pH gradient ranging from very acidic soils to those with high pH. Note that the availability of most nutrients is greatest at neutral pH values (Figure 9.7). By influencing mineral solubility, soil pH also affects the weathering rate of parent material, the formation of clay minerals, and soil development. Thus, the weathering process and soil acidification often go hand in hand. The activity of soil microorganisms also is influenced by soil acidity as we see in Figure 9.7.

There are several sources of H^+ that contribute to the lowering of soil pH, the weathering of parent material, and the removal of base cations. Perhaps the most important source of H^+ in soil results from the respiration of plant roots and soil microorganisms. Carbon dioxide produced during respiration dissolves in soil solution and forms carbonic acid (H_2CO_3). The prolonged exposure of soil minerals to this relatively weak acid results in their solubilization and the eventual removal of base cations over sustained periods of weathering. Much stronger organic acids (e.g., fulvic and humic acids) are produced as byproducts of the microbial decomposition of plant tissues. Plant roots also exude organic acids that similarly act on soil minerals, solubilizing them over time. Over extended time periods (i.e., 10 000–1 000 000 years), these sources of acidity facilitate the weathering of soil minerals and lower pH.

The industrial activities of humans also have influenced soil chemistry. Oxides of nitrogen (NO_x) and sulfur (SO_2), released during the burning of fossil fuels, can further oxidize in the atmosphere to produce nitric (HNO_3) and sulfuric (H_2SO_4) acids. The addition of these acids in precipitation has raised concerns in eastern North America and Central Europe. Because the constituents of "acidic deposition" are relatively strong acids, they have the potential to act on soil minerals in the same manner as the acids produced by the metabolism of plants and soil microorganisms. Some soils in eastern North America, particularly those with coarse textures where granitic bedrock is shallow, are sensitive to acidic deposition because of their low cation-exchange capacity, and low base saturation.

It is important to consider soil acidity, mineral weathering, and the removal of base cations simultaneously, because they are co-occurring processes in soil. As mentioned earlier, base saturation represents the ability of soil to buffer the input of H^+ from chemical and biological sources. Base cations buffer the soil reaction when they weather from soil minerals and replace H^+ adsorbed to exchange sites. The H^+ so released initially enters soil solution; however, it is easily leached from the soil resulting in a decrease in acidity. Clearly, over long time periods, the ability of soil minerals to relinquish base cations to weathering can be exhausted. In such a situation, pH declines to the point where the soil reaction is dominated by Al^{3+} and Al hydroxides. At a very low soil pH (<4.0), Al exists as Al^{3+} and, along with H^+, occupies the majority of cation-exchange sites. In soil solution, Al^{3+} can react with water in the following manner:

$$\text{Adsorbed Al}^{3+} \leftrightarrow \text{Al}^{3+}\text{in soil solution} \leftrightarrow \text{Al}^{3+} + \text{H}_2\text{O} \leftrightarrow \text{AlOH}^{2+} + \text{H}^+.$$

In studying this equation, note that the reaction of H^+ with bases in soil solution, or its leaching from the soil profile, will shift the equilibrium to the right, producing a "new" H^+ during formation of $AlOH^{2+}$. This mechanism constitutes a buffering system that maintains acidic soils at a low pH. Furthermore, Al^{3+} released in the reaction is toxic to plants and can greatly restrict root growth (Runge and Rode 1991). High concentrations of Al^{3+} in soil solution are known to reduce root elongation, kill root meristems, and interrupt the functioning of the plasmalemma in aboveground tissues (Foy et al. 1978).

Most forest soils range from extremely acid (pH 4.0) to slightly acid (pH 6.5). Where a particular forest soil lies along this gradient is substantially influenced by organic-matter additions (e.g., leaves, roots, twigs, reproductive structures, fine roots) from overstory trees and the acids produced during microbial decomposition. The general trend is for conifers such as pines, spruces, hemlock, and Douglas-fir to increase surface-soil acidity (i.e., decrease pH) to a greater extent than hardwoods or northern white-cedar.

An example of how individual trees influence soil acidity is provided by tulip tree and eastern hemlock in eastern Kentucky (Boettcher and Kalisz 1990). Although these trees co-occur on the same soil parent material, the soil pH under tulip tree (pH 4.7) is consistently greater than that beneath eastern hemlock (pH 4.0). In addition, quantities of the exchangeable bases Ca^{2+} and K^+ also are lower beneath eastern hemlock. In eastern Washington, organic-matter additions from western hemlock (pH 4.0) lower surface soil pH to a much greater extent than western red cedar (pH 5.9) when both species occur on the same soil parent material. These examples illustrate that differences in the litter biochemistry of forest trees can have a substantial influence on the chemical properties of soils.

SOIL ORGANIC MATTER

Although organic matter composes a relatively small fraction of most forest soils (e.g., <1–15%), it has a profound effect on a wide array of physical, chemical, and biological properties. As noted earlier, soil organic matter (i.e., aboveground and belowground plant litter) contributes to aggregate formation, which in turn influences the amount of soil water available to plants. It also functions as a "storehouse" of plant nutrients, supplying most of the nitrogen used in the annual growth of forest ecosystems as it is metabolized by soil microorganisms. Importantly, soil organic matter is the substrate for the growth and maintenance of heterotrophic microbial populations in soil. It is through the metabolic activities of these organisms that nitrogen and other plant nutrients bound in plant litter and soil organic matter are released into soil solution for plant use. Because the nutrients so released via microbial metabolism can be re-assimilated by plants, soil organic matter represents an important "weigh station" in the cycling and storage of nutrients within forest ecosystems (see Chapter 19).

The organic matter entering the soil originates from aboveground and belowground sources of plant litter. Aboveground sources consist of leaves, reproductive structures, twigs, and tree stems, whereas roots (fine and coarse) are the primary belowground source of litter. In most forests, belowground litter from fine roots equals or exceeds aboveground litter production (i.e., leaves, seeds, flowers). In general, plant litter contains approximately 15–60% cellulose, 10–30% hemicellulose, 5–30% lignin, and 2–5% protein (Paul and Clark 1996). In soil, these compounds are metabolized by microorganisms, producing energy, CO_2, H_2O, and humus as end products. The biochemical constituents of deal fungal and bacterial cells also are a component of humus, comprising approximately 25% of this material. Humus, which composes the Oa horizon, is a complex and chemically resistant material that gives surface soils their dark color and unique chemical properties. Due to its advanced state of decay, humus does not physically or chemically resemble the plant material from which it originated. Humus also is chemically resistant to microbial degradation and can remain in soil for periods of 100–3000 years (Paul and Clark 1996). The surface of humus can have a net negative charge, resulting from the dissociation of H^+ from hydroxyl (−OH), carboxylic (−COOH), or phenolic (C_6H_{11}–OH) groups. At high pH values, the cation-exchange capacity of humus (150–300 cmol kg^{-1}) can exceed that of many silicate clays. As such, cation- or anion-exchange reactions mediated by humus represent an important mechanism influencing nutrient availability in soil. In some soils, approximately 50% of the total cation-exchange capacity of soil can be attributed to humus. The relatively high cation-exchange capacities of the sandy soils of mixed mineralogy in Table 9.4 directly result from their high organic-matter contents.

The organic-matter content of soil reflects a balance between the addition of organic matter from plant production and its loss during microbial decomposition. Because forest harvesting can alter both rates of litter input and loss through decomposition, it also has the potential to alter the quantity of organic matter and associated plant nutrients stored in soil. In Chapter 19, we further consider the impact of forest harvesting on soil organic-matter dynamics and the cycling of plant nutrients.

The organic-matter content of soil exerts an important influence on the available water content of soil. Soil organic matter holds relatively large quantities of water at field capacity, but its permanent wilting point also is proportionally high, providing only small quantities of plant available water. However, organic-matter content is the primary factor influencing soil-aggregate formation. In turn, soil aggregation influences the proportion of micropore and macropore spaces, which directly controls the water-holding characteristics of soil. Consequently, soil organic matter exerts its main influence on the water-holding characteristics of soil through its influence on soil structure, rather than by how much water it can directly hold for plant use. Well-aggregated, fine-textured soils with ample organic-matter contents (5–10%) generally hold large quantities of available water, making them good substrates for plant growth.

SOIL CLASSIFICATION

A taxonomic system of soil classification, referred to as the soil taxonomy, is widely used throughout North America. It was developed by the soil survey staff of the United States Department of Agriculture to classify soils in regard to their potential for agricultural management (Soil Survey Staff 1975). This system is based on measurable morphological characteristics present within a particular soil profile. The primary advantage of this approach is that the soil profile itself, rather than the soil-forming process, is classified. Soil taxonomy is modeled after the plant taxonomic system, with categories ranging from order (broad grouping) to series (narrowest category).

The main soil orders found beneath North American forests are entisols, inceptisols, spodosols, alfisols, histosols, andisols, and ultisols; there are other soil orders, but they generally occur beneath desert, grassland, or agricultural soils.

Entisols (recent soils) beneath forests are mineral soil without, or with only, the beginnings, of horizon development. They often occur on talus slopes, floodplains, sand dunes, and where bedrock lies close to the land surface. **Inceptisols** (from Latin *inceptum*) are more weathered than entisols and contain a weakly developed B horizon. These soils have a wide geographic distribution, and, with the exception of arid climates, can be found in most regions in North America. Inceptisols are common forest soils in the Pacific Northwest, Rocky Mountains, and the eastern United States. They often occur in well-drained, upland landscape positions, but also can be found along river corridors.

Alfisols (from the chemical abbreviation for aluminum and iron) typically form in cool to hot humid areas and are common under deciduous forests in the eastern United States. They are characterized by shallow dark-surface horizons, medium to high base saturation, and the accumulation of silicate clay in the B horizon (i.e., Bt horizon; t denotes the accumulation of pedogenic clay). Alfisols are more weathered than inceptisols, but are less weathered than spodosols.

In cold and temperate climates, the process of leaching can give rise to the formation of **spodosols** (from Greek *spodos*, wood ash). These soils are characterized by the presence of a strongly developed E horizon and the accumulation of humus and oxides of Fe and Al in the B horizon (Bh or Bhs horizons). Spodosols often form beneath boreal forests occurring on sandy parent materials. In general, these soils are best developed beneath spruce and fir, species whose litter generally acidify the surface soil, and are common in the northeastern United States, northern Lake States region, and the Pacific Northwest. It should be noted that spodosols also can form beneath coniferous forest in warm climates. In Florida, spodosols are often encountered in landscape positions in which coniferous forests are seasonally flooded.

Andisols are derived from volcanic ejecta with the soil profile containing at least 60% volcanic glass. These soils form any place on Earth where volcanic activity has delivered a blanket of volcanic ash or related materials as surface deposits. Most often, these soils are poorly developed, but can be highly productive for forest growth as well as agriculture. In North America, andisols are located in the Pacific Northwest beneath highly productive forest often dominated by Douglas-fir.

Histosols form in poorly drained landscape positions and are characterized by a high organic-matter content ($\geq$20%). These soils can form anywhere the land surface is continually saturated with water, and thus occur in all climates and have a global distribution. Forest vegetation occurring on these organically derived soils includes: black spruce bogs in the northern Lake States, black ash–red maple swamps in the Northeastern United States, and the pocosin and cypress swamps of the Southeastern United States. Large expanses of forested histosols also can be found in Scandinavia, Siberia, and Canada.

Ultisols (form the Latin word *ultimus*) are a common forest soil on old land surfaces in warm, humid climates, such as those of the southeastern United States. These soils are characterized by an accumulation of silicate clay in the B horizon. However, ultisols are distinguished from alfisols by a low base saturation—the result of more intense weathering. These soils are widely distributed in the eastern US, extending southward from Maryland to Florida and westward from the East Coast to the Mississippi River Valley. They also occur in portions of the Pacific Northwest and eastern California and can be found on old land surfaces in Australia, Africa, India, southern China, and southern Brazil.

Oxisols (from the French *oxide* and the Latin word for soil, *solum*) support forest vegetation in the tropical and sub-tropical regions of Central and South America, Southeast Asia, and Africa. The subsoil of these highly weathered soils contains an accumulation of kaolinite and oxides of Al and Fe. The old land surfaces on which these soils occur, in combination with the intense weathering of humid tropical climates, can give rise to profiles exceeding 15 m in depth.

Forest and agricultural ecosystems differ in ways that limit the use of the soil taxonomy to classify forested landscapes. In the western United States, for example, a wide range of forest habitat types can be found on the same taxonomic unit of soil (Neiman 1988), making it difficult to use soil classification to predict the occurrence of forest vegetation as well as its productivity. The primary reason for such a disparity is that forest vegetation reflects a myriad of interacting physical and ecological factors such as physiography, harvesting frequency and intensity, and prior land uses that are not considered by soil taxonomy. Nevertheless, soil factors that reflect moisture and nutrient regimes, such as texture, aggregation, and coarse-fragment content, are often related to the occurrence of some forest types (Neiman 1988). Because the relationship between forest communities and the soil developing beneath them is multifactorial and dynamic, it is likely that any single-factor classification, such as the soil taxonomy, will be of limited use in predicting the distribution and growth of forest ecosystems. In Chapter 13, we further discuss the limitation of single-factor systems for classifying forested landscapes.

LANDFORM, SOIL, AND FOREST VEGETATION: LANDSCAPE RELATIONSHIPS

Changes in topography, vegetation, and parent material across a landscape give rise to marked differences in soil formation, even within a relatively small geographic area. In Michigan's Upper Peninsula, for example, forest ecosystems dominated by white pine, sugar maple, and red oak can occur in well-drained, upland landscape positions where a thin blanket of glacial drift overlies bedrock (Figure 9.8). Soils forming in these landscape positions are inceptisols, characterized by a shallow profile and the minimal development of a B horizon. The deeper deposits of glacial drift

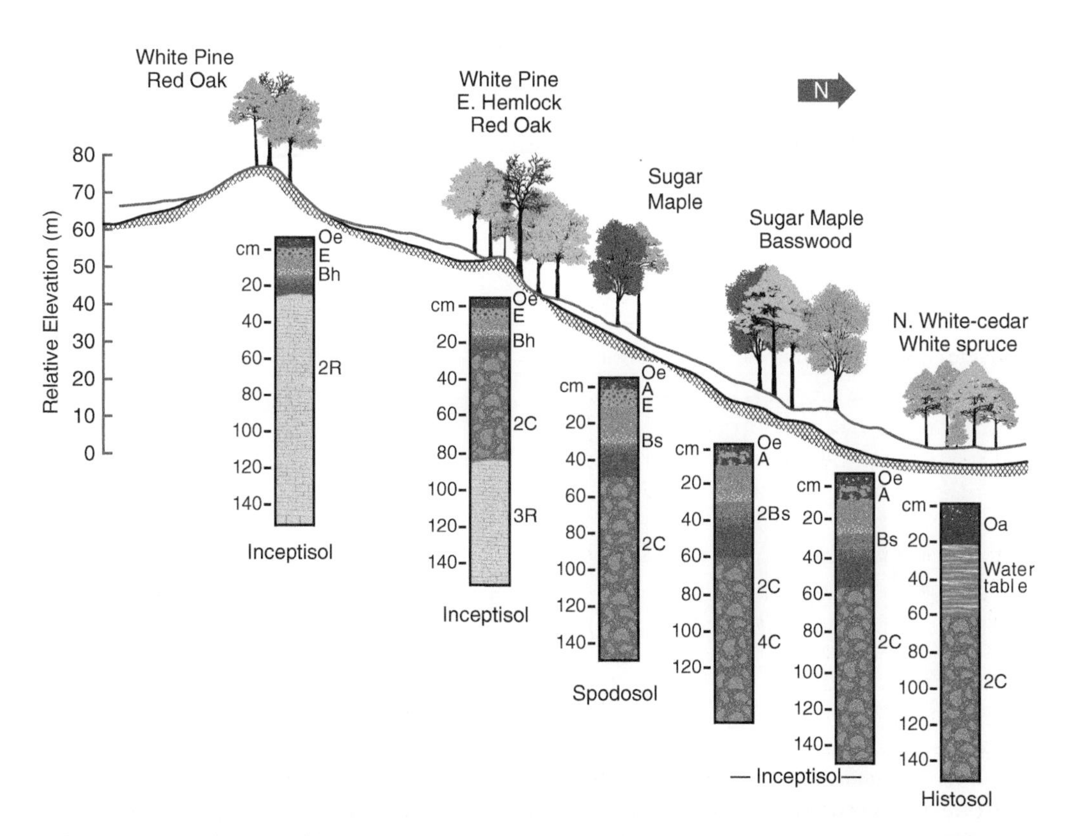

FIGURE 9.8 A physiographic cross section in the Upper Peninsula of Michigan illustrating the influence of topography and forest vegetation on soil profile development. *Source:* Pregitzer et al. (1983)/ John Wiley & Sons. Reprinted with the permission of the Soil Science Society of America, Segoe, Wisconsin, USA.

downslope give rise to spodosols, which occur beneath a canopy of sugar maple and basswood; these soils have well developed E and Bh horizons. In lowerslope positions also dominated by sugar maple, inceptisols form in relatively recent deposits of more recent colluvium. The formation of an inceptisol in this landscape position is related to the relatively short duration over which the parent material has weathered. Swamp forests dominated by black ash and northern white-cedar occur in the poorly drained bottomslope positions where the accumulation of organic matter has led to the formation of histosols (Figure 9.8). The patterns of soil formation described in the earlier text are repeatable features in the landscape, occurring in other locations with a similar set of soil-forming factors (i.e., parent material, vegetation, time).

In the southeastern United States, topography exerts a similar influence on the soil developing beneath the flatwood vegetation of the lower coastal plain (Figure 9.9). In this region, relatively small elevational differences differentiate well-drained from poorly drained landscape positions. This land surface is relatively old (100 000 years before present) compared to the relatively recent (approximately 8000 years before present) deposition of glacial materials in Michigan's Upper Peninsula. As a consequence, the parent material giving rise to this soil has been affected by climate, biota, and topography for a longer duration.

In uplands dominated by longleaf pine, relatively dry conditions give rise to inceptisols—poorly developed soils with minimal horizon formation. The warm, humid climate of this region, in combination with sandy parent material and coniferous vegetation, gives rise to spodosols in somewhat poorly drained landscape positions with a fluctuating water table. Histosols further form in very poorly drained landscape position in the Southeast United States, similar to the landscape distribution of these soils in other regions.

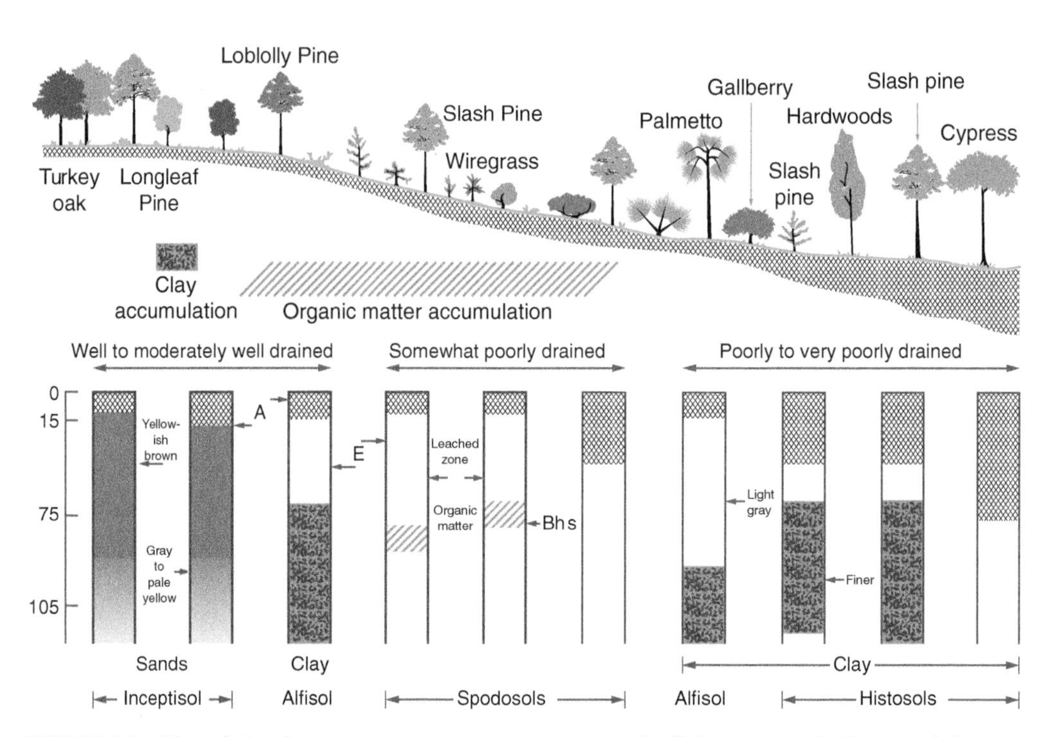

FIGURE 9.9 The relationship among topography, vegetation, and soil development in the coastal plain flatwoods of the southeastern US. Organic-matter accumulation in the surface horizon dramatically increases as one moves from the well-drained upslope positions to the poorly drained bottomslope positions. In this region of the United States, relatively small changes in elevation elicit large changes in soil drainage, profile development, and overstory composition. *Source:* Adapted from Pritchett and Smith 1970. Reprinted with permission from *Tree Growth and Forest Soils*, © 1970 by Oregon State University Press.

SUGGESTED READINGS

Binkley, D. (1995). The influence of tree species on forest soils: processes and patterns. In: *Proceedings of the Trees and Soils Workshop*, Agronomy Society of New Zealand Special Publication No. 10 (ed. D.J. Mead and I.S. Cornforth), 1–33. Canterbury, New Zealand: Lincoln University Press.

Paul, E.A. and Clark, F.E. (1996). *Soil Microbiology and Biochemistry*, 2e. New York: Academic Press 340 pp.

Sanchez, P.A. (1976). *Properties and Management of Tropical Soils*. New York: Wiley 618 pp.

Weil, R.R. and N.C. Brady 1996. Nature and Properties of Soil, 11 Prentice-Hall, New Jersey. 740 pp.

Fire

CHAPTER 10

In North America, fire has affected virtually all of the upland forests in the South, the Lake States and adjacent Canada, the West, and most of those in the Northeast, Appalachian Mountains, and central states have been burned more or less frequently. In the boreal forests of Alaska and Canada, fire has been a powerful natural factor affecting vegetation and wildlife. Even wetlands such as swamps (Cypert 1973; Ewel 1990), bogs, and marshes have burned, although less frequently, and their vegetation affected. Fire was extensively used by indigenous people prior to European colonization. Books by Kozlowski and Ahlgren (1974), Wein and MacLean (1983), Johnson (1992), Agee (1993), Whelan (1995), Bond and van Wilgen (1996), Johnson and Miyanishi (2001), and Pyne (1982, 2019) provide an entry into the voluminous literature on fire.

Fire has always been a critical process shaping the evolution of species and the functioning of ecosystems in which they reside. Prehistoric fire and the remarkable interactions of fire with humans and wildlife are considered by Schüle (1990). Fire regimes, which are defined by the frequency, intensity, severity, extent, and time of occurrence of fires, are characteristic of different regional and local landscape ecosystems. Fire is a principal influence on plant traits and life cycles as well as on diversity and ecosystem processes such as carbon, nutrient, and water cycling; biomass accumulation; and succession. Fire plays many major roles in landscape ecosystems around the world (Wright and Heinselman 1973; Swanson 1981). It influences:

Geomorphic and hydrologic processes of hillslopes and stream channels.

Physical and chemical properties of soils.

Nutrient loss.

Biomass accumulation.

Genetic adaptations of plants.

Plant composition and diversity, mortality, regeneration, growth, and succession.

Wildlife habitat and wildlife population dynamics.

Presence and abundance of forest insects, parasites, and fungi.

In this chapter, we consider fire as a physical site factor, examining its effects on forest species and forest site quality. The role of fire as a disturbance factor in forest ecosystems and their communities is considered in Chapter 16, and its interactions with climate change are discussed in Chapter 20.

Forest Ecology, Fifth Edition. Daniel M. Kashian, Donald R. Zak, Burton V. Barnes, and Stephen H. Spurr.

FIRE AND THE FOREST TREE

CAUSES

Fossil charcoal, which represents evidence of past natural fires, has been identified in the Carboniferous coal deposits of 400 million years ago and in the Tertiary deposits of brown coal (Harris 1958; Komarek 1973). Meteorites were a rare ignition source of fires, and falling igneous rock from volcanic eruptions undoubtedly caused local fires then as it does today. Primarily, however, fires were caused by lightning, which today causes about 50000 wildland fires worldwide each year (Taylor 1974). Notably, this estimate is less than 1% of the estimated 182 million cloud-to-ground discharges that occur in the forests and grasslands of the world each year. Between 70 and 100 lightning flashes are estimated to occur every second worldwide (Pearce 1997), but not all strike the ground. In the United States, about 10000 lightning-caused fires occur each year, and about 80% of these are in the Rocky Mountains and Pacific Coast states. When fuel and climatic conditions are conducive to ignition by lightning, a single lightning storm may start many small fires in these regions.

Humans have been the most significant cause of fires throughout much of the modern world. Prior to European colonization, indigenous people set many fires, and Europeans later followed suit. The high intensity and frequency of such fires, which are often associated with logging and land clearing, have in many cases altered the character of forests and affected site quality.

FIRE REGIME

Studies of fire history have helped to define the kind(s) of fire and the prominent immediate effects of fire that characterize an area. A **fire regime** is typically characterized by type, frequency, intensity, severity, size, and seasonality, with fire type, frequency, and severity the most important. Fire **type** includes **ground**, **surface**, and **crown** fires. Fire **frequency** refers to the recurrence of fire in a given area over time and may be expressed in a number of ways (Agee 1993). *Return interval*, or the average number of years between successive fires, may be expressed for a given point (for example, a single fire-scarred tree or a small group of trees) or for an area. Frequency is also expressed as *fire rotation* or *fire cycle*, which is the length of time required to burn over an area equal to that under consideration. These approaches and their computation are described and contrasted in detail by Agee (1993).

Fire **intensity** refers to the amount of heat released by the fire, often estimated by the length of the flames, whereas **severity** expresses the *effect* of fire on soil or vegetation (seed bank, mortality of plants). Specifically, severity is usually expressed as a function of tree mortality or soil damage. For example, an intense spring fire, when the soil is moist, may be low in severity, whereas a low-intensity summer fire during drought conditions may be severe in its effects on soil properties. Intensity is typically estimated in kilowatts per meter length of fire front ($kW m^{-1}$) and in flame length (m) along the fire front (Johnson 1992; Agee 1993). The **seasonality** of fire may have differential effects on vegetation and soil. For example, in the southeastern coastal plain of the United States, the proper proportion of winter fires to fire-free years is critical to the perpetuation of pure longleaf pine forest because pine regeneration is prevented by annual burning and summer fires (Marks and Harcombe 1981).

Three primary fire regimes are recognized (typically based on severity, or whether fires are nonlethal to the dominant aboveground vegetation or cause stand replacement): nonlethal understory fires (including frequent or infrequent surface fires), stand-replacing fires (short- or long-frequency crown fires), and mixed-severity fires. The latter regime applies to combinations of understory and stand-replacing fires that may occur in two ways: (i) a *variable* fire regime of frequent, low-intensity surface fires typically followed by an infrequent, stand-replacing fire, or (ii) a *mixed* fire regime of individual fires alternating between nonlethal understory burning and stand-replacing fires, creating a fine-scale pattern of young and older trees. The second type of mixed-severity regime is more often discussed in the literature.

It is important to understand that nearly all fires result in a mosaic of severities, ranging from unburned or hardly burned to completely scorched, mass mortality events within the same fire perimeter. As a result, most burned areas are a mosaic from unburned survivors, to low-severity areas with few trees killed, to high-severity areas with many or all trees killed. This variation is often ignored for simplicity in describing how fires affect forests, but this relative proportion of severities is often what implicitly defines a fire regime.

Fire Types, Frequency, and Severity Van Wagner (1983) identifies five main types of fire in the northern forest based on their physical behavior, including two kinds of surface and crown fires:

Ground fires that smolder in deep organic layers (less than 10 kW m^{-1}).

Surface backfires that burn against the wind (100–800 kW m^{-1}).

Surface headfires that burn with the wind (200–15 000 kW m^{-1}).

Crown fires advancing as a single front (8000–40 000 kW m^{-1}).

Crown fires that include high-intensity spotting fires (up to 150 000 kW m^{-1}).

Surface fire is the most common type of fire, which burns over the forest floor, consuming litter, killing aboveground parts of herbaceous plants, shrubs, and small trees, and typically scorching the bases and crowns of larger trees. Mortality of shrubs and trees increases with greater fuel accumulation on the surface. Surface fires are very sensitive to wind speed; surface headfires may attain quite high intensities in brush in leafless hardwood stands and in open forests where trees are sparse, or crowns are high above the ground. In addition to fire intensity, the amount of tree mortality depends on the species, the age of the tree, and rooting habit. Young pines may succumb to a surface fire, whereas older individuals of the same species survive because their thicker bark protects the cambium from heat damage and their crown height above the flames protects the canopy. A shallow rooting habit, whether due to the inherent nature of the species or site conditions (wetland or bedrock), increases susceptibility to fire injury compared to that of the deep rooting habit typical of upland species such as oaks and hickories.

Surface fires tend to kill young trees of all species (though often just the aboveground portion) and most trees of less fire-resistant species of all sizes. However, pole-size to mature trees of fire-resistant species survive light surface fires in varying proportions. Survival in a surface fire for most fire-resistant tree species is not typically dictated by damage to the stem cambium, but by their susceptibility to root injury and to crown scorch by hot gases rising above the flames. For example, Van Wagner (1970) reported that a light surface fire will leave mature crowns of red pine undamaged, whereas a hot surface fire will kill a red pine stand just as surely as a crown fire 10 times as intense. Observations of fire-damaged red and white pines suggest that the trees will die if more than 75% of the crown is killed.

Fires sweeping the forest floor may generate ground fires, which burn in thick accumulations of organic matter, such as peat, that overlies mineral soil. They are flameless when they burn below the surface, and they kill most plants with roots growing in organic matter. Ground fires burn slowly and generate very high temperatures. In moist organic matter, heat from the fire dries out material adjacent to the burning zone and perpetuates a zone of combustible fuel. Ground fires tend to be persistent and serve as reignition sources for surface fires. Recent examples from near-arctic regions have documented summer ground fires burning in peatlands that smolder beneath the snow throughout the winter! Such fires, sometimes called "zombie fires" by wildland firefighters, may reignite aboveground vegetation early in the spring.

Surface fires are often closely associated with crown fires. Surface fires, fueled by accumulations of organic matter and whipped by winds, may scorch and ignite crowns of trees. Crown fires travel from crown to crown, often in dense even-aged stands, killing most trees in their path. Notably, large, intense crown fires may leave unburned strips (sometimes called "stringers"; Kashian et al. 2012) due to powerful, downward air currents (Simard et al. 1983) or other mechanisms. Where the forest is patchy or consists of small groups of trees such as in much of the dry-climate, ponderosa pine forests of western North America, some individuals or patches of trees may *torch* or briefly carry a crown fire, but an extensive crown fire is highly unlikely. Conifers are more susceptible to crown fires because of the high flammability of their foliage and the greater likelihood of their occurrence in pure stands compared to broad-leaved species. Sparks and burning embers may be carried ahead of the main fire to start *spot fires*, or new surface fires often far away from the site of the main crown fire. Notably, especially in mountainous areas, surface fires are common, and their frequency is related to climatic factors of temperature and precipitation associated with elevation. However, these surface fires can lead to severe crown fires given the right climatic and fuel conditions.

Fire frequency, intensity, severity, and burning pattern are primarily controlled by climate, fuel accumulation and flammability, soil water, and especially topography. Fire frequency was probably greatest in grasslands prior to European colonization, where burns every 2–3 years were common (Wells 1970). In dry ponderosa pine forests of western North America, the average interval between fires varied from 2 to 18 years (Weaver 1974; Dieterich and Swetnam 1984; Savage and Swetnam 1990). Average intervals vary because of the method of determination (Agee 1993) and ecosystem type; pine forests of drier low-elevation ecosystems burned more frequently than those of moist slopes and higher elevations. For example, Arno (1976) reported fire frequency over an elevational gradient (1150–2600 m) for three watersheds in the Bitterroot Mountains of western Montana for the period 1735–1900. Low-elevation forests of ponderosa pine and Douglas-fir burned at about 9-year intervals (range 2–20 years). Intervals between burns increased with increasing elevation, approaching 35 years at high elevations (range 2–78 years).

Intense surface fires may damage the cambium and scar the tree. When an individual tree is scarred multiple times, the fire scar records the actual fire frequency of a single place (Spurr 1954). One ponderosa pine in Arno's study was scarred by 21 fires from 1659 to 1915, an average interval of 13 years (Figure 10.1). Nearby trees were scarred by additional fires, demonstrating that fires do not always burn hot enough to cause injury, particularly when they occur so frequently that only light accumulations of fuel have occurred. Similarly, Dieterich and Swetnam (1984) studied a small, suppressed ponderosa pine in Arizona containing 42 fire scars over the 178-year period from 1722 to 1900. A detailed historical fire chronology of the 97-ha area showed that whereas the mean fire interval for the tree was 4 years, the interval for the area was 2 years, that is, not every fire was recorded by this tree. Thus, records of fire frequency may reliably indicate the frequency of severe fires, but may underestimate the frequency of all fires. Light to moderate surface fires, although often undetected, are nevertheless important in regulating seedling composition and distribution and reducing the concentration of fuels on the forest floor.

In contrast to frequent fires in dry, low-elevation, submontane ponderosa pine and mixed conifer forests, fires occur infrequently but regularly in high-elevation subalpine ecosystems of the Pacific Northwest and northern Rocky Mountains (Agee 1993). Dominant trees are firs (often subalpine fir), spruces (often Engelmann spruce), mountain hemlock, and whitebark pine. Fire frequency is spatially heterogeneous because of local variation in ecosystem conditions and fuel accumulation. For example, fire-return intervals varied from over 1500 years for ecosystems dominated by mountain hemlock to differing intervals for whitebark pine ecosystems: from 50 to 300 years (Arno 1980) to a low of 29 years (Morgan and Bunting 1990).

In the Sierra Nevada Mountains of California, a frequency of 9 years between fires was found at one locality in the giant sequoia–mixed conifer forest during the period 1705–1873

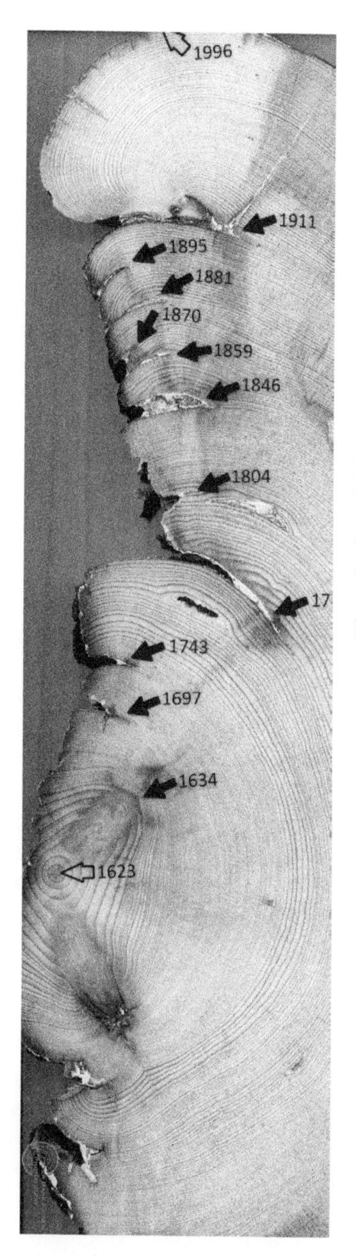

FIGURE 10.1 Multiple fire-scar cross section of a ponderosa pine showing 11 fire scars from 1623 through 1996.

Photo by Peter M. Brown, Rocky Mountain Tree-Ring Research

(Kilgore 1973), comparable to the overall 8-year frequency of fire reported for ponderosa pine forests (Weaver 1951). As is typical of many other western forests, moister east and north slopes do not burn as readily as drier south and west slopes, but they may burn more intensely than those that burn more frequently. A similar relationship develops along elevational gradients from warm and dry low elevations to the cool and moist slopes of high altitudes. Studies of redwood in California also illustrate that fire-return intervals are site-specific and not species-specific. Fire-return intervals for moist coastal redwood sites at the northern part of the range may be as long as 500–600 years (Viers 1980), but intervals of 33–50 years (Viers 1980) and 20–29 years are reported (Jacobs et al. 1985; Finney and Martin 1989) on drier, interior and more southerly sites. Similar variation in return interval by specific ecosystem conditions is also reported for subalpine forests.

In conifer forests of the Boundary Waters Canoe Area of northern Minnesota, detectable fires were relatively frequent (Heinselman 1973, 1981), burning at approximately 4-year intervals during the period 1727–1886. European colonization activities increased fire frequency, and the interval dropped to 2 years. Major fires that burned over large areas occurred at longer intervals of 21–28 years.

The emerging pattern from fire history studies is one of cyclic occurrence that is determined by the multiple and interrelated physical and biotic features of site-specific ecosystems. Fire regimes of regional and local ecosystems are affected by climate, physiography, vegetation type (affecting fuel accumulation and its flammability), and human activities. Variations in these factors have resulted in fires of different intensities and severities—from frequent, light surface fires, which kill few trees but reduce fuel accumulations and create patterns of seedling and sapling distribution, to infrequent, high-severity crown fires that kill most trees and regenerate entire stands. This ecosystem-dependent range of fire frequency, intensity, and severity was instrumental in the evolution of the life span and species characteristics and, more generally, the kind of vegetation present in each region (Chapter 16). There exist obvious spatial patterns of regional forest composition throughout North America that have resulted from variations in fire frequency and severity (Figure 10.2).

FIRE ADAPTATIONS AND KEY CHARACTERISTICS

Certain key characteristics and adaptive traits of forest trees and shrubs appear to have developed in response to fire. Many individual characteristics are typically cited, such as thick insulative bark and sprouting ability, but probably few of these are genetic adaptations that are exclusively fire-determined. Seasonally hot, dry sites favor fire and support species that not only are likely to have fire adaptations but also are physiologically adapted to live in such severe environments. Selective forces of the total environment affect the plant's evolutionary response. This point is confirmed by the discovery that many species of tropical rain forests persist in human-modified ecosystems now characterized by frequent fire. Such "fortuitous adaptations" allow species to survive fires or quickly establish following fire regardless of evolutionary derivation (Kauffman and Uhl 1990). Such adaptations include thick bark, vegetative sprouting, and seeds stored in the soil. Therefore, our emphasis in this section is on those traits of woody species that are suited to survive and reproduce in ecosystems characterized by fire, that is, fire-dependent ecosystems.

The characteristics probably most directly elicited by fire are closed cones (cone serotiny, i.e., delayed or heat-opening of cones); thick fire-resistant bark; buried or protected buds and subsequent sprouting; early flowering, fruiting, and dispersal; and the "grass stage" in longleaf and other pines. In addition, the relative life span of many forest species has been shaped by the frequency of stand-replacing fires (Gill et al. 1981). As a result, moderately long-lived, fire-dependent species such as Douglas-fir and eastern white pine are able to compete with long-lived, late-successional associates such as western hemlock and western redcedar, or sugar maple and American beech on mesic sites that burn infrequently but within their life span.

The key characteristics of forest species enabling their persistence in fire-prone ecosystems may be grouped into four general categories to emphasize major life history features related to fire and site: (i) **avoid** fire damage, (ii) **recover** following fire damage, (iii) quickly **colonize** sites after fire, and (iv) **promote** or **facilitate** fire. Species especially fire-resistant or susceptible typically exhibit the extreme expression of these characteristics. Examples of each category are cited in the following text:

(i) **Avoiding Fire Damage:**

Thick, insulated bark (many pines and oaks, western larch, giant sequoia, and redwood).

Buried buds protected by soil (aspens, sweetgum, sumacs, and many shrubs).

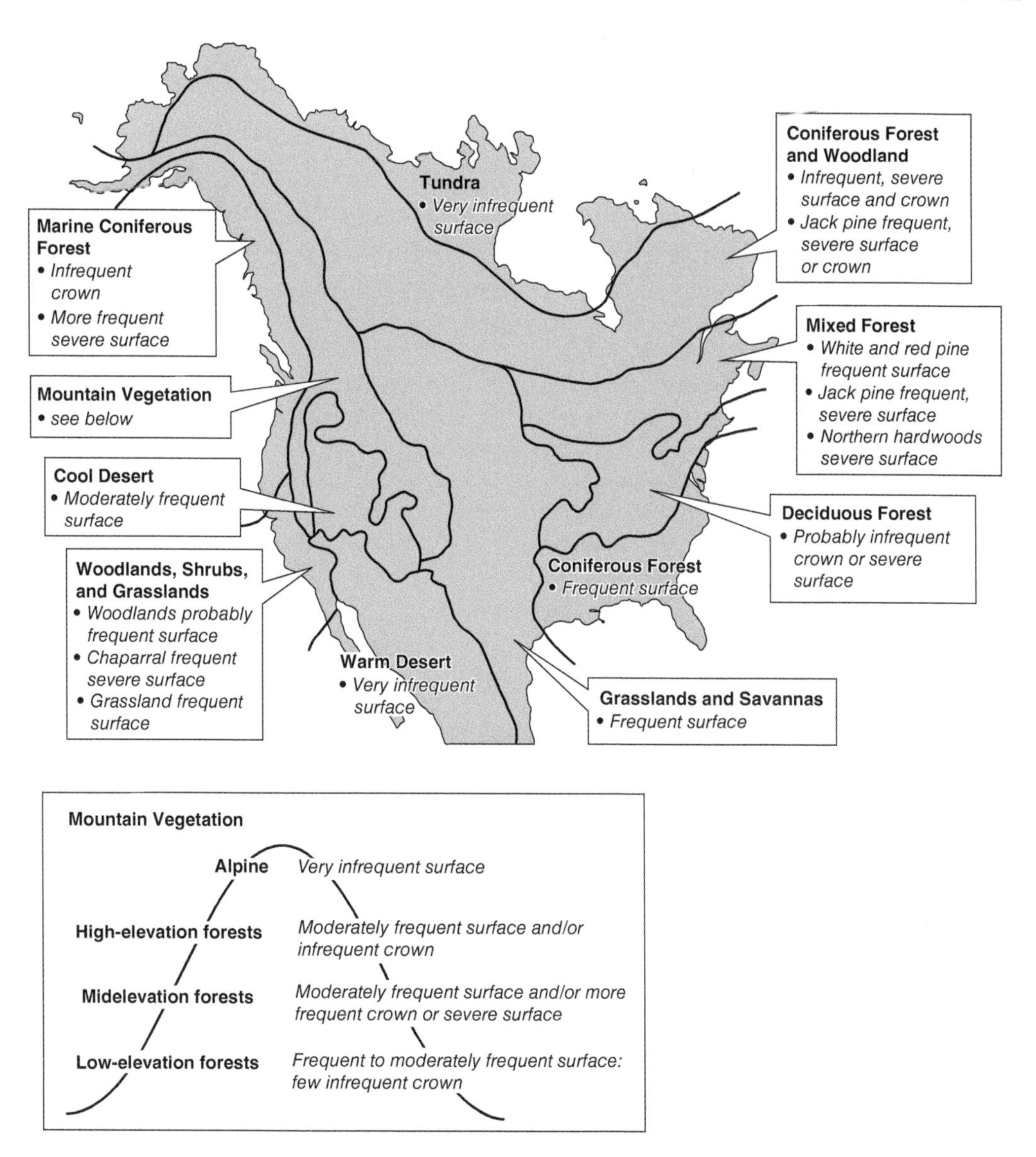

FIGURE 10.2 Pre-European colonization fire regimes of broad vegetation types. Based on broad ecoregions of Bailey (1995). *Source:* Vale (1982) / with permission of American Association of Geographers.

"Grass stage"—longleaf pine of the United States (see Chapter 16), and several pines of Mexico and the Caribbean region (Mirov 1967, p. 417), and chir and Merkus pines of Asia.

Deep rooting—taproot in young plants (upland oaks and hickories).

Rapid juvenile growth—crown grows above the surface fire zone and heat-resistant bark is formed (pines).

Basal crook—dormant buds on the lowermost stem are protected from fire by a crook of the stem that brings the buds in contact with mineral soil (pitch, shortleaf, pond, and other hard pines).

Branch habit and self-pruning ability—rapid self-pruning of branches decreases the likelihood of crown fire (larches, pines, and Douglas-fir in closed stand conditions), whereas low or drooping branching habit and poor self-pruning ability increase the likelihood of crown fire (true firs, hemlocks, and spruces).

Stand habit—open-grown stands decrease the probability of crown fire and also afford less fuel (western larch, ponderosa pine, and longleaf pine).

Fire-resistant live foliage—hardwoods are much less flammable than conifers; among conifers, larches have less flammable foliage than pines, Douglas-fir, spruces, and true firs.

Rapid foliage decomposition—slows fuel accumulation and reduces the opportunity for fire ignition and spread (ashes, elms, sugar maple, and basswood).

The differential resistance of tree species of the northern Rocky Mountains to fire damage and mortality was shown in a classic study by Flint (1930). He determined species' relative resistance using characteristics such as bark thickness of old trees, rooting habit, resin in old bark, branch and stand habit, relative flammability of foliage, and abundance of lichens on stems. Western larch, ponderosa pine, and Douglas-fir are very resistant, whereas western hemlock and subalpine fir are low to very low in resistance. Although species vary greatly in resistance, severe fires largely erase differences in resistance (Wellner 1970).

(ii) **Recovering from Fire Damage:**

Sprouting—new shoots arise from living stems or roots after fire damage:

Stems:

Rhizomes (many shrubs and herbaceous plants).

Root collar or crown (oaks, paper birch, black cherry, redwood, and chaparral species)—basal sprouting is rare in conifers, but in hard pines (shortleaf, pitch, Monterey, etc.) sprouts may occasionally arise from dormant buds formed in the axils of primary needles at the base of the stem (Stone and Stone 1954).

Lignotubers (burls)—relatively large stem swellings at or below ground incorporate many clusters of dormant buds, common in redwood and nearly all eucalypt species and Mediterranean shrubs (Gill et al. 1981; James 1984).

Bole—dormant buds along the bole initiate new shoots after the crown is scorched or killed (pitch pine, redwood, big-cone Douglas-fir, and eucalypt species).

Roots:

Important in some trees and shrubs (aspens, rock elm, sweet gum, sassafras, sumacs, and Acacia species).

Deep rooting—taproot or sinker roots provide food reserves for rapid regeneration of new shoots (upland oaks and hickories).

(iii) **Colonizing Burned-Over Areas:**

Early flowering and seed production—enable a species to reproduce itself sexually on a site where fires regularly or occasionally occur at short to moderate intervals (jack pine, lodgepole pine, pitch pine, and *Xanthorrhoea* species of Australia (Gill et al. 1981).

Light, wind-borne seeds—facilitate widespread dissemination of seeds, some of which may reach burned sites (birches, aspens, willows, hemlocks, and larches).

On-plant seed storage and fire-triggered dehiscence of fruits and cones—closed cones that contain viable seeds persisting on branches are typical of many species of the pine family (including jack, lodgepole, pitch, pond, sand, table mountain, knobcone, and Monterey pines) and cypress species of arid sites of southern California (Vogl et al. 1988). Black spruce exhibits semi-serotinous cones; the cone scales open upon drying and close when wet. Seed dispersal is therefore periodic, sometimes occurring over a 2-year period.

Shrubs of the Australian genus *Banksia* retain follicles on their branches that remain closed unless they have been in contact with flames (Gill et al. 1981). Many other taxa of the Australian flora also have this wood-fruit habit. Also, some eucalypt species of southeastern Australia retain their fruit 4 or more years with seed release following fire.

No dormancy—certain species of fire-prone ecosystems have seeds that do not enter true dormancy. They germinate readily at any favorable time following post-fire rains (jack, lodgepole, longleaf, and pitch pines).

Heat-induced germination—hard-coated seeds of certain species of *Arctostaphylos*, *Ceanothus*, and *Rhus* tend to lie dormant in the soil (in redstem ceanothus up to 150 years (Mutch 1976)). Germination is favored by fire, which cracks the seed coat and generates the heat needed to stimulate germination.

(iv) **Promoting Fire Occurrence:**

Traits that increase the likelihood of fire and thereby favor regeneration of the species over others:

Flammable foliage and bark—needles of pines, many other conifers, and the foliage of some angiosperms are highly flammable, decompose slowly, and form a ready fuel source for surface fires (Mount 1964; Mutch 1970; Gill 1977). The bark of certain species, such as paper birch and some eucalyptus species, is decay-resistant yet highly flammable, creating an excellent and persistent fuel source on the forest floor.

Retention of foliage near the ground—promotes crown fires (understory conifers, especially firs, northern white-cedar and western red cedar; juvenile oaks).

Short stature—brings foliage close to the ground where a surface fire may spread to the crown (young or slow growing trees with flammable foliage, jack pine).

Retention of dead, lower branches—serves as "ladder" fuel promoting fire spread to the crown (narrowleaf cottonwood, jack pine, and some oaks).

STRATEGIES OF SPECIES PERSISTENCE

Species persist in fire-prone ecosystems using one or more of these and other adaptations, regenerating themselves either by seed or by vegetative means. Rowe (1983) distinguished five kinds of "strategies," based on an understanding of fire adaptations, life cycle, and mode of persistence of plants in ecosystems of northern and boreal forests. Plants reproducing primarily by seeds are termed **invaders**, **evaders**, and **avoiders**. Plants reproducing primarily by horizontal and vertical extension of vegetative parts are the **resisters** and **endurers**. These groups were conceived for boreal conditions, but have wide application elsewhere, as Agee (1993) has shown for the species of western North America.

Invaders are the early arrivers that are successful because they produce large amounts of short-lived, wind-disseminated seeds. Once established, they flower and fruit profusely or spread vegetatively. They are typically the shade-intolerant, pioneering, "fireweed" plants that establish vigorously after fire regardless of its frequency or intensity. Many herbs exemplify this group; woody invaders include paper birch, aspens, and willows.

Evaders store seeds in the canopy, humus, or mineral soil, placing seeds to evade high temperatures, followed by rapid seed germination and establishment. The plants themselves may be killed by fire, but their seeds are protected. They include short-lived ephemerals as well as intolerant to shade-tolerant species that persist into later successional stages (Rowe 1983). Woody evaders of the boreal forest include jack and lodgepole pines and black spruce.

Avoiders arrive late in succession and prosper where fire cycles are long. They essentially lack direct adaptations to fire and are often said to occupy unburned areas or ecosystems relatively undisturbed by fire. They are typically mesophytic and shade-tolerant. Included in the group are many herbs and shade-requiring mosses and lichens; balsam fir and white spruce are woody examples in northern and boreal forests. Sugar maple, basswood, beech, and southern magnolia are examples of deciduous avoiders of eastern and southern forests.

Resisters are the few intolerant species whose adult stages can survive *low-severity* fires. They continue growing *vegetatively* in spite of fire (whereas evaders are likely to be killed by fire). Boreal examples include jack and lodgepole pines and cottongrass, a circumpolar sedge, whose dense tussock form resists fire. Giant sequoia; ponderosa, red, shortleaf, loblolly, and longleaf pines; western larch; and many thick-barked oaks are also resisters. This group is not mutually exclusive with evaders.

Endurers are composed of the large and diverse group of species, both shade-tolerant and shade-intolerant, that re-sprout following fire. They regenerate from roots, root collars, rhizomes, and other belowground organs whose vertical position in the insulating humus and mineral soil strongly determines these species' persistence and abundance. Aspens, birches, and many shrubs, including ericaceous bog species, are woody representatives of the group.

These groups of species are closely related to the site-specific ecosystem where they occur and its particular fire cycle and fire behavior. Ecosystems characterized by high-severity fires favor invaders, evaders, and endurers; those with low-severity surface fires favor resisters as well as evaders and endurers. Species may belong to more than one group. For example, jack and lodgepole pines may be invaders, evaders, and resisters because of their multiple adaptations in fire-prone ecosystems.

Closed-Cone Pines Many species' life-history traits are strongly related to fire frequency and severity in fire-prone ecosystems. Such is the case with the varied group of pines bearing serotinous cones: jack and lodgepole pines of northern and western North America; pitch pine and pond pine of the Coastal Plain of the eastern United States (Lutz 1934; Ledig and Fryer 1972; Givnish 1981; Ewel 1990); table mountain pine of the southern Appalachians (Zobel 1969); and knobcone, Bishop, and Monterey pines of California (Vogl 1973). Closed cones may persist on the tree for many decades and still bear viable seeds. In lodgepole pine, millions of seeds per hectare may be stored in serotinous cones. In such cones, the cone scales are bonded by resin that melts above 45 °C (113 °F), typically due to heat generated from fire. The scales open as they dry, and the seeds are disseminated. Amazingly, seeds of jack and lodgepole pines may remain viable even when cones in the tree crowns are exposed to temperatures of 900 °C (1652 °F) for 30 seconds.

Both jack and lodgepole pines may bear various proportions of serotinous cones and those that open readily (nonserotinous) under normal climatic conditions. These proportions vary both within an individual tree and among trees in a population. The range of closed cones per tree in jack pine is from 0 to 100%; the average for the species is about 78% (Schoenike 1976). Open-cone trees are found in the southern portion of the range, whereas trees with serotinous cones predominate in the boreal and western range (Figure 10.3). In the south, jack pine occurs in mixed stands with oaks and red pine, and fires are more often light surface fires than in the boreal forest. In the boreal forest, jack pine is regenerated by crown fires that open the cones and prepare the seedbed,

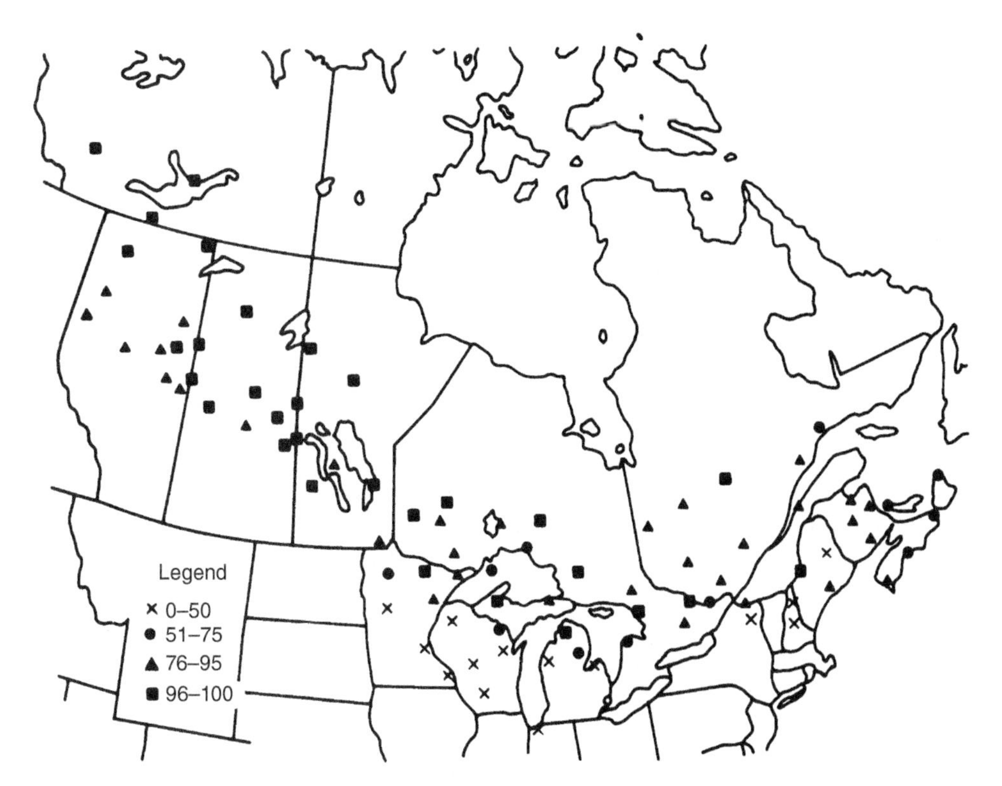

FIGURE 10.3 Variation in the percentage of closed cones per tree in jack pine. *Source:* Based on Schoenike (1976).

thereby perpetuating the predominance of the serotinous habit. Similarly, in lodgepole pine, stands originating after a severe burn typically produce trees with a high percentage of closed cones (Muir and Lotan 1985).

More recent studies of geographic variability in serotiny have examined the proportion of trees in a population that bears serotinous versus open cones. For example, the percentage of serotinous trees for jack pine in northwest Wisconsin was 83% for populations where forests were characterized by high-severity crown fires, but only 9% where forests burned more frequently by low-severity surface fires (Radeloff et al. 2004). In Yellowstone National Park in northwestern Wyoming, Tinker et al. (1994) found a higher percentage of trees with serotinous cones in lodgepole pine stands on xeric sites compared to mesic sites, and a significant negative correlation of closed cones with elevation. Expanding this work, Schoennagel et al. (2003) found that the proportion of serotinous trees was higher but varied with stand age at low elevations where stand-replacing fires burned every 135–185 years. At high elevations where fires burned every 280–310 years, the proportion of serotinous trees was low and did not vary with stand age. Moreover, younger stands at low elevations were least likely to have a high proportion of serotinous trees, but this proportion increased with stand age. Elsewhere, populations of a given species with a very high proportion of cone serotiny also typically occur in the most frequently burned ecosystems (Givnish 1981). For example, serotiny is common in pitch pine populations of the Coastal Plain, but rare in the Appalachian Mountains where it occurs with serotinous populations of table mountain pine that grow on more fire-prone sites. Taken together, these data suggest a very close relationship between fire regimes (especially frequency and severity) and expression of the serotinous trait across large landscapes.

Givnish (1981) emphasized the importance of natural selection by fire in maintaining the serotinous habit in pitch pine in a local portion of its distribution—the Pine Barrens—on the Coastal Plain of New Jersey. In the Pine Plains area of the barrens that topographically favor frequent fire every 6–8 years (Lutz 1934), 99% of the individuals are serotinous. Cone serotiny decreases with distance in all directions from the Pine Plains; infrequently burned lowland sites (fire frequency 16–28 years) averaged 84% serotinous individuals. Givnish and Lutz stressed that the flatness and physiographic location of the Pine Plains affected the incidence of fires and promoted high cone serotiny, in that fires from many directions and distances could carry fire uninterrupted into the Pine Plains.

FIRE AND THE FOREST SITE

Many writers once considered and described fire to be a destructive agent with few or no beneficial aspects. The success of prescribed fire in improving silviculture, reducing fuels, and managing wildfire risk has led to the realization that the effect of fire on forest ecosystems is complex and may often be entirely beneficial. The many important functions of fire and its beneficial effects in forest ecosystems are considered in Chapter 16. In this section, we examine fire effects on site conditions that affect the regeneration and growth of plants. We distinguish between the *indirect* effects of fire on site quality, through its effects on vegetation, organic matter, and soil organisms, and its *direct* effects on soil properties and microclimate.

INDIRECT EFFECTS

The indirect effects of fire on site quality depend upon changes in the vegetation, as discussed in detail in Chapters 16 and 17. A high-severity fire will kill most or all of the plants above the soil surface, and even a surface fire will kill smaller vegetation and affect dead organic matter. Soil's physical, chemical, and biological properties are heavily influenced by live vegetation and dead organic matter (Chapter 9), and thus fire may invoke major changes upon the soil via its effects on these two factors. When vegetation is killed, loss of shade affects soil temperature, ceased transpiration affects soil moisture, and the death of roots affects soil stability. In these examples, fire effects are on the organic components of the ecosystem which in turn affect the soil, and thus those effects are indirect.

Perhaps the most obvious way fire indirectly affects the site is by its role in the type of vegetation growing there. Vegetation re-establishing after fire tends to be made up of light-seeded species that colonize from outside the burned area, species with perennial root systems capable of vegetative reproduction, and species with dormant seeds stimulated by heat. Many legumes and chaparral species (*Ceanothus* spp.; Hanes 1988) fall in these categories, and the abundance of these and other nitrogen-fixing plants often increases after fire. In such a case, although previously accumulated nitrogen may be volatilized by the fire, a rapid increase in available nitrogen often occurs and the overall site quality may be temporarily improved. By contrast, in many parts of the world, recurrent fires favor the development of shrubby vegetation composed of sprouting species with characteristically tough foliage low in nutritional value and slow to decompose. Heather in northern Europe, blueberry and other ericaceous species and bracken in many countries, scrub oaks around the Northern Hemisphere, and many chaparral types (the broad-sclerophyll scrub or brush-land vegetation) of California (Hanes 1988) and the Southwest, all are plants that become dominant after heavy and repeated fires.

Fire that removes forest tree cover and forest floor organic matter may substantially change the temperature regime of soils. In general, soil temperatures generally increase following fires because of the removal of vegetation and organic matter and the blackening of the surface. Average maximum soil surface temperatures on burned sites may be from 3 to 16 °C higher than on comparable unburned areas (C. Ahlgren 1974a). Soils of openings in southern forests are colder than those under the canopy, whereas in northern forests the reverse is true (Komarek 1971). In the boreal forest, tree canopies promote the deepest frost beneath them, and in permafrost areas (ground that is permanently frozen) surface soil thaw is least beneath a canopy (Rowe and Scotter 1973). A large part of the Alaskan boreal forest is in the zone of discontinuous permafrost (Van Cleve et al. 1986). Forest canopies in winter intercept snow that would otherwise provide insulation from the cold on the ground, and organic matter built up under forest cover in summer provides insulation from heat. Fire therefore contributes both to increased heat flow into the soil in summer (via removal of tree cover, creating a blackened surface, and reducing surface organic cover) and decreased heat outflow in winter (increased snow cover). These factors cause permafrost to melt at increasing depths below the surface following fire, although the effects are generally temporary (Certini 2005). Viereck (1982) reported that thaw depth increased steadily for 9 years following fire. Such belowground changes significantly affect surface hydrology, vegetation composition, and site productivity.

Removal of live vegetation by fire affects soil moisture as well as temperature, although the actual effect is variable depending on fire severity. An immediate effect may be that the loss of vegetation that once intercepted precipitation before it reached the soil and transpired away soil water would increase soil moisture. At the same time, however, the loss of organic matter on the forest floor may increase evaporation from the soil surface. High-severity fires that consume soil organic matter may greatly reduce its porosity and infiltration (see in the following text), as well as its water-holding capacity, leaving it with significantly less soil moisture. For example, a modeling study of post-fire soil-water balance in New Mexico found higher soil moisture following light- or moderate-severity fires due to reductions in evapotranspiration when the vegetation was killed (Atchley et al. 2018). However, high-severity burns were more strongly characterized by larger increases in surface runoff of water compared to increased soil moisture, resulting in drier soils, suggesting that there is a threshold of fire intensity at which soil moisture is affected.

The removal of organic matter is probably the most obvious effect of fire on soils. Fire regulates dry-matter accumulation on the forest floor and in coarse woody debris, thereby affecting fuel loadings and ultimately controlling the severity of burning. As previously discussed, fire severity affects the density and composition of forest vegetation, which in turn influences site quality. Abnormally long intervals between fires, such as those caused by prolonged fire exclusion, often lead to large fuel accumulations and high concentrations of organic matter. For example, a study conducted in the oak-dominated forests of the Missouri Ozarks showed that in a regional ecosystem characterized by surface fires that burn every 8–15 years, 50% of the litter re-accumulated within 2 years after fire, 75% within 4 years, and 99% after about 12 years (Stambaugh et al. 2006), well within the period between fires. Under fire suppression in this and other forest types, uncharacteristically intense fires that eventually occur may preclude or delay the reestablishment of normal vegetation on the site or change the type of vegetation present, although indirect effects of change or lack of vegetation are much less than the direct effects of severe burning.

Soil microbes and fungi significantly influence site quality by decomposing organic matter, fixing nitrogen, and providing aeration (Chapter 19). The effects of burning on soil organisms are highly variable depending on fire intensity, depth into the soil profile, time elapsed following burning, the nature of the soil, and the vegetation of the site (I. Ahlgren 1974b; Rundel 1981; Dooley and Treseder 2012). Fire immediately reduces the biomass of microorganisms, mostly in the upper 5 cm of soil; fatal temperatures for many microbes may be less than 100 °C (DeBano et al. 1998), which is below the temperature achieved during many fires. Light surface burns may therefore have almost no effect on microbial biomass, but extreme high-severity fires have been demonstrated to completely sterilize the upper soil horizons (Certini 2005). In a meta-analysis of 42 studies of microbial and fungal responses to fire, Dooley and Treseder (2012) found that fires reduced microbial biomass on average by about 33%. In addition to lethal heat, post-fire declines in microbes have been attributed to declines in soil carbon following fires (Waldrop and Harden 2008). General trends for fungi following fires are less clear, although they are more heat-sensitive and thus may be more highly affected (although some fungi have been reported to be fire-adapted; Wicklow 1975). Fungal biomass decreases more so than microbial biomass after fire, estimated by Dooley and Treseder (2012) to be by about 48%.

Microbial abundance typically increases after fire, often multifold, apparently due to the sudden availability of organic substrate available in the soil, increases in soil pH, increases in soil temperature and moisture following removal of tree canopies by fire, and other soil chemical changes associated with burning. Deposition of ash increases soil pH and the availability of nitrogen (Wan et al. 2001), the latter of which may increase due to higher mineralization rates and stimulate further microbial growth. Fungi, which are tolerant of acidic soils, may decline after fire in favor of bacteria when ash deposits increase soil pH. Microbial and fungal growth may also benefit from reduced evapotranspiration that increases soil moisture and reduced shading that increases soil temperature when fire kills aboveground vegetation, although temperature increases are most beneficial in colder climates (Smith et al. 2008; Treseder et al. 2004). Where re-growth of fire-adapted vegetation is rapid, one would expect that such changes in soil temperature and moisture are relatively short-lived.

In their meta-analysis, Dooley and Treseder (2012) found that the negative effects of fire on microbial biomass lasted about 15 years, whereas aboveground net primary productivity (ANPP; Chapter 18) may recover as quickly as 4 years after a fire. Organic matter in the soil accumulates much more slowly than ANPP, suggesting that the post-fire recovery of microbial biomass depends more strongly on re-accumulation of soil organic matter than on re-vegetation. For example, Litton et al. (2003) found that microbial biomass was still lower in burned stands compared to unburned stands even 13 years after a high-severity fire in subalpine lodgepole pine forests, probably because of a lack of organic layer re-accumulation relative to rapid forest re-vegetation.

Other soil fauna such as beetles, spiders, mites, collembola, centipedes, and millipedes are typically reduced by burning but increase thereafter (I. Ahlgren 1974b). Invertebrates in general are more mobile than microbes and fungi, which allows them the potential to move deeper into the soil to escape the heat of a fire. Ants are less affected than other fauna because their behavior enables them to survive in lower soil layers and they are adapted to xeric conditions of post-fire topsoil. The social organization and rapid colonizing ability of ants also enable them to reestablish populations rapidly after fire. Notably, reduction of litter on the forest floor by fire has a negative effect on the diversity and abundance of soil invertebrates, such that re-accumulation of the litter layer is likely to be a limiting factor for them much as it is for microbial biomass (Certini 2005).

DIRECT EFFECTS

Fire directly affects site quality via (i) the burning of organic matter above and on the mineral soil, and (ii) the heating of the surface layers of the soil. The burning of organic matter releases carbon dioxide, nitrogenous gases, and ash to the atmosphere and deposits minerals as ash. Ash from wood and litter is more soluble than the organic matter from which it was formed. Thus, in creating ash, fire increases available minerals, at least temporarily, lessens soil acidity and increases base saturation, and decreases total soil nitrogen.

Two classic studies are indicative of many studies on fire effects. In the ponderosa pine region of Arizona, burning increased soluble nutrients when the surface layer of unincorporated organic matter was ashed (Fuller et al. 1955). This increase in turn raised pH, available phosphorus, exchangeable bases, and total soluble salts; caused a decrease in organic matter and nitrogen to a depth of 20–30 cm; and increased microbial activity. Unfortunately, the soil surface was compacted by rains when litter was removed in this study, reducing the rate of water infiltration. In the Douglas-fir region of eastern Washington, Tarrant (1956b) investigated the effects of slash burning on physical soil properties. Light burning increased the percolation rate of water within the upper 8 cm of soil, but severe burning seriously impeded water drainage by about 70%.

Minerals in ash are readily available to plants because of its small particle size and high solubility, but these same characteristics also make the ash susceptible to leaching by precipitation. If the ash is washed down into the soil so that the roots can absorb the nutrients dissolved from it, site quality is usually temporarily improved. Such a process is typical on level, coarse soils of sandy to loamy texture. However, site quality is lowered if the ash is leached down below the tree roots (or is washed off the surface by runoff), which is apt to happen on very coarse sands (leaching) or clayey soils with considerable slope (runoff).

Total nitrogen may be lost through volatilization, which is closely related to fire intensity (Rundel 1981). Maximum ground temperatures during forest fires typically range between 200 and 300 °C (Rundel 1983), but may reach 500–700 °C in heavy fuels such as slash (Dunn and DeBano 1977). Nitrogen volatilization begins when temperatures reach 200–400 °C (Neary et al. 1999). Working in the coastal Douglas-fir region, Knight (1966) found no nitrogen loss in soils heated to 200 °C, a 25% loss at 300 °C, and a 64% loss at 700 °C. Nitrogen loss is also proportional to the amount of fuel consumed, and considerable nitrogen may be lost when fire is intense. For example, a high-severity fire in a second-growth, mixed-conifer stand in northcentral Washington resulted in a loss of about 97% nitrogen originally in the forest floor and a loss of 66% of the nitrogen in the A horizon (Grier 1975). In this case, nitrogen re-accumulation is expected to occur relatively rapidly due to N fixation by symbionts associated with snowbrush (*Ceanothus velutinus*; see Chapter 19). Although major volatilization of nitrogen occurs with fire, available forms of nitrogen are commonly higher on burned than unburned sites (Rundel 1981).

Much of the nitrogen lost through burning of litter and vegetation is organic nitrogen and in a form unavailable to plants. The organic matter layer in coniferous forests in boreal regions, for example, contains a large amount of unavailable nitrogen (Viro 1974). Organic nitrogen that is not volatilized must be mineralized and converted into ammonia for vegetation to have access to it (Chapter 19). The degree to which fire affects site quality thus depends largely on the ability of the succeeding vegetation and soil bacteria to replace the available nitrogen lost in burning. Higher post-fire pH due to ash deposition can improve the soil environment for free-living, N-fixing bacteria and thus may initiate an increase in available nitrogen, albeit slowly. Nitrogen loss is often considered to be a deleterious effect of fire, but fire at the same time increases the abundance of ammonium (Neary et al. 1999). Ammonium is eventually transformed to nitrate and is lost to leaching if not taken up by plants, such that nitrogen availability lowers to pre-fire levels in only a few years (Certini 2005).

Importantly, heating and burning of organic matter are of much more consequence than heating of the mineral soil, as the mineral soil horizons rarely experience the same heat as the organic horizons above. Nevertheless, even low-severity fires may heat the surface soil to 50 and 100 °C (122 and 212 °F) at 5 cm depth (Agee 1973). Where organic horizons are very thick and are burned with hot fires, mineral soil at the surface may reach 275 °C (527 °F) (Sackett and Haase 1992). Temperatures of surface soil in very hot fires may approach 700 °C (1292 °F) beneath slash piles and in such cases may exceed 100 °C as deep as 22 cm below ground. Removal of the entire organic horizon by fire may expose mineral soil to direct irradiation by the sun, elevating soil temperatures for months or even years (Neary et al. 1999). Soil aggregates may be broken down in this heated zone, first by the heat and later by the direct striking action of raindrops, resulting in loss of soil structure and lowered infiltration capacity of the surface soil, increasing surface runoff. In many cases, burning heavy accumulations of slash in piles causes intensely hot fires and may alter the physical structure of the underlying soil. For example, lower tree densities and markedly slower conifer growth were documented on burned slash pile sites in northwestern Montana compared to adjacent unburned areas (Vogl and Ryder 1969). The reduction in growth was attributed to impaired water infiltration and other physical soil properties.

ORGANIC MATTER AND EROSION

Fire effects on the site are greatest where the soil is composed almost entirely of organic matter, and where the removal of organic matter by fire exposes highly erodible soils to heavy rain. Burning peat bogs after drainage or a series of dry seasons results in complete destruction of the soil and reproduces swamp conditions. The lost organic soils would take thousands of years of peat accumulation to replace them, and such areas can be virtually eliminated as potential sites for forest growth. In the United States, many bogs on the Atlantic Coastal Plain and in the recently glaciated parts of the Northeast and the Lake States have been destroyed as forest sites by fire. Similarly, burning organic matter located directly atop rock will eliminate the soil. In glaciated portions of Canada and the northeastern United States and in many mountainous regions, thin accumulation of humus provides the only nutrition for forest trees, and fire may completely remove this thin layer (Kelsall et al. 1977).

Removal of organic matter by fire that exposes highly erodible soil on steep slopes may result in catastrophic erosion or landslides (Rundel 1981; Swanson 1981). Post-fire landslides require brief (15 minutes or more) heavy rainstorms to trigger debris flows. Such events are natural and have occurred regularly in steep areas that burn; rainstorms that initiated 77% of debris flows in the southwestern United States recur approximately every 2 years (Staley et al. 2020). The classic example of this process is in southern California, where chaparral species and the organic matter from them protect the granitic or clayey soils lying at approximately the angle of repose in steep mountains. Swanson (1981) estimates that over 70% of the long-term sediment yield from steep chaparral slopes is the result of fire effects. Chaparral is highly susceptible to burning, especially when it reaches about 30 years of age, because of the dense, highly flammable fuel accumulations, shrubby growth habit, and seasonal dry, windy periods (Biswell 1974; Keeley and Keeley 1988). Following wildfire, Biswell (1974) reports that debris movement may increase dramatically, especially on south-facing slopes (Figure 10.4). In this instance, most of the movement is during the dry period. If the characteristic heavy rainstorms of this semiarid region strike before re-vegetation of the burn, whole slopes may wash downhill.

Fire-caused erosion and landslides occur in many different physiographic contexts. Fires can also cause serious erosion in the northern Rocky Mountains where erosive soils occur on steep

Photo courtesy of the Pacific Southwest Research Station, USDA Forest Service

FIGURE 10.4 Dry-creep erosion may be severe immediately after intense wildfire in old-growth chaparral on slopes above the angle of repose.

slopes, the fires tend to be of high severity, and multiple burns frequently occur. In general, the amount of sediment transported a year after fires ranges from very low in flat terrain and with dry post-fire weather to extreme in steep terrain with heavy thunderstorms. Soil erosion typically increases with fire severity. Notably, erosion is a temporary process and decreases as vegetation recovers following fires, but it can lead to significant losses of nutrients from ecosystems (Neary et al. 1999, 2005).

The burning of litter and organic matter in the soil may be significant in causing reduced infiltration, increased surface runoff, and erosion in many areas of the western United States where water-repellent soils have been reported (DeBano et al. 1967; Meeuwig 1971; Dyrness 1976). Soils can become resistant to wetting in multiple ways. Removal of vegetation and litter layers by fire allows the force of raindrops to be directly absorbed by the soil rather than by the organic matter, which degrades soil structure at the surface and greatly reduces infiltration. In such cases, droplets do not readily penetrate and infiltrate, but "ball up" and remain on the soil surface for variable periods of time. This phenomenon is widespread in sandy soils throughout much of the wildland areas of western North America supporting chaparral or coniferous vegetation, as well as many other parts of the world (Foggin and DeBano 1971). By increasing surface runoff, this process reduces soil moisture, allows nutrient-rich ash to be eroded away, and effectively dries the soil, reducing microbial activity, nutrient availability, and plant growth (Neary et al. 1999).

Fire can actually promote the development of water-repellent soils. In unburned areas, non-wettable or hydrophobic organic molecules produced by litter decomposition coat surface soil particles, creating a weak water-repellent layer between the litter layer and the mineral soil (Figure 10.5; DeBano 1969). High-intensity fires consume litter and volatilize the hydrophobic substances, which diffuse downward and attach to cooler soil particles lower in the soil profile (Figure 10.5). This deeper water-repellent layer forms as a result of fire when soil temperatures rise above 176 °C (349 °F), but is destroyed when temperatures exceed 288 °C (550 °F) (DeBano et al. 1976). As a result, water repellency is high after summer and fall wildfires such that rain

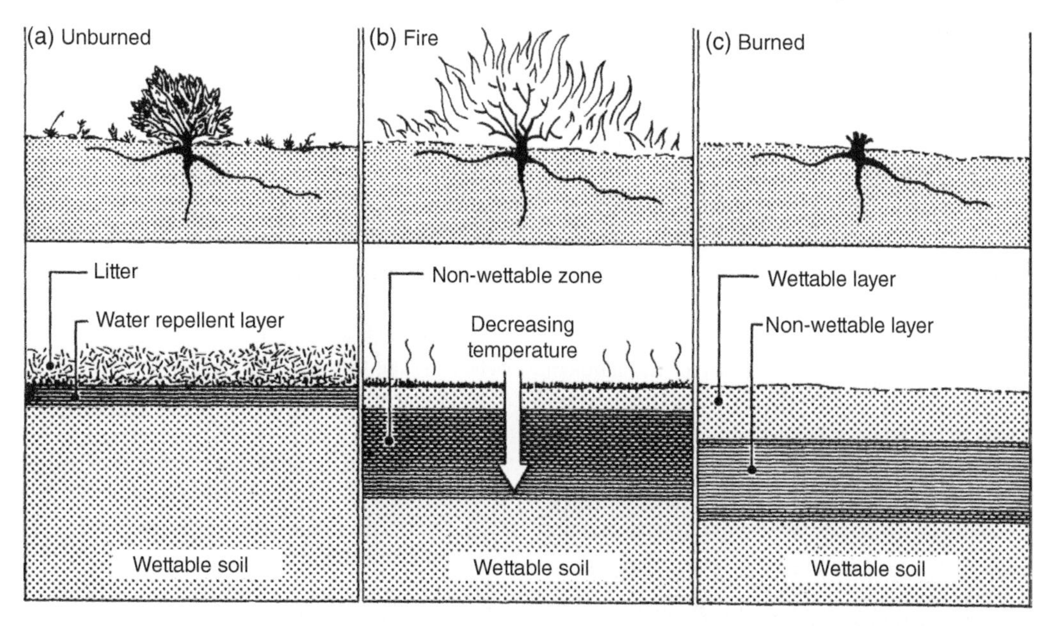

FIGURE 10.5 Soil non-wettability before, during, and after fire. (a) Before fire, the non-wettable substances accumulate in the litter layer and mineral soil immediately beneath it. (b) Fire burns vegetation and litter layer, causing non-wettable substances to move downward along temperature gradients. (c) After a fire, non-wettable substances are located below and parallel to the soil surface on the burned area. *Source:* DeBano et al. (1967) / United States Department of Agriculture / Public Domain.

falling on the soil surface infiltrates readily until impeded by the shallow non-wettable layer, facilitating surface runoff. In regions where the terrain is flat, this process contributes to drying of soils, but in steep regions can cause serious erosion (Neary et al. 1999).

BENEFICIAL EFFECTS OF FIRE

Thus far we have emphasized mainly the deleterious effects of fire on soils and site quality (other than the release of ammonium), but controlled burning has shown to benefit site quality in many forest ecosystems. On some sites, such as in the boreal forest where thick layers of humus tend to develop under coniferous vegetation, burning improves site quality because it enhances the post-fire thermal regime. Increased soil temperatures in burned areas hasten spring development of roots and shoots. Extremely high post-fire soil temperatures may cause seedling mortality and delay forest regeneration, although fire-dependent species on fire-prone sites are typically adapted to tolerate such extreme conditions.

As long demonstrated by many classic studies, the effect of fire on site quality is often relatively subtle. Metz et al. (1961) showed that repeated burning in the sand plains of the Southern Pine region at worst had no major detrimental effect on site quality and, at best, was locally beneficial to the soil. Likewise, nine successive annual surface burns under even-aged shortleaf and loblolly pines had little effect on the nutrient content or structure of the topsoil on loess soils on flat terrain in Arkansas (Moehring et al. 1966). Studies of soil biota after 20 years of annual prescribed burning on a very fine sandy loam on the Coastal Plain indicated little evidence that fire adversely affected metabolic processes (Jorgensen and Hodges 1970). Burning had no effect on total concentration of fungi, although it did reduce their abundance by decreasing the organic horizon. The number of bacteria in the organic layer was reduced by annual burning but not in mineral soil. Stark (1977) found that controlled burning in a Douglas-fir–larch forest on sandy loam soils in northwestern Montana generally reduced accumulated fuels without nutrient loss, runoff, or erosion, and estimated that burning could be conducted on this soil for a very long time with no problem of soil fertility. Notably, ecosystems such as the sand plains of the Lake States, where organic matter is the major source of nutrients for plants, may not benefit from burning, especially for species with relatively high nutrient requirements (Stoeckeler 1948, 1960).

SUGGESTED READINGS

Agee, J.K. (1993). *Fire Ecology of Pacific Northwest Forests.* Washington, D.C.: Island Press 505 pp. Chapters 1, 4, 5, and 6.

Arno, S.F. (1980). Forest fire history of the northern Rockies. *J. For.* 78: 460–465.

Baker, W.L. (2009). *Fire Ecology in Rocky Mountain Landscapes.* Washington, D.C.: Island Press 632 pp.

Johnson, E.A. (1992). *Fire and Vegetation Dynamics.* Cambridge University Press 129 pp. Chapters 4 and 6.

Johnson, E.A. and Miyanishi, K. (ed.) (2001). *Forest Fires: Behavior and Ecological Effects.* New York: Academic Press 594 pp.

Rundel, P.W. (1981). Fire as an ecological factor. In: *Physiological Plant Ecology I., Responses to the Physical Environment*, New Series, vol. 12A (ed. O.L. Lange, P.S. Nobel, C.B. Osmond and H. Ziegler). New York: Springer–Verlag.

Wein, R.W. and MacLean, D.A. (1983). *The Role of Fire in Northern Circumpolar Ecosystems.* New York: Wiley 322 pp.

Whelan, R.J. (1995). *The Ecology of Fire.* Cambridge: Cambridge University Press 346 pp.

Wright, H.E. Jr. and Heinselman, M.L. (1973). The ecological role of fire in natural conifer forests of western and northern North America: introduction. *Quat. Res.* 3: 319–328.

Site Quality and Ecosystem Evaluation and Classification

CHAPTER 11

Thus far we have considered separately the forest tree and the physical factors that affect it within a landscape ecosystem framework. In this chapter, we begin an important focus on "site" of the forest ecosystem—the often-complex combination of factors that not only influences tree and forest productivity but affects the entire range of management and conservation applications. Forest ecosystems consist of (i) geographic position in space and its associated physical factors (**site**), and (ii) the **biota** (forest trees and associated plants and animals) that occupy these places and that are supported by site factors. The physical factors of climate, physiography, soil, and disturbances determine the site conditions that affect the occurrence and diversity of organisms, their productivity, and how they change in space and time. The biota in turn modify the direct effects of the physical factors. Obvious examples of how biota affect physical factors include how forest canopy trees affect understory light, temperature, and soil development, or how uprooted or fallen trees in uplands and wetlands (Chapter 8) create microtopography that affects forest composition and dynamics. Therefore, the forest site is *the sum total of all the factors affecting the distribution and growth of forests or other vegetation* (see site diagram, p. 120).

For centuries humans have evaluated the site quality of forests for a variety of purposes. Both simple and complex methods have been attempted with varying degrees of success. We conclude this section of this book, covering the forest environment, with a discussion of approaches to site-quality evaluation. Forest managers were historically concerned mostly with trees and typically with the potential land quality for fiber production, and single-factor methods of evaluation tend to reflect these objectives. Emphasis has largely shifted from the stand (bioecosystem concept) to the land or ecosystem (geoecosystem concept), such that we no longer consider site-quality evaluation only in terms of tree or stand production. Instead, forest managers seek an understanding of whole ecosystems as an attempt to assure their sustainability (Rowe 1992b; Franklin et al. 2018; Palik et al. 2021).

Forest and land managers evaluate sites and ecosystems for recreation, aesthetic values, water and wildlife conservation and management, ecosystem and biological diversity, and maintenance of ecosystem processes in addition to, or to the exclusion of, fiber production. Approaches involving multiple factors and mapping of entire landscape ecosystems have increased in popularity because they provide both the basis for biomass production and the ecological framework for ecosystem conservation, management, and restoration at multiple scales. We also include in this chapter a review of single-factor site evaluation and conventional use of tree height to estimate site quality, as this process has produced a wealth of insights concerning plant growth and occurrence for over a century.

A forest scientist must integrate many physical site factors to effectively estimate forest site quality in terms of forest productivity. Site factors are interdependent, but are also dependent in part upon the forest, which is itself a major site-forming factor. Because of these interactions, estimating site quality from an evaluation of a few important site factors, important as it is in practical forest ecology, is only approximate. Nevertheless, estimating forest productivity is of the utmost

Forest Ecology, Fifth Edition. Daniel M. Kashian, Donald R. Zak, Burton V. Barnes, and Stephen H. Spurr.

importance in forest ecology and ecosystem management. Productivity, or actual site quality, may be measured directly for a few forests where accurate long-term records of stand development and growth have been maintained. Most often, it can only be estimated indirectly by one or more of the alternatives that are considered in this chapter:

Vegetation of the forest:

Tree height (site index method).

Ground vegetation (indicator species and species groups).

Overstory and ground-cover vegetation in combination.

Factors of the physical environment:

Climate.

Physiography.

Soil survey and soil-site methods.

Multiple-factor and multiple-scale approaches (using some or all the abovementioned factors, disturbance regime, and forest land-use history).

Multiple-factor methods have broader applicability than productivity estimation alone, and have reflected the shifting emphasis toward multiple forest values. Significant reviews of traditional forest site-quality estimation have been published by Coile (1952), Rennie (1962), Ralston (1964), Jones (1969), and Carmean (1970a, 1977, 1996), and comprehensive reviews of site-quality evaluation, its history, methods, and application, by Carmean (1975) and Hagglund (1981), also are available.

DIRECT MEASUREMENT OF FOREST PRODUCTIVITY

Actual forest productivity, at least in terms of fiber production, is typically measured in terms of the gross volume of bole wood per unit area per year over the normal rotation. This gross mean annual increment (m.a.i.) may be computed from long-term permanent sample plot data. Classic studies showed Douglas-fir on a pumice soil site on the North Island of New Zealand to yield a gross m.a.i. of 31 $m^3\ ha^{-1}\ yr^{-1}$, and yields of 36 $m^3\ ha^{-1}\ yr^{-1}$ may be expected from Monterey pine on similar sites (Spurr 1963). In the United States, gross m.a.i. may reach 15 $m^3\ ha^{-1}\ yr^{-1}$ on high-quality sites, but average productivity in temperate forests of North America and Europe is approximately 5 $m^3\ ha^{-1}\ yr^{-1}$. Note that growth in these values is presented as gross—the total amount of wood accumulated by all trees within a given unit of time without accounting for natural mortality, removal by humans, or decrease in wood volumes due to rot. Gross values allow a manager to obtain comparable increment measurements of aboveground production. In reality, actual gross productivity data are rare, and actual yield is conditioned by genetic factors, age or rotation, stand history, and stand density as well as site factors. Nevertheless, actual growth is the proven productivity of a site and therefore represents the closest available approximation of potential productivity.

Theoretically, on a given site, a stand of a given species and age will produce the same amount of wood per year at various densities as long as the site is fully occupied. Trees in unthinned stands will fully occupy a site as long as they retain good crown development and vigor. Likewise, trees in thinned stands or even in open stands will normally fully occupy the site, even when excess crown space is available in the stand, if they are able to fully utilize available soil water and nutrients. Thus, thinned and unthinned stands should provide the same total increment if mortality within a growth period, whether from cutting activity by humans or from natural causes, is included.

Modern forest ecologists no longer attempt to estimate forest productivity using the growth of the boles alone, but in terms of all components of the forest ecosystem. Ecosystem ecologists now measure not only the stems but also the branches, leaves, and roots; the organic matter in the forest floor; and even the animals inhabiting the forest, to provide a more exact appraisal of the entire forest ecosystem. A detailed discussion of forest biomass and carbon productivity and its measurement is provided in Chapter 18.

TREE HEIGHT AS A MEASURE OF SITE

The height of free-grown trees is well known to be the best indicator of a site's capacity to produce wood of that species. Free-grown trees are defined as individuals of a given species and age that have grown without suppression by an overstory canopy since their establishment. The height of free-grown trees is the tree dimension least influenced by stand density and may thus be used as an *index* of site quality in even-aged stands that vary in density and silvicultural history. Tree height is especially appropriate for shade-intolerant and midtolerant species compared to tolerant species whose individuals may be suppressed in the understory for variable periods of time.

Tree height of the majority of a forest stand at a specified age is commonly termed **site index**. In applying site index, a set of height/age curves are developed for a geographic area, such as the Coastal Plain and Piedmont physiographic regions of the southeastern United States (Figure 11.1). Using the mean total height and age of a representative sample of dominant "site trees," this set of curves can be used to determine the site index for any given loblolly pine stand in the area. In Figure 11.1, the site index is 50 for a site supporting dominant pines that are 55 ft tall at 60 years, and is 95 for a site supporting pines that are 85 ft at 40 years. In the United States, site index is defined as the average height that the dominant portion of an even-aged stand will have at a standard age. This standard age is generally 50 years in the eastern United States, and 100 years

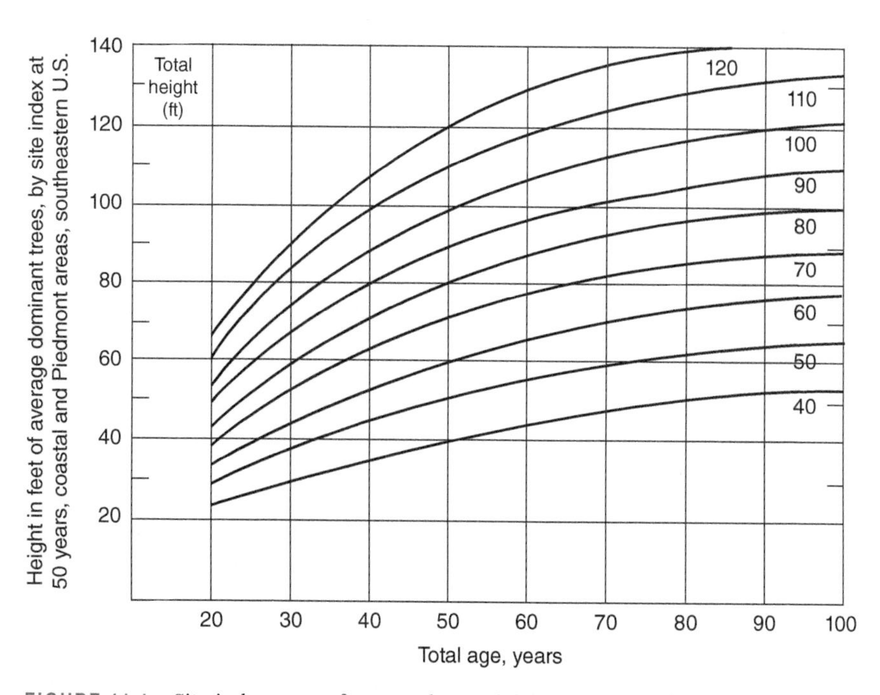

FIGURE 11.1 Site-index curves for second-growth loblolly pine in the Coastal Plain and Piedmont areas of the southeastern United States. *Source:* Hampf (1965) / United States Department of Agriculture / Public Domain.

for the longer-lived species of the West. Occasionally, other standard ages are specified for a particular species or region, for example, 25 years for pulpwood rotations in the South. Importantly, site index is simply a number that is used to indicate the relative productivity of a stand of a given species at one place compared to stands of the same species on other sites. While useful for such simple comparisons, site index is unable to indicate *why* a given site may be poor, medium, or good for growth and yield of the species currently growing there.

Determining and measuring dominant or codominant trees is often problematic, and thus the trees that should be measured for site determination are carefully selected. Often the height of an objectively determined sample of the larger trees in the stand is measured. In Britain and some other countries, the concept of **top height** is widely used, which refers to the arithmetic mean height of the 250 largest-diameter trees per hectare. It is often not feasible to measure so many trees per hectare for height, and heavy thinning practices commonly reduce the stand to fewer than 250 trees per hectare. In such cases, a mean height based on the largest 100 trees per hectare (sometimes defined as mean **predominant height)** is used. Top heights may also be based upon even fewer of the largest diameter trees when many are not available, such as the 60 largest-diameter trees per hectare, the 8–12 largest trees, or even upon the 1 largest tree per hectare.

Tree height is the single best measure related to the site productivity of a given species, but it is not necessarily unrelated to other factors, nor is there always a perfect correlation between stand top height and site productivity. Stand density, particularly extremes of stand density, may influence height growth, and under such circumstances site-index curves are developed separately for different stand density classes, as has been done for lodgepole pine in the Rocky Mountains (Alexander 1966; Alexander et al. 1967). Moreover, it is not uncommon for the same site index to be determined for a given species on two different sites because of differences in stand density rather than similarities in productivity (Assmann 1970; Curtis 1972). For example, for Douglas-fir in northern Idaho and northwestern Montana, the density of trees is greater in even-aged pure stands than in uneven-aged mixed-species stands of the same dominant height (Sterba and Monserud 1993). Although the site index for Douglas-fir for these two sites is the same, the productivity is different.

Within regional ecosystems, genetic factors likely are strong drivers of height growth, especially across an area heterogeneous in climate and landform (Monserud and Rehfeldt 1990). Genetic control of tree height growth within a species tends to be weaker at local scales, and site factors such as soil water and nutrients are more important. Naturally occurring aspen clones on sandy soils in northern Lower Michigan have been found to differ greatly in height on the same site, with some clones more than twice as tall as adjacent clones of the same age (Zahner and Crawford 1965). Such variation is much more likely in species that develop multi-stemmed clones than in species where each stem is a different genotype. Competition among clonal trees is primarily between stems of the same genotype, such that individual trees in slow-growing clones are more likely to survive than in stands where competing individuals are of different genotypes. In such a case, sampling the largest dominant trees within the same clone may overestimate site quality if the clone is genetically superior, and the mean total height from at least five different clones rather than five "site trees" is probably more appropriate.

A final point is that the condition of the site at the time of stand establishment, as well as competition from other vegetation in early years, may affect height growth. Both naturally reseeding and planted pines will usually show different growth trends on old fields, where soil nutrients, microorganisms, structure, and drainage have been altered, compared to cutover sites.

SITE-INDEX CURVES

Determining site index on the basis of tree height typically utilizes a height-over-age growth curve to estimate height at a standard age. In the past, such curves for North America were developed from a series of regression curves, based upon a single guiding curve and harmonized to have the same form and trend (Figure 11.1). This approach has many weaknesses that have been well described (Spurr 1952b). Most obviously, the technique is sound only if the average site quality does not change among age classes. For example, it is often the case that younger stands are found on generally better sites (perhaps because of early logging on these sites), while remaining old growth stands are concentrated on poorer sites. In such a situation, the average curve will be warped upward at younger ages and downward at older ages. The reverse situation can also occur.

The assumption that the shape of the height-growth curve is unchanging among sites is a second major weakness of the conventional technique. This generalization may provide poor results when differences in climate, landform, or soil conditions are present. For instance, if the depth of soil is limited by physical reasons, tree-height growth may be normal up to the point where the depth of the soil becomes a limiting factor (curve B; Figure 11.2). In another ecosystem and soil, the same species may grow slowly until roots reach an underlying enriched horizon or a deep-lying water supply, after which growth will be accelerated (curve C). The shape of these two growth curves may differ from the normal growth curve on a normal soil (curve A).

Similarly, harmonizing site-index curves as part of the standard technique assumes that site differences are apparent at early ages. Harmonizing site-index curves assumes that a site producing a taller tree at age 50 or 100 will also produce a taller tree at all other ages. However, many plantations and even-aged natural stands on marginal sites grow normally at young ages and exhibit sharply decreased growth in middle life. For example, planted black walnut trees on seven contrasting sites in southern Illinois (Figure 11.3) showed rapid early growth, even on the poorer sites, but slowed abruptly after 10 years (Carmean 1970b). These *polymorphic patterns* are closely related to soil conditions. In Figure 11.3, trees on plot 1 are growing on a deep, well-drained alluvial silt loam, whereas those on plots 4–7 are growing on a bottomland silt loam soil underlain at 1 m or less by a gravelly subsoil. Curves harmonized with a standard average curve would be unable to show such plant–soil relationships.

Height-growth patterns vary not only across the geographic range of a species but also among different local ecosystem types (local areas of contrasting soil and topography). Height-growth

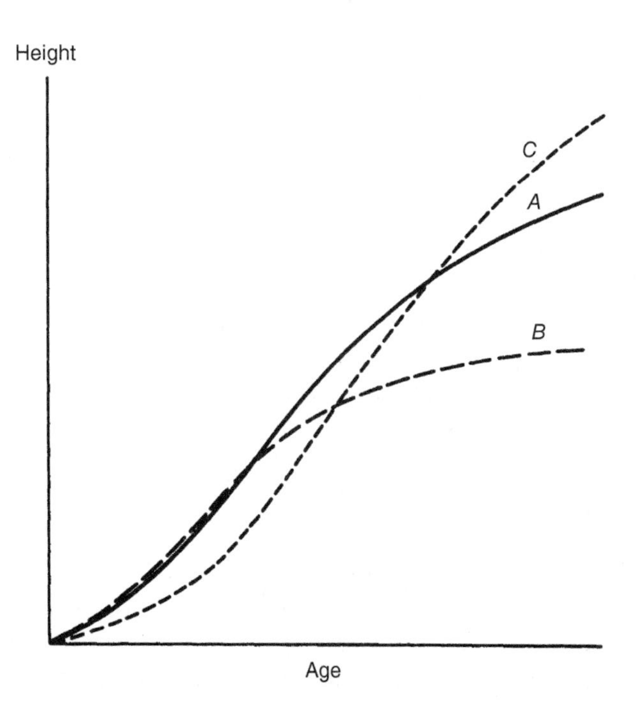

FIGURE 11.2 Theoretical effect of soil on height growth. A: normal height growth in homogeneous soil. B: height growth on good but shallow soil. C: height growth on soil poor at the surface but with a fertile horizon beneath.

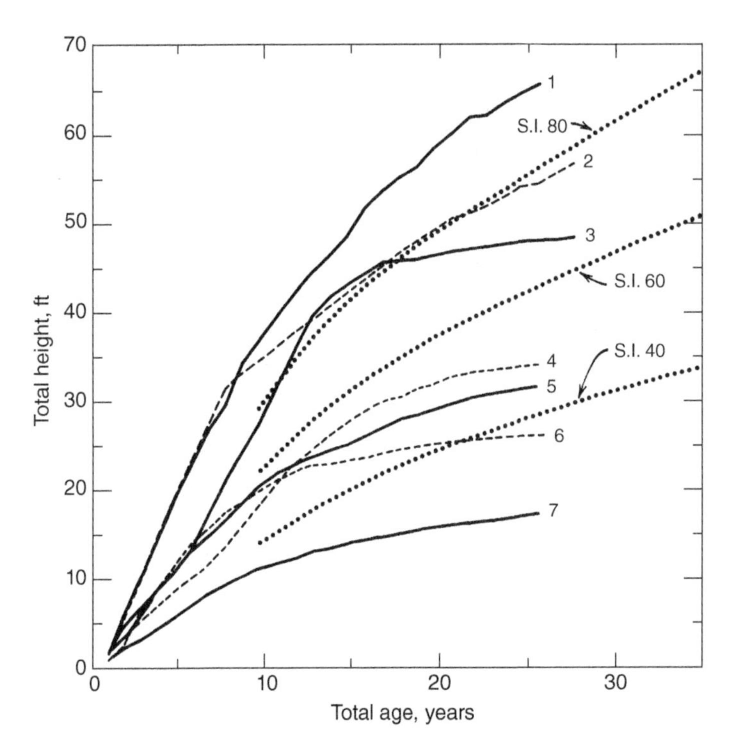

FIGURE 11.3 Polymorphic height-growth curves of dominant black walnut trees from plantations in southern Illinois. The height–age curves are averages from 36 sectioned trees growing in plots on 7 different sites. Plot 1: deep, well-drained alluvial silt loam; plot 3: similar to 1 but with restricted internal drainage; plots 4–6: bottomland silt loam underlain at 0.5–1 m by gravelly subsoil; plot 7: bottomland silt loam underlain at 0.3 m by a gravelly, cherty subsoil. *Source:* Carmean (1970b) / with permission of Oregon State University. Reprinted with permission from *Tree Growth and Forest Soils*, © 1970 by Oregon State University Press. Also shown are the regional harmonized site-index curves for black walnut (Kellogg 1939).

patterns of oak, for example, not only vary between different soil texture groups but also vary with aspect and slope within soil groups. Likewise, height-growth patterns of jack pine similarly vary across even minor differences in soil texture as well across landforms (Kashian and Barnes 2000). Indeed, polymorphic site-index curves have been repeatedly demonstrated to better characterize the variable height-growth patterns of forest trees compared to simple monomorphic pattern portrayed by regional harmonized curves (Carmean 1970b, 1975).

It is not uncommon for relatively small differences in curve shape to exist for a broad area. For example, in northern Ontario, jack pine exhibited only slight differences in average height-growth patterns among several different glacial landforms and soils, and only one set of curves were developed for this broad range (Carmean et al. 2001). However, pines on the poorest sites (8 m at 50 years) exhibited a relatively linear growth pattern, whereas those on the best sites (22–24 m at 50 years) were more curvilinear (rapid growth rate for the first 50 years and a gradually decreasing rate to 100 years). For black spruce, the most linear pattern was on the poorest sites where spruce grew on nutrient-poor, organic soils, and the pattern became increasingly curvilinear as site quality improved (Carmean et al. 2006). Polymorphic site-index curves have been prepared for many species (Carmean 1975, 1996), typically for broad regions. Polymorphic curves for specifically mapped ecosystem types were developed for Norway spruce and other species for intensive forest management in southwestern Germany (Barnes 1984), as well as for jack pine in northern Lower Michigan.

Weaknesses of standard site-index curve techniques suggest that height-growth curves should be based upon actual measured growth of trees for specific sites rather than averaging together height and age values from plots across a species' range. Doing so requires measuring actual tree growth among a series of soil-site conditions such that separate growth curves may be developed if necessary. These data may be collected from long-term permanent sample plots. Where permanent plot data are absent, **stem analysis** may be used to reconstruct past growth over time by sectioning trees from top to bottom. The pitfalls of this technique are more easily overcome than those of temporary plot techniques. The appropriate regression methods for estimating site index based on stem analyses are discussed by Curtis et al. (1974a,b).

A third method to reconstruct height growth with age is to measure the distance between nodes on the boles of trees that produce distinct annual whorls, such as Douglas-fir and many pines. Such measurements can be used to create accurate height-growth curves (Beck 1971; Carmean 1996). Height growth is fairly constant in the middle portion of the height-over-age growth curve, such that the average annual distance between whorls may be used to predict height growth over a future short period. The use of mean annual height increment over the middle period of height growth, the **growth-intercept method**, is well established as a measure of site quality. The 5-year intercept above breast height has proved better correlated with site quality than total height for loblolly, shortleaf, and slash pines (Wakeley and Marrero 1958). The technique has also been adapted to red pine, Douglas-fir, southern pines (Carmean 1975), black spruce, jack pine, and balsam fir (Mailly and Gaudreault 2005). Importantly, growth-intercept models have resulted in different site-index curves for jack pine growth in natural stands compared to plantations (Guo and Wang 2006).

COMPARISONS BETWEEN SPECIES

Different species growing on the same site will show variation in height response to a given forest site, and to a much greater degree than trees of differing genetic character within a species. Site-index curves may be quite different for two different species on the same site, because individual species differ in their maximum growth rates (Aber and Melillo 1991) and they respond differently to environmental conditions. Nevertheless, site index as predicted from height measurements of one species may sometimes be used to estimate the site index of the other. Carmean (1975, 1996) lists such studies, including examples from the southern pines, eastern hardwoods, northern and boreal trees, and western conifers; examples for northern Idaho conifers are given by Steele and Cooper (1986).

ADVANTAGES AND LIMITATIONS

Site index is the method for determining site quality that is most often employed in North American forests because it provides a convenient and reliable estimate of forest productivity. Site index is also useful in selecting the most productive tree species for specific sites. Developing site index curves for specific ecosystem types or groups of ecosystems with similar site conditions and genetic potential is generally useful, although it should be applied with great caution due to the various problems cited previously. Site index is not appropriate, however, for determining silvicultural practices geared toward regeneration and care of stands. Instead, site-quality studies and ecosystem classification using vegetation and physical site factors are more appropriate and have attracted universal attention. Such studies provide the ecological basis for determining silvicultural practices for various management goals as well as a basis for determining productivity. In the following text, we discuss how these other approaches used to estimate site quality are useful not only for understanding wood production potential, but also in understanding the broader ecological significance of site quality in managing ecosystems for wildlife, water, recreation, and wilderness.

VEGETATION AS AN INDICATOR OF SITE QUALITY

Tree species presence, relative abundance (also called "coverage"), and relative size reflect the nature of the local forest ecosystem of which they are a part and therefore may be useful indicators of site quality. Because of happenstance and elements unrelated to site quality, the correlation between site quality and these factors may not always be apparent. Factors such as plant competition and mutualisms, herbivory, low light conditions in the understory, or past events in the history of the vegetation such as drought, insect outbreaks, or human disturbance may disrupt the relationship. In many instances, however, vegetation sufficiently reflects site characteristics to make it a successful index of site quality. Plants themselves may be used as **phytometers**, or the measure of site, integrating the effects of many climatic and soil factors that are difficult to measure directly. They are not precise phytometers, however, and their indicator value changes with changes in regional macroclimate and physiography.

Tree species are long-lived, relatively unaffected by stand density, and easily identified in all seasons of the year, and as such are useful indicators. The occurrence of species with an extremely narrow ecological amplitude is indicative of a particular site. Demanding hardwoods such as black walnut, white ash, and tulip tree develop best only on moist, well-drained, protected sites rich in soil nutrients and characterized by well-developed organic soil horizons. Most trees, however, have a wider ecological amplitude and may occur and prosper on a wide variety of sites such that their presence is of little indicator value. For such species, relative abundance and size may be a better indicator than their presence. For example, in similar eastern hardwood forests to those described in the earlier text, the greater the proportion of northern red oak to black and white oak in the forest, the better will be the soil-water conditions and the general site quality. Therefore, the sizes of free-grown dominant oaks of these three species in even-aged stands of uniform density may be used as an index of site quality.

Plant species of the shrub, herb, and moss–lichen layers (termed groundflora or ground cover) of the forest understory are often more restricted in their ecological tolerances than tree species and may therefore be more useful as plant indicators. This relationship is particularly strong in the boreal forest, where the dominant tree species are spruces, firs, pines, and birches that are few in number and widespread in their distribution on various sites, and thus have relatively poor indicator value. Site classification methods based upon ground-cover indicator species have been successful under such circumstances. They are most easily applied where regional variation in altitude, precipitation, and soil conditions is high and where humans have not markedly altered the original vegetation. These methods have also been developed and used where such variation is more subtle and past human disturbances are nontrivial. Notably, ground-cover species are more apt to be influenced by stand density, past history, and the composition of the forest than are tree species. Plant indicators have been adapted successfully to many temperate forests of North America. Overstory dominants and physical ecosystem factors are increasingly used with ground-cover vegetation to estimate site quality and classify and map forest ecosystems.

The classic example of plant indicator species is Cajander's system of site types for Finland (1926), designed to distinguish site-quality classes in pure and mixed stands of spruce, pine, and birch. Pine height growth, for example, differed on four such site types (Figure 11.4); this system remains useful decades later for the site conditions where it was developed (Lahti 1995). The Cladina type is characterized by a lichen understory and occurs on dry, sandy, poor-quality heaths. Mosses and lichens are also present on the Calluna type, but heather (*Calluna vulgaris*) is typically the predominant species. The Vaccinium type is more dominated by low shrubs than mosses or lichens, particularly lingonberry (*Vaccinium vitis-idaea*), and occurs on moderately dry, sandy soils and on glacial ridges. Finally, the Myrtillus type, named for the predominant *Vaccinium myrtillus*, generally lacks lichens and is found on richer soils supporting late-successional spruce forest, but frequently converted to pine by fires, felling, and silvicultural control.

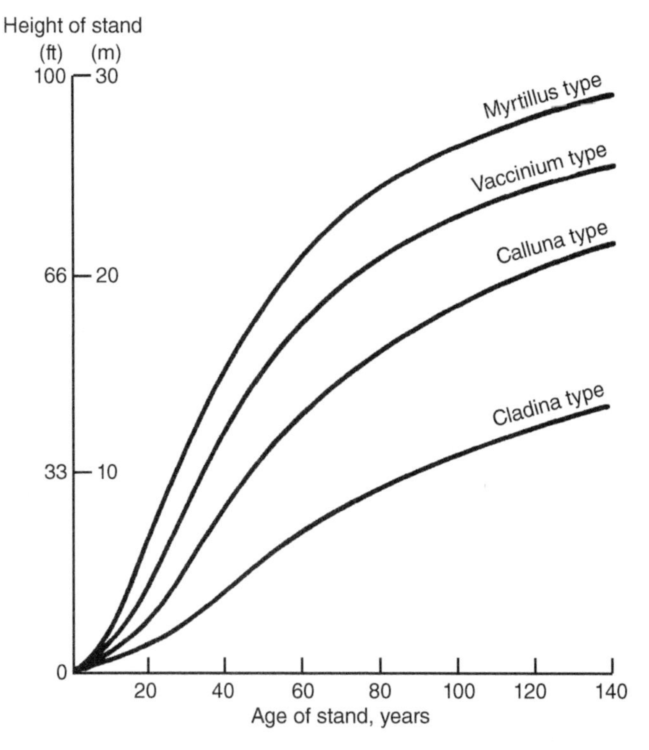

FIGURE 11.4 Height-over-age curves of Scots pine on four site types in Finland characterized by different indicator species.

It should be remembered that sites typically vary along a gradient rather than forming distinct and mutually exclusive site classes. For example, a sharp soil and microclimatic contrast may occur between an upland and adjacent wetland swamp or between a sandstone-derived residual soil and a limestone-derived residual soil, but otherwise site changes tend to be gradual. Segmenting this ecological gradient into classes or types is more useful in ecological comparisons and in land management. For example, Rowe (1956) identified five classes along a soil-water gradient from dry to wet in the mixed boreal forest in Manitoba and Saskatchewan. Understory vegetation was assigned to each class depending on their soil-water requirements, thereby forming groups of species that indicate distinctive positions along the soil-water continuum.

SPECIES GROUPS OF GROUND COVER

A site may be characterized in terms of very few species, but these key indicator plants may or may not be present at a given locality because of chance, past forest history, or present competitive conditions. For example, it is awkward to classify a site or ecosystem type as an Oxalis–Cornus type if both species are missing, yet other characteristics such as the sum total of the vegetation and the growth characteristics of the ecosystem may otherwise fit the classification well. These pitfalls may be overcome by using many ground-cover species grouped together using their similar environmental requirements as determined by intensive field study. **Ecological species groups** have been determined by a number of different approaches (Daubenmire 1952; Rowe 1956; Sebald 1964; Mueller-Dombois and Ellenberg 1974; Barnes et al. 1982; Ellenberg 1988; Klinka et al. 1989) depending on the nature of site and vegetation conditions of the study area and the purpose for which the method is to be used. The concept of ecological species groups is attributed to Duvigneaud (1946), but plant ecologists have long recognized the occurrence of plants with similar ecological distributions.

Ecological species groups are developed using one of two different general approaches. In the first, each species is assigned an indicator value for individual site factors (e.g., low nitrogen, high nitrogen). For example, Ellenberg (1974) listed the indicator value (in classes 1–9) of approximately 200 vascular plants for each of the gradients of light, temperature, continentality, soil water, pH, and soil nitrogen in the western part of central Europe. In the second approach, groups of species that reflect a similar combination of site factors are constituted. In both approaches, the "importance" or abundance of the groundflora species present on a site is quantitatively rated with percent **coverage**, using the aerial crown or foliage coverage of the species. Coverage for a given species is defined as the proportion of a ground area that is covered by a vertical projection of the foliage of all individuals of the species. This process is tedious, time-consuming, and subjective, such that ecologists typically estimate coverage on a scale from <1 to 100% using a series of cover classes whose number varies depending on user purpose, site conditions, and number and rarity of the species present (Mueller-Dombois and Ellenberg 1974; Spies and Barnes 1985b; Klinka et al. 1989).

Indicator Plants of Coastal British Columbia Over decades V. J. Krajina and his students characterized the plant–environment relationships of over 3000 species in British Columbia. As part of this effort, Klinka et al. (1989) described the indicator value of 416 species of coastal British Columbia using 4 site attributes: climate, soil water, soil nitrogen, and ground-surface material. The gradient of each attribute was segmented into classes (6 for climate, 6 for soil water, 3 for soil nitrogen, and 5 for ground-surface material), and each species was placed into a class for each attribute, forming 20 indicator species groups. For example, indicator values were assigned to 337 species in the 6 classes for soil water: (i) excessively dry to very dry (17 species), (ii) very dry to moderately dry (50 species), (iii) moderately dry to fresh (74 species), (iv) fresh to very moist (107 species), (v) very moist to wet (59 species), and (vi) wet to very wet (39 species). A given site or ecosystem is then characterized by the presence–absence and coverage (abundance) of different indicator species in groups of each of the four attributes.

Ecological Species Groups Ecological species groups were first used for forest ecosystem classification, mapping, and site-quality evaluation in the southwestern German state of Baden-Württemberg (Sebald 1964; Dieterich 1970; Barnes 1984). Plant species are grouped together that repeatedly co-occur in areas with similar combinations of soil water, nutrients, light, and other factors, as they are perceived to have similar ecological requirements or tolerances. Species groups consist of herbs, shrubs, and (less commonly) mosses and lichens of the groundflora, and each group is named for the most characteristic species. In North America, the approach has been applied in distinguishing and mapping landscape ecosystem types in old-growth forests in Michigan (Pregitzer and Barnes 1984; Spies and Barnes 1985b; Simpson et al. 1990) in highly disturbed oak–hardwood forests of southern Michigan (Archambault et al. 1990) and Wisconsin (Hix 1988), in jack pine forests of northern Michigan (Kashian et al. 2003), in forests of the southern Appalachian Mountains (Abella and Shelburne 2004), and in ponderosa pine forests of Arizona (Abella and Covington 2006a), among other systems. Methods used to develop and evaluate the groups are given by Spies (1983), Spies and Barnes (1985b), and Archambault et al. (1990). An example of 2 contrasting groups of the 16 groups used to map forest ecosystems in Upper Michigan (Spies and Barnes 1985b) illustrates their indicator value:

Clintonia borealis group (6 species): Most common on moist, very infertile soils supporting conifers. Dry to wet. Very infertile to infertile. Shade tolerant.

Caulophyllum thalictroides group (5 species): Characteristic of very fertile, moist to very moist soils. Moist to very moist. Fertile to very fertile. Shade tolerant.

Co-occurrence of plants in the field is due to many interacting physical and biotic factors, including mutualisms and competition with other plants, and the groups constituted represent the integrated effects of multiple-factor gradients. The absolute indicator value of a plant for any individual factor may therefore be difficult to determine. Once constituted, the groups may be used to help distinguish landscape ecosystems in the field by their presence or absence and by the relative coverage of plants in each group (see spectral format in Table 11.1). Importantly, ecological species groups are never used alone, but always with attributes of physiography, soil, microclimate, and the composition and vigor of overstory trees.

How widely can the perceived indicator value of an individual species or species group be extrapolated beyond its local use? Applications of this approach in Germany and Michigan show that the indicator value of a given species changes from one regional landscape ecosystem to another. As macroclimate, landform type and pattern, species genetics, and species competition and mutualisms change from one region to another, the relative indicator value of a species also changes. Among regions, the usefulness of certain species increases (especially as new species assert dominance) and that of other species declines. As such, a given species does not always indicate the same relative level of a given site factor (e.g., high soil fertility, low soil water) in different regions. For example, four different sets of ecological species groups were required to map landscape ecosystems in four macroclimatically different areas of northern Lower Michigan and Upper Michigan, suggesting that a regional landscape classification (e.g., Albert et al. 1986; Figure 2.10 in Chapter 2) is a useful framework to determine how widely to extrapolate a set of ecological species groups. In any case, it is clear that indicator species and groups should be used with great care.

PLANT ASSOCIATIONS AND HABITAT TYPES IN THE WESTERN UNITED STATES

Widespread deforestation in many parts of the world has made it difficult to evaluate or classify sites using native or pre-European colonization forest trees and groundflora. An exception to this trend occurs in the northern Rocky Mountains, where relatively pristine old-growth forests have facilitated the development of an indicator system employing both trees and understory plants. Rexford Daubenmire (1952; R. and J. Daubenmire 1968) used groups of understory species, termed **subordinate unions**, in combination with late-successional overstory species, termed **dominant unions**, to distinguish forest or plant associations. A **habitat type** (literally, the type of late-successional vegetation on a particular site or habitat) is the collective area of a given plant association. Different areas with the same habitat type may have similar environmental conditions, and thus the concept has value for land management.

Operational Site Classification Based on Vegetation Land classification using the habitat type method (Pfister et al. 1977; Pfister and Arno 1980) has been applied to millions of hectares of federal forests in the Rocky Mountains, the Intermountain Region, parts of the Southwest, California (Wellner 1989), and in the Pacific Northwest (Hall 1989). Habitat types are typically applied to parts of states (e.g., Montana, Pfister et al. 1977; northern Idaho, Cooper et al. 1991), to individual National Forests or parts thereof (Alexander et al. 1986), or a community type (trembling aspen, Mueggler and Campbell 1986). Habitat types are typically developed independent of a regional ecological framework.

The habitat type approach is based on the concept of the *natural potential climax* (polyclimax theory of Tansley 1935; R. and J. Daubenmire 1968; Pfister and Arno 1980), which assumes that overstory and understory vegetation integrate and express the environmental complex for a specified geographic area. A habitat type represents all parts of the landscape that support or may support the same primary climax vegetation (Alexander 1986). A climax community is defined as the final stage of forest succession for a given habitat; without natural or other disturbance,

Table 11.1 Representative ecological species groups and ecosystem types (site units) along moisture gradients, upper Neckar growth district, Baden-Württemberg, southwestern Germany[a].

Ecological species group	Ecosystem type (site unit)														
	Moderately fresh					→		Moderately dry					Fresh → wet		
	1	2	3	4	5	6	7	8	9	10	11	12	13	14	15
Vaccinium myrtillus group	●	●	●	●	·						●	·			
Leucobryum glaucum group	•	•	·												
Bazzania trilobata group	•	·									·				
Deschampsia flexuosa group	●	•	●	●	•	·	·				●	•	•		
Pirola secunda group				·	·	·	•	·	·	·		·			
Milium effusum group	·	·	•	•	●	●	•	·	·	•	·	●	●	●	●
Elymus europaeus group						•	•			·				•	
Aruncus sylvester group														·	●
Ajuga reptans group				·	·	•	·				·	•	●	•	•
Stachys silvatica group					·	·						·	·	●	●
Chrysanthemum corymb group							·	•	·	·					
Carex glauca group							•	●	●	•					
Molinia caerulea group											·		·		

Key: ●, species of the group abundant; •, species of the group moderately abundant; ·, species of the group rare.

[a]Thirteen of Sebald's 24 groups are shown. Major differences among the groups are indicated by space between the sets of groups. Two sets of site units (ecosystem types) are ordered along gradients from moderately fresh to moderately dry (units 1–10) and from fresh to wet and somewhat poorly drained (units 11–15).

Source: After Sebald (1964).

the climax community develops and perpetuates itself (Cooper et al. 1991, p. 134). Habitat types identify areas of similar climax vegetation (biota), not necessarily areas of the same ecosystem type. The climax vegetation upon which the classification is based is called a "plant association."

Classically in the Rocky Mountains, the habitat type system determines late-successional overstory dominants (termed **series**) that occur along an elevational gradient from grassland to alpine tundra (Figure 11.5). Understory species (shrubs and herbs) in each series are used to identify many different plant associations, and the type is named for the potential climax community type or plant association (e.g., the *Pseudotsuga menziesii/Calamagrostis rubescens* habitat type). The series level is typically denoted by the most shade-tolerant, late-successional tree species adapted to the site and a dominant or indicator understory species of the plant association. Series order habitat types along an elevational gradient that approximates climate. For example, ponderosa pine occupies areas that are warmer and drier than areas where Douglas-fir or Engelmann spruce and subalpine fir are dominant (Figure 11.5). A finer level, termed **phase**, is used to designate major within-type variation in understory vegetation associated with geographic, topographic, or edaphic features (Youngblood and Mauk 1985). The ubiquitous *Pseudotsuga menziesii/Calmagraostis rubescens* habitat type in Montana has four phases which range geographically from northwestern to southwestern Montana and are found at elevations from 823 to 2377 m (Pfister et al. 1977).

Habitat types and other classification systems based on vegetation are widely used in forest management on public lands for timber, wildlife, range, and watershed management (Layser 1974; Ferguson et al. 1989). Specifically, such systems have been used for growth and yield evaluation (Monserud 1984; Stage 1989), recreational use studies (Helgath 1975), forest protection (Arno 1976), fire effects (Fisher 1989), fire management (Arno and Fischer 1989), and

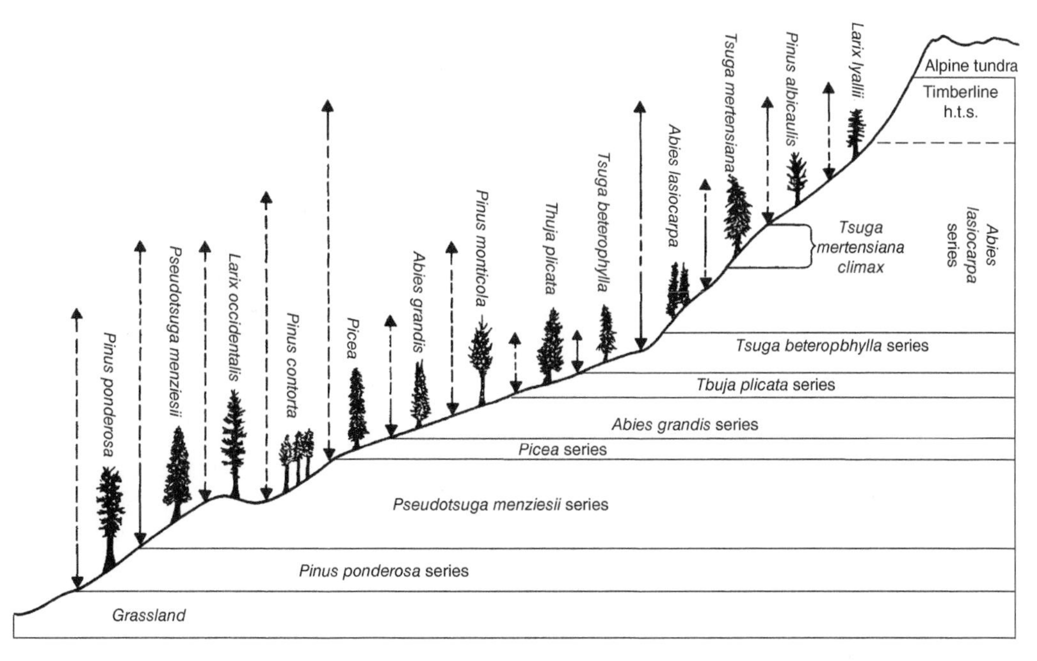

FIGURE 11.5 Distribution of forest trees along an elevational gradient in the Rocky Mountains of northwestern Montana. Arrows show the relative elevational range of each species; the solid portion of the arrow indicates where a species is the late-successional (climax) dominant, and the dashed portion shows where it is early successional (seral). *Source:* Pfister et al. (1977) / United States Department of Agriculture / Public Domain.

natural area preservation (Schmidt and Dufour 1975; Wellner 1989). Recent research has advocated the use of habitat types in forest conservation planning, particularly in Europe (Kovač et al. 2016, 2020; Culmsee et al. 2014). The use of habitat types and communities has provided an important framework for management and has contributed to inventorying, describing, and classifying forested lands. At least 127 classifications were developed primarily on US Forest Service lands from 1952 to 1987 (Wellner 1989), and 909 habitat types, community types, and plant communities were described for the Rocky Mountains alone by the mid-1980s (Alexander 1985).

The habitat type approach arose from a very old tradition of vegetation classification (Daubenmire 1989). Early foresters focused their attention narrowly upon forest cover types and their delineation, with an emphasis on stands rather than land such that a vegetation taxonomy was very useful. Habitat types are based on the phytosociological classification approach developed by Cajander (1926) and transplanted by his students to Canada (Pfister and Arno 1980). Cajander's approach is taxonomic, developing a classification of vegetation from plot samples and then applying the classification to other sites. Habitat types remain useful especially where forest land management problems concern particular cover types that are to be maintained, logged, and regenerated with the same species (Rowe 1984b). Using a taxonomic key to habitat types, and supplemented by selected topographic and soil properties, each forest stand can be classified and assigned an appropriate prescription for management.

Moreover, habitat types provide a framework for extensive management of large areas, although increasing the area in question naturally increases the number of taxonomic units. For example, a reworking of approximately 4.9 million hectares of northern Idaho, including the area of the Daubenmires' (1968) original study, yielded a five-fold increase in taxonomic units. It is notable that the large extent of a habitat type—including all land areas potentially capable of producing a similar plant association at climax—incorporates considerable variation in physiography and soils. Hanks et al. (1983), working in northern Arizona, report:

> *Two land areas with obvious differences in measurable environmental factors may fall within the same habitat type, if they are equivalent with respect to plant requirements. For example, a habitat type can occur on two sites with different soils and climatic regimes when greater moisture holding capacity of soils on one site compensates for a drier climate. Because of compensating factors a vegetation type may occur on different physical environments.*

Daubenmire (1976) also cites an example from the northern Rocky Mountains: ". . .the *Pseudotsuga/Physocarpus* forest occurs on steep north-facing slopes at its lowest altitudinal limits, moves onto zonal soils at intermediate elevations, then onto the shallow soils of steep south-facing slopes at its highest limits" (Daubenmire and Daubenmire 1968). The important point is that with its basis in vegetation alone, a given habitat type may not necessarily identify a landscape ecosystem type that is homogeneous in physical environment. Indeed, soil classifications have been found to poorly align with habitat types (Daubenmire and Daubenmire 1968; Neiman 1988), in part because separate disciplines have different goals, assumptions, and methods. As Daubenmire (1970) recognized, the soil-site properties that play key roles in patterning vegetation are not among those emphasized in soil classification. Neiman (1988) concluded in his broad study of northern Idaho that further delineation of habitat types, based on soil variation, would permit greater accuracy in predicting site capabilities and response to disturbance.

APPLICATIONS AND LIMITATIONS OF VEGETATION

In Site evaluation and classification, vegetation integrates the effects of many interacting factors, and key species may indicate specific site conditions. Like any single ecosystem component, however, the use of vegetation has limitations, as discussed by Rowe (1984b). First, vegetation is strongly controlled by macroclimate and microclimate, and thus its use should occur within regional ecosystem hierarchies that utilize climate (Albert et al. 1986; Bailey 2009). Second, vegetation is highly sensitive to disturbance and thus having information about historic disturbance regimes is important. Existing vegetation and potential natural vegetation need to be understood for land-management applications (Eshelman et al. 1989). Third, vegetation is floristically complex and may require the user to identify the entire complement of vascular plants and even mosses and lichens. Fourth, vegetation changes over time, and successional patterns should be understood, especially where original vegetation has been heavily disturbed by humans (Neiman 1988). Fifth, vegetation varies widely in occurrence, abundance, coverage, and biomass. Plants that are lacking may be due to chance or historic causes, such that reliable sampling methods are necessary in assessing plant indicator value. Finally, vegetation varies greatly in its vertical layering, which is of major significance in animal ecology and wildlife management.

ENVIRONMENTAL FACTORS AS A MEASURE OF SITE

Often trees or understory plants are less useful than site factors in assessing forest productivity. Such situations might include those where vegetation is lacking or heavily disturbed. Examples are non-forested or agricultural lands slated to be reforested; areas recently subjected to wildfire, logging, heavy grazing, or other severe disturbances; and areas to be converted between forest types. In these cases, site productivity must be estimated from an analysis of the physical environment rather than from the vegetation.

Physical factors, such as climate, physiography, and soil as discussed in previous chapters, may be used singly or in combination. The usefulness of any factor depends on its consistent, simple, and inexpensive measurement, as well as its high correlation to forest productivity. Many site factors are not useful as site indices because few data are available for the factors or because it is difficult or expensive to obtain these data. Other factors are not used because they are not sensitive enough as a measure of site quality. The most useful environmental factors are those that are "limiting," or in short supply regarding the demand by forest trees, so that small changes in the supply of the factor will result in measurable changes in tree growth. Thus, different environmental factors are useful in different regional and local forest ecosystems.

Climate factors often provide a rough index of productivity among adjacent forest regions or among altitudinal zones within a geographic region. Climate is a much stronger driver of genetic differences over a species' geographic and altitudinal range than is soil. Temperature and precipitation data are most often used to compare forest growth in various geographic regions or altitudinal zones, assuming similar soil conditions or at least assuming soil conditions closely related to climate. Tree growth will vary widely within a given climatic region depending on physiographic and soil conditions; soil factors are therefore particularly useful in local studies of site quality.

Increasing the number of factors used in creating an index of forest productivity will improve its assessment, because the site portion of a forest ecosystem is defined as the sum total of many factors. Unfortunately, increasing the number of factors typically increases the cost and the time needed to complete the analysis. Where site classifications are developed for application to a broad area, multiple factors should be used at both regional and local scales.

CLIMATIC FACTORS

The crowns and boles of trees live in the air and are affected by it, and thus climate has a strong effect on tree growth. Macroclimatic factors have long been used to distinguish differences among major forest regions (Schlenker 1960; Rowe 1972; Findlay 1976; Ecoregions Working Group 1989). However, long-term climatic data from forest sites are often very difficult to obtain.

In addition, data interpretation for individual interrelated variables is problematic, and vegetation itself is often a more easily measured and useful integrator of the complex of climatic factors. Most site-quality evaluations for management decisions take place at regional scales (Chapter 2) where the average climate may not vary widely, although local climate may vary significantly from place to place within the management unit. Local climate has seldom been used in site evaluation, in part because other variables such as land-use and forest management history have their own effects on forest growth. In addition, the control of local climate on local topography and soil suggests that an ecosystem classification based upon topography and soil implicitly assumes a local climate classification as well. For example, the same factors that make a soil very well drained are apt to insure good cold-air drainage, and a very poorly drained soil is apt to result from a topography that inhibits the drainage of cold air as well as of soil water. Thus, very poorly drained sites are also characterized by climatic factors such as temperature extremes and a short growing season.

PHYSIOGRAPHIC LAND CLASSIFICATION

Physiography directly or indirectly affects many other key ecosystem factors, and thus it provides a framework for managing ecosystems at both regional and local scales. As emphasized in Chapter 8, the shape, parent material, and location of specific landforms influence the reception of solar energy and the runoff and retention of precipitation (Rowe 1984b; Swanson et al. 1988). In the mid-south of the United States, Smalley (1991) developed and applied a landform-based taxonomic system for about 12 million hectares of the Cumberland Plateau and Highland Rim physiographic provinces of Virginia, Tennessee, Kentucky, and Georgia. Following the system of land stratification of Wertz and Arnold (1972, 1975), Smalley classified sites on these landscapes using five hierarchical levels (province, region, subregion, landtype association, landtype). Landtypes, the most detailed level (spatial scale 1:10 000 to 1:60 000), are described by nine factors of geography, geology, soil, and the most commonly occurring trees and shrubs. Notably, slope position and aspect were particularly important in distinguishing landtypes characteristic of highly dissected terrain in the Cumberland Mountains of Kentucky, Tennessee, and Virginia (Figure 11.6; Smalley 1984).

Land classification based on physiography is particularly appealing because of a general lack of soil survey information and a long history of human disturbances that have drastically changed how natural vegetation relates to site. As a result of human disturbances, existing forests vary widely in their stand conditions, productivity is reduced far below its potential, and few stands are appropriate for a direct measure of site productivity. In terrain with high topographic relief, landforms may correlate better to tree growth than soil series or the existing tree vegetation. In such a situation, landtypes become the basic unit of land management rather than existing forest stands whose boundaries may be artifacts of past land use. If necessary, this broad landform base facilitates subdivision into finer ecosystem types based on local topography, soil, and ground-cover vegetation.

PHYSIOGRAPHIC AND SOIL FACTORS: SOIL-SITE STUDIES

Many researchers have examined the potential of using physiography and soil to predict forest site quality. For areas where physiography (specific landforms, topographic factors; see Chapter 8), soil, and stand conditions are highly variable, such an approach would be very advantageous. Likewise, using physical factors rather than biota would be necessary where trees are lacking, the area is stocked with unwanted species, or the area is dominated by trees unsuited for site-index measurements. Carmean (1975, 1977, 1996) cites over 180 soil-site studies for the United States and Canada and discusses their applications and limitations; Burger (1972) cites many additional studies in Canada.

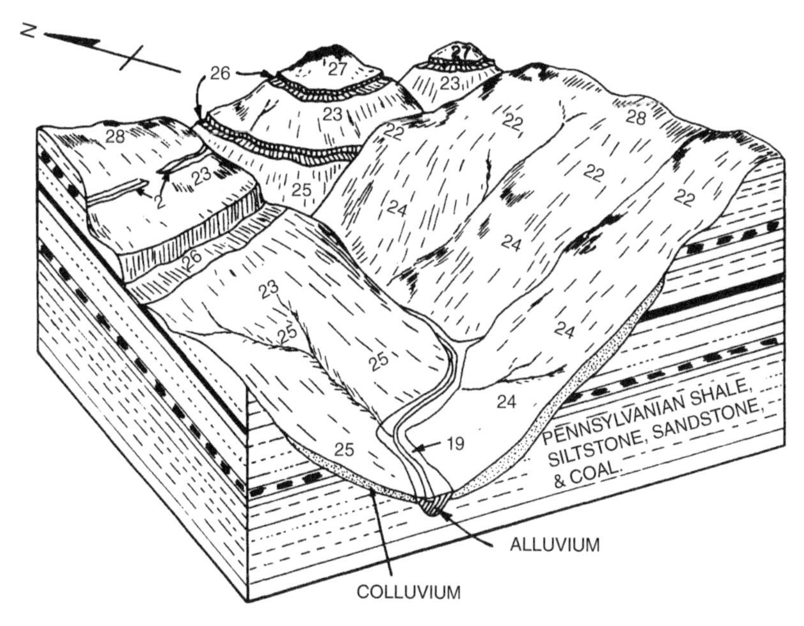

FIGURE 11.6 Land types characteristic of dissected forest terrain of the Middlesboro syncline and Wartburg Basin–Jellico Mountains of Kentucky, Tennessee, and Virginia. Legend: 2, shallow soils and sandstone outcrops; 19, mountain footslopes, fans, terraces, and stream bottoms; 22, upper mountain slopes–north aspect; 23, upper mountain slopes–south aspect; 24, colluvial slopes, benches, and coves–north aspect; 25, colluvial slopes, benches, and coves–south aspect; 26, surface mines; 27, narrow shale ridges, points, and convex upper slopes; 28, broad shale ridges and convex upper slopes. *Source:* Smalley (1984) / United States Department of Agriculture / Public Domain.

Depending upon the nature of the specific site, many individual soil and physiographic factors may be correlated with the site index of the desired forest species. **Soil-site** studies involve measuring many soil and site variables (e.g., soil depth, texture, and drainage class; slope position; aspect) and relating them with multiple regression analyses to tree height or site index. Combining these with other soil and topographic factors has produced useful formulas to estimate site index. In a classic study, Zahner (1958) restricted regressions to soil groups within a limited geographical region and related the site index of two southern pines to surface soil thickness, the proportion of clay and sand in the subsoil, and the slope percentage. Equations derived from such soil-site studies were then used to develop tables and graphs for estimating site index in the field. Similarly, combinations of soil and site variables in successful soil-site studies explained 65–85% of the variation in tree height or site index (Carmean 1975). A disadvantage of this approach, however, is that many soil and site variables are difficult or tedious to measure in the field, and thus somewhat less precise equations are often developed using variables that are most easily measured.

Two important caveats are applicable to soil-site studies. First, many investigators have read causal relationships into regressions based upon soil and topographic factors when the data only support correlation. Importantly, such correlations may reflect a causal relationship attributable to another unmeasured soil characteristic rather than the data at hand. Second, the dependent variable of the regression equations is often site index as read from harmonized site-index curves. As discussed previously, such values are suspect by the very nature of the method used to construct the site-index curves, and thus caution is necessary in using soil-site equations.

The growth potential of forest trees appears to be chiefly affected by the soil volume occupied by tree roots and the availability of water and nutrients in this soil space. Therefore, the

effective depth of the soil—the depth of the portion of the soil that is capable of being occupied by roots of the tree—is of primary importance. The effective soil depth may be limited by a layer that prevents further downward growth of roots, such as bedrock near the surface (Green and Grigal 1979, 1980), the position of the water table during the growing season, a coarse, dry soil stratum, or a highly compact and impervious stratum such as a well-developed hardpan. Soil factors have been correlated with site quality using many measures of effective soil depth (Coile 1952; Carmean 1975). The most important soil factors are the depth of the A horizon above a compact subsoil, the depth to the least permeable layer (usually the B horizon), the depth to mottling (indicative of the mean depth to restricted drainage), and thickness of the soil mantle over bedrock. All these measures are important when soils are shallow but are less important for deep soils where downward root development is unimpeded. In addition to effective soil depth, other soil profile characteristics that affect soil water, soil drainage, and soil aeration are also important for site quality, particularly soil texture and structure of the least permeable horizon (again usually the B).

The physical properties of the soil that govern soil-water and aeration are often closely related to topographic position of the site (slope position) and other landform factors, as is microclimate. Landform is the chief correlate of the patterns of soil and vegetation and thus site productivity, because it subsequently influences local climate near the ground, belowground climate induced by soil water, and nutrient regimes via parent materials and soil-water availability. When the relationships between landform and soil are well known, topographic site can be quickly recognized and evaluated using aerial photographs and topographic maps without the necessity of soil measurement. For example, an index of forest site quality was developed for use in the Ridge-and-Valley physiographic province of the central and southern Appalachian Mountains using only aspect, slope percentage, and slope position (Meiners et al. 1984). Subjectively ranking field sites from 1 to 5 for each of these three variables provided a simple index (3 = lowest quality; 16 = highest quality) that provides a rapid evaluation of relative site quality. This physiography-based index is highly correlated with site index of oak in the Allegheny Mountains of West Virginia and Maryland and with m.a.i. of trees on steep slopes in the Ridge-and-Valley province of Virginia (Ross et al. 1982).

Relative elevation, aspect, slope position, and degree of slope have repeatedly produced useful site relationships. Studies of site quality for oak forests in the Appalachian Mountains and the Appalachian Plateau (Trimble and Weitzman 1956; Doolittle 1957; Carmean 1967; McNab 1987) have shown close relationships between site quality and relative position between ridge top and cove, aspect, and degree of slope. Equations based on topographic features alone explained more than 75% of the variation in black oak height in southeastern Ohio (Carmean 1967). Likewise, topographic position was a better predictor of site quality than height–age site-index curves for mixed-oak forests of western Maryland (Sturtevant and Seagle 2004). Topography and site quality are closely related because topography is correlated with important soil features such as A horizon depth, subsoil texture, stone content, organic-matter content, and nutrient availability (Figure 11.7). The relationship of topography with microclimate is also critical, especially in hilly terrain. Northeast aspects and lower slopes usually have cool, moist microclimates and thus are more productive sites; southwest aspects and upper slopes and ridges have dry and warm microclimates and hence are usually less-productive sites. In the Appalachian Mountains, the relationship of site quality to aspect for mixed upland oak forests resembles a cosine curve (Figure 11.8; Lloyd and Lemmon 1970).

Notably, the predictive value of landform and topographic variables for tree height and site quality is fairly inconsistent among studies, partly because of different methods of scoring or measuring the variables. McNab (1987, 1991, 1993) developed indices to quantify slope position and local landform shape. Site index of tulip tree was significantly correlated to landform index at four sites in the Blue Ridge physiographic province (McNab 1989), suggesting that microsites may be an important source of spatial site variation. Such microsites represent local ecosystem types where the beneficial effects of leaf litter accumulation and decomposition (Stone 1977; Welbourn

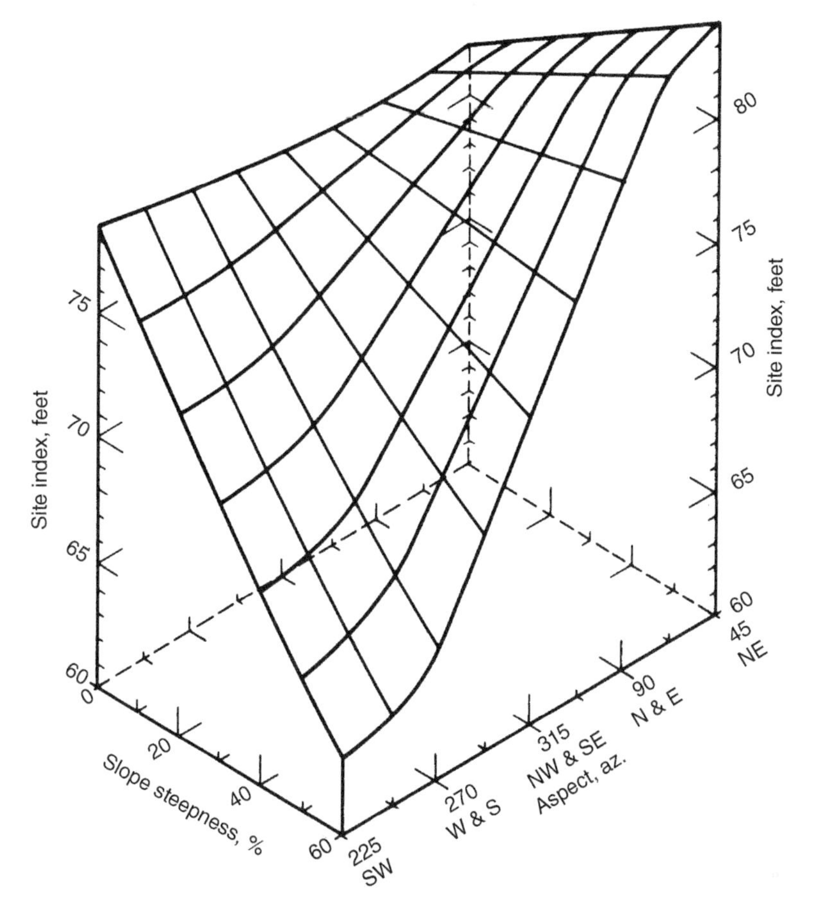

FIGURE 11.7 Relationship between aspect, slope steepness, and site index for black oak growing on medium-textured, well-drained soils. Site index increases from southwest-facing slopes to northeast-facing slopes. These increases are very pronounced for steep slopes, but site index increases related to aspect are relatively minor on gentle slopes. For southwest-facing slopes, site index decreases drastically with increased slope steepness, whereas it increases slightly on northeast slopes as slopes become steeper. *Source:* Carmean (1967) / with permission of John Wiley & Sons. Reprinted from *Soil Science Society of America Proceedings*, Vol. 31, p. 808, 1967 by permission of the Soil Science Society of America.

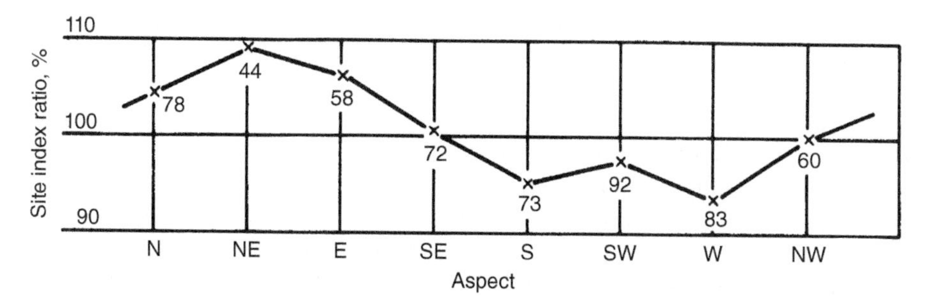

FIGURE 11.8 Productivity curve with a site-index ratio over aspect; based on 560 soil-site index plots on 27 soil series in mixed upland oak forests of the Appalachian Mountains. The site-index ratio is the ratio of the plot site index to the average site index of all plots of that soil series. *Source:* Lloyd and Lemmon (1970) / with permission of Oregon State University. Reprinted with permission from *Tree Growth and Forest Soils*, © 1970 by Oregon State University Press.

et al. 1981), combined with gravitational movement of subsurface water on mountain slopes (Hewlett and Hibbert 1963; Dwyer and Merriam 1981), create especially favorable nutrient and soil-water conditions.

Minor variations in topography may be highly important in very flat locations for various reasons depending on the nature of the site. On the dry, flat, sandy, nutrient-poor glacial outwash plains of northern Lower Michigan, small changes in elevation (as little as 15 m) or variations in landform shape create subtle changes in soil texture because of the way the parent material was deposited by fluvioglacial processes (Kashian et al. 2003). These subtle soil differences result in variations in nutrient and soil-water conditions that produce marked variation in the height growth of jack pine that is directly predictable by landforms across the region. In areas where the water table is close to the soil surface, small changes in topography reflect the effective depth of the soil over poorly aerated lower horizons. On southern river bottomland oak sites, topographic features varying only centimeters in elevation are correlated with differences in the silt and clay content of the soil, with flooding conditions, and with soil aeration, and thus are highly related to site quality (Beaufait 1956).

SOIL SURVEYS

Soil surveys, which emphasize soil series and phases, were developed for application in agriculture rather than forest ecosystem management (Chapter 9). Soil surveys provide maps and a taxonomic classification of forest soils, but have not been precise in estimating forest site quality (Grigal 1984). In part, these failures are attributed to the basic differences between soil taxonomic units and soil mapping units. Typically, soil taxonomic units include too much variation in forest productivity, as estimated by site index, to be useful (Rowe 1962; Jones 1969; Carmean 1970a, 1975; Grigal 1984; Neiman 1988). For example, loblolly pine site index ranged from 59 to 105 on soils of the Ruston series (Covell and McClurkin 1967). Similar variation in site index within soil taxonomic units also has been reported for numerous species in eastern hardwood forests (Carmean 1970a, 1975). Soil series prove more satisfactory when they incorporate specific soil, landform, and local topographic factors that are closely related to forest productivity (Richards and Stone 1964; Carmean 1967, 1970a; Grigal 1984).

MULTIPLE-FACTOR METHODS OF SITE AND ECOSYSTEM CLASSIFICATION

Evaluating site quality with single-factor approaches, such as site index and vegetation indicators, physiography, and soil, uses only individual components of the ecosystem. Such an approach is troublesome because site quality is the sum total of factors affecting the land's ability to produce forests. Site productivity is always better estimated by including more factors in the process, and a multifactor approach improves our ability to understand ecosystem processes as the basis for managing forests. This is why intensive multifactor methods have been developed and used successfully in Europe for over 85 years, and similar methods have more recently proliferated widely in Canada and the United States.

Many multifactor methodologies have been developed, resulting in a wide variety of terminology used to describe the ecosystem units within them. In practice, a common feature of multifactor methods is the consideration of a given ecosystem in two parts: the physical ecosystem factors of a specific geographic place (site) and the plants and animals occurring there (biota). The geographic space may be described with terms such as land type, site type, site unit, geosystem, physiographic site, or forest site. The biota is often described with terms such as vegetation type or cover type. Ecosystem units themselves (i.e., the volumetric land units integrating site and biota) are also described with various names, from the broadest (ecodomain) to the finest (ecosystem type). Notably, geographic site terms such as site unit, site type, and total site are often used to indicate landscape ecosystems because these spatial units have the same or similar properties and potential vegetation *regardless of present vegetation*. Despite the names of the various methods and their terminologies, the common focus among the examples presented in the following text is on geographic area or site.

ECOSYSTEM CLASSIFICATION AND MAPPING IN BADEN-WÜRTTEMBERG

The first operational and systematically applied multi-scale, multi-factor system of ecosystem classification and mapping was developed in the southwestern German state of Baden-Württemberg (Barnes 1984, 1996). The area is extremely diverse in climate and geology, including the Rhine River Basin, adjacent Black Forest, a limestone plateau (Swabian Alb), and glaciated terrain stretching nearly to the Swiss Alps. Much of the original forest of beech and oak was replaced by Norway spruce monocultures now often several generations old. An ecological approach integrating geology, physiography, climate, soil, vegetation, and forest history was developed to classify and map forest ecosystems as a way to diversify and improve multiple-use land management. This detailed and systematic approach is based upon classification and mapping at multiple scales (Figure 11.9). Once mapped, differences in local ecosystem types (termed "site units") are evaluated with the growth and biomass production of major tree species. Detailed silvicultural recommendations are then made for ecosystems at regional and local scales. This system, applied for almost 75 years, emphasizes natural regeneration of native species (Mühlhäusser and Müller 1995) and is used in long-term planning, forest and wildlife management, and for preservation of unique ecosystems (Dieterich et al. 1970; Arbeitsgruppe Biözonosenkunde 1995). Amazingly, the interdisciplinary team who developed this multi-scale and multi-factor approach integrated ecosystem components at a time in the 1940s when disciplinary fields of geology, climatology, soil science,

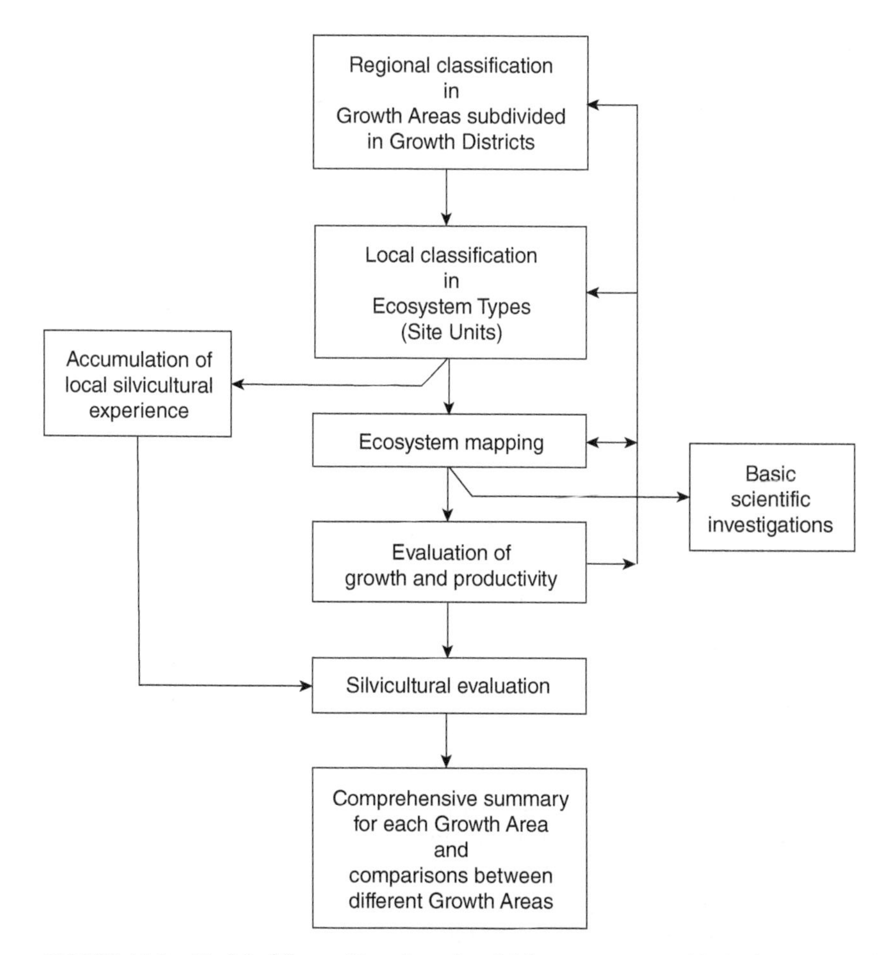

FIGURE 11.9 Model of the multi-scale and multi-factor system used in Baden-Württemberg, southwestern Germany to classify, map, and evaluate regional and local landscape ecosystems.

and vegetation science were pursued for their own sake. This team approached the practical application of these disciplines in forestry as the simultaneous integration of ecosystem components in classification and field mapping (Schlenker 1964; Mühlhäusser et al. 1983).

A major emphasis of this approach lies in **regionalization**, where a regional framework is provided that allows subdivision of broad ecosystems into successively smaller ecosystem units. In Baden-Württemberg, the synthesis of factors at the regional level leads to a division of the state into major landscapes (termed "growth areas"), which are in turn subdivided into minor landscapes ("growth districts") (Figure 11.9). This regional framework limits sweeping generalizations and prescriptions that were often made for species over wide areas having vastly different environments and histories. At the local level (within districts), individual ecosystem types (site units) are identified, mapped at a spatial scale of 1:10 000, and described. Ecosystem types are distinguished and mapped in the field using local differences in physiography, microclimate, soil factors of texture, structure, pH, depth, and water and nutrient status, and overstory and ground-cover vegetation. Ecosystem types include individual sites that are similar in physical site characteristics, silvicultural and management potential, disturbance regimes, incidence of disease and insect attack, and growth rates and tree biomass.

The German team also pioneered the use of ecological species groups (described in the earlier text) to distinguish and characterize local ecosystem types (Schlenker 1964; Sebald 1964; Dieterich 1970). In the Upper Neckar growth district of Baden-Württemberg (Sebald 1964), 24 ecological species groups were used to differentiate 30 ecosystem types along soil-water gradients (Table 11.1). Several plant species comprise each group, which indicate certain site factor complexes because of similar environmental requirements or tolerances. Some species groups have a wide ecological amplitude, such as the *Milium effusum* group, whereas others have a narrow amplitude, such as the *Aruncus sylvester* group (Table 11.1). Ecosystems are distinguished by the presence or absence of groups and the relative abundance of the species in the respective groups. Units are clearly differentiated along a gradual trend of differences in species groups when they are arranged along two soil-water gradients (from moderately fresh to moderately dry, units 1–10, and from fresh to wet, units 11–15). Importantly, the indicator value of each group is reliable only within the rooting zone of the species in the group.

In field mapping, ecological species groups are used simultaneously with physiographic and soil characteristics to delineate ecosystem boundaries. In part, this is because some units are well defined by vegetation in the species groups alone, such as those units at opposite ends of the gradients. On the other hand, adjacent site units (e.g., units 1 and 2 in Table 11.1) may have similar species groups and would need to be differentiated in the field using soil and topographic features. For example, unit 1 is a podzolized loamy sand on level terrain, whereas unit 2 is a podzolized sand on a moderately steep slope of south to southwest aspect. The combined technique using soil and physiography as well as vegetation is always faster and more reliable.

APPLICATIONS OF MULTI-FACTOR METHODS IN THE UNITED STATES AND CANADA

Ecosystem Classification and Mapping in Michigan The Baden-Württemberg approach has been modified and applied in Michigan at both regional and local levels. Regional ecosystems were classified based mainly on climate and physiography, which are the dominant physical factors driving ecosystems at regional scales. First, climates in Michigan were classified using separate classifications for growing season temperature, winter temperature, and growing season precipitation variables (Denton and Barnes 1988). The climate classifications were then combined with physiographic and soils data to produce a general hierarchical classification and map of regional landscape ecosystem regions and districts in the state at three spatial scales: regions, districts, and subdistricts (Albert et al. 1986; Figure 2.10 in Chapter 2). This approach has been extended to Wisconsin and Minnesota at the regional and district levels (Albert 1995).

Local landscape ecosystem types have been classified and mapped in Michigan for both old-growth forests (Barnes et al. 1982; Pregitzer and Barnes 1984; Spies and Barnes 1985a; Simpson et al. 1990) and highly disturbed landscapes (Archambault et al. 1990; Zou et al. 1992; Pearsall 1995; Walker et al. 2003). A physiographic diagram illustrating the correspondence of landform, soil, and vegetation for each ecosystem type in part of the McCormick Experimental Forest in Upper Michigan is presented in Figure 11.10. Ecosystem types in this old-growth forest are relatively homogeneous in site conditions and groundflora. A map of the area illustrates the fine-scale occurrence of ecosystems in formerly glaciated landscapes (Figure 11.11), where ecosystems recur in intricate but *predictable* patterns. For example, rocky and fire-prone ridges support white pine and northern red oak; outwash plains and slopes support northern-hardwood forests, and wetlands of various kinds are found along the Yellow Dog River. The major ecosystem types occupy over 95% of the area and

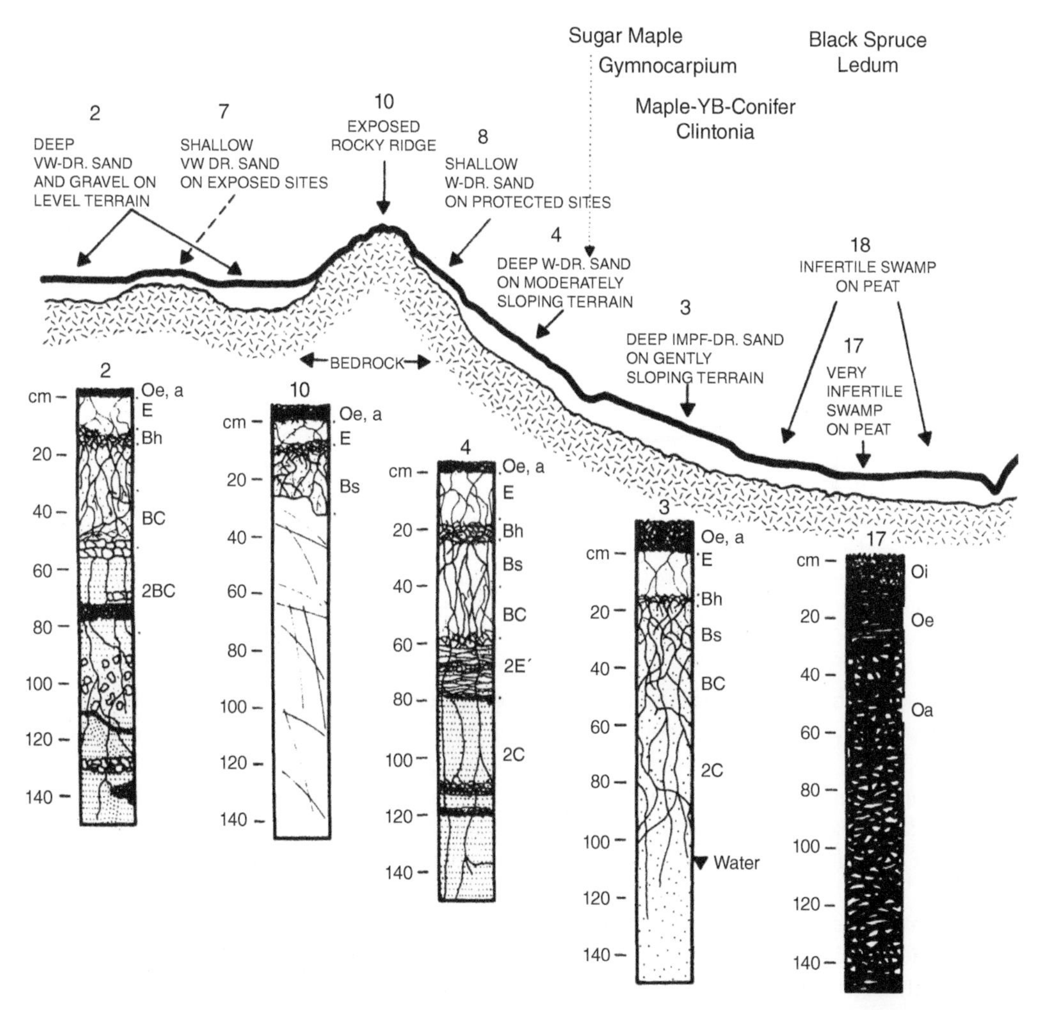

FIGURE 11.10 Physiographic diagram of selected landscape ecosystem types of the McCormick Experimental Forest, Ottawa National Forest, in Upper Michigan. Ecosystem types (identified by number) are characterized here by late-successional overstory species and characteristic ecological species groups, soil profile, and their physiographic position in and along the landscape gradient. *Source:* Barnes et al. (1982). Reprinted with the permission of the Society of American Foresters.

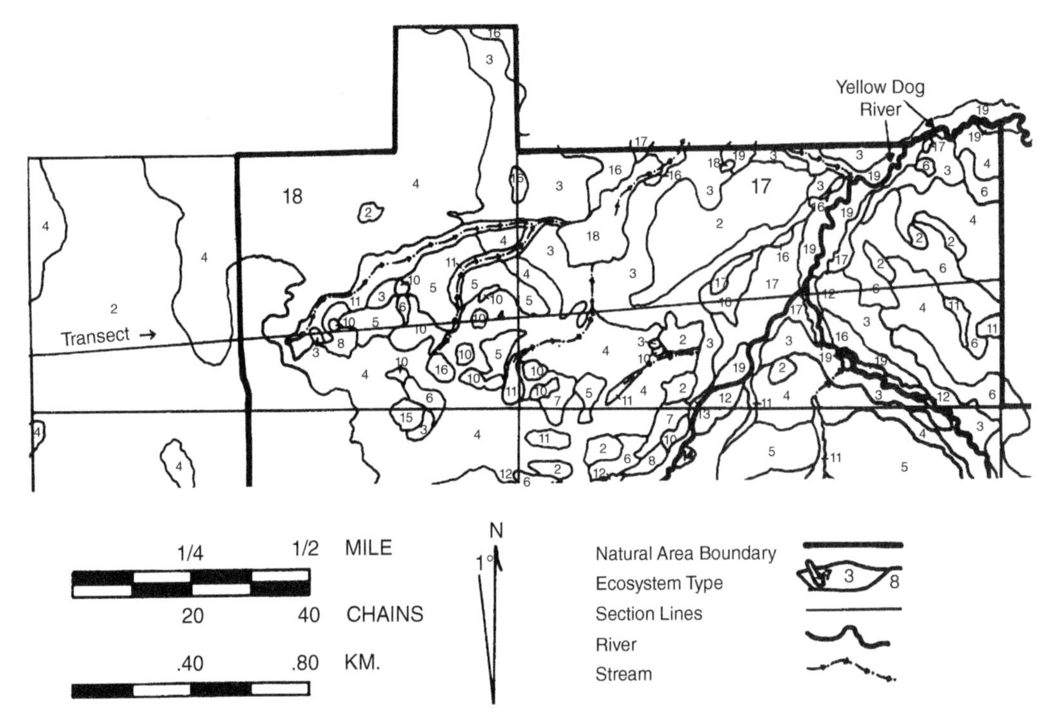

FIGURE 11.11 Map of local landscape ecosystem type for part of the McCormick Experimental Forest, Ottawa National Forest, in Upper Michigan. A transect line running roughly west to east illustrates the pattern of ecosystems; see also Figure 11.10 for their relative physiographic position to one another. On the west, an extensive flat, infertile outwash plain (type 2) is dominated by stunted sugar maple. Just to the east, a crystalline-rock ridge runs approximately northwest to southeast. Ecosystem type 10 occurs along the ridge top, and thin-soil types 7 and 8 are associated with it. On the northeastern slopes of the ridge, types 4 and 5 occupy mid and lower slope positions, respectively. Adjacent to the Yellow Dog River are wetland ecosystems of acid (type 18) and circumneutral (type 19) soils. On lower slopes, wet-mesic ecosystem type 3 often occurs adjacent to the swamps. Across the river on hilly, ice-contact terrain occur alternating steep sandy southwest slopes dominated by eastern white pine and hemlock (type 6), and steep, sandy northeast slopes dominated by sugar maple (type 4). One chain equals 66 ft. or 20.1 m. *Source:* Barnes et al. (1982) / with permission of Oxford University Press. Reprinted with permission of the Society of American Foresters.

can be grouped into four ecosystem groups for management purposes: sandy outwash and ice-contact terrain (types 2, 4, and 6), fertile lower slopes and valleys (types 5 and 11), rocky sites with shallow soils (types 7, 8, 9, and 10), and wetlands and wet-mesic adjacent slopes (types 3, 16, 17, 18, and 19).

Mapping and distinguishing ecosystems at the resolution shown in Figure 11.11 are often perceived as too expensive for development in North America, where management that might justify its cost is less intensive than in Europe. Long traditions of intensive land use are much different in Europe than those in North America. Nevertheless, this fine level of detail is also useful in understanding the ecological diversity of landscapes, to which the diversity of plants and animals (biodiversity) is closely related (Lapin and Barnes 1995; Pearsall 1995). As biodiversity is an increasingly important driver of forest management (Chapter 14), particularly as land use intensifies in North America, understanding the fine-scale resolution of landscape ecosystems will inevitably prove valuable. Notably, however, the ecosystem approach developed in southwestern Germany and applied in Michigan is applicable at several scales to meet appropriate management objectives (e.g., see Kashian et al. 2003; Walker et al. 2003), and fine-scale landscape ecosystems represent only one part of the classification system.

Ecosystem Classification in the Southeastern United States In the southeastern United States, landscape ecosystem classification is recognized as critical to managing ecosystems (Jones and Lloyd 1993). A multi-factor approach similar to that described in the earlier text has been applied in the Blue Ridge province of the southern Appalachian Mountains (McNab 1991) and in South Carolina (Jones 1991). Using a physiographic map of South Carolina (Meyers et al. 1986) as the regional hierarchical framework, landscape ecosystems were distinguished and mapped using a combination of landform, soil, and vegetation (Jones 1991). A physiographic diagram of the model for the Interior Plateau subregion of the Midlands Plateau region of the Piedmont province (Figure 11.12) illustrates five ecosystems named by their landscape position in relation to soil-water availability (xeric to mesic) and the vegetative associations occurring on these sites. Maps of these ecosystem types provide the basis for interpreting productivity and silvicultural management, but they also provide a framework for predicting potential habitat for endangered species, wetlands delineation, and ecological restoration.

A similar landscape ecosystem classification was developed for a portion of the Gulf Coastal Plain in southwestern Georgia (Goebel et al. 2001). Most of the upland forested area in the region is on karst topography dominated by fire-dependent longleaf pine, with mixed hardwoods along riparian areas. Seventeen ecosystems were identified and classified into six hierarchical levels, using physiographic systems (fluvial or karst) and specific landforms (topography, landforms, and terrain shape) at the highest three hierarchical levels and soil drainage class, soil texture, and vegetation at the lowest three levels. The classification was effectively tested by using it to map ecosystems in the field (Palik et al. 2000). Mapping the landscape ecosystems illustrated that less than half of the landscape supported plant communities that could be considered reference communities

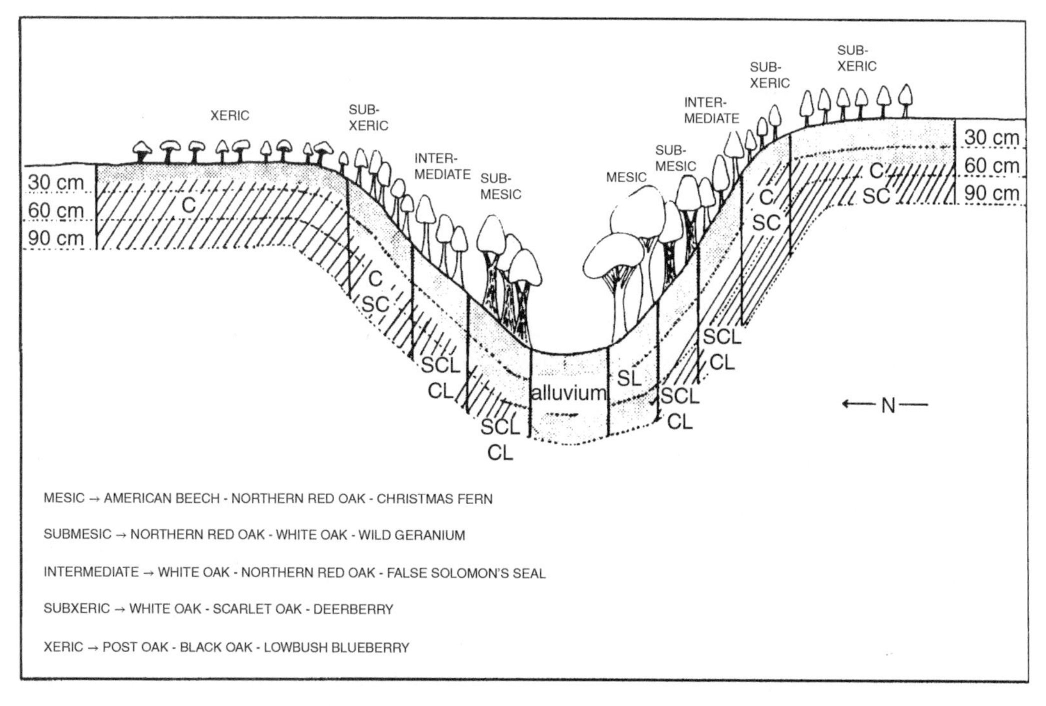

FIGURE 11.12 Physiographic diagram and landscape ecosystem classification model for the Interior Plateau subregion of the Midlands Plateau region of the Piedmont Province, South Carolina. Soil designations: C, clay; CL, clay loam; SC, sandy clay; SL, sandy loam; SCL, sandy clay loam. *Source:* Jones (1991) / United States Department of Agriculture / Public Domain.

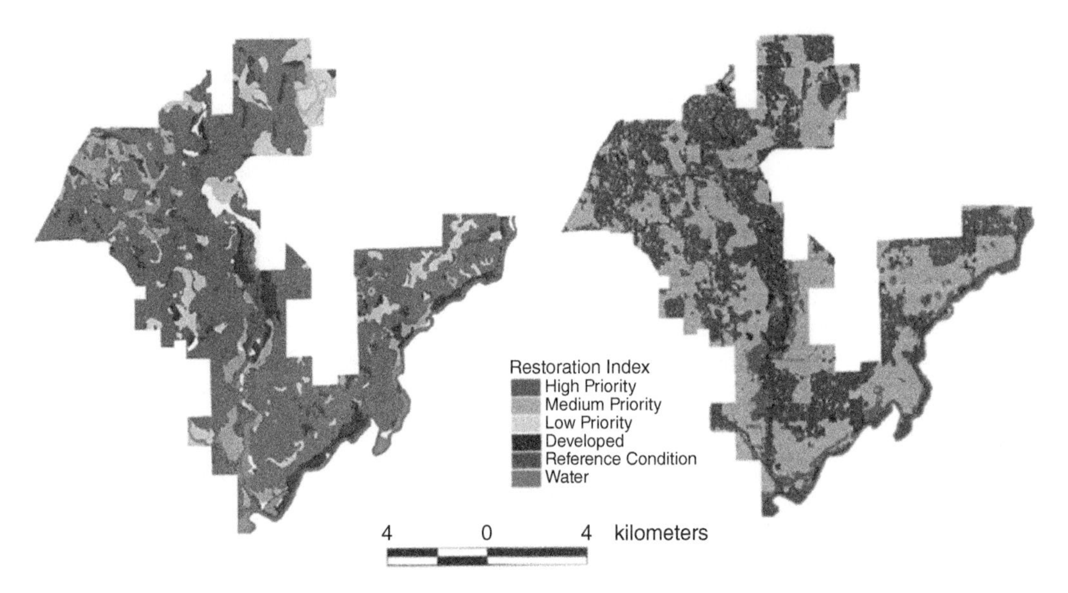

FIGURE 11.13 Landscape ecosystem map (left) and restoration priorities (right) developed for a portion of the Jones Ecological Research Center in southwestern Georgia. Despite the fine-grain occurrence of ecosystems, about 80% of the area found to be of highest priority for restoration occurred within 3 landscape ecosystems. *Source:* Palik et al. (2000).

for restoration, but it also provided an appropriate framework for prioritization of the landscape for restoration based on rarity and levels of disturbance (Figure 11.13). Notably, 80% of the high-priority sites occurred within only three of the identified ecosystems (Palik et al. 2000).

Ecosystem Classification in the Southwestern United States In portions of the southwestern United States, exclusion of historically frequent surface fires since European colonization in forests dominated by ponderosa pine has resulted in the accumulation of fuels and a subsequent shift to stand-replacing fires (Covington et al. 1994; Chapter 16). Fire exclusion has also led to the increase in dense stands of small-diameter trees and the reduction of the historically common open stands dominated by old trees, which has altered understory plant diversity and wildlife habitat. A study in northern Arizona ponderosa pine forests identified 10 landscape ecosystems in the region with the objective of providing an ecological framework for restoration (Abella and Covington 2006b). Most ecosystem types occurred across broad areas on this very flat landscape and were differentiated by differences in soil texture and carbon content rather than by physiographic variables, which nevertheless provides important guidance for restoration and conservation. Although not developed within a regional framework, this work further shows the value of identifying and understanding landscape ecosystems in a forested landscape heavily altered by humans.

National Classification Systems in the United States A national hierarchical system for ecological land classification (ELC) has been developed for the United States using many of the principles described in the earlier text. With the goal of developing a consistent framework for application nationwide, the National Hierarchical Framework of Ecological Units (NHFEU; Cleland et al. 1997) was devised to identify areas at multiple scales that are similar in their capabilities and potential for management. Similar to the Baden-Württemberg approach, the ecological units were devised based on climate, physiography, soils, and potential natural vegetation, with cooperation among the various disciplines representing each component encouraged as the system is utilized

(Avers and Schlatterer 1991). The system includes eight hierarchical ecological units based on decreasingly smaller scales of climate, physiography, soils, and potential natural vegetation: ecoregional units (domain, division, and province), subregional units (section and subsection), landscape units (landtype association, or LTA), and land units (landtype and landtype phase). Strategic planning and assessment are the main activities at the level of ecoregional units, statewide or multi-National Forest at the subregional unit level, individual National Forest planning at the landscape unit level, and specific project management at the land units. Thus, most planning and management activities within a given National Forest occur at the spatial scales of the LTA and land unit, within the regional framework of the broader-scale units. In addition to ecosystem mapping, land managers in the United States are encouraged to use the NHFEU, at the appropriate ecological scale, for activities such as resource assessment and management, environmental analyses, watershed analyses, determining desired future conditions (restoration goals), and monitoring (Cleland et al. 1997).

Ecological Land Classification in Canada Canada has been a leader in the use of multi-scale and multi-factor ecosystem classification. The Canadian versions of the approach, as in Europe, are characterized by many different systems and complex terminologies (e.g., Rubec 1992; Wicken 1986), but their broad scale of application sets them apart. Establishment of a common national system utilizing digital ecosystem mapping databases was completed in 1995 as part of the country's official ELC (Marshall et al. 1999). The classification was last updated in 2017, with hierarchical levels including ecozones (15 units across Canada), ecoprovinces (53 units), ecoregions (194 units), and ecodistricts (1027 units). For clarity, Canada also has an official classification for forests called the Canadian Forest Ecosystem Classification System (CFEC 2010), but it is a classification emphasizing plant associations and communities rather than landscape ecosystems. The CFEC is part of the broader Canadian National Vegetation Classification System, which is analogous to the US National Vegetation Classification System described in Chapter 2. Prior to publication of the ELC, the enormous diversity of Canada's forests had produced many different approaches and terminologies for its ecosystem classification and mapping. The diversity and detail of nine of these approaches, representing eight provinces and northern Canada, are presented in a single issue of the *Forestry Chronicle* (Canadian Institute of Forestry 1992). Despite Canada's adoption of an official national classification system, we focus in the following text on a few of the preceding systems to illustrate the diversity of approaches to ELC.

Hills' physiographic approach Cajander's method and that of the Zürich-Montpellier School of Plant Sociology (Lemieux 1965; Burger 1972) heavily influenced early attention to site quality in Canada, placing the main emphasis on ground vegetation. The pioneering work of Angus Hills in Ontario in the 1950s and 1960s shifted the focus toward physiography in evaluating site quality, in particular a *total site* or ecosystem approach integrating climate, soil, and vegetation on a landform basis. The idea that landform influenced the distribution of plants and animals as well as local climate, drainage, and soil formation was novel and unusual at the time (Rowe 1992a). This total site system (Hills 1952, 1960; Hills and Pierpoint 1960; Burger 1972; Hills 1977; Burger 1993) was undertaken originally to provide accurate, descriptive resource maps for land-use decisions. Hills provided a regional ecosystem framework by dividing Ontario hierarchically into 13 regional landscape ecosystems he called **site regions** (Figure 11.14; Burger 1993). A site region was further subdivided into **site districts** on the basis of relief and type of bedrock or parent material. Hills used physiographic features as a basic frame of reference because "they remain most easily recognizable in a world of constant change" (Hills 1952). Thus, **physiographic site types** and **forest types** (characterized by both overstory and groundflora) were combined to form total site types (Figure 11.14). Hills's basic approach continues to guide ecological classification today specifically in Ontario (Baldwin et al. 2000; Crins et al. 2009), and also at a national level (Marshall et al. 1999).

Other approaches used in Canada Development of a series of goal-oriented forest ecosystem classifications intensified beginning in the early 1980s for a broad geographic zone across north-central Ontario (Jones et al. 1983; Sims et al. 1989; Sims and Uhlig 1992). These classifications were developed for use as stand-level planning tools and for a specific and predefined range of uses. They are designed to help managers develop strategies and alternatives, such as prescribed burning guidelines, estimating susceptibility of a site to erosion, predicting productivity of specific tree species, and rating habitat for white-tailed deer. Other examples of application are given by Sims and Uhlig (1992).

Other Canadian systems focus primarily on vegetation types and are especially useful where human disturbances are minimal, for example, in Newfoundland and Labrador (Meades and Roberts 1992) and British Columbia (MacKinnon et al. 1992). The detailed floristic research of Damman (1964) facilitated forest site classification in Newfoundland. The phytosociological approach has remained strong in British Columbia, although Krajina and associates have developed a more inclusive biogeoclimatic approach that also considers climate, soil, and topography (MacKinnon et al. 1992).

In summary, multiscale and multifactor landscape ecosystem classification and mapping provide the appropriate units required for extensive and intensive management of forest ecosystems. German methods illustrate that ecosystem classification and mapping can meet the most intensive (and changing) management and conservation practices even in areas of highly variable landforms, soil, and climate, and for sites that have a long disturbance history. Today, remote sensing, geographic information systems, and computer-based techniques facilitate systematic ecosystem classification and mapping to provide a basic framework for land management and planning around the world.

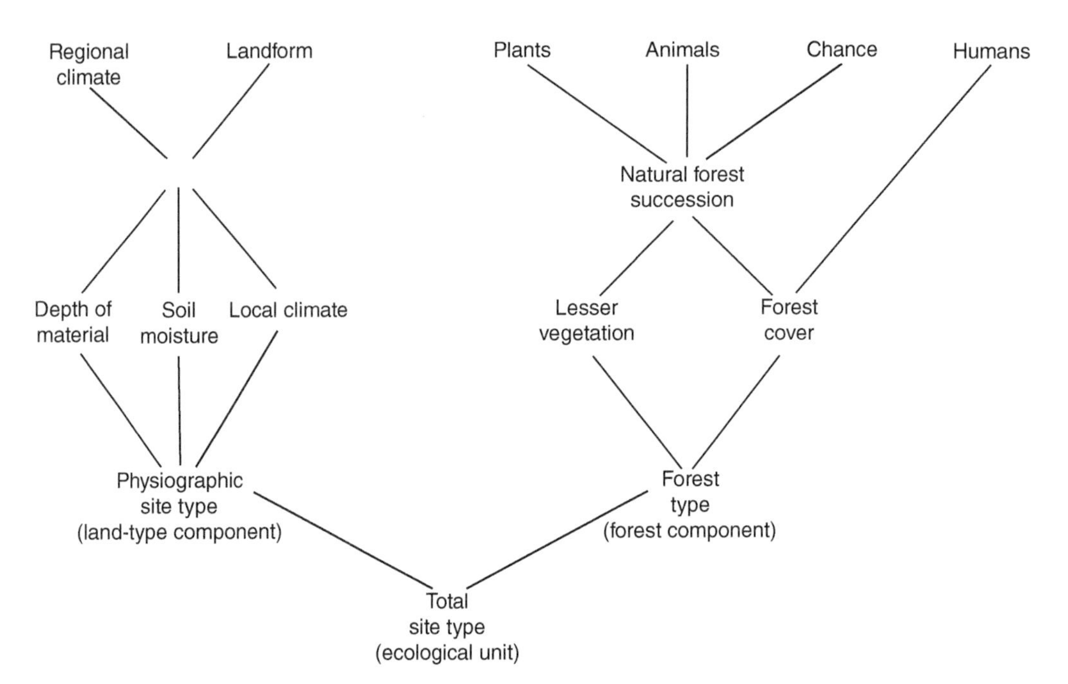

FIGURE 11.14 Model of the classification of total site types (landscape ecosystems) by Hills' method in Ontario, Canada. *Source:* Hills and Pierpoint (1960) / CAB International.

SUGGESTED READINGS

Barnes, B.V., Pregitzer, K.S., Spies, T.A., and Spooner, V.H. (1982). Ecological forest site classification. *J. For.* 80: 493–498.

Carmean, W.H. (1975). Forest site quality evaluation in the United Sates. *Adv. Agronomy* 27: 209–269.

Sims, R.A. 1992. Forest site classification issue. Forestry Chronicle, 68(1):21–120.

Grigal, D. F. (1984). Shortcomings of soil surveys for forest management. *In* J. G. Bockheim (ed.). Symp. Proc. *Forest land classification: Experience, problems, perspectives.* NCR-102 North Central For. Soils Com., Soc. Am. For., USDA For. Serv., and USDA Soil Cons. Serv.

Palik, B.J., Goebel, P.C., Kirkman, L.K., and West, L. (2000). Using landscape hierarchies to guide restoration of disturbed ecosystems. *Ecol. Appl.* 10: 189–202.

Pfister, R.D. and Arno, S.F. (1980). Classifying forest habitats based on potential climax vegetation. *For. Sci.* 26: 52–70.

Rowe, J. S. (1984). Forestland classification: limitations of the use of vegetation. *In* J. G. Bockheim (ed.), Symp. Proc. *Forest Land Classification: Experience, Problems, Perspectives.* NCR-102 North Central For. Soils Com., Soc. Am. For., USDA For. Serv., and USDA Soil Cons. Serv.

Rowe, J. S. (1991). Forests as landscape ecosystems: implications for their regionalization and classification. *In* D. L. Mengel and D. T. Tew (eds.), Symp. Proc. *Ecological land classification: applications to identify the productive potential of southern forests.* USDA For. Serv. Gen. Tech. Report SE-68. Southeastern For. Exp. Sta., Asheville, NC.

Rowe, J.S. and Sheard, J.W. (1981). Ecological land classification: a survey approach. *Environ. Manag.* 5: 451–464.

Forest Communities

PART 4

Forest ecosystems are characterized by a layered structure of functioning parts—forest biota sandwiched between physiography and soil below them and the atmosphere above them. Though they are often the main emphasis, forest communities of trees and associated plants and animals form only one key structural component of whole forest ecosystems.

Evolving from these structured, ecological environments in response to the abiotic factors therein, plants are integral parts of all ecosystem processes. Their individual forms and their community physiognomy, therefore, reflect the vertical stratification of the air above and the soil and physiography below. Plants of the forest community, individually and collectively, are instrumental in myriad of changes that occur in the soil and atmosphere of forest ecosystems.

Animals are also an indispensable part of forest ecosystems, and they are considered in Chapter 12. We emphasize the role of animals in affecting the life cycles of plants in this chapter. Animals, small and large, affect all aspects of the life history of plants from reproduction to death, and they even promote "life-after-death" in clonal plants. Because animals attack and consume plants at all stages in their life cycle, we consider in detail a forest plant's ability to defend itself from animals. We also consider the mutualisms that develop between animals and plants, including the critical processes of pollination, seed dispersal, germination, and establishment. Finally, the enormous effects of large animals associated with humans (livestock) on forest ecosystems receive special treatment.

The forest community is considered in Chapter 13 as an integral part of the forest ecosystem, but we work to ground communities in the context of their specific environments. Communities are aggregates or collections of organisms in a particular place at a given time. We review traditional views of the forest community, their development, and their shortcomings. Interactions among organisms are emphasized, including the concepts of mutualism, competition, and niche, and we use these concepts to discuss the classic processes of forest structural development. Finally, because of its importance in community dynamics, the concept of understory tolerance of forest trees is examined in detail.

This part also includes a chapter on the diversity of both ecosystems and organisms (Chapter 14). The diversity of whole ecosystems determines biodiversity itself, but

Forest Ecology, Fifth Edition. Daniel M. Kashian, Donald R. Zak, Burton V. Barnes, and Stephen H. Spurr.

as with many ecological characteristics, the biota is often a main focus. Therefore, in Chapter 14, both biodiversity and ecosystem diversity are considered at regional and local scales. The value of biodiversity and its measurement is treated in detail, along with its association with ecosystem function, and the maintenance and conservation of ecosystems and their biotic diversity. These considerations of forest biota together set the stage for the chapters in Part 5 on forest ecosystem dynamics.

Animals in Forest Ecosystems

CHAPTER 12

Animals of all sizes form indispensable parts of forest ecosystems, influencing forest community composition and ecosystem processes. Animals are likewise strongly affected both by the physical environment and by the plants with which they associate.

Plants provide shelter and food for animals. Green plants form the base of most **trophic systems**—the food webs that comprise plant and animal relationships within ecosystems. Trophic levels in a system vary but may consist of plants, the animals (**herbivores**) that eat them (including **browsers** and **grazers**), the animal predators (**carnivores**) and parasites that feed on the herbivores, and the scavengers and **detritivores** that eat animal remains and excrement. A trophic system is completed by **decomposers**, most typically fungi and bacteria but also some animals, that degrade and mineralize plant litter and animal residues. Trophic systems are complex, and a detailed treatment is beyond our scope. Instead, we emphasize here other interactions of animals and plants, including plant defense, the role of animals in regulating plant life history and production, and the effects of large animals on forest ecosystems.

We focus in this chapter on understanding the role of animals in forest ecosystems. Humans, of course, are animals with overwhelming effects on forest ecosystems, and those effects are discussed in detail in Chapters 20, 22, and 23, among others. A long history by humans of emphasizing timber production and wildlife habitat has focused much attention on the activities of forest animals perceived by humans to be destructive, such as insects that cause outbreaks or livestock and native herbivores that reduce tree species regeneration. We emphasize from an ecosystem perspective that animal activities are only destructive as a human construct, and that such activities may be critical and beneficial even when they appear to negatively impact forest resources. The ecological contributions of native forest animals in the evolution of plants and their indispensable roles in ecosystem processes are immensely significant but often underappreciated.

PLANT DEFENSE

There are many observations of woody species–animal relationships that we assume to be mutual adaptations, though often without rigorous demonstration of cause and effect. Perhaps the best examples of plant–animal co-adaptations are found in the area of plant defense to herbivory. The existence of plant defense is often evidenced by the fact that herbivores consume only a fraction of the plants available to them. All parts of woody plants are subject to herbivory at any stage of their life cycle, and it follows that they exhibit many defense mechanisms that presumably evolved due to the presence of herbivores and seed predators. In all cases, the nature of plant defense has developed within the particular site conditions of the ecosystem (Mattson and Haack 1987; Herms and Mattson 1992).

Plant defenses may be physical in nature or may involve specialized plant chemistry. Physical defenses employed by woody plants include structures or surface texture that may prevent herbivorous injury or destruction. Such defenses may include leaf toughness, the presence of trichomes

Forest Ecology, Fifth Edition. Daniel M. Kashian, Donald R. Zak, Burton V. Barnes, and Stephen H. Spurr.

(small hair-like outgrowths on shoots that deter herbivores, as in oaks; Hardin 1979), or the presence of specialized organs or tissues such as thorns or resin ducts. Many tree species, including black locust, honey locust, osage-orange, hawthorns, junipers, and some hard pines, as well as many woody shrubs and vines (roses, greenbriers, blackberries, raspberries), have prickles, spines, thorns, or sharp needle-leaves that deter browsing. Physical defenses in tree species tend to be concentrated in the juvenile phase when foliage and stem feeding by rodents, rabbits, and other herbivores is most likely. A review of physical defense is provided by Hanley et al. (2007).

Chemical defenses include the presence of **secondary chemicals**, compounds utilized for purposes other than metabolism, which tend to reduce the palatability or digestibility of plant tissues to herbivores. Qualitative defense chemicals, or **toxins**, disrupt herbivore metabolism or development even at low concentrations. Examples of toxins include alkaloids, terpenes, and cyanogenic glycosides, and they act as the main defense against mammalian herbivores (Bryant et al. 1991). Quantitative defense chemicals, or **digestibility reducers**, inhibit a herbivore's absorption of nutrition from the plant tissues they have consumed. Examples include resins, cellulose, and phenolics including tannins and lignins, all of which must be produced in high concentrations to be effective. An enormous number and diversity of secondary chemicals have been described in higher plants, including thousands of alkaloids alone. Overall, herbaceous plants have more qualitative than quantitative defense chemicals, and woody plants have a mix of both types. Early-successional tree species are likely to have fewer defensive chemicals in general than late-successional species, as evidenced by their higher susceptibility to herbivory (Coley et al. 1985; Coley 1980).

The investment into plant defense changes as plants age, mainly because of age-related changes in resource allocation (Boege and Marquis 2005). Secondary chemicals are costly for a plant to produce because the plant is expending energy (and carbon) on a process other than growth or reproduction. It is proposed that plants should therefore invest in defense only when the benefit of avoiding herbivory outweighs the cost of reduced growth or reproduction. If resource availability is a main control of plant defense, then plants newly emerged from a seed should have higher defense because they can allocate stored resources in seeds and cotyledons (Herms and Mattson 1992; Stowe et al. 2000). Once a seedling is well developed and established, defense is likely to be lowest because plants have a large root-to-shoot ratio and most resources are allocated to growth. By the sapling stage, plants can produce more photosynthate and allocate these resources to defense. At reproductive age, allocation to defense may decrease if the plant has heavy investment into reproduction (as in early-successional species), but it may continue if reproduction is delayed or intermittent (as in late-successional species). Finally, plant defense is likely to decrease with the reduction in most metabolic functions that occurs with old age (Figure 12.1). Empirical evidence for trends in defense with plant age is inconsistent, however. In a meta-analysis of 116 published studies, Barton and Koricheva (2010) found a general trend of increasing chemical defense during the seedling stage followed by an increase in physical defenses during the vegetative juvenile stage.

INVESTMENT IN PLANT DEFENSE

The general concept of resource allocation generally dictates why plant defense may be emphasized in some plant species more so than others. A series of hypotheses, all of which consider resource allocation, has developed to explain these differences (Stamp 2003). Critically, the mix of plant genotypes that survive on a given site depends on pressures exerted by biotic interactions including competition, mutualisms, and herbivory. The relative importance of these biotic factors, however, is determined by physical site factors, emphasizing the influence of the abiotic as well as the biotic components of landscape ecosystems.

The **apparency hypothesis** (Feeny 1976; Rhoades and Cates 1976) suggests that plants difficult for herbivores to find will experience less herbivory, and such plants should therefore

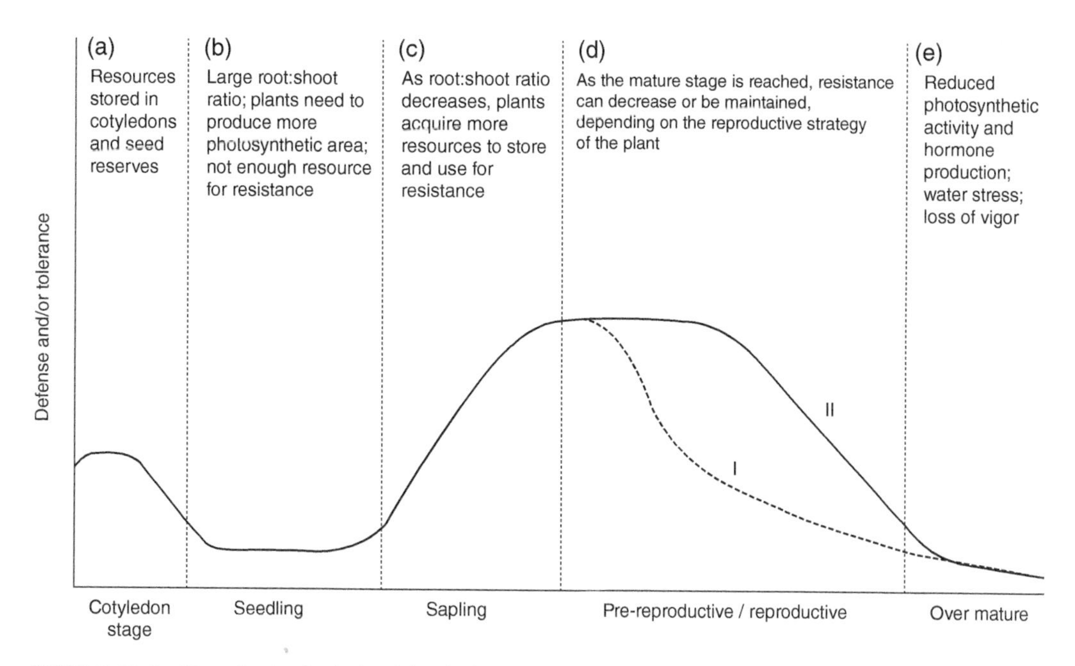

FIGURE 12.1 Hypothesized relationship of plant investment into defense at various life stages. Plant defense is moderate at (a) the cotyledon stage when the plant relies on stored resources, but decreases at (b) the seedling stage when the plant emphasizes growth over other functions. Defense reaches its highest point at (c) the sapling stage when excessive photosynthate is produced, then either continues (II) or decreases (I) at (d) the reproductive stage depending on the species' reproductive strategy. Plant defense decreases in (e) old age with other metabolic functions. *Source:* Boege and Marquis (2005) / with permission of Elsevier.

invest less into plant defense (typically qualitative defense chemicals). Likewise, plants—and plant tissues—that are more apparent to herbivores, such as the dominant tree species in a forest or the long-lived, evergreen needles on a conifer, should have more defense chemicals (quantitative defense chemicals). Thus, young leaves of trees tend to be lower in defense chemicals compared to older leaves. A study of downy birch in the boreal forest of Russia found that larger individuals of downy birch suffered higher insect herbivory; background losses of foliage to insects increased from small saplings to large saplings and then to mature trees (Zverev et al. 2017). Moreover, many of the small individuals escaped from herbivory altogether. Notably, the apparency hypothesis has been criticized for inconsistent support for its predictions (Smilanich et al. 2016).

The **resource-availability hypothesis** (Coley et al. 1985; Coley 1987) emphasizes plant growth rather than apparency as a determinant of investment in plant defense. This hypothesis suggests that fast-growing plants are better able to tolerate herbivory compared to slow-growing plants, and thus fast-growing plants invest less into plant defense. This is because slow-growing plants typically grow on sites where resources are limited, such that replacing tissue lost to herbivory would be more costly. There is much evidence for higher defense investment in slow-growing species (Endara and Coley 2011), although differences in defense investment are generally higher in tropical forests compared to temperate and boreal ecosystems (Van Zandt 2007).

The **growth–differentiation balance hypothesis** (Herms and Mattson 1992) suggests that resource allocation within plants balances growth with defense mechanisms that limit herbivory (Figure 12.2). On favorable sites, plants allocate photosynthate to growth rather than defense so that they may maximize their acquisition of readily available resources of light, moisture, and nutrients. Plants therefore tolerate and compensate for herbivores with vigorous growth

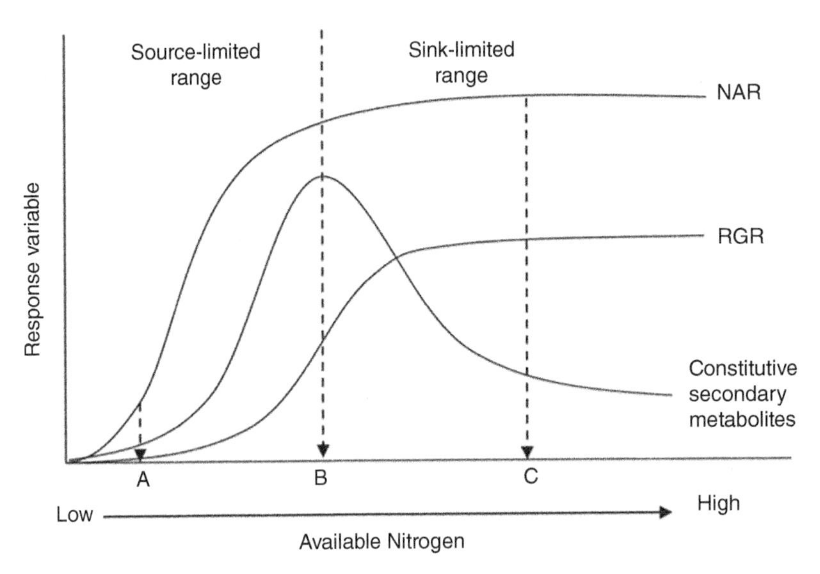

FIGURE 12.2 Relationship of allocation to net assimilation rate (NAR), relative growth rate (RGR), and secondary metabolism along a gradient of low- to high-nutrient availability. The vertical lines at A, B, and C represent high, moderate, and low levels of nitrogen, respectively. At B and C, net assimilation is constant, whereas the relative growth rate and secondary metabolism are inversely correlated. The physiological trade-off between growth and secondary metabolism is apparent. *Source:* From Hattas et al. (2017), as adapted from Herms and Mattson (1992).

that keeps them competitive. On poor sites, plants allocate photosynthate to structural and chemical traits that enhance the structure and function of existing cells, including those involved in plant defense, because the loss of growth (and the subsequent need for re-growth if plant tissues or organs are lost or damaged) is particularly costly where resources are scarce. Thus, plants must either outgrow herbivory or defend against it, but not both (Herms and Mattson 1992). The growth–differentiation hypothesis is difficult to test rigorously (Stamp 2004), and only limited evidence exists to support it (e.g., Glynn et al. 2007).

PLANT DEFENSE AGAINST INSECTS

Plants exhibit a variety of defense strategies against insects, from fully compensating for any and all injuries (complete tolerance) to chemical and physical defenses that provide complete or near-complete immunity (Mattson et al. 1988). Even immunity may be overcome during periods of stress or under outbreak conditions, however (Mattson and Haack 1987), such that most plants have evolved partial tolerance whereby they regulate and thus minimize damage with chemical and morphological defenses. The type of plant defense system depends on multiple factors, including the type of tissue targeted (meristematic tissue, phloem, etc.), timing of the attack (early versus late in the growing season), and the type of attack itself (single insect or mass attack). Mattson et al. (1988) ranked 13 insect-feeding guilds according to their potential injury to woody plants. Insects least damaging to trees and shrubs are leaf and twig gall-formers and those affecting late-season or prior year's leaves, and phloem and sapwood borers and root feeders are intermediate in their effects. Most damaging are phloem and cambium borers on main stem and roots because they can destroy the plant's conductive tissue faster than the plant is able to replace it.

Examples of Injury and Plant Defense In conifers, and particularly the pine family, oleoresin ("pitch") is a terpene-based compound exuded from resin ducts in needles, shoots, and bark to deter foliage feeders and bark beetles that excavate galleries in the inner bark tissues. Though used as a mechanism for defense, however, insects use the vapors from oleoresins to search for host trees. When a tree is attacked by insects and begins to increase its production of oleoresin, the resin vapor becomes concentrated and serves to attract additional bark beetles to the attacked tree. Thus, trees damaged by lightning or previous attack, or even freshly cut trees, tend to suffer more from mass attack by bark beetles than vigorous, standing trees (Hanover 1975). For example, western larch trees are strongly resistant to bark beetle attack while standing and healthy, but are immediately attacked after being felled (Furniss 1972). Likewise, moisture-stressed loblolly pines that abnormally lose needles and exude resin have been shown to attract insects (Heikkenen et al. 1986).

Plants require an unwavering defense against bark beetles that feed on the cambium because their girdling of the conducting phloem and xylem has devastating effects (Mattson and Haack 1987). When one insect becomes established, it uses pheromones to rapidly attract others, contributing to mass attack. Healthy trees are generally unharmed by insects at endemic population levels, but become attractive to beetles when the trees are stressed (Larsson 1989). Oleoresin pressure is high in young and vigorously growing pines, physically repelling beetles entering the bark ("pitching out") or rendering the beetles physiologically impaired by chemical properties of the resin, thus preventing them from reproducing (Figure 12.3). The terpenes myrcene and limonene in healthy ponderosa pine may actually kill western pine beetles feeding on its needles or bark (Smith 1966). Conversely, the severity of bark beetle attack is greatest when resin defenses are low due to the natural aging process or from major stresses caused by competition, drought,

FIGURE 12.3 Lower bole of a lodgepole pine in the initial stages of attack by mountain pine beetle in Grand Teton National Park, Wyoming. Resins in the tree have "pitched out," or physically expelled, beetles that have bored into the bark such that each bore hole is marked by "plugs" of hardened sap (six such plugs are denoted with white arrows). Additional and continued attack by beetles will eventually overcome the tree's defenses and kill it.

Photo by Dan Kashian

pollution, logging damage, disease, or defoliation. Bark beetles exhibit various degrees of tolerance to resin toxicity, and as a group are host-specific and more tolerant to resins of their own host species than those of other hosts.

Anti-herbivory resins and other defense chemicals are also found in hardwood species. Certain hardwood trees such as cottonwoods have resins that deter insect herbivory (Curtis and Lersten 1974), as do young leaves of the desert shrub creosote to avoid defoliation (Rhoades 1976). Juglone is a compound produced by walnuts and shagbark hickory that deters feeding by some bark beetle species. Tannins also tend to deter herbivory when present in high concentrations, especially in oaks. A study of acorns (oaks) showed a high concentration of tannins around the embryo at the apical end, thereby confining weevil activity to the less-protected basal end (Steele et al. 1993). Oak leaves are most susceptible to insect herbivory during their development in the spring when tannins are absent or scarce, but the late and rapid flushing of preformed leaves and shoots of oaks minimizes the time insects may feed and reproduce using these tissues. This rapid leaf-flushing trait in north temperate forests has apparently elicited reciprocal adaptations of insects and their hosts. For example, flushing time may differ by as much as 3 weeks among trembling aspen clones (Barnes 1969), and populations of tortricid caterpillars predominantly infest leaves of early flushing clones (Witter and Waisanen 1978). A similar relationship was also reported for larvae of tortrix moths on oaks in Russia (Sukachev and Dylis 1964) and Europe (DuMerle 1988). In sugar maple, early flushing buds suffered greater damage by pear thrips than trees with late budburst (Kolb and Teulon 1991).

There are many studies that provide examples of how site conditions may affect the intensity and nature of insect attack and reciprocal plant defenses. Pinyon pines growing on cinder fields in Sunset Crater of northern Arizona live in a highly stressful environment with limited moisture and low nutrient status. The pines suffer unusually high levels of chronic and severe herbivory by many insects, to the point that tree architecture is altered, growth is reduced, and female reproductive function is eliminated (Whitham and Mopper 1985; Mopper et al. 1991a). Pinyons growing on less-severe sandy loam soils adjacent to the cinder fields, however, are rarely attacked and exhibit normal reproduction and growth, despite producing significantly less resin than the Sunset Crater pines. Notably, insects at Sunset Crater not only reduced tree growth directly by consuming above-ground tissue but also indirectly by reducing the amount of ectomycorrhizal fungi associated with roots of susceptible individuals (Gehring and Whitham 1991; Del Vecchio et al. 1993). These studies therefore show that the incidence of herbivory may influence how photosynthate is differentially allocated to growth, defense, or to maintain mycorrhizae.

Nutrition Insects assimilate nutrients from plant tissue with very low efficiency, such that variations in the nutritional value of plants have large implications for insect herbivores. Plants growing on poor sites (low moisture and nutrients) are likely to produce lower-quality tissues, which may be detrimental to insect performance and thus effectively inhibit herbivory (Herms and Mattson 1992). The plant-stress theory of herbivory, however, predicts that plant stress induced by limited moisture or nutrients actually benefits insects by increasing the concentrations of usable substances in foliage and creating a more favorable balance of nutrients (Mody et al. 2009; Mattson and Addy 1975; White 1978). Evidence for both situations comes from experimental studies of sawflies on pinyon pines at Sunset Crater, Arizona (Mopper and Whitham 1992; see pp. 337–338). This seeming paradox, whereby abiotic stresses may decrease or increase performance, may be explained by the type of stress considered. Mopper and Whitham (1992) contrast sustained plant stress, such as poor soil-nutrient conditions or prolonged drought prevailing while insects are both active and inactive, with "simultaneous" plant stress, such as low precipitation while the insect is feeding or ovipositing (Figure 12.4). They predicted that under simultaneous stress, insect performance will be high when plant stress is low and drops rapidly as stress increases. At the highest level of plant stress, the plant becomes an inadequate food source. By contrast, under

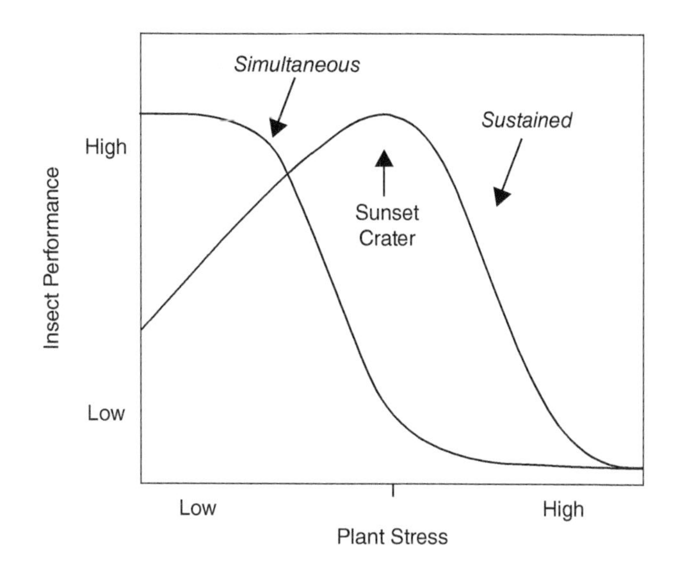

FIGURE 12.4 Hypothetical relationship between insect performance and simultaneous and sustained plant stress. Conditions at Sunset Crater, Arizona are indicated. *Source:* Mopper and Whitham (1992) / John Wiley & Sons. Reprinted with permission of the Ecological Society of America.

sustained stress, insects are relatively successful at sites of low chronic plant stress, even more successful at sites of intermediate stress, but unsuccessful when high stress makes the plant an inadequate food source. More recent studies have also suggested that intermittent stresses rather than sustained stresses improve plant quality for some herbivores (Huberty and Denno 2004; Sconiers and Eubanks 2017), particularly those that feed on sap rather than leaves or other plant tissues. For example, aphids were found to be more abundant on intermittently stressed Sitka spruce compared to those continuously stressed (Major 1990).

In summary, woody plants exhibit a wide variety of defenses against insect herbivores including those associated with different individual genotypes. Even the likelihood of somatic mutations in modular shoots of long-lived plants or clones may represent a mosaic of genotypes that might effectively prevent herbivores from evolving metabolic pathways to overcome plant defenses (Whitham and Slobodchikoff 1981). These new and diverse plant defenses force herbivores to co-evolve and find new ways to avoid or resist them. These intricate interrelationships among site conditions, vegetation, and herbivores further cement the important role of animals in the structure and function of ecosystems on the landscape.

Plant Hybrid Zones as Reservoirs for Insect Diversity Plant hybrid zones have been found to be critical centers of insect abundance and diversity (Kearsley and Whitham 1989; Whitham 1989; Floate et al. 1993; Whitham et al. 1994). In Weber Canyon, Utah, 85–100% of the *Pemphigus betae* gall aphid population occurs on less than 3% of its host population in a 13-km hybrid zone of Fremont and narrowleaf cottonwood (Figure 12.5b; Whitham 1989). Hybridization has altered the well-engineered defense of each parent species such that aphid populations are more viable on the hybrid. The concentration of aphids on such a small segment of the host population suggests that susceptible hybrid plants not only act as insect reservoirs in ecological time, but they may also have inhibited the aphids from adapting to the more numerous parent hosts in evolutionary time.

The same hybrid zone in Weber Canyon is also superior habitat for the free-feeding beetle, *Chrysomela confluens* (Figure 12.5a), in part because the early leaf flush of narrowleaf cottonwood and the hybrid provides the first source of abundant food for beetles in spring. In addition, staggered leaf phenology in the hybrid zone allows beetles to shift onto newly flushed Fremont cottonwoods as foliage of the hybrid and narrowleaf trees declines in quality. Thus, the hybrid zone is a phenological reservoir that increases beetle fecundity and leads to chronic

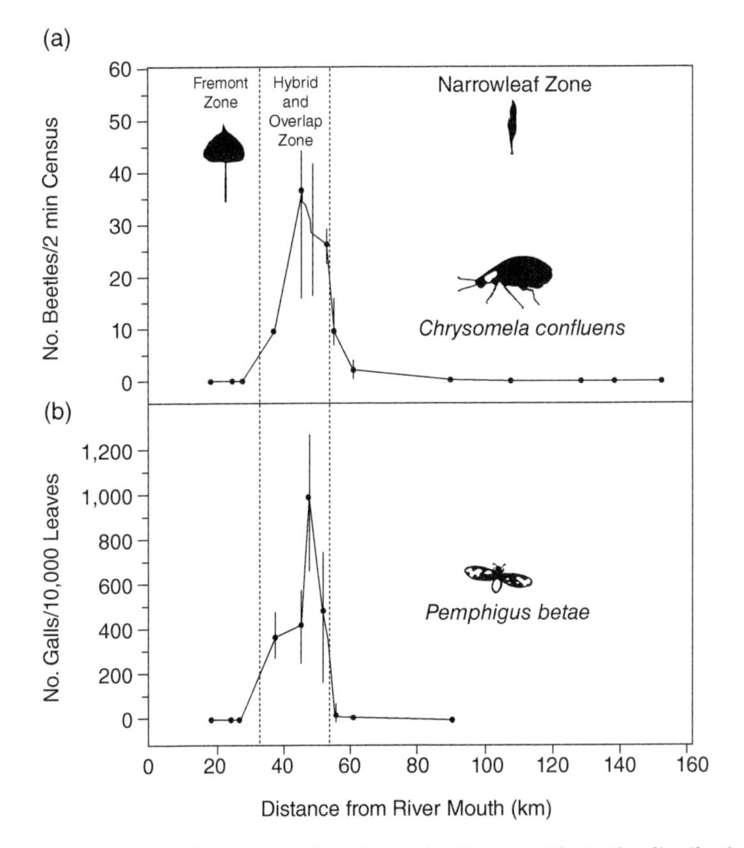

FIGURE 12.5 Occurrence of two insect herbivores with similar distributions in the hybrid zone between Fremont and narrowleaf cottonwoods along the Weber River, northern Utah. (a) Free-feeding beetle, *Chrysomela confluens*, values are 3-year means (±1 SE). (b) Galling aphid, *Pemphigus betae*. *Source:* Floate et al. (1993) / John Wiley & Sons. Reprinted with permission of the Ecological Society of America.

herbivory every year. This beetle species may also be an example of the **hybrid bridge hypothesis** (Floate and Whitham 1993), which suggests that hybrid plants help herbivores to switch or adopt additional plant host species, thus allowing them to evolve a larger range of host species when hybrids are present. A study of two red oak species and their hybrid in central Mexico found that 32% of insects in the study were specific to one of the parental oaks, 23% to the other parent, and 9% to the hybrid (Tovar-Sanchez and Oyama 2006), and that the hybrid oak supported intermediate levels of herbivory between the parents. The increased genetic diversity among the vegetation of the hybrid zone appears to support greater genetic diversity among the herbivores.

PLANT DEFENSE AGAINST MAMMALS

Plants must also defend against browsing mammals as well as insects, and plants employ chemical defenses against mammals in stems and leaves as well as physical defenses of prickles, spines, and thorns. Boreal mammals follow a well-studied pattern of woody plant food selection in browsing a variety of evergreen and deciduous species, growth phases, and plant parts, primarily as a strategy to avoid secondary chemicals (Bryant 1981; Bryant et al. 1985; Huntly 1991). These chemicals strongly influence the palatability and food value of plant tissues during the time they are dormant in the winter. Birches and willows are heavily browsed by voles and snowshoe and mountain hares in boreal ecosystems, with heavy preference for juvenile rather than adult growth. These herbivores have relatively little access to adult growth, such that their herbivory would be expected to select for heavy chemical defenses in juvenile growth (Barton and Koricheva 2010). Small droplets containing resin with secondary chemicals such as papyriferic acid are located on the surface of birch twigs and young stems. The palatability of birch seedlings and saplings (1–8 years old) to the mountain hare is negatively correlated with the number of resin droplets, and hares are keenly able to detect palatable plants (Rousi et al. 1989, 1991). Birches seem to be especially low-quality browse species. For example, when a previously multispecies site on Alaska's Kenai Peninsula became white birch-dominated after a forest fire, the moose population starved (Oldemeyer et al. 1977).

Various woody plants produce juvenile-phase sucker shoots heavy in secondary chemicals when browsing is severe. Even fast-growing boreal plants that emphasize compensatory growth to replace browsed tissues allocate significant resources to defend against winter browsing (Bryant et al. 1985). For example, juvenile sprouts of trembling aspen contain a chemical that deters browsing by beavers such that they avoid juvenile saplings in favor of large trees that have low concentrations of a specific phenolic compound (Basey et al. 1988, 1990). However, where juvenile trees are uncommon, as in areas newly occupied by beavers, beavers select smaller, *non-juvenile* trees to maximize their net energy intake. Woody plants found in boreal or subalpine regions with harsh winter conditions may be more likely to experience herbivory by mammals that preferentially browse adult over juvenile growth compared to those found in moderate climates (Swihart and Bryant 2001).

ROLES OF ANIMALS IN PLANT LIFE HISTORY

Invertebrates and vertebrates affect many ecosystem processes, including nutrient cycling, water cycling, and the regeneration and succession of forest trees. Animals influence these processes by dispersing pollen and seeds, decomposing organic matter, mixing the soil, and damaging and killing trees. In many instances, animals may even regulate forest composition and growth. In the following sections, we examine some of the key roles of animals in affecting plants from birth to death.

POLLINATION

Animals play a critical function in the life cycle of woody plants via their role in pollination. Temperate and boreal forests are dominated mostly by species that are wind-pollinated, but animal pollination is widespread among tropical woody species. Pollination is accomplished by insects (bees, wasps, flies, beetles, butterflies, and moths), birds (especially hummingbirds in the New World), and bats (Baker et al. 1983). Animal pollinators are mainly attracted to plants using nectar; fragrance; flower color, shape, and size; and, in the case of birds, insects visiting the blossoms. A comprehensive account of animal–plant interactions in pollination ecology is given by Willmer (2011).

In North America, many families and genera of woody species are dominantly or wholly insect-pollinated. Understory species are primarily insect-pollinated, whereas most upper tree canopy species are mainly wind-pollinated. Major insect-pollinated groups include most species of the Ericaceae, Fabaceae, and Rosaceae, as well as some species of the genera *Acer, Aesculus, Catalpa, Cornus, Magnolia, Liriodendron, Nyssa, Salix, Sassafras, Tilia*, and *Zanthoxylum*. In the case of tulip tree, insect pollination is inefficient (Boyce and Kaeiser 1961), and only about 10% of the seeds may be viable. Nevertheless, enough viable seeds are produced per tree that natural regeneration is not limited.

SEED DISPERSAL

Animals are instrumental in maintaining and spreading woody plant populations by connecting site and plant through seed dispersal. The body of literature on seed dispersal is enormous, presented most often in a biotic context of animal–plant mutualisms, co-adaptations, and co-evolution (Janzen 1983). We emphasize here that dispersal is part of plant regeneration, and plant regeneration is a process that links plants and animals to physical site factors where the success of dispersal is played out. Some woody species produce highly nutritious fruits adapted for a small group of specialized frugivores ("specialists") that reliably disseminate the seeds (Howe 1993). One example of a woody plant catering to specialists would be various parasitic mistletoes, which have nutritious, sticky fruits with seeds that tend to pass through the disperser's gut, have peak availability when other food sources are scare, need to be dispersed to locations at the tops of host trees, and are dispersed by birds from only eight families (Watson and Rawsthorne 2013). Other woody species offer less-nutritious fruits in very high abundance, relying on common and numerous frugivores that are individually less reliable ("generalists"), but collectively effective in dispersing the seeds (Howe 1993). Most animal-dispersed temperate woody species, with some exceptions, fall on the continuum between these extremes. Seed dispersers of woody plants tend to be vertebrate animals (birds, mammals, fish, and reptiles).

Three key features characterize an animal–plant–site dispersal system that is successful in the establishment and persistence of the plant species. First, the fruit or seed must be attractive to the disperser, either by sight, smell, or taste. Second, the fruit must be at its most attractive at the same time that the seed matures, because premature ingestion destroys the developing seed. Finally, a sufficient number of viable seeds must escape predation or digestion by the dispersal agent and be deposited where the chances of successful establishment are high. Some escape mechanisms include burying or regurgitation of the seeds by the disperser; hard, smooth, seed coatings that assure undamaged passage through digestive tracts; and darkly or inconspicuously colored fruits coupled with brightly colored accessory parts. In the latter case, the animal is attracted to the fruit by a red or orange aril, peduncle, or bract, and if the dark-colored fruit is dropped, it is not readily found (Janzen 1969). Regarding dispersal of seeds via vertebrate guts, Janzen (1983) concluded that virtually all traits of seeds and fruits have probably been modified for protection of the seed in passing through the guts of dispersing animals.

Fish and Reptiles Fish eat pulpy seeds of various woody species growing along rivers and in recent decades have been recognized as important dispersers. Many riparian trees and shrubs tend to have fruits and seeds dispersed by water, and thus dispersal by fish would allow upstream transport that is not otherwise possible. Most of the evidence for dispersal by fish is in tropical regions, although across wide biogeographical regions (Correa et al. 2007; Horn et al. 2011). Seed dispersal by fish has also been documented in North America for channel catfish in the Mississippi River (Chick et al. 2003). Catfish were observed to consume fruits of red mulberry and swamp privet during high water periods, and consumed seeds had higher rates of germination than unconsumed seeds when harvested from fish guts. It is assumed that fish defecation would eventually deposit the seeds on floodplain sites suitable for germination once floodwaters receded.

Reptiles, particularly turtles and tortoises but also alligators, lizards, and snakes, have a keen sense of smell and may eat fruits after they have dropped off trees or when borne close to the ground. The fruits of iguana hackberry are eaten by climbing iguanas. Although most modern reptiles are not vegetarians, an increasing number of cases of reptilian frugivory have been documented. For example, box turtles were found to be important dispersers of two palm species and one understory shrub in dry, fire-dependent pine forests of southern Florida (Liu et al. 2004).

Birds Birds are primary dispersal agents and have developed many adaptive interrelationships with plants. Birds disperse seeds either by disgorging fruits or seeds carried in the mouth, or by excreting seeds contained in fruits that have been eaten. Rarely do birds carry fruits on the outside of their body (except for sticky fruits of mistletoe, referenced in the earlier text).

Seeds are often destroyed when they are eaten and digested by birds, but certain dispersers do not immediately consume the fruits and seeds they collect. Many wood pigeons, thrushes, nutcrackers, crows, and waxwings can disgorge whole fruits and seeds from their beak. An important group of avian dispersers attempt to store or hide their food either above or below ground but then neglect to recover all or part of it (Pesendorfer et al. 2016). Birds that disperse and cache acorns and other nuts above ground, for example, usually do so in tree cavities and bark crevices, such as the acorn woodpecker of California (Stacey and Koenig 1984). The acorn woodpecker imbeds thousands of acorns, almonds, and hickory nuts into small holes it drills in the bark of communal storage trees (Figure 12.6) known as "granaries" (Pavlik et al. 1991). Squirrels and other rodents may then carry them to a germination site, completing dispersal. Birds that routinely cache nuts and seeds below ground include jays and nutcrackers. Placement of the seeds in the ground increases the probability of successful establishment for many woody angiosperms dispersed in this manner (oaks, beeches, hickories, chestnuts, hazelnuts). In the eastern deciduous forest, the blue jay disperses nuts more than several hundred meters and caches them in the ground (Johnson and Webb 1989); documented dispersal flights are up to 1.9 km in Virginia and up to 4 km in Wisconsin (Darley-Hill and Johnson 1981; Johnson and Adkisson 1985). From a single Wisconsin woodlot, jays were estimated to have dispersed 150 000 viable beechnuts in only 27 days (Johnson and Adkisson 1985).

The mutualism between wingless-seeded white pines of semi-arid and subalpine environments in North America, Asia, and Europe and nutcrackers, jays, and woodpeckers (Lanner 1981, 1990) may be one of the most striking. The birds disperse millions of pine seeds over

Photo by Lorraine Bruno, Sonoma County, CA

FIGURE 12.6 This California acorn woodpecker (Melanerpes formicivorus) has imbedded thousands of acorns in the bark of this tree near Sebastopol, California.

long distances and bury them in favorable microsites, but the seeds also are an abundant and highly nutritious food source for the birds. Germination and establishment of unrecovered seeds is likely the only way for species such as limber and whitebark pines to become systematically established (Lanner 1980, 1982). Dispersal is more than simply a bird–pine interaction; insects may indirectly affect the dispersal process by both birds and mammals, both of which avoid pines with reduced seed crops due to insect infestation (Christensen and Whitham 1991, 1993). Notably, specific site conditions, whether xeric woodlands or subalpine mountain sites, affect the life histories of both plant and animals and their mutualistic association.

Note that gathering seeds and caching them in distributed locations away from where they were collected, a behavior known as *scatter-hoarding*, only becomes an important method of seed dispersal when some of the seeds are left unrecovered by the disperser. Thus, plants have evolved various strategies that reduce the likelihood that cached seeds will be recovered (Vander Wall 2010). First, the production of attractive, large, nutritious seeds and fruits is useful when it stimulates dispersers to hoard or collect and cache more seeds than will be recovered for food. Second, if seeds are difficult to eat because of physical or chemical barriers that take handling time to overcome, seeds are more likely to be hoarded than immediately eaten. A third strategy is **masting**, where all the individuals of a population of trees or shrubs produce a large crop of fruit (usually nuts) at the same time (Chapter 4), which causes animals to collect and store more food than is needed. Finally, odorless seeds are less likely to be detected once cached. Each of these traits is common and may have evolved in response to scatter-hoarding (Vander Wall 2010).

The framework of **seed-dispersal effectiveness** describes the contribution of a disperser to the fitness of a plant species (Schupp et al. 2010). Components of seed-dispersal effectiveness include visitation rate and the number of seeds acquired per visit (dispersal quantity), how the seeds are treated in the mouth and gut, and factors of seed deposition that affect seedling emergence, survival, and growth (dispersal quality). In the southwestern United States, pinyon jays and Clark's nutcrackers disseminate pine seeds and bury them 2–3 cm deep in loose soil at communal caching areas. Caching usually occurs on south-facing slopes that are free of snow by late winter (Vander Wall and Balda 1977), such that cached seeds provide a critical food source to initiate the breeding season (Ligon 1978). When pinyon seeds are abundant, entire flocks of 200–300 pinyon jays gather and store them day after day over a period of months. Using conservative figures of 30 seeds per trip and 4 trips per day, Ligon (1978) estimated a flock of 250 jays would store 30 000 seeds per day and approximately 4.5 million seeds over a 5-month period. The Clark's nutcracker may carry an average of 55 and as many as 95 seeds of the smaller-seeded whitebark pine in its sublingual pouch per trip (Vander Wall and Balda 1977). A single Clark's nutcracker could disperse 1225 seeds per day for 80 days or 98 000 seeds per individual per year (Hutchins and Lanner 1982). Clearly, both nutcrackers and jays have very high seed-dispersal effectiveness in this system.

Though not explicitly considered within the seed-dispersal-effectiveness framework, long-distance dispersal (best exemplified by birds) is also considered to benefit plant fitness because it allows for colonization of new areas and eventually tree migration and the expansion of species' ranges (Johnson and Webb 1989; Cain et al. 2000). Pinyon jays can cache seeds up to 22 km from pinyon stands, and so their role of disseminating pinyon seeds over long distances is enormous. In a study of ponderosa pine, which has winged seeds for short-distance dispersal by wind but is also dispersed over long distances by Clark's nutcrackers and pinyon jays, four isolated Wyoming populations of ponderosa pine were found to have been established by long-distance dispersal (Lesser and Jackson 2013). Ages and genotypes of each individual tree in the populations showed that long-distance dispersal was responsible for initial colonization of each of the four sites. However, the initial populations were unable to reproduce via on-site reproduction while young and required additional long-distance dispersal events to continue population growth until they were able to expand on their own.

Species such as limber pine and the stone pine group (including whitebark pine of western North America, Swiss stone pine of Europe, and Korean, Japanese, and Siberian stone pines of Asia) are found at the highest forest elevations and at timberline in harsh environments. The integrated bird–plant–site dispersal system involving these species largely determines their morphology, successional status, population age structure, and tree spacing. The morphology of trees and cones appears to be influenced evolutionarily by nutcracker species (Lanner 1980). Cones are displayed in a highly visible position on steeply upswept limbs, rigidly attached and retained in the crown where seeds are visible but cannot fall out even when shaken or rotated—ideal for foraging by nutcrackers. Furthermore, the population genetic structure of pine stands is largely shaped by the nutcracker filling its sublingual pouch with seeds from only one or a few trees before caching. As a result, the multiple seedlings of whitebark pine from a single cache are more closely genetically related than individuals from distant clumps (Furnier et al. 1987). Thus, the local population structure and possibly the mating system of the pines are strongly related to avian behavior in seed harvest and caching.

In addition to the white pines described in the earlier text, *Juniperus* is the other exceptional group of bird-dispersed conifers. Eastern redcedar has cones that effectively resemble bluish berries, thereby enticing cedar waxwings, sparrows, robins, warblers, and mockingbirds to consume entire cones with seeds. A study of eastern redcedar in southwestern Virginia found that 65% of cones were dispersed away from the parent trees by birds, and 61% of cones were dispersed over long distances (Holthuijzen et al. 1987; Figure 12.7). As with the white pines, physical site factors supporting eastern redcedar are again a key part of the dispersal system. Birds dispersed redcedar

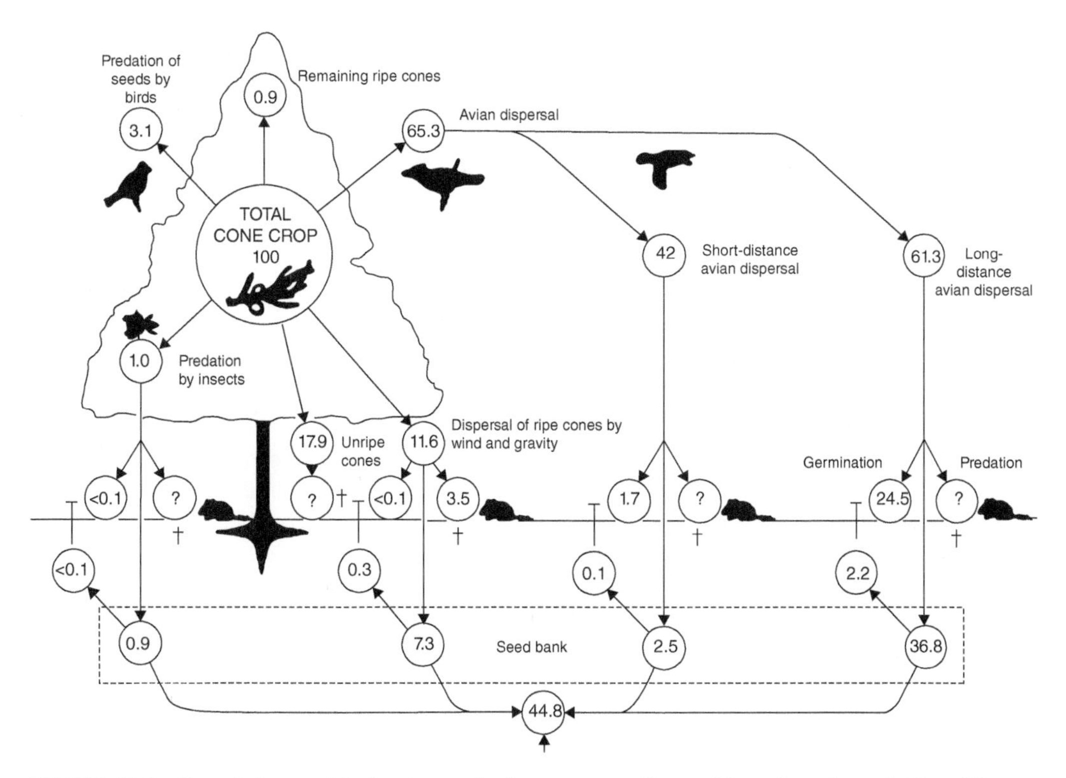

FIGURE 12.7 Descriptive model of eastern redcedar cone-crop dispersal from June through May of the following year. Numbers in circles are percentages of the total cone crop and are means of four sample trees in southwest Virginia. *Source:* Holthuijzen et al. (1987). / with permission of Canadian Science Publishing. Reprinted with permission of National Research Council Canada.

cones to barrens and rock outcrops that were open, dry, and relatively fire-free, forming "cedar glades" on the pre-European colonization landscapes of the Midsouth and Appalachian Mountains. Combined with the ability of redcedar to germinate quickly and survive on dry sites, the species was able to persist in regions otherwise overwhelmingly dominated by deciduous forests. These same attributes also make the species today a successful colonizer of old fields and abandoned pastures.

Mammals Mammals such as rodents, ungulates, bats, and certainly humans are also important dispersal agents. Tropical forests include a diverse suite of mammal dispersers because they have fruits adapted for mammal dispersal year-round, and many fruits are eaten by both mammals and birds. Rodents are the primary agents in temperate forests. Many deciduous woody plants in northern temperate forests have their seeds dispersed by scatter-hoarding rodents as well as birds. Many of these species produce nuts, such as oaks, hickories, walnuts, chestnuts, horse-chestnuts, hazelnuts, and beech (Vander Wall 1990). As with many birds, rodents often destroy the seeds, but many are cached and stupidly forgotten, allowing the unrecovered seeds to germinate and establish. Acorns and hickory nuts, for example, are dispersed by squirrels for distances up to 50 m from the parent tree, appreciably farther than what might be accomplished by wind or gravity. This longer-distance dispersal is important for hickory species that rely on gaps in the oak forest or the forest edge for their eventual development into the overstory.

Secondary chemicals in acorns are effective in deterring rodent feeding so that if other more palatable foods are available, acorns may be cached rather than eaten on the spot. Oaks are known to concentrate tannins in the apical part of the seed where the embryo is located, such that the apical part is less palatable and the probability of embryo survival of seeds discarded by the consumers is higher (Steele et al. 1993). Squirrels prefer acorns of the white oak group to those of the red oak group because tannins are three to four times higher in the red oak group, but tannins in both groups deter rodent feeding (Short 1976). Moreover, acorns of the white oak group germinate quickly after falling and diminish in palatability after sprouting (Smith and Follmer 1972). By contrast, acorns of the red oak group lie dormant over winter, with low palatability, and germinate the following spring. Compared to legumes and other seeds, or dried and fleshy fruits, acorns have relatively low levels of protein and phosphorus and insufficient nitrogen that do not supply adequate nutrition to squirrels (Short 1976). The relatively large size of the nuts of many temperate oaks and hickories is probably an adaptation to entice dispersers with a valuable energy source, although nut size is probably related to both latitude and site quality. Many oaks and hickories grow on dry or dry-mesic sites where energy from seed reserves is required for initial development of a tap root during the seedling phase to penetrate a surface mat of organic matter and reach subsurface moisture during the growing season.

As dispersal agents, mammals differ from birds in that they have a keener sense of smell and have teeth that masticate seeds. They are also typically larger, spend a significant portion of their lives on the ground, are most active between dusk and dawn, and are color-blind feeders (van der Pijl 1972). The characteristics of fruits eaten by mammals reflect these differences and include a favorable smell; a hard skin; protection of the seed itself against mechanical destruction (often a stone-like covering of the seeds as in all drupes), often assisted or replaced by the presence of secondary chemicals in the seed; non-essentiality of color; and in many cases large size that causes them to drop to the ground.

Also notable is the individual behavior of mammals and birds during the dispersal process. Scatter-hoarding requires decisions about seed selection, whether to consume the seed or cache it, how far to disperse it, and where to cache it (Lichti et al. 2017). Differences in these decisions among individuals within an animal species have been shown to affect many important ecological processes and have been examined as animal *personalities* (Wolf and Weissing 2012; Carere and Maestripieri 2013). A lab study of nearly 650 voles, mice, and shrews captured from an experimental forest in Maine showed that 90% of behaviors were repeated, indicating animal personality (Brehm et al. 2019). Personality affected the size of seeds selected, the distance seeds were dispersed, where the seeds were cached, and the probability of consuming a seed. For example, the distance of seed dispersal was influenced by differences in anxiety (indicated by time spent grooming) in mice and timidness (indicated by time spent in the open) in voles (Figure 12.8). Moreover, the proportion of mice and voles with a given personality trait differed among forests with different structure; mice, for example, were more active and less timid in even-aged forests compared to reference forests (Brehm et al. 2019). Reviews of the emerging field of animal personality are provided by Zwolak and Sih (2020), Carere and Maestripieri (2013), and Smith and Blumstein (2008).

Ungulates consume a wide variety of vegetation and thus are likely to disperse seeds, although their influence on seed dispersal is probably strongest in tropical forests and savannas. For example, many African savanna ruminants consume considerable amounts of tree fruits, particularly legumes. Many acacias provide leathery, nutritious pods containing extremely hard and smooth seeds, which evade or resist strong molars. Large grazing mammals extinct since the Pleistocene (such as extinct horses, gomphotheres, glyptodonts, and ground sloths) may have also been important in seed dispersal of certain plants in Central American lowland forests (Janzen and Martin 1982). Even some temperate deciduous species with large, sweet-fleshed fruit, for example, Kentucky coffeetree, osage-orange, honey locust, pawpaw, persimmon, and ginkgo, may have had denser populations and much wider ranges in the past compared to today because they were formerly dispersed by now-extinct megamammals. Delcourt and Delcourt (1991) speculated that osage-orange, which is a widely planted but relatively rare tree with fleshy fruits approximately the size of a softball, may have formerly been dispersed by mastodons!

Other mammals are also prominent but less-influential seed dispersers in temperate forests. Bats are important dispersal agents of woody species, primarily in tropical Asia and Africa

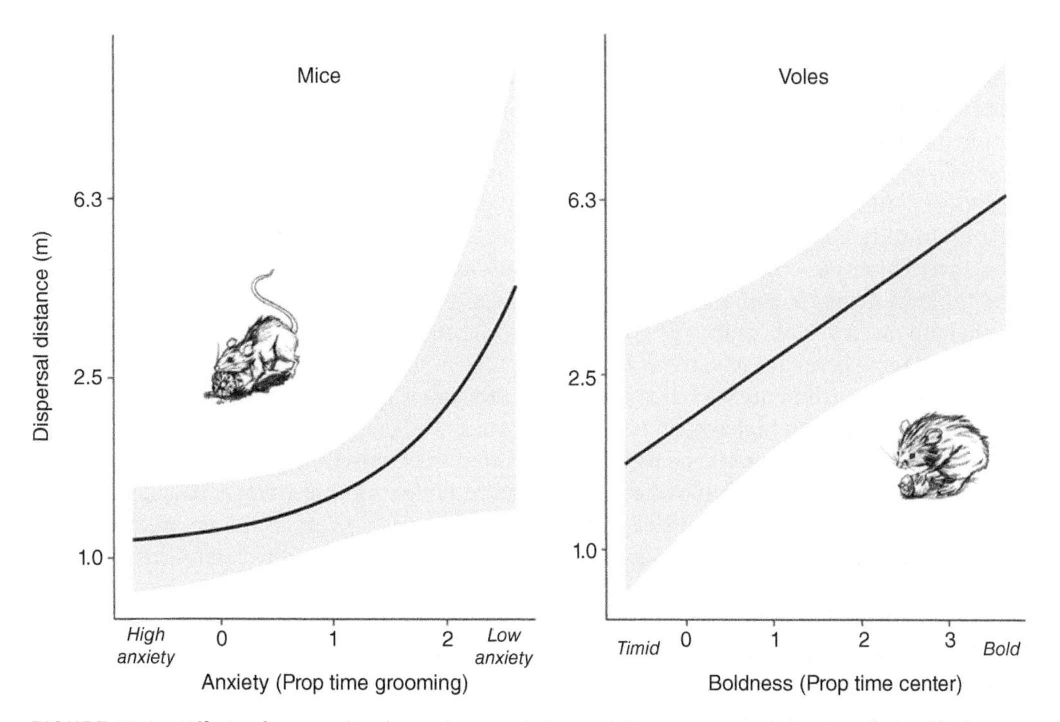

FIGURE 12.8 Effects of personality (x-axes) on seed dispersal distance (y-axes). Anxiety (quantified as the proportion of time spent grooming) decreases dispersal distance in mice (left) and boldness (quantified as the time spent in the open) increases dispersal distance in voles (right). Y-axes are on a log-10 scale. *Source:* Brehm et al. (2019).

(van der Pijl 1957), eating fruits that are of drab color, have a musky odor, and are often large and exposed outside the foliage. They transport seeds within about 200 m of the fruit source. Primates such as monkeys and apes are mostly destructive, eating everything edible, ripe or unripe, with an apparently limited dispersal role.

Finally, we cannot ignore the role of humans as significant seed-dispersal agents, spreading plants (as well as insects and pathogens) into diverse areas outside their native ecosystems, often as horticultural and forest introductions. The rate and distance of seed dispersal of many plant species by humans are likely unmatched by any other animal group. The role of humans in seed dispersal of many plants considered to be invasive species, as well as introduced and destructive insects and pathogens, is discussed in detail in Chapter 21.

GERMINATION AND ESTABLISHMENT

Many animals primarily influence germination of temperate tree species by caching seeds in the forest floor where dormancy requirements (if any) may be satisfied. Many seeds are adapted to pass through the digestive tracts of birds and mammals unharmed, and digestive juices may weaken the seed coat, favoring the absorption of water and increasing eventual germination.

Animals aid in plant establishment and plant growth, especially on nutrient-poor sites, by transporting and "fertilizing" the soil with mycorrhizae (see Chapter 19) that help the plant obtain water and nutrients (Maser et al. 1978; Maser and Maser 1988). Although some mycorrhizal fungi produce aboveground fruiting bodies with wind-dispersed spores, many produce belowground fruiting bodies that do not discharge spores and thus depend on animals for dispersal. When animals eat mycorrhizal fruiting bodies, they consume fungal tissue that contains nutrients, viable fungal spores, nitrogen-fixing bacteria, yeast, and water. The fruiting bodies are digested and the undigested material containing these components is excreted in the forest and in disturbed areas. Under favorable conditions, roots of seedlings or established trees in contact with fecal pellets may be inoculated with the mycorrhizal fungus when spores germinate. Fungal spores may also be dispersed by adhering externally to the disperser's body using spines, ridges, hooks, and other structures (Halbwachs et al. 2015). Dispersal of mycorrhizal spores in temperate forests has been documented for a diverse set of animal dispersers, including springtails, insects, mites, millipedes and centipedes, slugs, salamanders, rodents, birds, and ungulates (see review in Vašutová et al. 2019). Vertebrates and certain fungus-eating flies disperse mainly ectomycorrhizal fungi by ingesting fruiting bodies, but these fungi may be also transported externally by springtails and mites. Invertebrates are the main dispersers of arbuscular mycorrhizae, particularly earthworms and millipedes. Slugs, snails, and beetles are known to be important dispersers of fungal propagules in both mycorrhizal groups.

The digging activities of various animals turn up mineral soil that provides a seedbed suitable for plant establishment. Sukachev and Dylis (1964) reported that the germination of oak and maple seeds was twice as high where 2–3% of the surface was covered with molehills compared to undisturbed soil. Rooting by wild boars may also remove thick moss, organic matter, or other vegetation that inhibits germination and establishment, although rooting animals may also destroy young seedlings and cause widespread damage. Rooting by wild boars in Russia may cause the replacement of hardwoods by Norway spruce. Feral pigs in the eastern United States are huge consumers of mast crops, which may shift forest species composition and tree density by interfering with seed dispersal.

On the other hand, digging activities may either inhibit or favor tree species depending on their specific mode of regeneration. In northern Arizona, belowground herbivory by pocket gophers on the roots of trembling aspen prevented its colonization of mountain meadows (Cantor and Whitham 1989). In 32 aspen-meadow associations, the distributions of aspens and pocket gophers overlapped 93% of the time. However, in the mountain plateaus of central and northern Utah, gophers may favor aspen regeneration by stimulating vigorous sprouting via belowground herbivory. Notably, site conditions are key to the presence of this process: limestone-derived soils in the West support gopher burrowing and nest building, but extensive animal burrowing is limited on sandy Michigan sites such that aspen sprouting is less common.

DECOMPOSITION, MINERAL CYCLING, AND SOIL IMPROVEMENT

The relative contribution of soil animals to decomposition varies with climate, being greatest at mid-latitudes and decreasing toward the poles (Swift et al. 1979). Where they are important, soil mesofauna (springtails, mites, earthworms) play a significant role in organic matter decomposition by:

1. Physically crumbling and breaking up tissues; increasing the surface area available for bacterial and fungal action.
2. Selectively consuming material such as sugar, cellulose, and even lignin.
3. Converting plant residues into humus.
4. Mixing decomposed organic matter into the upper layer of soil.
5. Forming complete aggregates between organic matter and the mineral fractions of soil (Edwards 1974).

Notably, climate appears to have a very strong effect on the importance of animals in terrestrial decomposition. Soil animals increase the rate of decomposition in temperate and wet tropical climates, but have a lesser effect where temperature or moisture limits their activity (Wall et al. 2008). Other aspects of the physical environment, forest composition, and the amount and kind of organic matter present strongly affect the number and diversity of microfauna and mesofauna. Soil invertebrate community composition depends on the vegetation present, such that soil fauna will change with the sequential development of vegetation on a site.

Animals, particularly insects, may indirectly affect litter decomposition via their effects on the canopy. Severe insect defoliation of the overstory usually results in more light, warmth, and moisture reaching the forest floor, all of which increase organic-matter decomposition and hence the rate of nutrient cycling. Insect defoliation also may regulate nutrient cycling by increasing the rate of litter fall, influencing the rate of nutrient leaching from foliage, stimulating the redistribution of nutrients within plants, increasing light penetration through foliage, and stimulating the activity of decomposer organisms (Mattson and Addy 1975). Herbivores can also change litter chemistry, which in turn affects decomposition, by disproportionally removing high-quality forage (Ritchie et al. 1998; de Mazancourt and Loreau 2000) or by inducing changes in the nutrient or secondary-chemical composition of infested plants that carry over to the litter. The effect of herbivory on decomposition rate may depend on the type of canopy being defoliated. Chapman et al. (2006) hypothesized that herbivory should increase decomposition of evergreen litter fall because leaves would reach the ground prematurely before nutrients could be translocated into the plant, but should decrease decomposition of deciduous litter fall because the insects would induce secondary compounds in infested plants.

The physical and chemical properties of soils are altered and often improved by soil-dwelling invertebrates and burrowing mammals. In addition to mixing organic matter with mineral soil, soil mesofauna improve soil aeration, moisture-holding capacity, and nutrient content. Soil structure is altered by soil mesofauna (insects, centipedes and millipedes, isopods, earthworms) when they ingest soil particles and mix them with finely ground and digested organic matter, forming small but durable aggregates that improve soil structure when excreted. In addition, soil animals may return considerable amounts of nutrients to the soil through their excreta and carcasses. Small animals leave their feces and fragments of litter at different depths of the soil as they travel through it. They improve aeration, porosity, and moisture-holding capacity by mixing humus into the surface mineral soil.

Many of the same soil-animal processes that alter the physical properties of soil also influence soil water. The various actions of soil-dwelling invertebrates and burrowing vertebrates increase water percolation into the soil, and organic matter mixing into soil increases its water-holding capacity. Severe defoliation by canopy insects limits interception and evapotranspiration and thereby increases soil-water inputs from precipitation. Finally, the drainage system of lands adjacent to streams and small lakes may be significantly changed by dam-building activities of beavers (Hammerson 1994). Animal effects on soil and nutrient cycling is treated in Chapters 9 and 19, respectively. The effects of earthworms in particular, which are invasive species in North America, are discussed in Chapter 21.

DAMAGE AND DEATH

Despite their positive influences on forest ecosystem structure and function, nearly every animal group inhabiting forest land has been cited for some kind of damage to forests, be it by browsing, chewing, gnawing, stripping, debarking, girdling, felling, or trampling woody plants. Animals were natural thinning agents prior to European colonization whose populations fluctuated with climate, food supply, and the populations of predators and parasites. Managers became particularly concerned about damage by animals in North America in the early twentieth century in the midst of major reforestation efforts. High levels of damage to plantations and naturally regenerated stands typically followed European colonization and deforestation, which had disrupted the natural site–plant–animal interrelationships of forest ecosystems. Management devoted to artificial regeneration by planting must often contend with problems initiated by humans, such as lack of predators and resulting high populations of herbivores. For example, ungulate browsing is probably the most common source of animal damage to forest regeneration in Europe and North America. In many areas of eastern North America, white-tailed deer populations exploded following the height of logging in the nineteenth or early twentieth century (Figure 12.9). Under such conditions, stands may become park-like, a distinct browse line is maintained, and forest

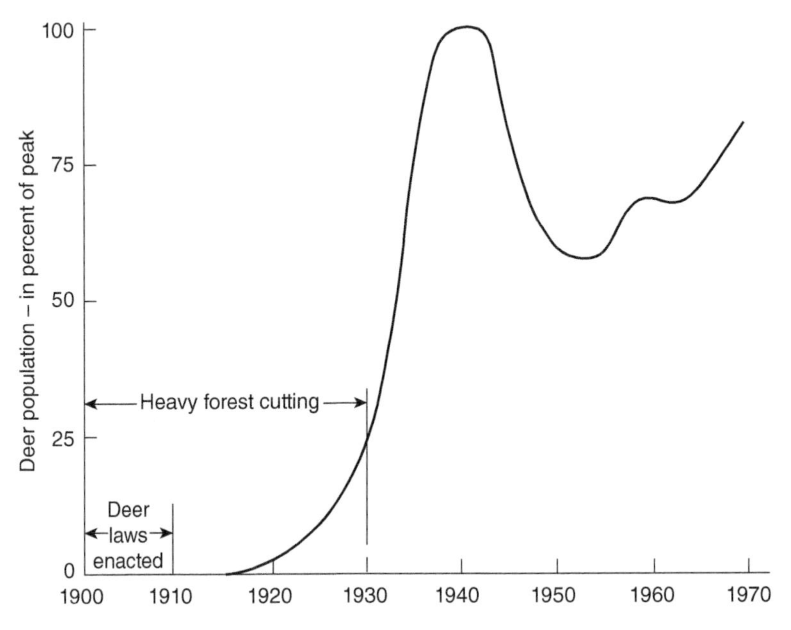

FIGURE 12.9 Change in white tailed deer populations in Pennsylvania from 1900 to 1970. *Source:* Marquis (1975) / U.S. Department of Agriculture / Public Domain.

FIGURE 12.10 Elk browsing has completely eliminated aspen regeneration outside a fenced exclosure on the National Elk Refuge, Jackson Hole, Wyoming.

regeneration is sparse or may fail entirely (Tilghman 1989). A similar situation exists in western North America with the predominance of elk, which have converted many forested areas in semi-arid climates to grass- or sagebrush-dominated vegetation, as shown by experimental ungulate exclosures (Figure 12.10). Regeneration failure of hardwood forests in Pennsylvania (Marquis 1975; Tilghman 1989), lack of sapling regeneration of eastern hemlock in the western Great Lakes region (Frelich and Lorimer 1985; Mladenoff and Steams 1993), and regeneration failure of trembling aspen clones in the Intermountain West (Kay 1997) are all attributed in part to high ungulate populations. In central Europe, high deer populations have severely limited forest regeneration and stand development for centuries. Most natural forest regeneration and plantings must be fenced from deer for several years during and after establishment. Several well-known examples of the drastic effects of ungulates on forest communities (often detailing the importance of predator control of ungulates) include deer on the Kaibab Plateau in Arizona (Binkley et al. 2006), elk in Yellowstone National Park in Wyoming and Montana (Ripple et al. 2015), and moose in Isle Royale National Park in Michigan (Wilmers et al. 2006).

Every forest tree is subject to animal damage, from seed and seedling to mature tree. Cones, fruits, and seeds are destroyed by various insects that may eliminate regeneration completely in non-mast years. Squirrels cut and collect conifer cones and cache them by the hundreds, storing them for 2 years or more. Mature seeds of many species may be consumed by birds and small mammals. In a classic study of a western Oregon clear-cut, birds and mammals caused a 63% loss of Douglas-fir seeds, whereas only a 16% loss was sustained by the smaller-seeded western hemlock, and seed loss for western red cedar was negligible (Gashwiler 1967). In a 2-year study of the fate of seeds of the relatively large-seeded ponderosa pine, only 4% of all seeds reaching maturity were available for germination (Table 12.1). Only the production of millions of seeds in excellent seed years may ensure that enough survive even heavy losses to germinate and live.

Once seedlings are established, the threat of seed predation changes to the threat of clipping and bark removal of seedlings by hares and rodents, stem and root girdling by beavers and various rodents (Crouch 1976; Teipner et al. 1983), and trampling and browsing by deer and other

Table 12.1 Effect of animal predation on seed availability of ponderosa pine in western Montana.

Fate of seeds (per 100 reaching maturity)	Number of seeds	Percent
Seeds used by animals before dispersal	66	66
Seeds dispersed	34	
Seeds used by animals after dispersal	30	88
Seeds available for germination	4	

Source: Schmidt and Shearer (1971) / U.S. Department of Agriculture / Public Domain.

ungulates. Saplings and pole-sized trees are subject to browsing by ungulates including deer, moose, and elk; girdling by beavers and porcupines; and defoliation by a variety of insects (Mattson et al. 1991). Larger trees are subject to damage by sapsuckers that drill large holes in the bark, killing or damaging a wide variety of orchard, shade, and forest trees.

Mature and overmature trees are subject to severe attack by a variety of insect feeders. Endemic levels of insect herbivory of 5–30% of annual foliage typically do not impair tree growth or total annual plant production (Franklin 1970; Mattson and Addy 1975). Under epidemic conditions, however, some insects may consume 100% of the foliage. In North America, about 85 species of free-feeding and leaf-mining insects (mostly caterpillars of butterflies and moths in the order Lepidoptera) can exhibit outbreaks and cause serious and widespread defoliation of forest trees (Mattson et al. 1991) exceeding 1000 contiguous hectares. Interestingly, these 85 species represent only 1–2% of forest Lepidoptera. Insect outbreaks are most likely in very old stands having low net primary production. For example, spruce-budworm outbreaks in the eastern and western United States tend to occur in overmature balsam fir and spruce stands (Mattson and Addy 1975), annually affecting 5.1 million hectares of commercial forests in the United States (Haack and Byler 1993). Insects typically act as one of many important disturbance agents (Chapter 16) and alone do not usually cause the regeneration of senescent forests. Instead, insect effects work in conjunction with fungi, fire, climatic stress, and windthrow to recycle aging forests, which then may be replaced by fast-growing, productive young stands.

In summary, animals interact with physical site characteristics and forest trees throughout their complete life cycle, beginning with pollination, seed dispersal, and establishment, through stand development and thinning, to the death of old trees and stands (and in clonal plants, life after death).

INFLUENCE OF LIVESTOCK ON FOREST ECOSYSTEMS

Humans have had the most far-reaching influence on forests of any animal through a variety of direct and indirect activities, and many of these are covered in detail in Chapters 20-23. One such activity, the introduction of hoofed grazing and browsing animals such as cattle, sheep, and other large animals has substantially changed forest ecosystems. At least 34% of all the forested area in the United States experiences grazing (Clason and Sharrow 2000). Around the world, severe grazing by livestock has probably been more important than any other factor in reducing the productive capacity of uncultivated land. Despite its human influence, we have chosen to include the effects of livestock in this chapter because of its pervasive influence on forests around the world and its similar effects as those of native grazers and browsers discussed in the earlier text.

Heavy grazing pressure from a single livestock species can cause entire changes in the structure of the plant community. Most broadly, these changes often alter the dominance of woody species compared to grasses. After European colonization of the Intermountain West in the United States in the nineteenth century, overgrazing from cattle and sheep depleted the bunch grass vegetation, reducing competition for woody species and the amount of fine fuels for fires critical to grassland persistence (Van Auken 2009). As a result, woody species such as big sagebrush, junipers, bitterbrush, mesquite, and serviceberry increased greatly and converted grassland into brushland (Wagner 1969). In the twentieth century, however, widespread increases of native mule deer placed heavy browsing pressure on these shrubs, often causing them to disappear and return the dominant vegetation to the original bunch grass type in many areas.

In the United States, grazing has greatly affected forest regeneration in heavily grazed woodlands of the East, central States, and the South, and throughout most of the open ponderosa pine forests of the West. Though forests are not typically the intended targets of grazing compared to rangelands, domestic ungulates inevitably wander into nearby or adjacent forests. In the West, severe overgrazing by livestock in the nineteenth century affected forest composition, fire regimes, and site conditions. For ponderosa pine forests of the Intermountain West, grazing altered both the composition and quantity of ground-cover vegetation. Removal of grassy and herbaceous competition and exposure of mineral soil by livestock helped prepare the ground for dense thickets of pine reproduction, and their removal reduced fine fuels that historically increased the likelihood of fire. Together with the lack of fires to thin dense stands, many ponderosa pine forests soon grew into almost impenetrable sapling thickets (Cooper 1960; Belsky and Blumenthal 1997). Grazing has had similar effects on forest regeneration in riparian areas of the West. For example, in southern Arizona, small seedlings of Fremont cottonwood were virtually eliminated by cattle grazing, threatening the future of the species' dominance in riparian areas (Glinski 1977). Riparian forests dominated by willow and black cottonwood in Oregon had little if any cottonwood or alder regeneration due to grazing, effectively altering future species composition (Kauffmann et al. 1983). A recent detailed treatment of livestock grazing effects on forest vegetation is given by Öllerer et al. (2019).

In part through its effects on plant communities, livestock grazing has had documented effects on animal biodiversity in many forest ecosystems, particularly in the West. For example, intensive cattle grazing in riparian areas of southcentral New Mexico affected populations of the acorn woodpecker (Ligon and Stacey 1995). The population decline was correlated with the loss of nearly all large granary trees of narrowleaf cottonwood and a lack of middle-aged trees to take their place. This "hole" in the tree-age distribution was attributed to a period of intensive cattle grazing during which regeneration of young cottonwoods was suppressed. Similar effects of grazing on animal populations are evident in small mammals, reptiles, fish, insects, and other birds (see review in Fleischner 1994).

Direct effects of heavy grazing on physical site factors result largely from the action of animal hoofs in compacting the surface soil. Consistent pounding from livestock hoofs breaks down soil aggregates, creating a crumb structure to the surface soil. As a result, the pore space of the surface soil is greatly reduced, greatly decreasing aeration and infiltration of rainwater, thereby causing surface runoff and potential erosion (Ohmart and Anderson 1982; Kauffman and Krueger 1984; Orodho et al. 1990). Erosion typically occurs by sheet erosion, although gullying has been documented for heavily grazed areas compared to ungrazed areas (e.g., Kauffman et al. 1983). For example, surface soil bulk density increased 7–17% after 26 years of livestock grazing in an oak woodland of the Sierra Nevada foothills in California (Tate et al. 2004) and by 6% for pine plantations after 7–8 years of grazing in British Columbia (Krzic et al. 1999). In a controlled grazing experiment of Douglas-fir forests and adjacent pastures in eastern Oregon, the pastures had higher bulk density and lower porosity compared to the forests, and water infiltration was 38% less (Sharrow 2007). Notably, these differences disappeared after 2 years when grazing was removed. Forests are therefore likely to better withstand the negative impacts of grazing than rangelands, though grazing effects in forests are still substantial.

Heavy livestock grazing is problematic for soil and water in the eastern United States just as it is in the West. For example, in a classic study in southern Wisconsin, highly compacted soils of heavily grazed woodlots had lower initial moisture content in the spring and dried out faster in the summer and late fall because of lowered soil permeability and increased runoff (Steinbrenner 1951). Water permeability of grazed soils averaged about only one-tenth of those of the ungrazed woodlots. However, the effects of livestock grazing on soil and water appear to be less in the eastern United States compared to the drier regions of the West. Despite the typical problems of heavy grazing pressures, livestock grazing *in moderation* can have negligible effects on forest soil and water (Patric and Helvey 1986), and there is less evidence that woodland grazing in the East, as typically practiced, has substantial adverse effects on water quality or on flooding in streams draining grazed woodlands.

A final point is that the effects of livestock grazing are not universally accepted as being negative (in contrast to overgrazing), and that many studies suggest that at least some of the often-accepted ideas may be inconsistently supported with empirical data. For example, a study of forest–grassland boundaries in Montana was not able to identify a relationship between livestock grazing and tree encroachment into the grasslands (Sankey et al. 2006). Moreover, some studies have argued that careful management of grazing may favor desirable vegetation structures and compositions (Darabant et al. 2007; Kaufmann et al. 2013), suppress woody invasive species (Chauchard et al. 2006; Mayerfeld et al. 2016), or aid in fuels reduction and fire mitigation (McEvoy et al. 2006; Varela et al. 2018). It is notable that many such studies are European, emphasizing potential cultural differences in attitudes and perceptions toward grazing, as well as how it is applied, that may be difficult to overcome and/or inapplicable in North America.

SUGGESTED READINGS

Chick, J.H., Cosgriff, R.J., and Gittinger, L.S. (2003). Fish as potential dispersal agents for floodplain plants: first evidence in North America. *Can. J. Fish. Aquat. Sci.* 60: 1437–1439.

Herms, D.A. and Mattson, W.J. (1992). The dilemma of plants: to grow or defend. *Q. Rev. Biol.* 67: 283–335.

Janzen, D.H. (1971). Seed predation by animals. *Annu. Rev. Ecol. Syst.* 2: 465–492.

Mattson, W.J. and Addy, N.D. (1975). Phytophagous insects as regulators of forest primary production. *Science* 190: 515–522.

Öllerer, K., Varga, A., Kirby, K. et al. (2019). Beyond the obvious impact of domestic livestock grazing on temperate forest vegetation–a global review. *Biol. Conserv.* 237: 209–219.

Stamp, N. (2003). Out of the quagmire of plant defense hypotheses. *Q. Rev. Biol.* 78: 23–55.

Vander Wall, S.B. (2010). How plants manipulate the scatter-hoarding behaviour of seed-dispersing animals. *Philos. Trans. R. Soc. B* 365: 989–997.

Whitham, T.G. and Mopper, S. (1985). Chronic herbivory: impacts on architecture and sex expression of pinyon pine. *Science* 228: 1089–1091.

Willmer, P. (2011). *Pollination and Floral Ecology*. Princeton University Press 832 pp.

Forest Communities

CHAPTER 13

In Chapter 1, we focused on organisms and the ecosystems to which they belong. Communities and populations, however, are categorically different from ecosystems. Communities are aggregates of organisms and are contained within ecosystems as one compositional component, but they themselves are not ecosystems. In this chapter, we consider aggregates of plants as one such ecosystem component. These collections of plants form the vegetative cover of forest ecosystems—from the overstory trees to the successively lower vertical layers of shrubs, herbs, ferns, and bryophytes, as well as the non-plant layers of lichens, fungi, and algae. We examine the community concept, communities as parts of landscape ecosystems, historical views of community, mutualisms and competitive relationships among plants, and the vertical structure of forests. These considerations lead to a discussion of disturbance in Chapter 16 and ecosystem succession in Chapter 17.

COMMUNITY CONCEPT

A **community**, simply stated, refers to the assemblage of organisms within a particular area at a given time. If only plants are considered, it is a plant community; if all organisms are considered, it is a biotic community which at broad scales is a **biome**. Community is a very generic term of convenience used to designate aggregates of organisms regardless of how broad or complicated (Cain and de Oliveira Castro 1959). Aggregates of plants over a broad area may be considered a community (the plant component of a biome), as may a local assemblage of plants associated with a specific site. Beyond the co-occurrence of plants that occupy space and have a spatial boundary, definitions of a community vary widely.

Early concepts of plant communities emphasized co-occurrence or mixtures (Clements 1905, p. 316) of species, physiognomy (life form), organism-like properties, and classification as idealized "types." In their detailed study of the community concept, Shrader-Frechette and McCoy (1993) observed that the community concept included many disparate ideas by the middle of the twentieth century, and focused on communities as "units of interacting species and habitats" in the second half of the century. Communities at this time were also defined as distinct ecological units, and units of dynamic stability. Whittaker (1975, p. 359) even defined plant communities as living systems. Nearly 70 years ago, Cragg (1953) observed that community had limitless meaning, ranging "from a piece of shorthand denoting an assemblage of organisms to something endowed with the attributes of organization which, in the absence of factual support, rivals the daydreams of the alchemists." McIntosh (1993) also observed that a clear definition of community has largely avoided many ecologists. Why is the concept of a plant community so contentious?

An important key to confusion about the nature of "communities" is that they are not ecosystems, ecological units, or living systems, but simply aggregations of organisms that happen to occupy a common geographic space. Communities are composed of individuals and species whose ecological amplitudes, mutualistic relationships, and competitive abilities allow them to

Forest Ecology, Fifth Edition. Daniel M. Kashian, Donald R. Zak, Burton V. Barnes, and Stephen H. Spurr.

coexist (Rowe 1984a). They are inseparable parts of landscape ecosystems, but they are not systems themselves as are organisms and ecosystems.

The problem of community definition is traceable to a primary focus on vegetation rather than landscape ecosystems. Plants and animals evolve together within site-specific environments (Chapter 3) as integral parts of ecosystems. Plant communities, therefore, appear to not only have plant properties but those of environmental and ecological conditions, past and present, in which the plants evolved. Community composition, physiognomy, vertical structure, and population fluctuations result as much from the effects of site factors on plants and their reciprocal interactions as they do from direct plant-to-plant interactions. We may regard plant life forms, patterns of species distribution, and competitive abilities as "plant" properties, but they have been determined in large part determined through interactions with the sites that support them.

It is well known that certain patterns of forest composition characterize extensive areas of the Earth. A regional forest ecosystem dominated by spruce and fir trees as well as pine and larch extends around much of Earth's boreal zone. This forest vegetation is heterogeneous, yet has a characteristic physiognomy and composition that make it immediately recognizable as a "spruce–fir" community or forest type. Furthermore, many of the smaller plants and many of the animals will be common to spruce–fir forests in different geographic locations. Notably, regional and local climates differ from place to place, and because site conditions are not always the same, the tree, shrub, and herb associates of spruce and fir will certainly differ from place to place, as may the relative proportions of spruce and fir and the races of these species (Chapter 3). Obviously, no two communities or ecosystems are exactly alike. Nevertheless, it is clear that an "acceptable likeness" exists (Rowe 1966), so that we convey an immediately recognizable concept and mental image to others when we speak of the boreal spruce–fir forest.

An important point is that this mental image is named by tree species, but is not simply vegetative cover. It includes regional ecosystems characterized by harsh climate and cold soil, supporting complicated biotic communities dominated by spruce and fir in the overstory. Both regional and local ecosystems are conventionally named by the dominant trees (oak–hickory forest, beech–maple forest), and thus we may lose sight of the more inclusive landscape ecosystem of which the community is only one part. Moreover, assigning a community to a class may suggest a uniformity that doesn't really exist on the ground.

GROUNDING COMMUNITIES

As communities are integral parts of ecosystems that form landscapes at broad and local scales, it is useful to examine community occurrence, composition, and structure in relation to their supporting ecosystems. These broad vegetative units may seem discrete and with sharp boundaries, but ecosystems and their communities form more or less continuous patterns or gradients in nature, typically following closely those of changing site conditions.

Florida Keys Regional ecosystems and their communities contain considerable heterogeneity, particularly viewable at local scales where variation is evident with the close match between site and the vegetation it sustains. In the Florida Keys, community structure and composition reflect the complex of factors and factor gradients, including physiographic position, hydrology, soil depth, salinity, frequency of tidal inundation, fire regime, and hurricane effects. The plant communities of recently undisturbed dry tropical forest, for example, are conceived as parts of mapped landscape ecosystems (here termed ecological site units [ESUs]) along a physiographic diagram (Figure 13.1; Ross et al. 1992). Such a map emphasizes that the pattern of communities is intimately related to the physical configuration of the island. Although this view shows spatially patterned ecosystems as if they were all discrete units, the concrete systems and their communities vary from place to place and are not homogeneous. In addition, each ecosystem and its community may grade gradually or

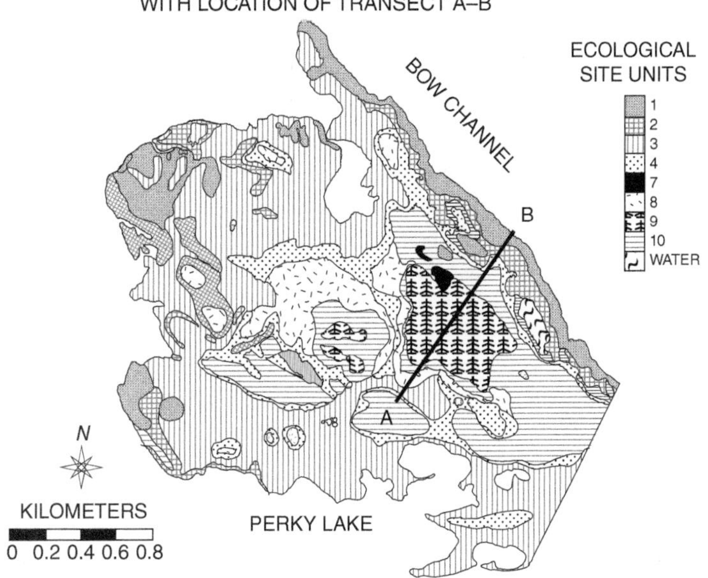

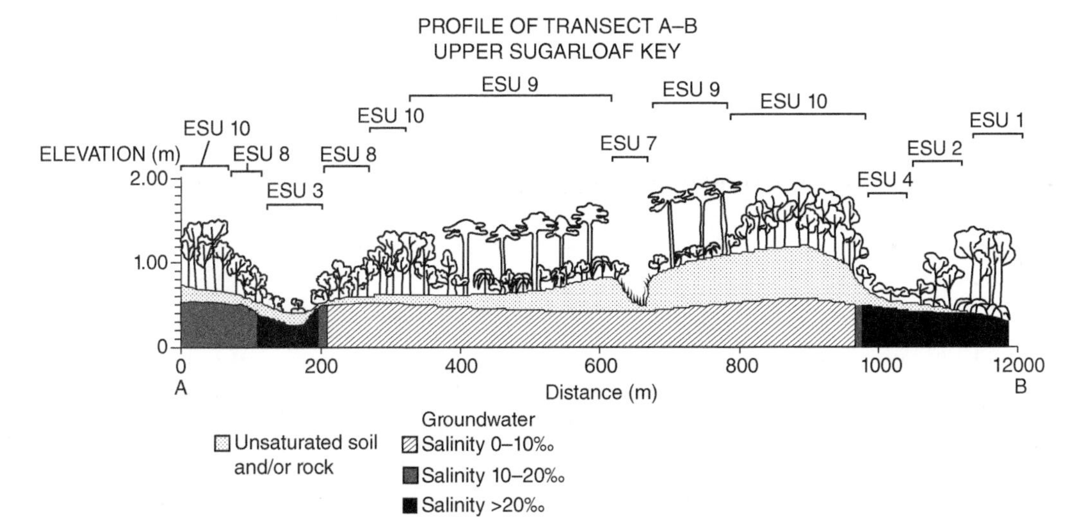

FIGURE 13.1 Ecological site units (landscape ecosystem types), geomorphology, and ground water characteristics for Upper Sugarloaf Key, Florida. Upper diagram is a map of the ecological site units, and the lower diagram is a cross section of part of the island along the transect line A–B shown on the map. *Source:* Ross et al. (1992) / Association for Tropical Biology and Conservation.

abruptly into those surrounding it. The nature of the transition zone or **ecotone** between ecosystems depends on site, disturbance, and community factors. In Figure 13.1, abrupt transitions are evident between ecosystems (ESUs 1 and 2, 7 and 9, 3 and 8 on the lower physiographic diagram) as well as gradual changes (ESUs 9 and 10, 8 and 10). Notably, the number of plant species in the respective communities varies greatly, from the relatively species-poor mangrove units (units 1–3) with only 5 species, to the species-rich pine rockland forest (unit 9) with 35 species.

Interior Alaska Viereck et al. (1984) studied communities and ecosystems in the taiga of interior Alaska (Figure 13.2). They first developed a vegetation classification, then an ecosystem type classification including physiography, soil, and other site factors associated with the vegetation. A group of communities near Fairbanks illustrates the diversity of early- and late-successional stands ranging from river floodplains and relatively warm southern slopes to the coldest north slopes at tree line. The communities are clearly related to physiographic (landforms of river valley and upland flats, and slopes, aspects, and elevation) and edaphic conditions (Figure 13.2). The warmest sites are occupied by deciduous angiosperms (balsam poplar, quaking aspen, paper birch) that form early-successional stands following fire. Aspen and birch communities (stands 15, 13, and 16 in Figure 13.2) eventually succeed to white spruce on well-drained, loess soil. Both lowland and upland black spruce-dominated ecosystems are evident. As in the Florida Keys, some but not all communities are abruptly separated due to site conditions (for example, balsam poplar on alluvial soil and the adjacent white spruce on permafrost; the black spruce muskeg versus the droughty bluff with quaking aspen). By contrast, the transition from forest to tree line and alpine tundra is more continuous and gradual.

Southern Illinois In a midcontinental region further different from Florida or Alaska in geology and macroclimate, local communities remain closely related to the interactions of geologic substrate, slope position, soil, microclimate, and disturbance regime. In the Shawnee Hills of

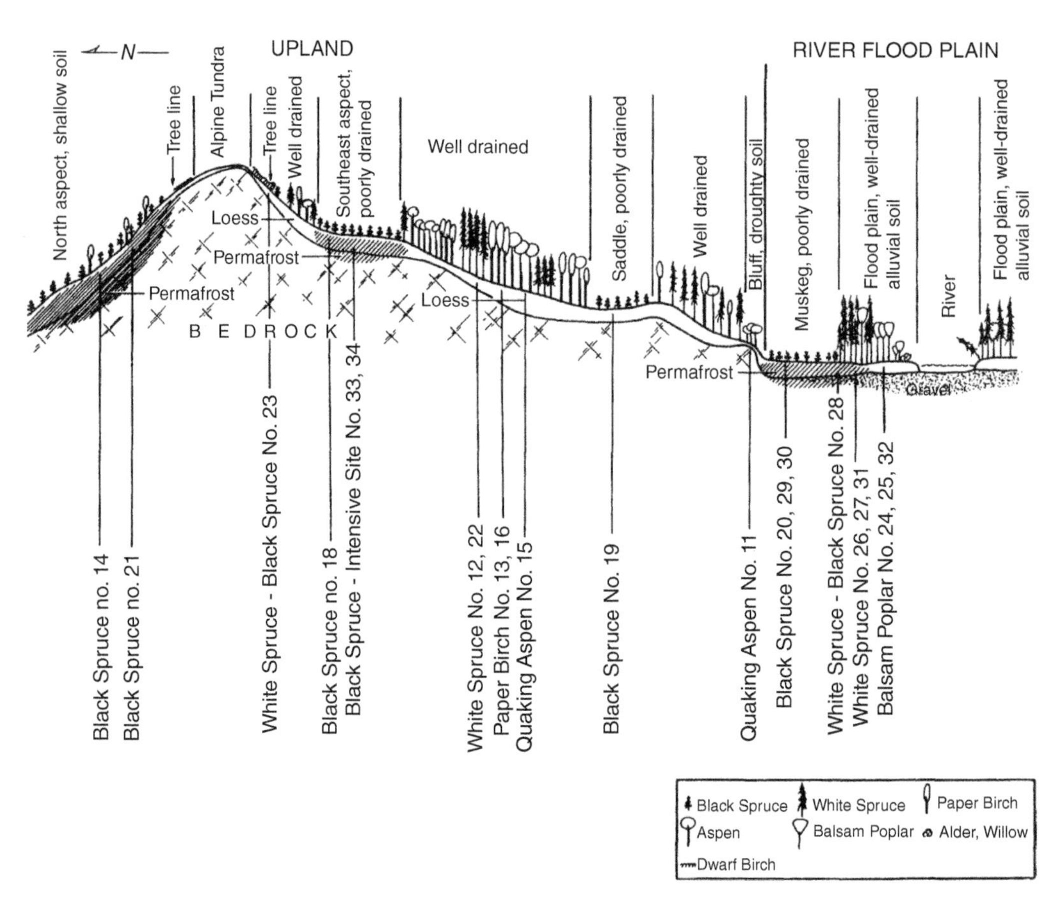

FIGURE 13.2 Forest communities of landscape ecosystems in interior Alaska. Shown is a physiographic diagram of the topography, landforms, vegetation, and parent material in the Fairbanks, Alaska area with locations of sampled stands. *Source:* Viereck et al. (1984) / U.S. Department of Agriculture / Public Domian.

southern Illinois, communities and local landscape ecosystems may be distinguished on a physiographic cross section of ridge and valley terrain (Figure 13.3; Fralish et al. 1991). Dominant forest cover types during both pre-European colonization (prior to 1800; Figure 13.3a) and "old-growth" forest cover types of 1988 that developed during European colonization characterized by fire suppression (Figure 13.3b) were examined in the same physiographic setting. The ecosystem types of the pre-European colonization period include: (i) shallow soil over bedrock, (ii) rocky upper slope, (iii) loess-covered ridge, (iv) upper north slope, (v) lower north slope, (vi) alluvial floodplain, (vii) terraces, and (viii) mid-south slope (Figure 13.3a). Geologic substrate, landform, and local topography of this area change relatively little over hundreds to several thousands of years and form the basis for distinguishing plant communities that reflect site conditions. Both gradual and abrupt changes are evident between ecosystems and their cover types. Eastern redcedar and post oak dominate the driest sites, white and black oaks dominate the fire-prone ridges and south slopes, and river birch (Bn in Figure 13.3) and American sycamore (Po) dominate alluvial floodplains. The white oak–black oak community was found in three ecosystems due to a relatively high fire frequency: loess-covered ridge, upper north slope, and lower north slope.

The old-growth forest cover types that exist today on the area (Figure 13.3b) are, for the most part, the same as those present prior to European colonization. The spatial positions of landform and soil are relatively unchanged such that ecosystem types have not changed. However, lack of fire has allowed more mesophytic species such as red oak and sugar maple to replace the white oak–black oak overstory on upper north slopes and lower north slopes, respectively (Figure 13.3b). Red oak has also replaced black oak in the loess-covered ridge ecosystem. Changes in these forest cover types illustrate

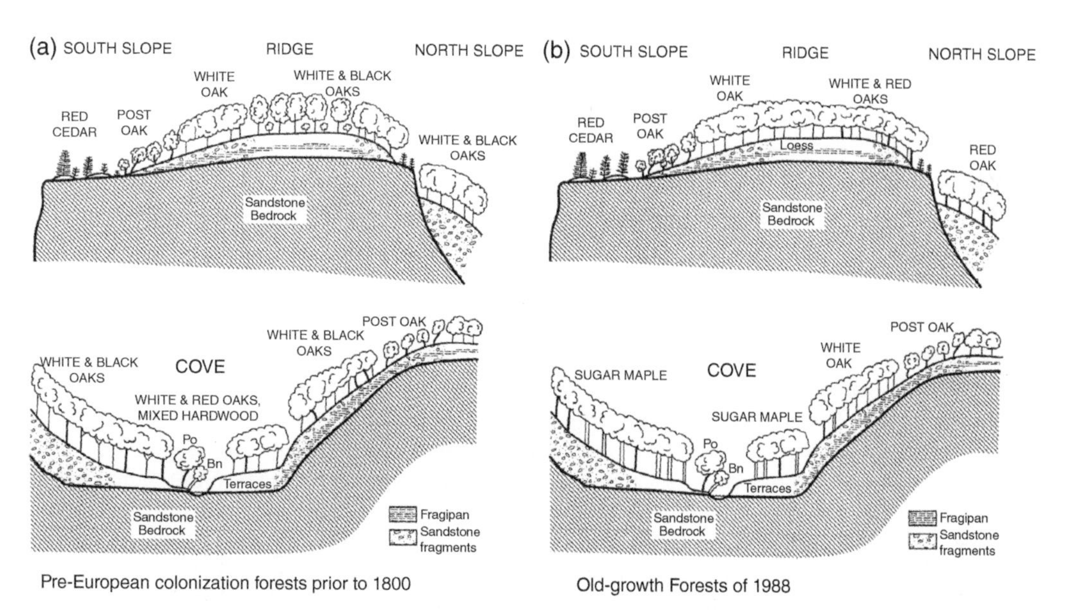

FIGURE 13.3 Physiographic diagram of the pattern of (a) pre-European colonization and (b) old-growth ecosystem types and forest cover types in the Shawnee Hills region of southern Illinois. Ecosystem types (cover types) include: (i) shallow soil over bedrock (eastern redcedar), (ii) rocky upper slope (post oak and white oak), (iii) loess-covered ridge (white and black oaks), (iv) upper north slope (white oak), (v) lower south slope (white oak), (vi) stream terraces (white and northern red oaks, sugar maple, basswood), (vii) alluvial floodplain (American sycamore and river birch), and (viii) mid-south slope (white and black oaks). Pre-European colonization communities before 1800; old-growth communities as of 1988. *Source:* Fralish et al. (1991) / with permission of The University of Notre Dame. © 1988, American Midland Naturalist. Reprinted with permission of the American Midland Naturalist.

the dynamic nature of vegetation in relation to disturbance, or lack thereof in the case of fire (Fralish et al. 1991). Notably, tree communities of some ecosystems are virtually unchanged. Thus, the composition of forest communities is highly dependent on the interaction of specific site conditions and disturbance regimes that influence the tree species' competitive ability on the Shawnee Hills landscape.

In summary, plant communities are visually distinctive features of the landscape. Understanding their composition, stand structure, dynamics, and spatial distribution for any landscape requires a solid understanding of the site conditions, disturbance regimes, history, and processes of the landscape ecosystems of which they are an inherent part.

VIEW FROM THE PAST: COMMUNITY CONCEPTS

The paradigm of the community as one component of landscape ecosystems differs from other community concepts because of its first-order focus on ecosystems. The historical focus of plant ecologists was on plant species and communities, their distribution, composition, and classification, and less on their relationship to site factors. Many of the early terms and underlying concepts of plant ecologists remain entrenched in the literature and used today.

Early plant geographers, such as Humboldt (see Botting 1973; McIntyre 1985), Schimper (1898), Gradmann (1898), and Warming (1909) emphasized the geographical distribution of plants (i.e., phytogeography) and classification. For example, Warming (1909), a critical figure in shaping modern ecology, developed a classification based on water plants and land plants, with the latter subdivided further into 12 primary groups. Henry Cowles, the noted American ecologist and a keen field ecologist and botanist, pioneered a physiographic classification of plant communities of the Chicago area (Cowles 1901). Cowles recognized that physiography influenced climate, soil, and geomorphic change (erosion, deposition), and therefore best accounted for the distribution of species and communities. Cowles purported to examine communities, but his emphasis was essentially on ecosystem units and their geomorphic change to which characteristic plants and communities were dynamically related.

SCHOOLS AND TERMINOLOGY

The field of community ecology was dominated by two schools in North America in the first half or two-thirds of the twentieth century. Frederic Clements (1905, 1916; Weaver and Clements 1929) first developed a school based on plant succession, followed by the Curtis–Whittaker group who emphasized an abstract vegetative continuum (Curtis 1959; Whittaker 1962, 1975; Fralish et al. 1993). Clements had a broad, comprehensive understanding of the historical development of the time, innovative ecological insights (especially regarding physiography and site–plant relations), and a complex emphasis on succession and vegetation dynamics despite the fact that his theories of vegetation as an organism, rigid classification, and monoclimax later became unacceptable (see Chapter 17).

Like many of his predecessors, Clements focused on the **plant formation** as the major unit of vegetation, characterized by a given physiognomic form that recurs on similar sites (i.e., grassland, temperate deciduous forest, tropical rain forest). Each formation was composed of various other distinctive community types, termed **associations**. Thus, the deciduous forest formation was composed of several different associations such as beech–maple, oak–hickory, and others. Braun (1950) mapped and described such associations as "forest regions" (Figure 2.6, Chapter 2). Küchler (1964) published a non-hierarchical classification of the "potential" vegetation units of the conterminous United States. In Canada, Rowe (1972) described forest regions, and Strong et al. (1990) summarized its vegetation zones. A worldwide classification of plant formations was published by Mueller-Dombois and Ellenberg (1974).

Europeans and Americans followed different courses in the use of the term "association" (Mueller-Dombois and Ellenberg 1974). Partly because the term was strongly identified with its rigid European definition and usage, and partly because Clements related it only to "climax" communities, "community" was soon used to replace "association" among American ecologists. The accuracy of "community" is still questioned today, and alternatives such as plant assemblage are often used. Its particular meaning in a given study depends on the context in which it is used and the modifiers applied to it—oak–pine community, early-successional community, etc. American ecologists often further subdivide the community or association into "layers." Thus, the spruce–fir formation of northeastern United States and eastern Canada includes the red spruce–balsam fir community, which may in turn be subdivided into an overstory or tree layer, an understory or shrub layer, and a groundcover layer. Notably, the additional term **forest type** (or **community type**) refers to a forest community defined only by overstory composition. Because the community is or should be defined by the total plant complement, its name sometimes takes into account characteristic groundcover plants.

Concepts of Clements and Gleason Plant ecology in the early twentieth century was conceptually dominated by the prevailing ideas of the plant association as a natural unit for study and classification. Clements's version of ecology was very tidy and orderly—far more orderly than nature itself (Egler 1968). Although today we understand that all populations have inherent heterogeneity, at that time populations were thought to be more or less uniform such that it was easy to believe in uniformity of plant formations and communities. Clements's extreme view likened the formation and association to a complex super-organism that arises, grows, matures, reproduces, and dies. The final, stable, self-maintaining, and self-reproducing state in the development of vegetation was the **climax** (climatic maximum). Clements believed that only one true climax occurred in a given climatic region, although common sense forced him to recognize and name the other units that actually occurred with even more terms such as pre-climax, post-climax, and dis-climax. A less extreme view held by plant ecologists such as Cooper (1913, 1923), Nichols (1923), and many European phytosociologists was that the association was in fact a series of separate similar units rather than a super-organism, variable in size but repeated in numerous examples, and analogous to a species.

By sharp contrast, Henry Gleason developed an "individualistic concept" of ecology and the plant association (Gleason 1917, 1926, 1939). Gleason objected to an organismal concept of vegetation as well as the classification of vegetation into rigid "pigeon-holes" of seemingly uniform types. He emphasized the variability of communities that were supposed to be of the same type, arguing that the variation occurred throughout both space and time due to chance and environmental effects. He described the vegetation unit as a temporary and fluctuating phenomenon, its origin, its structure, and disappearance dependent on the selective action of the physical environment and on the nature of the surrounding vegetation (Gleason 1939).

Gleason was a field botanist by training, and he advocated a floristic, individualistic approach to vegetation (Nicolson 1990). Gleason used "individualistic" at two levels of organization: (i) for individuals that make up a species, and (ii) for species as individualistic components of a community (Egler 1968). He emphasized that the mixture of species in a community results from both migration and environmental selection, and that a species' spatial distribution depends upon its individual peculiarities of these two major factors. Gleason (1936) noted that the true nature of the association is determined by "the physical environment, which decides what kinds of plants may exist in it, and the living plants themselves, which tend to control and to modify the physical environment. Individuals of a species grow wherever they find favorable conditions, disappear from areas where the environment is no longer endurable, and occur in company with any other species of similar environmental requirements." Notably, Gleason did not recognize or consider the many mutualisms pervasive in communities, which suggests that individuals and species aren't the laws unto themselves to the degree he may have assumed.

Gleason's argument for individuality of communities is strongest at regional scales where differences are most evident. He asserted that similar environments within a region result in similar floras that could be classified together and noted that abrupt and gradual changes in environmental factors were often what caused similar changes among different communities. Gleason's floristic approach was revolutionary in stressing variation, the importance of chance events in plant colonization, and the influence of vegetation on the site. However, Gleason's approach continued a primary focus on species and communities rather than the ecosystems to which these organisms belong. From a landscape ecosystem perspective, communities aren't necessarily individualistic because they are influenced by the communities that surround them. For example, forest communities of dry-mesic ecosystems protected from fire by adjacent swamps (Chapter 10) have different composition and structure compared to those on unprotected sites. Likewise, communities on mesic sites are often burned and changed by fires spreading from adjacent flammable communities. Communities of upper-slope ecosystems influence those on adjacent lower slopes through movement of wind-blown leaf litter and by water and nutrients flowing downslope. Many such exchanges, from one community to another through landform and soil-based linkages, provide a new perspective on the individualistic concept of the community.

Phytosociology in Europe About the same time Clementsian ideas of the community were taking root, the "association" was technically defined by ecologists at the Zürich-Montpellier School of Phytosociology and adopted at the International Botanical Congress of 1910 as "a plant community of definite floristic composition, presenting a uniform physiognomy, and growing in uniform habitat conditions." The school of phytosociology identified with Braun-Blanquet (1921, 1964) had a primary objective of hierarchical vegetation classification, and the association was the basic unit of this system. Braun's approach was established by 1921 when he urged a floristic rather than an ecological classification system (van der Maarel 1975), with plant communities conceived as vegetation types recognized by their floristic composition. In this system, the association has a type specimen complete with author, date, and description analogous to that of a plant species: for example, "*Abieti-Fagetum* Oberdorfer 38"; and the "*Galio-Carpinetum* (Buck-Feuct 37) Oberdorfer 57 em. Th. Müll. 66." Interestingly, Braun never adopted an organismal approach (van der Maarel 1975), and there is little justification for the naming of communities as if they were species. Such a floristic approach has made important contributions to the description and classification of vegetation, with hundreds of thousands of communities described (Ellenberg 1988). The approach is widely used in eastern and central Europe (Oberdorfer 1990) and by many plant sociologists throughout the world. Excellent descriptions and perspectives of the Braun-Blanquet approach and the Zürich-Montpellier School of Phytosociology are available (Poore 1955; Becking 1957; Whittaker 1962; Shimwell 1971; Westhoff and van der Maarel 1973; Mueller-Dombois and Ellenberg 1974; van der Maarel 1975; Podani 2006; Pott 2011; Mirkin et al. 2015; Guarino et al. 2018).

CONTINUUM CONCEPT

The **continuum concept** of vegetation is the second major school of plant ecology in the United States, including both the continuum approach of John Curtis and associates and the **gradient analysis** approach of Robert Whittaker (Curtis and McIntosh 1951; Curtis 1959; Whittaker 1962, 1967; Cottam and McIntosh 1966; McIntosh 1967, 1968, 1993). The continuum approach is an extension of Gleason's floristic–individualistic concept of the community, which was overshadowed by the Clementsian and European association concepts by the 1940s. The concept of a vegetation continuum was developed in reaction to the concept of the association as a relatively discrete unit. The basis for the continuum concept is that vegetation varies continuously across a landscape (Curtis 1959; McIntosh 1968). Notable plant ecologists such as Daubenmire (1966) accepted that vegetation varied continuously but struggled with whether the transitional areas (ecotones) between obvious communities could themselves be classified.

There is little question, even by continuum critics, that vegetation or floras are continua (Daubenmire 1966). Vegetation varies continuously because (i) changes in environmental conditions vary, thus affecting the establishment and composition of the vegetation; (ii) genecological variation of tree species is typically clinal, and a given genotype often has wide phenotypic plasticity to tolerate a number of environments (Chapter 3); (iii) historical and chance events may reinforce points 1 and 2 so that no communities are exactly alike; and (iv) a continuum of successional change in time is superimposed upon changes in composition in space. Furthermore, whether discontinuity or relative continuity may be demonstrated for a tract of vegetation in the field may depend on methods of field sampling or analysis.

The opposition between those who favored classification and those who favored continua was cast as a dichotomy of two extremes, suggesting that a middle ground would soon prevail. Logically, studies of communities in nature would reveal that both gradually changing species aggregations and relatively abrupt changes and acceptably similar compositional types exist (just as Gleason had observed). This was not the case, however, because the continuum concept emphasized the arrangement of communities in *abstract space*, as models, not necessarily in nature. Lambert and Dale (1964) recognized that many ecologists had become confused with continuum and continuity because the terms were applied both to the actual continuity on the ground and to the continuity in abstract models. Both Curtis (1959) and McIntosh (1967, 1993) recognized the discontinuities or abrupt changes of vegetation type but held them as irrelevant and not at odds with the abstract continuum.

In retrospect, the continuum approach changed the viewpoint of many plant ecologists, encouraged countless compositional studies of vegetation, and introduced increasingly sophisticated multivariate analyses to classify and ordinate species and communities. Importantly, the approach also generated controversy and confusion (Dansereau 1968; Austin 1985), and a subsequent loss of interest in generalizations about plant communities. The approach contributed to the notion that plant communities are too intricate and complicated to observe any general patterns (Noy-Meir and van der Maarel 1987). Another perspective, however, is that as a subject of study the plant community is unrewarding because it is incomplete, and, like climate, physiography, or soil, is only one part of the landscape ecosystem. Communities are not functional systems and therefore have no processes in isolation, but in fact function within and as parts of ecosystems from which they derive their resources. It is these communities as ecosystem parts that ecologists and managers seek to manage, conserve, and restore.

COMMUNITY AS A LANDSCAPE ECOSYSTEM PROPERTY

Austin and Smith (1989) reinforced the idea that communities are properties of ecosystems in reformulating the continuum concept. Figure 13.4 shows a landscape with four species present. Five species associations may be recognized due to their frequency of occurrence along a transect from low to high elevation: A, AB, B, C, and D (seen along x-axis in Figure 13.4). The combinations BC and CD are regarded as transitions or ecotones. The "communities" composed of coexisting species result from the site-specific conditions and spatial pattern of this landscape. However, the elevational gradient shows a continuum of species regularly replacing one another in the sequence A, B, C, and D with increasing elevation (right side of diagram, Figure 13.4). Thus, what appear as relatively distinct species associations horizontally along the x-axis of distance can be viewed as a continuum along the abstract elevational gradient.

If the transect in Figure 13.4 were taken through an adjacent area where the first bench or terrace were 30 m lower, then the combination AB would become rare because only species A would be present. Similarly, if the second bench was 30 m higher, the communities recognized would be A, B, BC, C, and D with ecotones AB and CD and the combination BC would become common. Thus, communities are a function of the whole landscape and specific landscape

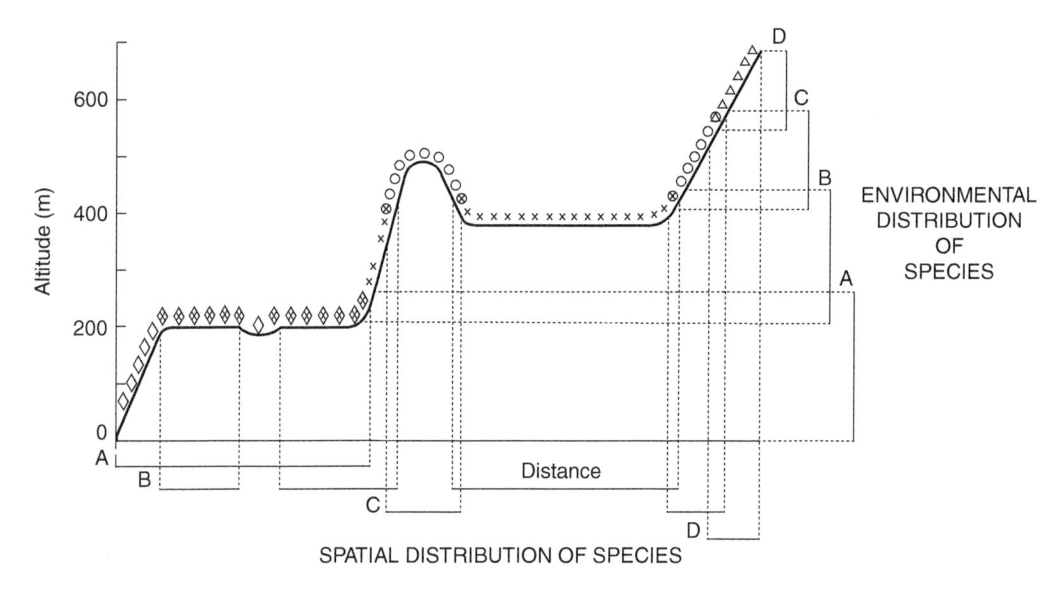

FIGURE 13.4 Pattern of co-occurrence of four species on a landscape along an indirect environmental gradient of elevation. Note the distribution of species along the x-axis (distance) illustrating distinctive species associations of A, AB, B, C, and D that occur at characteristic positions on the diagram. Then observe the continuous variation of composition (regular replacement of A, B, C, and D) along the elevational gradient on the right side of diagram. Elevation is an indirect gradient because its effect on growth is through the site factors of temperature, water, and nutrients. *Source:* Austin and Smith (1989) / Springer Nature. © 1989 Kluwer Academic Publishers. Reprinted by permission of Kluwer Academic Publishers.

ecosystems, and either abrupt changes or gradual transitions may occur depending on the pattern of recurring ecosystems in a given landscape. Co-occurring groups of species can be recognized for any particular area and may be recurring. Labeling or classifying these communities is useful for communication, management, and research, but extrapolation of these communities to other regions or landscapes will be accurate only if the regions have similar patterns of site factors (Austin and Smith 1989).

The elevation continuum in this example would be valid and applicable within similar transects of the same regional ecosystem, but not in a different region where the growth-influencing variables of climate (temperature, water) relate differently to elevation. Austin and Smith (1989) concluded that: (i) the concept of co-occurring species is only relevant to a particular landscape and its pattern of combinations of site factors, and (ii) the continuum concept applies only to abstract environmental space, not necessarily to any geographic distance on the ground or to any indirect environmental gradient (such as elevation). Further details of this model of the continuum concept are provided by Austin (1990).

EXAMPLES OF SPATIAL VARIATION IN FOREST COMMUNITIES

DISCRETE FOREST COMMUNITIES

Forests are probably most accurately considered in terms of gradient patterns. What appear to be sharp boundaries between forest communities are likely due to abrupt local changes in environmental gradients or to a completely different vegetational history of the two communities. Discrete communities provide evidence of the existence of discrete differences in past or present environmental conditions. The boundary between two communities is more commonly a belt or zone which may vary in width. In the forest–grassland transition, for example, there will always be an outer belt of forest that will be modified by the adjacent open areas, and an inner belt of grassland that will be modified by the adjacent forest. As mentioned in the earlier text, the transition zone between two communities is termed an **ecotone**, which usually embodies some of the ecological features of the two communities, but often has a specific site characteristic of its own.

Persisting, abrupt site differences often give rise to sharp forest type boundaries. Examples of such site differences include: (i) a sharp boundary between two geological formations producing two contrasting soil types that differ in water and nutrients available to vegetation; (ii) a sharp boundary between landforms that results in abrupt differences in soil drainage conditions, such as between a poorly drained swamp and a well-drained upland; (iii) a sharp boundary in topographic position that affects local climate, such as a knife-edged ridge separating a north and a south slope; and (iv) a sharp boundary in vegetation structure that affects the local climate and soil conditions, such as a forest edge-facing grassland, a shrub community impinging upon open rock surfaces, or a logging boundary between a clear-cut and a standing forest. Sharp boundaries between plant communities may also arise from historical events such as fires, tornadoes and other windstorms, salt spray from the sea, fumes from smelters, logging, and agricultural development of land.

Coastal California: Giant and Pygmy Forests Soil scientist Hans Jenny and others (1969, 1980) provided a remarkable example of diverse ecosystems and communities related to landforms along the Mendocino coast of north-central California. Rising seas during the Pleistocene cut terraces into the sandstone, and retreating seas covered them with beach sands, gravels, and clays. The terraces were elevated by tectonic forces and today occur at elevations of approximately 30, 53, 91, 130, and 198 m above sea level (Figure 13.5). Hillsides and steep canyon slopes are dominated by redwood and Douglas-fir, whereas the three upper, oldest terraces have distinctive hardpan soils and support a sparse, dwarf or pygmy forest of slender, stunted cypresses and dwarfed pines (bishop pine and Mendocino White Plains lodgepole pine). The soil is a spodosol (see Chapter 9)

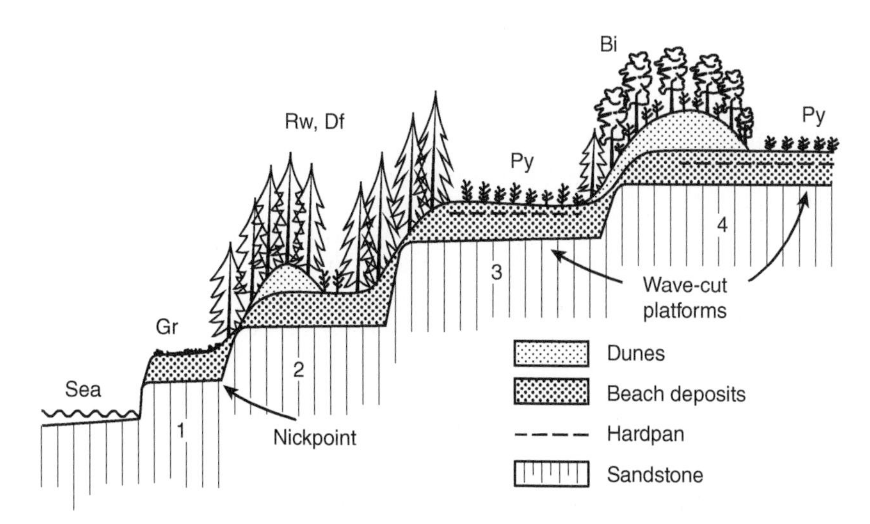

FIGURE 13.5 Physiographic arrangement of four marine terraces (1, 2, 3, and 4) in the Fort Bragg, California area, with a young dune on the second and a very old dune on the fourth terrace. Gr = grassland; Rw, Df = redwood–Douglas-fir forest; Bi = bishop pine forest; Py = pygmy forest. Horizontal distance is 4.8 km, and vertical distance is 152 m above sea level. *Source:* Jenny et al. (1969) / California Botanical Society.

characterized by extreme acidity and a thick hardpan layer of iron or clay. Water accumulates above the hardpan layer in late fall, forming a perched water table that may flood the entire surface. The surface water table disappears in late spring by evapotranspiration and seepage, and the hardpan dries out, hardens, and imparts extreme xeric conditions in summer. The distinctive vegetation probably results from many physical and chemical factors associated with soil development in the parent material of the terrace.

The lowermost terrace supports grass and pine forests, and the second terrace supports a pattern of different communities from redwood, Douglas-fir, and western hemlock to dwarf pines, depending on soil development. Neither of the lower two terraces has experienced soil development as advanced as the upper terraces, because they are of more recent origin. The communities of the terraces are therefore strongly dependent on physiography—the flat terrace form and its parent material on which soil development occurs.

Sand dunes are present on all terraces, exhibiting three relatively distinct plant communities depending on age and soil development. Recent dunes on the lowest terrace are still moving inland and are stabilized by lupines and lodgepole and bishop pines. Young dunes on the second terrace are thousands of years old and even older on the third terrace; these dunes have developed favorable water and nutrient conditions to support magnificent forests of redwood, Douglas-fir, and grand fir. Very old dunes of the fourth and fifth terraces are highly acidic and infertile, but have no hardpan formation. Very old dunes support bishop pines with a dense understory of ericaceous species, wax myrtle, and chinquapin. Thus, three kinds of dune ecosystem types and their communities are evident due to their age and soil development.

Forest–Grassland Ecotone Abrupt changes between forest and grassland in the tropical and temperate zones may or may not be associated with abrupt changes in Site. Once the forest edge has been established, such as by fire or land clearing, contrasting climatic and soil conditions between the forest and grassland may itself perpetuate the forest border. Grassland often originates in forested areas following fire that kills trees and thereby creates an environment at the ground more suitable for the development of grasses than for tree regeneration. The grassland persists once established because of the inability of the adjacent forest trees to colonize the site—whether due to the recurrent incidence of fires (Wells 1965, 1970; Rowe 1966; Veblen and Lorenz 1988); failure of tree seedlings to penetrate the sod and reach a suitable medium for establishment (Jakubos and Romme 1993); excessive root competition with grasses for soil water; or the absence of mycorrhizae (Langford and Buell 1969). Examples of fire-created grasslands that have persisted for hundreds of years include the alpine meadows of the western American mountains, the fingers of prairie extending into the Black Hills of South Dakota, and the extension of the Prairie Peninsula east into Indiana and Ohio, all under climates suitable for tree growth.

Importantly, forest–grassland borders are not static, and colonization of one type occurs by the other, as climatic conditions have fluctuated over geologic time. For example, subalpine meadows in Yellowstone National Park have been experiencing colonization by lodgepole pine from adjacent forests for over a century (Jakubos and Romme 1993), probably due to a regional climatic trend toward warmer and wetter growing seasons since about 1870. In many other parts of the world, grasslands within forested regions are being colonized by forest both because of more efficient fire suppression and because of changing climate.

Alpine Tree Lines The tree line of forests in mountain ranges, which results from the interaction of trees and the site over a long period of time, is an obvious example of an abrupt forest edge. Ecologists distinguish three kinds of altitudinal tree lines (Huggett 1995): the **forest line** or timberline, which is the upper limit of tall, erect trees growing at normal forest densities; the **tree line**, or the limit at which individuals recognized as trees (>2 m) grow; and the **tree-species line**, or the point up to which tree species will grow only in deformed habit (Krummholz or "crooked wood"). Tree lines are not always sharp (Armand 1992), but the discontinuity between trees and low-growing vegetation is indeed abrupt when two or three of these lines coincide.

Many explanations have been proposed for the formation of timberline (Daubenmire 1954; Wardle 1985; Huggett 1995). In the zone of stunted and recumbent trees, wind, wind-blown snow, snowpack, and other factors produce a harsh and exposed zone near the ground through which trees cannot grow normally. Timberline location (as contrasted with climatic factors that cause dwarfing and recumbent growth) is primarily determined by heat deficiency during the growing season at high altitudes, which limits growth and winter-hardening of shoots. Wardle (1968) found this to be the case for timberline in Colorado, where timberlines are among the highest in the world despite desiccating wind and low winter temperatures. He also found that wind blew Colorado sites free of snow, leaving seedlings unprotected during the winter and an absence of melt water in the spring to moisten the rocky, coarse-grained soils. The causes of tree lines are much debated, but most ecologists believe that climate is their main determinant (Huggett 1995).

MERGING FOREST COMMUNITIES

As much as distinct plant communities reflect abrupt changes in site, major disturbance, and land history, gradual changes in site or vegetational history are expressed as similarly gradual changes in forest composition. Forests of a generally similar history may exhibit such differences over a geographical stretch of many kilometers, or over gradual changes in elevation, aspect, soil water, and soil fertility. Climate also affects forest composition, and the relative continuity of macroclimate in space and time favors gradual change. In the absence of an abrupt change in site of vegetational history, the composition and structure of forests generally vary continuously.

Eastern Deciduous Forest—Southern Appalachians The forests of the eastern United States, characterized by deciduous hardwood species but also containing evergreen hardwoods and conifers (Barnes 1991), show many gradual changes at both broad and local scales. The greatest complexity and size of individual trees is found in the southern Appalachian region and Cumberland Plateau to the west in Kentucky and Tennessee, in which old, eroded slopes, valleys, and ridges provide many ecological gradients of forest composition.

Whittaker (1956) studied composition gradients with altitude and with "moisture gradients" within altitudinal belts in the Great Smoky Mountains. His "moisture gradient" referred to the complex gradient that occurs from valley bottoms to dry slopes, but not its causation. Along this gradient the dominant trees varied from mesic species in the valley bottoms to xeric species on the dry, exposed portions of slopes. For example, between 750 and 1050 m in elevation, eastern hemlock was most abundant in the bottoms, with silver bell, red maple, chestnut oak, scarlet oak, pitch pine, and table-mountain pine each becoming more numerous toward the drier end of the gradient. On mesic sites, hemlock and red maple were most abundant at low elevations (600–900 m) with silver bell, yellow birch, sugar maple, and basswood reaching maximum abundance between 900 and 1220 m, and yellow buckeye, mountain maple, and American beech most common at higher elevations. Whittaker thus found the broad forest pattern of the Great Smoky Mountains to be one of continuous gradation of stands along these generalized gradients, yet with certain relatively discontinuous types (Figure 13.6). Notably, he recognized the existence of sharp discontinuities between cove forests and beech stands on south-facing slopes, between grassy balds and forests, and between heath balds and spruce–fir forests. As described by Whittaker (1962):

> *The whole pattern was conceived to be a complex continuum of populations, with the relatively discontinuous types confined to "extreme" environments and forming a minor part of the whole. Allowing for discontinuities produced by disturbance and environmental discontinuity, the vegetation pattern could be regarded as a complex mixture of continuity and relative discontinuity.*

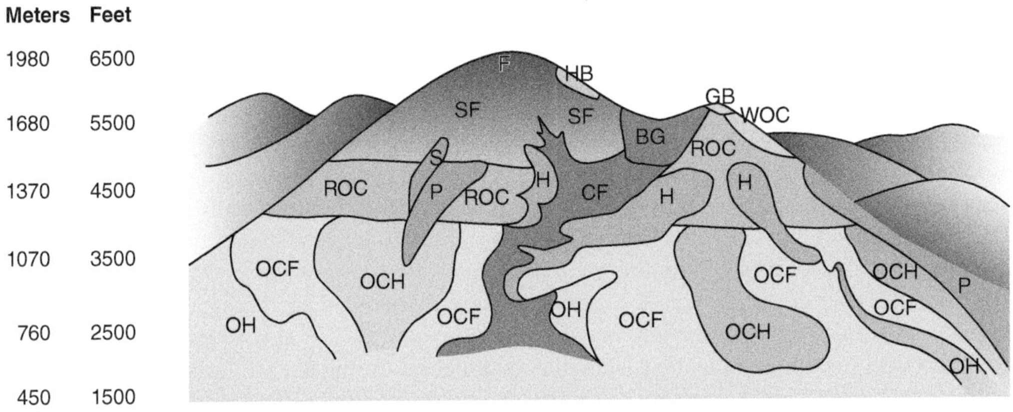

Redrawn, by permission, from R. H. Whittaker. 1956. Ecological Monographs 26: 1–80

FIGURE 13.6 Topographic distribution of vegetation types on an idealized west-facing mountain and valley in the Great Smoky Mountains. Vegetation types: BG, beech gap; CF, cove forest; F, Fraser fir forest; GB, grassy bald; H, hemlock forest; HB, heath bald; OCF, chestnut oak–chestnut forest; OCH, chestnut oak–chestnut heath; OH, oak–hickory forest; P, pine forest and pine heath; ROC, red oak–chestnut forest; S, spruce forest; SF, spruce–fir forest; WOC, white oak–chestnut forest. *Source:* Whittaker (1956) / John Wiley & Sons. Reprinted with permission of the Ecological Society of America.

In contrast to Whittaker's sampling approach along an elevational gradient, Hack and Goodlett (1960) emphasized local physiographic features in the central Appalachians (Chapter 8). In a similar region to that studied by Whittaker, they concluded that species assemblages generally coincided with landform units and often changed abruptly with changes in the form of the slope. Thus, at least part of how we conceive communities—whether they occur along gradients or with sharp boundaries—is heavily influenced by the scale and detail of a given study and by the sampling methods employed.

New England Much of the forest landscape in central New England is occupied by forests in different successional stages on old agricultural fields. Revegetation of old fields following widespread land clearing and farming has occurred on poorly to excessively drained soils of glacial origin. Composition of these forest communities therefore varies in response to a temporal gradient that includes the development of old-field succession, and a spatial gradient that includes the range of site conditions from wet to dry. Absolute elevation above sea level and aspect are less important in affecting the composition of the forest in this region of moderate elevations and rolling topography.

In Harvard Forest in central Massachusetts, Spurr (1956b) classified all existing stands according to their relative position on a successional gradient (early-successional, transitional, or late-successional) and their relative position on a soil-moisture gradient (from somewhat excessively drained to very poorly drained). Occurrence of individual tree species varied consistently with the two gradients. Northern red oak and red maple were prominent in all successional stages, and one or the other was prominent on all sites (see data for transitional stands in Table 13.1). Both species' occurrence and abundance were strongly related to soil water; northern red oak was most frequent on the well-drained and red maple on the very poorly drained sites. Other species were more specific in their site associations. White oak was most frequent on somewhat excessively drained sites, paper birch on well-drained sites, and white ash on somewhat poorly drained sites. In total, the forest communities of Harvard Forest appear to represent a continuous gradient correlated with successional stage and soil water.

Table 13.1 Occurrence of species as major components in transitional middle-aged stands in the Harvard Forest, central Massachusetts.

Site	Northern red oak	Red maple	Paper birch	White oak	White ash
		Frequency (percent)			
Somewhat excessively drained	67	0	0	33	0
Well drained	95	61	14	13	3
Somewhat poorly drained	81	81	9	6	19
Poorly drained	42	100	8	0	8
Very poorly drained	20	100	0	0	0

COMPETITION AND NICHE DIFFERENTIATION

Ecologists acknowledge that communities are not composed of mutually exclusive sets of species. Individual species have different genetically based physiological characteristics, such that different species are effective competitors on many different sites. A given species may be highly competitive and thus dominate in one community on a given site, but may be less competitive (though still present) in adjacent communities having different environmental conditions. Within a given climatic region, most forest species probably develop optimally under similar site conditions of favorable water and nutrients. Thus, two main processes in space and time, competition and mutualism, coupled with adaptation of species to particular site conditions, are what elicit differentiation of what we recognize as relatively distinct communities.

For example, many forest species in much of central Europe reach their optimal development under similar site conditions (Figure 13.7), but the superior competitive ability of European beech and two oak species restricts the other species to only portions of their potential range. Two significant factors, soil water and nutrients (typically correlated with acidity), are used as the axes in the "ecograms" of Figure 13.7. Tree growth is absolutely limited only by waterlogged soils and by extremely shallow soils over rock, both represented by dotted lines in the diagrams. The **physiological optimum range** of each species, or the conditions of maximum growth in the absence of competitors, is hatched more densely (Figure 13.7) than their potential tolerance range, or **physiological amplitude**. The optimum ranges of the species tend to coincide, but the most successful species are the best competitors, having a long life span, moderate growth rate, and ability to both tolerate and produce deep shade. Beech is able to exclude all or almost all other trees from most of its optimal range (Figure 13.7), birches dominate on wet soils only when these are very acid, and the proportion of oaks becomes greater the poorer and drier *or* the wetter the soil. In this humid, submontane climate, deciduous trees flourish on almost all soils and essentially exclude conifers except toward the limits of existence of forests where greater disturbance is likely. At such locations, Scots pine is able to dominate either on very dry sites, whether calcareous or acid, or at the edge of raised, oligotrophic bogs.

Plants having similar ecological requirements and tolerances occur together in recognizable communities because their genetic makeup and physiological responses allow them to be successful competitors in varying ways. Tree, shrub, and herb species with seemingly similar requirements and tolerances are able to exist in a given arid, cold, or wet environment because they occupy different niches. That is, the plants either occupy different microsites or they utilize a given site differently. This might be accomplished by differences in their morphological–physiological responses and life-history traits such as time of germination; amount and timing of shoot and root growth; depth of rooting; differential allocation of photosynthate to leaves, stems, and roots; and adaptive mechanisms for obtaining and retaining water. Thus, species become **niche-differentiated** when they are co-adapted in their physical and temporal environment.

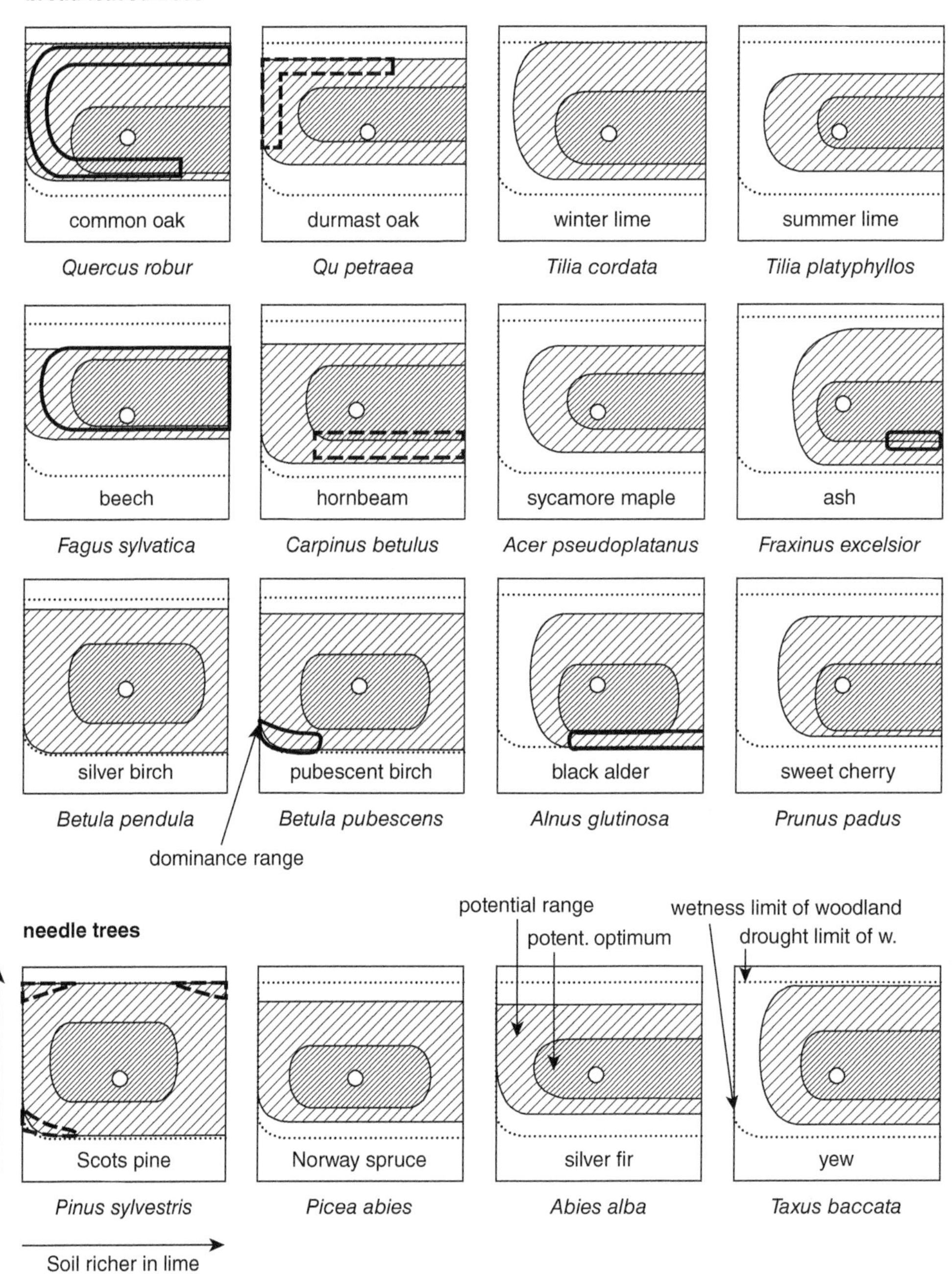

FIGURE 13.7 Ecograms showing the potential and optimum range of important tree species of central Europe in the submontane belt of a temperate suboceanic climate as related to soil-water and nutrient gradients. Broad-hatch = "physiological amplitude" or potential range of tolerance; narrow-hatch = "physiological optimum" range or potential optimum; area with thick black border = range where the species achieves a natural dominance under natural competition; broken border = species is co-dominant with others, or in the case of *Pinus*, this co-dominance applies only in the south and east of central Europe. In each of the ecograms, the y-axis represents the degree of wetness of the site. The x-axis covers the range from extremely acid to very basic soils. Above the upper dotted line it is too dry for tree growth; below the lower line it is too wet. The small circle in the center of each ecogram indicates average conditions. *Source:* Ellenberg (1988) / Cambridge University Press. Reprinted from *Vegetation Ecology of Central Europe*, © 1988 by Cambridge University Press. Reprinted with permission of Cambridge University Press.

INTERACTIONS AMONG ORGANISMS

Plant interactions, be they with animals or other plants, are ubiquitous and countless in nature. Such interactions are quite diverse and may be grouped by net effect (positive, negative, or neutral) that each species has on its associates (Table 13.2). The most important of these interactions in shaping the way organisms and evolution are perceived to be the antagonistic (negative) interactions of competition and predation. By contrast, interactions with no net effect on one or both partners (commensalisms, amensalisms, or neutralisms) are rarely studied. The remaining type of interaction—mutualism—benefits both species and is also referred to as symbiosis, cooperation, or facilitation. Mutualisms were once somewhat neglected in ecological studies relative to their importance in nature, but an enormous amount of information has now been accumulated (e.g., Bronstein 2015; Frederickson 2017; Schupp et al. 2017; Chomicki et al. 2019). Although competition is an important process, it is unlikely that the natural world is organized primarily around a single process. Rather, the influences of symbioses, cooperation, and mutualisms are also pervasive and underlie the evolution and ecology of most species. Ecological and evolutionary considerations of mutualisms are considered by Bronstein (1994, 2009, 2015) and Boucher (1985).

MUTUALISMS IN FOREST ECOSYSTEMS

Mutualisms occur throughout the entire life cycle of woody plants, and their importance cannot be underestimated (Janzen 1985). The four main types of mutualisms include: (i) nutritional, including the breakdown and supply of nutrients; (ii) protection from extreme site conditions or from enemies; (iii) dispersal of pollen, spores, or seeds; and (iv) supply of energy by plants to animal associates through photosynthesis.

Symbiotic Mutualisms—Mycorrhizae The association of roots with fungal hyphae may be the most important symbiotic mutualism for plants. Such an association is called a **mycorrhiza** and may increase the water and nutrient availability to the plant and protect roots from soil pathogens. Recent research estimates that there are about 50 000 fungal species that form mycorrhizal associations with about 340 000 plant species (van der Heijden et al. 2015; Genre et al. 2020). About 82% of angiosperms are mycorrhizal, as are most conifers except for a few pines (Brundrett 2002). The association between plant roots and mycorrhizal fungi is clearly the rule rather than the exception. Mycorrhizal relationships are discussed in detail in Chapter 19.

Nonsymbiotic Mutualisms Nonsymbiotic mutualisms between two species are those that occur in the absence of a physical connection, such as pollination, seed dispersal, protection, and decomposition of organic matter. Pollination of flowers by animals is critical to successful sexual reproduction in many woody plants, as is seed dispersal. The mutualisms of pollination and seed dispersal by animals are discussed in detail in Chapter 12. Protective mutualisms between woody plants and ants and other insects are common, whereby the plant benefits from protection from

Table 13.2 Grid of species interactions, grouped according to their net effect for each of the two interacting species.

		Effect of Species 2		
		Positive effect	**Negative effect**	**Neutral effect**
Effect of Species 1	**Positive effect**	Mutualism		
	Negative effect	Predation/parasitism	Competition	
	Neutral effect	Commensalism	Amensalism	Neutralism

predators, parasites, and diseases, and the insects benefit from shelter and sometimes food (Boucher et al. 1982). Ants are extremely important protective mutualists in the tropics and to a lesser extent in temperate forests (Huxley and Cutler 1991). Finally, an indispensable mutualism in all forests is the decomposition of organic matter on the forest floor and in the soil by many kinds of animals, fungi, and bacteria, thus ultimately making available nutrients and CO_2 to plants (Chapter 19).

COMPETITION

Competition is an antagonistic relationship between two species attempting to utilize the same resource when that resource is limited in supply. As such, competition is an important agent that shapes communities. For plants, competing species are capable of occupying the same geographic space, and competition is considered over the entire life cycle (Grubb 1985). Competition also occurs when the species interfere with each other's ability to utilize a resource that is not limited, such that the yield of one plant is reduced as a result of another plant being present (Harper 1961). Competition in plants occurs both above and below ground, and its outcome is influenced by the plants' inherent requirements or tolerances of light, temperature, water, nutrients, and other site resources. In the earlier text, we defined these inherent tolerances or requirements as a species' niche; the **principle of competitive exclusion** states that one of two species occupying the same niche will eventually be outcompeted by the other.

The eventual death of the most dominant individuals in a community and the constant demand of individual trees for crown and root space create changes in the structure and composition of forest communities. As the main canopy trees grow larger, competition for growing space increases, with a few individuals gaining space, a few maintaining their space, and an increasing majority losing space and eventually succumbing. As dominant individuals die due to lightning, fire, wind, insects, or diseases, a portion of the site is released from the main canopy for occupancy by a growing and developing understory. These types of competitive relationships occur in all forests. *Intraspecific competition* (between the trees of the same species) typically does not affect the composition of the forest community, but *interspecific competition* (among individuals of different species) results in a change in species composition and thus has a direct effect on forest succession.

Forest Community Structure and Composition Even-aged stand structure results when trees regenerate *en masse* at roughly the same time, as following fire or flooding. Almost all pines, Douglas-fir, black spruce, cottonwoods, and many other species may form natural even-aged stands. Individuals in such a stand are nearly the same age and are often of a similar height and diameter. By contrast, uneven-aged stands are formed by various long-lived hardwoods and conifers and exhibit great diversity in tree age and size. Uneven-aged stands are typical of naturally regenerated hemlock–northern hardwood forests (comprised of eastern hemlock, sugar maple, beech, yellow birch, and basswood, among others) and of some spruce–fir forests of central Europe. Here, the trees of a given species may range 300 years or more in age. At broad scales in many forests, a mosaic of small even-aged groups forms the uneven-aged forest; this structure was probably characteristic of upland oak forests of the East and ponderosa pine forests of the West prior to European colonization.

Competition for light, water, and nutrients in even-aged stands depends largely on stem density, or the number of stems per unit area. In time, **canopy closure** develops (i.e., the crowns of the trees come together to exploit all available growing space) and crowns of individual trees differentiate into classes (Figure 13.8). A combination of environmental and genetic factors allows some trees to develop rapidly and exhibit large, well-formed crowns, while others grow more slowly, and their crowns become more or less restricted. Eventually the slower-growing trees are gradually overtopped and suppressed. While crowns utilize growing space to compete for available light, roots utilize belowground growing space to compete for water and nutrients. Root competition is often severe in even-aged stands, though it is far less obvious than crown competition. The

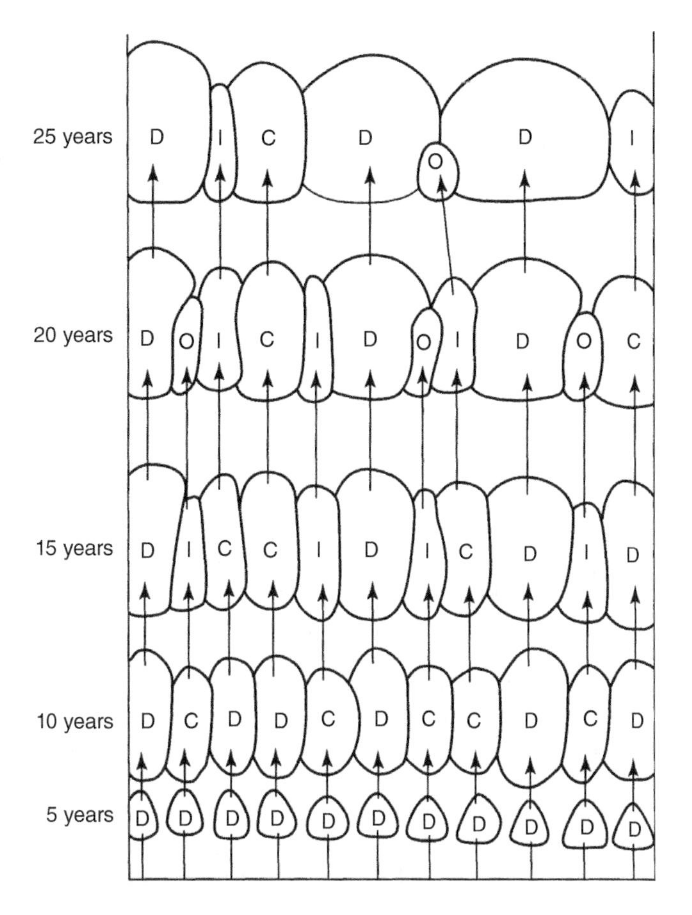

FIGURE 13.8 Differentiation of trees of a pure even-aged stand into crown classes. D, dominant; C, co-dominant; I, intermediate; O, overtopped. *Source:* Smith et al. (1997) / John Wiley & Sons. Reprinted from *The Practice of Silviculture* by D. M. Smith, B. C. Larson, M. J. Kelty, and P. K. S. Ashton, © 1997 by John Wiley & Sons, Inc. Reprinted with permission of John Wiley & Sons, Inc.

root system of a pine in an even-aged stand, for example, may compete with several hundred trees, whereas its crown competes only with a few adjacent trees. Parsons et al. (1994) found that significant loss of nitrogen to groundwater did not occur until root gaps resulting from the removal of clusters of 15–30 lodgepole pines were created in even-aged stands in southeastern Wyoming.

Intense competition in even-aged stands can be illustrated using a simple classification of crown classes (Figure 13.8). Crown classes are often used by ecologists and forest managers to judge the vigor of the stand and for determining and planning thinnings and other harvest operations. The major crown classes are:

Dominant: Trees with crowns extending above the general level of the canopy and receiving full light from above and partly from the sides. Dominant trees are larger and more vigorous than the average stems in the stand and have well-developed crowns.

Co-dominant: Trees with crowns forming the general level of the canopy or somewhat below. Co-dominant trees receive full light from above but only moderate or limited amounts from the sides. They typically have medium-sized crowns that are more or less crowded on the sides.

Intermediate: Trees shorter than the preceding classes but with crowns extending into the canopy formed by the dominant and co-dominant trees. Intermediate trees receive little direct light from above and virtually none from the sides, and typically have small crowns that are considerably crowded on the sides.

Overtopped: Trees with their crowns entirely below the general canopy level, receiving no direct light from above or from the sides.

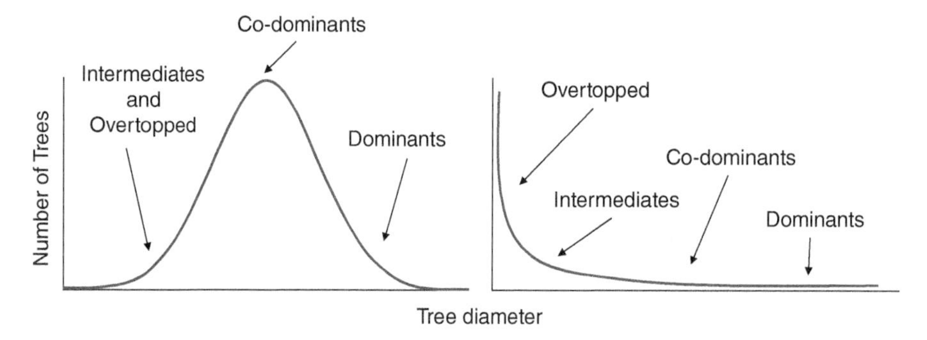

FIGURE 13.9 Typical tree diameter distributions for a classic even-aged stand (left) and an uneven-aged stand (right). Moderate-sized co-dominant trees are most abundant in even-aged stands, but small, overtopped trees are most abundant in uneven-aged stands.

Crown classes of individual trees can be combined with tree diameters to better understand the structure of a stand; structure is also illustrative of light competition. Even-aged stands typically have a tree diameter distribution resembling a bell-shaped curve, with many moderate-sized stems representing the co-dominant trees, and far fewer small (intermediate and suppressed trees) and large (dominant trees) stems (Figure 13.9). Uneven-aged stands have a diameter distribution that follows a negative exponential curve, with many small, often young, overtopped stems and only a few larger, older stems (co-dominants and dominants). These classical diameter distributions represent the two extremes of stand structure, and many variations exist between them. For example, the diameter distribution of even-aged stands may resemble that of an uneven-aged stand if they contain multiple species that differ in shade tolerance or a have well-established layer of regeneration beneath the canopy.

The distribution of tree ages may be used to further study competition among individuals within a species population in an even-aged stand or among different species in an uneven-aged stand. The distribution of tree ages in a stand may be used as a proxy for time, allowing ecologists to infer changes that have occurred or may happen in the future. For example, in an old-growth hardwood forest in New Hampshire, sugar maple and beech exhibit sharply declining populations, typical of species that periodically establish many seedlings on the shaded forest floor (Figure 13.10; persistent seedlings, Chapter 4). Many seedlings survive and eventually some replace the dominant overstory trees. By contrast, red spruce exhibits a lack of young seedlings and may diminish in abundance over time. Red spruce is not competitive in the younger age classes with beech and maple on this site in the absence of disturbance.

Vertical Structure Competition for light among forest species also results in the development of vertical vegetation structure. For example, a multistoried forest may have an upper layer of overstory trees, one or more subdominant layers of younger stems of the overstory trees and/or mature trees of subcanopy species, layers of high and low shrubs, and a ground layer of grasses, herbs, mosses, liverworts, and lichens. The species of each layer are effective competitors where they occur because they are genetically adapted to make the best use of the space, light, and microclimatic resources of their respective vertical positions.

Considerable structure occurs in the tree layer alone depending on species composition and site conditions. Generally, more vertical layers are found in the presence of more favorable site conditions, especially soil water. For example, cove forests in the southern Appalachian Mountains exhibit complex vertical structure. In addition to a diverse overstory, there may be several understory tree layers and a dense layer of tall ericaceous shrubs, primarily the rosebay rhododendron. Tropical rain forests are renowned for even more pronounced vertical structure, including huge

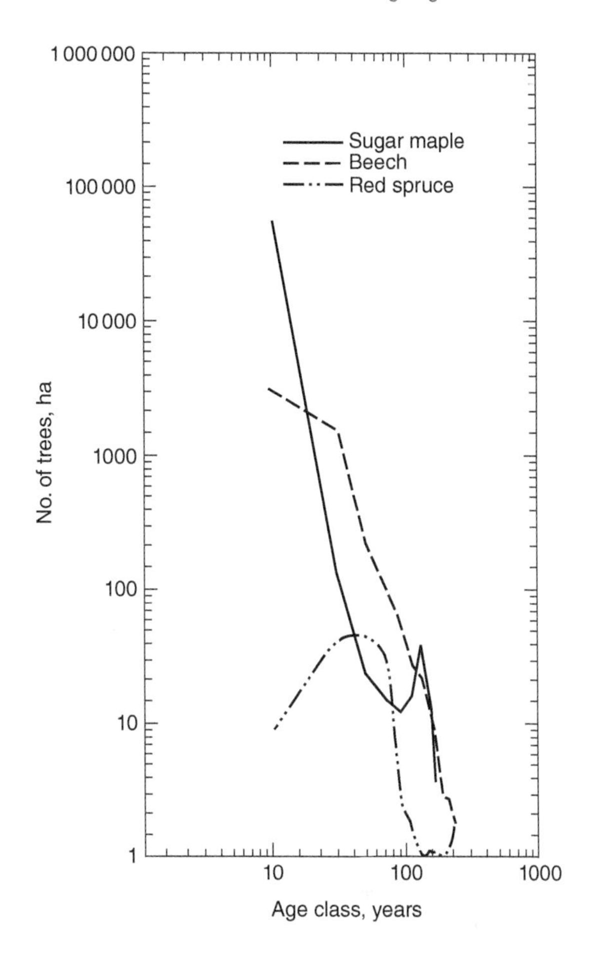

FIGURE 13.10 Numbers of trees per hectare (log scale) in an old-growth northern hardwood forest over midpoints of 20-year age classes (log scale) for sugar maple, beech, and red spruce. *Source:* Leak (1975) / John Wiley & Sons. Reprinted with permission of the Ecological Society of America.

emergent trees whose crowns extend 5–10 m above the general canopy level and are fully exposed to sunlight. In such multistoried forests, each vertical stratum has a different microclimate and often a distinct assemblage of insects and other animals.

The overstory canopy of forests intercepts much of the incoming irradiance, and thus species of the lower layers must be physiologically and structurally adapted to use the continuously decreasing amount of light available beneath the canopy and at the forest floor. The canopy is never completely unbroken over large stretches, however, and various light-requiring subdominant trees, shrubs, and herbs utilize the well-lighted microsites and growing space to extend their crowns into openings. The crown densities of overstory trees vary widely, so that considerably more light may penetrate one forest overstory compared to another. As such, the degree of development of vertical structure varies across forests of different composition. For example, the shrub layer is particularly well developed in oak forests of eastern North America compared to beech–sugar maple forests because oak crowns are less dense, and the canopy is more open than that of beech–sugar maple forests. The increased light, together with more frequent fires in oak forests, favors the growth, sprouting, and clonal spread of understory shrubs.

Many pioneer species that colonize areas following major stand-replacing disturbances such as fire initially exhibit only a single tree layer. Near-pure natural overstory monocultures may be formed by species such as lodgepole and jack pines, aspens, and the southern pines following stand-replacing fire. The image of a plantation-like expanse of even-aged and uniform stems is misleading, however. In the dense stands that arise following a crown fire, for example, trees are typically patchily distributed over tens or hundreds of hectares (Kashian et al. 2004; Turner et al. 2004; Kashian et al. 2017). Such stands may be dense and may lack a well-developed understory when young and middle-aged, but with increasing age, gaps form and regeneration begins to develop. A classic example is lodgepole pine forests of the central and northern Rocky Mountains that may exhibit even-aged stand structure over large areas but show increasing vertical structure in older stands (Kashian et al. 2005b).

Evidence of considerable variation in stand structure is accumulating and may be associated with physiography and soil conditions, variability in fire regimes, or insect and pathogen infestation patterns. Working in central Colorado, Parker and Parker (1994) identified two distinct kinds of stand structures in 120–140-year-old lodgepole pine: those with closed canopies, high basal areas and stem densities, and low pine sapling and seedling densities of pine; and those with open canopies, low tree densities, and significantly higher densities of pine saplings and seedlings. The structure of closed stands result from dense regeneration developing rapidly following crown fires, while nearly continuous recruitment of pine occurs in the open stands. Kashian et al. (2005b) confirmed the presence of both of these stand structures in the lodgepole pine forests of Yellowstone National Park and found that the structures tended to converge after about 200 years with age-related development.

Stand Density The abundance and size of trees occupying a site have important implications for the trees themselves, the site, and for a land manager responsible for controlling forest composition, growth, and reproduction. Stand density is typically expressed as the number of stems or basal area per unit area.

Except at very low stand densities, tree mortality in most stands is mostly caused by *density-dependent competition* (also called *self-thinning*) among trees and between tree seedlings and other vegetation for growing space, specifically light, water, and nutrients. Mortality is highest in the seedling stage when the number of seedlings per unit area is highest. In a sugar-maple-dominated forest in Wisconsin, Curtis (1959) reported 99% mortality of sugar maple seedlings over a 2-year period; this mortality rate is initially independent of age and declines as the seedlings mature (Hett 1971). Surviving individuals become larger as the outcompeted plants die, and the smaller plants are continually eliminated from the population. There is therefore a strong relationship between plant size and stand density, whereby larger size (or biomass) of the stems is associated with fewer stems per unit area, and vice versa.

The relationship between plant size and stand density has been extensively explored as a "fundamental truth" or natural law of plant populations. Reineke (1933) showed that plotting the logarithm of number of trees per acre against the logarithm of average diameter of fully stocked, even-aged forest stands typically resulted in a negatively sloping, linear relationship. Plant ecologists have since designated this as the "$-3/2$ law of self-thinning" because the slope of the line of mean volume or biomass of surviving plants versus stand density is approximately $-3/2$ for populations of many different kinds of plants (Drew and Flewelling 1977; White 1985; Silvertown and Lovett-Doust 1993).

This relationship is illustrated in Figure 13.11 for stand density and volume for five stands of loblolly pine of different initial densities over a 50-year period (Peet and Christensen 1987). Competition increases as stand development and self-thinning occur such that stand density decreases and mean tree volume increases. The slope of the thinning curve is -1 for a given stand when its net biomass production equals its loss by mortality, and the slope will be greater (less negative) if the rate of biomass loss exceeds the rate of net production. Competition is low in a young pine stand prior to canopy closure, and the slope is flatter than $-3/2$, but stand mortality rate converges on a $-3/2$ slope when the canopy closes and resources become limiting. The rate of convergence to the $-3/2$ line is clearly dependent on initial density, with stands at high initial densities thinning

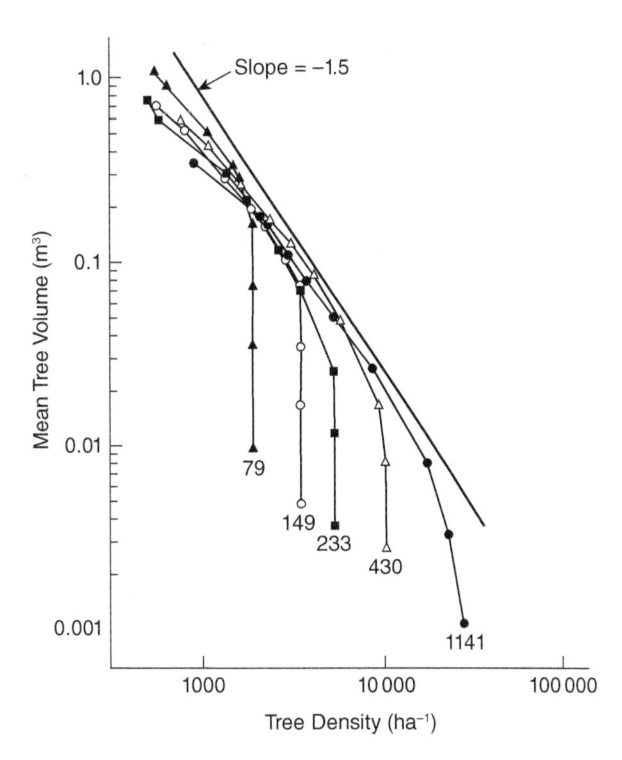

FIGURE 13.11 Relationship between tree density and mean tree volume for a series of loblolly pine stands over a 50-year period. The different initial stand densities are indicated by the number below each line (79–1141 stems originally in the 0.1-acre plots). The straight line has a slope of −1.5, as predicted by the law of thinning. The tree populations approach an upper limit of density defined by this −3/2 line, but no stand closely tracks the line. *Source:* Peet and Christensen (1987) / with permission of Oxford University Press. Reprinted from BioScience (vol. 37, no. 8, page 591), © 1987 by the American Institute of Biological Sciences.

more rapidly and those at low densities taking a longer time to reach the line. No single stand clearly defines the −3/2 line, yet all stands together do define such a line, which individual stands approach but never cross. The line therefore defines the maximum stocking level. The relationship holds for many stands of many species, of many ages on many sites (Long 1980), although some exceptions have been reported (Lonsdale 1990; Osawa and Allen 1993). Notably, the intercept of the line may vary widely depending on species and site conditions (Silvertown and Lovett-Doust 1993). In forestry, the relationship has been widely applied to stand-density management of Douglas-fir plantations (Drew and Flewelling 1979).

The relationship between tree diameter and stand density means that careful control of density is extremely important for forest managers. If large trees are the objective, wide initial spacing achieved via planting or by thinning is necessary to maximize diameter growth, yet still fully utilize site resources. Low stand density may negatively affect wood quality for some uses, however, because trees become branchy, the wood knotty, and/or growth rings may be too wide. In such cases, a balance must be reached for a given species on a given site through appropriate density control, depending on the end product desired. As long as the site is fully stocked, such that trees are making full use of available resources, a species will produce the same amount of wood per year over a relatively wide range of densities regardless of whether the trees are large or small. More trees can be grown per unit area and a greater volume of wood can be obtained as site quality improves. Thus, understanding site quality is of great importance in regulating stand density. Detailed considerations of stand density from the standpoint of growth and yield and silvicultural management of forests are available in several texts (Oliver and Larson 1996; Nyland et al. 2016; Ashton and Kelty 2018; Burkhart et al. 2018).

Competition and Overstory Composition Trees in the main canopy of a forest increase yearly in height, bole size, length of each growing branch, and number of leaves. A tree must grow to remain alive and vigorous, and thus it must increase its growing space—its utilization of the site. In doing so, competition among growing individuals in the overstory ultimately eliminates some trees, particularly those species genetically and physiologically less suited for survival under existing site conditions. The result is a gradual (or occasionally an abrupt) change in the composition of the overstory.

Table 13.3 summarizes 60 years of growth of a middle-aged, mixed hardwood stand on an island in the St. Mary's River between Michigan and Ontario. Relatively short-lived aspens and birches practically disappeared from the stand during this period, and sugar maple suffered steady mortality, with basal area slowly decreasing after 1978. Red maple declined greatly in abundance, but the growth of the remaining trees compensated for the loss of individuals to create a steady increase in basal area. Sugar and red maples made up 72% of the basal area by 1993. There were few northern red oaks on the island, but they grew relatively fast and made up a substantial proportion of the basal area. The better competitive ability of sugar maple, red maple, and northern red oak at the expense of aspen and birches brought about major changes in stand composition.

Notably, permanent plot data record how the overstory composition changed over time, but it does not chronicle the cause of tree mortality in the interval between measurements. It might be assumed that most of these trees simply were suppressed (leaf area and root feeding area were reduced) to the point that they died for their inability to maintain a positive carbon balance under competition for site-specific resources of light, water, and nutrients. However, death may have occurred from many causes. Even "natural mortality" of a tree often occurs during some period of extreme stress, such as in a hot and dry spell in late summer, in a severe unseasonable frost, or in an extremely cold winter. In other cases, the tree becomes weakened to the point that it becomes susceptible to insects or disease. Usually, several factors are interrelated and together account for death (Mueller-Dombois 1987).

Competition in the Understory Overstory competition involves relatively few individuals and is spread out over many years, but competition near the forest floor involves many shrubs,

Table 13.3 Changing composition over 60 years in a northern hardwood forest on Sugar Island, St. Mary's River, Upper Michigan.

Species	1933	1944	1955	1971	1978	1989	1993
			Stand density (trees/ha)				
Sugar maple	803	741	652	502	474	403	381
Red maple	314	210	173	158	151	133	133
Paper birch	89	40	22	7	7	7	5
Bigtooth aspen	49	7	7	7	7	5	5
Northern red oak	44	44	44	37	37	35	35
Yellow birch	35	17	15	10	10	5	2
All trees[a]	1364	1077	927	734	697	598	571
			Basal area (m^2/ha)				
Sugar maple	9.4	11.6	12.7	13.2	13.3	12.4	11.8
Red maple	7.1	7.5	7.9	9.0	9.7	10.1	10.7
Paper birch	2.8	1.6	1.3	0.5	0.5	0.6	0.4
Bigtooth aspen	1.4	0.4	0.5	0.8	0.9	0.8	0.9
Northern red oak	2.0	2.8	3.6	5.0	5.4	6.2	6.7
Yellow birch	1.0	0.7	0.7	0.6	0.6	0.2	0.1
All trees[a]	24.0	25.0	27.3	29.7	31.0	31.0	31.2

[a] Includes some individuals of species not listed in the earlier text.

herbaceous plants, and grasses as well as trees, often in very large numbers and changing on an annual basis. Many plants are adapted to life in the understory, such as herbs that complete a major portion of their annual growth before the canopy trees fully flush; herbs, many shrubs, and small trees that are generally tolerant of understory site conditions; and trees that can both thrive in the understory and also occupy the overstory itself when the opportunity occurs.

Wendel (1987) estimated the sheer numbers of potentially competing understory stems and plants in north-central West Virginia oak forests. Forest floor samples from each of four different sites were collected and the plants allowed to develop in a greenhouse during the following growing season. Nine species of tree seedlings and sprouts averaged 981 000 stems/ha, 8 species of woody shrubs averaged 243 000 stems/ha, and 25 species of herbs averaged 220 000 stems/ha. Most of the newly germinating seedlings in this study, grown free from competition, would not have survived in the forest understory, but the potential for competition in the forest understory is nevertheless enormous in these ecosystems.

Similar to overstory trees, several factors are usually at play in the death of understory plants. Unsuccessful competitors lose vigor and die in part due to starvation resulting from inadequate photosynthesis due to inadequate irradiation, resulting in insufficient carbon fixation for the plant to survive. Lack of sufficient photosynthate leads to a "death spiral," as the root system is weakened and unable to obtain adequate water to support a vigorous cambium and crown. Inadequate soil water is another factor that contributes to the death of outcompeted understory plants, as are insects and diseases, particularly when the plant has reached a weakened condition (Decker 1959). The site factors relating to understory survival are discussed in the following text.

UNDERSTORY TOLERANCE

An important factor shaping the composition of forest communities is the physiological capacity of plants to persist, or not, in the low light conditions of the forest understory. Those that can persist in these conditions have the potential to reach overstory status, whereas those that cannot are destined to die. These basic attributes therefore shape forest composition over time. Plants that become shaded in open habitats may avoid shade by growing taller or otherwise toward the light (called *shade avoidance*), but those found in the forest understory do not have such a luxury. Survival of understory plants is central to an understanding of forest succession, because forest trees able to both survive as understory plants and then respond to release to reach overstory size will eventually represent a major component of the forest community. A forest tree that can survive and thrive under a forest canopy is said to be **understory-tolerant** (or simply **tolerant),** whereas one that can thrive only in the main canopy or in the open is classified as **understory-intolerant** (or **intolerant**). There is a continuum of tolerance from one extreme to the other.

The general biological meaning of "tolerance" refers to the capacity of a plant to be genetically adapted and to have the physiological ability to be a good competitor under conditions generally unfavorable to other plants. A salt-tolerant plant is therefore one that is adapted to grow in soil with a high salt concentration, and a drought-tolerant plant is one that can grow at low soil-water levels that would be fatal to most other plants. Many pioneer species are intolerant of low light but tolerant of high irradiance, extreme temperature, drought, and low nutrient status often found on exposed, open sites. However, in this chapter, and elsewhere in this book, our unmodified use of the term "tolerance" refers to a plant's vigor in the forest understory due to its genetic adaptations and physiological attributes to this complex environment. Although this is often described in the literature as "shade tolerance" because light is typically a limiting factor in the forest understory, "understory tolerance" is the more acceptable term because of the prevalence of many other abiotic and biotic factors that contribute to understory conditions, such as soil-water stress, temperature, browsing, and disease. The term "shade tolerance" should be applied specifically to situations where light is clearly the dominant factor.

Notably, the concept of tolerance is somewhat misleading and has been described as inadequate (Grubb 1985). Many plants do not simply tolerate the conditions with which they are associated in the field; they may actually require them for maximum growth. For example, some plants associated with shade require it because they perform poorly in full sunlight. Likewise, flood-tolerant trees such as red ash, eastern cottonwood, sweetgum (Broadfoot and Williston 1973), and water tupelo (McKevlin et al. 1995) may actually require flooding to outcompete other plants in establishment and for optimum growth.

CHARACTERISTICS OF UNDERSTORY-TOLERANT AND -INTOLERANT SPECIES

A tolerant tree is one that grows and thrives under a forest canopy, and its vigor under such conditions may be characterized in various ways. An intolerant tree is usually characterized by opposite extremes of the same traits. Many characteristics have been cited as criteria or indicators for the determination of tolerance:

Tolerance of Suppression in the Understory: Tolerant trees may live for many years under dense forest canopies. Many species are able to germinate and survive a few years under such conditions, but only tolerant trees will persist and continue to grow for decades or centuries.

Response to Release: The vigor of tolerant trees is evident when they are able to initiate "release," or immediate and substantial growth following removal of the overstory. A tolerant tree may survive for years or even decades in the understory with very little growth, and yet grow extremely rapidly soon after release.

Crown and Bole Development: The lower branches on tolerant trees maintain foliage longer and to a greater extent than those on intolerant trees, such that tolerant trees will have deeper and denser crowns. Tolerant trees will therefore naturally prune at a slower rate, will maintain a greater number of leaf layers in the crown, and will maintain healthy and vigorous leaves deeper into the crown. The greater depth of crown of tolerant trees results in more tapered boles compared to those of intolerant trees (Larson 1963b).

The shoot growth and branching pattern of a species are adapted to provide the amount of light required of its foliage. The relatively fast branch growth and spreading form of American elm, sycamore, and silver maple, for example, create a well-lighted crown necessary to maintain the leaves of these species. By contrast, sugar maple and American beech leaves tolerate considerable shade and thrive in the interior of a crown that is more compact and has slower branch growth than that of intolerant species.

Stand Structure: Tolerant trees persist over long periods of time in natural mixed stands and tend to be successful when mixed with other species of equal size. Consequently, tolerant trees may form denser stands with more stems per unit area compared to intolerant trees.

Growth and Reproductive Characteristics: Tolerant trees inherently grow more slowly than intolerant trees, especially in moderate to high light conditions (in dense shade, a slow growing, tolerant species may appear to grow faster than a dying intolerant species). Tolerant deciduous angiosperms (beeches, maples) tend to have seasonally determinate shoots, whereas intolerants (willows, aspens, cottonwoods, birches) have seasonally indeterminate shoots (Marks 1975). Tolerant trees typically mature later, flower later and more irregularly, and live longer than intolerant trees.

Populations of the respective extreme types of tolerance have evolved under differing selection pressures of environment and plant competition such that they belong to complex and contrasting adaptation systems. Intolerant species are typically early-successional species that may colonize a wide variety of sites, successful because of two major types of adaptations. First, they are able to rapidly establish on disturbed sites, grow quickly in the open, produce seed early in their lives, and disseminate seeds widely. These characteristics have enabled intolerant species to perpetuate themselves wherever fire, windstorm, flooding, cultivation, or other disturbances have eliminated or reduced the existing vegetation, and they may depend on such disturbances for persistence in many climatic regions. Intolerant species also exhibit important adaptations to extreme site conditions, such as dry or infertile sites as well as to the temperature extremes of initially disturbed sites. They may form relatively permanent communities on extreme sites where they may outcompete more tolerant species. For example, willows in annually flooded bottomlands, jack pine on nutrient-poor sands of the Lake States, table-mountain pine on the driest and least fertile soils of the Appalachian Mountains, and sand live oak on sands of the southern Coastal Plain are all examples of intolerant species occupying extreme sites.

Tolerant species establish, grow, and have become adapted to conditions markedly different from those of intolerants, typically more mesic (moist, sheltered, and fertile sites). They replace intolerant species during succession and perpetuate themselves through adaptations favoring survival and growth in a shaded understory, establishment in relatively undisturbed litter layers, and long life spans. Their presence tends to reinforce the shaded, moist, humid, and relatively disturbance-free environment favorable to their own regeneration once they attain dominance in the canopy (Chapter 4). Notably, although it is instructive to compare the two extremes of tolerant and intolerant species, species typically have varying degrees of intermediacy from very intolerant to extremely tolerant.

TOLERANCE RATINGS OF TREE SPECIES

Tolerance varies with genetically different individuals and races of a given species, and with different regional climates, local site conditions, plant associates, vegetative conditions (seedlings versus sprouts), and especially with age. A given species may appear to be more tolerant in one part of its geographic range than another, one site in comparison to another, and/or in one forest type compared to another. For example, eastern white pine is thought to be mid-tolerant in Minnesota but intolerant in New England. In central New England, it is more tolerant on dry, sandy soils than on moister sandy loams. Like most species, eastern white pine is more tolerant as a juvenile compared to the adult phase. For those species that vigorously sprout, such as oaks, hickories, ashes, and tulip tree, the sprouts are more tolerant than seedlings of the same species in part because they are better able to absorb water and nutrients from the well-established root system of the parent. Thus, understory tolerance is not only a relative matter, as a species' tolerance rating will depend upon the regional ecosystem, specific site, age, and its associates. Tolerance ratings must be interpreted with great care and with these points in mind.

At the same time, the approximate tolerance rating of more common forest tree species should be known and understood by ecologists and forest managers as a general frame of reference. We present such a frame of reference as an estimate of the relative understory tolerance of tree species assuming representative site conditions in Table 13.4. Notably, such estimates can and do vary among observers and researchers. Niinemets and Valladares (2006) attempted to assign a numerical value for shade tolerance on 806 woody species across the Northern Hemisphere using the 5-class ranking scale of Baker (1949). While this protocol appears useful for many species, in some cases the attempt to generalize across all regional ecosystems and sites provided relative values contrary to those observed by some ecologists in the field. For example, in the forests of the Lake States, black oak is an intolerant species that grows on dry and nutrient-poor sites, forming savannahs prior to European colonization. Northern red oak is a mid-tolerant tree found on dry-mesic to mesic sites, often found as a minor component of mesic and nutrient-rich sites dominated

Table 13.4 Relative understory tolerance of selected North American forest trees[a].

Eastern North America		
Gymnosperms	**Angiosperms**	**Western North America**
	Very tolerant	
Eastern hemlock	Flowering dogwood	Western hemlock
Balsam fir	American beech	Pacific silver fir
	Sugar maple	Pacific yew
	Tolerant	
White spruce	Basswood	Spruces
Black spruce	Hop-hornbeam	Western redcedar
Red spruce		White fir
		Grand fir
		Subalpine fir
		Redwood
	Mid-tolerant[b]	
Eastern white pine	Yellow birch	Western white pine
Slash pine	Silver maple	Sugar pine
	Red maple	Douglas-fir
	White oak	Noble fir
	Northern red oak	
	Hickories	
	White ash	
	Elms	
	Intolerant[b]	
Red pine	Black cherry	Ponderosa pine
Shortleaf pine	Tulip tree	Junipers
Loblolly pine	Sweetgum	Red alder
Eastern redcedar	Sycamore	Madrone
	Black walnut	
	Black oak	
	Red ash	
	Scarlet oak	
	Sassafras	
	Very intolerant	
Jack pine	Paper birch	Lodgepole pine
Longleaf pine	Aspens	Whitebark pine
Virginia pine	Black locust	Digger pine
Tamarack	Eastern cottonwood	Western larch
	Pin cherry	Cottonwoods

Survival in the understory is related to light irradiance, moisture stress, and other factors. As a general guide to the light irradiance component, we estimate the range of *minimum* percentage of full sunlight for a species to survive in the understory at each of the five arbitrary levels of tolerance.

Very tolerant species may occur when light irradiance is as low as 1–3% of full sunlight. *Tolerant* species typically require 3–10% of full sunlight. *Intermediate* species 10–25%. *Intolerant* species, 25–50%. *Very intolerant* species, at least 50%. For example, an intolerant species competing in the understory is unlikely to survive with less than about 25% of full sunlight (unless other compensating factors are favorable).

[a] Based on representative site conditions for the respective species.

[b] Many *mid-tolerant* and *intolerant* species are observed to be more tolerant in the juvenile phase than in the adult phase. Notable examples include black cherry, white ash, and elms.

by sugar maple and beech. White oak is also a mid-tolerant tree, but is less tolerant than northern red oak and is found on dry to dry-mesic sites, rarely with mesic species (Barnes and Wagner 2004). By contrast, Niinemets and Valladares (2006) characterize black oak and northern red oak as equally shade-tolerant, but black oak as the more drought-tolerant, and white oak as the most shade- and drought-tolerant of the three species. We therefore emphasize that all forest ecologists must learn to understand the relative tolerance and competitive ability of trees using their own observations and field experiences and not from a textbook table. Nevertheless, a "standard condition" or general starting point is appropriate, given an understanding of the nature of understory tolerance.

EXAMPLES OF UNDERSTORY TOLERANCE IN FOREST ECOSYSTEMS

We provide four examples of understory tolerance using photographs from contrasting forest ecosystems. In a young, second-growth stand of intolerant tulip tree in a cove site of the southern Appalachian Mountains, sufficient light and water reach the understory to favor development of a conspicuous and diverse community of mid-tolerant and tolerant species, such as oaks, maples, and beech, as well as a rich shrub and herbaceous flora (Figure 13.12). A stand of mid-tolerant oaks in the Missouri Ozarks (Figure 13.13) illustrates a dominant oak overstory and an open, partially shaded understory of oaks and other vegetation. In contrast to the understory of the tulip tree forest, growth of the oak understory is curtailed by the shade cast by the overstory and low soil water during the growing season. In a stand of sugar maples in northern Michigan, dense shade from the overstory favors the very-tolerant sugar maple seedlings over all other species in the understory (Figure 13.14). In deep shade, sugar maple seedlings survive but grow slowly; in small openings where more light is available (background of Figure 13.14), they respond with accelerated growth and, barring disturbance, will perpetuate the dominance of sugar maple. Finally, the western white pines of a stand in northern Idaho originally colonized the site following a fire, forming a dense stand two centuries later with a canopy that intercepts most of the incoming light (Figure 13.15). The pines will gradually be replaced by the very tolerant species in the understory, such as western hemlock, grand fir, and western redcedar, in the absence of fires, windstorms, or logging.

NATURE OF UNDERSTORY TOLERANCE

The relative tolerance of a given species growing in a given ecosystem can be estimated reasonably accurately, but understanding the nature of its tolerance is much more difficult. The nature of tolerance may be examined at the species level in terms of adaptations, but the specific trait of survival in the understory is usually investigated by studying the site factors and the physiological processes involved.

Two major hypotheses have been proposed to investigate the traits determining a species' understory tolerance as it relates specifically to light, and understanding these perspectives is useful in framing further discussion of the nature of understory tolerance. The *carbon gain hypothesis* emphasizes that plants in low light maximize the capture of light while minimizing respiration costs needed for tissue maintenance (Givnish 1988). Thus, shade tolerance may be increased by any trait that helps a plant more efficiently use light and more efficiently increases photosynthesis and net carbon gain. Recent evidence has suggested that the better performance of shade-tolerant species in low light results more from minimizing carbon losses during dark respiration than by increasing carbon gain (Chapter 18; Walters and Reich 2000b; Craine and Reich 2005). A relationship arising from the carbon gain hypothesis is that plants with high growth rates in high light have low survivorship in low light, and vice versa (Kobe et al. 1995). A second hypothesis related

FIGURE 13.12 Young second-growth stand of tulip tree in a cove of the Appalachian Mountains, North Carolina. *Source:* U.S. Forest Service photo.

FIGURE 13.13 Uncut, old-growth oak stand in the Ozark Mountains of southeastern Missouri. *Source:* U.S. Forest Service photo.

FIGURE 13.14 Sugar maple stand in the Upper Peninsula of Michigan. Unbrowsed sugar maple seedling shown on left and browsed seedling on right. *Source:* U.S. Forest Service photo.

FIGURE 13.15 The old-growth western white pines will eventually be replaced, barring disturbance, by very tolerant conifers that have become established in the understory; northern Idaho. *Source:* U.S. Forest Service photo.

to shade tolerance is the *stress tolerance hypothesis*, which suggests that plant survival in low light is explained best by its capability to resist biotic and abiotic stresses in the understory (Kitajima 1994). One process for such stress tolerance may be a higher allocation to storage, such that a plant is able to tolerate periods of low light by utilizing stored photosynthate produced during periods of high light (Kobe 1997; Canham et al. 1999; Myers and Kitajima 2007). Many shade-tolerant species exhibit determinate growth and terminate annual growth earlier in the growing season compared to the indeterminate growth of intolerant species, allowing shade-tolerant species more time to allocate photosynthate toward storage rather than growth (Kikuzawa 2003). A review of the evidence supporting and refuting each of these hypotheses is presented in Valladares and Niinemets (2008).

Environmental Factors Relating to Understory Tolerance The most obvious ecological feature of the understory is low light irradiance, and as suggested in the earlier text, many ecologists associate a plant's ability to survive in the understory solely based on its tolerance to shade. Notably, abundant observational and experimental evidence has shown that light irradiance is sufficient under most forest canopies to permit most forest trees to photosynthesize at rates high enough to compensate for respiration losses. Moderate canopy cover, as in pine and oak forests, allows ample light to reach the forest floor, and even dense canopy cover like that formed by spruce–fir and sugar maple–beech forests permits sunflecks and light at the edges of small gaps that support plant survival and growth in the understory. In general, light is the single limiting factor in understory survival only where light on the forest floor is less than about 2% full sunlight (Chapter 6).

Light is not the only environmental factor greatly modified by the forest canopy. When a forest canopy is dense, almost all climatic and soil factors differ from those on similar but open sites. Soil water in particular is among the environmental factors affected. The importance of soil water in regulating understory establishment and growth was demonstrated in a classic study in 1904 by Fricke, who cut the roots of competing understory trees by trenching around small, poorly developed seedlings in a Scots pine stand in Germany. The seedlings responded vigorously, indicating that their growth had been inhibited by soil-water limitation created by the competing roots of the overstory trees rather than by low light. These experiments were repeated with similar results under eastern white pine in New Hampshire (Toumey and Kienholz 1931), loblolly pine in North Carolina (Korstian and Coile 1938), and others.

Understory tolerance should not be attributed solely to soil water just as it should not be attributed to light alone. Both soil water and light as well as nutrients (Chapter 19) contribute to understory survival. Often soil water and light availability occur on opposite gradients, with dry sites correlated to open areas and shaded areas more mesic, leading many ecologists to the idea that shade and drought tolerance are inversely related to one another (Niinemets and Valladares 2006; Figure 13.16). The logic behind such an idea is that tolerance to one environmental factor incurs a physiological cost to the plant, making it difficult or unable to also tolerate additional stressors. Trade-offs between shade tolerance and flood tolerance have also been documented in floodplains of the Southeast (Battaglia and Shari 2006). Given that other factors of the understory environment, such as temperature and humidity, are also likely to influence understory tolerance as they become limiting, it is clear that understory tolerance includes much more than the effects of light availability.

The intricate association between the crown and roots of understory plants illustrates how shade and drought tolerance are related. Loblolly pine in North Carolina and elsewhere, for example, has relatively intolerant seedlings that seem to photosynthesize more than enough to counterbalance respiration losses. However, seedlings beneath loblolly pine canopies do not photosynthesize enough to permit their root systems to expand into the deeper soil. Without a root system of sufficient extent and depth, the seedlings struggle to compete with the roots of overstory

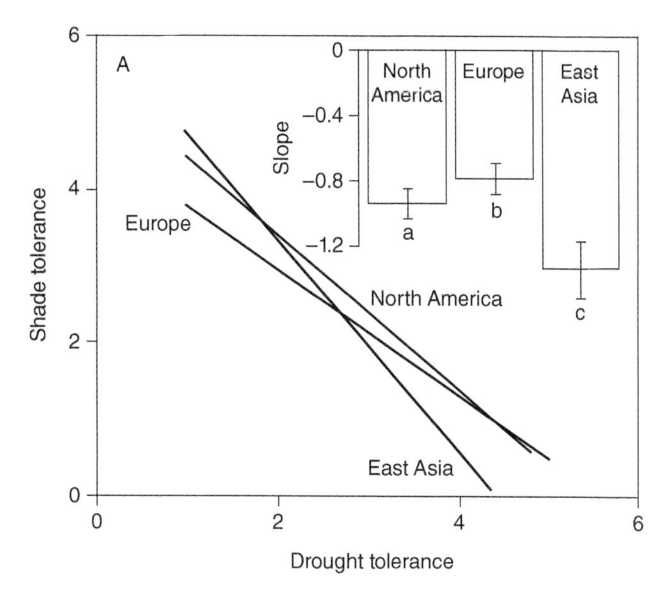

FIGURE 13.16 Inverse relationship of shade tolerance to drought tolerance for 806 woody species from three continents. Insets provide the slopes of the regression lines for each continent, with the steepest slope found for East Asian species. *Source:* Niinemets and Valladares (2006) / John Wiley & Sons.

trees and understory plants. As a result, they often develop somewhat weakened tops over time and may die during a period of hot, dry, midsummer conditions when soil-water stress becomes unusually severe. By contrast, mid-tolerant oaks and other hardwoods under similar conditions develop root systems extensive and deep enough for them to survive droughts.

While loblolly pine seedlings may require a rate of photosynthesis high enough to develop a sufficient root system in the moderate light conditions of the understory, either light or root competition may be relatively more important under other conditions. For example, light is at extremely low levels under dense Sitka spruce and western hemlock in temperate rain forests of the Pacific Northwest, but the site is almost always wet or at least moist such that light is obviously the more important factor. Under open oak woodland types or under dry ponderosa pine forests, light availability in the understory is usually well above any critical level but soil water is always in short supply, such that soil water is the more important factor. Always, it is the ecosystem-specific interaction of light, water, temperature, and other environmental factors as well that together determine understory survival and growth.

Physiological Processes Relating to Tolerance Basic genetic differences among species dictate that some understory plants can survive and grow under the site conditions of the understory while others cannot. Such genetic differences are expressed in species' physiological responses to these conditions. Based on the carbon gain hypothesis, any trait that improves a species' capability of capturing carbon will improve its understory (specifically, shade) tolerance (Givnish 1988), but the importance of many such traits in improving survival under low light is disputed. Valladares and Niinemets (2008) summarized 40 such traits related to the physiology, biochemistry, anatomy, and morphology of leaves; crown architecture; and whole plant characteristics. Notably, the expected values for shade-tolerant plants of 26 of the 40 traits have been challenged by at least one study. Logically, we would expect that in having adapted photosynthetic mechanisms to low light intensity, a shade-tolerant plant would exhibit traits to efficiently harvest and utilize limited available light (crown geometry and photosynthetic rate) and maintain low rates of respiration. For the purposes of this discussion, examples of characteristics of surviving plants compared to failing plants in the understory include: (i) greater photosynthesizing leaf area, (ii) more efficient photosynthesis per unit leaf area, (iii) lower rates of dark respiration and thus lower compensation points and light saturation points, and (iv) lower relative growth rates.

As discussed in the earlier text, tolerance to one environmental factor often leads to a trade-off in tolerance to another. A similar physiological trade-off is present in intolerant species, which have evolved highly adapted photosynthetic and respiration mechanisms to maximize their production in full sunlight at the cost of lowered efficiency under shaded conditions. The differences in rates of dark respiration between tolerant and intolerant species are probably the most important and least debated (Valladares and Niinemets 2008) determinants of success or failure in forest shade (Loach 1967; Walters and Reich 2000b; Craine and Reich 2005; Baltzer and Thomas 2007). Given that daily net carbon gain is the difference between photosynthesis and respiration, plants in a shaded understory may spend many more hours below than above even their lower light compensation point. Shade-tolerant plants are thus able to perform well in low light by minimizing their loss of carbon dioxide. By contrast, the high photosynthetic rates of intolerant species are offset to some extent by high rates of respiration (Chapter 18), and thus are less adapted to shaded conditions where photosynthetic rates are inevitably lower.

Photosynthetic rates have long been posited as an important mechanism for successful plant performance and survival in the shade (Logan and Krotkov 1969; Logan 1970). Tolerant species typically exhibit lower photosynthetic rates, presumably because of reduced photosynthetic constituents in the leaves such as RuBisCO, ATP synthase, and electron carrier per unit of leaf surface, which translates into lower carbon fixation (Givnish et al. 2004; Pallardy 2008). As such, shade-tolerant plants have a greater carbon gain and correspondingly better survival in low light, but tolerant plants tend to grow slowly relative to intolerant plants regardless of whether they are in the shade or in full sunlight. Notably, photosynthetic capacity (the maximum rate at which leaves are able to fix carbon during photosynthesis) in intolerant species is higher than that of tolerant species in both high and low light (Kitajima 1994; Walters and Reich 1996; Reich et al. 2003), but higher rates of respiration in intolerant plants reduce their net carbon gain. Thus, selection for a high rate of photosynthesis and high growth rate at high light irradiances may inevitably limit the plant in the shade.

Northern red oak was examined in a study of growth and CO_2 exchange of several northern hardwood species in controlled environments (Walters et al. 1993) that bear out the trade-offs between photosynthesis, growth rate, and respiration. The intolerant paper birch had a photosynthetic rate over 3 times that of the tolerant sugar maple, but also had respirations rates 1.5–2 times those of sugar maple (Table 13.5). Paper birch also had high relative growth rates compared to the tolerant sugar maple. Northern red oak is fire-dependent and mid-tolerant and may grow in both open and moderately shaded environments. Northern red oak was intermediate in relative growth rate, remarkably low in respiration, and had the highest ratio of root weight to total plant weight. Thus, the higher understory tolerance of red oak relative to paper birch incurs a cost of a lower relative growth rate. The relationship of photosynthesis and total plant respiration in determining the carbon balance of individual plants and whole ecosystems is considered in Chapter 18.

Table 13.5 Comparison of photosynthesis and respiration for tolerant and intolerant species.

	Photosynthesis (nmol CO_2 g^{-1} s^{-1})		Respiration (nmol CO_2 g^{-1} s^{-1})		
Species	**Leaf**		**Leaf**	**Stem**	**Root**
Paper birch	High light	225	22.9	23.2	37.5
	Low light	102	25.3	22.3	42.1
Northern red oak	High light	94	11.1	10.4	12.7
Sugar maple	High light	69	14.8	10.0	20.5

Photosynthesis is net leaf CO_2 assimilation expressed on a leaf mass (nmol CO_2 g^{-1} s^{-1}) base of seedlings grown in high and low light conditions in growth chambers. Respiration is in mass-based values (nmol CO_2 g^{-1} s^{-1}) as overall seedling means.
Source: Walters et al. (1993) / Springer Nature. Reprinted from Oecologia.

SUGGESTED READINGS

Austin, M.P. and Smith, T.M. (1989). A new model for the continuum concept. *Vegetatio* 83: 35–47.

Bronstein, J.L. (2009). The evolution of facilitation and mutualism. *J. Ecol.* 97: 1160–1170.

Gleason, H.A. (1926). The individualistic concept of the plant association. *Bull. Torrey. Bot. Club* 53: 7–26.

Grubb, P.J. (1985). Plant populations and vegetation in relation to habitat, disturbance and competition: problems of generalization. In: *The Population Structure of Vegetation*, Handbook of Vegetation Science, Part III (ed. J. White). Dordrecht: W. Junk.

Kitajima, K. (1994). Relative importance of photosynthetic traits and allocation patterns as correlates of seedling shade tolerance of 13 tropical trees. *Oecologia* 98: 419–428.

McIntosh, R.P. (1993). The continuum continued: John T. Curtis' influence on ecology. In: *Fifty Years of Wisconsin Plant Ecology* (ed. J.S. Fralish, R.P. McIntosh and O.L. Loucks). Madison: University of Wisconsin Press.

Niinemets, Ü. and Valladares, F. (2006). Tolerance to shade, drought, and waterlogging of temperate northern hemisphere trees and shrubs. *Ecol. Monogr.* 76: 521–547.

Diversity in Forests

CHAPTER 14

The diversity of organisms, how to measure it, and hypotheses about its causes have long interested researchers studying all types of ecological systems. Exploring natural forests, inventorying species, searching for rare species, and posing questions such as why a species is present, rare, or abundant at some sites but not at others are fundamental to forest ecologists. "Diversity" simply means "variety" in ecology, but the concept quickly becomes complex with attempts to make it quantitative or comparing it among areas. At its simplest level, studying diversity is a descriptive pursuit in which species are listed or counted.

Diversity may be studied at many different levels of organization, from genes to ecosystems. In this chapter, our treatment of diversity is primarily at organism and ecosystem levels. Diversity is closely intertwined with many interacting topics covered individually in this book, including paleogeology and ecology, ecosystem geography, plant physiology, and human activities. Moreover, we again emphasize the physical and ecological processes of forests whose ultimate result includes the diversity of ecosystems and organisms that inhabit them. Our overview of diversity is necessarily limited, and readers are encouraged to delve deeply into books that treat organismal diversity (Wilson 2010; Ricklefs and Schluter 1993; Huston 1994; Rosenzweig 1995; Gaston and Spicer 2014), measurement of species diversity (Krebs 1998; Stohlgren 2007; Magurran and McGill 2011; Magurran 2013), and conservation biology (Meffe and Carroll 1997; Groom et al. 2006; Primack 2010; Van Dyke and Lamb 2020; Hunter and Gibbs 2021).

CONCEPTS OF BIOLOGICAL AND ECOSYSTEM DIVERSITY

The natural world contains many different types of diversity. For example, the diversity of physical features of bedrock, landform, geomorphic processes, parent material, and climate are strong drivers of the growth and distribution of plants. The diversity of biota—plants and animals, their gene pools, and their aggregation in populations and communities—is of great interest and concern. We define **biological diversity (biodiversity)** as the types and numbers of organisms and their patterns of distribution. In addition to the organisms themselves, this definition also includes the genetic diversity of gene pools and other types of diversity related to groupings of biota, such as the diversity of families and genera, populations and communities, and "structural" properties of communities (vertical layering, plant density, patchiness, etc.).

Biota are not ecosystems, and thus biodiversity should be distinguished from ecosystem diversity, although landscapes and ecosystems are often included as part of "biodiversity." We define **ecosystem diversity** as the kind and number of ecosystems in an area, including the patterns of association of certain ecosystems with one another and the recurrence of these patterns on a given landscape. Ecosystem diversity can be estimated at multiple spatial scales using the same diversity indices used to measure species diversity (Romme 1982). Maintaining and conserving biodiversity depend on our understanding of ecosystem diversity because species belong to ecosystems.

Forest Ecology, Fifth Edition. Daniel M. Kashian, Donald R. Zak, Burton V. Barnes, and Stephen H. Spurr.

THE VALUE OF SPECIES DIVERSITY

Scientists today have relatively little knowledge of the Earth's biodiversity (Solbrig 1991), and estimates of the Earth's biota seem to be continually updated. For example, Wilson (2010) estimated that 10–100 million species existed, only 1.4 million of which had been described and identified. He estimated that 73% of the Earth's biota were animals, mostly insects (53% of the total), and 18% plants, mostly angiosperms. The fact that the combined diversity of insects and flowering plants is so large reflects the intricate symbioses that unite these groups (Wilson 2010), and in fact a large fraction of plants depend on insects for pollination and dispersal (Chapter 12). Wilson's estimates have been refined in recent decades; one well-cited study estimates that 8.7 million eukaryotic species exist on Earth, with animals representing an even higher proportion (91%) of the 1.2 million species identified, though still mostly insects (Mora et al. 2011). This estimate remarkably suggests that 86% of the Earth's species have yet to be discovered and described, which would require almost 500 years to complete at the current pace of scientific progress (May 2011). An even more recent estimate suggests that the world's biodiversity is dominated by bacteria (78%), with animals (mostly insects) representing only 7% and plants less than 1% (Figure 14.1; Larsen et al. 2017). These estimates were in part made based on estimates of the number of parasitic species, many of which are bacteria, likely to be found to be specific to each insect species. As the technology for genome sequencing refines the boundaries between species, new estimates of biodiversity will surely come to light.

Where does biodiversity exist in forests? Admittedly in this book we draw the most attention to plants, and in particular to woody plants. Biodiversity in forests is found mainly in understory plants (rather than the dominant trees) and animals that use forests as habitat. However, most of the diversity of forests is found in soil organisms. Soil biodiversity per unit area is often far greater than aboveground biodiversity (Nielsen et al. 2015). Soil microbial organisms (most fungi and bacteria) probably represent a pool of diversity of tens of thousands of different species per gram of soil (Roesch et al. 2007), depending on soil type, climate, and the researcher's definition of what makes a microbial species. Soil invertebrates such as insects, springtails, arachnids, nematodes, earthworms, and protists also add to the belowground biodiversity pool. Despite our focus mainly on plants and other aboveground diversity in this chapter, we acknowledge the enormous diversity of life below ground.

Despite estimates of a seemingly extraordinary number of organisms, the vast majority of plant and animal species that have ever lived on the Earth are extinct (Raup 1986). As many as four billion species of plants and animals alone are thought to have lived at some point in the geologic

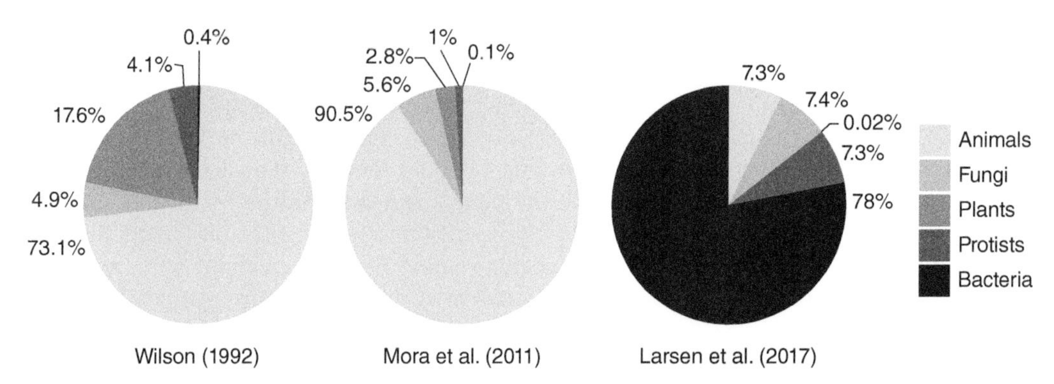

FIGURE 14.1 Three estimates of the relative proportion of different groups of organisms. Earlier estimates suggest that animals (mostly insects) dominate Earth's biodiversity, but recent estimates suggest that bacteria are the dominant group. *Source:* From Larsen et al. (2017) / with permission of University of Chicago Press.

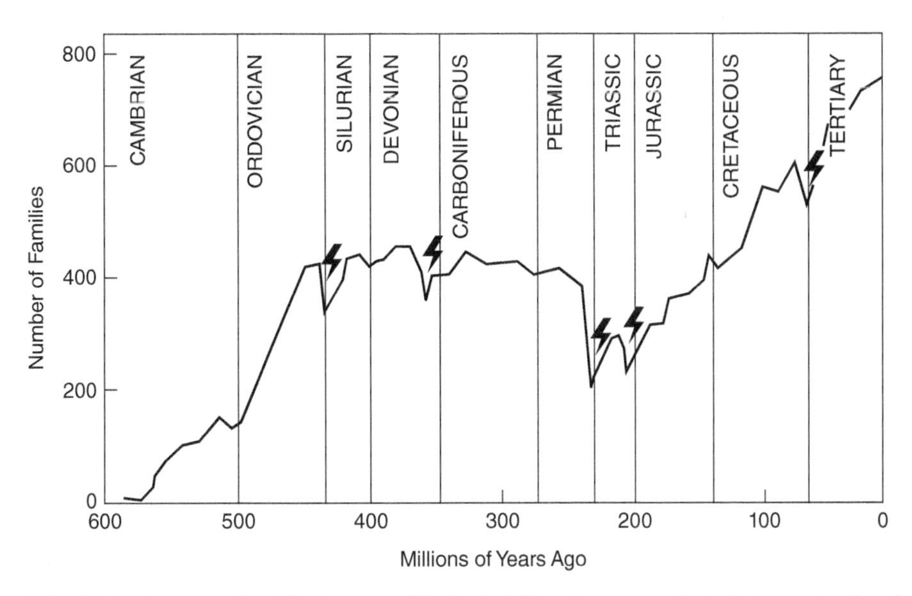

FIGURE 14.2 Example of the increase of biological diversity over geological time using data from families of marine organisms. A slow increase is seen, with occasional setbacks through mass global extinctions. There have been five such extinctions so far, indicated here by lightning flashes. A sixth major decline is now underway as the result of human activity. *Source:* THE DIVERSITY OF LIFE by Edward O. Wilson, Cambridge, Mass.: The Belknap Press of Harvard University Press, Copyright © 1992 by Edward O. Wilson. Used by permission. All rights reserved.

past (Simpson 1952), but far fewer are thought to exist today. Thus, species extinction has been almost as common as origination. Much evidence exists that major extinction events, like that at the end of the Cretaceous, are regularly spaced in geologic time (Raup 1986), but smaller turnovers occur as well. The trend for biodiversity has been upward for the past 600 million years even in the face of such extinctions (Figure 14.2). Global biodiversity of plants (Figure 14.3) peaked in the Cenozoic because: (i) the aerobic environment was created, (ii) land masses became fragmented, and (iii) species were increasingly pushed into regional and local ecosystems of the developing landscape (Wilson 2010). The number of plant species in local floras has more than tripled in the past 100 million years (Figure 14.3). Despite this seemingly exploding biodiversity, a sixth great extinction event is thought by many to be currently occurring (Wilson 2010). In this context, we consider the value of biodiversity, as well as the increasing role of biodiversity in shaping the sciences of ecosystem conservation, management, and restoration.

VALUE OF BIODIVERSITY

There are many reasons for conserving, promoting, and managing biodiversity (Berry et al. 2018; Burton et al. 1992), and a review of the extensive literature confirming the value of biodiversity is well beyond our scope here. Biodiversity maintains or provides important **ecosystem services**, or the benefits that ecosystems provide to humans (Cardinale et al. 2012; Chapter 23). Biodiversity has many ecological benefits for ecosystems and their functioning (Lefcheck et al. 2015; Cardinale et al. 2006; Chapin et al. 2000), some of which we discuss later in this chapter. A sample of such benefits might be renewable resources (Cardinale et al. 2012; Piotto 2008), reducing plant invasion (Quijas et al. 2010; Levine et al. 2004), reducing insect pests (Letourneau et al. 2009), improving soil quality (Quijas et al. 2010), increasing carbon sequestration (Cardinale et al. 2012), and indicating the loss of ecosystem integrity (Burton et al. 1992), among many others.

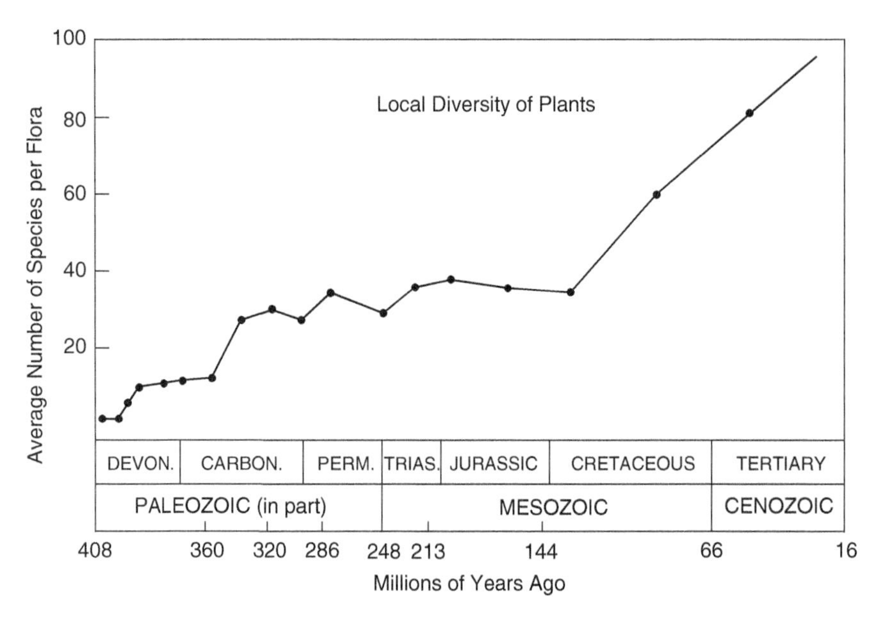

FIGURE 14.3 The average number of plant species found in local floras over geologic time. The number of plants found in local floras has risen steadily since the invasion of the land by plants 400 million years ago. The increase reflects a growing complexity in terrestrial ecosystems around the world. *Source:* THE DIVERSITY OF LIFE by Edward O. Wilson, Cambridge, Mass.: The Belknap Press of Harvard University Press, Copyright © 1992 by Edward O. Wilson. Used by permission. All rights reserved.

Biodiversity has been directly related to the critical ecosystem service of human health. Opportunities for identifying natural products for future pharmaceutical value increase with species diversity (Chivian and Bernstein 2008; Mendelsohn and Balick 1995). Pharmaceutical compounds derived from wild plants are often of global importance. The Madagascar periwinkle, for example, is used in the treatment of childhood leukemia and Hodgkin's disease, and is reportedly worth millions in sales each year (Shiva 1990). Likewise, the western yew tree is the source of taxol, a compound found to have strong activity against a number of cancers (Wani et al. 1971). High biodiversity is also thought to be associated with reduced transmission and emergence of diseases that affect humans (Keesing and Ostfeld 2015), such as zoonotic diseases (Keesing and Ostfeld 2021; Keesing et al. 2010) and lyme disease (Ostfeld and Keesing 2000). Moreover, green space and wilderness that support and maintain biological diversity can promote human well-being (Easley et al. 1990; Kaplan 1992; Thompson and Barton 1994; Gaston et al. 2007), an important part of human mental health.

Economic value may be placed on the ecosystem services that biodiversity provides (Paul et al. 2020; Hanley and Perrings 2019; Edwards and Abivardi 1998; Gowdy 1997), although the valuation of biodiversity remains an important discussion among environmental economists. Costanza et al. (1997) estimated the market value of the world's ecosystem services to be about US\$33 trillion/year (10^{12}), which was more than the global gross national product at the time (US\$18 trillion/year). At that time, forests contributed 14% of the \$33 trillion, and wetlands (in part forested swamps and floodplains) contributed another 15%. A revised estimate in 2011 showed a global loss in ecosystem services due to land use and management changes (Costanza et al. 2014). Moreover, the future value of ecosystem services may increase by as much as US\$30 trillion/year or decrease by as much as US\$51 trillion/year depending on land use and management scenarios (Kubiszewski et al. 2020). This bottom-line approach provides a powerful incentive for conservation of the natural capital stock. Notably, many ecologists and others would additionally, or alternatively, cite very real and intrinsic reasons for sustaining the Earth's ecosystems, reasoning that the existence of species and ecosystems has value irrespective of humans (e.g., Leopold 1949; Regan 1981; Norton 1982; Naess 1986; Rowe 1990; Ghilarov 2000; Alho 2008; Justus et al. 2009; Fearnside 2021).

COMMON THREATS TO DIVERSITY

Species extinction is the most well-known human impact upon biodiversity, and it has escalated with increasing human populations and their need to utilize natural ecosystems for agriculture, forest management, and urbanization. The main threat to diversity in North America and around the world is therefore the human population, particularly as it increases in size. All threats to diversity stem from humans either directly or indirectly. A detailed discussion of threats to diversity is well beyond our scope, but the loss of species has occurred largely because of four main human impacts:

Direct exploitation of species by humans, either by removing every last individual of a species or by reducing their populations to the point that they have little resilience to events that might cause their extinction. Trees or other woody plants are less commonly exploited to extinction by humans than are animals. However, many conservationists have raised concerns about fuelwood collection in developing countries whose increasing demand is likely to lead to species extinctions in areas where it is concentrated (Gaston and Spicer 2014).

Habitat loss and fragmentation is perhaps the most widespread threat to diversity in forests. Forest loss via conversion to agriculture and development is extensive worldwide, estimated as a 29% decrease since significant human disturbance at the turn of the twenty-first century (Klein Goldewijk 2001). Similar land-use changes are likely to occur in the future not only because of continued human population growth but also due to climate change (Chapter 20). Remaining forests are likely to support fewer species due to their fragmentation into small, isolated patches (Chapter 22).

Invasive species that have the potential to outcompete native species or alter the ecosystems in which native species occur are an increasingly important threat to diversity (Chapter 21). Both native and introduced species respond to similar site factors in successfully establishing in an ecosystem (Sax 2002), and thus ecosystems with higher native species richness also tend to have more invasive species. In a summary of the 100 worst invasive species across the world based on their impacts on biodiversity and human activities (Baskin 2002), 25 of the species were woody land plants and 4 were insects or pathogens of forest trees.

Climate change is a human impact likely to have enormous consequences for biodiversity in the next 50–100 years. Changing climate is predicted to alter the distribution of many species at a rate too high for species to adapt to the changing environmental conditions. Species extinctions due to climate change have been predicted to reach as high as 54% (Urban 2015). Although some species are likely able to disperse to new suitable habitat, a recent study of 538 plants and animals by Román-Palacios and Wiens (2020) suggested that 57–74% of the species would not disperse quickly enough to avoid extinction. However, they predicted that species adaptations were more likely to occur than previously thought, such that only 16–34% of the species would go extinct in the next 50 years. The effects of climate change on forests are described in detail in Chapter 20.

MEASURING DIVERSITY

LEVELS OF DIVERSITY

Diversity includes two important components: richness and evenness. Richness refers to the number of units (alleles, species, families, communities, ecosystems) per unit area, and evenness refers to their abundance or dominance relative to one another. These two concepts can be applied and measured at many levels of organization: ecosphere and continental-level ecosystems, regional ecosystems, local ecosystems, species, and the genetic diversity of allele frequencies within populations of a species. Species are typically the foci of biodiversity measurement because they are easily observed, and the most common taxonomic unit used in ecological studies of forests. The complexity of forest ecosystems is of note, however, because the patterns of diversity for trees may not reflect those of ground-cover vegetation, birds, or microorganisms.

Biodiversity is typically studied at the local level in areas of <25 ha, but may also be examined at broader spatial scales. Diversity at broad scales is often referred to in the literature as landscape or regional diversity. Here, the focus may be on relatively broad vegetative cover types such as those in Yellowstone National Park (Romme 1982; see Chapter 22) or the White Mountains of New Hampshire (Reiners and Lang 1979). At broad spatial scales, diversity results from the interaction of multiple patterns of: (i) species distributions along gradients of limiting factors, (ii) recurrent spatial patterns of local ecosystems, and/or (iii) communities in different stages of succession following disturbance. Diversity of taxonomic units above the species level (i.e., genera, family, order) are often examined in continental-level studies (Currie and Paquin 1987; Latham and Ricklefs 1993a, b). At this spatial scale, diversity is related to macroclimate, physiography, continental drift, and the historical development of plant floras. By contrast, observed patterns of richness and evenness at local scales (<25 ha) are determined by interspecific interactions and local site factors of microclimate, landform, and soil.

Whittaker (1960, 1972, 1975, 1977) was among the first ecologists to recognize that species diversity may be measured in several ways and at several spatial scales. First, he recognized four increasingly larger spatial scales at which **inventory diversity** could be measured:

Point diversity: at a microsite scale with samples taken from within a homogeneous site, that is, a microsite within a local landscape ecosystem type (e.g., 100–500 m^2).

Alpha diversity: at a homogeneous site, that is, for a given landscape ecosystem type or community in a homogeneous site (micro-ecosystem level, e.g., <1 to several hundred hectares). Simply, it is defined as the variety of organisms at a particular site. It is sometime referred to as *local diversity* in the literature.

Gamma diversity: at a larger landscape unit, that is, for an island or a regional ecosystem. Gamma diversity includes the diversity of all ecosystems within a region (e.g., 625–2500 ha).

Epsilon diversity: total diversity for an area encompassing a group of areas of gamma diversity, that is, macro-ecosystem level (e.g., >2500 ha).

As an analogy, Magurran (1988) described a leaf as point diversity, a plant as a unit of alpha diversity, a group of plants as gamma diversity, and the entire forest containing the group of plants as epsilon diversity.

Whittaker also recognized that species composition changes across the landscape, both along environmental gradients, and from one generally homogeneous site to the next. This change from one area to the next, regardless of scale, he called **beta diversity**. Beta diversity measures how different two or more areas are from each other based on the variety of species present. Beta diversity may compare the similarity of sites using species distinction, or it may relate species richness across different spatial or temporal scales. Jurasinski et al. (2009) proposed that the first of these versions of beta diversity be termed **differentiation diversity**, while the second should be termed **proportional diversity**.

Diversity may also be used to describe the structural diversity of communities as well as their composition. In forested systems, this analysis focuses on the spatial pattern of species occurrence (random, regular, or patchy), stand density, diameter class distribution, and the vertical layering of a forest. Structural diversity is of considerable importance to wildlife in that the number of animals capable of using a forested system depends on the presence of appropriate areas for nesting, feeding, resting, and hiding. Structural diversity for old-growth, mixed-age or -size, and mixed-species stands is typically high, whereas that for plantations or single-age, single-species stands is low. Storch et al. (2018) developed an index of structural diversity for forests in Baden-Württemberg, Germany using 11 variables relating to tree diameter, tree height, deadwood decay classes, deadwood diameter, bark diversity, diversity of flowering and fruiting, and tree species richness. Expressed as a value between zero and one, the index may be used to examine structural diversity across large-scale forest inventories to support biodiversity monitoring. Similarly, McElhinny et al. (2006) developed a structural diversity index for dry forests of Australia that included 13 variables related to vegetation cover of forest strata, perennial species richness, life-form richness, diameter and basal area of live trees, number of hollow or dead trees, length of deadwood, and litter dry weights. The index was then used to compare and differentiate among forest structures across different sites.

MEASUREMENT

Inventory Diversity: Alpha Diversity There are various ways to measure species diversity. Perhaps the simplest measurement is to count the number of species present in a designated area. This number is **species richness:** the oldest, most fundamental, and least ambiguous of the diversity measurements (Peet 1974). Richness varies with the area sampled, so it is often expressed as number of species per unit area. Richness will increase with the area sampled up to a point where most or all species have been captured in the sample. Increasing the sample area at this point does not increase the species number. In using sample plots to determine the richness for a given ecosystem, it is therefore important to use plots large or numerous enough to encompass most of the species. The total number of species for most major organism groups is difficult to determine for large, forested areas. The species–area relationship is consistent for homogeneous areas, such as a single ecosystem type with relatively little microtopographic variation, relatively small gaps or disturbance features, and uniform climate. Once the sampled area expands beyond a single local ecosystem, however, environmental heterogeneity increases, and different ecosystems with different species are included in the sample. It is therefore important to recognize ecosystem boundaries if reliable estimates of alpha diversity of species are to be obtained.

Diversity also depends on the distribution of species across an area, termed **evenness**, or the degree to which all species share dominance in an area. For example, forest stands A and B in Figure 14.4 both have eight species and thus the same richness. Stand A is dominated by one or two species, with the other species being relatively uncommon, whereas stand B exhibits a more equal abundance of all species and thus exhibits greater evenness. In this case, stand B is more diverse because it has the more equal abundance of the species with the same number of species. Two measures of diversity in Figure 14.4 that combine richness and evenness into one index—Shannon's and Simpson's diversity indices—are higher for stand B than for stand A.

Many diversity indices summarize richness and evenness into a single number, each with its theoretical advantage under specific circumstances. Two of the most commonly used indices for measuring species diversity are Simpson's index and the Shannon–Wiener (or Shannon's) index. The Shannon–Wiener index is probably the most commonly used diversity index, and is computed as:

$$H' = -\sum_{i=1}^{S} p_i \ln(p_i),$$

where S is the number of species in the sample, and p_i is the proportion of individuals that belong to species i. H' generally ranges from a low of around 1.5 to a high of 4.5, rarely exceeding this value (Margalef 1972). The Shannon–Wiener index is sensitive to the number of species in a

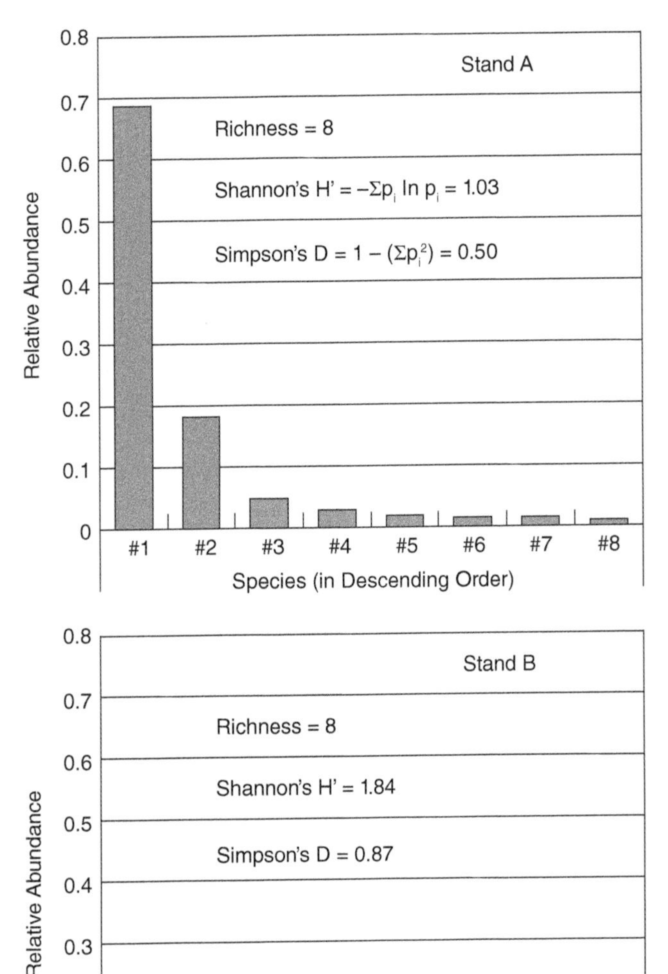

FIGURE 14.4 Comparison of species' alpha diversity exhibited by two hypothetical stands having the same number of tree species. Species richness for each stand is eight. Two diversity indices are given, where p_i is the proportional abundance of each species, *i*, and the summation is performed over all eight species. Because stand B exhibits greater evenness, with less dominance by any one species, it has greater overall diversity than stand A, as measured by the indices. *Source:* Reprinted with permission by the *Forestry Chronicle* from Burton et al. (1992).

sample, so it is biased toward measuring species richness (Figure 14.4). The evenness component of H' is computed as (Magurran 1988):

$$E = \frac{H'}{H'\max} = \frac{H'}{\ln S}.$$

In this computation, the Shannon–Wiener index is scaled by its maximum possible value (H' max), which would occur if all species had equal abundance and happens to be the natural log of S.

Simpson (1949) developed an index of diversity that is computed as:

$$D = \sum_{i=1}^{S} \left(\frac{n_i (n_i - 1)}{N(N-1)} \right),$$

where n_i is the number of individuals in species i and N is the total number of species in the sample. Diversity is inversely related to D, so Simpson's index is usually expressed as its complement, 1-D or 1/D. Simpson's index is sensitive to the most common or dominant species in a community and is relatively insensitive to rare species. For this reason, Simpson's index, or its complement (1-D), is sometimes used as a measure of evenness (Figure 14.4).

Given reasonable standardization, multiple diversity indices computed for the same data tend to be correlated, and sampled areas will rank similarly in diversity as measured by various indices. However, diversity indices vary in how they are computed and interpreted, so it is inappropriate to compare diversity measured by one index with that measured by another. Diversity values will also vary depending on which organisms or forest strata (overstory; ground cover, all layers) are sampled, the specific area sampled, and the method of sampling. Thus, the diversity index sensitive to the factors of interest in the study should be selected.

Diversity may also be assessed with species abundance models, which express commonly observed abundance distributions of species. Examples of such models include the log normal distribution, the geometric series, the logarithmic series, and MacArthur's broken-stick model (Magurran 1988). A common technique is to plot species abundance or dominance against their rank order of abundance, dominance, or coverage. These species-abundance plots can be used to compare not only the relative richness of communities but also to provide insight into evenness. Figure 14.5 shows species-abundance curves for ground-cover plants sampled in two very different ecosystem types on glaciated terrain in northern Lower Michigan: (i) an excessively drained, infertile, sandy outwash plain supporting a sparse, short overstory of small bigtooth and trembling aspens (ecosystem type 1), and (ii) a moderately well-drained, loamy, fertile moraine supporting a moderately dense canopy of tall bigtooth and trembling aspens, white ash, paper birch, and basswood (ecosystem type 116). In this example, species abundance is measured as percent coverage. Ecosystem type 1 shows lower ground-cover species richness (37 versus 69 species in ecosystem type 116) and a sharp decrease in dominance from the most dominant species (mean coverage = 44%) to the second most dominant species (mean coverage = 2%). Such a plot is characteristic of ecosystems extreme in factors such as microclimate, fire frequency or severity, light level, or fertility, and in an early-successional stage. By contrast, ecosystem type 116 has nearly twice the species richness as type 1 and its plot exhibits a more gradual drop in dominance, from about 3.3% to <1%, from the most to least dominant species (Figure 14.5b). This plot is more typical of ecosystems that are not at the extremes of microclimate, fertility, light, or disturbance, and are in a mid-successional stage.

Differentiation or Beta Diversity If the interest is in measuring how different or similar the variety of species is across a range of samples (Magurran 1988), then beta diversity is estimated. Communities of a given area that have very few species in common collectively have high beta diversity, and vice versa. This difference in species variety may be determined for ecosystems or communities that occur along an environmental gradient or within an area supporting diverse sites. Because the term "beta diversity" gives little clue to the spatial or ecological context of the diversity that is measured, beta diversity may be usefully understood as differentiation diversity (Whittaker 1977; Magurran 1988)—a measure of species diversity along a gradient containing different ecosystems or throughout an area where multiple ecosystems occur.

Beta diversity is typically measured by determining the similarity in species composition between pairs of communities or ecosystems. Six methods of measuring beta diversity are assessed and discussed by Wilson and Shmida (1984) and Magurran (1988). Two basic approaches are similarity coefficients and measuring change along gradients using a variety of indices. Percentage similarity (PS) and the Jaccard and Sorensen similarity coefficients are widely used where the presence or absence of species is known. These methods relate the number of shared species to the total number of species in two samples. For example, if two samples are taken and each has the same 10 species, the similarity is 100%, but if only five species are shared, then the similarity is only

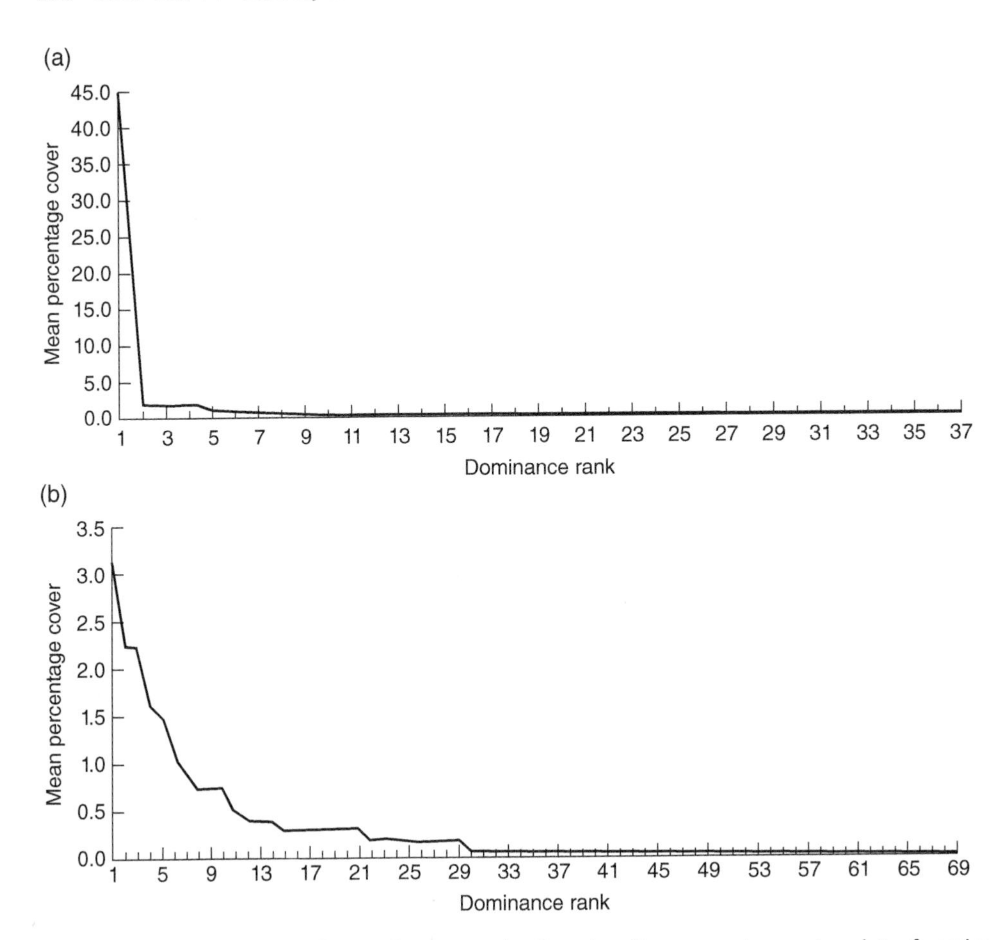

FIGURE 14.5 Comparison of ground-cover species diversity of two ecosystems using plots of species abundance: (a) Ecosystem type 1 occurs on an outwash plain landform with excessively drained, infertile, sandy soil. (b) Ecosystem type 116 occurs on a moraine with moderately well-drained, fertile, loamy soil. Abundance is measured by percentage cover. Ground-cover species for each ecosystem are ranked in order by their percent coverage along the x-axis (1 = species with greatest abundance). Richness is indicated by the number of species shown at the far right on the x-axis. Evenness is indicated by the shape of the curve, with ecosystem 116 being more "even" in distribution than ecosystem type 1. *Source:* Pearsall (1995) / Pearsall.

50%. This example of percentage similarity is a qualitative (presence/absence) estimate of beta diversity; the Sorensen index contains a quantitative estimate that can be used when abundance data are available. Another quantitative measure, the Morista–Horn coefficient, is independent of sample size and thus is preferred among all similarity measures (Wolda 1981).a

DIVERSITY OF LANDSCAPE ECOSYSTEMS

Most diversity studies have focused primarily on species, especially the alpha and beta diversities defined by Whittaker (1977). Rather than strictly biota, ecosystem diversity is concerned with local and regional landscape ecosystems. Diversity of landscape ecosystems can be a surrogate for biodiversity (Bailey et al. 2017), and conservation of "geodiversity" is considered as important as conservation of biodiversity (Shaffer 2015; Schrodt et al. 2019) as exemplified in the "Conserving the Stage" approach (Anderson and Ferree 2010) that is now the basis for The Nature Conservancy's conservation planning for the continental US (Anderson et al. 2016). Diversity exists and can be treated at a variety of levels of ecological organization, but applying the concepts of diversity to the hierarchically nested series of ecosystems or to vegetation types throughout a large landscape is relatively uncommon. Exceptions are Romme's (1982) study of forest ages across the landscape in Yellowstone National Park (Chapter 22), Loehle and Wein's (1994) study of forest vegetation types in western Tennessee, and Pearsall's (1995) study of ecosystem diversity in northern Lower Michigan. Recent work by Anderson et al. (2014, 2016) has incorporated ecosystem diversity—defined as "the variety of landforms created by an area's topography, together with the range of its elevation gradients"—into a landscape model of climate resilience. Distinguishing and mapping landscape ecosystems at multiple scales provide the spatial framework for determining ecosystem diversity at multiple spatial scales.

EXAMPLES OF DIVERSITY

Ground-Cover Species Diversity in Northern Lower Michigan In the following text, we examine measures of species diversity on a large landscape in northern Lower Michigan. The purpose is not only to compare various measures of alpha and beta diversity but also to examine diversity at several ecosystem spatial scales (major and minor landforms) and soil types, by groups of ecosystems, and by individual ecosystem types. We also consider ecosystem diversity.

Ecosystem Groups Landscape ecosystems of multiple scales were distinguished and mapped at the 4000-ha landscape of the University of Michigan Biological Station in northern Lower Michigan (Lapin and Barnes 1995; Pearsall 1995; Zogg and Barnes 1995). The terrain was shaped by two late-Wisconsin glacial advances and retreats between 15 000 and 13 000 years ago. The area was highly disturbed by logging and repeated post-logging fires in the late nineteenth and early twentieth centuries. Early-successional species, primarily bigtooth and trembling aspens, now dominate the overstory of upland ecosystems, having replaced many kinds of upland and lowland conifers and hemlock–northern hardwood forests common in the pre-European colonization forest over 200 years ago.

Landscape ecosystems were distinguished using physiography, microclimate, soil, and ground-cover vegetation, and mapped at four hierarchical scales: major landforms, minor landforms, ecosystem groups, and local ecosystem types. The three major landforms of outwash plain, ice-contact terrain, and interlobate moraine were initially subdivided into 21 ecosystem groups (Lapin 1990). Six of these 21 groups were used to analyze alpha and beta diversity of ground-cover vegetation (Figure 14.6). The key site features of the six ecosystem types are given in Table 14.1.

Ground-cover species richness ranged from 27 species to 16 per 100 m^2 plot, and Shannon–Wiener heterogeneity ranged from 2.21 to 0.61 (Table 14.2). Ecosystem group 20 (moist and nutrient-rich) was the richest and most heterogeneous, and ecosystem group 4 (dry and nutrient-poor) was the least diverse. Ecosystem groups were ranked in alpha diversity as follows, from more diverse to less diverse: (i) group 20 (moist and nutrient-rich), (ii) group 17 (moderately moist and moderately nutrient-rich), (iii) group 11 (dry and nutrient-rich), (iv) group 2 (moist, nutrient-poor, and climatically extreme), (v) group 1 (dry, nutrient-poor, and climatically extreme), and (vi) group 4 (dry and nutrient-poor) (Tables 14.1 and 14.2). Multiple-comparison methods often showed significant pairwise differences (Table 14.2).

Species richness and diversity tended to increase with moisture and nutrient availability, as long as factors such as temperature, light, and herbivory were not limiting. Fire-regenerated aspens that currently dominate the overstory of all ecosystem groups provide a favorable light environment for many vascular plants. Pre-European colonization forests—dominated by eastern white pine, American beech, and eastern hemlock (Kilburn 1960)—were probably denser than today's relatively open canopies of aspens. As such, ground-cover diversity measures applied to forests as they were prior to the arrival of Europeans are likely to yield very different results compared to the same areas today.

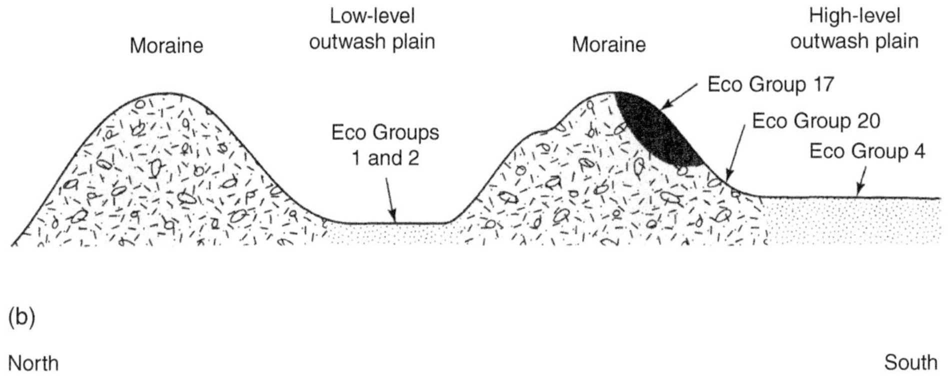

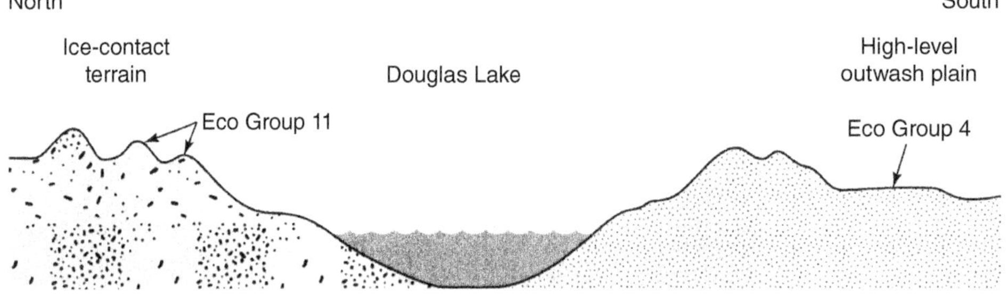

FIGURE 14.6 Physiographic diagrams illustrating the location of six ecosystem groups on major landforms (outwash plain, moraine, ice-contact terrain) of the University of Michigan Biological Station, Emmet and Cheboygan counties, northern Lower Michigan: (a) West–east transect south of Douglas Lake showing location of ecosystem groups 1, 2, 17, 20, and 4. The low-level outwash plain (Pellston Plain) is a huge frost pocket that is over 3 km wide between high interlobate moraines; vertical scale exaggerated. (b) North–south transect showing the location of ecosystem groups 11 and 4. *Source:* Lapin and Barnes (1995). Reprinted by permission of Blackwell Science, Inc.

Table 14.1 Comparative summary of site conditions of six landscape ecosystem groups of the University of Michigan Biological Station, Emmet and Cheboygan Counties, northern Lower Michigan.

Ecosystem group[a]	Site condition	Physiography	Soil and drainage
20	Moist, nutrient rich	Interlobate moraine; flat and moderate slopes	Loamy sand over clayey calcareous till; well drained
17	Moderately moist, moderately nutrient rich	Interlobate moraine; flat	Sandy soil with many heavy-textured bands; noncalcareous; well drained
11	Dry, nutrient rich	Ice-contact terrain	Calcareous, gravelly medium sand, somewhat excessively drained
2	Dry to seasonally moist; nutrient poor; wide daily and seasonal temperature extremes	Low-lying outwash plain between two moraines; flat	Deep, non-calcareous medium sand; moderately well drained
1	Dry, nutrient poor; wide daily and seasonal temperature extremes	Low-lying outwash plain between two moraines; flat	Deep, non-calcareous medium sand; excessively drained
4	Dry, nutrient poor	High-level outwash plain; flat	Deep, non-calcareous medium sand; excessively drained

[a]Pre-European colonization cover type—groups 20, 17, and 11: hemlock–northern hardwoods; groups 1, 2, and 4: eastern white pine, red pine, and northern red oak.

[a]Present cover type—groups 20, 17, and 4: bigtooth aspen; groups 1 and 2: trembling and bigtooth aspens; group 11: northern hardwood species and bigtooth aspen.

Source: Lapin and Barnes (1995). Reprinted with permission of Blackwell Science, Inc.

Table 14.2 Comparison of alpha diversity indices for ground-cover plant species among six ecosystem groups of the University of Michigan Biological Station, Emmet and Cheboygan Counties, northern Lower Michigan.

	Ecosystem group[a]					
Index[b]	**20**	**17**	**11**	**2**	**1**	**4**
S_{GC}	27.2	18.4	17.2	23.6	19.4	15.8
	a	bc	bc	ac	bc	b
S_H	14.2	7.8	9.0	10.2	9.4	7.2
	a	b	b	b	b	b
S_W	13.0	10.6	8.2	13.4	10.0	8.6
	a	abc	c	ab	ac	c
H'	2.21	1.66	1.15	0.99	0.70	0.61
			a	ab	b	b
1-D	0.827	0.704	0.537	0.420	0.268	0.241
	a	ab	bc	cd	cd	d
E	0.68	0.57	0.41	0.31	0.23	0.22
	a	a	b	bc	c	c

[a] Number of plots = 5 for all groups. There was no significant difference between groups that share a letter, alpha = 0.1.
[b] S_{GC} = ground-cover species richness, S_H = herbaceous species richness, S_W = woody species richness, H' = Shannon–Wiener heterogeneity, 1-D = Simpson's heterogeneity, E = Shannon–Wiener evenness.
Source: Lapin and Barnes (1995). Reprinted with permission of Blackwell Science, Inc.

Diversity measures based on ground-cover vegetation show a much higher level of diversity than would be seen by examining only the overstory vegetation. On this landscape, there are significant differences in ground-cover composition with an overstory dominated only by two closely related tree species—bigtooth and trembling aspens. This relationship emphasizes that alpha and beta species diversity to some extent are driven by how sample plots are defined and bounded. If plot boundaries are set by community type (i.e., by the bigtooth aspen cover type), it may be assumed that any one plot will be representative of the whole cover type. If plot boundaries are determined by ecosystem type (distinguished based on landform, soil, and climatic factors), then patterns of diversity emerge within the cover types from which the idea of representativeness can be better gauged.

Analysis of beta diversity further demonstrates the differences in ground-cover vegetation among ecosystem groups using Jaccard's coefficient and percentage similarity (PS; Table 14.3). Jaccard's coefficient is a qualitative measure based on species presence/absence, whereas PS is the similarity of the percentage coverage of the ground-cover species (Krebs 1989). Jaccard's coefficient shows that ecosystem groups 20 (moist, nutrient-rich) and 11 (dry, nutrient-rich) have the most species in common (C_j = 56%) and contain several species that are not present in other groups. Three nutrient-rich groups (20, 17, and 11) characterized by pre-European colonization vegetation of hemlock–northern hardwood forest are the most similar ecosystem groups based on presence–absence data, but the most similar two groups are only slightly over 50% alike. The least similar groups (less than one-third similar) are the various combinations of nutrient-rich (20, 17, and 11) and the nutrient-poor, climatically extreme ecosystem group (1).

PS indicates extreme differences in species coverage among the ecosystem groups (Table 14.3). Percentage similarity ranged from 2 to 92%, with most pairs of ecosystem groups ranging from 12 to 36% similar. Ecosystem group 11 (dry, nutrient-rich) is very dissimilar to the three nutrient-poor groups (2, 4, and 1; PS = 2–3%). The pairs with greatest PS (92) are groups 1 and 4, both dry and

Table 14.3 Comparison of beta diversity measures for ecosystem groups of the University of Michigan Biological Station, Emmet and Cheboygan counties, Northern Lower Michigan. Jaccard's coefficient[a] for all combination pairs of six ecosystem groups is shown outside parentheses, and the percentage similarity[b] (PS) is shown within parentheses.

Group	20	17	11	2	1	4
20	—					
17	53 (57)	—				
11	56 (28)	47 (36)	—			
2	42 (23)	36 (14)	31 (2)	—		
1	33 (20)	30 (12)	28 (2)	39 (56)	—	
4	36 (23)	46 (16)	34 (3)	39 (55)	37 (92)	—

[a] Jaccard's coefficient (species presence/absence): $C_j = j/(a + b - j)$, where j equals the number of species present in both samples, a equals the number of species present in sample 1, and b equals the number of species present in sample 2 (Magurran 1988).
[b] Percentage similarity, PS = minimum ($p1i$, $p2i$), based on the coverage of a species in a sample where $p1i$ equals the proportion of species i in sample 1 and $p2i$ equals the proportion of species i in sample 2 (Krebs 1989).
Source: Lapin and Barnes (1995). Reprinted with permission of Blackwell Science, Inc.

nutrient-poor, although they occur on different landforms and are not spatially adjacent (Figure 14.6). Notably, ecosystem groups may share many species, but the species' relative coverage may differ widely across the ecosystems. For example, ecosystem groups 20 and 11 share 56% of the species (Table 14.3), but their PS in coverage is only 28%. Conversely, groups 1 and 2 share only 39% of the same species, but are 56% similar in species coverage.

Ecosystem Types The classification and mapping of 125 local landscape ecosystem types (finer units within ecosystem groups) for the 4000-ha tract facilitated the study of alpha and beta diversity of ground-cover species within and among diverse major and minor landforms. For example, several measures of alpha diversity in Table 14.4 illustrate differences in diversity for major landforms (outwash plains and moraines), groups of ecosystem types within these landforms, and ecosystem types with calcareous versus non-calcareous soils. Notice also in Table 14.4 that the diversity values are presented in two ways: for all ecosystems and plots within a given unit and the average per unit area (within parentheses). For example, total species richness for outwash plains with 41 ecosystem types is 182, whereas the average per 150 m^2 plot is 20.2. For major landforms, moraines exhibit greater richness and heterogeneity of ground-cover species per unit area than outwash plains (26 versus 20, respectively) despite the greater number of ecosystem types in the outwash plains. Ecologically, this pattern relates to the greater soil-water and nutrient availability on moraines compared to the outwash plains.

Notably, the differences in ground-cover diversity between major landforms at the Biological Station have persisted for 25 years since the data shown in Table 14.4 were collected. Ricart et al. (2020) re-sampled ground-cover vegetation at the major landforms in 2015 and found that species richness and Shannon–Wiener heterogeneity remained higher on moraines relative to outwash plains. Values of both metrics declined over the 25-year period for both major landforms, but these results show the close relationship between diversity and physical site factors even after a quarter-century of forest succession.

Within moraines, diversity is higher in the bigtooth aspen-dominated interlobate moraine compared to the hemlock–northern hardwood-dominated Colonial Point moraine (Table 14.4). These landforms are very similar in their high soil-water and nutrient availability, but species richness is low on the Colonial Point moraine due to the dense overstory and understory of beech, sugar maple, hemlock, and other species that shade the forest floor. By contrast, open crowns of

Table 14.4 Alpha diversity measure of ground-cover diversity for upland ecosystem groups on outwash plain and moraine landforms of the University of Michigan Biological Station, Emmet and Cheboygan Counties, northern Lower Michigan.

Ecosystem group	Number of ecosystem types	Number of plots[a]	Index[b]			
			S	*H′*	1-*D*	*E*
Major landforms						
Outwash plain	41	175	182 (20.2)	1.77 (1.22)	0.54 (0.48)	0.34 (0.41)
Moraine	9	48	141 (25.8)	3.06 (1.84)	0.93 (0.71)	0.62 (0.58)
Minor landforms						
Interlobate moraine	3	26	123 (31.5)	3.05 (21.3)	0.93 (0.80)	0.63 (0.62)
Colonial Point moraine	6	22	76 (19.1)	2.29 (1.49)	0.81 (0.60)	0.53 (0.52)
Low-level outwash plain	11	34	132 (23.6)	1.53 (1.16)	0.46 (0.45)	0.31 (0.36)
High-level outwash plain	21	121	142 (18.9)	1.75 (1.17)	0.56 (0.47)	0.35 (0.40)
Soil type in outwash plains						
Calcareous outwash	18	70	142 (22.4)	1.89 (1.33)	0.60 (0.52)	0.38 (0.43)
Non-calcareous outwash	23	105	149 (18.3)	1.63 (1.05)	0.50 (0.43)	0.32 (0.36)

[a] Plot size = 150 m^2.

[b] *S* = ground-cover species richness, *H′* = Shannon–Wiener heterogeneity, 1-*D* = complement of Simpson's index, *E* = Shannon–Wiener evenness. Diversity values *outside parentheses* are based on all ecosystem types and plots for a given unit. Diversity values *inside parentheses* are expressed on a per unit area basis, in this study 150 m^2. For example, there is a total of 182 different ground-cover species in the 175 plots and 41 ecosystem types in outwash plain landform, whereas the average number of species per unit area (150 m^2) is 20.2. See text for discussion.

Source: Courtesy of Douglas Pearsall.

bigtooth aspens dominate ecosystems of the Interlobate moraine (Figure 14.6). In general, for these two landforms, richness is highest when canopy cover is intermediate, but richness is lowest where canopy cover is high and the low light at the forest floor is limiting to many species.

Two contrasts are shown in Table 14.4 for outwash plains—between (i) outwash plains of different elevation, where low-lying outwash plains are climatically extreme (giant frost pocket, Figure 14.6), and (ii) non-calcareous versus calcareous ecosystem types. The contrast of the colder low-level outwash plains with warmer high-level plains shows their similar values of *H′*. Note that the high-level outwash contains nearly twice as many ecosystem types as the low-level outwash, resulting in its higher diversity measures. The contrast of calcareous versus non-calcareous ecosystems in outwash plains (with similar overstory coverage of aspens) shows that the more nutrient rich calcareous outwash exhibits greater diversity. Calcareous ecosystems are able to support both nutrient-requiring and non-nutrient-requiring plants, whereas some plants requiring relatively high nutrient levels are excluded in the dry, non-calcareous outwash soils.

Beta diversity of ground-cover vegetation among paired comparisons of ecosystems in upland landforms is essentially the same for ecosystem types of outwash plains and moraines. Beta diversity is relatively low in the Interlobate and Colonial Point moraines where there are relatively few ecosystem types and species richness and coverage are similar.

ECOSYSTEM DIVERSITY

Thus far we have considered species diversity, but similar analyses of richness and evenness may be completed for ecosystems on landscapes where they have been distinguished and mapped at a given scale (see Figures 2.11 and 11.11 in Chapters 2 and 11, respectively). For example, the Pine River flows through a large expanse of Lake Superior beach in the Huron Mountains of Upper Michigan (Figure 14.7). Six riverine ecosystem types are tightly clustered along the river, and their abundance, size, and spatial distribution contrast with the single, flat, fire-prone beach ecosystem dominated by jack pine. The riverine ecosystems have diverse soil conditions ranging from extremely acid and infertile peat to very fertile muck, as well as a remarkable richness of plant species of different life forms (Simpson et al. 1990). Similar "hotspots" of ecosystem diversity were identified at the 4000-ha Biological Station tract described in the earlier text (Pearsall 1995). Importantly, high biodiversity is associated with these hotspots of ecosystem diversity in the glacial terrain of Michigan. Thus, land managers may efficiently prioritize areas of high biodiversity for preservation by identifying areas of high ecosystem diversity.

Conventional alpha and beta diversity indices may be used to quantitatively assess the diversity of landscape ecosystems and landforms. Pearsall (1995) used this approach to utilize the number of ecosystems, multiple characteristics of their physiography, soil, and vegetation, and their pattern in the landscape to determine the ecosystem diversity of landforms and ecosystems of the Biological Station tract. Ecosystem diversity was highest in outwash plains, and the centers of highest diversity apparent from ecosystem maps were confirmed by this multivariate diversity measure.

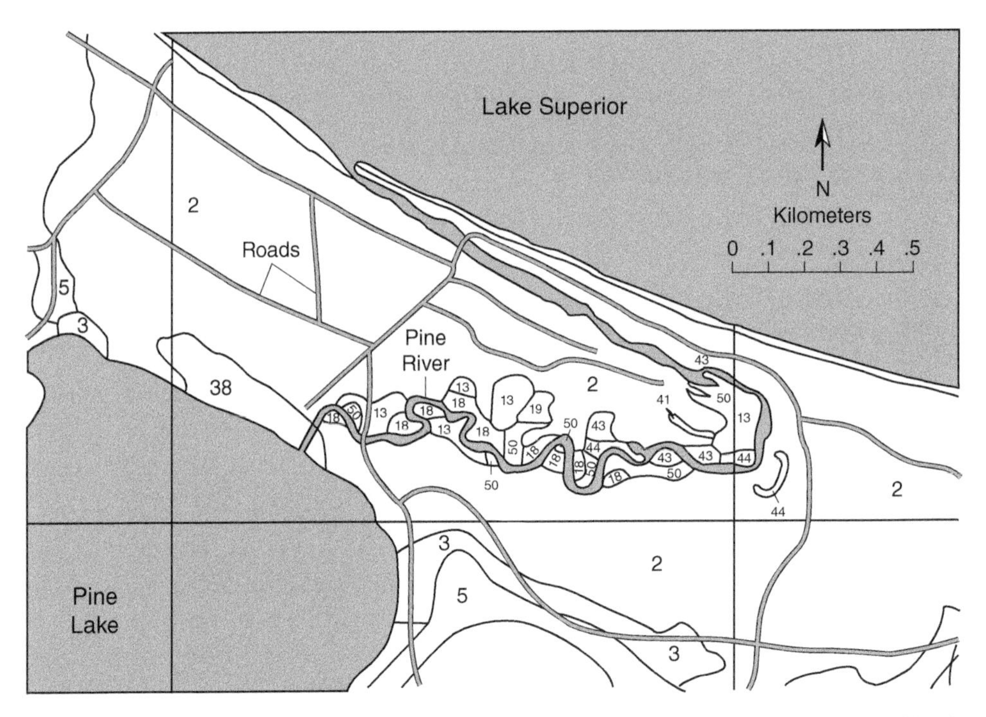

FIGURE 14.7 Landscape ecosystem diversity. Comparison of the number, size, and pattern of landscape ecosystem types of an area along the south shore of Lake Superior, Huron Mountain Club, Marquette County, Upper Michigan. In contrast to the extensive beach terrace (ecosystem type 2), six small ecosystem types (13, 18, 41, 43, 44, and 50) border the Pine River where it flows through the beach ecosystem. *Source:* Simpson et al. (1990) / Simpson.

CAUSES OF SPECIES DIVERSITY

In the following text, we examine biological diversity from the broad scale of continents to the fine scale of ecosystem types, with the objective of exploring the factors working together to create the patterns of diversity we observe today.

DIVERSITY AT CONTINENTAL AND SUBCONTINENTAL SCALES

Paleogeography and Continental Relationships The differences in tree species diversity that exist among regions of the Earth have interested biogeographers and ecologists for at least two centuries. Wet tropical forests have 10 times as many tree species as moist forests of the northern and southern temperate zones. No single mechanism has gained general acceptance as an explanation for this pattern, and several factors may contribute. Notably, ecologists have often, perhaps inappropriately, invoked local-scale deterministic processes of competition and other interactions to explain global-scale diversity patterns (Latham and Ricklefs 1993b). Nevertheless, differences in diversity among regions with similar climate suggest that local and regional processes have important implications for regional species richness, as do historical and evolutionary events.

Latham and Ricklefs (1993b) examined present-day taxonomic diversity among major moist temperate forest regions of the Northern Hemisphere: (i) northern, central, and eastern Europe, (ii) east-central Asia, (iii) Pacific slope of North America, and (iv) eastern North America. A total of 1166 species comprise the characteristic tree flora of these regions, approximately in the ratio 2:12:1:4 (Table 14.5); eastern Asia has by far the most species and genera. A total of 63 genera occur in either two or three of these regions (i.e., those neither endemic nor cosmopolitan), and 59 of these 63 genera (94%) occur in eastern Asia. Eastern Asia and eastern North America share the most genera (20), reflecting the well-known range disjunction of many plants inhabiting both regions (Boufford and Spongberg 1983). However, the temperate forests of eastern and western North America share few genera and none uniquely, despite their proximity.

The high diversity of temperate tree flora in eastern Asia compared to other regions of the Northern Hemisphere probably arose from differences in colonization history and perhaps in subsequent rates of proliferation of endemic taxa (Latham and Ricklefs 1993b). Colonization may have occurred more frequently in Asia, giving rise to new native taxa and their geographical spread among the temperate regions. Asia is characterized by many more species at the local level than found in temperate forests elsewhere, reflecting this species proliferation. Considerable site differentiation resulting from mountainous terrain and the relative lack of continental glaciation are reasons for such proliferation. Notably, most cosmopolitan taxa of temperate trees

Table 14.5 Summary of moist temperate forest trees in the Northern Hemisphere by taxonomic level and geographic region.

	Number of tree taxa characteristic of moist temperate forests in:				
Taxonomic level	Northern, central, and eastern Europe	East-central Asia	Pacific slope of North America	Eastern North America	Northern Hemisphere (total)
Subclasses	5	9	6	9	10
Orders	16	37	14	26	39
Families	21	67	19	46	74
Genera	43	177	37	90	213
Species	124	729	68	253	1166

Source: Latham and Ricklefs (1993b) / with permission of University of Chicago Press. Reprinted with permission from *Species Diversity in Ecological Communities*, Ricklefs and Schluter, eds., © 1993 by the University of Chicago. Reprinted with permission of the University of Chicago Press.

are likely to have originated in eastern Asia and dispersed to Europe and North America. Diversity differences between temperate and tropical flora reflect a physiological barrier to colonization of temperate zones that can only be overcome by the evolution of freezing tolerance. The important point is that latitudinal gradients in richness are best explained by historical and evolutionary factors affecting land masses and floras rather than present-day ecological interactions.

Glaciation Continental glaciation has importantly affected species diversity, and some of these effects are described in Chapter 15. Species displaced by glaciation may have gone extinct when they were unable to migrate to their former locations. In central Europe, the Alps prevented southerly migration of many species resulting in their extinction. Moreover, continental ice sheets altered the climate of areas even at great distances from the advancing or retreating ice fronts; these changes are discussed for North America in Chapter 15. Glaciation has also affected global sea levels; low sea levels during the Pleistocene Era formed a land bridge between Asia and North America, allowing immigration of humans into the Americas that facilitated elimination of many large North American mammals and altered disturbance regimes of fire and flooding. The extent of human influence on diversity in the Americas is unclear, but their introduction during the Pleistocene had a remarkable effect on the structure and diversity of North American forests.

Latitude and Elevation Plant species richness decreases with both increasing elevation and increasing latitude toward the poles (Billings 1973, 1995). This decrease results from evolutionary and geographic processes over the very long term as well as local ecological conditions such as decreasing mean annual and growing season temperatures, drought stress, and ultraviolet-B irradiation at high elevations (Billings 1995).

The latitudinal gradient in diversity of plants from the tropics to the poles has generated great scientific interest. The pattern has been demonstrated to have been present for flowering plants throughout most of the Cretaceous (Crane and Lidgard 1989). Overall diversity for organisms does not occur along a particularly smooth gradient from the poles to the equator, however; the diversity gradient is much steeper in the Northern Hemisphere than south of the equator, where diversity declines more slowly from the equator to the South Pole (Platnik 1991, 1992). The diversity gradient is complex and not well understood, but the reasons for its existence are related to (i) tropical areas having more land area than temperate and polar areas, such that speciation will occur at higher rates and extinction at lower rates (Rosenzweig 1992); (ii) areas nearer to the equator have higher available energy, providing more resources to therefore support more species (Turner et al. 1996; Wright et al. 1993), and (iii) evolutionary rates may have been higher in the tropics without interruption by glaciations and/or associated drying climates (Rohde 1992).

Clearly, factors other than latitude affect richness. For example, the geographical pattern of tree species richness for North America shows that richness is indeed greatest in the southeastern United States and decreases steadily northward (Figure 14.8). However, richness also decreases in the arid parts of the continental interior and is near its lowest values in North America at the same latitude as the southeastern United States. Tree species richness is strongly related to annual actual evapotranspiration and mean annual temperature using large geographic quadrats (Figure 14.8, Currie and Paquin 1987), but no significant relationship exists between actual evapotranspiration and local tree richness of broad-leaved deciduous trees at finer scales using 0.5–10-ha plots (Latham and Ricklefs 1993a,b). The tree richness pattern in North America is thought to reflect the evolutionary history of broad-leaved trees and the relative newness of continental arctic climates compared to the unglaciated areas of the southeastern United States (Latham and Ricklefs 1993a,b) as well as local ecological interactions.

Plant diversity is strongly related to elevational gradients as well as latitudinal gradients. Plant diversity generally decreases with elevation (Billings 1987), especially when elevation is

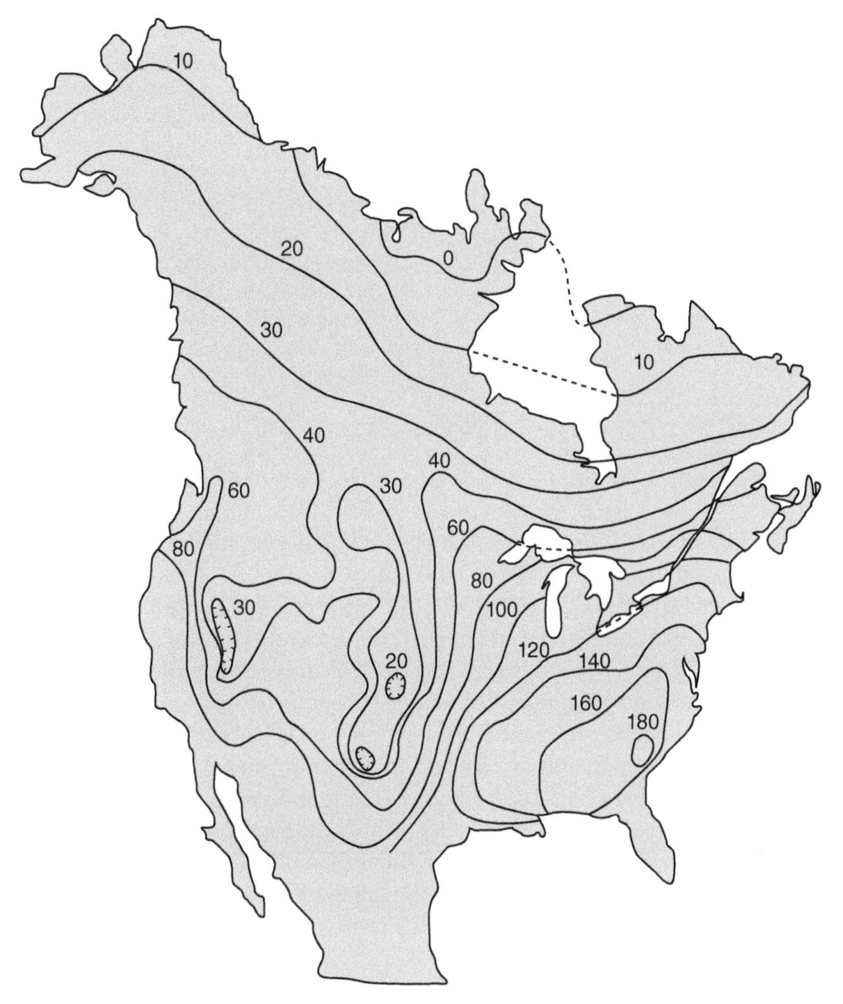

FIGURE 14.8 Species richness of North American trees north of the Mexican border. Contours connect points with the same approximate number of species per quadrat. Data are based on 620 native tree species in quadrats each 2 1/2° × 2 1/2° south of 50° N, and 2 1/2° × 5° (longitude) north of 50° N. *Source:* Currie and Paquin (1987) / Springer Nature. Reproduced with permission from *Nature* [Currie, D. J., and V. Paquin. Large-scale biogeographical patterns of species richness.], ©1987 Macmillan Magazines Limited.

combined with increasing latitude as illustrated in Figure 14.9 for high mountain ranges in Europe and Asia. Declining diversity is not always simple with increasing elevation; often diversity increases from low to mid elevations, then decreases toward high elevation, although diversity at low elevations typically exceeds that at high elevations (Grytnes and Vetaas 2002). Similar to latitude, diversity gradients related to elevation are typically explained by land area (reduced land area at higher elevations), and energy availability (peaking at mid elevations), but also by the isolation of high-elevation areas that limits immigration of species and thus speciation.

Whittaker (1977) contrasted patterns of plant species diversity in forests of two mountain ranges and two different regional ecosystems of the western United States (Figure 14.10). In the relatively humid, maritime climate of the Siskiyou Mountains of southwestern Oregon, richness of woody plant species decreased with increasing elevation (Whittaker 1960). In addition, herb diversity was highest at middle elevations and lower in dense stands with a sclerophyll tree stratum at

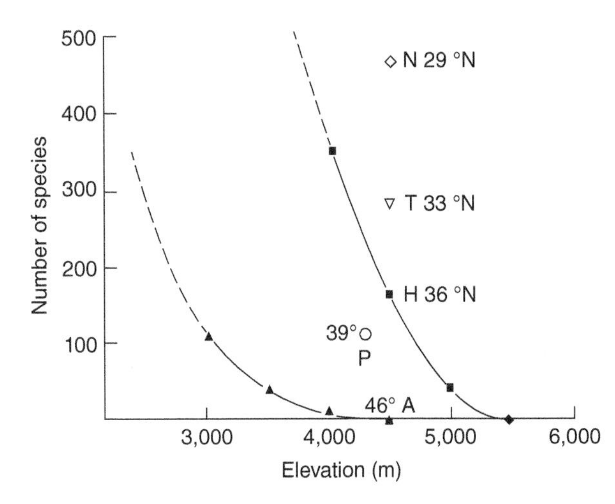

FIGURE 14.9 Relationship of the number of plant species with increasing elevation in the European Alps (left) compared with high mountain ranges in central Asia (right) having many peaks above 6000 m. The letters show the increasing number of species with decreasing latitude: A = Alps, P = Pamirs, H = Hindu Kush, T = Tibet, N = Nepal Himalaya. *Source:* Breckle (1974) / Nordic Publications in Botany. Reprinted with permission by the Council for Nordic Publications in Botany, Copenhagen, Denmark.

low elevations. Maximum richness was found in mesic forests at middle elevations (approximately 1200 m), with a secondary maximum at lowest elevations (Figure 14.10). At all elevations, richness increased from xeric (ridges and south–southwest aspects) to mesic and wet ravines. By contrast, the drier, continental climate of the Colorado Front Range of the Rocky Mountains supports more open forests without sclerophyllous trees. Forests give way to shrubland and grassland at low elevations and to alpine meadows at high elevations. Plant richness is low in forests of middle elevations and middle topographic positions of exposed, open slopes, but is higher in ravines, alpine meadow transitions, and the more open stands of driest topographic sites and low elevations (Figure 14.10). Again, richness increases from xeric aspects to mesic and wet ravines. These examples emphasize the importance of considering biodiversity through studies of regional ecosystems (southwestern Oregon mountains versus Colorado Rocky Mountains) and the local ecosystems (mesic ravines, open mid-elevation slopes, xeric ridges, subalpine woodland, etc.) that occur at positions along gradients of temperature and moisture in the respective mountain systems.

DIVERSITY AT LOCAL SCALES

It remains difficult to provide many generalities or principles regarding biodiversity, but two concepts seem reasonably clear: ecological heterogeneity promotes diversity, and extreme conditions constrain diversity. For example, a heterogeneous spatial pattern of ecosystems or their components (climate, physiography, soil, hydrology), disturbance regimes, and stand structures promote diversity. By contrast, the most extreme physical conditions of temperature, water, and/or soil chemistry limit biodiversity. Above all, biodiversity is closely related to the spatial pattern, diversity, and physical site characteristics of regional and local ecosystems. It is beyond the scope of this chapter to examine the enormous array of factors and multiple-factor determinants of species diversity, but we discuss selected ecological relationships in the following sections.

Physiography and Soil At regional and local scales, physiography and soil, as described in Chapters 8 and 9, influence the distribution and diversity of plants. Physiography in particular affects plant diversity by its effects on microclimate and soil development. Diversity in mountainous regions in the temperate zone and in unglaciated areas is often very high because of the dissected terrain and the resulting diverse sites and niches available for plant occupancy. Even in less dissected terrain, microtopography provides highly localized environmental heterogeneity (Chapter 8, Figure 8.16; Foster 1988a), thereby resulting in increased

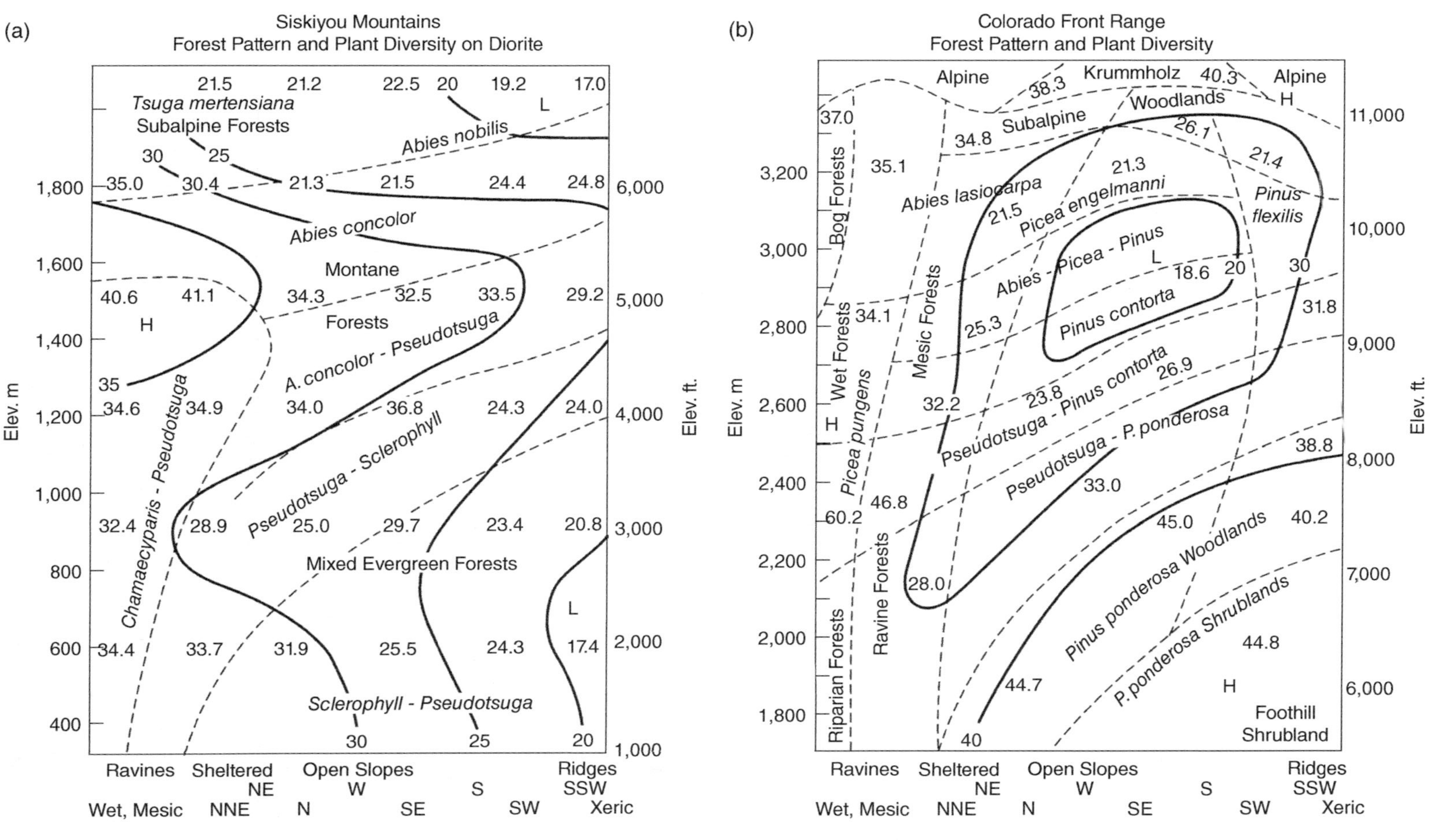

FIGURE 14.10 Patterns of plant species' richness in the Siskiyou Mountains of southwestern Oregon and the eastern slope of the Front Range of the central Rocky Mountains in Colorado. Numbers are for vascular plants species in 0.1-ha quadrats averaged for several stands representing a given combination of elevation and physiographic position. *Source:* Whittaker (1977) / Plenum Publishing Corp. Reprinted from *Environmental Biology* with permission of the Plenum Publishing Corp.

species diversity. Regeneration of some understory tolerant species depends on microtopography formed by treefall or coarse woody debris mounds. For example, yellow birch and eastern hemlock establish preferentially on moss-covered logs and mounds in hemlock–northern hardwood forests (Chapter 4), as do western hemlock and Sitka spruce in old-growth forests of the Pacific Northwest (Harmon 1987; Harmon and Franklin 1989).

Soil-water, nutrients, and oxygen availability also are responsible for many observed differences in diversity. Soils extreme in one or more attributes—low water or nutrient availability, high specific mineral content (e.g., soil derived from serpentine parent material), or very shallow—often support forests with low tree diversity (e.g., pine barrens, black spruce swamps). Notably, such soils may support many herbs or rare and unusual herbs that have adaptations making them competitive on such sites. Conversely, soils high in nutrient availability may support high tree species richness. For example, vascular plant richness was strongly related to soil nutrient availability for a broad range of ecosystems in the Piedmont Province of North Carolina (Figure 14.11). High tree species richness was also associated with abundant moisture, and moister sites had more species where soil fertility was low.

Deep, moist, or high-nutrient soils may support either high- or low-diversity forests depending on history, amount of disturbance, and successional stage. Dominance by one or a few species often occurs more rapidly under favorable site conditions, where all organisms can potentially grow rapidly (Huston 1993, 1994). Under such conditions, the best competitors are able to eliminate most other species by monopolizing an essential resource such as light. Under poorer soil-water and nutrient conditions, superior competitors are unable to dominate as rapidly, and more species are likely to coexist, leading to smaller and/or fewer individuals. For example, canopy coverage in late-successional stands may be very dense in mesic beech–sugar maple and hemlock–northern hardwood forests of the western Great Lakes region. Few species survive in the understory and ground-cover layers of these forests other than young individuals of the canopy species. Ecosystems on the fertile and moist moraine referenced in Figure 14.6, for example, exhibit relatively low ground-cover diversity (Table 14.4, Colonial Point moraine). However, as discussed in the earlier text, a bigtooth aspen community on a similarly fertile and mesic soil (Table 14.4, Interlobate moraine) has approximately 40% greater ground-cover richness due to favorable light

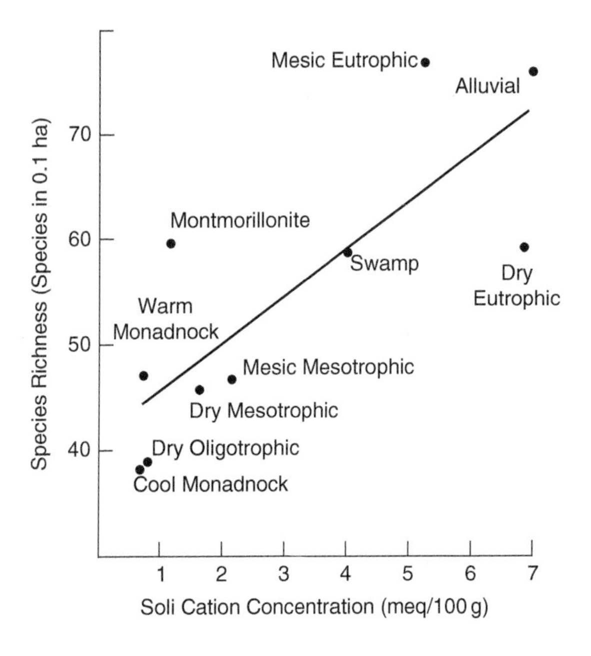

FIGURE 14.11 Relationship between species richness (average number of plant species in a 0.1-ha sample) and average exchangeable cation (Ca, Mg, K) for a diverse range of ecosystem types, representing wet, mesic, and dry sites of the North Carolina Piedmont. *Source:* Peet and Christensen (1980b) / Eidgenössische Technische Hochschule Zürich. Reprinted with permission of the ETH Zürich, Finanzdienste, CH8092 Zürich, Switzerland.

conditions resulting from the aspens' relatively open crowns. Ground-cover diversity observed in summer in ecosystems with a closed canopy may be somewhat misleading because of a relatively diverse flora of early spring ephemerals that complete much of their aboveground growth before the canopy trees leaf out in the spring.

Community Composition and Structure The diversity of birds, mammals, and other forest animals may be affected by: (i) forest species composition, and (ii) structure/age differences within and among communities, including vertical layering of vegetation, horizontal arrangement of trees and other vegetation within and among stands (e.g., random versus patchy stem distribution), and the distribution of standing and downed dead trees. A classic case is the relationship of bird species richness and structural complexity of forest vegetation. Bird species diversity has been shown to be positively related to foliage height diversity in northwestern North America (MacArthur and MacArthur 1961), but negatively related to it in northern Patagonia, Argentina (Ralph 1985; Huston 1994, p. 41). In the Pacific Northwest, the northern spotted owl inhabits old-growth forests, needing both large hollow trees for nesting and open subcanopy areas for foraging. Controversy over the spotted owl was a watershed event of American environmental policy in the 1980s and 1990s (Yaffee 1994). The controversy involved conserving structurally heterogeneous old-growth forests, which are lost following logging, to retain habitat of the owl and many other organisms of these unique forests (Franklin et al. 1981; Franklin et al. 1997). By contrast, the rare red-cockaded woodpecker of the southeastern United States breeds in old-growth longleaf, slash, or loblolly pine communities with low structural diversity. This bird nests in colonies in cavities carved over generations in large, living, fungal-infected, resin-rich pines (Lennartz 1988; McFarlane 1992). The birds preferentially reside in ecosystems with large trees and having little or no mid-story, few tall shrubs, and generally sparse ground cover. The woodpecker's preference for simplified vertical structure is probably related to protection from predators, hot fires, and availability of insect prey.

Disturbance and Succession It is difficult to provide generalized textbook examples about the effect of disturbance on species diversity given the wide range of ecosystem variation in composition, structure, and function. The concept most often cited relating to the association between diversity and disturbance is the **intermediate disturbance hypothesis** (Connell 1978). This hypothesis suggests that species richness is low at very high disturbance frequencies because only the most resistant species are able to survive, but also at very low disturbance frequencies because the most competitive species exclude other species. Richness is thereby theorized to be maximized at intermediate frequencies (Figure 14.12). The intermediate disturbance hypothesis has received mixed support based on empirical studies (Bendix et al. 2017; Huston 2014; Svensson et al. 2012, Mackey and Currie 2001), and the general validity of the model is still debated. Recent work suggests that disturbance intensity likely interacts with disturbance frequency to determine species diversity (Yeboah and Chen 2016; Miller et al. 2012; Hall et al. 2012).

Disturbance is ecosystem-specific and thus maintains a combination of trees, ground cover, and animals consistent with the geographic setting of regional and local ecosystems (Chapter 16). Disturbance will maintain either high or low species diversity depending on ecosystem type, and thus patterns of diversity in hurricane-prone ecosystems will differ from those in fire- or flood-prone ecosystems. For example, frequent fires prior to European colonization maintained jack pine as the only dominant species on some dry, outwash-plain ecosystems in the Lake States, but tree species richness sharply increases on adjacent mesic moraines where fires are virtually absent. Simultaneously, frequent fire in outwash-plain ecosystems maintains high herbaceous species richness by sustaining high light conditions via a lack of jack pine canopy closure.

Fire is the disturbance most commonly studied in North America for its effects on diversity (Thom and Seidl 2016). A universal example of fire disturbance *reducing* temporal species richness is observed in ecosystems where mesophytic, fire-sensitive species colonize the site when fire is

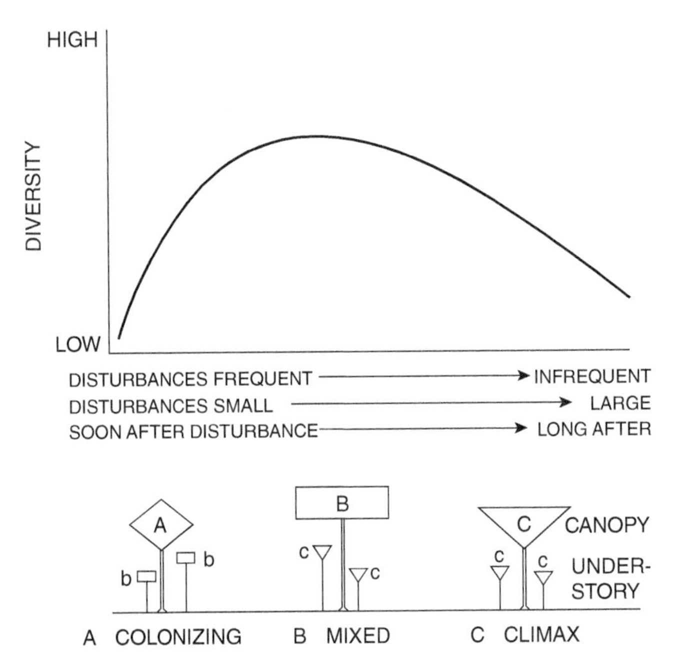

FIGURE 14.12 General relationship between tree species diversity and disturbance related to frequency of disturbance, size of disturbed area, and temporal scale corresponding to early- (colonizing), mid- (mixed), and late-successional stages of succession in the wet tropical Budongo rain forest of Uganda. The diagram represents the "intermediate disturbance hypothesis" where a mixture of early- and late-successional tree species exhibits the greatest tree-species diversity at an intermediate position in disturbance frequency and size and time after disturbance. *Source:* Connell (1978) / with permission of American Association for the Advancement of Science. Reprinted with permission from *Science* 199, p. 1303, © 1989 by the American Association for the Advancement of Science.

absent, increasing richness, but richness then decreases at the occurrence of the next fire. For example, dry-mesic oak forests of the Lake States and northeastern United States were characterized by frequent, light surface fires prior to European colonization. Shade-tolerant, mesophytic species that were able to invade during the fire-free interval were eliminated before they were recruited to the canopy. Fire exclusion in these forests following European colonization has today removed the disturbance necessary to eliminate mesophytic invaders such that open-canopy, oak-dominated forests are currently more species-rich. In the further absence of fire, however, oak forests will eventually be replaced by low-diversity mesic forests with dense canopies and a low-diversity, shade-tolerant understory as the intermediate-tolerant oak species are outcompeted (Nowacki and Abrams 2008).

Over longer temporal scales, however, frequent, less-severe disturbances are often necessary for the maintenance of diversity in ecosystems characterized by periodic fire. For example, lack of periodic fires in prairie or savanna ecosystems allows the encroachment of woody plants (Curtis 1959) and creates a reduction in herbaceous diversity. In an attempt to restore pre-European colonization of northern pin oak savanna structure and composition in central Minnesota, annual prescribed burning was begun in 1964 (White 1983, 1986). After 13 years of burning, oak basal area was significantly reduced, and shrub cover eliminated in burned versus unburned plots. The richness of herbs subsequently increased in burned plots by about 60%.

Disturbance also affects species diversity in successional sequences. Total plant species diversity is associated with the four general stages of secondary succession following disturbance (Chapter 17; Figure 17.4): (i) stand initiation (establishment), (ii) stem exclusion (self-thinning),

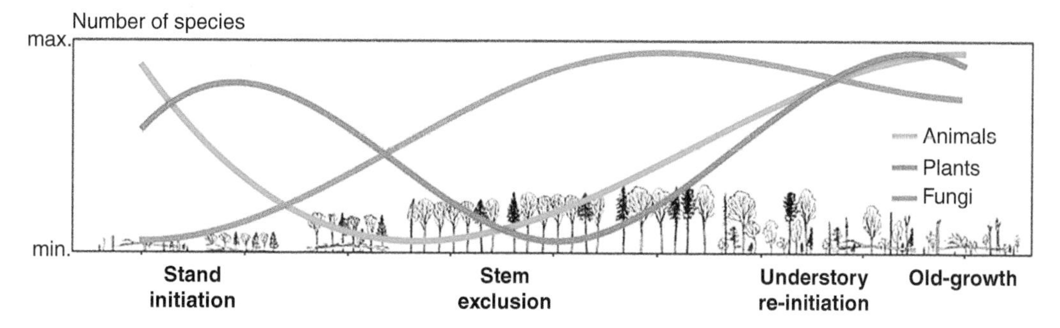

FIGURE 14.13 Predicted number of species along forest succession for animals, plants, and fungi for beech forests of central Europe. Successional stages along the x-axis approximate those described in the text and are not to scale. *Source:* Hilmers et al. (2018) / John Wiley & Sons.

(iii), understory re-initiation (transition), and (iv) old growth. Vascular plant diversity may be characterized for the respective stages as follows: (i) diversity is high in an open site soon after disturbance as species resprout or colonize by seeding; (ii) diversity is low once canopy closure occurs and the overstory "shades out" all but the most understory-tolerant species; (iii) diversity is low but increasing once gaps begin to appear in the overstory; and (iv) diversity ranges from low to high (depending on canopy density and number of gaps) as minor and major disturbances begin to grossly change the structure of the declining old-growth forest. This general pattern of diversity with successional stage has been demonstrated for Rocky Mountain forests (Peet 1978) and old-field mixed conifer–hardwood forests of the North Carolina Piedmont (Christensen and Peet 1981; Peet and Christensen 1987). In a study of European beech forests of central Europe, Hilmers et al. (2018) confirmed this "U-shaped" response of species richness for plants and animals over the course of forest succession, but an increasing trend of fungi that reached its maximum during understory re-initiation (Figure 14.13). Species assemblages were most similar to each other among early- and late-successional stages rather than in mid-succession. For plants, changes in species richness over forest succession were strongly driven by changes in abundance, that is, more individual stems—and thus more species—were present early and late in succession due to lower canopy cover. Obviously, disturbances occurring during succession would affect diversity in different ways depending on ecosystem type and disturbance characteristics—kind (wind, fire, insect, etc.), origin (natural or human), and severity.

FOCAL SPECIES IN CONSERVING DIVERSITY

Ecologists, land managers, and the public have increasingly focused on maintaining the local, regional, and global diversity of species, resulting in the emergence of an entire science of conservation biology within the last half-century. Maintaining diversity in forests requires understanding the status of all species that occur there, which is rarely realistic given available resources. **Focal species**, or those whose measurement provides information beyond that about its own status, are therefore often used in efforts to conserve diversity. There exist many types of focal species used for this purpose which may be distilled to five categories (Miller et al. 1999; Dayton 1972): (i) *keystone species*, or those with an ecological effect disproportionate to their abundance; (ii) *umbrella species*, or those requiring large areas whose protection would encompasses many other species; (iii) *flagship species*, or those with broad public appeal whose plight is likely to stimulate public support for conservation efforts; (iv) *indicator species*, or those associated with particular environmental conditions, and (v) *foundation species*, or those dominant in an ecosystem that influence structure and function through their abundance. We have treated indicator species in detail in Chapter 11; in the following sections, we discuss foundation and keystone species for their relevance to forests.

FOUNDATION SPECIES

Ecologists recognize that the most abundant species play major roles in controlling the rates and directions of many ecosystem processes. These species, termed **foundation species** (Dayton 1972), dominate an ecosystem in abundance and influence, and are often crucial for maintaining communities because they typically provide the major energy flow and the physical structure that supports and shelters other organisms (Gentry and Dodson 1987; Ashton 1992). Foundation species are typically plants and are usually trees in forested ecosystems. As we emphasize repeatedly throughout this book, trees in a forest shape its structure, microclimate, and light availability as well as its biomass and carbon dynamics and biogeochemistry. The potential loss of foundation tree species therefore will strongly affect the local environment on which a variety of other species depend (Ellison et al. 2005). The loss of foundation tree species is increasing across the world, which will inevitably influence diversity.

Several examples of the effects of foundation species loss on biodiversity have been documented and summarized by Ellison et al. (2005). For example, eastern hemlock grows in pure stands in the eastern United States, where its dense canopy intercepts light and precipitation, and its acidic, slowly decomposing litter slows nutrient cycling and creates infertile soils. Plant species richness in hemlock-dominated ecosystems is therefore typically very low. Hemlock is in rapid decline due to the hemlock woolly adelgid, an invasive insect that often results in the replacement of hemlock by deciduous species. Loss of hemlock may result in increases in plant species richness at local scales (e.g., Martin and Goebel 2013), but it is also likely to reduce broadscale variability in species richness (Ellison et al. 2016). Perhaps most importantly, hemlock in the southern Appalachian Mountains is notable for providing a shaded, cool environment along streams, such that its loss may impact the diversity of aquatic communities (Adkins and Rieske 2015; Webster et al. 2012).

Foundation species in forests are typically considered to be the dominant species, but recent research suggests that more than one species may be foundational in an ecosystem at some scale (Angelini et al. 2011). Specifically, Thomsen et al. (2018) reviewed 140 published studies examining **secondary foundation species**, or those species provided habitat by primary foundation species which themselves provide structurally complex habitat that alters environmental conditions for additional species in the community. They found that the presence of secondary foundation species increased species abundance and richness in many different ecosystems. For example, a study of Spanish moss, a flowering plant epiphytic on live oak in southeastern Georgia showed that oaks (primary foundation species) improved epiphyte survival by shading and reducing temperature on the Spanish moss (secondary foundation species). In turn, Spanish moss reduced drying effects and predation on insects, supporting communities that were larger and more species-rich than those found on live oaks without Spanish moss (Angelini and Silliman 2014). Studies such as these show that the biotic drivers of diversity in forests involve much more than just the dominant species, and that efforts to conserve diversity is therefore a very complex endeavor.

KEYSTONE SPECIES

In contrast to foundation species, a **keystone species** is one whose impact on its community or ecosystem is disproportionately large relative to its abundance (Power et al. 1996; Paine 1969). The concept of a keystone species is based on less abundant species that have major community and ecosystem effects, rather than those species that have high influence because they are very abundant or dominant. Identifying influential but less-obvious species is important for understanding how losing species will affect ecosystem function. Keystone species typically have high trophic status, and can exert effects through consumption (e.g., pocket gophers, elephants, wolves, trout, kangaroo rats, etc.) as well as by interactions and processes such as competition, mutualism, dispersal, pollination, disease, and by modifying abiotic factors (Bond 1993; Mills et al. 1993). One example of a keystone species is the badger, whose mounds maintain diversity in prairie floras (Platt 1975). Relatively few woody plants are identified as keystone species; fig trees that provide food for animals were the only woodland keystone species identified by Power et al. (1996).

The keystone species concept is helpful in identifying the most suitable areas for biodiversity preserves, understanding complex linkages among ecosystem biota and site–biota interactions, and in managing for single species. Power et al. (1996) gave three useful insights from the keystone species concept: (i) land managers should carefully consider the consequences of the loss of species for which no obvious role in ecosystems has been discovered, (ii) introduced alien species may, like keystone species, have potential strong effects disproportionate to their biomass, and (iii) there is no well-developed protocol of identifying keystone species; the field is littered with far too many untested anecdotal "keystone species."

ENDEMICS AND RARE AND ENDANGERED SPECIES

Much of the concern about diversity involves species considered rare, in danger of extinction, or endemic. Very often such species serve as flagship species for conserving diversity. **Endemic** species are those whose ranges are extremely limited in size, such as those restricted to an ecological region, a single mountain range, serpentine soils or other spatially limited sites, or to politically bounded units such as states or provinces. Endemics occur in both diverse and non-diverse areas, but they often contribute substantially to regional diversity.

Endemism is not uniform across North America. Areas of high endemism are usually correlated with unusual environmental conditions, relative isolation of an area, or a history of changing environmental conditions that have limited some species to a restricted area (Gaston and Spicer 2014). There are an estimated 186 endemic species of trees in the United States, representing about 29% of tree species richness (Jenkins et al. 2015; Figure 14.14). The highest numbers of endemic species occur in California, Texas, and Florida (Gentry 1986), many of which (though not all) occur in forest ecosystems. The highest concentration of endemic trees is found in the southeastern United States. Most of Florida's endemic plants are associated with swamps, flatwoods, and hardwood forests in northern and central Florida. A large number of Floridian species are

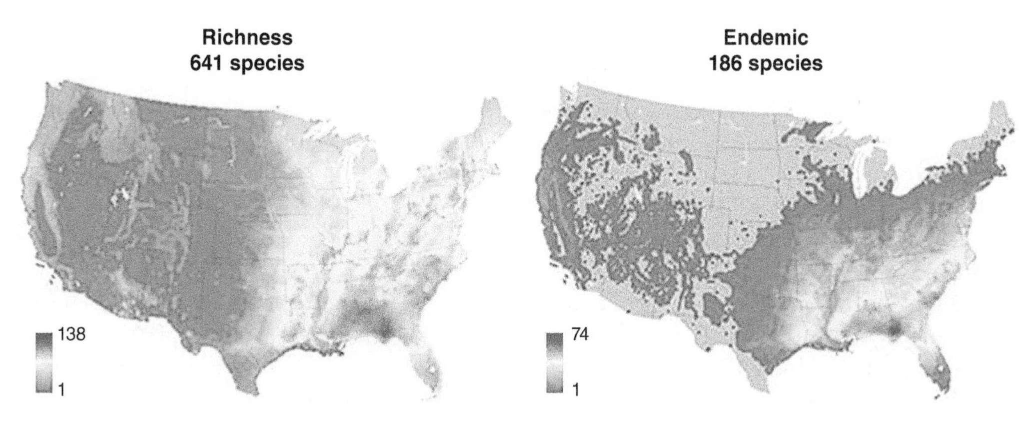

FIGURE 14.14 Tree species richness and endemic tree species of the lower continental United States. Total richness is the number of all tree species; endemics are species whose entire range is within the lower 48 states. *Source:* Jenkins et al. (2015).

associated with a small area known as "scrub" (Myers 1990). Many plants with very limited distributions apparently arose on isolated islands when most of the current peninsula was below sea level and thus persist today in isolated scrub "islands." Animals also evolved on these islands, including the Florida scrub jay and the sand skink, the latter of which is a lizard that "swims" just below the sand surface. These species are well adapted to survive on very specific sites and on an archipelago of scrub islands, but they are regionally rare, and many are considered to be threatened or endangered due to encroachment of agriculture and urbanization into the ecosystems that support them.

DIVERSITY AND THE FUNCTIONING OF ECOSYSTEMS

Ecologists have begun to uncover important roles for diversity in the way ecosystems function. That is, biodiversity has become increasingly recognized not only as a description of the biota in ecosystems, but as an important driver of the way ecosystems work. Biodiversity has been shown to be strongly correlated with several important attributes of ecosystem function (Lefcheck et al. 2015; Naeem et al. 2012; Cardinale et al. 2006; Chapin et al. 2000), although causal relationships remain the subject of much debate. In the following text, we briefly discuss two of these relationships with reference to forests: the relationship between biodiversity and productivity, and the importance of biodiversity for ecosystem stability.

BIODIVERSITY–PRODUCTIVITY RELATIONSHIP

Species richness tends to be correlated with primary productivity (Oehri et al. 2017; Balvanera et al. 2006; Waide et al. 1999). Most of the data supporting this relationship have surfaced from grasslands and other herbaceous communities rather than forests, but the relationship is most often described as positive but decelerating as species richness increases (Figure 14.15a; Liang et al. 2016b; Zhang et al. 2012). Notably, the form of the relationship may change as a result of many different factors, including the method of characterizing productivity (Sheil and Bongers 2020; Groner and Novoplansky 2003), the type of organisms studied (Mittelbach et al. 2001), the scale of the analysis (Mittelbach et al. 2001; Waide et al. 1999), and many other factors. For example, in a study of more than 115 000 forest plots across the United States, Fei et al. (2018) found an important effect of climate on the biodiversity–productivity relationship, with a hump-shaped relationship characterizing mesic climates and a linearly positive or non-significant relationship in dry climates. Meta-analyses have revealed strong biodiversity effects on productivity that may be as large as effects of other factors such as drought, fire, or herbivory (Tilman et al. 2012; Hooper et al. 2012).

The mechanisms driving the relationship between biodiversity and productivity are well studied but remain unclear. Both variables depend on climate and physical site factors, and therefore are likely to be positively correlated; that is, physical site characteristics tend to determine the growth (productivity) of organisms and the number of species populations (Allen et al. 2002). The positive effect of biodiversity on productivity is generally explained by two main mechanisms. First, communities with higher species richness typically have species that are able to differentiate their niches (Chapter 13) in a way that they become complementary to one another. This **complementarity** may provide for more efficient use of resources and nutrient retention, reduced competition (Hooper et al. 2005), and relief from herbivory and pathogens (Civitello et al. 2015) that may lead to increased productivity. Similarly, the positive effects of one species on another (**facilitation**; see Chapter 17) may increase the productivity of a community as a whole, and thus a diverse community is more likely to be more productive because it contains species that are able to facilitate others (McIntire and Fajardo 2014; Hooper et al. 2005). Second, more diverse communities are more likely to contain dominant species with high productivity—often called the "sampling effect" (Hooper et al. 2005). Complementarity has been shown to be less prevalent in

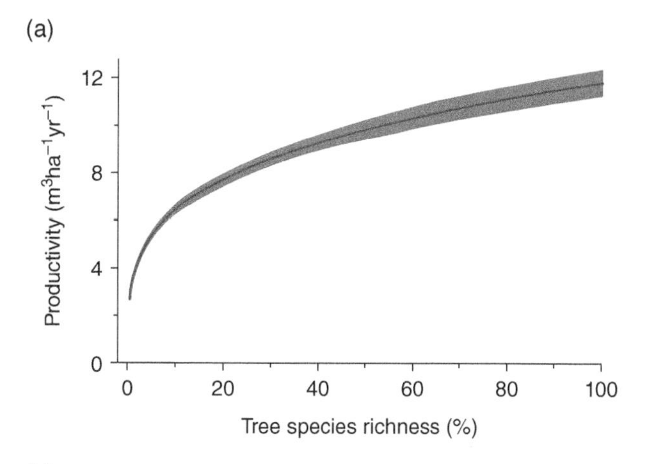

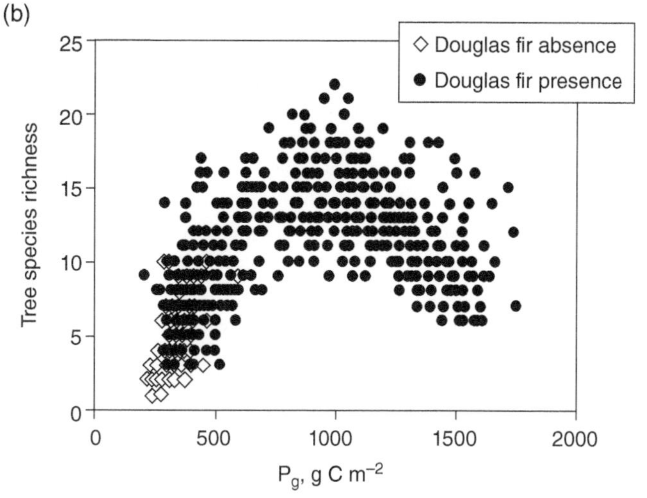

FIGURE 14.15 Relationship between tree species richness and productivity. (a) Effect of tree species richness on productivity using a global dataset from 44 countries. Productivity increases at a decreasing rate with species richness. *Source:* After Liang et al. (2016b) / with permission of American Association for the Advancement of Science. (b) Effect of productivity on tree species richness in the northwestern United States, using gross photosynthesis per square meter as a proxy for productivity. Species richness peaks at intermediate levels of productivity. *Source:* Swenson and Waring (2006) / John Wiley & Sons.

temperate forests, where species interactions tend to lead to competitive exclusion, compared to the harsher environments of boreal forests where facilitation among species may be more important (Paquette and Messier 2011).

A great deal of research has also examined how the biodiversity–productivity relationship may also work in the opposite direction; that is, whether productivity is an important driver of species richness. This relationship is commonly described as hump-shaped or unimodal, with richness peaking at intermediate levels of productivity (Figure 14.15b; Liang et al. 2016a; Waide et al. 1999). Competitive exclusion theory would explain this relationship as stress and a lack of resources dominating at low productivity, supporting few species, and competition dominating at high productivity, limiting species richness (Tilman 1982; Huston 1979; Grime 1979). High productivity could also simply translate into more individuals, which eventually translates into more species (Evans et al. 2005). Excellent reviews of this direction of the biodiversity–productivity relationship are given by Grace et al. (2016), Gillman and Wright (2006), and Loreau et al. (2001).

The importance of the biodiversity–productivity relationship for forests is not simply academic. If forest productivity truly is driven by biodiversity, then the negative effect of biodiversity loss could have serious economic implications for global forest resources. Moreover, decreases in productivity would reduce the ability of forests to absorb and sequester carbon, likely resulting in consequential effects on the global carbon cycle. The biodiversity–productivity relationship further emphasizes the "value" of biodiversity discussed at the beginning of this chapter; the economic value of biodiversity in maintaining commercial forest productivity across the globe is estimated to be as much as $490 billion per year (Liang et al. 2016b). It also suggests that biodiversity plays a key role in how ecosystems function.

THE ROLE OF BIODIVERSITY IN ECOSYSTEM STABILITY

Ecosystems with communities higher in species richness are thought to be more stable. There are many ways to define ecosystem stability (Ives and Carpenter 2007), but perhaps the simplest definition of a stable system is one changing relatively little from its average state, even in the face of environmental fluctuations (also called *ecosystem resistance*; Chapter 16). Stability may be considered from the perspective of species or from the perspective of ecosystems, and this frame of reference has implications for how the biodiversity–stability relationship is interpreted. From a species perspective, higher species richness may destabilize ecosystems because competition for resources among species will limit the number of individuals in a system, reducing species' population sizes and increasing the likelihood that some species will go extinct (May, 1974). From an ecosystem perspective, higher species richness creates a higher likelihood that an increased abundance of one species can compensate for the decreased abundance of another if the species vary in how they respond to environmental fluctuations—a process known as **compensatory dynamics**. Because both types of relationships are present in natural systems, diversity may stabilize ecosystems while simultaneously destabilizing the abundances of individual species (Hector et al. 2010; Tilman et al. 2006; Tilman 1996). As with the biodiversity–productivity hypothesis, the relationship between biodiversity and stability emerged from studies of grasslands, and fewer studies of the relationship have been conducted in forests.

Most recent attention to the biodiversity–stability relationship examines the apparent positive influence of biodiversity on ecosystem-level functions such as resource capture and biomass production (Cardinale et al. 2011). Indeed, a large body of research suggests that such functional attributes are more stable in diverse communities (Hector et al. 2010; Jiang and Pu 2009; Cottingham et al. 2001). Loreau and De Mazancourt (2013) synthesize the mechanisms driving the relationship between diversity and ecosystem stability as follows:

1. Differences in species' fundamental niches drive their differential preference of environmental factors such as temperature, rainfall, and resources. As a result, species' populations respond differently to environmental fluctuations, creating a more stable ecosystem as a whole. Differing environmental responses among species is always stabilizing and is thought to be a major factor causing compensatory dynamics between species.

2. Species populations differ in their intrinsic rates of natural increase, such that they respond at different speeds to environmental changes. Similar to the first mechanism mentioned in the earlier text, the differences in response time tend to generate asynchronous population dynamics and thereby promote ecosystem stability. This mechanism is also compensatory.

3. Reducing the strength of competition increases stability through the *overyielding effect*, where mean total biomass of a diverse community increases compared to a non-diverse community or monoculture. Increased abundance or biomass reduces the strength of demographic stochasticity in population dynamics, which again contributes to ecosystem stability.

As described in the earlier text, the biodiversity–stability relationship is well studied but not as well understood, particularly for forests. Although the literature continues to evolve rapidly, thorough reviews and syntheses of the relationship are provided by Loreau and De Mazancourt (2013), De Mazancourt et al. (2013), Ives and Carpenter (2007), and McCann (2000), and Mori et al. (2017) review the concept relative to applied research in forest ecosystems. An important point is that a stabilizing effect of biodiversity on ecosystem functions would imply yet another implicit value of biodiversity that would clearly justify its conservation.

FOREST MANAGEMENT AND DIVERSITY

Silviculture and forest management include many operations, such as regenerating and tending forests, that are very different from the ecological processes occurring in natural stands. Creation of even-aged, even-sized stands using clear-cutting and artificial reforestation (Oliver and Larson 1996; Smith et al. 1997) has been the most common management system for many decades. This system creates highly simplified stands that lack many aspects of community structure and complexity, such as snags, coarse woody debris, multiple canopy layers, irregular tree spacing, and gaps (Franklin 1995). Forest ecosystems may be simplified at the genetic, stand, and landscape levels, as well as temporally (succession). The economic criteria of efficient forest management (e.g., whole-tree harvesting) and high productivity of commercially important tree species have largely driven forest simplification because it was held that the structural complexity found in natural stands was not essential for sustained tree productivity of a site. Research on the importance of stand structure and biological legacies as they affect ecosystem function, biodiversity, and long-term site productivity has been conducted by Franklin (1995, 1997) and Franklin et al. (1997, 2018). In the following sections, we briefly discuss the impacts of traditional forest management techniques on diversity and present alternative silvicultural approaches designed specifically to accommodate diversity as much as possible.

EFFECTS OF TRADITIONAL FOREST MANAGEMENT ON DIVERSITY

Clear-cutting is the forest management method most extensively studied because of the enormous changes it brings to ecosystems and their biota (Boyle 1992). Despite common assumptions to the contrary, the effects of clear-cutting on diversity are not always clear-cut (so to speak). Diversity of animals and plants may increase or decrease depending on the specific sites, scales, and patterns of clear-cuts, as well as the methods employed, including post-harvest burning or other site preparation. The effects of clear-cutting typically affect species differently, and therefore do not usually create a simple reduction in species diversity. Species dependent on old-growth or near-natural forests are often lost, presumably until suitable environmental conditions are restored, whereas colonizing species or those dependent on disturbance may thrive.

In a study of natural and managed Douglas-fir forests of the Pacific Northwest, Halpern and Spies (1995) found that changes in understory diversity were fairly short-lived following clear-cutting and slash burning. Populations of most vascular plants recovered quickly to original levels prior to canopy closure, although diversity may remain depressed for decades on severely burned sites, and some species may experience local extinction (Halpern and Spies 1995). Notably, the effects of clear-cutting may vary by ecosystem, even within the same region. For example, plant species richness did not differ between cut and uncut plots 50 years after clear-cutting in a sugar-maple-dominated ecosystem in Upper Michigan (Albert and Barnes 1987), but clear-cutting an eastern-hemlock-dominated ecosystem resulted in higher richness in the cut plots after 50 years (Hix and Barnes 1984). The differences in the hemlock system reflect the sudden change from a low-moisture, acidic, nutrient-poor site beneath a dense hemlock canopy to an open, less acidic, more nutrient-rich environment facilitated by regeneration of hardwoods on the logged site (Albert and Barnes 1984). In any case, clear-cutting obviously creates the potential for strong effects on species richness, and thus silvicultural prescriptions that maintain diversity of stand structure and site conditions will be most effective in maintaining diversity. Likewise, any management application that precludes or delays the re-establishment of old-growth attributes (such as intensive, short-rotation plantation forestry) may result in long-term loss of diversity.

Diversity in forests is also affected by patterns of forest management at broad spatial scales. Fragmentation of forests—or the breakup of forests into progressively smaller and isolated patches—by roads, agricultural fields, and clear-cut patches (Chapter 22) also influences diversity. The consequences of ecosystem fragmentation are not only the creation of "islands" of various sizes but also include large changes in the physical environment that markedly influence species and gene pool diversity. In many ways, the effects of fragmentation resemble those of clear-cutting or other activities that reduce or eliminate forest interior conditions. The significance of fragmentation and edge effects has been considered by many (With 2019; Fahrig 2003; Harris 1984; Saunders et al. 1991, Ledig 1992), and we examine it in detail in Chapter 22.

Our main focus in this chapter has been on ecosystem and species diversity, but genetic diversity of forests (i.e., species diversity and gene diversity within species) has also been greatly influenced by human activities. Forest management may affect genetic diversity via its manipulation of the evolutionary processes of extinction, selection, drift, gene flow, and mutation (Ledig 1992). Clear-cutting, for example, decreases population size and connectivity and thus increases differentiation, genetic drift, and inbreeding in remaining adult trees, but not necessarily in regenerating trees (El-Kassaby et al. 2003). By contrast, management that retains some degree of stand structure and regeneration is less likely to affect genetic diversity of adult trees but could promote inbreeding and genetic drift of the regeneration (Sagnard et al. 2011). Because managers select the trees for harvesting and for regeneration, diversity may be increased (e.g., exposing recessive genes and increasing the occurrence of novelties and mutants), but these practices can also reduce diversity. Forest harvest practices that cause fragmentation, alter population sizes, or simplify age structure all affect the breeding systems of forest trees. Ledig (1992) describes how habitat alteration, environmental deterioration, and domestication of forest species may all lead to reduced biodiversity, and Ratnam et al. (2014) review the effects of forest management on genetic diversity in forests across the world.

PRESERVING DIVERSITY IN MANAGED FORESTS

The most common method of diversity preservation is the establishment of preserves—an area of land set aside to protect the species and ecosystems within it. The concept of biodiversity preserves has been heavily influenced by island biogeography theory of MacArthur and Wilson (1967; Chapter 22), whereby preserves are viewed as islands of suitable habitat within an ocean of unsuitable habitat. Based on island biogeography theory, preserves function better to maintain diversity when they are large such that core habitat remains; smaller preserves work better if they are connected. A great deal of research has been dedicated to optimal preserve design, much of which we discuss in Chapter 22.

The success of a preserve system for diversity protection is usually assessed based upon the proportion of area set aside. Worldwide, protected areas have higher abundances of individual species and higher species richness compared with areas having alternative land uses (Coetzee et al. 2014). In 2020, the International Union for Conservation of Nature (IUCN) estimated that 17% of the Earth's land area was set aside as preserves based on their own set of guidelines (UNEP-WCMC, IUCN, NGS 2020). For comparison, about 8% of the lower 48 states in the United States was within an IUCN categorized protected area as of 2015, mostly in the western half of the country where a large proportion of land is in public ownership (Jenkins et al. 2015). Relatively few preserves are available in the southeastern United States, where species richness and endemism are the highest in the country (Figure 4.14), and most land is in private ownership. These trends emphasize the limitations of preserves in conserving diversity: preserve design is often subject to political and economic pressures that minimize their effectiveness. Many researchers argue that the solution is to either simply increase the amount of area in preserves or set aside higher-quality areas into preserves, but these strategies are often not practical.

It became clear almost three decades ago that a preserve system alone is not sufficient to effectively sustain forest diversity (Hansen et al. 1991; Marcot 1997; Westman 1990; Wilcove 1989). The world's forested lands are dominated by semi-natural ecosystems and cut-over land such that conservation of biological and ecosystem diversity rests primarily on innovative management of the land outside of protected areas. To a large degree, the ecosystem services described in the earlier text and in Chapter 23 are provided by the biological and ecosystem diversity found on all landscapes regardless of their history of human activities, such that conserving diversity is a fundamental need in all forests (Lindenmayer and Franklin 2002). As such, a variety of silvicultural procedures have been developed for the conservation of biodiversity on managed lands. A detailed discussion of forest management techniques is beyond our scope, but we review the basics of forest management that incorporate diversity in the following sections.

ECOLOGICAL FORESTRY: INCORPORATING BIODIVERSITY INTO FOREST MANAGEMENT

Traditional silvicultural harvest and regeneration methods—clear-cut, seed-tree, and shelterwood in even-age management systems and selection used in uneven-aged systems (Smith et al. 1997)—were created solely for regeneration and subsequent growth of commercially important tree species. These traditional methods have typically been applied without concern for diversity and usually lack the structural complexity of natural stands. Lindenmayer and Franklin (2002) summarize that sustaining diversity in managed forests requires shifting the focus from what is *extracted* from a forest, such as timber or other resources, to what *remains* in place after the resources are harvested. As biodiversity depends on factors such as structural complexity, special habitats or niches that might result from that complexity, keystone and foundation species, and biological legacies, retaining these factors in a post-harvest forest is critical for maintaining diversity. Hunter and Schmiegelow (2011) simplify this process into a "micro approach" that provides structures and complexity necessary for habitat (e.g., snags or downed logs, or microclimatic conditions) within forest stands, and a "macro approach" that provides a diversity of stand conditions on the landscape (e.g., mixtures of open- and closed-canopy forests, or forests at different successional stages). This multi-scale process may be relatively easily incorporated into modern silvicultural systems that manage forests for multiple values, including wood production as well as diversity (Franklin et al. 2018).

Broadly speaking, managing forests from the perspective of conserving biodiversity and ecological productivity as guided by natural disturbance regimes is known as **ecological forestry** (Palik et al. 2021; Franklin et al. 2018; Seymour and Hunter 1999). General components of ecological forestry include retaining portions or features of the pre-harvest stand in the managed stand, using knowledge of forest succession to manipulate established stands, using whatever silvicultural return intervals are necessary to achieve desired structure or ecological processes, and using knowledge of large-scale processes to plan management on landscapes (Palik et al. 2021). Compared to traditional forestry focused on timber production, ecological forestry maintains a diversity of structures and processes, values complexity and heterogeneity, and emphasizes ecosystem diversity. By contrast, production forestry utilizes only the subset of structures and processes that support economic goals, values simplicity and homogeneity, and emphasizes optimizing tree growth (Franklin et al. 2018).

The structural complexity of old-growth Douglas-fir–western hemlock forests of the western Cascades that range in age from 250 to over 600 years has been studied in detail (Franklin et al. 1981; Harmon et al. 1986; Franklin et al. 1987; Maser et al. 1988; Spies and Franklin 1991; Franklin 1993; Spies 1997). These forests provide an effective model from which the principles of ecological forestry have been developed. Aspects of structural complexity here include varied sizes and conditions of live trees, multiple canopy layers, and presence of canopy gaps, as well as snags and downed logs and woody debris. These structural features markedly influence biotic composition of the forest, ecosystem processes such as biomass production, nutrient cycling, and succession, and regeneration following natural disturbances. In particular, biological legacies—surviving organisms and organically derived structures such as snags, logs, and organic soil horizons (Chapter 17)—are essential for the rapid reestablishment of forests that have highly diverse composition, function, and stand structure (Franklin and MacMahon 2000; Franklin et al. 2000). Creating or maintaining structural complexity in managed stands in this region is an effective example of managing forests for multiple and complex objectives, including both wood production and biodiversity (Lindenmayer and Franklin 2002; Franklin et al. 1997). Palik et al. (2021) and Franklin et al. (2018) provide excellent overviews of the variety of methodologies in ecological forestry; we provide an example of one such method in the following text.

Variable-Retention Harvest System The variable-retention harvest system attempts to retain elements of the rich structural, compositional, and functional diversity in managed stands that are characteristic of old-growth forests. Variable-retention harvest prescriptions focus upon providing the structures and complexity necessary for habitat within stands as well as enhancing connectivity in the managed landscape (Franklin et al. 1997). The term "variable retention" is used because the structures retained may vary widely in amount (percent of area covered), kind, and spatial pattern (Figure 14.16). Structural retention focuses largely on biological legacies—both living and dead organic structures, including living trees of various species, sizes, and conditions and their derivatives such as snags and logs on the forest floor. Such structures provide habitats for many species (Carey and Johnson 1995) and ecosystem functions (Franklin et al. 1981; Spies 1997). Retaining biological legacies acts to (i) perpetuate species and genotypes (lifeboating); (ii) enrich

Photo courtesy of Jerry F. Franklin

FIGURE 14.16 Variable-retention harvest system in old-growth Douglas-fir forest, western Oregon. Silvicultural prescription is for dispersed structural retention of large live trees (135-year old), snags, and logs (20–25 trees per hectare) to meet minimal goals for coarse woody debris. H. J. Andrews Experimental Forest, western Oregon.

structure; and (iii) reduce the contrast in structure between managed and unmanaged areas ("softening the matrix"; Palik et al. 2021).

Lifeboating provides a way for genetic and species diversity to be carried from the pre-harvest stand to the post-harvest stand. Plant species may be specifically lifeboated by retaining mature individuals as seed sources or potential asexual reproducers by sprouting. Retention of biological legacies may perpetuate biota in general by (i) providing habitat or refugia that might otherwise be lost from the harvested area; (ii) ameliorating microclimatic conditions in relation to those that would be encountered without retention, and (iii) providing energy and nutrients to the post-harvest stand (Palik et al. 2021). For example, Figure 14.16 shows dispersed retention of living trees and woody debris following harvest in a 135-year-old Douglas-fir forest. Retention of trees in this manner retained microclimatic conditions critical for survival of some organisms. In addition, retention of live trees and shrubs provides critical habitat to maintain populations of soil organisms including mycorrhizae (Louma et al. 2006; Amaranthus et al. 1994).

Besides serving as refugia immediately after logging, structural retention enriches the complexity of managed forest stands for an entire rotation. Suitable conditions for species can thereby be re-established much faster than would otherwise be possible. Retention may provide suitable habitat for species that are generally rare or absent in young stands of simple and homogeneous structure. For example, many forest stands 80–200 years old in the Douglas-fir region provide suitable nesting and foraging habitat for the northern spotted owl and other species associated with late-successional forests. By retaining some old-growth Douglas-fir trees, managed stands may provide suitable nesting and foraging habitat for spotted owls within 50 or 60 years of harvest. Without retention, it may take 120 years or more to create the necessary structural elements (Franklin et al. 1997). Notably, lifeboating and structural enrichment may be accomplished simultaneously using the same biological legacies (e.g., live trees that both perpetuate plant species but also provide habitat).

Facilitating the movement of organisms within a managed landscape is a third value of structural retention. Connectivity is enhanced in the harvest unit by retaining structures in the harvested area that facilitate the dispersion of some organisms, in addition to creating corridors between intact stands. The objective here is to make the traditionally non-habitat, managed stands ("the matrix"; see Chapter 22) more hospitable for dispersion by retaining, for example, well-spaced logs, trees, and shrub patches for protective cover or habitat.

Designing a Variable-Retention Harvest System Developing a variable-retention harvest system requires that managers identify which and how much structures to retain, and the spatial pattern for the retention (i.e., *dispersed* or *aggregated*, or some combination of these). As described in the earlier text, structures to retain may include live trees, especially large-diameter trees, snags in varying states of decay, logs and other debris on the forest floor, forest understory, and undisturbed layers of forest floor. Dispersed retention creates an even distribution of legacies (Figure 14.16), which effectively provides wildlife habitats and future sources of energy across a large area. Aggregated retention retains small patches of forest ("green trees"), which tends to provide more microclimatic modification and retain more diversity by canopy structure and species composition. For example, in one of the first trials of aggregated retention (Figure 14.17), foresters of the Plum Creek Timber Company retained 15% of the green trees as aggregates in a clear-cut unit. Similarly, the City of Seattle experimented with aggregated retention using teardrop-shaped islands of green trees left between radiating clear-cut strips (Figure 14.18). In both cases, these compact aggregates are better at modifying microclimate and may be less susceptible to windthrow, whereas linear aggregates can provide better visual screening, wildlife corridors, and stream protection.

How Well Does Variable Retention Conserve Biodiversity? Approximately three decades have passed since the initial variable-retention systems were established, providing ample opportunity for assessing their effects on diversity. Meta-analyses of variable-retention studies have suggested

Photo courtesy of Jerry F. Franklin

FIGURE 14.17 One of the first examples of aggregated retention in which 15% of green trees were left in the clear-cut area which lies in the white polygon shown. Aggregated retention is often advantageous from standpoints of logging costs, safety concerns, and management efficiency. Plum Creek Timber Company lands in southwestern Washington State.

Photo courtesy of Jerry F. Franklin

FIGURE 14.18 Variable-retention harvest system in Douglas-fir forest illustrating the aggregated method of retaining live trees following harvest. Teardrop-shaped patches of trees are left, providing corridor-like connections of live-tree groups to the intact stand. Note individual trees left between the islands of green trees. City of Seattle, Cedar River Watershed, western Washington State.

that the system is a valuable way to conserve biodiversity while simultaneously achieving timber production (Beese et al. 2019a). An analysis of 39 studies that compared variable-retention harvesting to clear-cutting in North America and Europe found that retaining structures in the harvested area reduced species loss in 72% of the studies, and typically improved habitat for early-successional insects, birds, and woody plants (Rosenvald and Lõhmus 2008). Over the 39 studies, species richness and abundance of birds and ectomycorrhizal fungi and the abundance of woody plants increased with tree retention. Notably, the coverage of grasses and herbs generally decreased in the presence of retained trees, but total species richness did not decrease because of the higher presence and survival of woody plants (Rosenvald and Lõhmus 2008). A more recent analysis of 78 studies showed that species richness and abundance were higher in retention areas compared to clear-cuts; the richness and abundance of species preferring open habitat were also higher in retention areas compared to unharvested areas (Fedrowitz et al. 2014). Retaining a higher proportion of trees generally increased species richness, but diversity does not appear to respond to the spatial arrangement of retained trees (aggregated or dispersed).

Effects of variable-retention harvesting are not all positive, although most negative impacts are on tree regeneration and growth rather than species richness and abundance. Some forest-interior species found in unharvested forests are unable to tolerate harvesting even when trees and other structures are retained, and some species that prefer open habitats benefit more from clear-cuts than from variable-retention harvests (Fedrowitz et al. 2014). Several studies also suggest that fungi and lichens, having many species associated with old forests, are particularly susceptible to losses from harvesting (Trofymow et al. 2003; Sillett et al. 2000). Negative effects for timber production are also non-trivial, as regeneration is often reduced in retention harvest units compared to clear-cuts. For example, increasing the proportion of retained trees resulted in a decline in aspen sucker density and volume 9 years after harvest in boreal mixed forests in Alberta, Canada (Gradowski et al. 2010); sucker density declined by 50% even when only 20% of the original basal area was left in the stand. In the Pacific Northwest, growth of 5–12-year-old seedlings of Douglas-fir and western white pine was reduced at higher levels of tree retention, and natural regeneration density was higher in dispersed treatments compared to aggregated treatments (Urgenson et al. 2013). Retained trees exposed to wind when adjacent trees are harvested may be damaged by wind at rates as high as 50% and increase with reduced retention and aggregate size (Beese et al. 2019a, b). These results suggest that managers need to be opportunistic with the flexibility provided by variable retention harvesting in order to best balance the management objectives of timber production and biodiversity conservation.

In general, ecological forestry is likely to play an important role in sustaining biodiversity where timber production is an important objective (Beese et al. 2019a). Importantly, both variable-retention harvest systems and preserves in combination are likely to be critical for conserving biodiversity rather than either approach alone (Lindenmayer and Franklin 2002). Regardless of the improvement of variable-retention harvesting over clear-cutting, some species will require forest interior conditions that cannot be provided in a harvested landscape regardless of the proportion of trees retained. Nevertheless, variable-retention harvesting and other methods encompassed by ecological forestry (such as *variable density thinning*; Palik et al. 2021; Churchill et al. 2013) are important innovations in sustaining biodiversity in forests on human-dominated landscapes.

EPILOG: CONSERVING ECOSYSTEM AND BIOLOGICAL DIVERSITY

The number of species in danger of extinction continues to increase as landscapes become increasingly dominated by humans (Vitousek et al. 1997). Currently, more than 1600 species or subspecies are considered to be endangered or threatened in the United States, 57% of which are plants, representing an increase of 78% since the last edition of this book. Concern for diversity stretches beyond endangered and threatened species, however; it is for the immense number of unknown species and for ecosystem processes that are basic to long-term sustainability of ecosystems. As such, targeting whole ecosystems for biodiversity conservation rather than individual species may be an appropriate approach (Rowe 1990, 1997; Anderson and Ferree 2010; Shaffer 2015).

It is obvious from our discussion of diversity that species tend to be the focus of conservation, which typically proceeds on a species-by-species basis. Ecologists have long realized that such an approach is not sufficient, and that a species-by-species approach will fail because it will quickly exhaust available time, financial resources, societal patience, and scientific knowledge. To even come close to attaining the goal of preserving biodiversity, broadscale approaches—at the levels of regional and local ecosystems—are the only way to conserve the overwhelming mass—millions of species—of existing biodiversity (Franklin 1993). Most forethinking biodiversity conservation today combines a "coarse-filter" approach of protecting broadscale or regional ecosystem processes and structures with a "fine-filter," species-by-species approach (Franklin et al. 2018). As we close this portion of the textbook that has focused largely on the biota of forests, we re-emphasize a focus on the *spaces*—ecosystems that provide the support system for the biota they contain—to facilitate protection of the *species* (Barnes 1993).

SUGGESTED READINGS

Franklin, J.F., Johnson, K.N., and Johnson, D.L. (2018). *Ecological Forest Management*. Long Grove, IL: Waveland Press 646 pp.

Huston, M.A. (1994). *Biological Diversity*. Cambridge University Press 681 pp.

Latham, R.E. and Ricklefs, R.E. (1993). Continental comparisons of temperate-zone tree species diversity. In: *Species Diversity in Ecological Communities: Historical and Geographical Perspectives* (ed. R. Ricklefs and D. Schluter). Chicago: University of Chicago Press.

Lindenmayer, D.B. and Franklin, J.F. (2002). *Conserving Forest Biodiversity: A Comprehensive Multiscaled Approach*. Washington, D.C.: Island Press 352 pp.

Magurran, A.E. (2013). *Measuring Biological Diversity*. New York: Wiley 272 pp.

Palik, B.J., D'Amato, A.W., Franklin, J.F., and Johnson, K.N. (2021). *Ecological Silviculture*. Long Grove, IL: Waveland Press 343 pp.

Wilson, E.O. (1992). *The Diversity of Life*. Cambridge, MA: Belknap Press, Harvard University Press 424 pp.

Forest Ecosystem Dynamics

PART 5

All ecosystems are dynamic entities that change through time and over space, be they aquatic or terrestrial. In the previous sections of this book, we have considered how the physical environment, in combination with life-history traits of forest trees and associated plants and animals, influences the composition and structure of forest ecosystems from place to place at a given time. In particular, we have discussed how climate, physiography, and soil influence the geographic distribution of forest ecosystems at local, regional, and global scales. In this part, we consider the mechanisms and ecological implications of changes that occur over time in ecosystems as influenced by the physical environment.

Changes that occur in ecosystems over very long temporal scales reflect the relationship of biota with physical factors that form ecosystems after changes in vegetation distribution and genetics occur over millennia in the form of plant migration, speciation, and evolution. In Chapter 15, we discuss quaternary biogeography and paleoecology to study changes in tree and forest distributions and environmental changes, respectively, as an attempt to understand the ecological foundations of the ecosystems we see today. Current ecosystems are impacted by disturbances of many kinds and in many ways. Fire, flooding, insects, windstorms, and harvesting can remove one complement of plants and animals and, over time, replace it with another. In Chapter 16, we treat disturbance as an ecosystem process, recognizing that the events that disrupt forest organisms are fundamental to ecosystems. The treatment of disturbance leads naturally to the consideration of forest ecosystem succession. In Chapter 17, we discuss the concepts, causes, mechanisms, and models of succession and provide examples to illustrate how the composition and structure of different forest ecosystems change following a particular type of disturbance.

Disturbance and subsequent changes in forest communities fundamentally direct the pattern in which carbon and plant nutrients are cycled and stored within forest ecosystems. In Chapter 18, we discuss the physiological processes controlling the capture, storage, and loss of carbon (C) by forest trees, a set of processes that directly control growth. We then build upon these principles to understand the capture, storage, and loss of C by entire forest ecosystems, and the extent to which disturbance alters this ecosystem

Forest Ecology, Fifth Edition. Daniel M. Kashian, Donald R. Zak, Burton V. Barnes, and Stephen H. Spurr.

process. Changes in the growth of forest ecosystems following a disturbance also changes the demand for nutrients. Nutrient cycling is considered in Chapter 19, in which we trace the flow of growth-limiting nutrients into, within, and out of forest ecosystems. We place particular emphasis on the extent to which both natural and human disturbances alter the pattern in which nutrients are cycled and stored within forests.

Long-Term Forest Ecosystem and Vegetation Change

CHAPTER 15

Early in this book, we described how landscape ecosystems exist at multiple spatial scales as nested geographic and volumetric segments of the ecosphere. In this chapter, we consider how forest ecosystems change over long temporal scales. We can easily observe important changes in some ecosystem properties and processes at scales of days, seasons, and years, but noticeable changes in landforms, soils, and biota occur over time frames of decades, centuries, and millennia. Succession is the forest ecosystem change most commonly considered by forest ecologists. It occurs over decades to centuries and creates differing vegetation composition, structure, and biomass (Figure 2.2 of Chapter 2; Chapter 17). At even longer temporal scales, changes in vegetation distribution and genetics occur over thousands and millions of years in the form of plant migration, speciation, and evolution (Figure 2.2 of Chapter 2), also within an ecosystem context. Here, we emphasize long-term distributional changes in north temperate forest trees over the last 20 000 years as they reflect associated climatic, physiographic, and soil changes in landscape ecosystems.

What we know about long-term ecosystem change comes from geological studies in formerly ice-covered areas, from plant fossils, and from the forest trees themselves. Plants are among the best indicators of historical landscape ecosystems for which there lacks a written record of climate, physiography, and soil conditions. In other words, we are able to track and interpret the migration of a tree species, but much less is known about the changes in regional and local climate, physiography, soil, and disturbance regimes that largely controlled these migrations. We lean heavily on using forest tree migration to provide insights into ecosystem change related to macroclimate. In turn, we also expect major changes in tree migration to occur with future changes in climate (Chapter 20).

Plants migrate with changing environments and may establish totally new spatial ranges, all the while leaving evidence of extinction in the fossil pollen record at places of former dominance. The genus *Sequoia*, for example, is today restricted to a single species distributed locally in California, but fossil records show former distributions in Europe, central Asia, North America, Greenland, Spitsbergen (midway between Norway and the North Pole), and the Canadian Arctic. Tundra and boreal species in North America no longer exist where they once did 20 000 years ago, except for small pockets at high elevations in areas such as the southern Appalachian Mountains. When species persist in a given geographic site for several thousands of years, the genetic makeup of those populations change as they adapt to regional and local climate (Chapter 3). For example, the American beech of a given locality in eastern North America 9000 years ago differs genetically from the beech population occupying the same place today.

CHANGE BEFORE THE PLEISTOCENE AGE

Flowering plants (angiosperms) and mammals have dominated the Earth since the beginning of the Cenozoic (66 million years ago or 66 Ma). Long-term variations in global temperatures and displacement of the land masses relative to their modern locations altered the prevailing climate,

Forest Ecology, Fifth Edition. Daniel M. Kashian, Donald R. Zak, Burton V. Barnes, and Stephen H. Spurr.

which in turn shaped the geographical distribution of forests in ways different than today. Tree taxa ancestral to many modern-day taxa were present in these forests, but it is highly unlikely that trees were associated in regional-level and local-level ecosystems in the same communities that we observe today. Nevertheless, plants occurred in spatial patterns that corresponded to climatic, physiographic, and soil patterns of the time. For example, tropical rainforests covered much of the southeastern United States 55–65 Ma, but high-elevation forests in the southern Appalachian Mountains supported temperate communities of alders, birches, and hickories. Elsewhere, broad-leaved evergreen forest dominated coastal regions, while inland forests at northern latitudes and high-elevation forests consisted of more temperate, deciduous woody plants such as ginkgos, viburnums, sycamores, birches, and elms (Graham 1999). By 16 Ma, as the global climate had dried and cooled, the Southeast was dominated by a dry tropical forest with scrubby pines on sandy sites and deciduous forests elsewhere (oak–chestnut, oak–hickory, and southern mixed hardwoods). Montane coniferous forests of spruces and hemlocks developed at high elevations in the Appalachians, and mixed hardwood–coniferous forests formed to the north featuring maples, hickories, chestnuts, and beeches, with cooler forests of larches, pines, hemlocks, and birches. Western forests had developed as woodland–savanna, mixed hardwood–conifer, and western coniferous forests at high elevations (Graham 1999).

We do not expect that historical plant communities would exactly resemble those of modern forests, nor do we expect that past species–site relationships would remain today. However, where the physical components of physiography, microclimate, and soil were generally similar to those today, certain ancestral genera of the canopy and groundflora layers may have been present in combinations similar to those that exist today. Such appears to be the case at least by the end of the Cenozoic (1.6 Ma), with various genera grouped together into associations that are familiar today. Among many important events at this time are included the development of a familiar deciduous formation of maple–basswood; a refinement of an Appalachian montane forest that featured firs, spruces, and hemlocks; and the appearance of a near-modern boreal coniferous forest (Graham 1999). Similarly, remnants of the wide-ranging dawn redwood forests of the Tertiary Period forests 50 Ma still exist today, discovered only in 1949 in a remote valley in the northern Hubei Province of China. Many genera of the Tertiary forests of northeastern Oregon (maples, alders, *Ailanthus*, hackberries, beeches, hornbeams, pines, oaks, hemlocks, elms, etc.), where the dawn redwood once grew, are also found today in the location in China where the dawn redwood presently grows (Tallis 1991).

PLEISTOCENE GLACIATIONS

Ice accumulates on the Earth's surface as the climate becomes progressively colder. The "Pleistocene Ice Age," beginning about 2.6 Ma, consisted of multiple glacial–interglacial cycles (currently estimated at about 17) of about 100 000 years duration. Data from a series of ice cores at Vostok, East Antarctica dating into the late Pleistocene suggest that cold glacial stages lasted 70 000–90 000 years, as glacial ice sheets built up on continents. These glacial stages were directly followed by a short and warm stage of about 10 000 years, though interglacial periods also varied in length (Figure 15.1). The last continental cold–warm cycle in North America, the Wisconsin Glaciation, reached its maximum extension 25 000–21 000 years ago. Thereafter, the climate warmed, and the sudden warming about 11 650 years B.P. (Before Present) marks the beginning of the present Holocene Epoch.

Previous glaciations eliminated forests when ice sheets overrode the land, with the exception of a few small refugia (Figure 15.2; see discussion in the following text). The present forest vegetation of these now-deglaciated lands therefore dates from the last continental retreat of Pleistocene ice sheets. Importantly, major changes in whole regional and local landscape ecosystems occurred throughout the Pleistocene. New landscapes were created by the dynamics of the glaciers and

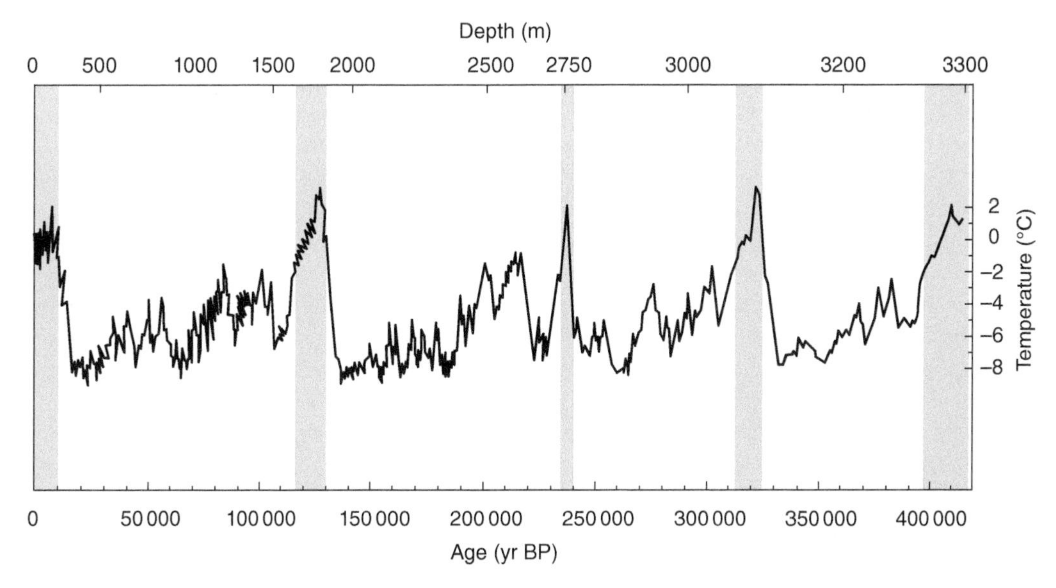

FIGURE 15.1 Temperature data determined from the Vostok ice core for the last 400 000 years incorporating five interglacial periods (shown as shaded columns). *Source:* Petit et al. (1999).

FIGURE 15.2 Maximum extent of continental glaciers in North America during the last ice age. Major refugia are marked by letters: A (southern refugium), B (Beringia), and C (Canadian Arctic Archipelago). *Source:* Levsen and Mort (2009) / with permission of Canadian Science Publishing.

through the action of waters as the ice disintegrated. These new features affected climate at regional and local scales in ice-covered and adjacent lands. The detailed consideration of these features is beyond our scope, but we discuss the significance of physiography in glacial landscapes for forests in Chapter 8.

ECOSYSTEM AND VEGETATIONAL CHANGE SINCE THE LAST GLACIAL MAXIMUM

As glacial ice receded, new physiographic features and their soil parent materials were created and colonized by organisms. Sediment-transporting meltwater formed new outwash plains, and lakes formed and reformed, reworking the lands they covered. Forests migrated onto the deglaciated land as the ice melted, and tree genera and certain species with distinctive pollen morphology in the prehistoric record clearly indicate dramatic changes that occurred across landscapes and waterscapes over the last 25 000 years. Pollen has been used to track and map the directions and rates of tree population migrations using fossil pollen and plant parts and by reconstructing forest composition (Delcourt and Delcourt 1987). The study of prehistoric plants is known as **Quaternary biogeography** if the emphasis is on geographical distribution of the flora, and as **Quaternary paleoecology** if environmental changes are of primary concern.

The composition of forests in times past was determined by the combination of geologic processes and interrelated climatic events. Climatic cooling and mountain building in the Tertiary Period were major factors in western North America (Graham 1999), but Pleistocene glaciation and postglacial climate have been the primary influences on ecosystem change and forest distribution and composition in eastern North America. Vegetation in temperate regions has been in an almost constant state of instability and adjustment due in part to an almost continuously changing climate over the past 20 000 years and even more fundamental changes in the configuration of the land and the composition of its surficial materials (Davis 1986). Overall, vegetation changed slowly during the glacial maximum until 17 000 years B.P., quickly between 16 000 and 8000 years B.P., then slowly again after 7000 years B.P. until vegetation change accelerated at about 500 years B.P. due to the impacts of European activities (Williams et al. 2004). Importantly, however, climate varied at every temporal scale important to forests, from years to decades to centuries to millenia (Jackson 2004). Present vegetation patterns are closely related to events in recent geological history, of which climatic change is one important part.

EASTERN NORTH AMERICA

The major features of climatic history in eastern North America during the end of the Pleistocene were summarized by Webb et al. (1993) and confirmed with the Vostok ice cores (Petit et al. 1999): (i) climate changed continuously from 18 000 years B.P. to the present, (ii) a major increase in mean annual temperature occurred between 15 000 and 9000 years B.P., (iii) the greatest seasonal contrast between cold winters and hot summers took place between 12 000 and 9000 years B.P., (iv) the modern spatial gradients for July temperature and annual precipitation first appeared at 9000 years B.P., (v) the time of maximum warmth was about 6000 years B.P. when temperatures were not more than 1–2 °C higher than today; and (vi) the time of maximum dryness in the last 12 000 years was 6000 years B.P. in the Midwest and 9000 years B.P. in the northeastern United States.

In addition to the prevailing climate, key characteristics of tree species affect their migration patterns and community composition. The potential range of environments in which trees can colonize and grow is far greater than what we see in their present-day natural occurrences for several reasons. Tree species persisting today have evolved a balance of genetic and non-genetic mechanisms to maintain fitness in a given environment or the flexibility to change with changing conditions (Chapter 3). Trees that have persisted for millions of years in temperate and boreal environments have evolved mechanisms for long-distance dispersal using wind, water, and mutualisms with birds and mammals. Competition may severely restrict forest species to a relatively small part of their potential geographic range, that is, their realized niche (Chapter 13). Finally, many forest trees have adaptations enabling them to colonize ecosystems with already-established communities and maintain themselves, depending on the specific site conditions (Chapter 21).

Overall Migration Sequence and Patterns An overview sequence of the location of various broad forest types that followed the last glaciation is presented in Figure 15.3. In eastern North America, the climate during glacial times was sufficiently cool even to middle latitudes, and forests were displaced to the south of a narrow band of tundra vegetation at the ice margin. During the late Wisconsin Glaciation, 18 000 years B.P., a boreal-like coniferous forest dominated by white spruce occurred on both sides of the Mississippi River Valley south to about 31° N latitude (Figure 15.3a). Just as today, structure and composition of the full-glacial boreal forest varied across the region, and forest structure resembled an open woodland moving westward (Adams 1997). Boreal species such as jack pine and white spruce were displaced as far to the south as the coastal plain between northcentral Georgia and northern Florida (Watts 1970; Watts and Stuiver 1980; Graham 1999; Jackson et al. 2000). Temperate genera such as walnuts, maples, oaks, and beeches were likely locally abundant on moist soils in this region (Delcourt 1980; Jackson et al. 1997). Temperate hardwoods grew locally in the Lower Mississippi Valley as far north as 35° N latitude (Jackson et al. 2000), but white spruce, fir, and tamarack extended southward to nearly 31° N latitude, occupying extensive sand flats of braided streams that carried glacial meltwater sediments into the Gulf of Mexico (Delcourt and Delcourt 1977; Royall et al. 1991). Evidence of shifting sand dunes east of the Appalachians also suggests forests in this area were much more open than those today (Wells 1992; Adams 1997). Forests dominated by pine (mainly jack pine with local red and white pines) were found in this region, possibly as far south as the Florida Panhandle (Jackson et al. 2000).

The Gulf Coastal Plain was more forested than areas to the north, with a higher proportion of deciduous trees (mainly oaks and hickories) mixed with southern pines (Watts and Stuiver 1980; Delcourt and Delcourt 1981). Oaks and hickories may have been favored by dry summers, particularly on fire-prone sites in the area (Adams 1997). Some researchers have speculated that a narrow transition zone of cool-temperate conifers and northern hardwoods existed from northern Mississippi to South Carolina between the spruce and jack-pine-dominated forest to the north and this oak–hickory–southern pine region on the Gulf Coastal Plain (Delcourt and Delcourt 1981). Mixed mesophytic forest may have existed along the bluffs east of the Mississippi Valley and within protected valleys of major river systems of the Southeast. These bluffs are composed of loess soil (fertile, mesic, often calcareous wind-deposited silt), which is readily eroded to form steep ravines and banks that would have provided protected sites for the species-rich mesophytic communities (Delcourt and Delcourt 1979).

Tree populations began to migrate northward as the climate slowly became warmer and wetter beginning about 16 500 years ago. By 14 000 years B.P., spruce–jack pine forest had colonized lands formerly occupied by tundra as the ice sheet retreated, and mixed hardwoods–conifer forest, now dominated by deciduous species, expanded northward and eastward across the southeastern United States (Figure 15.3b; Watts 1979; Delcourt 1980; Whitehead 1981; Adams 1997). Oaks, in particular, expanded quickly northward from the southeastern United States sometime before 14 000 years B.P. (Williams et al. 2004). Mixed mesophytic forest also expanded to the north and east into the Appalachian Highlands at this time (Delcourt and Delcourt 1981). A period of cooling and drying at about 13 000 years B.P. caused severe winters in the Northeast and a reversion from a mixed hardwood–conifer forest (oak, ash, spruce, and fir) to one more closely resembling a boreal community (spruce, fir, larch, birch, and alder). Pines rapidly increased in abundance during this period in the Great Lakes region (Williams et al. 2004), largely replacing spruces probably due to a regional decrease in moisture availability (Shuman et al. 2002). The colder and drier climatic phase ended sometime near 11 500 years B.P. resulting in a decline in boreal vegetation in the northeastern United States toward a forest dominated by white pine and oak, and deciduous genera continued to increase in abundance and range elsewhere in eastern North America (Adams 1997). Red spruce and Fraser fir populations were stranded at high elevations in the southern Appalachians and persist today on some of the higher montane peaks.

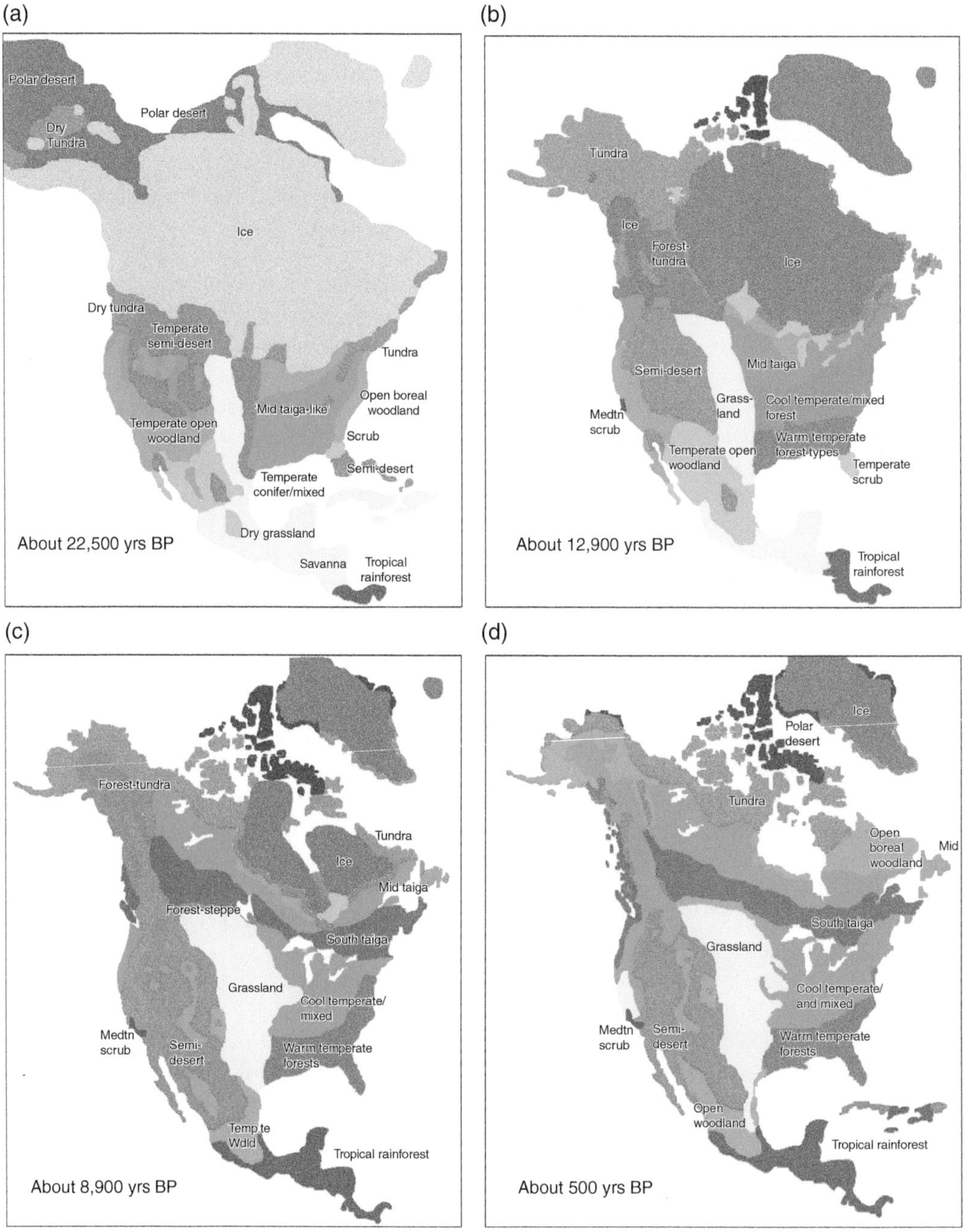

FIGURE 15.3 Maps of vegetation change for North America. (a) Map of vegetation for 22 500 years B.P. (b) Map of vegetation for 12 900 years B.P. (c) Map of vegetation for 8900 years B.P. (d) Map of vegetation for 500 years B.P. *Source:* Adams (1997) / Adams.

Major changes in vegetation continued throughout the Holocene, most of which was slightly warmer than at present. With increasing warmth and aridity in the Great Plains and a general increase in the prevailing westerly winds, oak–hickory forest, oak savanna, and prairie shifted eastward between 9000 and 7000 years B.P. (Figure 15.3c). Fire, fanned by the westerly winds across the predominantly flat landscape, was a major factor in pushing back and maintaining the prairie--

forest border. Reaching into Minnesota, Illinois, Indiana, southern Michigan, and Ohio, prairie openings gradually replaced mixed pine–deciduous forests to the north and elsewhere deciduous forests and oak savannas. This **Prairie Peninsula** (Transeau 1935; Stuckey 1981) is evidenced by the widespread occurrence of prairie soils, remnants of oak savannas, and by the persistent fragments of the prairie until the initiation of farming that occurred with European colonization (Whitney 1994). Pines and temperate hardwoods gradually replaced prairies, although prairie openings probably survived in eastcentral Minnesota until 4000 years B.P. (Wright 1971). The Prairie Peninsula posed a barrier to the northward migration of upland mesophytic species such as beech, hemlock, and tulip tree. These species migrated into the Midwest from the east.

After 7000 years B.P., mixed conifer–hardwoods forest had moved into southern Canada, the mixed mesophytic forest species were restricted to moist coves and slopes of the Cumberland and Allegheny Plateaus, and oak–chestnut communities dominated the southern and central Appalachians. Southern pines replaced oak–hickory forests on sandy uplands of the Gulf and Atlantic Coastal Plains, and oak–hickory–pine forests were restricted to the Piedmont Province and the Ozark and Ouachita Mountains. Throughout this interglacial period, regional physiographic features of plains, plateaus, and mountains were important in shaping the diverse vegetation types of modern regional ecosystems.

The present potential (pre-European colonization; Figure 15.3d) vegetation is similar to that at 5000 years except for a minor southward shift of the boreal forest and a retreat to the west of the prairie-forest boundary. These changes were caused by a long-term cooling trend at high latitudes and increased precipitation across the Midwest that began at about 6000 years B.P. In general, the present potential vegetation boundaries are similar to those of Braun's (1950) forest regions based on broad physiography (Figure 2.6, Chapter 2). This correspondence illustrates the importance of regional physiographic features in understanding the distribution of broad vegetation types. Macroclimate is an important but not isolated driver of vegetation distribution since the last glacial maximum. Moreover, multiple geologic and soil factors as well as disturbances such as fire, flooding, and windstorms acted at all spatial and temporal scales to shape the vegetation patterns of regional and local ecosystems.

Ecosystem Change in the Southern Appalachians Delcourt and Delcourt (1987) describe ecosystem change in both time and space for Mt. LeConte (elevation 2000 m) in Great Smoky Mountains National Park, eastern Tennessee to western North Carolina (Figure 15.4). At 20 000–16 500 years B.P., mean annual temperatures at elevations above 1500 m in the southern Appalachians were well below 0 °C, creating perennially frozen ground (permafrost), and permanent snowpacks. Intensive freeze–thaw churning of soil prevented establishment of trees, but the montane environments were suited for alpine tundra. Coniferous krummholz dominated lower elevations, at and below 1500 m. From 16 500 to 12 500 years B.P., rising temperatures to near 0 °C led to the development of a patchwork cover of tundra and krummholz above 1500 m. Boreal and cool-temperate forests became established at intermediate and lower elevations, respectively.

Climatic amelioration after 12 000 years B.P. made temperate stream processes more influential than other geomorphological processes. By 10 000 years B.P., conifers and northern hardwoods occupied mountain summits and upper slopes, and temperate deciduous hardwoods (including mixed mesophytic forest) dominated the lower elevations. Today, a great diversity of ecosystems comprises the landscapes of Mt. LeConte at all elevations (Figure 15.4), determined largely by the physiographic diversity in elevation, aspect, and slope characters of the mountain landforms.

WESTERN NORTH AMERICA

In western North America, the continental ice sheet did not extend nearly as far south as it did in eastern North America (Figure 15.2). Species were nevertheless displaced to the south, and mountain glaciation was widespread in the Rocky Mountains and other ranges south of the continental ice border. This mountain glaciation complicated the postglacial history in the Pacific

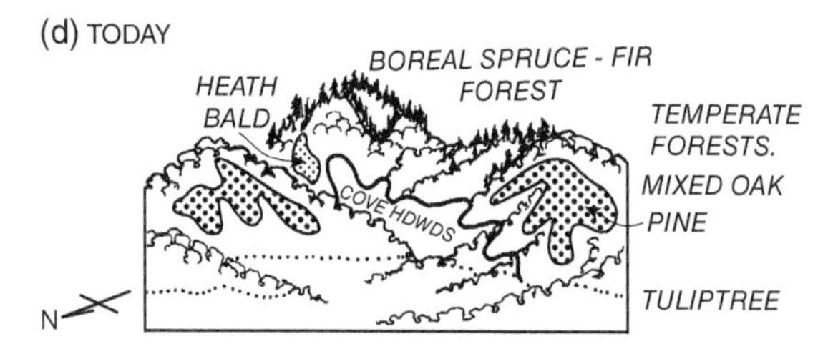

(c) 12,500 to 10,000 yr B.P.

KRUMMHOLZ

BOREAL FOREST

TEMPERATE DECIDUOUS FOREST

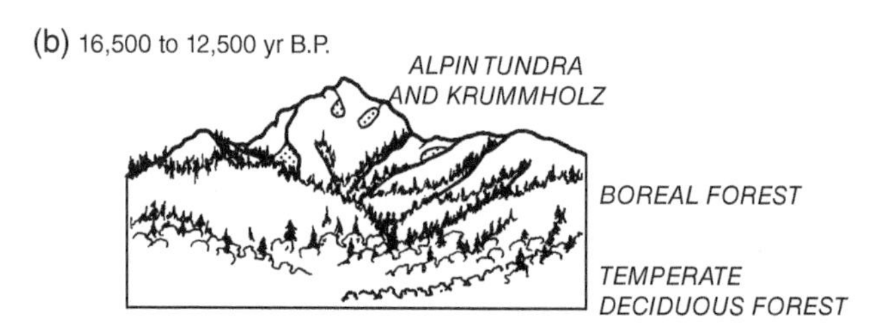

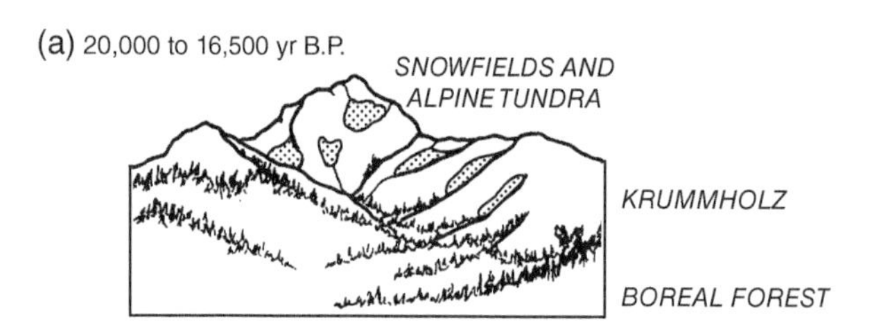

FIGURE 15.4 Diagrammatic illustration of vegetational changes on Mt. LeConte, Great Smoky Mountains National Park: (a) full-glacial interval, 20 000–16 500 years B.P.; (b) late-glacial interval, 16 500–12 500 years B.P.; (c) early-Holocene interval, 12 500–10 000 years B.P.; (d) today. *Source:* After Delcourt and Delcourt (1987) / Springer Nature. Reprinted from *Long-term Dynamics of the Temperate Zone*, © 1987 by Springer-Verlag New York, Inc. Reprinted by permission of Springer-Verlag New York, Inc.

Northwest (from Oregon north to Alaska) because the ice radiated east and west from the north–south-oriented mountain ranges rather than simply moving southward, as in eastern North America. In addition, maritime climatic conditions originating from the Pacific Ocean moderated at least some of the postglacial climatic fluctuations compared to those found further east. Therefore, refugia for various plants and animals along the Pacific Coast, including Sitka spruce and western hemlock, served to restock much of the glaciated terrain rather than migration from the south as in eastern North America (Hansen 1947, 1955; Heusser 1960, 1965, 1983; Barnosky 1987).

Near the last glacial maximum (22 500 years B.P.), most of the Pacific Northwest was alpine tundra with only scattered areas of spruce (Whitlock and Bartlein 1997). During this period, the central Rockies included a mosaic of open conifer forests and dry shrubland, depending on altitude, and sparse temperate forest in the Southwest and along the Pacific Coast (Figure 15.3a; Adams 1997). A late-glacial cold period about 14 500 years ago was followed by a warmer and drier trend that ultimately gave way to a cooler, more humid climate. This period was also important for the retreat of the ice such that a narrow ice-free corridor linked Alaska to the continental

United States (Figure 15.3b). Boreal forest was present in the lowlands of Washington State, and birches became common in the tundra regions of Alaska and northwest Canada. As the ice-free corridor widened by about 13 000 years ago, spruce parkland and forest developed in western Canada. This conifer-dominated area spread to the eastern foothills of the Canadian Rockies by about 11 000 years ago. Some studies have suggested that a lodgepole pine parkland developed in the Pacific Northwest and northwestward as far as southeastern Alaska (MacDonald and Cwynar 1985). Lodgepole pine reached its current most northward locality possibly a century ago (Cwynar and MacDonald 1987) and migrated at an average speed of about 180 m per year over a distance of 200 km in about 12 000 years (18.3 km per century). As the climate ameliorated, Douglas-fir became the dominant species of the Pacific Northwest during the warmer and drier times of the early-to-mid Holocene. Where rainfall was heavy, western hemlock replaced pine and mountain hemlock, and Sitka spruce became well represented from British Columbia to southeastern Alaska. In the Willamette Valley, the succession was to Douglas-fir and Oregon oak, and grassland achieved dominance on the dry plateaus east of the Cascades. As in the East, the approximate boundary dates of this period of maximum warmth and dryness are 8000–4000 years B.P.

PATTERNS OF TREE GENERA AND SPECIES MIGRATIONS

Pollen analyses have been used to describe the individual migration pathways, directions, and rates of various genera of forest trees of eastern North America rather than the location of broad vegetation types. Classic studies by Davis (1983a, b) examined migration routes and arrival times of spruces, white pine, beech, and chestnut (Figure 15.5), among others. Many trees migrated generally from southern refuges northward along a broad front as seen for the spruces (Figure 15.5a). In contrast, white pine (Figure 15.5b) and hemlock migrated north and west from refugia on the central coastal plain or on the exposed continental shelf (Davis 1983a, b). American beech (Figure 15.5c) migrated northward, and after about 9000 years ago expanded westward through the Great Lakes region. American chestnut (Figure 15.5d) migrated through the Appalachian Mountains in a northeasterly direction and arrived in New England after most of the other tree species with which it was associated.

Early-successional species with many tiny wind- and water-dispersed seeds migrated at the highest rate (Godwin 1956). For example, aspens expanding into tundra areas averaged a rate of advance of 263 m per year from 14 000 to 8000 years B.P. The comparable migration rates for heavy-seeded species such as oaks and hickories were 126 and 119 m per year, respectively. In general, availability of suitable sites for establishment and subsequent growth was more limiting than dispersal rate for migration rates for boreal and temperate trees in eastern North America (Delcourt and Delcourt 1987). Physiography and soil therefore played important local roles in glacial–interglacial range expansion and late-Quaternary community dynamics.

Birds and other animals were important seed dispersal agents and aided in establishment for many plant species during post-glacial periods. These mutualisms are significant in tree species of the beech family, especially for oaks, American beech, and American chestnut (Johnson and Webb 1989), all of which are known to be dispersed and cached by blue jays. Blue jay acorn dispersal may explain the rapid expansion of the range of oaks in late glacial times relative to beech and chestnut. The slow migration rate of chestnut may be partially explained by the preference by jays for small nuts relative to chestnut. Beech migration was probably limited more by climate than by dispersal (Johnson and Webb 1989), as well as by its narrow site requirements for establishment and late seed-bearing age. When dispersal occurs over long distances (typically by birds), plants may reach suitable sites for establishment far ahead of the leading edge of their main advance. Similarly, strong winds can place small seeds in outlying sites far ahead of the advancing front. For example, by 12 000 years ago aspen populations had established in Maine,

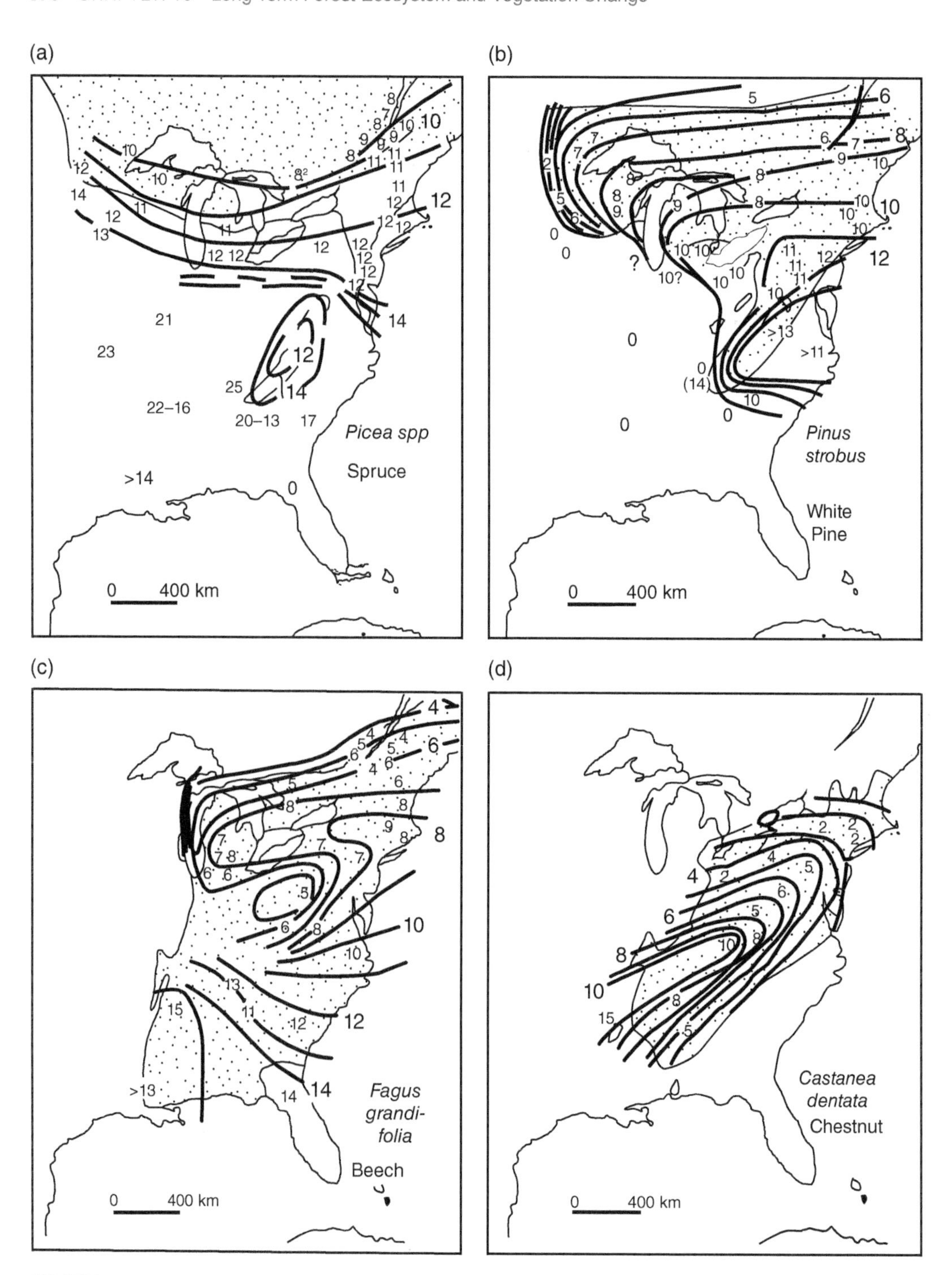

FIGURE 15.5 Migration routes of tree species in eastern North America during the late-glacial and early Holocene. Small numbers on the map indicate the time of arrival at individual sites. Contours show the leading edge of population advance at 1000-year intervals. Shaded area is the range for the genus or species prior to European colonization. *Source:* After Davis (1983b) / Missouri Botanical Garden Press. Reprinted by permission of Margaret B. Davis and the Missouri Botanical Gardens.

then 700km east of the continuous range margin for aspen in the Lake States (Davis and Jacobson 1985; Delcourt and Delcourt 1987). In the western Great Lakes region, hemlock was advanced by long-distance dispersal of seeds, forming isolated populations ahead of the main front (Davis 1987).

MIGRATION IRREGULARITIES AND DISTURBANCE

In imagining the re-vegetation of recently deglaciated terrain, we envision smooth waves of trees and other organisms establishing and migrating over the newly available sites to reach their present areas of distribution. Importantly, migration appears "smooth" only at the temporal scale of millennia, and migration over centuries would appear to respond more directly to barriers or corridors such as major rivers or mountain ranges. The pollen record has recognized numerous irregularities from smooth advances even at long temporal scales, such as the migration of American beech around or across Lake Michigan (Davis et al. 1986; Webb 1986; Davis 1987; Delcourt and Delcourt 1991) and patterns of hemlock expansion in the western Great Lakes region (Davis et al. 1986; Davis 1987). In New England, early colonization of sites by elms is noted 300–700years earlier than the main "explosion" in elm pollen values (Webb et al. 1983; Davis and Jacobson 1985).

Once initial re-vegetation occurred, disturbances by wind, flooding, and fire were undoubtedly very important in providing open sites for establishment of migrating species. Both severe and minor disturbances were important in incorporating late-arriving species into existing forest communities. Fire, in particular, provided suitable seedbeds, temporarily reduced competition, and generally favored the migration and maintenance of all boreal species, pines, oaks, and many other temperate species. Even a species such as eastern hemlock, although perceived as fire-sensitive, establishes readily on fire-prepared seedbeds. For example, eastern hemlock colonizes burned areas at the same time as trembling aspen, paper birch, or white pine, but grows far more slowly, reaching the overstory only after 100years or more. This pattern has been recognized in the pollen record (Swain 1973, 1978) as initial high proportions of aspen and birch pollen, and increases in white pine pollen after 50years, and hemlock after 100years (Swain 1973, 1978). The relationship of climate change and fire regimes and their collective influence on temperate forest composition have been demonstrated in several paleoecological studies (Green 1981; Clark 1989, 1990) and reviewed by Patterson and Backman (1988) and Delcourt and Delcourt (1991). Again, macroclimate is a main driver of ecosystem and vegetation change, but geomorphic processes also change in time and interact with climate at regional and local levels. These geomorphic processes also interact and drive disturbances. Thus, long-term vegetation change is a complex ecosystem process involving many interrelated factors.

MIGRATION FROM GLACIAL MICROREFUGIA

When considering migration of species during Pleistocene deglaciation, pollen studies are unlikely to detect small or outlying populations. Reconstruction of the late-glacial distribution of vegetation relies on the amount of pollen produced and deposited in sediments. A high proportion of pollen of a species in a sediment suggests that the species was not only present, but present at an abundance high enough to produce most of the pollen found in the sample. The converse is not necessarily true. When pollen of a species is lacking within a sample, it is only be concluded that there was not enough pollen produced by the species to leave a substantial record in the sample. Often the conclusion is that the species was absent at that location, but it might also mean that the species was present in small pockets distant from where the sample was taken, or that the species was present but not abundant enough to leave a pollen record. In short, pollen data are very good at indicating the distribution of species that reach higher densities, but they are far less effective at identifying small populations, or even widespread, sparser populations (McLachlan and Clark 2004).

Understanding the details of paleoecological studies is important for how we estimate the rates of migrating taxa. Accurately assessing the migration rates of forest taxa during the rapid changes in climate during deglaciation has become increasingly important in the current era of climate change, because it provides some perspective on how quickly these taxa will adapt as the contemporary climate continues to change (Chapter 20). As described earlier in this chapter, many temperate species and genera present in eastern North America today are thought to have originated mostly from the southeastern United States during the last glacial maximum, with those species having moved northward as the ice sheet retreated (Figures 15.3 and 15.5). This notion of Pleistocene vegetation, first advocated by Edward Deevey (1949), reasoned that the colder climate that accompanied the ice sheet displaced temperate species far south of their current distributions. These regions of **refugia** generally include central Texas, the Lower Mississippi River Valley, the Florida Peninsula, the Atlantic and Gulf Coastal Plains, and the southern Appalachians (Soltis et al. 2006; Jaramillo-Correa et al. 2009). An alternate argument is that many temperate eastern taxa maintained smaller populations closer to the ice margin during the last glacial maximum (Jackson et al. 2000; Delcourt 2003), as initially evidenced by the presence of plant fossil specimens discovered farther north. As a counter to Deevey's hypothesis of southern refugia, E. Lucy Braun (1950) argued that more northern regions, particularly the Cumberland Plateau of western Kentucky and Tennessee, served as a smaller refugium for temperate taxa (today recognized as those of the mixed mesophytic forest), although the surrounding region was dominated by boreal vegetation. These northern **microrefugia** would have been small populations surviving on suitable sites in an otherwise unsuitable climate. If microrefugia, particularly those located closer to modern species' ranges, contributed to post-glacial re-vegetation, then ignoring them may overestimate the species' migration rates (McLachlan et al. 2005). Such populations would have had to persist throughout the last glacial maximum to be likely sources of re-vegetation, so their importance is often ambiguous, and the focus remains on the displacement of taxa into southern geographical areas.

The advent of molecular and genomic methodologies in the early twenty-first century provided strong evidence for refugia that pollen studies were unable to detect. These techniques can be used to compare the genetic structures of living trees from across their geographic range and thus reconstruct the migration paths of populations as they expanded. In a classic study, chloroplast DNA was used to examine the migration routes and rates of two temperate deciduous species in eastern North America: red maple and American beech (McLachlan et al. 2005). Pollen-based studies had concluded that red maple originated from a refugium near the Lower Mississippi River Valley, migrating northward and westward into the Great Lakes region and the Northeast. Molecular evidence, however, suggested that more northerly populations from the Appalachian Mountains contributed most to the spread of red maple to the north. Likewise, American beech was assumed to have originated from a refugium on the Gulf Coastal Plain based on pollen evidence (Davis 1981; Delcourt and Delcourt 1987), but the genetic structure of most beech populations in the Upper Great Lakes region did not resemble that of beech found in traditional southern refugia. Thus, low-density populations of red maple and American beech likely persisted near the ice sheet rather than in southern refugia, and at least some of these populations were influential in post-glacial re-vegetation. It is possible that additional small populations that made little or no contribution to re-vegetation also existed during this period, which are not detectable using these methodologies.

Modern genomic evidence does not suggest that every species existed in microrefugia north of their southern refugia during the last glacial maximum. For example, hickory species are thought to have originated from the Gulf Coastal Plain based on reconstructions from pollen data (Watts and Stuiver 1980; Delcourt and Delcourt 1981). Genomic evidence, however, suggests that bitternut hickory expanded from a northern microrefugium near the confluence of the Mississippi and Ohio Rivers where habitat was otherwise inhospitable (Figure 15.6a), whereas pignut hickory indeed expanded from a main southern refugium in central Mississippi on the eastern Gulf Coastal

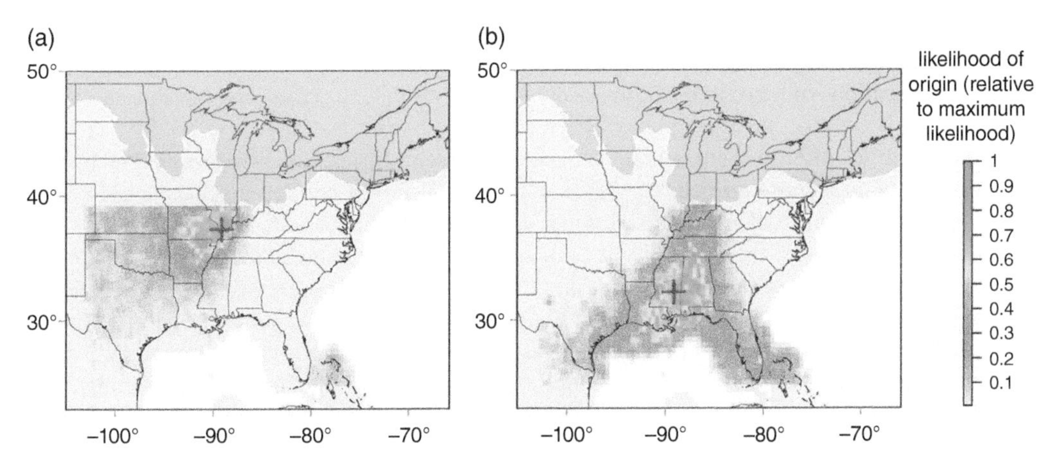

FIGURE 15.6 Expansion origins for (a) bitternut hickory and (b) pignut hickory in eastern North America during the last glacial period based on genomic evidence. The red cross represents the maximum likelihood of the point of origin, with the color ramp indicating the probability of alternate points of origin. Blue shaded area represents glaciated regions. *Source:* After Bemmels et al. (2019) / with permission of PNAS.

Plain (Figure 15.6b; Bemmels et al. 2019). Increasing molecular evidence will undoubtedly show that both northern microrefugia and traditional southern refugia were important in the re-vegetation by temperate taxa of deglaciated terrain during the late Pleistocene and early to mid Holocene (Soltis et al. 2006; Jaramillo-Correa et al. 2009).

INDEPENDENT MIGRATION AND SIMILARITY OF COMMUNITIES THROUGH TIME

Paleoecologists have made enormous advances over the years in determining the general patterns of response of a variety of tree genera to the changing conditions of the last 20 000 years. Most paleoecological literatures of the past two decades acknowledge that plant taxa have migrated independently of each other in response to climatic changes, and that plant communities arise fortuitously (e.g., Graham 1999). This individualistic nature of plant taxa migration includes not just the rate of migration, but also the timing of migration and its direction of range expansion and taxa abundance (Williams et al. 2004). As a result, many of the tree genera and species that we observe growing together today did not co-occur in the past, and many associations or communities recognized during the Holocene have no modern analog (Williams et al. 2004; Williams and Jackson 2007). An example of an "extinct" community is one that featured spruces, sedges, ash, and hornbeams. This community was very common in eastern North America between 17 000 and 12 000 years ago, but today the geographic ranges as well as the local site requirements of these taxa overlap far less, if at all.

By contrast, many modern forest communities date at least to the early Holocene, such as several boreal assemblages, oaks and hickories, and American beech and eastern hemlock. In some cases, such as with oaks and hickories, the historical formation can be detailed only to genera by the pollen record, and it remains unclear whether the same species were associated together during the Holocene as are associated today. Modern genomic data have the capability to phylogenetically differentiate among species within genera (e.g., Bemmels et al. 2019) and should soon address many of these issues. Other taxa associated today had similar geographic ranges in the past and had the opportunity to co-occur geographically and ecologically. Notably, the range expansion of some tree species, such as hemlock and beech, lagged behind those of other temperate species with which they are associated today, and did not migrate at the same rate or along the same paths during the Pleistocene (Davis 1983a). Using pre-European colonization ecosystems of 300 years ago as a baseline, ecosystem and community similarity would tend to be least similar in the early stages of migration and become progressively similar as these species colonized established communities.

Although we often recognize plant formations present during the glacial period as resembling modern forest communities (e.g., spruce–jack pine, mixed conifer–northern hardwoods, mixed mesophytic, oak–hickory–southern pine), their occurrence does not necessarily imply that the local ecosystems and their communities of the past are the same as those of today. This is because the similarity of communities, past and present, depends on the similarity of the physical environment and disturbance regimes of the site where the plants grow. Some species are similar enough in their site requirements and tolerances that they respond similarly to disturbances of fire, flood, windstorm, and specific site conditions of soil water and nutrient availability. For example, we would expect combinations of "fire species" such as black spruce, tamarack, and leatherleaf; jack pine and red pine; eastern white pine and eastern hemlock; and loblolly and shortleaf pines to be more often associated together (with characteristic groundflora species) in fire-prone ecosystems of their respective regional ecosystems rather than growing in floodplain ecosystems with willows and cottonwoods or elms and silver maples. Thus, where regional gradients or local environments of the past were similar to those of today, it is probable that certain species occurred in characteristic site-species patterns not unlike those of today. There is no exact correspondence between present and past communities, nor can entire intact communities migrate together because they are structurally and functionally linked to physiographic and soil features of landscape ecosystems that do not migrate.

SUGGESTED READINGS

Davis, M.B. (1983). Holocene vegetational history of the eastern United States. In: *The Late Quaternary Environments of the United States, The Holocene*, vol. 2 (ed. H.E. Wright and S. Porter). Minneapolis: University of Minnesota Press.

Delcourt, P.A. and Delcourt, H.R. (1987). *Long-Term Forest Dynamics of the Temperate Zone*. Springer-Verlag 439 pp.

Delcourt, H.R. and Delcourt, P.A. (1991). *Quaternary Ecology, a Paleoecological Perspective*. New York: Chapman and Hall 242 pp.

Graham, A. (1999). *Late Cretaceous and Cenozoic History of North American Vegetation*. New York: Oxford University Press 350 pp.

McLachlan, J.S., Clark, J.S., and Manos, P.S. (2005). Molecular indicators of tree migration capacity under rapid climate change. *Ecology* 86: 2088–2098.

Williams, J.W. and Jackson, S.T. (2007). Novel climates, no-analog communities and ecological surprises. *Front. Ecol. Environ.* 5: 475–482.

Williams, J.W., Shuman, B.N., Webb, T. III et al. (2004). Late-quaternary vegetation dynamics in North America: scaling from taxa to biomes. *Ecol. Monogr.* 74: 309–334.

Disturbance

CHAPTER 16

Disturbances affect and often drive ecosystem composition, structure, and function by killing organisms and changing site conditions. Disturbances selectively disrupt biota and initiate succession, the course and rate of which are often directed by disturbance-caused changes in composition, species interactions, and spatial patterns of organisms. These changes in species composition occur in the context of disturbance effects on parent material, soil, and hydrology. All natural ecosystems have been shaped by disturbances in some form, from plate tectonics and volcanism to local avalanches, floods, and fires. Disturbances are important functional components and a vital part of all ecosystems over long periods of time.

Particular kinds of natural disturbance are characteristic of particular kinds of landscape ecosystems—such as avalanches in mountains, fires in dry plains, hurricanes along coastal regions, or floods in river valleys. Disturbances influence changes difficult to observe, such as the genetic differentiation of plants of these ecosystems, as well as the obvious vegetation changes in ecosystem structure and composition. Many kinds of natural disturbances were prevalent in the forest ecosystems of North America long before the first migrants arrived from eastern Asia. Humans thereafter significantly disturbed environments in new ways as they developed cultural landscapes. For many decades, researchers assumed pre-European colonization forests to be undisturbed old-growth, Clemenstian "climax" forests (Whitney 1994). Moreover, most forest ecologists agree that widespread and localized disturbance patterns of wind, fire, and other disturbance types associated with regional ecosystems and landforms were commonplace (Frelich 2002). Whitney (1994) observes: "If the primeval forest did not consist of stagnant stands of immense trees stretching with little change over vast areas (Cline and Spurr 1942), neither was it an amalgamation of pioneer species recovering from one form of disturbance or another."

Human-caused disturbances are superimposed on those of natural ecosystems, and human alteration of original landscape ecosystems has been enormous throughout the world. Humans significantly affect the soil, microclimate, and biota wherever they occupy a region as they develop a culture. Trees have been the principal fuel and building material of almost every society for over 5000 years, accompanied by deforestation (Perlin 1989), and specific evidence of widespread human disturbance has been documented for North America as early as the sixteenth century (Denevan 1992). Working in the eastern United States, Abrams and Nowacki (2019) used tree pollen and charcoal records to compare pre-European colonization and modern forests, finding that modern forests contain tree species more shade-tolerant and drought-intolerant and less fire-adapted than early forests. As such changes could not be explained by climate alone, they concluded that vegetation changes were likely driven by reduced burning, associated with shifts in human population and land use well prior to European colonization. Whitney (1994) has described disturbances of all kinds, natural and human, and their consequences from 1500 AD to the present for eastern North America. Overall perspective and details of disturbance are provided by Pickett and White (1985), Oliver and Larson (1996), Coutts and Grace (1995), and Johnson and Miyanishi (2001). Here, we consider the concept of disturbance and the kinds of disturbances that bring

Forest Ecology, Fifth Edition. Daniel M. Kashian, Donald R. Zak, Burton V. Barnes, and Stephen H. Spurr.

about secondary succession—fire, wind, insects and pathogens, logging, land clearing, biotic changes—as an introduction to a detailed examination of succession as an ecological process in Chapter 17.

Forest ecologists and managers are concerned mainly with how disturbances initiate ecosystem change, specifically regeneration of forest species and changes over time in composition, stand structure, and biomass. These changes to the existing forest are often described as forest dynamics or secondary succession, in contrast to primary succession, which occurs on previously unvegetated sites such as water or bare soil or rock. Disturbances of one sort or another in any forest region are constantly altering the course of forest dynamics and initiating or restarting secondary succession. Understanding how disturbances shape ecosystems remains one of the most exciting fields of forest ecology.

CONCEPTS OF DISTURBANCE

As humans increasingly alter natural systems in novel and unprecedented ways, the concept of disturbance in ecology has become somewhat more complex than it appears at first consideration.

DEFINING A DISTURBANCE

The word "disturbance" in ecology is generally interpreted to mean *any relatively discrete event in time that disrupts the composition, structure, and function of ecosystems.* Disturbances typically kill some but not all of the organisms in an ecosystem, and in the process resources such as nutrients, water, growing space, or light are made available to new or surviving organisms (Pickett and White 1985). When disturbances are natural (not caused by humans), such as wildfires, windstorms, hurricanes, or insect outbreaks, they are typically part of a well-defined historical and repeating disturbance regime (Pickett and White 1985; Turner 2010). That is, natural disturbances are usually neither unique nor novel events, but part of a recurring series of events that have long shaped the ecosystems and landscapes where they occur. Human disturbances on the other hand, such as logging or land clearing, may be particularly unique to a given ecosystem in terms of their severity, frequency, or seasonality, and therefore may elicit very different ecological responses.

An important component in defining disturbances is that they are discrete in time whether natural or human-caused, meaning they have a beginning and an end, and thus are absolute and not relative (Pickett and White 1985; White and Jentsch 2001). Non-discrete events that affect ecosystems are more aptly categorized as **perturbations** rather than disturbances, where some defining parameter of the system changes and the system departs from its normal behavior or trajectory. Perturbations are typically gradual changes and are often caused by humans; climate change is perhaps the best example of a perturbation. Keeley and Pausas (2019) identified common perturbations to fire regimes (see description in the following text) as human-caused changes in specific fire parameters, such as alterations in the frequency of fire, the pattern of burning across landscapes, the amount and types of fuels accumulated in some ecosystems, or the intensity of fires. Ecosystems adapted to fire are not necessarily threatened by fire itself but by changes in the way fire burns in ecosystems.

The definition of disturbance given in the earlier text is not perfect and is not synchronous with all other definitions of disturbance. For example, disturbances have also been defined primarily based upon their duration. Many ecologists today think about disturbances in terms of "pulse" and "press" disturbances (Bender et al. 1984). **Pulse disturbances** are typically short-term, abrupt, easily delineated disturbances, such as fires, floods, or even timber harvest, after which the ecosystem responds and recovers. By contrast, **press disturbances** are disturbances of a prolonged intensity that have long-term impacts on an ecosystem, often resulting from changes to its physical structure. Examples of press disturbances in forests might be clearing of forests for agriculture, chronic browsing by deer or elk, or climate change. This categorization of disturbances is simple, and it clearly incorporates long-term stressors and gradual changes into the definition. The duration of a disturbance has important implications for how an ecosystem recovers. Understanding the length of time that has passed since the disturbance ended is equally important for understanding the structure and function of ecosystems.

Generally, an individual natural disturbance event occurs within a sequence of similar events over space and time for a particular ecosystem, known as a **disturbance regime** (White and Jentsch 2001). Disturbance regimes are typically characterized using descriptors of average disturbance frequency (how often the disturbance occurs in the ecosystem), size (the typical area disturbed by an individual disturbance event), intensity (the physical energy of the event, such as wind speed or the heat released by a fire), and severity (the effect of the event on the ecosystem). Often disturbance regimes are described in terms of one type of disturbance (e.g., a fire regime or flood regime), and this is characteristically the most pervasive disturbance type for a particular site. However, a disturbance regime may include different types of disturbance agents. For example, the fire regime of Yellowstone National Park is characterized by large (>5000 ha), infrequent (occurring every 100–300 years), stand-replacing wildfires (most or all trees are killed in a burned patch; Romme and Despain 1989). However, small (1–5 ha) stand-replacing fires occur between large fires almost every year (Despain 1990), widespread bark beetle outbreaks occur every 20–40 years (Cole and Amman 1980), and individual trees are uprooted by wind at least weekly. Each of these disturbances occurs at its own spatial scale (from a single tree to a forest >5000 ha) and temporal scale (from weekly or daily events to those that occur every few centuries). Notably, these different disturbance types may interact and affect the frequency and intensity of the others, often in novel ways. For example, bark beetle outbreaks have been shown to affect the probability of subsequent burning by large, infrequent fires in Yellowstone (Lynch et al. 2006).

Subjectively, the most important effect of disturbances on ecosystems is that they alter available energy, substrate, and community composition, creating spatial and temporal variability within and among ecosystems. The way ecosystems respond to disturbances has two primary components. **Resistance** is the ability of an ecosystem to remain unchanged when subjected to some driver of disturbance; generally, a more resistant system requires a stronger force to change it (Carpenter et al. 2001). For example, Sánchez-Pinillos et al. (2019) measured resistance in the boreal forest of Quebec, Canada as the similarity in forest structure and composition between the pre-disturbance state and the state immediately after disturbance. Black spruce forests were highly resistant to spruce budworm outbreaks because spruce bud burst and budworm larval emergence were not synchronized. Outbreaks that were severe or of long duration, however, had the potential to overcome this resistance and cause an acute disturbance event, leading to collapse of black spruce forests into treeless systems. By contrast, balsam fir forests showed low resistance to spruce budworm outbreaks, easily shifting toward dominance of paper birch, but the disturbed forests recover to the pre-disturbance composition quickly (Sánchez-Pinillos et al. 2019). This ability of an ecosystem to recover from a disturbance and return to its pre-disturbance state is called its **resilience**. Resistance and resilience together determine an ecosystem's **stability**, a term which has become ambiguous and cumbersome in the ecological literature (Grimm and Wissel 1997; May and McLean 2007).

The conceptual bases of resistance and resilience are often illustrated using "ball-in-cup" diagrams (Figure 16.1; Holling and Meffe 1996; Lamothe et al. 2019). Ecosystems—particularly biota—can persist in a **stable state**, a condition defined by the combination of the biotic community and the disturbance regime. For example, a lodgepole pine forest can be maintained through time by fire; the forest and disturbance regimes together constitute a stable state, though at any point some areas may be burned or in an intermediate stage of recovery.

Many ecosystems contain multiple stable states toward which the system may move after a disturbance (Beisner et al. 2003), with unique species compositions and disturbance processes. In

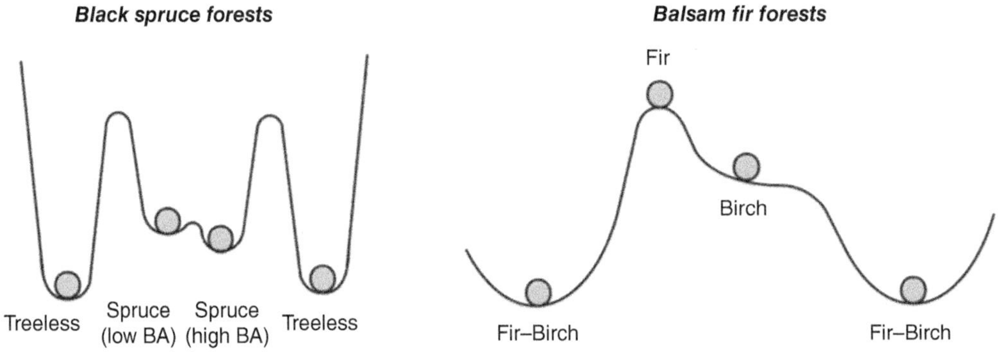

FIGURE 16.1 Ball-in-cup diagrams representing resistance, resilience, and multiple stable states in the boreal forest of Quebec. Black spruce forest has high resistance, represented by narrow deep cups, implying that a significant disturbance would be necessary to move spruce forests to treeless systems (though only a small disturbance would be needed to shift between high- and low-density forests). Balsam fir forests have low resistance but high resilience, represented by broad, shallow cups that would require less disturbance for movement among stable states. Notably, pure fir forests are extremely easy to shift toward a different species composition, thus the "ball" is located at the top of a narrow peak. *Source:* Sánchez-Pinillos et al. (2019).

a ball-in-cup diagram, the bottom of the cup represents a given stable state. In the black spruce forest example described in the earlier text, alternate stable states include high-density spruce forest, low-density spruce forest, and treeless former forests. Having high resistance, each stable state in the black spruce forest is represented as a ball in a steep-sided cup, because it would take a large force, or a severe disturbance, to move the ball out of such a position (Figure 16.1). By contrast, the balsam fir forest, with high resilience but low resistance, is represented by balls in broader cups with shallower walls, where transitions between stable states (in this case fir forest, birch forest, or mixed forest) occur easily (Figure 16.1).

DISTURBANCE AS AN ECOSYSTEM PROCESS

Importantly, any truly ecological definition of disturbance encompasses the disruption of all ecosystem components, such as landforms, soil, vegetation, and any of the processes that link these components. However, disturbance definitions often have an organismal focus, such as "a force that kills at least one canopy tree" (Runkle 1985), "the mechanisms which limit the plant biomass by causing its partial or total destruction" (Grubb 1988), or events that make growing space and other resources available to survivors or new colonists (Glenn-Lewin and van der Maarel 1992). These are certainly disturbances, but the concept does not end here. Disturbance is often conceived as representing death and destruction when organisms are the primary focus, but they are in fact fundamental natural events that occur in all ecosystems and, in many cases, are responsible for the stability of those organisms through time. Even recent recognition of the importance of disturbances for ecosystem structure and function tends to have an organismal focus in terms of recovery.

Flood, fire, or wind that kills forest organisms allows the study of recovery mechanisms and the sequence of organisms that replace those destroyed. Doing so, however, requires at least an understanding of the underlying ecosystem processes that mediate the changes following disturbance, and such an understanding results from conceiving disturbance itself as an ecosystem process. In doing so, we find that certain disturbances are characteristic of physiographic features (Chapter 8), for example, and disturbances may be expressed by rates of occurrence per unit of time (flooding and avalanche frequency, fire return interval) just as the rates of change of other ecosystem processes are expressed.

An ecosystem perspective of disturbance brings several insights worth considering:

Disturbance types are specific to regional ecosystems. Most natural disturbances are not ubiquitous; that is, their occurrence is not randomly located but rather is shaped by macroclimate and physiography at broad spatial scales. For example, some parts of North America are more susceptible to hurricanes (Gulf and Atlantic coasts) than other areas because of their location relative to the ocean and the movement of tropical storms. Severe thunderstorms and tornadoes occur in the mid-continent because of the relatively flat topography and their location where dry cold air from the north meets warm humid air from the south. Lightning-caused fires occur in the Intermountain West because of a dry climate resulting from Pacific storm precipitation blocked by the Cascade and Sierra Nevada ranges to the west. At finer spatial scales within these regions, specific ecosystem characteristics (physiography, soil, vegetation) and position of an ecosystem in relation to other ecosystems strongly influence the kind, frequency, and severity of disturbance (Chapter 8).

Disturbances affect sites as well as biota. The dramatic effects of disturbance on individual plants and communities are often the focus, but disturbances have many major effects on forest sites that impact subsequent regeneration and forest succession. Certain disturbances create new landforms, rearrange parent material, or otherwise change soil properties. These include volcanism, earthquakes, and glaciation at broad spatial scales and land movement (debris flow, mudslide, soil creep), floodwater, wind, and fire at finer scales. Wind creates new landforms such as dunes and loess deposits, and floodwater creates new landforms by cutting riverbanks and forming point bars, levees, and terraces. Wind also rearranges soils and creates diverse microtopography by uprooting trees (Chapter 8). Removal or killing of vegetation also changes the light and temperature regimes of the site. Thus, disturbances have far-reaching effects on regeneration and forest succession beyond their direct effects on biota through both severe and subtle site changes.

Disturbances vary in their effects. Different disturbances, such as fire, wind, and flooding, each affect site and biota differently, resulting in different processes of regeneration and succession. For example, wind primarily damages large, old, overstory trees rather than young stands or the understory of old stands. By contrast, fire almost always affects the understory—whether it is crown fire that kills the overstory and kills or even consumes the understory, or surface fire that kills or severely damages only the understory. Unlike wind and fire, flooding has more significant direct effects on landform and soil and more indirectly affects the biota, although understory plants are much more vulnerable to flooding than overstory trees. Fire more commonly maintains and regenerates early-successional communities compared to wind, whereas wind disturbance may hasten succession to more tolerant species if they are well established in the understory. Flooding maintains those plants tolerating or requiring its effects.

Disturbances are vital to ecosystems. Ecosystems well-adapted to the frequency and intensity of a disturbance regime may actually require the disturbance in order to maintain a normal, resilient system (Waring 1989). Perturbation of such a system then typically results from the *suppression or exclusion* of disturbances such as fire and flooding, often with new and unexpected consequences. Fire, for example, is a fundamental feature of many ecosystems, from southern pine forests to subarctic black spruce–jack pine forests. Fire suppression or exclusion in these systems becomes the perturbation to those systems with accompanying changes in ecosystem structure, composition, and function. In many fire-dependent systems, fire-dependent species, including many oaks and pines, fail to regenerate, allowing more fire-sensitive species to take advantage of the changed site conditions.

Disturbances leave both biological and physical legacies that shape subsequent forest dynamics. Natural disturbances such as fires, floods, windstorms, avalanches, or insect and disease outbreaks kill trees and other organisms, but they leave behind **biological legacies** in the form of snags (standing dead trees) and downed logs. The patchy nature of disturbances also leaves behind large living legacies of the pre-disturbance forest, including trees. Biological legacies help to initiate and shape secondary succession following the disturbance. Studies of post-disturbance regeneration have emphasized the importance of biological legacies for the rapid reestablishment of forests with high diversity of composition, function, and stand structure. By contrast, old agricultural fields and large clear-cuts offer few legacies. **Physical legacies** following disturbance include bare or rearranged mineral soil (after fires, floods, or landslides), new microtopography resulting from tree uprooting by wind, soil movement via debris flows, and the cutting and depositing actions of floodwater.

SOURCES OF DISTURBANCE

Where do disturbances originate? For decades, ecologists assumed that disturbances are external and are initiated from outside the plant community (e.g., Grubb 1988). Lightning strikes a tree; floodwaters kill plants intolerant of low O_2; spongy moths (formerly gypsy moths) or spruce budworms arrive *en masse*. In each of these cases, disturbances appear to originate from the outside because of an organism or community focus. When the focus is on ecosystems rather than organisms, disturbance frequency, extent, severity, and consequences are each closely related to the physical site conditions where the plants are growing rather than simply the plants themselves: the lightning-struck tree is on a ridgetop, not in the protected valley; the flooded plants are on the first bottom of the floodplain rather than the elevated second bottom. An ecosystem context is therefore paramount in understanding disturbance effects.

Examples are innumerable. In determining the frequency and severity of fire, site and biota are inseparable: dry, nutrient-poor sites favor conifers with flammable foliage, whereas moist, fertile sites favor species with rapidly decomposing leaves and reduced fuel accumulation. Windthrow is prevalent on sites where rooting depth is limited but where trees may grow tall and are susceptible to wind bursts, such as sites with high water tables or hardpan development. The interrelationship of water table and tree crowns is revealed whenever trees in or near wetlands and swamps are cut or die from disease: the water table rises due to lessened transpiration, risking injury or death to the root systems of other tree species. In Lower Michigan, the death of American elms in swamp ecosystems due to Dutch elm disease in the 1960s and 70s facilitated a rise of the water table and the subsequent death of basswood and red maple trees, initiating an even further raising of the water table (Barnes 1976). The roots of northern white-cedar were eventually overwhelmed as well, and only the most tolerant tree species of a high water table, black ash, remained—until it was mostly killed by emerald ash borer in the 2000s (Kashian and Witter 2011).

Many disturbances caused by humans are more impactful to site or biota than natural disturbances that occurred before human occupation of North America (about 12 000 years ago). Clearing of forests for agriculture drastically changed both site and biota. Widespread logging and the repeated severe fires that followed it destroyed forest understories of many areas of the upper Great Lakes in the late nineteenth and early twentieth centuries. These disturbances drastically changed conditions for regeneration and hence the structure and composition of forests. Modern commercial clear-cutting and timber extraction, followed by planting, generate simplified forests in structure and composition, and different in function from naturally regenerated ecosystems (Franklin 1995). Ongoing recovery of herbaceous layers in Appalachian cove forests following clear-cutting may require centuries (Duffy and Meier 1992). Fire and flood exclusions since the early 1900s have dramatically changed forest ecosystems, as have increasing urbanization and fragmentation of landscapes by roads, fields, and power lines and the introduction of invasive exotic species.

In summary, disturbance is ecosystem-specific, and its kind, source, frequency, severity, and consequences are best understood in this context. In the sections that follow, we examine several major disturbances that affect the structure, composition, and processes of forest ecosystems.

MAJOR DISTURBANCES IN FOREST ECOSYSTEMS

The most obvious disturbances in forest ecosystems are those that alter sites or affect the forest community by killing either the trees in the overstory or the trees and other plants in the understory.

FIRE

It was stated in the first edition of this textbook that *fire is the dominant fact of forest history* (Spurr 1964). Most forest ecosystems of the world—excepting only the perpetually wet rainforests such as those in southeastern Alaska, the coast of northwestern Europe, and the wettest belts of the tropics—have burned at fairly regular intervals for many thousands of years. Fire continues to be a major disturbance factor in most North American forests even under present-day conditions when awareness of forest fires is high and forest fragmentation interrupts the continuity of fuels across most landscapes. Organized fire suppression activities were mounted beginning in about 1900, becoming increasingly effective after 1945 and thereby greatly reducing the number and size of fires.

The incidence of fire was quite different prior to the twentieth century. Humans from many different societies and cultures throughout the world had until relatively recently no compunction about burning forests, and no desire, intent, or ability to put out either human- or lightning-ignited fires. On the contrary, fires have been set deliberately around the world for thousands of years to clear underbrush, improve grazing, drive game, improve game habitat, combat insects, without thought, or just for the hell of it. As more research is conducted into the ecological history of fire, the more it is obvious that frequent burning has been the rule, not the exception, for the vast majority of the forests of the world as far back as we have any evidence (Chapter 10).

Fires have been influential in the development of heathlands and moors of western Europe and the British Isles, for many savannas within tropical forest belts, for high-elevation meadows within the mountainous forests of the Americas, and, in general, for the persistence of grassland areas on upland sites within forest regions. Dominance of many pine and oak forests around the world is due to fire, as are the vast areas of Douglas-fir in the Rocky Mountains and Pacific Northwest and eucalyptus in Australia. The extensive areas of spruce in the boreal forest of North America and Eurasia are also structured to a great extent by past fires (Bloomberg 1950; Sirén 1955; Viereck 1973). Even hemlock–northern hardwood forests of Upper Michigan, mesic and dominated by sugar maple, have been regularly burned by fires, though very infrequently (Frelich and Lorimer 1991).

Though sweeping and perhaps overstated, these statements reflect the feeling of many forest ecologists who have increasingly realized the great importance of fires and subsequent succession in framing forest composition and structure (Cooper 1961; Agee 1993; Whelan 1995). Fire is clearly a disturbance that alters the composition and structure of an existing stand, but it is also a natural ecosystem factor whose effects have long been incorporated in species' adaptations and ecosystem dynamics (Chapter 10). In the following text, we highlight the importance of fire in forest ecosystems by examining the similar roles it plays in many different fire-dependent systems.

Roles of Fire in Forest Ecosystems Fire has played important and similar roles in fire-dependent conifer and hardwood forests around the world prior to significant human colonization. Many ecosystems depend strongly on fire; existence for many forest species literally begins and ends with fire, and thus the spatial mosaic of species and structures created by fire is important to maintaining many forest systems. The following major functions and processes are regulated by fire:

Regeneration and Reproduction: Severe fires kill all or most of the trees in existing stands (often called "stand-replacing" fires) and begin the process of regeneration for the forest that follows. As a selective force, fire also elicits the following reproductive characteristics of forest trees:

1. Asexual reproduction, mainly via sprouting, occurs in all fire-dependent angiosperms and in some conifers (Chapter 10).
2. Sexual reproduction by light, wind-blown seeds is characteristic of many fire-dependent tree species. Moreover, self-fertility in pines may have evolved as a result of fire, although only a small percentage of self-fertilized seeds are viable in many pines. Red pine is the exception and is highly self-fertile, producing viable seeds and seedlings showing no growth depression, unlike most other tree species (Fowler 1965a,b). A single isolated red pine that survives a fire may therefore self-fertilize and perpetuate itself by establishing a colony of seedlings.

Preparation of Seedbeds and Dry-Matter Accumulation: Fire sharply reduces the amount of litter, sometimes baring mineral soil, and thus greatly enhances seedling establishment. Seed germination is favored when seeds are partially buried in the ash layer. Fire also regulates dry-matter accumulation, influencing the severity of burning.

Reduction of Competing Vegetation: The elimination of trees, shrubs, and herbs by severe fires favors the reestablishment of a new stand. Light surface fires at regular intervals diminish encroaching vegetation, reducing competition for light, soil water, and nutrients. This effect is typical in many mixed conifer–hardwood stands in the southern United States in which understory hardwood regeneration is reduced by periodic burning. Similarly, in oak forests, periodic surface fires kill seedlings of shade-tolerant species that continuously establish on the forest floor and grow into the understory.

Nutrition: Recurrent surface fires in a forest reduce organic matter to its basic components of water, CO_2, and mineral nutrients that are otherwise unavailable to forest vegetation (Chapter 19). Nutrients made available by fire are particularly important in nutrient-poor conifer forests and wherever site conditions are unfavorable for decomposition. Although nitrogen is lost from soil after burning, post-fire soil conditions often favor nitrogen-fixing soil organisms.

Thinning: Periodic surface fires reduce the density of pure stands of seedlings that establish after fire, particularly in conifer-dominated ecosystems. Larger, faster growing seedlings with thicker bark are favored, temporarily reducing competition for soil water and nutrients and increasing resistance to mortality that might occur due to drought conditions that often accompany fires.

Sanitation: Dense, even-aged stand conditions may be conducive to disease and insect outbreaks. Outbreaks that kill many trees increase fuel concentrations that in turn lead to intense fires that terminate the outbreaks (such as those of bark beetles and the spruce budworm) and create conditions for the establishment of a new even-aged stand that is initially resistant to disease and insect attacks. The maturing stand will eventually be again susceptible to outbreaks, and a self-perpetuating cycle is established. Fires also eliminate plant parasites such as mistletoes on ponderosa pine, lodgepole pine, and black spruce.

Succession: Depending on site conditions and the frequency and intensity of recurring fires, fire tends to retain fire-dependent species on an area as long as the species are adapted to these fire characteristics. Ponderosa pine, the southern pines, red and white pines, upland oaks, and many other species are often regenerated generation after generation, even when they are not the most shade-tolerant or late-successional species of their respective regions. Changes (particularly increases) in the fire frequency or severity may result in the replacement of a given species by a species more able to take advantage of post-fire conditions. For example, in the Rocky Mountains, Engelmann spruce is replaced by Douglas-fir, Douglas-fir is replaced by ponderosa pine, and ponderosa pine is replaced by grasses with decreasing elevation, due to differences in water availability as well as the frequency of fire.

In addition to differences in species composition that varied with fire frequency, a mosaic of different age classes in each forest type was typical of many North American upland forests prior to European colonization. This mosaic resulted in part from the erratic burning patterns of fires and the resulting mosaic of fire severities, which in turn was controlled by heterogeneous soil-site conditions and fuel buildup in the different ecosystems (Williamson and Black 1981). All or most fires create a spatial mosaic of severities, from unburned or hardly burned areas to areas where all trees are killed, resulting in a corresponding mosaic of unburned surviving trees to a gradient of low to high tree mortality, and an eventual mosaic of tree age classes across the burned area.

Very shade-tolerant species of both hardwoods and conifers, such as sugar maple, red maple, beech, hemlocks, firs, northern white-cedar, and western redcedar, tend to be poorly adapted to fire. These species occupy protected or moist sites that are least susceptible to burning and did not develop adaptations to cope with fire. During fire-free intervals, these species colonize adjacent areas, establishing and eventually replacing fire-dependent species in the absence of fires intense enough to kill them. Exceptions to this process exist, however: although its shallow roots make eastern hemlock susceptible to fire, it regenerates vigorously following burns due to highly suitable post-fire seedbeds for its tiny, wind-dispersed seeds.

Wildlife Habitat: Fire universally creates a mosaic of habitats and niches for wildlife of many kinds (Chapter 10). Depending on how they affect and shape a particular species' habitat, fires clearly benefit some wildlife species more than others. However, overall animal species diversity, as with plants, tends to increase following fire until crown closure occurs, and then declines.

In the following text, we present a few of the many examples that demonstrate these roles and significance of fire. This selection represents important North American trees and regions as an effort to illustrate the importance of fire in forest ecosystems.

Pines in New England and the Lake States In the northeastern United States and adjacent Canada, the occurrence of the two- and three-needled pines (red, jack, and pitch pines) as well as that of pure even-aged stands of white pine (not to be confused with individual white pine in mixed stands) is strongly related to the past occurrence of forest fires. During the early colonization period, many of the fires resulted from burning by European colonists in land-clearing operations, but much evidence suggests that fires were commonly set by indigenous people for many hundreds of years before this period (Little 1974; Denevan 1992; Munoz et al. 2010; Pinter et al. 2011; Marlon et al. 2013; Abrams and Nowacki 2015, 2019). Lightning apparently played a lesser role in causing fires in eastern forests compared to the West, probably because of the heavy precipitation that often accompanies summer thunderstorms in the East.

Many ecological studies of relict old-growth pine stands have shown that fire was important in their formation. In both northwestern Pennsylvania (Lutz 1930; Hough and Forbes 1943) and southwestern New Hampshire (Cline and Spurr 1942), more or less pure even-aged stands of old-growth eastern white pine originated from past forest fires. By contrast, nearby mixed forest types with occasional dominant white pine were relatively free from evidence of past burns (Abrams 2001). Recent studies suggest that white pine is favored by a mean interval between severe fires of 150–300 years (Frelich 1992); longer fire intervals allow mesic hardwoods to invade such that fires are less common, and shorter fire intervals shift the dominant vegetation toward paper birch or aspen (Frelich 2002). In northwestern Minnesota, extensive even-aged stands of old-growth red pine clearly date from a series of forest fires, many of which antedate the advent of the colonists (Spurr 1954; Frissell 1973). In Canada, the abundance of red pine was found to be directly related to fire intensity and was most favored by moderate-intensity fires rather than severe crown fires (Flannigan 1993). In many regions of the northern Lake States, frequent (25–35 years), low-intensity fires supported even-aged stands of red pine (Drobyshev et al. 2008).

Jack pine in the Lake States and Canada and pitch pine on sandy soils near the mid-Atlantic coast (Little 1974) are virtually completely fire-dependent (Chapter 10). Jack pine grows in nearly pure stands on dry, sandy soils, forming a highly flammable vegetation type. Severe crown fires burn jack pine forests during hot, dry periods at intervals of a few decades, killing all or most of the vegetation in a patchy distribution. Following fires, the subsequent generation of trees arises from multiple sources. Jack pine regenerates from seeds stored within many years' serotinous cones in the tree crowns, and thereby benefits from severe crown fires. Red pine regenerates from seeds of trees with bark of sufficient thickness and lacking enough lower branches to permit them to survive surface fires. Hardwoods such as bigtooth and trembling aspen, northern red oak, and red maple regenerate by sprouts arising either from roots or the lower part of the stem not killed by the fire. Some pioneer hardwoods regenerate by the dissemination of seeds into the area from afar either by wind for aspens and birches, or by birds for the cherries. Infrequently, some hardwoods such as pin cherry and black cherry regenerate from seeds stored in the seed bank of the forest floor (Marks 1974).

Western Pines and Trembling Aspen The relationship between fire and pines in the northeastern United States is applicable to most of the other pine species in the United States. Lodgepole pine in the Intermountain West plays an analogous role to that of its close relative, jack pine, in the Lake States. It regenerates on recent burns, largely from seeds stored in serotinous cones of trees killed by crown fires or severe surface fires, to form dense, even-aged, post-fire stands (Figure 16.2). In 1988, fires burned

Photos by Dan Kashian

(a)

(b)

FIGURE 16.2 Lodgepole pine regeneration following fire in Yellowstone National Park. (a) Lodgepole pine revegetation in a burned patch approximately 20 years after fire. Seedling density in this patch is nearly 500 000 stems per hectare, with height averaging only 1–1.5 m. (b) Lodgepole pine thickets develop following fire when initial seedling regeneration is dense. This 130-year-old stand of about 5400 stems per hectare in Yellowstone averages only 10 m high and 9 cm in diameter at breast height. The numerous naturally thinned dead trees suggest the stand density was at least 11 000 stems per hectare within the last few decades.

570 000 ha of the Yellowstone Plateau in northwestern Wyoming, including approximately 240 000 ha of forested land in Yellowstone National Park (Schullery 1989). The fires began in mid-July and were essentially uncontrollable until snow and cool temperatures arrived in mid-September (Ellis et al. 1994). Regeneration of lodgepole pine at varying levels of density blanketed the area in succeeding years in patterns related to fire severity, pre-fire stand density and cone serotiny, and site conditions (Kashian et al. 2004; Turner et al. 2004).

The Yellowstone fires of 1988 raised important questions as to whether the Park's policy of fire suppression from 1872 to 1972 led to an abnormal buildup of fuels and therefore to abnormal fire spread, intensity, and severity. Romme and Despain (1989) concluded that the 1988 fires should not be viewed as an abnormal event for this area, and were in fact similar in total area burned, rate of spread, and severity to those that occurred in the early 1700s. Crown fires similar in extent and behavior probably occurred on this landscape every 100–300 years and are driven primarily by weather conditions rather than fuels. Although a century of fire suppression may have had some influence on size and behavior of the 1988 fires, suppression is difficult and generally ineffective in forests characterized by crown fires, and the 1988 fires had undergone normal successional dynamics and fuel buildup following the last major fires approximately 280 years ago. Notably, Yellowstone National Park has experienced several years of large area burned (>15 000 ha) since 1988, including 1994, 2003, 2007, and 2016 (https://famit.nwcg.gov/applications/FireAndWeatherData, accessed August 19, 2021).

Depending on site conditions, lodgepole pine forests are gradually replaced by shade-tolerant Engelmann spruce and subalpine fir if the period between fires reaches several hundreds of years. White and black spruce also play a part as late-successional species in the northern range of lodgepole pine in Alberta, and Douglas-fir and other more shade-tolerant western conifers become prominent toward the Pacific Northwest. In subalpine and high foothills in Alberta, Horton (1956) estimated that succession from pine to spruce and fir will occur if free from fire from 225 to 375 years, but it takes much longer on drier southern slopes, if indeed it ever takes place.

Pure stands of ponderosa pine result from a long and complex fire history at lower elevations and in warmer, drier portions of western forests (Figure 16.3; Cooper 1960; Fischer and Bradley 1987). Ponderosa pine is a pioneer species following fire in the cooler and moister portions of its range and is gradually replaced by more shade-tolerant conifers such as Douglas-fir, incense cedar, and white fir. In the warmer and drier parts of its range, those species cannot generally outcompete ponderosa pine and it dominates the forest. Fire regimes and stand development of ponderosa pine are discussed for a variety of regional and local ecosystems by Fischer and Bradley (1987), Agee (1993), Covington and Moore (1994), Arno et al. (1995), Veblen et al. (2000), and Baker and Ehle (2001), and are reviewed in detail in Chapter 10.

Trembling aspen is one of the few hardwood species able to coexist with conifers in the fire-dominated western mountains. In the northern Rocky Mountains and in western Canada, it competes as a pioneer via wind-disseminated seeds on higher, cooler, and wetter sites. In warmer and drier regions such as the Great Basin and the central and southern Rocky Mountains, aspen persists mainly via root suckers and may be maintained as a late-successional type by fire in some localities (Mueggler 1985). In contrast to the eastern and northern parts of its range, aspen clones of the central and southern Rockies may become very large; a single aspen clone in Utah is 43 ha in extent and contains thousands of genetically identical stems (Kemperman and Barnes 1976; Mock et al. 2008; Chapter 4, Figure 4.14). Some researchers have speculated that individual clones may have been perpetuated for thousands of years following rare events of seedling establishment by a combination of fire and low ungulate browsing intensity (Barnes 1975; Kay 1997). However, recent molecular analyses suggest that many very small, even single-stem clones exist in and among the larger clones, meaning that sexual reproduction (by seed rather than root suckers) is not a negligible process and likely contributes to aspen persistence (Mock et al. 2008).

FIGURE 16.3 Fire and ponderosa pine. (a) A surface fire burns grass and litter of this ponderosa pine stand in central Idaho. (b) Fires prior to European colonization time maintained open, grassy, parklike stands of ponderosa pine such as this one in western Montana. *Source:* U.S. Forest Service photos.

In the moister northern Rocky Mountains, aspen seedlings may establish readily following fire in appropriate microsites. Seedling establishment is also possible, although rare, further south in the species' range given a very narrow window of conditions dictated by sustained moist weather conditions immediately following the fire. For example, following the major fires in Yellowstone Park in 1988, thousands of aspen seedlings were observed over a broad range of vegetation types (Renkin et al. 1994; Turner et al. 2003a; Romme et al. 2005), in part because the growing season in 1989 was unusually cool and wet and allowed extensive seedling establishment.

Many aspen clones of the central and southern Rocky Mountains are now deteriorating. Concern about aspen decline in the western United States, which features clones characterized by aging mature trees without regeneration, began in the 1970s (Schier 1975) and reached its height beginning in the 1990s (Kay 1997). Aspen decline was most evident at local scales (Kashian et al. 2007), and was attributed primarily to fire suppression and excessive ungulate browsing (Baker et al. 1997; Kaye et al. 2005; Kulakowski et al. 2013). In the 2000s, aspen in the West began to experience a phenomenon termed **sudden aspen decline**, characterized by an episode of unusually high overstory crown thinning and mortality (in addition to the previously recognized lack of regeneration) that occurs at broad scales over several years or more (Worall et al. 2008, 2013). Sudden aspen decline has been linked to drought-related moisture stress but is also associated with multiple interacting factors including climate, land-use history, and successional dynamics (Singer et al. 2019).

Southern Pines The dependence of pine forests upon recurring fires is most evident in the southern pine belt of the southeastern United States. Of the four most common southern pines, longleaf pine depends the most upon recurring surface fires for its persistence (Chapman 1932). Longleaf pine dominated much of the Coastal Plain prior to European colonization (Christensen 1988a) but is now considered an endangered forest type in the United States (Noss et al. 1995). Pre-European colonization longleaf pine forests occurred on three fairly different kinds of sites—dry sandhills, intermediate areas, and wetlands that included shrub-bogs, savannas, and flatwoods—each with different fire effects and successional pathways (Marks and Harcombe 1981). Longleaf pine regeneration is prevented by annual burning and summer fires, and thus the proper ratio of winter fires to fire-free years is critical if pure longleaf pine forests are to persist. Furthermore, longleaf pine stands require burning during the "grass stage" to control brown spot disease, as well as repeated burning following canopy formation to avoid the establishment of and replacement by understory hardwoods in the absence of fire. After many attempts at complete fire exclusion by southern foresters in the first half of the twentieth century, prescribed burning has been heavily used to achieve these ecological objectives.

Longleaf pine is extremely shade-intolerant and regenerates most successfully in full sunlight on mineral seedbeds created by forest fires that destroy and/or consume all but scattered longleaf seed trees. It is one of the most fire-resistant of all pines. Seedlings appear like dense bunches of grass (Figure 16.4), and this stage is known as the **grass stage**. Longleaf pine seedlings, unlike all other North American pines, remain in the grass stage for about 6 years (range 3–12 years) while they develop a deep tap root. A surrounding sheath of dense needles in the grass stage protects the bud from fire, and if burned away are replaced by a new set from stored carbohydrates and nutrients in the large tap root. Rapid stem elongation, called "bolting," eventually occurs, quickly elevating the crown above the level of periodic surface fires.

The other common southern pines—loblolly pine, slash pine, and shortleaf pine, as well as sand pine, Virginia pine, pitch pine, and other species of local occurrence—are pioneer species that establish after fires, giving way to shade-tolerant hardwood mixtures in their absence (Little and Moore 1949; Wahlenberg 1949; Campbell 1955; Myers 1990). The dependence of these species upon fire has given rise to practical use of fires as a silvicultural tool in these systems that will maintain natural succession in the pioneer pine stage or to restore forests after long periods of fire suppression (Stanturf et al. 2002; Ryan et al. 2013).

FIGURE 16.4 A dense stand of natural longleaf pine seedlings in the Coastal Plain of South Carolina following a prescribed burn to prepare the seedbed. *Source:* U.S. Forest Service photo.

Douglas-Fir in the Pacific Northwest Characteristic summer droughts in the Douglas-fir region of northern California, western Oregon and Washington, and southern British Columbia create highly flammable conditions during the hottest period of the year. As a result, extensive forest fires in the region have been common since pre-European colonization. Douglas-fir is a pioneer species following fires on the west side of the Cascades and over much of the Coastal Range, so long as adjacent unburned Douglas-fir stands are able to supply a source of seed. More shade-tolerant conifers, such as western hemlock, Sitka spruce, and western redcedar, later colonize the understory of Douglas-fir forests and may eventually dominate after 500 or more years when the dominant Douglas-fir matures and senesces (Figure 16.5; Franklin et al. 1981). Many fire-history studies of Douglas-fir forests suggest that stand-replacing fires occurred every 200 to 400+ years, but others reconstruct a regime of multiple patchy fires of varying severity recurring every 20–100 years (Morrison and Swanson 1990). Fire history and stand development of Douglas-fir throughout its Pacific Northwest range are discussed by Agee (1993). An excellent example of a fire history study in Douglas-fir forests is provided by Taylor and Skinner (1998), and Perry et al. (2011) provide a treatment of mixed-severity fire regimes in the region.

Giant Sequoia The native giant sequoia-mixed conifer forests of the Sierra Nevada of California are also fire-dependent (Kilgore 1973; Agee 1993; Swetnam 1993). Fire in sequoia forests prepares a soft, friable, mineral seedbed upon which the tiny seeds of sequoia fall and are lightly buried, favoring germination (Hartesveldt and Harvey 1967). Seedling establishment is limited by water availability (Rundel 1972), and litter that is partially burned may contain more available water than unburned litter (Stark 1968).

Periodic surface fires, some of which were presumably set intentionally by indigenous people, maintained an open and parklike setting prior to European colonization (Figure 16.6a; Biswell 1961). Fires in these ecosystems probably burned every 15 years (Swetnam et al. 2009). Such fires eliminated small shade-tolerant trees, particularly white fir, which continually colonized the understory. Burning by indigenous people was gradually eliminated, and fire suppression became increasingly efficient

FIGURE 16.5 The dominant, old-growth Douglas-fir individual (left) in this western Washington stand will eventually be replaced in the absence of fire by more tolerant firs and western hemlocks in the understory. *Source:* U.S. Forest Service photo.

upon European colonization. As a result, surface fuels built up, and white fir and other shade-tolerant species often formed thickets under the sequoias (Figure 16.6b). A vertical sequence of "ladder fuel" was thus formed from ground level to low-hanging fir branches to the top of understory crowns, 10–30 m above the surface (Kilgore and Sando 1975), such that fire in surface fuels was more likely to pass through understory crowns and reach sequoia crowns. Unlike redwood, giant sequoia is a poor sprouter and will die once the crown is killed. Fires have occurred more severely and more frequently in these forests since 2015 (Stephenson and Brigham 2021); high-severity fires in 2020 and 2021 are estimated to have killed up to 15% of all existing large sequoias across its range (Stephenson and Brigham 2021). Prescribed burning every 5–8 years is needed to reduce surface fuels, remove small understory trees and the lower crowns of large understory trees, and prevent continued encroachment of a shade-tolerant understory.

Fire History and Behavior Fire-history studies in forests increase our awareness of fire as a dominant force over virtually all of the northern and western landscapes. Such studies are fantastically abundant in the literature. Three classic studies illustrate fire history, behavior, and effects.

Northern Lake States Heinselman's (1973) classic study in the Boundary Waters Canoe Area in northern Minnesota documents 400 years of fire occurrence and effects on a landscape where fire has driven forest composition and structure for nearly 10 000 years. Recurrent fires had maintained nearly three-fourths of the region in recent burns and stands of small-diameter trees prior to the arrival of logging in about 1895. Thus, relatively little of the 215 000 ha tract, now reserved as wilderness, was subjected to the predominant harvesting method of the time: "high-grading" (cut the best and leave the rest). Effective fire control began about 1911.

Heinselman developed a fire chronology based on the age of existing stands, tree-ring counts from sections or cores from fire-scarred trees, and historical records. Nearly all of the forest burned one to several times between 1595 and 1972, but much of the burned area is accounted for by just a few major fires. In the European colonization period, fires in 1875 and 1894 accounted for 80% of

(a)

(b)

FIGURE 16.6 1890–1970: 80 years of fire exclusion in the giant sequoia-mixed conifer forests of Yosemite National Park (Confederate Group, Mariposa Grove). (a) 1890: A parklike stand as the result of periodic fires. *Source:* Kilgore 1972; Historical documentation by Mary and Bill Hood. Photo by George Reichel. Courtesy of Mrs. Dorothy Whitener. (b) 1970: The parklike stand was invaded by thickets of white fir. By 1970 firs obscured all but the fire-scarred sequoia on the left. Such thickets provide ladder fuels that could support a crown fire fatal even to mature sequoias. *Source:* National Park Service photo by Dan Taylor.

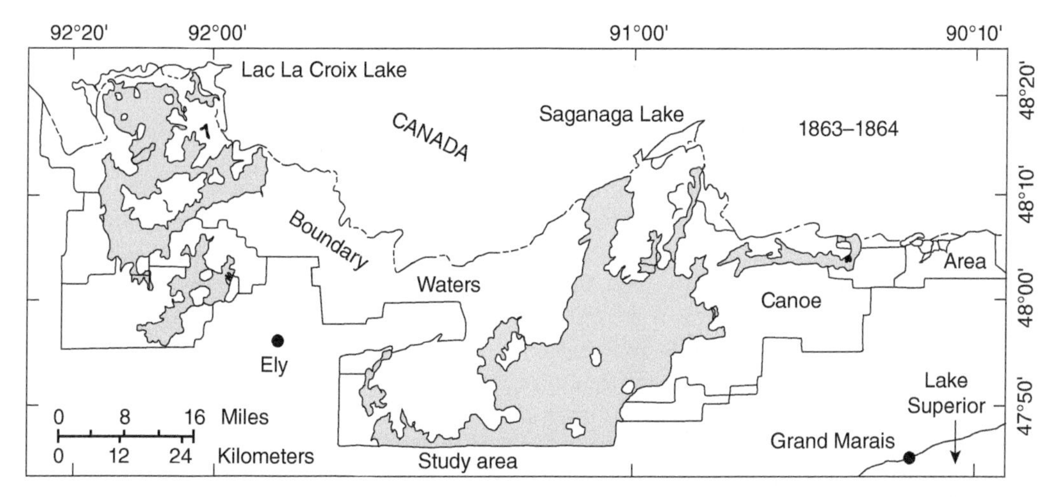

FIGURE 16.7 Severe fires of 1863 and 1864 burned (shaded area) over 1800 km² (700 mi²) of the Boundary Waters Canoe Area, northern Minnesota. *Source:* Heinselman (1973) with permission of Cambridge University Press. Reprinted with permission of the Quaternary Research Center, University of Washington, Seattle.

the total area burned; but just five brief fire periods accounted for 84% of the total prior to European colonization. These major fire periods coincided with prolonged, subcontinental-scale summer droughts. Climate combined with fuel accumulations (driven by dry matter accumulation, spruce budworm outbreaks, and blowdowns) to create optimum burning conditions at long intervals. The fires of 1863 and 1864, associated with the major droughts of those years, burned 44% of the area (Figure 16.7). Fire in 1864 also burned over 400 000 ha in Wisconsin.

Heinselman estimated the natural fire rotation (the average number of years necessary to burn the entire area) since 1595 to be about 100 years, although the natural rotation for different ecosystems varied. For example, ecosystems dominated by aspen–birch forests burned more frequently (50 years or less) than those dominated by red and white pines (150–350 years). In red and white pine ecosystems, fire typically occurs as light surface fires that reduce the organic layer and understory competition, with occasional severe fires that kill the overstory and provide openings for regeneration. A second severe fire that occurred only a few years after the first could eliminate many conifers, being too small to withstand the heat and needle scorch and too young to bear cones. Regeneration in that scenario comes from hardwood sprouts of trembling aspen, paper birch, red maple, and northern red oak.

Fires over the entire area produce a mosaic of stands of different composition and ages driven by the complex of site factors and fire size and severity, generating diversity in stand structure and species composition. Both plants and animals are adapted to these ecosystems and community patterns and are directly or indirectly adapted to fire. Moose, beaver, black bear, snowshoe hare, woodland caribou, small mammals, and birds are all adapted and dependent on the mosaic of different habitat conditions created by fire. Notably, widespread fire exclusion is now restructuring the entire regional ecosystem, gradually creating changes in forest composition and eliminating the niches of many wildlife species. Restoring the natural vegetation mosaic in this now human-dominated wilderness will be difficult but will surely depend on the use and management of fires.

Boreal Forest and Taiga Wildfires burning over diverse sites create a vegetation mosaic in the boreal forest and taiga of North America and have a major influence on the region's plant and animal diversity (Lutz 1956; Slaughter et al. 1971; Rowe and Scotter 1973; Viereck 1973, 1983; Bonan and Shugart 1989). The patchy pattern of vegetation is maintained by fires and provides diverse wildlife populations. Seven of the ten major boreal tree species are pioneers on disturbed sites and have adaptations that favor rapid invasion after fires (Chapter 10). Only balsam and subalpine fir, whose cones disintegrate at maturity, and whose seeds are not retained in the crown, are not well adapted for immediate post-fire regeneration. The understory and groundcover layers of the boreal forest are dominated by clonal species whose underground stems and roots are protected from fire and regenerate rapidly by sprouting.

The taiga lies north of the boreal forest and is a broad, open area of slow-growing spruces interspersed with scattered, well-developed stands of trees and treeless bogs, locally called muskeg (Viereck 1973; Van Cleve et al. 1986). Vegetation pattern is closely related to fire history and **permafrost**, or permanently frozen ground with water often incorporated as lenses of pure ice. Permafrost perches the water table and thus cools the ground by creating a deep, long-lasting ice layer on undisturbed sites (Bonan and Shugart 1989). Thick layers of moss insulate the forest floor during summer months, limiting soil thawing to depths of less than 1 m (termed the *active layer*). Fire may increase this annual depth of thaw via removal of this insulating layer and other organic matter; historically the upper permafrost layers recover as coniferous vegetation regrows. In many areas, permafrost and vegetation are in delicate balance and altering the depth of summer thaw can significantly affect the presence and composition of the vegetation. Increasing the depth of thaw following fire may cause the areas over ice lenses to subside, creating a polygonal mound-and-ditch pattern. This pattern has been documented in paper birch stands at least 40–50 years after fire. These sites may succeed to black spruce and stabilize as permafrost reforms, or small ponds may develop and alternate in the cycle with black spruce.

Northern Rocky Mountains Fire has a major ecological influence in the semi-arid, mountainous, coniferous forests of northern Idaho, western Wyoming and Montana, eastern Washington, and adjacent portions of Canada (Habeck and Mutch 1973; Arno 1976; Agee 1998; Baker 2009; McKenzie et al. 2011; Pyne 2019). Fire in this region (as well as insect outbreaks, discussed in the following text) was second only to precipitation as the major factor shaping forests until fire suppression and exclusion became effective in the 1940s. Large stand-replacing fires were common due to the extremely dry summer climate and the buildup of fuels in areas that were densely forested. Fuel buildup was exacerbated by forest insects and diseases, which killed many trees in dense stands over extensive areas. Between 1 and 2 million hectares burned in 1910 alone following prolonged drought.

Many forests in the regions may be "reburned," or burned multiple times at short intervals. In some cases, the first fire will reduce the severity of a second fire, if it occurs, by consuming fuels and reducing the abundance of fire-sensitive vegetation (Parks et al. 2014). However, where the first fire results in abundant growth of flammable vegetation or accumulation of hazardous fuels, the probability and severity of the second fire may increase. Famously, parts of the 1910 fires reburned in either 1917 or 1919, or both, and even again in subsequent years. Snags were particularly important for ignition by lightning, and wind-blown embers from tall snags could start new fires far in advance of the burning front. A recent study by Harvey et al. (2016a) in northern Idaho and western Montana and Wyoming found that initial high-severity fires facilitated high-severity reburns when the interval between fires was longer than a decade. Successive stand-replacing fires were most common in high-elevation forests where abundant tree regeneration tends to follow burning. Coastal systems in the West tend to reburn at shorter intervals between fires when conditions are wetter, and interior forests when conditions are drier, suggesting that burn intervals are largely determined by vegetation productivity and the accumulation of fine fuels (Buma et al. 2020).

Unstoppable, severe fires occur in regional ecosystems that are characterized by the climatic and vegetational features conducive to crown fire. An excellent example with a long, documented fire history is the Yellowstone Plateau that includes Yellowstone National Park mentioned in the earlier text (Romme 1982; see also Chapter 22). Most of the Yellowstone Plateau is above 2300 m (7500 ft) in elevation, and most of the forested area is dominated by relatively even-aged lodgepole pine forests between 100 and 350 years old. The 1988 Yellowstone fires burned in one of the driest summers on record in the Rocky Mountains with extremely dry fuel conditions. In addition, twice as many lightning storms were recorded in 1988 than in the average year (Wuerthner 1988). Following the fires, lodgepole pine regeneration density varied across the burned area from zero to 535 000 stems per hectare (Turner et al. 1994), with variation explained mostly by elevation and, secondarily, distance to unburned areas. An earlier study of lodgepole pine regeneration in moderately burned, severely burned, and unburned stands found the highest seedling densities in moderately burned areas (Ellis et al. 1994). Where burn severity was high, seeds stored in serotinous cones in the tree crowns were likely consumed by the fire, resulting in relatively low post-fire seedling densities. The varying or patchy pattern of burn severity allowed seeds from lightly or moderately burned areas to supply much of the seed source for adjacent severely burned areas. Local levels of cone serotiny, in addition to burn severity, were also likely to have affected post-fire lodgepole pine regeneration (Turner et al. 1997, 1999).

Fire is also a key factor in other parts of the Rocky Mountains and has been examined in detail by Peet (1981), Veblen (Veblen and Lorenz 1986, 1991; Veblen et al. 1991a; Veblen 2000), Baker (2009), McKenzie et al. (2011), and Knight et al. (2014), among many others.

Fire Suppression and Exclusion At a temporal scale of centuries, fire is an integral part of the evolution and perpetuation of forest species as parts of landscape ecosystems. It exists as an internal factor to ecosystems rather than an external disruptive force that ecologists once considered to be a "disturbance." From this perspective, increasingly efficient alteration of the timing, frequency, intensity, or severity of fire via its suppression or exclusion by humans, if continued indefinitely, constitutes a perturbation that will greatly change ecosystems (Smith et al. 1997). Many such changes are already evident.

Fires were common before European colonization, set by lightning and indigenous people, and remained common during colonization, even increasing in frequency in many places. Unless they threatened human life, livestock, or other property, fires were little regarded except as a local nuisance. Public opinion arose against fire in the early 1900s, fueled by unrestrained burning and the large destructive fires of the late 1800s and 1910. Such public opinion led to the development of rigid suppression policies whereby fire was considered a disastrous threat to the forest and to human life and property rather than a natural phenomenon. Even today the public is exposed to verbiage about fires in forests too often described as "catastrophic," "destructive," or "devastating" without context of its ecological importance. Fire damage can be seen immediately and thus encouraged anti-fire "propaganda;" its beneficial effects, on the other hand, might take years to materialize (Harper 1962).

Today, technological advances and a century of experience have led to efficient fire control, making it increasingly possible to extinguish fires very early in their development. Effective suppression is not possible in all forest types, particularly where crown fires dominate the fire regime such as the boreal forest and high-elevation forests in the Rocky Mountains, because fires in these regions are generally driven by extreme weather conditions. Even crown fire-dominated forests have experienced a significant reduction in the area burned since the middle of the twentieth century, however, often increasing the incidence of wind damage and insect and disease attacks as the proportion of mature and late-successional forest increases and spatial connectivity of those forests is high (due to a loss of the fire-created mosaic of the past). Where surface fires are dominant, fire suppression is more easily accomplished, increasing the intervals between fires and permitting a buildup of fuel and a decline in animal and plant diversity. Prolonged buildup of fuel over longer fire-free intervals in these systems often leads to severe fires, even shifting surface-fire regimes to crown fire regimes, resulting in fires more impactful to the ecosystem than what occurred in the original forest, and at intensities for which the ecosystem may not be adapted.

Many federal agencies and other organizations in the United States, such as the National Park Service, the United States Forest Service, and The Nature Conservancy, have instituted fire management programs in recent decades to allow fires to resume their natural role in forest ecosystems, where possible. Prescribed burning is an important tool in restoring fire to ecosystems having experienced fire suppression. Fire suppression and exclusion permit some thin-barked, slow-growing species to grow into size classes that are fire-resistant (Harmon 1984), such that restoration of the original fire interval alone will not restore the original forest composition and structure. Thus, prescribed burning is often combined with efforts such as thinning or other harvesting methods to reduce stand density in restoring formerly fire-prone forest ecosystems. Fire management and prescribed burning have been considered from many standpoints and in relation to many different forest ecosystems (Slaughter et al. 1971; Wright and Heinselman 1973; Kayll 1974; Kilgore 1975, 1976a,b,c; Mutch 1976; Martin et al. 1977; Arno et al. 1995; Knight et al. 2014; North et al. 2015; Thompson et al. 2018).

Because fire is an important ecosystem process, it is also an ecosystem-specific process and thus fire suppression or exclusion has varying effects depending on the physical site factors and forest species that make up the ecosystem. Perhaps the most well-known example of the negative effects of fire suppression is found in low-elevation ponderosa pine forests in the Rocky Mountains (Covington and Moore 1994; Veblen et al. 2000). These forests were historically characterized by short-interval surface fires that maintained low fuel availability, minimal undergrowth, and low stand densities. However, the relatively light surface fires were easily and successfully suppressed following European colonization. As a result, fire suppression and exclusion had major effects upon these forests, allowing fuel buildup, increased tree density, and altered tree species composition, as well as high-intensity fires when fires do occur. Suppression-induced increases in fuels and tree density were blamed for several major fires in the West in the 2000s. In response, some recent forest management in the West—often regardless of forest type—has focused upon proactively reducing fuels and stand density as an attempt to reduce fire intensity. However, increasing evidence suggests that ponderosa pine forests varied in structure prior to European colonization, with some naturally dense and burned by crown fires rather than light surface fires (Williams and Baker 2012). These data further support the idea that most fires consist of a mosaic of burn severities ranging from low to high, making proactive fuels management difficult to carry out appropriately. Moreover, high-elevation forests in the Rocky Mountains dominated by lodgepole pine or spruce and fir were historically characterized by dense stands, long fire intervals, and crown fires that generally occur during prolonged dry periods. Such crown fires were limited by weather conditions rather than by fuel accumulations and are very difficult to suppress, such that the effect of suppression in these forests is restricted primarily to reducing the area burned. It therefore makes little ecological or practical sense to attempt to reduce fire severity by altering forest structure, reducing fuels, or incorporating prescribed burning in high-elevation forests as would be appropriate in ponderosa pine forests. These differences again highlight the site- and species-specific nature of fire as an ecosystem process (Schoennagel et al. 2004). Effects of fire suppression on western forests are summarized in recent reviews and studies by Hagmann et al. (2021), Haugo et al. (2019), Keeling et al. (2006), Keane et al. (2002), and Arno et al. (2000), among others.

WIND

Wind disturbances have enormous consequences in forest ecosystems, affecting forest structure, composition, and function at regional and local scales. The occurrence of wind disturbances is much more difficult to identify and track compared to more dramatic disturbances such as fire, and thus its prevalence in forests across the world is probably understated. Like fire, wind disturbances occur within a particular regime at a frequency, severity, and size specific to a given ecosystem or landscape. Varying regimes have varying effects as wind encounters forested landscapes. In this section, we examine the ecological effects of wind in forest ecosystems.

Widespread and Local Effects Certain regional ecosystems of North America and the world are characterized by distinctive wind patterns. *Hurricanes* occur predictably along the Gulf and Atlantic coasts of North America, where sustained winds may reach velocities of over 200 km h^{-1} (1989 hurricane Hugo, 216 km h^{-1}; 1992 hurricane Andrew, 242 km h^{-1}; 2005 hurricane Katrina, 280 km h^{-1}). Hurricanes may cause forest destruction over huge swaths across a landscape (50-km wide for Andrew; Pimm et al. 1994). *Tornadoes* are most prevalent in the central part of the United States with "tornado alleys" extending from Texas to Nebraska and Kansas east across central Indiana. Tornadoes have higher wind velocities than hurricanes, but their damage is more localized (typical width is 100–200 m, maximum width 2 km). Wind speeds of >430 km h^{-1} were reported for the 1993 Kane tornado in northwestern Pennsylvania (Peterson and Pickett 1995), although only about 3% of tornadoes have winds that exceed 320 km h^{-1}. Straight-line winds in the form of downbursts during severe thunderstorms are common in the Lake States (Canham and Loucks 1984). A single thunderstorm may produce multiple downbursts as it moves across the landscape; this family of downbursts is known as a *derecho*. A derecho that occurred in northern Wisconsin in 1977 included 26 separate downbursts and created a damage path 20-km wide and 200-km long across the forested landscape (Frelich 2002).

As opposed to wind events that affect large portions of a landscape, local wind gusts (90–150 km h^{-1}), associated with cyclonic storms that break off or uproot one or more trees, may affect only small areas (20–100 m^2) in the forest. These canopy gaps or exposed patches on the ground occur over vast areas in North America, and their accumulated ecological significance may be equal to or greater than that caused by large wind events (Bormann and Likens 1979). Notably, winds capable of causing forest disturbances occur over a broad range of velocities, and they encounter regional and local ecosystems with many different site conditions and vegetation at different times of the year. Despite the many gradients, the nature and amount of wind disturbance are somewhat predictable by regional ecosystem, specific landform–soil drainage conditions, site exposure, and tree species and height.

Principles of Wind Damage Wind is turbulent near the Earth's surface due to the nature of the ground over which the air is moving, with higher turbulence along rougher surfaces. Forests form a particularly rough surface such that wind over them is much more turbulent than over a field. Turbulence is organized into coherent **gusts** that move widely across the forest (Figure 16.8). Wind becomes unstable as it moves across a forest, and each gust consists of a rapid increase in wind speed and a downward movement of air into the canopy (see #5 in Figure 16.8). The strongest gusts exert a force on trees up to 10 times larger than does the mean wind velocity (Quine et al. 1995), making them the most consistent cause of windthrow in forests.

Changes in vegetation height or composition induce additional turbulence and wind acceleration. Forest edges, roads, and other open areas that cause wind to deflect upward tend to increase damage to trees. In addition, the force of the wind on trees nearer to a forest edge is substantially greater than that on interior trees, and trees that grow on edges have significantly higher physical resistance to wind. By contrast, trees suddenly exposed by the formation of a new edge or the opening of a patch lack such a physical resistance and are more vulnerable to wind damage than those that grow on a forest edge. The edges between cutover and uncut areas tend to deflect wind currents, particularly at high elevations in mountainous areas, producing increased wind velocities where the deflected currents join together (Figure 16.9). Wind damage in mountainous forested country is often due to local acceleration of wind by either topography or forest edges (Gratkowski 1956; Alexander 1964).

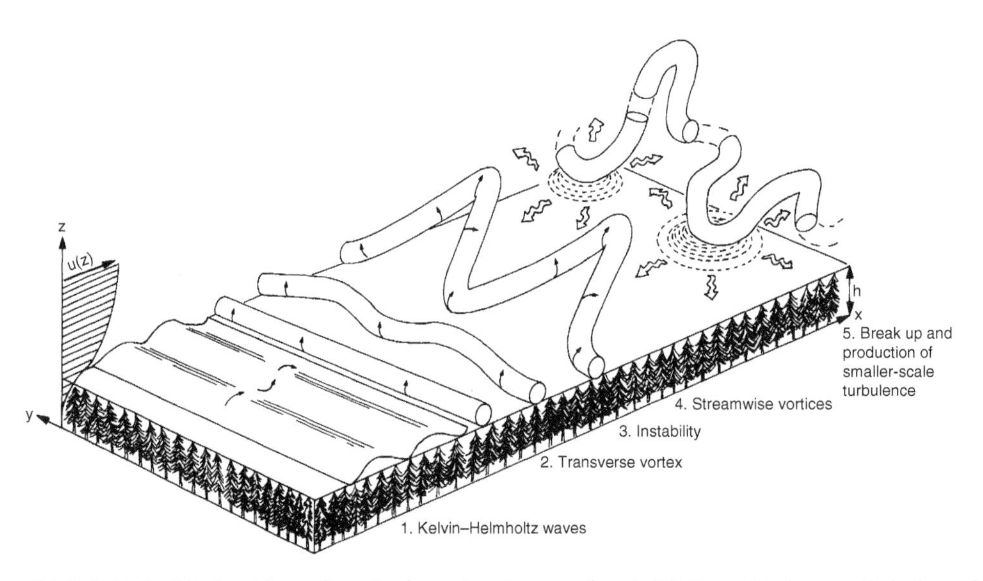

FIGURE 16.8 Idealized formation of coherent gusts over a forest. (1) The rapid change of wind speed (u) at the top of the canopy (z = h) is unstable and leads to the emergence of Kelvin–Helmholtz waves. (2) The waves become transformed into across-wind vortices. (3) These vortices are unstable and begin to distort. (4) The distortion produces coherent gusts aligned in the direction of the wind. The gusts propagate across the forest and if strong enough, lead to wind damage. (5) Eventually the gusts become distorted and break up. *Source:* Finnigan and Brunet (1995) / Cambridge University Press. Reproduced with permission of the British Forestry Commission.

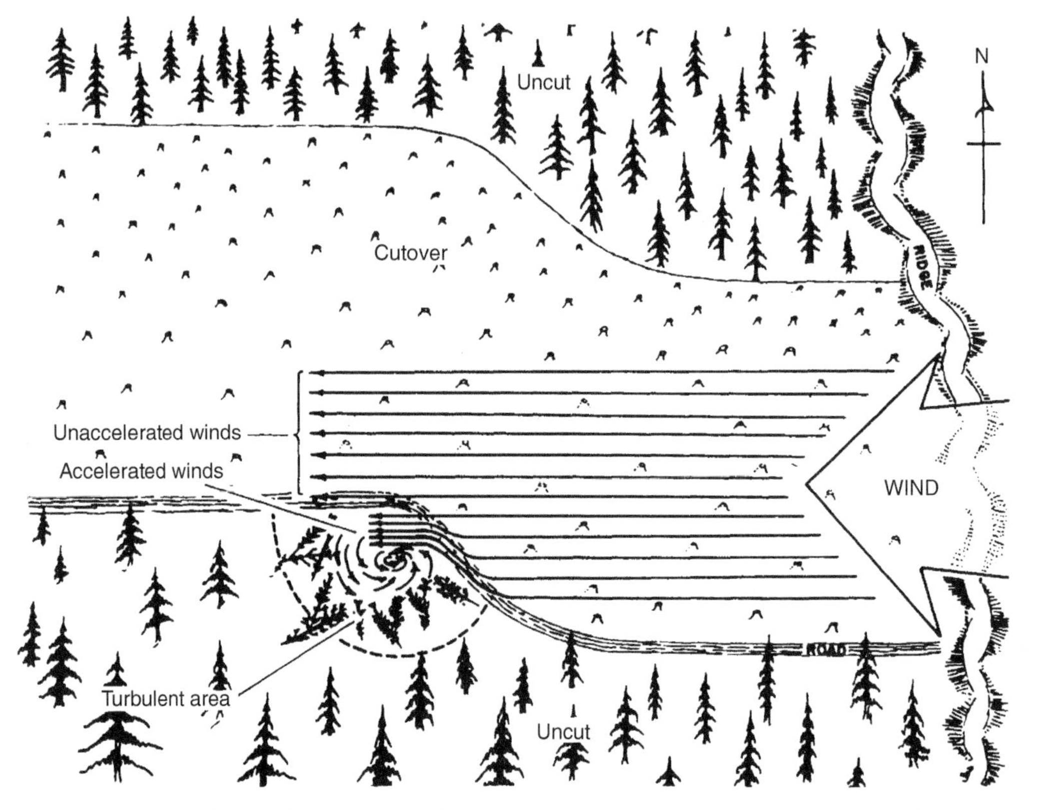

FIGURE 16.9 Windthrow of standing trees following a harvest cut. A change in the direction of the cutting boundary that was parallel to the prevailing winds acts to funnel wind into standing timber, causing a pocket of blowdown. Routt National Forest, northwestern Colorado. *Source:* Alexander (1964) / with permission of Oxford University Press. Reprinted with the permission of the Society of American Foresters.

Topography has a strong influence on wind damage to forests. Wind velocity is often assumed to be at its highest near the tops of slopes and ridges, but this is the case only when the wind direction is perpendicular to the slope. When the wind direction is at a sharp angle to the slope, the wind velocity is highest at the midslope position, and at the valley floor when the wind is parallel to the slope (Frelich 2002). Local wind acceleration can also take place on gradual and smooth lee slopes during severe windstorms, in gaps and saddles of main ridges, and in narrow valleys or V-shaped openings in the forest that constrict the wind flow. Forests subject to hurricanes or other windstorms where wind direction is lateral and in the same direction often experience damage related to slope aspect (Chapter 22; Boose et al. 1994). Where damaging winds result from vertical downbursts, however, there may be little correspondence between wind damage and topography (Frelich and Lorimer 1991).

Windthrow (uprooting of trees) and snapping is most likely where air currents are concentrated to form high wind velocities at a particular spot. Windthrow most often occurs in shallow-rooted species on shallow or poorly drained soils. Soil provides tree stability, and thus a tree whose root system is in contact with more soil mass is less likely to be uprooted by wind. Rooting depth, rather than lateral root spread, is strongly determined by soil conditions. Rooting depth is most often restricted by inadequate oxygen supply (poorly drained soils) or impenetrable layers (bedrock, ortstein, etc.). Tall trees with large crowns are especially susceptible to wind damage, and conifers are more vulnerable than deciduous trees in winter, when deciduous trees are leafless. It is possible to identify species at high risk for windthrow—black spruce, white pine, and Norway spruce, for example—and high-risk sites.

Broadscale Disturbance by Hurricanes Occasional severe storms, particularly hurricanes, may destroy the overstory canopy across hundreds or thousands of hectares of forest, initiating secondary forest succession over large areas.

Gulf and Southern Atlantic Coasts More than 50 hurricanes have made landfall along the southeastern Louisiana coast since 1851 (Blake and Gibney 2011). Hurricane Katrina made landfall in southeast Louisiana and the Mississippi coastline on August 29, 2005, with hurricane-force winds that extended at least 100 km inland. It was at the time the third most intense storm on record to make landfall in the United States, killing or severely damaging 320 million trees over 84 000 km^2 (Chambers et al. 2007). Wind damage to forests in the region was spatially variable, however, with 60% severely disturbed, 35% moderately disturbed, and 7% only lightly disturbed in the Lower Pearl River Valley and its surrounding area (Wang and Xu 2009). Overall, post-hurricane damage patterns were closely related to stand conditions and site characteristics (Chapman et al. 2008; Kupfer et al. 2008). Forests closest to rivers and streams in the region were more severely disturbed than areas further away from bottomlands, related to the increased susceptibility of trees to wind on poorly drained sites. In addition, older stands with a higher percentage of hardwoods were more susceptible to wind damage. In general, wetland forests were more susceptible to damage than other forest types with the exception of those having high proportions of cypress and/or tupelo, which were resistant to wind damage in wetlands because of their buttressed roots. Oaks and sweetgum were more susceptible to strong winds because of their shallow rooting. On uplands, evergreen pines (especially longleaf pine) were more resistant to wind damage compared to oaks. Shorter trees with smaller diameters in the shrub/scrub ecosystems of the region were the most resistant to wind damage.

Hurricane Andrew struck on August 24, 1992, in south Florida and severely affected ecosystems over an area 50-km wide by 100-km long (Loope et al. 1994; Pimm et al. 1994). From 25–40% of overstory trees in pinelands were uprooted or snapped at heights of 1–6 m; 2–3 times as many pines were broken as were uprooted. The understory was relatively unaffected. Tall, large-diameter hardwood trees of hammock ecosystems (small islands of trees within broad wetland areas) were extensively damaged (20–30% downed or large branches broken off), but forests dominated by bald cypress showed only modest damage. Mangroves suffered heavy damage, reaching 80–95% trunk snapping and uprooting in the most vulnerable sites (Smith et al. 1994). Interestingly, small patches of surviving mangroves were saplings that had previously regenerated in gaps caused by lightning-induced fires. These small gaps provided nuclei for recolonization of destroyed forests, given that the Florida mangroves only reproduce viable propagules on stems less than 1 m in height. A primary concern during post-hurricane succession is the acceleration of introduced plant species spread that was already occurring before Hurricane Andrew. Propagule dispersal and facilitation of establishment and further spread in new canopy gaps of exotic tree species, such as the Chinese tallow, have proven to be problematic over the two decades since the event occurred (Conner et al. 2002; Henkel et al. 2016; Smith et al. 2020).

New England—1938 Hurricane Hurricanes also regularly and historically affect forests of the northern Atlantic coast and New England region. Seven major hurricanes along different tracks from 1620 to 1950 have inflicted light to severe damage (Chapter 22, Figure 22.7) at intervals of a few to more than a 100 years. A great hurricane in 1938 severely damaged more than 240 000 ha of central New England's forests along a 100-km swath from central Connecticut and Massachusetts to the northwest corner of Vermont (Foster and Boose 1992; Whitney 1994). Foster (1988b) found that damage to species and stands from this severe windstorm occurred quite predictably and specifically; damage was not heavy everywhere. Similar to patterns found after Hurricane Katrina, post-hurricane damage was a mosaic of differentially damaged stands controlled by the physiography, wind direction, soil type, and nature of the pre-hurricane vegetation, especially species composition, tree height, and density (Spurr 1956a; Foster 1988a,b; Foster and Boose 1992). A case study of disturbance across the New England landscape is presented in Chapter 22.

Wave-Regenerated Fir Species Wave-regeneration of subalpine fir species is an example of specific broadscale, wind-induced disturbance patterns. In high-elevation balsam fir forests of the northeastern United States (Sprugel 1976; Sprugel and Bormann 1981) and fir forests of central Japan (Sato and Iwasa 1993), persistent wind (rather than wind velocity per se) and other factors maintain fir regeneration and other ecosystem processes over long time periods. The fir canopy is broken by numerous crescent-shaped strips of standing dead trees, with mature forest on the leeward side of the dead trees and young, vigorously regenerating forest on the windward side (Figure 16.10). This "wave" of death and regeneration moves slowly through the forest, with seedlings replacing the dying trees at the wave's leading edge. The waves move in the general direction of the prevailing wind at constant speeds of 0.37–3.3 m per year. The effects of prevailing winds, including winter desiccation, summer cooling, increased rime deposition, and breakage of branches and roots, are presumably the causes of tree mortality of the dieback zone. Shallow, rocky soils of these ecosystems also contribute to death of the fir trees, as strong winds induce tree sway and cause root breakage on sharp rock surfaces, thereby reducing water conducting tissue and encouraging root disease fungi (Harrington 1986). Seedlings established under the canopy trees are released upon canopy tree death and eventually become canopy trees themselves in 60–80 years (Figure 16.10). The wave-regenerated forest is unusually orderly due to an omnipresent and highly predictable nature of the prevailing wind disturbance (Sprugel and Bormann 1981).

FLOODWATER AND ICE STORMS

Floodwaters have shaped and rearranged physiographic features of streams of all sizes for tens or even hundreds of thousands of years (Chapter 8). Through inundation, sedimentation, and influence on local hydrology, floodwaters control the regeneration of riverine species and thus shape the distinctive pattern of forest vegetation associated with riverine landforms. Flood

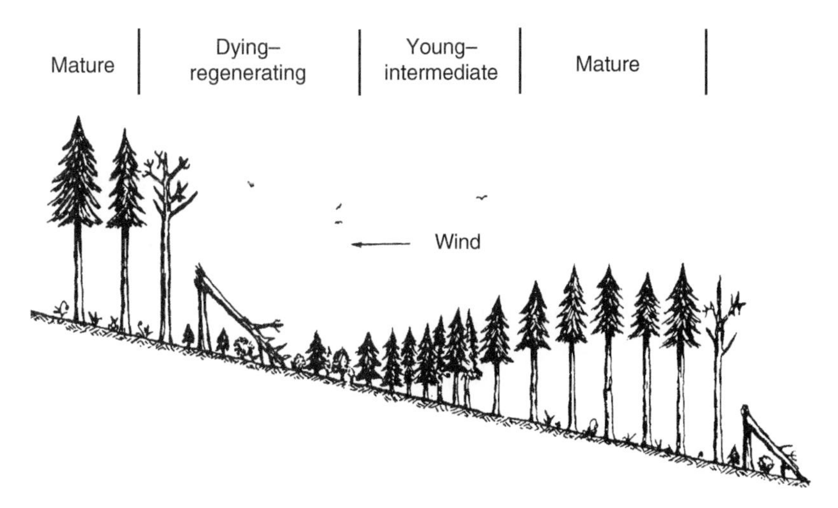

FIGURE 16.10 Diagrammatic cross section through a regeneration wave. Fir regeneration is initiated in and along the edges of the mature stand and released as the trees die. *Source:* Sprugel and Bormann (1981) / American Association for the Advancement of Science. Reprinted with permission from *Science* 211 : 390–393, © 1981 by the American Association for the Advancement of Science.

regimes—the long-term duration, frequency, and temperature of floodwater—have elicited genetic differentiation of river valley species. The many different physiological adaptations of river valley species to lack of oxygen for root metabolism enable particular species and races to dominate diverse riverine ecosystems (Gill 1970; McKnight et al. 1981; Hook 1984).

Wind rapidly and visually impacts the overstory, but floodwaters slowly and pervasively change the forest site itself by bank cutting and soil deposition. Topographic sites available for establishment of vegetation are constantly created and destroyed by the flux of sedimentary deposits (Malanson 1993). Flooding also controls seedling regeneration by continuously and subtly eliminating upland invaders, favoring physiologically adapted plants and providing water and nutrients, and thus significantly determines the composition and dynamics of riverine vegetation. By altering ecosystem processes, flood control measures and human development of floodplains have changed the successional patterns associated with flood regimes. Many upland species are currently found in bottomland sites where flooding is excluded, emphasizing the enormous selective force of floodwaters on seedling establishment. Damage to overstory and understory trees alike by ice floes and woody debris moved downstream by floodwater can be considerable (Lindsey et al. 1961).

Ice or glaze storms occur when rainfall freezes on contact with leaves, branches, or boles of trees, and affects tree growth when branches break under the weight of the accumulating ice. Ice is particularly damaging when ice-coated branches are subsequently affected by wind. Damaging ice storms occur regularly throughout the eastern United States; Kashian and Barnes (2021), for example, noted ice storm damage in southeastern Michigan that occurred at least twice per decade over the last 60 years. As with many disturbances, some species are more susceptible to ice damage than others (Downs 1938; Whitney and Johnson 1984), particularly fast-growing species such as black cherry, aspen, willow, and basswood. Oaks, sugar maple, white ash, and hickories are less likely to suffer ice damage, and conifers are particularly resistant to damage even with a constant presence of foliage. Severe ice storms may create canopy gaps, providing an opportunity for susceptible species such as black cherry to grow into the canopy. In addition to direct effects of ice storms on the overstory by severe branch or stem breakage, ice damage may indirectly affect forest tree species by providing entry for pathogens or weakening trees, so they are more subject to insect attack.

INSECTS AND DISEASE

Insects and disease-causing pathogens that kill or reduce the growth of trees affect millions of hectares of forests each year in North America and are major regulators of forest ecosystems (Chapter 12; Romme et al. 1986; Haack and Byler 1993; Castello et al. 1995). Insects and pathogens create widespread disturbances and influence succession by killing or weakening trees singly or in patches. These disturbances reciprocally affect insect populations in many ways (Schowalter 1989). Unlike fire, wind, or even logging (see in the following text), insects and disease affect overstory trees directly without changing the understory, forest floor, or soil. Root fungi, bark beetles, and inner-bark borers are the primary tree killers, tending to attack one particular species or genus of trees, and strongly control the rate and direction of succession (Franklin et al. 1987). Small gaps resulting from the death of overstory trees killed by pathogens or insects often favor shade-tolerant species already established in the understory. Larger areas of overstory tree mortality have been shown to result in increases of understory plant diversity and productivity (Pec et al. 2015).

Epidemic levels of insects or disease may occur, especially those due to bark beetles, and create a much larger effect on forested landscapes than endemic or background herbivory or disease. For example, the return interval for mountain pine beetle outbreaks in lodgepole pine forests of the northern Rocky Mountains is about 20–40 years, with an average duration of about 6–10 years (Cole and Amman 1980; Taylor et al. 2006; Alfaro et al. 2010). These components of the disturbance regime vary with climate and tree density, such that outbreak severity will vary based on these factors. An unprecedented outbreak of mountain pine beetle occurred in British Columbia between the mid-1990s and about 2017. Peaking in 2005, the rate of spread and severity of the outbreak were the largest recorded in Canadian history, resulting in the loss of 58% of the province's merchantable pine volume over >18 million hectares (Figure 16.11; Canadian Forest Service 2021). The size and duration of the outbreak have been attributed to a

Photo courtesy of Colorado State Forest Service

FIGURE 16.11 Lodgepole pine trees killed by mountain pine beetles near Granby, Colorado. The red-needled trees in the photo are those recently killed by the beetle.

combination of factors, primarily the simultaneous maturation of lodgepole pine stands across the landscape (due to past logging and effective fire suppression) and warming winter temperatures. Other native insect species have a similar influence on forests. In some Colorado subalpine forests, the effects of spruce beetle outbreaks appear to be as great as those of fire (Veblen et al. 1991b). Having coevolved with the natural disturbance agents that affect them, forest ecosystems are generally resilient to insect and disease outbreaks, but exotic or invasive insects and disease may have more drastic effects on forests. The effects of some exotic insects and diseases are described later in this chapter and in Chapter 21. Finally, insects and disease often interact with other natural disturbances, such as fire and wind; these disturbance interactions are also discussed in the following sections.

CATASTROPHIC AND LOCAL LAND MOVEMENTS

Catastrophic events such as volcanism, earthquakes, and landslides create new ecosystems and shape species occurrence and distribution. Volcanic activity on several continents has provided new substrate for forest development, such as in the Cascades of the Pacific Northwest, the Andes in Chile (Veblen 1987), the Hawaiian Islands, and the Changbai mountains of northeastern China (Barnes et al. 1992). The 1980 eruption of Mt. St. Helens has provided abundant evidence of the overwhelming effects on existing vegetation and the rapidity of succession on newly formed substrates as a result of biological legacies of diverse kinds (Franklin et al. 1985; Franklin et al. 1988; Franklin 1990; Dale et al. 2005c; Del Moral and Chang 2015). Landslides in mountainous terrain provide new substrates for plant occupancy (Pabst and Spies 2001). For example, a 1960 earthquake in the southern Chilean lake district triggered volcanic activity and thousands of landslides that provided excellent regeneration sites for southern beeches (*Nothofagus* spp.) (Veblen 1987). Such broadscale disturbances occur frequently enough to influence forest composition and structure over long temporal scales. Landslides and rock avalanches initiated by prehistoric earthquakes also shaped landscapes and changed forests in mountains of the Puget Sound area of Washington State about 1000 years ago (Jacoby et al. 1992; Schuster et al. 1992). Catastrophic and local debris avalanches—rapid downhill flows of soil, rock, and vegetation almost always preceded by heavy rains—are major disruptive and land-shaping forces in the Appalachian Mountains (Hack and Goodlett 1960; Eschner and Patric 1982; Jacobson et al. 1989).

LOGGING

Lumbering was a major factor in the development of the United States, and major changes in forest composition followed it. Post-logging fires, fueled by woody debris left after logging, also contributed significantly to change in forest composition (Whitney 1994). Aside from large-scale clear-cutting, logging as a disturbance may be similar to wind in that tree harvesting removes the overstory and releases the understory. Many understory hardwood species cut in logging-related operations resprout to form vigorous and fast-growing stems that then compete for overstory space.

As a general rule, the competitive ability of post-harvest regenerating trees will depend on the intensity and pattern of the cut. Light, partial cuts such as thinning and improvement cutting favor tolerant species, particularly those already established in the understory, and tend to accelerate forest succession rather than return it to an earlier stage (Figure 16.12). Moderately heavy partial cuts or cutting of small groups of trees to mimic the formation of canopy gaps (group selection cutting) favor mid-tolerant species. Clear-cutting or the creation of large gaps with a diameter of at least twice the height of the stand favors the invasion of pioneer species, particularly if mineral soil is exposed by the harvest operation. A forest manager can therefore greatly influence subsequent forest composition and the rate of succession by regulating the intensity and pattern of cutting.

FIGURE 16.12 Partial cutting in this eastern white pine stand in Maine has favored advanced regeneration of tolerant red spruce. *Source:* U.S. Forest Service photo.

The differential effect of partial cutting versus clear-cutting on forest succession is clear in the Douglas-fir region of Washington and Oregon. Older stands that are cut are more or less even-aged Douglas-fir dating to forest fires that occurred 150–500 years ago (Franklin et al. 1981). Clear-cutting in such stands during or immediately after a seed year provides the reestablishment of even-aged Douglas-fir. Clear-cut patches ranging from 1 to 15 ha in size will normally regenerate to Douglas-fir within a few years with seed dispersed from surrounding uncut forest. On any site, the new forest that follows a clear-cut is composed of pioneer species that range from shade intolerant to mid-tolerant in their competitive ability in the understory in that region. On clear-cuts on the lower and wetter sites in the Coastal Range, red alder colonizes prolifically as a pioneer species. Although large-area clear-cutting and planting of Douglas-fir was a conventional forestry practice for several decades, an increasingly common view is that Douglas-fir can be regenerated and grown under a partial overstory created with a variable retention harvest system or shelterwood (Chapter 14, Figures 14.16–14.18).

Logging also affects forest succession by the differential removal of one species over another—thus changing the composition of the forest. Such effects are exemplified in the mixed-wood forests of Maine where logging has occurred nearly continuously for almost 150 years (Whitney 1994). Early logging focused on large white pines suitable for masts for wooden sailing ships and house construction, but later shifted to the large-scale cutting of red spruce to build and supply sulfite and groundwood pulp mills. Removal of the red spruce from a community that included tolerant balsam fir, sugar maple, and beech created a mixed forest of these species. Balsam fir was more utilized as the supply of red spruce decreased. Partial cutting was soon replaced by clear-cutting, resulting in a rising water table that precluded forest in many of the wetter flats; pioneer species such as trembling aspen and paper birch colonized the drier sites. Thus, forest composition of central and northern Maine has been markedly shaped by successive waves of logging, each concentrating upon different species.

LAND CLEARING

Many forests occur in humid regions that are suitable for the raising of agricultural crops. The great majority of the world's population is found in forested regions, and thus much of the world's forests have inevitably been cut and the land cleared for agriculture. The massive forest clearance that occurred in the eastern United States and its effects are described by Whitney (1994). In the tropics, forest land use for agriculture is often (but not always) transitory, with fields being allowed to revert to forest again after a few years. Even in the temperate zone, much land cleared and formerly used for farming has been found unsuitable for continued cropping or has been supplanted by bringing better lands into production. This land has been planted to forest or allowed to revert naturally to forest; many currently forested lands belonging to State and National Forests in the Upper Midwest have followed this process. Secondary forest succession following land abandonment is therefore an important process, occurring over many hundreds of thousands of hectares in well-settled forest areas.

DISTURBANCE INTERACTIONS

As described in the earlier text, most ecosystems are subjected to more than one type of disturbance. The result of those interactions may be cumulative (increasing impact) or negative (e.g., the second disturbance is less impactful or less likely to occur), or the disturbances may interact to produce a novel or unexpected result (Paine et al. 1998). Interacting disturbances—specifically that the effects of one disturbance may either change the likelihood or alter the effects of another—have long been recognized by forest managers and ecologists. For example, extensive blowdown of subalpine forest in 1939 led to an epidemic of spruce beetles in the 1940s that devastated 290 000 ha of the White River National Forest alone (Hinds et al. 1965). The largest and/or most severe wildfires typically occur during periods of prolonged drought, as in the infamous Peshtigo Fire of Wisconsin in 1871, the massive western fires of 1910, and the 1988 fires in and around Yellowstone National Park, all of which burned through prior burn locations. Fire-scorched or drought-stressed trees are often sentinels to successful attack by bark beetles and other insects (Mattson and Haack 1987). Mudslides often follow fire in steep-slope, chaparral-dominated ecosystems of California and in other ecosystems of the western United States. The pattern and nature of multiple disturbances are characteristic of regional ecosystems with their distinctive climates (Mediterranean, northern continental, coastal/marine), physiographic features (mountains, plains), and soil/hydrology (excessively drained, poorly drained).

Interacting disturbances have been studied extensively over the last two decades and may be characterized as two types of interactions (Buma 2015). Disturbances that alter ecosystem resistance, by changing the likelihood, intensity, severity, or size of the subsequent disturbance, are called **linked disturbances** (Simard et al. 2011). Linked disturbances occur when the first disturbance creates a legacy, such as reduced availability of host trees for an insect or altered distribution of fuels, that mechanistically affects the second. In a study of a spruce beetle outbreak in the 1940s in western Colorado, Kulakowski et al. (2016) found that young stands of Engelmann spruce and subalpine fir that regenerated from fires in the late-nineteenth century were not susceptible to the spruce beetle outbreak. Thus, the location and severity of the spruce beetle outbreak were altered by legacies of the fires in the late 1800s (in this case, lack of mature, susceptible trees). Other examples of linked disturbances include those presented by Kulakowski and Jarvis (2013), Kulakowski et al. (2012), Simard et al. (2011), Lynch et al. (2006), Bigler et al. (2005), and Veblen et al. (1994).

When disturbances alter the resilience of an ecosystem to a subsequent disturbance such that the interacting disturbance changes the likelihood or rate of recovery of the ecosystem, the interaction is the result of **compounded disturbances**. In a study of American beech dominance in coastal New England, Busby et al. (2008) found that intensive forest harvesting in the nineteenth century favored regeneration of oaks and beech, but repeated hurricanes thereafter favored beech persistence in the understory over oaks. A severe hurricane in 1944 and heavy deer browsing released beech saplings (typically avoided by deer) while preventing further oak establishment. The reduced resilience of oaks as a result of compounded disturbances in this example has created long-term changes in forest composition and diversity (Busby et al. 2008). Other examples of compounded disturbances include those presented by Andrus et al. (2021), Carlson et al. (2017), Kulakowski et al. (2013), and Girard et al. (2009).

Interestingly, the same events can simultaneously result in both linked and compounded disturbance interactions (Buma 2015), as illustrated by two independent studies that examined the interaction of a severe wind event with subsequent fire in the subalpine forests of northwestern Colorado (Figure 16.13). Kulakowski and Veblen (2007) found that the changes in fuels resulting from the blowdown altered the severity (but not the size) of the fire that followed, suggesting that wind and fire in this case were linked disturbances. In a separate study in the same area, Buma and Wessman (2012) found that the blowdown affected resilience of the system by altering seed dispersal into the areas disturbed by wind and severe fires. In particular, wind-damaged areas were too large to be revegetated by spruce and fir seeds dispersed from adjacent unburned forests, and

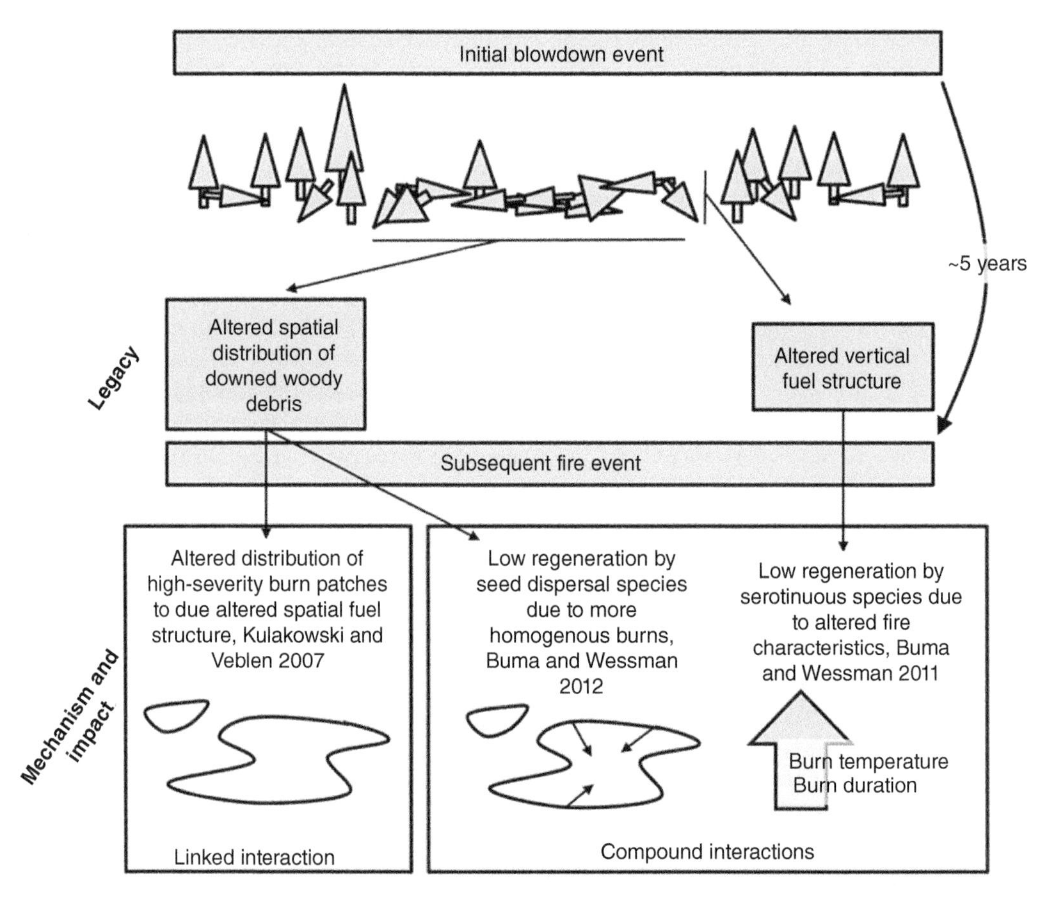

FIGURE 16.13 The interaction of a large wind event and subsequent fire in Colorado show that linked and compound interactions may occur from the same events. Burn severity was higher in the areas affected by wind (linked disturbance interaction), but the blowdown reduced post-fire resilience by limiting seed dispersal into the large and severely burned patches from adjacent unburned areas and from serotinous cones within the area (a compound disturbance interaction). *Source:* Buma (2015) / John Wiley & Sons.

the increased severity of the fires in the wind-affected areas consumed the serotinous cones of lodgepole pine and the seeds therein. These mechanisms reduced the ability of the ecosystems to recover, suggesting a compounded disturbance interaction (Figure 16.13). An excellent review of disturbance interactions and their ecological implications is provided by Buma (2015).

BIOTIC COMPOSITION CHANGES

Secondary succession originates primarily from fire, windstorms, logging, and land clearing, but a second major kind of disturbance—the addition or subtraction of plant or animal species—will alter the trajectory of forest succession. These changes in the flora or fauna may create considerable disturbance to existing ecosystems and thus may well be considered as initiating secondary succession in a very real sense.

ELIMINATION OF SPECIES

When a species is eliminated or greatly reduced in abundance, its place in an ecosystem is typically taken by other organisms. In many instances, species loss is the result of the introduction of exotic insects or pathogens to which native tree species have not adapted. There are numerous examples of species' elimination or near-elimination that have occurred over the last 100 years as a result of exotic insect or pathogen introduction. The American chestnut, once a major overstory dominant in eastern North America (Braun 1950), was virtually eliminated from the eastern deciduous forest by the 1940s by a pathogen introduced from Asia in 1904. American chestnut was replaced primarily by its major associates in New England, the mid-Atlantic states, and the southern Appalachians (Keever 1953; Woods and Shanks 1959; Good 1968; Stephenson 1974). Dutch elm disease, caused by two related pathogens introduced from Europe in the early-twentieth century, drastically reduced the abundance of mature American elm trees in swamp and river floodplain forests of the eastern and midwestern United States, replaced by a variety of different species depending on site conditions. In southeastern Michigan, American elm has not been eliminated from deciduous swamps but remains as an understory and subcanopy overstory species rather than a dominant overstory species (Barnes 1976). It will likely persist by seeds from young elm trees with an average life span that is drastically reduced. Emerald ash borer is an invasive insect introduced into the metropolitan Detroit area from Asia in the 1990s. Specific to the ash genus, the insect had spread to 36 states and five Canadian provinces by 2022, killing nearly all ash trees larger than about 1 in. in diameter once it reaches high population levels. Where ash was found in mixed stands, succession has mainly featured replacement by associates depending on the species and site conditions (Kashian and Witter 2011). However, pure stands of black ash, typically found on poorly drained soils, may be potentially converted to shrub wetlands or other cover types if ash mortality is high (Kolka et al. 2018; Palik et al. 2021). Like American elm, most ash species are likely to persist, but only as seedlings, saplings, and sprouts from top-killed overstory trees and the seeds they produce (Kashian 2016). Effects on forests of invasive species such as chestnut blight, Dutch elm disease, and emerald ash borer are discussed in detail in Chapter 21.

Reduction of tree species in a community is not always the result of the introduction of exotic species. For example, in the spruce–fir forests of eastern Canada and the northeastern United States, spruce budworm epidemics have greatly reduced the abundance of overmature and mature balsam fir (together with lesser amounts of black and white spruces) over large areas. The budworm apparently has played a major role in these mixed softwood forests in suppressing the proportion of the very-tolerant balsam fir compared to that of the somewhat less-tolerant spruces (Bognounou et al. 2017). Notably, the response of balsam fir in this case is a local reduction in abundance rather than a widespread elimination because it has coevolved and adapted to the long-term presence of the native spruce budworm.

In the animal portion of forest ecosystems, the virtual elimination of many predators from most forests in North America—particularly wolves, bears, cougars, and lynx—has played a role in increasing the populations of their prey: deer, rabbits, and other herbivores. This in turn results in greater browsing of the understory, including tree regeneration. Such disturbance precludes the regeneration of black cherry in areas of Pennsylvania (Marquis 1974, 1975, 1981; Tilghman 1989) and hemlock in many places in the northern United States (Frelich and Lorimer 1985; Peterson and Pickett 1995). Reintroduction of wolves into Yellowstone National Park in 1995 had significant effects on elk populations, both in their population size (reduced by predation) as well as their foraging behavior. As a result, browsing was reduced and recruitment to the overstory was increased for trembling aspen and various species of cottonwoods and willows (Ripple and Beschta 2003, 2012; Beschta and Ripple 2016).

ADDITION OF SPECIES

The species occupying a given site may simply be those that have access to that site at the time of its availability rather than those best adapted to compete and grow on that site. Invasion of the site by better competitors often results in substantial changes in forest succession. The chestnut blight fungus, Dutch elm disease fungus, and emerald ash borer cited in the previous section are examples of accidental introductions that have greatly modified forests. Other organisms may be trees, other higher plants, fungi, bacteria, and animals of all kinds (Chapter 21). Important deliberate forest-tree introductions in Europe include Douglas-fir, eastern white pine, northern red oak, and black locust; in the northeastern and north-central United States: Scots pine, Norway spruce, Norway maple, black pine, white mulberry, Chinese tree of heaven, Siberian elm, and European larch; and in temperate zones of the Southern Hemisphere: Monterey pine, patula pine, and slash pine. These and many others have become vigorous and often dominant species in the local flora. The effects of additions of such species to forests is the subject discussed in detail in Chapter 21.

SUGGESTED READINGS

Agee, J.K. (1993). *Fire Ecology of Pacific Northwest Forests*. Washington D.C.: Island Press 493 pp. Chapters 5 and 7–12.

Covington, W.W. and Moore, M.M. (1994). Postsettlement changes in natural fire regimes and forest structure: ecological restoration of old-growth ponderosa pine forests. *J. Sustain. For.* 2: 153–181.

Heinselman, M.L. (1973). Fire in the virgin forests of the boundary waters canoe area, Minnesota. *Quat. Res.* 3: 329–382.

Paine, R.T., Tegner, M.J., Johnson, E.A., and E.A. (1998). Compounded perturbations yield ecological surprises. *Ecosystems* 1: 535–545.

Pickett, S.T.A. and White, P.S. (1985). *Natural Disturbance and Patch Dynamics*. New York: Academic Press 472 pp.

Sprugel, D.G. and Bormann, F.H. (1981). Natural disturbance and the steady state in high-altitude balsam fir forests. *Science* 211: 390–393.

Whitney, G.G. (1994). *From Coastal Wilderness to Fruited Plain*. Cambridge: Cambridge Univ. Press 451 pp.

Forest Succession

CHAPTER 17

All components of ecosystems change over time, including climate, landforms, soil, and biota, and we have discussed the basis of dynamic change in forest ecosystems in previous chapters. Change may be gradual or rapid depending on the particular disturbances that characterize regions and local sites (Chapter 16). Ecosystem development and change are most easily observed in vegetation, especially the forest trees, which may change dramatically over time following disturbances such as fires or windstorms.

One manifestation of ecosystem change is **forest succession**, which is often characterized as the change in species composition or the replacement of the biota of a site by one of a different nature. It is notable that other aspects of ecosystem change, such as soil development, have also been documented (Olson 1958; Miles 1985; Matthews 1992), and changes involving biomass accumulation and nutrient cycling are described in Chapters 18 and 19. In this chapter, we focus upon changes in species composition. Importantly, forest succession is what happens in one place over an extended period of time, measured in tens to thousands of years. In other words, succession happens with space fixed and time changing (Rowe 1961b). It is characterized by a sequential change in the relative structure, kind, and abundance of the dominant species. We focus here on the intervals between disturbances and within a time period of the same order of magnitude as the life span of the longest-lived organisms in the successional sequence (tens to thousands of years).

Forest succession progresses in nearly infinite ways and is driven by many different factors, and many other processes occur simultaneously with it. However, succession can only have meaning in the context of a particular geographic framework defined by regional and local ecosystems. The heart of the matter is that *location-specific ecosystems* undergo development, change, and succession. The way a given ecosystem changes when disturbed is displayed more or less visibly according to the resilience and inertia of its component parts. Our attention is quickly drawn to the visibly changing vegetation, but attempts to explain and predict its successional path separate from its geographic context are unlikely to be rewarding.

Because vegetation is most easily observed, plant succession has historically been defined in the context of plant species or communities. Definitions include those of Grime (1977): "a progressive alteration in the structure and species composition of the vegetation" and Finegan (1984): "the directional change with time of the species composition and vegetation physiognomy of a single site." Forest succession obviously deals with changes in biota, but the way we conceive its context (plant or ecosystem, part or whole) is critical in understanding the patterns and processes of succession over the complexity of regional and local ecosystems. Troll (1963b) was the first to emphasize and explicitly describe succession as a landscape process involving all ecosystem components, including climate, physiography, and soil as well as vegetation.

Adopting vegetation as a primary focus of succession contributed to the development of an elaborate and often criticized concept, with literally thousands of published papers introducing additional terminology and controversy. A perceptive review by Pickett et al. (1987) clears up much

Forest Ecology, Fifth Edition. Daniel M. Kashian, Donald R. Zak, Burton V. Barnes, and Stephen H. Spurr.

confusion, and a landscape ecosystem perspective provides the spatial framework for understanding how succession works. All successions have these key characteristics that affect their course and rate: (i) a sequence of concomitant environmental and vegetational change that may be characterized by stages for convenience and practical applications; (ii) disturbance regimes; and (iii) mechanisms (processes) occurring at different times in the sequence. Above all, the possible successional sequences of interacting organisms, mechanisms, and disturbances are controlled or mediated by the regional and local landscape ecosystems where succession takes place.

In this chapter, we examine successional concepts, the evolution of the concept of forest succession, causes of ecosystem and forest change, and the mechanisms involved. The focus is of course on vegetation change but placed in an ecosystem context as far as possible. Examples of primary and secondary succession are presented to illustrate succession in diverse settings. For details beyond the scope of our treatment, the following books and reviews provide diverse viewpoints: Clements (1916, 1936); Tansley (1929, 1935); Drury and Nisbet (1973); Connell and Slatyer (1977); Golley (1977); Miles (1979, 1987); White (1979); West et al. (1981); Finegan (1984); Pickett et al. (1987); Glenn-Lewin et al. (1992); Matthews (1992); Peet (1992); Worster (1994); Schmitz et al. (2006); Swanson et al. (2011); Walker and Wardle (2014); and Prach and Walker (2020).

BASIC CONCEPTS OF SUCCESSION

PRIMARY AND SECONDARY SUCCESSION

Primary succession is ecosystem change that occurs on previously unvegetated terrain, uninitiated by a disturbance. For example, primary succession on recently deglaciated terrain in Alaska eventually supports spruce forest after first being colonized by communities of willow or alder. **Secondary succession** follows a disturbance to an existing forest that disrupts ecosystem processes and destroys or disrupts existing biota. The distinction between primary and secondary succession is often more arbitrary rather than real, but it is followed here as a matter of convenience in understanding the dynamics of forest ecosystems.

BIOLOGICAL LEGACIES

Biological legacies that follow disturbances such as fires, floods, and windstorms distinguish the course of ecosystem change during secondary succession. Biological legacies are biologically generated elements of a pre-disturbance ecosystem that persist in some form after a disturbance occurs (Franklin et al. 2018). Examples of biological legacies include: (i) living organisms or portions of organisms that survive a disturbance, (ii) organic debris, and (iii) biotically derived patterns in soils and understories (Franklin 1995). Living legacies include intact plants and animals, rhizomes, and dormant spores and seeds. Important biotically derived structures include standing dead trees (snags) and fallen logs (coarse woody debris), large soil aggregates, and dense mats of fungal hyphae. These diverse biological legacies, remaining from pre-disturbance ecosystems, strongly influence the paths and rates of forest recovery and ecosystem function by shaping post-disturbance community structure and/or providing substrate for new colonizers. In fact, the types and abundances of biological legacies are among the most important ecological outcomes of disturbances (Franklin et al. 2018), although they differ sharply among different kinds of disturbances (Table 17.1).

SUCCESSIONAL PATHWAYS, MECHANISMS, AND MODELS

Pickett et al. (1987) define the terms successional pathway, mechanism, and model. A **successional pathway** is the temporal pattern of vegetation change, which typically shows change in community

Table 17.1 Biological legacies generated by different types of severe disturbances.

	Disturbance type				
Biological legacy	**Severe wildfire**	**Wind**	**Insect outbreaks**	**Volcanoes**	**Clearcut harvesting**
Live trees	Few	Few or absent	Depends on forest composition	Few, confined to margins	Absent
Snags	Abundant	Few	Abundant	Abundant	Absent
Downed woody debris	Common	Abundant	Eventually abundant	Abundant	Absent
Uproots	Absent	Abundant	Absent	Few	Absent
Intact understory	Rare	Abundant	Abundant	Absent	Very little
Mineral seedbed	Abundant	Abundant with uproots	Absent	Absent if buried	Abundant

Source: Adapted from Franklin et al. (2018) and Swanson et al. (2011).

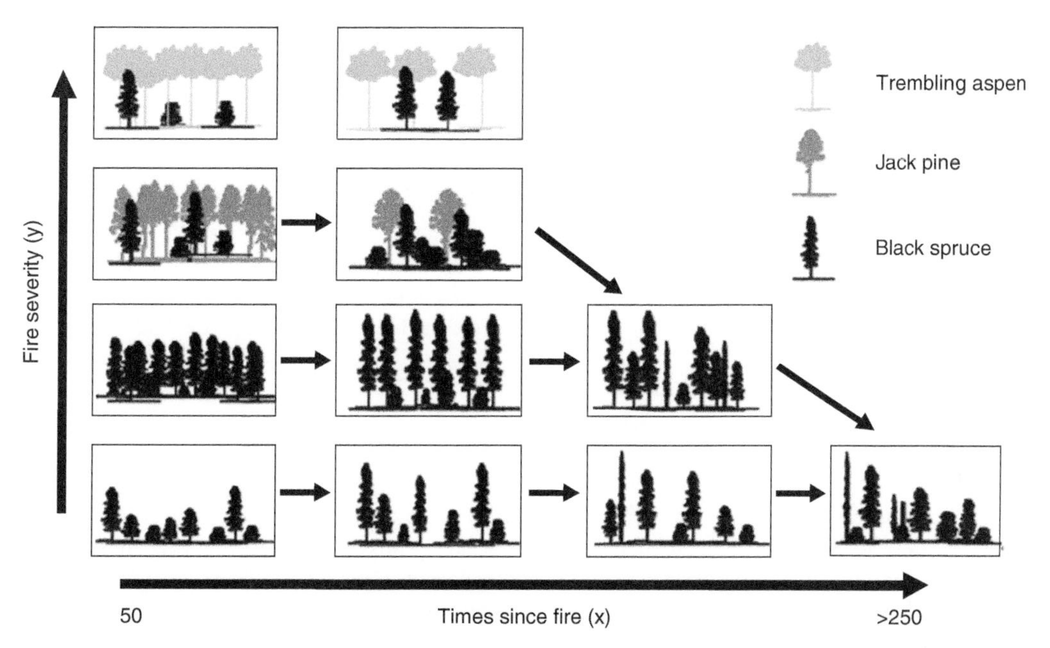

FIGURE 17.1 Alternative successional pathways for aspen–conifer forests in the clay belt of northwestern Quebec, Canada. Forest composition and structure are strongly determined by the severity of the last fire (y-axis) and the time elapsed since the last fire (x-axis). High-severity fires initially result in forests ranging from aspen-dominated to a mixture of jack pine and black spruce. Low-severity fires result in black spruce-dominated forests across a range of density. With time, forests generally converge toward open, multi-aged black spruce forests. The rate and direction of forest succession on this landscape are strongly influenced by site factors, especially soil type. *Source:* From Franklin et al. (2018).

types and may describe the decrease of particular species populations. Figure 17.1 is an excellent example of successional pathways because the initial site conditions in conjunction with disturbance are major factors in determining ecosystem change. A **successional mechanism** is an interaction or process that contributes to successional change. Examples include general ecological processes such as dispersal, competition, and establishment, and physiological processes such as biomass allocation (Chapter 18) and nutrient uptake (Chapter 19) that affect plant form and reproductive ability. A **successional model** is a conceptual map that explains a successional pathway by identifying and specifying the relationship among the mechanisms and the various stages of the pathway.

AUTOGENIC AND ALLOGENIC SUCCESSION

Successional change has traditionally been characterized as either **autogenic** (by endogenous factors) or **allogenic** (by exogenous factors) depending whether successive changes are brought about by the action of the plants themselves on the site or by external factors (Tansley 1929). Primary succession is an autogenic process in the classical view because it is plant-driven, and secondary succession is allogenic because it is driven by periodic disturbances. The dichotomy between autogenic and allogenic factors is actually quite artificial: the causal factors don't fit strictly into one category or the other (White 1979; Matthews 1992). Tansley (1929, p. 678) recognized this problem when he proposed the concept, cautioning that real successions show a mixture of the two classes of factors.

Today, the ecosystem rather than the plant is or should be the focus, because change is elicited by a combination of factors both within the ecosystem and external to it. Allogenic factors (physical processes and disturbances such as fires and windstorms) may be initiated outside a given ecosystem, but their effects often strongly depend upon site conditions (e.g., topography, aspect, soil depth, drainage) and the associated vegetation within a given ecosystem. For example, the aspect, topography, soil properties, and vegetation of an ecosystem, and its position in relation to fire breaks, strongly affect fire frequency and severity. Windstorm effects are significantly related to soil depth, drainage, rooting ability, and the height of tree species of the affected ecosystem. Furthermore, autogenic factors of litter accumulation and plant-induced changes in microclimate, soil water, and nutrients all affect the outcome of allogenic disturbances. In turn, autogenesis is almost always influenced by allogenic factors. Thus, allogenic and autogenic factors are closely interrelated.

HOW IS SUCCESSION DETERMINED?

In either primary or secondary succession, ecosystem and vegetation change should ideally be tracked at one place over time such that an *actual* successional pathway for a given ecosystem might be determined. However, such monitoring is rarely possible due to the long life span of forest trees, and much of what we know about succession therefore comes from short-term observations at different places. The typical approach used to characterize long-term succession is to *infer* successional trends from stands of different ages, all of which are assumed to have developed on the same site and under the same disturbance conditions. This approach has come to be known as a **chronosequence** (Major 1951) or **space-for-time substitution** (Pickett 1988). If constructed carefully and replicated, chronosequences can be indispensable for inferring how forest ecosystem structure and function change over centuries (e.g., Kashian et al. 2013). A chronosequence approach is not without its problems, however (Matthews 1992; Johnson and Miyanishi 2008). Chronosequences are generally most appropriate when successional pathways are likely to converge, biodiversity is low, and disturbance is infrequent and/or not severe (Walker et al. 2010). Chronosequences are less suitable when successional pathways are likely to diverge, or the ecosystems are species-rich, highly disturbed, or arrested in time. The actual chronosequence of ecosystem and forest change is one that is site-specific over time, and ideally this should be the process that is observed. Several examples of chronosequences are described later in this chapter.

EVOLUTION OF THE CONCEPT OF FOREST SUCCESSION

The dynamic nature of forests was recognized by naturalists in North America and Europe as early as the eighteenth century in a form that approximates a plant-based (bioecosystem) concept of forest succession (Spurr 1952a). Such ideas gained significant traction in the nineteenth century. In Europe, beginning with Hundeshagen in 1830, observed changes in forest composition were the subject of specific articles by professional foresters and botanists. Hundeshagen pointed out instances of spruce replacing beech and other hardwoods in Switzerland and Germany, and of spruce and other species taking the place of birch, aspen, and Scots pine. The first detailed North American report of species composition changes was apparently that of Dawson in 1847. Working in the Maritime Provinces of eastern Canada, Dawson recognized the effects of windthrow and fire in the forests found by European colonists and distinguished between successional trends in small clearings following cutting, following a single fire, following repeated fires, and as a result of agricultural use of land.

As early as 1860, the writer Henry David Thoreau recognized that pine stands on upland soils in central New England were succeeded after logging by even-aged hardwood stands that today constitute the principal forest type of the region. He named this trend "forest succession" and published an essay on "The Succession of Forest Trees." His late natural history writings, published long after his death (Thoreau 1993), are remarkably filled with insights on ecological succession: pines are "pioneers" to oaks and "lusty oaken carrots" that are a "remarkable special provision for the succession of forest trees." In articles published in 1875 and 1888, Douglas discussed at some length the concepts of forest succession and pioneer species, and presented an explanation of how short-lived, light-seeded pioneer species formed the first forest types on burned-over pine land.

FORMAL ECOLOGICAL THEORY

A general and formal theory of plant succession evolved slowly but was well established by the beginning of the twentieth century when several American ecologists systematized its study. The foundations of plant ecology as a study of vegetation dynamics and succession were initiated by Henry C. Cowles (1899, 1901), who analyzed succession on sand dunes of Lake Michigan, beginning with uncolonized sand and ending with a mature forest. Although dunes were formed and destroyed by wind, he found that "vegetation profoundly modifies the topography." In 1911, Cowles noted that the original plants in any habitat give way "in a somewhat definite fashion to those that come after." Furthermore, he recognized that changes in vegetation caused by climatic, topographic, and biotic changes occurred concurrently, at different rates, and at times in different directions.

A contemporary, Frederic E. Clements, next developed an elaborate philosophical structure of plant succession (1916, 1949) that attempted to categorize and formalize all eventualities of plant community change. Specific examples of forest succession were documented by William S. Cooper, with his studies of Isle Royale in Michigan (1913) and the colonization by plants following glacial retreat in Glacier Bay, Alaska (1923, 1931, 1939). Clements, in particular, developed a complex nomenclature to describe plant succession in a meticulous, tidy organization of nature that has both facilitated and greatly complicated the efforts of his successors. Some of his terms have taken a permanent place in the vocabulary of ecologists (vegetation dynamics, biome), others have persisted but with broadened and changed meanings (climax), while many have been dropped from general usage (cyriodoche, consocies). The work of Clements is described in detail in the following text.

HOW DOES SUCCESSION WORK?

Succession involves many factors and processes operating within a geographic framework at multiple scales (Table 17.2). Understanding the key characteristics of ecosystem change and what they might tell us about succession is critical for examining how succession works. Key characteristics determining actual vegetation change (with space fixed and time changing) include the following:

The physical site conditions of the Earth's regional and local ecosystems provide a geographic framework that determines the organisms involved and the rate and trajectory of succession.

Disturbance factors and processes associated with particular regional and local ecosystems provide substrates for primary and secondary succession and periodically affect the rate and trajectory of succession.

Colonizing organisms have an enormous range of life-history traits (genetic adaptations of structural architecture, longevity, and physiological processes) and regeneration strategies that are tightly linked to site conditions and disturbance regimes.

Changes in site conditions and organisms occur reciprocally, with one affecting the other and vice versa.

Organisms interact through competition and mutualisms.

With these characteristics we can expect that: (i) successional pathways are likely to be unique, and no two pathways will be the same; (ii) different successional pathways will occur where regional and local ecosystems are also diverse; (iii) there are a great number of different and simultaneously operating mechanisms that control the rate and trajectory of succession; (iv) distinguishable patterns and stages of succession are evident in ecosystems not severely affected by humans; (v) species occurring early in succession modify the environment in ways that may limit their own regeneration and may favor or inhibit the regeneration and growth of species occurring late in

Table 17.2 Causes of succession: Hierarchical summary of factors and processes.

- **Landscape ecosystem characteristics**
  - Regional climate, physiography, vegetation patterns
    - Local climate, physiography, landform, soil, species distribution
      - **Site availability**
        - Disturbance (kind, periodicity, size, severity, dispersion, disturbance history of target site and adjacent ecosystems)
        - **Differential species dispersal and establishment**
          - Landscape position in relation to seed source availability
          - Reproductive mode of species (sexual, asexual) and fecundity
            - Dispersal (dispersal agents, landscape configuration)
            - Propagule pool for establishment (time since land disturbance, land use treatment)
          - **Differential species performance** (in regeneration and post-establishment development)
            - Genetic adaptations and plasticity (germination and establishment traits, understory tolerance, assimilation and growth rates, biomass allocation, plant defense)
            - Competition and mutualism (presence, kind, amount)
            - Environmental stress (climatic cycles, site history, prior occupants)
            - Allelopathy (soil chemistry, neighboring species)

succession; and (vi) although succession proceeds at various rates, there is no end point of the process.

With these key characteristics, what are the significant features of how succession works? We may summarize the *general trends* of how succession works as follows:

1. Following a disturbance that creates either (i) new land (unvegetated substrate for primary succession) or (ii) a disrupted forest with a vegetated site having remnants of the pre-disturbance forest (with secondary succession following), substrate is available for recolonization or continued growth by organisms.
2. Species from surrounding areas or within the disturbed area itself occupy the disturbed site in varied patterns and temporal sequences. Where a disrupted forest undergoes secondary succession, new colonists may be dispersed into the site or are already present in surviving vegetation, the soil seed bank, or the vegetative propagule bank (Figures 4.2 and 4.13, Chapter 4).
3. The colonizing plants develop on the site and either become established or die.
4. Plants interact through competition and mutualisms and thereby affect further plant establishment and development over time.
5. Plants change site conditions, such as light, temperature, or soil properties, and processes which may or may not affect them as they continue to interact with other plants and animals.
6. All of the abovementioned processes are modified by multiple disturbances that are ecosystem-specific in type, frequency, and severity.
7. Some species persist, thrive, and dominate at their time during succession, whereas others drop out or diminish in abundance.

The nature and rate of change of vegetation depend on the spatial and temporal scales under consideration, the site conditions of regional and local ecosystems, species' life-history traits and regeneration strategies, disturbance regimes, chance effects, and other processes (Table 17.1). If it is known which species are present, which are in the vicinity of a particular land area, and the ecology of the species in the context of the various land-vegetation patterns, then it is possible to make an educated guess as to what is *likely* to happen in the future (Rowe 1961b).

CLEMENTSIAN SUCCESSION

It is instructive to compare the scenario just described with Clements' model of succession, the first to describe the vegetative approach and a legacy for our current understanding and the basis for opposing ideas. Clements (1905; Weaver and Clements 1929) declared what he saw as the rules of succession in no uncertain terms. The basic processes were clear cut, and stages proceeded in a systematic, stepwise, directional, and predictable process (Figure 17.2). In this model, *ecesis* (germination, establishment, and growth) occurs immediately after plants reach the site (migration). Dominant plants change the environment (*reaction*) to favor or "facilitate" the dominance of the next community that can better compete at the site. The end point of a given succession is reached as the **climax** formation or association is attained. The plants that dominate at the climax stage modify their environment such that they perpetuate themselves, resist change, and therefore create the community of maximum stability. Clements also assumed that all primary successions in a climatic region eventually converge to the same climax community from multiple starting points (**monoclimax**). A critical flaw of Clements' was that in emphasizing the influence of plants upon

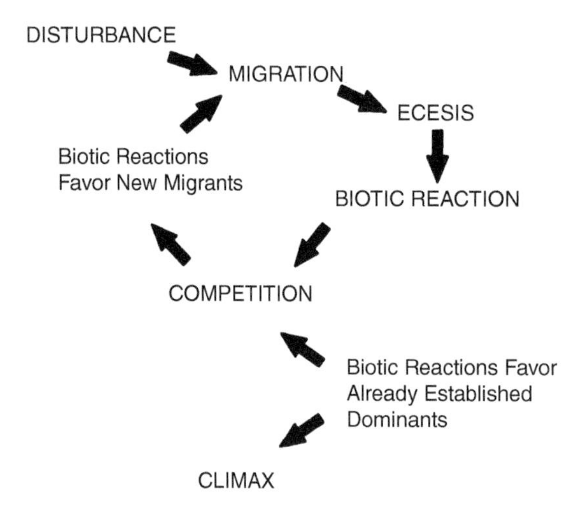

FIGURE 17.2 Classical (Clementsian) model of autogenic plant succession. *Disturbance* brings about nudation followed by arrival of propagules (*migration*) that germinate and grow (*ecesis*) and modify the site (*biotic reaction*) in *competition* with one another. Ecesis, reaction, and competition often occur simultaneously. Competition and reaction continue in cycles (left) that favor new immigrants and hence new communities; ultimately (lower right) the reactions lead to the climax or end stage. *Source:* From Christensen (1988b) / University of Washington Press.

the habitat (i.e., biotic reaction in Figure 17.2); he saw the driving force of succession to be changes caused by internal factors (i.e., autogenic succession). It is therefore important to consider the stages of succession as a basis for examining insights, problems, and misconceptions about orderly succession.

STAGES OF SUCCESSION

In describing plant succession, we begin with a hypothetical unvegetated substrate or disturbed forest and describe the successive plant communities that will occupy this site. Doing so requires assumptions that (i) the regional climate will remain unchanged, and (ii) disturbances such as severe windstorms, fires, or insect outbreaks are absent. Such assumptions are unrealistic given the hundreds or thousands of years involved in forest successions. However, these assumptions provide the view of succession as a more or less orderly and predictable sequence that depends upon the character of the physical site and associated biota. The assumptions also provide the framework to examine some problems of this perspective.

The recognition of stages is a matter of convenience rather than of their actual occurrence. At best, stages are simply wave-like replacements of species with similar ecologies, and at worst, they are arbitrary divisions in a continuum (Matthews 1992). It is notable that depending on site and species availability, plant communities vary in their rate of change as new species colonize the site and existing species either reproduce or disappear. The arbitrary classification of this continuum into stages, characterized by the dominance or presence of certain species and plant life-forms, is certainly useful and a convenience worth maintaining.

Primary Succession Unvegetated sites range from pure mineral material (rock or soil) to water; plant colonization and growth are most favored by mixtures of soil and water (i.e., moist, well-drained mineral soil). Primary plant succession beginning with dry rock material (either as rock or as mineral soil) is termed a **xerarch** succession; that beginning with water is termed a **hydrarch** succession; while that beginning with moist, aerated soil is a **mesarch** succession.

The major stages of primary succession are generally consistent, but many types of primary successions exist (Matthews 1992). Both the specific successions and the vegetational stages within each are arbitrarily chosen. In Table 17.3, a series of 10 stages is given for each representative type of primary succession. Notably, some stages are omitted under conditions where the next successional life-form (tree, shrub, herb) is capable of directly colonizing an earlier vegetational

Table 17.3 Stages of primary succession.

Stage	Xerarch	Mesarch	Hydrarch
1	Dry rock or soil	Moist rock or soil	Water
2	Crustose lichens	(Usually omitted)	Submerged water plants
3	Foliose lichens and mosses	(Usually omitted)	Floating or partly floating plants
4	Mosses and annuals	Mostly annuals	Emergents
5	Perennial forbs and grasses	Perennial forbs and grasses	Sedges, sphagnum, and mat plants
6	Mixed herbaceous	Mixed herbaceous	Mixed herbaceous
7	Shrubs	Shrubs	Shrubs
8	Intolerant trees	Intolerant trees	Intolerant trees
9	Mid-tolerant trees	Mid-tolerant trees	Mid-tolerant trees
10	Tolerant trees	Tolerant trees	Tolerant trees

type. In the Clementsian paradigm, the series of stages at a given site is termed a **sere** and each stage a **seral stage**.

There exist several potential misconceptions in relating the possible stages to real-world succession. First, succession does not necessarily begin with stage 1, proceeding neatly through each successive stage unidirectionally. Second, stages do not typically proceed discretely one after another in relay fashion, but instead exhibit considerable overlap between them. Third, there is no predetermined time period for each stage to begin and end; depending on the site, one stage may occupy the site for a long time and seemingly terminate succession until site conditions change. In sum, real-world succession typically has stages that are far less uniform and rigid than those listed in Table 17.3.

The various life-forms and developmental stages in succession may be characteristic of more than one stage, and indeed, many persist through many stages. Some mosses, for instance, may colonize a site early in succession and persist until the later stages characterized by tolerant trees. Tree seedlings often establish themselves very early in succession along with annuals and grasses or beneath shrubs, particularly in secondary succession when tree seed sources surround the disturbed area. Thus, all plants of each stage do not necessarily appear and die out abruptly at an appointed time; rather they may overlap in various sequences according to the life-history traits of each species. For example, in the Appalachian Mountains many clonal shrubs and alder, rhododendron, and mountain laurel thickets persist for 20–40 years or more, preventing the establishment of tree seedlings within them (Niering and Goodwin 1974; Damman 1975; Vose et al. 1994).

The actual species composition of the different stages depends upon those species that have access to the site in question, either in terms of the relative location of seed sources or their ability to be dispersed to the site (Clements 1905; Abrams et al. 1985; McCune and Allen 1985; McClanahan 1986; Matthews 1992; Fastie 1995; Buma et al. 2017). Thus, the actual plant communities present on a given site depend upon the available plants as well as upon the site itself. Such plants may be already on the site as buried seeds and propagules (roots, rhizomes) or may be transported by wind, water, or animals from adjacent or distant ecosystems.

The importance of species availability and dispersal in succession was clearly borne out in a study by Buma et al. (2017), who resampled Cooper's original permanent plots in Glacier Bay, Alaska after 100 years of succession. The plots were located by Cooper at particular distances from a retreating glacier to represent time since glaciation, and thus various stages along the successional sequence. The first 50 years of succession seemed to confirm the basic hypotheses of plant succession that plant coverage, in general, and willow coverage, particular to this location, would be higher in "older" plots. The second 50 years of succession, however, suggested that the spatial location of seed sources and species dispersal into the plots began to become more important than time since glaciation. Moreover, the plots were highly variable in canopy coverage, species composition, and soil development. In particular, predicted late-successional dominant species (Sitka spruce) did not establish in plots where willow arrived early and dominated. These results highlight that, in some cases, early species arrival has an enormous influence on successional pathways (Buma et al. 2017).

Egler (1954) described the dichotomy of two models, "relay floristics" and "initial floristic composition" (IFC), that contrast the strategies of species as they colonize old fields. Relay floristics (Figure 17.3a) refers to the appearance and disappearance of stages of species at a site. This model suggests that one group of plants favors or *facilitates* the establishment of the next

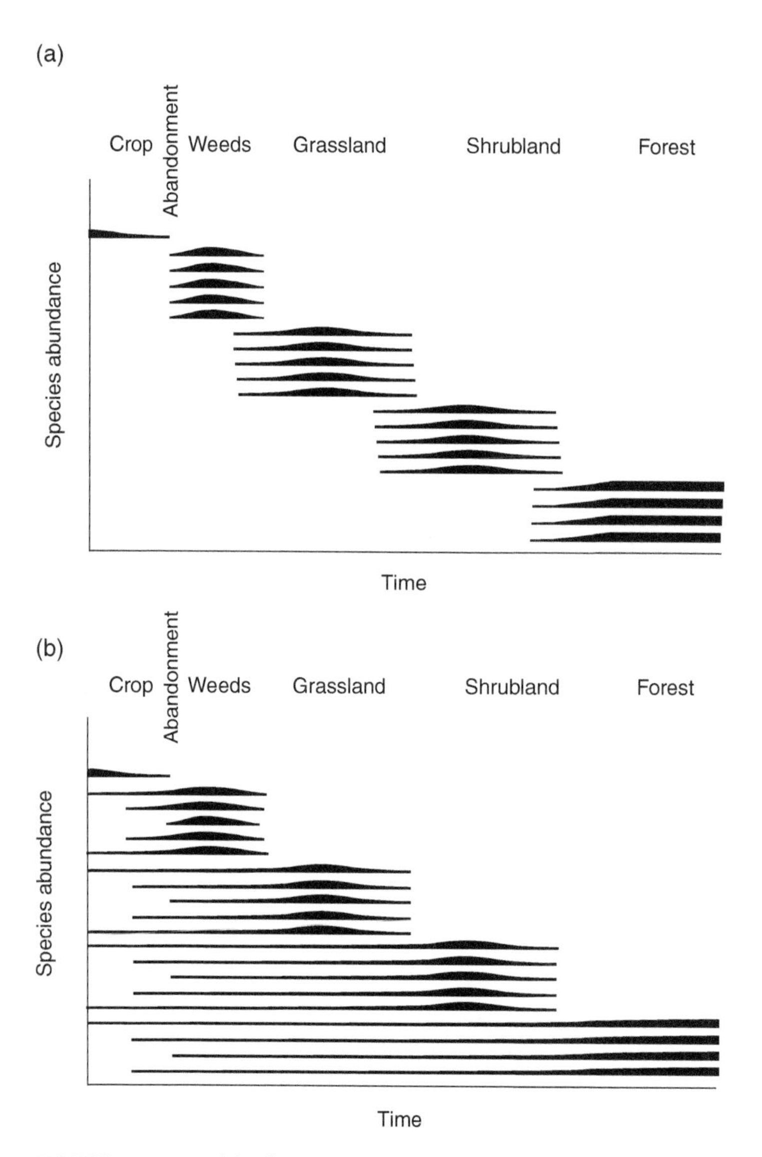

FIGURE 17.3 Models of succession from studies of abandoned agricultural fields (old-field succession). (a) Relay floristics—colonization of successive groups of species occurs in sequential phases. (b) Initial floristic composition (IFC) —all species establish at or shortly after initiation of succession but still dominate sequentially in stages. *Source:* Egler (1954) / Springer Nature. Vegetation science concepts I. Initial floristic composition, a factor in old-field vegetation development. *Vegetatio* 4:414, © 1954 by Uitgeverij Dr. W. Juink, Den Haag. Reprinted with kind permission from Kluwer Academic Publishers.

group in the sequence by modifying the environment. Some attribute this rigid scheme to Clements, but, in fact, he acknowledged the less discrete nature of the stages as they appeared. In 1905, Clements observed that many migrants appeared into a new, denuded, or greatly modified habitat to then be sorted into three groups by the establishment process: (i) those that arrive but are unable to germinate or grow and soon die, (ii) those that grow normally under the conditions present, and (iii) those that arrive but remain dormant through one or more of the earlier stages to appear at a later stage. Matthews (1992, p. 294) also observed that the relay floristics model, as shown here, does not realistically represent the lack of discreteness of the species groups in primary succession.

In contrast to the relay floristics model, the IFC model (Figure 17.3b) suggests that all species are present at or shortly after succession begins, as acknowledged by Clements, and then occur as separate stages over time. The IFC model incorporates the idea that succession is a function of species' life-history traits rather than simply facilitation by those species arriving earliest. Perhaps the most questionable assumption in Egler's IFC model is that all or most species are present at the start of succession and additional species arrivals do not occur. For example, Finegan (1984) argued that shrubs and small trees (cherries, juniper), some pines, and red maple are early colonizers in old fields in the eastern United States, but oaks and hickories colonize significantly later and are not present at the start. Finegan also demonstrates many other situations where succession does and does not resemble the IFC model. Matthews (1992) examined the IFC model in studies of primary succession on recently deglaciated terrain in Alaska and found an appreciable lag between colonization by pioneer species and that of loose groupings of later colonizers. These differential patterns of establishment draw the IFC model into question, as more sequences resemble the relay floristic model without the discreteness shown in Egler's diagram (Figure 17.3a). In reality, there simply lacks a model of forest succession that is universally applicable to all sites and situations.

Secondary Succession To some degree, secondary succession mimics primary succession in its basic principles. Most forest ecosystems regenerate rapidly following disturbances, and so secondary succession is often described in terms of the redevelopment of forest tree populations. Ecologists working in even vastly different forest regions have produced similar conceptual models of post-disturbance forest succession featuring four phases or stages (Table 17.4). In European beech forests, Watt (1947) described "upgrading" and "downgrading" periods that he divided into four phases, of which the **gap phase** has had the most application elsewhere. In general, we recognize two basic stages, known as *aggrading* (also upgrading or building) versus *senescing* (also downgrading, degenerating, or declining). Among the best-known models in North America are those featuring stand development (Oliver 1981; Table 17.4) and populations (Peet and Christensen 1980a, b; Table 17.4). The stand development model was developed in New England and the Pacific Northwest (Figure 17.4a). The population model applies particularly to the

Table 17.4 Stages of secondary succession.

Forest maturation model (Franklin et al. 2018)	Stand development model (Oliver 1981)	Population model (Peet and Christensen 1980a, b)	Beechwood model (Watt 1947)
Disturbance and legacy creation			Upgrading → Gap
Preforest	Stand initiation	Establishment	Bare
Forest canopy closure			↓
Young forest	Stem exclusion	Thinning	Downgrading → Oxalis
Mature forest	Understory reinitiation	Transition	↓
Old forest	Old growth	Steady state	Rubus

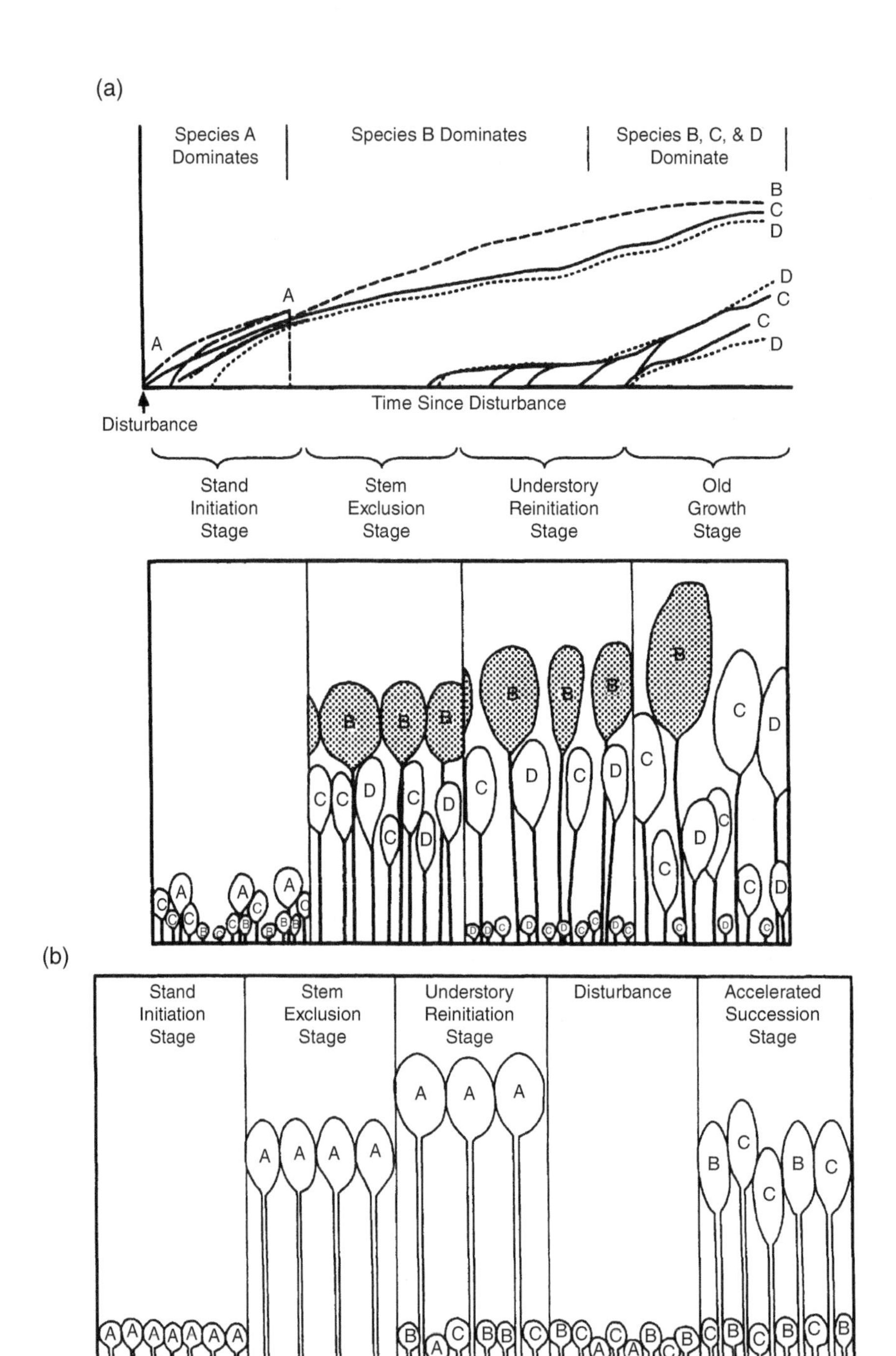

FIGURE 17.4 Stages of stand development following major disturbance. (a) All trees forming the forest are already present in the stand or colonize soon after disturbance. However, the dominant overstory tree species changes as stem number decreases and vertical stratification of the species progresses. Height attained and duration of each stand varies with species, site conditions, and disturbances. Barring intervening disturbances, the "old growth" stage may be reached in less than 200 to over 500 years. *Source:* After Oliver (1981) / with permission of Elsevier. (b) Alternative diagram of disturbance-mediated accelerated succession in a pioneer forest community. Species A is a pioneer tree, whereas B and C are later successional species. Disturbances may include logging, windthrow, ice storm, fire, and insect/disease epidemic. The old-growth stage of species B and C is not shown. *Source:* After Abrams and Scott 1989. Part (a) is reprinted from Oliver, C. D. 1981. Forest development in North America following major disturbances. *Forest Ecology and Management* 3(3): 156, © 1981 with kind permission of Elsevier Science-NL, Sara Burgerhartstrast 25, 1055 KV Amsterdam, The Netherlands. Part (b) after Abrams and Scott 1989. Reprinted with the permission of the Society of American Foresters.

Piedmont Plateau in the southeastern United States, where abandoned agricultural land reverted to pine-mixed hardwood forest. Both models also apply widely to forests in other regions with the following characteristics: (i) developing following a major disturbance, (ii) having single or several age classes, and (iii) having stems that regenerate during a relatively short period following disturbance. These models are described in the following text. By contrast, Watt's beechwood example was developed for forests on mesic sites and is most applicable in forest regions where small gaps are characteristic. Franklin et al. (2018) developed a model we term the forest maturation model that focuses heavily on the importance for biodiversity of early-successional forests, old forests, and biological legacies (Table 17.4). It was developed for forests that experience infrequent, high-severity disturbances such as those in the Pacific Northwest but is broadly applicable to forests disturbed more frequently and/or less severely.

All models of secondary succession are initiated by a disturbance that occurs at fine to broad scales. Such disturbances create marked structural changes in forests by killing trees, but their ecological function is to free resources such as light, nutrients, and soil moisture as well as to create a suite of biological legacies that will direct the rate and course of succession and forest redevelopment. Described as the **disturbance and legacy creation event** by Franklin et al. (2018), such an event is transitional between the undisturbed forest and the beginning of secondary succession (Table 17.4).

In the **stand initiation** or **establishment stage** (also called **reorganization**), plants regenerate the area, and new individuals and species continue to appear for several years. Regenerating plants may develop from (i) newly dispersed or buried seeds (e.g., species A in Figure 17.4a), (ii) sprouts from stems and roots remaining from disturbed plants, and (iii) **advanced regeneration** (species B, C, D)—saplings and seedlings of the previous forest understory that accelerate growth when released by disturbance. Stand initiation may last from 5 to 100 years depending on the type and severity of disturbance, species' regeneration strategies, weather, herbivory, and many other factors (Oliver and Larson 1996). During this period, invasion continues until resources (particularly growing space) become limiting, and plant species diversity is often maximized.

Stand initiation is analogous to the **preforest stage** (Franklin et al. 2018), which precedes the redevelopment of a closed-canopy forest. At this point in succession, diversity of both plants and animals is relatively high, biological legacies are common, habitat niches and food sources are numerous, and food webs and trophic relationships are complex relative to later points in succession. Trees are typically present and are developing in the new forest, but do not yet dominate the ecosystem. Incorporating such a non-forest stage into a conceptual model of forest succession is critical because the lack of trees results in an open, structurally complex stage that is often the most species-rich of the successional process (Swanson et al. 2011). The ecological importance of early-successional forests is often overlooked (Swanson et al. 2011; King and Schlossberg 2014), in part because such forests are perceived as recently disturbed ecosystems in the midst of a "recovery" to a more recognizable, familiar forested condition.

The preforest or stand initiation stage is followed by another important transitional stage, the **forest canopy closure event**, which creates sudden and dramatic changes in environmental conditions (Franklin et al. 2018). This event represents the reassertion of strong dominance of a site by trees, although the degree of dominance (and the speed at which canopy closure occurs) varies among forest types depending on the density of the canopy. The most obvious environmental changes that occur with canopy closure include the limitation of light at the forest floor (Chapter 6) and the moderation of the microclimate (Chapter 7). However, canopy closure also has significant influence on other ecosystem factors, such as competition and diversity. Ecosystem processes such as biomass accumulation and carbon storage are also affected, because canopy leaf area approaches its maximum at about the time of canopy closure, thus determining the rate of photosynthesis, carbon allocation, and growth (Chapter 18). For example, in lodgepole pine forests in Yellowstone National Park, Kashian et al. (2013) found that tree density strongly controlled carbon storage only until canopy closure occurred, after which tree size and stand age became more important because leaf area had been maximized.

Once canopy closure occurs, the processes that occur in the next stages are generally related to competition or biomass accumulation. In the **stem exclusion** or **thinning stage** (also called **aggradation** or **young forest**), virtually all growing space is utilized, crown closure occurs, overtopped seedlings die, and tree regeneration is limited or ceases (Figure 17.4a), although herbs and shrubs are typically still present in the understory. In addition, more vigorous trees grow larger and capture the space from weaker trees, and severe self-thinning of the initial cohort takes place (Peet and Christensen 1987; Figure 13.11, Chapter 13). Figure 17.4a illustrates that the pioneer species A has been excluded and other species (B, C, D) assort themselves vertically depending on growth rate and biomass allocation. Many examples of development in single- and multiple-cohort stands are described in detail by Oliver and Larson (1996). In general, these competitive effects result in a forest with relatively uniform structure. Biological legacies (mostly snags and fallen logs) remain from the pre-disturbance forest, and many trees killed due to thinning are also present. Plant species diversity is often minimized in this stage.

In the **understory reinitiation**, **transition,** or **mature forest stage** (Figure 17.4a), the overstory begins to thin out or "break up" due to insects, pathogens, or other disturbances, thereby increasing the light reaching the understory. Gaps may begin to form in the canopy, releasing some of the smaller trees previously suppressed in the understory or subcanopy layers. Most overstory trees have maximized their height growth and crown spread. Tree regeneration begins, and individuals of advanced regeneration may live for a few years or decades and then die. Longevity of regeneration less than a meter in height in the understory varies substantially, for example, from 3 or 4 years in cherry bark oak to over 100 years for Pacific silver fir (Oliver and Larson 1996). Together, these features significantly increase the structural complexity of the stand relative to the stem exclusion stage. Pre-disturbance biological legacies have significantly decayed and declined, but snags and logs from the post-disturbance forest begin to accumulate during this stage.

The **old-growth**, **steady-state**, or **old forest stage** (i.e., the late-successional forest) develops as overstory trees age and increase in height and biomass (Figure 17.4a) while invasion and regeneration continue. Senescence, chronic insects and pathogens, and minor and major disturbances cause tree crowns to further disintegrate or entire trees to die and disintegrate singly and in groups. As a result, some advanced regeneration and understory trees are able to reach the overstory. Second- and even third-generation trees grow into the canopy, and thus canopy trees begin to vary in their size. Very old forests may lack any representative trees from the original post-disturbance cohort, such that a 500-year-old forest may lack trees older than 300 years! Structural diversity is therefore maximized in the old-growth stage, both in terms of the abundance of diverse tree structures as well as the patchiness of structure (Franklin et al. 2018). For example, in old-field Piedmont forests, hardwoods dominate over declining loblolly pines, and relatively even-aged patches resulting from previous gaps now undergo a miniature version of the three previous stages. Importantly, the old-growth stage is defined not only by tree age and size but also by abundant biological legacies (such as snags and coarse woody debris) and ecological processes that are very different from those in earlier stages (Franklin et al. 1981; Harmon et al. 1986; Oliver and Larson 1996).

The importance of the old-growth stage for forest ecosystems has been emphasized by many researchers and is reviewed extensively by Franklin et al. (2018). Old-growth forests are ecologically unique in North America in large part because twentieth-century forest management deemed them unsuitable or "overmature" for timber production and other resources, resulting in their harvesting and conversion to earlier successional stages. The structural diversity provided by the old-growth stage is critical for its provision of many habitats and ecological niches, as well as for providing critical living space for rare or endangered species. The old trees that persist within many old-growth forests have ecological significance of their own, for their provision of habitats in their cavities, large branches, and complex crowns; resistance to future disturbance because of their thick bark; and their contributions to genetic diversity (Franklin and Johnson 2012). Old trees that die continue to have ecological value as snags, logs, and coarse woody debris, all of which are common in the old-growth stage. In general, the ecological importance of individual trees could be summarized as increasing slowly as they age and increase in complexity, and then gradually declining once they die (Franklin et al. 2018).

There are many possible variations of the four-stage models presented in Table 17.4 depending on the site, vegetation, and disturbance characteristics specific to the ecosystem. One alternative is illustrated in Figure 17.4b, where pioneer species A dominates in the first two stages only to be disturbed in the midst of the understory reinitiation stage, causing accelerated succession of species B and C. Thereafter, a B–C old-growth stage would be characterized either by areas lacking disturbance, or perhaps by minor or major disturbance which, in turn, might lead to replacement of B by C, C by B, or B and C by A! Logging or windthrow of the overstory when advanced regeneration is present (Figure 16.12, Chapter 16) is one example of the Figure 17.4b alternative, and other specific examples are given by Abrams and Scott (1989).

Again, for both primary and secondary succession, complete, stage-by-stage successions rarely, if ever, occur in nature. Disturbances constantly disrupt the gradual internal changes that occur in an ecosystem and may set back, accelerate, or permanently change successional trajectories. Moreover, we emphasize that succession is ecosystem-dependent, and a predictable sequence of vegetation stages can be expected within the context of most ecosystems. Disturbances often only temporarily alter the sequence unless they act to restart succession or permanently change the site, as with erosion, deposition, fire, and windstorm. In such cases, the disturbance may initiate an entirely new successional sequence. For example, on well-drained sand soils, nutrients are retained primarily in the accumulated soil organic matter. A severe fire that destroys the stand and the organic matter may actually change soil conditions so drastically that a new succession is initiated (Damman 1975).

Even a vegetation stage itself can change the site to the degree that there is a resulting change in succession. In Newfoundland, for example, Damman (1975) reported a relatively predictable succession following fire on most sites unless a *Kalmia* heath became established. If such a heath became firmly established after fire, it initiated soil changes leading to thin, iron-pan formation, water logging, and peat bog formation, preventing the recovery of forest vegetation and creating a whole new successional sequence. Although succession does not always follow predicted patterns, disturbance- or vegetation-mediated changes in site conditions may strongly determine the trajectory of the new pattern.

SUCCESSIONAL CAUSES, MECHANISMS, AND MODELS

This section describes some of the key characteristics, causes, mechanisms, and models of succession. These features gained attention as the role of disturbance in forest ecosystems was increasingly appreciated, as the number of ecosystems in which succession was observed increased, and in reaction to the classical model.

Key Characteristics and Regeneration Strategies Succession functions as the interaction of species with different genetically determined life histories and regeneration strategies. There are likely limitless possibilities for differences between species in their regeneration requirements, allowing replacement of the plants of one generation by those of the next (Grubb 1977). Ecologists often attribute these traits and strategies to the species themselves, but they are intimately related to the site conditions that elicited them (Chapter 3). The characteristics of forest trees paramount to succession include: understory tolerance; seed production, size, and dispersal; growth rate and longevity; tree crown architecture; resistance to insects and pathogens;

biomass production; allocation of carbohydrates; and nutrient requirements. It has been hypothesized that the competitive ability of species is regulated during succession by the changing ratio between nutrient availability and light (Tilman 1985, 1988), and that community composition should therefore change whenever the relative availability of the two limiting resources changes. Known as the *resource-ratio hypothesis*, this idea has been criticized for its lack of general applicability (Glenn-Lewin and van der Maarel 1992, p. 31) and because its predictions have not been robustly tested (Miller et al. 2005, 2007).

The *vital attributes* concept (Noble and Slatyer 1977, 1980) identifies the characteristics of a species that are vital in determining its role in successional sequences. Such attributes include means of recovery from disturbance, ability to establish in the face of competition, and time needed to reach critical life-history stages. The basic trait of differential longevity is also critical for some successional sequences (Egler 1954, 1977; Drury and Nisbet 1973). For example, short-lived pioneer species may be replaced by long-lived, late-successional species that simply outlive them, and they are not regenerated even by subsequent periodic disturbances.

Each of these characteristics (e.g., seed size, nutrient requirements, vital attributes, longevity, etc.) evolved in response to site-specific ecosystem conditions (Chapter 3), and broad generalizations without an ecosystem context can be misleading. For example, in certain ecosystems small seed size and a short-lived pioneer role are key factors in explaining successional pathways where such species are replaced by heavier-seeded, more understory-tolerant, long-lived species. However, not all small-seeded species are short-lived pioneers. For example, hemlocks (*Tsuga*) and cedars (*Thuja*) are small-seeded, very tolerant, long-lived groups that persist in certain kinds of ecosystems whose seedbed conditions (e.g., moist to wet sites; bare mineral soil) and disturbance regimes (e.g., fire, erosion, and deposition) favor them. By contrast, walnuts are heavy-seeded but intolerant and fast-growing; likewise, their vital attributes evolved and are expressed in specific kinds of ecosystems wherever they persist today in North America and Asia. In summary, the key life-history attributes and regeneration strategies are inherent parts of many overlapping, sequential, ecosystem-specific processes, such as regeneration (Chapter 4), competition and mutualisms (Chapter 13), and biomass accumulation and growth (Chapter 18). These processes, involving both site and plants, explain succession on an ecosystem-by-ecosystem basis. Walker and Chapin (1987) describe this process model for primary succession in Alaska, and details are available in a number of sources (Walker and Chapin 1986; Walker et al. 1986; Matthews 1992).

Availability and Arrival Sequence of Species Successional trajectories are strongly affected by the availability of species and their arrival at a disturbed site. The timing and magnitude of species invasion vary depending on their occurrence in adjacent ecosystems, the spatial position and distance of the disturbed site in relation to seed sources, the site conditions and disturbance history of the disturbed area, and species life-history traits. Differential arrival often explains why different successional pathways occur on similar sites. In secondary successions following windstorms or fires, surviving understory plants, clonal plants, and those germinating from the soil seed bank have immediate access to the site. For example, the differential arrival of species largely determined three pathways of primary succession in Alaska where a 220-year glacial retreat progressively exposed parent material for colonization (Fastie 1995). The variability of species arrival was related to the distance from each site to the closest seed source of Sitka spruce at the time of deglaciation. In secondary succession following logging and fire in the Great Lakes region, some eastern white pine seed trees were able to survive, and the variable successional pathways on a given area today demonstrate the influence of these remnant seed sources (Palik and Pregitzer 1994).

Facilitation, Tolerance, and Inhibition Connell and Slatyer (1977) proposed three alternative models of succession based on the mechanisms of facilitation, tolerance, and inhibition

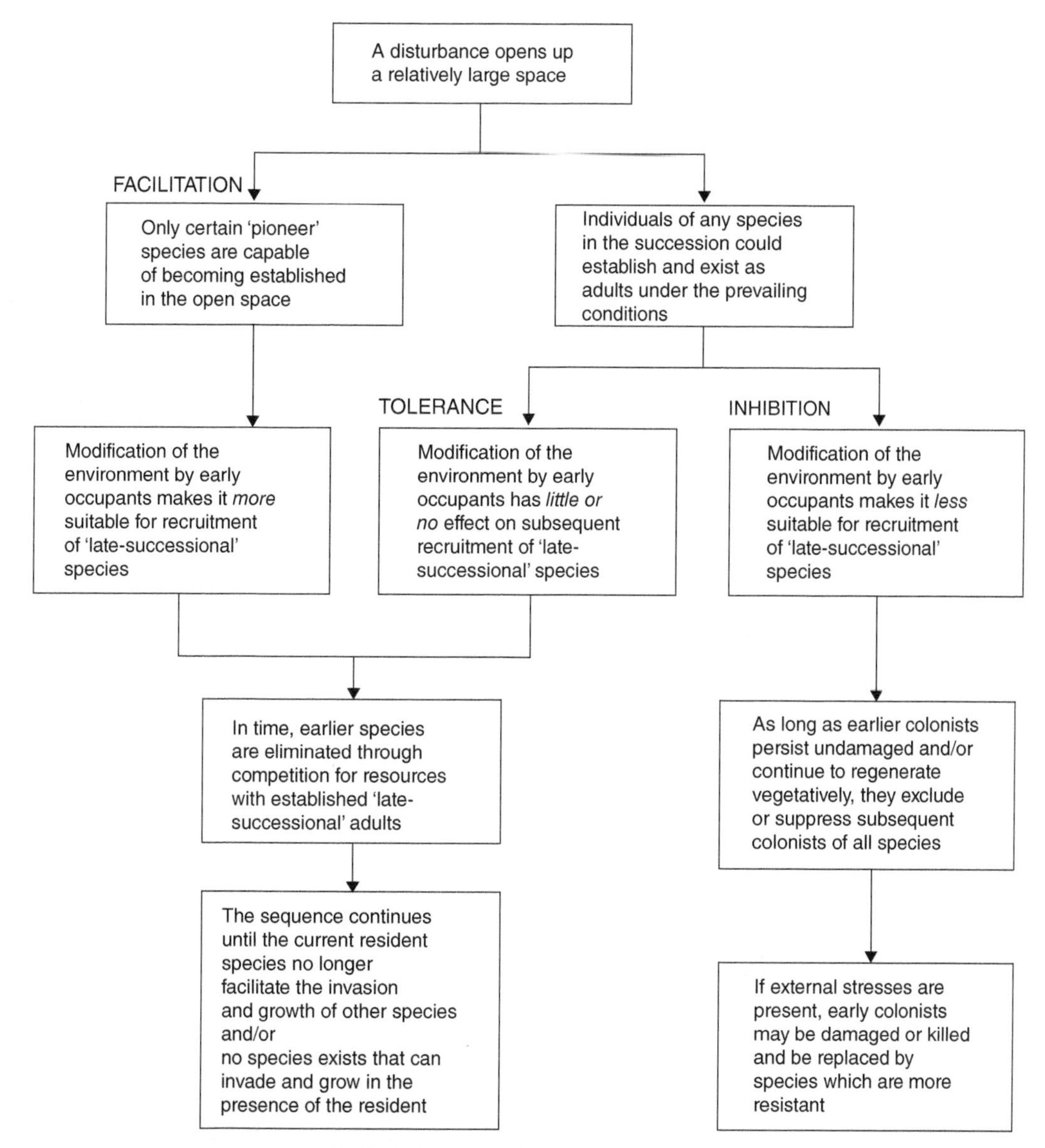

FIGURE 17.5 Three autogenic mechanisms—facilitation, tolerance, and inhibition—that determine successional trends. *Source:* After Connell and Slatyer (1977) / University of Chicago Press, as modified by Begon et al. 1990. Reprinted with permission from *American Naturalist*, © 1977 by the University of Chicago. Modification reprinted with permission of Blackwell Science.

(Figure 17.5). Notably, many ecologists have found these models restrictive (Huston and Smith 1987) and oversimplified (Christensen and Peet 1981) because each was developed to represent a single, opposing, alternative pathway (Pickett et al. 1987). In reality, nature features an enormous variety of pathways and mechanisms that may simultaneously influence successional change in complex ways.

Facilitation is analogous to the model of relay floristics (Figure 17.3a), whereby early successional species modify the environment to favor later successional species. In addition to the amelioration of environmental stress, facilitation may operate by increasing the availability of resources and enhancing colonization ability (Pickett et al. 1987). Thus, a species with one combination of life-history traits may establish and modify environmental conditions in a manner that favors individuals with other combinations (Huston and Smith 1987). Facilitation in this sense has been reported in many instances (e.g., Cooper 1923, 1931, 1939; Jenny 1941; Christensen and Peet 1981; Walker and Chapin 1986; Wood and del Moral 1987; Matthews 1992; Fastie 1995; Baumeister and Callaway 2006; Calder and St. Clair 2012). Chapin et al. (1994) investigated facilitation during succession following deglaciation at Glacier Bay, Alaska (Figure 17.6). Life-history traits determined the pattern of succession, and initial site conditions and facilitation (where present) influenced the rate of succession as well as the composition and productivity of the late-successional community. Facilitative effects were strongly related to the increase in nitrogen that is enhanced by all plants but especially by alder.

Inhibition, where species established early inhibit subsequent colonization of other species by pre-empting space and other resources, was also an important process at Glacier Bay (Figure 17.6). A common example of inhibition is when shrubs resist colonization by trees (Niering and Egler 1955; Webb et al. 1972; Niering and Goodwin 1974; Damman 1975; Chapin et al. 1994).

The mechanism of **tolerance** relates to the ability of a species to tolerate low resource levels. In such situations, the initial species modify the environment, but this change has little or no effect on subsequent recruitment and growth of later successional species (Figure 17.5). Instead, late-successional species replace early-successional species either by active competition or by simply being longer-lived. Active and passive concepts of tolerance are discussed by Pickett et al. (1987).

Importantly, facilitation, inhibition, and tolerance describe relative, rather than absolute, processes (Huston and Smith 1987). The three processes almost always occur simultaneously with varying degrees of importance and are likely to occur in most primary and secondary successions. Various combinations of the three processes are reported for primary succession on glacier forelands (Matthews 1992; Chapin et al. 1994; Fastie 1995), in old-field succession in New York (Gill and Marks 1991), and on the North Carolina Piedmont. On Alaskan floodplains,

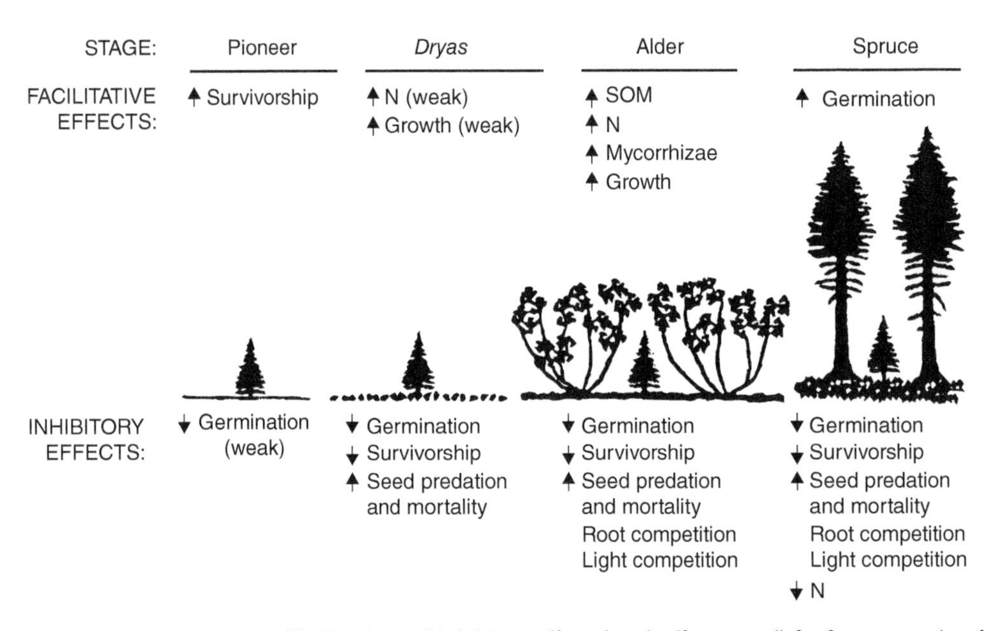

FIGURE 17.6 Summary of facilitative and inhibitory effects (weak effects noted) for four successional stages on establishment and growth of spruce seedlings at Glacier Bay, Alaska, as determined from field observations, field experiments, and greenhouse studies. SOM = Soil organic matter. *Source:* After Chapin et al. (1994) / John Wiley & Sons. Reprinted with permission of the Ecological Society of America.

many different processes and factors (e.g., life history, chance, facilitation, competition, herbivory, etc.) affect how alder and spruce interact during succession, and no single successional process or model adequately describes successional change in these ecosystems (Walker and Chapin 1987).

CHANGE IN ECOSYSTEMS

Ecosystems are constantly changing over time and space: from night to day, seasonally, and year to year as climate and soil change, and as the cycle of organisms' activity changes. As such, a regional or local ecosystem never can and never does reach a complete balance or permanence except for something arbitrarily chosen by humans for their own understanding.

END POINT OF SUCCESSION?

Forest succession has been recognized from the earliest days of natural history study, but whether there is a last stage of succession has been debated for decades. Cowles observed in 1901: "The condition of equilibrium is never reached, and when we say that there is an approach to the mesophytic forest, we speak only roughly and approximately. As a matter of fact, we have a variable approaching a variable rather than a constant." On the other hand, Clements defined the term "climax" relatively rigidly as the end point of succession and used it as the driver of his vegetation classification. Over decades many descriptors were added to the term to improve it conceptually, such as "polyclimax," "polyclimatic climax," "site climax," "climax pattern," or position of relative stability. Regardless of these modifications, the term implies a terminus of succession, and epitomizes succession within a vegetation context rather than an ecosystem context. In fact, the term "climax" is inappropriate within an ecosystem context. It is more useful to recognize pioneer or early-successional, mid-successional, and late-successional species or groups for position in characterizing vegetation along the temporal scale for site-specific ecosystems. These terms (early, mid, late) are not absolute but are relative to the specific ecosystem in the landscape.

In any case, climax terminology permeates the literature such that it is useful to examine how the terms are used. Macroclimate is the dominant factor forming communities in Clements' monoclimax theory, whereas other factors (physiography, soil, fire, biota) are of secondary importance. When other factors are of equal importance, a different climax community is possible or even likely for different physiographic settings, soil types, and disturbance factors. This is the **polyclimax** theory (Tansley 1939), which holds that biotic succession will move toward a climax for any combination of environment and organisms, but that the specific nature of the climax will vary with environmental factors and biotic conditions. Cowles (1899, 1901), Moss (1913), and especially Nichols (1923) contributed heavily to polyclimax theory. In fact, Nichols argued for a different **physiographic climax** on each site that differed more or less from the regional **climatic climax**.

For example, in his classic studies of plant succession on the sand dunes of lower Lake Michigan, Cowles at first thought that succession on all sand dune sites would eventually reach the same forest stage of mixed mesophytic hardwoods. In a later study of the same area, Olson (1958) concluded that the drier, lower fertility, and more exposed sites would never support such a community but would instead be occupied more or less permanently with a black oak–blueberry type. Furthermore, Olson suggested that although vegetation changes would continue to become progressively slower with time, but they would never reach complete stability. Over the years, the accepted concept of climax became broader, less rigid, and more closely approximated the polyclimax approach. Tansley (1939) was instrumental in broadening the concept by adopting a dynamic viewpoint in recognizing that "positions of relative equilibrium are reached in which the conditions and composition of the vegetation remain approximately constant for a longer or shorter time."

Polyclimax theory is also mired in its own terminology: climatic, edaphic, topographic, topoedaphic, fire, zootic, salt, etc. (see descriptions in Oosting 1956 and Daubenmire 1968). Many of these terms approximate or are merely more specific distinctions of Clements' climaxes. For example, in the polyclimax approach the Clementsian climax was renamed "climatic climax;" Clements' subclimax may be either an "edaphic" or "topographic climax;" Clements' disclimax becomes a "fire" or a "zootic climax." In addition, like Clements, only one climatic climax was recognized. A **polyclimatic climax** theory was later developed (see Meeker and Merkel 1984) that recognized more than one climatic climax for a macroclimatic region.

Whittaker (1953, 1975) recognized a gradational **climax pattern** of vegetation corresponding to environmental gradients—a continuum of climaxes. Whittaker (1953) wrote that climax composition was determined by

> *all "factors" of the mature ecosystem—properties of each of the species involved, climate, soil and other aspects of site, biotic interrelations, floristic and faunistic availability, chances of dispersal and interaction, etc. There is no absolute climax for any area, and climax composition has meaning only relative to position along environmental gradients and to other factors.*

The "climax pattern" continuum has clear problems in application to management, and so Dyksterhuis (1949, 1958) proposed the use of site units to characterize succession and climax for range classification. This approach is termed **site climax** by Meeker and Merkel (1984). Other viewpoints and details of the climax concept have been presented by many authors (Phillips 1931, 1934–35; Tansley 1935; Clements 1936; Cain 1939; Whittaker 1953, 1975; Daubenmire 1968; Langford and Buell 1969; White 1979; Matthews 1992).

SUCCESSION AS AN ECOSYSTEM PROCESS

A consistent thread throughout the abovementioned discussion is that a general model of succession based on species or communities alone is unrewarding. It is somewhat useful to develop a framework of stages based on the development of stands or populations, but specific site conditions are the primary determinant of the duration, distinctness, and predictability of those stages. Species' life-history traits are inseparably linked to the sites and disturbance regimes to which they are adapted. An ecosystem approach to forest succession provides a functional and geographical framework for understanding these multiple mechanisms and processes that drive succession.

An ecosystem approach to forest succession requires that the first-order focus is the regional and local ecosystem rather than component parts, such as vegetation or soil. Succession is a multidimensional process with many factors and processes occurring at multiple scales (Table 17.2) that can be assessed for specific regional and local ecosystems. A "top-down" approach that begins with an understanding of the geographic hierarchy of regional and local ecosystems, their physical factors, biotic interrelationships, and past histories (Chapter 2) is often most appropriate in this regard. Each regional ecosystem will exhibit unique successional pathways and mechanisms, but their character and pattern will be further different among climatic–physiographic regions.

For example, forest succession in the southern Appalachian Mountains is different than that of the Yellowstone Plateau, the slopes of the Sierra Nevada, or the southeastern Coastal Plain. Likewise, the four ecosystem regions in Michigan (Figure 2.10, Chapter 2) exhibit different patterns of succession due to differences in climate, physiography, soil, and vegetation. Within a given region in Michigan, ecosystem districts (Figure 2.10, Chapter 2) have their own characteristic successional patterns for local ecosystems occurring as sand plains, acid swamps, calcareous swamps, rocky ridges, or moist and fertile lower slopes. Variation in succession may even be expected for local ecosystems (Simpson 1990) due to past disturbance or management (logging, drainage), disturbance frequency and severity, seed source availability, and microsite variation, among other

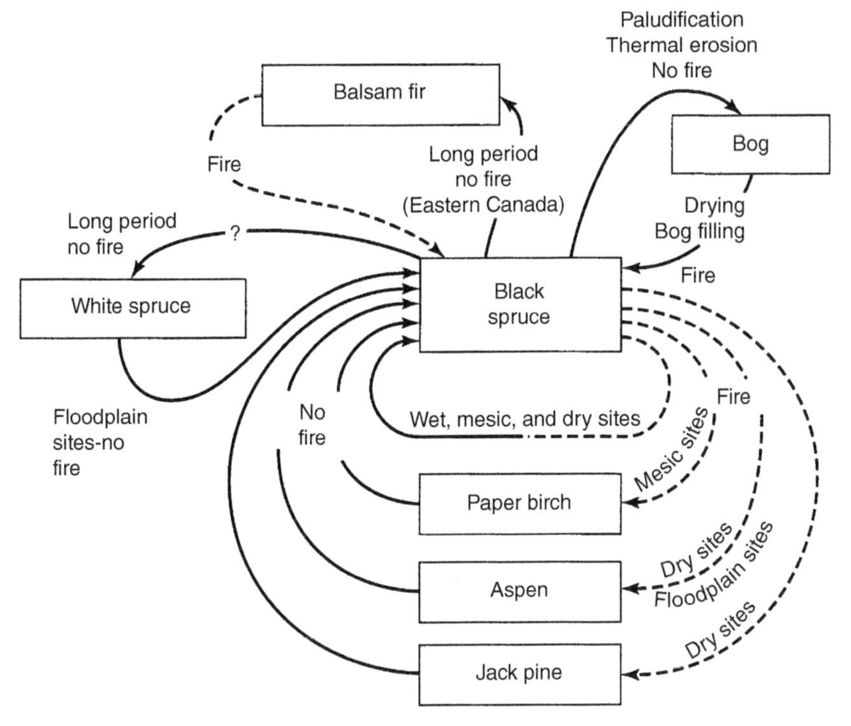

FIGURE 17.7 General example of successional pathways involving black spruce on different ecosystems, defined by their site conditions, of the boreal forest in Alaska and Canada. *Source:* After Viereck (1983) / John Wiley & Sons. Reprinted with permission from SCOPE 18, The Effects of Fire in Northern Circumpolar Ecosystems, edited by Wein, R. W. and D. A. MacLean 1983 © SCOPE, John Wiley & Sons, Ltd., Chichester, UK.

factors. The factors that obscure successional trends are significantly reduced by specifically defining the geographic framework and working with more or less homogeneous site conditions.

A general example of multiple, ecosystem-specific successional pathways is presented in Figure 17.7 based on the site characteristics and disturbance regimes of the boreal forest of Alaska and Canada. On many wet, mesic, and dry sites, black spruce replaces itself with or without fire (Viereck 1983). In the southern regions of the boreal forest, colonization of burned black spruce forest by birch and aspen is more common than in the north. Long fire-free periods in eastern Canada enable balsam fir to replace black spruce (Figure 17.7). Many other ecosystem-specific pathways are described by Viereck (1983).

A specific example of an ecosystem approach to succession comes from research on biomass accumulation and nutrient cycling in northwestern lower Michigan (Figure 17.8; Zak et al. 1986; Host et al. 1987; Host et al. 1988). Site-specific ecosystems were first distinguished at two levels: *landforms* (outwash plain, ice-contact drift, and moraine) and *ecosystem types* within landforms (named by overstory and ground-cover species). Successional trends for landforms and ecosystem types were then estimated based on an understanding of site conditions and life-history traits of the species in the understory and ground-cover layers (Figure 17.8). Importantly, the oak cover types failed to distinguish the landforms and ecosystem types as well as the predicted successional pathways, suggesting that an understanding of entire ecosystems—not just communities defined by dominant species—is necessary to ascertain successional pathways. Moreover, different pathways were predicted for ecosystem types within each landform, and these pathways, based on multiple ecological factors, are the best estimate of successional trends available for making

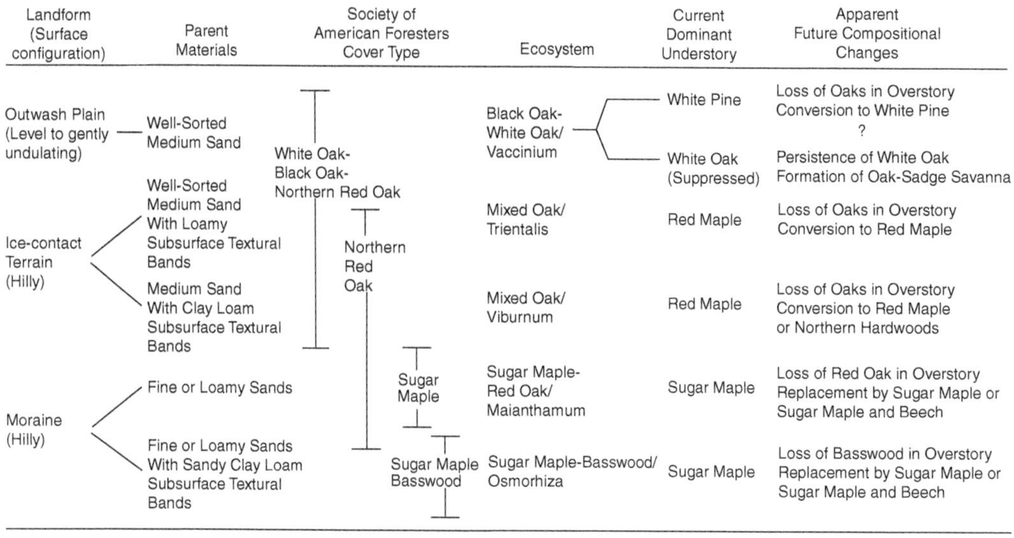

Landform (Surface configuration)	Parent Materials	Society of American Foresters Cover Type	Ecosystem	Current Dominant Understory	Apparent Future Compositional Changes
Outwash Plain (Level to gently undulating)	Well-Sorted Medium Sand	White Oak-Black Oak-Northern Red Oak	Black Oak-White Oak/ Vaccinium	White Pine	Loss of Oaks in Overstory Conversion to White Pine ?
				White Oak (Suppressed)	Persistence of White Oak Formation of Oak-Sadge Savanna
Ice-contact Terrain (Hilly)	Well-Sorted Medium Sand With Loamy Subsurface Textural Bands	White Oak-Black Oak-Northern Red Oak; Northern Red Oak	Mixed Oak/ Trientalis	Red Maple	Loss of Oaks in Overstory Conversion to Red Maple
	Medium Sand With Clay Loam Subsurface Textural Bands	White Oak-Black Oak-Northern Red Oak; Northern Red Oak	Mixed Oak/ Viburnum	Red Maple	Loss of Oaks in Overstory Conversion to Red Maple or Northern Hardwoods
Moraine (Hilly)	Fine or Loamy Sands	Northern Red Oak; Sugar Maple	Sugar Maple-Red Oak/ Maianthamum	Sugar Maple	Loss of Red Oak in Overstory Replacement by Sugar Maple or Sugar Maple and Beech
	Fine or Loamy Sands With Sandy Clay Loam Subsurface Textural Bands	Northern Red Oak; Sugar Maple Basswood	Sugar Maple-Basswood/ Osmorhiza	Sugar Maple	Loss of Basswood in Overstory Replacement by Sugar Maple or Sugar Maple and Beech

FIGURE 17.8 Estimated future compositional change associated with two levels of landscape ecosystems (landforms and ecosystem types) in northwestern lower Michigan. *Source:* After Host et al. (1987) / Oxford University Press. Reprinted with the permission of the Society of American Foresters. Site conditions, overstory cover type (Eyre 1980), and current dominant understory species are also shown.

management decisions. Being specific to ecosystems that recur throughout the landscape allows the proposed trends to be monitored and tested over time. Such an approach is applicable for improved management, conservation, and restoration of ecosystems on many or most public and large private lands in North America.

EXAMPLES OF FOREST SUCCESSION

PRIMARY SUCCESSION ON RECENTLY DEGLACIATED TERRAIN

Primary succession is perhaps best studied on the land newly formed in front of a glacier, and such detailed studies of the glacier foreland have been underway since the late 1800s. Matthews (1992) provides a comprehensive review of primary succession on recently deglaciated terrain, examining physical and ecological processes and successional stages. From his geoecology viewpoint, succession in front of the retreating glacier is a process of the developing landscape when all parts of the ecosystem develop simultaneously. The most important visible features of the landscape in the glacier foreland are the landforms and their associated parent materials, and the developing vegetation is a function of the dominant factor *terrain age*. Other factors are subordinate: climate, physiography, parent material, organisms, as well as physical site change and disturbance. Matthews found that vegetation chronosequences are generally limited by differences among sites, both the initial environment and their subsequent environmental histories, but their careful use can provide approximations of succession at particular sites. For example, the chronosequence at Glacier Bay, Alaska is a series of eight stages that occur along a continuum:

I. Early pioneer stage: mountain avens *(Dryas drummondii)* and prostrate willows, mosses, and herbs.

II. *Dryas* mat: coalescing mats of *Dryas*.

III. Late pioneer stage (after 15–20 years): young Sitka alder and black cottonwood shorter than 2–3 m.

IV. Open thicket stage (after 20–25 years): clumps of alder.

V. Closed thicket stage (after 25–30 years): dense alder with individual erect specimens of balsam poplar, shrubby willows, and Sitka spruce.

VI. Balsam poplar line stage (after 30–40 years) where poplars reach a height of 4–5 m and form a canopy and visible line above the alder.

VII. Spruce forest stage (at 75–90 years) when spruce replaces alder.

VIII. Spruce–western hemlock stage (speculative because the forest may remain dominated by spruce even after 200 years).

The horizontal and vertical structure of vegetation on the glacier foreland of the Grand Glacier d'Aletsch in the Swiss Alps has long been studied with permanent plots (Figure 17.9). The sequence of herbs to shrubs to trees typically reflects the sequence of plant size, longevity, and competitive power, but trees may colonize much earlier in the sequence depending on seedbed conditions and seed availability. In the Canadian Rocky Mountains, establishment time ranges from 10 to 15 years to 80 years (Luckman 1986).

Matthews' review of primary succession in the glacier foreland reveals that succession may proceed in trajectories that are parallel, but also in those that are strongly divergent or convergent. Generally, divergent pathways are most clear during early stages and are associated with relatively severe physical environments; convergence is favored by relatively strong biotic controls. Biomass and species richness tend to increase during primary succession, although patterns of diversity vary considerably from site to site (Matthews 1992).

Matthews notes that most of the general successional models (as previously described) are not well applied to recently deglaciated terrain. Such land instead is better described by a geoecological model that emphasizes physical environmental processes as well as biological processes. In addition, models of succession as a landscape process must incorporate spatial variation in site

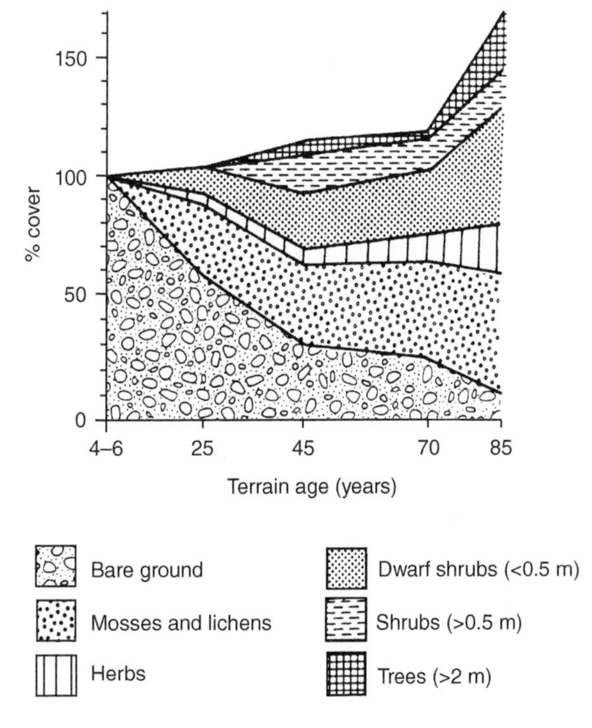

FIGURE 17.9 Coverage of bare ground and five vegetative strata on recently deglaciated terrain, which is on terrain of increasing age, Grand Glacier d'Aletsch, Swiss Alps. *Source:* After Lüdi (1945). Reprinted with permission of the 17, Stiftung Rübel, Zurich, Switzerland.

characteristics through time. In sum, Matthews strongly emphasizes the properties (landforms and glacial sediments, climate, soil) and processes (erosion by water; frost weathering, heaving, and sorting; solifluction; wind deposition and erosion; and soil development) of the physical landscape.

Matthews' summary points concerning succession on recently deglaciated terrain provide real-world insights into the mechanisms of succession discussed previously in this chapter:

Allogenic and autogenic processes interact during succession and are often inseparable.

Establishment is more important than migration (species arrival) in determining species composition in the harsh environment of glacial forelands. Here, it is likely that pioneers require some special physical, chemical, or microbial conditions as well as lack of competition.

Multiple processes—including physical as well as biological processes—are characteristic in replacement of pioneers by later colonizers. Pioneer species may decline more due to physiochemical changes associated with leaching (nutrient depletion, balance of available nutrients, toxicity associated with pH changes) and other physical changes than to the rise of species of the next successional stage.

The action of allogenic factors (physical processes and disturbances) as compared with autogenic action of the plants (the importance of life-history traits) influences both the nature and rate of succession in the harsh physical environment of the glacier forelands. In favorable environments (high in resources; low in stresses), there is a rapid increase in the importance of autogenic processes through time (Figure 17.10, see "1"). This increase may reflect the rapid accumulation of biomass that occurs as soil water and nutrients become more favorable in a favorable climate or a relatively undisturbed site. By contrast, allogenesis, as a determinant of vegetational change, is reduced sharply (1′) as the intensity of physical processes and disturbances decrease plant biomass and vegetative control of the environment increases. As the environment becomes less severe, biological production is constrained, allogenic processes become relatively more important in any successional stage, and the control of succession by allogenic processes becomes longer. The relative importance of allogenesis in Figure 17.10 (curves 2′ and 3′) is highest in more severe environments. That is, on very severe sites, plants themselves modify the environment to a lesser extent than on favorable sites. This relationship can be applied at both the broad scale of regional

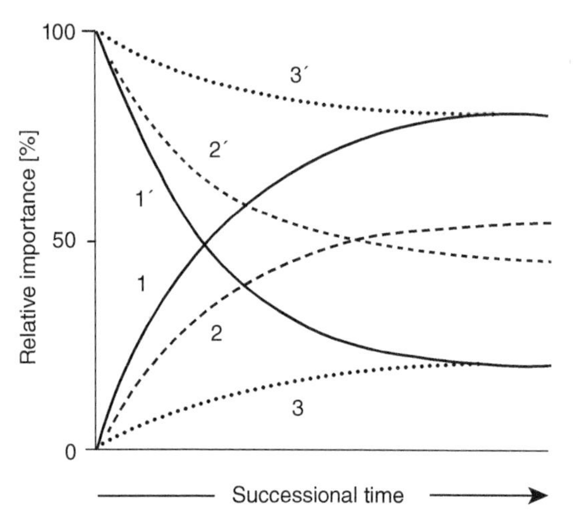

FIGURE 17.10 Relative importance of autogenic (1, 2, 3) and allogenic (1′, 2′, 3′) processes under increasing environmental severity (1 and 1′ with solid line = least severe; 2 and 2′ with dashed line = moderately severe; 3 and 3′ with dotted line = most severe) in the context of recently deglaciated terrain. The model assumes species/community context with autogenic processes relating only to vegetational processes; allogenic processes are a combination of physical processes and disturbance. *Source:* After Matthews (1992) / Cambridge University Press. Reprinted from *The Ecology of Recently Deglaciated Terrain,* © 1992 by Cambridge University Press, Reprinted with the permission of Cambridge University Press.

ecosystems (increasing latitude, for example, boreal forest, tundra, polar desert) and the local scale within the different landforms of the glacier foreland. As described in the earlier text, however, autogenic and allogenic processes are not mutually exclusive, and their relationship to site severity applies only to the extent that these processes can be distinguished. It appears likely that they are most distinct in the harshest environments.

SUCCESSION FOLLOWING THE ERUPTION OF MOUNT ST. HELENS

On May 8, 1980, Mount St. Helens erupted in the Cascade Mountains of Washington in the Pacific Northwest. The eruption produced a catastrophic disturbance to approximately 600 km^2 of old-growth and planted forests dominated by western hemlock, Douglas-fir, and Pacific silver fir. The disturbance created by the eruption was extremely heterogeneous due to the diversity of geological processes that occurred. These processes included pyroclastic flows (a landslide and flows of hot gases and pumice fragments) that eliminated ecosystems and landforms over about 60 km^2, a lateral blast zone that blew down or scorched forests over 500 km^2, and the deposition of more than 5 cm of ash over 1000 km^2 (Dale et al. 2005a). Whether organisms survived the eruption depended heavily on which of these zones they resided in prior to the disturbance, with little survival in the area of pyroclastic flows, some survival of underground plant parts and burrowing animals or organisms protected by snow in the blast zone, and greater survival of taller organisms in the ashfall zone. The presence of snow was a major factor in organism survival and the subsequent rate and trajectories of succession (Figure 17.11).

Much like we observe in areas of severe fires or windstorms, the heterogeneity of disturbance intensity resulting from the eruption had important consequences for post-eruption succession. Unlike fires or windstorms, the pyroclastic zone generally buried or scoured away any organic material toward its terminus (Dale et al. 2005b), and succession there was therefore mostly primary (Dale and Denton 2018). However, secondary succession characterized the blast zone and the ashfall areas because there was survival of some to many organisms. Remarkably, some individuals of most plant species present before the eruption survived the disturbance in refuges of various forms (Del Moral 1983; Halpern and Harmon 1983; Adams et al. 1987; Zobel and Antos 1992), and survivors represented most life-forms and successional stages. For example, snowbanks that had persisted into early May of 1980 protected the tree seedlings and shrubs that they had buried, allowing them to immediately resume growth upon snowmelt (Franklin and MacMahon 2000). Together with populations dispersed into the disturbed area, these survivors facilitated and shaped secondary succession (Dale and Crisafulli 2018), and succession occurred most rapidly from areas with surviving organisms (Franklin and MacMahon 2000).

Succession following the eruption of Mount St. Helens has been extensively studied, and a series of important lessons have emerged. The concept of biological legacies emerged from the aftermath of the Mount St. Helens eruption, and there they were found to be extremely important in directing the rates and trajectories of succession (Dale et al. 2005a). Pre-eruption legacies included dead organic material (mainly trees blown over or left standing but scorched in the blast zone), but also organisms that survived the blast but died before they were able to reproduce (Dale and Crisafulli 2018). Ecologists had largely ignored the importance of such legacies for ecosystem structure prior to 1980, but they were found to be important in shaping post-eruption erosion and deposition processes as well as supporting many organisms with protective cover, food, habitat, and nutrients (Franklin and MacMahon 2000).

An additional important lesson provided by succession at Mount St. Helens is that succession is extremely complex and does not always follow ecological theory. Heterogeneity in organism survival and in the distribution of biological legacies created many different points of initiation of succession across the landscape that resulted in many different pathways of succession that did not necessarily converge. In some cases, adjacent sites undergoing primary succession supported very

FIGURE 17.11 Top: Photo chronosequence taken from 1983 to 2004 showing changes in plant community development in the blowdown zone of Mount St. Helens at a site that was snow-covered at the time of the eruption. Snow cover facilitated the survival of numerous plants, including conifer saplings. Bottom: Photo chronosequence showing changes in plant community development from 1983 to 2016 in the blowdown zone at a site that was not snow-covered at the time of the eruption. *Source:* From Dale and Crisafulli (2018); photos by Peter Frenzen, USDA Forest Service.

different plant communities, emphasizing the importance of random or chance events in the successional process (Dale and Crisafulli 2018). Specific to forests, several late-successional coniferous species (e.g., western hemlock, Pacific silver fir) were found to colonize the primary succession sequence on the debris flow area within the first 3 years after the eruption, although seedling density greatly decreased by 5 years post-disturbance (Dale et al. 2005b). These results highlight the differences between ecological theory and the reality of ongoing succession in real-world ecosystems following complex disturbance events. Other succession research completed at Mount St. Helens through 2015 is summarized in detail by Crisafulli and Dale (2018).

SECONDARY SUCCESSION FOLLOWING FIRE IN PONDEROSA PINE FORESTS OF WESTERN MONTANA

Fire leads to ecosystem change throughout forests of the western United States (Chapters 10 and 16). Fischer and Bradley (1987) described the major role of fire in determining successional pathways in grasslands and forests in western Montana. For example, ponderosa pine of the submontane zone in western Montana competes with grasses in dry ecosystems at low elevations and with Douglas-fir in the cooler and moister ecosystems of the montane zone. In this submontane zone, fire acts to (i) maintain grasslands, (ii) maintain open ponderosa pine stands, and (iii) encourage ponderosa pine regeneration. Frequent fires maintain the grassland community by killing pine seedlings (stage 1; Figure 17.12a). Over occasional long fire-free periods, ponderosa pine seedlings become established gradually as the result of a seedbed-preparing fire (stage 2). The seedlings develop into saplings in the absence of further burning, and fires during this period may kill young trees (stage 3) or thin them (stage 4). In time, light surface fires produce an open stand of mature trees (stage 5) that may accumulate enough fuel to produce a stand-replacing fire (stage 6) which is followed again by grassland (Figure 17.12a).

Multiple successional pathways are evident in these forests using a starting point of open, park-like, old-growth ponderosa pine forest (Figure 17.12b). The interaction or absence of fire (timing and severity) and the life-history traits of pine and grasses result in various states of pine stands (e.g., A, B2, C3, E2, etc.) and pathways. For example, starting with an open, park-like, old-growth pine stand (A), frequent fires (pathway 33) maintain grassland (K), but moderate-severity fire (pathway 5) leads to a closed-canopy, multistoried pine stand (stage D1). A dense stand (El) occurs in the absence of fires and is either maintained by low-severity fire (pathway 7) or succeeds to grassland (F) after severe fire (pathway 9). This example illustrates the complexity of succession and the lack of any end point of the successional process. Similar examples of fire ecology for the diverse range of ecosystems of eastern Idaho and western Wyoming are also available (Bradley et al. 1992).

Quantitative succession models are beyond the scope of our treatment, but many simulation models have been developed for succession (Huston et al. 1988; Botkin 1993). A mechanistic, ecological process model simulating fire succession of coniferous forests of the northern Rocky Mountains is also available (Keane et al. 1996), as is a commonly used forest succession simulator with versions specialized for individual forest regions of the United States (Crookston and Dixon 2005). Recent model development also includes a spatially explicit landscape-scale succession model capable of incorporating climate change and additional disturbances (Scheller et al. 2007).

SECONDARY SUCCESSION AND GAP DYNAMICS

Forest trees may establish and successfully reach the overstory canopy via three modes (Veblen 1992): (i) following severe disturbance, (ii) by establishing in overstory gaps, or (iii) rarely, by continuous regeneration in a disturbance-free understory and recruitment into an overstory without discrete gaps. Understory intolerant species typically reach the overstory following severe disturbances, whereas mid-tolerant and tolerant species often utilize relatively small treefall gaps. Tolerant species may recruit without disturbance if the canopy is not too dense, but they usually establish in a disturbance-free understory (Lorimer et al. 1988; Busing 1994) and then require at least periodic episodes of overstory crown thinning. We examine regeneration in canopy gaps in this section.

When an opening or gap occurs in the forest overstory, advanced regeneration and new stems colonizing the gap from the understory form a patch of vegetation that responds to the newly changed environmental conditions—leading to a type of secondary succession termed **gap dynamics** (Pickett and White 1985). Watt (1925, 1947) described four phases (Table 17.4) in the development of pure European beech forests in England. One of these four phases was the "gap phase" that occurred in old-growth beech stands. Gaps are initiated by major and minor

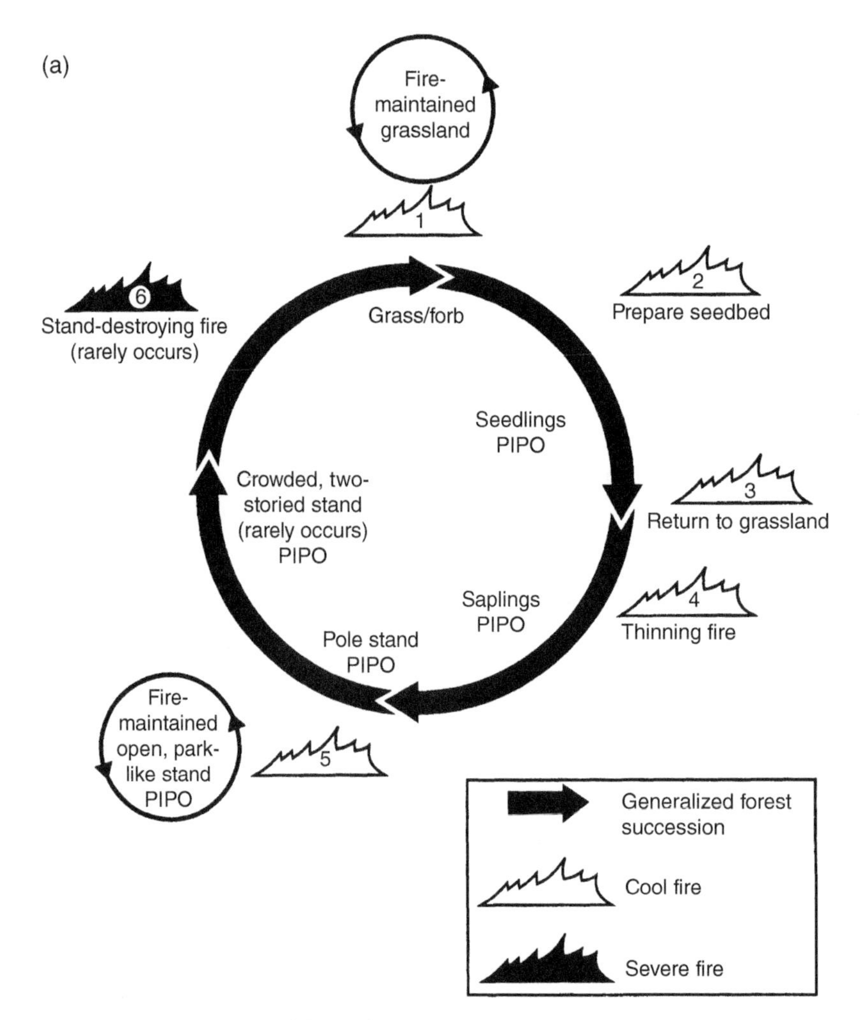

FIGURE 17.12 The role of fire in forest succession in warm and dry ponderosa pine (PIPO) ecosystems. (a) Generalized diagram of succession illustrating situations where grassland is maintained or where ponderosa pine is favored by a less-frequent fire regime. (b) Specific diagram of multiple successional pathways starting from an open, park-like old growth ponderosa pine stand. Numbers (1, 2, etc.) indicate successional pathways; letters (A, B) indicate states of pine forest or grassland. *Source:* After Fischer and Bradley (1987) / U.S. Department of Agriculture / Public Domain.

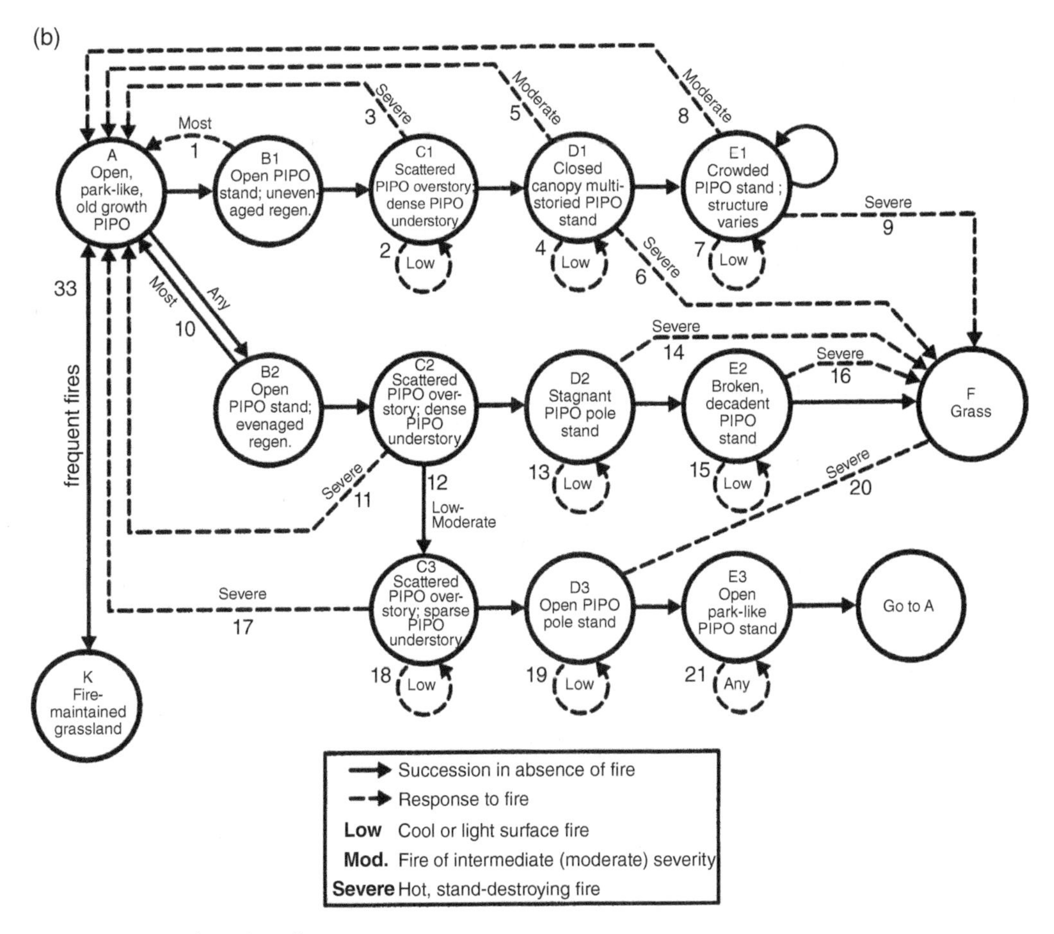

FIGURE 17.12 (Continued)

disturbances, and most emphasis has been placed on severe disturbances such as windstorms, fires, flooding, and landslides (Chapter 16). However, fine-scale or tree-fall gaps also significantly affect the structure and composition of mesic forests around the world, especially those dominated by understory tolerant trees.

Small canopy gaps are characteristic of many mesic forests of eastern and western North America. Larger, major disturbances occur at very long intervals in these forests, such that small gaps and slight but chronic disturbances to overstory crowns necessarily play a major role in tree regeneration. Lorimer (1977, 1980) found that mesic pre-European colonization forests in eastern North America and the Appalachian Mountains were characterized by all-aged stands—stands with an inverse J-shaped size-class distribution (Figure 13.9, Chapter 13). In mesic old-growth forests of eastern North America, Runkle (1982) estimated average annual canopy disturbance rates of less than 1% per year, suggesting that it requires 110–125 years on average for the entire canopy to be disturbed. For both tropical and temperate forests, Veblen (1992) estimated an average canopy rotation time of approximately 100 years and a range of 50–575 years.

Late-successional species can perpetuate themselves more or less indefinitely in and near small canopy gaps in the absence of larger or more severe disturbances. Regeneration of species such as beeches, sugar maple, western and eastern hemlocks, firs, and basswoods occurs when persistent seedlings in the understory fill the gaps, excluding less-tolerant species that need to first establish in the newly formed gap. A canopy profile through a single-tree gap in a sugar maple–beech forest is illustrated in Figure 17.13a. The gap is 50 m² in the tree canopy and affects 250 m² at ground level and formed from the breakage of a single stem 13 m above the ground about 3 years before the profile was made. Such small gaps may be filled by (i) adjacent canopy trees expanding branches into it or (ii) already-established saplings filling it from below. This process takes place in temperate forests in gaps smaller than approximately 0.04 ha; the abundance of intolerant species sharply increases in larger gaps (Runkle 1985; Stewart et al. 1991; Busing 1994). Many authors report that two or more gaps are required for most saplings of tolerant species to reach the overstory. Mean residence time in the understory for tolerant species is often long; for sugar maple, it is often over 100 years and can be as long as 230 years (Barden 1981; Canham 1985).

Mesic Douglas-fir–western hemlock forests of the Pacific Northwest appear to exhibit slower gap dynamics than other forests (Spies et al. 1990). Gap formation is typically less than the range of 0.5–2.0% reported for temperate deciduous forests (Runkle 1985). Nevertheless, vertical forest structure provides an important context for gap dynamics in these forests. In Figure 17.13b, a small gap occurs in an old-growth Douglas-fir–western hemlock forest whose overstory trees are over twice as tall as those in the beech–sugar maple forest in Figure 17.13a. Because of their great height and narrow crowns, the death of one or a few conifers in such forests may not transmit enough light for the regeneration of mid- or intolerant species (Canham et al. 1990; Spies et al. 1990). Gaps may remain important for the regeneration of Douglas-fir and other early-successional species where trees are shorter and canopies less dense, such as mixed-conifer forests (Spies and Franklin 1989).

Larger gaps may allow mid-tolerant and intolerant species (opportunists or gap-phase species) to establish and reach the overstory. Such species have light, wind-disseminated seeds with the ability to germinate on small patches of exposed mineral soil, such as those created by the uptorn roots of windthrown trees. Juvenile growth under conditions of partial shade and partial root competition is rapid so that at least some seedlings can outgrow and overtop already-established tolerant species in the openings. Typical gap-phase replacement species include yellow birch and white ash, and with increasingly larger gaps include tulip tree, northern red oak, black cherry, and sweet birch. For example, Webster and Lorimer (2005) found that yellow birch was able to take advantage of gaps as small as 100–400 m² in hemlock–northern hardwood forests in northern Wisconsin, although those most successful in reaching the canopy were already present at the time of gap formation.

Small gaps create a shifting mosaic during the transition and old-growth stages of mesic forests (see Table 17.4), but the forest canopy is not a simple closed-canopy matrix punctuated by gaps (Lieberman et al. 1989). Tree crowns thin with age as their branches die, are sheared off by wind, or are broken by colliding with the branches of adjacent trees. The intensity of light reaching the forest floor therefore changes over time, and gaps represent an extreme condition where light may travel unimpeded to the forest floor. Latitude and the height of trees determine how much light penetrates beyond the perimeter of the gap itself and under the branches of surrounding overstory trees (Canham et al. 1990). In addition, succession in canopy gaps depends on many factors in addition to light, because of spatial variability in the amount of vegetative cover, soil water and nutrients, occurrence of coarse woody debris, and bare soil. Other factors include those associated with different regional and local ecosystems: latitude, physiography, drainage, nutrient cycling, tree species composition, disturbance regime, mutualisms with animals, and mycorrhizae. Of primary consideration in assessing gap dynamics is the disturbance type that creates the gap. For example, the specific type of wind damage can affect succession; trees uprooted by wind may form larger gaps and provide bare soil for colonizing species, whereas stem breakage may favor existing advanced regeneration in a smaller, more shaded gap.

Runkle (1981, 1982, 1985, 1990) discovered in a variety of mesic forests of the eastern United States that small gaps 50–200 m² favor tolerant species but allow opportunists to persist in low densities. Gaps caused by the death of one species that favor seedlings of another species (i.e., *reciprocal*

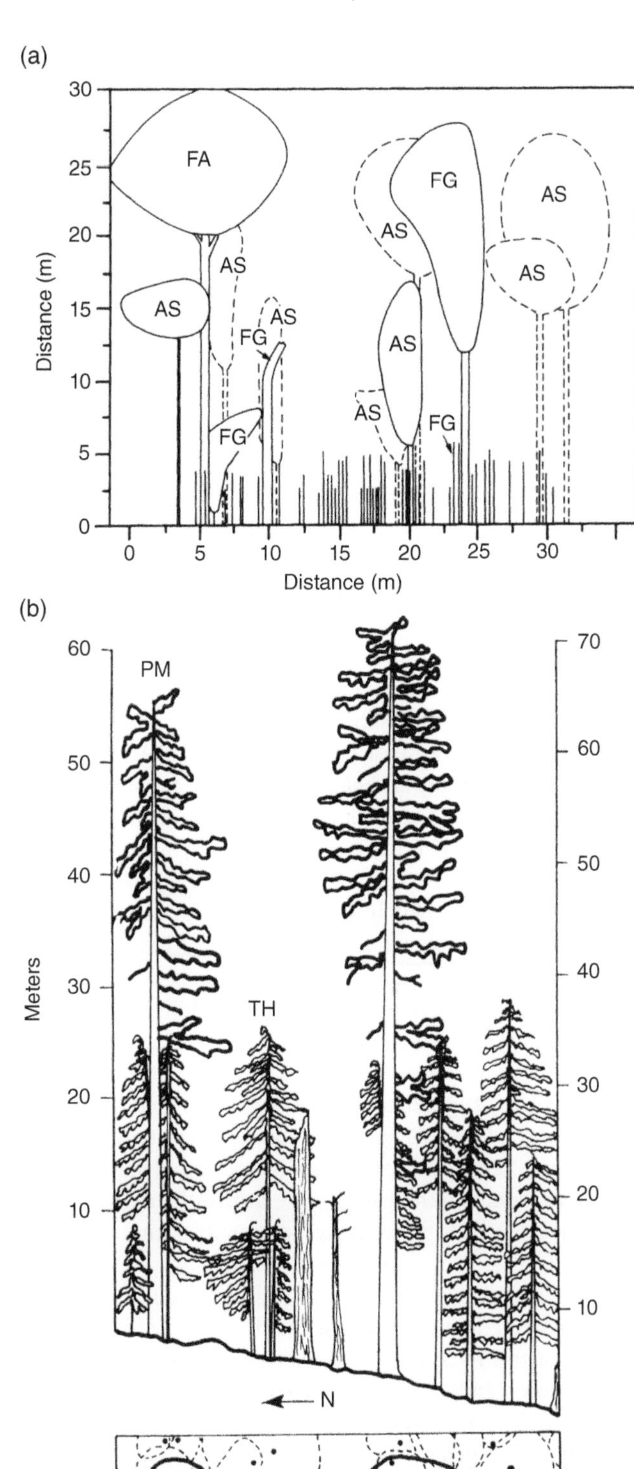

FIGURE 17.13 Examples of tree-fall gaps in eastern and western North American ecosystems. (a) Overstory canopy profile through a single-tree gap. The gap, 50 m^2 in the tree canopy and affecting 250 m^2 at ground level, was formed when one stem broke at a height of 13 m. Crowns with solid lines are within 5 m of the observer, and crowns with broken lines are between 5 and 10 m of observer. FA, *Fraxinus americana* (white ash); FG, *Fagus grandifolia* (American beech); AS, *Acer saccharum* (sugar maple). The vertical lines in the understory are FG; thick lines are AS. (b) Vertical and horizontal profile through a 450-year-old Douglas-fir/western hemlock stand in western Oregon. *Source:* After Spies et al. (1990) / Canadian Science Publishing. Reprinted with permission of National Research Council Canada. The small gap in the transect center (8–10 years old) was created when a large Douglas-fir broke off and fell on a smaller hemlock, breaking its crown. Tree regeneration consists entirely of western hemlock seedlings less than 0.5 m tall. PM, *Pseudotsuga menziesii* (Douglas-fir); TH, *Tsuga heterophylla* (western hemlock); G, canopy gap. *Source:* After Runkle (1990). Reprinted with permission of National Research Council Canada.

replacement; Fox 1977; Woods 1979, 1984; Runkle 1981) are more common in low-diversity forests such as those dominated by beech and sugar maple stands, although exceptions exist. In the southern Appalachian Mountains, for example, four mid-tolerant to intolerant species—northern red oak, white ash, black cherry, and tulip tree—occupied 3% of the canopy area via infrequent success in gaps, especially those formed by multiple trees (Barden 1981).

Gap Specialists: American Beech and Sugar Maple American beech and sugar maple, two of the most common species in eastern North America, are long-lived, very tolerant, late-successional species of mesic forests where severe disturbances are very infrequent. These two species are the most likely of all deciduous species to perpetuate themselves over many centuries because they are able to regenerate in shaded understories and utilize treefall gaps to gain the canopy. Where the two species co-occur, beech occurs over a wider variety of sites because it is less nutrient-demanding, tolerates wetter sites (Crankshaw et al. 1965; Leak 1978), and is more shade tolerant than sugar maple. For example, beech occurred on 256 soil types in pre-European colonization forests of Indiana, second only in distribution to white oak (Crankshaw et al. 1965). However, many investigators have sought to explain the co-existence of these late-successional species in mesic forests by their performance in relation to the light relations of forest understories.

A long-term study of beech and sugar maple coexistence is a 16-ha beech–sugar maple-dominated ecosystem that is part of Warren Woods in southwestern Michigan. Beech was the main overstory species in 1933 as it was prior to European colonization in the surrounding area (Kenoyer 1933). At that time, beech had 10 times the number of trees >15 cm diameter at breast height (DBH) and represented 82% of the basal area (Cain 1935). Three-fourths of the treefalls from 1946 to 1976 were of beech (Brewer and Merritt 1978). Meanwhile, sugar maple was more common as seedlings and saplings. Long-term studies of the understory show that both species maintained themselves over the entire area without marked replacement of one by the other, a relationship termed *allogenic coexistence* (Poulson and Platt 1996). In this case, "allogenic" is emphasized because wind creates the treefall gaps that appear to explain the observed pattern of species interaction. Notably, autogenic ecosystem factors, such as old (300 years), tall (30–40 m) trees with broad crowns, hollow trunks, brittle wood, and shallow rooting, also are significant in gap formation, which has increased markedly since 1975 in this aging stand.

Beech and sugar maple understory individuals are able to coexist in this ecosystem due to their differences in understory tolerance, tree architecture, and how they respond to light levels both in the understory and in treefall gaps (Runkle 1985; Canham 1988; Poulson and Platt 1996). Beech replaces sugar maple in the deeply shaded understory and in small gaps where its saplings exhibit horizontal (plagiotropic) growth to obtain light. Sugar maple is favored where multiple treefall gaps occur; sugar maple has more rapid vertical (orthotropic) extension growth enabling it to quickly colonize the gap. Where a mix of treefall gaps of different sizes occurs at appropriate frequencies, both species may coexist at near-equal abundances over the forest as a whole (Figure 17.14).

Despite the importance of species trade-offs based on light relations, the Warren Woods example is incomplete without consideration of soil properties and nutrient and water relations specific to this local ecosystem. Soils at Warren Woods are sandy and strongly acid, and the high spring water table tends to restrict rooting to the acidic topsoil. The less nutrient-demanding understory beech should be favored over sugar maple on the low nutrient status of this site and contribute significantly to beech dominance with minimum canopy disturbance. This relationship was reflected by the overwhelming dominance of beech overstory in 1933 (Cain 1935). As windstorm disturbances increase in frequency within the forest as they have over the last several decades, understory dominance may shift to sugar maple (Figure 17.14, panel 4). The proportion of the species in the understory and overstory may reflect soil and nutrient differences in other ecosystems where sugar maple is favored, provided sufficient light is available for sugar maple regeneration. For example, sugar maple dominates only in two favorable ecosystems in the White

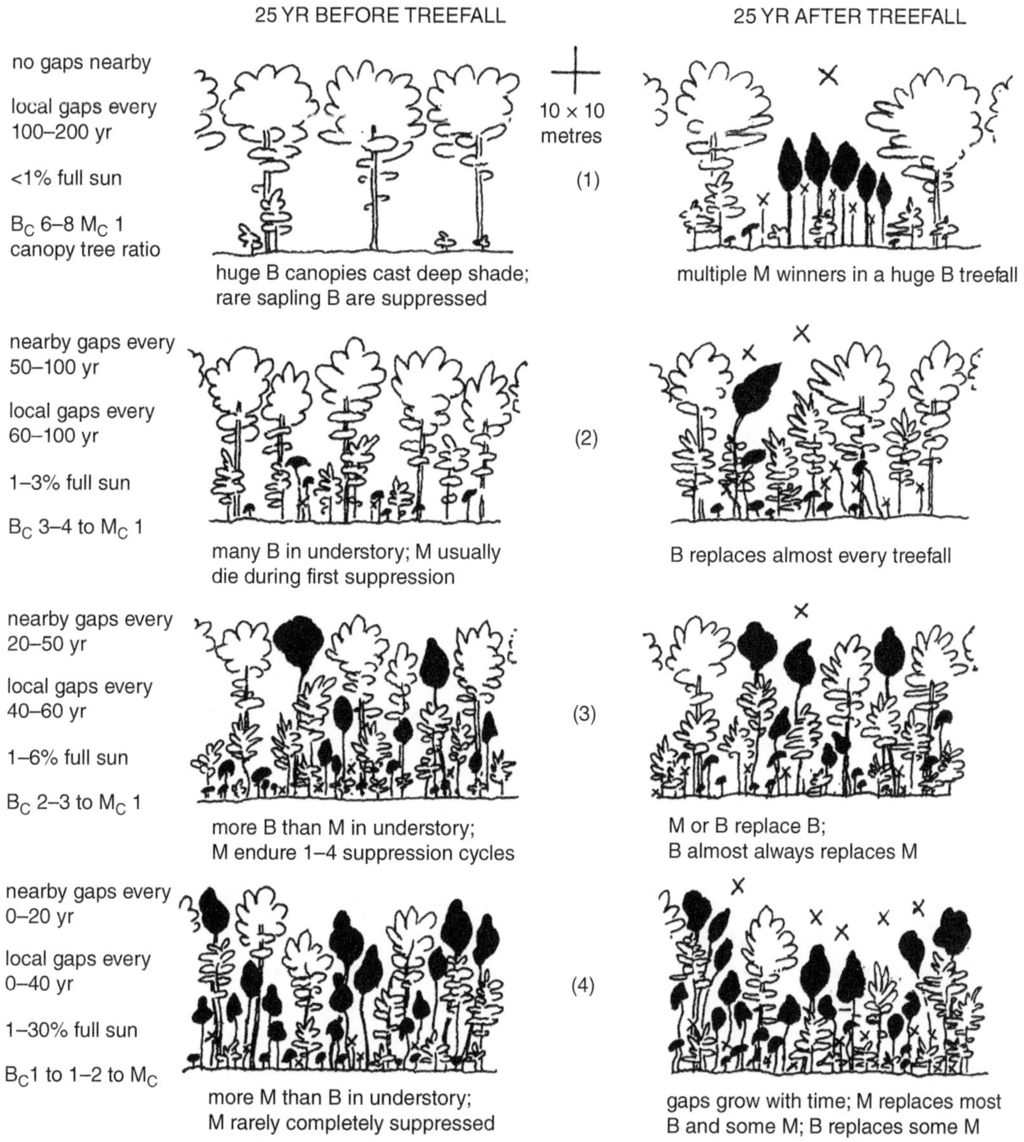

25 YR BEFORE TREEFALL

25 YR AFTER TREEFALL

10 × 10 metres

(1)

no gaps nearby

local gaps every 100–200 yr

<1% full sun

B_C 6–8 M_C 1 canopy tree ratio

huge B canopies cast deep shade; rare sapling B are suppressed

multiple M winners in a huge B treefall

(2)

nearby gaps every 50–100 yr

local gaps every 60–100 yr

1–3% full sun

B_C 3–4 to M_C 1

many B in understory; M usually die during first suppression

B replaces almost every treefall

(3)

nearby gaps every 20–50 yr

local gaps every 40–60 yr

1–6% full sun

B_C 2–3 to M_C 1

more B than M in understory; M endure 1–4 suppression cycles

M or B replace B; B almost always replaces M

(4)

nearby gaps every 0–20 yr

local gaps every 0–40 yr

1–30% full sun

B_C1 to 1–2 to M_C

more M than B in understory; M rarely completely suppressed

gaps grow with time; M replaces most B and some M; B replaces some M

FIGURE 17.14 A conceptual model of coexistence of beech (B) and sugar maple (M) at Warren Woods, southwestern Michigan. The sequence of four panels shows that infrequent and small treefall gaps (panel 1) favor beech (open silhouettes), whereas frequent and large gaps (panel 4) favor sugar maple (closed silhouettes) in the understory. The left panel of each pair, 25 years before treefall, shows how the understory percentages of full sunlight levels, due to branch gaps and nearby treefall gaps, affect forest structure and the relative success of beech and sugar maple. The right panel, 25 years after a treefall (x marks the former position of the crown top of a fallen tree), shows responses of species to gaps. Vertical and horizontal axes are drawn to a 10-m scale, and only individuals >1 cm DBH and >1–2 m tall are shown. B_c = beech canopy tree; M_c = sugar maple canopy tree. *Source:* After Poulson and Platt (1996) / John Wiley & Sons. Reprinted with permission of the Ecological Society of America.

Mountains of New Hampshire, whereas beech is more abundant in eight (Leak 1978). In addition, in the heavy-textured, calcareous morainal soils that supported beech–sugar maple forest prior to European colonization in central and southeastern Michigan (Quick 1923; Veatch 1953, 1959), sugar maple appears to be much more vigorous than beech and the tree replacement patterns differ from those at Warren Woods. For example, in 1997 in an old-growth forest on a clay-rich, calcareous, nutrient-rich moraine in southeastern Michigan, sugar maple dominated the overstory (49% sugar maple to 23% beech for basal area), and its saplings overwhelmingly dominated the understory (62–13% for number of stems). Notably, sugar maple has become even more dominant on this site over the last 20 years in the understory (now 66–14% for number of stems). Experimental evidence demonstrates that growth of sugar maple seedlings under low light is enhanced 2 to 4 times on rich sites in part because of higher nitrogen and water availability (Walters and Reich 1997).

FIRE AND OAK DOMINANCE—OAKS AT RISK

Oaks dominate the overstory of dry and dry-mesic sites in much of eastern North America today, but the current generation may be their last on all but the most extremely dry sites. Succession of oak forests to more mesophytic and understory tolerant species has been documented for many different regions and ecosystems (Christensen 1977; Lorimer 1984; McCune and Cottam 1985; Host et al. 1987; Pallardy et al. 1988; Hammitt and Barnes 1989; Abrams and Downs 1990; Nowacki et al. 1990; Nowacki and Abrams 1991; Abrams 1992; Shotola et al. 1992; Abrams 2003; Aldrich et al. 2005; Nowacki and Abrams 2008; Johnson et al. 2019). Although oaks dominate in the overstory, seedlings are relatively few and there is virtually no recruitment to the sapling understory. Such a trend, if continued, would likely create major differences in ecosystem structure and function, including shifts in both animal and plant communities.

Upland oaks evolved in dry, fire-prone ecosystems and are characterized by a suite of fire and dry-site adaptations (Chapter 10; Abrams 1990). Depending on regional and local site conditions, fire seasonality, frequency, intensity, and severity are instrumental in maintaining oak regeneration and determining the distribution and occurrence of oak species. Widespread fire exclusion throughout much of the East is perceived as the major reason for lack of oak regeneration, as is herbivory by deer in many areas (Chapter 12). Red maple and sugar maple are the understory tolerant, mesophytic species replacing oaks in the understory on all but the driest sites. Opportunistic species such as black cherry, sweet birch, blackgum, and sassafras can also establish and develop in the absence of fire where oak advance regeneration fails and logging or wind disturbance to the oak overstory provides sufficient light.

Successful oak regeneration depends on an abundance of oak saplings in the forest understory before disturbance, that is, the presence of advanced regeneration. There are many factors that drive oak regeneration, such as seed production and dispersal or animal damage to seeds and seedling (Lorimer 1993). However, fire is critically important for successful oak regeneration because it kills competitors and promotes resprouting of young oaks. In addition, fire provides a source of readily available nutrients for developing oak sprouts when it consumes organic matter. As mid-tolerant species, young upland oaks are at a competitive disadvantage from many early-successional species in open understories as well as from the species that tolerate shade. Thus, mesophytic species are always invading the understories of oak forests by wind and animal dispersal, but oak dominance is maintained by fire.

Light and moderate surface fires historically occurred at all stages of stand development in oak forests, indiscriminately killing tree seedlings, saplings, and shrubs. Few species have the resprouting ability exhibited by oaks (and their usual associates, the hickories) in the seedling or sapling stage, such that repeated fires act to purge the understory of mesophytic competitors in oak–hickory forests. Young oak stems killed by fire resprout vigorously from well-established root systems, although the thick bark of many young saplings allows them to survive surface fires. The density of the oak overstory may be high enough to deter or significantly slow the invasion of mesophytic species into the understory during stand initiation and stem exclusion. Fire is necessary to kill invaders and reduce shrub competition from the understory reinitiation stage onward, when oaks are themselves becoming established and forming advanced regeneration. When crown breakage or gaps form during the old-growth stage, the advanced regeneration of oaks and hickories is well positioned to dominate the understory (as in Figure 13.13, Chapter 13). On the driest sites of ridges and fire-prone barrens, oaks (blackjack, post, chestnut, black, northern pin) may be self-replacing or only very slowly replaced by other species (Abrams 1992). In these ecosystems, fire, wind, soil movement, or other disturbances, together with other species adaptations to extreme sites, may maintain the intolerant oak species and associates of hickory or pine.

Understanding a complex process such as forest succession is a tall order. Almost all ecological research (at least on secondary succession) has been conducted within the life span of existing forests, and therefore successional pathways must always be inferred or modeled rather than observed. It is notable that macroclimate has been considered to be the grand driver of succession since the concept was first articulated, and thus nearly everything we think we know about how forests are likely to change over time is likely to be altered in the current era of rapid climate change. From this perspective, despite the fact that forest succession has been studied in detail for well over a century, ongoing and future research on successional processes should remain at the forefront of forest ecology. We examine climate change and its effects on forests in the final part of this book.

SUGGESTED READINGS

Abrams, M.D. (1992). Fire and the development of oak forests. *BioScience* 42: 346–353.

Buma, B., Bisbing, S., Krapek, J., and Wright, G. (2017). A foundation of ecology rediscovered: 100 years of succession on the William S. Cooper plots in Glacier Bay, Alaska. *Ecology* 98: 1513–1523.

Chapin, F.S. III, Walker, L.R., Fastie, C.L., and Sharman, L.C. (1994). Mechanisms of primary succession following deglaciation at Glacier Bay, Alaska. *Ecol. Monogr.* 64: 149–175.

Christensen, N.L. (1988). Succession and natural disturbance: paradigms, problems, and preservation of natural ecosystems. In: *Ecosystem Management for Parks and Wilderness* (ed. J.K. Agee and D.R. Johnson). Seattle: University of Washington Press.

Crisafulli, C. and Dale, V. (ed.) (2018). *Ecological Responses at Mount St. Helens: Revisited 35 Years After the 1980 Eruption*. New York: Springer 351 pp.

Glenn-Lewin, D.C. and van der Maarel, E. (1992). Patterns and processes of vegetation dynamics. In: *Plant Succession: Theory and Prediction* (ed. D.C. Glenn-Lewin, R.K. Peet and T.T. Veblen). London: Chapman and Hall.

Matthews, J.A. (1992). Chapter 6. Plant succession: processes and models. In: *The Ecology of Recently-Deglaciated Terrain*, 250–316. Cambridge: Cambridge University Press.

Pickett, S.T.A., Collins, S.L., and Armesto, J.J. (1987). Models, mechanisms and pathways of succession. *Bot. Rev.* 53: 335–371.

Carbon Balance of Trees and Ecosystems

CHAPTER 18

Although forest ecosystems cover only 21% of the Earth's land surface, they constitute a disproportionately large share of terrestrial plant mass (75%) and its annual growth (37%). Because plants are largely constructed of carbon (47%), the extent to which forests are altered by human activities (i.e., harvesting or conversion to other types of vegetation) has a substantial influence on the pattern in which carbon (C) is cycled and stored at local, regional, and global scales. Recent attention has focused on factors influencing the fixation and release of C by trees and forest ecosystems (i.e., their C balance), because the rising CO_2 concentration of the Earth's atmosphere and projected changes in global climate have the potential to alter the present-day geographic distribution of forests and the rate at which they sequester CO_2 from the atmosphere (Pastor and Post 1986; Melillo et al. 1993). Such a change is of both ecological and economic importance, because the production of plant matter in forest ecosystems not only influences the global cycling of C but also represents an indispensable source of food, fuel, fodder, and fiber for human populations around the Earth. Globally, forests also are home to countless species of animals, as well as vascular and non-vascular plants. Importantly, aspects of climate change, such as rising atmospheric CO_2, altered temperature and precipitation regimes, and the deposition of human-derived nitrogen (N), have the potential to alter the C balance of forest trees, forest ecosystems, and also that of the entire Earth.

The C balance of plants (i.e., growth) is controlled by the quantity of CO_2 fixed through photosynthesis and the rate at which photosynthetically fixed C is returned to the atmosphere by respiring plant tissues. Light, temperature, and the availability of soil water and nutrients constrain these physiological processes and consequently influence the productivity of plants and the ecosystems in which they occur. One needs to only examine desert, grassland, and forest ecosystems to understand that the processes controlling the productivity of individual plants also have a substantial influence on the productivity of terrestrial ecosystems. This situation is particularly true in forests, in which the growth and metabolism of plants far outweigh that of any other organism.

From an ecological perspective, the flows of C and energy are somewhat interchangeable; that is, plants transduce solar energy into C-based biochemical energy via photosynthesis. Further, it is plant-derived biochemical energy that drives the subsequent flow of energy through entire terrestrial ecosystems. In this chapter, we focus on the physiological processes of plants that control the C balance (i.e., the flow of energy) of terrestrial ecosystems, especially forests. Importantly, the principles we present here apply equally to the C dynamics of all terrestrial ecosystems, whether they are tundra, desert, or grassland. We discuss how climate and soil nutrient availability influence the C balance of plants, and hence their growth. We then build upon this information to understand how environmental factors influence the C balance of ecosystems at local, regional, and global scales.

Forest Ecology, Fifth Edition. Daniel M. Kashian, Donald R. Zak, Burton V. Barnes, and Stephen H. Spurr.

CARBON BALANCE OF TREES

Only a small fraction of the radiant energy reaching the Earth's surface (2%) is used by green plants to assimilate atmospheric CO_2 into organic compounds via photosynthesis. These compounds are used to construct new plant tissue, maintain existing tissue, create storage reserves, or provide defense against insects and pathogens. Simply stated: trees grow (or gain C) when the amount of CO_2 fixed through photosynthesis exceeds the amount of CO_2 lost from respiring leaves, branches, stems, and roots. Understanding the factors influencing the balance between photosynthesis and respiration is of ecological importance, because plants with the greatest net C gain under a specific set of environmental conditions (i.e., light, water, and nutrient availability) are often the best competitors. It follows that fast-growing plants require rapid rates of photosynthesis, but the converse is not always true. As you will learn, the net C gain of plants is not influenced by photosynthetic rate alone (Ceulemans and Saugier 1991). In the following section, we trace the flow of photosynthetically fixed CO_2 to the construction and maintenance of plant tissue, focusing on how environmental factors influence the net C gain of plants (i.e., growth).

PHOTOSYNTHESIS, DARK RESPIRATION, AND LEAF C GAIN

The processes of photosynthesis and respiration occur in the leaves of all plants, and the balance of these more or less opposing physiological processes controls the net C gain of leaves. **Gross photosynthesis** is the *total* amount of C plants assimilate from the atmosphere. However, a portion of that fixed C is returned to the atmosphere from leaves as CO_2 during **dark respiration.** This process is so termed because leaf respiration can only be determined in the absence of C fixation and is hence measured on unilluminated leaves. In leaves, the process of dark respiration provides the biochemical energy (i.e., ATP) to repair proteins and membranes mediating photosynthesis. **Net photosynthesis** is the balance between gross photosynthesis and leaf dark respiration, and it represents the amount of photosynthate available for the growth and maintenance of non-photosynthetic tissue, as well as that allocated to storage and defense. Light intensity, temperature, and the availability of water and soil nutrients (especially N) all influence photosynthesis, respiration, and the C gain of leaves.

Rates of net photosynthesis measured under saturating light, optimum air temperature, low vapor pressure deficit, and ambient atmospheric CO_2 represent the **maximum photosynthetic capacity** of plants (see Table 18.1). Care should be taken in interpreting Table 18.1, because maximum photosynthetic capacity varies with leaf canopy position (shade versus sun leaves), as trees develop from a juvenile to an adult phase (see ponderosa pine), soil nutrient availability, and measurement technique. Also be aware that maximum photosynthetic capacities are often poorly correlated with the C gain of forest trees (Ledig and Perry 1969), because environmental factors (light, temperature, water, and soil nutrients) constrain photosynthesis under field conditions, often well below its maximum. Nevertheless, Table 18.1 illustrates several important physiological differences among forest trees that have important ecological implications.

The maximum photosynthetic capacity of forest trees ranges from 2 to 25 $\mu mol\ m^{-2}\ s^{-1}$ (Table 18.1), generally lower than that of most agronomic crops (Ceulemans and Saugier 1991). Shade-intolerant deciduous species have some of the highest photosynthetic capacities of all trees (up to 25 $\mu mol\ m^{-2}\ s^{-1}$; Table 18.1), whereas those of shade-tolerant deciduous species are much lower (3–6 $\mu mol\ m^{-2}\ s^{-1}$; Table 18.1). Recognize that shade tolerance is only one of several characteristics enabling a species to persist in the understory (i.e., understory tolerance; Chapter 13). Although some conifers have high photosynthetic capacities (see Scots and Monterey pine), most are modest when compared to those of deciduous trees.

Canopy architectures efficient at intercepting solar radiation can outweigh moderate photosynthetic rates, allowing some trees to attain large C gains despite the relatively low net

Table 18.1 The photosynthetic capacity of trees under conditions of saturating light, optimum temperature and water regimes, and ambient atmospheric CO_2 (350 μmol mol^{-1}). Rates are expressed in micromoles of CO_2 assimilated by 1 square meter of leaf during 1 second.

Shade-understory tolerance	Maximum photosynthetic capacity μmol CO_2 m^{-2} s^{-1}
Tolerant	
Deciduous trees	
sugar maple	3–4
American beech	4–5
hop-hornbeam	6
American basswood	3
flowering dogwood	4–6
Coniferous trees	
European silver fir	2–4
white fir	8
grand fir	4–5
balsam fir seedling	2
Sitka spruce	3–9
Norway spruce	2–7
subalpine fir	9
Mid-tolerant	
Deciduous trees	
white oak	2
northern red oak	4–5
red maple	3–5
white ash	4–5
yellow birch	9
Coniferous trees	
Engelmann spruce	3–5
Douglas-fir	2–6
Intolerant	
Deciduous trees	
black oak	7
sassafras	6
black cherry	4
sweet gum	6–11
tulip tree	7–17
paper birch	10
black locust	13–17
red ash	20–25
bigtooth aspen	14
cottonwood	15–19

(*Continued*)

Table 18.1 (Continued)

Shade-understory tolerance	Maximum photosynthetic capacity μmol CO_2 m^{-2} s^{-1}
trembling aspen	20–22
black poplar hybrid	20–25
Coniferous trees	
lodgepole pine	3–9
Ponderosa pine	
seedlings	25
mature trees	5–14
old-growth	4
Scots pine	10–16
Monterey pine	17
loblolly pine	3–6

Source: All values are expressed on a one-sided leaf area basis and are summarized from Bazzaz (1979), Ceulemans and Saugier (1991), Jurik et al. (1988), Larcher (1969), Wallace and Dunn (1980), and Walters et al. (1993).

photosynthetic rates of individual leaves. The rapid growth and modest net photosynthetic capacities (2–10 μmol m^{-2} s^{-1}) of some conifers (e.g., Douglas-fir) illustrate this point (Table 18.1). Coniferous canopies disperse incoming solar radiation over a larger number of leaves than broad-leaved canopies, enabling them to capture more solar energy at high light intensities. The orientation of coniferous leaves and branches also produces less self-shading of leaves lower in the canopy, enabling them to photosynthesize at a relatively high rate. Also note that the photosynthetic capacities in Table 18.1 are expressed on an area basis (μmol CO_2 per square meter of leaf surface per second), a factor that can be compensated for by canopy architecture as described in the earlier text. Additionally, coniferous species often maintain higher leaf areas than deciduous species, creating a larger surface area to capture solar radiation. For example, Douglas-fir and many true firs maintain leaf areas 2–6 times greater than that of some broad-leaved species. As such, low net photosynthetic rates in some conifers are offset by an efficient light-gathering canopy architecture, enabling them to achieve relatively large C gains, despite relatively low rates of maximum net photosynthesis.

LIGHT AND LEAF C GAIN

Net photosynthesis varies substantially with light intensity, and the response of trembling aspen (shade intolerant) and northern red oak (mid-tolerant) leaves to increasing light intensity illustrates this important point (Figure 18.1). Sun leaves of trembling aspen attain a greater light-saturated rate of net photosynthesis (i.e., maximum photosynthetic capacity; see right-hand portion of Figure 18.1) than do the sun leaves of red oak. The shade leaves of trembling aspen also have greater light-saturated rates of net photosynthesis than do the shade leaves of northern red oak, but rates in the shade leaves of both species are lower than those of sun leaves. Although shade leaves photosynthesize at a lower rate than sun leaves, they can contribute as much as 40% of the C assimilated by tree canopies (Schulze et al. 1977).

Plants attaining high rates of net photosynthesis, such as trembling aspen and other shade intolerant species, typically contain large amounts of chlorophyll and the enzymes used for CO_2 fixation (Field and Mooney 1986), which represent the biochemical machinery plants use to convert light energy into biochemical energy. High respiration rates are associated with plant tissues

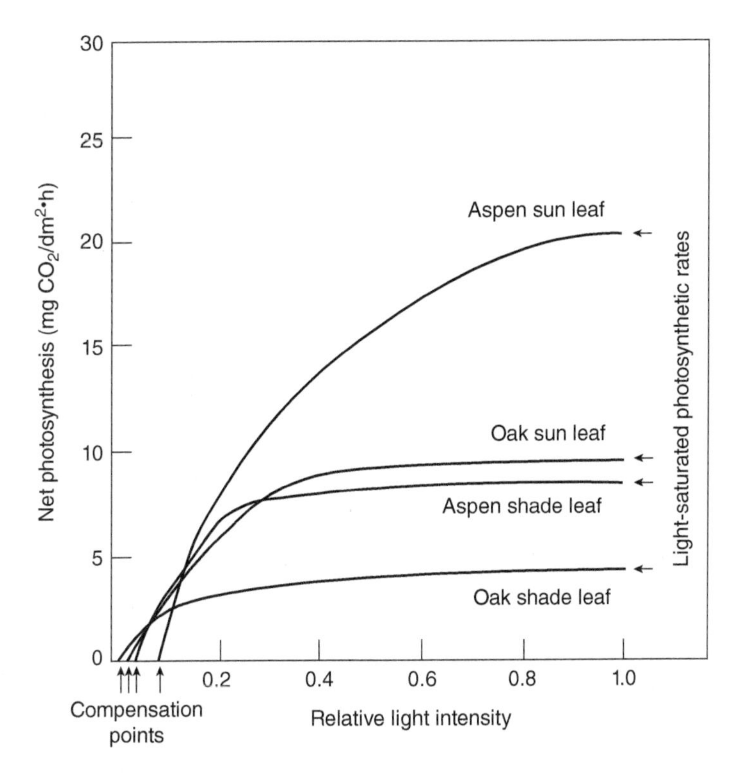

FIGURE 18.1 The photosynthetic response to changing light intensity for trees of contrasting tolerance. Trembling aspen is an intolerant species, whereas northern red oak is a mid-tolerant species that can persist in the forest understory. *Source:* After Loach (1967) / John Wiley & Sons.

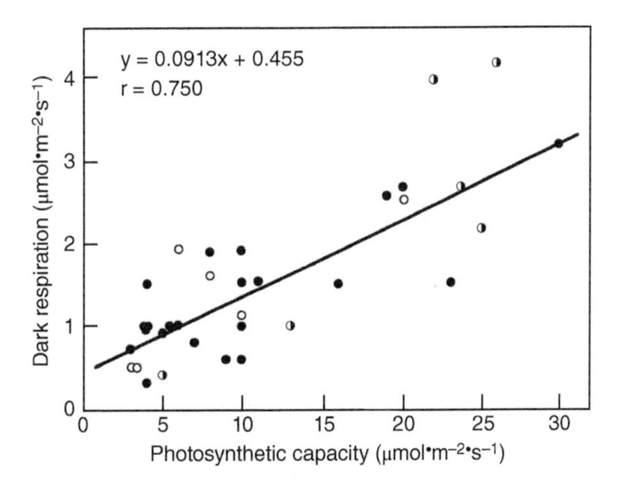

FIGURE 18.2 The relationship between photosynthetic capacity and leaf dark respiration for deciduous broad-leaved (closed circles), evergreen broad-leaved (half-open circles), and coniferous (open circles) tree species. *Source:* After Ceulemans and Saugier (1991) / John Wiley & Sons. Reprinted from *Physiology of Trees,* © 1991 by John Wiley & Sons, Inc. Reprinted with permission of John Wiley & Sons, Inc.

containing large amounts of metabolically active enzymes (Amthor 1984); such high rates of respiration are needed to replace and repair enzymes and membranes involved in the biochemical processes mediating photosynthesis. Consequently, leaves with high net photosynthetic rates also respire rapidly. Leaf dark respiration represents a relatively constant proportion (7–10%) of the maximum photosynthetic capacity of many forest trees (Figure 18.2; Ceulemans and Saugier 1991). However, rates of net photosynthesis rarely attain their maximum under field conditions, because light, soil nutrients, or water are often in short supply. Under these conditions, dark respiration can account for 10–20% of annual net photosynthesis in a wide range of coniferous and deciduous tree canopies (Ryan et al. 1994).

Because rates of net photosynthesis vary with light intensity, whereas dark respiration remains constant across a range of light intensities, the ability of a species to persist in low light environments is related to the balance of these physiological processes. In Figure 18.1, the light intensities at which net photosynthesis is zero (i.e., light compensation point) for the sun and shade leaves of trembling aspen are substantially higher than those of northern red oak. This relationship results from high dark respiration rates that maintain rapid rates of photosynthesis in aspen leaves. Dark respiration is lower in shade tolerant species, thereby enabling them to maintain positive rates of net photosynthesis (i.e., positive leaf C gain) at much lower light intensities. This ability illustrates the point that high photosynthetic rates alone do not insure a high C gain in all light environments.

TEMPERATURE AND LEAF C GAIN

Air temperature directly controls leaf C gain by influencing both rates of gross photosynthesis and dark respiration (Figure 18.3). The difference between these two physiological processes (i.e., net photosynthesis) reaches a maximum between 15 and 25 °C for most temperate trees (Kozlowski and Keller 1966; Mooney 1972). Rates above and below this range rapidly decline because low and high temperatures limit the process of CO_2 fixation. In Figure 18.3 we see that gross photosynthesis increases with rising temperature to an optimum and then rapidly declines with a further increase in temperature. Dark respiration also increases with rising temperature, attains a maximum, and then declines markedly as temperature continues to rise. However, the response of dark respiration to temperature differs from that of gross photosynthesis in several important ways (Figure 18.3). First, the maximum rate of dark respiration occurs at a much higher temperature than does the maximum rate of gross photosynthesis. Also, the slope of the gross photosynthesis and respiration curves differs at high temperatures (i.e., 35–50 °C). In this range of temperatures, gross photosynthesis declines rapidly, whereas dark respiration continues to increase. Note also that leaf C gain approaches zero as temperature rises between 35 and 43 °C. At temperatures above 43 °C, rapid rates of dark respiration combined with slow rates of gross photosynthesis result in a negative C

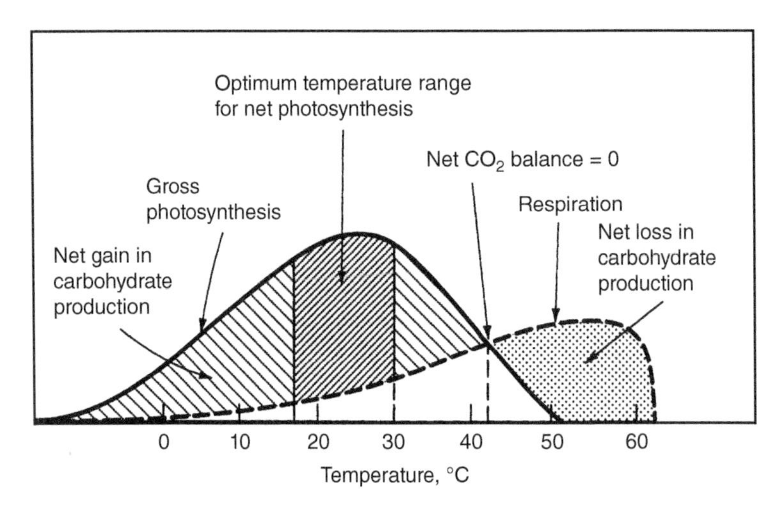

FIGURE 18.3 The temperature response of gross photosynthesis, dark respiration, and leaf C gain. Gross photosynthesis and dark respiration in leaves are temperature-dependent processes, and leaf C gain is maximized at the temperature where the difference between these processes is greatest. *Source:* After Daniel et al. (1979) / McGraw-Hill. Reprinted from Daniel, T., J. A. Helms, and F. S. Baker, *Principles of Silviculture,* 2nd ed., © 1979 by McGraw-Hill, Inc. Reprinted with permission of The McGraw-Hill Companies.

balance and the net loss of C from the leaf. Thus, you can see that changes in rates of gross photosynthesis and dark respiration in response to temperature have a great impact on leaf C gain (Berry and Bjorkman 1980; Berry and Downton 1982), controlling whether it is positive (net C gain) or negative (net C loss).

WATER AND LEAF C GAIN

Water is important for plant growth and its availability directly influences photosynthesis and leaf C gain. Plants are faced with a fundamental trade-off because CO_2 fixation and water loss occur simultaneously in leaves. Trees open their stomata allowing CO_2 to diffuse from the atmosphere to enzymatic sites of CO_2 fixation in the leaf mesophyll. At the same time, water vapor diffuses out of stomata in response to differences in water potential between the leaf and atmosphere (see Chapter 9). Declines in leaf water content from 5 to 10% often have little influence on photosynthesis (Hanson and Hitz 1982), but a drop in leaf water potential beyond this level can sharply reduce CO_2 fixation. Low leaf water potentials cause stomata to close, restricting the loss of water to the atmosphere and the flow of CO_2 into the leaf. At leaf water potentials of −1.0 to −2.5 MPa (Figure 18.4), stomatal closure begins to limit the diffusion of CO_2 into the leaf, thereby slowing rates of photosynthesis. As leaf water potentials fall below a critical level, stomata completely close and photosynthesis ceases.

Although the species in Figure 18.4 inhabit a wide range of sites (dry-mesic uplands for white pine and northern red oak; wet lowlands for speckled alder, red ash), leaf water potential limits photosynthesis over a surprisingly narrow range of values. Plants occurring in dry sites have several physiological and morphological adaptations enabling them to maintain high leaf water potentials at relatively low leaf water contents (percent water). Physiological changes in cell wall elasticity, membrane permeability, and the concentration of solutes within plant cells can maintain high leaf water potentials (i.e., less negative), allowing photosynthesis to occur at relatively low leaf water contents (Bradford and Hsiao 1982). Trees inhabiting dry environments also have a higher rate of gas exchange, lose turgor at a lower leaf water potential, and have greater root:shoot ratios, leaf thicknesses, guard cell lengths, and stomatal densities than trees from mesic sites (Abrams et al. 1994). Such adaptations provide dry-site species with increased water-use efficiency (μmol CO_2 fixed/mmol of H_2O transpired), allowing them to fix more CO_2 per unit of water lost.

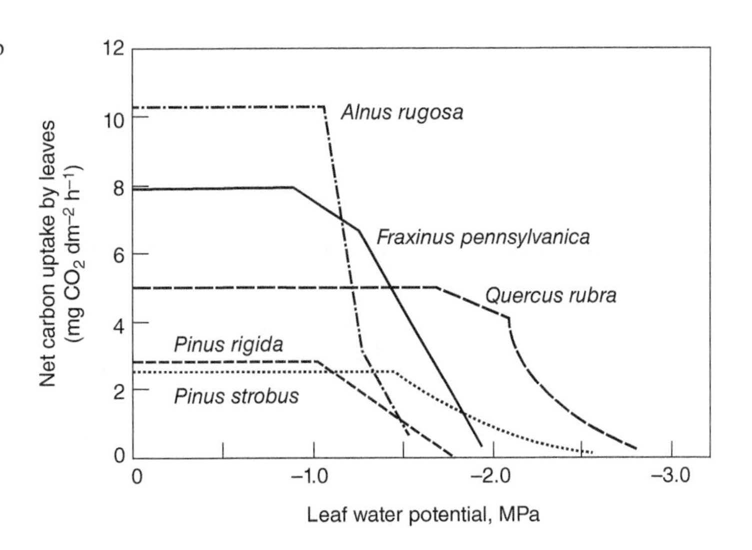

FIGURE 18.4 The relationship between leaf water potential and net photosynthesis for deciduous and coniferous trees occurring in different habitats. *Source:* After Hinckley et al. (1981) / with permission of Elsevier. Reprinted from *Water Deficits and Plant Growth*, © 1981 by Academic Press, Inc. Reprinted with permission of Academic Press, Inc.

SOIL NITROGEN AVAILABILITY AND LEAF C GAIN

Photosynthesis as well as other biochemical processes in the leaf are influenced by the supply of soil nutrients. Nitrogen is a constituent of chlorophyll, other light-harvesting pigments, photosynthetic membranes, and the enzymes mediating photosynthesis. Their presence makes leaves rich in N compared to most other plant tissues. Thus, the supply of N can have a large influence on the ability of a leaf to fix CO_2 from the atmosphere. For example, leaf N concentration is directly related to the photosynthetic rate of a wide array of herbaceous plants (Field and Mooney 1986). Net photosynthesis in forest trees also increases with leaf N concentration, but coniferous and deciduous species respond much differently (Brix 1971; Linder et al. 1981). In deciduous species, net photosynthesis is particularly responsive to leaf N concentration (Linder et al. 1981). For example, the highest rates of net photosynthesis for European white birch occur at the greatest leaf N concentration (5%), and rates sharply drop as leaf N concentrations decline (Figure 18.5). Coniferous species, on the other hand, lack the ability to increase photosynthetic enzyme contents to the same extent as deciduous trees, making photosynthesis much less responsive to changes in leaf N concentration, which often reflects soil N availability. Photosynthetic rates in conifers, such as Douglas-fir, initially increase up to 25% following N fertilization, but rates rapidly decline with time (Brix 1971). However, conifers often increase whole-tree C gain following N fertilization by producing more foliage, thus creating a larger photosynthetic surface area to capture CO_2 (Brix and Ebell 1969; Brix 1971). In deciduous trees, N fertilization often increases both the photosynthetic rate of individual leaves as well as leaf production, resulting in a larger and more photosynthetically active canopy.

CONSTRUCTION AND MAINTENANCE RESPIRATION

Forest trees are unique compared to other types of terrestrial vegetation due to their size and longevity. As a consequence of their large mass, forest trees require substantial amounts of photosynthate (i.e., carbohydrates produced by photosynthesis) to maintain the function of living cells located in the non-photosynthetic tissues of branches, stems, roots, and reproductive structures. Photosynthate also is required for the construction of new leaves to capture light, roots to forage for water and nutrients, and reproductive structures to insure regeneration. The construction and maintenance of these tissues require the products of photosynthesis to generate cellular energy (i.e., ATP). Respiration during the biosynthesis of new tissue is **construction respiration**, whereas respiration used to biochemically sustain already existing living tissue is **maintenance**

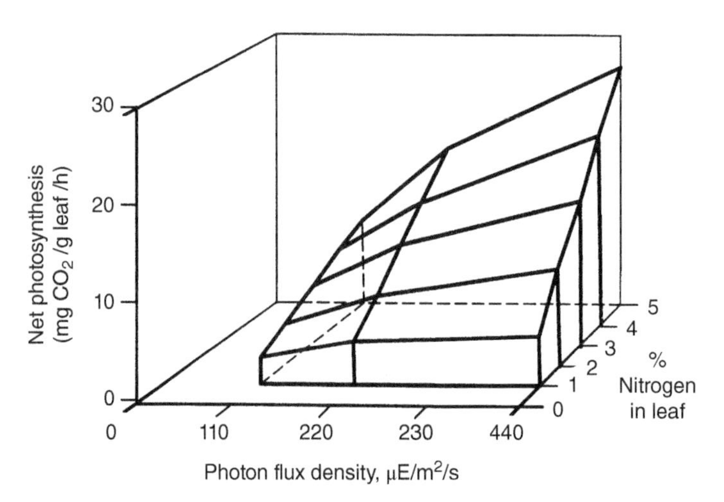

FIGURE 18.5 Leaf N concentration is an important factor controlling the rate of net photosynthesis in deciduous species. The highest rates of net photosynthesis for European white birch occur at saturating light intensities and high leaf N concentrations. *Source:* Linder et al. (1981) /U.S. Department of Energy / Public Domain.

respiration. These physiological processes, which return fixed CO_2 to the atmosphere, have an important influence on the C balance of all plants.

Construction Respiration: Respiration provides the cellular energy (ATP) used to convert the products of photosynthesis into the biochemical constituents of new plant tissue. Thus, photosynthetically fixed CO_2 is lost to the atmosphere (as CO_2) during the synthesis of new plant tissue. This loss (i.e., construction respiration) is calculated knowing the biochemical constituents of a particular plant tissue and the biosynthetic pathways by which they are formed. This information can be used to determine the amount of CO_2 produced during the biosynthesis of a particular plant compound or organ. Glucose, a product of photosynthesis, is often considered the substrate for biosynthesis when calculating the cost of construction respiration during the biosynthesis of a plant compound or a tissue (Penning de Vries 1975; Chung and Barnes 1977).

Look at Table 18.2 and note that the needles and shoots of loblolly pine contain large proportions (%) of carbohydrates, lignin, and phenolic compounds, whereas nitrogenous compounds, lipids, and organic acids are present in smaller amounts. The biosynthetic pathways giving rise to these compounds require different amounts of substrate (i.e., glucose) and produce different amounts of CO_2 as they are biochemically synthesized. For example, the synthesis of 1 g of carbohydrate during cell wall construction requires 1.2 g of glucose as substrate, of which 11% is lost as CO_2 (Table 18.2). Although lipids are present in relatively small amounts, their construction produces relatively large amounts of CO_2 compared to other biosynthetic pathways.

By knowing the constituents of a particular plant tissue and the amount of CO_2 lost during their biosynthesis, one can calculate construction respiration for plant tissues. For example, during the construction of 1 g of loblolly pine needles, approximately 19% of substrate glucose is respired during biosynthesis. Similarly, construction respiration consumes approximately 16, 18, and 17% of the glucose used to synthesize loblolly pine shoots, cambium, and roots, respectively (Chung and Barnes 1977). Clearly, accounting for these losses of C is essential for understanding the C balance of whole plants, as well as the C balance of forest ecosystems. It is important to realize that construction respiration is a fixed cost to the plant. Regardless of temperature or other environmental conditions that a plant faces, the amount of photosynthate used to synthesize a protein, carbohydrate, or for that matter a leaf, is invariant. The process of construction respiration may occur at a faster or slower pace depending on temperature, but the amount of photosynthate used to construct a compound or tissue remains the same. This differs dramatically from maintenance respiration, which is highly response to environmental conditions.

Table 18.2 The biochemical constituents of the needles and shoots of loblolly pine, and the proportion of carbon respired during their biosynthesis.

	Needles	Shoots	Construction respiration[a]
Compound	---------------Percent	-------------------------	-------------------------
Carbohydrates	35.7	38.0	11
Lignin	23.1	23.3	14
Phenolic compounds	21.0	20.0	29
Nitrogenous compounds	9.3	8.4	25
Lipids	5.6	5.3	49
Organic acids	3.8	3.5	32

[a] Calculated as the percent of glucose lost as CO_2 during the synthesis of a compound in the left-hand column.

Source: After Chung and Barnes (1977) / Canadian Science Publishing. Reprinted with permission of National Research Council Canada.

Maintenance Respiration: Carbohydrates produced by photosynthesis are used to sustain the function of living plant tissues during the process of maintenance respiration. Protein synthesis and replacement, membrane repair, and the maintenance of ion gradients across cell membranes all require the expenditure of energy initially derived through photosynthesis (Ryan 1991). The rate of maintenance respiration, like all other enzymatic processes, increases predictably with temperature. For most plants, maintenance respiration roughly doubles for every 10 °C increase in ambient temperature. That is, maintenance respiration increases exponentially with a linear increase in temperature. This fundamental fact has important implications for the C balance of plants, as well as entire ecosystems, as the Earth's climate continues to warm from fossil fuel burning.

Maintenance respiration also is strongly influenced by tissue N concentration (Waring et al. 1985), wherein tissues with high N concentrations also have high rates of maintenance respiration. In fact, leaf maintenance respiration for a wide range of plant species can be predicted from N concentration using the following equation (Ryan 1991):

$$R_M = 0.0106\,N$$

where R_M is leaf maintenance respiration (mmol of CO_2 mol tissue C^{-1} h^{-1}) and N is the total nitrogen concentration of that tissue (mmol N mol tissue C^{-1}). This relationship arises because the majority of N present in plant tissue occurs in proteins, and a large proportion (60%) of maintenance respiration is used for their repair and replacement (Penning de Vries 1975). Such a relationship also contributes to the high rates of leaf dark respiration that accompany rapid photosynthetic rates (see Figure 18.2). Herein lies an important link between soil N availability, leaf N concentrations, and the physiological processes that control plant C gain. Greater soil N availability can increase the N concentration of plant tissues. This can lead to greater rates of photosynthesis in leaves and more rapid rates of maintenance respiration in non-photosynthetic tissue. However, increases in maintenance respiration of non-photosynthetic tissues (i.e., plant roots) are surpassed by increases in net photosynthesis in leaves with higher N concentrations. In combination, these responses often lead to higher plant C gains when soil N availability increases.

As trees develop from juvenile to adult phases, the ratio of photosynthetic to non-photosynthetic tissues decreases, a factor that substantially influences maintenance respiration and the C balance of whole trees. Canopy leaf area reaches a maximum relatively early in tree development, thereby imposing a limit on canopy photosynthesis and the supply of photosynthate for tissue construction and maintenance. Although the supply of carbohydrate is constrained by canopy photosynthesis, the demand for carbohydrate rises as the proportion of non-photosynthetic tissue continues to increase during tree ontogeny. For example, stemwood accounts for a large proportion of total tree mass (approximately 50–70% in mature trees) and contains both living (sapwood) and dead tissue (heartwood). Although the proportion of living, non-photosynthetic tissue in stems decreases as trees mature, the absolute mass of living non-photosynthetic tissue continues to increase. Consequently, the amount of photosynthate allocated to the maintenance of living cells in stemwood dramatically increases as trees develop from a juvenile to an adult (Agren et al. 1980; Waring and Schlesinger 1985; Ryan 1988). Remember that the product of photosynthate used to maintain the mass of live, non-photosynthetic tissue in mature trees cannot be used for the construction of new tissue. Consequently, the greater cost of tissue maintenance in mature trees results in a decline in annual growth. Annual growth also can be slowed by reductions in the net photosynthetic rate of adult trees. As you will read later in this chapter, increases in maintenance respiration and declines in growth as trees mature have important implications for the C balance of forest ecosystems.

ALLOCATION TO STRUCTURE, STORAGE, AND DEFENSE

In the previous paragraphs, we have discussed physiological processes influencing the fixation and loss of C by forest trees. Accounting for these processes, one can construct a C budget illustrating how plants partition C into the growth and maintenance of leaves, stem, roots, reproductive structures, as well as the production of defense and storage compounds. Such an approach can shed light onto how light, water, and nutrients influence the allocation of photosynthate into organs that aid in gathering light or the acquisition of water and soil nutrients.

Summarized in Table 18.3 is the annual C budget for a young Scots pine tree (Agren et al. 1980). This budget quantifies the partitioning of photosynthetically fixed CO_2 into the growth of various plant tissues and accounts for their respiration during construction and maintenance. Approximately 88% of fixed C in this young tree was used to produce new tissue, whereas only 10% was consumed during construction and maintenance respiration. The production of new roots to forage for water and nutrients in soil represented a large investment of photosynthate, accounting for 56% of annual net photosynthesis. In combination, root growth and respiration comprised almost 62% of annual net photosynthesis, indicating a substantial proportion of fixed C was allocated to belowground growth and metabolism to acquire soil resources.

The amount of photosynthate plants allocate to growth, storage, and the production of defense compounds is under strong genetic control. Patterns of C allocation vary among species, within a species seasonally as well as with increasing age. Environmental factors also play an important role in influencing the allocation of C into structure, storage, and defense. Figure 18.6 illustrates the C allocation priorities in lodgepole pine. Allocation of photosynthate to new foliage and buds represents the highest priority for carbohydrate, followed by the production of new roots. Tree growth often is limited by light, water, and nutrients, making it necessary for the plant to first meet the carbohydrate demand of tree crowns and roots. The storage of carbohydrate in leaves, stem, and coarse roots has a lower priority than allocation to new leaves and roots. Allocation of

Table 18.3 The carbon budget of a 14-year-old Scots pine tree.

	Assimilation	Allocation	Percent of total
		g C year-1	
Net photosynthesis	1723		
Growth			
Current needles		286	16.6
Branch axes		132	7.7
Stem		145	8.4
Roots		960	55.6
Total growth		1523	88.4
Construction and maintenance respiration			
Stem		49	2.8
Branch axes		15	0.9
Roots		109	6.3
Total respiration		173	10.0
Growth + respiration		1696	98.5
Unaccounted net photosynthesis		27	1.5
Total		1723	100.0

Source: After Agren et al. (1980).

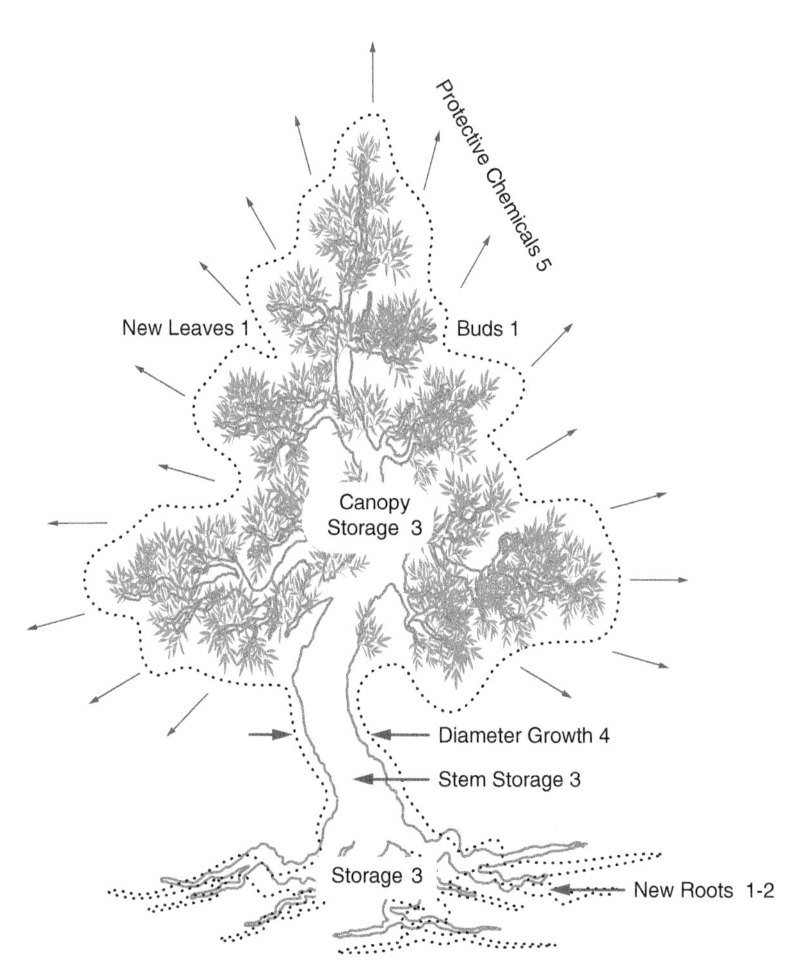

FIGURE 18.6 The allocation of photosynthate into structure, storage, and defense in lodgepole pine. Tree components with low numbers (e.g., one) represent strong priorities for allocation, whereas components with high numbers indicate a low priority for allocation. *Source:* Waring and Pitman (1985) / John Wiley & Sons. Reprinted with permission of the Ecological Society of America.

carbohydrate to stem growth follows that to storage and the production of chemicals to defend against herbivory, which is the lowest priority.

Patterns of C allocation vary substantially over the growing season, wherein certain tissues function as sinks or sources for carbohydrate. Nowhere is this more evident than during the phenological development of deciduous trees. Storage reserves in buds, branches, stems, and roots function as a source of carbohydrate to build new leaves early in the growing season. Once the expanding canopy attains a positive C balance, the priorities of allocation shift. In sugar maple seedlings, for example, large amounts of photosynthate are allocated to the production of new leaves early in the growing season (Table 18.4). The proportion of C allocated to stems and coarse roots increases following canopy development (i.e., mid growing season), and represents the greatest sink for photosynthate late in the growing season. The amount of carbohydrate stored in stems, coarse roots, and fine roots also increases over the growing season, attaining the highest priority late in the growing season. These carbohydrate stores are then used to construct new leaves, as well as new fine roots, at the start of the next growing season.

Table 18.4 The seasonal C allocation to leaves, stem, and roots in sugar maple seedlings.

	Percent of growth allocated to:			
	Leaves	Stem	Coarse roots[a]	Fine roots[b]
Early growing season	85	5	0	10
Mid growing season	30	30	25	15
Late growing season	0	5	60	35

[a] Coarse roots are 2–5 mm.

[b] Fine roots are <2 mm.

Source: After Burke et al. (1991). / Canadian Science Publishing. Reprinted with permission of National Research Council Canada.

LIGHT AND C ALLOCATION

By knowing the C allocation pattern of trees, one can begin to understand how the availability of light, water, and nutrients influences allocation to structure, storage, and defense. Carbohydrate produced during photosynthesis is first allocated to new leaves and then new fine roots (<2 mm in diameter; Figure 18.6). These are the energy, water and nutrient gathering structures of trees, and they have the highest priority for photosynthate. If the carbohydrate demand of new leaves and fine roots is satisfied, then remaining carbohydrate can be allocated to storage, the growth of stemwood, and the production of chemicals to defend against herbivory. Such a pattern has been observed for both coniferous (Waring and Pitman 1985) and deciduous species (Burke et al. 1991).

In very dense stands of lodgepole pine, low light levels limit photosynthesis and carbohydrate production, even when soil resources are abundant. In such a situation, carbohydrate allocated to meet the demand of new leaves and roots occurs at the expense of allocation to lower priorities of storage, stem growth, and the production of defense chemicals. As a consequence, stem diameter growth declines and the susceptibility to mountain pine beetle attack increases; trees with low diameter growth are those most susceptible to beetle attack (Waring and Pitman 1985). Thinning these forests increases light availability, decreases intraspecific competition for light, and increases the photosynthesis of remaining trees. Therefore, carbohydrate limitation within the remaining trees lessens, leading to increases in diameter growth and a reduction in susceptibility to mountain pine beetle attack. Nitrogen fertilization produces a similar response, wherein greater soil N availability increases the production of new foliage, which, in turn, increases canopy photosynthesis and the supply of carbohydrate within the tree. This is an interesting link between environmental conditions, changes in the allocation of photosynthate, and the susceptibility of a tree to an insect attack.

SOIL NITROGEN AVAILABILITY AND C ALLOCATION

The availability of soil nutrients is a potent modifier of C allocation in forest trees, substantially influencing the proportion of carbohydrate allocated to leaves, stems, and roots. Fine roots (< 1 mm in diameter) are the water- and nutrient-absorbing organs of forest trees, and their biomass usually declines as water or nutrient availability increases. By contrast, the amount of foliage often increases when soil water and nutrients are in abundant supply, suggesting that a trade-off occurs between aboveground and belowground plant growth as soil resource availability increases. That is, the root to shoot ratio (root:shoot) of plants declines along a gradient of increasing soil nutrient availability. In sugar maple, for example, N fertilization increased the mass of leaves and stem by 12%, whereas root mass (coarse + fine) declined by an equivalent proportion (calculated from Burke et al. 1991). Scots pine and a black poplar hybrid respond in a similar manner to N fertilization, but these species exhibit an even more dramatic shift from roots to leaves (Linder and Axelsson 1982; Pregitzer et al. 1995). Herbaceous species also apparently decrease root and increase leaf mass along gradients of increasing soil N availability (Tilman 1988). Such a pattern suggests that the plants "invest" smaller amount of carbohydrate into roots when soil resources are relatively abundant. As a result, carbohydrate is available to produce relatively more foliage, a response that allows the plant to capture more light energy on N-rich soils.

Nevertheless, there is considerable debate regarding the extent to which soil N availability alters the *total* allocation of carbohydrate to the production of belowground plant tissues. This debate stems from our limited understanding of fine-root production (i.e., birth) and mortality (i.e., death), factors that greatly influence the total amount of carbohydrate that plants allocate to belowground growth. Unlike the leaves of deciduous trees, which are initiated in spring and abscise during autumn, the birth and death of fine roots occurs simultaneously throughout the growing season (Hendrick and Pregitzer 1992). Consequently, relatively little can be inferred about patterns of C allocation by comparing the standing crops (i.e., biomass) of leaves and roots across gradients of soil nutrient availability.

Theoretically, the proportion (i.e., percent) of plant mass composed by fine roots could decline in N-rich soil by several alternative responses: (i) fine-root production could remain constant while mortality increases (Figure 18.7a), (ii) production could decline while mortality remained constant (Figure 18.7b), or (iii) both production and mortality could increase with rates

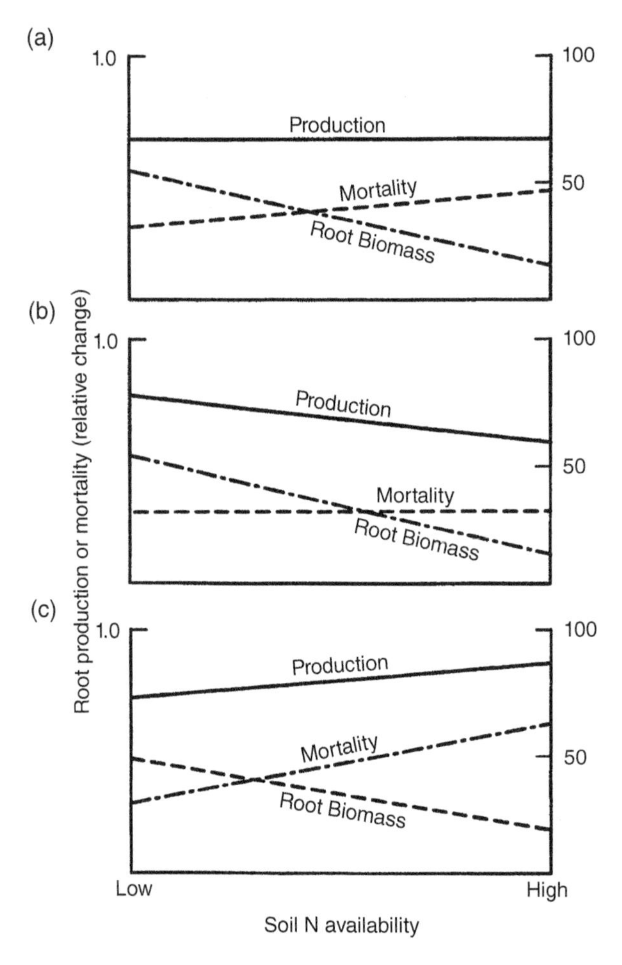

FIGURE 18.7 Three alternative mechanisms by which allocation to fine roots could decline with increasing levels of soil N availability. Each alternative has very different implications for the total amount of carbohydrate allocated to fine-root production.

of mortality surpassing those of production (Figure 18.7c). The first mechanism suggests that the life span of fine roots decreases, but root production (i.e., birth) remains constant; such a response would not alter the total amount of carbohydrate allocated to fine roots. The second mechanism argues for a decline in the total amount of carbohydrate allocated to fine roots, wherein fewer roots are produced with the same life span (i.e., no change in mortality). In the third alternative, fine-root production increases, but greater N availability decreases the life span of fine roots, which thereby increases the total amount of carbohydrate "invested" into these structures. In other words, fine roots "live fast and die young" in the third alternative. Clearly, determining which of these alternative responses is correct will rest on our understanding of the processes controlling fine-root production and mortality.

Attaining such an understanding has been difficult, largely because of the numerous problems associated with studying live plant roots in soil. Most methods of studying the production and mortality of fine roots modify the soil in ways that may invalidate the observations. However, technology has enabled ecologists to actually view and record the birth, growth, and death of individual tree roots. Clear plastic tubes (mini-rhizotrons) can be placed in the soil of a forest stand, and a miniaturized video camera can be used to capture images of individual roots from their birth to death (Figures 18.8 and 18.9; Hendrick and Pregitzer 1992; Iversen et al. 2012). The root images can be digitized, processed by computer, and production can be

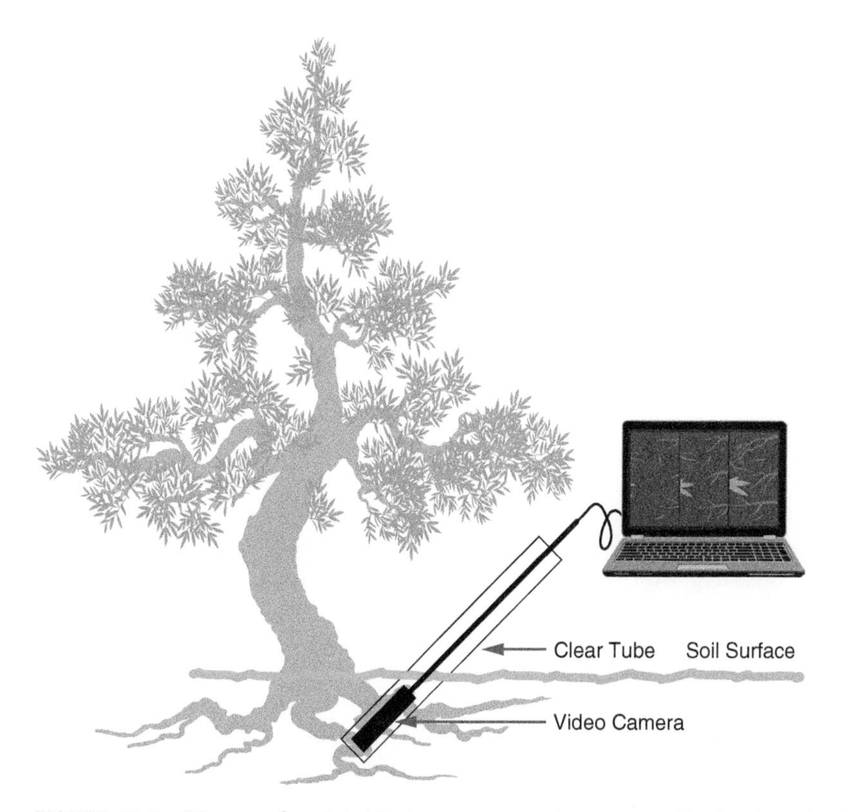

FIGURE 18.8 Diagram of a mini-rhizotron system used to capture video images of actively growing plant roots. The clear plastic mini-rhizotron tube (A) is placed into the soil using an auger, and the micro-video camera (B) is passed down the tube to record video images of the plant roots with the use of a video cassette recorder (VCR). The controller allows the operator to focus the camera and adjust light levels in the mini-rhizotron tube. Recorded images are subsequently digitized with the use of computer software. *Source:* After Hendrick and Pregitzer (1992) / John Wiley & Sons. Reprinted with permission of the Ecological Society of America.

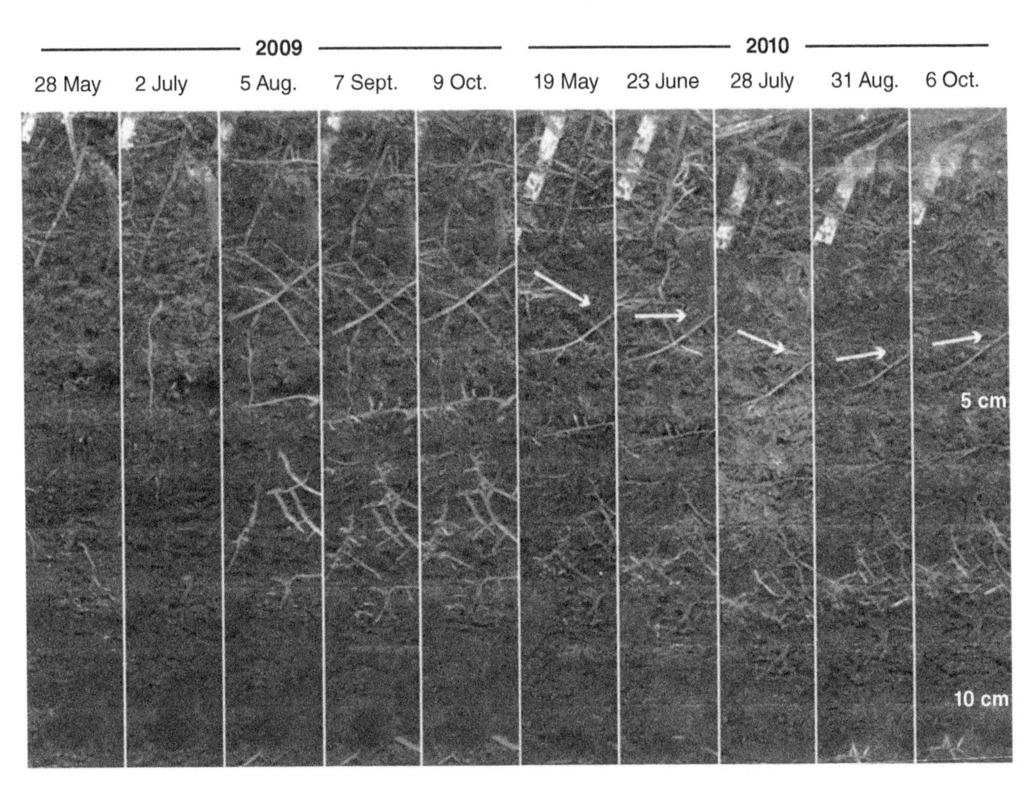

FIGURE 18.9 Images of the Scots pine fine-root growth over two growing seasons in surface soil (0–10 cm). Notice that fine roots are produced throughout the growing season and experience a pulse of mortality from autumn to spring (9 Oct 2009 to 19 May 2010). *Source:* Image obtained from Iversen et al. (2012).

calculated as the cumulative increase in root length over one growing season. Similarly, mortality can be determined as disappearance of roots that had previously appeared in the video image (i.e., a negative change in length). The results of such a study are shown in Figure 18.10, which illustrates the cumulative change in root production and mortality for a black poplar hybrid (a fast-growing intolerant tree) growing in soils of low and high N availability (Pregitzer et al. 1995).

The leaf mass of black poplar hybrid in the high N soil of this experiment increased by 35%, and fine-root mass decreased by 25%, relative to plants in low N soil. The decline in fine roots and increase in leaves are consistent with traditional views of changes in plant C allocation and with increases in soil N availability (a decline in root:shoot ratio). In the high-N soil, however, fine-root production increased, whereas fine-root life span decreased (i.e., increase in mortality; Figure 18.10a,b), supporting the idea that more, not less, photosynthate was actually allocated to fine-root growth in N-rich soil (Pregitzer et al. 1995).

The underlying mechanisms for an increase in fine-root mortality can be found in the relationships among soil N availability, the N concentration of fine roots, maintenance respiration, and the longevity of plant tissue. Tissue N concentrations in the black poplar cultivar also increased in response to greater soil N availability. Conceptually, fine roots growing in N-rich soil should have relatively high tissue N concentrations (i.e., ion uptake proteins) and high rates of maintenance respiration, compared to fine roots growing in N-poor soil (Ryan 1991). As a consequence, plants growing on N-rich soil should allocate proportionately more carbohydrate to fine roots, which have a greater maintenance "cost." Although we know little about the influence of tissue N concentration on the life span of fine roots, leaf life span is inversely related to photosynthetic rate,

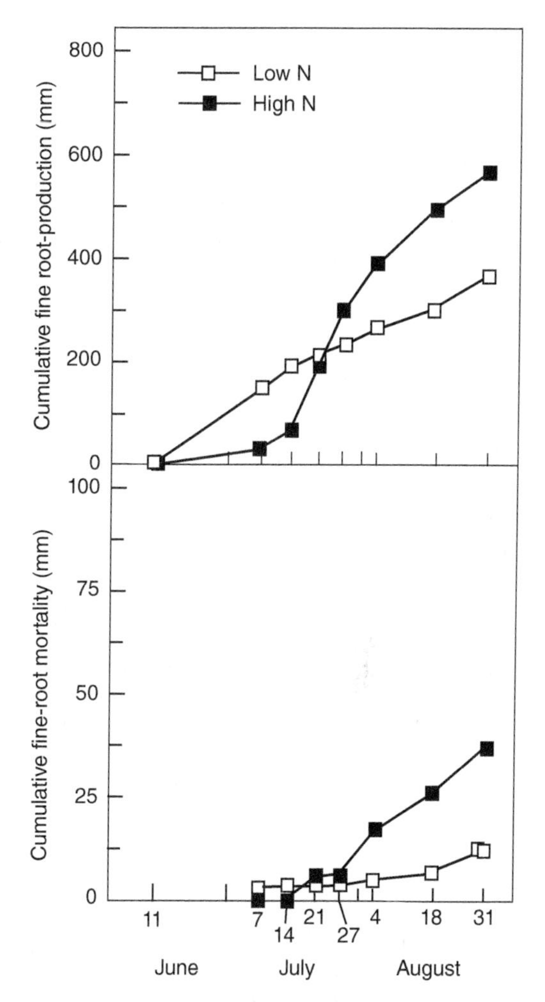

FIGURE 18.10 The influence of soil N availability on fine-root production and mortality of a black poplar hybrid. Note that both fine-root production and mortality increase with greater levels of soil N availability. Although production increases in the high-N soil, the standing crop of fine roots is kept small by rapid rates of fine-root mortality. Such a pattern suggests that the total amount of carbohydrate allocated to fine roots increases in N-rich soil. *Source:* After Pregitzer et al. (1995) / John Wiley & Sons.

tissue N concentration, and metabolic activity (Larcher 1969; Chabot and Hicks 1982; Reich et al. 1991, 1992). If the same relationship holds for fine roots, then one would expect fine-root life span to decline as soil N availability increases. Taken together, these results support the view that the allocation of C to fine-root growth and maintenance increases in N-rich soil, but greater rates of fine-root mortality (i.e., decreased life span) produce a proportionately smaller mass of fine roots.

We still have a great deal to learn about the C allocation patterns and fine-root dynamics of forest trees. Certainly, the processes controlling fine-root production and life span are of great importance for understanding the C allocation patterns of plants. Without a clear understanding of plant C allocation, we cannot fully understand how changes in light and soil resources influence the C balance of ecosystems. In the following section, we will build on the processes controlling the C balance of plants to understand how climate, soil resources, and disturbance influence the C balance of entire ecosystems.

CARBON BALANCE OF ECOSYSTEMS

The physiological processes by which plants photosynthetically fix CO_2 from the atmosphere and how plants allocate photosynthate to growth, storage and defense are important at a global scale. For example, approximately 810 Pg of C presently reside in the Earth's atmosphere, but 2000 Pg of C are stored in the vegetation and soil of Earth's terrestrial ecosystems, the majority

(60%) of which occurs in forests. Because of this fact, the establishment and management of forests have been looked to as a means to decrease the accumulation of anthropogenic CO_2 in the atmosphere. However, to do this effectively, we need to understand the physiological processes by which CO_2 fixed by forest ecosystems is stored within them and the processes that release this stored C back to the atmosphere. To do this, we must consider the activities of organisms other than trees to understand the flow of C into, within, and out of forest ecosystems. For example, most soil microorganisms use dead plant tissue (i.e., leaves, branches, stems, and roots) as an energy source. They incorporate a portion of the plant-derived C into microbial cells, release microbially derived organic compounds that aid in humus formation (i.e., soil C storage), and return the remainder to the atmosphere as CO_2 during microbial respiration. Leaf feeding (phytophagous) insects annually consume from 5 to 30% of the canopy leaves in most forests. Although such a loss of photosynthetic surface area often has little influence on tree growth (Mattson and Addy 1975), it represents a transfer of C from one ecosystem component to another. That is, C once residing in plants has been transferred to insects, portions of which are lost during their respiration and in excrement. In the sections that follow, we consider how climate, soil nutrient availability, and disturbance influence the processes controlling the cycling and storage of C in terrestrial ecosystems. Collectively, these processes are of global importance, due to the large amount of C stored in the Earth's forest ecosystems. We will later draw on this information to understand the cycling and storage of plant nutrients within forests, a topic covered in Chapter 19.

BIOMASS AND PRODUCTIVITY OF FOREST ECOSYSTEMS

Several concepts are important in understanding the C balance of ecosystems, whether they are dominated by forest trees, prairie grasses, desert shrubs, or arctic tundra. **Biomass** is the dry mass of living organisms and dead organic matter contained in a defined area, usually one square meter or a hectare ($g\,m^{-2}$ or $kg\,ha^{-1}$, respectively). In some instances, biomass is reported in its C or energy equivalent (i.e., $g\,C\,m^{-2}$ or $cal\,m^{-2}$), a necessary tool for understanding the flow of C (or energy) within ecosystems. In forest ecosystems, biomass is located in 5 major pools: (i) the aboveground and belowground tissues of overstory and understory plants, (ii) woody debris consisting of dead, fallen tree stems, (iii) forest floor, (iv) mineral soil, and (v) the tissues of heterotrophic organisms (decomposers and consumers). Although biomass in woody debris, forest floor, and mineral soil is composed of dead organic matter, it is convention to consider these compartments when discussing the distribution of biomass within forest or other terrestrial ecosystems.

Compare the amount and distribution of biomass among boreal, temperate, and tropical forest ecosystems in Table 18.5. Notice that the total biomass of a tropical rain forest is almost five times that of a black-spruce-dominated boreal forest; the temperate coniferous and deciduous forests lay between these two extremes. Differences in the proportion of total biomass located in overstory trees, forest floor, and mineral soil also occur among the forests in Table 18.5. For example, forest floor and mineral soil compose the largest pool of biomass in the boreal forest, whereas the majority of biomass in the tropical forest resides in overstory trees. The temperate coniferous and deciduous forests in Table 18.5 also differ in biomass distribution, wherein overstory trees compose a larger proportion of total biomass in the coniferous forest.

Caution should be used in interpreting Table 18.5, because there is substantial variation in the amount and distribution of biomass within forests from boreal, temperate, and tropical regions. For example, overstory biomass in tropical forests ranges from $10\,Mg\,ha^{-1}$ in dry forests to $540\,Mg\,ha^{-1}$ in moist forests (Brown and Lugo 1982), similar to the range of overstory biomass found in temperate deciduous (2–578 $Mg\,ha^{-1}$; Grier et al. 1989) and coniferous (5–810 $Mg\,ha^{-1}$; Grier et al. 1989) forests.

Table 18.5 The distribution of biomass in selected boreal, temperate, and tropical forest ecosystems.

	Boreal	Temperate		Wet tropical
	Alaska	Washington	New Hampshire	Amazon
Location	USA	USA	USA	Brazil
Overstory				
Dominant species	black spruce	Douglas-fir	sugar maple–beech	mixed species[a]
Age (years)	95	60	55	mature
Biomass pools		(Mg ha^{-1})		
Overstory	50	410	165	990
Woody debris	–	9	29	18
Forest floor	76	15	48	7
Mineral soil	152	119	173	250
Heterotroph	<1	<1	<4	<1
Total	278	553	419	1265

[a] The overstory of this tropical rain forest is composed of 50 species, but the total number of plant species exceeds 600 species per hectare. Species in the Leguminosae, Euphorbiaceae, and Sapotaceae families compose the majority of plant biomass.

Source: Data have been summarized from Bormann and Likens (1979), Harmon et al. (1990), Klinge et al. (1975), Schowalter (1989), Van Cleve et al. (1983), and Zak et al. (1994).

Primary production is the annual growth of all living plants within an ecosystem, and it can be subdivided into several basic components. **Gross primary production** (GPP) is the total amount of C fixed in an ecosystem during the process of photosynthesis (Figure 18.11). In all ways, GPP is the ecosystem equivalent of the gross photosynthesis of an individual plant; it is the summed gross photosynthesis of all plants in an ecosystem. As with an individual plant, a portion of the total C fixed by an ecosystem is lost to the atmosphere via leaf dark respiration and the construction and maintenance respiration of non-photosynthetic plant tissue (Figure 18.11). The difference between gross primary production and plant respiration is termed **net primary production** ($NPP = GPP - R_A$, where R_A = total plant respiration consisting of leaf dark respiration, construction respiration = R_C, and maintenance respiration = R_M). Gross and net primary productivity can vary by an order of magnitude in forest and grassland ecosystems. However, the proportion of GPP lost via plant respiration (R_A) is surprisingly constant among the ecosystems in Table 18.6, varying from 50 to 60%. The majority of that loss arises from the maintenance respiration (R_M) of non-photosynthetic tissues (75–88% of R_A), whereas construction respiration (R_C) represents a much smaller proportion of R_A (12–24%). Much as the rate of net photosynthesis controls the C gain of an individual plant, the relationship between GPP and R_A directly influences NPP, and hence the rate at which C accumulates within or is lost from terrestrial ecosystems.

The growth and metabolism of decomposing organisms in soil and that of animals foraging within forest ecosystems are fueled by organic matter produced through NPP (Figure 18.11). The proportion of NPP remaining after accounting for losses to herbivory and litterfall is the **live biomass accumulation** of plants (Figure 18.11), and it can be either a positive or negative value. For example, a forest that has suffered a devastating windstorm, an insect outbreak, or has been harvested can experience a negative live biomass accumulation.

Net secondary production is the annual increase in the biomass of all organisms obtaining their energy from net primary production, including decomposers and consumers (i.e., heterotrophic organisms). Realize that not all organic matter assimilated by decomposers or consumers is

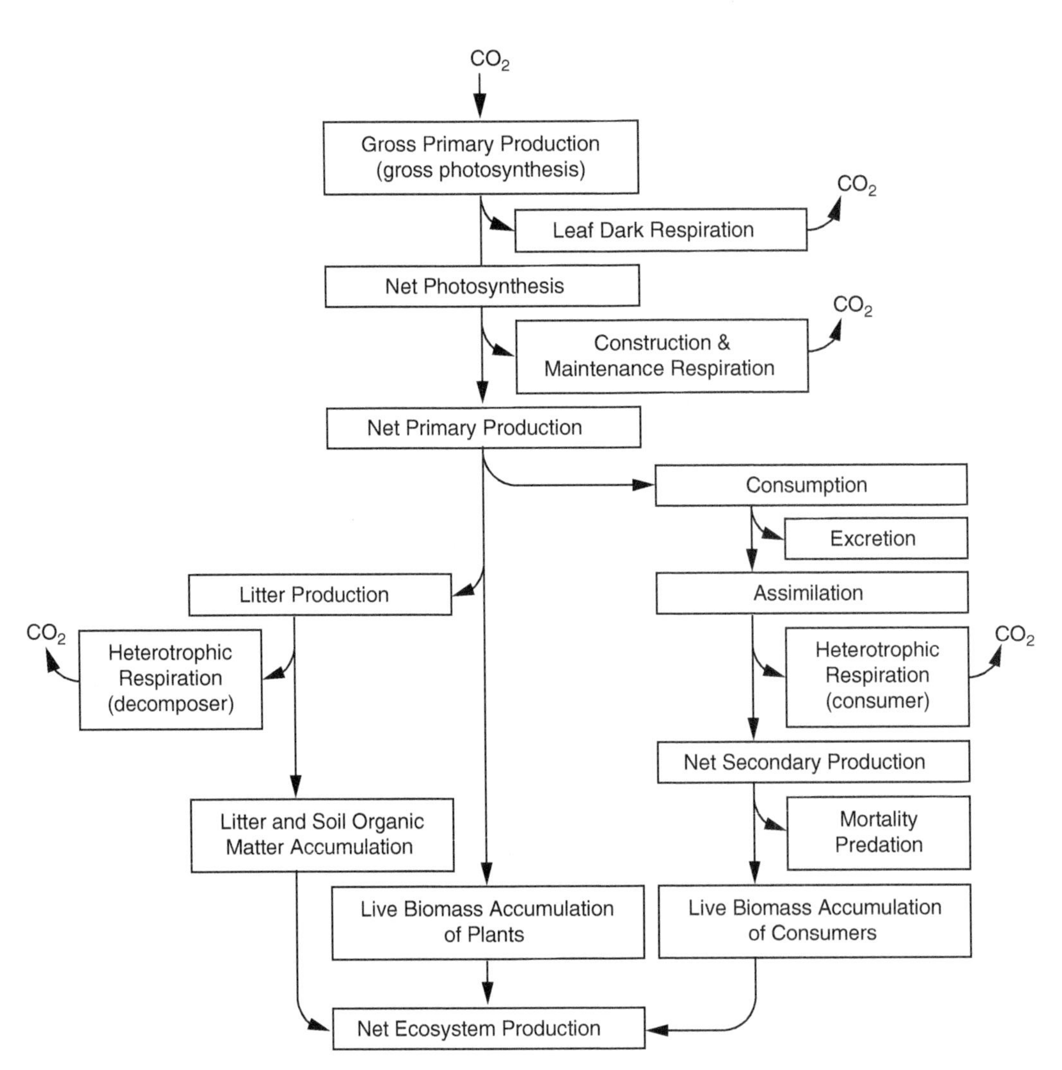

FIGURE 18.11 The processes controlling the C balance of terrestrial ecosystems. Gross primary production is the process responsible for C entering terrestrial ecosystems, whereas the construction and maintenance respiration of plants and animals is responsible for returning the C fixed through photosynthesis back to the atmosphere as CO_2. Also note that decomposing organisms in soil ultimately return C fixed by plants to the atmosphere. *Source:* After Aber and Melillo (1991) / Saunders College Publishing. Figure from *Terrestrial Ecosystems* by John D. Aber and Jerry M. Melillo, copyright © 1991 by Saunders College Publishing, reprinted with permission of the publisher.

Table 18.6 Estimates of total plant (R_A), construction (R_C), and maintenance (R_M) respiration in relationship to gross primary production (GPP) and net primary productivity (NPP) in forest and grassland ecosystems.

Ecosystem	Plant biomass (Mg C ha^{-1})	GPP	NPP	R_A	R_C	R_M	R_A/GPP	RM/RA
		Mg C ha^{-1} y^{-1}					Percent	
Forest								
Deciduous	88	21.7	7.3	14.4	1.8	12.6	66	88
Oak–pine	71	12.8	6.0	6.8	1.5	5.3	53	78
Grassland								
Tallgrass	8	9.5	3.7	5.8	0.9	4.9	61	84
Shortgrass	–	1.5	0.7	0.8	0.2	0.6	53	75

Recall that NPP = GPP–R_A and $R_A = R_C + R_M$; estimates of GPP do not include leaf dark respiration.
Source: After Ryan (1991) / John Wiley & Sons. Reprinted with permission of the Ecological Society of America.

used to build biomass; portions are lost during heterotrophic respiration (R_H) and in excrement (Figure 18.11). The live biomass of consumers accumulates only when the rate of net secondary production is greater than the rate at which consumer biomass is lost via mortality and predation.

Litter production, both aboveground and belowground, and its decomposition directly control organic matter or biomass accumulation in the forest floor and mineral soil. Organic matter accumulates in these ecosystems pools when its rate of production (leaves and roots) exceeds the rate at which it is decomposed by soil microorganisms and returned to the atmosphere as CO_2 (Figure 18.11). Clearly, changes in either the rate of litter production or decomposition have the potential to alter the accumulation of organic matter in soil, a topic we will consider later in this section.

The annual rate of biomass accumulation in live plants, live animals, and soil organic matter is termed **net ecosystem production** (NEP), and it represents the summed change in all ecosystem biomass pools (Figure 18.10), and it is defined in the following expression:

$$NEP = GPP - (R_A + R_H).$$

Net ecosystem production, or the C gain of ecosystems, can be either positive or negative depending on the change in the biomass of live plants, live animals, and soil organic matter. Biomass accumulates within terrestrial ecosystems only when the amount of C fixed during GPP exceeds the rate at which C is lost via the respiration of plants (R_A) and heterotrophic organisms (R_H). That is, NEP > 0 when GPP > ($R_A + R_H$).

In forests, the biomass and respiration of organisms that consume plant material and their predators are relatively small compared to that of the overstory trees, forest floor, and mineral soil. Moreover, the amount of NPP consumed by herbivores, primarily insects, in forests is relatively small (3%), compared to grasslands, such as the Serengeti, in which large herds of grazing animals (i.e., wildebeest, zebra, antelope) annually consume a significant proportion of aboveground NPP (ANPP, 66%). Consequently, NEP in forests is largely controlled by the difference between NPP and the rate at which CO_2 is returned to the atmosphere during organic matter decomposition (R_H). Any forest disturbance that reduces or eliminates the live biomass accumulation of plants and increases rates of soil organic matter decomposition could lead to a net loss of C from an ecosystem. Under such a condition, NEP is negative and total ecosystem biomass declines. Ecosystems in such a state are a source of CO_2 to the atmosphere, rather than a sink, which has global implications for the amount of C stored by terrestrial ecosystems.

Understanding the C balance of forest ecosystems has clear management implications—any management practice that tilts the balance toward negative NEP has the potential to alter the amount of C sequestered by forest ecosystems. Nevertheless, it is important to realize that the C balance of forest ecosystems is dynamic and can rapidly become positive following destruction of plants by harvest, fire, disease, or insect outbreak.

MEASUREMENT OF BIOMASS AND PRODUCTIVITY

Ecologists have long been interested in understanding how and why the productivity of ecosystems differs from one geographic location to another. Clearly, biomass and productivity differ markedly among forest, agricultural, grassland, or desert ecosystems. Nonetheless, measuring the biomass and productivity of an ecosystem is not an easy task, particularly in forests in which tremendous amounts of biomass reside in the aboveground and belowground portions of overstory trees.

Three approaches provide ecologists with insight into the biomass and productivity of forest ecosystems. One frequently used approach is based on relationships that exist between the dimensions of forest trees (diameter and height) and their weight. These species-specific, or allometric, relationships are typically expressed in the form of a mathematical equation in which the diameter at breast height (DBH) and height of a particular tree is used to estimate the weight of leaves, branches, stem, and sometimes roots. Development of **allometric biomass equations** is a very labor-intensive task in which individual trees of a particular species which span a range of DBH and height are harvested, divided into components, dried, and weighed. Allometric equations have been developed for most North American forest trees and a significant number of shrub species (Whittaker 1966; Tritton and Hornbeck 1982; Smith and Brand 1983). Because of the great difficulty of collecting tree roots from the soil, most allometric biomass equations are used to predict the aboveground biomass and productivity of forest ecosystems.

With the use of allometric biomass equations, aboveground net primary productivity can be estimated with the following expression:

$$\text{ANPP} = \Delta B + \text{L}$$

In this equation, the annual change in biomass (ΔB) of all aboveground plant tissue plus the biomass of senesced leaves, seeds, and flowers (L) produced in that year is equal to the aboveground net primary productivity of the ecosystem. The biomass of senesced leaves, seeds, and flowers is estimated by collecting this material in a fixed area as they are shed by overstory trees. Estimates of root biomass and root litter are often unavailable and hence are often absent from estimates of net primary productivity.

Aboveground net primary productivity is estimated by making measurements of tree or shrub heights and diameters in successive years, computing the biomass for each year using allometric equations, and subtracting the biomass estimate in year 1 from that estimated in year 2. Estimates of net aboveground primary productivity obtained using this technique are readily available for a wide array of forest ecosystems distributed throughout North America and Europe (see Reichle 1981; Grier et al. 1989).

A second approach estimates net primary productivity by directly measuring the physiological processes controlling the C balance of ecosystems. This approach entails measuring photosynthesis and respiration for representative ecosystem components (i.e., leaves, branches, stems, soil) and then extrapolating their CO_2 flux to the entire ecosystem. Ecologists have used this approach to obtain estimates of gross primary productivity, ecosystem respiration, and net primary productivity for a late-successional oak–pine ecosystem at the Brookhaven National Laboratory on Long Island, New York (Botkin et al. 1970; Woodwell and Botkin 1970).

In this study, leaves, branches, stems, and the soil surface were enclosed within small chambers that were connected to gas analyzers which measured CO_2 concentration over time (Figure 18.12). Because the entire forest canopy cannot be practically enclosed within a chamber, estimates of photosynthesis and dark respiration from a known quantity of leaves must be scaled to the entire forest canopy. Care must be taken to obtain estimates of photosynthesis and dark respiration for sun and shade leaves, thereby accounting for differences in photosynthesis from the top to the bottom of the canopy. Similarly, estimates of branch, stem, and soil respiration must be extrapolated to a flux from the entire ecosystem. Given that rates of CO_2 exchange from a small portion of the entire ecosystem are actually measured, great care must be taken to ensure that data are collected on representative samples of each ecosystem component. Otherwise, estimates could be greatly in error when they are extrapolated to the entire ecosystem.

Using this approach, gross primary productivity at the Brookhaven forest was estimated to be 14.0 megagrams of C per hectare (Mg C ha^{-1}) during the growing season, whereas leaf dark respiration (4.0 Mg C ha^{-1}), stem respiration (5.5 Mg C ha^{-1}), and soil respiration (4.5 Mg C ha^{-1}) totaled 14.0 Mg C ha^{-1}. These results indicate that net ecosystem productivity was 0 Mg C ha^{-1}, suggesting that this relatively late-successional forest is neither gaining nor losing C (i.e., GPP = $R_A + R_H$).

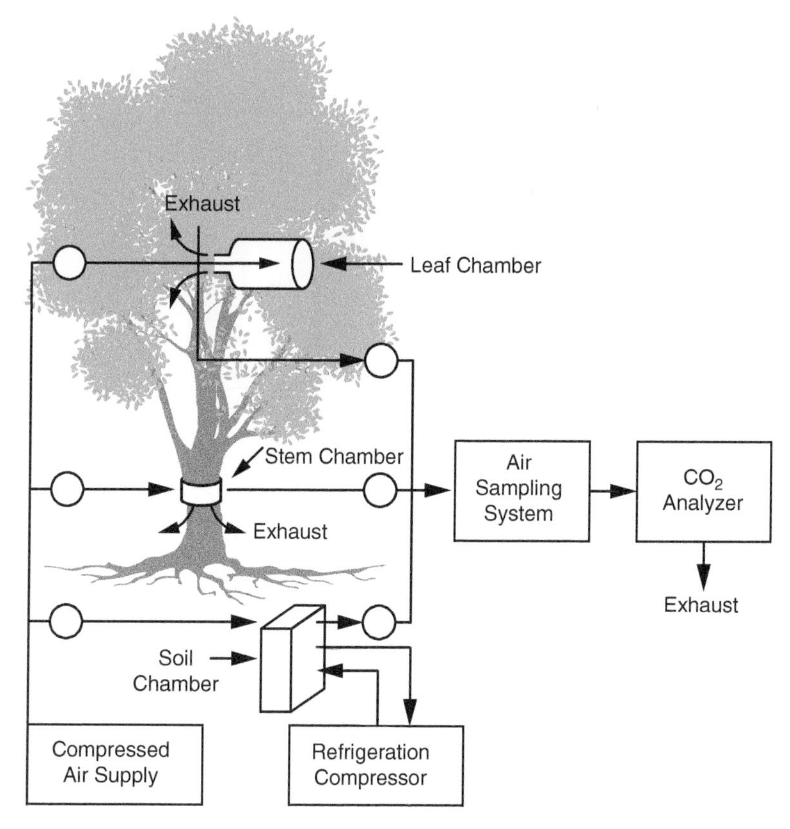

FIGURE 18.12 A CO_2 exchange system used to measure the C balance of forest ecosystems based on quantifying the physiological processes responsible for C fixation and loss. *Source:* After Woodwell and Botkin (1970) / Springer Nature. Reprinted from *Analysis of Temperate Forest Ecosystems,* © by Springer-Verlag Berlin, Heidelberg 1970. Reprinted with permission of Springer-Verlag New York, Inc.

One drawback of the small chamber approach is that, despite great effort, estimates of NPP are often inaccurate because root respiration cannot be separated from the respiration of soil microorganisms (i.e., soil respiration = root maintenance, root construction, and microbial respiration). This currently remains a challenge for ecologists because it places uncertainty on the actual belowground C budget of forest ecosystems, as well as other terrestrial ecosystems across the Earth.

Recently, micrometeorological techniques have provided a third approach to study the exchange of CO_2 between terrestrial ecosystems and the atmosphere. The **eddy covariance** method quantifies the net exchange of CO_2 by measuring vertical gradients of CO_2 from the forest floor to above the canopy. In combination with wind speed, wind direction, and temperature, the vertical CO_2 profile can be used to estimate the flux of C into and out of ecosystems on hourly, daily, and yearly time intervals (Figure 18.13). This technique has been used to study the C balance of a wide range of terrestrial and aquatic ecosystems distributed around the Earth, including forests, savanna, grassland, desert, tundra, and lakes. This micro-meteorological approach has enabled scientists to gain understanding of the degree to which the Earth's ecosystems function as a sink or source of CO_2 to the atmosphere, thereby improving our understanding of the Earth's C balance. In a second-growth northern hardwood ecosystem at the Harvard Forest in Massachusetts, for example, net ecosystem production estimated using the eddy correlation was 3.7 Mg C ha^{-1} y^{-1}. Ecosystem respiration, calculated from the CO_2 flux during nighttime

(a)

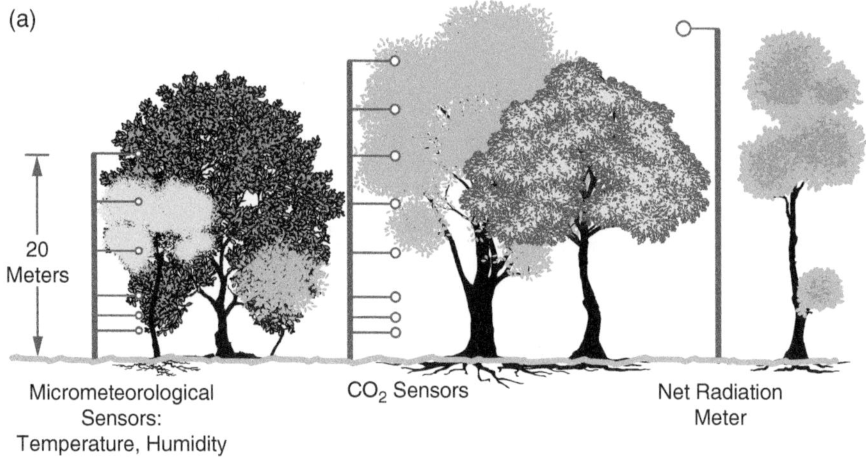

(b)

FIGURE 18.13 Instrumentation of a forest ecosystem to measure the net exchange of C with the atmosphere. (a) The eddy covariance approach uses the vertical gradient of CO_2 from above the canopy to the forest floor to estimate the net uptake and release of CO_2 by terrestrial ecosystems. (b) Photograph of the eddy covariance tower located at the Harvard Forest in Massachusetts, United States. *Source:* J. William Munger.

hours, was 7.4 Mg C ha^{-1} y^{-1}, suggesting that gross primary productivity was 11.1 Mg C ha^{-1} y^{-1} (Wofsy et al. 1993). In contrast to the late-successional forest at Brookhaven, New York, this second-growth northern hardwood ecosystem is accumulating C based on the fact that NEP is positive, that is, $GPP > R_A + R_H$.

It is important to realize that the eddy correlation technique also suffers from the fact that soil respiration cannot be broken into separate contributions of plant roots and soil microorganisms, making it difficult to estimate the proportion of net ecosystem production and net primary production occurring below the soil surface. One cannot simply accomplish this task by separating roots from soil, and then measuring the respiration of each component; root respiration is higher in the presence of rhizosphere microorganisms, and the respiration of rhizosphere microorganisms is higher in the presence of roots. This reciprocal relationship confounds our ability to discern the separate contribution of roots and microorganisms to the flux of CO_2 from soil. Globally, this flux is 8 times greater than the annual combustion of fossil fuels by humans, making it a significant component of the Earth's C balance. Clearly, the greatest uncertainty associated with estimating the C balance of ecosystems, and consequently the Earth, occurs in measuring rates of root production and the amounts of CO_2 separately respired by plant roots and soil microorganisms. This remains a frontier in the study of ecosystem C balance.

CLIMATE AND PRODUCTIVITY

Although there is still much to learn about the C budget of forests and other terrestrial ecosystems, striking patterns and predictable differences in ANPP occur over the Earth's surface. Geographic differences in ANPP are not random and are strongly linked to patterns of climate. It is no surprise that the productivity of desert or shrub steppe ecosystems is much less than that of prairie or forests.

Differences in temperature and precipitation over the Earth's surface influence the distribution of ecosystems and the processes controlling ANPP. Figure 18.14 summarizes the relationship between temperature, precipitation, and the ANPP of a wide array of terrestrial ecosystems, including tundra, deserts, grasslands, and forests from tropical, temperate, and boreal regions. Aboveground net primary productivity is relatively low in cold (Figure 18.14a) and dry (Figure 18.14b) climates, and it rapidly rises as both temperature and precipitation increase. This relationship is

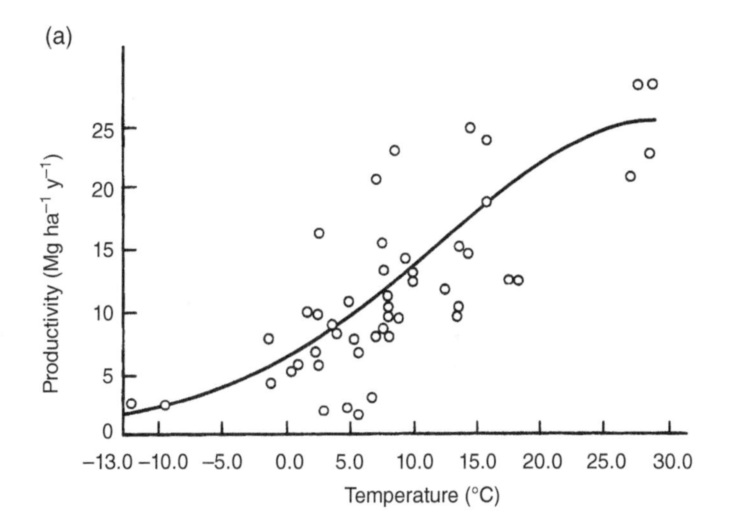

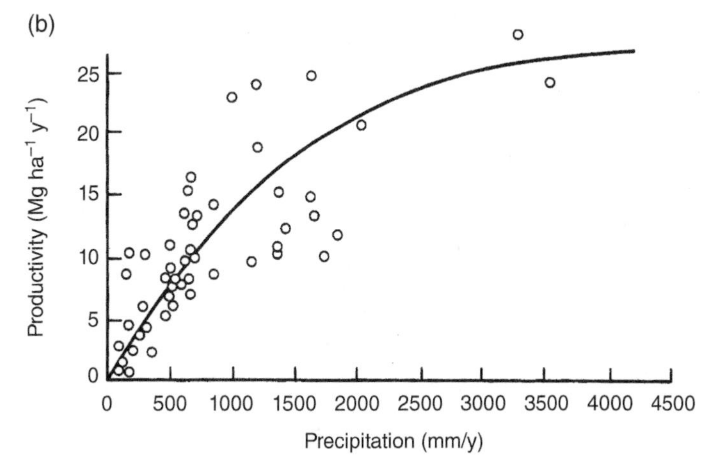

FIGURE 18.14 (a) Mean annual temperature, (b) mean annual precipitation, and the aboveground net primary productivity (ANPP) of terrestrial ecosystems. Data used to construct these relationships were collected from a wide array of terrestrial ecosystems including deserts, grasslands, forests, and tundra. Increasing temperature and precipitation generally result in greater amounts of ANPP. *Source:* After Lieth (1973) / Springer Nature.

striking, given the wide diversity of plants and ecosystems used to compile this relationship. Why are global-scale patterns of ANPP so closely related to climatic factors, such as temperature and precipitation?

The connection between climate and ANPP arises due to a fundamental trade-off during photosynthesis—plants lose water when they open their stomates to fix CO_2 from the atmosphere. At a global scale, the amount of CO_2 fixed by plants and terrestrial ecosystems is limited by the quantity of water available for transpiration. Actual evapotranspiration (AET) is an ecologically important climatic attribute that incorporates precipitation, temperature, and the water-holding capacity of soil to estimate the actual amount of water lost to the atmosphere from evaporation (soil surface) and plant transpiration. AET often is compared with potential evapotranspiration (PET), which estimates the amount of evapotranspiration that would occur if the supply of soil water was unlimited. In regions where temperatures are high and precipitation is infrequent (i.e., deserts), PET greatly exceeds AET because low amounts of precipitation limit evapotranspiration even when ample energy is available to evaporate water from the surfaces of soil and leaves.

Figure 18.15 illustrates the relationship between the ANPP of a wide array of terrestrial ecosystems from different climatic regimes and AET. Desert ecosystems occur in regions where little water is available for evapotranspiration and hence can be found at the low end of the AET axis in Figure 18.15. Forests, by contrast, occur where annual amounts of precipitation exceed AET (upper end of the axis); a situation in which the availability of water does not often limit evapotranspiration. Notice that a positive relation exists between ANPP and AET—an increase in the amount of water returned to the atmosphere via AET results in greater ANPP. Again, this relationship exists because plants must open their stomates to fix CO_2 from the atmosphere and simultaneously lose water during transpiration. At a global scale, the amount of water available for transpiration sets the upper bounds on plant C gain, and hence is responsible for differences in ANPP among ecosystems from different climatic regions.

Mountainous regions exhibit dramatic temperature and precipitation gradients that create differences in ecosystem productivity equivalent to those depicted in Figure 18.15, albeit at a much smaller spatial scale. The Coastal and Cascade Mountain ranges in the Pacific Northwest help create marked changes in climate, vegetation, and productivity across relatively small distances

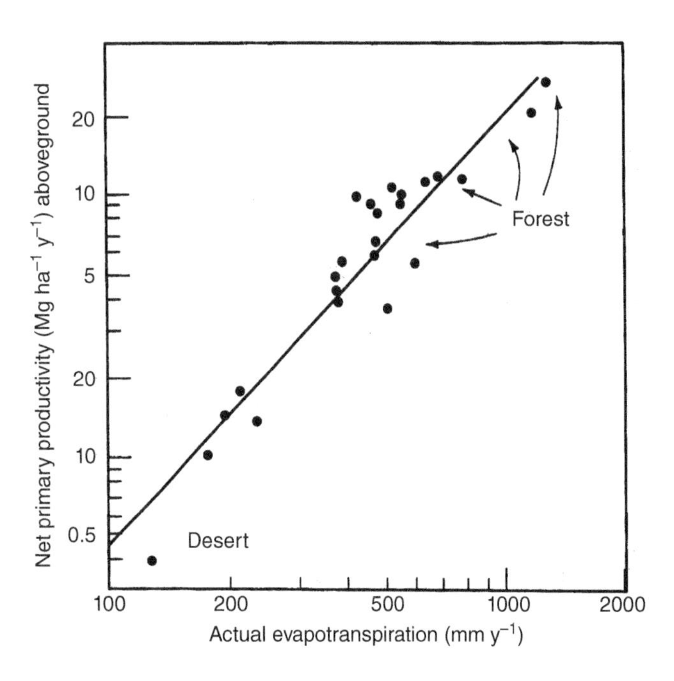

FIGURE 18.15 Actual evapotranspiration (AET) and aboveground net primary productivity (ANPP) for alpine, boreal, temperate, and tropical terrestrial ecosystems. The positive relationship between AET and ANPP emphasizes that climate factors impose important constraints on global patterns of ANPP. *Source:* After Rosenzweig (1968) / University of Chicago Press. Reprinted with permission from *The American Naturalist*, © 1968 by the University of Chicago. Reprinted with permission of the University of Chicago Press.

(Chapter 8). At low elevations in the Coastal Range (200 m) near the Pacific Ocean, relatively warm and wet (246 cm y^{-1}) conditions give rise to tall-statured forests dominated by Sitka spruce and western hemlock (Franklin and Dyrness 1973). Inland lies the Cascade Range where relatively warmer and drier conditions at lower elevation give rise to forests ecosystems dominated by Douglas-fir and grand fir. Moving eastward and up in elevation, orographic cooling produces progressively colder and wetter climatic conditions in which Douglas-fir, mountain hemlock, and Pacific silver fir dominate the forest overstory. Rain shadow conditions occur on the east slope of the Cascade Range, creating progressively warmer and drier climates as one moves further eastward and down in elevation. Ponderosa pine ecosystems occur on the east slope (870 m) of the Cascade Range in regions receiving 40 cm y^{-1} of precipitation; western juniper (25 cm y^{-1}) and sagebrush (20 cm y^{-1}) occur in the drier climates at lower elevations further on the east slope.

Along this relatively broad climatic gradient in the Pacific Northwest, ANPP ranges from 0.3 Mg ha^{-1} y^{-1} in dry, sagebrush ecosystems to 15 Mg ha^{-1} y^{-1} in wet, coastal western hemlock ecosystem (Gholz 1982), a range of values equivalent to that in Figure 18.15. In this mountainous region, patterns of ANPP are strongly related to differences in temperature, precipitation, and soil water storage. Gholz (1982) combined these physical site factors to estimate the water availability in Pacific Northwest ecosystems using the following expression:

$$WB = P - E + SWC.$$

In this equation, P is monthly precipitation (cm), E is monthly potential evaporation (cm), SWC is the average monthly soil water content (cm), and WB is the monthly water balance. Aboveground annual net primary productivity is relatively low in juniper and sagebrush ecosystems in which potential transpiration greatly exceeds precipitation and soil water content (i.e., a negative water balance; Figure 18.16). As either precipitation increases or evaporation decreases, the annual water balance in this region becomes less negative, which relates to a linear increase in ANPP (Figure 18.16). Again, this reinforces the idea that water availability sets a fundamental limit on ecosystem productivity by controlling the amount of CO_2 fixed through photosynthesis.

Although evapotranspiration is strongly related to global gradients of productivity, it is not the sole determinant of ecosystem productivity. In Figure 18.15, study the variation in ANPP at the point where AET equals 500 mm; note that ANPP ranges from 3 to 10 Mg ha^{-1} y^{-1}at this point in Figure 18.15. Such variation is not uncommon in temperate forests in which ANPP differs by a factor of 10 (Grier et al. 1989), even though AET varies to a lesser extent (500–700 mm). Also realize that ecosystems with very low productivity can be found in warm, tropical regions receiving relatively large amounts of precipitation, something that would not be predicted from the relationships depicted in Figures 18.14 and 18.15. What factors cause productivity to differ so widely among ecosystems within a particular climatic region? In the following sections, we explore the extent to which soil properties, species composition, and stage of ecosystem development modify NPP and NEP within a particular climatic region, somewhat independently of the amount of water available to plants.

SOIL PROPERTIES, FOREST BIOMASS, AND ANPP

Within a particular climatic region, soil parent material can exert a substantial influence on the productivity of forest ecosystems through its influence on soil water-holding capacity. Recall that sandy-textured soils hold relatively lower amounts of plant available water compared to soils of loam or clay loam texture (Chapter 9). In the Lake States, glacial activity (10 000–14 000 years BP) created local landscapes in which parent materials range from coarse outwash sands to relatively fine-textured glacial till; such changes occur over relatively short distances, for example, 100 m to

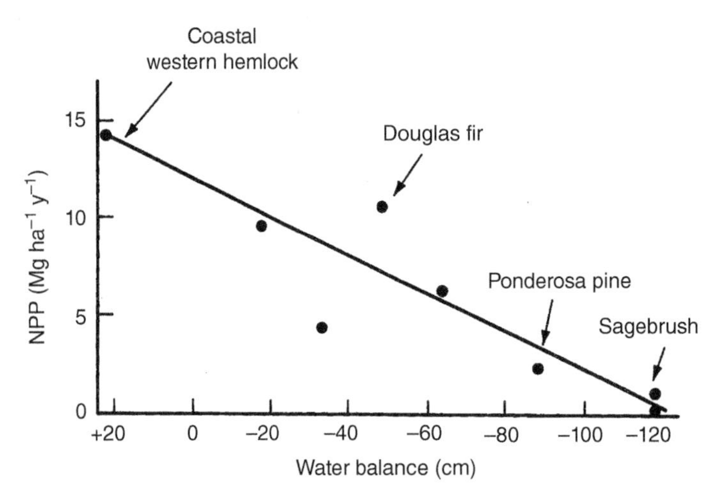

FIGURE 18.16 The water balance and aboveground net primary productivity (ANPP) of ecosystems in the Pacific Northwest. The relationship between water balance and ANPP emphasizes the idea that increases in water availability along climatic gradients in mountainous regions relate to an increase in net primary productivity. *Source:* After Gholz (1982) / John Wiley & Sons. Reprinted with permission of the Ecological Society of America.

1 km. Because these contrasting parent materials occur within a particular climatic regime, they present the opportunity to examine how landform-mediated differences in soil parent material (i.e., plant available water) modify the productivity of forest ecosystems.

One example occurs in northern Lower Michigan, where productive northern hardwood forests growing on loamy glacial till occur in the same landscape as lower productivity oak-dominated (black and white) ecosystems of the sandy outwash plains. Overstory biomass (208 Mg ha^{-1}) and productivity (3.2 Mg ha^{-1} y^{-1}) of a northern hardwood ecosystem are twice that of xeric oak ecosystem (105 Mg ha^{-1} and 1.5 Mg ha^{-1} y^{-1}; Host et al. 1988). Clearly, soil texture can greatly modify the influence of regional climate on the productivity of forest ecosystems.

Differences in soil parent material also give rise to differences in soil N availability, which, at a regional climatic scale, controls the productivity of many boreal, temperate, and a some tropical forests. Unlike agricultural ecosystems that receive N additions from chemical fertilizer, soil N availability in forests and other less-intensively managed ecosystems is controlled by the activity of saprotrophic soil microorganisms. Most soil microorganisms obtain energy and essential nutrients during the decomposition of plant litter (senesced leaves and fine roots). During this process, only a portion of the N contained in litter is used by soil microorganisms to maintain or build biomass. The remainder is released into soil solution as ammonium (NH_4^+), where it can be assimilated by plant roots (i.e., N mineralization; see Chapter 19 for details). In the Upper Lake States, variation in ANPP among forest ecosystems is related to differences in soil N availability, wherein relatively high rates of ANPP (overstory only) occur in soils with high rates of N mineralization (Figure 18.17). It is important to note that fine-textured soils, which hold relatively large amounts of plant available water, also release the greatest amounts of N for plant growth. The relationship between ANPP and soil N availability is not unique to Lake States forests, because ANPP also is related to soil N availability in grassland (Schimel et al. 1985), shrub steppe (Burke 1989), and desert (Fisher et al. 1987) ecosystems.

Soil texture can modify the influence of climate on ANPP by controlling the amount of water and N that is available for plant growth. In other words, climate sets the regional potential for productivity through patterns of temperature and precipitation, but that potential is constrained, or realized, by differences in soil texture which in turn control the availability of water and N within the local landscape.

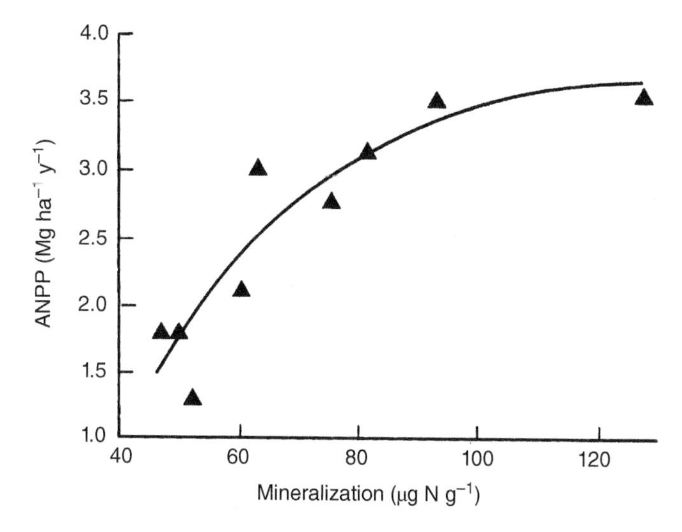

FIGURE 18.17 In the Lake States region, aboveground productivity of forest ecosystems is well correlated with the rate at which soil microorganisms release N from soil organic matter. Illustrated in this figure is the ANPP (overstory only) and N mineralization rate of nine upland forest ecosystems in northern Lower Michigan. The positive relationship between these variables suggests that soil N availability imposes an important constraint on ecosystem productivity within the climate of the Great Lakes region. *Source:* After Zak et al. (1989) / Canadian Science Publishing. Reprinted with permission of National Research Council Canada.

BIOMASS ACCUMULATION DURING ECOSYSTEM DEVELOPMENT

Following an event that destroys vegetation, living plant biomass increases over time from near zero to that of a mature forest ecosystem (Figure 18.18). Concomitant changes also occur in the activities of decomposing organisms and in the biomass stored in forest floor and mineral soil. The rate at which biomass accumulates in forest ecosystems is controlled by the amount of CO_2 fixed through photosynthesis (i.e., GPP) and the amount of C returned to the atmosphere via maintenance and construction respiration of plants and saprotrophic soil microorganisms (i.e., $R_A + R_H$). These processes are illustrated for an idealized forest following the destruction of living plant biomass in Figure 18.18.

In the newly established forest, living plant biomass ($Mg\,ha^{-1}$) is nearly zero and it rapidly increases during the first 20–25 years of ecosystem development. After this period of time, the increase in living plant biomass begins to slow and it eventually attains a maximum value, which is near year 50 in Figure 18.18b. The biomass of leaves initially increases and attains a maximum value after approximately 20 years (Figure 18.18b), the point at which canopy closure occurs. Recall that net canopy photosynthesis, and hence the amount of photosynthate available for plant growth and maintenance, is a product of leaf photosynthetic rate ($\mu mol\,m^{-2}\,s^{-1}$) and total canopy leaf area, as well as canopy geometry. Leaf biomass ($Mg\,ha^{-1}$) and area ($m^2\,ha^{-1}$) are well correlated in most forest trees, indicating that the amount of leaf area (or biomass) at canopy closure sets the upper limit on canopy photosynthesis and GPP. It follows that this relationship also sets a limit on the amount of C that can be allocated to growth and maintenance. Notice that total biomass (non-photosynthetic tissue) continues to increase well after leaf biomass has attained its maximum (Figure 18.18b). This means that net photosynthesis attains a maximum set by environmental conditions early in ecosystem development, but the further accumulation of non-photosynthetic biomass causes losses from respiration to increase. This dynamic causes NPP to slow over time, the ecological implications of which we discuss in the following paragraphs.

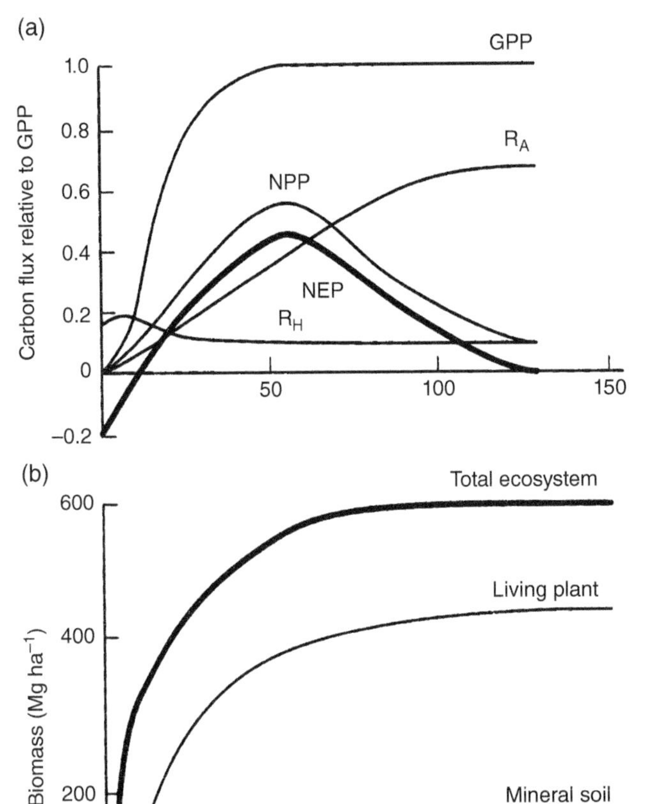

FIGURE 18.18 Changes in the processes controlling (a) the C balance of forest ecosystems and (b) the rate of biomass accumulation following a disturbance that destroys living plant biomass. Such a disturbance includes intensive harvesting of overstory trees, fire, or windstorm. Change in net secondary production is not illustrated and accounts for a relatively minor flux of C in forest ecosystems.

The pattern of total biomass accumulation in this forest ecosystem indicates the balance between GPP and $R_A + R_H$ changes during ecosystem development. In Figure 18.18a, GPP is relatively consistent after approximately 20 years, whereas R_A continues to increase for a longer period of time. This pattern suggests that an increasingly greater amount of fixed C is allocated to tissue maintenance, diminishing the amount allocated to the construction of new tissue. This is consistent with the increase in non-photosynthetic biomass (total) long after the biomass of photosynthetic tissue has reached its maximum at canopy closure. The initial increase and subsequent decline in NPP (i.e., $GPP - R_A$) reflects this fundamental trade-off: C allocated to maintenance cannot be allocated to growth. Clearly, the C balance of plants exerts direct control on the C balance of entire ecosystems.

Changes in forest floor biomass following a disturbance reflect a balance between litter production (i.e., abscised leaves, fine roots, twigs, reproductive structures) and the amount of litter consumed by decomposing organisms (Figure 18.18b). Early in ecosystem development (0–15 years), plant litter production is low, but activities of decomposing organisms are enhanced by the relatively warmer, moister conditions at the soil surface, resulting from the absence transpiration by overstory trees. Note that R_H increases during the first several years following disturbance.

Low litter production in combination with enhanced decomposition rates results in an initial decline in forest floor biomass. Biomass begins to accumulate in the forest floor when litter production exceeds decomposition, a situation that occurs between 20 and 50 years after disturbance in the example illustrated in Figure 18.17. The relatively constant biomass of the forest floor late in ecosystem development suggests that litter inputs to forest floor are balanced by rates of decomposition. That is, the forest floor is neither gaining nor losing biomass late in ecosystem development.

The amount of organic matter stored in mineral soil reflects the long-term (1000–1 000 000 years) balance between plant litter production and the rate at which decomposing organisms return it to the atmosphere as CO_2. Biomass or organic matter stored in mineral soil often changes little following a disturbance that destroys a forest overstory (Figure 18.18b; Bormann and Likens 1979). In some situations, the amount of organic matter lost from the forest floor and mineral soil is quickly regained by inputs from rapidly developing vegetation (Gholz and Fisher 1982). However, soil erosion following disturbance can result in a substantial reduction in soil organic matter, particularly in areas of high rainfall and steep topography.

Net ecosystem production measures the C balance of terrestrial ecosystems, and it can dramatically change during ecosystem development. Immediately following a disturbance, high rates of heterotrophic respiration (R_H) and low rates of NPP (GPP – R_A) result in a net flux of C from the ecosystem, as depicted in Figure 18.18a. This negative C balance is marked by a reduction in total ecosystem biomass, which results in a negative NEP. Recall that any disturbance that diminishes NPP and accelerates rates of organic matter decomposition (R_H) has the potential to reduce NEP below 0. In such a situation, ecosystems are sources of CO_2 to the atmosphere, rather than functioning as a sink. Additionally, total ecosystem biomass increases only when GPP exceeds the respiratory loss of C from plants and heterotrophic organisms, a situation occurring from 5 to 75 years following disturbance in our example. Note that NEP declines to 0 late in ecosystem development (130 years), as a result of the relationship $GPP = R_A + R_H$. Because the fixation and loss of C on an ecosystem basis are in balance, total ecosystem biomass reaches a relatively constant value late in ecosystem development (Figure 18.18b).

The generalized patterns illustrated in Figure 18.18 occur in all terrestrial ecosystems, albeit at different magnitudes and over different periods of time. After an initial decline, for example, total ecosystem biomass rapidly accumulates in young slash pine plantations in Florida (Figure 18.19; Gholz and Fisher 1982). Maximum leaf biomass and canopy closure occur after 5 years, which sets an upper limit on canopy leaf area and the amount of photosynthate available for growth and maintenance. Notice that total biomass continues to accumulate long after canopy closure, indicating that greater amounts of GPP are allocated to maintain the increasing amount of non-photosynthetic tissue (Figure 18.19b). This trade-off in C allocation from growth to maintenance is further illustrated by the peak in tree and understory biomass at 25 years ($GPP \approx R_A$; Figure 18.19a). After 25 years of growth, overstory trees are commercially harvested and seedlings are planted to reestablish a new plantation. These activities again reduce living plant biomass to near zero and restart the process of biomass accumulation in the newly developing plantation.

A substantial amount of biomass is also found in dead trees either killed by a disturbance or subject to long-term mortality as forests age. Severe wildfires, windstorms, hurricanes, and insect and pathogen outbreaks kill trees but leave behind a majority of their biomass as deadwood, either standing or fallen onto the forest floor. One notable exception to this pattern is harvesting, which actively removes most large wood (note that Figure 18.17b lacks a pattern for deadwood and is therefore representative of forest recovery following harvesting). Older forests produce less deadwood via chronic mortality than do disturbances, but mortality of individual old trees is an important and long-lasting source of deadwood, particularly when the trees are large (Harmon and Hua 1991). Deadwood is therefore typically most abundant immediately after a disturbance,

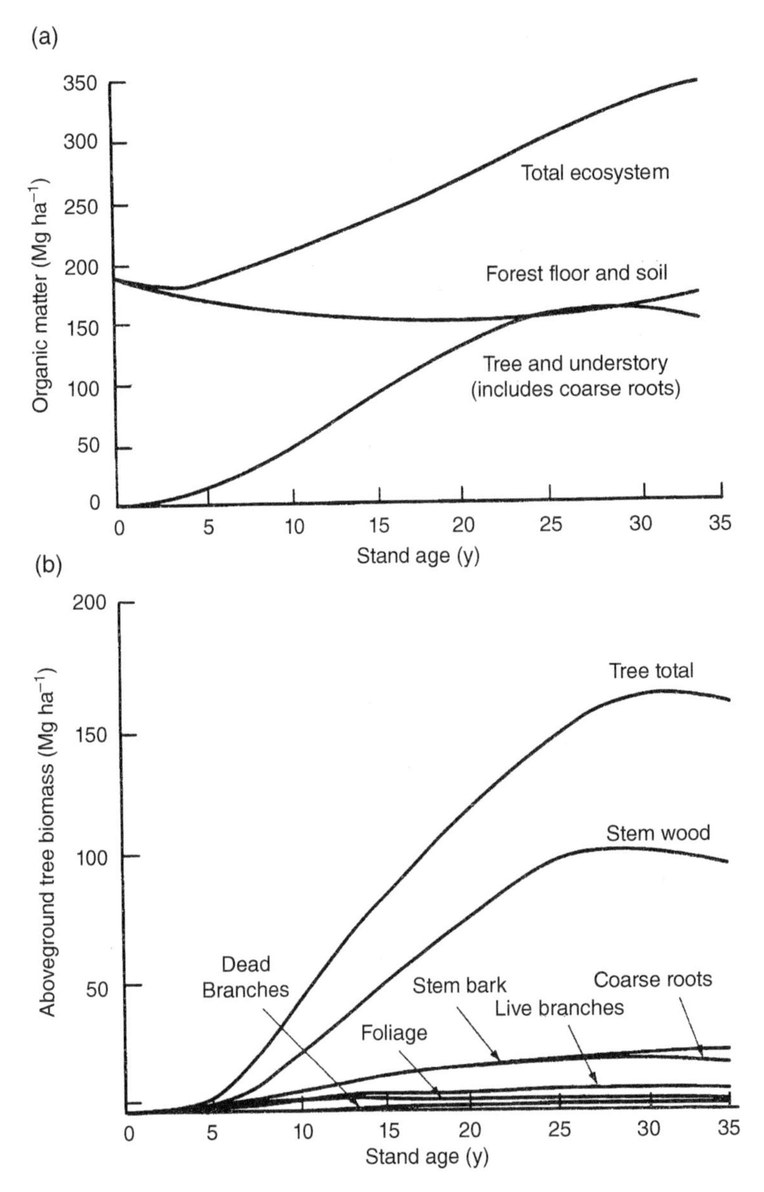

FIGURE 18.19 Changes in (a) ecosystem biomass and (b) the components of living plant biomass in a developing slash pine ecosystem in Florida. *Source:* After Gholz and Fisher (1982) / John Wiley & Sons. Reprinted with permission of the Ecological Society of America.

declines to a minimum in middle stand ages when the wood decays with only minor additional inputs from the reestablished forest, and then increases again in old age due to mortality of older trees (Sturtevant et al. 1997; Kashian et al. 2013).

The dynamics of deadwood in forests are often critical to NEP. When disturbances kill many or all trees and create a substantial amount of deadwood, NEP is strongly shaped by the balance between (i) the amount of carbon lost via heterotrophic respiration that occurs with the decomposition of deadwood, and (ii) the amount of carbon gained in reestablishing vegetation. The high rates of heterotrophic respiration and corresponding negative NEP immediately following a disturbance, as described in the earlier text, largely occur because of the decomposition of deadwood, which continues for several decades (Crutzen and Goldhammer 1993). Biomass accumulation in growing trees eventually compensates for C lost through decomposition as the forest reestablishes and R_H slows, and NEP becomes positive. In old forests, NEP eventually approaches zero as NPP is reduced in older trees and tree mortality produces additional large deadwood. Deadwood is therefore an important, though often overlooked, component of ecosystem C dynamics. For example, deadwood represented 36% of total ecosystem biomass in lodgepole pine forests in Yellowstone National Park, ranging from 2% in a middle-aged stand with little deadwood to 80% immediately after a stand-replacing fire (Kashian et al. 2013).

Natural disturbance regimes of fire and windstorms also alter the C balance of forest ecosystems. Prior to European settlement, fire was an important component of ecosystem development in the hardwood forests of southern Wisconsin (Figure 18.20; Loucks 1970). Repeated, random fires altered the C balance of these ecosystems by destroying living plant and reducing GPP. Biomass oxidized by fire to CO_2 also represents a loss of C and causes a decline in total ecosystem biomass. In combination, these responses cause NEP to fluctuate in a cyclic manner that reflects the magnitude and frequency of fire. Notice that the period between some fires in Figure 18.19 is sufficient for R_A and R_H to equal GPP, producing an NEP that is near 0 late in ecosystem development. In some ecosystems, fire may occur during a period of total biomass accumulation ($GPP > R_A + R_H$), thus preventing total biomass from attaining a "steady state" that might occur late in ecosystem development.

Natural disturbance that destroys living plant biomass can occur over relatively long periods of time (i.e., 100s to 1000s years) in some ecosystems, thus allowing them to reach a "steady state" in total biomass. In the Pacific Northwest, old-growth ecosystems dominated by Douglas-fir and western hemlock can attain ages of 450 years or more. Although NEP is low (or 0) in these old-growth forests, they have accumulated large amounts of C relative to second-growth forests (Table 18.7). Total ecosystem biomass of old-growth forest is over twice that of the second-growth forest, suggesting that GPP exceeds $R_A + R_H$ for a substantial period of time between 60 and 450 years after disturbance. The accumulation of biomass in stemwood and coarse woody debris accounts for 77% of the increase in total biomass between the second- and old-growth forests.

Although NPP and NEP decline through time, substantial amounts of biomass can accumulate in old-growth forests. Clearly, the conversion of old-growth to second-growth forests has the potential to greatly alter the cycling and storage of C in the Pacific Northwest.

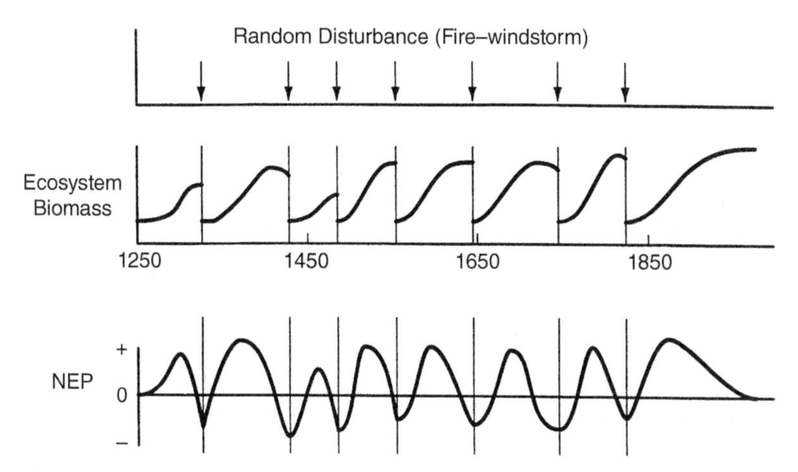

FIGURE 18.20 The relationship between fire frequency, total ecosystem biomass, and net ecosystem productivity for forests in southern Wisconsin. Fire was an important disturbance prior to European settlement that functioned to alter ecosystem C balance in a cyclic manner. The present suppression of fire in Lake States forests has greatly altered this relationship. Note the lack of fire following 1850 and the relatively long period of time over which NEP is greater than 0. *Source:* Modified from Loucks (1970).

Table 18.7 Biomass pools in second- and old-growth ecosystems dominated by Douglas-fir and western hemlock in the Pacific Northwest.

	60-year-old forest	450-year-old forest
Component	$Mg\,ha^{-1}$	
Foliage	12	13
Branch	15	56
Stem	308	687
Coarse roots	62	151
Fine roots	12	12
Forest floor	15	55
Coarse woody debris	8	206
Mineral soil	119	119
Total	551	1299

Source: Based on Harmon et al. (1990). Effects on carbon storage of conversion of old-growth forests to young forests. Reprinted with permission from *Science*, p. 700, © 1990 by the American Association for the Advancement of Science.

SOIL N AVAILABILITY AND BELOWGROUND NET PRIMARY PRODUCTIVITY

Allocation of C to belowground plant growth accounts for a substantial proportion of NPP in forest ecosystems, as well as others. Studies conducted in coniferous and deciduous forests suggest 50% of NPP can be allocated to the growth and maintenance of roots. Earlier in this chapter (see C Balance of Plants), we observed that soil N availability substantially alters the proportion of carbohydrate allocated to the leaves, stem, and roots of individual plants. In this section, we extend this idea to an ecosystem level and summarize patterns of belowground NPP in forest ecosystems. We focus on the extent to which soil N availability modifies the proportion of NPP allocated to fine roots and associated mycorrhizae, which are the nutrient and water absorbing organs of forest trees (Chapter 19).

Quantifying fine-root production in forest ecosystems remains one of the most important, but difficult, challenges to ecologists. Unlike leaves that senesce and abscise during a defined time of the year, root production and mortality are continuous processes occurring throughout the entire growing season. Most often, belowground NPP is estimated by measuring the monthly change in root biomass over 1 year. Other techniques include the use of mini-rhizotrons and several methods that employ a nutrient budget. These methods of quantifying belowground production have led ecologists to very different conclusions regarding the influence of soil N availability on belowground NPP. Regardless of this uncertainty, it is clear that the production of plant tissue below the soil surface in forest ecosystems is a substantial component of NPP.

Two alternative views regarding the influence of soil N availability on fine-root dynamics have emerged (Figure 18.21). One hypothesis argues that fine-root life span (i.e., mortality) does not change along gradients of soil N availability. It also predicts that fine-root production declines in N-rich soil, enabling the plant to allocate proportionately more carbohydrate to foliage and stems (Hypothesis 1 in Figure 18.21). Such a view contends that fewer roots are needed to forage for soil resources when they are in abundant supply. Evidence for this view comes from a comparison of two 40-year-old Douglas-fir ecosystems occurring on soils that differ in N availability (Table 18.8). Sequential monthly measurements of fine-root biomass were used to estimate fine-root production and allometric equations were used to estimate the change in coarse root biomass. The proportion of NPP allocated to fine-root production declined from 36.4% in the N-poor soil to 7.9% in the N-rich soil, but rates of fine-root mortality were unchanged (Keyes and Grier 1981). At the same time, allocation to leaves rose from 33.1% in the N-poor soil to 55.5% in the

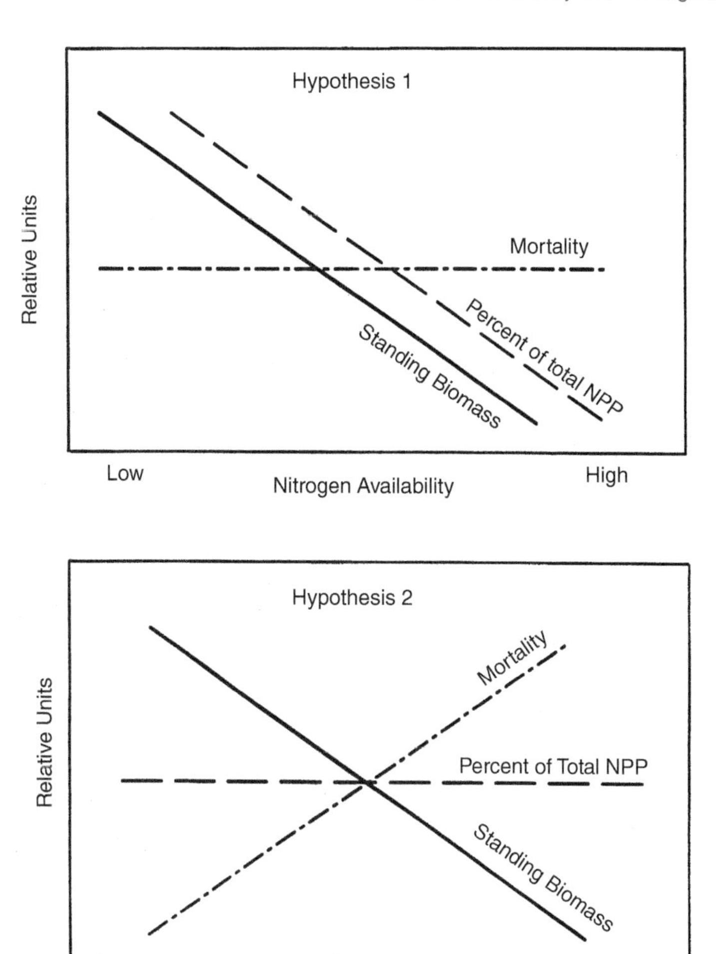

FIGURE 18.21 Two alternative hypotheses describing the allocation of NPP to the production of fine roots in forest ecosystems. *Source:* Reprinted from Hendricks et al. (1993) / with permission of Elsevier. With kind permission of Elsevier Science—NL, Sara Burgerhart-straat 25, 1055 KV Amsterdam, The Netherlands.

N-rich soil (Table 18.8). In combination, these observations suggest that the proportion of NPP allocated belowground declined in N-rich soil, facilitating a decline in fine-root biomass.

Another view contends that allocation to fine roots remains constant along a gradient of N availability, but the life span of fine roots declines when soil N availability is relatively high (Figure 18.21; Hypothesis 2). Therefore, the proportion of NPP allocated to fine-root growth remains constant, but an increase in mortality facilitates a reduction in fine-root biomass when soil N availability is high. This view of fine-root dynamics has developed from the use of an N budget to estimate the amount forest trees allocate to the annual production of fine roots. It assumes that the allocation of N to fine roots equals the difference between the amount of N assimilated by forest trees and the amount of N allocated to aboveground growth. This approach has been applied to a series of hardwood and coniferous forests in Wisconsin that occur along a gradient of soil N availability (Nadelhoffer et al. 1985; Figure 18.22) and has yielded results in direct contrast to the aforementioned study in Douglas-fir. Using the N budget approach, the biomass N content ($g N m^{-2}$) of fine roots declined along an increasing gradient of soil N availability, but the total amount of N allocated to

Table 18.8 The allocation of net primary productivity to aboveground and belowground plant tissues in 40-year-old Douglas-fir stands growing on soils of low and high fertility.

Component	Low N soil ($Mg\ ha^{-1}\ y^{-1}$)	Percent of total	High N soil ($Mg\ ha^{-1}\ y^{-1}$)	Percent of total
Stem	5.1	33.1	9.9	55.5
Branch	0.2	1.3	0.6	3.4
Leaves	2.0	13.0	3.2	18.0
Coarse roots	2.5	16.2	2.7	15.2
Fine roots	5.6	36.4	1.4	7.9
Total	15.4	100	17.8	100

Source: After Keyes and Grier (1981) / Canadian Science Publishing. Reprinted with permission of National Research Council Canada.

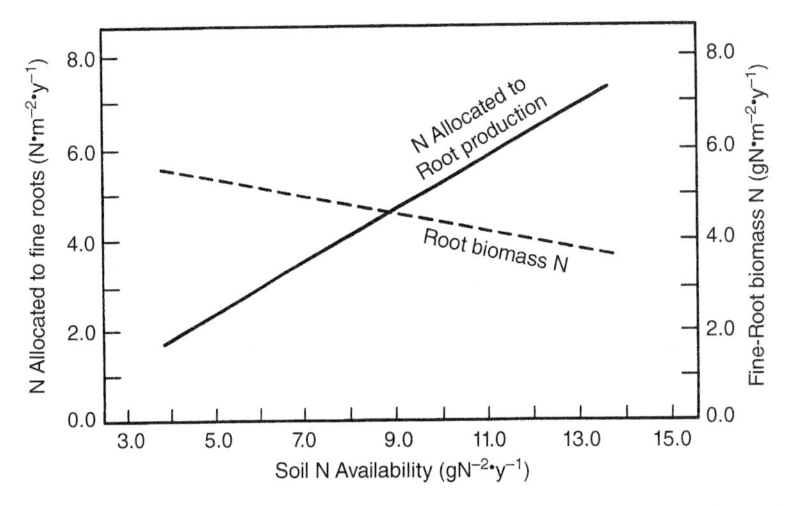

FIGURE 18.22 Changes in the N content of fine roots (dashed line) and the total amount of N allocated to fine-root production (solid line) along a gradient of increasing soil N availability. *Source:* After Nadelhoffer et al. (1985) / John Wiley & Sons. Reprinted with permission of the Ecological Society of America.

fine-root production increased; NPP allocated to fine roots varied little and averaged 27%. The inverse relationship between fine-root biomass N and the total amount of N allocated to fine-root production indicates that increased mortality at high soil N availability maintained a relatively low fine-root biomass. Recall that plant tissues with high N concentrations often have high maintenance respiration costs and a short life span. These attributes may partially explain the increase in fine-root mortality, because roots in high-N-availability soil often have a high N concentration (i.e., proteins mediating ion uptake), which results in more rapid rates of maintenance respiration.

A third approach, based on a short-term C budget, has also produced evidence for a positive relationship between aboveground and belowground NPP (Raich and Nadelhoffer 1989). It is based on the principle that all aboveground and belowground litter production is respired by decomposing organisms on an annual basis. Under that assumption, the allocation of C to root growth and maintenance is proportional to the flux of C from soil respiration minus the C in aboveground leaf litter production (for details, see Raich and Nadelhoffer 1989). Soil respiration and leaf litter production have been quantified in many forest ecosystems, and Figure 18.23 summarizes the relationship between C allocated to leaf litter and fine roots (fine-root production, construction respiration, and maintenance respiration) for 30 forest ecosystems occurring in tropical, temperate, and boreal regions. Notice that leaf litter and the amount of C allocated to fine

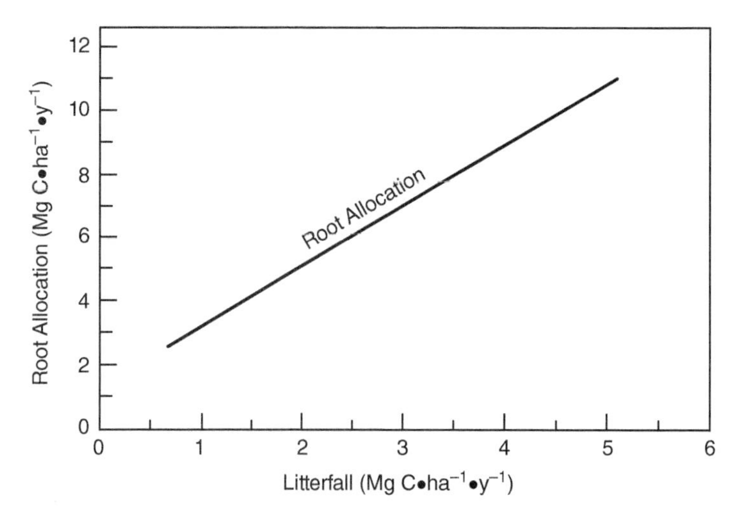

FIGURE 18.23 Leaf litter and fine-root production in tropical, temperate, and boreal forests. Estimates of the total amount of C allocated to fine roots (production + respiration) were derived using a C budget approach. *Source:* After Raich and Nadelhoffer (1989) / John Wiley & Sons. Reprinted with permission of the Ecological Society of America.

roots appear to be positively related on a global basis. Such a relationship provides further evidence that a relatively constant proportion of NPP is allocated belowground, which provides additional support for Hypothesis 2 in Figure 18.21.

All the approaches discussed use indirect methods of quantifying fine-root dynamics. That is, none directly observe the production and death of individual roots, which is the only approach that has the potential to resolve our uncertainty of the belowground C budget of forest ecosystems. In the next chapter, we provide further discussion on how the architecture, production, and mortality of roots mediate the acquisition of soil resources. As you will learn, gaining insight into the belowground dynamics of plant growth remains a difficult endeavor for ecologists. Consequently, we still have a great deal to learn regarding the ecosystem-level allocation of C to fine-root production in forest ecosystems before we can resolve the disparity in observations of fine-root production and mortality. This fact clearly limits the ability of ecologists to accurately estimate the NPP of forest ecosystems, and to fully understand the function of forest ecosystem in the global C cycle. Despite this limitation, it is clear that the growth and maintenance of plant tissues below the soil surface account for a significant proportion of C fixed via photosynthesis in many forest ecosystems, and therefore is a globally significant component of the cycling and storage of C in terrestrial ecosystems.

SUGGESTED READINGS

Bormann, F.H. and Likens, G.E. (1994). *Pattern and Process in a Forested Ecosystem*. New York: Springer-Verlag 266 pp.

Givnish, T.J. (1988). Adaptation to sun and shade: a whole-plant perspective. *Aust. J. Plant Physiol.* 15: 63–92.

Mooney, H.A. (1972). The carbon balance of plants. *Annu. Rev. Ecol. Syst.* 3: 315–346.

Pearcy, P.W., Ehleringer, J., Mooney, H.A., and Rundel, P.W. (ed.) (1991). *Plant Physiological Ecology: Field Methods and Instrumentation*. New York: Chapman and Hall 457 pp.

Smith, W.R. and Hinkley, T.M. (ed.) (1995). *Resource Physiology of Conifers: Aquisition, Allocation, and Utilization*. New York: Academic Press 396 pp.

Raghavendra, A.S. (ed.) (1991). *Physiology of Trees*. New York: John Wiley 509 pp.

Nutrient Cycling

CHAPTER 19

Although carbon (C), hydrogen (H), and oxygen (O) form the basic building blocks of all biological tissue, plants require a suite of 14 other elements, termed nutrients, in order to maintain existing tissue and build new biomass (see Table 9.3, Chapter 9). These nutrients mostly enter terrestrial ecosystems from the atmosphere (wet and dry precipitation) or through the weathering of soil minerals. Plants assimilate nutrients from soil, incorporate them with photosynthetically fixed CO_2 to form living biomass, and eventually return these nutrients to soil in dead leaves, roots, branches, and stems (i.e., plant litter). After dead plant material enters the forest floor (leaves, twigs, etc.) and surface mineral soil (fine roots), it is subject to decomposition, a microbially mediated process that releases organically bound nutrients into inorganic forms that can again be assimilated by plant roots. The uptake of nutrients by plant roots, their incorporation into living tissue, and the release of nutrients during organic matter decomposition cause nutrients to flow or *cycle* within terrestrial ecosystems. Nutrient cycles are biogeochemical processes, so named because they are controlled by the physiological activities of plants and soil microorganisms, as well as by the geochemical processes in soil that control nutrient supply (Chapter 9).

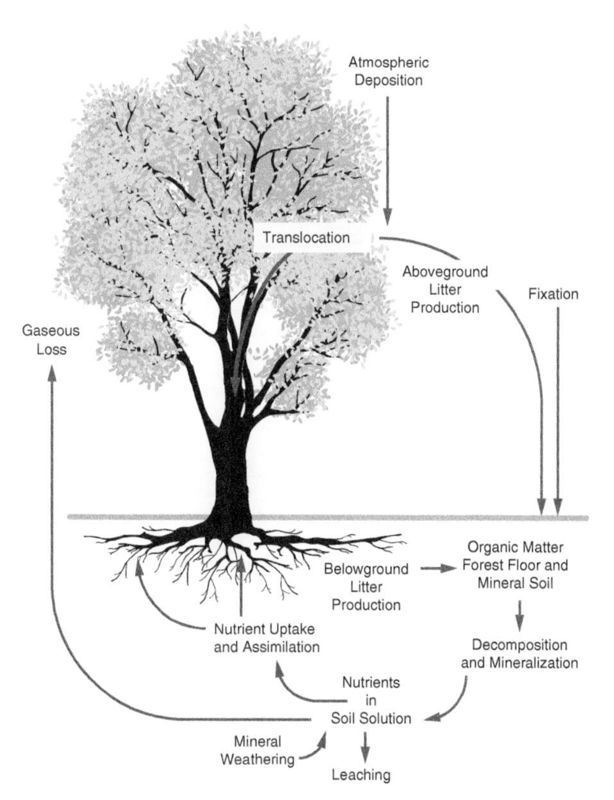

FIGURE 19.1 A conceptual diagram of the processes controlling the flow of nutrients into, within, and out of forest ecosystems. Nutrients enter forest ecosystems through atmospheric deposition, N_2 fixation, and mineral weathering. The flow of nutrients within forest ecosystems is controlled by nutrient uptake, the translocation of nutrients from senescent tissue, the return of nutrients in leaf and root litter, the decomposition of this litter, and the microbial release (i.e., mineralization) of nutrients from organic matter. Leaching and gaseous losses (i.e., denitrification) are processes by which nutrients are lost from forest ecosystems.

Forest Ecology, Fifth Edition. Daniel M. Kashian, Donald R. Zak, Burton V. Barnes, and Stephen H. Spurr.

Recall that net primary productivity (NPP) differs among boreal, temperate, and tropical forests (Chapter 18). Because nutrients are assimilated along with CO_2 to form living biomass, differences in net primary productivity among ecosystems suggest that rates of nutrient cycling also must differ. In this chapter, we discuss processes controlling the input of nutrients to forest ecosystems, the redistribution of nutrients within forest ecosystems by the metabolic activities of plants and soil microorganisms, and the loss of nutrients from forest ecosystems to streams, groundwater, and the atmosphere. These processes are illustrated in Figure 19.1, which outlines the organization of this chapter. Nutrient uptake by plant roots is central to the cycling of nutrients in forest ecosystems, and we place emphasis on the physiological and morphological mechanisms by which plants forage for nutrients in soil. We also discuss the extent to which the C balance of terrestrial ecosystems influences nutrient retention and loss within forests, particularly during forest development. Our treatment of nutrient cycling focuses on nitrogen (N), because the supply of this nutrient in soil most often limits the productivity of boreal, temperate, and several types of tropical forests (i.e., dry and montane tropical forests).

NUTRIENT ADDITIONS TO FOREST ECOSYSTEMS

Nutrient additions to forest ecosystems are relatively small compared to the annual requirements of actively growing overstory trees, typically contributing 5–30% of the annual need for most plant nutrients. Nevertheless, input processes serve as important conduits by which nutrients flow into forests and other terrestrial ecosystems, and over relatively long periods of time (100–200 years), they contribute significantly to the nutrient capital of an ecosystem. Nutrients enter terrestrial ecosystems through geological, hydrological, and biological processes. These are highlighted in the reduced version of Figure 19.1, which lies next to this paragraph. In the paragraphs that follow, we discuss the magnitude of nutrient inputs to forest ecosystems from mineral weathering, atmospheric deposition, and biological fixation, the primary mechanisms by which nutrients flow into forest ecosystems.

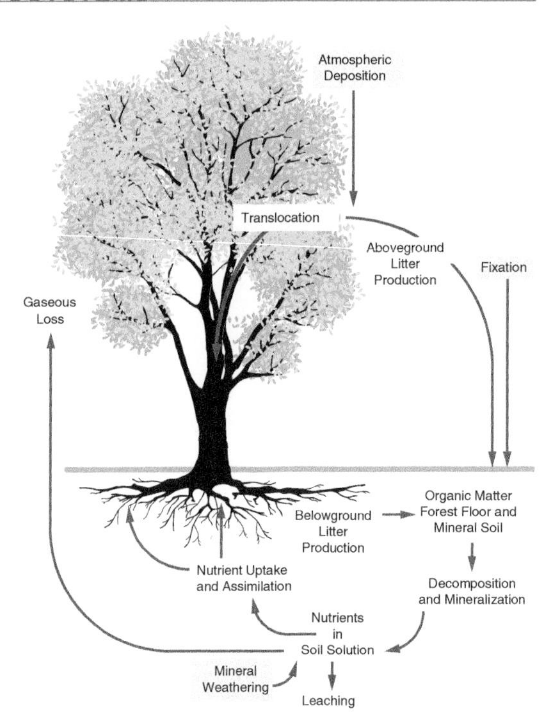

MINERAL WEATHERING

Plant nutrients contained in the chemical structure of rocks and soil minerals are largely unavailable to plants. Through physical abrasion and chemical dissolution, nutrients are slowly released from geologic materials into forms that plants take up from soil solution. The rate at which geologic materials weather to yield plant nutrients is strongly influenced by climate, temperature, and precipitation, which directly control rates of chemical reactions in soil, especially the solubilization of plant nutrients from minerals. Rates of mineral weathering are quite slow, and because not all geologic materials are chemically similar, they weather at different rates and yield different plant nutrients. All plant nutrients are found in the chemical structure of rocks, and therefore can be supplied to the plant from mineral weathering, albeit at an annual rate that does not meet the annual nutritional needs of plants. In Chapter 9, we discussed mineral weathering as it pertains to soil horizon development, soil acidity, and base saturation. Here, we focus on aspects of mineral weathering that control the input of plant nutrients to forest ecosystems.

The most well-known attempt at quantifying mineral weathering in ecosystems began in 1963, when a team of scientists quantified the nutrient cycles of northern hardwood forests in the White Mountains of New Hampshire (Likens et al. 1977). In this region, a number of comparable watersheds (i.e., similar in climate, size, geology, topography, soil parent material, and vegetation) are underlain by a layer of impermeable bedrock, preventing rainfall from reaching groundwater. Because these watersheds are sealed, streamflow represents the main process by which nutrients can exit. Therefore, the difference between the atmospheric deposition of nutrients and streamwater loss of nutrients should approximate the annual release of nutrients from mineral weathering. Using this approach, a large number of watershed-level studies of mineral weathering have been completed, allowing a comparison of weathering rates for different climates and parent materials. In areas dominated by silicate rocks, plant nutrients weather from soil parent materials in the following order:

$$Ca > Na > Mg > K > Fe$$

This sequence partially reflects mineral solubility and partially the degree to which these elements participate in the formation of secondary minerals, such as phyllosilicate clays.

Table 19.1 summarizes nutrient inputs from mineral weathering for forest ecosystems occurring in different climates and on different parent materials. Notice that the weathering of dolomitic parent materials ($CaCO_3$ and $MgCO_3$) beneath limber and bristlecone pines yields relatively large amounts of Ca compared to weathering of adamellite (SiO_2:$NaAlSi_3O_8$:$CaAl_2Si_2O_8$). The weathering of Ca- and Mg-bearing minerals represents an important input of these nutrients in many terrestrial ecosystems, satisfying 35% of Ca and Mg required for the annual growth of some temperate

Table 19.1 Rates of mineral weathering for forest ecosystems occurring in different climates and on different soil parent materials.

Ecosystem	Parent material	Location	P	K	Ca	Mg
			kg ha^{-1} y^{-1}			
Wet tropical	Alluvium	Venezuela	–	–	6	1
Douglas-fir	Volcanic tuff	Oregon	–	2	47	12
Limber and	Dolomite	California	–	4	86	52
Bristlecone pines	Adamellite	California	–	8	17	2
Trembling aspen	Glacial till	Wisconsin	1	4	7	–
Northern hardwood	Glacial till	New Hampshire	13	7	21	4

Source: Data have been summarized from Boyle and Ek (1973), Fredriksen (1972), Hase and Foelster (1983), Likens et al. (1977), Marchand (1971), and Wood et al. (1984).

forests (Table 19.1; Likens et al. 1977). Similarly, the weathering Ca-, Fe-, and Al-phosphate minerals provides an important source of P for plant growth. It had been thought that mineral weathering was not a source of N to plants, because few rocks contained N in their chemical structure. Recently, a global analysis of rock chemical composition has revealed that more N resides within them than previously thought. This realization suggests that N released from rock weathering could compose 8–26% of annual N inputs to terrestrial ecosystems, prior to the anthropogenic input of N from atmospheric deposition and fertilizer use (Houlton et al. 2008). Figure 19.2 summarizes the amounts of N contained in surface rock (Figure 19.2a), the amounts annually released through weathering (Figure 19.2b), and the percent contribution of annual N inputs supplied via mineral weathering. This summary has dramatically revised our thinking of how N enters terrestrial ecosystems.

ATMOSPHERIC DEPOSITION

Over the past several decades, ecologists have become increasingly interested in understanding the geographic pattern and amount of nutrients entering terrestrial ecosystems from the atmosphere. In many portions of the Earth, human activity has altered rainfall chemistry and hence the atmospheric deposition of nutrients in dramatic ways. In the northeastern United States, central Europe and broadly across Asia, terrestrial and aquatic ecosystems receive enhanced inputs of NO_3^-, SO_4^{2-}, and H^+ (i.e., acidic deposition or acid rain) from fossil fuel burning. Both NO_3^- and SO_4^{2-} contain essential plant nutrients, and elevated inputs of these ions have altered the rate and pattern by which N and S are cycled and stored within terrestrial ecosystems.

Atmospheric deposition occurs through three processes: (i) **wet deposition**—the addition of nutrients contained in rain or snow, (ii) **dry deposition**—the direct deposition of atmospheric particles and gases to vegetation, soil, or water surfaces, and (iii) **cloud deposition**—the input of small, nonprecipitating water droplets (in clouds and fog) to terrestrial surfaces (Fowler 1980; Lovett and Kinsmann 1990; Lovett 1994). Wet and dry deposition occur in all terrestrial ecosystems, but their importance as a source of nutrients varies greatly from place to place. By contrast, cloud deposition is generally restricted to coastal and mountainous regions that are frequently immersed in clouds or fog (Lovett 1994), like coastal giant redwood forests of northern California, United States.

Wet deposition occurs when solid particles (0.2–2 mm diameter) and gases in the atmosphere dissolve in water droplets that fall to Earth as rain or snow (Lovett 1994). This process is controlled by several factors that vary markedly from one region to another and result in a broad range of nutrient inputs from wet deposition. In North America, continental-scale patterns of wet deposition are relatively well understood (Figure 19.3). For example, anthropogenic emissions of N and S in the Midwest, Southeast, and Northeast result in widespread patterns of increased NO_3^- and SO_4^{2-} deposition over the eastern United States. By contrast, ammonium (NH_4^+) and Ca^{2+} inputs from wet deposition are relatively low in the eastern United States but are elevated in the eastern Great Plains region and the Midwest. The plowing of agricultural fields and traffic on unpaved roads cause Ca^{2+} associated with dust particles to be carried eastward from the Great Plains region by the prevailing winds. Ammonium released into the air from heavily fertilized agricultural fields and volatilized from animal manure in concentrated feeding operations is transported eastward in a similar manner.

Wet deposition entering forest ecosystems can pass through the canopy as throughfall or portions can flow down tree stems in stemflow. The chemical composition of wet deposition can be greatly altered by the "wash off" of dry material accumulated on plant surfaces, the leaching of nutrients from plant tissues, and the direct assimilation of nutrients into leaves. Potassium, Ca^{2+},

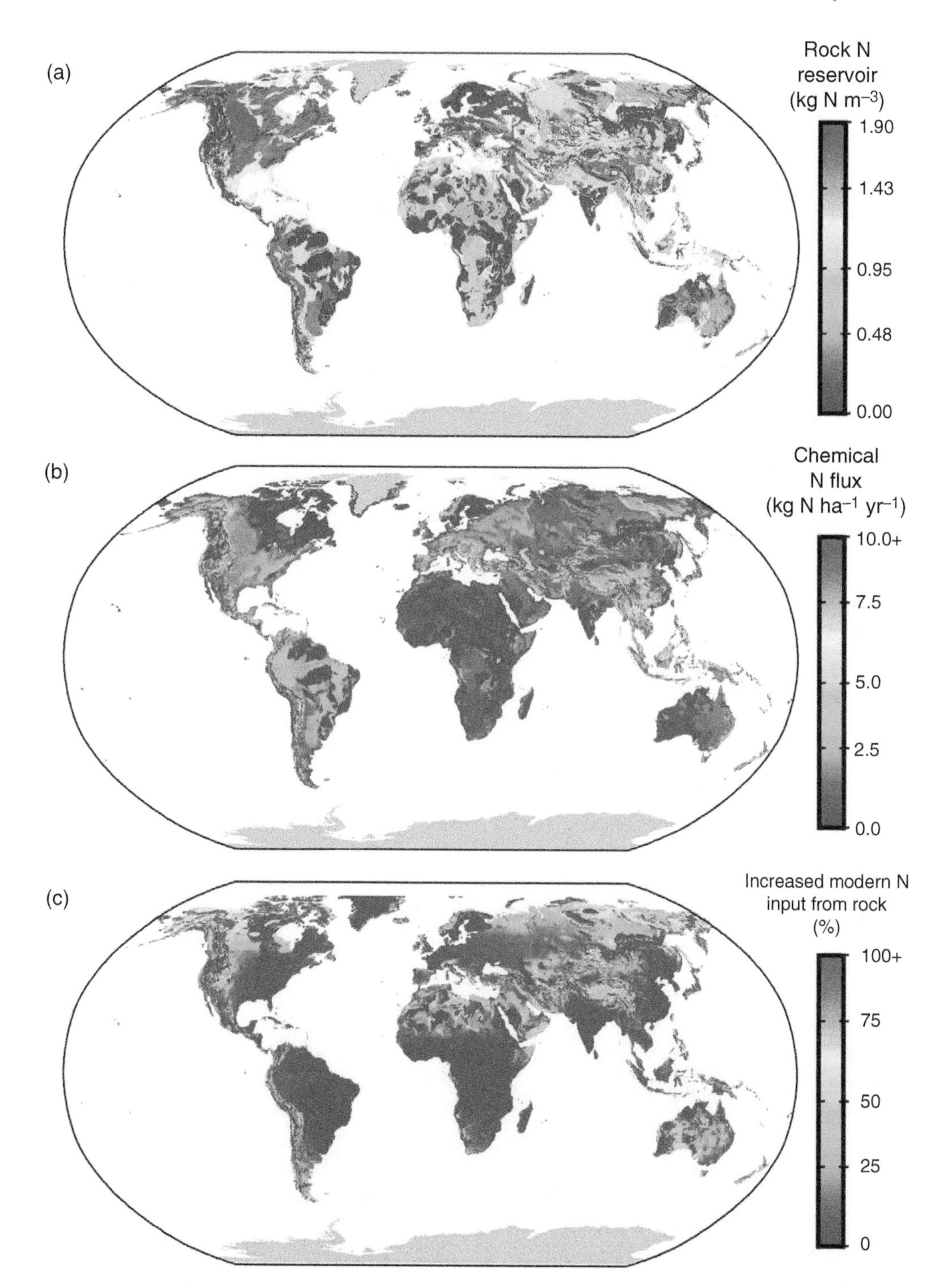

FIGURE 19.2 (a) The amount of N contained in surface minerals, (b) the annual release of N from minerals, and (c) the percent of annual ecosystem N inputs supplied by the weathering of N from surface minerals. Cool colors (blue) indicate low inputs and amounts, whereas warm colors (red) indicate the opposite. *Source:* Houlton et al. (2008) / American Association for the Advancement of Science.

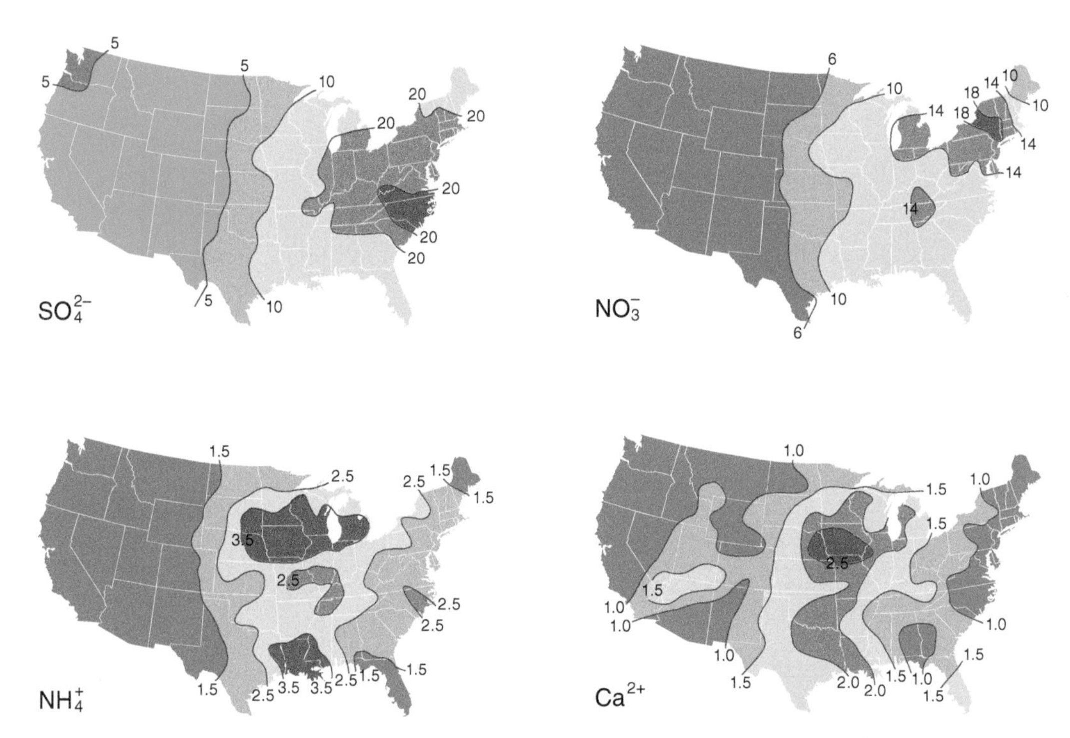

FIGURE 19.3 Geographic patterns of sulfate (SO_4^{2-}), nitrate (NO_3^-), ammonium (NH_4^+), and calcium (Ca^{2+}) in wet deposition across the continental United States. Values are in kilograms per hectare per year (kg ha^{-1} y^{-1}). *Source:* Lovett (1994). Reprinted with permission by the Ecological Society of America.

and Mg^{2+} are easily leached from leaves (Tukey 1970), and their concentrations in throughfall can be greater than that of the original wet deposition. This may be particularly important in tropical forests in which species-specific differences in canopy leaching are related to the abundance of epiphytes (Schlesinger and Marks 1977). In some cases, however, forest canopies directly assimilate water-soluble nutrients (i.e., N) from wet precipitation, thereby lowering their concentration in throughfall (Olson et al. 1981). Stemflow is significant in that it returns a relatively concentrated supply of nutrients directly to the base of the tree (Gersper and Holowaychuk 1971), where they may accumulate and be taken up by the tree.

Determining the extent to which the "wash off" of dry material and leaching influence the nutrient concentration of throughfall and stemflow can be difficult. However, it appears that 85% of the SO_4^{2-} in throughfall originates from the "wash off" of dry material deposited on leaves (Lindberg and Garten 1988), making the dry deposition of nutrients an important process in forest ecosystems. The high amount of leaf surface in forests makes them effective collectors of dry materials suspended in the atmosphere.

Dry deposition is a complex process involving atmospheric chemistry, wind velocity, and canopy characteristics like leaf shape, orientation, spatial arrangement, and area (Lovett 1994). Solid particles (< 5 μm diameter) enter terrestrial ecosystems through gravitational sedimentation, the process by which the force of gravity pulls particles suspended in the atmosphere toward the Earth's surface. Some particles are small enough to be kept aloft by wind turbulence and enter terrestrial ecosystems only when they impact the surface of vegetation. Even smaller dry particles and gases directly enter the plant through stomata and lenticels.

The movement of atmospheric particles toward leaf surfaces is determined by the change in particle concentration from the atmosphere to the leaf surface and the resistance to flow along that path (Garland 1977). Atmospheric scientists estimate the input of particles to terrestrial ecosystems by determining their deposition velocity:

$$\text{Deposition velocity} = \frac{\text{Rate of particle dryfall}\left(\text{mg cm}^{-2}\text{ of leat s}^{-1}\right)}{\text{Concentration in air}\left(\text{mg cm}^{-3}\right)}.$$

With knowledge of deposition velocity (cm sec^{-1}), atmospheric concentration (mg cm^{-3}), and leaf area ($\text{m}^2\ \text{m}^{-2}$), atmospheric scientists estimate the amount of nutrient-containing dry particles entering a particular forest ecosystem. Total dry deposition (mg m^{-2}) is simply the mathematical product of atmospheric concentration, deposition velocity, and canopy leaf area.

Dry deposition of nutrients is somewhat more variable than wet deposition for several reasons (Figure 19.3). The ratio of wet to dry deposition increases with distance from the source, because readily dry-deposited materials are depleted from the air as it travels downwind (Ollinger et al. 1993). Second, differences in canopy structure and leaf morphology among trees cause nutrients in dry material to be captured with different efficiencies, depending on overstory composition. Nonetheless, dry deposition can bring significant amounts of nutrients into forest ecosystems relative to those contained in wet deposition.

Figure 19.4 summarizes wet, dry, and cloud depositions for a series of forest, prairie, and agricultural ecosystems distributed across North America. Notice that the ratio of dry to wet deposition is highly variable among these ecosystems, reflecting differences in the distance to nutrient sources and canopy characteristics. The prairie ecosystem (AR) located downwind from Chicago, Illinois, United States receives relatively greater inputs of S and N in dry deposition than would be expected for an ecosystem in the Midwest (Figure 19.4) because it is relatively close to a major source of anthropogenic NO_3^- and SO_4^{2-} production.

Cloud deposition results when small, nonprecipitating water droplets containing dissolved nutrients come in contact with terrestrial surfaces, such as tree canopies. In contrast to wet deposition, tree canopies exert a substantial influence on the amount of nutrients entering forest ecosystems from cloud deposition. This process is a particularly important nutrient input in ecosystems that lie in coastal and mountainous regions (WF and CD in Figure 19.4), whereas in continental areas (e.g., Midwest), cloud deposition contributes relatively minor amounts of nutrients to terrestrial ecosystems. For example, relatively large quantities of SO_4^{2+}, NO_3^-, and NH_4^+ enter spruce/fir ecosystems in the northeastern United States and red spruce ecosystems at high elevation in the Appalachian Mountains through cloud deposition (Figure 19.4). Over 50% of the atmospheric deposition of NH_4^+ in these ecosystems occurs through cloud deposition.

BIOLOGICAL FIXATION OF NITROGEN

Biological inputs of N to forest ecosystems occur through symbiotic associations of soil bacteria and tree roots, free-living bacteria in litter and mineral soil, as well as epiphytic lichens living on the external surfaces of trees and decaying wood (Melillo 1981; Waughman et al. 1981; Boring et al. 1988). Several genera of soil bacteria, actinobacteria (filamentous bacteria), and cyanobacteria have the ability to reduce or "fix" atmospheric N_2 into NH_4^+. Depending on the fixation process, NH_4^+ can be directly used by the microbial cell or it can be transported from the microorganism to the host plant as N-containing organic compounds. The process of N_2 fixation is catalyzed by the nitrogenase enzyme system, which requires substantial quantities of energy (i.e., ATP). The following equation summarizes this process:

$$N_2 + 16\,ATP + 8e^- + 10H^+ \xrightarrow[\text{Nitrogenase}]{} 2NH_4^+ + H_2 + 16\,ADP + P_i.$$

The high energy cost (i.e., ATP) associated with N_2 fixation has important ecological implications and gives rise to widely different rates of fixation depending on the fixation process and the microorganisms involved. Soil organic matter provides small amounts of energy for fixation by free-living soil bacteria, whereas carbohydrates from photosynthesis can fuel rapid rates of N_2 fixation by symbiotic bacteria and actinobacteria inhabiting plant roots. As a result, quantities of N entering terrestrial ecosystems from N_2 fixation are highly variable and range from as little as 1 kg N ha^{-1} y^{-1} for free-living soil bacteria to 200 kg N ha^{-1} y^{-1} from symbiotic fixation in the roots of legumes, alder, and other higher plants (Cole and Rapp 1981).

Rates of free-living N_2 fixation by free-living organisms are low and equivalent to rates of atmospheric deposition (Figures 19.2 and 19.3) in most terrestrial ecosystems. Free-living heterotrophic bacteria of the genera *Azotobacter* and *Clostridium* fix atmospheric N_2 into forms that plants can assimilate after their cells die and cellular constituents are decomposed by other soil microorganisms. Because soil organic matter supplies the energy for free-living fixation, this process is greatest in soils with relatively high organic matter contents (Granhall 1981). In the forests of the Pacific Northwest and northeastern United States, free-living N_2 fixation has been observed in decaying logs (Roskoski 1980; Silvester et al. 1982), a relatively rich-energy substrate for heterotrophic bacteria. In most cases, the input of N to forest ecosystems from free-living N_2 fixation is relatively small, approximately 1–5 kg N ha^{-1} y^{-1} (Boring et al. 1988).

Free-living cyanobacteria (blue-green algae) are common soil microorganisms that can, in some instances, be a source of N in some terrestrial ecosystems. Fixation by these photosynthetic microorganisms is associated with high-light environments, and in most forests, fixation by cyanobacteria is limited by low light levels at the soil surface. However, algal crusts that form on the surface of some desert soils can have exceptionally high rates of N_2 fixation (Rychert et al. 1978). Nitrogen-fixing cyanobacteria also enter into mutualistic associations with some fungi to form N_2-fixing lichens that bring N into forest ecosystems. *Nostoc*, an N_2-fixing cyanobacteria, is one of two species of algae present in the lichen *Lobaria oregana* that inhabits the crowns of old-growth Douglas-fir. *Lobaria* is estimated to fix approximately 8–10 kg N ha^{-1} y^{-1}, an amount comparable to the input of N via atmospheric deposition (see Figure 19.4).

Nitrogen-fixing actinobacteria and unicellular bacteria can infect the fine roots of certain higher plants and induce the formation of root nodules, which are the location of symbiotic N_2 fixation. These small ($<$ 5 mm), round, hollow structures are formed on individual roots and enclose large numbers of bacterial or actinobacterial cells, the agents of N_2 fixation. Nitrogen fixed by the microbial symbiont is transferred to the plant, and carbohydrate supplied by the host in turn provides energy for the microbial fixation of N_2 in the root nodule. Because the cost of N_2 fixation is subsidized by a supply of carbohydrate (i.e., energy) from the plant, symbiotic fixation can bring substantial quantities of N into terrestrial ecosystems, far surpassing fixation by free-living bacteria and cyanobacteria.

Legumes are perhaps the most important plants with N_2-fixing bacteria inhabiting their roots. They occur as overstory species in both temperate and tropical forests. However, not all legumes have the ability to fix N_2 in association with *Rhizobium*, their bacterial symbiont. Black locust, honey locust, acacias, and mesquite are trees of the legume family that occur in temperate forests; there are hundreds of similar species in the tropics. Also, many herbs and shrubs of this family inhabit the forest floor throughout much of the Earth, many becoming dominant following forest harvesting or fire. Soil bacteria belonging to the *Rhizobium* genus exclusively infect the roots of legumes and enter into a symbiotic N-fixing relationship.

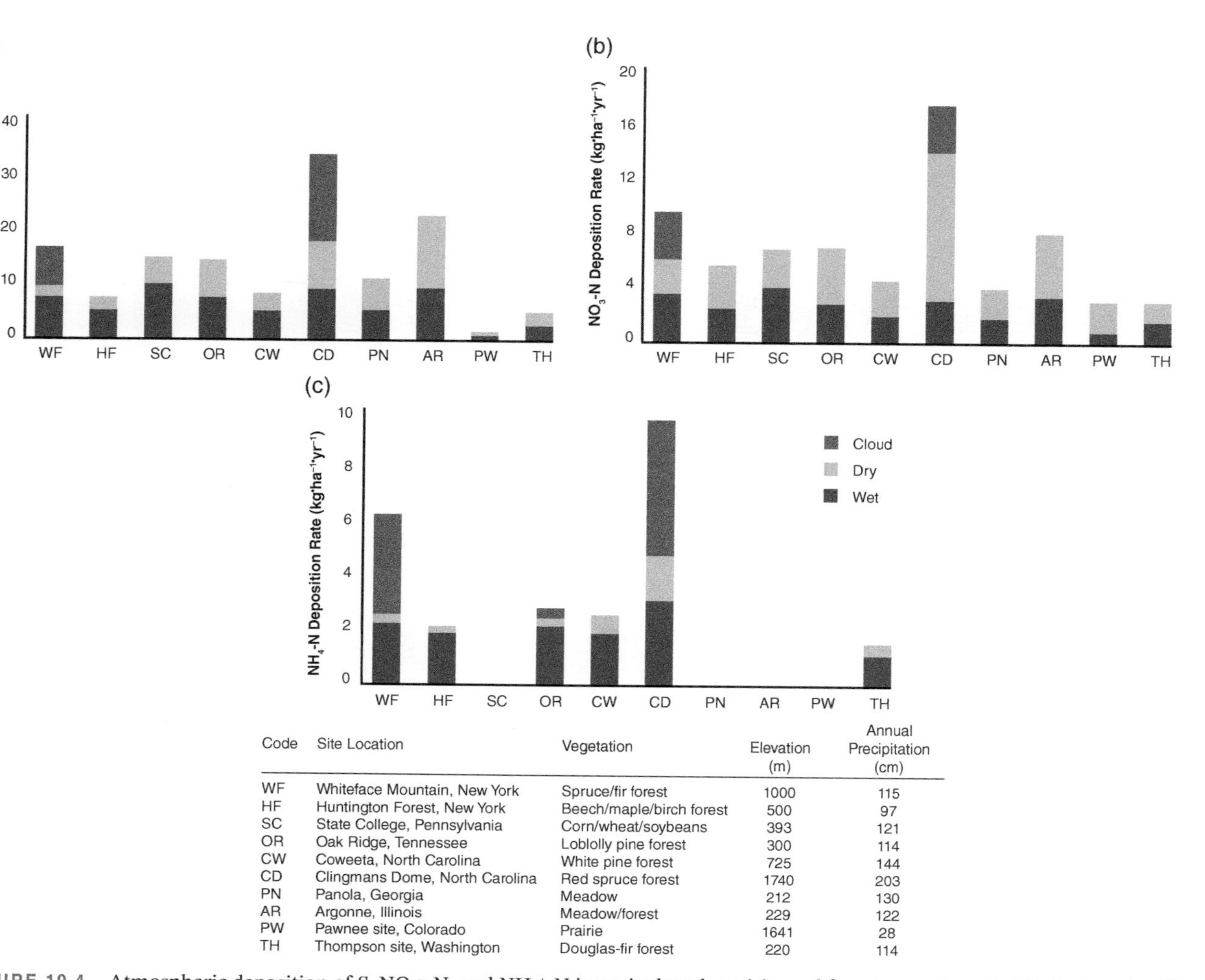

Code	Site Location	Vegetation	Elevation (m)	Annual Precipitation (cm)
WF	Whiteface Mountain, New York	Spruce/fir forest	1000	115
HF	Huntington Forest, New York	Beech/maple/birch forest	500	97
SC	State College, Pennsylvania	Corn/wheat/soybeans	393	121
OR	Oak Ridge, Tennessee	Loblolly pine forest	300	114
CW	Coweeta, North Carolina	White pine forest	725	144
CD	Clingmans Dome, North Carolina	Red spruce forest	1740	203
PN	Panola, Georgia	Meadow	212	130
AR	Argonne, Illinois	Meadow/forest	229	122
PW	Pawnee site, Colorado	Prairie	1641	28
TH	Thompson site, Washington	Douglas-fir forest	220	114

FIGURE 19.4 Atmospheric deposition of S, NO_3^--N, and NH_4^+-N in agricultural, prairie, and forest ecosystems in North America. The relative contribution of wet, dry, and cloud deposition is depicted. Note that cloud deposition is an important process in coastal areas and at high elevations. *Source:* Lovett (1994). Reprinted with permission of the Ecological Society of America.

In the southeastern United States, black locust readily establishes following a major disturbance, such as fire or clear-cut harvest. Fixation in the root nodules of this leguminous tree is a substantial input of N to the forests of the southern Appalachians, particularly during the early to intermediate stages of ecosystem development (Waide et al. 1988). Four years following clear-cut harvest, rates of N_2 fixation equaled 48 kg N ha^{-1} y^{-1}, a substantial input of N relative to atmospheric inputs (see Figures 19.2 and 19.3). Rates subsequently increased to 75 kg N ha^{-1} y^{-1} at 17 years following harvest, and after 38 years of ecosystem development rates declined to 33 kg N ha^{-1} y^{-1} (Waide et al. 1988). Compare these rates with mineral weathering and atmospheric deposition, which are orders of magnitude slower.

Actinobacteria (filamentous bacteria) of the genus *Frankia* form a symbiotic relationship with the roots of *Alnus*, *Ceanothus*, *Casuarina*, *Elaeagnus*, *Comptonia*, and *Myrica*. This symbiotic relationship is particularly important in the Pacific Northwest, where N_2 fixation by red alder and *Frankia* can substantially increase the amount of N entering forest ecosystems. One example comes from a comparison of 38-year-old red alder and Douglas-fir stands that initially established adjacent to one another on the same soil type. Differences in N pools between them resulted directly from N_2 fixation by the alder–*Frankia* symbiosis (Table 19.2). An additional 85 kg N ha^{-1} y^{-1} have accumulated under red alder, an input of N far exceeding atmospheric deposition (see Figure 19.3). Much of the N_2 fixed by the red alder–*Frankia* resides in forest floor and mineral soil, where it is eventually released during organic matter decomposition. The shrub *Ceanothus velutinus* also occurs in the Pacific and Inland Northwest and enters into an N_2-fixing symbiosis with *Frankia* that can contribute as much as 100 kg N ha^{-1} y^{-1} following forest fire or harvest (Youngberg and Wollum 1976).

Greater N availability resulting from N_2 fixation on N-poor soils can dramatically increase rates of biomass accumulation and nutrient cycling during ecosystem development. In the Hawaiian Islands, *Myrica faya* is an exotic N_2-fixing tree that has greatly altered the N cycle of native forests (Vitousek et al. 1987). This invasive, early-successional species establishes on volcanic ash flows that were formerly colonized by *Metrosideros polymorpha*, a native, non-N_2 fixing tree. In the absence of *M. faya*, N accumulates in ash-flow soils at a very slow rate, constraining rates of ecosystem development and productivity (approximately 5 kg N ha^{-1} y^{-1} from free-living fixation and atmospheric deposition). Nitrogen fixation by *M. faya* brings an additional 18 kg N ha^{-1} y^{-1} into the ecosystem, increasing the annual rate of N accumulation by a factor of three. The N_2 fixed by *Myrica* eventually enters the soil via leaf and root litter, and during decomposition, relatively greater amounts of N are released into soil solution where they are assimilated by plant roots.

Table 19.2 The accumulation of N in forest ecosystems dominated by nitrogen-fixing (red alder) and non-nitrogen-fixing (Douglas-fir) tree species. The overstory of each ecosystem is 38 years old and differences in the amount of N contained in each ecosystem pool reflect the fixation of nitrogen by red alder.

Ecosystem pool	Douglas- fir	Nitrogen kg N ha^{-1}	Red alder	Annual accumulation kg N ha^{-1} y^{-1}
Overstory	320		590	7.1
Understory	10		100	2.4
Forest floor	180		880	18.4
Mineral soil	3270		5450	57.4
Total ecosystem	3780		7020	85.3

Source: Cole and Rapp (1981) / Cambridge University Press. Reprinted from *Dynamic Properties of Forest Ecosystems*,

NUTRIENT CYCLING WITHIN FOREST ECOSYSTEMS

Nutrients entering forest ecosystems from mineral weathering, atmospheric deposition, and biological fixation can enter soil solution where they are absorbed by plant roots. Within the plant, nutrients taken up from soil solution participate in a wide array of physiological processes, and in some cases, growth-limiting nutrients are removed (i.e., translocated) prior to the shedding of some plant tissues. The production of plant litter above and below the soil surface eventually returns nutrients to forest floor and mineral soil, where they are released from litter by soil microorganisms. During the process of litter decomposition, soil microorganisms incorporate organically bound nutrients (and C) into their biomass and release excess nutrients into soil solution where they can again be taken up by plant roots. Thus, the rate at which nutrients flow within forest ecosystems is controlled by the physiological activities of plants and soil microorganisms, and their requirement for growth-limiting nutrients.

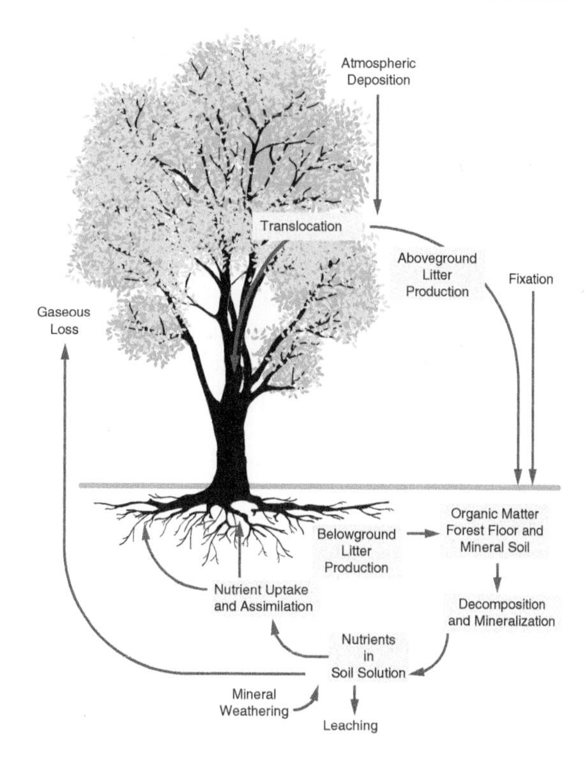

In the sections that follow, we trace the flow of nutrients within forest ecosystems. The cycling of nutrients within ecosystems occurs through: (i) root nutrient absorption via the processes of uptake and assimilation, (ii) nutrient allocation to biomass construction and maintenance, (iii) nutrient translocation from senescent tissue, (iv) the return of nutrients in aboveground and belowground litter, and (v) the microbially mediated release of inorganic nutrients into soils solution (i.e., mineralization) during organic matter decomposition. These processes are highlighted in the reduction of Figure 19.1, which is located above. Central to this discussion is the ability of plants to forage for nutrients in soil solution, a process mediated by physiological and morphological adaptations of plant roots. Also of importance is the growth dynamics of soil microorganisms during litter decomposition, because they control the release of organically bound nutrients from plant litter and soil organic matter. As you will learn later in this chapter, soil microorganisms are the "gate keepers" of N availability within forest ecosystems.

NUTRIENT TRANSPORT TO ROOTS

Nutrients must diffuse to the root surface from soil solution before they can be taken up and incorporated into biologically active compounds. The uptake of nutrients by plant roots is initially constrained by rates of organic matter decomposition, mineral solubility, cation/anion exchange reactions, as well as diffusion through soil solution. Although these processes control nutrient availability in soil, the fact that nutrients are present in soil solution or on exchange sites does not ensure uptake by plant roots. In the paragraphs that follow, we discuss the processes controlling the flow of nutrients to the root surface, uptake from soil solution, and their incorporation into biologically active compounds within the plant.

Nutrients in soil solution move toward root surfaces in response to two processes: **mass flow** and **diffusion**. Mass flow is a passive process in which ions move with the flow of water for transpiration; they are physically carried along from soil solution to the root surface. Diffusion occurs when ions move from a region of high concentration to one of low concentration, for example, from soil solution to the root surface. In most soils, the concentration of nutrients in solution is low enough that mass flow does not meet the demand of actively growing plants. Rapid uptake by plant roots also causes some nutrients to be virtually absent near their surface, thereby forming zones of depletion around individual roots. Consequently, the supply of nutrients to roots is constrained by the rate at which ions diffuse toward root surfaces from areas of higher concentration in soil solution (Nye 1977). Nutrients such as PO_4^{3-}, K^+, and NH_4^+ diffuse slowly in water and rapid uptake depletes their concentration near root surfaces so that concentrations increase as the distance from the root surface increases. Although NO_3^- is very mobile in soil solution and diffuses rapidly toward roots, a high demand by most plants maintains low NO_3^- concentrations throughout soil solution. Because of its high rate of diffusion, mass flow may meet the NO_3^- demand of some forest trees.

The size of the zone of depletion surrounding an individual root is directly proportional to its radius and the rate at which a particular ion diffuses in water. The radius of the zone of depletion ($R_{depletion}$ in cm) for any ion is

$$R_{depletion} = a + 2\sqrt{Dt},$$

where a is the root radius (cm), D is the diffusion coefficient for an ion (cm^2 sec^{-1}), and t is time in seconds (Nye and Tinker 1977). One also can compare root size (cm^3) to the zone of depletion (cm^3) it creates using the following expression:

$$\left[a + 2\sqrt{Dt}\right]^2 / a^2$$

Given this relationship, smaller roots exploit a greater volume of soil than larger roots (Fitter 1987). Recall that large roots have higher construction costs than small roots, indicating that plants "invest" relatively greater amounts of photosynthate into the production of large-diameter roots (Chapter 18). By forming small-diameter roots, plants increase the volume of soil exploited for water and nutrients, while reducing the amount of photosynthate allocated to root production. It is likely that this is why roots generally less than 1 mm in diameter are the nutrient- and water-absorbing organs of forest trees.

Once nutrients arrive at the root surface, they must be taken up from soil solution and incorporated into biologically active compounds. In some soils, the supply of some nutrients is excessive, and they are actively excluded from uptake. It is not uncommon to observe accumulations of $CaCO_3$ adjacent to the roots of desert shrubs growing on highly calcareous soils (Klappa 1980). In most instances, however, nutrient supply is low (especially for N and P), and plants actively forage for nutrients in soil. Nutrient foraging occurs through the physiological mechanisms of nutrient uptake and through morphological changes in root architecture that increase the absorptive area of plant root systems.

NUTRIENT UPTAKE AND ASSIMILATION BY ROOTS

Nutrients are incorporated into biologically active compounds (i.e., amino acids, nucleic acids) in a two-step process consisting of uptake and assimilation. **Uptake** is the physiological process by which nutrients in soil solution are actively transported across cell membranes into roots. **Assimilation** occurs when inorganic nutrients transported into plant cells are biochemically incorporated into organic compounds such as amino acids, nucleic acids, lipids, or other biologically active compounds.

High rates of enzymatic uptake are one physiological means by which plants can maximize acquisition of growth-limiting nutrients. For example, the uptake of nutrients from soil solution is mediated by enzymes associated with the membranes of very-fine roots (< 0.5 mm diameter). The rate of enzymatic activity mediating nutrient uptake increases with increasing nutrient concentrations in soil solution until the capacity of the enzyme system is saturated. Uptake capacity for nutrients varies widely among tree species, and this process can be highly responsive to soil temperature (Figure 19.5). In a comparison of trees of the taiga, rapidly growing species such as

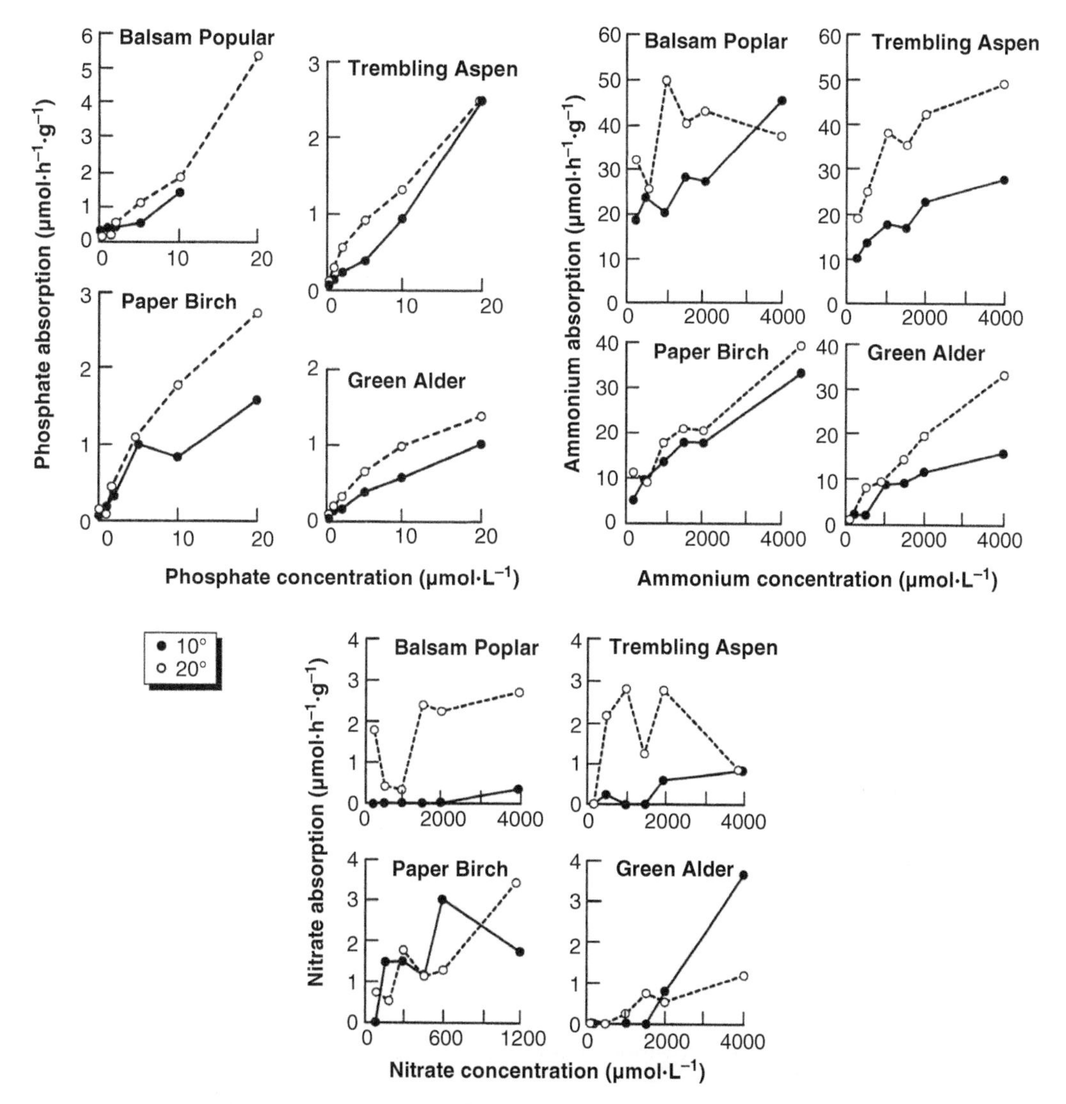

FIGURE 19.5 Nutrient uptake by four taiga trees in response to nutrient concentration and soil temperature. Uptake rates of NH_4^+, NO_3^-, and PO_4^{3-} were measured on excised roots of seedling growing under laboratory conditions. *Source:* Chapin et al. (1986). Reprinted from *Oecologia,* © Springer-Verlag Berlin, Heidelberg 1986. Reprinted with permission of Springer-Verlag, New York, Inc.

balsam poplar and trembling aspen have relatively high uptake capacities for NH_4^+, NO_3^-, and PO_4^{3-}, compared to the slower-growing paper birch and green alder (Chapin et al. 1986). Balsam poplar and trembling aspen also occur on relatively warm soils in the taiga and exhibit rapid increases in nutrient uptake with rising soil temperature (Figure 19.5).

Although the productivity of many temperate forests is constrained by soil N availability, we understand relatively little regarding the ecological importance of NH_4^+ versus NO_3^- uptake by overstory trees. Ammonium (NH_4^+) often is the dominant form of N in forest soils, and some trees exhibit a physiological preference for NH_4^+ over NO_3^-. This relationship can be observed in Figure 19.5 for taiga trees, and it also has been documented for several overstory trees in temperate forests (Figure 19.6). In the fine roots of Douglas-fir and sugar maple, rates of NH_4^+ uptake far exceed rates of NO_3^- uptake, suggesting that these widely distributed trees have a

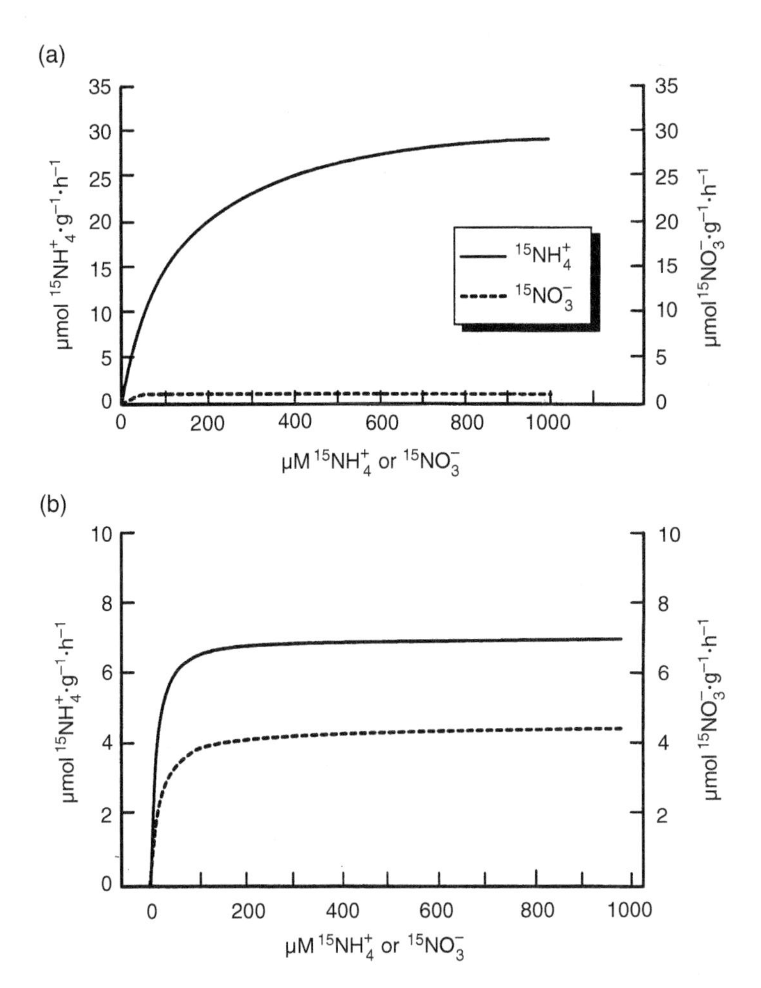

FIGURE 19.6 Ammonium and NO_3^- uptake in the fine roots of (a) sugar maple and (b) Douglas-fir. At any given concentration, rates of NH_4^+ uptake exceed those of NO_3^-, suggesting that these overstory species have a physiological preference for NH_4^+ over NO_3^-. *Source:* Kamminga-Van Wijk and Prins (1993) / Springer Nature. After Kamminga-Van Wijk and Prins 1993 and Rothstein et al. 1996. Panel (a) reprinted from *Oecologia,* © Springer-Verlag Berlin, Heidelberg 1996. Reprinted with permission of Springer-Verlag, New York, Inc. Panel (b) reprinted from *Plant and Soil,* 1993, Vol. 151, "The kinetics of NH_4^+ and NO_3^- uptake by Douglas-fir from simple N-solutions and from solutions containing both NH_4^+ and NO_3^-" by C. Kamminga-Van Wijk and H. Prins, pp. 91–96, © 1993 by Kluwer Academic Publishers. Reprinted with kind permission from Kluwer Academic Publishers.

physiological preference for NH_4^+, especially for sugar maple which has very low maximum rates of NO_3^- uptake.

Western hemlock and jack pine commonly occur on acidic soils with extremely low NO_3^- availability. These overstory species also have a limited capacity for NO_3^- uptake and appear to satisfy their demand for N through rapid NH_4^+ uptake (Lavoie et al. 1992; Knoepp et al. 1993). In jack pine, NO_3^- uptake enzymes are present at very low levels in fine roots (Lavoie et al. 1992), and when seedlings are supplied with NH_4^+ or NO_3^- as a sole source of N, those supplied with NH_4^+ attain twice the biomass of seedling grown solely on NO_3^-. (Lavoie et al. 1992). This observation suggests that jack pine has a low physiological capacity for NO_3^- uptake, even when concentrations in soil solution are relatively high. The inability of this species to use NO_3^- suggests that it would be a poor competitor in NO_3^--rich soils, relative to plants that have rapid rates of NO_3^- uptake and assimilation.

There are several reasons why NH_4^+ uptake is more rapid than NO_3^- uptake in many forest trees. First, NH_4^+ is often the dominant form of N in soil solution and small amounts can strongly inhibit enzymes involved with NO_3^- uptake and assimilation. Second, plants must reduce NO_3^- to NH_4^+ before it can be assimilated into biologically active compounds, a process that requires substantial amounts of energy (ATP). When this reaction occurs in leaves, it is subsidized by energy from the light-harvesting reactions of photosynthesis. However, many woody plants have the ability to reduce NO_3^- in roots, which requires the transport of reducing compounds from the leaf. Most often, plants with rapid rates of NO_3^- uptake and assimilation occur in high-light habitats in which excess energy from photosynthesis can be allocated to NO_3^- reduction.

A second physiological mechanism for maximizing the uptake of limiting nutrients is via the production of root enzymes that release nutrients from soil organic matter. This is particularly true of PO_4^{3-}, which diffuses slowly in soil solution. Phosphatases are enzymes produced by plants and microorganisms that act upon soil organic matter to yield PO_4^{3-} from organically bound P. The release of these enzymes into soil appears to be an important mechanism for increasing PO_4^{3-} supply to roots in PO_4^{3-}-poor soils. Phosphatase activity associated with the roots of arctic tundra plants can supply up to 65% of their annual PO_4^{3-} demand (Kroehler and Linkins 1988).

ROOT ARCHITECTURE, MYCORRHIZAE, AND NUTRIENT ACQUISITION

Root Architecture In addition to physiological mechanisms, plants also forage for nutrients by altering root architecture and growth. Ecologists have become increasingly interested in the structure and function of plant root systems, because plants appear to alter root branching patterns, or architecture, in response to soil nutrient availability (Fitter 1987; Fitter and Stickland 1991). Several analyses suggest that fine-root systems consisting of only a main axis and primarily laterals (herringbone architecture; Figure 19.7) are the most efficient at nutrient acquisition (Fitter and Stickland 1991; Berntson and Woodward 1992). The herringbone arrangement maximizes the volume of soil exploited by roots, while minimizing the overlap between the zone of depletion created by adjacent fine roots; however, the construction cost of this type of root system is high. It requires the production of larger diameter roots (e.g., higher construction), each of which transports a greater proportion of the total flow of water and nutrients into the aboveground portion of the plant (Figure 19.7). By contrast, the mean root diameter of a dichotomously branched root system is much smaller than that of a herringbone architecture, resulting in relatively lower construction costs. This spatial arrangement of roots allows the depletion zones of adjacent roots to overlap, and thus it is less efficient at foraging for nutrients than the herringbone architecture.

Differences in the cost of constructing and maintaining these root systems, and the efficiency with which they explore soil for nutrients, suggest that root-system architecture should vary in response to soil nutrient availability. Plants growing in nutrient-poor soils should allocate relatively more photosynthate to a highly efficient root system for nutrient uptake (i.e., herringbone

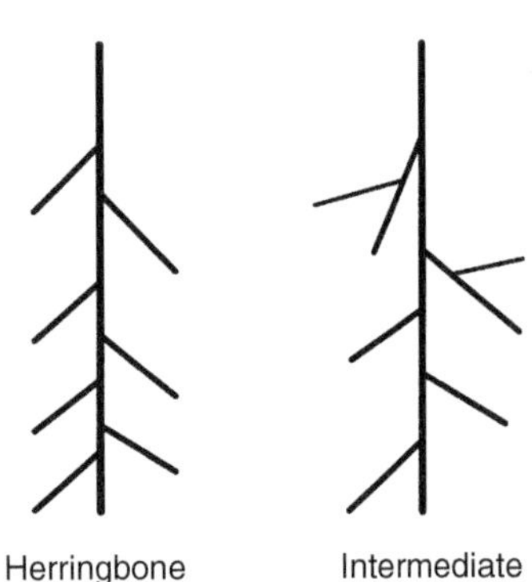

FIGURE 19.7 Examples of plant root systems with different degrees of branching. *Source:* Fitter (1987) / John Wiley & Sons.

branching pattern), whereas those growing in nutrient-rich soil should invest relatively less C into a dichotomously branched root architecture. These predictions hold for different herbaceous plants (dicots) from nutrient-rich and nutrient-poor soils, and also hold for the same species of herbaceous plants grown in soils of different fertility (Fitter and Stickland 1991; Berntson and Woodward 1992; Taub and Goldberg 1996). The fine-root system of forest trees appears to respond in a similar manner, but unfortunately, few studies have focused on these organisms. The fine roots of pin cherry rapidly proliferate within experimental patches ($125\,cm^3$) of N-rich soil by increasing rates of production and the degree of branching, that is, their fine-root system becomes more dichotomous (Pregitzer et al. 1992). Although there are clear differences in the construction cost of fine-root systems, it is also important to consider the maintenance cost of these structures, because maintenance costs could exceed the cost of their construction. This is especially true for fine roots with high N concentrations (enzymes for ion uptake), but, unfortunately, we presently do not have sufficient observations to understand the relationships among construction costs, maintenance costs, and fine-root longevity for a large number contrasting plant species.

Importantly, we still are learning how form and function of plant root systems are related, as well as how this might be coordinated with aboveground plant form and function (i.e., canopy geometry, maximum rates of net photosynthesis). Recent global analyses of plant root systems indicate that plants effective in nutrient-rich soil have a lower specific root length (i.e., smaller diameter per unit length) and higher N concentration than plants which are good competitors for nutrients in nutrient-poor soils (Weigelt et al. 2021). These root traits correspond to aboveground form and function, wherein leaves of plants in nutrient-rich soil are thinner, have high N concentrations, and more rapid rates of maximum net photosynthesis than those from N-poor soils. Like leaves, fine roots with high N concentrations likely have high maintenance costs to fuel the rapid uptake of nutrients from soil solution. Although this is an intriguing parallel, we still lack a comprehensive understanding of how fine root form (architecture) and function (ion uptake) are coordinated among forest trees or any other plant species. This remains an important gap in our understanding of the belowground physiology and natural history of plants as well as how they are coordinated, or not, with the aboveground plant attributes.

Mycorrhizae Forest trees increase their ability to forage for nutrients in soil by entering into a symbiotic relationship with soil fungi. This form of mutualism, termed a **mycorrhiza** or literally a *root-fungus,* increases the volume of soil exploited by an individual plant. Mycorrhizal fungi are important for the successful establishment of some tree seedlings. Mycorrhizae also facilitate an important nutrient cycling process—plant nutrient uptake and, to some extent, the decay of soil organic matter. In this section, we further explore the extent to which mycorrhizal fungi increase the ability of plant roots to forage for nutrients within soil.

Root hairs are not common in nature, because tree roots are almost universally colonized by mycorrhizal fungi. In pines, spruces, firs (and all other genera of the Pinaceae), birches, beeches,

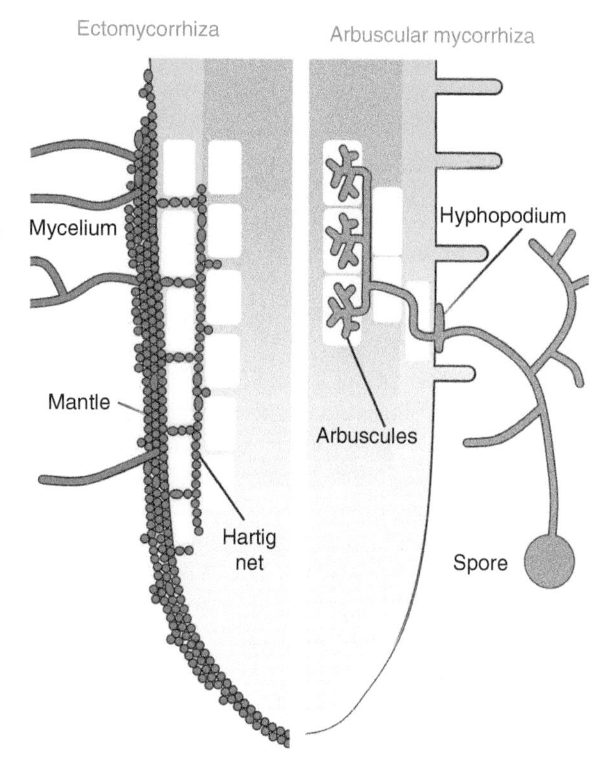

FIGURE 19.8 Shown are the manner in which the hyphae of ectomycorrhizal fungi (ECM) and arbuscular mycorrhizae (AM) fungi penetrate plant roots. A notable difference between is the fact that ECM hyphae grow throughout root cell wall, whereas AM hyphae oppress plant cell membranes and form arbuscules, which are the site of photosynthate and nutrient exchange.

oak, basswoods, and willows, ectomycorrhizal fungi (ECM; *ecto* meaning outside) form a sheath or mantle surrounding fine roots giving them a characteristic swollen appearance (Figure 19.8). Fungal hyphae penetrate the space between the outer cortical cells, but do not enter individual root cells. Over 2400 species of fungi are known to form ECM on North American trees (Marx and Beattie 1977). A less conspicuous group, arbuscular mycorrhizae (AM), forms no sheath, but individual hyphae grow within and between epidermal and cortical cells of roots. Many plant species form AM, including cultivated crops, grasses, and most tropical tree species (Janos 1987). Temperate species forming AM include redwood and many hardwoods: maples, ashes, tuliptree, sweet gum, sycamore, black walnut, and black cherry (Marx and Beattie 1977; Harley and Smith 1983).

Ectomycorrhizae and AM are effective accumulators of nutrients (and water) because they increase the absorbing surface area of tree roots, thereby allowing the plant to forage for nutrients in a greater volume of soil. Recall that small fine roots (approximately 0.1 mm in diameter) more effectively exploit soil for nutrients than larger roots. Fungal hyphae (10 μm in diameter) are much smaller than fine roots, which further increases the ability of plants to exploit soil for nutrients. Additionally, the hyphae of mycorrhizae can extend great distances into the forest floor and mineral soil. For example, 1 mm^3 of soil can contain 4 m of ECM hyphae that transport water and nutrients back to the host plant. In a 450-year-old Douglas-fir ecosystem in Oregon, 5000 kg ha^{-1} of ECM occur in the surface soil (0–10 cm), composing over 11% of the total root biomass (Trappe and Fogel 1977). Moreover, a single Douglas-fir–*Cenococcum* ectomycorrhizae can form 200–2000 individual hyphae, some of which extend over 2 m into soil and form more than 120 lateral branches or fusions with other hyphae.

As compared to non-mycorrhizal roots, those infected with mycorrhizal fungi tend to be more metabolically active, exhibiting rapid rates of respiration and ion uptake. Mycorrhizae are widely known to increase the PO_4^{3-} uptake of forest trees growing in P-poor soils; however, mycorrhizae also facilitate the uptake of other plant nutrients such as N (Bowen and Smith 1981). The mycorrhizae formed by the fungus *Paxillus involutus* are able to take up substantial amounts of NH_4^+ from soil solution, assimilate the NH_4^+ into amino acids, and transfer the newly formed amino acids to European beech (Finlay et al. 1989). *Paxillus involutus* also is able to assimilate NO_3^-, but it does so at a substantially lower rate. In addition to the direct uptake of ions from soil solution, mycorrhizae have the ability to produce extracellular enzymes which facilitate organic matter decomposition (Antibus et al. 1981; Dodd et al. 1987) and also produce organic acids which release plant nutrients from soil minerals (Bolan et al. 1984).

Ectomycorrhizae and AM fungi differ from one another in many important ways that have implications for plant nutrient supply. Foremost, AM fungi are evolutionary ancient and are thought to have mediated the transition of aquatic plants to those on land. They represent an evolutionary narrow group of fungi (i.e., monophyletic), which assist plants with the acquisition of both water and phosphorus. These organisms produce phosphatase enzymes which release PO_4^{3-} from organic matter, as well as organic acids that solubilize PO_4^{3-} from soil minerals. By contrast, ECM have evolved more recently and have arisen from multiple saprotrophic fungal ancestors (i.e., polypheletic; approximately 80 independent times). While some ECM have retained genes that encode enzymes mediating organic matter decay, others have lost them during their evolution into root symbionts (Pellitier and Zak 2018). Isotopic evidence suggests that ECM who have retained genes for saprotrophic function can modify soil organic matter and provide the N contained within it to a host plant (Pellitier et al. 2021). This appears particularly important where soil N supply is slow and large quantities of N are locked with soil organic matter. It is interesting to note that a large number of boreal trees associate with ECM whose genomes retain genes with decay function, likely a response to slow rates of organic matter decay in these cold ecosystems.

Because mycorrhizal fungi are heterotrophic microorganisms, they require a supply of carbohydrate for growth, for uptake of nutrients from soil, and for translocation of nutrients to the host plant. The cost of forming mycorrhizae is not inconsequential. Carbohydrate formed by the plant that could otherwise be used for growth, maintenance, or other physiological functions must be transferred to the mycorrhizal fungi in return for a greater nutrient supply. In a Pacific silver fir ecosystem, ECM compose only 1% of total ecosystem biomass, but their growth and maintenance consume over 15% of net primary production (Vogt et al. 1983). Because of the high cost of forming mycorrhizae, plants growing in nutrient-poor soil tend to form more mycorrhizae than those growing in nutrient rich-soils, because they are more dependent on them for nutrient acquisition.

PLANT LITTER AND THE RETURN OF NUTRIENTS TO FOREST FLOOR AND SOIL

Plant litter production above and below the soil surface directly controls the amount of nutrients returned to the forest floor and mineral soil, and therefore constitutes important processes controlling the cycling of nutrients within forest ecosystems. Nutrients absorbed by roots and mycorrhizae are used for a wide variety of physiological functions, including the growth of new tissue, the maintenance of existing tissue, storage, and the production of defense compounds—fates similar to that of photosynthetically fixed CO_2 (Chapter 18). Prior to abscission, both coniferous and deciduous trees withdraw nutrients from leaf tissue and transport them into the nearby branches for storage. These materials are removed from storage when growth commences in spring and are used to form new leaves and construct a canopy. However, it is not clear whether nutrients are withdrawn from fine roots prior to their death. Consequently, both the production of litter and the concentration of nutrients contained within it control the rate at which nutrients absorbed by plants are returned to the soil from which they came. In the paragraphs that follow, we explore geographic patterns of aboveground and belowground plant litter production and the extent to which plants withdraw nutrients prior to leaf abscission, a process controlling the nutrient concentration of leaf litter.

Leaf and Root Litter Production On a global basis, leaves account for approximately 60–75% of total aboveground litterfall in forest ecosystems; the remaining proportion is composed of woody material (approximately 30%) and reproductive structures (1–20%). In Chapter 18, we found that global patterns of aboveground net primary production (ANPP) were strongly related to climate, particularly temperature and actual evapotranspiration. Leaf litter production in forest ecosystems also is strongly related to global patterns of climate, because of the aforementioned relationship. In Figure 19.9, leaf litter production is low at high latitudes where short growing seasons limit plant growth; it increases toward the equator where plant growth can occur throughout the entire year. Also notice that substantial variation occurs at any particular latitude (Figure 19.9). This regional-scale variability in leaf litter production undoubtedly results from the modification of climate by physiography (i.e., slope and aspect), differences in soil water and nutrient availability, or disturbance.

Recall that leaf biomass (or area) places the upper limit on the amount of photosynthetically fixed CO_2 available for NPP (Chapter 18). As a result, leaf biomass and ANPP are positively related across a wide array of terrestrial ecosystems, including deserts, grasslands, and forests (Webb et al. 1983). Because leaves are temporary plant tissues, even in evergreen coniferous forests, one would expect that patterns of leaf litterfall should be related to ANPP. The solid line in Figure 19.10 indicates a 1:1 relationship between ANPP and leaf litter production, wherein all ANPP is allocated to leaf production. At low values of ANPP (i.e., $< 7.5\,\text{Mg}\,\text{ha}^{-1}\,\text{y}^{-1}$), note that leaf litter production represents a large proportion of ANPP. At higher rates of ANPP, leaf litter constitutes a smaller proportion, reflecting a greater allocation of photosynthate to other plant tissues and functions.

Although we have a clear understanding of global patterns of leaf litterfall, our knowledge of the production of plant litter below the soil surface is incomplete. It is clear, however, that the death of fine roots represents a significant proportion of NPP in many forest ecosystems, often exceeding leaf litter production. Notice that fine-root litter represents 40–330% of leaf litter in the temperate coniferous and deciduous forests summarized in Table 19.3. In the Pacific Northwest, the production and death of roots can account for 59–67% of NPP in Pacific silver fir ecosystems, whereas leaf litter production ranges from 6 to 8% of NPP (Vogt et al. 1983). Fine roots often have nutrient contents (N and P) greater than or equivalent to leaf litter, suggesting that fine-root litter represents a substantial transfer of nutrients from plant tissue to forest floor and mineral soil.

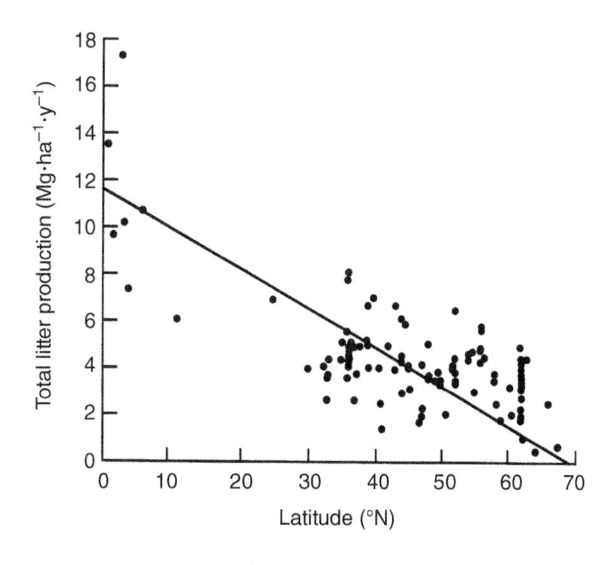

FIGURE 19.9 Global patterns of leaf litter production in forest ecosystems. *Source:* Modified from Bray and Gorham (1964) and O'Neill and DeAngelis 1981. Modified from *Advances in Ecological Research,* ©, 1964 by Academic Press, Ltd. London. Reprinted with permission of Academic Press, Ltd. London and from *Dynamic Properties of Forest Ecosystems,* © Cambridge University Press 1981. Reprinted with permission of Cambridge University Press.

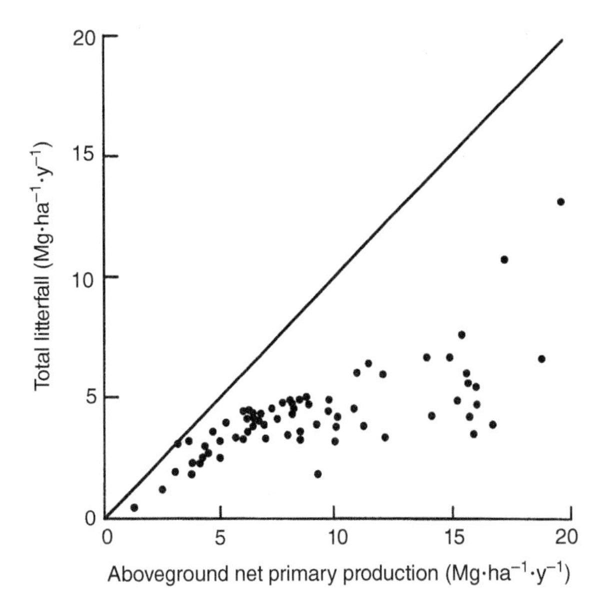

FIGURE 19.10 The relationship between aboveground net primary production (ANPP) and aboveground litter production in temperate and boreal forest ecosystems. The solid line indicates a 1:1 relationship between ANPP and the production of aboveground litter. At low levels of ANPP, virtually all production is allocated to leaves, and the departure from the 1:1 line at high levels of ANPP indicates a greater allocation of production to non-photosynthetic tissues. *Source:* After O'Neill and DeAngelis (1981) / Cambridge University Press. Reprinted from *Dynamic Properties of Forest Ecosystems,* © Cambridge University Press 1981. Reprinted with permission of Cambridge University Press.

Table 19.3 Leaf litter and root litter production in forest ecosystems from different geographic locations in temperate North America. The ratios of fine-root litter to leaf litter indicate that the production of fine-root litter in forest ecosystems ranges from 40 to 330% of leaf litter.

	Litter		
Ecosystem	Leaf	Fine-root	Root:leaf
	Mg ha^{-1} y^{-1}		
Temperate coniferous			
Pacific silver fir	1.5–2.2	4.4–7.4	2.9–3.3
Douglas-fir	2.8–3.0	2.4–2.7	0.9
Scots pine	2.1	2.0	1.0
White pine	2.9	2.6	0.9
Mixed pine	3.1	2.6	0.8
White spruce	2.7	1.6	0.6
Red pine	2.5–5.3	1.3–2.0	0.4–0.5
Temperate deciduous			
Black oak	4.1	5.9	1.4
Northern red oak	4.2	5.2	1.2
White oak	3.6	4.1	1.1
Sugar maple	2.9	4.0	1.4
Paper birch	2.8	3.2	1.1
Mixed hardwoods	4.4	2.7	0.6
American beech	3.2	1.3	0.4

Source: Vogt et al. (1986) / with permission of Elsevier and Nadelhoffer et al. 1985. Modified from *Advances in Ecological Research,* © 1986 by Academic Press, Inc. Reprinted with permission of Academic Press, Inc. and from *Ecology* by the permission of the Ecological Society of America.

Recall that the return of nutrients to the forest floor and mineral soil is controlled by the production of plant litter and its nutrient concentration. From the previous paragraphs, it should be clear that forest ecosystems dramatically differ in the production of litter, both above and below the soil surface. In the following paragraphs, we discuss the withdrawal, or **retranslocation**, of nutrients prior to the shedding of leaf litter. This process directly influences the concentration (i.e., percent) of nutrients in litter, which together with litter production controls the total return of nutrients in plant litterfall.

Nutrient Retranslocation The biochemical CO_2-fixing machinery of leaves requires substantial quantities of N and P to form photosynthetic proteins, as do the nutrient uptake and assimilation mechanisms of fine roots. As a result, green leaves and live fine roots contain relatively high concentrations of nutrients compared to other plant tissues, such as stems and branches. Because low quantities of N and P in soil often limit plant growth, plants have evolved mechanisms to conserve growth-limiting nutrients within their tissues. As noted earlier in the text, in both coniferous and deciduous trees substantial quantities of N and P (and other nutrients) are retranslocated from leaves prior to abscission and are mobilized into storage. Much of the N and P removed from leaves resides within adjacent branches, stem, and large structural roots, where they can be remobilized to meet the demand of newly forming leaves when growth resumes. However, the extent to which forest trees retranslocate nutrients from senescent leaves is highly variable (Killingbeck 1996); it changes from year to year within an individual species and changes among species from different habitats.

The retranslocation of many nutrients reflects their availability in soil and their demand within the plant for subsequent growth. In Brazilian rainforest trees, 17–73% of N and 41–83% of P are translocated from leaves prior to abscission. Although retranslocation efficiency (i.e., percent removed) varies among rainforest species, they all reduce P to similar concentrations in abscised leaves (i.e., 0.04–0.06% P; Scott et al. 1992). The retranslocation of N is not as complete and relatively large amounts of N (>1.0%) remain in abscised leaves (Scott et al. 1992). Differences in the amount of P and N remaining in abscised leaves of rainforest trees reflect the fact that P availability in soil, not N, limits the productivity of wet tropical forests (Vitousek 1984).

The extent to which forest trees retranslocate nutrients from senescing leaves directly influences the **nutrient-use efficiency** of litter, which is defined by the following expression:

$$\text{Nutrient use-efficiency} = \frac{\text{Dry weight of leaf litterfall}\left(\text{kg ha}^{-1}\text{y}^{-1}\right)}{\text{Nutrient content of leaf litterfall}\left(\text{kg ha}^{-1}\text{y}^{-1}\right)}.$$

Plants that retranslocate large quantities of nutrients prior to leaf abscission have high nutrient-use efficiencies and return relatively few nutrients (per unit biomass) to the forest floor in litterfall.

On a global basis, there is a remarkable relationship between the N- and P-use efficiency and nutrient availability in tropical, temperate, and boreal forests. In Figure 19.11, the amount of nutrients returned in litterfall (x-axis) is used as a surrogate for nutrient availability—large amounts of nutrients residing in litterfall occurs where their availability in soil is high. Notice that N-use efficiency is inversely related to the amount of N contained in litterfall, with the exception of tropical forests (Figure 19.9a). The inverse relationship between N-use efficiency and litterfall N in temperate coniferous and deciduous forests indicates that N-use efficiency is high in ecosystems in which relatively small quantities of N are annually cycled (i.e., those with low litterfall N contents).

In tropical ecosystems, litterfall N varies widely (10–180 kg N ha^{-1} y^{-1}), but N-use efficiency is relatively constant (approximately 80; Figure 19.11a). The fact that N-use efficiency changes little, whereas the amount of N annually cycled to the forest floor varies by a factor of three,

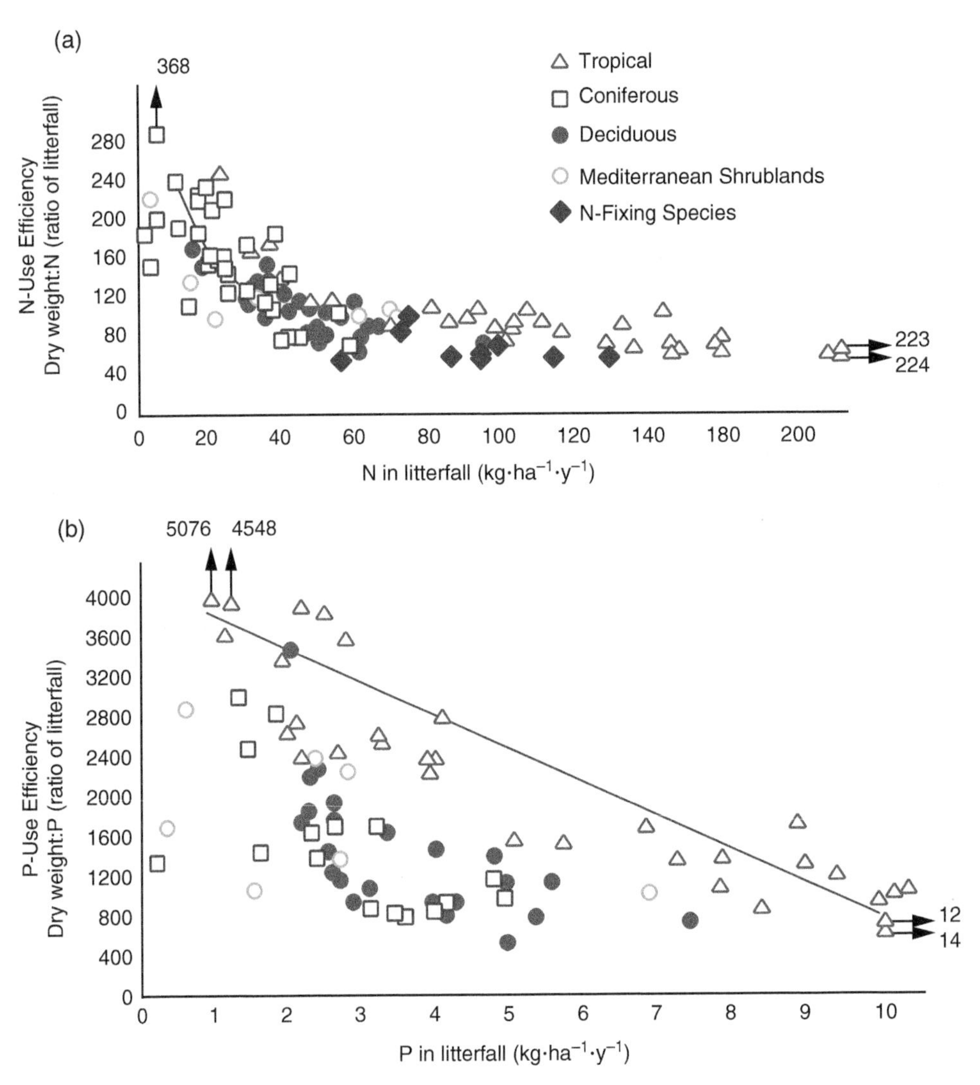

FIGURE 19.11 The relationship between plant nutrient-use efficiency and the amount of nutrients returned in leaf litterfall for tropical, coniferous, temperate deciduous, and Mediterranean shrublands. Forests dominated by N-fixing plants are denoted with a diamond. Plant nutrient-use efficiency is estimated as the ratio of litterfall mass to its (a) N or (b) P content. The amount of N and P returned in litterfall (kg N or P ha^{-1} y^{-1}) is a relative index of soil N or P availability. *Source:* After Vitousek (1982) / John Wiley & Sons. Reprinted with permission from *American Naturalist,* © 1982 by the University of Chicago. Reprinted with permission of the Chicago University Press.

indicates that N availability does not influence the N-use efficiency of tropical forest trees. In contrast to this pattern, P-use efficiency and the amount of P returned in litterfall are inversely related in tropical forests (Figure 19.9b). This pattern suggests that P translocation from senescent leaves is high (i.e., high P-use efficiency) where P availability is relatively low within the ecosystem. Also notice that there is a weak relationship between P-use efficiency and litterfall P in other forest ecosystems. These patterns are consistent with the observation that N generally limits the productivity of boreal and temperate forests, whereas a limited supply of P constrains productivity in wet tropical forests.

Taken together, these observations suggest that limited supplies of N or P result in greater translocation prior to leaf abscission, thereby increasing N- or P-nutrient-use efficiency of leaf litter. This is likely an evolved mechanism that permits plants growing in low N or P soils to conserve these growth-limiting nutrients. Forest trees that conserve nutrients through a high nutrient-use efficiency should have a competitive advantage on nutrient poor sites, and it is not uncommon to observe species replacement along gradients of nutrient availability (Pastor et al. 1984; Zak and Pregitzer 1990). Individual species also are able to alter nutrient-use efficiency in response to nutrient supply, which may allow them to persist in soils of markedly different fertility. Northern red oak, for example, occurs on soils with a wide range of N availability in the northern Lake States; N-use efficiency in this species drops as soil N availability increases; similar responses have been observed in sugar maple, basswood, and American beech (Zak et al. 1986). Although fine-root mortality represents a significant input of litter to forest floor and mineral soil, we do not yet understand the extent to which soil nutrient availability influences the nutrient-use efficiency of fine roots. While it is relatively easy to collect leaves prior to and following abscission, fine roots present a unique challenge to gain this same insight—they remain buried in soil and are difficult to observe and analyze for nutrient content.

The combined influence of plant litter production and nutrient-use efficiency on the amount of N returned to forest floor and mineral soil is summarized in Table 19.4. Notice that N inputs from fine-root litter are equivalent to or greater than the amount of N contained in leaf litter, regardless of whether forests occur in tropical, temperate, and boreal regions (i.e., root:leaf N ratio >1). The combined amounts of N entering forest floor and mineral soil are generally greatest in tropical forests, in which leaf decay is rapid. Recall rates of N input from atmospheric deposition (see Figure 19.3) and free-living fixation (approximately 1–5 kg N ha^{-1} y^{-1}) and realize that leaves and fine roots annually contribute 10 times more N to forest floor and mineral soil. Clearly, aboveground and belowground litter production and its nutrient-use efficiency represent an important process influencing the internal cycling of nutrients within terrestrial ecosystems.

NUTRIENTS IN THE FOREST FLOOR

Large amounts of organic matter and nutrients can accumulate on the soil surface in boreal and cool temperate forests, emphasizing the importance of the forest floor in controlling the flow of nutrients within these ecosystems (Table 19.5). Nevertheless, forest floor organic matter and nutrient content do not vary in a predictable manner among tropical, subtropical, and warm-temperate forests. In fact, the quantity of organic matter and nutrients in the forest floor of dry and montane tropical forests differ greatly from wet tropical forests and are more similar to that in temperate forests (Anderson and Swift 1983). Nevertheless, several important generalizations can be drawn from the information in Table 19.5. First, the accumulation of organic matter and nutrients in the forest floor of cold-temperate and boreal evergreen forests is much greater than that in warm-temperate, subtropical, or tropical forests. In temperate and tropical regions, deciduous forest floors contain half the organic matter of evergreen forest floors, regardless of whether they are dominated by needleleaf or broadleaf evergreen species. Although the greatest forest floor N and P contents occur in boreal evergreen forests, there is substantial variation among the N and P contents of tropical, subtropical, and temperate forest floors.

The quantity of organic matter and nutrients contained in the forest floor of tropical, temperate, and boreal forests reflect two opposing processes: the production and the decomposition of plant litter. Consequently, it is not surprising that patterns of leaf and root litter production alone do not well reflect differences in forest floor mass among different forest ecosystems (compare Tables 19.4 and 19.5). Temperature, precipitation, and the biochemical constituents of plant

Table 19.4 The nitrogen content of leaf and fine-root litter in forest ecosystems from different geographic locations.

Ecosystem	Litter		Root:leaf N
	Leaf	Fine root	
	kg N $ha^{-1}y^{-1}$		
Tropical			
Broadleaf evergreen	119	255	1.9
Warm temperate			
Broadleaf deciduous	36	44	1.2
Cold temperate			
Needleleaf evergreen			
White pine	21	40	2.0
Mixed pine	16	36	2.2
white spruce	28	22	0.8
red pine	12	19	1.6
Broadleaf deciduous			
black oak	31	79	2.5
Northern red oak	30	62	2.0
white oak	26	47	1.8
sugar maple	23	47	2.0
paper birch	25	43	1.7
Boreal			
Needleleaf evergreen	24	26	1.1

Compare the magnitude of nitrogen entering the soil from leaf and fine-root litter and notice that the N content of fine-root litter is equivalent to or greater than that of leaf litter.

Source: Vogt et al. (1986) / with permission of Elsevier and Nadelhoffer et al. 1985. Modified from *Advances in Ecological Research*, © 1986 by Academic Press. Inc. Reprinted with permission of Academic Press, Inc. and from *Ecology* by the permission of the Ecological Society of America.

litter all influence the rate at which leaf and root litter is decomposed by soil microorganisms; differences in these factors contribute to the variation in forest floor accumulation among ecosystems in Table 19.5. Earthworm activity is also of importance, because these organisms incorporate fresh leaf litter into surface mineral soil horizons which accelerates the decomposition process (Edwards and Bohlen 1996). In the absence of earthworms, relatively thick, distinct organic horizons develop over thin A horizons. By contrast, discontinuous and thin organic horizons form over thick, organic-matter-rich A horizons in the presence of earthworms.

A simple model considering plant litter production and decomposition has been used to gain insight into forest floor dynamics (Olson 1963). This mass-balance approach assumes that the annual rate of litter decomposition equals the annual rate of plant litter production, such that the amount of organic matter in the forest floor is unchanged. This condition only occurs very late in forest ecosystem development, and it is where this approach has been used to estimate the rate of decomposition. Under this assumption, a constant proportion (k) of forest floor organic matter decomposes on an annual basis:

$$\text{Litter Production} = k\left(\text{Forest Floor Mass}\right).$$

Table 19.5 Forest floor mass and nutrient contents in tropical, temperate, and boreal forest ecosystems. These values have been summarized for a wide range of forest ecosystems occurring in tropical, temperate, and boreal regions of the Earth.

Ecosystem	Forest floor		
	Organic matter ($Mg\,ha^{-1}$)	N	P
		$kg\,ha^{-1}$	
Tropical			
Broadleaf evergreen	22.6	325	8
Broadleaf semideciduous	2.2	35	–
Broadleaf deciduous	8.8	–	14
Subtropical			
Broadleaf evergreen	22.1	121	5
Broadleaf deciduous	8.1	–	–
Warm temperate			
Broadleaf evergreen	19.1	60	4
Needleleaf evergreen	20.0	362	25
Broadleaf deciduous	11.5	163	12
Cold temperate			
Needleleaf evergreen	44.6	200	10
Broadleaf deciduous	32.2	624	50
Boreal			
Needleleaf evergreen	44.6	875	81

Source: Vogt et al. (1986) / with permission of Elsevier. Modified from *Advances in Ecological Research,* © 1986 by Academic Press, Inc. Reprinted with permission of Academic Press. Inc.

In this equation, k represents the decomposition rate constant for the entire forest floor. With knowledge of litter production and forest floor mass, values of k are derived using the following equation:

$$k\left(\mathrm{y}^{-1}\right)=\frac{\text{Litter production}\left(\mathrm{Mg\,ha^{-1}y^{-1}}\right)}{\text{Forest floor mass}\left(\mathrm{Mg\,ha^{-1}}\right)}.$$

When decomposition is rapid, relatively small amounts of organic matter accumulate in the forest floor, and values for k are typically greater than $1\,\mathrm{y}^{-1}$. This situation occurs in late-successional tropical forests in which microbial activity has the potential to respire more organic matter than that contained in aboveground and belowground litter production (Cuevas and Medina 1988). By contrast, decomposition is slow in cold, boreal forests where substantial amounts of organic matter and nutrients accumulate in the forest floor; values of k are approximately $0.01\,\mathrm{y}^{-1}$.

If the forest floor is neither gaining nor losing organic matter (i.e., it is in equilibrium), then k can be used to estimate the mean residence of time of organic matter or nutrients (l/k), which is the average time organic matter or nutrients reside in the forest floor. Values range widely among the forest ecosystems listed in Table 19.6, primarily reflecting differences in temperature, precipitation, and plant litter biochemistry. The mean residence time for organic matter, N, and P in the forest floor of tropical forests is less than 1 year (Table 19.6), whereas these constituents can reside within the forest floor of boreal forests for tens to hundreds of years.

Table 19.6 The mean residence time of organic matter, nitrogen and phosphorus, in the forest floor of tropical, temperate, and boreal forest ecosystems. Mean residence time of forest floor organic matter is calculated by dividing the forest floor mass (kg ha^{-1}) by the annual leaf litterfall mass (kg ha^{-1} y^{-1}); residence times for nutrients are calculated in the same manner.

	Mean residence time in years[a]		
Ecosystem	Organic matter	N	P
Tropical			
Broadleaf evergreen	2	2	1
Broadleaf semideciduous	0.4	0.2	–
Broadleaf deciduous	0.9	–	2
Subtropical			
Broadleaf evergreen	6	3	2
Broadleaf deciduous	2	–	–
Warm temperate			
Broadleaf evergreen	3	1	2
Needleleaf evergreen	5	14	11
Broadleaf deciduous	3	5	4
Cold temperate			
Needleleaf evergreen	18	33	22
Broadleaf deciduous	10	19	11
Boreal			
Needleleaf evergreen	60	138	225

[a]Mean residence times are $1/k$.

Source: Vogt et al. (1986) / with permission of Elsevier. Modified from *Advances in Ecological Research.* © 1986 by Academic Press, Inc. Reprinted with permission of Academic Press, Inc.

Relatively large amounts of N and P (Table 19.6) in the forest floor of cool temperate and boreal forests, combined with long residence times ($1/k$), emphasize the importance of the forest floor as a site for nutrient storage in these forest ecosystems. Also, compare the mean residence times of coniferous (i.e., needleleaf) and deciduous forests, and notice that organic matter and nutrients reside in the forest floor of coniferous forests for relatively longer periods of time. This pattern reflects the fact that conifers contain greater proportions of organic compounds that are not easily metabolized by soil microorganisms, thus they have slower rates of litter decomposition.

Mean residence times in Table 19.6 are calculated using rates of aboveground litter production alone, which comprise only 28–56% of total litter input to forest floor (see Table 19.4). Using leaf litter production as the primary input of plant litter, the mean residence time of organic matter in the forest floor of cool temperate forests ranges from 8 to 67 years. The range of values is much lower when fine-root litter is included in the calculation of mean residence time (e.g., 5–15 years; Vogt et al. 1986), further emphasizing the importance of belowground litter production in studying the storage and cycling of nutrients in forest ecosystems.

ORGANIC MATTER DECOMPOSITION AND NUTRIENT MINERALIZATION

To fully understand the process of decomposition and the release of nutrients from plant litter and organic matter, one needs to consider the factors that control microbial growth and maintenance in soil. Once abscised leaves, dead roots, and other plant tissues enter forest floor and mineral soil, they become a growth substrate for saprotrophic microorganisms. During the decomposition process, soil bacteria, actinobacteria, and fungi assimilate the organic compounds contained in plant litter into their cells for biosynthesis (i.e., growth and maintenance), albeit at different rates depending on the types of compounds contained in plant litter. Whether soil microorganisms release nutrients into, or assimilate nutrients from, soil solution depends on the biochemical constituents of plant litter and their suitability for microbial growth and maintenance. In the paragraphs that follow, we discuss how plant litter functions as a substrate for microbial growth and maintenance in soil, and, in turn, how these microbial processes influence the supply of growth-limiting nutrients to plants. As you will read later in this section, the amount of energy that soil microorganisms derive from plant litter controls the processes of decomposition and soil N availability.

Biochemical Constituents of Plant Litter The leaves, stems, and roots of plants are constructed of a remarkably small set of organic compounds, regardless of differences in phylogeny and growth form. Plant litter is largely composed of cellulose (15–60%), hemicellulose (10–30%), lignin (5–30%), protein (2–15%), fats (1%), and soluble compounds such as sugars, amino acids, nucleic acids, and organic acids (10%). These organic compounds carry out essential metabolic functions, provide physical support, and defend plants against herbivores and pathogens. Upon entering soil, however, they become substrates that yield different amounts of energy for microbial growth and maintenance.

The process of decomposition is mediated by the microbial production of enzymes that harvest the biochemical energy contained in compounds formed during the biosynthesis of plant tissue. In most terrestrial ecosystems, microbial growth in soil is limited by usable forms of energy, and amounts contained in annual litter production are only sufficient to maintain (i.e., no growth) microbial populations in soil. The microbial production of enzymes that make the energy contained in plant compounds available is energetically expensive. Consequently, energy produced during the microbial metabolism of a particular plant compound must surpass the cost of the enzymes used to harvest its energy in order for microorganisms to use it as a growth substrate.

The use of a plant-derived organic compound as a substrate for microbial growth is determined by: (i) the amount of energy released by breaking different types of chemical bonds, (ii) the size and three-dimensional complexity of molecules, and (iii) its nutrient content (N and P). Look at Figure 19.12 and notice that the complexity of chemical bonds, three-dimensional structure, and nutrient content varies dramatically among the primary constituents of plant litter. Simple sugars (i.e., carbohydrates), such as glucose, are among some of the first products of photosynthesis, and although their concentration in fresh litter is low (<5%), these water-soluble compounds yield substantial amounts of energy for microbial growth. In soil, glucose and other simple sugars can be directly taken up and used to fuel energy-producing biochemical reactions (i.e., glycolysis and tricarboxylic acid cycle which generate ATP) within microbial cells. Given these characteristics, it should not be surprising that simple sugars are rapidly decomposed in plant litter.

Amino acids are the building blocks of proteins and molecules that carry out all biochemical reactions. A single protein contains thousands of amino acid subunits, making them too large to be directly taken up by microbial cells. Proteases released by microbial cells enzymatically cleave proteins into smaller polypeptides and eventually into their amino acid subunits. These relatively small, high-energy-yielding molecules can then be taken up by microbial cells and used for protein synthesis or energy production. Notice that proteins and amino acids are the only plant compounds in Figure 19.12 that contain N, an essential nutrient also required for microbial growth and maintenance. This fact has important implications for the release of N from plant litter during the process of mineralization, which we discuss later in this section.

Starch is a storage carbohydrate (Figure 19.12) and its synthesis occurs when the supply of photosynthate exceeds the carbohydrate requirements for the construction and maintenance of plant tissues. The concentration of this energy-storing glucose polymer is typically low in many types of plant litter, because starch is used for maintenance respiration prior to the senescence of leaves and

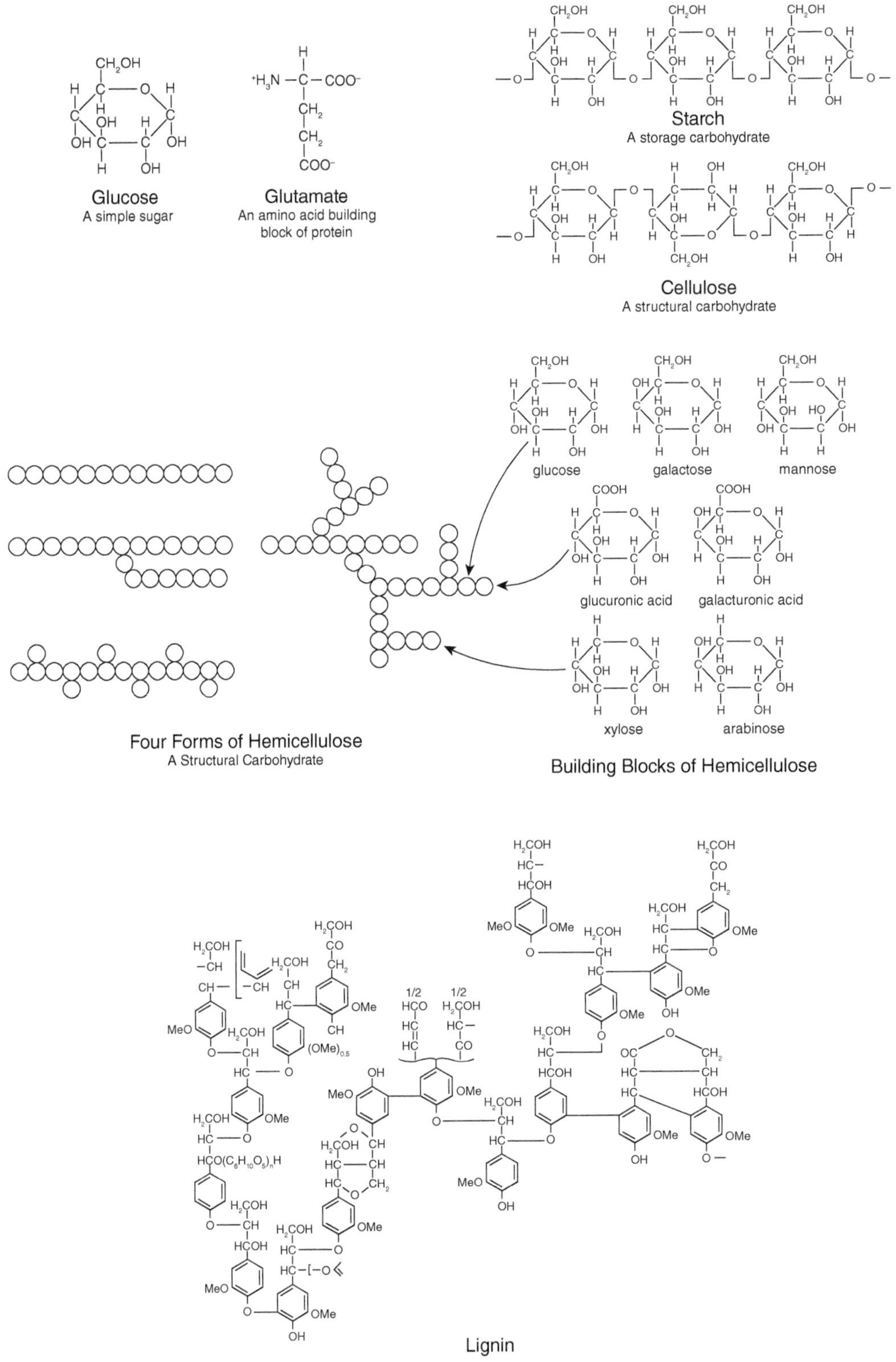

Glucose
A simple sugar

Glutamate
An amino acid building block of protein

Starch
A storage carbohydrate

Cellulose
A structural carbohydrate

glucose galactose mannose

glucuronic acid galacturonic acid

xylose arabinose

Four Forms of Hemicellulose
A Structural Carbohydrate

Building Blocks of Hemicellulose

Lignin

FIGURE 19.12 The chemical constituents of plant litter that influence the rate of organic matter decomposition in forest floor and mineral soil. Notice that nitrogen is a component of amino acids and proteins, but it is not found in the chemical structure of the other plant compounds illustrated in this figure.

fine roots. Starch is broken down by many soil bacteria, actinobacteria and fungi, which synthesize extracellular enzymes (i.e., amalyses) that cleave starch into subunits consisting of several glucose molecules; these subunits can then be taken up and assimilated by microbial cells where they are used to produce energy. Although the decomposition of starch is somewhat slower than that of simple sugars, its metabolism yields relatively large amounts of energy (ATP) for microbial growth.

Cellulose is the primary component of plant cell wall (Figure 19.12) and is the most abundant compound in plant litter. It is often associated with hemicellulose and lignin, which also are present in the plant cell wall. Cellulose is formed by linking glucose molecules into an unbranched, semicrystalline polymer. Notice that starch and cellulose differ in the type of bond joining adjacent glucose subunits (Figure 19.10). Unlike starch, the glucose subunits of cellulose cannot be mobilized by plants for use in energy production once it is formed. Cellulose is a structural polymer, whereas starch is an energy-storage polymer.

The decomposition of cellulose occurs as a two-step process. First, the bacterial production of the enzyme cellobiohydrolase repeatedly cleaves off two glucose subunits from the end of the molecule. Second, the resulting disaccharide, cellobiose, is degraded into glucose by the enzyme β glucosidase. As with starch degradation, the glucose subunits can be directly assimilated by the decomposing microorganism. Unlike proteins and amino acids that contain N, both starch and cellulose are exclusively composed of C, H, and O.

Hemicellulose is a heterogeneous group of polymers composed of several types of plant sugars (Figure 19.12). The linking of these sugar subunits produces branched as well as unbranched molecules, which in pure state are often broken down quite rapidly. In nature, however, the association of hemicellulose with other substances such as lignin in the plant cell wall makes their breakdown more difficult. Pectin (polygalacturonic acid) is a type of hemicellulose molecule found in the middle lamella of plant cell walls, and it is broken down by several enzymes collectively known as pectinases. These enzymes appear to be primarily produced by soil fungi and actinobacteria, which use the sugar subunits as a source of energy. The initial entry of mycorrhizal fungi into plant roots is thought to be facilitated by the production of pectinases. There is little difference between cellulose and hemicellulose in energy yield and hence their rate of decay by microorganisms.

Lignin is by far the most biochemically complex constituent of plant litter, containing a variety of chemical bonds and three-dimensional conformations (Figure 19.12). It is an abundant component of leaf litter, and high concentrations occur in most woody tissues. Lignin molecules surround cellulose and hemicellulose microfibrils in the plant cell wall, providing rigidity and protecting them against pathogenic fungi and bacteria that break down plant cell walls. Because of its biochemical complexity and the fact that most soil microorganisms cannot enzymatically harvest the energy contained within its chemical bonds, lignin decomposition occurs at a very slow rate. White-rot (*Pleurotus*, *Phanerochaete*) and brown-rot fungi (*Poria*, *Gloeophyllum*) are primarily responsible for the degradation of lignin in plant litter. In order to degrade lignin, these organisms require the presence of an alternative energy source, which functions as the primary growth substrate. As a result, only small portions of the C contained in lignin are actually incorporated into fungal cells; the majority enters into reactions that eventually form humus.

Dynamics of Decomposition The chemical constituents of plant litter and their use for microbial biosynthesis directly control the rate at which plant litter decays on and in the soil. The microbial breakdown of the biochemical constituents in plant litter can be described using the following first-order rate equation:

$$A_t = A_0 e^{-kt}.$$

In this equation, the amount of substrate remaining (A_t in $\mu g g^{-1}$) at any point during the decomposition process is proportional to its initial concentration (A_0 in $\mu g g^{-1}$) and its first-order rate constant for decomposition (k in days^{-1}, months^{-1}, or years^{-1}). First-order decomposition rate constants are experimentally derived values obtained by measuring the decline in a substrate

Table 19.7 Rate constants for the decomposition of organic compounds contained in plant litter.

Compound	Rate constant for decay k (day^{-1})
Glucose	0.500–1.000
Protein and simple sugars	0.200
Cellulose	0.036–0.080
Hemicellulose	0.030–0.080
Lignin	0.003–0.010

Values were determined by decomposing purified compounds under laboratory conditions.
Source: Adapted from Alexander (1977), Paul and Clark (1996), and van Veen et al. (1984).

during its decomposition, often under laboratory conditions. As such, they directly reflect how rapidly a particular compound can be used for microbial biosynthesis. Rapidly decomposed substrates have a high k, whereas substrates that provide little energy for microbial growth and maintenance have low values. Decomposition rate (k) constants also are sensitive to temperature and soil water potential, doubling for a 10 °C increase in temperature and attaining maximum values near field capacity.

In Table 19.7, we summarize decomposition rate constants for the primary constituents of plant litter. Notice that glucose, a small, high-energy-yielding molecule has a very-rapid decomposition rate constant, as do other simple carbohydrates and proteins. Contrast these values with the low-rate constant of lignin. Complex chemical bonds, subunits that provide little energy for microbial growth, and a folded three-dimensional structure that protects the inner portion of the lignin molecule against enzymatic attack all contribute toward its slow rate of microbial degradation. Also notice that cellulose and hemicellulose have similar decomposition rate constants, reflecting the fact that these molecules provide equivalent amounts of energy for microbial growth (Table 19.7).

The biochemical attributes of plant-derived molecules that influence their use for microbial growth are clearly reflected in the decomposition rate constants listed in Table 19.7. However, plant litter is not composed of a single organic compound, but instead contains varying proportions of simple sugars, protein, cellulose, hemicellulose, and lignin. How do different proportions of these compounds in plant litter influence the rate at which it is metabolized by soil microorganisms?

Consider the decomposition of an abscised leaf composed of 10% protein and simple carbohydrates, 25% lignin, and 65% cellulose and hemicellulose (Figure 19.13a). The overall decline in leaf mass seen in Figure 19.11a results from microbial respiration for growth and maintenance, which returns this photosynthetically fixed CO_2 to the atmosphere. Notice that the decomposition of this leaf is initially rapid (0–20 days), but that rate slows down during the later stages of decay. During the initial stages of decomposition, the leaching of water-soluble compounds and the metabolism of proteins and simple carbohydrates (high decomposition rate constants) contribute to a rapid loss of C from the leaf. In this example, proteins and simple carbohydrates are almost totally degraded after only 15 days of microbial decay. Metabolism of cellulose and hemicellulose influences the intermediate stages of decomposition (10–50 days), while the latter stages (50–100 days) of decomposition are dominated by lignin breakdown. The concentration and decomposition rate constants of these compounds additively yield the overall pattern of leaf decomposition in Figure 19.13a.

The situation is much different for an abscised leaf containing lower amounts of lignin (5%) and higher proportions of protein and simple sugars (25%), and cellulose and hemicellulose (70%; Figure 19.13b). Compared to the leaf in Figure 19.13a, the leaf containing 5% lignin (Figure 19.13b) exhibits a more rapid overall decline in C than the leaf containing 25% lignin (Figure 19.13a).

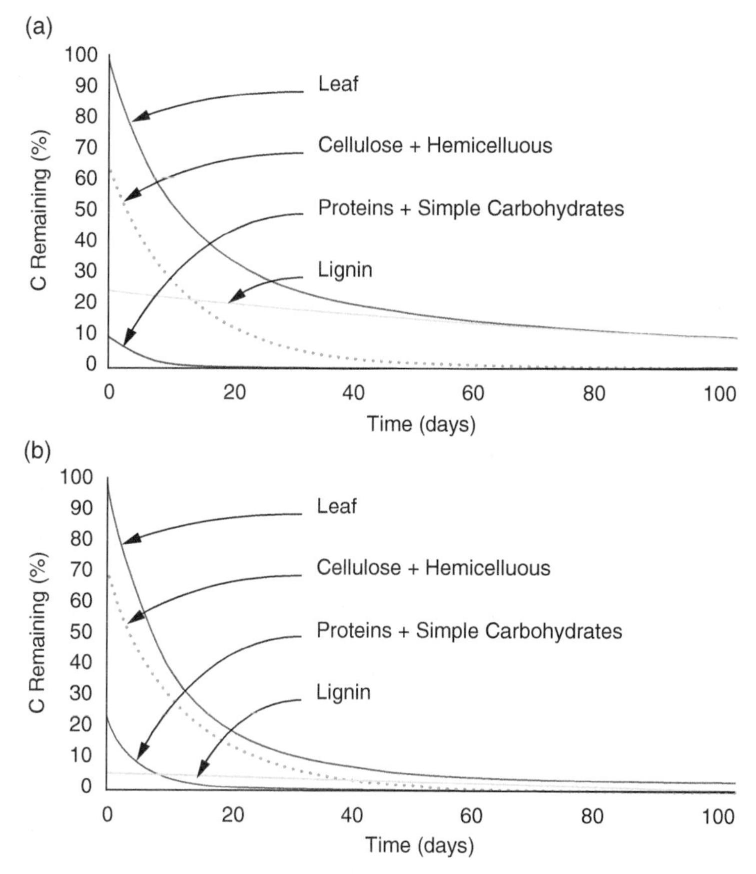

FIGURE 19.13 The decomposition of an abscised leaf in relationship to initial concentrations of protein, simple carbohydrates, hemicellulose, cellulose, and lignin. The amount of leaf C remaining is controlled by the concentration and decomposition rate constant of protein and simple sugars, cellulose and hemicellulose, and lignin. The amounts of leaf C remaining in panels (a) and (b) were calculated using the following series of first-order decay equations (percent leaf C $= A_1e^{-k1t} + A_2e^{-k2t} + A_3e^{-k3t}$, where A_1 = percent protein and simple carbohydrates, A_2 = percent cellulose and hemicellulose, and A_3 = percent lignin). The decomposition rate constant values (k) are 0.2 days^{-1} for protein and simple carbohydrates, 0.08 days^{-1} for cellulose and hemicellulose, and 0.01 days^{-1} for lignin. Note that differences in initial concentrations have a profound influence on leaf decomposition. High lignin concentrations and low concentrations of other constituents result in relatively slow decomposition rates.

Moreover, a higher proportion of protein and simple carbohydrates (Figure 19.13b) results in a rapid initial decline in leaf C, much greater than that of the leaf material lower in protein and simple carbohydrates (Figure 19.13a). Clearly, the relative proportion of easily metabolized, high-energy-yielding plant compounds, such as simple sugars, and of biochemically complex molecules that function as poorer substrates (i.e., lignin) for microbial growth has a profound influence on the overall decay of plant litter entering forest floor and mineral soil.

Decomposition is not solely controlled by differences in the biochemical constituents of plant litter. It also is controlled by the quantity of N and other nutrients available for microbial biosynthesis. In the Alaskan taiga, N and lignin concentrations in leaf litter have an important influence on its decomposition rate constant and the rate at which it decomposes in the forest floor. High concentrations of lignin, which reduce the overall energy yield of litter for microbial biosynthesis, result in substantial declines in decomposition rate constants (Figure 19.14a). In contrast to

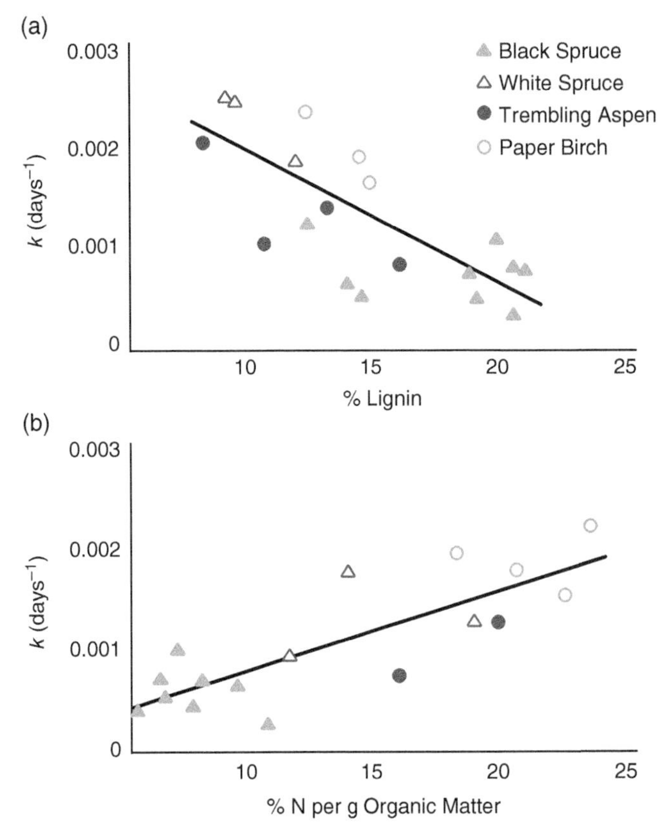

FIGURE 19.14 The relationship between decomposition rate constant (k) and (a) the lignin and (b) nitrogen concentration of leaf litter produced in the Alaskan taiga. *Source:* After Flanagan and Van Cleve (1983) / Canadian Science Publishing. Reprinted with permission of National Research Council Canada.

this relationship, decomposition rate constants are rapid in leaf litter with high concentrations of N (Figure 19.14b). A similar relationship can be observed in the northeastern United States, where high lignin:N ratios in the leaf litter of deciduous trees are reflected in slow rates of decomposition (Melillo et al. 1982).

The relationship between decomposition rate and litter N concentration results from the fact that soil microorganisms, like plants, require N to synthesize the biochemical constituents that form new cells and maintain existing functions. Fungal biosynthesis generally requires 1 atom of N for every 5–15 atoms of C assimilated during the decomposition of a particular substrate. Bacteria, on the other hand, require 1 atom of N for every 3–5 atoms of C. As a result, soil fungi have a C:N ratio ranging from 15:1 to 5:1, whereas those of soil bacteria are much lower (i.e., 5:1 to 3:1; Paul and Clark 1996). Look again at Figure 19.12 and realize that proteins and amino acids are the primary N-containing constituents of plant litter. Because protein and amino acid concentrations are low (2–15%) in most plant tissues, newly abscised leaves and dead fine roots are energy-rich and N-poor substrates for microbial metabolism.

Once litter enters forest floor and mineral soil, its organic constituents can only be used for microbial growth if there is a sufficient source of N and other nutrients. Leaf litter always has a C:N ratio wider (e.g., 50:1) than that of microbial cells, and its decomposition requires the presence of an additional source of N. Otherwise, soil microorganisms are unable to use the organic compounds contained in litter for biosynthesis. The positive relationship between litter N concentration and decomposition rate constants illustrates this point (Figure 19.14b); low concentrations of N in plant litter limit the rate at which organic compounds are used for microbial growth and hence slow its rate of decomposition. The inverse relationship between lignin and decomposition rate indicates that increasing proportions of poor-energy-yielding compounds in plant litter also slow the process of decomposition (Figure 19.14a).

In summary, decomposition is controlled by the energy yield of organic compounds (e.g., proportion of cellulose vs. lignin) in plant litter and the amount of N and other nutrients available for microbial growth and maintenance. Herein lies an important ecological link between N-use efficiency, litter N concentration, and microbial activity during the process of decomposition. Plants with a high N-use efficiency withdraw a relatively large proportion of N prior to leaf abscission, thereby lowering the N concentration of leaf litter. The production of leaf litter with a low N concentration slows decomposition and the subsequent release of inorganic N and other nutrients. Thus, nutrient-use efficiency and plant litter biochemistry can exert an important feedback on the rate at which nutrients cycle between plants and soil. In the following paragraphs, we further explore the microbially mediated release of N during organic matter decomposition—processes that influence the productivity of many forest ecosystems.

Nitrogen Immobilization and Mineralization Although forest floor and mineral soil represent the largest pools of N in forest ecosystems, the majority (90%) of N contained within them resides in organically bound forms that are largely unavailable to plants. Saprotrophic soil microorganisms mediate soil N availability within terrestrial ecosystems, because their growth and maintenance control the amount of inorganic N (i.e., NH_4^+) that is released during the decay of litter and soil organic matter. The balance of microbial NH_4^+ assimilation and release supplies approximately 90% of the N that is annually taken up and assimilated by plants in unmanaged ecosystems and is well correlated with NPP in many forest ecosystems.

In the previous section, we learned that litter decomposition was controlled by the types of organic compounds and the N required to use them for microbial growth. In the following paragraphs, we further explore the extent to which these factors control microbial requirements for N and the release of NH_4^+ from soil organic matter, the rate-limiting step in soil controlling the supply of N for plant growth in unmanaged terrestrial ecosystems.

When plant litter enters the soil, it is colonized by saprotrophic microorganisms, and high-energy-yielding constituents (i.e., organic acids, simple sugars, and carbohydrates) are preferentially used for microbial biosynthesis. Because this material is an energy-rich and N-poor substrate for microbial growth, NH_4^+ must be assimilated from soil solution to form new N-containing compounds within microbial cells. **Nitrogen immobilization** is the microbial uptake and assimilation of NH_4^+ into organic compounds. This process is illustrated by the following reaction in which an organic acid combines with NH_4^+ to produce an amino acid:

$$\begin{array}{c} COO^- \\ | \\ C=O + NH_4^+ \\ | \\ CH_2 \\ | \\ CH_2 \\ | \\ COO^- \\ | \\ \text{Organic Acid} \\ (\alpha\text{-ketoglutarate}) \end{array} \xleftarrow[\text{Glutamate dehydrogenase}]{} \begin{array}{c} COO^- \\ | \\ H-C-NH_2 + H_2O \\ | \\ CH_2 \\ | \\ CH_2 \\ | \\ COO^- \\ | \\ \text{Amino Acid} \\ (\text{L-glutamate}) \end{array}$$

Amino acid synthesis, facilitated by the glutamate dehydrogenase enzyme, is the first step in the production of proteins needed for the growth and maintenance of microorganisms. Nitrate (NO_3^-) also can be used by soil microorganisms, but it first must be reduced to NH_4^+ before it can form an amino acid. Nitrogen immobilization is the result of microbial growth on an energy-rich and N-poor substrate, and it characterizes the initial stages of litter decomposition. The hyphal networks of fungi, which extend from forest floor into mineral soil, are particularly effective at transporting NH_4^+ into freshly decaying litter.

Nitrogen mineralization is the release of NH_4^+ during the microbial breakdown of proteins, amino acids, and other N-containing organic compounds. Consequently, N mineralization is the reverse of the N immobilization reaction illustrated earlier in the text, and it is carried out by *all* microorganisms involved with organic matter decomposition. Whether an amino acid produced during protein degradation is used as an energy source or as a building block for microbial protein synthesis depends on a series of feedback controls. For example, carbohydrates contained in litter are preferentially used to generate energy in microbial cells, thereby preserving any amino acids present for microbial protein synthesis. Nonetheless, this relationship can change depending on the availability and energy yield of organic compounds remaining in plant litter.

During the initial stages of decomposition, carbohydrate availability is relatively high and amino acids contained in litter are directly assimilated for microbial protein synthesis. However, carbohydrate supply (i.e., energy) begins to limit microbial growth during the latter stages of litter decomposition (Figure 19.13). When quantities are insufficient to meet the maintenance energy requirements of microbial cells, mortality occurs and the constituents of the dead cells begin to serve as substrates for surviving microorganisms. Under energy-limited conditions, the C-skeletons (i.e., organic acids) of amino acids are used by surviving microbial cells to generate energy. During this process, NH_4^+ is released into soil solution where it can be assimilated by plant roots, participate in cation exchange reactions (Chapter 9), or enter into other microbially mediated processes. Because of this, virtually all N molecules in plants have first cycled microbial cells.

Nitrogen mineralization and immobilization occur simultaneously during the process of litter decomposition, albeit at different rates depending on the organic compounds present and the amount of N available to use those compounds for microbial metabolism. At what stage during litter decomposition is there a net release of NH_4^+ into soil solution where it can be taken up and assimilated by plant roots? Nitrogen is released from decomposing litter only when the gross rate of N mineralization exceeds the gross rate at which NH_4^+ is incorporated into microbial biomass (i.e., gross N immobilization).

Figure 19.15 illustrates the relationship between the C and N contents of leaf litter and changes in microbial metabolism that influences the gross immobilization and mineralization of N. During the initial phases of decomposition, plant litter is relatively rich in energy-yielding compounds, poor in N, and has a C:N ratio (50:1; Figure 19.13a) wider than that of microbial cells (approximately <8:1). In order to use litter constituents for growth or maintenance, soil microorganisms need to assimilate NH_4^+ or NO_3^- from soil solution. The initial demand for N for microbial biosynthesis causes high rates of gross N immobilization and an initial accumulation of N in decomposing litter (compare panels a and c of Figure 19.15). Comparing the decline in leaf C over time with rates of microbial respiration illustrates that litter mass loss results from the return of leaf C to the atmosphere as CO_2 during microbial respiration (Figure 19.15a,b). The initial increase in litter N from gross N immobilization, in combination with the loss of C during microbial respiration, causes leaf C:N to decline over time.

Later in the decomposition process, decomposing microorganisms are limited by substrates that provide little energy (i.e., energy- or C-limited phase), and microbial respiration and biosynthesis subsequently decline (Figure 19.15b). Proteins and other N-containing compounds released from cell death during this energy-limited phase of decomposition are then metabolized by surviving microbes. These organisms release NH_4^+ and use the resulting organic compounds for

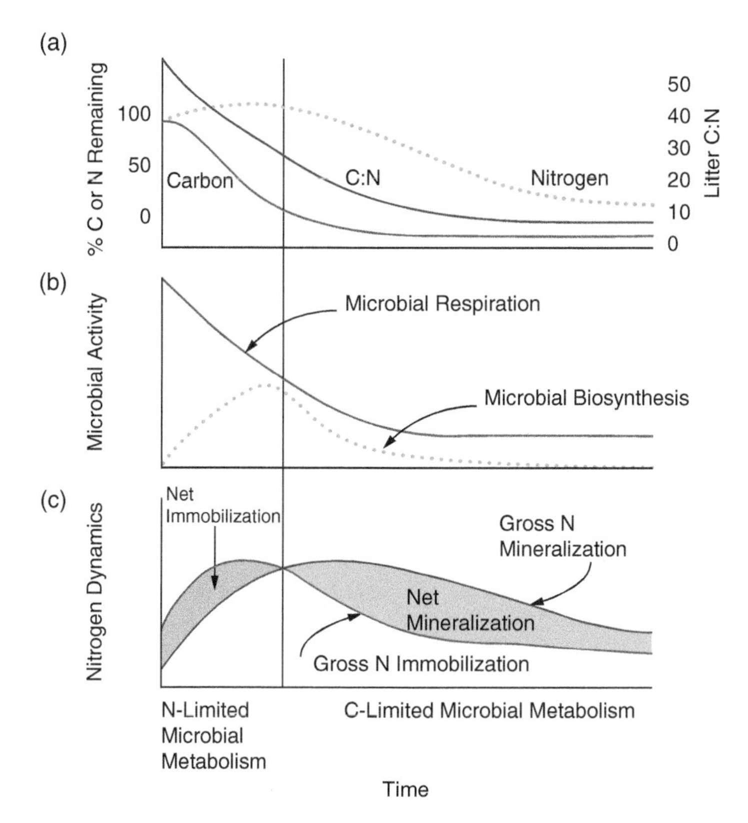

FIGURE 19.15 Changes in the C and N contents of leaf litter during the process of decomposition. Leaves enter the forest floor at time = 0, and (a) the initial decline in C results from (b) microbial respiration and biosynthesis. (c) Gross rates of immobilization exceed those of mineralization, due to the high-energy–low N stratus of leaf litter, which creates a microbial demand for N. The amount of N available for microorganisms to harvest the energy contained in plant litter initially limits decomposition. Following the use of high-energy-yielding substrates in litter, microbial biosynthesis slows and the demand for N declines. Gross mineralization exceeds gross immobilization during the period of C limitation, in which soil microorganisms are releasing NH_4^+ from N-containing organic compounds in order to harvest the energy contained within them.

energy production. As a result, gross rates of N immobilization fall below gross rates of N mineralization, and there is a net release of NH_4^+ from decomposing litter (net N mineralization). These dynamics illustrate an important point—the microbially mediated processes of gross N mineralization and gross N immobilization control the transfer of N from soil organic matter to soil solution where it can be taken up by plant roots. Although the biomass of soil microorganisms contains only 1.5% of the C and 3.0% of the N within forest ecosystems (Wardle 1992), their metabolic activities truly make them the "gate keepers" of soil N availability. We direct the reader to Staaf and Berg (1982) for a discussion of P, S, K, Ca, Mg, and Mn mineralization during plant litter decomposition.

Nitrogen Availability in Forest Ecosystems It is important to realize that net N immobilization only characterizes the initial phases of plant litter decomposition (Figure 19.15c). Soil organic matter or humus is the end product of litter decomposition; it has a low C:N (10:1) and contains few compounds that can be used for microbial biosynthesis. It is characterized by the net release of NH_4^+, which supplies the majority of N that is taken up and assimilated by plants on an annual basis. Generally, net N mineralization rates increase with the organic matter content of mineral soil (Marion and Black 1988). We direct the reader to Binkley and Hart (1989) and Hart et al. (1994) for a review of current techniques for measuring soil N transformations.

In North American temperate forests, rates of net N mineralization range from 30 to 120 kg N ha^{-1} y^{-1} (Pastor et al. 1984; Zak and Pregitzer 1990; Binkley 1995). This process supplies approximately 90% of the N available for plant uptake on an annual basis in forest ecosystems. For example, compare rates of net N mineralization with inputs from atmospheric deposition (Figures 19.2 and 19.3) and free-living N_2 fixation (*see* Biological Fixation of Nitrogen). Notice that rates of net N mineralization in most forest ecosystems are approximately 10 times greater than the combined annual inputs from atmospheric deposition and free-living N_2 fixation.

Nitrification The NH_4^+ produced during the process of net N mineralization can be retained by cation-exchange reactions, taken up by plant roots, or it can be oxidized to NO_3^- during the process of **nitrification**. Nitrification is an ecologically important N-cycling process because, in the absence of plant uptake or microbial immobilization, NO_3^- can rapidly move in the downward flow of water through the soil profile. The high mobility of this ion, in combination with the low anion-exchange capacity of most temperate soils, makes NO_3^- the form of N most often lost from terrestrial ecosystems. Once NO_3^-, or any other nutrient, has passed below the majority of plant roots, it is lost from the ecosystem and eventually enters groundwater, streams, or lakes. Consequently, nitrification is of great importance for understanding patterns of N loss from forest ecosystems.

Nitrification is the two-step oxidation of NH_4^+ to NO_3^- by chemoautotrophic bacteria and archaea. For well over 100 years, NH_4^+ oxidation was thought to be carried out exclusively by chemoautotrophic bacteria. With the advent of molecular tools, we now understand that chemoautotrophic archaea can be dominant agents of NH_4^+ oxidation in soil. They were previously unknown, because they will not grow on laboratory media, as do NH_4^+ oxidizing bacteria. Both NH_4^+ oxidizing bacteria and archaea use NH_4^+ as an energy source to fix CO_2 from the soil atmosphere. Nitrifying bacteria are sensitive to pH, water potential, and aeration; nitrifying archaea appear to be more important where NH_4^+ concentrations in soil are very low. In general, bacterial nitrification diminishes below pH 6.0, is most rapid at matric potentials of −0.01 to −1.0 MPa, and requires the presence of O_2 (Paul and Clark 1996). However, nitrification has been observed in some acidic forest soils. There is some evidence to suggest that several heterotrophic bacteria (*Arthrobacter* spp.) and fungi (*Aspergillus* spp.) can produce small amounts of NO_3^- from NH_4^+, but their ecological significance in forest soils is largely unknown (Schimel et al. 1984).

The first step in the process of nitrification in soil is mediated by chemoautotrophic bacteria (*Nitrosomonas*, *Nitrosospira*, *Nitrosococcus*, and *Nitrosovibrio*), as well as chemoautotrophic archaea (*Nitrososphaera* and *Nitrosopumilus*) in the following manner:

$$NH_4^+ + 1.5O_2 \rightarrow NO_2^- + 2H^+ + H_2O.$$

The NO_2^- produced in the above reaction is used by *Nitrobacter winogradskyi* and *Nitrobacter agilis* to obtain energy in the second reaction in the nitrification process:

$$NO_2^- + H_2O \rightarrow H_2O \cdot NO_2^- \rightarrow NO_3^- + 2H$$
$$2H + 0.5O_2 \rightarrow H_2O$$

Notice that NH_4^+ and NO_2^- oxidation require O_2, making nitrification unlikely in very poorly drained soil. The production of H^+ during NH_4^+ oxidation (first reaction) can lower the pH of soil with high NH_4^+ concentrations and rapid nitrification rates.

The potential for NO_3^- loss following nitrification has drawn the attention of ecologists and soil microbiologists, who have sought to understand the factors controlling the activity of nitrifying bacteria and archaea. Competition appears to be an important controlling factor, because plants, decomposing microorganisms, and nitrifying bacteria all require NH_4^+ as either a biosynthetic building block or as an energy source. Most plant roots and decomposing microorganisms have NH_4^+ uptake rates more rapid than those of NH_4^+-oxidizing bacteria, making plant uptake and microbial immobilization important controls constraining nitrification in many ecosystems (Robertson and Vitousek 1981; Robertson 1982; Zak et al. 1990). Nitrification rates can greatly increase in the absence of plant uptake, further suggesting that plant-microbe competition can limit nitrification. There is some evidence that plant compounds synthesized to deter herbivory (tannins and terpenoides) can inhibit nitrifying bacteria (White 1986, 1988).

Nitrification rates dramatically differ among intact forest ecosystems, ranging from 0 to 100% of net N mineralization (Pastor et al. 1984; Zak and Pregitzer 1990; Binkley 1995). The highest rates of nitrification often occur in forest soils with rapid rates of net N mineralization, ample soil water, and neutral pH. Because of its high rate of diffusion and low retention by anion-exchange reactions in soil, one would expect that NO_3^- loss following disturbance should vary according to initial nitrification rate and the extent to which it increases following the removal of plant uptake. We build on our understanding of nutrient uptake by plants, organic matter decomposition, and nitrification to understand patterns of N loss prior to and following disturbance later in this chapter.

NUTRIENT LOSS FROM FOREST ECOSYSTEMS

In the previous sections, we saw that relatively large quantities of nutrients annually circulate between overstory trees, forest floor, and mineral soil. In intact forest ecosystems, a small proportion of the nutrients that annually cycle among these ecosystem pools are lost through physical export via the hydrologic cycle or biological export to the atmosphere. Highlighted in the reduced version of Figure 19.1 (to the right) are leaching and gaseous loss, major pathways of nutrient loss.

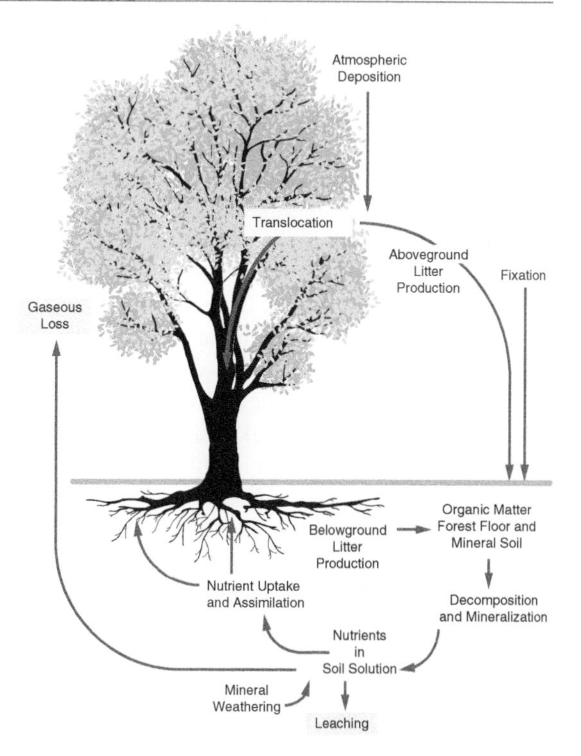

Nutrients in soil solution that exceed plant and microbial demands can be transported below the rooting zone by the downward flux of water through the soil profile. Nutrients moving in this way eventually enter ground or surface waters where they then become a nutrient input for aquatic ecosystems. Therefore, the nutrient cycles of terrestrial and aquatic ecosystems are intimately linked through the flow of water in the landscape. Nutrient loss from forest ecosystems also influences water quality and the nutrient dynamics of aquatic ecosystems. In the paragraphs that follow, we discuss the physical and biological loss of nutrients from intact forest ecosystems, thus completing the journey of nutrients into, through, and out of forest ecosystems. We later build upon this foundation to understand the extent to which natural disturbance and human activities influence nutrient loss from forest ecosystems.

NUTRIENT LEACHING FROM FOREST ECOSYSTEMS

Leaching is the physical process by which nutrients exit terrestrial ecosystems in the downward flow of water through the soil profile. This process can occur when precipitation exceeds the amount of water lost through transpiration and evaporation. Water moving downward through the soil profile has the potential to entrain and transport nutrients dissolved in soil solution. Nutrients are subject to leaching *only* when water is moving downward through the soil profile. Therefore, the hydrologic cycle provides the transport mechanism for the leaching of nutrients from terrestrial ecosystems. Once below the rooting zone, nutrients are lost from the forest ecosystem and are headed for groundwater or surface water.

The high demand for growth-limiting nutrients (N or P) by overstory trees typically maintains low concentrations in soil solution (*also see* Nutrient Uptake and Assimilation) and thus lessens the potential for nutrient leaching in forest ecosystems. The leaching loss of N is often much less than N inputs via atmospheric deposition and free-living N_2 fixation (Vitousek and Melillo 1979). On the other hand, non-limiting nutrients (i.e., Ca or K) can accumulate in soil solution, because the rate of supply from weathering or the atmosphere exceeds their biological demand. Under these conditions, leaching can export relatively large quantities of non-limiting nutrients from terrestrial ecosystems.

It should be remembered, however, that excess water must be present to serve as the transport mechanism for nutrient leaching. In temperate deciduous forests, the greatest potential for nutrient leaching occurs in early spring prior to canopy development and in late autumn after leaf fall. During these periods, a low biological demand for nutrients and water, and the resulting ample amounts of water moving through the soil profile due to low transpiration rates, provide the potential for leaching.

The potential for nutrient leaching, particularly the loss of limiting nutrients such as N and P, has drawn the attention of forest ecologists over the past 50 years. In 1963, one of the first attempts to quantify nutrient leaching from intact forests began at the Hubbard Brook Experimental Forest in New Hampshire (Likens et al. 1977). Using small, bedrock-sealed watersheds drained by a single stream, a team of ecologists began measuring nutrient leaching from northern hardwood forests (*also see* Mineral Weathering). In this unique situation, leached nutrients could only exit the bedrock-sealed watershed in streamflow. Many studies have been conducted subsequent to this pioneering research; the reader is directed to Henderson et al. (1978), Vitousek and Melillo (1979), and Swank and Crossley (1988).

Table 19.8 summarizes leaching losses of limiting (N, P, S), non-limiting (K, Ca), and nonessential (Si) plant nutrients from the second-growth northern hardwood forests at the Hubbard Brook Experimental Forest; also provided are nutrient inputs from atmospheric deposition and biological N_2 fixation. Notice that leaching losses of N, P, and S are much lower than their input, suggesting that these nutrients are being taken up and assimilated by forest vegetation. Contrast this pattern to the losses of K, Ca, and Si, which exceed inputs from atmospheric deposition. These nutrients weather from the K-, Ca-, and Si-rich bedrock of the Hubbard Brook Experimental Forest at a rate surpassing plant demand, giving rise to their net export from the forest. In Table 19.8, export or retention is the difference between leaching loss and input.

Similarly, Henderson et al. (1978) compared leaching losses of N, K, and Ca from watersheds dominated by old-growth Douglas-fir in the Pacific Northwest with those dominated by oaks and hickories in the southeastern United States. Despite ample amounts of precipitation in these widely different forests (158–233 cm y^{-1}), leaching losses of N were low (<0.5 kg N ha^{-1} y^{-1}) and

Table 19.8 Nutrient leaching from an intact northern hardwood-dominated watershed in the Hubbard Brook Experimental Forest, New Hampshire.

Nutrient	Leaching loss	Input	Export (-) or retention (+)
	kg ha^{-1} y^{-1}		
N	4.0	20.7	16.7
P	0.019	0.036	0.017
S	17.6	18.8	1.2
K	2.4	0.9	-1.5
Ca	13.9	2.2	-11.7
Si	23.8	0	-23.8

Nutrient inputs via atmospheric deposition and biological fixation are also summarized. Nutrients required for plant growth in large quantities are accumulating within these second-growth forests. Leaching losses are greater than additions for nutrients in excess of plant demand (i.e., Ca) or those not required for plant growth (i.e., Si).

Source: Likens et al. (1977) / Springer Nature. Reprinted from *Biogeochemistry of a Forested Ecosystem*, © 1977 by Springer-Verlag New York, Inc. Reprinted with permission from Springer-Verlag.

were much less than the annual input of N from atmospheric deposition (1–6 kg N ha^{-1} y^{-1}). These ecosystems also leached higher amounts of K and Ca than they received in atmospheric deposition, again reflecting mineral weathering rates in excess of plant demand. In summary, these examples illustrate that leaching is controlled by the biological demand for nutrients and the availability of excess water to transport nutrients below the rooting zone.

In the next section, we will consider how these factors change following natural and human-induced disturbances which initiate the process of ecosystem development (i.e., secondary succession). Overstory harvesting, which also constitutes a loss of nutrients, will be treated in our discussion of forest harvesting and nutrient loss later in this chapter. In the following section, we discuss the biological loss of N from forest ecosystems through the process of denitrification.

DENITRIFICATION

Nitrogen is the only plant nutrient that enters terrestrial ecosystems through a biological process. It is also one of the few plant nutrients that leaves terrestrial ecosystems via a biological process. **Denitrification** is the microbially mediated reduction of NO_3^- to nitrous oxide (N_2O) or N_2, which returns N to the atmosphere (Tiedje 1988). Denitrification results in the loss of a limiting nutrient, potentially influencing the productivity of terrestrial ecosystems. Moreover, N_2O produced during denitrification is a "greenhouse gas" and its concentration in the atmosphere continues to rise (Elkins and Rosen 1989). Consequently, atmospheric scientists, ecosystem ecologists, and microbial ecologists have become increasingly interested in understanding the environmental and biological controls on this process (Davidson and Swank 1987).

The denitrification pathway is used by soil bacteria (facultative anaerobes) as an alternative respiratory pathway to generate ATP when O_2 is absent in the soil atmosphere. Species within 13 genera of bacteria have the ability to use NO_3^- for respiration. Those common in soil (*Pseudomonas*, *Bacillus*, and *Alcaligenes*) carry out the following overall reaction:

$$5CH_2O + 4H^+ + 4NO_3^- \rightarrow 2N_2 + 5CO_2 + 7H_2O.$$

The reduction of NO_3^- is accomplished in several steps ($NO_3^- \rightarrow NO_2^- \rightarrow NO \rightarrow N_2O \rightarrow N_2$) that are mediated by a series of enzymes collectively known as reductases (Knowles 1981). Some organisms lack the enzymatic capacity to carry out the final reductive step ($N_2O \rightarrow N_2$) and produce N_2O

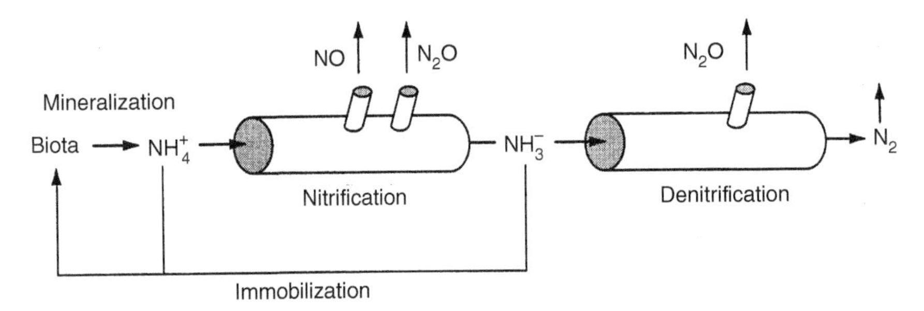

FIGURE 19.16 Transformation of N by soil microorganisms resulting in the gaseous loss of N from forest ecosystems. Gaseous loss of N can occur during the process of nitrification as well as during denitrification. *Source:* Firestone and Davidson (1989) / John Wiley & Sons. Reprinted from "Microbial basis of NO and N_2O production and consumption in soil" by M. K. Firestone and E. A. Davidson, In M. O. Andreae and D. S. Schimel (eds.), *Exchange of Trace Gasses between Terrestrial Ecosystems and the Atmosphere,* © 1989 by John Wiley & Sons Limited. Reproduced with permission.

as the end product of denitrification. In some situations, N_2O can account for one-third of N lost during denitrification (Robertson and Tiedje 1988). Whether the end product of denitrification is N_2O or N_2, the majority of denitrifying bacteria require organic compounds (CH_2O) in order to generate energy. Field techniques for measuring denitrification in terrestrial ecosystems are summarized by Tiedje et al. (1989).

Summarized in Figure 19.16 are the combined sources of gaseous N loss during denitrification and nitrification. There is a substantial body of evidence suggesting that gaseous N loss can also occur during nitrification. Nitric oxide (NO) and N_2O are gaseous by-products that are released by nitrifying bacteria and archaea as they oxidize NH_4^+ and NO_2^- (Firestone and Davidson 1989). Apparently, NO and N_2O are chemical intermediates that can dissociate from the NH_4^+- and NO_3^--oxidizing enzyme systems—processes that may be an important source of gaseous N loss in some forest soils (Robertson and Tiedje 1984).

The physiological requirements for denitrification place considerable restrictions on where this process occurs in the landscape. Most often, denitrification is associated with poorly drained soils in which the water table is near or at the soil surface for a considerable portion of the year. Soils in these landscape positions are often rich in organic matter and experience periodic low water levels during which nitrifying bacteria can produce NO_3^- when O_2 is present in the soil atmosphere. Once the soil becomes inundated, O_2 diffusion slows and the soil becomes anaerobic. Denitrification can ensue, provided that organic substrates and NO_3^- are present.

Groffman and Tiedje (1989) studied denitrification in northern temperate forests occurring on soils of sand, loam, and clay loam texture. Soils of each textural class were also located in well-drained, moderately well-drained, and poorly drained landscape positions. Denitrification increased from well-drained to poorly drained landscape positions and was generally higher on finer-textured soils in which anaerobic conditions are present within soil aggregates. For example, losses were substantial (40 kg N ha^{-1} y^{-1}) in lowland forests occurring on poorly drained clay loam soil, whereas denitrification was minimal in upland forests located on well-drained sandy soils (<1 kg N ha^{-1} y^{-1}). The lack of available NO_3^-, the result of rapid plant uptake, was the primary factor limiting denitrification during midsummer.

Nitrate uptake by plants also appears to be an important constraint on denitrification in wet tropical forests. Denitrification is rapid in lowland tropical rainforests that were recently cleared of vegetation, but only for the first 6 months following harvesting (Robertson and Tiedje 1988). After 6 months, denitrification was substantially lower than that in intact rainforests, apparently the result of N uptake by rapidly regrowing forest and accelerated decomposition, which lessened organic substrate availability to denitrifying bacteria. In combination, these results suggest that physical factors, such as soil drainage and aeration, directly control the activity of denitrifying bacteria and archaea. However, denitrification is also controlled by the activities of plants and heterotrophic microorganisms that influence NO_3^- availability in soil, which can dramatically change following the removal of overstory trees.

THE CYCLING AND STORAGE OF NUTRIENTS IN FOREST ECOSYSTEMS

In the preceding sections, we have discussed the physical, chemical, and biological processes controlling the flow of nutrients into, within, and out of forest ecosystems. Recall that nutrients enter forest ecosystems from mineral weathering, atmospheric deposition, and biological N_2 fixation. They cycle within forest ecosystems due to root uptake, aboveground and belowground litter production, decomposition, and nutrient mineralization. And nutrients are lost from forest ecosystems through leaching and denitrification. Here, we synthesize the specific processes of nutrient input, internal flow, and loss into overall patterns of nutrient cycling and storage in forest ecosystems.

NUTRIENT STORAGE IN BOREAL, TEMPERATE, AND TROPICAL FORESTS

The uptake of nutrients by roots, the return of nutrients in aboveground and belowground litter, and the release of nutrients during decomposition result in the continual movement of nutrients between plants and soil, albeit at different rates depending on the particular ecosystem. It is a widely held misconception that overstory trees contain the majority of nutrients within wet tropical forests, and that the soils beneath them are relatively poor in organic matter and nutrients. The origin of this misconception is obscure, but our current understanding of tropical forests clearly refutes this notion. For example, the organic matter and nutrient content of many tropical forest soils are equivalent to that of temperate forest soils (Klinge and Rodrigues 1968a,b; Sanchez 1976). In some instances, as much as 60% of N and P in wet tropical forests resides in the organic matter-rich surface and subsurface horizons of Spodosols (Klinge and Rodrigues 1968a,b). What differs between temperate and tropical forests is the rate at which nutrients cycle from soil to plants and back to soil again. Recall the rapid mean residence time of organic matter and nutrients in the forest floor of some tropical forests (Table 19.6). Declines in soil fertility following the harvesting of tropical forests are more likely related to a rapid decomposition and loss of nutrients than to the removal of nutrients in the harvested trees (Cole and Johnson 1980).

In Table 19.9, we summarize the distribution of N within boreal, temperate, and tropical forests. Notice that overstory trees in the wet tropical forest contain 10 times more N than those of the boreal forest and 3–5 times more N than the overstory of temperate forests. Regardless of this large difference, comparable quantities of N reside within the soil beneath tropical, temperate, and boreal forests. For example, mineral soil contains 62% of the N in the wet tropical forest and 69–84% of the N within the temperate and boreal forests in Table 19.9. When forest floor and mineral soil are combined, these proportions increase to 70% of total ecosystem N in the wet tropical forest, and 90% of total ecosystem N in the boreal and temperate forests.

From this analysis, it should be clear that forest management practices that erode or remove (i.e., windrowing or scalping) nutrient-rich O and A soil horizons have the potential to greatly reduce the nutrient capital of forest ecosystems, regardless of whether they occur in boreal, temperate, or tropical regions. Recall the rates of nutrient input from atmospheric deposition and mineral weathering earlier in this chapter (Figure 19.3 and Table 19.1) and realize that substantial reductions in forest floor and mineral soil nutrient pools can only be replenished over relatively long periods of time. Also realize that the loss of organic matter-rich surface horizons reduces rates of nutrient mineralization.

Table 19.9 Distribution of N in tropical temperate, and boreal forest ecosystems.

	Boreal		Temperate		Wet tropical
Location	Alaska, USA	Washington, USA	New Hampshire, USA	Tennessee, USA	Amazon, Venezuela
Dominant species	Black spruce	Douglas-fir	sugar maple–beech	Oak–hickory	Mixed species
Age (y)	55	42	55	30–80	Mature
Nitrogen			$kg\,N\,ha^{-1}$		
Overstory	134	316	491	497	1670
Understory	51	21	9	–	–
Forest floor	657	233	1100	334	406
Mineral soil	2200	2476	3600	4500	3507
Total ecosystem	3042	3046	5200	5331	5583

The amount of N in the overstory (leaves, branches, stems, large roots) increases from boreal to tropical forests, but the largest amount of N resides within the forest floor and mineral soil of these markedly different forest ecosystems.
Source: Adapted from Cole and Rapp (1981), Jordan et al. (1982), and Likens et al. (1977).

THE NITROGEN AND CALCIUM CYCLE OF A TEMPERATE FOREST ECOSYSTEM

The complete N cycle of a second-growth northern hardwood forest is depicted in Figure 19.17a. In this diagram, boxes represent the storage of N in ecosystem pools ($kg\,N\,ha^{-1}$) and the arrows depict processes ($kg\,N\,ha^{-1}\,y^{-1}$) by which nutrients flow from one ecosystem pool to another. The N cycle of this forest ecosystem is dominated by storage within soil ($3600\,kg\,N\,ha^{-1}$), forest floor ($1100\,kg\,N\,ha^{-1}$), and overstory trees ($532\,kg\,N\,ha^{-1}$). Look at Figure 19.17a and compare nutrients stored in overstory trees, forest floor, and mineral soil with rates of aboveground litter production ($54.2\,kg\,N\,ha^{-1}\,y^{-1}$), belowground ($6.2\,kg\,N\,ha^{-1}\,y^{-1}$) litter production, and net N mineralization ($69.6\,kg\,N\,ha^{-1}\,y^{-1}$). Notice that relatively small amounts of N are annually transferred from one ecosystem pool to another, compared to amounts stored in trees, forest floor, and mineral soil. Also notice that the total input of N is greater than the total loss of N, indicating an overall retention of N in this rapidly growing forest ecosystem.

The Ca cycle of the same northern hardwood ecosystem is depicted in Figure 19.17b; it is similar to and different from the N cycle in several important aspects. The Ca cycle is dominated by storage in mineral soil, forest floor, and overstory biomass with relatively small amounts of Ca annually flowing between these pools, similar to the pattern of N cycling and storage. However, the relatively large amounts of Ca stored in soil occur in the chemical structure of inorganic minerals, whereas N is stored in soil organic matter, not soil minerals. The large mineral pool of Ca provides weathering inputs ($21.1\,kg\,Ca\,ha^{-1}\,y^{-1}$) that are relatively large. Also note that Ca translocation (arrow from canopy to stem) during leaf senescence is not an important process; in fact, the Ca increases in leaves prior to abscission. Contrast this with the large amount of N translocated from leaves prior to senescence. Also contrast patterns of nutrient input and loss for the Ca cycle with that of the N cycle where losses are much less than inputs. Calcium losses are six times greater than Ca inputs from atmospheric deposition, suggesting that Ca is not being retained by this ecosystem (compare panels a and b of Figure 19.17).

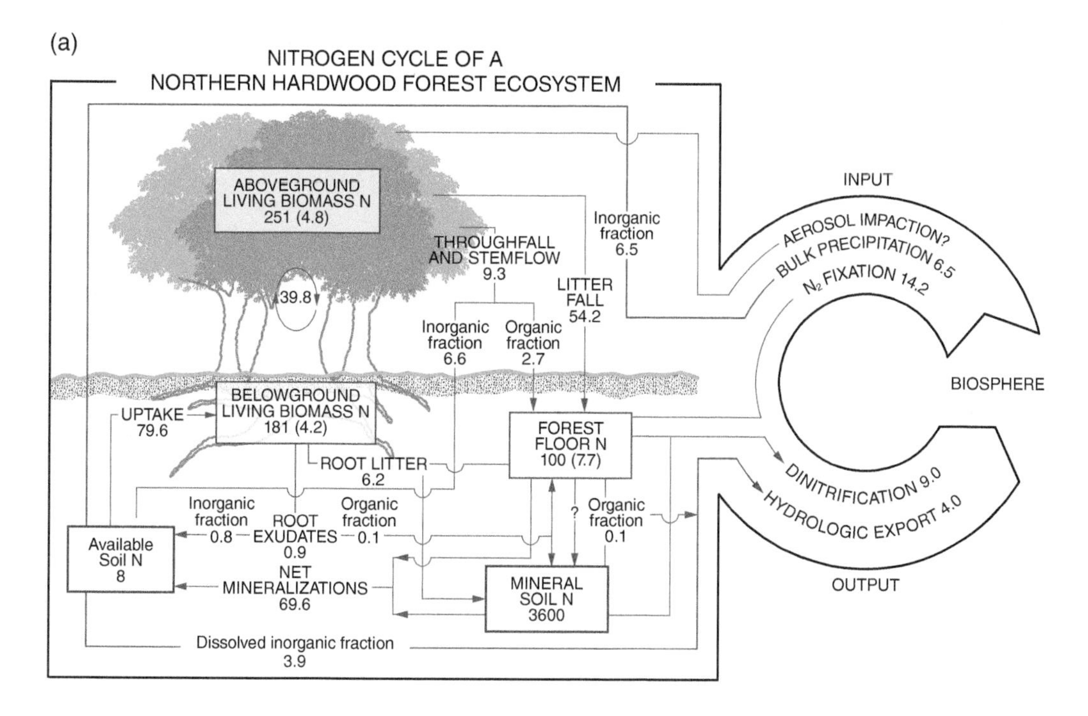

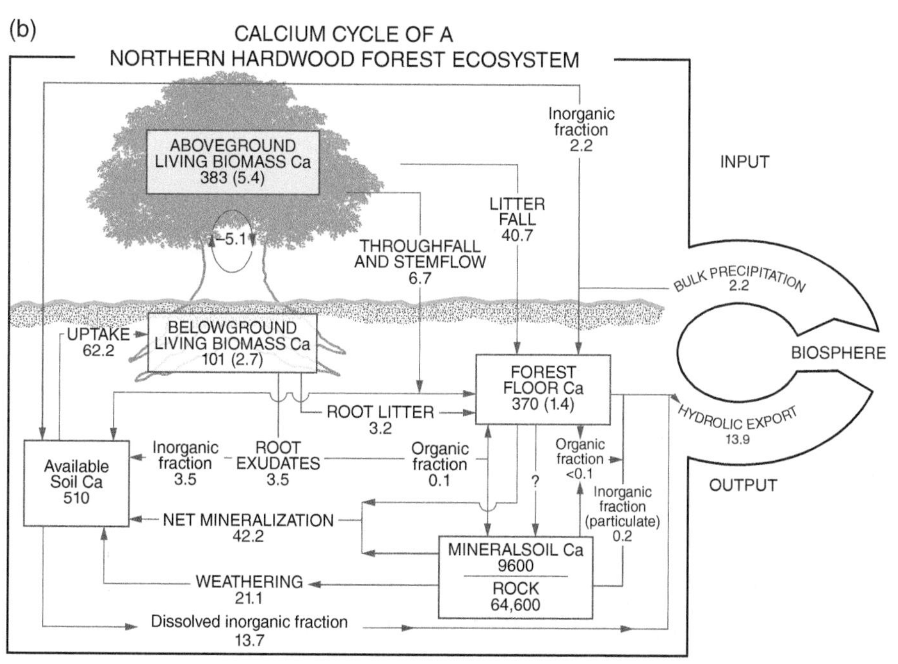

FIGURE 19.17 (a) The nitrogen and (b) calcium cycles of a second-growth northern hardwood forest at the Hubbard Brook Experimental Forest. Boxes in each diagram represent ecosystem pools in which N or Ca is stored, and the arrows represent processes by which nutrients are transferred from one ecosystem pool to another. Nutrient pools are measured in $kg\,ha^{-1}$ and the processes of transfer between pools are presented in $kg\,ha^{-1}\,y^{-1}$. *Source:* (a) Bormann et al. (1977) / American Association for the Advancement of Science. Reprinted with permission from *Science* 196, p. 982. © 1977 by American Association for the Advancement of Science. (b) Likens et al. (1977) / Springer Nature, *Biogeochemistry of a Forested Ecosystem,* © Springer-Verlag Berlin, Heidelberg 1977. Reprinted with permission of Springer-Verlag, New York, Inc.

Why is N retained by this forest ecosystem, and why is there a net export of Ca? Recall that biomass accumulates early in ecosystem development, because net ecosystem productivity (NEP) is positive (NEP > 0, when GPP > $R_A + R_H$; Chapter 18). Growth-limiting nutrients such as N are rapidly taken up to build new biomass, whereas non-limiting nutrients such as Ca can accumulate in soil solution—their supply exceeding the biological demand to build new biomass. A positive NEP and the resulting accumulation of biomass within this forest ecosystem (Gosz et al. 1978), combined with the fact that N limits the rate of biomass accumulation and Ca does not, cause the net retention of N and a net export of Ca from this developing forest ecosystem. In the following section, we build upon these principles to understand patterns of nutrient retention and loss during ecosystem development.

ECOSYSTEM C BALANCE AND THE RETENTION AND LOSS OF NUTRIENTS

In 1975, Peter Vitousek and William Reiners proposed and tested a hypothesis integrating patterns of nutrient loss and retention with the C balance of forest ecosystems (Vitousek and Reiners 1975). It was based on the idea that patterns of NEP control nutrient loss and retention in forest ecosystems. Notice in Figure 19.18a that net ecosystem production is low or negative during the initial stages of succession, increases during the intermediate stages of succession, and declines in the

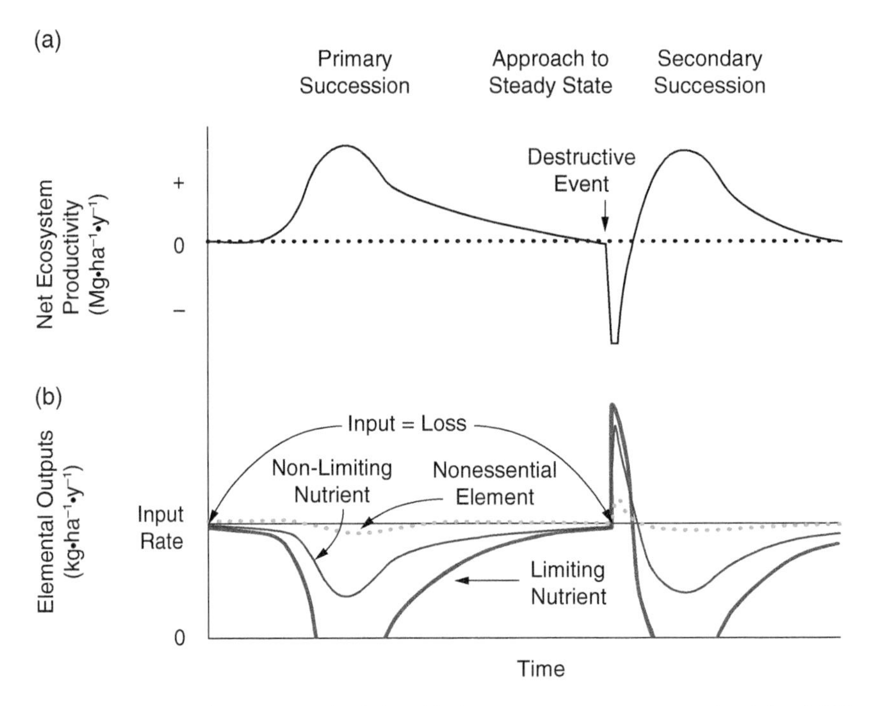

FIGURE 19.18 The conceptual relationship between net ecosystem productivity (NEP) and nutrient loss from a terrestrial ecosystem during primary and secondary successions. The upper panel depicts changes in NEP over time and the lower panel illustrates the corresponding changes in nutrient input and loss. Notice that nutrient inputs are constant over time, whereas the loss of nonessential, essential non-limiting, and essential limiting nutrients mirrors changes in NEP. *Source:* Vitousek and Reiners (1975) / Oxford University Press. Reprinted from *BioScience* (Vol. 25, No. 6, page 377), © 1975 by the American Institute of Biological Sciences.

latter stages of succession (i.e., ecosystem development). In Chapter 18, we saw that patterns of NEP during ecosystem development result from changes in gross primary productivity (GPP), the respiration of plants (R_A), and heterotrophic organisms (R_H). When net ecosystem production is positive ($GPP > R_A + R_H$), biomass and nutrients should accumulate within ecosystems, because they are needed to build the accumulating biomass.

Vitousek and Reiners (1975) reasoned that the loss of growth-limiting nutrients should be greatest early or late in succession when NEP is low and there is little or no demand for nutrients. Examine Figure 19.18a,b and observe the proposed inverse relationship between NEP and the loss of limiting nutrients. When NEP is less than zero, total ecosystem biomass declines (not shown in Figure 19.18), and the loss of limiting nutrients should exceed their input. When NEP is zero, total ecosystem biomass is at an equilibrium, and the loss of limiting nutrients should equal their input. And, when NEP is greater than zero, total ecosystem biomass increases, and the loss of limiting nutrients should be less than their input. That is, limiting nutrients should be retained in ecosystems in which NEP is positive and biomass is accumulating. They also predicted that these same patterns should hold for non-limiting nutrients, but to a much smaller extent (Figure 19.18a,b). Because nonessential nutrients are not required to build plant biomass, changes in NEP during ecosystem development should have little influence on their retention or loss (Figure 19.18a,b).

To test this hypothesis, Vitousek and Reiners (1975) located a series of bedrock-sealed watersheds that were entirely dominated by mid- (high NEP) or late-successional (low or 0 NEP) forests. If their hypothesis was correct, differences in NEP between mid- and late-successional forests should be reflected in the loss of limiting, non-limiting, and nonessential nutrients. These watersheds were located in close proximity to one another and received an equivalent input of nutrients in precipitation. Because each was sealed by underlying bedrock, nutrients that leached below the rooting zone could only exit in streamwater. Over a 2-year period, they compared the concentration of NO_3^- (limiting nutrient), K^+, Mg^{2+}, Ca^{2+} (non-limiting), and Na^+ (nonessential) in the streams draining the watersheds dominated by the mid- and late-successional forests.

By comparing nutrient concentrations in streamwater, it was clear that NO_3^- loss from late-successional forests was much greater than NO_3^- loss from mid-successional forests (Table 19.10 and Figure 19.19), supporting the contention that loss of a growth-limiting nutrient is low in ecosystems where NEP is positive. The export of non-limiting nutrients (K^+, Mg^{2+}, Ca^{2+}) was slightly greater from late-successional forests, whereas the loss of Na^+, a nonessential nutrient, did not differ between mid- and late-successional forests. These results support the idea that the C balance of terrestrial ecosystems exerts a substantial influence on the cycling and storage of nutrients.

Table 19.10 The mean concentration of limiting, non-limiting, and nonessential plant nutrients in the streamwater draining watersheds dominated by mid- and late-successional forests in New Hampshire.

Nutrient	Streamwater concentration μeq L^{-1}		
	Late-successional	Mid-successional	Late:mid ratio
NO_3^-	53	8	6.6
K^+	13	7	1.8
Mg^{2+}	40	24	1.7
Ca^{2+}	56	36	1.6
Na^+	29	28	1.0

Source: Vitousek and Reiners (1975) / Oxford University Press. Reprinted from *BioScience* (Vol. 25, No. 6, page 378), 1975 by the American Institute of Biological Sciences.

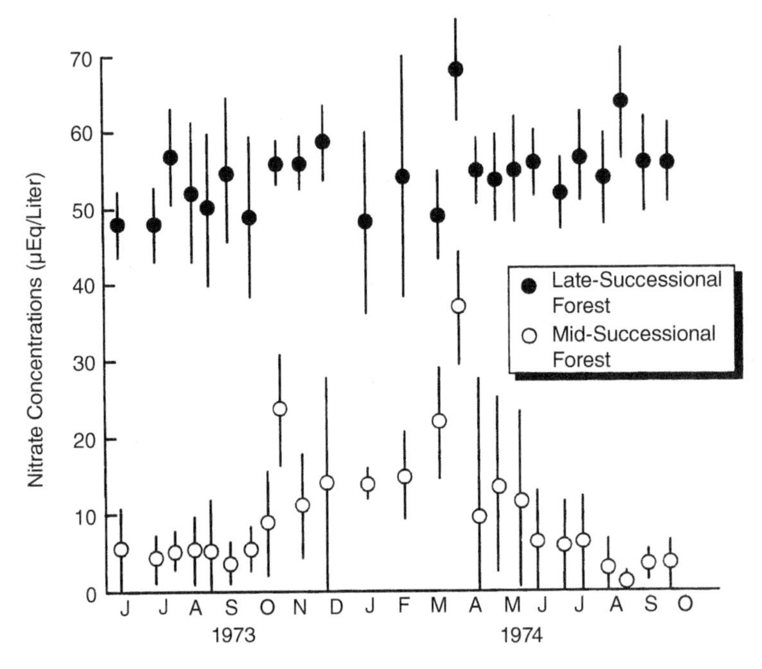

FIGURE 19.19 Streamwater nitrate (NO_3^-) concentrations from watersheds dominated by mid- and late-successional forests in the White Mountains of New Hampshire. *Source:* After Vitousek and Reiners (1975) / Oxford University Press. Reprinted from *BioScience* (Vol. 25, No. 6, page 379), © 1975 by the American Institute of Biological Sciences.

FOREST HARVESTING AND NUTRIENT LOSS

Following the removal of overstory trees, there are several fundamental changes that drastically alter the C balance of forest ecosystems and, consequently, patterns of nutrient loss. Gross primary productivity immediately declines after overstory removal, because there is little leaf area to convert solar radiation into plant biomass. Greater amounts of solar radiation reaching the soil warm its surface and increase the rates of organic matter decomposition. Large declines in transpiration increase soil water contents, which also increases organic matter decomposition and the return of CO_2 to the atmosphere. These changes in ecosystem C balance are illustrated in Figure 19.18 as the negative values of NEP (i.e., $GPP < R_A + R_H$) following a disturbance that initiates secondary succession. Low rates of nutrient uptake by overstory trees, in combination with a greater nutrient supply from accelerated rates of organic matter decomposition, may lead to the loss of nutrients from forest ecosystems.

One of the first and probably most well-known attempts to quantify the influence of forest harvesting on nutrient loss occurred at the Hubbard Brook Experimental Forest in the White Mountains of New Hampshire (Likens et al. 1977; Bormann and Likens 1979). To quantify nutrient loss, scientists located a series of similar watersheds (i.e., size, topography, geology, forest vegetation, and harvesting history) that were drained by a single stream. Nutrients leached below the rooting zone encounter shallow bedrock and flow laterally toward the stream draining each watershed. For a period of several years (1957–1965), researchers measured the volume and nutrient concentration of streamwater draining two watersheds dominated by 55-year-old northern hardwood forests. Notice in Figure 19.20 that precipitation, streamflow, evapotranspiration, and nutrient concentrations are nearly identical during this period. In 1965, however, the overstory of one watershed was clear-cut and herbicide was applied for 3 subsequent years to preclude revegetation (Figure 19.20). This treatment resulted in a decrease in evapotranspiration, an increase in streamflow, and the greater export of NO_3^-, K^+, and Ca^{2+} (compare control versus clear-cut in

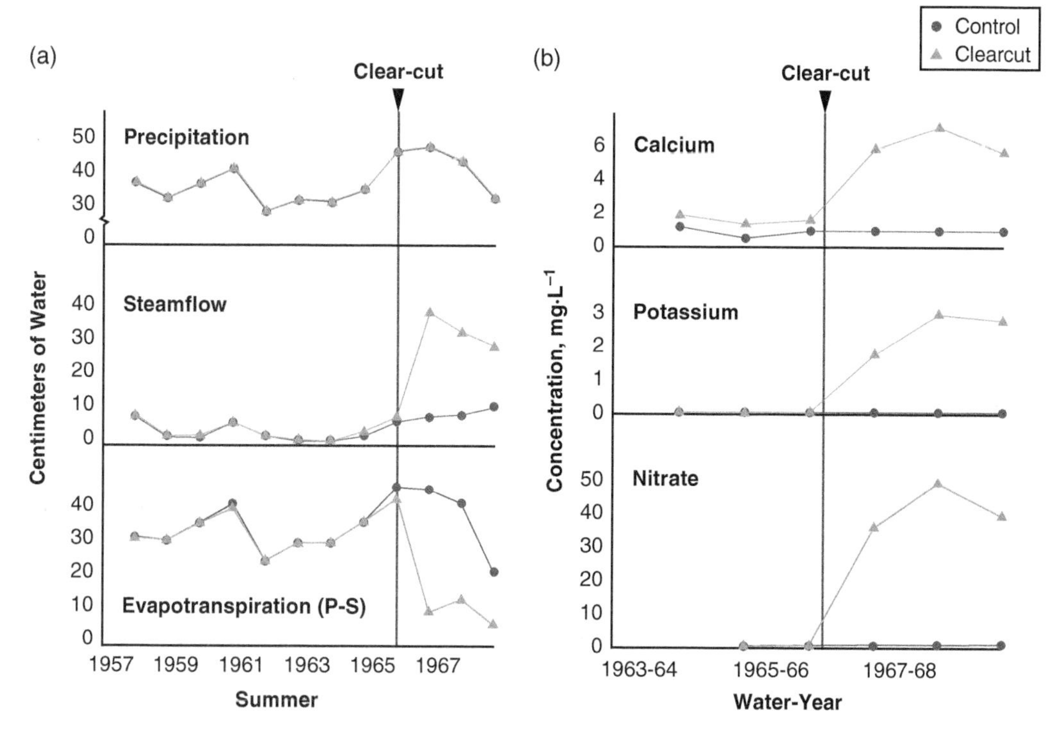

FIGURE 19.20 Short-term changes in the flow of water and nutrients through intact and clear-cut northern hardwood forest ecosystems at the Hubbard Brook Experimental Forest in New Hampshire. The shaded portion of the figure indicates the period of time during which regeneration was prevented with the use of herbicide. Notice that harvesting increases streamflow and the export of nutrients from the harvested watershed. *Source:* Bormann and Likens (1979) / Springer Nature. Reprinted from *Pattern and Process in a Forested Ecosystem,* © 1979 by Springer-Verlag New York Inc. Reprinted with permission of Springer-Verlag, New York, Inc.

Figure 19.20). Related studies demonstrated that clear-cut harvest also increased denitrification (Melillo et al. 1983). As plants recolonized the clear-cut site, NEP began to increase, biomass accumulated on the site, and streamwater nutrient concentrations returned to preharvest concentrations (Figure 19.21). This example illustrates the importance of NEP in regulating the loss of nutrients from clear-cut forest ecosystems.

It is important to point out, however, that not all forest management is conducted in such a manner nor are these northern hardwood forests typical of all temperate forests. Silvicultural systems ranging from single-tree selection to whole-tree harvest, soils of broadly different fertility, and differences of the balance between precipitation and evapotranspiration (i.e., excess water) all contribute to variation in nutrient loss from managed forest ecosystems. Moreover, not all temperate forests experience large losses of NO_3^- following disturbance. In the southeastern United States, for example, NO_3^- loss from intensively harvested loblolly pine plantations was reduced by the addition of logging slash to the forest floor, which increased net N immobilization by soil microbial communities and fostered N retention (Vitousek and Matson 1984). Temperate forests prone to NO_3^- loss are generally those with high rates of net N mineralization and nitrification prior to disturbance (Vitousek et al. 1982).

In addition to the export of nutrients to surface waters and the atmosphere, nutrients can also be lost from forest ecosystems in harvested biomass. One approach for quantifying the impact for forest harvesting on nutrient cycles uses a "balance sheet" to keep track of all nutrient inputs and losses (Silkworth and Grigal 1982). Using this technique, one can estimate the influence of

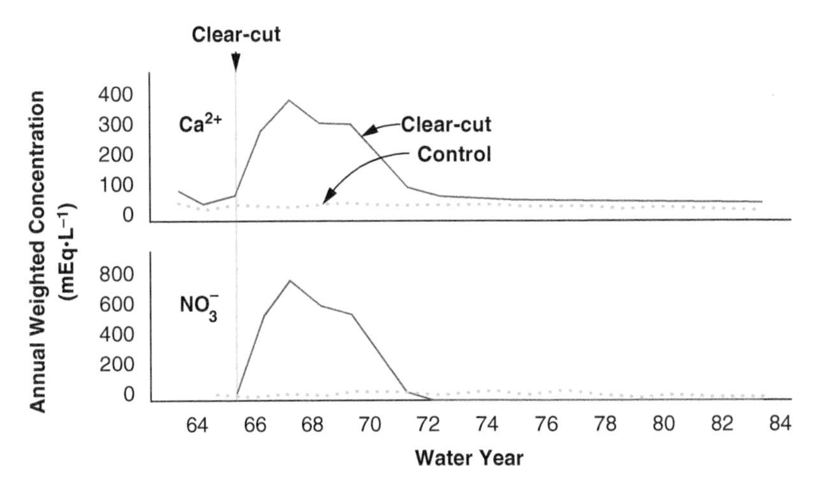

FIGURE 19.21 Long-term trends in NO_3^- and Ca^{2+} loss from intact and clear-cut northern hardwood forest ecosystems. The shaded portion of the figure indicates the period of time during which regeneration was prevented with the use of herbicide. The decline in nutrient loss from the clear-cut forest coincides with the accumulation of biomass (i.e., $GPP > R_A + R_H$, therefore $NEP > 0$) and a renewed demand for nutrients. *Source:* Likens et al. (1978) / American Association for the Advancement of Science. Reprinted with permission from *Science* 199, p. 493, © 1978 by American Association for the Advancement of Science.

repeated harvesting on nutrient storage in forest ecosystems. If the loss of nutrients during harvest is not met by nutrient additions during the next rotation, then forest harvesting can cause an overall decline in nutrient storage. The time it would take to replenish harvest-associated nutrient losses can be greatly reduced for some nutrients by altering the amount and type of biomass removed from forests. The leaves of trembling aspen are relatively rich in N and P and leaving them on the site dramatically reduces harvest losses. Similarly, trembling aspen has Ca-rich bark, and leaving this biomass component on site can reduce the net export of Ca from harvesting. This example illustrates the need to assess nutrient inputs and harvest-associated exports to assess the long-term influence of harvesting on the storage of nutrients within forest ecosystems.

SUGGESTED READINGS

Anderson, J.M. and Swift, M.J. (1983). Decomposition in tropical forests. In: *Tropical Rain Forest: Ecology and Management* (ed. S.L. Sutton, T.C. Whitmore and A.C. Chadwick). Oxford: Blackwell 498 pp.

Bormann, F.H. and Likens, G.E. (1979). *Pattern and Process in a Forested Ecosystem*. New York: Springer-Verlag 253 pp.

Hedin, L.O., Armesto, J.J., and Johnson, A.H. (1995). Patterns of nutrient loss from unpolluted, old-growth temperate forests: evaluation of biogeochemical theory. *Ecology* 76: 493–509.

Likens, G.E., Bormann, F.G., Pierce, R.S. et al. (1977). *Biogeochemistry of a Forested Watershed*. New York: Springer-Verlag 146 pp.

Paul, E.A. and Clark, F.E. (1996). *Soil Microbiology and Biochemistry*, 2e. New York: Academic Press 340 pp.

Roy, J. and Gamier, E. (1994). *A Whole Plant Perspective on Carbon-Nitrogen Interactions*. The Hague, The Netherlands: SPB Academic Pub. bv 313 pp.

Schlesinger, W.H. (1991). *Biogeochemistry: An Analysis of Global Change*. New York: Academic Press 443 pp.

Staaf, H. and Berg, B. (1982). Accumulation and release of plant nutrients in decomposing scots pine needle litter: long-term decomposition in a scots pine forest II. *Can. J. Bot.* 60: 1561–1568.

Vitousek, P.M. and Sanford, R.L. Jr. (1986). Nutrient cycling in moist tropical forests. *Ann. Rev. Ecol. Syst.* 17: 137–167.

Vogt, K.A., Grier, C.C., and Vogt, D.J. (1986). Production, turnover, and nutrient dynamics of above- and belowground detritus in world forests. *Adv. Ecol. Res.* 15: 303–377.

Forests of the Future

PART 6

The intentional and unintentional influences of humans on forest ecosystems cannot be denied. These influences only intensify as the human population continues to increase. Notably, many of the issues faced by forests today are unprecedented in human history, such as climate change, or the broad scale loss of diversity, or wholesale reconfiguration of landscapes, or the addition of new species to areas where they become invasive. Earlier editions of this textbook focused on renewable forest resources and the dangers of overexploitation or unwise management as the most pressing issues faced by forests. Over the course of only 60 years since the first edition of this book, the issues, though still largely anthropogenic in nature, have become far larger and more daunting.

In this part, we explore the ecological concepts related to some of the major human influences on forests in the twenty-first century. Climate change is perhaps the most prevailing environmental issue faced by forests and most other ecosystems. We examine the effects of climate change on forests in Chapter 20 by first summarizing the changes in temperature and precipitation patterns over the last century, and then exploring the potential effects of these changes on tree growth, mortality, phenology, and regeneration. The potential for changes in tree species and forest type distributions is studied, as are the effects of climate change on disturbances and forest carbon storage. Finally, we address climate change adaptation for forests.

A second enormous human influence on forests, the introduction of invasive plants and animals, is addressed in Chapter 21. We first examine the traits characteristic of invasive species. We then consider the potential impacts of invasive species on forests, including their effects on competition, fire regimes, and carbon and nutrient cycling. Finally, a short description of basic invasive species mitigation is presented.

The final two chapters of this book are not focused on specific environmental issues for forests, but rather review ecological subdisciplines that explore such issues in various ways. In Chapter 22, we summarize landscape ecology—the study of spatial patterns and heterogeneity of landscape features—as it pertains to forests. Landscape ecology is rooted in broad-scale, human-centered environmental issues and has developed its own ecological theory and analyses to study them. A selection of landscape ecological concepts is

Forest Ecology, Fifth Edition. Daniel M. Kashian, Donald R. Zak, Burton V. Barnes, and Stephen H. Spurr.

presented for forests, including forest fragmentation and connectivity, the interactions of disturbance and heterogeneity, historical range of variability, and the interactions of landscape pattern and ecological processes. As a conclusion, the basis for understanding and addressing sustainability in forests is presented in Chapter 23, including a review of the concept of ecosystem services. We end the chapter with a perspective for humans and their relationship to ecosystems, with the hope that this age-old question has become somewhat clearer (or at least more compelling) throughout the pages of this book.

Climate Change and Forest Ecosystems

CHAPTER 20

As we have seen throughout the chapters of this book, climate is the overarching driver of forest ecosystems on the Earth. Because it acts on specific physiological mechanisms of plants and other organisms, climate is a major factor determining genetic differentiation and speciation (Chapter 3), species distributions (Chapters 7 and 15), competition (Chapter 13), disturbance regimes (Chapters 10 and 16), and growth rates and carbon balance (Chapter 18). The distribution of vegetation at multiple scales is largely determined by variation in climate with latitude, elevation, and proximity to large water bodies and mountain ranges (Chapter 2). Characteristic temperature and precipitation patterns, as they interact with vegetation, parent materials, and physiographic position, are important in determining soil processes and soil development (Chapter 9). It therefore stands to reason that changes in climate, whether natural or anthropogenic, will alter the distribution of forests and their future productivity, much as they have altered the spatial distribution and species composition of forest ecosystems in the past (Chapter 15). In the last edition of this textbook, we posited that we did not know the likely biological consequences of changes in climate, and instead suggested likely changes in plant distributions and species migrations based on forest responses to past glaciations. Now that nearly 25 years have passed, we have sufficient data and improved technology to predict those consequences with more confidence.

Climate change is an enormously broad topic, and the field of climate change biology is both emerging and constantly evolving. We present in this chapter selected elements of climate change biology relevant to forest ecosystems. We first provide an overview of climate history and the natural and anthropogenic causes of variability, and we summarize our current understanding of climate trends. We then proceed with a summary of climate change impacts on tree physiology, growth, and phenology; potential tree population responses including range shifts; effects on disturbances and succession; and impacts on carbon storage and its feedback to climate change. It is challenging to present even an adequate sampling of climate change impacts on forests in a single chapter; recent textbooks such as those by Hannah (2021), Post (2013), and Newman et al. (2011) provide a comprehensive background on climate effects on biota and ecosystems. We provide many examples in the following text from the United States and North America, rather than at broader spatial scales, as an attempt to distill the extensive data now available.

CLIMATE CHANGE CONCEPTS

In describing climate change we must first distinguish between "climate" and "weather." Climate is the long-term trend in weather conditions of a particular region of the Earth. A relatively long timescale—the rule of thumb is typically a minimum of 30 years—is necessary to reliably determine a climate trend, because shorter timescales are subject to enough variability that trends are difficult to ascertain. This is why a single, very dry summer is no more indicative of climate change, or lack thereof, than a single, very warm winter. Such conditions are more appropriately referred to as weather conditions rather than climate.

Forest Ecology, Fifth Edition. Daniel M. Kashian, Donald R. Zak, Burton V. Barnes, and Stephen H. Spurr.

The basis of climate change is abiotic, the physics of which hinge on the prevalence of water vapor and carbon dioxide (CO_2) in the Earth's atmosphere. A process known popularly as the "**greenhouse effect**" occurs because both gases are transparent to visible light from the sun and also trap heat. As sunlight warms the surface of the Earth, long-wave radiation is emitted back from the surface, some of which is absorbed by water vapor and CO_2. The absorbed radiation is reemitted by the gas molecules, and some of this reemitted radiation is directed back toward the Earth, warming the lower atmosphere. The relationship between these "greenhouse gases" and global temperature is no longer hypothesized but observed in retrospect using ice-core data (see Chapter 15 for description of ice-core data). Carbon dioxide is the best-studied greenhouse gas and is our focus in this chapter, although others, such as methane (CH_4), also exist and have important relationships to land use and ecological systems. There has been a consistent and repeated occurrence of warm periods of climate over at least the last 800000 years that correspond with high levels of atmospheric CO_2 (Figure 20.1). Temperature and atmospheric CO_2 are intricately related in a positive feedback in the Earth's climate system: increasing CO_2 causes rising temperatures due to the greenhouse effect, and increasing temperature releases stored CO_2 from the oceans into the atmosphere.

The long-term variability in temperature and CO_2 over the past 800000 years is certainly not the result of human activities, but due to natural phenomena such as variations in the Earth's tilt and wobble on its axis (Milankovitch cycles), solar intensity, volcanic eruptions, and natural variations in greenhouse gas concentrations, all of which have occurred over very long temporal scales. Many of these natural phenomena are thought to be the drivers of past glacial and interglacial periods. Virtually all climate scientists today acknowledge that modern atmospheric CO_2 concentrations and global temperatures have increased at unprecedented rates since the middle of the nineteenth century, and that natural causes alone cannot explain these changes. Instead, it is human activity—primarily the burning of fossil fuels and deforestation, both of which transform sequestered carbon at or beneath the Earth's surface into atmospheric carbon—that best explains the rise of surface temperatures in the last 150 years (Figure 20.2). Meehl et al. (2004) used hindcasting climate models to show that only models that incorporated both natural and anthropogenic forcings could reproduce the rate and degree of global temperature change over the twentieth century. The 2021 report of the Intergovernmental Panel on Climate Change has

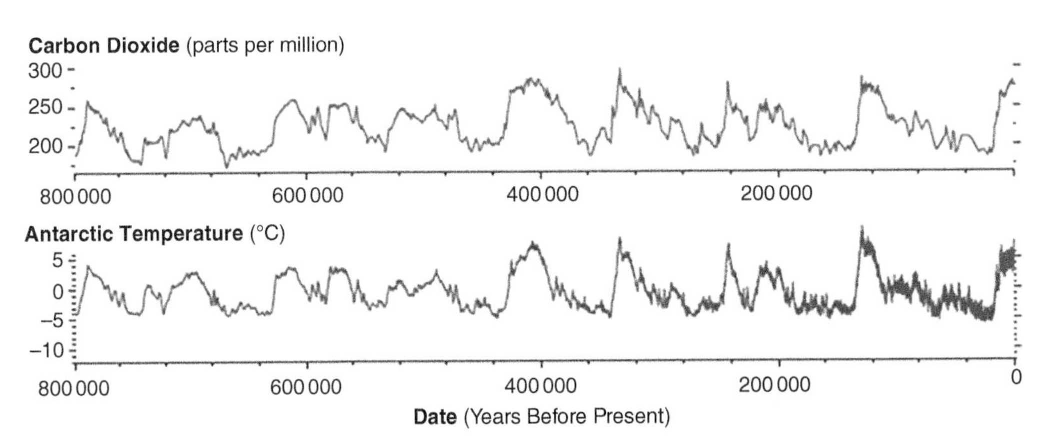

FIGURE 20.1 Variation in temperature and atmospheric CO_2 concentration over the last 800000 years in Antarctica. Temperature and CO_2 have varied in concert from at least 800 millennia. Data are from the European Project for Ice Coring in Antarctica (EPICA) Dome C ice core in Antarctica; CO_2 was determined from air bubbles in the ice core. *Source:* Data from Jouzel et al. (2007) and Lüthi et al. (2008); graphic by Robert Simmon, NASA Earth Observatory.

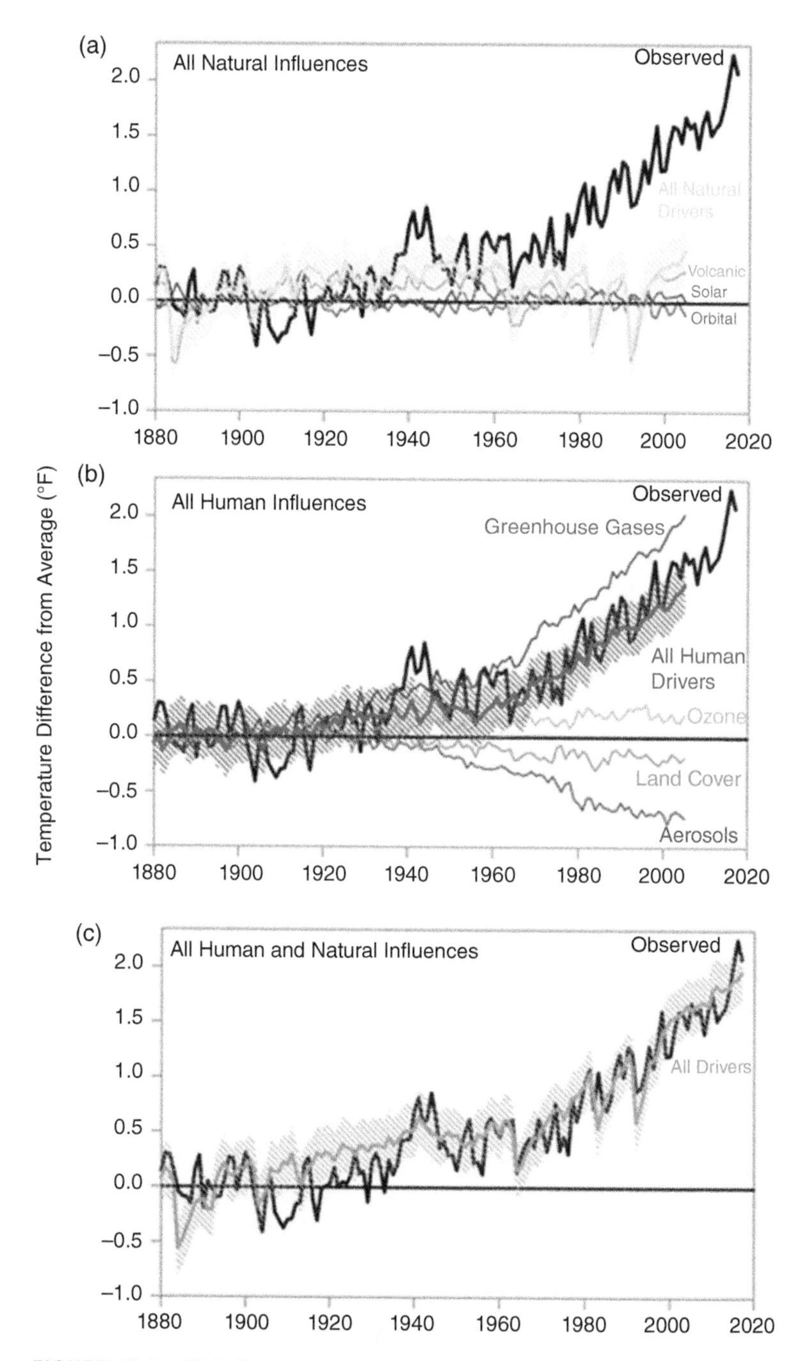

FIGURE 20.2 Global temperature changes due to (a) natural forcing, (b) human forcing, and (c) both natural and human forcings. Using climate models (colored lines) to reproduce observed changes in global temperatures (black lines), natural phenomena such as volcanic eruptions, solar activity, and variations in the Earth's orbit show no long-term trend in global temperatures over the last 140 years. While human influences such as aerosols and changes in land cover have had a net cooling effect over the last 80 years, most of the long-term warming trend in global temperatures is explained by increases in greenhouse gases. Combining both natural and human factors in the climate model matches the warming trend closely since 1950, suggesting the dominant role of human influences on global temperature changes. *Source:* From Hayhoe et al. (2018) / U.S. Government Printing Office / Public Domain.

declared that the increase of CO_2 and other greenhouse gases since 1850 is the result of human activities (IPCC 2021). Notably, atmospheric CO_2 concentration, at 421 ppm at the time of this writing, only in 1950 exceeded 300 ppm after being less than that for hundreds of thousands of years.

A discussion of climate change is often distilled to one of temperature change because it is among the most easily measured and observed factors of the Earth's climate system. Though the popular term "global warming" is often used interchangeably with climate change, it oversimplifies a changing climate in a complex system such as the Earth because it suggests that temperature is the only factor affected and that it only changes in a specific way (Post 2013). Instead, climate change is most usefully considered to include all of the abiotic changes that have occurred over the last 150 years. Obvious indicators of climate change include rising and more variable surface temperatures, changes in precipitation and snow cover, and changes in ice cover on sea and land. The first two of these indicators, changes in temperature and precipitation and their corollaries, are the main emphasis of this chapter.

EFFECTS ON TEMPERATURE

Temperature has many direct effects on trees, and many of these effects were reviewed in Chapter 7. Thus, changes in climate that affect temperature are extremely relevant for forests. The surface of the Earth has increased in temperature by 1.09 °C between about 1850 and 2020 (IPCC 2021). This estimate suggests a rate of increase in temperature higher than even that suggested in the last edition of this textbook, which reported a warming of 0.5 °C since 1800 and an increase of 0.27–0.39 °C since 1900 based on estimates by Henry et al. (1994). Remarkably, the period 2010–2019 was likely the warmest decade over the last 125 000 years, and the period of 2016–2020 was the hottest 5-year period since 1850. The majority of the warming is attributed to human activities, primarily greenhouse gas emissions, with only a small fraction (−0.1 °C to +0.1 °C) due to natural forcing (IPCC 2021). Hansen et al. (2006) reported a temperature increase of 0.2 °C per decade since 1980; despite a slower period of warming between 1998 and 2012 relative to the periods preceding and following it, the Earth's surface temperature appears to be warming at an increasing rate. In the United States, annual average temperature increased by 0.7 °C between 1901 and 2016. This warming accelerated between 1976 and 2016, with recent decades the warmest in 1500 years (Hayhoe et al. 2018).

The Earth's warming contains a great deal of variation, among years, decades, centuries, and millennia (Figure 20.2), as well as spatially. Some parts of the United States have experienced general warming trends while others have not (Figure 20.3). Between 1901 and 2016, the largest increases in temperature occurred in the western half of the United States, particularly in Alaska, the Northwest, the Southwest, and the northern Great Plains, where average temperature increased by more than 0.8 °C (1.5 °F; Hayhoe et al. 2018). However, much of the southeastern United States has had a neutral or a slight cooling trend during this period, apparently due to a winter shift in the position of the polar jet stream around 1950 that allowed cold air masses to more frequently penetrate the region. It is notable, however, that the Southeast has been warming at an accelerated rate since the early 1960s. This spatial variability in warming trends presents a significant challenge to understanding how forests will respond to climate change over broad regional scales (Post 2013).

In addition to long-term trends in surface temperature, there are several other temperature-related indicators of climate change that are relevant to forest ecosystems. As the degree of warming varies across years and decades, it also varies seasonally, with some seasons warming more than others in some regions. In the United States, minimum temperatures have increased faster than maximum temperatures, and a trend toward warmer winters has been evident since 1896, particularly at night and in the northern parts of the lower 48 states (USGCRP 2017). Average winter

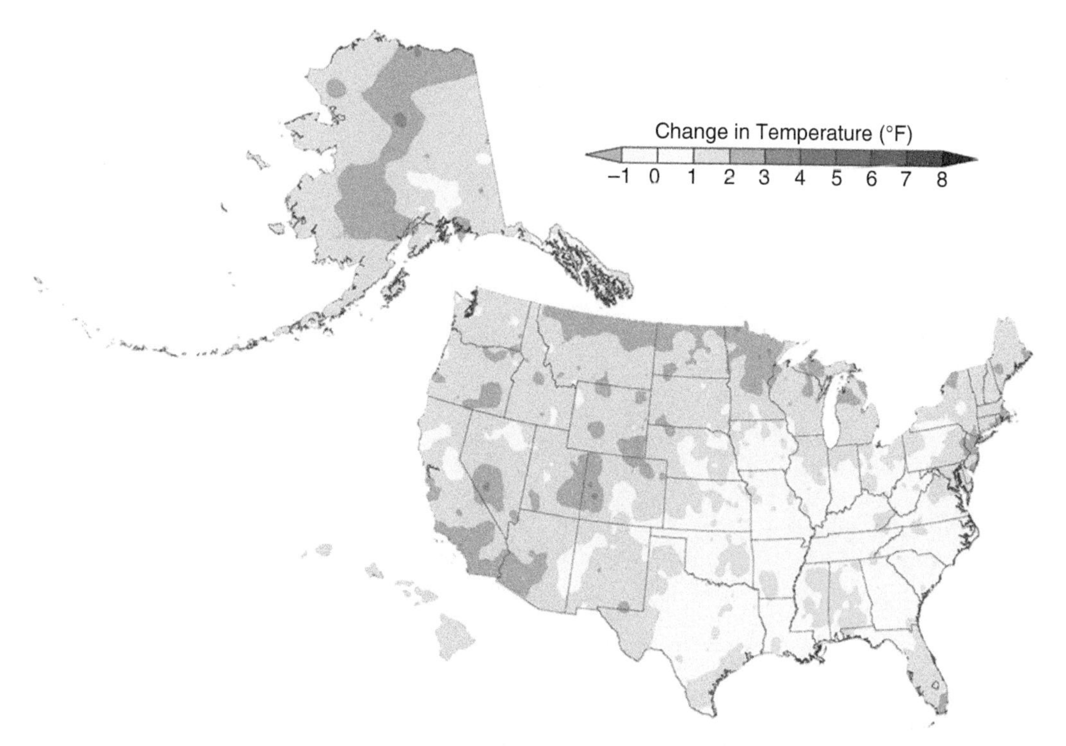

FIGURE 20.3 Observed changes in annual average temperature across the United States for the period 1986–2016 relative to 1901–1960 for the contiguous United States and 1925–1960 for Alaska, Hawaii, Puerto Rico, and the US Virgin Islands. Significant warming has occurred across much of the western United States and Alaska, though moderate cooling has occurred in the southeastern United States. Note that scale is in degrees Fahrenheit; for reference, 1° of change in Fahrenheit degrees = 0.56° of change in Celsius degrees. *Source:* From Hayhoe et al. (2018) / U.S. Government Printing Office / Public Domain.

temperatures have increased by nearly 1.7 °C (3 °F) across the contiguous 48 states, spring temperatures by 1.1 °C (2 °F), and fall temperatures by 0.8 °C (1.4 °F). These seasonal changes also vary spatially. The largest winter increases occur in the northern states (the Northeast, Lake States, Montana, and Dakotas) with additional increases in the Intermountain West; winter increases were smaller or neutral in the Southeast. Most states warmed in the spring, summer, and fall, except for a few in the Southeast that had little to no overall change or cooled slightly (NOAA 2021a). As discussed in the following sections, seasonal changes in temperatures may have important implications for plants in that they affect water supply and usage by plants, and may affect growing season length and heat sums, which are important in setting limits of some tree distributions.

An additional factor to consider regarding effects of climate change on temperature is its effect on temperature extremes. Unusually hot temperatures in the summer can affect water usage by trees, as well as lead to stressed conditions causing increased susceptibility to attack by insects and diseases, increased frequency or extent of fire or other disturbances, or death. Extreme or uncharacteristic hot or cold periods are a natural part of weather patterns in most regions, but the frequency and intensity of such periods are likely to change as overall surface temperatures warm. Across the United States, an increasing trend of unusually hot summer days (defined as those above the 90th percentile) has occurred since about 1980 (NOAA 2021b). Notably, the frequency of unusually hot summer nights, indicating a lack of nighttime cooling, has increased at an even faster rate than that for daytime temperatures, and the occurrence of unusually cold temperatures (especially nighttime extreme lows) has decreased (Figure 20.4). Unusually hot days have increased

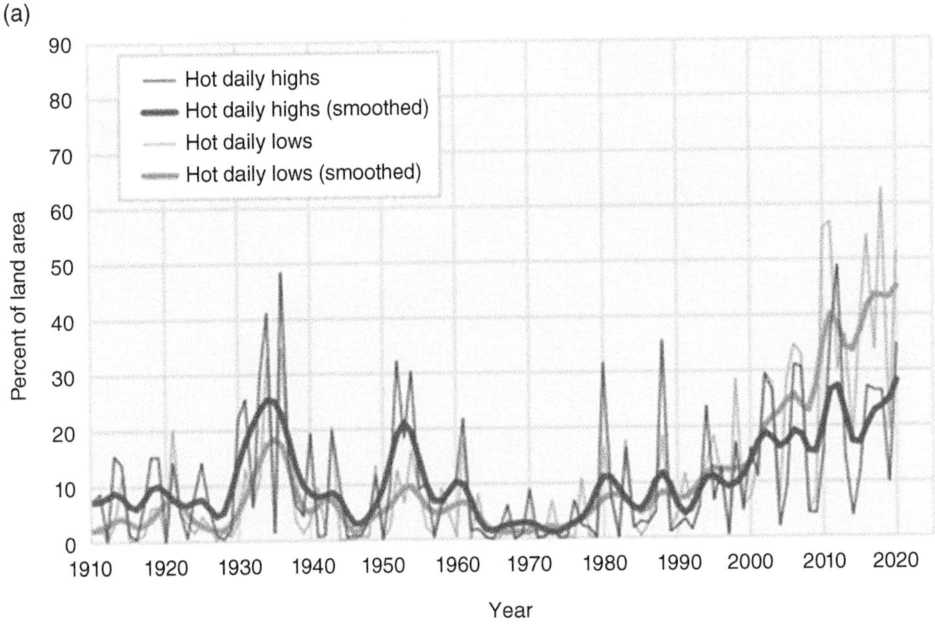

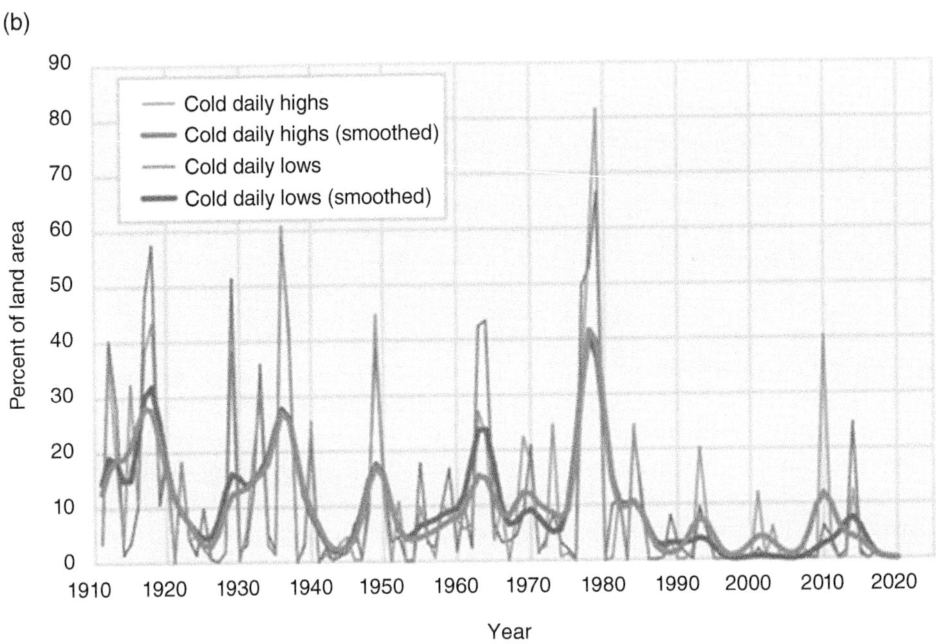

FIGURE 20.4 Percentage of the land area of the contiguous United States with (a) unusually hot daily high and low temperatures during June, July, and August, and (b) unusually cold daily high and low temperatures during December, January, and February. Unusually hot summer days and nights have increased since the 1980s, whereas unusually low winter temperatures (particularly at night) have become less common. *Source:* Data from NOAA (2021b); graphic from https://www.epa.gov/climate-indicators.

in frequency along the Gulf and Atlantic Coasts and in the Southwest (including southern California) since 1948 but decreased in frequency in the Upper and Lower Midwest and along the Mississippi River (NOAA 2021a). Unusually cold days have become less common across most of the contiguous United States, particularly in the West.

EFFECTS ON PRECIPITATION

Precipitation is an additional abiotic component vastly affected by climate change, but the effects of climate change on precipitation are somewhat more difficult to interpret from an ecologically relevant perspective compared to temperature. Precipitation patterns are strongly influenced by temperature, as well as the interrelationships among air currents, large water bodies, and topography. Precipitation alone does not correlate well with tree distributions or the boundaries of most major forest communities because evaporation and transpiration are heavily influenced by temperature as well as water. **Potential evapotranspiration**, or the amount of water that could be evaporated from land, water, and plant surfaces if soil water was limitless, has been shown to be more closely related to forest distribution than temperature in many regions (Patric and Black 1968). Although we review in the following text the changes and spatial variation that occur in precipitation with climate change, we caution that such changes are closely linked to temperature.

As with air temperature (Chapter 7), precipitation varies with latitude and altitude. Along the Pacific Coast, mean annual precipitation increases sharply with altitude, at rates varying from 13–17 mm per 100 m in the coastal range of Washington to 7–8 mm in the Sierra Nevada. Maximum precipitation occurs at the middle elevations—ranging from perhaps 900 m in the Olympics to 1500 m in northern California and 2400 m in the southern Sierra Nevada. Above these elevations, air currents become depleted of moisture and precipitation decreases with further altitude. The effect of elevation on rainfall is much more predictable in the interior mountains of western North America, being approximately 3–4 mm per 100 m rise in elevation (range 900–1500 m).

Precipitation also varies longitudinally across North America. The mountain ranges become progressively drier moving eastward from the Pacific Coast. Globally, forests tend to occur where precipitation exceeds transpiration, and precipitation is generally too low in the major rain shadow that characterizes the ranges and prairie regions east of the Rocky Mountains. Precipitation gradually increases further east due to the influence of maritime moisture from the Gulf of Mexico. High precipitation is found in the Southeast, particularly in the southern Appalachian Mountains, and in the Northeast, especially in the White Mountains.

Compared to temperature, there are far fewer data and thus more uncertainty in the historical levels of global precipitation, but average precipitation over land appears to have increased since 1950, with a faster rate of increase since the 1980s (IPCC 2021). Across the United States, average annual precipitation has increased by about 4% since the beginning of the twentieth century with marked regional differences (Figure 20.5). Much of the increase in precipitation has occurred in the northern, eastern, and central parts of the country (including the Great Plains), while the southeastern and western (particularly southwestern) regions of the country have experienced notable decreases, particularly in southern California and Arizona. Precipitation has increased at an average rate of 0.25 cm (0.1 in.) per decade globally since 1901; precipitation in the contiguous 48 states has increased at a rate of 0.51 (0.2 in.) per decade (Hayhoe et al. 2018).

As with temperature, trends in precipitation vary seasonally. Comparing the present day (1986–2016) to the early twentieth century (1901–1960), the largest increase in precipitation occurred in the fall for the contiguous United States, where the change is greater than 15% in the northern Great Plains and most of the eastern portion of the country (Easterling et al. 2017). Increases in precipitation are smaller in the spring months (3.5%), with the northern half of the country wetter and the southern half drier. Summer precipitation also increased across the country (3.5%), with patterns including increases in precipitation in the Upper Midwest, Northeast, southern Great Plains, and southern California, but drying in the Intermountain West, Southwest, Southeast, and Mid-Atlantic. Winter precipitation increased the least (2%), with drying over much of the western United States and Southeast but wetter conditions in the southern Great Plains and parts of the Lower Midwest (Easterling et al. 2017). Winter drying in the western United States

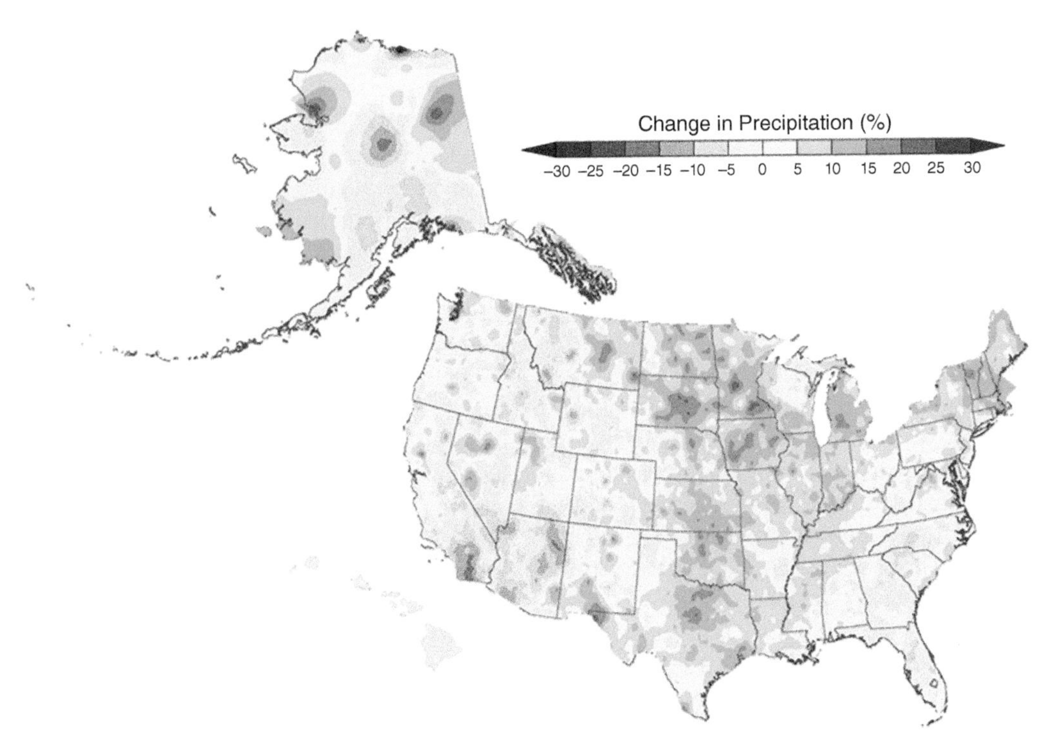

FIGURE 20.5 Observed changes in annual average precipitation across the United States for the period 1986–2016 relative to 1901–1960 for the contiguous United States and 1925–1960 for Alaska, Hawaii, Puerto Rico, and the US Virgin Islands. Increased precipitation has occurred in the Great Plains and Northeast, but the Southwest has experienced a notable decrease. *Source:* From Hayhoe et al. (2018) / U.S. Government Printing Office / Public Domain.

suggests declines in snowpack, particularly in middle to high elevations in mountainous areas and shifts to more winter precipitation falling as rain rather than snow.

Changes in the type of precipitation falling on some regional ecosystems can have important implications for the water balance of forests and other ecosystems. If climate change causes a greater proportion of winter-season precipitation to fall as rain rather than snow in such ecosystems, it will affect the extent and depth of **snowpack**, which is the amount or thickness of accumulated snow on the ground. In mountainous areas, snowpack is essential to the water balance of forest ecosystems, as it slowly releases stored water as runoff throughout much of the spring and summer as the snow melts. Snowpack is also important for insulating trees from freezing temperatures and winter wind at high elevations. Earlier melting of snowpack due to warmer winters and springs has the potential to contribute to soil-water deficits in mid and late summer because of the short duration of summer rainfall events that are ineffective at recharging the soil profile for uptake by plants. As such, earlier snowmelt also accelerates weather conditions conducive to wildfire, particularly in the western United States.

The average amount of snowfall across the country has decreased by about 0.2% each year since 1930, in part because the proportion of precipitation falling as rain instead of snow has also increased across a large portion of the contiguous United States. The proportion of rainfall has increased especially in the Pacific Northwest and the Midwest, though some areas of the Lake States have experienced a higher proportion of snowfall (Kunkel et al. 2009). April snowpack has decreased across much of the western United States, especially in the Pacific Northwest and northern Rocky Mountains. Snowpack has peaked earlier in the year across much of the West, averaging 8 days earlier since 1982. These patterns are particularly evident in states such as Colorado, New Mexico, and Utah (USDA-NRCS 2020).

Given the importance of precipitation for forest ecosystems, changes in the variability in precipitation within and among years are an additionally important aspect of climate change. Changes in the intensity and frequency of precipitation are characteristic of changing climate during the last 150 years. Higher surface temperatures, particularly those of the oceans, increase evaporation, creating heavier rain or snow events when the more moisture-laden air converges into storm systems. Heavy precipitation suggests that precipitation is occurring in more intense events and may not necessarily mean that the total amount of precipitation at a location has increased. The relevance to forests of heavy precipitation events is higher runoff and reduced infiltration of water into the soil, resulting in lower soil-moisture levels and potentially higher soil-water stress. Heavy precipitation events also lead to increases in soil erosion and flooding, the latter of which can be devastating depending on its timing and duration. Unusually heavy snow events may result in breakage and damage to trees, particularly where heavy snow is uncharacteristic of local or regional macroclimate and species lack adaptations to it.

Globally, heavy precipitation events have increased in frequency and intensity over a majority of land regions since 1950, with expectations that such increases will continue with further surface warming (IPCC 2021). In the United States, the frequency and intensity of heavy precipitation events have increased more than average precipitation (Hayhoe et al. 2018), and the proportion of precipitation that falls as intense, single-day events has increased since the 1980s and especially since 1996 (Easterling et al. 2017). Since 1901, the Northeast and the Midwest have increased 38% and 42%, respectively, in the proportion of total annual precipitation falling in the heaviest 1% of events. The value for the Northeast increases to 55% when measured from 1958 (Hayhoe et al. 2018). Increases for the Pacific Northwest and the Southeast were 22% and 18%, respectively, since 1901 and 9% and 27% since 1958.

At the other end of the spectrum from heavy precipitation events are droughts, or prolonged and abnormal periods when precipitation is lacking, creating water deficiencies for ecological systems. Droughts are often but not always accompanied by unusually warm conditions. Drought may affect trees directly by reducing growth during the growing season, or by reducing flowering, seed production, seed germination, and seedling survival. Indirectly, drought may affect trees by creating dry conditions that facilitate more frequent ignition, higher intensity, and faster spread of fires. Drought is also associated with insect attack and diseases in trees because it creates an additional stressor to trees that may weaken their resistance to these agents. Recurrence of adverse climatic patterns at a given location may increase the probability of disease or insect attack and would favor more drought-resistant species on sites with high soil-water stress.

The frequency and intensity of droughts have increased in many areas around the world, particularly in those that are already drought-prone. It remains difficult to characterize the role of climate change in the occurrence of droughts, however; some regions of the world, such as southern Europe and West Africa, have experienced longer and more intense droughts, but others, such as central North America, have experienced less frequent, less intense, or shorter droughts since the 1950s (IPCC 2013). This pattern holds true in the United States, which has experienced wetter than average conditions over the last half-century (Wehner et al. 2017). The most consistent and common droughts over the last 50 years have been observed in southwestern states (California, Arizona, and New Mexico), but the eastern United States has been generally wetter. Over the last two decades (2000–2020), as much as 70% of the land area of the United States experienced abnormally dry conditions at some point; more than half of the country experienced a moderate to severe drought in 2012 (National Drought Mitigation Center 2021).

CLIMATE CHANGE EFFECTS ON THE FOREST TREE

TREE GROWTH AND MORTALITY

Tree growth is closely linked to climate, and the direct effects of climate change on tree growth depend on several factors, including water and nutrient availability, increasing atmospheric CO_2, the timing of increases in temperature, the adaptability of species to new climatic conditions, the trees' thermal optimum for growth, and potential changes in trees' competitive ability (Way and Oren 2010). The temperature regime of a given regional or local ecosystem has particularly strong implications for plant growth (Chapter 7). The simplest evidence of this relationship comes from tree rings, which widely show increased growth during warmer growing seasons, especially in northern forests or at high elevations (Bunn et al. 2005; D'Arrigo et al. 2008). Notably, however, the relationship between warming and growth quickly degrades after 1950 for many species (Briffa et al. 1998; Lloyd and Fastie 2002; D'Arrigo et al. 2004). This weakening relationship is thought to be explained by increasing physiological stress and reduced growth that occurs when trees experience an increase in temperature beyond a threshold without also increasing moisture availability (i.e., a drought; D'Arrigo et al. 2008). Examining white spruce at tree line in northwestern Canada, D'Arrigo et al. (2004) found a positive correlation between tree growth and temperature between 1901 and 1965, but a negative correlation after 1965. A threshold summer temperature of 7.8 °C was identified as repeatedly occurring after 1965, suggesting that increasing temperatures associated with climate change below a threshold may promote tree growth, but those beyond the threshold may reduce it. Positive effects of increasing temperatures on tree growth are thought to be limited to species found in temperate and boreal regions with cooler climates and do not appear to benefit those found in warmer environments such as the tropics (Way and Oren 2010).

In a meta-analysis of studies that examined effects of temperature on tree growth, Way and Oren (2010) identified the following trends for trees experiencing higher temperatures: (i) increased leaf biomass and decreased root mass. This shift to a lower root-to-shoot ratio at higher temperatures likely increases susceptibility to drought stress; (ii) increased height growth and shoot growth affected mainly by daytime temperatures; (iii) moderate increases in stem growth, affected mainly by nighttime temperatures; (iv) stronger responses by deciduous species compared to conifers. Coniferous trees, found often on poor-nutrient sites and whose life history trade-offs include balancing productivity with nutrient retention, may be less able to increase their growth under favorable conditions; and (v) higher sensitivity of trees from warmer locales to warming temperatures compared to those from cooler locales, probably because trees at warmer locales typically persist closer to their temperature threshold or optimum.

The general pattern of tree growth described in the earlier text—an increase in growth with increasing temperatures to a maximum, followed by a decrease—mirrors those of plants' physiological processes. Photosynthesis and respiration generally increase to an optimum temperature before they decline, due mainly to quicker enzyme function, until lethal high temperatures are reached. The same processes that lead to increased growth in a plant may also lead to its being stressed. For example, the rate of photosynthesis theoretically increases with increasing temperature (Chapter 18), but the greater vapor pressure deficit of warmer air may cause stomata to close in response to greater transpiration (Mott and Parkhurst 1991) and/or due to soils dried faster by increased soil-water uptake by roots (Will et al. 2013), reducing photosynthesis. Despite increasing photosynthetic rates, rising temperatures associated with limited precipitation or soil moisture may therefore eventually lead to tree mortality (Allen et al. 2010, 2015). Thus, warming alone without substantial increases in precipitation may induce tree mortality if warming is severe. This is because drought in combination with warming only requires a moderate temperature increase to induce mortality (Figure 20.6) because a lack of (or even small increases in) precipitation does not compensate for increased vapor pressure deficits. Overall, a warmer climate will decrease tree growth in forests where water is limiting.

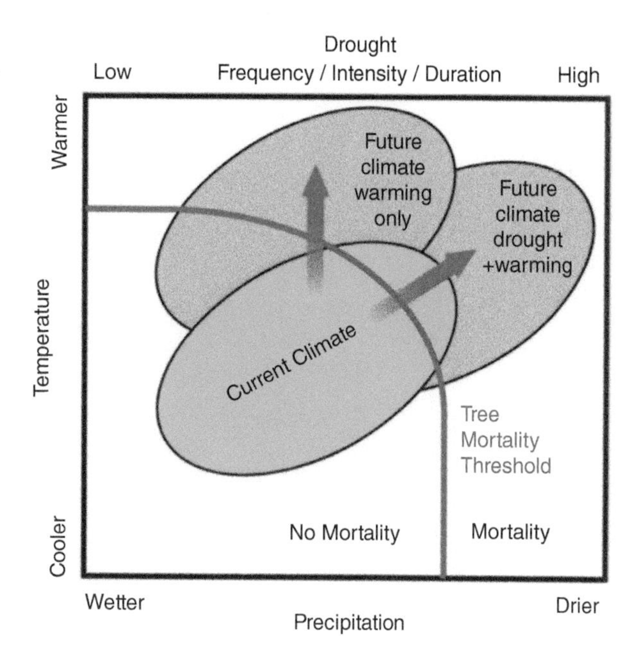

FIGURE 20.6 Conceptual diagram showing potential shifts of current climate and results on tree mortality. Only a small portion of the current climate exceeds the upper or lower threshold for temperature and precipitation to support trees, respectively, but more extreme temperatures alone or more droughty conditions together with warming may produce greater risk of mortality for current tree populations. *Source:* From Allen et al. (2015) / John Wiley & Sons.

At the same time, photosynthesis and ultimately tree growth increase when trees are grown in an atmosphere enriched with CO_2, as is present in most scenarios of climate change (Ainsworth and Rogers 2007; Norby and Zak 2011; Dusenge et al. 2019). Notably, trees use water more efficiently as atmospheric CO_2 increases because they reduce water loss via stomatal conductance (Keenan et al. 2013). There may therefore be a "fertilization effect" of increasing CO_2 that would increase tree growth in the presence of elevated CO_2. Depending on the strength of such an effect and the ability of trees to acclimate to rising temperature and CO_2, the higher photosynthetic rates and water use efficiency associated with CO_2 fertilization could theoretically counteract the decreased growth that occurs with warming temperatures (Sperry et al. 2019). It appears likely that tree growth will increase with future increases in CO_2, if only because rising atmospheric carbon appears to be particularly well correlated with increased water-use efficiency (Adams et al. 2020).

The concept of CO_2 fertilization is appealing because it suggests an important negative feedback. Rising concentrations of CO_2 in the atmosphere may increase tree growth and sequestration of carbon in forest biomass, thereby dampening future rates of increase in atmospheric CO_2. However, experimental tests of CO_2 fertilization, most of which have been conducted at free-air CO_2 enrichment (FACE) facilities or within open-top experimental chambers, have often raised more questions about the wider implications of such an effect. Experimental data have shown that CO_2 fertilization is a short-lived phenomenon when soil nutrients (especially nitrogen) are limiting, particularly in older forests where its effect may otherwise be weak (Reich et al. 2006; Norby et al. 2010). For example, in a FACE experiment in a sweetgum forest in Tennessee beginning in 1997, increases in growth under elevated CO_2 declined from 24% in 2001 to 9% in 2008 (Norby et al. 2010). When plots were fertilized with nitrogen, tree growth immediately responded positively and in a sustained manner, even as the CO_2-fertilized plots (and non-nitrogen–fertilized plots) were declining (Figure 20.7). These results suggest that nitrogen availability was limiting to tree growth and declining over time in a maturing forest whose growth was initially enhanced by elevated CO_2. A similar trend was documented globally by Wang et al. (2020), who used satellite and ground-based data sets to show that the CO_2 fertilization effect declined across the world from 1982 to 2015, likely due to changing nutrient concentrations and soil-water availability.

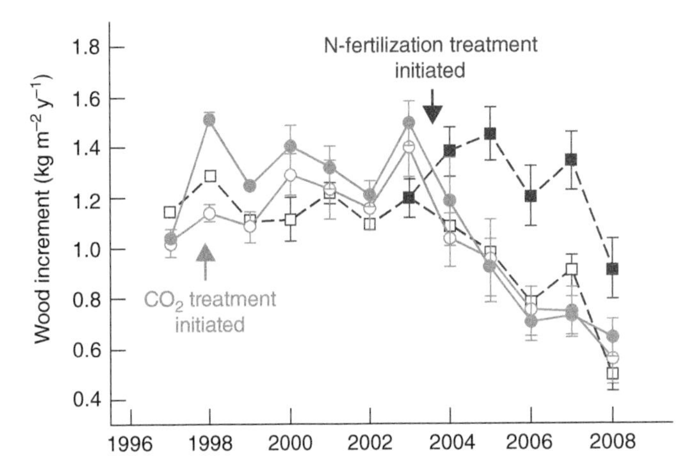

FIGURE 20.7 Tree growth response to CO_2 enrichment and nitrogen addition in a sweetgum forest in Tennessee. Elevated CO_2 (solid circles) initially increased growth but declined thereafter and did not differ from non-enriched plots. Nitrogen fertilization (closed squares) caused an immediate and sustained increase in growth. *Source:* From Norby et al. (2010) / with permission of PNAS.

An additional issue with CO_2 fertilization effects on tree growth is its potential trade-off with tree longevity. Utilizing 539 permanent plots sampled between 1960 and 2009 in the boreal forest of Alberta, Canada, Searle and Chen (2018) found an increase in tree mortality over 50 years, as well as a correlation between high lifetime growth rate and higher probability of mortality compared to slower growing trees. Reduced longevity was explained largely by decreasing water availability, to which faster growing or larger trees apparently are more sensitive (Hember et al. 2017). This trend has been corroborated for many tree species and growing environments across North and South America, Europe, Australia, and Asia (Brienen et al. 2020). Given that the rate of tree mortality has increased in recent decades (van Mantgem et al. 2009), the potential trade-off between rising atmospheric CO_2 and tree longevity is compelling for carbon dynamics in a changing climate. Fast-growing trees contribute to rapid carbon uptake from the atmosphere, but large, old trees are critical for maintaining the carbon in biomass rather than the atmosphere. Tree mortality returns stored carbon to the atmosphere as dead trees decompose over decades, offsetting any CO_2 fertilization effect (Bugmann and Bigler 2011), and fertilization-related gains in carbon storage from faster tree growth will therefore be short-lived if the trees die before they are large enough to store large amounts of carbon (Körner 2017). Forest productivity and carbon storage are discussed in detail in Chapter 18, and their role in climate change adaptation is discussed further later in this chapter.

PHENOLOGY

Trees, like all plants, vary in their seasonal rhythms, and important events such as flowering, spring bud burst, fall bud set, and seed maturation and dispersal have very regular and distinct cycles during and among years. Seasonal variations in temperature are important cues for the timing of these events, known as a species' **phenology**, and climate change has the potential to alter this timing (Cleland et al. 2007). Temperature-related factors such as growing season length, heat sum, and last spring and first fall frosts direct many phenological activities (Chapter 7). Many of the same ecological factors and relationships that affect tree species distribution, such as the relationship between physiography and temperature (Chapter 8) or the spatial relationships of landscape ecosystems, have similar, observable effects on phenology. For example, Fisher et al. (2006) used satellite data to detect leaf flush phenology in deciduous forests in southern New England. Satellite data were able to quantify variations in phenology with local topography, and in one example detected a gradient of leaf flush over 2 weeks over a 500-m distance and a 30-m elevational change on the side of a shallow valley (Figure 20.8). As discussed in Chapters 7 and 8, cold-air drainage into the valley significantly alters its microclimate, reducing temperatures to the point that phenology is also affected. Moreover, several studies have documented that vegetation in urban areas experiences earlier springs leading to earlier tree flowering and leaf flushing, as well as longer growing seasons, compared to adjacent non-urban areas due to the warmer microclimate characteristic of cities (Jochner and Menzel 2015). Such long-standing examples of the link between phenology and temperature should provide a great deal of foresight into the effects of rising global temperatures on tree phenology that are likely to occur with climate change.

Many physiological events that occur in the spring have been linked to warmer spring temperatures (IPCC 2014), and thus phenological events have begun to serve as important and effective indicators of climate change. Perhaps the most well-cited example of climate change effects on phenology is a study by Schwartz et al. (2006), who studied the first bloom and first leaf dates of lilacs and honeysuckles around the world as a means of detecting changes in spring onset. First leaf and first bloom (approximating the onset of bud burst in deciduous forest trees) dates in the contiguous United States, for example, have occurred earlier in the spring in the last few decades in the northern and western parts of the country where temperature increases have occurred, but later in the South where cooling trends have been prevalent (see temperature trends discussed in the earlier text). First leaf dates averaged 1.2 days earlier per decade across the world for the period 1955–2002, and first bloom dates averaged 1 day earlier per decade. Notably, the winter chill date—the date at which lilacs and honeysuckles are ready to respond to spring warmth after satisfying their winter chilling requirement—is occurring 1.5 days earlier per decade in central and eastern North America, increasing the length of the growing season (Schwartz et al. 2006). Studies linking altered tree phenology to altered temperature regimes have become increasingly common, especially for forests in seasonally cold climates (Gunderson et al. 2012; Gill et al. 2015; Richardson et al. 2018).

The phenology of species and individuals—even those co-occurring—does not necessarily respond to rising temperatures in the same way. For example, leaf flushing and flowering of co-occurring species appear to become less synchronous in warm, early springs (Zohner et al. 2018; Montgomery et al. 2020), probably because species differ in how they respond to early onset of warm spring temperatures. Co-occurring species with differing heat sums (Chapter 7) exhibit different sensitivities to warm springs and thus their timing of bud burst respond differently to earlier spring onset. There is also evidence that bud burst in some tree species is particularly cued by their winter chilling requirement (Zohner et al. 2016; Nanninga et al. 2017) or by photoperiod (Basler and Körner 2012) even more so than heat sums, and these species may respond more sluggishly to earlier springs (Montgomery et al. 2020). Cooler, later springs allow most species to meet their chilling and/or photoperiodic requirements, and such species are therefore more likely to be synchronized in their bud burst. Synchrony in leaf phenology is important for forest trees because those that exhibit early bud burst have a higher risk of damage due to late frosts (Kollas et al. 2014; Vitasse et al. 2014), which could facilitate shifts in species composition in a changing climate. Likewise, asynchronous flowering, especially among individuals of the same species, could have negative effects on pollination success (Zohner et al. 2018).

Phenological events are not restricted to overstory trees, of course, but also occur in the shrubs, seedlings, and saplings of the understory. Leaf phenology of many plants in the understory

FIGURE 20.8 Varying leaf phenology related to topography and microclimate in Rhode Island. The study location included a 30-m elevational change over 500 m of horizontal distance. Leaf flush is phenologically later at lower points in the landscape (the phenological gradient encompassed a 2-week duration) because of cold-air flow and a colder microclimate. This difference in phenology was detectable with satellite imagery. *Source:* From Fisher et al. (2006) / with permission of Elsevier.

is often characterized by leaf flush and expansion days or even weeks before the overstory trees in order to uptake sufficient carbon prior to canopy closure (Heberling et al. 2019). This shade-avoidance strategy, known as **phenological escape**, not only enhances the growth of understory plants but is critical for their growth and survival (Augspurger 2008; Lee and Ibáñez 2021a) because it provides the majority of the seedlings' annual carbon assimilation (Figure 20.9). At issue is that warming spring temperatures are currently driving earlier canopy leaf flush (Piao et al. 2019), such that the mismatched phenology between overstory and understory may diminish. However, using phenology modeling and field experiments of sugar maple and northern red oak seedlings in southern Michigan, Lee and Ibáñez (2021b) predicted that seedling leaf phenology would be more sensitive to spring warming relative to the nearby overstory trees that shade them. As a result, the duration of phenological escape by tree seedlings should increase under warmer springs, thus

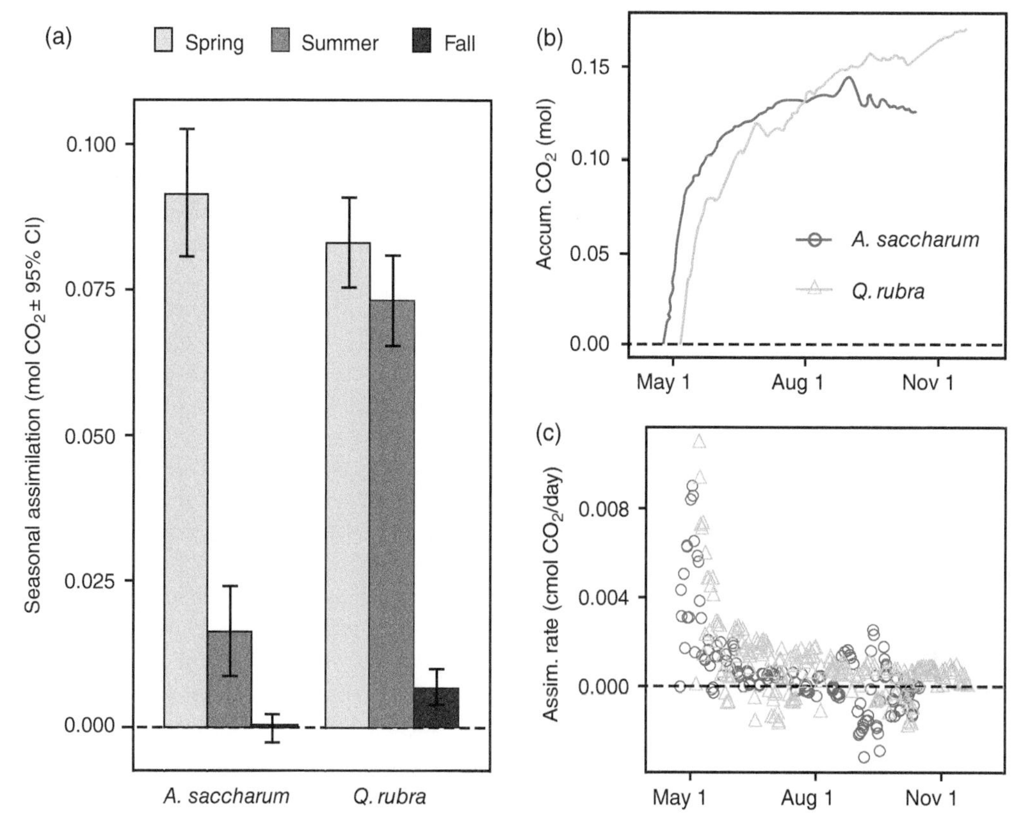

FIGURE 20.9 Seedling carbon assimilation for sugar maple and northern red oak resulting from phenological escape in southern Michigan. The majority of carbon is assimilated by tree seedlings in early spring, when light is most available in the understory prior to leaf flushing of the canopy trees. (a) Seasonal foliar carbon assimilation estimates. (b) Additive carbon assimilation over the growing season. (c) Daily assimilation rates for individual sugar maple and northern red oak seedlings. *Source:* From Lee and Ibáñez (2021a) / John Wiley & Sons.

increasing seedling access to early spring light. In the end, however, increased phenological escape is not likely to increase seedling performance in the understory because of likely increases in growing season respiration resulting from reduced water availability and warmer temperatures under climate change (Lee and Ibáñez 2021b). Such increased respiration would largely offset or exceed additional carbon gained by tree seedlings via earlier leaf flush in warmer springs. Notably, phenological escape by tree seedlings has also been documented in the fall, where seedlings maintain their leaves beyond canopy tree leaf drop (Gill et al. 1998), but late-season carbon uptake via this phenomenon is minimal (Lee and Ibáñez 2021a).

REGENERATION

Forest ecologists often focus on the response of mature trees to rising temperatures and changes in moisture availability, but tree regeneration deserves equal attention. Even prior to anthropogenic climate change, forests are likely to have adjusted to changes in environmental conditions primarily by changes in seedling occurrence and abundance. Regeneration trajectories are also useful in predicting future forest structure, distribution, and condition. Regeneration is directly affected by changes in temperature and precipitation, and these effects apply at multiple points of the reproductive cycle—during flowering and cone production, seed germination, seedling establishment, and seedling growth and survival. Increased atmospheric CO_2 concentrations are also likely to affect seedling growth, as described in the earlier text, though our focus here is on changes in temperature and moisture availability.

Importantly, effective tree regeneration depends on success at each stage of reproduction (Chapter 4), and specific changes in temperature and/or precipitation may be beneficial at one regeneration stage but detrimental at another. Overall, higher growing season temperatures benefit the early stages of regeneration, including flowering, pollination, and germination, because these processes accelerate under warmer spring temperatures (Boucher et al. 2019) and may begin sooner in early springs (Classen et al. 2010; Prevéy et al. 2018). Moderately higher temperatures may also benefit seedling growth and survival so long as adequate moisture is available (Fisichelli et al. 2014a), but increasingly higher temperatures eliminate most benefits of warming (Boucher et al. 2019). For example, a study of ponderosa pine in the western United States compared historical regeneration potential (1910–2014) to future potential using a climate–water balance model (Petrie et al. 2017). The study found that higher temperatures and lower moisture availability under climate change reduced seedling growth and survival relative to the historical period, but higher temperatures favored "flowering" (the pollination period for conifers), seed production, and germination. The models predicted higher future regeneration for 2020–2059, but also that increasingly higher temperatures after 2059 would reduce seedling growth and survival enough that the likelihood of regeneration failure in ponderosa pine would approach 60% (Petrie et al. 2017). Similarly, Ibáñez et al. (2017) found in Michigan that warmer temperatures benefitted seed production of red and sugar maples, but were detrimental to seedling establishment and survival, which were highest in cooler years. Moreover, increasing temperatures were more detrimental to seedling establishment and survival in southern locations compared to those three degrees latitude further north.

Of course, temperature changes do not occur in isolation, and moisture availability is important at every stage of regeneration. We differentiate changes in "moisture availability" from those in precipitation, given that flowers and seeds, germination, establishment, growth, and survival all depend specifically on soil moisture (Vose et al. 2016; Boucher et al. 2019). Soil-moisture availability may decrease even where precipitation is increasing if warming temperatures increase transpiration rates and evaporation from the soil surface. Seedlings are particularly susceptible to decreases in soil moisture because of their relatively small root system that limits their contact with, and thus moisture uptake from, soil water (Will et al. 2013). Seedling mortality and reduced growth associated with reductions in moisture availability have been documented for many regional ecosystems (Erickson et al. 2015; Rother et al. 2015). In eastern North America, seedlings that persist in the understory are typically shade-tolerant, and many shade-tolerant species lack tolerance of drought (Chapter 13), potentially making seedlings of shade-tolerant species particularly susceptible to mortality under reduced soil-moisture conditions. Projected soil-moisture deficits under climate change are particularly concerning where regeneration occurs only episodically in wetter years, as in western North America. For example, ponderosa pine forests of the western United States typically have a regeneration niche limited to periods of high moisture availability that are likely to decrease in frequency with climate change (Petrie et al. 2017; Davis et al. 2019).

Given the likelihood for increasing occurrence of wildfires (see discussion in the following text), rising temperatures, and increasing drought under climate change, as well as questions about the resilience of forests under such conditions, much attention has been given to tree regeneration following wildfires in western North America. Overall, tree regeneration has thus far declined following wildfires in the twenty-first century compared to the end of the twentieth century due to warmer and drier conditions (Stevens-Rumann et al. 2018; Stevens-Rumann and Morgan 2019; Rodman et al. 2020), suggesting that either longer recovery periods will be necessary for burned areas or a conversion of dry forests to non-forest cover types will occur. Using a series of models to examine tree seedling regeneration of Douglas-fir and ponderosa pine in 21 wildfire areas of the northern Rocky Mountains, Kemp et al. (2019) found that average summer temperature best explained post-fire seedling densities for both species. A temperature-tolerance threshold for Douglas-fir seedlings was determined to be 17 °C, above which Douglas-fir would not regenerate. Notably, over 80% of the 177 sampled sites were projected to exceed the threshold by the middle of the twenty-first century, which would sharply decrease Douglas-fir regeneration (Figure 20.10).

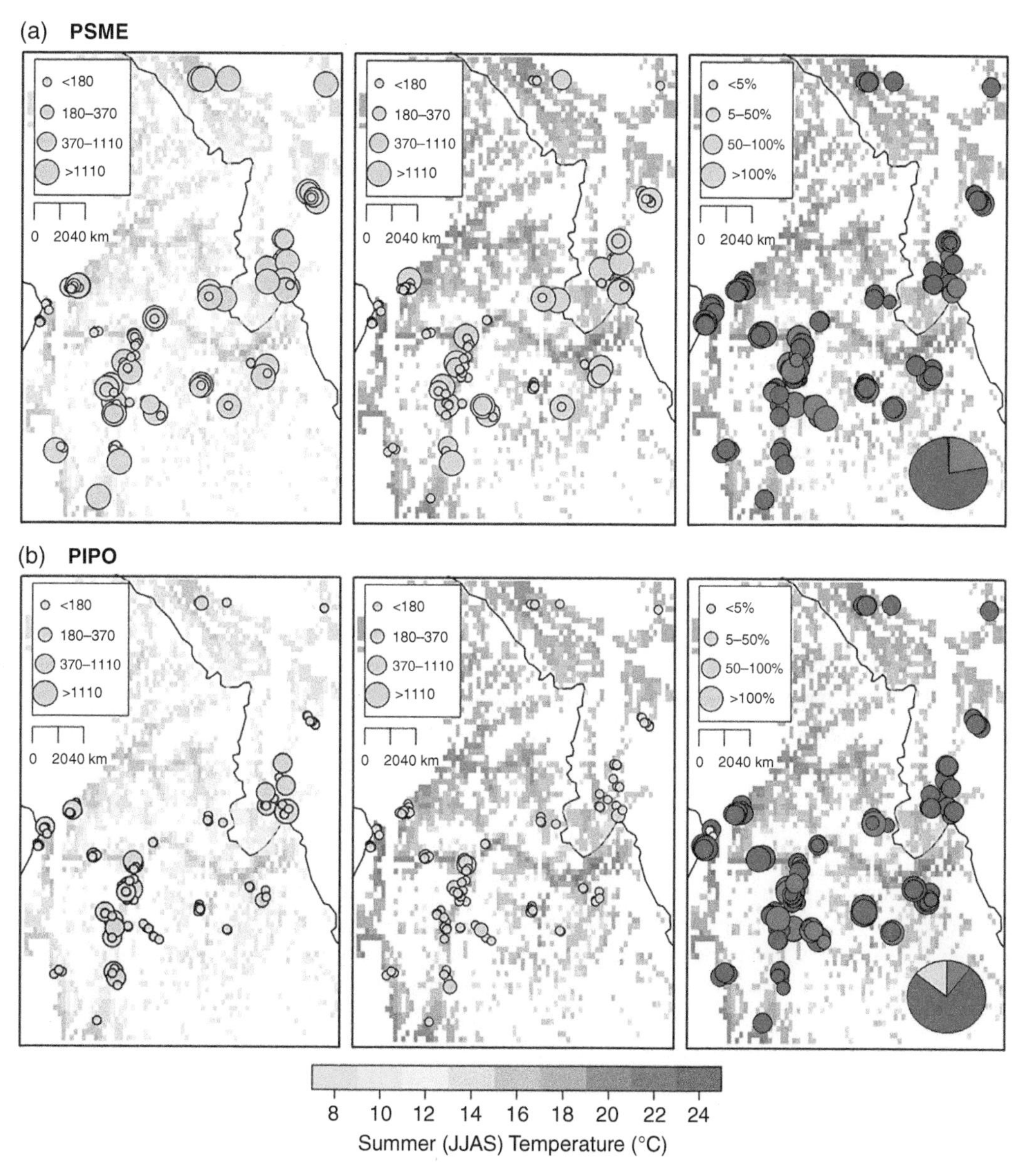

FIGURE 20.10 Map of current (left), predicted future (center), and change in (right) (a) Douglas-fir (PSME) and (b) ponderosa pine (PIPO) seedling densities. Map backgrounds are the average summer temperature of dry mixed-conifer forest currently (1981–2010; left) and predicted for mid-century (2041–2070; center and right). Seedling densities are considered to have changed if the density increased or decreased by at least 5% over the predicted period. Pie charts display the percent of sites with increase (blue), decrease (red), and no change (gray) for each species. *Source:* Kemp et al. (2019) / John Wiley & Sons.

The threshold for ponderosa pine was slightly higher at 19 °C, but future regeneration potentials were also projected to decrease. In subalpine forests, post-fire subalpine fir and Engelmann spruce regeneration declined with drought severity and distance to seed sources but was higher on cooler and wetter aspects (Harvey et al. 2016b). Post-fire regeneration may also be limited indirectly by climate change if wildfire severity, frequency, and/or extent are increased by increasing temperatures and decreasing precipitation; these scenarios are discussed in the Disturbance section in the following text.

CLIMATE CHANGE EFFECTS ON TREE SPECIES DISTRIBUTIONS

Perhaps of longest interest to ecologists concerned about climate change is the potential for changes in tree species' distributions. Range shifts and species migrations in relation to climate were considered well before modern climate change issues were ever examined as part of ecological research. Predicting how tree species may shift in their distributions is largely based on the idea that populations at the margins of a species' range occur on less favorable sites and in suboptimal climates. Such populations are therefore more susceptible to decline or to go locally extinct, leading to gradual changes in their geographical ranges (Lawton 1993). Given that climate change is particularly acute at northern latitudes, species in those locales might be expected to migrate toward more suitable growing conditions. Studies of migrations and shifting tree distributions may be categorized as those that (i) reconstruct the migration and re-distribution of tree species following glacial retreat at the end of the last ice age (Chapter 15); (ii) examine how species' ranges have relatively recently shifted since the presumed onset of anthropogenic climate change (about 1850); or (iii) project future distributions using modeling techniques. As paleoecological reconstructions of immediate post-glacial tree distributions were discussed in detail in Chapter 15, observed changes over the last 150 years and projected changes are our focus here.

OBSERVED RANGE SHIFTS

Given the hypothesis that species' distributions will change mostly at their margins, ecologists have expected that species would migrate toward the poles (northward in the Northern Hemisphere) and upward in elevation as temperatures warm. Most studies searching for evidence of such movement have depended on large networks of repeatedly measured forest inventory plots to examine changes in species occurrence across large areas over time (e.g., Woodall et al. 2009; Zhu et al. 2014; Fei et al. 2017; Sittaro et al. 2017; Boisvert-Marsh et al. 2019). In general, tree species in the Northern Hemisphere have been found to be moving higher in elevation (Holzinger et al. 2008; Lenoir et al. 2008), but northward migration has been somewhat less consistent. In one of the first studies to examine climate-related tree species migration across broad areas of the eastern United States, Woodall et al. (2009) compared the average latitude of large adult trees to the average latitude of seedling density for 40 species. In comparing seedlings to trees, those species having seedlings both further to the north and south of adult trees at the extent of the species' range would be expected to be expanding their range; those with seedlings further to the north but not to the south to be expanding northward; those with seedlings further to the south but not the north to be expanding southward; and those with adults further to the north and south than seedlings to be contracting in their range (Zhu et al. 2012). The average latitude for seedlings was more than 20 km further north than that of tree biomass for northern species (Figure 20.11), revealing tree regeneration in the northern portion of these species' ranges and suggesting northward tree migration by as much as 100 km per century for some species. Notably, migration in the southern portion of tree species distributions did not appear to occur (i.e., ranges expanded to the north but did not contract from the south; Woodall et al. 2009). Additional studies have corroborated similar northward shifts

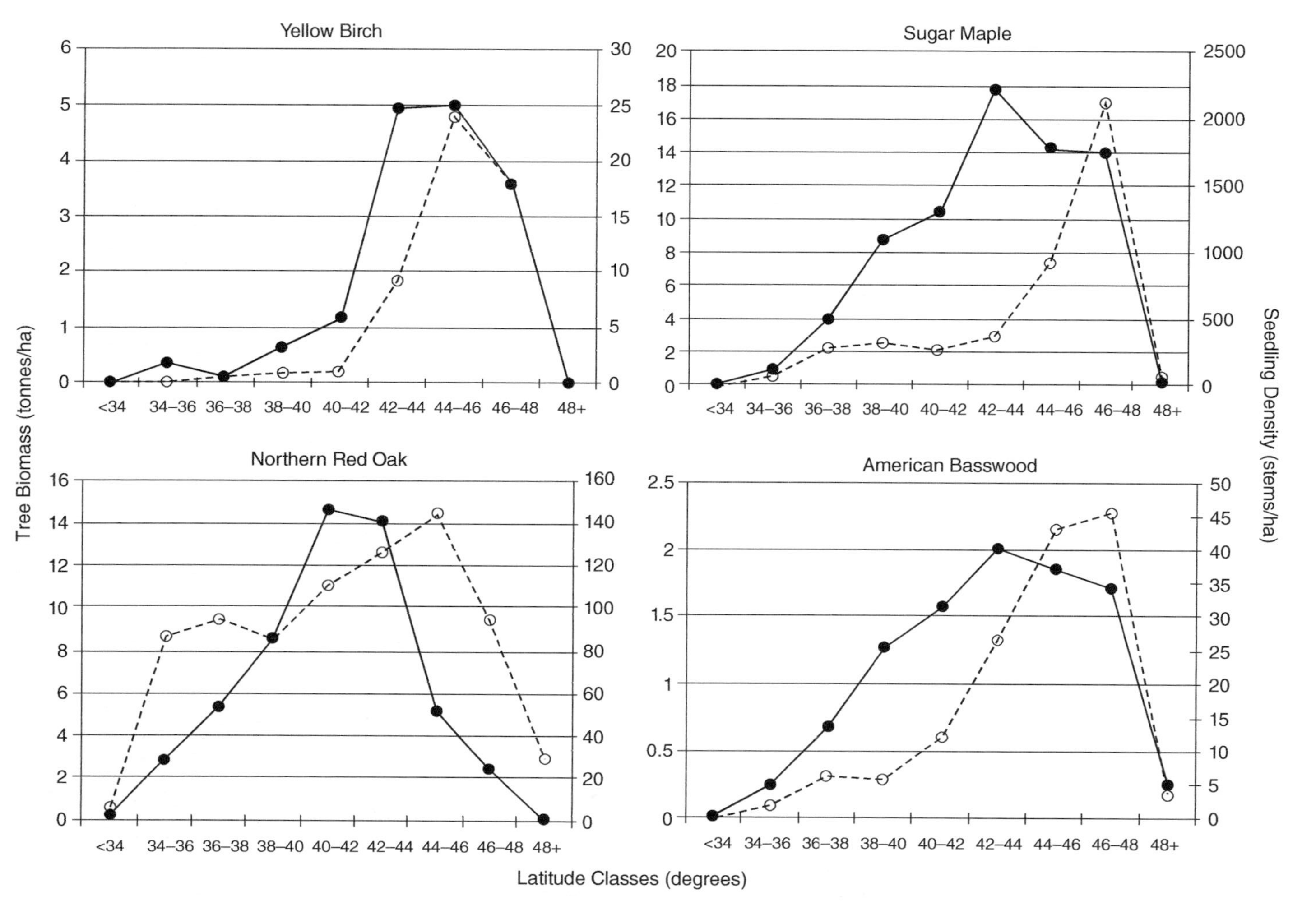

FIGURE 20.11 Comparison of tree biomass (solid lines) and seedling densities (dotted lines) across 2° latitude classes for four northern species in the eastern United States. Average and peak seedling densities are found further north than average and peak tree biomass for each species, suggesting a northward expansion of each species' range. *Source:* Woodall et al. (2009).

for portions of species' ranges in Canada (Boisvert-Marsh et al. 2014, 2019; Sittaro et al. 2017). For example, rising temperatures increased seedling recruitment for sugar maple, American beech, and red maple in the northern portions of their range in Quebec, Canada, but decreased it in the south, driving range shifts northward (Boisvert-Marsh et al. 2019).

Subsequent studies have found less evidence for northward migration in North America. Using the same data set as Woodall et al. (2009) to examine potential range shifts for 92 species, Zhu et al. (2012) emphasized that such analyses must be conducted at the boundaries of species' distributions rather than simply comparing average latitudes for seedlings and adult trees, as geographic range is realistically defined by boundaries rather than central tendencies. In examining adult trees and seedlings only within the 95th percentile of latitude for each species, they found no evidence that migration was greatest where climate change is the strongest. Instead, they found evidence for range contraction rather than northward expansion in 59% of the species studied; only 21% of the species showed a northward shift, and 16% showed a southward range shift. Additional studies using this methodology have suggested that most species' ranges in eastern North America are actually stable or possibly contracting rather than expanding northward (Woodall et al. 2013, 2018; Zhu et al. 2014). The concern with this "failure" to migrate northward of many species is the potential that tree species are unable to migrate at a pace similar to that of climate change occurring at northern latitudes (Neilson et al. 2005; Sittaro et al. 2017).

An important consideration is whether range shifts could occur in an eastward–westward shift in addition to (or instead of) a northward–southward shift. Tree species ranges in the eastern United States have been shown to have shifted to the west (median rate 15.4 km per decade) more than to the north (11 km per decade) in response to recent increased precipitation and moisture availability (Fei et al. 2017). For 86 tree species distributions (defined by abundance) over 35 years, important regional differences also exist in species migrations. Most species in the northern hardwood region shifted north at 20.1 km per decade, those in the central hardwood region shifted west at 18.9 km per decade, those in the southern pine–hardwood region shifted west at 24.7 km per decade, and those in the forest–prairie transition moved west at 30 km per decade. The faster western shift relative to the northern shift suggests that tree species distributions were more sensitive to precipitation compared to temperature between 1980 and 2015 (Fei et al. 2017), and emphasizes that change in precipitation is an equally important factor to consider in climate change effects on range shifts.

PROJECTED CHANGES IN TREE SPECIES DISTRIBUTIONS

Projecting future tree distributions requires some sort of species distribution modeling or forecasting of both climate and climate–species relationships. Models are often criticized by scientists and laypersons because they are by definition a simplification of what happens in the natural world and depend on a modeler's understanding of how the natural world works. Nevertheless, predicting something that has not yet happened requires such models, and projecting future tree distributions requires accepting some uncertainties in the data. A recent review of forest modeling challenges under climate change is presented by Scheller (2018). We also emphasize that projections of tree species are almost continuously changing with improvements in climate model projections and tree sampling data, such that those presented in the following text are for illustrative purposes rather than precise predictions of where a given tree species may be distributed under various climate change scenarios.

One method of predicting future distributions is the use of *climate envelopes*, which identify the values of a series of climate variables at each geographic location where the species occurs, then define the species' range using the maximum and minimum values of those variables (McKenney et al. 2007a). Using projections of future climate variables from one of several climate models with scenarios of high or low CO_2 emissions, a climate envelope may then be established for future climate scenarios and compared to the current envelope (Figure 20.12). Such a method thereby predicts future potential tree species habitat—in other words, where a given tree species could grow, ignoring the influence of biotic interactions such as competition, dispersal, or genetic adaptation to growing conditions (McKenney et al. 2007b). A second method uses statistical methodology called regression tree analysis (RTA) to project future tree species distribution from predictor variables that include climate variables as well as soil, elevation, land use, and indices of landscape fragmentation (Iverson et al. 2008, 2019a). Predictor variables are used to estimate the *importance value* (an index integrating the frequency, size, and density of a given species) of tree species under current and future climate scenarios. The importance values are mapped in contiguous 400-m and 100-m plots across the landscape, such that current and projected distributions can be compared (Figure 20.12). Notably, the RTA method also predicts only potential tree species habitat in the absence of biotic interactions. Neither method necessarily predicts species migration or even the potential distribution of tree species, but instead

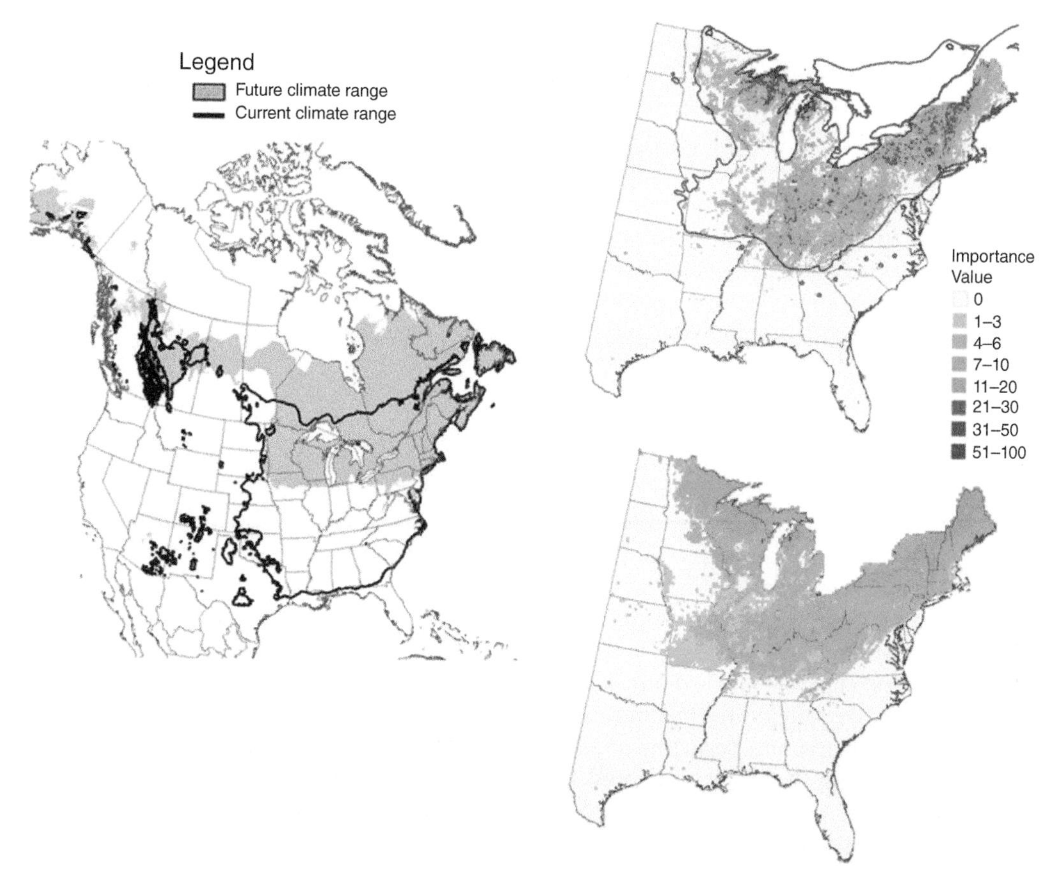

FIGURE 20.12 Future projections of the distribution of sugar maple in the eastern United States using climate envelopes (left) and regression tree analysis or RTA (right). Climate envelopes show a northward movement of potential habitat under future climate for the period 2071–2100, suggesting that much of the southern portion of the current climate envelope will become unsuitable due to warming temperatures. The RTA analysis suggests that sugar maple will become a less important species on the landscape by 2100 (bottom right) compared to its current distribution (top right). Future climate projections are based on different climate models for the two methods. *Source:* Climate envelopes map from McKenney et al. (2007b) / with permission of Oxford University Press; RTA map from Peters et al. (2020).

presents the area that would be suitable to colonize under climate change. Other modeling techniques have also been used to project future tree distributions (e.g., He et al. 2017; Wang et al. 2017).

Results of species distribution modeling suggest rather drastic changes for many tree species by the year 2100. Using a model that assumed any of the 130 species studied could disperse beyond their current climate envelope, McKenney et al. (2007a) found an overall average of a 12% decrease in envelope size and a northward shift of about 700 km. Notably, 72 of the 130 species exhibited a shrinking envelope, and 11 of the 72 species, most currently located in the southeastern United States, had envelopes that decreased in size by over 60%. Compared to current climate envelopes, future envelopes of 58 species were projected to increase in size up to 43%, mostly those located in the eastern United States and along the West Coast. These results suggest two important points. First, climate envelopes can shift rather large distances quite quickly, and it is unlikely that trees will migrate at a similar rate. Relatively slow migration rates have been corroborated by Iverson et al. (2004, 2019b) in a modeling context. In such a case, it is likely that many tree species will be mismatched with climate by the end of the twenty-first century. Second, species with currently limited distributions, such as many in the southeastern United States, are likely to be in danger of extinction because climate change drastically reduces the size of their climate envelope.

Results were even more drastic when the models were run under the assumption that tree species would only persist within current envelopes (i.e., they were unable to disperse beyond their current envelopes). In this scenario, future envelopes decreased by 58% and shifted northward only 330 km—probably decreasing more in size due to a more limited northward shift. Again, species with the largest decreases in envelope size were coastal with limited current distributions, particularly in the southeastern United States. A compelling prediction to emerge from this work was that by 2100 global change will have altered the climate of the southern United States beyond what any of the 130 species studied can currently tolerate, and that the species whose climate envelope showed the greatest latitudinal shift northward were from the southeastern portion of North America (McKenney et al. 2007a). When model runs were completed again using updated climate models (McKenney et al. 2011), climate envelopes were found to be as much as 10% larger in size but shifted up to 2.4° latitude further (Figure 20.13).

Similar results surface using RTA modeling. Iverson et al. (2019a) examined 125 tree species in the eastern United States and found that 72% of the species would gain suitable habitat under climate change, 21% of the species would lose suitable habitat, and 8% would remain unchanged, although model reliability varies widely among species. Those species gaining the most habitat were either (i) currently found in a warm and dry region (i.e., the southwestern portion of the United States). These species' distributions are heavily driven by temperature and expand greatly when provided much warmer temperatures; (ii) relatively rare and able to expand their range under climate change; or (iii) currently distributed in a southern location and expected to expand northward by 2100. About three to four times as many species moved northward instead of southward, with further distances occurring under more severe climate change scenarios (Figure 20.14). In recent RTA work that combined forest tree data from the United States and Canada for 25 species, Prasad et al. (2020) found that species projected to lose habitat did so mainly along the southern limits of their range, while habitat gains occurred mainly along the northern limits of their range.

The use of importance values in the RTA methodology provides the added advantage of examining the changes in species occurrence that might take place with climate change. Iverson et al. (2019a) identified the top three species for each state in the eastern United States as determined by their importance value and found that most of these species decreased in suitable habitat under climate change. For example, in Michigan, the species with the top three importance values are red maple, sugar maple, and trembling aspen, which decline to 98%, 82%, and 89% of their

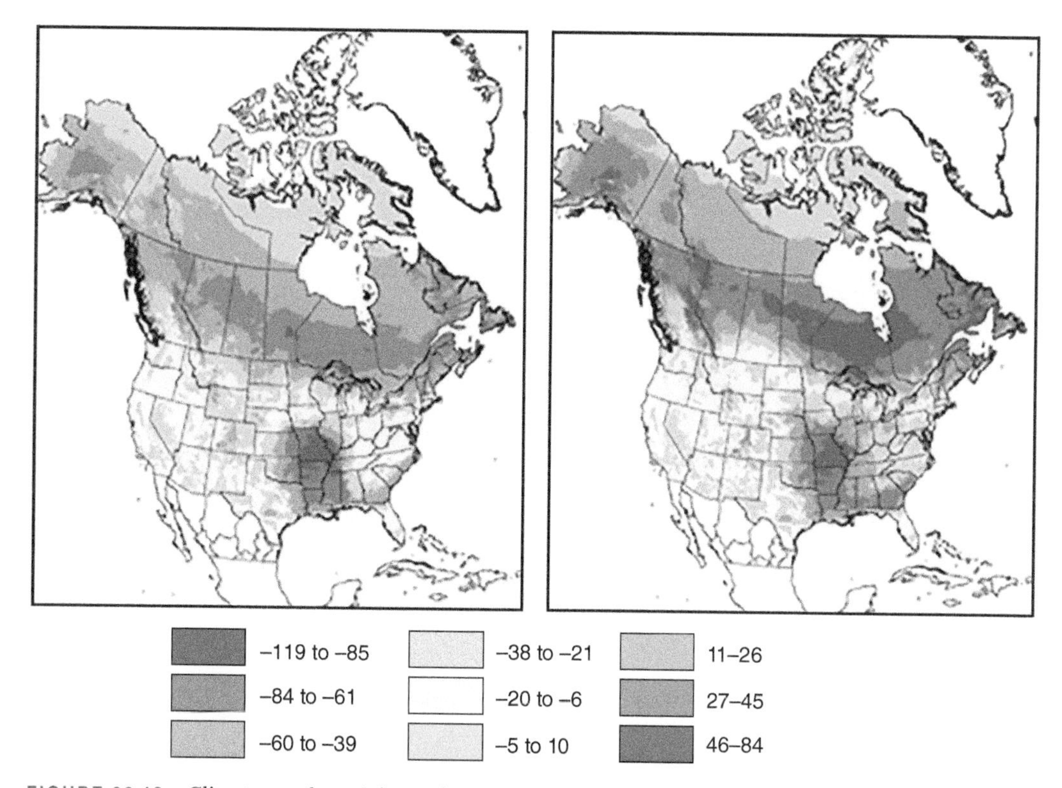

FIGURE 20.13 Climate envelope richness (i.e., number of tree species) differences between current (1971–2000) and future (2071–2100) periods for North America using two different climate models. Larger numbers (warmer colors) represent decreases in the number of tree species. Northern gains with southern losses (particularly from the southeast quarter of the continent) suggest northward migrations of many tree species. Left panel utilized the Parallel Climate Model (PCM); right panel utilized CCSM3.0 climate model. *Source:* McKenney et al. (2011).

current importance values under a low carbon emissions scenario and 87%, 74%, and 80% under a high emissions scenario. Over all states and regions in the eastern United States, importance values of the first (loblolly pine) and third (sweetgum) most common species increased under climate change by 21% and 48% of current importance values under a low emission scenario and by 32% and 73% under a high emissions scenario. The second most common species, red maple, decreased to 99% and 92% of its current importance values under low and high emissions scenarios, respectively (Iverson et al. 2019a).

PROJECTED CHANGES IN FOREST TYPE DISTRIBUTIONS

Notably, efforts to predict future tree distributions are focused primarily on species shifts rather than shifts in communities or forest types. Evidence from the pollen record and post-glacial reconstructions of tree species distributions clearly shows that post-glacial northward migration of species occurred independent of each other and not as forest types (Chapter 15). Thus, many forest types that existed during the period of glacial retreat have no modern analog, and likewise many modern types did not exist 12000–15000 years ago. Most ecologists therefore assume that such independent migration is likely to again occur as anthropogenic climate changes take place, and that tree species' range shifts are most likely to result in community disassembly and re-assembly,

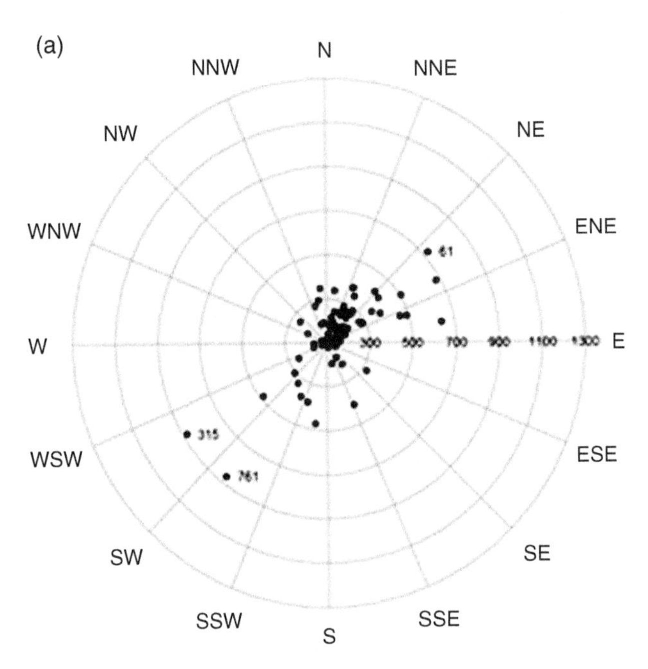

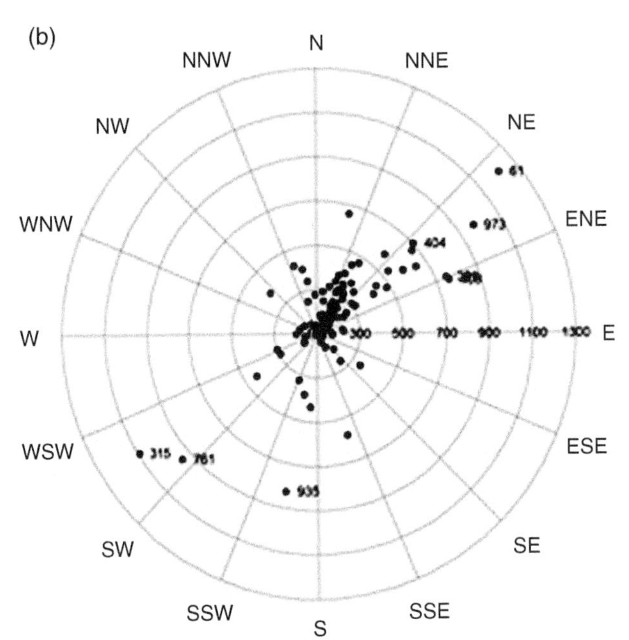

FIGURE 20.14 Modeled changes in distance and direction of mean centers of suitable habitat for 125 species utilizing different climate models. (a) Average of three climate models using low-emission scenario. (b) Average of three climate models using current emission scenarios. Both models suggest a northeasterly migration for most species, with further distances predicted using the high-emission scenarios. *Source:* Iverson et al. (2019a).

creating new forest types over the long term. Three important differences exist between current species shifts and historical post-glacial migrations that create uncertainties, however. First, the rate of anthropogenic climate change is higher than what occurred during glacial retreat, and it is unclear how a higher rate of change will affect tree migrations relative to those that occurred historically. Second, human influences that may affect migration rates via their impacts on dispersal—such as landscape and forest fragmentation, alteration of disturbance regimes that may encourage or inhibit tree regeneration, and invasive species that may compete with tree species—did not exist during post-glacial migration, and the presence of such barriers to migration on the landscape may affect tree species associations. Finally, current range shifts are occurring in the presence of well-established communities rather than recently de-glaciated terrain. These "occupied landscapes" may affect dispersal and competitive outcomes, and thus how future associations are assembled.

Despite these caveats, several researchers have attempted to model how forest types may be affected by climate changes, at least in the short term. A study by Iverson et al. (2008) using previous climate models that have since been updated suggested that changes in forest type distribution in the United States are most likely at northern latitudes. Spruce-fir forests, already limited to the far northern areas of the eastern United States, are likely to be essentially eliminated south of the Canadian border (Figure 20.15), as are the white-red-jack pine forests currently common in the Lake States. Refugia for some of these boreal species may include high elevations in the northern Appalachian Mountains or north-facing slopes where microclimates mitigate rising temperatures. Aspen–birch forests may be retained only in northern Minnesota under low-emissions scenarios and are eliminated under high-emissions scenarios. By contrast, oak–hickory and oak–pine forests were projected to increase to the north; maple–beech–birch forests were stable under low-emissions scenarios but reduced under high emissions, often replaced by oak–hickory forests. There is some evidence that red maple has already moved into the southern edge of the boreal forest due to warming temperatures at its northern extent (Fisichelli et al. 2014b). Other forest types, such as elm–ash–cottonwood and the southern pines, were stable in the analysis. The northern forests lost under climate change currently provide significant economic values to the region in the form of commercial products and tourism (Iverson et al. 2008).

In considering the potential migration of forest types, it is important to reiterate that communities are grounded within sites, and sites do not migrate with climate change. Important site factors other than climate also shape the composition of species within ecosystems, and enormous variation in these factors exists within a given species' climate envelope or suitable habitat that is not and cannot be easily modeled. For example, in the northeastern United States, upland oak species, with or without warming temperatures, will require periodic fires to open the canopy and remove competitors if they are to persist. Warming temperatures and reduced precipitation in the eastern United States would facilitate such disturbances and perhaps will accompany the northward migration of oak-dominated forests, but oaks probably will still be limited to drier sites with droughty soils, at least in the short term. Floodplain forests dominated by silver maple are likely to be favored by increased spring precipitation and flooding regardless of warming temperatures. Mesic forests dominated by species such as sugar maple, basswood, and American beech, which prefer moist but not wet sites, may find refugia from warmer temperatures and decreasing soil moisture in cool, low wetlands commonly nearby their current sites as they dry out. There will likely come a time when the severity of climate change overrides current site–species relationships, but short-term changes in forest type distributions are likely to remain closely tied to physical site factors over the next several decades.

CLIMATE CHANGE EFFECTS ON FOREST DISTURBANCES

The effects of climate change on forests involve much more than the simple ability of tree species to tolerate changes in temperature and precipitation. We have discussed throughout this book the central role of disturbances in shaping forest structure and function (Chapters 10 and 16) as well as landscapes (Chapter 22). Many disturbance regimes are strongly driven by climate, so naturally climate change is likely to have significant impacts upon where disturbances occur, how often they occur, their severity, and their extent. Changes in disturbance regimes, in turn, impact how well

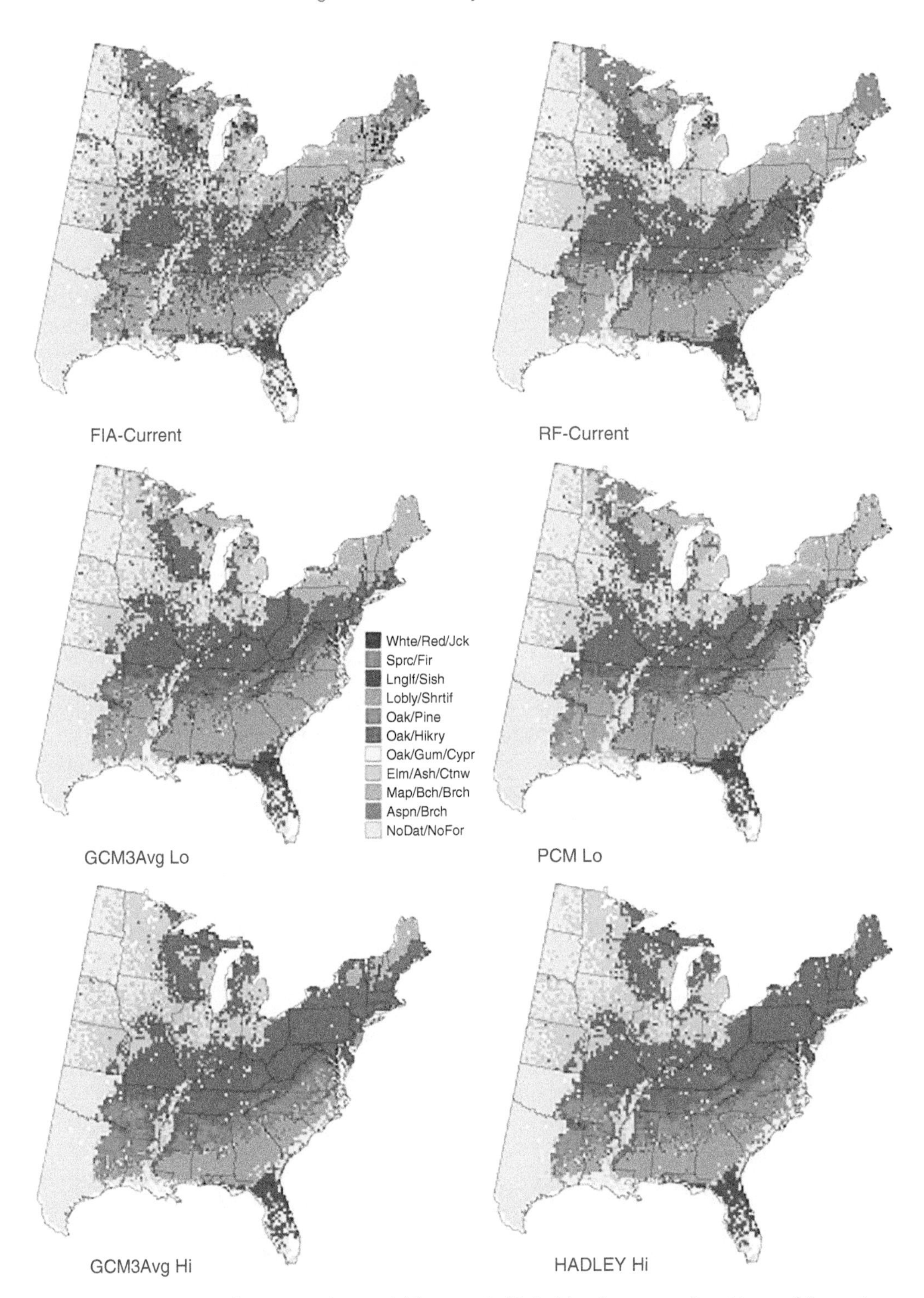

FIGURE 20.15 Maps of current and potential future suitable habitat for current forest types of the eastern United States. Maps include current inventory estimates of current distribution of abundance (FIA current), the modeled current map (RF current), two low-emission model scenarios, and two high-emission model scenarios. *Source:* From Iverson et al. (2008) / Elsevier.

forests are able to provide ecosystem services (Chapter 23) such as carbon storage, biodiversity, and water resources. The ecological resilience of forests is also at issue, as changing disturbance regimes have the capacity to move ecological systems beyond tipping points that would result in the loss of forests (Reyer et al. 2015; Johnstone et al. 2016). In sum, changes in forests due to climate change are most likely to be expressed over the next century as extensive changes in forest composition and structure due to altered disturbance regimes rather than slow-paced species migrations that reflect physiological responses to changes in temperature, precipitation, and CO_2 (Dale et al. 2001). This is especially the case because disturbances create rapid changes, whereas long-lived organisms such as trees would express changes more slowly—especially if changes are expressed through regeneration (Johnstone et al. 2010).

The onset of climate change and observable changes in disturbances have greatly increased attention toward forest disturbances in the last few decades, and general trends across all disturbance types may be gleaned from the literature. In a review of 647 studies investigating climate change effects on disturbances, Seidl et al. (2017) found that 42% of the studies found temperature-related factors to be the most important climatic driver of changes in disturbance regimes, particularly in northern latitudes and especially in the boreal forest. About 38% focused on precipitation, primarily water availability, mostly toward southern latitudes and especially in the tropics. These two broad climatic factors are as impactful in their effects on disturbance regimes as they are on growth, regeneration, phenology, and species distributions.

Climate change is likely to affect the frequency and severity of many kinds of disturbances, such as fire, drought, windstorms, ice storms, hurricanes, invasive species, insect and pathogen outbreaks, and landslides (Dale et al. 2001). In the following sections we focus on fires, insect and pathogen activity, and wind. In their review, Seidl et al. (2017) noted that where climate change is projected to produce warmer and drier conditions, over 82% of the studies identified an increase in fires and 78% in insect activity. Where climate was projected to be warmer but wetter, only 55% identified an increase in fires and 65% in insect activity. Wetter climates also translated to an increase in wind disturbance (89%) and diseases (69%) (Seidl et al. 2017). The importance of these disturbances for forests is described in Chapters 10, 12, and 16; in the following text we focus mainly upon how they may be altered by climate change.

FIRES

With much of the climate across the globe predicted to become warmer and drier and the dramatic effects of fires on forests, ecologists have paid great attention to the effects of climate change on fire regimes across the world. Fire frequency, extent, intensity, and seasonality are closely linked to weather and climate as well as to forest species composition, stand structure, and site characteristics (Chapter 10). Many decades of fire suppression in many forests have altered forest structure to such an extent that it can sometimes be difficult to differentiate the effects of climate change on fire behavior and occurrence from the effects of fire suppression. Climate change directly affects fire regimes by influencing the moisture content of fuel, which in turn affects fire initiation and spread (Williams and Abatzoglou 2016), ignition sources such as lightning (associated with storms), and the way fire spreads (via influences on wind speed and other weather factors). Williams and Abatzoglou (2016) estimated that human-caused climate change in the western United States resulted in extremely dry fuels in 75% more forested area for an additional 9 days during the fire season. They also attributed 4.2 million ha more burned area in the West to climate change between 1984 and 2015, which was almost twice the area expected otherwise. Climate change can also influence fire regimes indirectly by affecting the productivity of vegetation and the decomposition of dead vegetation, both of which impact the availability of fuels (Pausas and Ribeiro 2013), by driving forest composition thereby increasing the flammability of forests, and by altering forest structure which influences fuel continuity. The effects of fire on forests are discussed in detail in Chapter 10.

Changes in climate have already affected fire regimes in many forests in observable ways. In dry, fire-prone forests such as those of the boreal region in Canada and those in the western United States, warmer temperatures and lower precipitation over the last 20 years have increased the area burned by fires (Gillett et al. 2004; Abatzoglou and Kolden 2013). Moreover, the likelihood of fires >5000 ha in size is projected to increase across much of the United States by the mid-twenty-first century, especially in the northern Rocky Mountains, the northern Lake States, and the southeastern Coastal Plain (Figure 20.16; Barbero et al. 2015). Some researchers have estimated that the area burned each year in the western United States could increase as much as sixfold by the middle of the twenty-first century with continued climate change (Litschert et al. 2012). The western United States has also experienced an increase in how often fires burn, due to earlier snowmelt and warmer spring and summer temperatures that lead to longer fire seasons (Westerling et al. 2006). The extraordinary fire season of 2020 in the western United States alone effectively doubled the area burned in the Rocky Mountains since 1984, and as a result the rate of burning in subalpine forests is now double that of the last 2000 years (Higuera et al. 2021).

The potential for significant effects of climate change on fire regimes in the future has been predicted for several decades (Dale et al. 2001). One of the most compelling examples of potential changes in fire regimes due to climate change is found in the Greater Yellowstone Ecosystem in the central Rocky Mountains, where the climate is predicted to increase temperatures and reduce precipitation to an extent that the disturbance regime will move well out of its historical range of variability (Chapter 22). Fires larger than 200 ha have burned every 100–300 years in the region for as many as 10 000 years, but the fire rotation is projected to shorten to less than 30 years by the mid-twenty-first century with continued warming, and the frequency of years with synchronized large fires across the region is likely to increase (Westerling et al. 2011). Altered fire frequency and extent to this degree are incompatible with the tree species that currently occupy the landscape and will almost certainly have consequences for the persistence of forests and carbon storage.

The important interactions between climate change and fire are not limited to how fire regimes are altered. An important effect only recently documented is how well forests recover following fires in an altered climate—that is, how might climate change affect the resilience of forests when the fires that burn them are more frequent, larger, or more intense? The important effect in such a scenario is the change to the fire regime as well as to the post-fire climate under which the forest must recover. One of the easiest ways to envision such a scenario is to consider post-fire regeneration. When fire burns through a forest, the severity of the fire determines how the forest recovers (Chapter 10). A low-severity fire that does relatively little harm to mature trees, such as historical fires in ponderosa pine forests, may recover rather easily and leave the forest appearing relatively unchanged. However, recovery of forests following high-severity fires that kill most of the mature trees, such as those that occur in boreal and subalpine forests, depends heavily on post-fire regeneration that will eventually replace the killed mature trees. Such regeneration is likely to gain more attention from ecologists because (i) climate change is causing larger, more frequent, and/or more severe fires, such that more burned area will probably rely on post-fire regeneration for recovery, and (ii) the tree seedlings that regenerate following fires will be subject to novel and in many cases harsher growing conditions under which they have to survive and grow. Reduced or failed regeneration, even when numerous fire adaptations and recovery mechanisms are present, strongly affects the ability of a forest to be resilient to fires. Forests that successfully regenerate will be able to recover from fires as they have in the past, but those that do not may be converted to a non-forested state (Coop et al. 2020). Thus, changing fire regimes and/or decreasing post-fire regeneration have the ability to reduce the amount of forest cover across the world (McDowell et al. 2020).

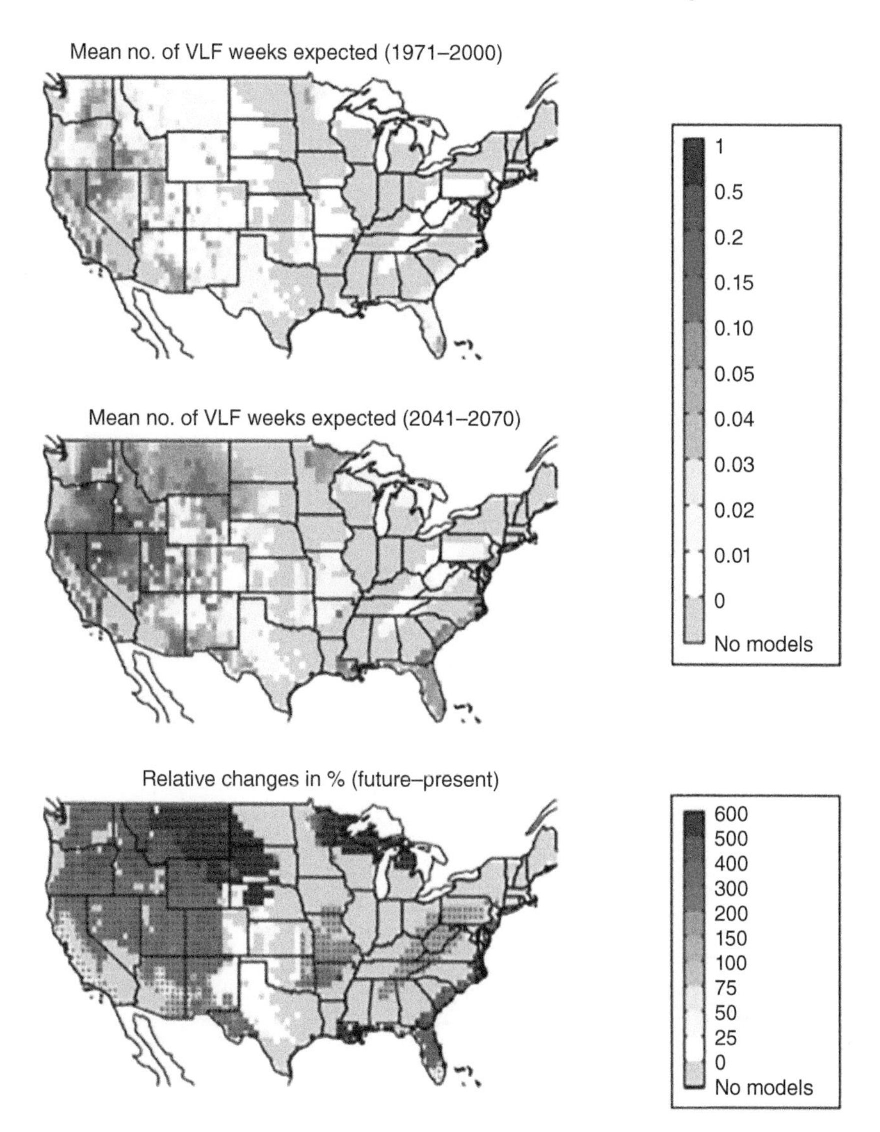

FIGURE 20.16 Mean annual number of very large fire (VLF; >5000 ha) weeks per surface unit for 1971–2000, 2041–2070, and the relative change between the two time periods. These large disturbances are projected to increase by mid-century, particularly in the northern Rocky Mountains, the Lake States, and the southeastern Coastal Plain. *Source:* From Barbero et al. (2015) / CSIRO Publishing.

As climate change progresses with increasing fire activity, rising temperatures, and reduced soil moisture, forest conversion has become increasingly common, especially in western North America. In general, the mechanisms behind such conversion can be summarized as (Coop et al. 2020):

Increased fire activity that kills all or most trees: Many current and historical fire regimes are stand-replacing and kill many or most trees, such as in many high-elevation forests dominated by lodgepole pine, spruce, and fir. However, increasing area burned, fire size, number of fires, and proportion of area burned at high severity are likely to reduce the resilience of such forests historically burned by large, infrequent crown fires as well as those burned by smaller, frequent surface fires (e.g., ponderosa pine forests) or mixed-severity fires (e.g., mixed conifer forests).

Such changes in fire regimes have several important effects. First, a growing number of large fires that burn at high severity will raise the number of large openings on the landscape (Figure 20.17a). Unless the forests are dominated by serotinous species (e.g., lodgepole pine), reseeding of the burned area would need to occur via live tree seed sources whose distance from many portions of a large patch may be beyond the trees' dispersal limit (Figure 20.17b; Johnstone et al. 2016). Moreover, many such landscapes become extremely harsh environments for seedling survival after they are burned, which may lead to forest conversion to non-forest or, at the very least, delay its recovery (Harvey et al. 2016a,b). For example, almost 40% of ponderosa pine forests burned by

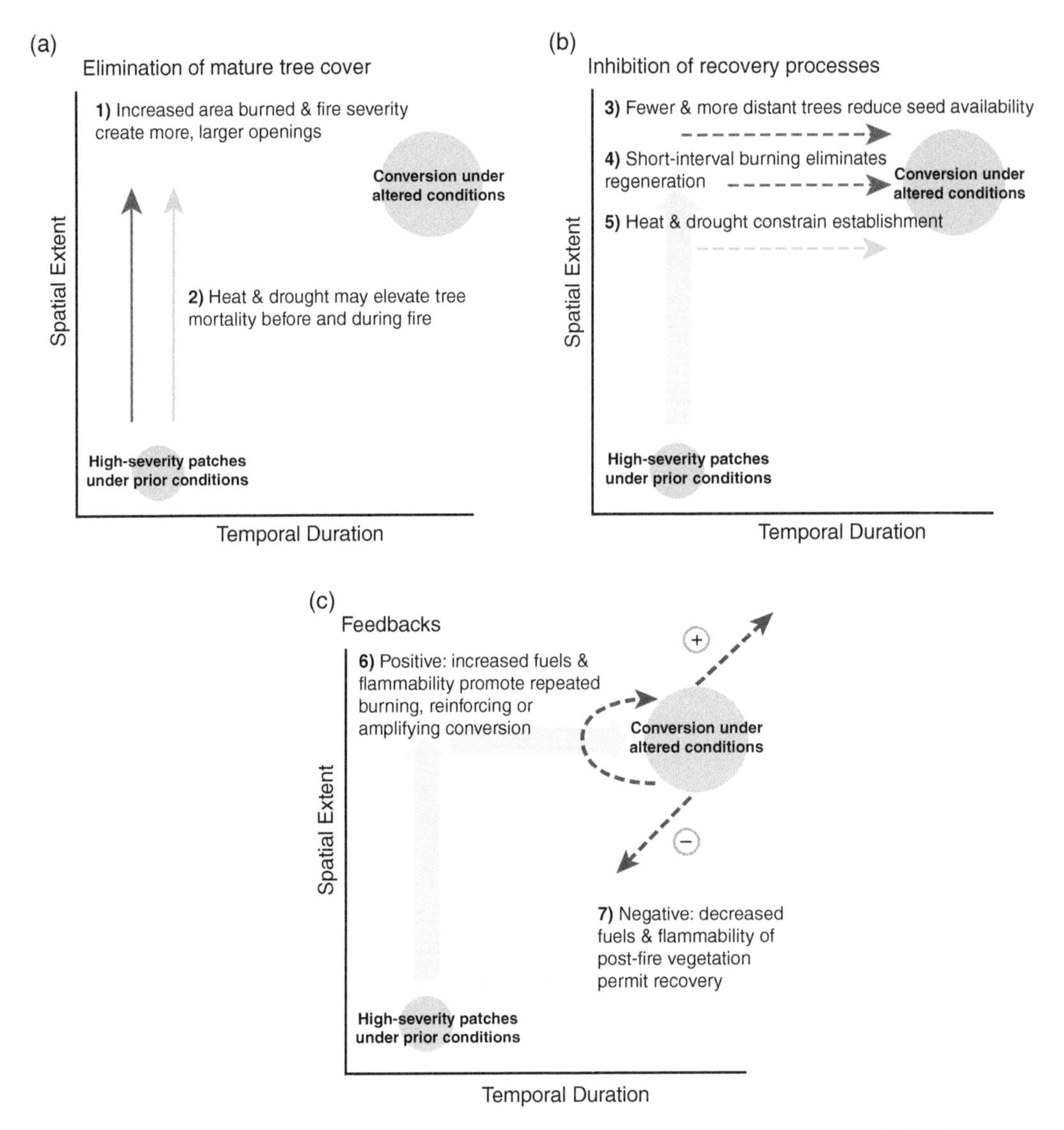

FIGURE 20.17 Processes directing forest conversion due to fire. (a) Conversion is initiated by fire (red arrow) or climate (yellow arrow) that kills extensive forested areas. (b) Conversion is maintained by failed or reduced post-fire regeneration of pre-fire tree species, prolonging forest recovery time. (c) Repeated or short-interval burning may further reinforce conversion by further limiting regeneration and recovery. *Source:* From Coop et al. (2020) / with permission of Oxford University Press.

high-severity fire as part of the 2002 Hayman Fire near Denver, Colorado still lacks significant tree regeneration nearly two decades after they were burned, due to a lack of seed sources and harsh post-fire environment (Rodman et al. 2020).

Increasing the area burned each year also increases the chances that an area recently burned will be burned again with relatively little time for the forest to recover between the two burns. Such *short-interval fires* may remove remaining live tree seed sources, reduce or eliminate any established tree regeneration, or shift vegetation composition toward species with superior post-fire resprouting ability (Coop et al. 2016). For example, short-interval fires in the Greater Yellowstone Ecosystem, where historical high-severity fires burned lodgepole pine forests on a 100–300-year rotation, combusted almost all pre-fire biomass in some reburns, sharply reduced post-fire regeneration relative to the previous (long-interval) fire, and converted many high-density stands on the landscape to low-density stands (Turner et al. 2019). Notably, forest resilience in this case was reduced by short-interval fires, even for a highly fire-adapted tree species with serotinous cones, because a second fire occurred before the regenerating trees from the first fire were able to begin cone production. Similar studies of black spruce, a semi-serotinous species in the boreal forest, have documented a switch between forest types (e.g., Hart et al. 2019) or a conversion of forest to grassland (Brown and Johnstone 2012) due to short-interval fires when trees are burned before they are able to produce cones.

Warmer and drier climate making survival of post-fire regeneration more difficult: Climate change may directly influence regeneration failure and potential forest conversion via the impacts of warming temperatures and reduced precipitation on tree and seedling survival and growth. As discussed in the earlier text, these changes in climate have resulted in significant tree mortality in dry forests across North America. Such mortality is unrelated to fires but may contribute to reduced seed sources critical to revegetating burned areas and may increase fuels for future fires (Figure 20.17a,b).

Perhaps more critical to potential forest conversion is the potential inability of trees to regenerate in a warmer and drier climate following fire, even when seed sources are available. Such post-fire regeneration failure attributed to climate change and potentially leading to forest conversion has already been observed in the western United States. For example, regeneration of ponderosa pine and Douglas-fir was low 8–15 years after fire at lower elevations in the Colorado Front Range, especially on dry sites and at greater distances from post-fire seed sources (Rother et al. 2015). Similarly, at lower tree line in Yellowstone National Park, Douglas-fir regeneration and forest replacement were low on dry sites, far from seed sources, and at the lowest elevations in areas of high-severity burning (Donato et al. 2016). These trends contrasted with rigorous Douglas-fir regeneration on mesic sites, possibly signaling effects of warming temperatures and reduced precipitation. At high elevations in Yellowstone, Rammer et al. (2021) used simulation models to predict widespread regeneration failure (between 28% and 59% of the forested area) in highly fire-adapted lodgepole pine forests by 2100 and particularly in non-fire-adapted forests dominated by Engelmann spruce and subalpine fir. Finally, Stevens-Rumann et al. (2018) attributed sharp decreases in twenty-first century post-fire tree regeneration at 52 wildfires in the Rocky Mountains to greater annual moisture deficits that occurred from 2000 to 2015. Importantly, it is not known whether such regeneration failures will truly lead to permanent forest conversion, or whether recovery to the pre-fire forest will simply be delayed, because post-fire regeneration in dry forests of the region occurs in pulses for many forest types such as ponderosa pine and mixed-conifer forests (Coop et al. 2020).

Disruption of fire–vegetation feedbacks: Fire regimes generally include feedbacks that promote the persistence of that regime, and alteration of fire regimes by climate change may change these feedbacks (Figure 20.17c). For example, fire managers have long known that young forests are less susceptible to fire than older forests because fuels are consumed by the previous fire; flammability increases as fuels re-accumulate. This *negative feedback* between fire and vegetation limits or delays future burning and thus typically provides an extended window during which post-fire regeneration and forest recovery may occur (Coop et al. 2020). However, such a feedback may be overridden by weather that promotes severe burning, and such weather is likely to increase under current climate trends. For example, windy conditions and dramatic declines in precipitation preceding and during the 1988 fires in Yellowstone National Park caused the fires to spread irrespective of forest age or fuel type (Christensen et al. 1989), essentially negating any negative feedback. Fire regimes can also be bolstered by *positive feedbacks*, whereby fires promote future fires, particularly when they are severe. In some systems, severe burning quickly creates high levels of fuels from killed vegetation and dense re-vegetation through sprouting that promote additional burning. Although such positive feedbacks perpetuate a fire regime under a given climate, a changing climate that results in even small changes in fire activity may result in relatively abrupt forest conversion (Figure 20.17c). For example, Tepley et al. (2017) documented conversion of dry montane conifer forests in California and Oregon to highly flammable shrubland under drier post-fire conditions and at greater distances from seed sources. In this system, recruitment of Douglas-fir occurs only in the first few years after severe fires, facilitating rapid forest recovery, because competing broadleaf trees and shrubs prevent a longer duration of seedling establishment. Warmer and drier temperatures that limit or reduce this immediate seedling recruitment are likely to shift the system toward a prolonged recovery time, and the onset of shorter-interval fires in a highly flammable landscape is likely to perpetuate shrub cover rather than forest cover (Tepley et al. 2017).

The potential for forest conversion due to changing fire regimes is at the forefront of forest ecology in an era of rapid climate change. Excellent reviews of this potential are provided by Buma et al. (2020), Coop et al. (2020), Hart et al. (2019), Hessburg et al. (2019), Stevens-Rumann et al. (2018), Tepley et al. (2017), Johnstone et al. (2016), and Anderson-Teixeira et al. (2013).

INSECTS AND PATHOGENS

Climate change, particularly the effect of warming temperatures, is likely to influence insect and pathogen effects on forests nearly as much as it does for fires. Insect outbreaks are already a major source of tree mortality across the United States; between 1984 and 2012, bark beetles alone killed more forest area than wildfires (Hicke et al. 2016). Many current and recent bark beetle outbreaks have been associated with rising temperatures and decreasing precipitation (Bentz et al. 2010), leading many ecologists to speculate that climate change is likely to further increase such outbreaks in the future. Pathogens are similarly expected to have large effects on forests because of climate change (Sturrock et al. 2011; Kolb et al. 2016). Climate change may facilitate outbreaks of insects and disease by (i) affecting the development and survival of insects and pathogens, and (ii) changing the susceptibility of the trees that they attack.

Insect and pathogen activity is strongly controlled by effects of climate. For insects, the most obvious effect is that warmer temperatures accelerate the rates at which they proceed through their life cycles. In some cases, warming temperatures have increased developmental rates to the extent that species with a 2- or 3-year life cycle have been able to complete their development within a single year (Figure 20.18; Bentz et al. 2010), which would create quicker population growth that potentially leads to outbreaks. Moreover, increased developmental rates may also lead to higher synchrony among populations, allowing mass attack that overwhelms tree defenses and leads to outbreaks (Logan and Powell 2001). Some insects have also been documented to increase their reproductive output by producing multiple generations per year instead of one. For example, Mitton and Ferrenberg (2012) observed that 20 years of rising temperatures in the Colorado Front

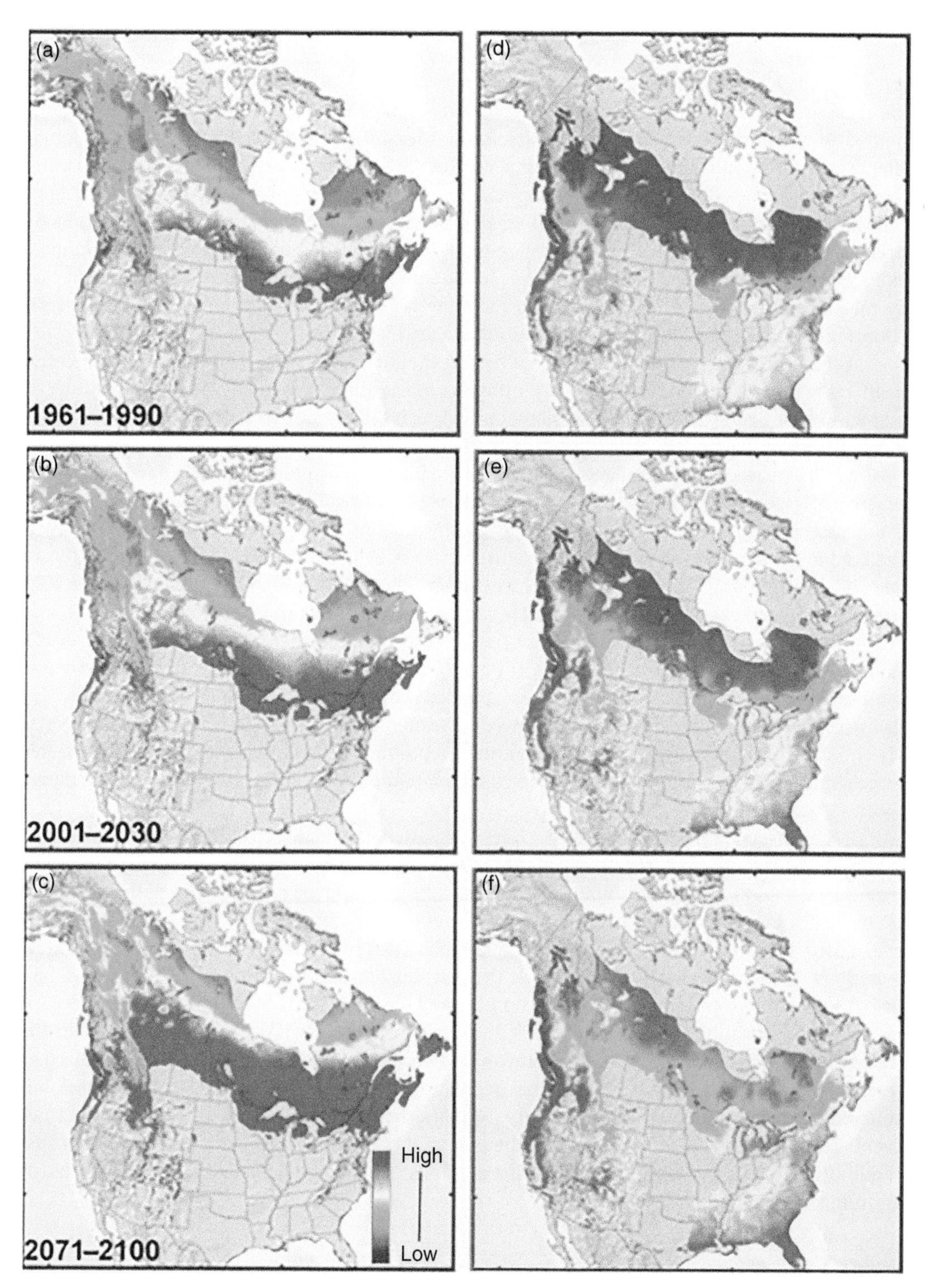

FIGURE 20.18 (a)–(c) Predicted probability of spruce beetle offspring developing in a single year in spruce forests across the range of this insect in North America during 1961–1990, 2001–2030, and 2071–2100. Forests with a higher probability of 1-year life-cycle duration have a higher probability of population outbreak and increased levels of spruce-beetle-caused tree mortality. (d)–(f) Predicted probability of mountain pine beetle cold survival across the range of pine species in the United States and Canada during the same three climate periods. Based on the response of mountain pine beetle to temperature, results suggest a low to moderate probability of range expansion across Canada and into central and eastern US forests by the end of the twenty-first century. *Source:* From Bentz et al. (2010) / with permission of Oxford University Press.

Range resulted in an earlier and longer dispersal period for mountain pine beetle and increased the life cycle from one to two generations per year. The potential for two generations is probably limited to warmer forests in North America until at least the middle of the twenty-first century (Bentz et al. 2019), but continued rising temperatures may eventually expand the potential to other forests as well. The effect of warming temperatures on insect development is not always straightforward, because some species have a winter-dormancy requirement that may become less likely to be met, which could upset insect development (Ayres and Lombardero 2000). Thus, climate change is not likely to affect all insect species in the same way. Pathogens tend to be more affected by moisture than by temperature because moisture is required for reproduction and dispersal (Desprez-Loustau et al. 2006), although the effects vary by species.

Climate change is also important in affecting the survival of insects and pathogens because it affects winter temperatures that may influence overwinter survival. Winter temperature is known to be a strong determinant of many insect species' ranges, such that small rises in winter temperature could permit such species to expand into previously unsuitable forests (Figure 20.18). Northward expansion of several insect species due to increased winter temperatures has been documented. For example, the southern pine beetle was historically limited to pine forests of the southeastern United States, but has moved northward into New Jersey, New York, and New England following a rise of winter temperatures there of 3–4 °C (Dodds et al. 2018). Similarly, an enormous mountain pine beetle outbreak in British Columbia between the mid-1990s and 2017 (Chapter 16) is likely related to consistently warmer temperatures there that began in 2003, and even included tree species that are not normally susceptible to this insect (Logan et al. 2010). Finally, the hemlock woolly adelgid, an invasive insect that attacks eastern hemlock and was limited to the mid-Atlantic in the 1970s has expanded northward into central New England, where its further spread has been significantly slowed by winter temperatures (Paradis et al. 2008). Notably, the short life cycle, sensitivity to temperatures, reproductive potential, and mobility of invasive insects and pathogens will likely result in rapid adaptation to climate change relative to native species (Newman et al. 2011).

Many insects and pathogens most successfully attack weakened trees, and a warmer climate is likely to have marked effects on tree vigor. Drought, characterized by reduced precipitation and increased temperatures, is the climate effect most commonly linked to insect and pathogen attack of stressed or low-vigor trees because it reduces the efficacy of tree defense (Raffa et al. 2008; Bentz et al. 2010). In turn, reduced tree defense across broad areas may result in outbreaks. For example, a study by Gaylord et al. (2013) examined the relationship between drought, tree defense, and insect attack in forests dominated by pinyon pine and one-seed juniper in New Mexico by experimentally manipulating precipitation. They found pinyon pine mortality to be the highest in the treatment that simulated drought, with pinyon ips bark beetle, which generally attacks stressed or recently dead trees, present in 92% of dead trees. Juniper was not susceptible to beetle attack but suffered canopy loss after 3 years of drought. Notably, the increased incidence of extreme weather events related to climate change also has the potential of reducing tree vigor and increasing susceptibility to insect and pathogen attack, for example, via damage by heavy snow, ice storms, or lightning (Ayres and Lombardero 2000).

WIND

Hurricanes drive the structure and composition of forests of Atlantic and Gulf coastal areas, but smaller-scale wind events such as downbursts, derechos, and tornadoes are a fundamental component of the disturbance regime of most deciduous forests in eastern North America (Peterson 2000). As discussed in Chapter 16, wind events of varying severity kill trees and disturb forest canopies, thereby initiating gap dynamics and altering stand structure and greatly impacting tree regeneration, establishment, and growth. It remains difficult to predict the general response of forests to wind events because forest types and tree species vary widely in their susceptibility to wind damage, and susceptibility also changes with forest age and density. However, it is certainly sound speculation that increasing intensity, duration, or frequency of wind events would increase disturbance to forests. Heavy precipitation events, often associated with thunderstorms, have become more frequent and intense in the United States since the 1990s (Easterling et al. 2017), particularly in eastern deciduous forests in the Midwest and Northeast. Given that downbursts and derechos arise from thunderstorms and thunderstorm conditions contribute to tornado formation, it is likely that such events will continue to increase in frequency and severity. Gregow et al. (2017) found increased windstorm damage to European forests between 1951 and 2010, with a threefold increase in average intensity of windstorms after 1990 attributed to climate change.

Perhaps equally important as direct effects, however, are indirect effects of climate change-induced increases in wind damage to forests, such as the effects of changing temperature and precipitation on the resistance of individual trees and forests to wind. Where winter temperatures are increasing, normally frozen soils that provide stable anchorage for trees during the winter months may thaw, leaving them susceptible to windthrow. Such effects have been a strong emphasis of forest research in Europe. For example, Usbeck et al. (2010) found that windstorm damage to forests increased by 22 times since 1858 in Switzerland, and that the most severe wind damage occurred when soil conditions prior to storms were thawed and wet. In this region, weather conditions attributed to climate change included much warmer and wetter winter conditions and higher maximum gust wind speeds. Blennow et al. (2010) modeled wind effects on forests in Sweden with particular attention to changes in frozen soils, finding that a changed wind climate and soil thawing increased the probability of wind damage especially in southern compared to northern forests. Climate change may also indirectly influence wind damage in forests by changing tree growth and forest dynamics. A modeling study of radiata pine plantations in New Zealand found that increased tree growth under climate change scenarios increased susceptibility to wind damage (Moore and Watt 2015). In addition to experiencing higher wind speeds, trees grown under increased CO_2 emissions grew taller but not larger in diameter, leaving them more susceptible to breakage by wind, particularly when grown at high density.

An important point about climate change effects on forest disturbances is that disturbances interact, such that changes in the frequency, intensity, duration, extent, or seasonality of one disturbance may affect the same characteristics of the next (Seidl et al. 2017, Chapter 16). For example, increases in wind damage or insect outbreaks brought about by climate change may increase fuel availability, leading to larger and more severe fires. Likewise, increases in pathogens may reduce root or stem stability, heightening the risk of wind damage. Climate change will affect all disturbances within any disturbance regime in both obvious and unforeseen ways, such that precisely predicting future forests will remain very difficult.

CLIMATE CHANGE EFFECTS ON FOREST CARBON

Climate change is often discussed today in terms of its impacts on ecosystem services provided by forests. Ecosystem services affected by climate change span a wide range of ecological processes, such as diversity (Chapter 14), forest resilience to disturbances (this chapter), and carbon storage. Forests are an enormous pool of carbon in the global carbon cycle, which is important because carbon stored in biomass is carbon *not* found in the atmosphere acting as a greenhouse gas (Pan et al. 2011). In the context of climate change, there is therefore great interest among ecologists in **carbon sequestration** in forests, or their ability to uptake carbon from the atmosphere, fix it into biomass, and thereby store it within an ecosystem. Greatly simplified, the amount of carbon stored in forest ecosystems is largely the result of the balance between the amount of carbon taken up into ecosystems via photosynthesis (net primary productivity or NPP) and the amount lost from ecosystems via decomposition or heterotrophic respiration. When NPP is greater than decomposition, an ecosystem acts as a sink of carbon, but it acts as a source of carbon to the atmosphere when decomposition is greater than NPP. In determining the location of carbon, this carbon balance is extremely important for the global carbon cycle, and anything that changes the balance, such as disturbance, human activities, climate change, or their interactions, can alter atmospheric CO_2. The carbon balance of forest ecosystems is discussed in detail in Chapter 18, and understanding how climate change may affect the components of carbon balance in forests is the brief focus here.

CLIMATE CHANGE EFFECTS ON CARBON GAIN: PRIMARY PRODUCTIVITY

In the context of climate change, we discussed in the earlier text that tree growth generally decreases with warming temperatures but increases in the presence of elevated CO_2. Individual tree growth generally decreases with warming temperatures where water is limiting, but rising CO_2 and the increased water use efficiency associated with it may outweigh the decrease. These trends in tree growth are scalable to ecosystems, in that NPP increases with rising temperatures when water is available but decreases with drought or reduced precipitation. Likewise, the rate of nutrient cycling and availability may potentially increase with warming temperatures depending on moisture availability, sustaining positive growth responses to elevated CO_2 (Norby et al. 2010). These physiological mechanisms are coupled to phenology, in that lengthened growing seasons resulting from the onset of earlier springs and later falls are likely to increase forest productivity (Myneni et al. 1997; Keenan et al. 2014). Productivity responses to climate change are notoriously variable among regions, however. In general, forest prod--uctivity in cooler and colder climates tends to respond favorably to warming (Rustad et al. 2001), but that in warmer and drier climates often does not (Peñuelas et al. 2007). However, Charney et al. (2016) used a modeling study to project changes in forest productivity with rising CO_2 in North America, finding negative impacts in the interior West but positive impacts on coastal forests in the West, Southeast, and Northeast. Studies of tree rings at the end of the twentieth century had also shown variable importance of warming temperatures, rising CO_2, and reduced precipitation among regions in the United States, with some factors more important in some regions than others (e.g., Graumlich et al. 1989; Peterson et al. 1990). Overall, primary productivity will probably respond favorably to climate change in the near future.

An important point is that productivity is not affected independently by warming temperatures, rising CO_2, or reduced precipitation. In fact, each of these climate change effects occur concurrently, and their interactions may have different implications for forest productivity (Newman et al. 2011). Warming temperatures, for example, might increase rates of nutrient cycling in cool and cold climates that are nutrient-limited, but they may simultaneously reduce soil moisture via evaporation. At the same time, even while warming temperatures may reduce soil-moisture availability, rising CO_2 increases water-use efficiency such that moisture stress is less acute or unimportant to growth. Site conditions, of course, will also play an important role in providing the context for potential changes in productivity with climate change. Cool, moist ecosystems such as those in coves or northern and eastern aspects are less likely to experience soil-moisture deficits in the presence of warming temperatures, and warmer, drier ecosystems or those on droughty soils are more vulnerable to rising temperatures, heavily influencing their primary productivity.

CLIMATE CHANGE EFFECTS ON CARBON LOSS: HETEROTROPHIC RESPIRATION

Primary productivity is only part of the carbon cycle, and the impacts of climate change on heterotrophic respiration are equally important for forest carbon storage. Enormous carbon pools are found in dead organic matter in soils, litter, and deadwood, and decomposition of this organic matter constitutes a release of carbon from ecosystems back into the atmosphere as CO_2. Increases in decomposition due to climate change could therefore create a positive feedback, as it would contribute greenhouse gases that would further exacerbate the process (Davidson and Janssens 2006). Decomposition is critical in making nutrients available for tree growth and thus for increasing primary productivity, but such increases in productivity are meaningless for carbon sequestration if the amount of carbon lost through decomposition equals or exceeds that gained through primary productivity. The process of decomposition and its role in nutrient cycling are detailed in Chapter 19. Here, we briefly review potential effects of climate change on decomposition, as it affects the carbon cycle.

Carbon will accumulate in the soil if the rate of decomposition is slower than the rate of carbon inputs from litter and dead vegetation. If primary productivity increases as a result of climate change, the amount of carbon available for decomposition will increase. Thus, the way climate change affects decomposition—through changes in temperature and soil-moisture availability—will greatly affect how forests store and cycle carbon. Microbial activity is clearly affected by temperature, and many ecologists have therefore assumed that warming temperatures will, in turn, increase decomposition rates. Microbial activity is also sensitive to soil moisture, however, and warming temperatures that lead to higher evaporation and thus drier soils may slow decomposition rates, depending on the soil-moisture conditions that result from the drying (i.e., decomposition in poorly drained soils may increase as they dry). Litter decomposition rates are usually low when mean annual temperatures are lower than about 10 °C, but much more variable when warmer than 10 °C (Figure 20.19), suggesting that factors in addition to temperature and precipitation are also impor-

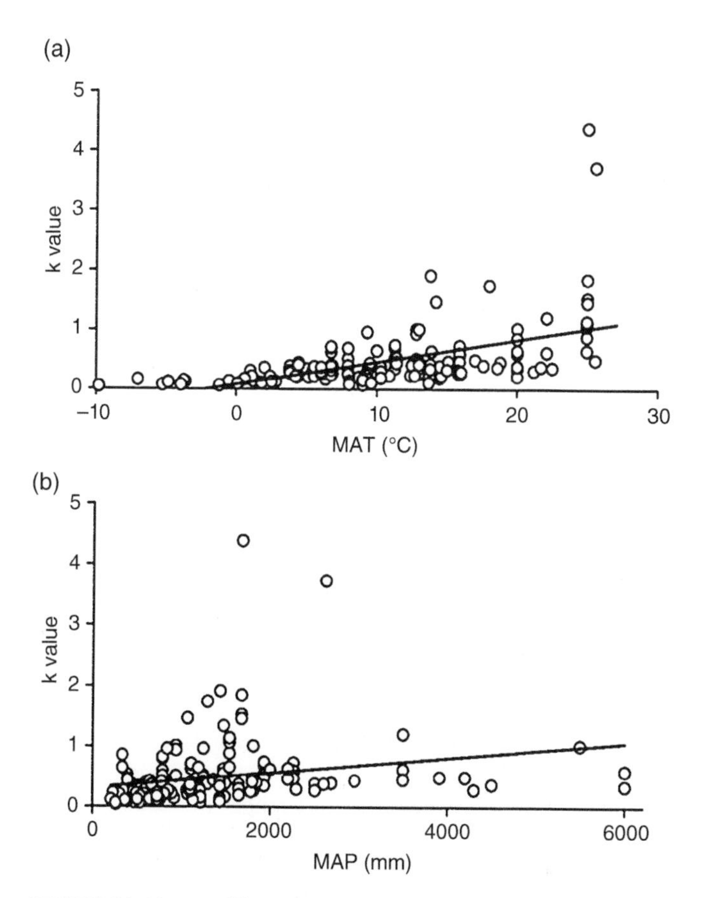

FIGURE 20.19 Leaf litter decomposition rates (k value) with (a) mean annual temperature, and (b) mean annual precipitation for an analysis of 70 publications of litter decomposition. *Source:* From Zhang et al. (2008) / with permission of Oxford University Press.

tant for decomposition (Zhang et al. 2008). Overall, much like primary productivity, the rate of decomposition is likely to rise with warming temperatures, but only if soil moisture remains available (Aerts 2006), and in concert with other factors such as litter quality (see the following text). Notably, several soil-warming experiments in forests have shown that increased decomposition rates may be limited in duration if litter input rates do not also increase, because available soil carbon will be effectively respired away (e.g., Bradford et al. 2008b; Bronson et al. 2008).

The effect of warming temperatures on decomposition is likely to be especially critical near the poles in permafrost areas—both forested and non-forested—where soils remain frozen year-round. The world's permafrost is extremely rich in soil carbon, estimated to hold twice as much carbon as is currently present in the atmosphere (Schuur et al. 2015), but most of it is unavailable to microbial activity and decomposition when frozen. As warming temperatures cause permafrost to thaw, decomposition of formerly unavailable carbon in the soil and its release into the atmosphere are likely to create an enormous positive feedback to climate change (Turetsky et al. 2019). Similarly, peatlands, many of which are forested and also in boreal and polar regions, contain an enormous amount of soil carbon that decomposes very slowly because of the anaerobic conditions found in waterlogged soils. If warming temperatures are accompanied by drier conditions, drying of peatlands could aerate soils and make carbon available for decomposition. In the case of both permafrost and peatlands, it is unknown how much of the carbon could be released to the atmosphere through decomposition, how quickly the release may occur, or how much primary productivity might compensate for the release (Turetsky et al. 2019), but these examples of temperature-induced changes on decomposition may demand the most attention from ecologists in the coming decades.

In considering leaf litter, the rate of decomposition also depends on the compound being decomposed, and thus substrate quality is likely to influence climate change effects on decomposition (Davidson and Janssens 2006). Global rates of litter decomposition are primarily driven by litter quality (Cornwell et al. 2008); plant species producing litter with high nutrient concentrations and lower concentrations of compounds more difficult to decompose (e.g., lignin) tend to decompose more rapidly (Chapter 19). Therefore, the potential for climate change to alter the quality of litter produced by tree species could be important for decomposition rates and loss of carbon from ecosystems. Relatively few studies to date have examined this relationship experimentally. In central Spain, experimentally increased temperature, reduced precipitation, and a combination of the two factors each reduced decomposition of a dry shrub species, attributed to reduced leaf quality (Prieto et al. 2019). Where temperature was increased, leaf litter had reduced concentrations of phosphorous and iron and higher C:P and C:N ratios—the latter of which stimulated nitrogen immobilization by microbes, thus decreasing decomposition rates. If such trends are consistent among other woody species across regional ecosystems, such reduction in litter quality could significantly reduce carbon losses from soil in a changing climate.

Despite the possibility that climate could alter the quality of litter in a given species, the changes in forest species composition brought about by climate change are most likely to have the strongest effects on litter quality within an ecosystem (Newman et al. 2011). It is well known, for example, that deciduous tree litter (such as that of maples) is more rapidly decomposed than coniferous litter (such as that of pines), at least in the ecosystems where each type of tree is found (Gholz et al. 2000). Likewise, deciduous litter decomposition rates vary by species (e.g., sugar maple leaves decompose faster than northern red oak leaves). Therefore, any changes in tree species composition on a given site might be expected to bring about changes in decomposition. For example, Alexander and Arthur (2014) examined changes in decomposition rates within Kentucky oak forests invaded by red maple due to fire suppression by manipulating leaf litter in field plots. Over only 18 months, red maple leaf litter initially decomposed much faster than litter from scarlet or chestnut oak, decreased nitrogen immobilization, and slowed mineralization. Although this experiment did not directly examine the effects of climate change and was conducted under current climate conditions, it reiterates a tight link between decomposition rates and species composition and suggests that decomposition rates are likely to respond quickly to climate-induced changes in species composition.

Finally, in addition to leaf litter, coarse woody debris is a major carbon pool representing 10–20% of heterotrophic respiration in mature forests (Bond-Lamberty et al. 2004), and the carbon it contains has a longer residence time before it is lost to the atmosphere. There is relatively little information about wood decomposition compared to that for leaf litter in forests (Harmon et al. 2011), and thus even less evidence for an influence of climate on the decomposition of wood. Overall, most studies find that the specific characteristics of the wood being decomposed—such as nutrient content, size, and tree species—are more important in explaining variations in the rate of decomposition than are variations in climate. In general, only 20–30% of the variation in wood decomposition is explained by climate both regionally (Bradford et al. 2014) and globally (Hu et al. 2018), and the majority of the variation is explained by wood characteristics. Understanding climate-induced changes in wood decomposition as it affects carbon loss from ecosystems is clearly an important area of additional research.

FEEDBACKS AMONG DISTURBANCE, CLIMATE CHANGE, AND CARBON IN FORESTS

Disturbances in forests kill trees and, in most cases (except for harvesting), do not remove the dead biomass, such that they are likely to alter the balance between primary productivity and decomposition and alter an ecosystem's carbon storage. Depending on the disturbance type, carbon may be immediately lost from the ecosystem; for example, fire oxidizes biomass and immediately removes carbon from the system, but insect outbreaks more slowly kill trees and do not create an immediate loss. Regardless of disturbance type, however, an ecosystem's carbon balance (net ecosystem productivity or NEP; Chapter 18) will shift from sink to source and back to sink as the forest recovers as a result of the balance between primary productivity and decomposition (Figure 18.18, Chapter 18). For a simple example, a stand-replacing fire in a mature conifer forest will shift the ecosystem from a carbon sink prior to the fire to a source after the fire (Kashian et al. 2006). Immediately following the fire, live vegetation is lacking such that primary production is low, but dead vegetation is prominent such that decomposition is high, and NEP is negative (i.e., the system is a carbon source; Figure 20.20). As post-fire succession proceeds, carbon sequestered by young, vigorous vegetation eventually exceeds that lost from decomposing dead vegetation, and NEP is positive (i.e., a carbon sink). Primary production decreases over decades as the forest ages, most or all of the post-fire dead vegetation decomposes away, and NEP approaches zero until the next disturbance. Thus, the landscape acts as a small carbon sink over many years, but as a strong carbon source for a few decades after it has burned. Provided that there is enough time for forests to recover between disturbances, most forests recover most of the carbon loss resulting from a given disturbance (Kashian et al. 2013). Post-disturbance forest carbon dynamics such as these are discussed in detail in Chapter 18.

Changes in forest carbon storage with succession and associated shifts between carbon sources and sinks have probably occurred at least since the Pleistocene. In the context of anthropogenic climate change, however, carbon storage is a larger concern because of its implications for atmospheric CO_2 and because of potential effects of climate change on disturbance regimes. This is especially true when the effects of changing disturbance regimes are considered across large, forested landscapes. In the example cited in the earlier text, more frequent fires would cause more of the landscape to act as a carbon source, which cumulatively would release more carbon into the atmosphere at a given point in time. Perhaps more concerning regarding carbon storage are the effects of climate change on post-disturbance recovery of forests. If, for example, more frequent or severe fires converted many high-density forests on a landscape to low-density forests, as has been

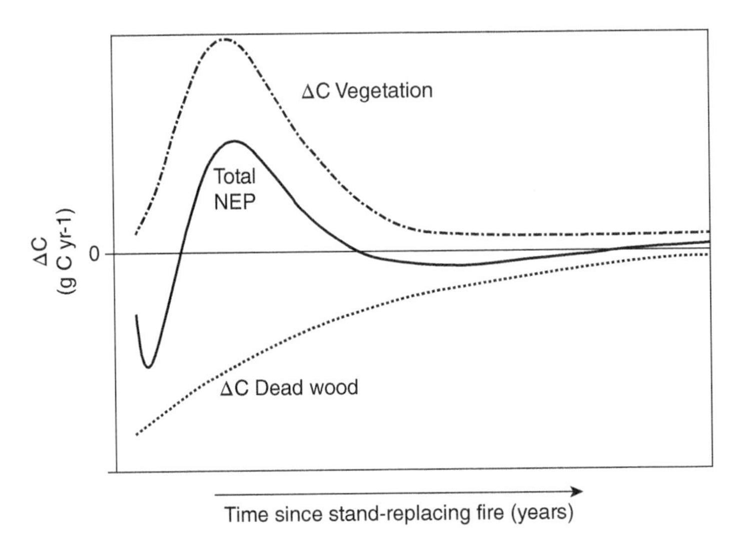

FIGURE 20.20 Hypothetical trajectory for forest carbon storage (net ecosystem production; NEP) after a stand-replacing fire. NEP is negative immediately after the fire because sparse vegetation has little carbon uptake while large amounts of dead vegetation have high carbon loss (ΔC is strongly negative) from decomposition. NEP soon increases, reflecting high carbon uptake in live vegetation (ΔC is strongly positive) as trees and other vegetation re-establish. NEP eventually decreases toward zero as low carbon uptake by vegetation is balanced by low carbon loss from dead wood. *Source:* From Kashian et al. (2006) / with permission of Oxford University Press.

documented for lodgepole pine in the Greater Yellowstone Ecosystem (Turner et al. 2019), carbon loss will occur for the landscape because the fires will create much dead vegetation to decompose but reduce the amount of vegetation available to uptake carbon. An even more extreme example is the potential for conversion of forest to non-forest, due to increased fire size or severity, short-interval fires, or less-suitable post-fire growing conditions. Although forests eventually recover all of their lost carbon after a disturbance, such recovery requires that forests remain forests; conversion to other cover types would result in a net loss of carbon from the landscape. Such dramatic cover type conversions were once considered only remotely possible (Kashian et al. 2006), but have become considerably more common as climate change progresses.

Because of their ability to affect carbon storage, insect outbreaks are also an important disturbance affected by climate change. Insect outbreaks are second only to wildfires as the largest source of tree mortality in western North America (Samman and Logan 2000). We have described in the earlier text that insect outbreaks are likely to increase in frequency, extent, and severity for some insect species as development accelerates with rising temperatures, overwinter survival increases with rising winter temperatures, and host trees become stressed by moisture deficits. The response of forest carbon storage to insects is not dissimilar to its response to fire, although there exists tremendous variability in carbon storage response across different types of insects (Hicke et al. 2012). In general, much like fire, insect outbreaks reduce primary productivity and shift carbon pools from live vegetation to deadwood as they kill trees, in some cases switching the ecosystem to a carbon source. Depending on the severity of the outbreak (proportion of trees killed), forest recovery following insect outbreaks is typically much quicker, and thus carbon sources are generally much shorter-lived. For example, Romme et al. (1986) found that lodgepole pine forests returned to pre-disturbance levels of primary productivity after only 5–15 years following mountain pine beetle outbreaks in Yellowstone National Park. Similarly, Kashian et al. (2013) were unable to detect differences in ecosystem carbon between mature lodgepole pine stands attacked by mountain pine beetle and unattacked stands by 25–30 years after the outbreak. For comparison, they found that the same ecosystems required about 250 years to recover most of the carbon following stand-replacing fires. The rate of recovery of carbon reflects the severity of the outbreak, with stronger sources and longer recovery times associated with more severe outbreaks. Even severe outbreaks rarely kill every tree or even most trees. The surviving trees are critical as seed sources for post-disturbance regeneration, and to partially compensate for carbon losses with rapid growth following release from competition with trees killed by the outbreak (Hicke et al. 2012; Hansen 2014). Compensatory contributions of surviving trees to post-outbreak primary productivity are a main reason that outbreaks have a reduced impact on carbon storage compared to fires.

Despite reduced impact compared to fires, insect outbreak-induced carbon losses from forests are certainly not trivial. For example, in its worst year, the enormous bark beetle outbreak that occurred in British Columbia in the early 2000s produced carbon emissions equal to about 75% of those from forest fires across all of Canada over a 40-year period and reduced primary production by the same amount that it increased in the 1980s and 1990s due to climate change (Kurz et al. 2008a). Notably, Thom et al. (2017) used a simulation model to find that bark beetle activity in an Austrian national park increased with climate change in the twenty-first century, but that changing tree species composition reduced host tree availability and decreased insect activity in the long run. Despite these results and the bevy of unknown factors that may shape future biotic disturbances, it is likely that continued and increasing insect outbreaks are likely to decrease forest carbon for at least the next several decades (Williams et al. 2016).

Fire, Carbon, and Climate Change in Forests of Yellowstone National Park The lodgepole pine forests of Yellowstone National Park in northwest Wyoming are an excellent example of how climate change may interact with disturbance regimes and carbon storage. Large, stand-replacing fires, comparable to those that burned 3200 km^2 of the park in 1988, have infrequently burned this landscape on a rotation of 100–300 years. Kashian et al. (2013) used a replicated chronosequence approach (Chapter 17) to measure all the carbon pools in 77 lodgepole pine forests aged 12 to >350 years having varied structures (tree densities) to characterize changes in carbon storage following large fires. They were unable to detect a carbon source (negative NEP), although the youngest stands in their analysis were 12 years old, suggesting that any carbon source is relatively short-lived on this landscape. This carbon source, if it exists, is probably short-lived because of the rapid regeneration of the burned area by lodgepole pine via serotinous cones and slow decomposition of deadwood in the cold, dry subalpine landscape. The duration of a carbon source would likely be longer on a landscape dominated by non-serotinous species or where decomposition was more rapid. The recovery of ecosystem carbon was driven mainly by primary productivity, as carbon gain in live vegetation (7780 g C/m^2) greatly exceeded carbon lost through decomposition (1960 g C/m^2) over the first 100 years. Most carbon accumulation occurred early; NEP was not statistically different from zero for forests older than 70 years (Kashian et al. 2013).

As discussed in the earlier text, most forests are able to recover carbon lost as a result of disturbance so long as forests regenerate rather than undergo a significant state change to another cover type (Kashian et al. 2006). However, the ability of a forest to recover after a disturbance requires a disturbance-free interval to which a particular forest type is adapted. In Yellowstone, Kashian et al. (2013) found that 90% of total ecosystem carbon in lodgepole pine forests was recovered in the first 100 years after fire, making landscape carbon well buffered for fires that recur every 100–300 years (Figure 20.21). Likewise, coniferous subalpine forests in Colorado were found to recover most of the carbon lost to disturbance after 100 years for fires that recur every 150–400 years (Bradford et al. 2008a), and jack pine forests in Michigan after about 20 years for fires that recur every 30–80 years (Rothstein et al. 2004). In these forests, carbon lost to disturbance is recovered prior to the next disturbance, and there is no net carbon lost to the atmosphere. Of issue, however,

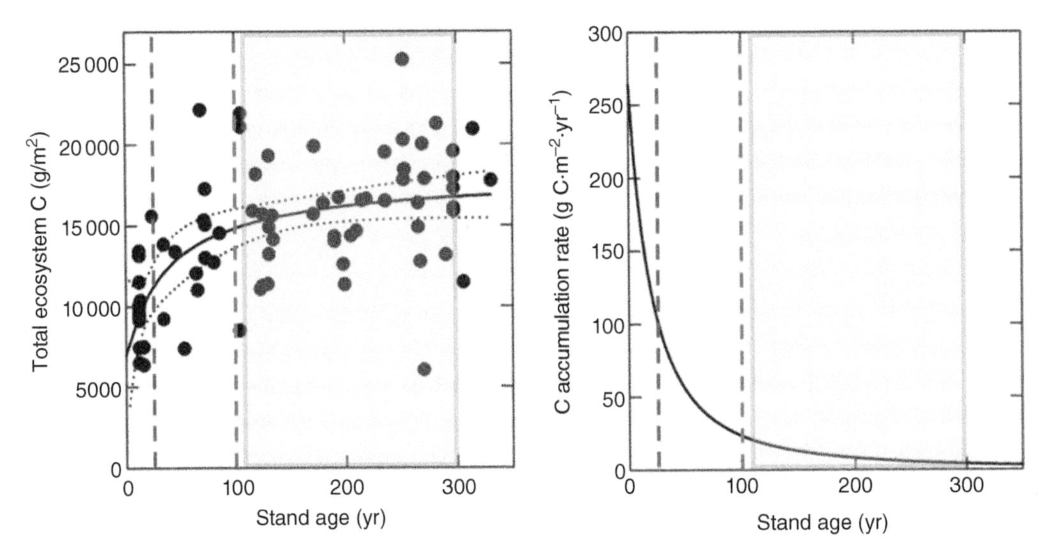

FIGURE 20.21 Total ecosystem carbon (left) and carbon accumulation rate (right) for 77 lodgepole pine stands in Yellowstone National Park. Total ecosystem carbon increases with stand age, but 90% of the increase occurs before age 100 years (blue dashed line). Carbon in the ecosystem was resilient to fire because fire frequency was historically 100–300 years (shaded box). However, fire rotation is expected to increase to 30 years (red dashed line) by 2050, which would prevent the system from recovering more than 75% of its carbon after a large stand-replacing fire. *Source:* From Kashian et al. (2013) / John Wiley & Sons.

is the possibility that climate change could shorten the fire rotation (or the rotation of non-fire disturbances on other landscapes), such that subsequent disturbances occur before carbon is able to recover. In Yellowstone, current projections allow for fire rotations as short as 30 years in the next several decades (Westerling et al. 2011), which would significantly reduce the amount of forest carbon on the landscape. Such short-interval fires would almost certainly lead to forest conversion to non-forest cover types (e.g., Harvey et al. 2016b; Turner et al. 2019), a significant net loss of carbon from the landscape, and potential feedback to further climate change.

Although our discussion of climate-disturbance–carbon interactions here is limited to fire and insect outbreaks, we emphasize that other disturbance types could also produce significant carbon losses from forested landscapes if altered from their current regimes by climate change. Wind and drought are two such examples. A third disturbance type, timber harvest, is extremely important in its potential to reduce carbon storage in forests despite its anthropogenic nature, and it leaves a long legacy of effects. We discuss timber harvesting in the following text; more detailed treatments of climate change and disturbance interactions for carbon storage are presented in Williams et al. (2012, 2016), Seidl et al. (2014), Vanderwel et al. (2013), and Kurz et al. (2008b), among others.

ADAPTING TO CLIMATE CHANGE EFFECTS ON FORESTS

There are several reasons to expect human responses to climate change. First, modern climate change is largely anthropogenic, and it is human nature to want to correct past mistakes. Second, forests provide a wide variety of values (such as timber resources but also clean water, wildlife habitat, recreation, scenic beauty, cultural values, etc.) that many ecologists and managers feel strongly about preserving for future generations. Responses to climate change might be categorized as either climate change *mitigation*, where emphasis is placed upon reducing carbon emissions from fossil fuels and attempting to stabilize or reduce the concentrations of greenhouse gases in the atmosphere, or *adaptation*, where ecologists and land managers develop strategies that assist ecosystems in coping with current and future climate change impacts. These responses have importantly different goals. Mitigation attempts to reduce or avoid human interference with climate to lessen the rate of climate change so that ecosystems may naturally adapt while maintaining food production and economic development (IPCC 2014). By contrast, adaptation has as its goal reducing vulnerability to the harmful effects of climate change. Thus, adaptation involves adjustments by humans to current or expected changes in climate and is our major emphasis in this section. As we have discussed, forests are a major player in the global carbon cycle, such that the ultimate and most obvious adaptation to climate change in forests is to avoid deforestation at a global scale. Our emphasis is not upon the obvious, but instead on a few major strategies of adaptation to climate change effects in forests: assisted migration, the use of climate refugia, and carbon management. There include a host of other equally important adaptation strategies that are not discussed in this chapter. Detailed treatments of such strategies are presented in Halofsky et al. (2018), Swanston et al. (2016), Millar et al. (2014), and Janowiak et al. (2014), among many other resources.

ASSISTED MIGRATION

The process of intentional, human-assisted movement of species or genotypes in response to climate change is called **assisted migration**, also called assisted colonization or managed relocation (Williams and Dumroese 2013). The ecological basis for assisted migration is the notion that the rate of climate change is likely to exceed the rate and extent of tree dispersal to newly suitable habitat, and the speed and distance at which climate zones will change will not allow species the time to relocate to suitable environments. As a result, many tree species are likely to be overtaken by an unsuitable climate and potentially be outcompeted by more quickly dispersing species (Minteer and Collins 2010, Vitt et al. 2010). Impediments to movement of tree species may include gradual rates of adaptation, but also may include physical barriers, such as fragmentation or geographic features, within a species' current range. Because it may lifeboat (Chapter 22) at-risk species or populations at risk due to climate change, assisted migration is considered by many to be an important tool for endangered species conservation (Hoegh-Guldberg et al. 2008) and building resilience when restoring forests (Dumroese et al. 2015).

In practice, assisted migration has been efficiently categorized into three forms, depending on the distance over which movement occurs (Figure 20.22; Handler et al. 2018). *Assisted population migration*, also called assisted genetic migration or assisted gene flow, involves within-range movement of seed sources or populations to new locations. Assisted population migration is an important tool for helping species to overcome physical barriers to movement within their current range. *Assisted range expansion* is defined as the movement of seed sources or populations from their current range to suitable areas just beyond the historical species range. Assisted range expansion has the goal of facilitating or mimicking natural dispersal, particularly when human activities have created barriers to such dispersal. *Assisted species migration*, also called species rescue, managed relocation, or assisted long-distance migration, moves seed sources or populations to locations far outside the historical species range, beyond locations accessible by natural dispersal. All three forms of assisted migration are meant to ensure that a species occurs in many redundant locations or across a range of conditions such that risks of unforeseen climate impacts are reduced (Handler et al. 2018).

The idea of assisted migration—particularly assisted species migration—has not been without controversy. Some ecologists have argued that anthropogenic-based migration of species may facilitate the effects of invasive species on other plants and ecosystems when the assisted species become established in new areas (Ricciardi and Simberloff 2009). Other potential challenges to

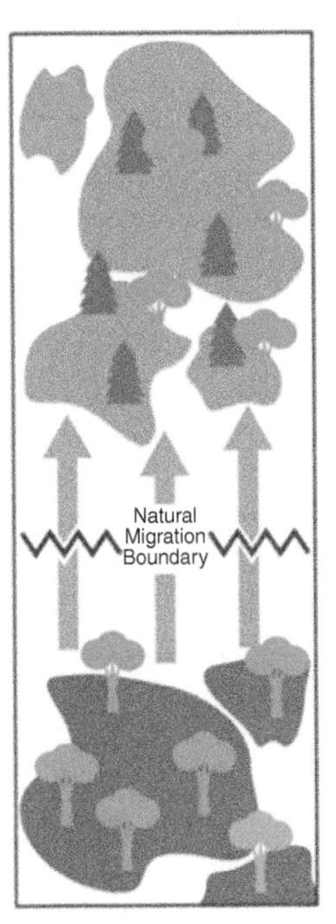

FIGURE 20.22 The three forms of assisted migration. Assisted population migration transfers seed sources of populations of a species to new locations within its current range. Assisted range expansion moves seed sources or populations from their current range to suitable areas just beyond the historical species range to mimic natural dispersal. Assisted species migration transfers seed sources or populations to locations far outside the historical species range, inaccessible by natural dispersal. *Source:* From Handler et al. (2018) / U.S. Department of Agriculture / Public Domain.

assisted migrations are also notable. For example, species established in new areas may hybridize with local species if they are capable of doing so, as in the spruces, pines, aspens, and oaks (Aitken et al. 2008), increasing the influence of humans on genetic diversity. The introductions of new species into an area may introduce or attract previously absent insects or pathogens. Pragmatically, moving species long distances outside of their current range, even when climate projections predict that such locations will provide suitable growing conditions by the end of the century, may not be successful if current growing conditions at those locations are not yet suitable (Handler et al. 2018).

Assisted migration can also be designed to maintain high levels of productivity and diversity in commercial forests (termed *forestry-assisted migration*; Pedlar et al. 2012). Forestry-assisted migration emphasizes moving populations (usually seed sources) within the current range of the species (assisted population migration) or expanding ranges along the margins of current geographical limits (assisted range expansion; Gray et al. 2011; Williams and Dumroese 2013). Assisted population migration has been incorporated into forestry practice in parts of Canada (Pedlar et al. 2011), led by British Columbia, and in the United Kingdom (Whittet et al. 2016).

Perhaps the best example of the complexities of assisted migration is that of whitebark pine, whose seeds were tested for germination and establishment far north of their current native range (McLane and Aitken 2012). Whitebark pine is a high-elevation conifer of the Rocky Mountains that is an important food source for wildlife and is currently threatened by both warming temperatures and an exotic pathogen (white pine blister rust). Seeds were planted from 7 populations in 8 locations bracketing the current northern (800 km north of the current boundary) and southern (600 km south of the current boundary) boundaries, and data were collected after 10 years. Germination occurred at all locations, but germination, establishment, and seedling growth were affected by seed size, winter snow cover, and the timing of snowmelt much more than latitude. In fact, the seedlings at the northernmost site were twice as tall on average as those on any other site in the study. Seedlings on sites with snowpacks at one of the extremes (i.e., either little or no protective snow cover in winter, or deep snowpacks that melted late in the year) had poorer rates of survival than on sites where snowpacks were continuous in winter and melted in April or May. Notably, the northernmost site with the tallest seedlings was also near the Pacific Coast and had relatively mild mean annual temperatures, the warmest average summer temperature of all other sites, and had intermediate snowfall (McLane and Aitken 2012). These results suggest that climate is not likely to be the sole driver of successful assisted migration, and that local site conditions are clearly important.

The information required for high confidence of success with assisted migration includes much more than simple changes in climate envelopes. Climatic extremes outside of the historical range of variability may be more important to assisted migration than averages; phenological relationships with temperature, photoperiod, and precipitation are likely to be extremely important in determining the establishment and survival of relocated species; and local effects of microclimate, herbivory, and competition of a species in a new environment cannot be ignored (Park and Talbot 2018). If eventual adoption of the approach is widespread, assisted migration will likely be complemented with other approaches designed to increase forest resilience in the face of climate change (see Janowiak et al. 2014) rather than serving as a stand-alone strategy. Detailed discussions of assisted migration are presented in Park and Talbot (2018), Handler et al. (2018), Park et al. (2014), Williams and Dumroese (2013), and Pedlar et al. (2012). A review of available assisted migration experiments is provided by Sáenz-Romero et al. (2021).

REFUGIA

Glacial **refugia** have long been discussed as pockets of species distinct from their surroundings that resulted from some sort of protection from the intense cold and fluctuations of glacial climates (Chapter 15). During glaciations, these areas harbored populations during periods of unfavorable climate in regions where the ice displaced suitable climates to the south or to lower elevations. Glacial refugia also allow the species they contained to re-colonize the areas around them as the climate warmed and the ice retreated. Species that thrive in warmer climates shrink into refugia during glacial periods, but those that thrive in cooler climates are often restricted to refugia between glaciations (Stewart et al. 2010). Today, as the climate continues to change, the latter refugia concept has been applied to small areas that are relatively buffered from those changes, and thus refugia may be effective in protecting especially sensitive species and ecosystems from climate change over the short term (Morelli et al. 2016). These refugia are considered to be potentially important foci of biodiversity conservation in an altered climate. They are not completely analogous to the interglacial refugia described in the earlier text because the climate will be warmer and the landscape will be fragmented, limiting dispersal (Ashcroft et al. 2012). Nevertheless, climate change refugia have been described as a "slow lane" for species diversity because they safeguard species for longer periods of time than those existing within the faster climatic changes of the broader region (Morelli et al. 2020), although these relative differences may often be short-lived (McLaughlin et al. 2017). However, even transient delays of climate change effects may be important in allowing species additional time to adapt to the new climate, or, perhaps more importantly and likely, to disperse.

Modern refugia often occur because of specific physical site conditions that locally influence resources available to species. Climate change refugia generally have locally stable and persistent climatic conditions despite changes that may occur around them (Ashcroft et al. 2012). Species within the refugia do not experience the climate conditions in the broader area, at least in the same way as the species outside of the refugia. The classic example of refugia are low areas in the landscape such as depressions and valleys that pool cold air (see Chapter 8; Figure 7.2 (Chapter 7) and Figure 20.8), particularly at night, resulting in a markedly different microclimate and often markedly different species composition and phenology. Other climate change refugia may buffer against moisture deficits, such as wetlands, riparian areas, or groundwater-fed springs. In mountainous areas of the Northern Hemisphere, north- and east-facing slopes and draws are typically shaded and are thereby cooler and wetter than their surroundings. Additional examples of climate change refugia are presented in Figure 20.23. These refugia are often considered microrefugia; macrorefugia have also been identified as broader areas occurring due to elevation and proximity to coastal areas (Stralberg et al. 2018).

In addition to protection from rising temperatures and reduced moisture resulting from climate change, other factors such as extreme events, streamflow timing and rate, and disturbances have also been associated with refugia (Figure 20.23; Krawchuk et al. 2020). In particular, *disturbance refugia* may include areas protected from disturbances that may increase in frequency or severity under climate change. For example, in a landscape study of mixed-conifer forests burned frequently by fires in the Sierra Nevada of California, Wilkin et al. (2016) found that cold-air pools were not only important microrefugia from rising temperatures, but they burned less frequently and at lower severity than the greater region. These reduced effects of fire are likely to have a positive effect on the protective capacity of the refugia and suggest that potential disturbances to refugia may override their buffering of rising temperatures or reduced moisture availability.

Much like assessing assisted migration, potential climate change refugia may be identified but are truly only working hypotheses as climate change is ongoing. Like most other projections

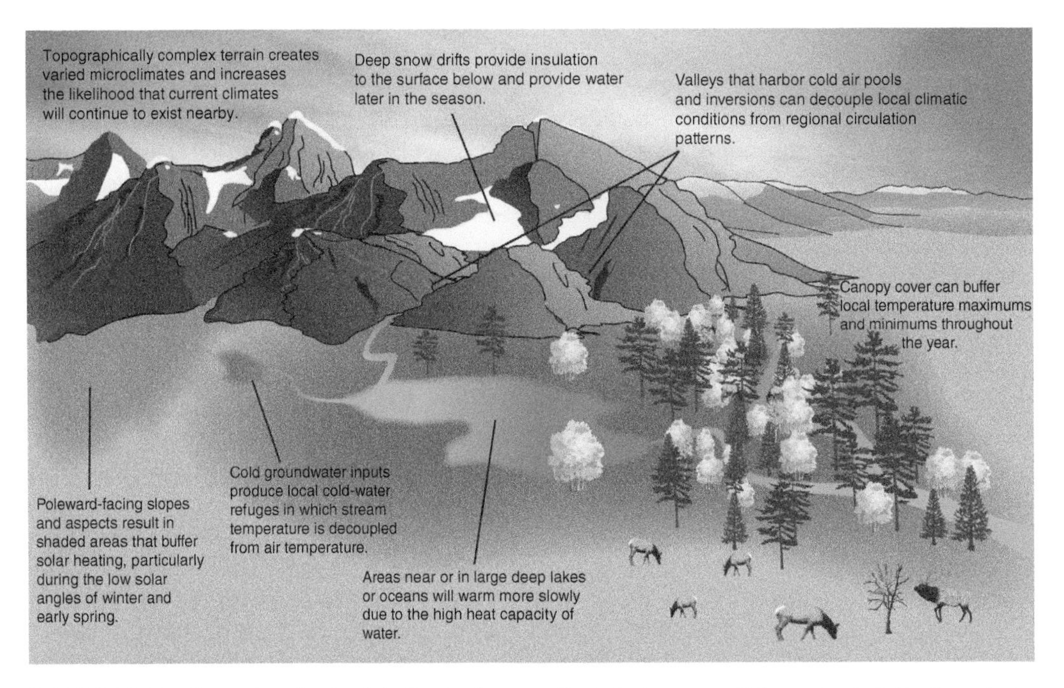

FIGURE 20.23 Examples of potential climate change refugia that may result from physical site characteristics on a landscape. *Source:* From Morelli et al. (2016) / Morelli TL et al / Public Domain CC BY 4.0.

associated with climate change, identification of potential refugia has had to rely on simulation modeling of climate and species distributions. Michalak et al. (2018) identified potential future climate change refugia across North America using projected changes in the size and distributions of current climate envelopes over time and different dispersal differences. They found that only 12% of North America could function as climate refugia, mostly at high elevations and in mountainous areas. Existing protected areas on the continent disproportionately included potential refugia, and refugia at lower latitudes and lower elevations were less likely to be included within protected areas. The refugia were also projected to shrink in size over time as climate change progresses, reiterating their vulnerability to climate change themselves.

The maintenance or creation of refugia is increasingly included as a strategy in management portfolios for climate change adaptation (e.g., Swanston et al. 2016). Approaches to such a strategy may include: (i) identifying unique sites and prioritizing them for protection. Such sites might include those with unique or diverse geology, physiography, soils, or biodiversity, particularly those relatively undisturbed by humans. They might also be identified as potential sites for assisted migration species that may be vulnerable elsewhere on the landscape; (ii) identifying and protecting sensitive or threatened species and ecosystems, such as those near the southern boundaries of their current range or restricted to a narrow range of site conditions. Methods to protect such species are not unlike those typical of species conservation but may also include protection of species and ecosystems across a range of environmental conditions and monitoring for potential evidence of migration; and (iii) establishing artificial refugia. The development of highly controlled environments such as nurseries, arboretums, or botanical gardens may be used to temporarily harbor rare or at-risk species until they can be moved to a new suitable habitat. Such an approach mirrors that of assisted migration but uses "refugia" to maintain species before they are relocated. Tactics include seed collection, seedling cultivation, and planting in a natural setting where protection from climate change can be assured. Much like establishing general protocols for conservation planning, effectively incorporating refugia into management, especially quickly enough to keep pace with a changing climate, remains an important challenge.

FOREST CARBON MANAGEMENT

The largest driver of climate change today is the combustion of fossil fuels, but forests have continuously been cited as important buffers in the global carbon cycle as it relates to atmospheric CO_2 (Harris et al. 2021). Management of forests has a strong influence on how they store carbon because it involves manipulation of tree species composition, competition within and between species, and stand structure in ways that shape primary productivity and, to a lesser extent, decomposition. Centuries-old forest management decisions—mainly past timber-harvesting activities—have left important legacies for carbon storage today. In the United States, forests currently act as a major sink of carbon and have done so since about the 1950s (Figure 20.24). By 1915, forests in the United States, most of them harvested mostly in the 1800s, were losing about 760 Tg of carbon each year, but by 2010 were gaining about 200 Tg each year (Birdsey et al. 2006). The nature of this increased forest growth, and the switch between carbon source and sink, is critical for future carbon storage and thus for climate change. Factors such as nitrogen deposition and CO_2 fertilization could explain the increased forest growth and would probably result in a stronger carbon sink in the short term. More likely, however, is that forests in the United States are responding to past and current land use such as reforestation of old agricultural land or regrowth of harvested forests (Ryan et al. 2010). In that case, the carbon sink will likely weaken as regrowth concludes, and the ability of forests to buffer fossil fuel combustion will be reduced. Current projections of forest carbon storage appear to show a weakening sink but suggest that effective forest carbon management may be able to reduce the weakening (Figure 20.24). Thus, the size of the carbon sink is only one part of the story, because the potential saturation of the sink is also fundamental to future carbon storage.

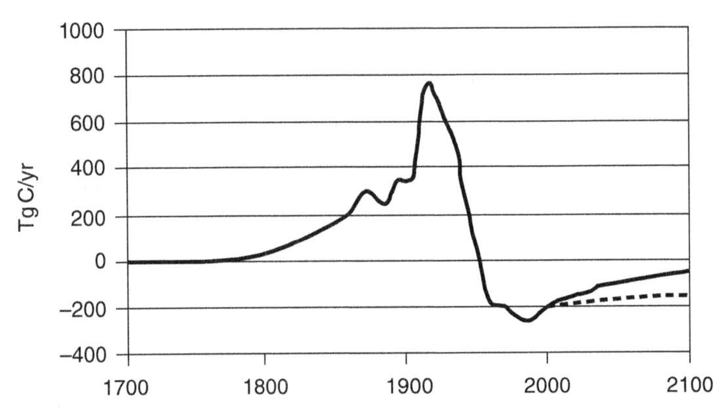

FIGURE 20.24 Carbon emissions for forests in the United States for 1700–2100. Values greater than zero represent a carbon source, whereas negative values represent a carbon sink. Projections beyond 2000 suggest a weakening sink but could be stabilized with forest management for carbon (dashed line). *Source:* From Birdsey et al. (2006).

Over the last two decades, forest managers have developed several silvicultural strategies to increase carbon stores. A detailed discussion of these strategies is beyond our scope; examples of such strategies as summarized by Ontl et al. (2020), Ryan et al. (2010), and Canadell and Raupach (2008) include:

Avoid the loss of forests: As we have described in the earlier text, forests are one of the largest pools of carbon in the terrestrial biosphere, such that deforestation results in an obvious release of carbon to the atmosphere. Of particular concern is conversion of forests to non-forest cover types, which not only results in the loss of carbon storage in trees but also losses through increased decomposition in the forest floor and soil. Avoiding deforestation may be achieved with conservation easements or best management practices on forested land.

Increase the extent of forested land: Just as avoiding forest loss will avoid loss of carbon to the atmosphere, adding new forest to unforested lands will increase carbon uptake and storage and remove significant amounts of carbon from the atmosphere. Such afforestation might occur where partial regeneration failure has resulted in very sparse forests, on marginal sites for agriculture or where forests have been cleared for crop land, or as riparian buffers.

Manage forests to reduce disturbances: Disturbances inevitably cause carbon losses from forests, and enhanced disturbances caused by climate change will exacerbate this process. Fire is the most influential disturbance in this regard, and several approaches may be used to reduce the incidence of fire in forests. These include using prescribed fire to reduce fuel loads and tree density or establishing an effective system of fuel breaks and fire lines to slow fire spread or reduce its intensity. Altering forest structure by thinning is also an important approach to reduce the threat of fire, lower the density of host species for a given insect or pathogen, reduce the risk of drought-induced mortality, and favoring species less susceptible to wind or ice damage. Thinning with an objective of increasing carbon storage should be conducted thoughtfully, as thinning nearly always reduces forest carbon at landscape scales (Ryan et al. 2010). Note that reducing disturbances in forests may be a competing objective with managing to mimic or incorporate the natural disturbance regime (see Chapters 10 and 16).

Facilitate post-disturbance recovery: Regeneration failure following disturbance, especially fires, is likely to be a major driver of forest conversion and carbon loss in warmer and drier climates. Examples of approaches that enhance rapid post-disturbance carbon uptake include quickly revegetating disturbed areas with seeding or planting, using native species likely to be well adapted to an altered climate, and protecting established seedlings and sapling from ungulate browsing or tree harvesting activities.

Manage forests to increase forest growth: Increasing forest productivity has always been an objective of forest managers, but in many cases current forest structure may require significant alterations to increase carbon uptake and storage. In doing so, managers might plant or favor with silviculture species that are better adapted to future conditions, creating large openings to encourage regeneration of shade-intolerant species that grow quickly, thinning to enhance age- or size-class diversity, or promoting species with higher wood density (and thus higher carbon density), selecting species with high growth rates, and controlling competing vegetation.

Many specific silvicultural techniques are available in managing for forest carbon storage. Some of these are discussed by Franklin et al. (2018) and Ashton et al. (2012). D'Amato and Palik (2021) place some of these strategies in the context of ecological forestry (Chapter 14). As the climate continues to change, management techniques will likely need to continually adapt to additional changes and conditions.

SUGGESTED READINGS

Coop, J.D., Parks, S.A., Stevens-Rumann, C.S. et al. (2020). Wildfire-driven forest conversion in western North American landscapes. *Bioscience* 70: 659–673.

Hicke, J.A., Allen, C.D., Desai, A.R. et al. (2012). Effects of biotic disturbances on forest carbon cycling in the United States and Canada. *Glob. Chang. Biol.* 18: 7–34.

Iverson, L.R., Prasad, A.M., Matthews, S.N., and Peters, M. (2008). Estimating potential habitat for 134 eastern US tree species under six climate scenarios. *For. Ecol. Manag.* 254: 390–406.

McKenney, D.W., Pedlar, J.H., Lawrence, K. et al. (2007). Potential impacts of climate change on the distribution of North American trees. *Bioscience* 57: 939–948.

Ontl, T.A., Janowiak, M.K., Swanston, C.W. et al. (2020). Forest management for carbon sequestration and climate adaptation. *J. For.* 118: 86–101.

Seidl, R., Thom, D., Kautz, M. et al. (2017). Forest disturbances under climate change. *Nat. Clim. Chang.* 7: 395–402.

Westerling, A.L., Hidalgo, H.G., Cayan, D.R., and Swetnam, T.W. (2006). Warming and earlier spring increase western US forest wildfire activity. *Science* 313: 940–943.

Williams, M.I. and Dumroese, R.K. (2013). Preparing for climate change: forestry and assisted migration. *J. For.* 111: 287–297.

Invasive Species in Forest Ecosystems

CHAPTER 21

Throughout this book we have emphasized the relationship between plant species and physical site factors, and that plants occur in ecosystems where they best compete and are adapted to conditions of climate, physiography, soil moisture and nutrients, and microclimate. However, the species occupying a given site are not necessarily those best adapted to grow on that site, but merely those that have access to that site at the time of its availability and are superior competitors once established. Invasion of a site by better competitors often results in substantial changes following natural or human disturbances. As human effects on the ecosphere increase and become more widespread, many of these better competitors are those introduced by humans, whose ecological adaptations are to sites and ecosystems from other parts of the continent or world.

The distribution of biota of the ecosphere has been restricted by oceans and other natural barriers for millions of years. During the last 125 years, human activities, especially international travel and trade, have broken these barriers, allowing introduced or "exotic" species to invade new continents at an increasing rate. Such **invasive species**—those species that arrive with human assistance, establish populations, and spread (Simberloff 2013)—have dramatically altered forest ecosystem diversity, function, and productivity. Many ecologists agree that the spread of invasive species is now one of the most serious ecological threats to the sustainability and productivity of native ecosystems (Mack et al. 2000; Pyšek et al. 2012). In a study of stressors to 1055 threatened or endangered plants in the United States conducted almost 25 years ago, Wilcove et al. (1998) found that 57% were impacted by invasive species; this proportion has almost certainly risen over the last three decades. Economic losses to society due to invasive species in the United States are estimated to be $120 billion each year, and losses of forest products alone are at least $2 billion annually (Pimentel et al. 2005). Together with habitat loss and fragmentation (Chapter 22), reductions in biodiversity (Chapter 14), and climate change (Chapter 20), the threat of invasive species is one of the most pressing issues faced by forest ecologists and managers in the twenty-first century.

In this chapter, we present multiple examples of the effects of invasive species on forest ecosystems. We first examine the characteristics that make a plant species particularly invasive and those of the ecosystems that appear to be particularly subject to invasion. We then discuss the impacts of invasive species on forests, with special emphasis on the effects of invasive plants, insects and pathogens, and animals on biota, soils, and disturbance regimes. The general threat to forest ecosystems by invasive plants is summarized by Poland et al. (2021) and Liebhold et al. (1995, 2017).

CONCEPTS OF INVASIVE SPECIES

DEFINITION OF INVASIVE SPECIES

Humans have caused many species to be introduced to new environments, but not all introductions are deliberate, and not all introduced species become invasive. Many or even most new plant species are introduced for horticultural or landscape uses, while many (not all) new insects are

Forest Ecology, Fifth Edition. Daniel M. Kashian, Donald R. Zak, Burton V. Barnes, and Stephen H. Spurr.

introduced accidentally through cargo, packing materials, or even through transported plants or soil. Invasive species occur along a general three-stage continuum: introduction, naturalization, and invasion and spread. An *introduced species* is simply one that has arrived with either purposeful or accidental human assistance, whether or not it becomes established and develops populations in a new environment (Simberloff 2013). To become invasive, an introduced species must (i) establish new populations, and (ii) spread widely from that point of establishment. Introduced species that establish but do not spread widely are *naturalized species*. Naturalized species are perhaps most common in highly disturbed areas or in heavily human-modified landscapes. Relatively few woody species are considered naturalized but not invasive; one example is the common lilac, which is hardy enough to survive for centuries once established but fails to spread widely because it lacks viable seeds for dispersal.

Williamson (1996) suggested that the development of invasive species follows the "rule of tens," which predicts that roughly 10% of introduced species become established (naturalized), and 10% of those established then spread and become invasive (meaning that, ultimately, approximately 1% of introduced species become invasive). A general lack of data makes it difficult to truly quantify the proportion of introduced species that follow the rule of tens. In California, Florida, and Tennessee, roughly 6, 10, and 13% of introduced plant species, respectively, have become invasive (Simberloff 2013).

Invasive species are most often considered to be non-native, suggesting that they were moved with human assistance to a new environment where they did not evolve. These non-native species may also be called *alien*, *exotic*, or *non-indigenous*, although the first two terms have fallen out of favor for their value-laden and somewhat judgmental use. Importantly, again, such species are not always invasive. It is also noteworthy that even native species may behave like invasive species if human disturbance allows them to colonize, establish, and spread into ecosystems normally inaccessible to them. For example, in the eastern United States, oak-dominated forests maintained by relatively frequent surface fires prior to European colonization have been invaded by native red maples and other mesophytic species due to fire suppression (Nowacki and Abrams 2008). With fire suppression or exclusion in these ecosystems, oak forests developed into closed-canopy forests that reduced the ability of oak to regenerate under the diminished light conditions and allowed shade-tolerant species such as red maple to establish in the understory. As the abundance of shade-tolerant species increased, the probability of fire decreased with the development of moist and cool microclimates and the accumulation of less-flammable fuels, reducing oak dominance. This process of "mesophication"—facilitated by the invasion of dry and dry-mesic sites by native, shade-tolerant, mesic species—has threatened to convert many oak forests to those dominated by maple and other mesic species (Nowacki and Abrams 2008).

A final point about invasive species is that their significance is heavily value-laden and varies from location to location (Coates 2007). That is, a species invasive and perceived as an ecological nuisance in one location is often heavily valued within its native range. There are numerous examples of tree species in this regard (Table 21.1). For example, northern red oak is a late-successional canopy dominant native to eastern North America that is valued for its wood quality, whose persistence is a major emphasis of forest management in the region. In forests of western and central Europe, however, introduced northern red oak is an invasive species that establishes and spreads in native woodlands, limiting regeneration of native tree species and severely reducing native diversity. Likewise, Douglas-fir is an iconic, long-lived canopy dominant of the old-growth forests of the Pacific Northwest of the United States and Canada whose protection from logging has been a major environmental issue in North America over the last four decades. In the Southern Hemisphere, however, particularly in New Zealand and in Patagonia, introduced Douglas-fir aggressively colonizes grasslands and forest understories, altering forest composition and reducing diversity. The perception of these two tree species valued in North America but abhorred elsewhere does not differ from that of the tree-of-heaven, one of the most notoriously invasive trees in

Table 21.1 Selected list of trees invasive in North America and those native to North America but invasive elsewhere.

Species	Origin	Invaded	Habit in ecosystems of origin	Habit in invaded ecosystems
Tree-of-heaven	China	Widespread in the United States except extreme north	Widespread, probably as an early-successional species	Dense, clonal thickets on forest edges even on poor sites; displaces natives
Norway maple	Europe, northern Turkey and Iran	Northeastern and northwestern United States	Generally, a scattered understory tree; rarely a dominant	Forms closed-canopy forests and casts deep shade
Russian olive	Europe and Asia	Mostly western United States	Shrublands and woodlands in riparian and moist areas	Grows in riparian areas and outcompetes native vegetation there
Common buckthorn	Europe and Asia	Northern United States and California	Dense thickets to scattered plants on open sites	Invades forests, prairies, and savannas and forms dense thickets
Box elder	North America	Central Europe, Australia	Scattered short-lived tree in riparian areas and colonizes uplands	Dominant in lowlands, riparian areas, and disturbed areas
Douglas-fir	Pacific Northwest and Intermountain West of North America	New Zealand, Chile, and Argentina	Long-lived canopy dominant in pure and mixed stands	Invades open grasslands and forest understories
Black cherry	Eastern North America	Western and central Europe	Early-successional and edge species dominant; common in understory	Invades understories of oak and pine forests, impedes regeneration of other tree species
Northern red oak	Eastern North America	Western and central Europe	Late-successional canopy dominant	Establishes and competes with native oaks, impedes regeneration, and reduces diversity

North America but highly valued in China for its medicinal, ornamental, and commercial value as a host plant for silk moths. Much like their invasiveness itself (see the following text), the perception of such species is highly context-dependent.

CHARACTERISTIC TRAITS OF INVASIVE PLANT SPECIES

What is it about invasive species that makes them particularly able to establish and spread beyond their point of introduction? A classic explanation is that species become successful in new locations in the absence of their coevolved natural enemies. Commonly called the *enemy release hypothesis*, the idea is that some species become invasive when introduced to a new environment that does not contain herbivores or pathogens that fully replace the natural enemies present in the native environment. As a result, an increase in abundance and distribution of the invasive plant may proceed unchecked in the new environment. Although criticized for its oversimplicity in explaining why some plants become invasive (Colautti et al. 2004), there indeed exists evidence for the enemy release hypothesis when the plant species is strongly regulated by pathogens in its native environment. For example, Norway maple is a popular street tree introduced from Europe in the eighteenth century that has invaded both urban woodlots and undisturbed forests of the northeastern United States and Canada (Martin 1999). Tar spot disease is a pathogen that affects Norway maple in Europe, and it was introduced to North America near the end of the nineteenth century. Webster et al. (2005) attributed a decline in stem growth of overstory Norway maples in Michigan to an outbreak of the disease. Moreover, studies near Montreal, Quebec have shown that the growth of Norway maple saplings and trees declined sharply and mortality increased as a result of an outbreak of tar spot disease, while the disease's effects on native sugar maple remained far less detrimental (Lapointe and Brisson 2011). Although the interactions between introduced herbivores/pathogens and invasive plants may differ in locations beyond those of the ecosystems in which they evolved, these "reunion" studies suggest that in at least some cases plants may become invasive when freed from natural enemies.

Despite the potential importance of natural enemies, humans have preconceptions about what allows invasive plants to establish and spread in new environments. One of the most common traits attributed to invasive plants is an extremely high growth rate that allows them to grow faster than native plants, thus outcompeting them or overgrowing them, often to the point of local extinction. Perhaps the best example of a fast-growing invasive plant is kudzu, a semi-woody vine native to Asia and introduced to the United States in the mid-1870s. Kudzu may grow as fast as *a foot per day* once established, may develop vines as long as 100 feet, and is well known for overtaking and overgrowing anything in its path, particularly in the southeastern United States (Figure 21.1). The other major characteristic of invasive plants, including many invasive trees, is their ability to form dense, monotypic (single-species) stands that shade out native plants. For example, in eastern North America, Norway maple forms dense stands of overstory trees that cast

FIGURE 21.1 Aggressive invasion of kudzu overgrowing trees, shrubs, and virtually all other features on a landscape in the southeastern United States.

deep shade, limiting or eliminating the growth of native herbaceous plants and the regeneration of native trees. Norway maple regeneration is much more successful beneath the dense shade cast by its own overstory than is regeneration of native trees, and its saplings also appear to outcompete native saplings in the understory (Galbraith-Kent and Handel 2008).

The traits that make some plant species invasive are more complex than simply a fast growth rate and an ability to exclude other species with shade, however. In general, invasive species are superior to native species in some functional traits that allow them to opportunistically acquire resources that help them to excel at growth, reproduction, and spread (Blumenthal 2005). Such traits include high relative growth rates, but also short generation time; light, small, and widely disseminated seeds; and fruits attractive to vertebrate dispersers (Rejmánek and Richardson 1996). Notably, a wide variety of fast-spreading plant species fit this description, including many of those classically considered to be pioneer or early successional species that are not invasive in their native environments.

Traits most likely related to invasiveness are often determined from meta-analyses of many studies that compare co-occurring native and naturalized species to those that are invasive, and such studies often indicate that many presumptions about what makes a species invasive are not necessarily correct. For example, Daehler (2003) examined 79 comparisons of co-occurring native and invasive plants and found that invaders were more likely to have higher leaf area and lower cost of construction for leaf tissue. Higher leaf area and lower tissue construction costs are together likely to provide a growth advantage to invasive species. However, most other traits did not differ between native and invasive plants (Figure 21.2); invasive plants did not necessarily grow faster, were not better competitors, nor were they better dispersers. In a study of 146 comparisons of naturalized and invasive plants of the same genus in Australia, Gallagher et al. (2015) found that invasive plant species had larger specific leaf area (the amount of leaf area per unit biomass), longer flowering periods, and were taller than naturalized species. Specific leaf area was especially important in distinguishing invasive from naturalized species and was associated with faster growth rates, higher leaf turnover, and shorter life spans. Invaders were also more tolerant of a wider range of rainfall and temperature conditions in the native range. Finally, invasive plants tend to show greater phenotypic plasticity than non-invasive plants (Davidson et al. 2011; Daehler 2003). Theoretically, higher plasticity should allow an invading species to better adapt to novel environmental conditions to which it is introduced, and to take advantage of environmental fluctuations common to disturbed ecosystems where invasive species commonly establish and spread.

In a comprehensive study of 117 trait comparisons between invasive and non-invasive plant species, Van Kleunen et al. (2010) identified multiple trends attributable to species' invasiveness. Invasive species had higher values for traits such as photosynthetic rate, transpiration, tissue nitrogen content, and lower leaf construction costs; allocation to leaf area and shoots, growth rate, and size (biomass of roots, shoots, and whole plants, and plant height); reproduction (number of flowers or seeds, germination); and survival. In other words, invasive species had higher values for traits related to high performance and faster growth, and ultimately higher fitness. Importantly, traits differed more between invasive and native plants than between invasive and naturalized non-native plants, and the traits of invasive plants generally did not differ from native plants when the native plants were invasive in other environments (Van Kleunen et al. 2010).

The traits associated specifically with invasive woody plants include a subset of those typically mentioned as being characteristic of invasive plants in general. In a study of 28 woody species known to be invasive in California, Grotkopp and Rejmánek (2007) compared the traits of invasive and naturalized (non-invasive) species within the same genus or family. They found that high relative growth rate in the seedlings of trees and shrubs and high specific leaf area were the best indicators of woody plant invasiveness, probably because these traits are related to a higher ability to opportunistically capture available resources. High growth rate was also a major predictor of

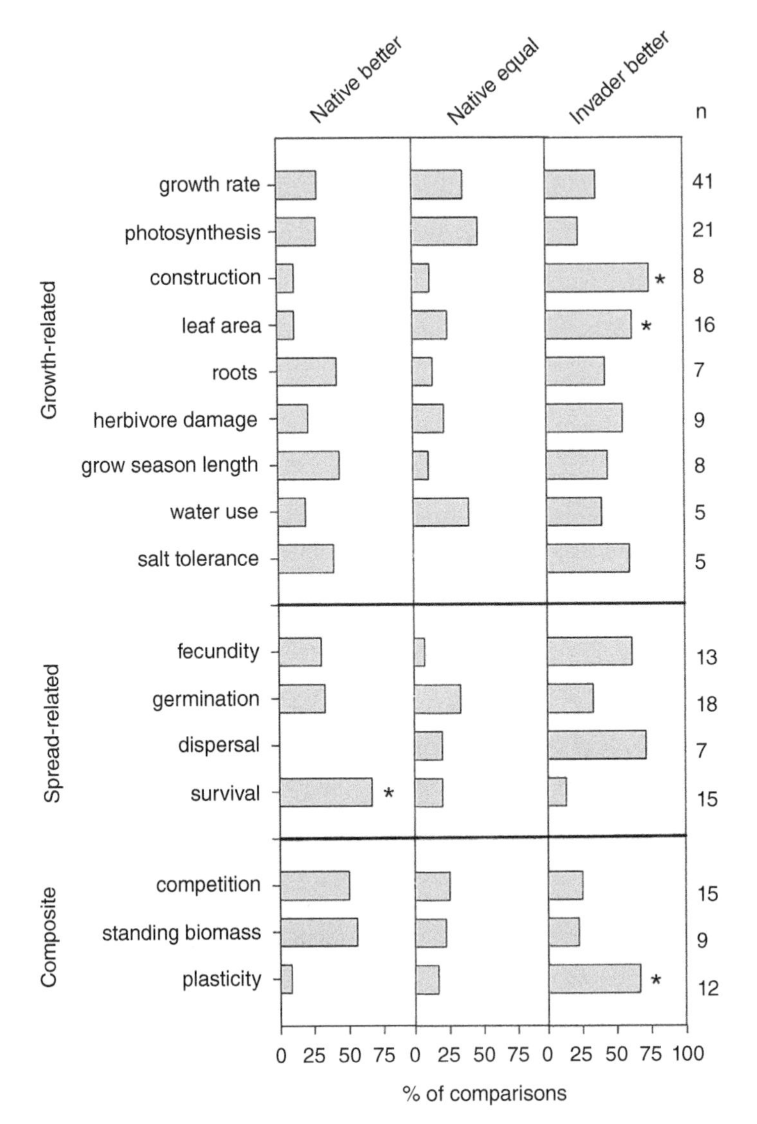

FIGURE 21.2 Performance of native versus invasive plants using 16 growth-related, spread-related, and composite traits. The first column shows the number of comparisons between co-occurring native and invasive plants in which the native performed better; the center column shows equal performance between native and invasive plants; and the third column shows better performance by the invasive plant in the comparison. Differences between invasive and native plants were relatively few; invaders had higher leaf area, lower construction costs, and higher phenotypic plasticity, and natives had higher survival.
Source: Daehler (2003) / Annual Reviews Inc.

invasiveness for trees in a literature review by Lamarque et al. (2011), who noted that several mechanisms related to invasiveness were useful in explaining tree invasions, and that no single factor was sufficient to predict whether an introduced tree would become invasive. This lack of a "silver bullet" in explaining tree invasions is well in line with invasive plants in general.

The physical site factors of the ecosystems being invaded seem to also heavily influence the incidence of plant invasions. For example, invasive species in regional ecosystems having a Mediterranean climate (long, dry growing seasons and mild, wet winters) tend to allocate more biomass to roots than non-invasive plants, allowing them to better survive the characteristic dry summers (Grotkopp and Rejmánek 2007). These patterns suggest that invasive species must not only be able to grow quickly and excel at acquiring resources, but also must be able to tolerate certain growing conditions or **environmental filters** that determine where they may be effective invaders. Such environmental filters appear to be important regardless of the traits possessed by the invader. Apparently, the site factors of the ecosystems being invaded select for similar traits as those found in native species, such that introduced plants that become naturalized typically have similar traits as native plants (Divíšek et al. 2018; Kraft et al. 2015). In some cases, these similarities occur because the introduced plants are closely related to the native plants (Strauss et al. 2006). To proceed from naturalized to invasive species, however, plants must also have traits that are different enough from those found in the community in which they have been introduced to give them an advantage over native species (Ordonez 2014).

Adding to the complexity of invasiveness are the growing conditions of the new environment (Daehler 2003). An advantage of invasives over native species depends on the availability of resources in a given ecosystem. In general, invasive plant performance increases under conditions of high resource availability (light, water, or nutrients), but native plants outperform invasives when resources are low or when natives are favored by a specific disturbance regime (Figure 21.3). As a result, very few species are able to be invasive everywhere they are able to establish (Daehler 2003). At the same time, these trends explain why ecosystems with high resource availability (such as open, urban, or other ecosystems highly disturbed by humans) or altered disturbance regimes (such as ecosystems affected by fire exclusion) are particularly susceptible to plant invasions. Exceptions to this trend, whereby invasive plants perform well when resources are limited, are not uncommon and are often related to the high resource-use efficiency of some invasive species (Funk and Vitousek 2007; Heberling and Fridley 2013; Jo et al. 2015, 2017).

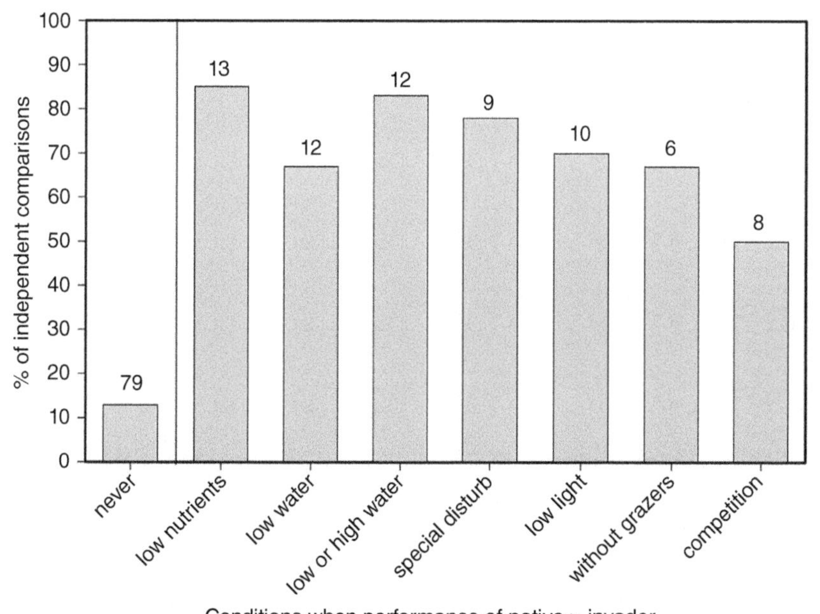

FIGURE 21.3 Growing conditions under which the performance of the native species was equal to or better than that of the invasive species in a pairwise comparison. Numbers over each bar are the total number of studies that manipulated each condition. *Source:* Daehler (2003) / Annual Reviews Inc.

NON-PLANT INVASIVE SPECIES IN FORESTS

Thus far we have described invasive species in forests as if they only included plants, but exotic insects, pathogens, and other animals are also critical in affecting forest ecosystems and deserve mention here. Phytophagous (plant-feeding) insects and pathogens are among the most widespread and impactful agents to forests in North America. Like invasive plants, insects and pathogens are often introduced unintentionally with their host plants during shipping of nursery stock, produce, or related products, or simply when the species hitchhike on any cargo transported from their native range to a new environment (McCullough et al. 2006; Westphal et al. 2008). Also, much like invasive plants, only a fraction of new introductions become established, and only a fraction of established populations begin to spread to become invasive.

More than 450 non-indigenous insect species and at least 16 pathogens, introduced between 1635 and 2006, are documented to attack one or more tree species in the United States (Aukema et al. 2010). The majority of these introductions have occurred since 1860 (Figure 21.4a). About 14% of these species are high-impact insects and pathogens, in that they kill trees at a rate above background levels; these high-impact species may cause tree mortality over broad areas. A selected list of some of these high-impact insects and pathogens is presented in Table 21.2. Notably, of the high-impact species, only two insects and one pathogen were introduced before 1860; between 1860 and 2006, a new damaging insect or pathogen was detected in the United States about every two years. Accumulation of non-indigenous species in the United States has been dominated by those that feed on sap (e.g., adelgids, aphids, and scales) or on foliage, although a significant increase in the number of introductions that feed on phloem and wood has occurred since about 1970 (Figure 21.4b).

Much like invasive plants, many, if not most, invasive insects are less ecologically troublesome in their native range. For example, the emerald ash borer feeds on stressed and dying ash trees in its native range in east Asia and is fairly uncommon at endemic levels. In North America, however, the borer was able to colonize and kill healthy ash trees, reproduce, and spread quickly across the continent (Herms and McCullough 2014), and has emerged as one of the most destructive invasive insects ever introduced there. In fact, North American ash used as ornamental trees in

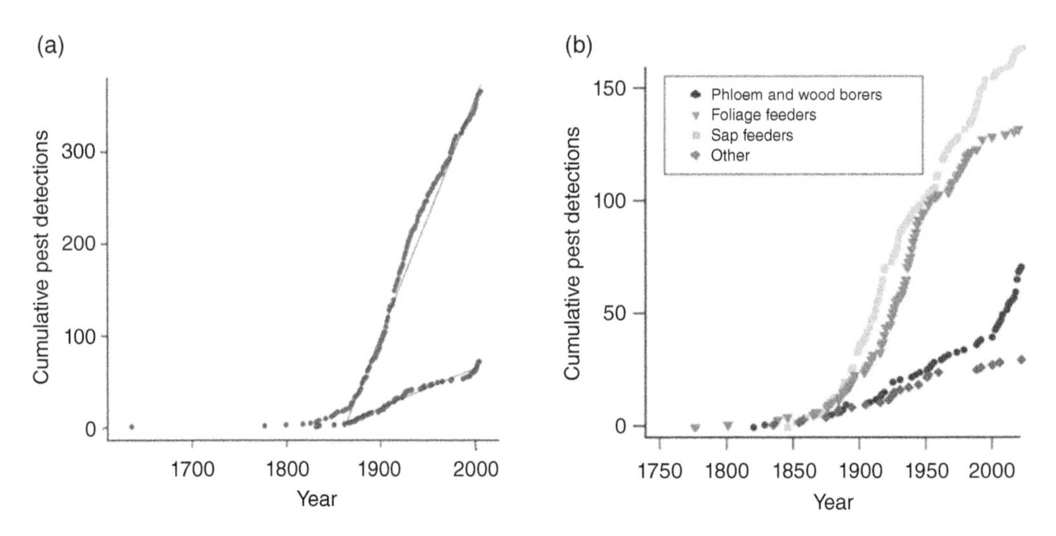

FIGURE 21.4 Cumulative detections of established forest pests over time for (a) total non-indigenous insects with line fitted for the years 1860–2006 (blue) and high-impact insects and pathogens for the years 1860–2006 (red); and (b) cumulative detections of non-indigenous forest insects by guild over time. *Source:* Aukema et al. (2010) / with permission of Oxford University Press.

Table 21.2 Selected list of important forest insects and diseases introduced into North America that have severely disrupted native ecosystems and decimated or displaced native species.

Agent	Latin name	Origin	Detection year	Hosts	Tissue attacked
Insects					
Spongy (formerly gypsy) moth	*Lymantria dispar*	Europe, Asia	1869	Hardwoods	Foliage
Winter moth	*Operophtera brumata*	Europe	1930	Hardwoods	Foliage
European pine sawfly	*Neodiprion sertifer*	Europe	1925	Pines	Foliage
Hemlock woolly adelgid	*Adelges tsugae*	Asia	1952	Hemlocks	Foliage
Balsam woolly adelgid	*Adelges piceae*	Europe	1908	True firs	Phloem
Asian longhorned beetle	*Anoplophora glabripennis*	Asia	1996	Hardwoods	Cambium, phloem
Emerald ash borer	*Agrilus planipennis*	Asia	2002	Ashes	Phloem
Beech scale[a]	*Cryptococcus fagisuga*	Europe	1890	Beech	Bark, cambium
Pathogens					
Chestnut blight	*Cryphonectria parasitica*	Asia	1876	Chestnuts	Cambium, phloem
White pine blister rust	*Cronartium ribicola*	Europe, Asia	1900	White pines	Needles, stems
Dutch elm disease	*Ophiostoma ulmi*	Europe	1927	Elms	Xylem, phloem
Dutch elm disease[b]	*Ophiostoma novo-ulmi*	Europe	1945	Elms	Xylem, phloem
Oak wilt	*Bretziella fagacearum*	Central or South America	1944	Oaks	Xylem
Laurel wilt	*Raffaelea lauricola*	Asia	2002	Redbay, sassafras	Xylem

[a] Beech scale is associated with the spread of the pathogen that causes beech bark disease (*Nectria coccinea var. faginata*). Trees remain free of the disease until they are infested with beech scale.

[b] Dutch elm disease is caused by two distinct species of a pathogen successively introduced into North America.

Source: Data from Aukema et al. (2010).

Asia have allowed local emerald ash borer populations to approach outbreak levels (Liu et al. 2003). Lack of resistance in North American ash relative to Asian ash species is a likely explanation for the rapid increase in emerald ash borer population in a new environment and its consideration as a highly destructive invasive species there (Liebhold et al. 2017).

The presence of invasive insects appears to vary widely across North America. There are far more non-indigenous, high-impact forest insects and pathogens present in eastern North America compared to western North America (Figure 21.5). There are likely multiple mechanisms driving this spatial pattern. First, the mostly deciduous forests of eastern North America contain a higher number of tree genera that could serve as hosts for newly established non-indigenous insects compared to the mostly coniferous forests of the West (Liebhold et al. 2013). Of the 450 non-indigenous insect species identified by Aukema et al. (2010), about half of the species feed on host species within a single genus, about 18% attack more than one genus within a single plant family, and about a third have a broader range of hosts that may span multiple plant families. As such, it makes sense that most insects are more likely to invade and become established where genera are more diverse, which might also explain why invasive insects from Europe are much more pervasive in North America than North American insects are in Europe (Mattson et al. 2007). In addition, given that international trade is a main vector of new introductions (Westphal et al. 2008), the longer history of such trade in eastern North America compared to the West is also likely an important factor driving the pattern of insect invasions across the continent.

Invasive tree pathogens in many ways have similar effects as invasive insects. Invasive pathogens may also cause widespread tree mortality. Classic examples include chestnut blight, which was introduced from Asia into North America shortly after the Civil War and killed the majority of American chestnut trees in eastern forests by the 1940s (Paillet 2002); and Dutch elm disease, which was first introduced in the 1920s from Europe, killing most large elms across the eastern United States in natural forests as well as those widely planted and prized as street trees (Hubbes 1999). Important differences exist between insects and pathogens, however. First, invasive pathogens are often extremely difficult to identify, such that pathogenic establishment and spread and resulting tree mortality may proceed for many years unchecked. In addition, pathogens

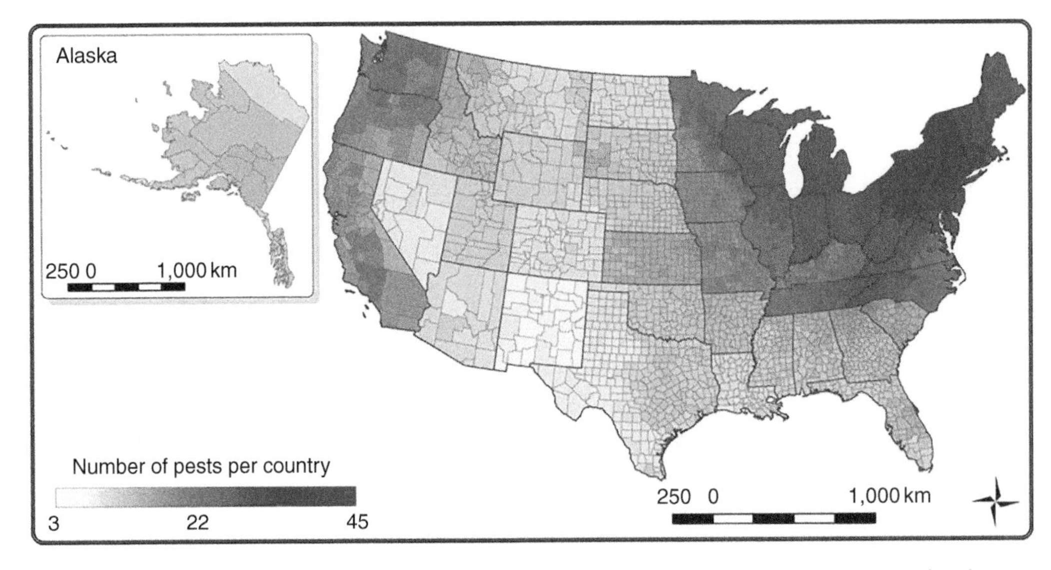

FIGURE 21.5 Geographical variation of the numbers of invasive, high-impact forest insects and pathogens across the United States. There are far more invasive insects and pathogens in the eastern United States with a clear area of concentration in the Northeast. *Source:* Liebhold et al. (2013) / John Wiley & Sons.

often exist in several strains or even species that may be introduced in succession and, as a result, increase the severity of a particular epidemic. For example, Dutch elm disease was caused by two distinct pathogen species that were introduced to North America in succession (Table 21.2), the second two decades after the first, resulting in prolonged and more severe effects of the disease (Brasier 2001).

Tree pathogens—including those invasive to a particular region—often work in tandem with either invasive or native insects that serve as vectors for infection or as dispersal agents (Wingfield et al. 2016). There are many examples of such mutualisms that have developed between invasive species. Dutch elm disease results from a pathogen introduced with timber from Asia; this timber was infested by both native and introduced bark beetles that facilitated the spread of the pathogen to native North American elms. More recently, beech bark disease has begun to cause extensive tree mortality of American beech across the eastern United States. The pathogen that causes beech bark disease does not attack trees until the trees have been heavily infested by the beech scale, an invasive insect introduced from Europe around 1890. The pathogen presumably enters the tree through the puncture wounds caused by the extensive feeding by the beech scale. The association between the beech bark disease pathogen and the beech scale is also present in the native range of each agent in Europe. Although the pathogen in North America was once thought to be the same as that causing beech mortality in Europe, recent evidence suggests that closely related pathogens might cause the disease on both continents, and the pathogen causing disease in North America is actually native (Wingfield et al. 2016).

Non-plant invasive species in forests are not limited to insects and pathogens, but also include soil invertebrates and large vertebrates (Liebhold et al. 2017). Invasive soil invertebrates include many that affect decomposition, such as some beetles and particularly earthworms. Earthworms have been documented to have a large effect on microbial and invertebrate soil communities and the cycling and storage of soil nutrients (Frelich et al. 2006). Introduced large vertebrates, such as feral swine, also have large impacts on forests. Extensive rooting and digging by feral swine have been documented to destroy tree regeneration and damage the litter layer in the southeastern United States (Campbell and Long 2009). Impacts to forest ecosystems of these and other invasive species are discussed in the following sections.

IMPACTS OF INVASIVE SPECIES ON FORESTS

Separate from understanding the *invasiveness* of a species—that is, the traits that allow an introduced species to be effective at establishing and spreading from its point of introduction—is understanding the *impact* of an invasive species, or its effect on native plants or ecosystems. The impact of an invasive species varies depending on the type of taxa (e.g., plant, insect, pathogen, or vertebrate) and ranges from changing abundance of an individual species to altering ecosystem function (Parker et al. 1999). Impacts of an invasive species depend largely on its distribution and abundance, which, in turn, depend on the stage of invasion (Mayfield et al. 2021). In this section, we focus on the ecological impacts of the major taxonomic groups of invasive species in forests and provide examples therein. A full treatment of invasive species impacts is given by Mayfield et al. (2021), Miniat et al. (2021), and Liebhold et al. (2017).

IMPACTS OF INVASIVE PLANTS ON FORESTS

Invasive plants may impact forests in several different ways, including changing the abundance and distribution of native species (for native plants, through competition; for other native taxa, by altering habitat structure and resource availability), altering fire regimes, and affecting processes that govern nutrient cycling and storage. Although plants are but one taxonomic group of invasive species in forests, they have a disproportionate effect on ecosystems for their ability to shape important factors such as vegetation structure, successional trajectories, primary productivity, nutrient cycling, and species habitats.

Competition One of the most obvious ways invasive plants affect forests is by outcompeting native plants for resources such as light, water, and nutrients. As we have seen, invaders tend to be favored where such resources are abundant, but natives are favored where resources are limited (Daehler 2003). Once plants invade a forest, they can drastically change its composition by competing with native tree regeneration and, in the case of invasive trees, by shading out (e.g., Norway maple) or crowding out (e.g., tree-of-heaven) native woody plants. In the case of invasive woody vines, such as kudzu or Asian bittersweet, the invasive plants may climb, overgrow, or smother native trees and other plants (Figure 21.1). For example, infestation by Asian bittersweet sharply reduced the growth of red oak and trembling aspen in a Massachusetts forest 14 years after establishment (Delisle and Parshall 2018). Although local changes in species abundance and compositions are well documented to be associated with invasive plants, invasive plants rarely push native plants fully to extinction (Gurevitch and Padilla 2004). By affecting native plants, however, invasives may also affect native invertebrate and vertebrate species that utilize or specialize on the native plants (Gandhi and Herms 2010).

Altered Fire Regimes In addition to altering the vegetation structure or the abundance of species in forests, some invasive plants cause system-level effects that create new disturbance regimes under which both native and invasive plants must persist (Asner and Vitousek 2005). Perhaps the best examples of such system-level effects include invasive plants that alter the fire regimes of forests—recognized as some of the most important ecosystem-altering species on Earth (Brooks et al. 2004; Mayfield et al. 2021). The best-studied of such plants are grasses. When an invasive grass establishes in a forest of a dry regional ecosystem, it provides the fine fuel needed to initiate and carry a fire, resulting in increased fire frequency and intensity. Following such fires, the invasive grasses recover more quickly than native grasses and plants, further increasing the susceptibility of the forest to additional fires (D'Antonio and Vitousek 1992).

The most notorious example of the invasive grass–fire cycle is that of cheatgrass in the Intermountain West. Cheatgrass was introduced from Europe or Asia prior to the Civil War and is now widespread across the United States, though it is most prominent and problematic in the West. Though its major effects are on rangelands and sagebrush steppe ecosystems, cheatgrass is commonly found in pinyon–juniper forests of the Great Basin, ponderosa pine forests across the West (Figure 21.6), and interior Douglas-fir forests of British Columbia. It can outcompete native plant species through soil water depletion (Melgoza et al. 1990) and may displace them as fire frequency increases. Repeated and frequent burning due to cheatgrass invasion has converted millions of acres of sagebrush steppe to annual grasslands (Monsen and Shaw 2000), and it has had similar effects on the forest types described in the earlier text. Cheatgrass is very difficult to control once it achieves dominance; those areas where perennial plants have been nearly completely displaced by cheatgrass and fires occur every five or fewer years are likely to become permanently dominated by cheatgrass over the long term (Mosely et al. 1999).

A similar impact of invasive grasses on forests is found in the southeastern United States, where cogongrass was introduced from Japan in 1912. Cogongrass is a tall, dense, and aggressive grass that outcompetes native plants for nutrients, crowds out native species, reduces light levels at the forest floor, and alters decomposition rates in southern pine forests (Brewer 2008; Holly et al. 2009). In a study of longleaf pine forests on the Gulf Coastal Plain in Mississippi, Brewer (2008) found that cogongrass invasion over 3 years displaced shorter plants that were endemic to longleaf pine forests and sharply reduced species richness. Tall saplings, shrubs, and vines, most of which were not endemic, were not displaced. In addition to displacing native plants, cogongrass also alters fire regimes by increasing fine fuel loads and fire intensity, resulting in higher tree

Left photo by Becky Kerns, USDA Forest Service; right photo by John M. Randall, The Nature Conservancy, Bugwood.org.

FIGURE 21.6 Cheatgrass invasion into ponderosa pine forest in the Blue Mountains, Oregon (left) and cogongrass invasion into slash pine forest in the southeastern United States (right).

mortality relative to historical fire regimes. For example, Platt and Gottschalk (2001) found higher aboveground biomass of fine fuels and litter biomass in slash pine forests where cogongrass was present than in nearby areas without it. Likewise, Lippincott (2000) found higher fuel continuity in longleaf pine forests invaded by cogongrass that led to more intense fires and higher young tree mortality, creating the potential that such forests could convert to non-indigenous grasslands.

Carbon and Nutrient Cycling Invasive plants may create system-level changes beyond altering fire regimes. On the whole, invasive plants appear to increase ecosystem productivity because they affect carbon uptake, storage, and cycling. For example, in a meta-analysis of 94 experimental studies, Liao et al. (2008) found that aboveground net primary production increased by 50–120% in ecosystems with significant plant invasions—particularly by woody plants—compared with uninvaded ecosystems. Such effects are greatest when invasive plants have significantly higher light and nutrient use efficiencies and growth rates than the native plants they are replacing. Effects on carbon cycling, in general, increase with the dominance of such plants. Interestingly, woody plants tend to increase carbon stocks in aboveground and belowground biomass and soil organic matter when they invade grasslands, but annual grasses tend to decrease soil carbon stocks when they invade forests (Miniat et al. 2021). Given that invasive plants are generally more competitive when limiting resources are more abundant (Daehler 2003), the strength of the effects of invasive plants on carbon cycling is likely to vary widely across ecosystems differing in resource availability and is difficult to generalize.

The differences in growth rates and tissue chemistry of invasive plants often increase their litter production and decomposition rates (Wardle and Peltzer 2017), respectively, and these differences can facilitate changes in fire regimes. In turn, altered fire regimes, particularly when fire frequency or intensity is sharply changed, will affect carbon and nutrient cycling in invaded ecosystems. For example, cover-type conversion of forests to grasslands due to invasion by cheatgrass or cogongrass, as described in the earlier sections, will quickly result in a loss of carbon storage (see Chapter 20). Studies of cheatgrass invasion in the Great Basin of the western United States have shown that an estimated 8 Tg of carbon has been released to the atmosphere due to increased fire frequency and intensity and land cover change from forest to grassland (Bradley et al. 2006). Continued invasion by cheatgrass will likely release another 50 Tg of carbon in the coming decades, converting large portions of the landscape from carbon sinks to carbon sources.

Invasive plants also appear to have a strong effect on the cycling of nutrients in forests. Plant invasions generally increase the amount of nitrogen found in aboveground plant tissues (Liao et al. 2008), increase nitrogen availability in soils, and stimulate microbial communities. In comparing five pairs of native and non-native buckthorn, honeysuckles, and bittersweets in deciduous forests of the eastern United States, Jo et al. (2017) found that invaders had greater litter production and litter nitrogen content that enhanced the flow of nitrogen from litter to the soil after 3 years. Moreover, available soil nitrogen uptake was increased in invaded forests because invaders had higher fine root production and longer roots. Increased rates of soil nitrogen mineralization and uptake by invasive plants therefore accelerate soil nitrogen cycling, which, in turn, is likely to support further increased productivity of invaders.

An additional, and perhaps most pronounced, mechanism of impacting nutrient cycling by invaders is by nitrogen fixation. Many invasive plants can house bacterial symbionts that are able to fix nitrogen and therefore provide the plant a significant competitive advantage on sites where nutrients are limiting. For example, black locust is a nitrogen-fixing tree native to the Ozarks and southern Appalachian Mountains, but widely introduced and naturalized across much of the United States. Black locust is considered invasive in Europe as well as in the western United States, the Midwest, New England, and northern California, even though the species is native to North America. Rice et al. (2004) found that black locust invasion in pitch pine-scrub oak forests of New York increased soil nitrogen concentrations and net nitrification rates relative to non-invaded sites. Net mineralization rates also greatly increased relative to non-invaded sites and were attributed to leaf litter with high nitrogen content and high turnover. The longevity of invasive effects on nutrient cycling remains an area of active research, and the magnitude of impacts is likely to be highly dependent on the ecological characteristics of a given ecosystem (Kumschick et al. 2015).

IMPACTS OF INVASIVE INSECTS AND PATHOGENS ON FORESTS

Most invasive insects and pathogens have relatively little effect on forests, but a small fraction have gained notoriety for killing or damaging millions of trees across very wide areas. The type of damage caused by invasive insects depends largely on the type of tissue they feed upon. For example, insects that feed on seeds and flowers reduce the ability of host trees to reproduce, whereas those that feed on foliage, sap, or roots generally reduce the vigor of trees. The spongy moth (formerly known as gypsy moth) is a defoliator of a wide variety of deciduous tree species in eastern North America. Severe defoliation by larval spongy moths creates a deficit of carbohydrate allocated to roots and shoots, thereby reducing nitrogen uptake and resulting in loss of growth and dieback in the crown (Kosola et al. 2001). The hemlock woolly adelgid is an aphid-like insect that feeds on the storage cells of hemlocks, mainly in eastern North America. Such feeding not only reduces carbohydrates in the tree but may cause xylem to form abnormalities that affect its ability to uptake water (Domec et al. 2013) and may slowly kill the tree. Widespread mortality of hemlocks, considered to be foundational species whose loss is likely to cause significant ecological shifts in habitat of both aquatic and terrestrial taxa, has occurred across much of the southern Appalachians.

Extensive tree mortality is most often caused by invasive bark and wood-boring insects. Emerald ash borer is a wood borer introduced in Michigan from China in the early 1990s that has killed millions of ash trees in eastern North America and is considered the most destructive invasive insect ever to be introduced in the United States (Aukema et al. 2011). Larval emerald ash borers feed at the nexus of phloem and xylem and rapidly girdle and kill the tree. Generally, invasive insects having the most severe and long-term impacts are those most likely to spread quickly, are specific to one or a few host species, and kill hosts that are abundant, dominant, and ecologically unique (Lovett et al. 2006). As such, emerald ash borer and hemlock woolly adelgid are likely to be more impactful than spongy moth, as are pathogens such as chestnut blight and Dutch elm disease.

By killing trees, invasive insects or pathogens create important changes in vegetation structure or species composition of a forest that may affect the rest of an ecosystem. When invasive insects or pathogens kill trees, they reduce canopy coverage or create canopy gaps that increase light availability and alter temperature and moisture regimes at the forest floor. In the short term, such disturbances are likely to affect tree regeneration, density, and basal area as well as the establishment and growth of both native and invasive plants. In killing or damaging trees, most invasive insects and pathogens will also increase organic inputs to the forest floor as leaves, branches, insect excrement and biomass, and coarse woody debris (Figure 21.7; Gandhi and Herms 2010). Over longer time periods, tree mortality caused by invasive insects and pathogens has the potential to alter tree species composition and forest structure, especially if a large proportion of the dominant overstory trees are killed by host-specific invasives (Figure 21.7), and thus sharply influence succession (e.g., Morin and Liebhold 2016). When a host species largely defines forest structure and/or stabilizes the functioning of an ecosystem as a foundational species (Chapter 14), removal of such a host may have dramatic effects (Ellison et al. 2005; Wilson et al. 2018).

Chestnut Blight, Dutch Elm Disease, and Forest Succession A host species that is eliminated or greatly reduced in abundance in forests is eventually replaced by other tree species. An outstanding example is the virtual elimination of American chestnut from eastern deciduous forests by chestnut blight. This invasive fungus was introduced from Asia about 1904, killing most of the mature chestnuts in New England within 20 years, and most in the southern Appalachians by the 1940s. It is unlikely in recorded history that a major forest tree has been so nearly eradicated. American chestnut was a major dominant of many dry and dry-mesic ecosystems in the southern Appalachian Mountains and in deciduous forests of eastern North America (Braun 1950), growing to enormous sizes in old-growth forests of the southern Appalachians (Figure 21.8). Surviving chestnuts infected with the blight typically produce non-viable fruit. Although we still observe sprouts from root systems (Griffin 1992), few trees are vigorous or survive long enough to reach the overstory.

Succession following the elimination of American chestnut often resulted in the simple replacement of that species by its former associates (Illick 1914, 1921; Korstian and Stickel 1927; Augenbaugh 1935; Keever 1953; Woods 1953; Woods and Shanks 1959; Day and Monk 1974; Stephenson 1974). Succeeding trees in the southern Appalachians are mainly oaks (especially chestnut, red, and white oaks), along with various hickories, and tuliptree on more mesic sites. Thus, eradication of chestnut in this region has caused the replacement of the former oak–chestnut type with an oak or oak–hickory type. Similar changes, but involving some different species, previously occurred in the middle Atlantic states and southern New England (Good 1968).

In the Allegheny Mountains of western Pennsylvania, logging and fire following the death of chestnut trees created open sites for invasion by early and mid-successional species (black cherry, black birch, black oak, sassafras, and black gum) together with species already present on the site such as sugar maple, red maple, white ash, and beech (Mackey and Sivec 1973). Without fire or other disturbances in the former chestnut forest, succession gradually proceeds to more tolerant and mesophytic species such as hemlock, sugar maple, and beech.

Photo by Dan Kashian

Photo by Kayla Perry, Kent State University

FIGURE 21.7 Impacts of emerald ash borer on forests in southeastern Michigan. Top: Extensive mortality of overstory ash trees. Note host-specific mortality in this mixed deciduous forest. Bottom: Dead overstory ash trees left as coarse woody debris on the forest floor following emerald ash borer infestation. Note the amount of sunlight now reaching the forest floor through canopy gaps created by killed ash trees.

Later in the twentieth century, Dutch elm disease, caused by a fungus first introduced into Ohio from Europe around 1927, virtually eliminated mature American elm trees from swamp and river floodplain forests of the eastern and midwestern United States within about 50 years. Studies of succession following the elimination of overstory elms in the Midwest indicate that elms are

FIGURE 21.8 American chestnut trees in an old-growth forest of the Great Smoky Mountains, North Carolina. On favorable sites, American chestnut trees could grow over 2 m in diameter and 36 m in height. *Source:* Photo courtesy of the American History Society.

replaced by a number of different species, depending on regional and local site conditions. In Illinois woodlands, sugar maple is the species most likely to increase in dominance where soils are not too poorly drained (Boggess 1964; Boggess and Bailey 1964; Boggess and Geis 1966). In southeastern Iowa, hackberry and box elder were the most frequent trees replacing elm (McBride 1973).

In southeastern Michigan, American elm was a late-successional dominant in deciduous swamp forests with black ash, red maple, and yellow birch. The latter two species now dominate the swamps in different proportions depending on site conditions (Barnes 1976). In contrast with chestnut in dry-mesic forests, American elm today remains in deciduous swamps, making up about 10–15% of the understory. However, old fields and other open upland areas are now much more important sites for elm regeneration than swamps. Unlike American chestnut, American elm typically reaches reproductive maturity before it succumbs to the invasive pathogen, such that it will likely be perpetuated for generations by seeds from young elm trees but the average life span will be drastically reduced. Resistant elm clones continue to hold promise for horticultural planting (Townsend et al. 1995).

The death of host trees due to invasive insects and pathogens has obvious effects on carbon and nutrient cycling in the forests that they affect. In general, invasive insects and pathogens reduce the productivity of forests, at least in the short term (Miniat et al. 2021), and the input of organic material to the forest floor and to dead wood carbon pools affects decomposition, soil respiration, and annual carbon budgets. Invasive species that extensively damage but do not necessarily kill their host species, such as spongy moth, are still likely to reduce tree growth and thus net ecosystem production (Clark et al. 2010). Invasive insects or pathogens that kill trees may also have indirect effects on carbon cycling when tree mortality results in replacement of the host species by new species with different rates of nutrient uptake, growth, or litter quality. Such replacements in some cases may have long-term implications for carbon cycling in affected forests (Lovett et al. 2006), depending on the specific successional dynamics. For example, the loss of ash in the United States due to emerald ash borer is likely to result in a significant loss of aboveground carbon in the short term, but much of this carbon is likely to be recovered by the increased growth of other tree species already present that will succeed ash (Flower et al. 2013).

Invasive insects and pathogens can also cause changes in nutrient cycling, often dramatically (Crowley et al. 2016). Again, the pulse of organic matter to the forest floor following invasive insect outbreaks or pathogen infestations, as well as the decrease in nitrogen uptake by trees immediately following outbreaks or infestations, can result in short-term increases in soil nitrogen availability (e.g., Orwig et al. 2008). Moreover, leaching and loss of nitrogen from soil may also result. For example, Eshleman et al. (1998) found that nitrogen "leakage" from forested watersheds in the Chesapeake Bay region occurred several months to a year after spongy moth defoliation, probably due to reduced nitrogen uptake from recovering trees and increased water discharge from soil resulting from reduced transpiration following defoliation. Nutrient cycling may also be affected over the long term when invasive-caused tree mortality results in tree species replacement. For example, hemlock woolly adelgid infestations have shifted dominance from eastern hemlock toward black birch in New England, increasing aboveground productivity and nitrogen uptake (Finzi et al. 2014). Likewise, beech bark disease in beech–sugar maple forests of the eastern United States has resulted in a shift toward sugar maple-dominated forests, in turn increasing litter decomposition, nitrogen cycling, and nitrogen leaching from soils in the region (Lovett et al. 2010).

IMPACTS OF INVASIVE ANIMALS ON FORESTS

A large number of animals other than insects have been introduced and become invasive in forests across the world. In the United States, over 260 vertebrates, including mammals, birds, amphibians, and reptiles, could be considered invasive (Fall et al. 2011). The effects of many of these species are local or regional in scale, such as those of goats or burros whose selective grazing impacts native vegetation; or are difficult to isolate, such as various bird species that aid in the dispersal of invasive plants. In this section, we describe the notorious impacts of one invertebrate (earthworms) and one vertebrate (feral swine) on forests in North America.

In northern regions of the United States that were glaciated during the last ice age, non-native earthworms have invaded many forests that had been free of such fauna for more than 12 000 years. North American earthworms found in unglaciated areas in the southeastern United States and the Pacific Northwest have not expanded their range into this region, either because of slow dispersal or because they do not tolerate relatively harsh winters. Where forests evolved without earthworms, litter decomposes more slowly on the forest floor, depending on site conditions. When earthworms are introduced to these areas, typically from non-native plants and soils and discarded fishing bait, the invasive worms voraciously eat the litter layer and may eliminate it completely. In doing so, earthworms may redistribute carbon throughout the soil profile, dramatically reducing carbon stocks on the forest floor (Bohlen et al. 2004). This redistribution may alter the distribution and composition of microbial and microarthropod communities in the soil, as well as the seedbed conditions for forest herbs and woody plant regeneration. In some forests, invasive earthworms also contribute to soil erosion, nutrient leaching, and acceleration of nutrient cycling to the extent that nutrient uptake by native plants cannot keep pace with available nutrients (Frelich et al. 2006). These impacts are likely to alter the abundance and diversity of tree regeneration and could create large, long-term changes in forest composition (Bohlen et al. 2004).

The presence of earthworms in agricultural settings was once universally considered to be beneficial (for example, see the third edition of this textbook), but the potential negative impacts of invasive earthworms on forests have only recently been recognized by ecologists. The leading edge of earthworm invasion was examined in a series of studies of four forests dominated by sugar maple in north-central Minnesota. Hale et al. (2006) found that high earthworm biomass was associated with lower herbaceous plant abundance and diversity as well as lower abundance and density of tree seedlings. High earthworm biomass was also associated with rapid disappearance of organic horizons, and increases in the thickness, density, and total soil organic matter content of the A horizon. Moreover, availability of nitrogen and phosphorus was lower where earthworm biomass was high (Hale et al. 2005a). Hale et al. (2005b) found a clear succession of invasive earthworm species, where species that live and feed exclusively in the litter layer may facilitate the establishment of species of earthworms that live and/or feed in the mineral soil, the latter of which prevent recovery of the forest floor. Thus, not only do earthworms invade forest ecosystems, but they also develop entire communities with species that compete and succeed one another as if they were native.

The first swine in the United States was introduced to North America from the domestic stock of early European explorers and colonists. Eurasian wild boars were later introduced for hunting purposes, and they interbred with domestic pigs where their occurrences overlapped. Today, feral swine cause extensive damage to forests and other ecosystems due to their digging and rooting activities, which has been shown to influence plant succession and species composition (Engeman et al. 2007) and decrease tree diversity and regeneration (Siemann et al. 2009). In altering species composition, feral swine indirectly impact the species by disrupting the ecosystems they inhabit. Feral swine also have direct effects on other animal species, in that they prey on many native invertebrates, birds, amphibians, and reptiles (Jolley et al. 2010) and compete with native fauna for food resources. In addition, feral swine consume the acorns of oaks and the nuts of hickories, representing an important source of competition with native wildlife for a critical food source as well as important seed predation for native tree species (Elston and Hewitt 2010). The degree of soil disturbance provided by feral swine makes it likely that they substantially affect forest ecosystem nutrient cycles, although empirical data are not definitive in this regard (Barrios-Garcia and Ballari 2012).

A PRIMER OF INVASIVE SPECIES MANAGEMENT IN FORESTS

Prevention, management, and control of invasive species are within the enormous and complex fields of applied and restoration ecology with textbooks dedicated to those subjects alone. There are literally hundreds of strategies available to control and manage invasive species, and a detailed treatment of such strategies is well beyond our scope. However, a series of common concepts, caveats, and methods characterize the management and control of all invasive species regardless of origin, taxonomy, or ecosystem, and we attempt in this section to provide a brief summary in the context of forest ecosystems. Recent exhaustive and comprehensive reviews of invasive species management are provided by Venette et al. (2021), Poland et al. (2021), Koch et al. (2021), Green and Grosholz (2021), Simberloff (2013), and Clout and Williams (2009), among many others.

A discussion of invasive species management first requires an understanding of the stage of the invasive process being addressed for a particular species. Regardless of the invasive species in question, management needs and approaches vary widely at different points along this invasion

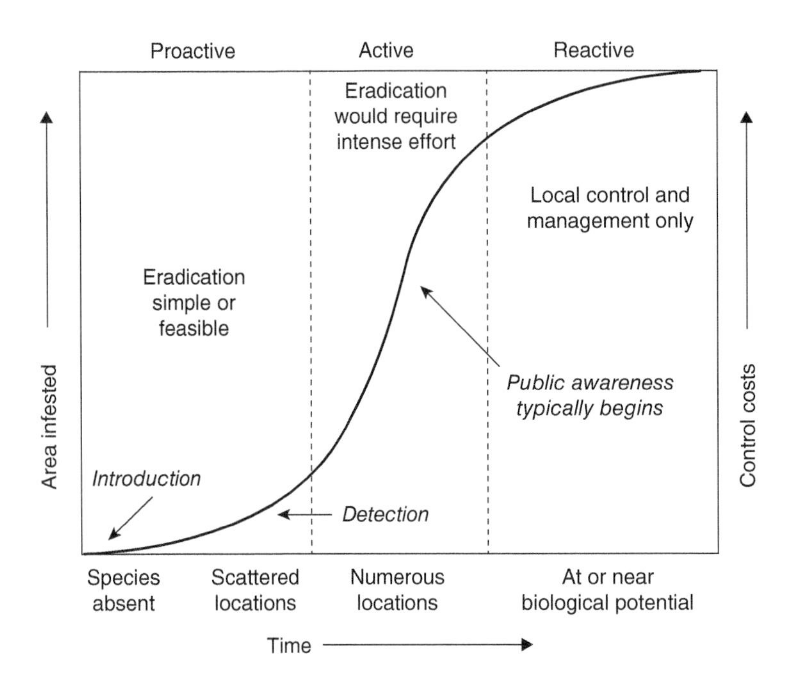

FIGURE 21.9 Population growth of a hypothetical invasive species and the timeline of associated management activities to control it. Eradication is most likely at the first sign of detection, which would require proactive management. Once the species has invaded numerous locations, eradication is more difficult, and reactive management when the invasive species approaches its biological potential is usually successful only locally. *Source:* Adapted from Hobbs and Humphries (1995).

process. The timeline for invasion by a hypothetical invasive species is illustrated in Figure 21.9. If we assume that the area infested by an invasive species can be represented by a logistic growth curve, with low initial area infested followed by rapid infestation of a landscape and finally a long-term stabilization of the area infested, then management effort (and thus activities) clearly must differ over time. Eradication of an invasive species is fairly simple or feasible, as well as inexpensive, when the species is present only in scattered locations across limited areas. Management at this stage would be considered proactive and would depend heavily on predicting which species might become invasive, preventing their further establishment and spread, and early detection if they do begin to spread (Epanchin-Niell and Liebhold 2015; Rout et al. 2014).

Unfortunately, most invasive species are not detected on a landscape until they have already begun to increase their populations, typically just before they experience the fastest increase in the amount of area infested and are detected at many locations (represented by the steep portion of the logistic growth curve; Figure 21.9). Because invasives exhibit rapid increases during this period, management is necessarily active and typically requires intense effort, which is accompanied by rapidly increasing cost of control. Even more unfortunate is that public awareness of the issues surrounding an invasive species often is not substantial until the invasive species is already rapidly increasing. When invasives infest all or most of what they are able on a landscape, their increasing distribution slows. Management at this point is reactive and is too late for eradication, resorting to control of the invasive in small, local areas at high cost. Among the important lessons illustrated in Figure 21.9 are that (i) early intervention is crucial in invasive species management, and (ii) a suite of intensive management strategies is necessary for active management of invasive species.

EARLY INTERVENTION STRATEGIES

Early intervention includes prediction, prevention, early detection, eradication, and other rapid responses (Venette et al. 2021). Prediction occurs by forecasting the likelihood that a species will become invasive and its potential consequences, a process called **risk assessment**. Prediction also depends heavily on **pathway analysis**, which is a standard process used to consider the potential avenues for a species to enter a given ecosystem. Prevention typically involves regulations and quarantines, which, together with pathway analysis, have been considered extremely valuable and inexpensive in their ability to avoid new invasions (Hulme 2009). Early detection depends heavily on surveillance, inventory, and monitoring of potential invasives in a new area of concern; specific methodologies are described in Venette et al. (2021) and Oswalt et al. (2021). Detection of a new invasive species or the spread of an invasive species to a new area then requires rapid responses such as eradication, containment, or suppression (Beric and MacIsaac 2015). Management at the early intervention stage may occur for recently introduced species that are not yet spreading, species present at low levels but not yet considered to be threatening, or invasive species threatening an area of interest (Venette et al. 2021).

MANAGEMENT APPROACHES FOR ESTABLISHED INVASIVE SPECIES

The priorities and strategies of invasive species management change quickly once an invasive species becomes established and has spread. These strategies turn from goals of eradication and containment to management strategies ranging in spatial scale from local approaches to landscape- or region-wide programs. Important principles of effective invasive species management at this stage, as reviewed by Poland et al. (2021), include the following: (i) it is necessary to understand the potential impacts of an invasive species as well as the impacts of its management; (ii) objectives should be specific to a given ecosystem or landscape; (iii) knowledge of thresholds is necessary for decision-making; and (iv) resources are inherently limited for managing invasive species and must be prioritized. Prioritization of resources is generally based on the degree of negative impact by the invasive species, the value or rarity of the ecosystems at risk for invasion, the size of the area potentially invaded and the stage of invasion, expected outcomes of management, and the availability of management tools.

Management approaches for invasive species have been categorized by Poland et al. (2021) to include the following specific to forests:

Regulatory control consists of measures to monitor or eliminate pathways that might allow introductions of invasives into a given ecosystem, such as seed certification programs, quarantines, and best management practices. Regulatory control is most common for recently introduced invasive species rather than long-established, widespread species. It is typically used for invasive insects, pathogens, and terrestrial wildlife, but not for plants.

Physical control involves the physical removal of invasive species or the construction of physical barriers to their invasion or spread. Examples include the pulling of invasive plants, trapping and removing invasive insects and other animals, or fencing to exclude large invasive animals. Physical control is expensive and generally limited to local areas. It is used in the control of all invasive taxa.

Cultural control includes activities that humans conduct in culturing or managing resources that make the establishment or spread of invasive species less likely. Avoiding the use of a host species susceptible to an invasive species in urban areas or using silvicultural techniques that increase health and vigor such as thinning or increasing diversity in forest stands are examples of cultural control. Cultural control is especially used for the control of invasive insects and pathogens.

Chemical control employs natural or human-made chemicals or microbial agents to directly affect invasive species in an effort to prevent infestation, eradicate populations, reduce impacts, or slow down spread by reducing the population. Chemical control is very commonly used for invasive insects and pathogens but is generally used for all invasive taxa.

Biological control uses living organisms to reduce the numbers of targeted invasive species such that the invasive species can be better managed with other strategies. The most common biological control agents are insects, usually predators or parasitoids, but may also include fungi, bacteria, and viruses. Biological control is generally highly successful but requires a long period of development to ensure its effectiveness against the invasive species but its harmlessness against non-target species. Biological control has been most successful with invasive insects but also holds much promise for the control of invasive plants.

Host resistance involves the propagation of host species populations that have genetic traits that leave them protected against (or less susceptible to) attack by invasive species. Host resistance is largely used as a strategy to reduce the impacts of invasive insects and pathogens, particularly those long established, on native plants. Identifying trees resistant to invasive pathogens is generally most successful in genetically diverse populations.

Reproductive control refers to the prevention of mating of invasive species or the development of their offspring using genetic or chemical manipulation. Classic examples include using pheromones to disrupt the mating of invasive insects or controlling fertility in other invasive animals. Reproductive fertility is effective but is currently restricted to a relatively small number of invasive species.

As suggested in the earlier text, a given invasive species will demand its own set of management approaches; a list of the management approaches used for the invasive species discussed in this chapter is presented in Table 21.3. The simultaneous application of several control methods in a program emphasizing sound ecology and economics is called **integrated pest management** (IPM). IPM programs have been developed for several of the invasive species discussed in this chapter. For example, such a program has been developed to control the spread of spongy moth, integrating all seven of the management approaches discussed in the earlier text (Sharov et al. 2002). Other examples of IPM programs include those for hemlock woolly adelgid and emerald ash borer, as well as several invasive plants and tree pathogens (Poland et al. 2021).

NOVEL ECOSYSTEMS AND INVASIVE SPECIES

In many ecosystems, humans have altered species composition, disturbance regimes, and ecosystems processes well beyond their natural range of variability (Chapter 22), and together with climate change, have created an entirely new ecological context in which ecosystems occur. In such cases, non-indigenous, introduced species may become better adapted to the new conditions compared to native species, resulting in new combinations of species. Moreover, the new ecological situation may be extremely difficult to reverse despite intensive management efforts. Such ecosystems are termed **novel ecosystems** (Hobbs et al. 2006), and may include many introduced species that represent the "new normal" in that they develop as unique, long-term stable states that feature the dominance or abundance of one or more invasive species. Because they are unlikely to be restored to a condition resembling that prior to human alteration, management of novel ecosystems may shift toward those aspects that best meet the needs of humans—such as for pollination, recreation, or carbon uptake—rather than toward the most natural condition possible (Hobbs 2007).

Table 21.3 Management approaches employed for the invasive species discussed in this chapter.

Species	Control method for established species						
	Regulatory	Physical	Cultural	Chemical	Biological	Resistance	Reproductive
Plants							
Tree-of-heaven		X		X			
Norway maple		X		X			
Common buckthorn		X		X			
Amur honeysuckle		X		X			
Kudzu		X	X	X			
Asian bittersweet		X		X			
Cheatgrass		X		X			
Cogongrass		X		X			
Insects							
Emerald ash borer	X	X	X	X	X	X	
Beech scale		X	X			X	
Spongy (gypsy) moth	X	X	X	X	X	X	X
Hemlock woolly adelgid	X	X	X	X	X	X	
Pathogens							
Dutch elm disease		X	X	X		X	
Chestnut blight		X				X	
Animals							
Earthworms	X	X		X	X		
Feral swine		X		X			X

Source: Data from Poland et al. (2021).

Several important examples of invasive species already described in this chapter fit the definition of novel ecosystems outlined in the earlier text. For example, ponderosa pine forests heavily invaded by cheatgrass in the western United States are systems that include new combinations of species that have altered fire regimes. The altered fire regimes are likely to perpetuate the plant communities that now include invasive plants at the expense of native species, including ponderosa pine itself. Initiated by human activities, the new ecological conditions are likely to be irreversible, and more frequent and intense fires that accompany cheatgrass invasion are likely to maintain a new stable state of cheatgrass dominance. As a result, management is likely to emphasize land use options viable under the new conditions rather than maintaining some historical benchmark of ecological integrity; in other words, future management may emphasize the *novel* conditions rather than *historical* conditions.

Related to the concept of novel ecosystems is the opinion that the impact, rather than whether a species is native or non-indigenous, should shape how humans value species. For example, Davis et al. (2011) present the example of tamarisk, a shrub genus from Europe, Asia, and Africa, that has invaded riparian areas of the dry American Southwest. Tamarisk has a reputation for high levels of evapotranspiration and thus its ability to alter hydrologic regimes in ecosystems where water is limiting, and has been targeted for removal or suppression since the 1930s. However, tamarisks have recently been documented to have similar rates of water use compared to native shrubs of the region (Stromberg et al. 2009), and today provide the preferred habitat for at least one endangered bird. Thus, despite it being a heavily human-modified riparian system of the arid Southwest, this novel ecosystem, together with its dominant, non-indigenous species, may actually be a desirable element on this landscape that performs valuable ecosystem functions (Aukema et al. 2010).

The concept of novel ecosystems as it pertains to invasive species has not been without controversy. Several ecologists have argued, often confrontationally, that the concept of novel ecosystems is seriously flawed because it enables conservationists and managers to ignore the enormous ecological effects of invasive species that we have discussed in this chapter (Simberloff et al. 2015). Similarly, they argue that equating the ecological value of invasive species to that of native species ignores the impacts of invasives (Simberloff 2015). Moreover, the central concept of novel ecosystems relating to non-indigenous species—that it might not be possible to remove all invasive or non-native species from an ecosystem—is argued by some ecologists to be a dangerous avenue for managers to forgo attempts at invasive species eradication and management when such attempts may be completely feasible (Aronson et al. 2014).

As with all controversies, the most useful answer is probably the one lying squarely in the middle. Most forest ecologists would probably rather see a near-natural forest ecosystem dominated by native species than a highly altered forest dominated by non-native and invasive trees and shrubs. At the same time, many ecologists would rather see a highly altered forest ecosystem on a landscape than to see no wooded areas at all. Perhaps the value of the novel ecosystem concept is that it labels non-natural ecosystems as something other than "degraded," thus providing managers and ecologists a new framework in which to define new objectives or paradigms for a seemingly different, "novel" system that is not likely to be restored to its original composition, structure, or function (Miller and Bestelmeyer 2016). If there is ecological value in such forest systems—and there may not always be—then it is probably beneficial to emphasize that value rather than abandon any concern because of its dissimilarity to pre-European colonization conditions.

SUGGESTED READINGS

Aukema, J.E., McCullough, D.G., Von Holle, B. et al. (2010). Historical accumulation of nonindigenous forest pests in the continental United States. *Bioscience* 60: 886–897.

Daehler, C.C. (2003). Performance comparisons of co-occurring native and alien invasive plants: implications for conservation and restoration. *Ann. Rev. Ecol. Evol. Syst.* 34: 183–211.

Lamarque, L.J., Delzon, S., and Lortie, C.J. (2011). Tree invasions: a comparative test of the dominant hypotheses and functional traits. *Biol. Invasions* 13: 1969–1989.

Liebhold, A.M., Brockerhoff, E.G., Kalisz, S. et al. (2017). Biological invasions in forest ecosystems. *Biol. Invasions* 1911: 3437–3458.

Poland, T.M., Patel-Weynand, T., Finch, D.M. et al. (2021). *Invasive Species in Forests and Rangelands of the United States: A Comprehensive Science Synthesis for the United States Forest Sector*. New York: Springer 500 pp.

Simberloff, D. (2013). *Invasive Species: What Everyone Needs to Know*. Oxford, UK: Oxford University Press 352 pp.

Williamson, M. (1996). *Biological Invasions*. New York: Springer 256 pp.

Forest Landscape Ecology

CHAPTER 22

Landscape ecology is a subdiscipline of ecology that, above all else, emphasizes **spatial heterogeneity** (variation from place to place) and patterns in ecological systems. In most (not all) cases, the heterogeneity of concern is that of vegetation, in part because it is most easily observed. For a long time, much of the field of ecology (including much of forest ecology) ignored heterogeneity and assumed, for the sake of simplicity, everything to be non-spatial. However, we casually see spatial heterogeneity in the natural world every day, by observing that one location is a better place to see wildflowers than another, or perhaps that some tree species are likely to be more productive on one site compared to another. Ecology is concerned with interactions, and it is clear that the strength of interactions changes with distance (i.e., objects closer to each other interact more strongly than those further apart). This is because *distance implies spatial location*, and therefore the patterns we see across landscapes and within ecosystems are likely to be important for ecological processes.

The discipline of landscape ecology is fairly new, having been developed only in the early 1980s, though its broad conception arose much earlier from a combination of geography and plant ecology, especially through the work of the German geographer Carl Troll (1939). Landscape ecology is well known for having two well-defined "schools" depending on where the discipline is centered. In Europe, the field is strongly human-centered (Wu and Hobbs 2007; Brandt et al. 2009) and deals mainly with the "built" environments of cities and densely populated areas of the region. There, landscape ecology focuses on the interrelationship between humans and their open and built-up landscapes (Naveh and Lieberman 1994). Together with biology and ecology, it emphasizes human-centered fields, such as the social, economic, geographic, and cultural sciences connected with modern land uses, and is used as the scientific basis for landscape planning, management, conservation, and restoration. In North America, the concepts of the European school of landscape ecology are often emphasized in the fields of landscape architecture, urban planning, and landscape design.

In North America, the framework for a new discipline in landscape ecology was defined by a small group of ecologists and geographers in 1983. Specifically, the field was defined as the study of relationships between spatial patterns and ecological processes unrestricted by a particular scale (Risser et al. 1984). This definition deviated from that of the European school, probably because of the relatively small number of participants and their particular interests and expertise (With 2019). Its emphasis on spatial patterns and ecological processes centered its stronger focus on natural systems, quantitative analyses using statistics and modeling, and the development of ecological theory. In contrast to a focus in Europe on cultural landscapes, the North American school was also more able to stress management of its still-remaining natural resources and large tracts of unbuilt land. Forest and other natural resource management has played a major role in how landscape ecology has developed as a field in North America over the last four decades (With 2019). As such, landscape ecology is much more commonly taught and studied in biology, ecology, and natural resource departments in North American colleges and universities.

Landscape ecology is commonly considered to be the study of features called **patches** or **elements** situated in a landscape, where a "landscape" occurs at spatial scales broader than

Forest Ecology, Fifth Edition. Daniel M. Kashian, Donald R. Zak, Burton V. Barnes, and Stephen H. Spurr.

individual organisms or ecosystem types. It considers the spatial pattern and heterogeneity of these features, their development and dynamics, exchanges between them, and their management (Turner and Gardner 1991). The focus on broad spatial scales is an artifact of the human-centered focus of the European school, but landscape ecology does not need to concentrate only on large spatial extents. Indeed, technological advances such as satellite and other remotely sensed imagery, the maturity of geographical information systems, and rapid improvements in the speed and capacity of computer processing have allowed ecologists to more easily make observations and collect data over large areas of land (Turner and Gardner 1991). However, landscape ecology does not dictate specific scales that should be studied, instead emphasizing the identification of the scales most appropriate for characterizing the relationships of spatial patterns and the ecological process of interest (Turner and Gardner 2015). For example, With et al. (1999) examined the response of crickets to habitat abundance and fragmentation using varying spatial patterns mowed into 25 m^2 plots in a field, because this was the appropriate "landscape" perceived by the crickets.

Landscape ecology encompasses an enormous diversity of possible topics that cannot possibly be covered well in a single chapter. Therefore, we emphasize some of the most important topics with particular importance to forests, including the ecological importance of landscape fragmentation, disturbances on landscapes, and the effects of landscape pattern on ecological processes. A full treatment of landscape ecology is provided in the definitive texts by Turner and Gardner (2015) and With (2019), as well as works by Wu and Hobbs (2007), Wiens and Moss (2005), Forman (1995), Naveh and Lieberman (1994), and Forman and Godron (1986).

CONCEPTS OF LANDSCAPE ECOLOGY

A **landscape** is something with which most of us have some level of familiarity. As described by Turner and Gardner (2015), we intuitively think of a landscape as a large area that we can observe from a single place at a single point in time. It could be a forested landscape, or a prairie landscape, or an agricultural or urban landscape, all of which contain patches of different types. For example, a forested landscape might include lakes and ponds as well as forests, and the forested areas might include patches of different forest types (e.g., oak forests and beech–sugar maple forests) or forests of different ages. An urban landscape might include patches of industrial areas, residential areas, and green areas such as parks or golf courses. As described in the earlier text, the concept of a landscape could be broadened to include something more than the human perspective, to include patches of things important to a particular organism of interest. Thus, a cricket's landscape might include patches of water, thick vegetation, and open areas, all in a small area as perceived by humans. Therefore, important is the presence of a **mosaic** of different patch types that may interact with each other—the presence of heterogeneity. Although many definitions exist for a landscape (Wiens and Moss 2005), simply defining it as an area exhibiting spatial heterogeneity in at least one factor of interest (Turner and Gardner 2015) seems most appropriate.

All landscapes may be broadly described with a few simple attributes. **Landscape structure** describes both the composition (different kinds or diversity of patches) and the configuration (spatial arrangement of the patches) of the landscape. Composition might include metrics such as the proportion of the landscape represented by each patch type, or the number of different patch types (patch richness). Configuration metrics might include the distribution of sizes of patches of a certain type, the isolation or connectedness of patches on a landscape, or the complexity of patch shapes. Landscape structure represents the patterns that have been quantified by landscape ecologists since the birth of the discipline. For example, Kashian et al. (2017) quantified the spatial pattern of jack pine seedling densities across a recently burned landscape in northern Michigan as an attempt to create a template for managers to mimic post-fire jack pine regeneration in plantations (Figure 22.1). Within the burned area, seedlings varied in density from fewer than 100 seedlings per hectare to more than 5000 per hectare, and were classified into five patch types

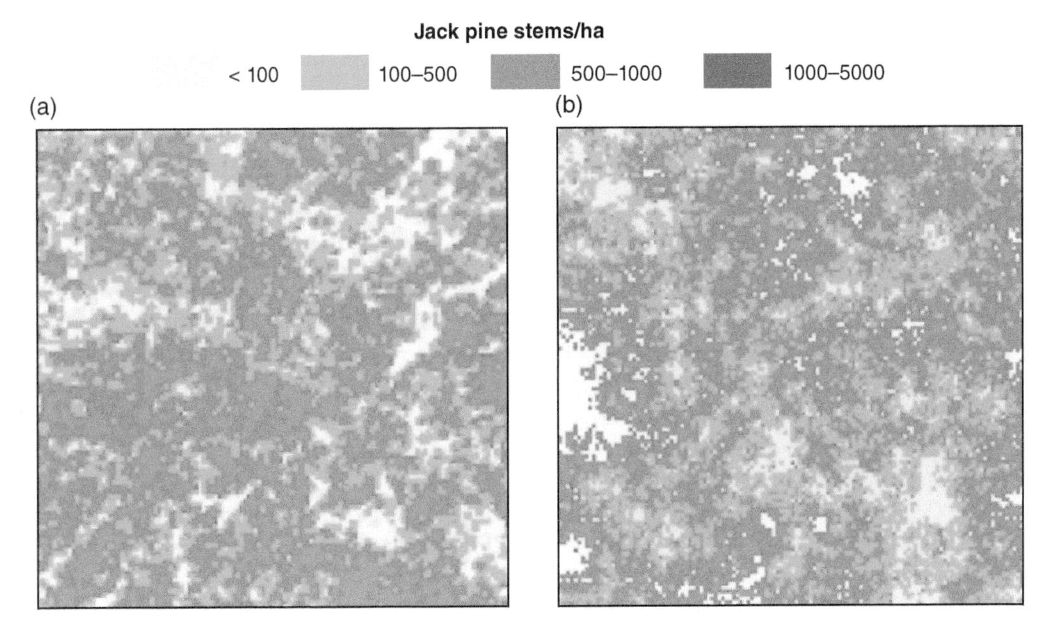

FIGURE 22.1 Characterization of post-fire landscape structure using jack pine seedling densities across a burned landscape in northern Michigan. Seedling densities were classified into five density classes on the burned landscape and the pattern quantified using a few indices of landscape structure. These indices were then used to simulate potential planting configurations that would best emulate the post-fire landscape. (a) Seedling densities on the actual landscape; (b) simulated pattern of seedling densities. Areas shown represent 164-ha regions of the burned area. *Source:* Kashian et al. (2017).

based on their density. Landscape structure was quantified using the proportion of landscape, mean patch area, patch density (number of patches per 100 ha), and distance to the closest patch of the same type for each patch type in the burned area. These indices were then used to simulate similar landscape patterns that could be used as planting configurations that would emulate a naturally regenerated landscape (Figure 22.1).

Landscape ecologists are concerned not just about the patterns present on a landscape, but also about what the patterns mean for ecological processes. **Landscape function** describes how the patches on a landscape interact with each other, often by examining flows of energy, materials, and species within and among them. A familiar example of interacting patches is the concept of *riparian buffers*, which are regions of intact vegetation (often forests) around streams and wetlands that act to increase water quality despite adjacent land uses. On agricultural landscapes, riparian buffers intercept sediment, nutrients, pesticides, and other materials in surface runoff from fields into the stream or wetland, as well as reduce nutrients and other pollutants moving with water flowing just below the soil surface. Riparian buffers are particularly important in removing nitrogen, the excess of which may originate from fertilizers or animal waste on agricultural landscapes and is a pollutant in surface and ground water (Carpenter et al. 1998). Nitrogen removal from water in riparian buffers may occur via uptake by vegetation or microbial activity, storage in the soil, or denitrification, the latter of which represents a loss of nitrogen from the system in gaseous form. Patch size—in this case buffer width—plays an important role in the effectiveness of buffers in removing nitrogen, with wider buffers more effective (Mayer et al. 2007). The spatial arrangement of riparian buffers on a landscape is also important, as maintaining buffers around stream headwaters appears to be most effective in sustaining water quality (Peterson et al. 2001). Thus, the structure of the landscape (the abundance and arrangement of patches of vegetation adjacent to streams) has a clear effect on the ecological process of nitrogen movement across the landscape.

A third major attribute of all landscapes is the process of **landscape change**, which examines how landscape structure and function change over time. Landscape change represents a major component of landscape ecology in forests because it incorporates the temporal effects of disturbances, succession, land use, and management. If the relationship between landscape structure and function is well understood, comparisons of how landscape patterns change over time allow explanations of functional changes as well as predictions of how those functions may change in the future. In a study in the midwestern United States, Radeloff et al. (2005) found that the number of housing units grew by 146% between 1940 and 2000, particularly around the suburbs of major metropolitan areas in the 1950s and 1960s and in rural forested areas separated from metropolitan areas in the 1970s and 1980s (Figure 22.2). Together these trends of landscape change have conservation concern for forests. Suburban sprawl intensively affects smaller forested areas with higher housing density, but rural sprawl probably has a stronger negative impact because it affects larger, less-altered forested areas, albeit at lower intensity. If continued, these changes have potential to impact breeding habitat for bird populations, as well as large mammals. Thus, understanding landscape change in this region provides important opportunities for using ecological principles in land use planning and policy development for growth management.

Landscape ecology can thus be summarized as being concerned with spatial heterogeneity at a variety of scales. The nature of this heterogeneity has specific implications for the way ecological flows of energy, nutrients, and species occur across the landscape, and those flows rely somewhat on how well patches are connected to one another. Finally, landscapes are dynamic, and the heterogeneity they contain is likely to change over time (Wiens 2005). We described the importance of ecological scale in Chapter 2; many of the other aspects of landscape ecology we cover in the following text with a series of examples specific to forests.

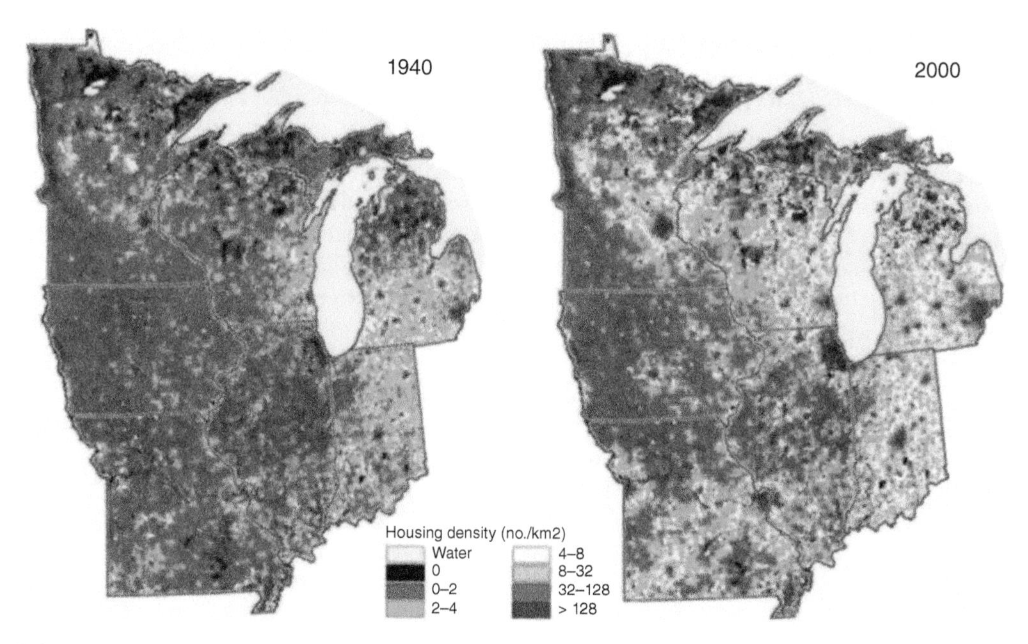

FIGURE 22.2 Housing density in 1940 and 2000 directing landscape change in the midwestern United States. Housing growth occurred both as suburban sprawl along the fringes of metropolitan areas and rural sprawl in previously unaltered forested areas. *Source:* Modified from Radeloff et al. (2005).

FOREST FRAGMENTATION AND CONNECTIVITY

Among the first concerns of landscape ecology was the idea of landscapes changing as a result of human activities. The most concerning of these changes was the idea that human land use such as forest clearing for agriculture, colonization, or forest harvesting was acting to break up large, contiguous tracts of forests with deleterious ecological effects (Harris 1984). Given its occurrence at broad spatial scales and its emphasis on spatial heterogeneity, **forest fragmentation** was a natural focus for landscape ecologists. The classic example of forest fragmentation was shown in a township in southern Wisconsin where land clearing and farming created an increasingly fragmented forest from 1882 (when only 30% of the original forest remained) to 1950 (Figure 22.3). By 1954,

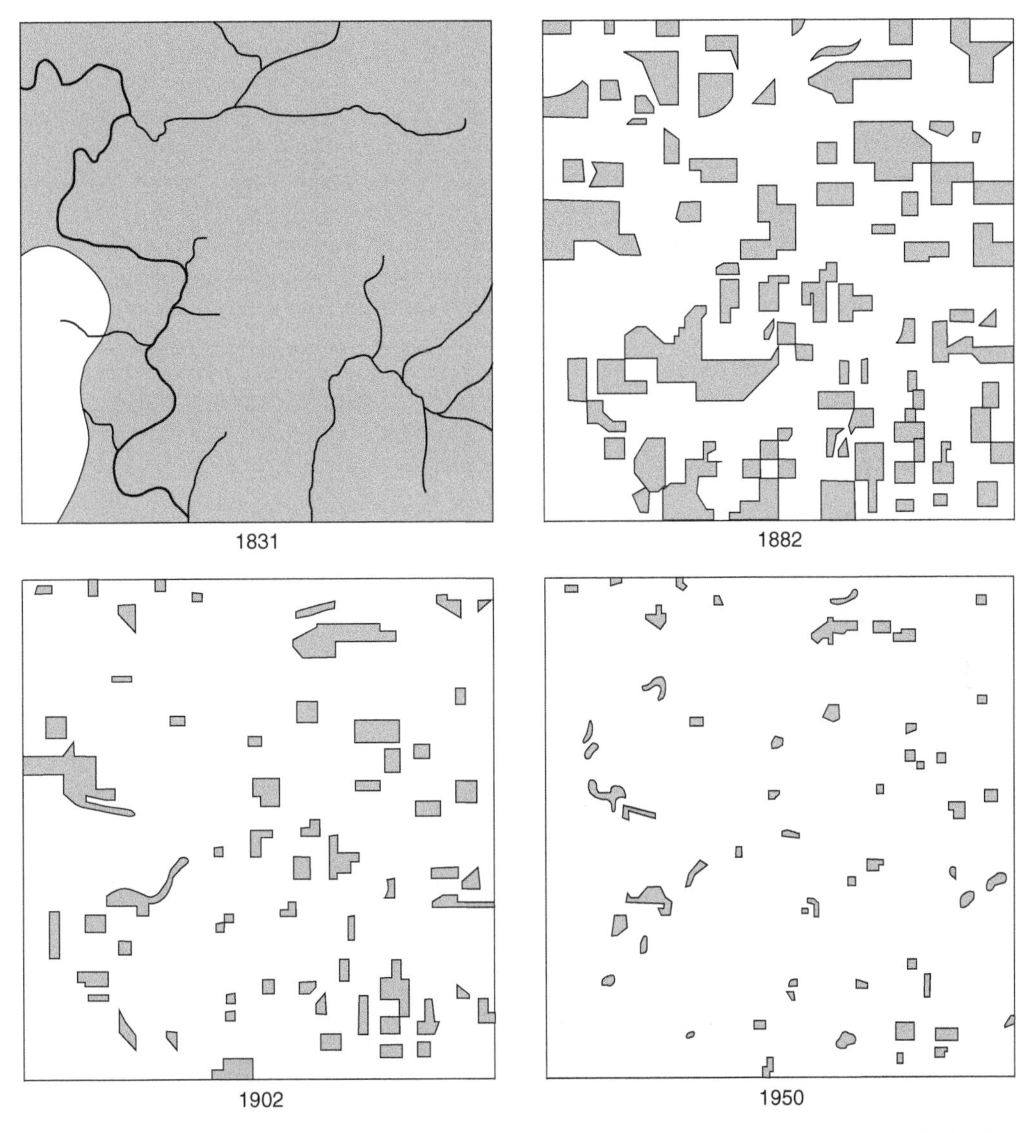

FIGURE 22.3 Fragmentation of a forested area of Cadiz Township, Green County, Wisconsin, during the period of European colonization. The township is 6 miles on a side. The shaded areas represent land remaining in, or reverting to, forest in 1882, 1902, and 1950. *Source:* Curtis (1956) / with permission University of Chicago Press.

Curtis (1956) found only 3.6% of the original forest remained, and 77% of what remained was so heavily grazed that tree regeneration was absent. The forest fragments are therefore functionally isolated from other and larger areas of forest. Much of our understanding of forest fragmentation and its ecological effects is based in patch-based ecological theory, as described in the following text.

PATCHES IN FOREST ECOLOGY

Many aspects of landscape ecology today are based on ecologists' conceptualization of landscapes as a mosaic of patches of habitats, ecosystems, or cover types. The patch–corridor–matrix model, as described by Forman (1995), suggests that the extent and configuration of patches on a landscape is what defines landscape pattern. As described in the earlier text, patches are to some extent human constructs and are defined based on the phenomenon under consideration, such as forest ages, forest types, or vegetation type (e.g., forest vs. field). Patches are considered to be distinct areas at a given point in time, with boundaries different enough from what surrounds them that they can be identified as a relatively homogenous area at a given spatial scale. **Corridors** are narrow, linear patches of a patch type that differs from those on either side (Forman 1995) and typically connect similar patch types on a landscape. Corridors may function as breeding habitat for organisms that enables gene flow across a landscape, may allow dispersal or migration across a landscape, or may act as a barrier or a filter to the movement of organisms. The most common or connected landscape element is known as the **matrix**, which is often considered to be an unusable space to the organisms or phenomenon of interest. Thus, the matrix in Figure 22.3 is non-forested area beginning in 1882.

Patch-based ecological theory is heavily influenced by the theories of island biogeography (MacArthur and Wilson 1967) and metapopulations (Levins 1969, 1970). **Island biogeography theory** was originally developed to explain community dynamics on oceanic islands and suggests that the number of species on island communities is affected by the size of the island and its distance from the mainland from which organisms disperse. Larger islands support more species because they have more complex habitat and large populations less threatened by extinction. In addition, islands further from the mainland support fewer species because they are colonized less frequently. Therefore, large islands closest to the mainland are likely to have the most species, while small islands far from the mainland are likely to have the fewest species. Island biogeography theory is rather easy to transpose onto fragmented landscapes, allowing assumptions about biodiversity in small, isolated patches of forest compared to larger, well-connected fragments.

Metapopulation theory adds an additional layer of complexity to patch-based theory. Metapopulations include groups of populations that are spatially separated, as they would be when occupying fragments of habitat on an altered landscape. Though separated in space, the populations interact through dispersal among the patches, forming a "population of populations" (Levins 1969). Metapopulation theory predicts that a specific population in the metapopulation may go extinct, leaving some patches on the landscape suitable but unoccupied, but the patch will eventually be re-colonized by immigration from an occupied patch. Thus, the metapopulation as a whole remains stable even if some of the populations within it do not.

Patch-based theory is one of the foundational aspects of forest landscape ecology and still permeates the field today. In general, most forest ecologists have realized that the natural world is not simply a series of patches surrounded by a featureless, unhabitable matrix. In such a scenario, every species—including trees and other plants—would "perceive" a very different landscape, making management or conservation nearly impossible. Instead, the matrix itself is made up of a mosaic of different patch types that may be suitable for dispersal or other activities of life history but not for breeding (Figure 22.4). As such, the matrix surrounding patches may not be completely analogous to an inhospitable ocean surrounding islands. Forest fragmentation such as that shown in Figure 22.3 may indeed create functionally isolated forests and woodlots for trees because the

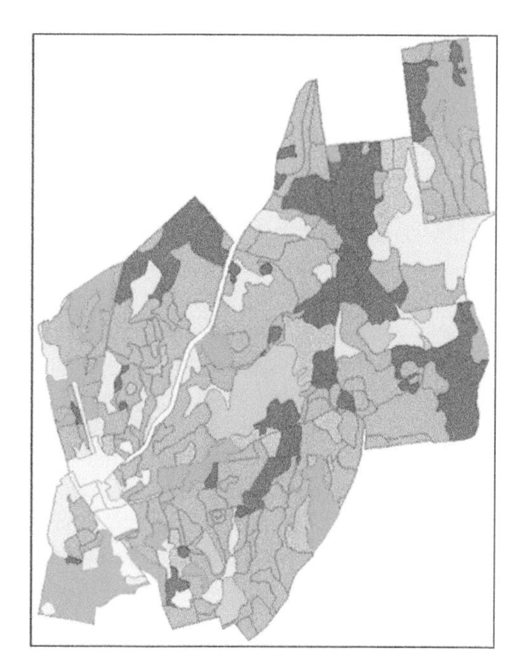

FIGURE 22.4 Distribution of eastern white pine on a portion of Harvard Forest, Massachusetts. White pine forest is highly fragmented when only the white pine patch type and the matrix are considered (left), but in reality the matrix forms a complex mosaic (right) of patches of other conifers (pink), mixed conifer stands (lighter greens), open areas (yellow), hardwood stands (orange), mixed hardwood and conifer (brown), and wetlands (blue).

nature of the matrix may prevent dispersal, regeneration, and/or establishment. We cannot assume, however, that some organisms, particularly animals, may not be able to traverse the matrix even as the patches within it become smaller and more isolated. This **landscape-mosaic** view is probably a more accurate representation of the natural world than that of the patch–corridor–matrix view, and its use for understanding and characterizing landscapes is at the forefront of forest landscape ecology.

FOREST FRAGMENTATION

As described in the earlier text, human activities have reduced total forest area and have fragmented remaining forests into progressively smaller patches isolated by adjacent plantations, roads, or agricultural and urban development (Harris 1984; Saunders et al. 1991). Forested land in the 48 contiguous United States is estimated to have occupied 400 million ha in the 1500s (Harrington 1991), and it was reduced by about 53% to approximately 188 million ha by the 1920s. By the early 1990s only 3–5% of pristine old-growth forest remained, principally in the Pacific Northwest (Miller 1992). Forest fragmentation is widespread in the eastern United States, where remaining forests date only from the 1920s and 1930s (Rudis 1995; Vogelmann 1995) and small, isolated woodlots interspersed in farmlands and suburbia comprise about 40% of the deciduous forest (Terborgh 1989). We discuss in the following text the ecological effects of fragmentation (and associated forest loss).

Ecological Effects Forests have always been naturally fragmented by large disturbances even prior to European colonization (Chapter 16). Disturbances such as fires, windstorms, floods, and avalanches characteristic of regional and local ecosystems created openings of various sizes on the landscape, providing the heterogeneity necessary for a diverse array of animals and plants. Pattern on modern landscapes differs from that prior to European colonization, however, at least in part because changes in land use, conversion of cover types, or other drivers of current fragmentation often prevent or reduce regeneration in disturbed patches.

An important consideration regarding forest fragmentation is that many of its effects are actually due to the loss of forest cover rather than simply its subdivision into relatively small, isolated patches (Fahrig 2003). Most landscape ecologists agree that fragmentation affects forests on a landscape by (i) reducing the amount of forest, (ii) increasing the number of forest patches, (iii) reducing the size of the forest patches, and (iv) increasing the isolation of the forest patches. However, these aspects are not independent of each other. For example, removal of some patches via forest loss would reduce the number of patches but increase the isolation of those that remain (Figure 22.5a), but subdivision of patches would increase the number of patches and reduce their size while decreasing their isolation (Figure 22.5b). Thus, forest fragmentation and loss are closely related to each other but are not equivalent. An increase in the number of patches on a landscape is clearly caused by fragmentation rather than loss (With 2019), but both patch size and isolation may be at least as well explained by loss as by fragmentation (Bender et al. 2003; Fahrig 2003). The important assertion, therefore, is that forest fragmentation changes the function of the patches of forest that remain as well as decreasing their size and increasing their isolation (van den Berg et al. 2001).

Fragmentation effects may be classified into those resulting from patch size and isolation, and those resulting from edge effects. Predictions about the effects of decreasing patch size on the diversity and abundance of species (most often animals) lean heavily on island biogeography theory: larger patches are expected to favor larger populations compared to smaller patches, and thus

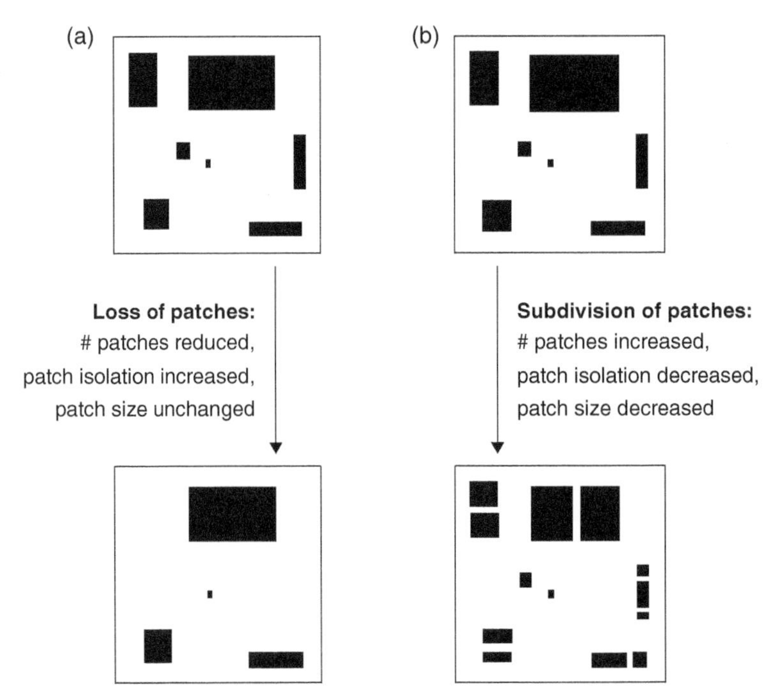

FIGURE 22.5 Potential effects of forest loss on landscape pattern. Fragmentation is expected to increase the number of small, isolated patches as it proceeds on a landscape, but these effects are not always consistent and some may be achieved by forest loss alone. (a) Forest loss may increase patch isolation yet reduce the number of patches, and (b) subdivision may increase the number of patches and patch size yet decrease patch isolation.

extinction rates are expected to be lower in larger patches. Most species have a minimum patch size requirement (Diaz et al. 2000), and thus fragmentation of forests may result in patches too small to provide adequate heterogeneity for territory size, food supply (Whitcomb et al. 1981), or other required features such as streams or wetlands. As described in the earlier text, patch isolation reflects the amount of forest in the landscape more so than its fragmentation, but isolated patches are less likely to be recolonized if a population within them goes extinct, depending on how well dispersers are able to move through the matrix among the patches (Ricketts 2001). For example, small mammals are more likely to move among forest patches through open habitat if the patches are close together and exposure to predation in the open areas will be limited.

The most unambiguous indicator of fragmentation is the amount of edge that is increased as the process takes place. The amount of edge increases proportionately as patch size decreases, often represented as a ratio of the length of the patch perimeter to the patch area. **Edge effects** are the ecological consequences of sharp boundaries between two patch types, and they tend to be stronger than effects of patch area (With 2019). In forests, edge effects are often represented by adjacent forests and disturbed areas or openings such as those created by harvesting; edges thus ring the perimeter of each forest patch. The forest edge is characterized by relatively sharp boundaries with microclimatic, vegetational, and biotic conditions markedly different than in the interior of the forest. In a series of studies of clear-cut harvesting in old-growth Douglas-fir forests of the Pacific Northwest, Chen et al. (1992, 1995) found warmer air and soil temperatures during the day and cooler temperatures at night at the edge compared to the interior forest, as well as lower relative humidity, higher short-wave radiation, and exponentially higher wind speed. These effects extended 30 to over 240 m into the forest from the edge. Although not tested explicitly for an edge effect, canopy cover, basal area, and trees per unit area decreased in the forest nearest the edge, dominant tree growth increased, tree mortality increased, and tree regeneration was impacted in a species-specific way.

The abrupt, human-created sharp boundaries along forest edges differ from those to which many forest species have adapted. Edges favor species adapted to them and select against those requiring interior conditions (Whitcomb et al. 1981; Yahner 1988, 1995; Saunders et al. 1991), potentially changing species richness. Such edge effects are evident with many forest animals but are particularly obvious with birds. **Edge species** of birds that forage and breed near forest edges are relatively independent of forest patch size and may be unaffected or even positively affected by fragmentation. By contrast, **forest interior species**—those that depend on large forest tracts and avoid edges and their effects—may decline in numbers with fragmentation. For example, Robbins et al. (1989) found that encountering a breeding pair of an area-dependent bird species such as the scarlet tanager is much more likely (>70% probability) in a 100-ha forest compared to a tract smaller than 10 ha (<50%).

Edge effects also favor certain edge-dwelling game species (white-tailed deer, red fox), raptors (great horned owl, red-tailed hawk), and songbirds (gray catbird, brown-headed cowbird; Yahner 1995). However, edges clearly have a negative effect on many other fauna and flora such as neotropical migrant birds, often because of higher predation and parasitism in edge habitat (Wilcove et al. 1986; Robinson et al. 1995; Yahner 1995). For example, nest predation and parasitism by the brown-headed cowbird increased with forest fragmentation in nine landscapes in Missouri, Indiana, Illinois, and Wisconsin. Cowbirds are nest parasites that lay their eggs in the nests of other "host" species, which then raise cowbirds at the expense of their own. Cowbird parasitism significantly increases with decreasing amount of forest cover. In some cases, nest parasitism so decreased reproductive success for some host species that they required immigration from source populations in heavily forested, less-fragmented landscapes for persistence (Gustafson and Crow 1994).

A classic and severe example of forest fragmentation occurs in Florida, in large part due to the influence of highways. Florida's native species evolved in a setting surrounded by the sea and not in the presence of high-density, high-speed traffic (Harris and Silva-Lopez 1992). Hard-surface highways in Florida were built at a rate of over 6 km/day between the 1940s and 1990s to accommodate human population growth rate and tourism. In this case, highways act as barrier corridors to animal movement, and automobile–animal collisions are the primary source of mortality for most of Florida's endangered large wildlife species, including the panther, black bear, key deer, and American crocodile. Harris and Silva-Lopez (1992) conclude that roads are perhaps the most important driver of forest fragmentation in the United States.

CONNECTIVITY

An important aspect of fragmentation is **connectivity**, which measures how well the structure of a landscape enables or hinders movement or flows among its patches. Connectivity provides an explicit measure of landscape function, in that it describes how pattern influences ecological processes on a landscape (With 2019). As it relates to fragmentation, connectivity is generally considered to mitigate patch isolation, even in a subdivided landscape, if movement among the patches is possible. The connections among patches on a landscape are typically *functional*, and as with many aspects of landscape ecology, what constitutes a functional connection between patches depends on the application or process of interest. Patches that are connected for birds, for example, may not be connected for hydrologic flow. Connectivity might be defined for one species or process as strict adjacency (i.e., patches that are touching or physically connected by a corridor), but by a threshold distance (such as a maximum dispersal distance) for another. For example, in a study of seed dispersal of hardwood species among forest fragments in southern Ontario, breaks in forest cover of 15–30 m, composed of open fields or roads, were not sufficient enough to limit seed dispersal among forest patches (Hewitt and Kellman 2002). Therefore, the landscape remained connected even when fragmented, presumably because the movements of birds and rodents were not restricted. Small patches of habitat that are physically isolated but within the threshold distance may exhibit connectivity by acting as stepping-stones for species' movement across a landscape that appears highly fragmented. Simulation modeling has shown that a landscape generally must be about 30% occupied by a given patch type for that type to be functionally connected (Fahrig 2003).

Examples of connectivity affecting the movement of organisms are many. A study examining the colonization of planted windbreaks in agricultural landscapes in Costa Rica found that forest connectivity increased tree recruitment (Harvey 2000). When windbreaks were connected to intact forests, tree seedlings were greater per unit area and tree species richness was higher compared to windbreaks not connected to forests. In particular, bird-dispersed tree species were more common and abundant in connected windbreaks because bird species were more likely to move through forested areas than open areas, and dispersal was therefore more likely in the connected windbreaks (Harvey 2000). D'Eon et al. (2002) examined connectivity between patches of old-growth forest, recent harvest, and recent wildfire in British Columbia as it varies among organisms' ability to move freely across its range. As described earlier in this chapter, old-growth patches were increasingly connected as the amount of total old-growth increased, but decreased where harvesting was more common. Carnivorous bird species, which were most able to move freely, perceived the landscape as connected and were able to access all patches, but smaller, less mobile species appeared to perceive a lack of connectivity. The degree of connectivity on landscapes has also been proposed to have importance for the ability of species to shift their ranges under changing climates. Fragmented landscapes are likely to slow but not prevent range shifts unless connectivity is reduced below the level at which metapopulations may persist (Opdam and Wascher 2004), although some empirical studies have suggested that connectivity is only one of several factors that may affect changes in species' distributions (e.g., Melles et al. 2011).

Although most often considered for its effects on organisms' ability to move across a landscape, connectivity also has important implications for ecological flows and processes. Wildfire spread and behavior are perhaps the most obvious examples of how ecological processes may be influenced by forest connectivity. The spatial arrangement of fuels on a landscape is a major driver of how fires spread, at least when weather conditions are not severe enough to override its importance (Turner and Romme 1994; Turner et al. 1994). Miller and Urban (2000) used a simulation model to examine the interaction between surface fire frequency and the connectivity of burnable area for forests in the Sierra Nevada in California. Connectivity of fuels was higher where fire was less frequent, probably because fuel is able to accumulate during longer fire intervals. However, fuel moisture also plays a critical role in fuel connectivity. Surface fuels become very connected when fuel moisture is very low and very disconnected when it is very high (Figure 22.6). Thus, the

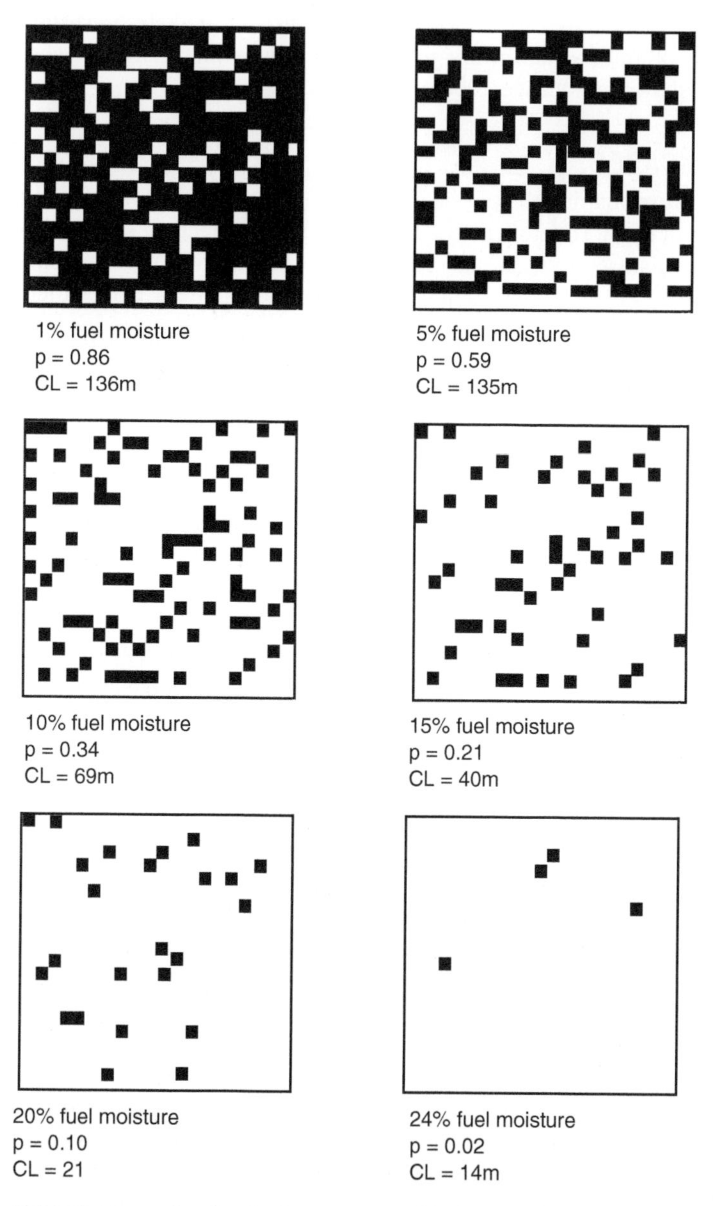

FIGURE 22.6 Simulated effects of fuel moisture on surface fuel connectivity in the Sierra Nevada of California. The proportion of the map that is burnable (p) and the correlation length (CL) are shown for each map. Correlation length is an index of connectivity that reports the average distance a fire can spread without leaving a patch of burnable area (higher correlation length suggests higher connectivity). Surface fuels become far more connected as fuel moisture decreases. *Source:* Miller and Urban (2000) Miller and Urban, 2000 / with permission Springer Nature.

spatial arrangement of fuels probably determines fire behavior and spread prior to fire suppression at intermediate fuel moisture levels. Fuel connectivity likely has no effect during wet periods or during droughts (Miller and Urban 2000). Fire suppression likely increases fuel connectivity by allowing fuel loads to accumulate and may result in more widespread fires.

Forest fragmentation is one result of a suite of human activities that have intensified concerns of ecologists and the public at large over diminishing forest lands, declining native plant and animal diversity, concomitant increases in exotic and invasive species, and sustaining natural ecological processes. Our understanding of forest fragmentation effects has led to an examination of forest and resource management practices and basic changes in policies and practices to sustain ecosystems. Detailed treatments of fragmentation are provided by Lindenmayer and Fischer (2013), Collinge (2009), and Bradshaw and Marquet (2002).

DISTURBANCES ON LANDSCAPES

Disturbances have been significant in shaping the physical features of landscape ecosystems and are instrumental in the evolution of biota. Heterogeneity on landscapes may affect the susceptibility and spread of disturbances, but disturbances may also act as an important driver of heterogeneity. The fact that forest disturbances both respond to and create landscape pattern makes them conventional objects of study in landscape ecology (Turner and Gardner 2015). We examine both of these interactions in the following text using examples of major disturbance events.

EFFECTS OF HETEROGENEITY ON DISTURBANCES

We discussed in Chapter 16 that specific regional ecosystems and certain local ecosystems are more vulnerable to some types of disturbances because of the unique combinations of characteristic physical site factors and biota. The susceptibility of certain ecosystems to disturbances can be assessed by examining the likelihood that the disturbance will occur, typically using historical occurrences of the disturbance to calculate a probability. When the historical record of disturbances is long, it is possible to identify the regional ecosystems most likely to experience a given disturbance as well as the physiographic sites within the regional ecosystem where disturbance is most common. Long-term records of disturbances in forests in New England have shown differential disturbance effects that correspond to these sites.

Hurricanes in New England The forests of central New England have been repeatedly affected by many types of disturbances, but the ecological effects of hurricanes have been most influential and studied at several spatial scales (Foster 1988a, b; Boose et al. 1994; Foster and Boose 1992, 1995). Hurricanes have historically struck southern and central New England every 20–40 years, with the most destructive storms in 1625, 1788, 1815, and 1938 moving over fairly generalized pathways (Figure 22.7). The 1938 hurricane included winds exceeding 200 km h^{-1} and affected forests along a 100-km^2 path, leaving a mosaic of differentially damaged forest controlled by physiography, wind direction, and the nature of the pre-hurricane vegetation. Southeastern winds were the strongest, and the most severe damage to forests occurred on south- to east-facing slopes and the northwestern shores of lakes. By contrast, damage was far less to forests in the lee of broad hills (northwest exposure), on the southeast and east shores of lakes, or in valleys. The severity of vegetation damage was related to physiography and forest type. For example, damage was most severe where exposure was greatest, such as immediately adjacent to boundaries of a pond and a large field (Figure 22.8).

Regionally, damage severity was also related to historical factors, such as the time since the last major storm and the previous pattern of human clearing and agriculture which, in turn, affected the heterogeneity of the pre-hurricane vegetation. Landscapes of central New England were especially prone to disturbance in 1938, because 30- to 100-year-old white pines dominated old fields and

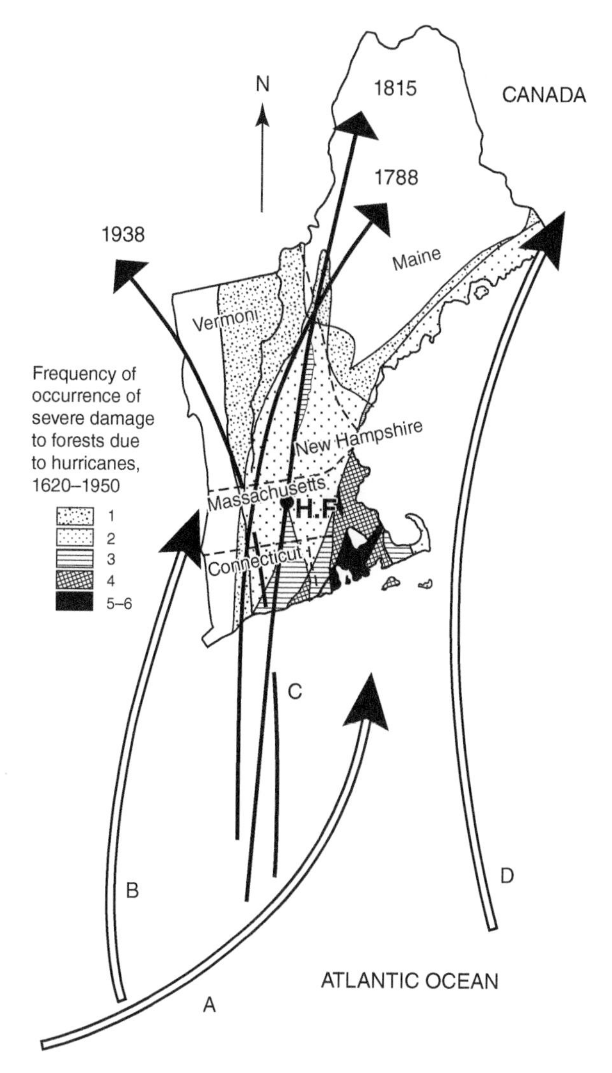

FIGURE 22.7 Four generalized pathways (A–D) of hurricanes in the New England region superimposed over a map of the frequency of occurrence of the major hurricanes from 1620 to 1950. In addition, the historical tracks of the hurricanes of 1788, 1815, and 1938 are shown. Hurricane damage ratings range from: 1 = light to 5–6 = severe. HF = location of the Harvard Forest, Petersham, MA. *Source:* Foster and Boose (1992) and Whitney (1994). Reprinted from the *Journal of Ecology* with permission of the British Ecological Society.

were highly susceptible to wind damage. A similar hurricane occurred in 1815 but caused much less damage because it occurred at the height of the agricultural period prior to reforestation. A legacy of the 1938 storm was that it destroyed older conifers and tall pioneer species, leaving more storm-resistant hardwoods and young understory conifers (Figure 22.9).

Despite an expectation that the 1938 hurricane would create large openings across a region, whereas smaller storms create small gaps, the hurricane produced a continuum of damage from individual trees to uprooting of entire stands. Entire windthrown stands are spectacular, but the hurricane actually created a fine-grained heterogeneous mosaic of damage, with most patches smaller than 2 ha (Foster and Boose 1992). This heterogeneity was due in part to broadscale physiography, land-use history, and the interrelated factors of stand structure and species composition. Stand-scale damage was closely related to topography and increasing stand age and height. Higher

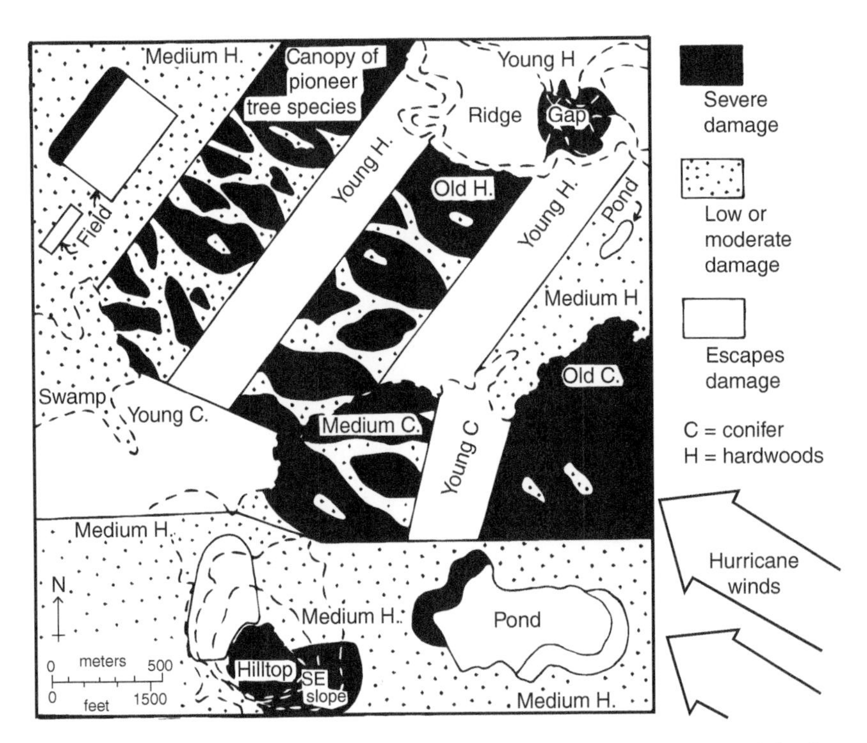

FIGURE 22.8 Diagram of post-hurricane damage classes from a characteristic landscape in southern New England. Young, medium-age, and old stands of conifers (C) or hardwoods (H) represent a hypothetical pre-hurricane vegetation mosaic. Conifer = white pine; hardwoods = oaks, red maple, birch, beech, hickories (and hemlock); pioneers = pines, aspens, white birch. Predicted damage distribution is primarily based on studies of Foster (1988a,b) and Foster and Boose (1992). *Source:* After Forman (1995) / Cambridge University Press.

Photo by Stephen H. Spurr. Reprinted courtesy of the Harvard Forest, Harvard University

FIGURE 22.9 The Harvard Tract of the Pisgah National Forest in 1942, 4 years after the 1938 hurricane.

stand density also reduced damage, apparently because stand opening increased air turbulence during the storm. White-pine-dominated conifer forests were significantly more susceptible than hardwood forests, as were fast-growing species that tended to occupy dominant canopy positions such as red pine, aspens, and white birch. Slower-growing oaks, hickories, red maple, and hemlock occupied subordinate canopy positions and were more protected from wind. The 1938 hurricane was also notable for the catastrophic uprooting of trees, especially white pine, attributed to soil saturation and reduced rooting strength resulting from the 15–35 cm of rain that preceded the storm.

Although the damage to forests caused by hurricanes in New England is intimately related to exposure as mediated by physiography and associated ecosystem characters, not all disturbances are influenced by landscape pattern. For example, in the Upper Peninsula of Michigan, topographic relief is far more subtle than in New England, and most widespread wind damage in this region is caused by vertical downbursts from thunderstorms rather than horizontal winds from hurricanes. In studying this system, Frelich and Lorimer (1991) were unable to find any effect of physiography (including slope position, aspect, or slope inclination) on patterns of wind disturbance, reflecting stark differences in regional weather patterns and broadscale physiography compared to hurricane-prone areas. Likewise, Hurricane Katrina struck the forests of the central Gulf Coast in southern Mississippi in 2005, an area also lacking appreciable topography. Stand age rather than physiographic characteristics best predicted damage severity in forests affected by this storm; aspect reflects gently sloping topography and was only marginally important in predicting damage (Kupfer et al. 2008). These studies reiterate that although landscape pattern may affect disturbances, these ecological effects are firmly rooted in the regional and local landscape ecosystems in which they occur.

Landscape Pattern Effects on Disturbance Spread The spread of a disturbance across the landscape may also be affected by heterogeneity, although studies are somewhat inconsistent in showing this relationship. Generally, disturbances spreading within a cover type are dampened when the landscape is more heterogeneous, and disturbances spreading among cover types are enhanced (Turner et al. 1989). For example, many coniferous forests in western North America were impacted by widespread and severe outbreaks of bark beetles in the early 2000s. Given that most bark beetles are host-tree-specific, outbreak spread generally occurs within a cover type. Though many factors govern the occurrence of outbreaks, a lack of heterogeneity in forest type and age likely enhanced the spread of outbreaks, especially where forests were widespread and well connected (Raffa et al. 2008). A study of the astonishing mountain pine beetle outbreaks in western Canada in the early 2000s found that almost 70% of the forest area dominated by the beetle's primary host, lodgepole pine, had simultaneously reached the preferred age and size for beetle attack (Taylor and Carroll 2004). This extremely low landscape heterogeneity clearly translated to high susceptibility to this disturbance.

By contrast, disturbances that spread among multiple cover types may be enhanced by heterogeneity. In studying outbreaks of the forest tent caterpillar in northern Ontario, Roland (1993) found that increasing amounts of edge on a landscape had strong effects on outbreak duration, and fragmented landscapes in general also experienced longer outbreak duration. Fragmentation may favor the caterpillar by restricting dispersal of the pathogens that limit its populations, and the warmer microclimate of edge habitat may favor caterpillar development (Roland 1993). In this case, heterogeneity represented by fragmentation enhanced the spread of the outbreak as it moved between forested and non-forested patches.

EFFECTS OF DISTURBANCES ON HETEROGENEITY

In describing how landscape heterogeneity affected disturbances in New England, we referenced the idea that the 1938 hurricane created a fine-grained mosaic of damage severities to forests. This mosaic of disturbance severities imparts heterogeneity in vegetation structure and composition as the forests recover. This heterogeneity is most obvious following large disturbances, and Foster et al. (1998) compared the patterns left by a diverse series of five large disturbance events in the 1980s and 1990s (Figure 22.10). Hurricane events in New England, as described in the earlier text, create a fine-grained mosaic of damage severities that feature uprooted or broken trees, damaged trees left alive, and undamaged forests, all varying in structural characteristics. In contrast to hurricanes, wind-related damage patterns to forests due to *tornadoes* are very complex due to their extreme variability and the potential for multiple touchdowns. An F4 tornado in the Tionesta Scenic Area in northwestern Pennsylvania in 1985 created a path of damage 19 km long and 1 km wide, with extremely sharp boundaries between damaged and undamaged areas but little influence of physiography on the patterns of damage severity (Figure 22.10).

Flooding of riparian areas occurs regularly, but extreme flooding such as the 1993 Mississippi River flood was notable for its effects on landscape pattern. Long periods of inundation in lower-lying areas of the floodplain, as well as flooding of areas of floodplain occupied by less flood-tolerant species, caused widespread tree mortality across the region. Landscape patterns linked to exceptional flooding events are therefore strongly influenced by the topography of river floodplains. Landscape patterns created by *volcanic eruptions*, such as that of Mount St. Helens in 1980, are often a combination of broad area of damage with an epicenter in the blast zone and "fingers" of damaged areas resulting from lava or pyroclastic flows (Figure 22.10). Forests within the blast zone at Mount St. Helens were subject to an explosive blast, a thermal wave, and the deposition of rock and ash. Landscape patterns were very complex within the blast zone due to influences of pre-eruption vegetation patterns, physiography, and multiple disturbances, and the degree of heterogeneity is likely to differ depending on the season of eruptions (Foster et al. 1998). Finally, *large wildfires* such as those that burned in Yellowstone National Park in 1988 very clearly create heterogeneity across the landscape as variability in forest age, structure, or species

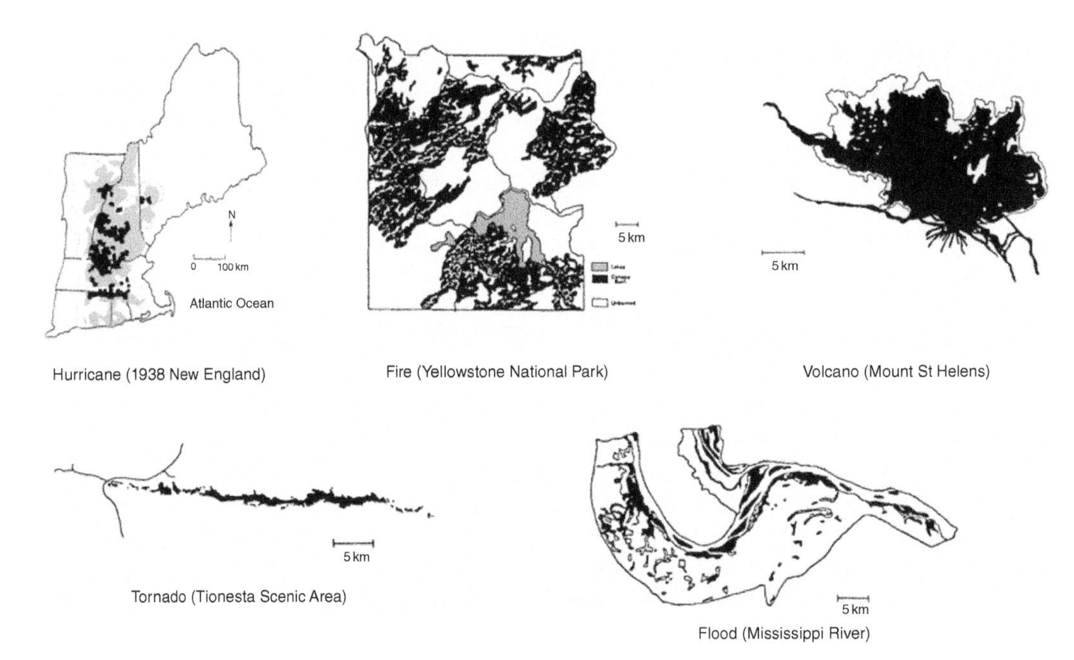

FIGURE 22.10 Comparison of landscape patterns of five contrasting large disturbances in forests across North America. For each disturbance, high-severity damage is shown in black, and lesser damage in gray. Disturbances include the 1938 hurricane in New England, the 1988 fires in Yellowstone National Park, the 1980 eruption of Mount St. Helens, an F4 tornado at the Tionesta Scenic Area in Pennsylvania in 1985, and the 1993 Mississippi River floods. *Source:* From Foster et al. (1998) / Springer Nature.

composition. Next, we use Yellowstone as an ideal example to illustrate the role of wildfires in creating significant landscape pattern.

Stand-Replacing Wildfires in Yellowstone National Park Yellowstone National Park is a large wilderness area in the central Rocky Mountains ideal for examining landscape changes that occurred for millennia prior to European colonization. Romme (1982) and Romme and Knight (1982) studied the fire history of lodgepole-pine-dominated forests in the 73-km^2 Little Firehole River watershed prior to the extraordinary Yellowstone fires of 1988. Fire history during a 350-year period was determined using tree-ring and fire-scar analyses (Chapter 10). Major wildfires in 1739, 1755, and 1795 burned more than half of the watershed. Large, stand-replacing fires (>1000 ha) were generally absent until 1988; only three fires >4 ha burned from 1795 to 1978, all of which were <100 ha. Fire suppression was not effective until the late 1940s, so the absence of large fires was largely due to lack of suitable fuel conditions over most of the watershed. Large fires occur infrequently in Yellowstone (every 100–300 years) because of very slow forest regrowth and fuel accumulation after fire (Romme 1982). Subalpine plateaus feature such large, infrequent fires in part because physical site factors (geology, physiography, and soils) and vegetation are relatively homogeneous over much of the area; large, contiguous forests over broad areas grow and accumulate fuel at approximately at the same rates; and fire spreads easily because the plateau topography has low relief and few natural barriers.

Romme and Knight (1982) identified three main stages of succession following fire on upland sites (early, 0–40 years; middle, 40–250 years; late, >250 years). The even-aged pine canopy that developed after fire deteriorates as the stands age and is replaced by trees able to survive in the understory and seed into gaps. Most of the subalpine plateaus in Yellowstone are dry and dominated by lodgepole pine, but subalpine fir and Engelmann spruce are common and may co-dominate with pines on occasional moist sites, particularly in older forests. This relatively homogeneous site-vegetation pattern contrasts with the diverse mosaic of physiography, human-use history, and disturbance regime of central New England. As a result, the forest landscape mosaic in Yellowstone is largely based on forest age and structure rather than species composition. Large fires in 1739, 1755, and 1795 reduced the high proportion of old-growth forest present in the early 1700s and replaced it with early-successional forests (Figure 22.11). As the forests burned by these fires aged, middle-successional stages became most abundant around 1800 and gradually transitioned to late-successional forest.

After creating a map of the mosaic resulting from multiple fires over 200 years, Romme (1982) applied species diversity indices (Chapter 14) to the mosaic to characterize landscape diversity from 1778 to 1978. Changes in landscape diversity were quantified using (i) a weighted average of richness (number of stand ages), evenness (proportion of the landscape occupied by each stand age), and patchiness; and (ii) the Shannon–Wiener index. Both measures show that the large fires of 1739, 1755, and 1795 created high diversity in the late 1700s and early 1800s, and a lack of large fires in the late 1800s during a 70-year period with no major fires resulted in low diversity (Figure 22.12). Two small fires and a series of mountain pine beetle outbreaks in the twentieth century led to differential rates of succession that increased landscape diversity (Romme 1982). These differences in landscape diversity over time in turn had important effects on bird and ungulate populations, which are described in detail by Romme and Knight (1982).

To this point, we have discussed heterogeneity in Yellowstone as having resulted from a series of stand-replacing fires over several centuries. As described in the earlier text, single disturbance events also may create complex landscape patterns, particularly if they affect broad spatial extents (Foster et al. 1998), as illustrated by the extensive Yellowstone fires of 1988. As described in the earlier text, very large, stand-replacing fires were a dominant factor on the Yellowstone landscape prior to European colonization, with such fires occurring at intervals of 100–300 years (Romme and Despain 1989). Due to prolonged, extreme drought and high winds that occurred

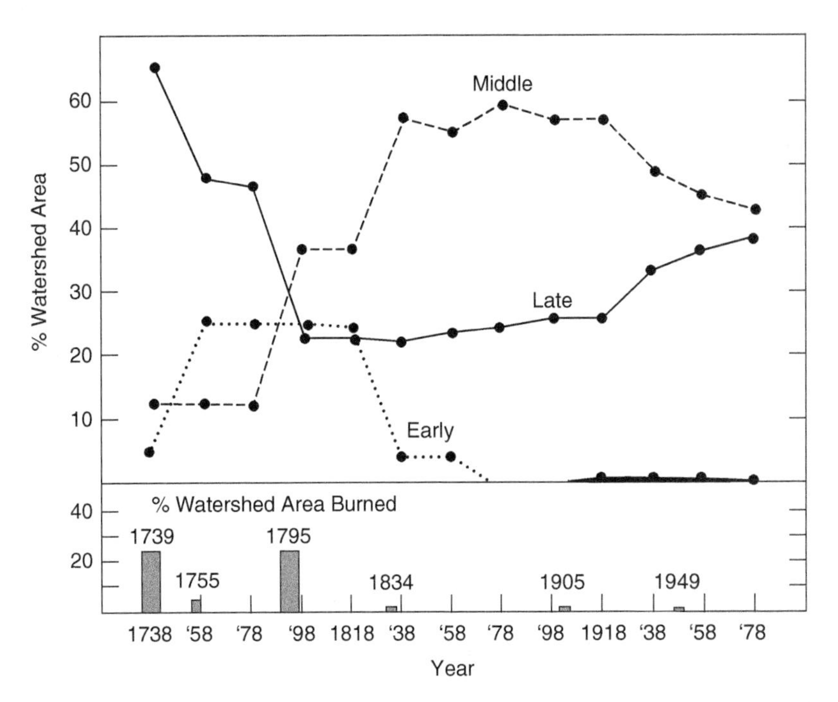

FIGURE 22.11 Percentage of the 73 km² Little Firehole River watershed occupied by early, middle, and late stages of forest succession from 1738 to 1878 in Yellowstone National Park, northwestern Wyoming.
Source: Romme and Knight (1982) / with permission of Oxford University Press.

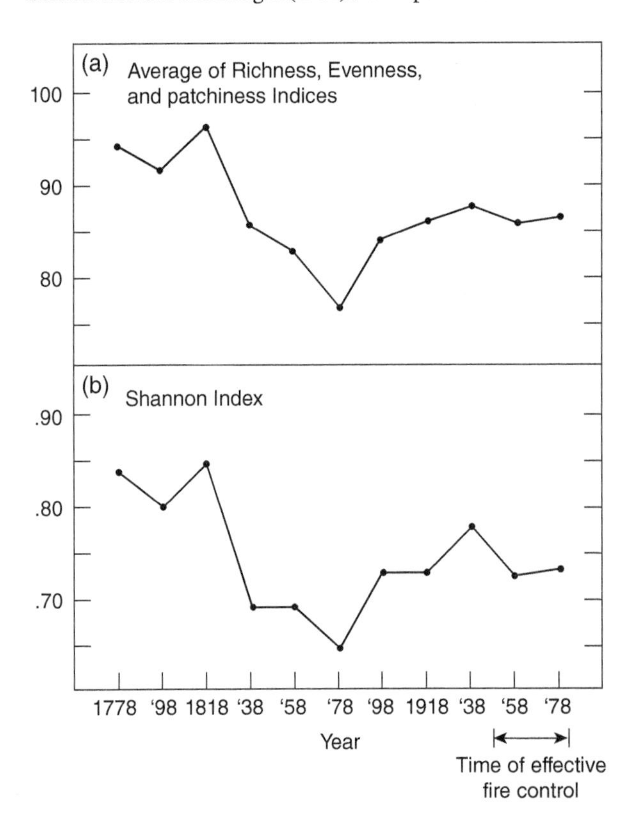

FIGURE 22.12 Changes in landscape diversity in the Little Firehole River watershed from 1778 to 1987 in Yellowstone National Park. (a) Average of richness, evenness, and patchiness indices; Y axis = percentage of maximum possible average of richness, evenness, and patchiness indices. (b) Shannon–Wiener diversity index; Y axis = percentage of maximum possible Shannon–Wiener diversity index. Maximum possible Shannon–Wiener index would occur where all stand ages have equal coverage in the landscape.
Source: Romme and Knight (1982) / with permission of Oxford University Press.

during the summer of 1988, fires burned approximately 250 000 ha of lodgepole pine forest. In contrast to expectations at the time, what resulted was not a homogenous landscape, but a markedly heterogeneous one, including a mosaic of burned and unburned patches with varying burn severities (Turner et al. 1994).

Sooner than 5 years after the 1988 Yellowstone fires, lodgepole pine regeneration varied substantially across the landscape, from areas with no seedlings to areas with >500 000 seedlings per hectare (Turner et al. 2004). Seedling densities truly varied in a fine-grained mosaic, with mean patch size 1.5 ha, 68 patches occurring per 100 ha, and similar patches separated on average by only 150 m (Kashian et al. 2004). Overall, small, dense patches of seedlings occurred within a matrix of large, sparser patches across the landscape (Figure 22.13). Variation in burn severity was important in determining post-fire seedling density; areas burned by severe surface fires had higher seedling densities than areas burned by crown fire or light surface fires (Turner et al. 1994). Lodgepole pine has serotinous cones (Chapter 10) that require heat to open, but may be incinerated by high-intensity fires. While crown fires may have incinerated many cones and light surface fires may have been insufficient to open them, severe surface fires likely provided the appropriate burn temperatures to maximize post-fire regeneration (Turner et al. 1994). The expression of the serotinous trait by lodgepole pine also varies across the landscape and is not well understood, but higher proportions of trees producing serotinous cones in a stand are correlated with shorter fire intervals and moderately aged stands (Tinker et al. 1994; Schoennagel et al. 2003).

The heterogeneity created by the 1988 fires in Yellowstone has many implications for the future landscape. Although succession and stand dynamics will proceed as the post-fire forests age,

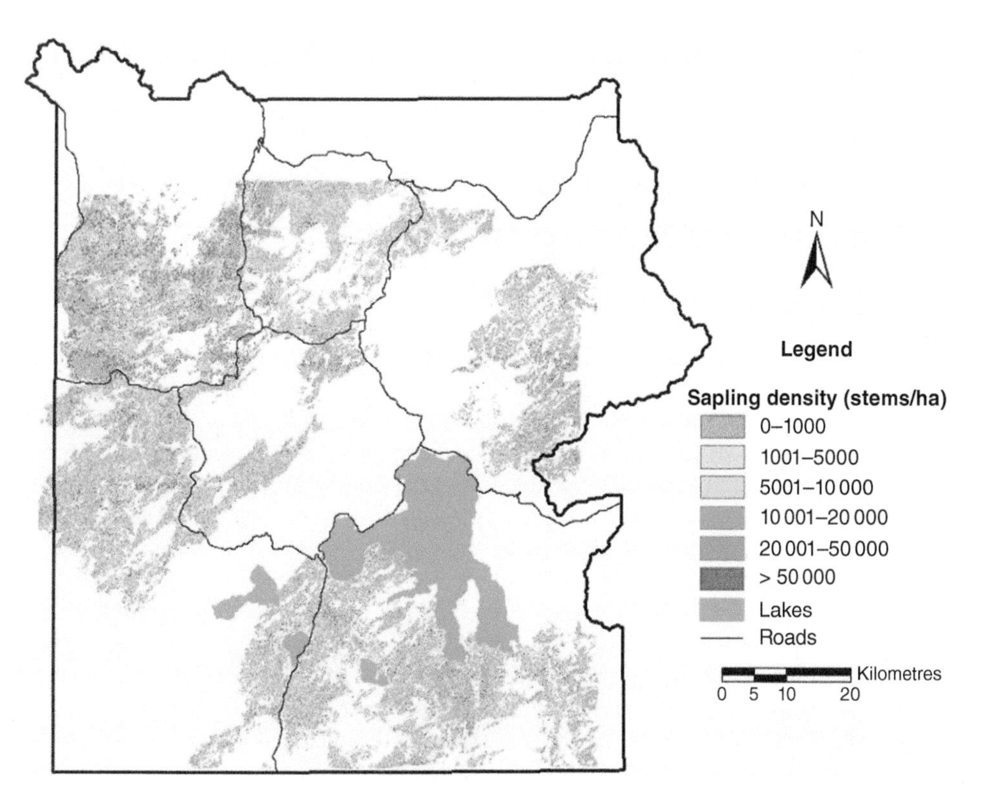

FIGURE 22.13 Heterogeneity of lodgepole pine regeneration density across the area burned in Yellowstone National Park in 1988. Regeneration density varied over six orders of magnitude in a fine-grained, complex spatial mosaic. *Source:* Kashian et al. (2004) / with permission of Canadian Science Publishing.

the landscape pattern created in 1988 is likely to leave a long-lasting legacy of the fires for decades to centuries. Studies of stand structural dynamics on the unburned portion of the Yellowstone landscape found high variability in stand density that decreased over time (Kashian et al. 2005b). Most stands on the landscape exhibited classic dynamics of a fire-regenerated, shade-intolerant species, showing evidence that they had initially been relatively dense and their density had decreased over time due to self-thinning (Chapter 17). However, several stands 50–150 years old had clearly regenerated sparsely after the fire that last burned them, allowing gradual and continuous recruitment to occur and density to remain stable or increase over time. Younger stands were not only more dense, but their density varied more across the landscape; stand density was generally similar among stands 200 years and older (Kashian et al. 2005b). The initial patterns of regeneration density produced by the 1988 fires largely remained in 2012 such that convergence had not yet occurred, but the mechanisms that would likely cause convergence across the landscape were observed (Turner et al. 2016). Notably, smaller fires (<5000 ha) that burn during the interval between large fires (Despain 1990), such as those examined by Romme (1982), become the main source of heterogeneity once stand structures converge following larger fires.

HISTORICAL RANGE OF VARIABILITY

Related to the concept of disturbance and the successional processes that follow is the idea that landscape patterns are not static, as illustrated in the earlier text by the studies of convergence in stand structure and function in Yellowstone National Park. The fact that landscape patterns change over time suggests that a single ecological description of a landscape would serve as a poor benchmark because it is not likely to capture the full range of natural conditions that occur over time. The **historical range of variability** (HRV; also called *natural range of variability*) is the variation in ecological conditions that occurs over time and space within a specified time frame and area without human intervention (Landres et al. 1999). The benefit of identifying an HRV is that ecologists and managers may assess the degree of alteration of current landscapes using a range of historical conditions rather than a specific condition at a single point in time. The HRV concept is useful in restoring current landscapes to historical conditions as well as designing management, such as forest harvesting, in such a way that it mimics historical conditions.

Two of the most important disturbance types that affect forests are fire and harvesting, particularly because humans have heavily influenced the extent and severity of each. As such, many (but not all) HRV studies have occurred in forests of western North America where these disturbances are particularly prevalent. A comparison of the landscape pattern of the wilderness area of Yellowstone National Park to that of the adjacent Targhee National Forest in Wyoming is an illustrative example of the use of HRV. Using a map of historical fires in Yellowstone as a reference landscape, Tinker et al. (2003) calculated a series of landscape metrics at 20-year intervals for the period 1705–1995 (Figure 22.14). Metrics included the total number of patches, average size of patches, the variation in patch size, and the amount of edge. The HRV for each metric was estimated using the range of values for each metric during the study period. The same metrics were then calculated for the Targhee National Forest, which had undergone clear-cut harvesting since the 1970s, prior to and after the onset of harvesting. The pre-harvest landscape of the National Forest fell within the HRV for each of the four metrics, but the post-harvest landscape did not (Figure 22.14). The post-harvest landscape featured more patches that were smaller and less variable in size compared to the HRV, indicating a heavily altered landscape pattern due to harvesting.

An example of the use of HRV in forests of eastern North America is found in jack pine forests of northern Michigan. Jack pine forests of this region are the breeding grounds for the rare Kirtland's warbler, a once-endangered songbird that nests beneath stands approximately 5–25 years old. Like lodgepole pine, jack pine is a species with serotinous cones (Chapter 4) and is dependent on fire for its regeneration, and thus the warbler's need for young jack pine stands leaves it also

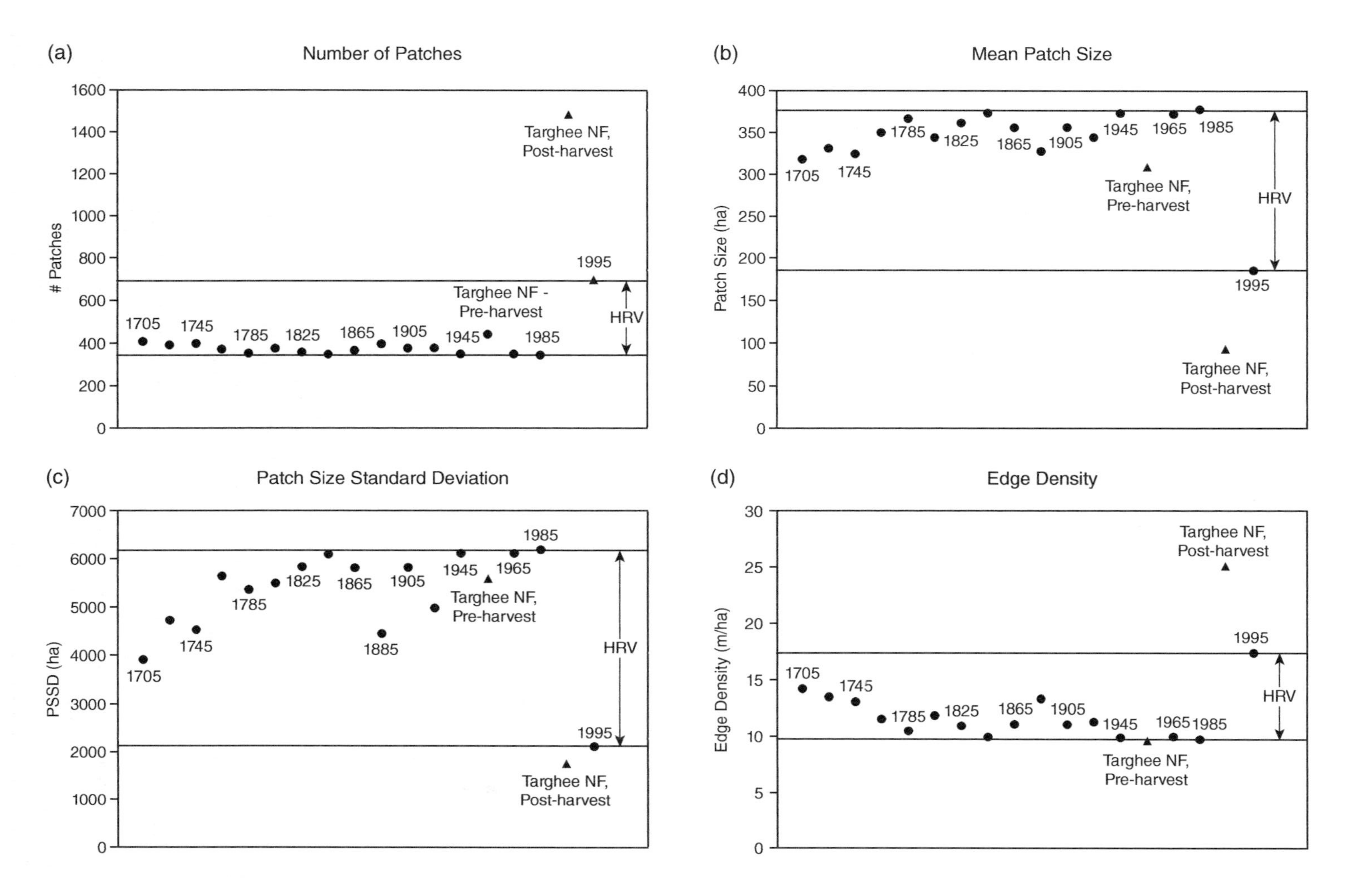

FIGURE 22.14 (a)–(d) Landscape metrics illustrating the deviation of the harvested landscape of the Targhee National Forest from the historical range of variability of the reference landscape of Yellowstone National Park. *Source:* Tinker et al. (2003) / with permission of Springer Nature.

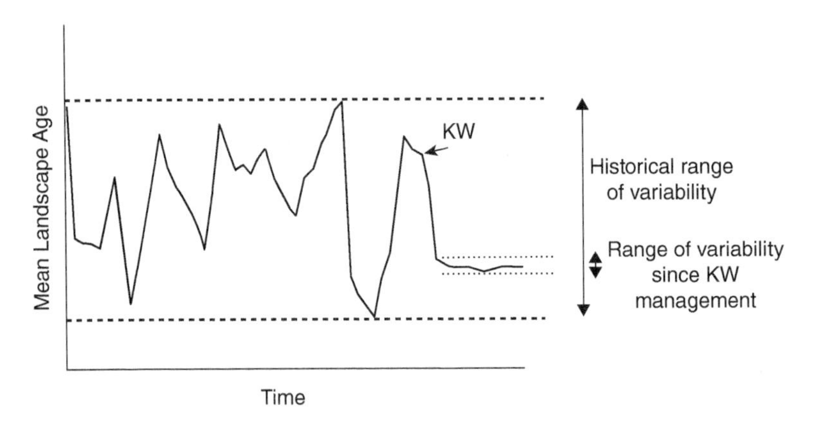

FIGURE 22.15 Conceptual representation of the historical range of variability of jack pine stand ages across the landscape in northern Lower Michigan. Prior to European colonization, the landscape was likely to be both younger and older than the 2015 landscape at various points in time, but current Kirtland's warbler habitat management has restricted the variability compared to its historical range. *Source:* Tucker et al. (2016) / with permission of Springer Nature.

heavily dependent upon fire. Fire suppression in the region in the twentieth century left the warbler reliant on forest management that provided young jack pine habitat using extensive plantations, typically at the expense of mature jack pine forests on the landscape. Historical fire return intervals in jack pine forests of the region were 30–60 years, and the conversion of older forests to young plantations was thus assumed to be an accurate mimicry of historical landscape conditions. Using original General Land Office survey notes collected in 1858, Tucker et al. (2016) reconstructed the distribution of historical age classes of jack pine forests across northern Lower Michigan. They found that the landscape under warbler management in 2015 was much younger than that in 1858; historical coverage of stands <20 years old was only 5% compared to 31% in 2015 and stands >50 years old occupied 76% of the landscape compared to 30% in 2015. This study did not specifically estimate an HRV (as did Tinker et al. 2003), and in fact the 2015 landscape likely fell within the HRV when one considers the historical tendency for this landscape to be burned by large fires. However, the marked difference in the age structure of the landscape prior to European colonization in 1858 suggests that the historical landscape was at least occasionally older during fire-free periods. Thus, forest management for the Kirtland's warbler, which maintained a much higher proportion of the forests on the landscape at a young age, had drastically reduced the variability of landscape age over time and may eventually eliminate it (Figure 22.15).

INTERACTIONS OF LANDSCAPE PATTERNS AND ECOLOGICAL PROCESSES

At the beginning of this chapter, we defined landscape ecology as the study of the interaction between pattern and process, and we might consider this to be a bi-directional relationship (Turner and Gardner 2015). For example, in our references to landscape ecosystems throughout this book, we have illustrated how ecological processes influence vegetation patterns across the landscape. In forming ecosystems, physical site factors including climate, physiography, soil, and hydrology influence ecological processes such as nutrient cycling (Chapter 19) or ecosystem productivity (Chapter 18), and patterns of species composition and cover type in turn reflect this heterogeneity. The opposite is also true, in that heterogeneity affects ecosystem processes and how they occur at broad spatial scales.

Much of our understanding of ecosystem processes in forests is articulated at a stand or site level from studies whose design ensured homogeneity of study sites to the extent possible. When ecosystem processes are considered at broad scales, heterogeneity creates an additional layer of complexity that must be accounted for, but it also provides additional insights (and potentially different answers) about how ecosystem function occurs. Turner and Chapin (2005) developed a framework distinguishing *point processes*, which generally include ecological rates, from *lateral transfers*, which include transport processes of matter among patches on a landscape. Point processes are typically measured at a specific place on the landscape, such as net primary production, carbon storage, or nutrient cycling, and highlight variability in the rates and the potential causes of that variability. Lateral transfers focus upon the flow of materials from one place to another, such as the flow of nitrogen across an agricultural landscape in the context of riparian buffers. We focus mostly on point processes in this section to briefly illustrate the interaction between heterogeneity and ecological processes. Detailed treatments on the effects of heterogeneity on ecological processes are presented by Turner and Gardner (2015), Lovett et al. (2005), and Hutchings et al. (2000).

LEAF AREA AND PRODUCTIVITY

The prevalence of climate change has spawned many recent studies that have examined carbon storage and ecosystem productivity over large extents. One of the earliest attempts at understanding how carbon dynamics vary at broad spatial scales examined evapotranspiration (ET) and photosynthesis for a 1540 km^2 coniferous forest landscape in northwestern Montana (Running et al. 1989). These processes represent important interactions between vegetation at the Earth's surface and the atmosphere and are closely related to net primary productivity (Chapter 18). Within a geographic information system, Running et al. (1989) determined leaf area index (LAI) from satellite imagery for the forest in 1200 grid cells, and ran an ecosystem model in each cell using LAI as an input variable. The model estimated annual ET and net photosynthesis for the landscape and demonstrated the enormous degree of spatial variability in these ecosystem processes and its association with physiography. For example, leaf area was highest where precipitation was the highest, which happened to be on the western slopes of the study area. Soil water holding capacity was also an important determinant of LAI and was lower on steep slopes or shallow soils often in spite of higher precipitation (Figure 22.16). Variation in ET was in part due to land use, where low ET was associated with young post-logging regeneration. ET dried out soils earlier in the season where snowmelt occurred earlier, and those regions of the landscape experienced a longer period of soil moisture stress. Photosynthesis reflected LAI but varied with temperature, decreasing at low elevation due to warmer temperatures and at high elevations due to cooler temperatures, and varied over the study area with soil moisture (Figure 22.16). Overall, this work showed the importance of quantifying the variability of ecosystem processes in attempting to understand and predict carbon storage and flux over large landscapes.

Heterogeneity in ecological processes that affect carbon dynamics may result from disturbances as well as physical site factors. Carbon dynamics, particularly when heavily influenced by aboveground productivity, are closely tied to disturbances and the recovery and succession of forests that follow them (Bradford et al. 2008a; Kurz et al. 2008b). For example, the heterogeneity of lodgepole pine regeneration following the 1988 Yellowstone fires (Figure 22.13) had important implications for leaf area and aboveground net primary productivity (ANPP) approximately 10 years after the disturbance (Turner et al. 2004). In fact, ANPP was closely related to seedling density and varied in a fine-grained mosaic much like post-fire regeneration. About 68% of the burned landscape had low ANPP ($<2\,Mg\,ha^{-1}\,year^{-1}$), reflecting a matrix of lower-density patches of seedlings, and 10% had high ANPP ($>4\,Mg\,ha^{-1}\,year^{-1}$). Similar to seedling density, Kashian et al. (2005a) predicted that the variation in productivity would decrease over decades or centuries during the fire-free interval as contrasting stand structures began to converge, and the initiation of this convergence has already been observed 24 years after the fire (Turner et al. 2016).

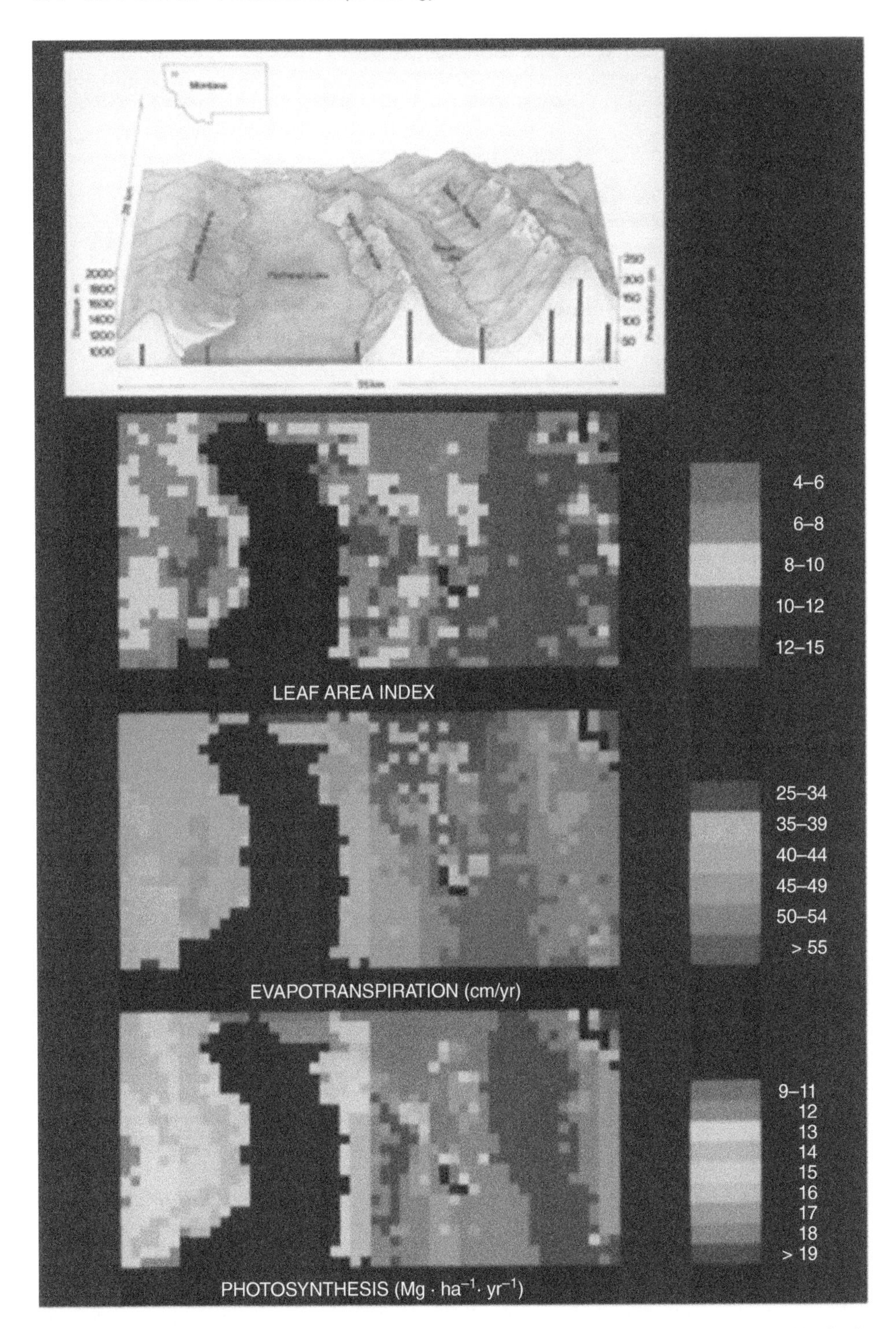

FIGURE 22.16 Heterogeneity of evapotranspiration and net photosynthesis as determined by leaf area index and an ecosystem model for a 1540 km^2 landscape in northwestern Montana. Heterogeneity of the ecosystem process closely shows physiographic differences of the forested landscape. *Source:* Running et al. (1989) / John Wiley & Sons.

FOREST CARBON DYNAMICS

Incorporating heterogeneity is particularly important for creating carbon budgets and for understanding the carbon balance on landscapes because of the variability of forest species composition, age, and other factors. A carbon balance may be determined from leaf to global scales, but net ecosystem productivity (NEP; also termed net ecosystem carbon balance, or NECB) in a forest essentially represents the balance between carbon fixed into live biomass via photosynthesis and that lost from dead biomass as it decomposes (Chapter 18). The importance of NEP for large landscapes relates to whether a region acts as a *carbon source* (losing more carbon to the atmosphere than it gains) or a *carbon sink* (gaining more carbon than it is loses to the atmosphere). Because forests are a major pool of carbon on the earth's surface, and they act to remove carbon from the atmosphere, disturbances that impact forests are especially important in determining whether a landscape is a source or sink of carbon and therefore how it contributes to climate change (Chapter 20).

As discussed in Chapter 18, forests are typically carbon sources immediately after a disturbance when small, live trees with relatively low ANPP do not compensate for carbon lost from high levels of dead, decomposing biomass killed by the disturbance. Depending on the forest type, NEP reaches its maximum several years or decades after the disturbance when tree growth is most vigorous (primary productivity is highest) and most dead biomass has decayed away, and forests then act as strong carbon sinks. By contrast, older forests have accumulated large stocks of carbon even as primary productivity declines and dead biomass begins to re-accumulate, and NEP approaches zero. Many studies employ a chronosequence approach to determine how carbon storage changes after a disturbance, whereby a series of stands are sampled that vary in age, and all differences among the stands are assumed to occur due to the effects of succession (Chapter 17). For example, Rothstein et al. (2004) sampled 11 jack pine stands ranging in age between 1 and 72 years in northern Michigan. They found that jack pine-dominated forests are initially a carbon source for 6 years after wildfires because decomposition exceeds primary productivity, then shift to a strong carbon sink until stands matured and primary productivity declined.

The patterns described in the earlier text are very common for systems where disturbances are stand-replacing and may also vary with stand structure (Figure 22.17; Kashian et al. 2006). Stand density is important because it affects the rate of carbon uptake and storage by vegetation. Because they have higher leaf area and photosynthate accumulates more quickly, dense stands should accumulate more carbon quicker to form a strong sink after the disturbance. Sparse stands should take longer to shift to positive NEP and achieve a weaker sink, but the duration of the sink is likely to be prolonged compared to dense stands (Kashian et al. 2006). Notably, Kashian et al. (2013) found that stand density was important for carbon storage in forests in Yellowstone National Park that had not yet reached canopy closure, but that tree size and age were most important after this point. Canopy closure represents the point at which stands reach their maximum leaf area, and thus changes in leaf area may be a critical driver of NEP after disturbances.

Given the heterogeneity of both stand age and structure across landscapes, estimates of NEP at broad spatial scales are extremely useful because they account for the variability that directly affects carbon storage. Landscapes dominated by young or sparse forests may be carbon sources while those dominated by older or dense forests may be carbon sinks, or any combination therein. In addition to the influence of physiography as described by Running et al. (1989), species composition also imposes spatial heterogeneity in carbon dynamics across landscapes. For example, boreal forests with mixed conifer and hardwood composition were found to store 47% more carbon in the aboveground vegetation compared to black spruce forests and 44% more than jack pine forests (Martin et al. 2005). In part, mixed stands were able to accumulate carbon more quickly because they contain species with contrasting shade tolerances and thus are able to form multilayered canopies, supporting a greater leaf area. Landscape carbon studies able to incorporate all factors that are spatially heterogeneous are very difficult and remain elusive, but clearly accounting for such heterogeneity is critical in understanding carbon storage at broad spatial scales.

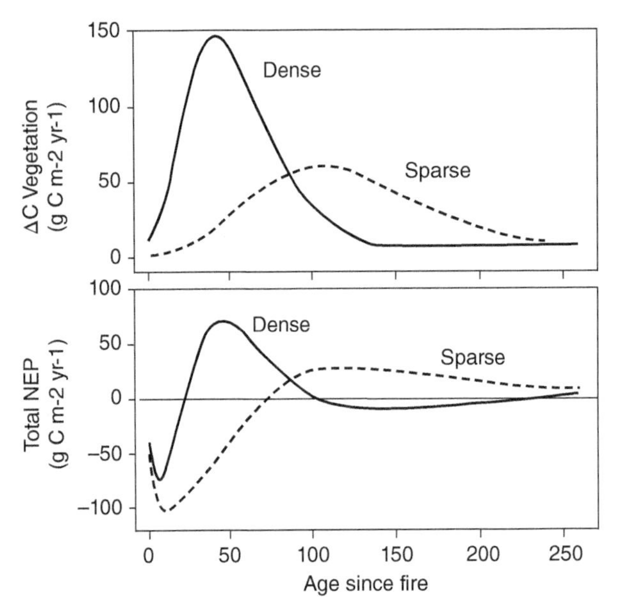

FIGURE 22.17 Carbon balance following disturbances for lodgepole pine forests of contrasting structure in the central Rocky Mountains. Carbon accumulates quickly and to a higher maximum for a short duration in dense stands compared to sparse stands, which are slower to accumulate carbon. These differences in vegetation drive similar changes in net ecosystem productivity (NEP). *Source:* Kashian et al. (2006).

Incorporating the heterogeneity present on landscapes into our understanding of carbon dynamics has implications for predicting how forested regions may contribute to atmospheric carbon and climate change. Any disturbance to a forest will release carbon to the atmosphere if it kills trees and will therefore temporarily shift the landscape to a carbon source. The amount of carbon released, and the strength of the source, will vary with the type and severity of the disturbance and the heterogeneity of these factors that may occur across a landscape. For example, stand-replacing wildfires will create a strong and immediate carbon source due to combustion and widespread tree mortality and subsequent decomposition, but less-severe surface fires may create only a weak source if relatively few trees are killed. Likewise, moderate insect outbreaks or windstorms will create a weaker, more prolonged carbon source because a higher proportion of trees survive the disturbance, and variation in their severity may affect the strength or duration of the source.

NUTRIENT DYNAMICS

Although it is clear that landscape heterogeneity in nutrient pools and fluxes is the rule rather than the exception for most systems, fewer researchers have been successful in both quantifying and explaining such patterns (Turner and Gardner 2015). Much of this work in temperate forests has focused on the nitrogen cycle, given that nitrogen is often the limiting nutrient for forest productivity (Chapter 19). A series of studies by Zak et al. (1986, 1989) and Zak and Pregitzer (1990) related the spatial variation in nitrogen-related processes to the patterning of landscape ecosystems in northern Lower Michigan. In comparing nine ecosystems located on various landforms and soil textures, net mineralization was lowest in a dry oak ecosystem and highest in a mesic northern hardwood ecosystem (Table 22.1). Nitrification was also highest in the northern hardwood ecosystem and far higher than in any other ecosystem sampled. Overstory biomass was asymptotically related to differences in mineralization, but was unrelated to nitrification rates. Thus, the spatial heterogeneity in at least some nutrient dynamics closely tracks the spatial distribution of ecosystems across the landscape (Zak and Pregitzer 1990).

Table 22.1 Association of mineralization and nitrification rates with landscape ecosystems in northern Michigan.

Dominant overstory species	Landform	Percent coarse and medium sand	Mineralization (μg N/g)	Nitrification (μg N/g)
Black and white oak	Outwash plain	72	52.0	0.9
Black and white oak	Outwash plain	70	49.6	0.3
Black, white, and red oak	Kame and moraine	67	47.2	0.2
Red, white, and black oak	Kame and moraine	63	60.2	0.2
Red and white oak	Kame and moraine	55	63.0	0.6
Red oak and red maple	Kame and moraine	41	81.8	0.2
Sugar maple and red oak	Moraine	61	75.5	4.1
Sugar maple and red oak	Moraine	56	93.4	1.2
Sugar maple, basswood, and white ash	Moraine	55	127.8	45.5

Source: Zak et al. (1989) / with permission of Canadian Science Publishing.

At broad spatial scales, spatial variability in some nitrogen transformations may be explained by the heterogeneity of soil texture. For example, Reich et al. (1997) found a strong influence of soils on mineralization for forests in Wisconsin and Minnesota, with higher rates of mineralization on finer-textured soils. Likewise, Groffman and Tiedje (1989) and Groffman et al. (1992) examined rates of denitrification, which tends to occur on wet sites in the absence of oxygen, in forest soils of southern Michigan. Rates of denitrification across the landscape varied with soil texture and associated drainage, and were differentially responsible for regional estimates of denitrification. Loamy soils represented 47% of the forest but 73% of the denitrification, and clay soils represented 9% of the forest but 22% of the denitrification. Meanwhile, sandy soils represented 44% of the forest but only 5% of the denitrification. Together, these studies are examples of ecosystem processes whose heterogeneity may be predicted by the heterogeneity of physical site factors across the landscape.

Heterogeneity of nutrient dynamics at broad spatial scales appears to be most easily related to physical site factors, but a general framework that associates nutrient dynamics with more general characteristics of landscape pattern has been more fleeting. However, an experimental study of forest fragmentation in Kansas may have begun to uncover a relationship between patch size and soil nitrogen dynamics (Billings and Gaydess 2008). The study was designed with the assumption that edge effects in fragmented forests result in changes in dominant tree species that in turn may affect soil nutrient cycling. The experimental site contained small ($4 \times 8\,m^2$) and large ($50 \times 100\,m^2$) patches of successional forest that were established in 1984. Large patches had greater rates of woody plant colonization, higher plant species richness, and likely higher aboveground net primary productivity (ANPP) per unit area than smaller patches, all of which became evident after about 15 years of succession in the experimental fragments. These differences in vegetation apparently resulted in greater rates of mineralization and nitrification in soils of the small patches compared to the large patches, associated with greater fine root biomass and root nitrogen concentration. Soils from small patches also had higher potential denitrification rates compared to large patches, which is often associated with faster nitrogen cycling rates (Venterea et al. 2003). Although these patterns have yet to be rigorously tested in other landscapes, the potential that forest patch size—or landscape configuration in general—may govern the spatial heterogeneity in nutrient cycling is intriguing.

Much like the landscape ecosystem approach we reference throughout this book, landscape ecology focuses on the importance of scale, is concerned with the effects on a specific place on a landscape by the ecosystems that surround it, and emphasizes physical site factors to some degree. The landscape ecosystem approach involves mostly a focus on natural systems, with perhaps some acknowledgement that humans are a significant disturbance to those systems. Additively, landscape ecology explicitly includes humans as an important variable in understanding spatial heterogeneity, including how their perceptions, values, and decisions shape landscapes. We have not addressed this human dimension in our brief overview of forest landscape ecology, but we do so in detail in our treatment of sustainability in the chapter that follows.

SUGGESTED READINGS

Fahrig, L. (2003). Effects of habitat fragmentation on biodiversity. *Annu. Rev. Ecol. Evol. Syst.* 34: 487–515.

Forman, R.T.T. (1995). *Land Mosaics, the Ecology of Landscapes and Regions.* Cambridge University Press 632 pp.

Foster, D.R. and Boose, E.R. (1992). Patterns of forest damage resulting from catastrophic wind in central New England, USA. *J. Ecol.* 80: 79–98.

Landres, P.B., Morgan, P., and Swanson, F.J. (1999). Overview of the use of natural variability concepts in managing ecological systems. *Ecol. Appl.* 9: 1179–1188.

Romme, W.H. (1982). Fire and landscape diversity in subalpine forests of Yellowstone National Park. *Ecol. Monogr.* 52: 199–221.

Turner, M.G. (1989). Landscape ecology: the effect of pattern on process. *Annu. Rev. Ecol. Evol. Syst.* 20: 171–197.

Turner, M.G. and Gardner, R.H. (2015). *Landscape Ecology in Theory and Practice*, 2e. New York: Springer 502 pp.

With, K.A. (2019). *Essentials of Landscape Ecology.* Oxford, UK: Oxford University Press 656 pp.

Sustainability of Forest Ecosystems

CHAPTER 23

At many instances in this book, we have suggested that humans have had far-reaching influences on forests, both directly and indirectly. Their influences have been discussed throughout the first 22 chapters of this book, including impacts on the overriding effects of climate (Chapter 20); extensive alteration of disturbance regimes (Chapter 16); intentional and accidental introduction of non-native plants, animals, and pathogens that eventually became invasive (Chapter 21); widespread reductions in biological and ecological diversity (Chapter 14); and fragmentation of forest landscapes (Chapter 22), among many others. Forest ecology does not exist in a vacuum, in that every forest ecosystem on Earth has in some way been influenced by humans. Although we struggle to understand how the natural world works, we cannot completely study forest ecosystems independent of human influence, especially when we attempt to consider the future. This final chapter grapples with the concept of sustainability: its meaning, its role in ecology, and its applications to forest ecology.

CONCEPTS OF SUSTAINABILITY

Without question, humans are part of ecosystems and are not external to them. Although sometimes difficult to imagine because of their wide-ranging influence, humans are supported by the physical site factors of landscape ecosystems as much as any other organism. At the same time, however, humans have a disproportionate influence on the ecosphere and have profoundly changed regional and local ecosystems in both dramatic and subtle ways. Perlin (1989) provides a remarkably powerful historical record of human destruction of forests that accompanied the development of civilizations from the Bronze Age in Mesopotamia to nineteenth-century America. In several papers, Kay (1994) speculated that the role of indigenous people is probably underestimated in its importance of changing the biota in western North America. He concluded that the modern concept of wilderness as areas without human influence is a myth because indigenous people changed the composition of entire plant and animal communities by limiting ungulate numbers and purposefully modifying the vegetation with fire. Of course, the effects of European colonizers in North America have included forests cleared for agriculture, logging, livestock grazing, long-term changes in dominant tree species, forest fragmentation, climate change, introduction of invasive species, and changes in disturbance regimes. The piercing story of human interventions in the complexity of ecosystems is compellingly described for the Blue Mountains of Oregon by Langston (1995).

THE PREVALENCE OF HUMAN VALUES IN FOREST ECOLOGY

The widespread decline of ecosystems has led to a great interest in the idea of **sustainability** since the 1970s (Du Pisani 2006). A general notion of sustainability is that current desirable conditions of an object or process—be they species composition, structure, recreational opportunities, a perception of wildness, etc.—continue to exist for the benefit of future generations. In most cases, sustainability considers the *maintenance* into the future of something currently present. For example, Chapin et al. (1996) defined a sustainable ecosystem as one that "over the normal cycle of disturbance events maintains its characteristic diversity of major functional groups, productivity, soil fertility, and rates of biogeochemical cycling." Similarly, Turner et al. (2013) defined sustainability as "use of the environment and resources to meet current needs without compromising the ability of a system to provide for future generations." One definition is strictly ecological and the other incorporates social and economic structures, but both definitions are sufficiently broad and ambiguous to place human values on real units of nature. Because humans differ in the way they perceive the world, placing their values onto specific components of landscape ecosystems inevitably results in differences of opinion.

In the first edition of this textbook, Stephen H. Spurr elegantly described how the ecosystems we see today are but a snapshot in time and space, and the values humans place on them may also be temporary. In scrutinizing the concepts of native and introduced species, he reasoned (Spurr 1964, p. 157):

> *. . . there is no meaning to the concepts of native and introduced species from a biocentric viewpoint. Characterizing a plant or animal as being exotic or endemic characterizes it only from the standpoint of man's relationship to it. Actually, all plants and all animals are introduced or exotic from a biocentric standpoint except at the very point in space where the particular gene combination was first put together. Whether the subsequent migratory pattern of that organism took place independent of man or with the help of man is important to man but not to the wilderness ecosystem itself.*
>
> *It is an interest of man to know whether he carried the coconut to a given tropical island or not. To the coconut it is of little importance as to whether it floated by itself in an ocean current or was lodged in the hull of a native dugout canoe which in turn floated on the ocean current. To a maple growing in given spot it is immaterial whether its seed flew there on its own wings or whether it was aided and abetted by the wings of an airplane. A wild cherry is unaffected by concern as to whether its seed was deposited by a seagull who spotted a target below or by a human who brought it thither in a paper bag . . . In short, from the viewpoint of the forest, there is no distinction between native and introduced species . . . All were migrants there.*

Given what is now known about invasive species (Chapter 21), imagine how most naturalists and ecologists would react if such a sentiment was written as fact in a modern textbook! Indeed, the passage was heavily edited for the second edition (Spurr and Barnes 1973) and eliminated by the third (Spurr and Barnes 1980), primarily because the values of ecologists (and society) began to change. Ecological data from the last 58 years suggest that introduced species can be bad because those that become invasive may outcompete native species, alter nutrient dynamics and/or soil communities, have negative economic implications, or may even lead to wholesale changes in a system. And yet relatively few of these data directly contradict Spurr's initial sentiments; instead, it is implicit in modern ecological studies that humans value native species over introduced species, find the nutrient dynamics of a near-natural ecosystem to be more important than of a heavily altered ecosystem, and abhor disruptions to a well-functioning economy. Data over the last six decades have also gradually moved toward acknowledging the importance and inevitability of change in ecological systems (Chapter 17), the major point that Spurr was attempting to make. Spurr might be amused today to see that we have finally accepted ecological change—as long as systems change as we think they should. Human values are typically not scientific, yet they are often overlain onto scientific inquiry, including that undertaken by forest ecologists. These values are what challenge scientists in incorporating notions of sustainability into scientific understanding.

HISTORICAL PERSPECTIVE OF SUSTAINABILITY IN FORESTS

Ours is not a textbook about the management of forests, but a brief discussion here is of use to illustrate how human values are an intrinsic part of sustainability. In fact, if sustainability's root concern is providing for future generations, then those concerned with forests may be among the first who explicitly integrated sustainability into their operating models (Westoby 1989), even as those models evolved and changed over decades and centuries. First and foremost, the concept of sustaining natural resources over the long term was present in the traditions of many ancient cultures, such as those of the indigenous peoples of New Zealand, Indonesia, Mali, India, and British Columbia (Gadgil and Berkes 1991). Regarding modern times, a review by Wiersum (1995) notes that sustainability was well expressed in forest management systems in Germany in the eighteenth century, and the philosophy of wise use consistent with future availability of forest resources was well articulated in the 1800s. Notably, the creation of the forestry profession in the United States resulted from concerns about sustainable use of forests in the 1900s. Sustainability in some form therefore has a very long association with forests and forest ecology (Franklin et al. 2018).

In North America, one aspect of timber management has been deeply associated with at least one form of sustainability for over a century: the idea of sustained yield. Sustained yield is a Western concept that because harvested trees can be regrown, forests are a renewable resource and can be managed for a continuous flow of resources for future generations. Notably, the term "resource" itself automatically invokes human values, both social and economic. Two general problems, among many, are inherent with the idea of sustained yield. First, one cannot assume that all forests are equally renewable, and as such their yields are not always sustainable. For example, aspen in the Lake States is typically clear-cut for pulpwood, sometimes as often as every 20–30 years, because the trees regenerate quickly from root sprouts after cutting and exhibit growth rapid enough to replace themselves within the time frame of our current social and economic systems. Likewise, Douglas-fir forests of the Pacific Northwest are also considered quite renewable after harvesting because they may produce sawlogs within a 50-year time frame on suitable sites, again within an appropriate social and economic time frame. These models place the emphasis on the potential product (in this case timber) as the resource of interest, such that clear-cutting a 500-year-old ("old-growth") Douglas-fir forest and placing it into a rotation of harvesting every 50 years are appropriate to achieve a sustained yield.

A second problem with sustained yield is its focus on a single resource provided by the forest. Using the abovementioned example, outside of the timber resource, transitioning an old-growth Douglas-fir forest to one that produces 50-year-old sawlogs does not consider the potential value of the old growth itself, which would not be considered to be renewable. The old-growth forest may provide a limited habitat for certain species (both plants and animals), or it may provide recreational opportunities or aesthetic values, or it may create an important protection for water resources not found in younger forests. Forest management in North America broadened to multiple-use forest management in the 1960s and 1970s in recognition of the idea that sustainability might also target resources other than timber (Clawson 1978). It also more firmly emphasized the importance of human values in its treatment of sustainability. Multiple-use management continued to include sustained yield as one of its multiple uses, but transitioned again in the 1990s to ecosystem management (Grumbine 1994; Brussard et al. 1998). Ecosystem management suffered somewhat from varying definitions and a somewhat ambiguous framework, but generally considered the balance between maintenance of biodiversity and ecosystem processes (including disturbances and the changes they invoked in ecosystems) and appropriate human uses. Its emphasis on "whole systems" included social and economic factors as well as ecological ones and hoped to integrate ecological conservation into socioeconomic models. Social factors, in particular, included concepts crucial to sustainability, such as freedom, justice, and equity. Ecosystem management therefore represented a further emphasis of the human factor into sustainability, and many of its basic tenets are recognizable today in concepts such as ecological forestry (Chapter 14).

ECOSYSTEM SERVICES

Central to the concept of sustainability is a clear definition of what exactly is being sustained. Simply put, **ecosystem services** are the objects and processes of natural ecosystems that are beneficial to humans, and they are typically the implicit or explicit targets of sustainability. The idea of ecosystem services was described by Daily et al. (2000), codified by the United Nations' Millennium Ecosystem Assessment (MEA 2005), and has since been discussed by many authors (e.g., Nesbitt et al. 2017; Alix-Garcia and Wolff 2014; Levin 2012; Perrings 2007; and many others). Notably, the focus of ecosystem services by design is human well-being; the ecosystems supplying the services benefit from protection or conservation only because of what they provide to humans rather than from any intrinsic value they have. It is often argued, however, that protecting ecosystem services contributes to the overall function of the ecosphere and in turn the regional and local ecosystems within it. Indeed, it has been recognized that many ecosystem services are irreplaceable, and their loss would somehow harm the integrity of ecosystems, regardless of direct human benefit (Ekins et al. 2003; Farley 2012; Neumayer 2013). Thus, ecosystem services are considered to be a vital bridge between ecosystems and society, and they justify the actions needed to undergo sustainability (Wu 2013, and references therein).

Ecosystem services are often grouped into four broad categories (MEA 2005; Carpenter et al. 2009): provisioning, regulating, supporting, and cultural. *Provisioning services*, originally defined as "ecosystem goods" by Daily (1997), include the potential products extractable from an ecosystem such as timber, food, water, or minerals, but also genetic and medicinal resources. *Regulating services* are less tangible than provisioning services because they describe more abstract benefits humans obtain when they maintain important ecosystems processes. Examples of regulating services might include pollination, which is important for fruit production and the persistence of plant species, or trophic dynamics, which are important for control of pests and diseases. *Supporting services* are those that are vital for the functioning of ecosystems and are necessary for most other ecosystem services to exist, such as soil formation, nutrient cycling, or oxygen production. Turner et al. (2013) also treat primary productivity and post-disturbance forest regeneration as major supporting services on specific landscapes. Supporting services are sometimes difficult to distinguish from regulating services because many regulating services may be supportive at large spatial or temporal scales. Finally, *cultural services* are non-material services obtained from ecosystems such as spiritual enrichment, aesthetic encounters, reflection, or recreation. A detailed treatment of ecosystem services in forests is provided by Solórzano and Páez-Acosta (2009); a truncated list of examples is shown in Table 23.1.

If ecosystem services are a main target of sustainability, then maximizing those services is typically its goal. However, not all ecosystem services of a forest are able to be maximized at the same time, and trade-offs ensue. Perhaps the best example of trade-offs in ecosystem services would be plantations or tree farms under a sustained yield scenario, where the provisioning service of timber would be maximized at the expense of biodiversity, wildlife habitat, or aesthetics. Likewise, wilderness areas might maximize cultural services such as recreation, spiritual uses, or aesthetics at the expense of provisioning services when potential resources are not extracted (Kinzig 2009). The point is that most services require their own unique set of ecological attributes such that it is difficult to maximize all or even most simultaneously.

Interestingly, many ecologists have argued that biodiversity preservation is the keystone of sustainability, because it augments the maintenance of most other ecosystem services through its heavy influence on ecosystem function (e.g., Cardinale et al. 2012). For example, in one of the earlier explorations of sustainability in forests, Lindenmayer and Franklin (2003) edited a book with chapters written by 13 forest ecologists and managers from around the world, many of whom defined sustainability as the preservation of forest biodiversity and one who defined it as "the

Table 23.1 Non-exhaustive list of examples from forest ecosystems of the four categories of ecosystem services.

Type of ecosystem service	Ecosystem good or function	Examples from forest ecosystems
Provisioning	Forest products	Timber, fiber
	Food	Fish, fruits, graze or browse for livestock, wildlife
	Genetic resources	Test and assay organisms, bioprospecting
	Biochemical/ medicinal resources	Drugs and pharmaceuticals, dyes, latex, and other resins
	Ornamental resources	Diversity of organisms with ornamental potential
Regulating	Climate regulation	Altering temperature, albedo, evapotranspiration
	Hydrological regulation	Regulating volume, quality, and timing of water flow
	Disturbance moderation	Dampening of wind disturbance, flood control
	Carbon sequestration and storage	Cycling and storage of carbon
	Pollination	Pollination of timber/fiber species, fruit production
	Trophic dynamics	Pest or disease control, reduction of herbivory on crops
Supporting	Soil formation and support	Physical support for plants, nutrient cycling, water
	Post-fire forest regeneration	Maintenance of cover types and habitats
	Site productivity	Framework for species composition, soil processes
Cultural	Recreation activities	Tourism, camping, hunting, hiking
	Spiritual inspiration	Use of forests for religious or individual experiences
	Aesthetics	Appreciation of scenery
	Cultural uses	Background for books, music, film, folklore

grand attempt at forest biodiversity rehabilitation." Although biodiversity is certainly an important attribute of a sustainable forest, its increase may necessarily be accompanied by trade-offs with some other services such as food or timber production and may have less or nothing to do with other services such as carbon storage or erosion control (Kinzig 2009; Reyers et al. 2012). Biodiversity appears to typically be most at odds with provisioning services. Naeem (2009) suggested that the ecosystem functions most closely related to biodiversity are generally supporting services, such as net primary production, nutrient cycling, and other biogeochemical processes, but that these services are not consumed and are least understood by people. Others have argued that biodiversity contributes more to regulating services such as invasion resistance and ecosystem resilience, as well as cultural services (Mace et al. 2012). In any case, some have argued that biodiversity is best conserved when ecosystem services are the target rather than biodiversity itself, because ecosystem services provide the economic and social justification that biodiversity has an instrumental value (Skroch and López-Hoffman 2009; Mace et al. 2012; Balvanera et al. 2014).

TOWARD A DEFINITION OF SUSTAINABILITY

If sustainability is built upon a balance between ecological factors and human well-being, and ecosystems services are its target, then how exactly might we define the term? Unfortunately, inconsistencies in its definition have plagued an already complex concept while its study has increased dramatically over the last three decades. Marshall and Toffel (2005) tabulated at least 100 different definitions of the term. As a matter of comparison, we present three such definitions as the most common and familiar to those who think about sustainability.

Sustainability has often been described as a "three-legged stool," with equal-sized environmental (integrity of ecosystems), social (justice, equity, and freedom), and economic (material goods and services) legs. In this model, also called the "triple bottom line" or the "people–planet–profit" model (Elkington 2004), all three aspects must be met simultaneously; an emphasis on any one more so than the others will not achieve sustainability. These three components have been described as separate but intersecting, best depicted with a Venn diagram where sustainability represents the limited area where all three components intersect, representing their achievement at the same time (Figure 23.1a). In natural resource management, the idea is to properly account for all three components and the interactions among them. The triple bottom line is increasingly used as a way to plan for sustainability and sustainable development in the business world, with the model often used as a metric of a corporation's commitment to sustainability.

If we accept that the triple bottom line describes a basic model of sustainability, then the likelihood that all three components are not equally emphasized must be considered. If the focus is on the entire system rather than any of its separate components, achievable sustainability is considered to be weak. *Weak sustainability* allows one component to substitute for another in the triple-bottom-line model, as long as the entire system increases or sustains its total capital (Figure 23.1b). Further, the model assumes there is no difference between natural capital (ecosystems or environment) and human-made capital when providing for the well-being of humans, and that natural capital lacks unique properties that cannot be substituted by human-made capital. Thus, weak sustainability could be achieved on a heavily logged forest landscape if the harvesting supported economic growth of the region and an appropriate standard of living for its people, because the social and economic spheres of the model would substitute for the environment. Wu (2013) contrasts weak sustainability with *absurdly strong sustainability*, where the environmental

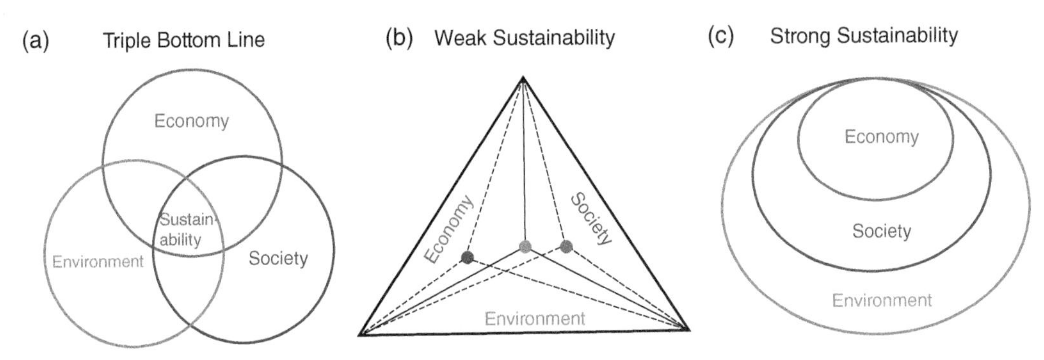

FIGURE 23.1 Three models relevant to sustainability: (a) the triple-bottom-line model outlines the basis of sustainability as the intersections of the three spheres of environment, economy, and society, assuming that all three must be met simultaneously to achieve sustainability; (b) weak sustainability assumes that any of the three components is substitutable for another, so long as the entire system increases or maintains its capital; (c) strong sustainability assumes that the three components are nested within each other such that environment constrains society and economy, and society constrains economy. *Source:* Wu (2013) / Springer Nature.

sphere is never substituted even when it could be or perhaps should be. Under such a model, species and ecosystems are protected above all else, even at the cost of human well-being (e.g., never clearing another forest for agriculture or cutting trees for fuel even if it means people would starve or go without heat in a local village). Thus, weak sustainability assumes that ecosystems and economic products or resources are equivalent in value and therefore may substitute for one another, while absurdly strong sustainability assumes they are polar opposites. A model of sustainability considered strong therefore lies somewhere between these two ends of the sustainability spectrum. Rather than assuming any of the three components are substitutable by the others as in weak sustainability, *strong sustainability* assumes that each of the three components compliments the others, because natural capital may not necessarily have an analogous human-made substitute. Rather than viewed separate from each other, strong sustainability assumes that the three components are viewed hierarchically, with economy nested within society and society nested within the environment (Figure 23.1c). In this model, social and economic activities exist within the context of fundamental ecological processes such as nutrient cycling, climate regulation, and water purification, and economy functions within the context of a specific sociological regime (Franklin et al. 2018). Strong sustainability thus firmly places the ecological/environmental aspect at the center of the paradigm.

An example of the differences between weak, strong, and absurdly strong sustainability can be illustrated using the susceptibility of western landscapes to fire. The Colorado Front Range is an area undergoing rapid economic development and extensive exurban sprawl into forested, fire-prone areas. This expansion of the wildland–urban interface (WUI) poses great problems for federal agencies responsible for protecting private landowners from wildfires burning on adjacent National Forests. Moreover, fire suppression in the twentieth century in the area has made the ponderosa pine and mixed-conifer forests overly dense and particularly susceptible to large, severe fires with the potential to heavily impact lives, property, and the landscape itself. This context has led to a great many discussions about the sustainability of continued development and the expansion of the WUI, and arguments have varied precisely based on the type of sustainability model under consideration. Using a weak sustainability model, the continued economic growth and provision of property might substitute for the ecological integrity of the landscape, such that appropriate management would include substantial fire suppression in the wake of further expansion of the WUI. On the other hand, an absurdly strong sustainability model might argue that the environment is not substitutable regardless of the consequences for humans, such that any fires that burn on the landscape should be allowed to burn without care for lives or property as a means of allowing altered forests to heal themselves. A strong sustainability model would probably propose a carefully planned development strategy that minimized environmental impact and mitigated risk to private landowners while allowing the natural disturbance regime to persist. Although it would take region-wide cooperation and collaboration, extensive regulation and legislation, and corporate support to achieve strong sustainable development in this manner, it is clear that a strong sustainability model is the most appropriate.

Different sustainability models will often lead to different landscapes that vary in the ecosystem services they provide (Wu 2013; Figure 23.2). Wilderness areas, for example, that emphasize the preservation of biodiversity and natural ecosystems provide high natural capital and are most appropriately viewed through an absurdly strong sustainability model. Regulating ecosystem services would dominate in such a landscape, and provisioning services would effectively be absent. At the other end of the spectrum, urban landscapes would constitute a weak sustainability model most closely, having low natural capital and providing mostly cultural and regulating services. Agricultural and semi-natural landscapes might follow more of a strong sustainability model because they clearly treat social and economic factors within the context of ecological factors. Agricultural landscapes are generally more altered and thus have less natural capital than semi-natural landscapes with a higher proportion of provisioning services, but neither landscape fully substitutes natural capital with human-made capital.

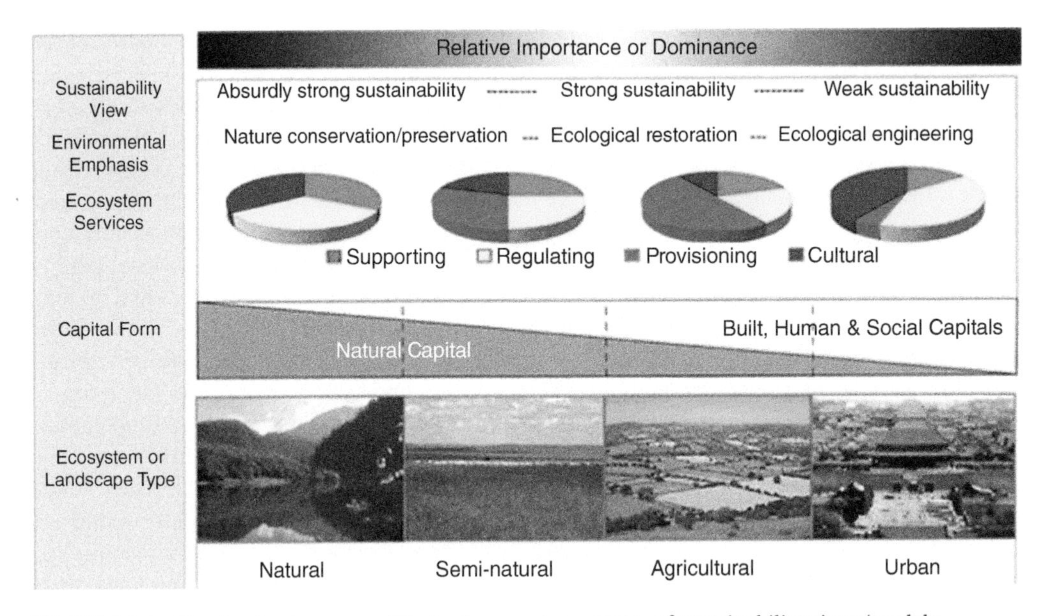

FIGURE 23.2 Varying landscapes and their placement on spectra of sustainability views/models, proportion of ecosystem services categories, and relative amount of natural capital. *Source:* Wu (2013) / Springer Nature.

The idea that different landscapes follow different sustainability models and provide different ecosystem services suggests that scale and spatial heterogeneity have strong implications for this concept. A mosaic of land use types is always present on modern landscapes, such that achieving strong sustainability at broad scales may require targets of weak sustainability (for example, in urban areas) or absurdly strong sustainability (for example, in wilderness areas) at smaller spatial scales (Wu 2013). Moreover, if landscape pattern influences ecological processes (Chapter 22) and biodiversity, then it likely also influences the provision of ecosystem services, making them patchy across space (and potentially across time). If true, then some patterns are more likely to favor the maintenance of ecosystem services, and thus some patterns are more likely to lead to sustainable landscapes (Wu 2013).

Notably, the patchiness, or spatial heterogeneity, of some ecosystem services may be important for sustaining them. In a comparison of two forested landscapes in the western United States (the Greater Yellowstone Ecosystem, a subalpine landscape of the central Rocky Mountains; and the coastal temperate rainforest of the Pacific Northwest), Turner et al. (2013) found that post-disturbance spatial heterogeneity was important for sustaining the ecosystem services of forest regeneration, primary production, carbon storage, natural hazard regulation, insect and pathogen regulation, timber production, and wildlife habitat. Spatial heterogeneity helps to sustain the supporting service of forest regeneration by determining the location of surviving pre-disturbance trees that will provide a seed source for regeneration, and this supporting service will in turn also affect other services such as primary production and carbon storage. Regulation of forest insects and pathogens is influenced by heterogeneity because it determines the spatial patterns of stands of susceptible age and size for insect or pathogen attack, as well as how their placement on a landscape might dampen or enhance an outbreak. The provisioning service of timber is clearly affected by site heterogeneity as determined by physiography, soils, and microclimate (Turner et al. 2013).

WHERE DO WE GO FROM HERE?

It is not arguable that we are in an age of ecological crisis. Nearly 20 years ago, Franklin (2003) noted that humans have extensively altered the physical and ecological contexts of forests to the point that they no longer resemble those from which the forests evolved. These include altered disturbance regimes, especially fire; changes in global and regional climate and chemical regimes (i.e., climate change and acid rain); introduction of invasive species; and fragmentation of landscapes that have changed the amount and spatial arrangement of various forest conditions. Each of these topics has been covered in detail in other chapters of this book (Chapters 10 and 20–22) and will not be revisited here, but their increased pervasiveness in modern forests has warranted each a detailed treatment that was not presented in this book's last edition. Franklin (2003) predicted that these four problems would intensify in the twenty-first century, and it seems as though he was correct.

So how do we ensure sustainability in forests? In as much as our living within an ecological crisis, this is a question still of great debate whose answer is not an easy or unambiguous one, but one that is clearly related to many of the concepts in this book. In a sense, sustainability of forests is subject to the "land sparing versus land sharing" debate (Phalan et al. 2011). One could simply keep humans out of at least some forests and allow the forests to function without human interference. This *land-sparing* argument suggests that intensively managed planation forests (sometimes called "fiber farms," usually in temperate regions) be used to supply necessary wood products, such that native forests could be left unaltered by humans and thus sustained (e.g., Bremer and Farley 2010; Paquette and Messier 2010). The adoption of fiber farms serves as a kind of ecological offset that allows larger reserve areas elsewhere on a landscape (Green et al. 2005; Fischer et al. 2008). This type of forest allocation is viewed by some as the easiest and most efficient movement toward forest sustainability because often-opposing objectives of stakeholders will conflict less often, such that substitution of human-made capital for natural capital will be less common. However, many forest managers counter that such a physical separation of forests into provisioning areas and preserves is likely to result in a failure of forests to fulfill the expectations and needs of society as well as undesirable outcomes for ecosystem function and biodiversity (Franklin 2003). These failures are likely because many forest landscapes have already been heavily altered by humans if only indirectly, and thus there are few or no candidate remnants suitable for preservation. Moreover, the lofty expectations humans have for forests in addition to wood products—such as biodiversity preservation and watershed protection—are probably not attainable without active management. Critics of land-sparing often follow the *land-sharing* model, which integrates different goals (such as biodiversity and/or ecosystem services and resource extraction) in the same area. Proponents of land-sharing in forests are primarily those who favor ecological forestry (Puettmann et al. 2008; Lindenmayer et al. 2012; Franklin et al. 2018), discussed in Chapter 14, where complex silvicultural systems (harvesting and regeneration) are developed that address more than simply the need for wood products. Of course, critics of land sharing counter that a focus on multiple goals simultaneously is less effective than a focus on either goal individually. In addition, decreased resource extraction that would accompany ecological forestry would necessitate increased extraction elsewhere (Gabriel et al. 2010).

There are certainly no easy answers for forest sustainability today, in part because sustainability is achieved by following very different pathways depending on the landscape. Lindenmayer and Cunningham (2013) summarized key principles for sustainable forests. First, there is an important need for improved planning and monitoring to avoid overcommitment of natural resources. To do so, managers and scientists must be able to identify the critical thresholds beyond which resource use will begin to degrade ecosystem services and biodiversity (Carpenter et al. 2011). Second, landscapes vary in their importance to ecosystem services and biodiversity, and thus it is critical to identify those places with disproportionate contribution to species persistence and ecological processes, such as breeding areas, refugia, or biodiversity hotspots. Research begun immediately after major natural disturbance events is often useful in this regard, particularly in understanding how refugia contribute to post-disturbance recovery (e.g., Turner et al. 2003a). Third, it is important to maintain connectivity, but only when connectivity is desirable for specific processes. Given that connectivity could refer to connected habitat for a species or group of species of interest or to connected ecological processes, and that some connectivity could lead to the spread of undesirable disturbances or species (Lindenmayer and Fischer 2013), identifying the type of connectivity important for sustainable forest management leaves much room for future research. Fourth, land-use practices may accumulate over space and time in such a way that threatens forest sustainability, and such land uses should be avoided. A classic example would be the great eastern white pine forests logged at the end of the nineteenth century in Michigan and other areas of the Lake States. Wholesale forest clearing and subsequent, repeated slash fires converted a widespread dry-mesic forest landscape to one dominated by xeric species such as jack pine and xeric oaks much more likely to burn. Finally, land uses typically fall upon a gradient between land sharing and land sparing rather than either end of the spectrum and are extremely context-dependent.

A final point about the prospects of sustainability draws on our original treatment of forest ecology as a study of landscape ecosystems, which integrates the biotic and physical aspects of whole, volumetric ecosystems. In his treatment of sustainable planning and design for large landscapes, Bailey (2002) reasons that sustaining ecosystems must occur at large spatial scales, and that local management must occur within a strong contextual knowledge of regional ecosystems. The ecological crises we described in the earlier text have at least partially resulted from a mismatch of human development and the limits of the regional ecosystems in which they occur. The sense of place provided by regional ecosystems provides patterns and processes that place bounds or limits on the social and economic components of the triple bottom line of that particular place, and thus recognizing the sense of place is necessary to conserve resources (Bailey 2002). Returning to our retrospective example of eastern white pine harvesting in Michigan, a lack of understanding and recognition of the cold winter climate, relatively flat topography, and dry, sandy soils that promote fires limited the level of post-harvesting forest regeneration and recovery that might have been expected in that regional ecosystem. Unfortunately, human activities exceeded the regional ecosystem's ecological limits in a way that has threatened its long-term sustainability.

EPILOG: EARTH AS A METAPHOR FOR LIFE

Thus far our discussion of sustainability has detailed the problems without suggesting solutions (an action Theodore Roosevelt called "whining"). We have been able to discuss in detail what sustainability is, but have been less able to describe how to achieve it in contemporary forests without a detailed discussion of forest management. Again, the fact that sustainability is defined in part by human values makes it extremely difficult—and undesirable—to simply stick to the science. As such, we take the opportunity to end this chapter (and this book) with a final abstract, rather philosophical consideration.

In thinking about sustainability, perhaps absurdly strong sustainability in particular, certain things such as biodiversity or ecological integrity have intrinsic value, independent of their value or contribution to human well-being. Rare or endangered species should be protected because they have value, regardless of how humans may benefit from them. Wilderness areas are not just valuable for the recreation, aesthetic, or spiritual values they provide humans, but also because they are an end in themselves. Ecosystem services, the major target of sustainability, contrast with this notion of intrinsic value because they are human-centered (Colyvan et al. 2009) and have value only based on what they provide for humans (Reyers et al. 2012). Focusing on the ultimate service that ecosystems provide humans is likely to oversimplify how ecosystems function (Ghazoul 2007). In addition, humans need to adapt to ecosystem function because ecosystems lack intentionality in serving humans (Salles 2011). Sustainability by definition is therefore a human-centered construct, illustrated by the simple fact that we need to label ecosystems with social or economic value to justify protecting them.

As described in the earlier text, humans have had an enormous, disproportionate influence on the ecosphere and on ecosystems at every scale. We could easily venture into a discussion of understanding humans' place in the world. Are we part of ecosystems, or external to them? Should we be stewards of ecosystems, regardless of whether or not we consider ourselves as members? Many (not all) people interested in forest ecology are likely to answer "yes" to both of these questions, but do humans deserve to be the center of focus? Rowe (1998) described "Earth" rather than "organism" as a metaphor for "life". A majority of ecologists and non-scientists alike consider organisms to represent living things, but Rowe argues that this is a reductionist view that turns the vantage point of humans inward toward themselves instead of outward toward their surroundings. Throughout this book, we have emphasized that organisms are but one part of ecosystems, and they cannot realistically be separated from physical factors of the Earth (air, water, soil, and energy). Organisms are parts of wholes, and it is the whole ecosystem that provides life to the organisms. In the past (and to some degree the present), ecological research often tended to gravitate toward the charismatic biota, either as the survival physiology of organisms or as the interactions between them. An ecological viewpoint that places organisms at its center, only later turning to what is around them, lends itself to a human-centered focus. When the whole ecosphere, rather than just the biota, is considered to be life, we put spaces at the center of our ecological viewpoint and look at its contents (Rowe 1998). Thus, humans become an integral part of ecosystems, but are not central to them.

A viewpoint alternative to one that is human-centered is not new; many have articulated the dangers of placing humans at the center of the world. In fact, many of the world's ecological problems have resulted almost exclusively from an anthropogenic view of the world. Aldo Leopold's land ethic famously moved the focus from humans to the land, with the idea that land was not a commodity owned by humans but a community that included them (Leopold 1949). Sigmund Freud (not an ecologist) suggested that humans make their biggest scientific strides when they realize they are not the center of the universe. For example, physical science leapt forward when Copernicus speculated that the Earth was not the center of the universe; biology advanced greatly when Darwin suggested that humans actually descended from other animals; and psychology developed quickly from Freud's theory that a personality is driven in part by the unconscious mind, out of a person's control (Freud 1917). It is certainly true that an additional scientific leap forward will be needed to find new solutions in this age of ecological crises, perhaps one facilitated by encouraging humans to consider that they are a tenant rather than the steward of ecosystems.

How does this philosophy translate to sustainability? It suggests that ecosystems have organisms, not the reverse; thus, ecosystems have humans, not the reverse. Rowe (2000) uses an analogy of the heart in a human body to describe the relationship of humans to ecosystems. The role of the heart is to keep the body healthy, not to oversee the body. Likewise, perhaps the role of humans is to keep ecosystems healthy, but as parts of those wholes rather than as their stewards. Transposing this to ecosystem services, ecosystems do not serve humans; in fact, the reverse should be true. We have no illusions to the fact that this view of the world is awkward and would be difficult if not impossible to instill within our current models of sustainability in forests. At the very least, however, we return to the notion of understanding forest ecology, and of ecology in general, as the study of whole ecosystems that include more than organisms (including humans) alone. If nothing else, that resembles a good start.

SUGGESTED READINGS

Bailey, R.G. (2002). *Ecoregion-Based Design for Sustainability*. New York: Springer 223 pp.

Balvanera, P., Siddique, I., Dee, L. et al. (2014). Linking biodiversity and ecosystem services: current uncertainties and the necessary next steps. *Bioscience* 64: 49–57.

Daily, G.C. (ed.) (1997). *Nature's Services: Societal Dependence on Natural Ecosystems*. Washington, DC: Island Press 412 pp.

Lindenmayer, D.B. and Franklin, J.F. (ed.) (2003). *Towards Forest Sustainability*. Washington, D.C: Island Press 212 pp.

Mace, G.M., Norris, K., and Fitter, A.H. (2012). Biodiversity and ecosystem services: a multilayered relationship. *Trends Ecol. Evol.* 27: 19–26.

Turner, M.G., Donato, D.C., and Romme, W.H. (2013). Consequences of spatial heterogeneity for ecosystem services in changing forest landscapes: priorities for future research. *Landsc. Ecol.* 28: 1081–1097.

Wu, J. (2013). Landscape sustainability science: ecosystem services and human well-being in changing landscapes. *Landsc. Ecol.* 28: 999–1023.

References

Abatzoglou, J.T. and Kolden, C.A. (2013). Relationships between climate and macroscale area burned in the western United States. *Int. J. Wildland Fire* 22: 1003–1020.

Abbey, E.P. (1968). *Desert Solitaire*. New York: McGraw-Hill 336 pp.

Abella, S.R. and Covington, W.W. (2006a). Vegetation–environment relationships and ecological species groups of an Arizona *Pinus ponderosa* landscape, USA. *Plant Ecol.* 185: 255–268.

Abella, S.R. and Covington, W.W. (2006b). Forest ecosystems of an Arizona Pinus ponderosa landscape: multifactor classification and implications for ecological restoration. *J. Biogeogr.* 33: 1368–1383.

Abella, S. and Shelburne, V.B. (2004). Ecological species groups of South Carolina's Jocassee Gorges, southern Appalachian Mountains. *J. Torry Bot. Soc.* 131: 220–231.

Abella, S.R., Shelburne, V.B., and Macdonald, N.W. (2003). Multifactor classification of forest landscape ecosystems of Jocassee Gorges, southern Appalachian Mountains, South Carolina. *Can. J. For. Res.* 33: 1933–1946.

Aber, J.D. and Melillo, J.M. (1991). *Terrestrial Ecosystems*. Philadelphia: Saunders 429 pp.

Abrams, M.D. (1990). Adaptations and responses to drought in *Quercus* species of North America. *Tree Physiol.* 7: 227–238.

Abrams, M.D. (1992). Fire and the development of oak forests. *Bioscience* 42: 346–353.

Abrams, M.D. (2001). Eastern white pine versatility in the presettlement forest. *Bioscience* 51: 967–979.

Abrams, M.D. (2003). Where has all the white oak gone? *Bioscience* 53: 927–939.

Abrams, M.D. and Downs, J.A. (1990). Successional replacement of old-growth white oak by mixed-mesophytic hardwoods in southwest Pennsylvania. *Can. J. For. Res.* 20: 1864–1870.

Abrams, M.D. and Nowacki, G.J. (2015). Exploring the early Anthropocene burning hypothesis and climate-fire anomalies for the eastern US. *J. Sustain. For.* 34: 30–48.

Abrams, M.D. and Nowacki, G.J. (2019). Global change impacts on forest and fire dynamics using paleoecology and tree census data for eastern North America. *Ann. For. Sci.* 76: 8.

Abrams, M.D. and Scott, M.L. (1989). Disturbance-mediated accelerated succession in two Michigan forest types. *For. Sci.* 35: 42–49.

Abrams, M.D., Sprugel, D.G., and Dickmann, D.I. (1985). Multiple successional pathways on recently disturbed jack pine sites in Michigan. *For. Ecol. Manag.* 10: 31–48.

Abrams, M.D., Kubiske, M.E., and Mostoller, S.A. (1994). Relating wet and dry year ecophysiology to leaf structure in contrasting temperate tree species. *Ecology* 75: 123–133.

Adams, W.T. (1992). Gene dispersal within forest tree populations. In: *Population Genetics of Forest Trees* (ed. W.T. Adams, S.H. Strauss, D.L. Copes and A.R. Griffin), 217–240. Boston, MA: Kluwer.

Adams J.M. (1997). Global land environments since the last interglacial. Oak Ridge National Laboratory, TN, USA. https://www.esd.ornl.gov/projects/qen/adams1.html. Accessed 02-17-2021.

Adams, A.B., Dale, V.H., Kruckeberg, A.R., and Smith, E. (1987). Plant survival, growth form and regeneration following the May 18, 1980, eruption of Mount St. Helens, Washington. *Northwest Sci.* 61: 160–170.

Adams, M.A., Buckley, T.N., and Turnbull, T.L. (2020). Diminishing CO_2-driven gains in water-use efficiency of global forests. *Nat. Clim. Chang.* 10: 466–471.

Adkins, J.K. and Rieske, L.K. (2015). A terrestrial invader threatens a benthic community: potential effects of hemlock woolly adelgid-induced loss of eastern hemlock on invertebrate shredders in headwater streams. *Biol. Invasions* 17: 1163–1179.

Aerts, R. (2006). The freezer defrosting: global warming and litter decomposition rates in cold biomes. *J. Ecol.* 94: 713–724.

Agee, J.K. (1973). Prescribed fire effects on physical and hydrologic properties of mixed-conifer forest floor and soil. Report 143, Univ. California Resources Center, Davis, CA.

Agee, J.K. (1993). *Fire Ecology of Pacific Northwest Forests*. Washington D.C: Island Press 493 pp.

Forest Ecology, Fifth Edition. Daniel M. Kashian, Donald R. Zak, Burton V. Barnes, and Stephen H. Spurr.

Agee, J.K. (1998). The landscape ecology of western forest fire regimes. *Northwest Sci.* 72: 24.

Agren, G.I., Axelsson, B., Flower-Ellis, J.G.K. et al. (1980). Annual carbon budget for a young Scots pine. *Ecological Bulletins* 32: 307–313.

Ahlgren, C. (1974a). Effects of fires on temperate forests: north central United States. In: *Fire and Ecosystems* (ed. T.T. Kozlowski and C.E. Ahlgren), 46–72. New York: Academic Press.

Ahlgren, I.F. (1974b). The effect of fire on soil organisms. In: *Fire and Ecosystems* (ed. T.T. Kozlowski and C.E. Ahlgren). New York: Academic Press.

Ahuja, M.R. (2005). Polyploidy in gymnosperms: revisited. *Silvae Genet.* 54: 59–69.

Ahuja, M.R. and Libby, W.J. (ed.) (1993). *Clonal Forestry. Vol. I, Genetics and Biotechnology*, 277 pp. Vol II, Conservation and Application, 240 pp. New York: Springer-Verlag.

Ainsworth, E.A. and Rogers, A. (2007). The response of photosynthesis and stomatal conductance to rising [CO_2]: mechanisms and environmental interactions. *Plant Cell Environ.* 30: 258–270.

Aitken, S.N., Yeaman, S., Holliday, J.A. et al. (2008). Adaptation, migration or extirpation: climate change outcomes for tree populations. *Evol. Appl.* 1: 95–111.

Albert, D.A. (1995). Regional landscape ecosystems of Michigan, Minnesota, and Wisconsin: a working map and classification. USDA For. Serv. Gen. Tech. Report NC-178. North Central For. Exp. Sta., St. Paul, MN. 250 pp. + map.

Albert, D.A. and Barnes, B.V. (1987). Effects of clearcutting on the vegetation and soil of a sugar maple-dominated ecosystem, western Upper Michigan. *For. Ecol. Manag.* 18: 283–298.

Albert, D.A., Denton, S.R., and Barnes, B.V. (1986). *Regional Landscape Ecosystems of Michigan*. Ann Arbor: School of Natural Resources, University of Michigan 32 pp.

Aldrich, P.R., Parker, G.R., Severson, J.R., and Michler, C.H. (2005). Confirmation of oak recruitment failure in Indiana old growth forest: 75 years of data. *For. Sci.* 51: 406–416.

Alexander, R.R. (1964). Minimizing windfall around clear cuttings in spruce-fir forests. *For. Sci.* 10: 130–142.

Alexander, R.R. (1966). Site indexes for lodgepole pine, with corrections for stand density; instructions for field use. USDA For. Serv. Res. Paper RM-24. Rocky Mountain For. and Rge. Exp. Sta., Fort Collins, CO. 7 pp.

Alexander, M. (1977). *Introduction to Soil Microbiology*, 2e. New York: Wiley 467 pp.

Alexander, R.R. (1985). Major habitat types, community types, and plant communities in the Rocky Mountains. USDA For. Serv. Gen. Tech. Report RM-123. Fort Collins, CO. 105 pp.

Alexander, R.R. (1986). Classification of the forest vegetation of Wyoming. USDA. For. Serv. Res. Note RM-466. Rocky Mountain For. and Rge. Exp. Sta., Fort Collins, CO. 10 pp.

Alexander, H.D. and Arthur, M.A. (2014). Increasing red maple leaf litter alters decomposition rates and nitrogen cycling in historically oak-dominated forests of the eastern U.S. *Ecosystems* 17: 1371–1383.

Alexander, R.R., Tackle, D., and Dahms, W. (1967). Site indexes for lodgepole pine, with corrections for stand density; methodology. USDA For. Serv. Res. Paper RM-29. Rocky Mountain For. and Rge. Exp. Sta., Fort Collins, CO. 18 pp.

Alexander, N.L., Flint, H.L., and Hammer, P.A. (1984). Variation in cold hardiness of *Fraxinus americana* stem tissue according to geographic origin. *Ecology* 65: 1087–1092.

Alexander, R.R., Hoffman, G.R., and Wirsing, J.M. (1986). Forest vegetation of the Medicine Bow National Forest in southeastern Wyoming: a habitat type classification. USDA For. Serv. Res. Paper RM-271. Rocky Mountain For. and Rge. Res. Sta. Fort Collins, CO. 39 pp.

Alfaro, R., Campbell, E., and Hawkes, B. (2010). *Historical Frequency, Intensity, and Extent of Mountain Pine Beetle Disturbance in Landscapes of British Columbia*. Victoria, B.C.: Canadian Forest Service, Pacific Forestry Centre 52 pp.

Alho, C.J.R. (2008). The value of biodiversity. *Brazilian Journal of Biology* 68: 1115–1118.

Alix-Garcia, J. and Wolff, H. (2014). Payment for ecosystem services from forests. *Annu. Rev. Resour. Econ.* 6: 361–380.

Allen, G.S. and Owens, J.N. (1972). *The Life History of Douglas-fir*. Ottawa: Environment Canada, For. Serv 139 pp.

Allen, R.B. and Platt, K.H. (1990). Annual seedfall variation in *Nothofagus solandri* (Fagaceae), Canterbury, New Zealand. *Oikos* 57: 199–206.

Allen, A.P., Brown, J.H., and Gillooly, J.F. (2002). Global biodiversity, biochemical kinetics, and the energetic-equivalence rule. *Science* 297: 1545–1548.

Allen, C.D., Macalady, A.K., Chenchouni, H. et al. (2010). A global overview of drought and heat-induced tree mortality reveals emerging climate change risks for forests. *For. Ecol. Manag.* 259: 660–684.

Allen, C.D., Breshears, D.D., and McDowell, N.G. (2015). On underestimation of global vulnerability to tree mortality and forest die-off from hotter drought in the Anthropocene. *Ecosphere* 6: 1–55.

Allen, R.B., Millard, P., and Richardson, S.J. (2017). A resource centric view of climate and mast seeding in trees. *Prog. Bot.* 79: 233–268.

Allendorf, F.W., Luikart, G.H., and Aitkan, S.N. (2012). *Conservation and the Genetics of Populations*, 2e. Hoboken, NJ: Wiley Blackwell 624 pp.

Alvarez-Uria, P. and Körner, C. (2007). Low temperature limits of root growth in deciduous and evergreen temperate tree species. *Funct. Ecol.* 21: 211–218.

Amaral, J., Ribeyre, Z., Vigneaud, J. et al. (2020). Advances and promises of epigenetics for forest trees. *Forests* 11: 976.

Amaranthus, M.P., Trappe, J.M., Bednar, L., and Arthur, D. (1994). Hypogeous fungal production in mature Douglas-fir forest fragments and surrounding plantations and its relation to coarse woody debris and animal mycophagy. *Can. J. For. Res.* 24: 2157–2165.

Ambroise, V., Legay, S., Guerriero, G. et al. (2020). The roots of plant frost hardiness and tolerance. *Plant Cell Physiol.* 61: 3–20.

Amoroso, M.M., Daniels, L.D., Baker, P.J., and Camarero, J.J. (ed.) (2017). *Dendroecology: Tree-Ring Analyses Applied to Ecological Studies*. Cham, Switzerland: Springer 420 pp.

Amthor, J.S. (1984). The role of maintenance respiration in plant growth. *Plant Cell Environ.* 7: 561–569.

An, M., Deng, M., Zheng, S.S. et al. (2017). Introgression threatens the genetic diversity of *Quercus austrocochinchinensis* (Fagaceae), an endangered oak: a case inferred by molecular markers. *Front. Plant Sci.* 8: 229.

Anderson, E. (1948). Hybridization of the habitat. *Evolution* 2: 1–9.

Anderson, M.G. and Ferree, C.E. (2010). Conserving the stage: Climate change and the geophysical underpinnings of species diversity. *PLoS One* 5: e11554.

Anderson, J.M. and Swift, M.J. (1983). Decomposition in tropical forests. In: *Tropical Rain Forest: Ecology and Management* (ed. S.L. Sutton, T.C. Whitmore and A.C. Chadwick), 287–308. Oxford: Blackwell.

Anderson, M.G., Clark, M., and Sheldon, A.O. (2014). Estimating climate resilience for conservation across geophysical settings. *Conserv. Biol.* 28: 959–970.

Anderson, M. G., Barnett, A., Clark, M. (2016). Resilient and connected landscapes for terrestrial conservation. http://easterndivision.s3.amazonaws.com/ResilientandConnectedLandscapesForTerrestialConservation.pdf. Accessed 01-10-2022.

Anderson-Teixeira, K.J., Miller, A.D., Mohan, J.E. et al. (2013). Altered dynamics of forest recovery under a changing climate. *Glob. Chang. Biol.* 19: 2001–2021.

Andersson, E. (1963). Seed stands and seed orchards in the breeding of conifers, World Consult. For. Gen. and For. Tree Imp. Proc. II FAO-FORGEN 63- 8/1:1–18.

Andrus, R.A., Hart, S.J., Tutland, N., and Veblen, T.T. (2021). Future dominance by quaking aspen expected following short-interval, compounded disturbance interaction. *Ecosphere* 12: e03345.

Angelini, C. and Silliman, B.R. (2014). Secondary foundation species as drivers of trophic and functional diversity: evidence from a tree–epiphyte system. *Ecology* 95: 185–196.

Angelini, C., Altieri, A.H., Silliman, B.R., and Bertness, M.D. (2011). Interactions among foundation species and their consequences for community organization, biodiversity, and conservation. *Bioscience* 61: 782–789.

Antibus, R.K., Croxdale, J.G., Miller, O.K., and Linkins, A.E. (1981). Ectomycorrhizal fungi of *Salix rotundifolia*. III. Resynthesized mycorrhizal complexes and their surface phosphatase activities. *Can. J. Bot.* 59: 2458–2465.

Antonovics, J. (1971). The effects of a heterogeneous environment on the genetics of natural populations. *Am. Sci.* 59: 593–599.

Arbeitsgruppe Biözonosenkunde (1995). *Ansätze für eine Regionale Biotop-und Biozönosenkunde von Baden-Wurttemberg*. Baden-Wurttemberg. Freiburg, Germany: Mitt. Forst. Versuchs u. Forschungsanst 166 pp.

Archambault, L., Barnes, B.V., and Witter, J.A. (1989). Ecological species groups of oak ecosystems of southeastern Michigan, USA. *For. Sci.* 35: 1058–1074.

Archambault, L., Barnes, B.V., and Witter, J.A. (1990). Landscape ecosystems of disturbed oak forests of southeastern Michigan, USA. *Can. J. For. Res.* 20: 1570–1582.

Arduini, G., Chemel, C., and Staquet, C. (2020). Local and non-local controls on a persistent cold-air pool in the Arve River Valley. *Q. J. R. Meteorol. Soc.* 146: 2497–2521.

Armand, A.D. (1992). Sharp and gradual mountain timberlines as a result of species interaction. In: *Landscape Boundaries: Consequences for Biotic Diversity and Ecological Flows* (ed. A.J. Hansen and F. di Castri), 360–379. New York: Springer.

Arno, S.F. (1976). The historical role of fire on the Bitterroot National Forest. USDA For. Serv. Res. Paper INT-187. Intermountain For. and Rge. Exp. Sta., Ogden, UT. 29 pp.

Arno, S.F. (1980). Forest fire history of the northern Rockies. *J. For.* 78: 460–465.

Arno, S.F. and Fischer, W.C. (1989). Using vegetation classifications to guide fire management. *In* D. E. Ferguson, P. Morgan, and F. D. Johnson (comps.), *Proceedings—Land Classifications Based on Vegetation: Applications for Resource Management*. USDA For. Serv. Gen. Tech. Report INT-257, Intermountain Res. Sta, Ogden, UT.

Arno, S.E, Scott, J.H., and Hartwell, M.G. (1995). Age-class structure of old growth ponderosa pine/ Douglas-fir stands and its relationship to fire history. USDA, For. Serv. Res. Paper INT-RP-481, Intermountain Res. Sta., Odgen, UT. 25 pp.

Arno, S.F., Parsons, D.J., and Keane, R.E. (2000). Mixed-severity fire regimes in the northern Rocky Mountains: consequences of fire exclusion and options for the future. In: *Wilderness Science in a Time of Change Conference* (ed. D.N. Cole, S.F. McCool, W.A. Freimund and J. O'Loughlin) (comps.). 1999 May 23–27; Missoula, MT. USDA For. Serv. Rocky Mountain Res. Sta. Proceedings RMRS-P-15-VOL-1. Ogden, UT.

Arnold, M.L. (1994). Natural hybridization and Louisiana irises. *Bioscience* 44: 141–147.

Arnold, M.L. (2015). *Divergence with Genetic Exchange*. Oxford, UK: Oxford University Press 272 pp.

Aronson, J., Murcia, C., Kattan, G.H. et al. (2014). The road to confusion is paved with novel ecosystem labels: a reply to Hobbs et al. *Trends Ecol. Evol.* 29: 646–647.

Ashcroft, M.B., Gollan, J.R., Warton, D.I., and Ramp, D. (2012). A novel approach to quantify and locate potential microrefugia using topoclimate, climate stability, and isolation from the matrix. *Glob. Chang. Biol.* 18: 1866–1879.

Ashton, P.S. (1992). Species richness in plant communities. In: *Conservation Biology* (ed. P.L. Fiedler and K.S. Jain), 3–22. London: Chapman and Hall.

Ashton, P.M.S. and Berlyn, G.P. (1994). A comparison of leaf physiology and anatomy of Quercus (section of Erythrobalanus-Fagaceae) species in different light environments. *Am. J. Bot.* 81: 589–597.

Ashton, M.S. and Kelty, M.J. (2018). *The Practice of Silviculture: Applied Forest Ecology*, 10e. Hoboken, New Jersey: Wiley 776 pp.

Ashton, M.S., Tyrrell, M.L., Spalding, D., and Gentry, B. (ed.) (2012). *Managing Forest Carbon in a Changing Climate*. New York: Springer 424 pp.

Asner, G.P. and Vitousek, P.M. (2005). Remote analysis of biological invasion and biogeochemical change. *Proc. Natl. Acad. Sci.* 102: 4383–4386.

Assmann, E. (1970). *The Principles of Forest Yield Studies*. Oxford: Pergamon Press 506 pp.

Aston, J.L. and Bradshaw, A.D. (1966). Evolution in closely adjacent plant populations. II. *Agrostis stolonifera* in maritime habitats. *Heredity* 21: 649–664.

Atchley, A.L., Kinoshita, A.M., Lopez, S.R. et al. (2018). Simulating surface and subsurface water balance changes due to burn severity. *Vadose Zone J.* 17: 180099.

Atkins, J.W., Bohrer, G., Fahey, R.T. et al. (2018a). Quantifying vegetation and canopy structural complexity from terrestrial LiDAR data using the FORESTR R package. *Methods Ecol. Evol.* 9: 2057–2066.

Atkins, J.W., Fahey, R.T., Hardiman, B.S., and Gough, C.M. (2018b). Forest canopy structural complexity and light absorption relationships at the subcontinental scale. *J. Geophys. Res. Biogeosci.* 123: 1387–1405.

Augenbaugh, J.E. (1935). Replacement of chestnut in Pennsylvania. Pa. Dept. For. Waters. *Water Resour. Bull.* 54: 1–38.

Augspurger, C.K. (2008). Early spring leaf out enhances growth and survival of saplings in a temperate deciduous forest. *Oecologia* 156: 281–286.

Augspurger, C.K. (2009). Spring 2007 warmth and frost: phenology, damage and refoliation in a temperate deciduous forest. *Funct. Ecol.* 23: 1031–1039.

Aukema, J.E., McCullough, D.G., Von Holle, B. et al. (2010). Historical accumulation of nonindigenous forest pests in the continental United States. *Bioscience* 60: 886–897.

Aukema, J.E., Leung, B., Kovacs, K. et al. (2011). Economic impacts of non-native forest insects in the continental United States. *PLoS One* 6: e24587.

Austerlitz, F., Mariette, S., Machon, N. et al. (2000). Effects of colonization processes on genetic diversity: differences between annual plants and tree species. *Genetics* 154: 1309–1321.

Austin, M.P. (1985). Continuum concept, ordination methods, and niche theory. *Annu. Rev. Ecol. Syst.* 16: 39–61.

Austin, M.P. (1990). Community theory and competition in vegetation. In: *Perspectives on Plant Competition* (ed. J.B. Grace and D. Tilman), 215–238. New York: Academic Press.

Austin, M.P. and Smith, T.M. (1989). A new model for the continuum concept. *Vegetatio* 83: 35–47.

Avers, P.E. and Schlatterer, E.F. (1991). Ecosystem classification and management on National Forests. In: *Proceedings of the 1991 Symposium on Systems Analysis in Forest Resources*. March 3–6, 1991, Charleston South Carolina. USDA For. Serv. Southeastern Exp. Sta, Asheville, NC. 423 pp.

Ayres, M.P. and Lombardero, M.J. (2000). Assessing the consequences of global change for forest disturbance from herbivores and pathogens. *Sci. Total Environ.* 262: 263–286.

Azizi Jalilian, M., Shayesteh, K., Danehkar, A., and Salmanmahiny, A. (2020). A new ecosystem-based land classification of Iran for conservation goals. *Environ. Monit. Assess.* 192: 1–17.

Baack, E., Melo, M.C., Rieseberg, L.H., and Ortiz-Barrientos, D. (2015). The origins of reproductive isolation in plants. *New Phytol.* 207: 968–984.

Bacles, C.F.E., Lowe, A.J., and Ennos, R.A. (2006). Effective seed dispersal across a fragmented landscape. *Science* 311: 628.

Bailey, R.G. (1983). Delineation of ecosystem regions. *Environ. Manag.* 7: 365–373.

Bailey, R.G. (1988). Ecogeographic analysis: a guide to the ecological division of land for resource management. USDA For. Serv. Misc. Publ. No. 1465. Washington, D.C. 18 pp.

Bailey, R.G. (1994). Ecoregions of the United States (map). USDA For. Serv., Washington, D.C.

Bailey, R.G. (1995). Description of the ecoregions of the United States. USDA For. Serv. Misc. Publ. 1391. Washington, D.C. 108 pp. + map

Bailey, R.G. (2002). *Ecoregion-Based Design for Sustainability*, 2e. New York: Springer 236 pp.

Bailey, R.G. (2009). *Ecosystem Geography*, 2e. New York: Springer-Verlag 251 pp + 1 map.

Bailey, R.G. and Hogg, H.C. (1986). A world ecoregions map for resource reporting. *Environ. Conserv.* 13: 195–202.

Bailey, R.G., Zoltai, S.C., and Wiken, E.B. (1985). Ecological regionalization in Canada and the United States. *Geoforum* 16: 265–275.

Bailey, J.J., Boyd, D.S., Hjort, J. et al. (2017). Modelling native and alien vascular plant species richness: At which scales is geodiversity most relevant? *Glob. Ecol. Biogeogr.* 26: 763–776.

Baker, F.S. (1949). A revised tolerance table. *J. For.* 47: 179–181.

Baker, W.L. (2009). *Fire Ecology in Rocky Mountain Landscapes*. Washington, DC: Island Press 632 pp.

Baker, M.E. and Barnes, B.V. (1998). Landscape ecosystem diversity of river floodplains in northwestern Lower Michigan, USA. *Can. J. For. Res.* 28: 1405–1418.

Baker, W.L. and Ehle, D. (2001). Uncertainty in surface-fire history: the case of ponderosa pine forests in the western United States. *Can. J. For. Res.* 31: 1205–1226.

Baker, H.G., Bawa, K.S., Frankie, G.W., and Opler, P.A. (1983). Reproductive biology of plants in tropical forests. In: *Tropical Rainforest Ecosystems* (ed. F.B. Golley), 183–215. New York: Elsevier.

Baker, W.L., Monroe, J.A., and Hessl, A.E. (1997). The effects of elk on aspen in the winter range in Rocky Mountain National Park. *Ecography* 20: 155–165.

Baldwin, D.J.B., Desloges, J.R., and Band, L.E. (2000). Physical geography of Ontario. In: *Ecology of a Managed Terrestrial Landscape: Patterns and Processes of Forest Landscapes in Ontario* (ed. A.H. Perera, D.L. Euler and I.D. Thompson), 12–29. Vancouver, British Columbia: UBC Press.

Balisky, A.C. and Burton, P.J. (1995). Root-zone soil temperature variation associated with microsite characteristics in high-elevation forest openings in the interior of British Columbia. *Agric. For. Meteorol.* 77: 31–54.

Baltzer, J.L. and Thomas, S.C. (2007). Determinants of whole-plant light requirements in Bornean rain forest tree saplings. *J. Ecol.* 95: 1208–1221.

Baluška, F., Čiamporová, M., Gašparíková, O., and Barlow, P.W. (1995). *Structure and Function of Roots*. Norwell, MA: Kluwer Academic Publishers 364 pp.

Balvanera, P., Pfisterer, A.B., Buchmann, N. et al. (2006). Quantifying the evidence for biodiversity effects on ecosystem functioning and services. *Ecol. Lett.* 9: 1146–1156.

Balvanera, P., Siddique, I., Dee, L. et al. (2014). Linking biodiversity and ecosystem services: current uncertainties and the necessary next steps. *Bioscience* 64: 49–57.

Bannister, M.H. (1965). Variation in the breeding system of *Pinus radiata*. In: *The Genetics of Colonizing Species* (ed. H.G. Baker and G.L. Stebbins), 353–372. New York: Academic Press.

Bao, Y. and Nilsen, E.T. (1988). The ecophysiological significance of thermotropic leaf movements in *Rhododendron maximum*. *Ecology* 69: 1578–1587.

Barber, H.N. (1955). Adaptive gene substitutions in Tasmanian eucalyptus: I. genes controlling the development of glaucousness. *Evolution* 9: 1–14.

Barber, H.N. and Jackson, W.D. (1957). Natural selection in action in Eucalyptus. *Nature* 179: 1267–1269.

Barbero, R., Abatzoglou, J.T., Larkin, N.K. et al. (2015). Climate change presents increased potential for very large fires in the contiguous United States. *Int. J. Wildland Fire* 24: 892–899.

Barbour, M.G. and Billings, W.D. (1988). *North American Terrestrial Vegetation*. Cambridge: Cambridge Univ. Press 434 pp.

Barbour, M.G. and Christensen, N.L. (1993). Vegetation. In: *Flora of North America, North of Mexico*, vol. 1 (ed. R.R. Morin) (conv. ed.), 97–131. New York: Oxford Univ. Press.

Barden, L.S. (1981). Forest development in canopy gaps of a diverse hardwood forest of the southern Appalachian Mountains. *Oikos* 37: 205–209.

Barnes, B.V. (1967). The clonal growth habit of American aspens. *Ecology* 47: 439–447.

Barnes, B.V. (1969). Natural variation and delineation of clones of *Populus tremuloides* and *P. grandidentata* in northern Lower Michigan. *Silvae Genet.* 18: 130–142.

Barnes, B.V. (1975). Phenotypic variation of trembling aspen in western North America. *For. Sci.* 21: 319–328.

Barnes, B.V. (1976). Succession in deciduous swamp communities of southeastern Michigan, formerly dominated by American elm. *Can. J. Bot.* 54: 19–24.

Barnes, B.V. (1977). The International Larch Provenance Test in southeastern Michigan, USA. *Silvae Genet.* 26: 145–148.

Barnes, B.V. (1984). Forest ecosystem classification and mapping in Baden-Württemberg, Germany. In: *Symp. Proc. Forest Land Classification: Experience, Problems, Perspectives* (ed. J.G. Bockheim). NCR-102 North Central For. Soils Com., Soc. Am. For., USDA For. Serv., and USDA Soil Cons. Serv. Madison, WI.

Barnes, B.V. (1991). Deciduous forests of North America. In: *Ecosystems of the World, Temperate Deciduous Forests*, vol. 7 (ed. E. Röhrig and B. Ulrich), 219–344. New York: Elsevier.

Barnes, B.V. (1993). The landscape ecosystem approach and conservation of endangered spaces. *Endanger. Species Update* 10: 13–19.

Barnes, B.V. (1996). Silviculture, landscape ecosystems, and the iron law of the site. *Forstarchiv* 67: 226–235.

Barnes, B.V. and Dancik, B.P. (1985). Characteristics and origin of a new birch species, *Betula murrayana*, from southeastern Michigan. *Can. J. Bot.* 63: 223–226.

Barnes, B.V. and Wagner, W.H. (2004). *Michigan Trees*. Ann Arbor, MI: University of Michigan Press 456 pp.

Barnes, B.V., Pregitzer, K.S., Spies, T.A., and Spooner, V. (1982). Ecological forest site classification. *J. For.* 80: 493–498.

Barnes, B.V., Xü, Z., and Zhao, S. (1992). Forest ecosystems in an old-growth pine—mixed hardwood forest of the Changbai Shan Preserve in northeastern China. *Can. J. For. Res.* 22: 144–160.

Barnes, B.V., Zak, D.R., Denton, S.R., and Spurr, S.H. (1998). *Forest Ecology*, 4e. New York: Wiley 774 pp.

Barnosky, C.W. (1987). Response of vegetation to climatic changes of different duration in the Late Neocene. *Trends Ecol. Evol.* 2: 247–250.

Barrios-Garcia, M.N. and Ballari, S.A. (2012). Impact of wild boar (*Sus scrofa*) in its introduced and native range: a review. *Biol. Invasions* 14: 2283–2300.

Barry, R.C. (1967). Seasonal location of the Arctic Front over North America. *Geogr. Bull.* 9: 79–95.

Barthelemy, D., Edelin, C., and Hallé, F. (1991). Canopy architecture. In: *Physiology of Trees* (ed. A.S. Raghavendra), 1–20. New York: Wiley.

Barton, N.H. (2001). The role of hybridization in evolution. *Mol. Ecol.* 10: 551–568.

Barton, K.E. and Koricheva, J. (2010). The ontogeny of plant defense and herbivory: characterizing general patterns using meta-analysis. *Am. Nat.* 175: 481–493.

Basey, J.M., Jenkins, S.H., and Busher, P.E. (1988). Optimal central-place foraging by beavers: tree-size selection in relation to defensive chemicals of quaking aspen. *Oecologia* 76: 278–282.

Basey, J.M., Jenkins, S.H., and Miller, G.C. (1990). Food selection by beavers in relation to inducible defenses of *Populus tremuloides*. *Oikos* 59: 57–62.

Baskin, Y. (2002). *A Plague of Rats and Rubber Vines: The Growing Threat of Species Invasions*. Washington, D.C.: Island Press 330 pp.

Basler, D. and Körner, C. (2012). Photoperiod sensitivity of bud burst in 14 temperate forest tree species. *Agric. Meteorol.* 165: 73–81.

Battaglia, L.L. and Shari, R.R. (2006). Responses of floodplain forest species to spatially condensed gradients: a test of the flood–shade tolerance trade-off hypothesis. *Oecologia* 147: 108–118.

Bauerle, W.L., Bowden, J.D., Wang, G.G., and Shahba, M.A. (2009). Exploring the importance of within-canopy spatial temperature variation on transpiration predictions. *J. Exp. Bot.* 60: 3665–3676.

Baumeister, D. and Callaway, R.M. (2006). Facilitation by *Pinus flexilis* during succession: a hierarchy of mechanisms benefits other plant species. *Ecology* 87: 1816–1830.

Bazzaz, F.A. (1979). The physiological ecology of plant succession. *Annu. Rev. Ecol. Syst.* 10: 351–372.

Beaufait, W.R. (1956). Influence of soil and topography on willow oak sites. U.S. For. Sen., Southern For. Exp. Sta. Occ. Paper 148. 12 pp.

Beck, D.E. (1971). Polymorphic site index curves for white pine in the southern Appalachians. USDA For. Serv. Res. Note SE-80. Southeastern For. Exp. Sta., Asheville, NC. 8 pp.

Beck, D.E. (1977). Twelve-year acorn yield in southern Appalachian oaks. USDA For. Serv. Res. Note SE-244, Southeastern For. Exp. Sta., Asheville, NC. 8 pp.

Beck, E.H., Heim, R., and Hansen, J. (2004). Plant resistance to cold stress: mechanisms and environmental signals triggering frost hardening and dehardening. *J. Biosci.* 29: 449–459.

Becking, R.W. (1957). The Zürich-Montpellier school of phytosociology. *Bot. Rev.* 23: 411–488.

Beese, W.J., Deal, J., Dunsworth, B.G. et al. (2019a). Two decades of variable retention in British Columbia: a review of its implementation and effectiveness for biodiversity conservation. *Ecol. Process.* 8: 1–22.

Beese, W.J., Rollerson, T.P., and Peters, C.M. (2019b). Quantifying wind damage associated with variable retention harvesting in coastal British Columbia. *For. Ecol. Manag.* 443: 117–131.

Begon, M., Harper, J.L., and Townsend, C.R. (1990). *Ecology, Individuals, Populations, and Communities*, 945. Boston: Blackwell Sci. Publ.

Beisner, B.E., Haydon, D.T., and Cuddington, K. (2003). Alternative stable states in ecology. *Front. Ecol. Environ.* 1: 376–382.

Belahbib, N., Pemonge, M.H., Ouassou, A. et al. (2001). Frequent cytoplasmic exchanges between oak species that are not closely related: *Quercus suber* and *Q. ilex* in Morocco. *Mol. Ecol.* 10: 2003–2012.

Bell, A.D. (1991). *Plant Form, An Illustrated Guide to Flowering Plant Morphology*. New York: Oxford Univ. Press 341 pp.

Belleau, A., Leduc, A., Lecomte, N., and Bergeron, Y. (2011). Forest succession rate and pathways on different surface deposit types in the boreal forest of northwestern Quebec. *Ecoscience* 18: 329–340.

Belsky, A.J. and Blumenthal, D.M. (1997). Effects of livestock grazing on stand dynamics and soils in upland forests of the interior West. *Conserv. Biol.* 11: 315–327.

Bemmels, J.B., Knowles, L.L., and Dick, C.W. (2019). Genomic evidence of survival near ice sheet margins for some, but not all, North American trees. *Proc. Natl. Acad. Sci.* 116: 8431–8436.

Bender, E.A., Case, T.J., and Gilpin, M.E. (1984). Perturbation experiments in ecology: theory and practice. *Ecology* 65: 1–13.

Bender, M.H., Baskin, J.M., and Baskin, C.C. (2000). Age of maturity and life span in herbaceous polycarpic perennials. *Bot. Rev.* 66: 311–349.

Bender, D.J., Tischendorf, L., and Fahrig, L. (2003). Evaluation of patch isolation metrics for predicting animal movement in binary landscapes. *Landsc. Ecol.* 18: 17–39.

Bendix, J., Wiley, J.J. Jr., and Commons, M.G. (2017). Intermediate disturbance and patterns of species richness. *Phys. Geogr.* 38: 393–403.

Bentz, B.J., Régnière, J., Fettig, C.J. et al. (2010). Climate change and bark beetles of the western United States and Canada: Direct and indirect effects. *Bioscience* 60: 602–613.

Bentz, B.J., Jönsson, A.M., Schroeder, M. et al. (2019). *Ips typographus* and *Dendroctonus ponderosae* models project thermal suitability for intra-and inter-continental establishment in a changing climate. *Front. For. Glob. Change* 2: 1.

Berdanier, A.B. and Clark, J.S. (2016). Divergent reproductive allocation trade-offs with canopy exposure across tree species in temperate forests. *Ecosphere* 7: e01313.

Beric, B. and MacIsaac, H.J. (2015). Determinants of rapid response success for alien invasive species in aquatic ecosystems. *Biol. Invasions* 17: 3327–3335.

Berntson, G.M. and Woodward, F.I. (1992). The root system architecture and development of *Senecio vulgaris* in elevated CO_2 and drought. *Funct. Ecol.* 6: 324–333.

Berry, J. and Bjorkman, O. (1980). Photosynthetic response and adaptation to temperature in higher plants. *Ann. Rev. Plant Physiol.* 31: 491–543.

Berry, J. and Downton, W.J.S. (1982). Environmental regulation of photosynthesis. In: *Photosynthesis, Development, Carbon Metabolism, and Plant Productivity*, vol. 2 (ed. R. Govindjee). New York: Academic Press 580 pp.

Berry, P.M., Fabók, V., Blicharska, M. et al. (2018). Why conserve biodiversity? A multi-national exploration of stakeholders' views on the arguments for biodiversity conservation. *Biodivers. Conserv.* 27: 1741–1762.

Beschta, R.L. and Ripple, W.J. (2016). Riparian vegetation recovery in Yellowstone: the first two decades after wolf reintroduction. *Biol. Conserv.* 198: 93–103.

Besnard, G., Acheré, V., Jeandroz, S. et al. (2008). Does maternal environmental condition during reproductive development induce genotypic selection in *Picea abies*? *Ann. For. Sci.* 65: 109.

Bidwell, R.G.S. (1974). *Plant Physiology*. New York: Macmillan 643 pp.

Bigler, C., Kulakowski, D., and Veblen, T.T. (2005). Multiple disturbance interactions and drought influence fire severity in Rocky Mountain subalpine forests. *Ecology* 86: 3018–3029.

Billings, W.D. (1973). Arctic and alpine vegetations: Similarities, differences, and susceptibility to disturbance. *Bioscience* 23: 697–704.

Billings, W.D. (1987). Constraints to plant growth, reproduction, and establishment in Arctic environments. *Arctic and Alpine Res.* 19: 357–365.

Billings, W.D. (1995). The effects of global and regional environmental changes on mountain ecosystems. *In* D. G. Despain (ed.), *Plants and their Environment: Proceedings of the First Biennial Scientific Conference on the Greater Yellowstone Ecosystem.* Tech. Report NPS/NRYELL/NRTR-93XX. US Dept. Interior, Nat. Park Serv., Nat. Res. Publ. Office, Denver. CO.

Billings, S.A. and Gaydess, E.A. (2008). Soil nitrogen and carbon dynamics in a fragmented landscape experiencing forest succession. *Landsc. Ecol.* 23: 581–593.

Binkley, D. (1995). The influence of tree species on forest soils: processes and patterns. *In* Mead, D. J., and I. S. Cornforth (eds.), *Proc. Trees and Soils Workshop*. Agronomy Society of New Zealand Special Publication No. 10. Lincoln Univ. Press, Canterbury.

Binkley, D. and Hart, S. (1989). The components of nitrogen availability assessments in forest soils. *Adv. Soil Sci.* 10: 57–115.

Binkley, D., Moore, M.M., Romme, W.H., and Brown, P.M. (2006). Was Aldo Leopold right about the Kaibab deer herd? *Ecosystems* 9: 227–241.

Bird, A. (2007). Perceptions of epigenetics. *Nature* 447: 396–398.

Birdsey, R.A., Pregitzer, K.S., and Lucier, A. (2006). Forest carbon management in the United States: 1600-2100. *J. Environ. Qual.* 35: 1461–1469.

Birkeland, P.W. (1974). *Pedology, Weathering and Geomorphological Research*. New York: Oxford Univ. Press 285 pp.

Biswell, H.H. (1961). The big trees and fires. *Natl. Parks Mag.* 35: 11–14.

Biswell, H.H. (1974). Effects of fire on chaparral. In: *Fire and Ecosystems* (ed. T.T. Kozlowski and C.E. Ahlgren), 321–364. New York: Academic Press.

Björklund, J., Seftigen, K., Schweingruber, F. et al. (2017). Cell size and wall dimensions drive distinct variability of earlywood and latewood density in Northern Hemisphere conifers. *New Phytol.* 216: 728–740.

Blake, E.S. and Gibney, E.J. (2011). *The Deadliest, Costliest, and Most Intense United States Tropical Cyclones from 1851 to 2010 (and other Frequently Requested Hurricane Facts)*. NOAA Technical Memorandum NWS NHC – 6. 47 pp.

Blasi, C., Carranza, M.L., Frondoni, R., and Rosati, L. (2000). Ecosystem classification and mapping: a proposal for Italian landscapes. *Appl. Veg. Sci.* 3: 233–242.

Blennow, K., Andersson, M., Sallnäs, O., and Olofsson, E. (2010). Climate change and the probability of wind damage in two Swedish forests. *For. Ecol. Manag.* 259: 818–830.

Bliss, L.C. and Cantlon, J.E. (1957). Succession on river alluvium in northern Alaska. *Am. Midi. Nat.* 58: 452–469.

Blonder, B., Graae, B.J., Greer, B. et al. (2020). Remote sensing of ploidy level in quaking aspen (*Populus tremuloides* Michx.). *J. Ecol.* 108: 175–188.

Bloomberg, W.J. (1950). Fire and spruce. *For. Chron.* 26: 157–161.

Blumenthal, D. (2005). Interrelated causes of plant invasion. *Science* 310: 243–244.

Boege, K. and Marquis, R.J. (2005). Facing herbivory as you grow up: the ontogeny of resistance in plants. *Trends Ecol. Evol.* 20: 441–448.

Boettcher, S.E. and Kalisz, P.J. (1990). Single-tree influence on soil properties in the mountains of eastern Kentucky. *Ecology* 71: 1365–1372.

Boggess, W.R. (1964). Trelease Woods, Champaign County, Illinois: woody vegetation and stand composition. *Trans. Ill. Acad. Sci.* 57: 261–271.

Boggess, W.R. and Bailey, L.W. (1964). Brownfield Woods, Illinois: woody vegetation and changes since 1925. *Am. Midi. Nat.* 71: 392–401.

Boggess, W.R. and Geis, J.W. (1966). The Funk Forest Natural Area, McLean County, Illinois: woody vegetation and ecological trends. *Trans. Ill. Acad. Sci.* 59: 123–133.

Bognounou, F., De Grandpré, L., Pureswaran, D.S., and Kneeshaw, D. (2017). Temporal variation in plant neighborhood effects on the defoliation of primary and secondary hosts by an insect pest. *Ecosphere* 8: 1–15.

Bohlen, P.J., Scheu, S., Hale, C.M. et al. (2004). Non-native invasive earthworms as agents of change in northern temperate forests. *Front. Ecol. Environ.* 2: 427–435.

Boisvert-Marsh, L., Périé, C., and de Blois, S. (2014). Shifting with climate? Evidence for recent changes in tree species distribution at high latitudes. *Ecosphere* 5: 1–33.

Boisvert-Marsh, L., Périé, C., and de Blois, S. (2019). Divergent responses to climate change and disturbance drive recruitment patterns underlying latitudinal shifts of tree species. *J. Ecol.* 107: 1956–1969.

Bolan, N.S., Robson, A.D., Barrow, N.J., and Aylmore, L.A.G. (1984). Specific activity of phosphorus in mycorrhizal and non-mycorrhizal plants in relation to the availability of phosphorus to plants. *Soil Biol. Biochem.* 6: 299–304.

Bolstad, P.V., Swank, W., and Vose, J. (1998). Predicting southern Appalachian overstory vegetation with digital terrain data. *Landsc. Ecol.* 13: 271–283.

Bonan, G.B. and Shugart, H.H. (1989). Environmental factors and ecological processes in boreal forests. *Annu. Rev. Ecol. Syst.* 20: 1–28.

Bond, W.J. (1993). Keystone species. In: *Ecosystem Function and Biodiversity* (ed. E.D. Schulze and H.A. Mooney), 237–253. New York: Springer-Verlag.

Bond, W.J. and van Wilgen, B.W. (1996). *Fire and Plants*. London: Chapman & Hall 263 pp.

Bond-Lamberty, B., Wang, C., and Gower, S.T. (2004). Net primary production and net ecosystem production of a boreal black spruce fire chronosequence. *Glob. Chang. Biol.* 10: 473–487.

Bonga, J.M. and von Aderkas, P. (1993). Rejuvenation of tissues from mature conifers and its implications for propagation in vitro. In: *Clonal Forestry I, Genetics and Biotechnology* (ed. M.R. Ahuja and W.J. Libby), 182–199. Berlin: Springer-Verlag.

Boose, E.R., Foster, D.R., and Fluet, M. (1994). Hurricane impacts to tropical and temperate forest landscapes. *Ecol. Monogr.* 64: 369–400.

Borchert, R. (1991). Growth periodicity and dormancy. In: *Physiology of Trees* (ed. A.S. Raghavendra), 221–245. New York: Wiley.

Boring, L.R., Swank, W.T., Waide, J.B., and Henderson, G.S. (1988). Sources, fates, and impacts of nitrogen inputs to terrestrial ecosystems: review and synthesis. *Biogeochemistry* 6: 119–159.

Bormann, F.H. and Likens, G.E. (1979). *Pattern and Process in a Forested Ecosystem*. New York: Springer-Verlag 253 pp.

Bormann, F.H., Likens, G.E., and Melillo, J.M. (1977). Nitrogen budget for an aggrading northern hardwood forest ecosystem. *Science* 196: 981–982.

Boshier, D., Broadhurst, L., Cornelius, J. et al. (2015). Is local best? Examining the evidence for local adaptation in trees and its scale. *Environ. Evid.* 4: 1–10.

Bossdorf, O., Richards, C.L., and Pigliucci, M. (2008). Epigenetics for ecologists. *Ecol. Lett.* 11: 106–115.

Botkin, D.B. (1993). *Forest Dynamics: An Ecological Model*. New York: Oxford Univ. Press 309 pp.

Botkin, D.B., Woodwell, G.M., and Tempel, N. (1970). Forest productivity estimated from carbon dioxide uptake. *Ecology* 51: 1057–1060.

Botting, D. (1973). *Humboldt and the Cosmos*. London: Harper and Row 295 pp.

Boucher, D.H. (1985). *The Biology of Mutualism*. London: Croom Helm 388 pp.

Boucher, D.H., James, S., and Keeler, K.H. (1982). The ecology of mutualism. *Annu. Rev. Ecol. Syst.* 13: 315–347.

Boucher, D., Gauthier, S., Thiffault, N. et al. (2019). How climate change might affect tree regeneration following fire at northern latitudes: a review. *New For.* 51: 543–571.

Boufford, D.E. and Spongberg, S.A. (1983). Eastern Asian–eastern North American phytogeographical relationships—a history from the time of Linnaeus to the twentieth century. *Ann. Missouri Bot. Gard.* 70: 423–439.

Bowen, G.D. and Smith, S.E. (1981). The effects of mycorrhizas on nitrogen uptake by plants. *Ecol. Bull.* 33: 237–247. Swedish Natural Sci. Res. Council, Stockholm.

Boyce, S.G. and Kaeiser, M. (1961). Why yellow-poplar seeds have low viability. USDA For. Serv. Central States For. Exp. Sta., Tech. Paper 186. 16 pp.

Boyle, T.J.B. (1992). Biodiversity of Canadian forests: current status and future challenges. *For. Chron.* 68: 444–453.

Boyle, J.R. and Ek, A.R. (1973). Whole tree harvesting: nutrient budget evaluation. *J. For.* 71: 760–762.

Boyle, T., Liengsiri, C., and Piewluang, C. (1990). Genetic structure of black spruce on two contrasting sites. *Heredity* 65: 393–399.

Bradford, K.J. and Hsiao, T.C. (1982). Physiological responses to moderate water stress. In: *Physiological Plant Ecology II: Water Relations and Carbon Assimilation*. Encyclopedia of Plant Physiology, vol. 12B (ed. O.L. Lange, P.S. Nobel, C.B. Osmond and H. Ziegler), 263–324. New York: Springer-Verlag.

Bradford, J.B., Birdsey, R.A., Joyce, L.A., and Ryan, M.G. (2008a). Tree age, disturbance history, and carbon stocks and fluxes in subalpine Rocky Mountain forests. *Glob. Chang. Biol.* 14: 2882–2897.

Bradford, M.A., Davies, C.A., Frey, S.D. et al. (2008b). Thermal adaptation of soil microbial respiration to elevated temperature. *Ecol. Lett.* 11: 1316–1327.

Bradford, M.A., Warren, R.J. II, Baldrian, P. et al. (2014). Climate fails to predict wood decomposition at regional scales. *Nat. Clim. Chang.* 4: 625–630.

Bradley, A.F., Fischer, W.C., and Noste, N.V. (1992). Fire ecology of the forest habitat types of eastern Idaho and western Wyoming. USDA For. Serv. Gen. Tech. Report INT-290. Intermountain Res. Sta., Ogden, UT. 92 pp.

Bradley, B.A., Houghton, R.A., Mustard, J.F., and Hamburg, S.P. (2006). Invasive grass reduces aboveground carbon stocks in shrublands of the western US. *Glob. Chang. Biol.* 12: 1815–1822.

Bradshaw, K.E. (1965). Soil use and management in the National Forest of California. In: *Forest–Soil Relationships in North America* (ed. C.T. Young-berg), 413–424. Corvallis: Oregon State Univ. Press.

Bradshaw, A.D. (1971). Plant evolution in extreme environments. In: *Ecological Genetics and Evolution* (ed. R. Creed), 20–50. Oxford: Blackwell Press.

Bradshaw, G.A. and Marquet, P.A. (ed.) (2002). *How Landscapes Change: Human Disturbance and Fragmentation in the Americas*. Ecological Studies 162. New York: Springer 383 pp.

Brady, N.C. (1974). *The Nature and Properties of Soils*, 8e. New York: MacMillan 639 pp.

Brady, N.C. (1990). *The Nature and Properties of Soils*, 10e. New York: Macmillan 621 pp.

Brandt, J., Antrop, M., de Blust, G. et al. (2009). Why a European chapter of IALE? *IALE Bull.* 27: 1–5.

Brasier, C.M. (2001). Rapid evolution of introduced plant pathogens via interspecific hybridization: hybridization is leading to rapid evolution of Dutch elm disease and other fungal plant pathogens. *Bioscience* 51: 123–133.

Braun, E.L. (1950). *Deciduous Forests of Eastern North America*. New York: McGraw-Hill 596 pp.

Braun-Blanquet, J. (1921). Prinzipien einer Systematik der Pflanzengellschaften auf floristischer Grundlage. *Jahrb. St. Gllen Naturw. Ges.* 57: 305–351.

Braun-Blanquet, J. (1964). *Pflanzensoziologie, Grundzüge der Vegetationskunde*, 3e. New York: Springer-Verlag 865 pp.

Bray, J.R. and Gorham, E. (1964). Litter production in forests of the world. *Adv. Ecol. Res.* 2: 101–157.

Brayshaw, T.C. (1965). *Native Poplars of Southern Alberta and Their Hybrids*. Canada: Dept. of Forestry Publ No. 1109. 40 pp.

Brayton, R. and Mooney, G.A. (1966). Population variability of *Cercocarpus* in the White Mountains of California as related to habitat. *Evolution* 20: 383–391.

Breckle, S.-W. (1974). Notes on alpine and nival flora of the Hindu Kush, East Afganistan. *Bot. Not.* 127: 278–284.

Breckle, S.-W. (2002). *Walter's Vegetation of the Earth: The Ecological Systems of the Geo-biosphere*, 4e. Berlin: Springer-Verlag 527 pp.

Brehm, A.M., Mortelliti, A., Maynard, G.A., and Zydlewski, J. (2019). Land-use change and the ecological consequences of personality in small mammals. *Ecol. Lett.* 22: 1387–1395.

Bremer, L.L. and Farley, K.A. (2010). Does plantation forestry restore biodiversity or create green deserts? A synthesis of the effects of land-use transitions on plant species richness. *Biodivers. Conserv.* 19: 3893–3915.

Brewer, S. (2008). Declines in plant species richness and endemic plant species in longleaf pine savannas invaded by *Imperata cylindrica*. *Biol. Invasions* 10: 1257–1264.

Brewer, R. and Merritt, P.G. (1978). Wind throw and tree replacement in a climax beech-maple forest. *Oikos* 30: 149–152.

Brienen, R.J., Caldwell, L., Duchesne, L. et al. (2020). Forest carbon sink neutralized by pervasive growth-lifespan trade-offs. *Nat. Commun.* 11: 1–10.

Briffa, K.R., Schweingruber, F.H., Jones, P.D. et al. (1998). Reduced sensitivity of recent tree-growth to temperature at high northern latitudes. *Nature* 391: 678–682.

Brinson, M.M. (1990). Riverine forests. In: *Ecosystems of the World, Forested Wetlands*, vol. 15 (ed. D. Goodall, A. Lugo, M. Brinson and S. Brown), 87–141. New York: Elsevier.

Brix, H. (1971). Effects of nitrogen fertilization on photosynthesis and respiration in Douglas-fir. *For. Sci.* 17: 407–414.

Brix, H. and Ebell, L.F. (1969). Effects of nitrogen fertilization on growth, leaf area, and photosynthesis rate of Douglas-fir. *For. Sci.* 15: 189–196.

Broadfoot, W.M. and Williston, H.L. (1973). Flooding effects on southern forests. *J. For.* 71: 584–587.

Bronson, D.R., Gower, S.T., Tanner, M. et al. (2008). Response of soil surface CO_2 flux in a boreal forest to ecosystem warming. *Glob. Chang. Biol.* 14: 856–867.

Bronstein, J.L. (1994). Our current understanding of mutualism. *Quarterly Review of Biology* 69 (1): 31–51.

Bronstein, J.L. (2009). The evolution of facilitation and mutualism. *J. Ecol.* 97: 1160–1170.

Bronstein, J.L. (2015). The study of mutualism. In: *Mutualism* (ed. J.L. Bronstein). Oxford, UK: Oxford University Press 320 pp.

Brooks, M.L., D'Antonio, C.M., Richardson, D.M. et al. (2004). Effects of invasive alien plants on fire regimes. *Bioscience* 54: 677–688.

Brouillet, L. and Whetstone, R.D. (1993). Climate and physiography. In: *Flora of North America, North of Mexico*, vol. 1 (ed. R.R. Morin) conv. ed. New York: Oxford Univ. Press.

Brown, C.D. and Johnstone, J.F. (2012). Once burned, twice shy: Repeat fires reduce seed availability and alter substrate constraints on *Picea mariana* regeneration. *For. Ecol. Manag.* 266: 34–41.

Brown, S. and Lugo, A.E. (1982). The storage and production of organic matter in tropical forests and their role in the global carbon cycle. *Biotropica* 14: 161–187.

Brown, C.L., McAlpine, R.G., and Kormanik, P.P. (1967). Apical dominance and form in woody plants: a reappraisal. *Am. J. Bot.* 54: 153–162.

Brundrett, M.C. (2002). Coevolution of roots and mycorrhizas of land plants. *New Phytol.* 154: 275–304.

Brussard, P.F., Reed, J.M., and Tracy, C.R. (1998). Ecosystem management: what is it really? *Landsc. Urban Plan.* 40: 9–20.

Bryant, J.M. (1981). Phytochemical deterrence of snowshoe hare browsing by adventitious shoots of four Alaskan trees. *Science* 213: 889–890.

Bryant, J.P., Chapin, F.S. III, Reichardt, P., and Clausen, T. (1985). Adaptation to resource availability as a determinant of chemical defense strategies in woody plants. In: *Chemically Mediated Interactions Between Plants and Other Organisms*. Recent Adv. Phytochemistry 19 (ed. G.A. Cooper-Driver, T. Swain and E.E. Conn), 219–237. New York: Plenum Press.

Bryant, J.P., Provenze, F.D., Pastor, J. et al. (1991). Interactions between woody plants and browsing mammals mediated by secondary metabolites. *Annu. Rev. Ecol. Syst.* 22: 431–446.

Bryson, R.A. (1966). Air masses, streamlines, and the boreal forest. *Geogr. Bull.* 8: 228–269.

Bryson, R.A. and Hare, F.K. (1974). The climates of North America. In: *Climates of North America* (World Survey of Climatology, Vol. 11). (ed. R.A. Bryson and F.K. Hare), 1–47. New York: Elsevier.

Brzostek, E.R., Dragoni, D., Schmid, H.P. et al. (2014). Chronic water stress reduces tree growth and the carbon sink of deciduous hardwood forests. *Glob. Chang. Biol.* 20: 2531–2539.

Budde, K.B., Heuertz, M., Hernández-Serrano, A. et al. (2014). In situ genetic association for serotiny, a fire-related trait, in Mediterranean maritime pine (*Pinus pinaster*). *New Phytol.* 201: 230–241.

Buechling, A., Martin, P.H., Canham, C.D. et al. (2016). Climate drivers of seed production in *Picea engelmannii* and response to warming temperatures in the southern Rocky Mountains. *J. Ecol.* 104: 1051–1062.

Bugmann, H. and Bigler, C. (2011). Will the CO_2 fertilization effect in forests be offset by reduced tree longevity? *Oecologia* 165: 533–544.

Buma, B. (2015). Disturbance interactions: characterization, prediction, and the potential for cascading effects. *Ecosphere* 6: 70.

Buma, B. and Wessman, C.A. (2012). Differential species responses to compounded perturbations and implications for landscape heterogeneity and resilience. *For. Ecol. Manag.* 266: 25–33.

Buma, B., Bisbing, S., Krapek, J., and Wright, G. (2017). A foundation of ecology rediscovered: 100 years of succession on the William S. Cooper plots in Glacier Bay, Alaska. *Ecology* 98: 1513–1523.

Buma, B., Weiss, S., Hayes, K., and Lucash, M. (2020). Wildland fire reburning trends across the US West suggest only short-term negative feedback and differing climatic effects. *Environ. Res. Lett.* 15: 034026.

Bunn, A.G., Graumlich, L.J., and Urban, D.L. (2005). Trends in twentieth-century tree growth at high elevations in the Sierra Nevada and White Mountains, USA. *The Holocene* 15: 481–488.

Buol, S.W., Hole, F.D., and McCracken, R.J. (1980). *Soil Genesis and Classification*. Ames: Iowa State Univ. Press 404 pp.

Burger, D. (1972). Forest site classification in Canada. *Mitt. Vereins forstl. Standortsk. Forstpflz.* 21: 20–36.

Burger, D. (1993). Revised site regions of Ontario: concepts, methodology and utility. Ontario For. Res. Inst., Sault Ste. Marie, For. Res. Report No. 129. 24 pp.

Burke, I.C. (1989). Control of nitrogen mineralization in a sagebrush steppe landscape. *Ecology* 70: 1115–1126.

Burke, M.K., Raynal, D., and Mitchell, M.J. (1991). Soil nitrogen availability influences seasonal carbon allocation patterns in sugar maple (*Acer saccharum*). *Can. J. For. Res.* 22: 447–456.

Burkhart, H.E., Avery, T.E., and Bullock, B.P. (2018). *Forest Measurements*, 6e. Long Grove. IL: Waveland Press 434 pp.

Burley, J. (1966). Provenance variation in growth of seedling apices of Sitka spruce. *For. Sci.* 12: 170–175.

Burns, G. P. (1923). Studies in tolerance of New England forest trees IV. Minimum light requirements referred to a definite standard. *Univ. Vermont Agr. Exp. Sta. Bull* 235.

Burns, G.R. (1942). Photosynthesis and absorption in blue radiation. *Am. J. Bot.* 29: 381–387.

Burns, R.M. and Honkala, B.H. (1990). *Silvics of North America, Vol. 2, Hardwoods*. Washington, DC: U.S.D.A. For. Serv. Agr. Handbook 65 886 pp.

Burton, P.J., Balisky, A.C., Coward, L.P. et al. (1992). The value of managing for biodiversity. *For. Chron.* 68: 225–237.

Busby, P.E., Motzkin, G., and Foster, D.R. (2008). Multiple and interacting disturbances lead to *Fagus grandifolia* dominance in coastal New England. *J. Torrey Bot. Soc.* 135: 346–359.

Büsgen, M. and Münch, E. (1929). *The Structure and Life of Forest Trees*, 3e (trans. T. Thomson). London: Chapman & Hall 436 pp.

Busing, R.T. (1994). Canopy cover and tree regeneration in old-growth cove forests of the Appalachian Mountains. *Vegetatio* 115: 19–27.

Buttò, V., Deslauriers, A., Rossi, S. et al. (2020). The role of plant hormones in tree-ring formation. *Trees* 34: 315–335.

Byram, G.M. and Jemison, G.M. (1943). Solar radiation and forest fuel moisture. *J. Agric. Res.* 67: 149–176.

Cain, S.A. (1935). Studies on virgin hardwood forests: III. Warren's Woods, a beech-maple climax forest in Berrien County, Michigan. *Ecology* 16: 500–513.

Cain, S.A. (1939). The climax and its complexities. *Am. Midi. Nat.* 21: 146–181.

Cain, S.A. and de Oliveira Castro, G.M. (1959). *Manual of Vegetation Analysis*. New York: Harper & Row 325 pp.

Cain, M.L., Milligan, B.G., and Strand, A.E. (2000). Long-distance seed dispersal in plant populations. *Am. J. Bot.* 87: 1217–1227.

Cajander, A. K. (1926). The theory of forest types. *Ada For. Fenn.* 29. 108 pp.

Calder, W.J., and St. Clair, S.B. (2012). Facilitation drives mortality patterns along succession gradients of aspen-conifer forests. *Ecosphere* 3: 1–11.

Callaham, R.Z. (1962). Geographic variability in growth of forest trees. In: *Tree Growth* (ed. T.T. Kozlowski), 311–325. New York: Ronald Press.

Cameron, R.P. and Williams, D. (2011). Completing an ecosystem classification system for Nova Scotia. *Nat. Areas J.* 31: 92–96.

Campbell, R.S. (1955). Vegetational changes and management in the cutover longleaf pine-slash pine area of the Gulf Coast. *Ecology* 36: 29–34.

Campbell, R.K. (1979). Genecology of Douglas-fir in a watershed in the Oregon Cascades. *Ecology* 60: 1036–1050.

Campbell, R.K. (1986). Mapped genetic variation of Douglas-fir to guide seed transfer in southwest Oregon. *Silvae Genet.* 35: 85–96.

Campbell, R.K. (1991). Soils, seed-zone maps, and physiography: guidelines for seed transfer in Douglas-fir in southwestern Oregon. *For. Sci.* 37: 973–986.

Campbell, T.A. and Long, D.B. (2009). Feral swine damage and damage management in forested ecosystems. *For. Ecol. Manag.* 257: 2319–2326.

Campbell, R.K. and Sorensen, F.C. (1978). Effect of test environment on expression of clines and on delimitation of seed zones in Douglas-fir. *Theor. Appl. Genet.* 51: 233–246.

Campbell, R.K. and Sugano, A.I. (1979). Genecology of bud-burst phenology in Douglas-fir: response to flushing temperature and chilling. *Bot. Gaz.* 140: 223–231.

Campbell, R.K., Pawuk, W.A., and Harris, A.S. (1989). Microgeographic genetic variation of Sitka spruce in southeastern Alaska. *Can. J. For. Res.* 19: 1004–1013.

Campbell, C.R., Poelstra, J.W., and Yoder, A.D. (2018). What is speciation genomics? The roles of ecology, gene flow, and genomic architecture in the formation of species. *Biol. J. Linn. Soc.* 124: 561–583.

Canada Inst. Forestry. (1992). *The Forestry Chronicle* 68(1):21–120.

Canadell, J.G. and Raupach, M.R. (2008). Managing forests for climate change mitigation. *Science* 320: 1456–1457.

Canadian Forest Ecosystem Classification System (CFEC) (2010). Natural Resources Canada. Canadian Forest Service. Great Lakes Forestry Centre, Sault Ste Marie, Ontario. Frontline Express 38. 2pp.

Canadian Forest Service (2021). Mountain Pine Beetle factsheet. https://www.nrcan.gc.ca/forests/fire-insects-disturbances/top-insects/13397#shr-pg0. Accessed August 25, 2021.

Canham, C.D. (1985). Suppression and release during canopy recruitment in *Acer saccharum*. *Bull. Torrey Bot. Club.* 112: 134–145.

Canham, C.D. (1988). Growth and canopy architecture of shade-tolerant trees: response to canopy gaps. *Ecology* 69: 786–795.

Canham, C.D. and Loucks, O.L. (1984). Catastrophic windthrow in the presettlement forests of Wisconsin. *Ecology* 65: 803–809.

Canham, C.D., Denslow, J.S., Platt, W.J. et al. (1990). Light regimes beneath closed canopies and tree-fall gaps in temperate and tropical forests. *Can. J. For. Res.* 20: 620–631.

Canham, C.D., Finzi, A.C., Pacala, S.W., and Burbank, D.H. (1994). Causes and consequences of resource heterogeneity in forests: interspecific variation in light transmission by canopy trees. *Can. J. For. Res.* 24: 337–349.

Canham, C.D., Kobe, R.K., Latty, E.F., and Chazdon, R.L. (1999). Interspecific and intraspecific variation in tree seedling survival: effects of allocation to roots versus carbohydrate reserves. *Oecologia* 121: 1–11.

Cannell, M.G.R. and Smith, R.I. (1986). Climatic warming, spring budburst, and forest damage on trees. *J. Appl. Ecol.* 23: 177–191.

Cannon, H.L. (1960). The development of botanical methods of prospecting for uranium on the Colorado Plateau. *U.S. Geol. Surv. Bull.* 1085-A: 1–50.

Cantor, L.F. and Whitham, T.G. (1989). Importance of belowground herbivory: pocket gophers may limit aspen to rock outcrop refugia. *Ecology* 70: 962–970.

Carbó, M., Iturra, C., Correia, B. et al. (2019). Epigenetics in forest trees: keep calm and carry on. In: *Epigenetics in Plants of Agronomic Importance: Fundamentals and Applications* (ed. R. Alvarez-Venegas, C. De la Peña and J.A. Casas-Mollano), 381–403. New York: Springer.

Cardinale, B.J., Srivastava, D.S., Duffy, J.E. et al. (2006). Effects of biodiversity on the functioning of trophic groups and ecosystems. *Nature* 443: 989–992.

Cardinale, B.J., Matulich, K.L., Hooper, D.U. et al. (2011). The functional role of producer diversity in ecosystems. *Am. J. Bot.* 98: 572–592.

Cardinale, B.J., Duffy, J.E., Gonzalez, A. et al. (2012). Biodiversity loss and its impact on humanity. *Nature* 486: 59–67.

Carere, C. and Maestripieri, D. (2013). *Animal Personalities: Behavior, Physiology and Evolution*. Chicago, IL: University of Chicago Press 507 pp.

Carey, A.B. and Johnson, M.L. (1995). Small mammals in managed, naturally young, and old-growth forests. *Ecol. Appl.* 5: 336–352.

Carlson, A.R., Sibold, J.S., Assal, T.J., and Negrón, J.F. (2017). Evidence of compounded disturbance effects on vegetation recovery following high-severity wildfire and spruce beetle outbreak. *PLoS One* 12: e0181778.

Carmean, W.H. (1967). Soil refinements for predicting black oak site quality in southeastern Ohio. *Proc. Soil Sci. Soc. Am.* 31: 805–810.

Carmean, W.H. (1970a). Site quality for eastern hardwoods. *In: The Silviculture of Oaks and Associated Species*. USDA For. Serv. Res. Paper NE-144. Northeastern For. Exp. Sta., Upper Darby, PA.

Carmean, W.H. (1970b). Tree height-growth patterns in relation to soil and site. In: *Tree Growth and Forest Soils* (ed. C.T. Youngberg and C.B. Davey), 499–512. Corvallis: Oregon State Univ. Press.

Carmean, W.H. (1975). Forest site quality evaluation in the United States. *Adv. Agron.* 27: 209–269.

Carmean, W.H. (1977). Site classification for northern forest species. *In* Proc. Symp *Intensive Culture of Northern Forest Types,* USDA For. Serv. Gen. Tech. Report NE-29. Northeastern For. Exp. Sta., Upper Darby, PA.

Carmean, W.H. (1996). Site-quality evaluation, site-quality maintenance, and site-specific management for forest land in northwest Ontario. Ontario Ministry Nat. Res., Northwest Sci. and Technology Unit, NWST Tech. Report TR-105, Thunder Bay, ON. 121 pp.

Carmean, W.H., Hazenberg, G., and Niznowski, G.P. (2001). Polymorphic site index curves for jack pine in Northern Ontario. *For. Chron.* 77: 141–150.

Carmean, W.H., Hazenberg, G., and Deschamps, K.C. (2006). Polymorphic site index curves for black spruce and trembling aspen in northwest Ontario. *For. Chron.* 82: 231–242.

Carneros, E., Yakovlev, I., Viejo, M. et al. (2017). The epigenetic memory of temperature during embryogenesis modifies the expression of bud burst-related genes in Norway spruce epitypes. *Planta* 246: 553–566.

Carpenter, S., Caraco, N.F., Correll, D.L. et al. (1998). Nonpoint pollution of surface waters with phosphorus and nitrogen. *Issues Ecol.* 3: 1–12.

Carpenter, S., Walker, B., Andries, J.M., and Ablel, N. (2001). From metaphor to measurement: Resilience of what to what? *Ecosystems* 4: 765–781.

Carpenter, S.R., Mooney, H.A., Agard, J. et al. (2009). Science for managing ecosystem services: beyond the millennium ecosystem assessment. *Proc. Natl. Acad. Sci.* 106: 1305–1312.

Carpenter, S.R., Cole, J.J., Pace, M.L. et al. (2011). Early warnings of regime shifts: a whole-ecosystem experiment. *Science* 332: 1079–1082.

Carter, G.A. and Smith, W.K. (1985). Influence of shoot structure on light interception and photosynthesis in conifers. *Plant Physiol.* 79: 1038–1043.

Carter, R.E., Mackenzie, M.D., and Gjerstad, D.H. (1999). Ecological land classification in the Southern Loam Hills of south Alabama. *For. Ecol. Manag.* 114: 395–404.

Casal, J.J. (2012). Shade avoidance. *Arabidopsis Book* 10: e0157.

Castello, J.D., Leopold, D.J., and Samllidge, P.J. (1995). Pathogens, patterns and processes in forest ecosystems. *BioScience* 45: 16–24.

Certini, G. (2005). Effects of fire on properties of forest soils: a review. *Oecologia* 143: 1–10.

Ceulemans, R.J. and Saugier, B. (1991). Photosynthesis. In: *Physiology of Trees* (ed. A.S. Raghavendra), 21–50. New York: Wiley.

Chabot, B.F. and Hicks, D.J. (1982). The ecology of leaf life spans. *Annu. Rev. Ecol. Syst.* 11: 233–257.

Chambers, J.Q., Fisher, J.I., Zeng, H. et al. (2007). Hurricane Katrina's carbon footprint on Gulf Coast forests. *Science* 318: 1107.

Chambless, L.F. and Nixon, E.S. (1975). Woody vegetation—soil relations in a bottomland forest of west Texas. *Tex. J. Sci.* 26: 407–416.

Chapin, F.S. III, Van Cleve, K., and Tryon, P.R. (1986). Relationship of ion absorption to growth rate in taiga trees. *Oecologia* 69: 238–242.

Chapin, F.S., III, Bloom, A.J., and Field, C.B. (1987). Plant response to multiple environmental factors. *BioScience* 37: 49–57.

Chapin, F.S. III, Walker, L.R., Fastie, C.L., and Sharman, L.C. (1994). Mechanisms of primary succession following deglaciation at Glacier Bay, Alaska. *Ecol. Monogr.* 64: 149–175.

Chapin, F.S. III, Torn, M.S., and Tateno, M. (1996). Principles of ecosystem sustainability. *Am. Nat.* 148: 1016–1037.

Chapin, F.S. III, Zavaleta, E.S., Eviner, V.T. et al. (2000). Consequences of changing biodiversity. *Nature* 405: 234–242.

Chapin, F.S. III, Matson, P.A., and Vitousek, P.M. (2012). *Principles of Terrestrial Ecosystem Ecology*, 2e. New York: Springer 544 pp.

Chapman, H.H. (1932). Is the longleaf pine a climax? *Ecology* 13: 328–335.

Chapman, S.K., Whitham, T.G., and Powell, M. (2006). Herbivory differentially alters plant litter dynamics of evergreen and deciduous trees. *Oikos* 114: 566–574.

Chapman, E.L., Chambers, J.Q., Ribbeck, K.F. et al. (2008). Hurricane Katrina impacts on forest trees of Louisiana's Pearl River basin. *For. Ecol. Manag.* 256: 883–889.

Charlesworth, B. and Charlesworth, D. (2010). *Elements of Evolutionary Genetics*. New York: W.H. Freeman and *Company* 768 pp.

Charney, N.D., Babst, F., Poulter, B. et al. (2016). Observed forest sensitivity to climate implies large changes in 21st century North American forest growth. *Ecol. Lett.* 19: 1119–1128.peterson.

Chauchard, S., Pille, G., and Carcaillet, C. (2006). Large herbivores control the invasive potential of non-native Austrian black pine in a mixed deciduous Mediterranean forest. *Can. J. For. Res.* 36: 1047–1053.

Chazdon, R. (1988). Sunflecks and their importance to forest understory plants. *Adv. Ecol. Res.* 18: 1–63.

Chen, J., Franklin, J.F., and Spies, T.A. (1992). Vegetation responses to edge environments in old-growth Douglas-fir forests. *Ecol. Appl.* 2: 387–396.

Chen, J., Franklin, J.F., and Spies, T.A. (1995). Growing-season microclimatic gradients from clearcut edges into old-growth Douglas-fir forests. *Ecol. Appl.* 5: 74–86.

Chick, J.H., Cosgriff, R.J., and Gittinger, L.S. (2003). Fish as potential dispersal agents for floodplain plants: first evidence in North America. *Can. J. Fish. Aquat. Sci.* 60: 1437–1439.

Chivian, E. and Bernstein, A. (ed.) (2008). *Sustaining Life: How Human Health Depends on Biodiversity*. Oxford, UK: Oxford University Press 568 pp.

Chomicki, G., Weber, M., Antonelli, A. et al. (2019). The impact of mutualisms on species richness. *Trends Ecol. Evol.* 34: 698–711.

Christensen, N.L. (1977). Changes in structure, pattern and diversity associated with climax forest maturation in Piedmont, North Carolina. *Am. Midl. Nat.* 97: 176–188.

Christensen, N.L. (1988a). Vegetation of the southeastern Coastal Plain. In: *North American Terrestrial Vegetation* (ed. M.G. Barbour and W.D. Billings), 317–363. New York: Cambridge Univ. Press.

Christensen, N.L. (1988b). Succession and natural disturbance: paradigms, problems, and preservation of natural ecosystems. In: *Ecosystem Management for Parks and Wilderness* (ed. J.K. Agee and D.R. Johnson), 62–86. Seattle: Univ. Washington Press.

Christensen, N.L. and Peet, R.K. (1981). Secondary forest succession on the North Carolina piedmont. In: *Forest Succession: Concepts and Applications* (ed. D.C. West, H.H. Shugart and D.B. Botkin), 230–245. New York: Springer-Verlag.

Christensen, K.M. and Whitham, T.G. (1991). Indirect herbivore mediation of avian seed dispersal in pinyon pine. *Ecology* 72: 536–542.

Christensen, K.M. and Whitham, T.G. (1993). Impact of insect herbivores on competition between birds and mammals for pinyon pine seeds. *Ecology* 74: 2270–2278.

Christensen, N.L., Agee, J.K., Brussard, P.F. et al. (1989). Interpreting the Yellowstone fires of 1988. *Bioscience* 39: 678–685.

Christensen, N.L., Bartuska, A.M., Brown, J.H. et al. (1996). The report of the Ecological Society of America committee on the scientific basis for ecosystem management. *Ecol. Appl.* 6: 665–691.

Christy, H.R. (1952). Vertical temperature gradients in a beech forest in central Ohio. *Ohio J. Sci.* 52: 199–209.

Chung, H.H. and Barnes, R.L. (1977). Photosynthate allocation in *Pinus taeda*. I. Substrate requirements for synthesis of shoot biomass. *Can. J. For. Res.* 7: 106–111.

Chung, M.Y., Son, S., Herrando-Moraira, S. et al. (2020). Incorporating differences between genetic diversity of trees and herbaceous plants in conservation strategies. *Conserv. Biol.* 34: 1142–1151.

Churchill, D.J., Larson, A.J., Dahlgreen, M.C. et al. (2013). Restoring forest resilience: from reference spatial patterns to silvicultural prescriptions and monitoring. *For. Ecol. Manag.* 291: 442–457.

Cicatelli, A., Todeschini, V., Lingua, G. et al. (2013). Epigenetic control of heavy metal stress response in mycorrhizal versus non-mycorrhizal poplar plants. *Environ. Sci. Pollut. Res.* 21: 1723–1737.

Cieslar, A. (1895). Über die Erblichkeit des Zuwachsvermogens bei den Waldbäumen. *Centralbl. gesam. Forstw.* 21:7–29.

Cieslar, A. (1899). Neues aus dem Gebiete der fostlichen Zuchtwahl. *Centbl. gesam. Forstw.* 25:99–117.

Civitello, D.J., Cohen, J., Fatima, H. et al. (2015). Biodiversity inhibits parasites: broad evidence for the dilution effect. *Proc. Natl. Acad. Sci.* 112: 8667–8671.

Clark, J.S. (1989). Ecological disturbance as a renewal process: theory and application to fire history. *Oikos* 56: 17–30.

Clark, J.S. (1990). Fire and climate change during the last 750 years in northwestern Minnesota. *Ecol. Monogr.* 60: 135–169.

Clark, K.L., Skowronski, N., and Hom, J. (2010). Invasive insects impact forest carbon dynamics. *Glob. Chang. Biol.* 16: 88–101.

Clason, T.R. and Sharrow, S.H. (2000). Silvopastoral practices. In: *North American Agroforestry: An Integrated Science and Practice* (ed. H.E. Garrett, W.J. Rietveld and R.F. Fisher), 119–147. Madison, Wisconsin, USA: American Society of Agronomy.

Classen, A.T., Norby, R.J., Sides, K.E., and Weltzin, J.F. (2010). Climate change alters seedling emergence and establishment in an old-field ecosystem. *PLoS One* 5: e13476.

Clawson, M. (1978). The concept of multiple use forestry. *Environ. Law* 8: 281–308.

Cleland, D.T., Avers, P.E., McNab, W.H. et al. (1997). National hierarchical framework of ecological units. In: *Ecosystem Management Applications for Sustainable Forest and Wildlife Resources* (ed. M.S. Boyce and A. Haney), 181–200. New Haven, CT: Yale University Press.

Cleland, E.E., Chuine, I., Menzel, A. et al. (2007). Shifting plant phenology in response to global change. *Trends Ecol. Evol.* 22: 357–365.

Clements, F.E. (1905). *Research Methods in Ecology*. Univ. Publ. Co., Lincoln, NB. 334 pp. (reprinted 1977, Arno Press, New York).

Clements, F.E. (1916). Plant succession: an analysis of the development of vegetation. *Carneg. Inst. Wash. Publ.* 242. 512 pp.

Clements, F.E. (1936). Nature and structure of the climax. *J. Ecol.* 24: 252–284.

Clements, F.E. (1949). *Dynamics of Vegetation: Selections from the Writings of Frederic E. Clements, Ph. D.* The H. W. Wilson Co., New York. 296 pp.

Cline, M. (1997). Concepts and terminology of apical dominance. *Am. J. Bot.* 84: 1064.

Cline, A.C. and Spurr, S.H. (1942). The virgin upland forest of central New England. *Harvard For. Bull.* 21.51 pp.

Clinton, B.D. and Baker, C.R. (2000). Catastrophic windthrow in the southern Appalachians: characteristics of pits and mounds and initial vegetation responses. *For. Ecol. Manag.* 126: 51–60.

Clout, M.N. and Williams, P.A. (ed.) (2009). *Invasive Species Management: A Handbook of Principles and Techniques*. Oxford, UK: Oxford University Press 320 pp.

Coates, P. (2007). *American Perceptions of Immigrant and Invasive Species*. Berkeley, CA: University of California Press 266 pp.

Coetzee, B.W.T., Gaston, K.J., and Chown, S.L. (2014). Local scale comparisons of biodiversity as a test for global protected area ecological performance: A meta-analysis. *PLoS One* 9: e105824.

Coile, T.S. (1937). Distribution of forest tree roots in North Carolina Piedmont soils. *J. For.* 35: 247–257.

Coile, T.S. (1952). Soil and the growth of forests. *Adv. Agron.* 4: 330–398.

Colautti, R.I., Ricciardi, A., Grigorovich, I.A., and MacIsaac, H.J. (2004). Is invasion success explained by the enemy release hypothesis? *Ecol. Lett.* 7: 721–733.

Cole, L.C. (1958). The ecosphere. *Sci. Am.* 198: 83–92.

Cole, W.E. and Amman, G.D. (1980). *Mountain Pine Beetle Dynamics in Lodgepole Pine Forests, Part 1: Course of An Infestation*. USDA For. Serv. Gen. Tech. Rep. INT-89. Intermountain Res. Sta., Ogden, UT. 56 pp.

Cole, D.W. and Johnson, D.W. (1980). Mineral cycling in tropical forests. *In* C. T. Youngberg (ed.), *Forest Soils and Land Use*. Proc. 5th North American Forest Soils Conf. Colorado State Univ., Fort Collins.

Cole, D.W. and Rapp, M. (1981). Elemental cycling in forest ecosystems. In: *Dynamic Properties of Forest Ecosystems*, Int. Biol. Programme 23 (ed. D.E. Reichle), 341–409. Cambridge: Cambridge Univ. Press.

Coley, P.D. (1980). Effects of leaf age and plant life history patterns on herbivory. *Nature* 284: 545–546.

Coley, P.D. (1987). Interspecific variation in plant anti-herbivore properties: the role of habitat quality and rate of disturbance. *New Phytol.* 106: 251–263.

Coley, P.D., Bryant, J.P., and Chapin, F.S. III (1985). Resource availability and plant anti-herbivore defense. *Science* 230: 895–899.

Collinge, S.K. (2009). *Ecology of Fragmented Landscapes*. Baltimore, MD: Johns Hopkins University Press 360 pp.

Colyvan, M., Linquist, S., Grey, W. et al. (2009). Philosophical issues in ecology: recent trends and future directions. *Ecol. Soc.* 14: 22.

Conkle, M.T. (1973). Growth data for 29 years from the California elevational transect study of ponderosa pine. *For. Sci.* 19: 31–39.

Conkle, M.T. and Critchfield, W.B. (1988). Genetic variation and hybridization of ponderosa pine. *In* D. Baumgartner and J.E. Lotan (eds.), Symp. Proc. *Ponderosa Pine, The Species and Its Management*. Washington State Univ., Dept. Nat. Res. Sciences. Pullman.

Connell, J.H. (1978). Diversity in tropical rain forests and coral reefs. *Science* 199: 1302–1310.

Connell, J.H. and Slatyer, R.O. (1977). Mechanisms of succession in natural communities and their roles in community stability and organization. *Am. Nat.* 111: 1119–1144.

Conner, W.H., Mihalia, I., and Wolfe, J. (2002). Tree community structure and changes from 1987 to 1999 in three Louisiana and three South Carolina forested wetlands. *Wetlands* 22: 58–70.

Cook, R.E. (1979). Asexual reproduction: a further consideration. *Am. Nat.* 113: 769–772.

Cook, E.R. and Kairiukstis, L.A. (ed.) (1990). *Methods of Dendrochronology: Applications in the Environmental Sciences*. New York: Springer 406 pp.

Coomes, D.A. and Grubb, P.J. (2000). Impacts of root competition in forests and woodlands: a theoretical framework and review of experiments. *Ecol. Monogr.* 70: 171–207.

Coop, J.D., Parks, S.A., McClernan, S.R., and Holsinger, L.M. (2016). Influences of prior wildfires on vegetation response to subsequent fire in a reburned Southwestern landscape. *Ecol. Appl.* 26: 346–354.

Coop, J.D., Parks, S.A., Stevens-Rumann, C.S. et al. (2020). Wildfire-driven forest conversion in western North American landscapes. *Bioscience* 70: 659–673.

Cooper, W.S. (1913). The climax forest of Isle Royale, Lake Superior, and it development. *Bot. Gaz.* 55 (1–14): 189–235.

Cooper, W.S. (1923). The recent ecological history of Glacier Bay, Alaska. *Ecology* 4: 93–128, 223–246, 355–365.

Cooper, W.S. (1931). A third expedition to Glacier Bay, Alaska. *Ecology* 12: 61–95.

Cooper, W.S. (1939). A fourth expedition to Glacier Bay, Alaska. *Ecology* 20: 130–155.

Cooper, C.F. (1960). Changes in vegetation, structure, and growth of southwestern pine forests since white settlement. *Ecol. Monogr.* 30: 129–164.

Cooper, C.F. (1961). The ecology of fire. *Sci. Am.* 204 (4): 150–160.

Cooper, S.V., Neiman, K.E., and Roberts, D.W. (1991). Forest habitat types of northern Idaho: a second approximation. USDA For. Serv. Gen. Tech. Report INT-236. Int. Res. Sta., Ogden, UT. 143 pp.

Corenblit, D., Tabacchi, E., Steiger, J., and Gurnell, A.M. (2007). Reciprocal interactions and adjustments between fluvial landforms and vegetation dynamics in river corridors: a review of complementary approaches. *Earth-Sci. Rev.* 84: 56–86.

Corenblit, D., Steiger, J., Gurnell, A.M. et al. (2009). Control of sediment dynamics by vegetation as a key function driving biogeomorphic succession within fluvial corridors. *Earth Surf. Process. Landf.* 34: 1790–1810.

Corenblit, D., Steiger, J., Charrier, G. et al. (2016). *Populus nigra* L. establishment and fluvial landform construction: biogeomorphic dynamics within a channelized river. *Earth Surf. Process. Landf.* 41: 1276–1292.

Cornwell, W.K., Cornelissen, J.H., Amatangelo, K. et al. (2008). Plant species traits are the predominant control on litter decomposition rates within biomes worldwide. *Ecol. Lett.* 11: 1065–1071.

Correa, S.B., Winemiller, K.O., Lopez-Fernandez, H., and Galetti, M. (2007). Evolutionary perspectives on seed consumption and dispersal by fishes. *Bioscience* 57: 748–756.

Costanza, R., d'Arge, R., De Groot, R. et al. (1997). The value of the world's ecosystem services and natural capital. *Nature* 387: 253–260.

Costanza, R., De Groot, R., Sutton, P. et al. (2014). Changes in the global value of ecosystem services. *Glob. Environ. Chang.* 26: 152–158.

Cottam, G. and McIntosh, R.P. (1966). Vegetational continuum. *Science* 152: 546–547.

Cottingham, K.L., Brown, B.L., and Lennon, T.J. (2001). Biodiversity may regulate the temporal variability of ecological systems. *Ecol. Lett.* 4: 72–85.

Coutts, M.P. and Grace, J. (ed.) (1995). *Wind and Trees*. Cambridge: Cambridge Univ. Press 485 pp.

Coutts, M.P. and Nicoll, B.C. (1991). Development of the surface roots of trees. In: *L'Arbre, Biologie et Development* (ed. C. Edelin), 61–70. Montpellier: Naturalia Monspeliensia.

Covell, R.R. and McClurkin, D.C. (1967). Site index of loblolly pine on Ruston soils in the southern Coastal Plain. *J. For.* 65: 263–264.

Covington, W.W. and Moore, M.M. (1994). Postsettlement changes in natural fire regimes and forest structure: ecological restoration of old-growth ponderosa pine forests. *J. Sustain. For.* 2: 153–181.

Covington, W.W., Everett, R.L., Steele, R. et al. (1994). Historical and anticipated changes in forest ecosystems of the inland west of the United States. *J. Sustain. For.* 2: 13–63.

Cowles, H.C. (1899). The ecological relations of the vegetation on the sand dunes of Lake Michigan. *Bot. Gaz.* 27: 95–117, 167–202, 281–308, 361–391.

Cowles, H.C. (1901). The physiographic ecology of Chicago and vicinity; a study of the origin, development, and classification of plant societies. *Bot. Gaz.* 31: 73–108, 145–182.

Cowles, H.C. (1911). The causes of vegetative cycles. *Bot. Gaz.* 51: 161–183.

Cragg, J.B. (1953). Book review of natural communities by LR. Dice. *Bull. Inst. Biol.* 1: 3.

Craine, J.M. and Reich, P.B. (2005). Leaf-level light compensation points in shade-tolerant woody seedlings. *New Phytol.* 166: 710–713.

Crane, P.R. and Lidgard, S. (1989). Angiosperm diversification and paleolatitudinal gradients in Cretaceous floristic diversity. *Science* 246: 675–678.

Crankshaw, W.B., Qadir, S.A., and Lindsey, A.A. (1965). Edaphic controls of tree species in presettlement Indiana. *Ecology* 46: 688–698.

Crawford, D.J. (1974). A morphological and chemical study of *Populus acuminata* Rydberg. *Brittonia* 26: 74–89.

Crins, W. J., Gray, P. A., Uhlig, P. W. C., and Wester, M. C. (2009). *The Ecosystems of Ontario, Part I: Ecozones and Ecoregions*. Ontario Ministry of Natural Resources, Peterborough Ontario, Inventory, Monitoring and Assessment, SIB TER IMA TR- 01, 71pp.

Crisafulli, C.M. and Dale, V.H. (ed.) (2018). *Ecological Responses at Mount St. Helens: Revisited 35 years after the 1980 Eruption*. New York, NY: Springer 351 pp.

Critchfield, W. B. (1957). Geographic variation in *Pinus contorta*. Maria Moors Cabot Found. Publ. No. 3. Harvard Univ., Cambridge, MA. 118 pp.

Critchfield, W.B. (1985). The late Quaternary history of lodgepole and jack pines. *Can. J. For. Res.* 15: 749–772.

Crone, E.E. and Rapp, J.M. (2014). Resource depletion, pollen coupling, and the ecology of mast seeding. *Ann. NY Acad. Sci.* 1322: 21–34.

Crookston, N.L. and Dixon, G.E. (2005). The Forest Vegetation Simulator: A review of its structure, content, and applications. *Comput. Electron. Agric.* 49: 60–80.

Crouch, G.L. (1976). Wild animal damage to forests in the United States and Canada. In: *Proc. XVI IUFRO World Congress, Div. II*, 468–478. Oslo, Norway.

Crowley, K.F., Lovett, G.M., Arthur, M.A., and Weathers, K.C. (2016). Long-term effects of pest-induced tree species change on carbon and nitrogen cycling in northeastern US forests: A modeling analysis. *For. Ecol. Manag.* 372: 269–290.

Crutzen, P.J. and Goldhammer, J.G. (ed.) (1993). *Fire in the Environment: The Ecological, Atmospheric, and Climatic Importance of Vegetation Fires*. New York: Wiley 416 pp.

Csilléry, K., Lalagüe, H., Vendramin, G.G. et al. (2014). Detecting short spatial scale local adap tation and epistatic selection in climate-related candidate genes in European beech (*Fagus sylvatica*) populations. *Mol. Ecol.* 23: 4696–4708.

Cuevas, E. and Medina, E. (1988). Nutrient dynamics within Amazonian forest ecosystems. I. Nutrient flux in fine litter fall and efficiency of nutrient utilization. *Oecologia* 68: 466–472.

Culmsee, H., Schmidt, M., Schmiedel, I. et al. (2014). Predicting the distribution of forest habitat types using indicator species to facilitate systematic conservation planning. *Ecol. Indic.* 37: 131–144.

Cuny, H.E., Rathgeber, C.B., Frank, D. et al. (2014). Kinetics of tracheid development explain conifer tree-ring structure. *New Phytol.* 203: 1231–1241.

Currie, D.J. and Paquin, V. (1987). Large-scale biogeographical patterns of species richness of trees. *Nature* 329: 326–327.

Curtis, J.T. (1956). The modification of mid-latitude grasslands and forests by man. In: *Man's Role in Changing the Face of the Earth* (ed. W.L. Thomas Jr.), 721–736. Chicago: Univ. Chicago Press.

Curtis, J.T. (1959). *The Vegetation of Wisconsin*. Madison: Univ. Wisconsin Press 657 pp.

Curtis, R.O. (1972). Yield tables past and present. *J. For.* 70: 28–32.

Curtis, J.D. and Lersten, N.R. (1974). Morphology, seasonal variation and function of resin glands on buds and leaves of *Populus deltoides* (Salicaceae). *Am. J. Bot.* 61: 835–845.

Curtis, J.T. and Mclntosh, R.P. (1951). An upland forest continuum in the prairie-forest border region of Wisconsin. *Ecology* 32: 476–496.

Curtis, R.O., DeMars, D.J., and Herman, F.R. (1974a). Which dependent variable in site index—height-age regressions. *For. Sci.* 20: 74–87.

Curtis, R.O., Herman, F.R., and DeMars, D.J. (1974b). Height growth and site index for Douglas-fir in high-elevation forests of the Oregon-Washington Cascades. *For. Sci.* 20: 307–315.

Cutter, A. (2019). *A Primer of Molecular Population Genetics*. Oxford, UK: Oxford University Press 272 pp.

Cwynar, L.C. and MacDonald, G.M. (1987). Geographical variation of lodgepole pine in relation to population history. *Am. Nat.* 129: 463–469.

Cypert, E. (1973). Plant succession on burned areas in Okefenokee Swamp following the fires of 1954 and 1955. *In Proc. Annual Tall Timbers Fire Ecology Conf.* 12:199–217. Tall Timbers Res. Sta., Tallahassee, FL.

Daehler, C.C. (2003). Performance comparisons of co-occurring native and alien invasive plants: implications for conservation and restoration. *Annu. Rev. Ecol. Evol. Syst.* 34: 183–211.

Daily, G.C. (ed.) (1997). *Nature's Services: Societal Dependence on Natural Ecosystems*. Washington, DC: Island Press 412 pp.

Daily, G.C., Soderquist, S., Aniyar, S. et al. (2000). The value of nature and the nature of value. *Science* 289: 395–396.

Dale, V.H. and Crisafulli, C.M. (2018). Ecological responses to the 1980 eruption of Mount St. Helens: key lessons and remaining questions. In: *Ecological Responses at Mount St. Helens: Revisited 35 Years after the 1980 Eruption* (ed. C. Crisafulli and V. Dale), 1–18. New York, NY: Springer.

Dale, V.H. and Denton, E.M. (2018). Plant succession on the Mount St. Helens debris-avalanche deposit and the role of non-native species. In: *Ecological Responses at Mount St. Helens: Revisited 35 Years after the 1980 Eruption* (ed. C. Crisafulli and V. Dale), 149–164. New York, NY: Springer.

Dale, V.H., Joyce, L.A., McNulty, S. et al. (2001). Climate change and forest disturbances. *Bioscience* 51: 723–734.

Dale, V.H., Crisafulli, C.M., and Swanson, F.J. (2005a). 25 years of ecological change at Mount St. Helens. *Science* 308: 961–962.

Dale, V.H., Campbell, D.R., Adams, W.M. et al. (2005b). Plant succession on the Mount St. Helens debris-avalanche deposit. In: *Ecological Responses to the 1980 Eruption of Mount St. Helens* (ed. V.H. Dale, F.J. Swanson and C.M. Crisafulli), 59–73. Springer. New York.

Dale, V.H., Swanson, F.J., and Crisafulli, C.M. (2005c). Disturbance, survival, and succession: understanding ecological responses to the 1980 eruption of Mount St. Helens. In: *Ecological Responses to the 1980 Eruption of Mount St. Helens* (ed. V.H. Dale, F.J. Swanson and C.M. Crisafulli). New York: Springer.

D'Amato, A.W. and Palik, B.J. (2021). Building on the last "new" thing: exploring the compatibility of ecological and adaptation silviculture. *Can. J. For. Res.* 51: 172–180.

Damman, A. W. H. (1964). Some forest types of central Newfoundland and their relation to environmental factors. *For. Sci. Monogr.* 8. 62 pp.

Damman, A. W. H. (1975). Permanent changes in the chronosequence of a boreal forest habitat induced by natural disturbances. *In* W. Schmidt (ed.), *Int. Proc. Symp. Sukzessionsforschung. Int. Verein. Vegetationsk.* J. Cramer, Vaduz.

Daniel, T.W., Helms, J.A., and Baker, F.S. (1979). *Principles of Silviculture.* New York: McGraw-Hill 500 pp.

Dansereau, P. (1968). The continuum concept of vegetation: responses. *Bot. Rev.* 34: 253–332.

D'Antonio, C.M. and Vitousek, P.M. (1992). Biological invasions by exotic grasses, the grass/fire cycle, and global change. *Annu. Rev. Ecol. Syst.* 23: 63–87.

Darabant, A., Rai, P.B., Tenzin, K. et al. (2007). Cattle grazing facilitates tree regeneration in a conifer forest with palatable bamboo understory. *For. Ecol. Manag.* 252: 73–83.

Darley-Hill, S. and Johnson, W.C. (1981). Dispersal of acorns by blue jays (*Cyanocita cristata*). *Oecologia* 50: 231–232.

D'Arrigo, R.D., Kaufmann, R.K., Davi, N. et al. (2004). Thresholds for warming-induced growth decline at elevational tree line in the Yukon territory. Canada. *Glob. Biogeochem. Cycles* 18: GB3021.

D'Arrigo, R., Wilson, R., Liepert, B., and Cherubini, P. (2008). On the "Divergence Problem" in northern forests: a review of the tree-ring evidence and possible causes. *Glob. Planet. Chang.* 60: 289–305.

Dasmann, R.F. (1984). *Environmental Conservation*, 5e. New York: Wiley 486 pp.

Daubenmire, R.F. (1936). The "Big Woods" of Minnesota: its structure, and relation to climate, fire, and soils. *Ecol. Monogr.* 6: 223–268.

Daubenmire, R.F. (1952). Forest vegetation of northern Idaho and adjacent Washington, and its bearing on concepts of vegetation classification. *Ecol. Monogr.* 22: 301–330.

Daubenmire, R.F. (1954). Alpine timberlines in the Americas and their interpretation. *Butl. Univ. Bot. Stud.* 11: 119–136.

Daubenmire, R.F. (1966). Vegetation: identification of typal communities. *Science* 151: 291–298.

Daubenmire, R.F. (1968). *Plant Communities.* New York: Harper & Row 300 pp.

Daubenmire, R. F. (1970). Steppe vegetation of Washington. Tech. Bull 62. Wash. State Agr. Exp. Sta., Pullman, WA. 131 pp.

Daubenmire, R.F. (1974). *Plants and Environment: A Textbook of Plant Autecology.* New York: Wiley 422 pp.

Daubenmire, R.F. (1976). The use of vegetation in assessing the productivity of forest lands. *Bot. Rev.* 42: 115–143.

Daubenmire, R. F. (1989). The roots of a concept. *In* D. E. Ferguson, P. Morgan, F. D. Johnson (comp.), *Proceedings—Land Classifications Based on Vegetation: Applications for Resource Management.* USDA For. Serv. Gen. Tech. Report INT-257, Int. Res. Sta., Ogden, UT.

Daubenmire, R.F., and Daubenmire, J.B. (1968). Forest vegetation of eastern Washington and northern Idaho. Washington Agric. Exp. Sta., Tech. Bull. 60. 104 pp.

Davidson, E.A. and Janssens, I.A. (2006). Temperature sensitivity of soil carbon decomposition and feedbacks to climate change. *Nature* 440: 165–173.

Davidson, E.A. and Swank, W.T. (1987). Factors limiting denitrification in soils from mature and disturbed southeastern hardwood forests. *For. Sci.* 33: 135–144.

Davidson, E.A., Belk, E., and Boone, R.D. (1998). Soil water content and temperature as independent or confounded factors controlling soil respiration in a temperate mixed hardwood forest. *Glob. Chang. Biol.* 4: 217–227.

Davidson, A.M., Jennions, M., and Nicotra, A.B. (2011). Do invasive species show higher phenotypic plasticity than native species and, if so, is it adaptive? A meta-analysis. *Ecol. Lett.* 14: 419–431.

Davies, W.J. and Kozlowski, T.T. (1974). Stomatal responses of five woody angiosperms to light intensity and humidity. *Can. J. Bot.* 52: 1525–1534.

Davis, M.B. (1981). Quaternary history and the stability of forest communities. In: *Forest Succession: Concepts and Application* (ed. E.C. West, H.H. Shugart and D.B. Botkin), 132–153. New York: Springer-Verlag.

Davis, M.B. (1983a). Holocene vegetational history of the eastern United States. In: *The Late Quaternary Environments of the United States. Vol. 2: The Holocene* (ed. H.E. Wright and S. Porter), 166–181. Minneapolis: Univ. Minnesota Press.

Davis, M.B. (1983b). Quaternary history of deciduous forests of eastern North America and Europe. *Ann. Missouri Bot. Card.* 70: 550–563.

Davis, M.B. (1986). Climatic instability, time lags, and community disequilibrium. In: *Community Ecology* (ed. J. Diamond and T.J. Case), 269–284. New York: Harper and Row.

Davis, M.B. (1987). Invasions of forest communities during the Holocene: beech and hemlock in the Great Lakes region. In: *Colonization, Succession, and Stability* (ed. A.J. Gray, M.L. Crawley and P.J. Edwards), 373–393. London: Blackwell Sci. Publ.

Davis, P.H. and Heywood, V.H. (1963). *Principles of Angiosperm Taxonomy.* New York: D. Van Nostrand 558 pp.

Davis, R.B. and Jacobson, G.L. Jr. (1985). Late glacial and early Holocene landscapes in northern New England and adjacent areas of Canada. *Quat. Res.* 23: 341–368.

Davis, M.B., Woods, K.D., Webb, S.L., and Futyma, R.P. (1986). Dispersal versus climate: expansion of *Fagus* and *Tsuga* into the Upper Great Lakes region. *Vegetatio* 67: 93–103.

Davis, M.A., Chew, M.K., Hobbs, R.J. et al. (2011). Don't judge species on their origins. *Nature* 474: 153–154.

Davis, K.T., Dobrowski, S.Z., Higuera, P.E. et al. (2019). Wildfires and climate change push low-elevation forests across a critical climate threshold for tree regeneration. *Proc. Natl. Acad. Sci.* 116: 6193–6198.

Day, F.P. Jr. and Monk, C.D. (1974). Vegetation patterns on a southern Appalachian watershed. *Ecology* 55: 1064–1074.

Dayton, P.K. (1972). Toward an understanding of community resilience and the potential effects of enrichments to the benthos at McMurdo Sound, Antarctica. In: *Proceedings of the Colloquium on Conservation Problems in Antarctica* (ed. B.C. Parker), 81–96. Lawrence, KS: Allen Press.

De Bodt, S., Maere, S., and Van de Peer, Y. (2005). Genome duplication and the origin of angiosperms. *Trends Ecol. Evol.* 20: 591–597.

De Mazancourt, C. and Loreau, M. (2000). Effect of herbivory and plant species replacement on primary production. *Am. Nat.* 155: 735–754.

De Mazancourt, C., Isbell, F., Larocque, A. et al. (2013). Predicting ecosystem stability from community composition and biodiversity. *Ecol. Lett.* 16: 617–625.

De Villemereuil, P., Gaggiotti, O.E., Mouterde, M., and Till-Bottraud, I. (2016). Common garden experiments in the genomic era: new perspectives and opportunities. *Heredity* 116: 249–254.

DeBano, L.F. (1969). The relationship between heat treatment and water repellency in soils. In: *Water-Repellent Soils* (ed. L.F. DeBano and J. Letey), 265–279. Riverside: Univ. Calif.

DeBano, L.F., Osborn, J.F., Krammes, J.S., and Letey, J. Jr. 1967. Soil wettability and wetting agents . . . our current knowledge of the problem. USDA For. Serv. Res. Paper PSW-43. Pacific Southwest For. and Rge. Exp. Sta., Berkeley, CA 13 pp.

DeBano, L.F., Savage, S.M., and Hamilton, D.A. (1976). The transfer of heat and hydrophobic substances during burning. *Soil Sci. Soc. Am. J.* 40: 779–782.

DeBano, L.F., Neary, D.G., and Ffolliott, P.F. (1998). *Fire Effects on Ecosystems*. New York: Wiley 352 pp.

Decker, J.P. (1959). A system for analysis of forest succession. *For. Sci.* 5: 154–157.

Deevey, E.S. Jr. (1949). Biogeography of the Pleistocene: part I: Europe and North America. *Geol. Soc. Am. Bull.* 60: 1315–1416.

Del Moral, R. (1983). Initial recovery of subalpine vegetation on Mount St. Helens. *Am. Midl. Nat.* 109: 72–80.

Del Moral, R. and Chang, C.C. (2015). Multiple assessments of succession rates on Mount St. Helens. *Plant Ecol.* 216: 165–176.

Del Vecchio, T.A., Gehring, C.A., Coob, N.S., and Whitham, T.G. (1993). Negative effects of scale insect herbivory on the ectomoycorrhizae of juvenile pinyon pine. *Ecology* 74: 2297–2302.

Delcourt, H.R. (1980). Late quaternary vegetation history of the eastern Highland Rim and adjacent Cumberland Plateau of Tennessee. *Ecol. Monogr.* 49: 225–280.

Delcourt, H.R. (2003). *Forests in Peril: Tracking Deciduous Trees from Ice-Age Refuges into the Greenhouse World*. Lincoln: University of Nebraska Press 244 pp.

Delcourt, H.R. and Delcourt, P.A. (1974). Primeval magnolia—holly—beech climax in Louisiana. *Ecology* 55: 638–644.

Delcourt, P.A. and Delcourt, H.R. (1977). The Tunica Hills, Louisiana—Mississippi: late-glacial locality for spruce and deciduous forest species. *Quat. Res.* 7: 218–237.

Delcourt, P.A. and Delcourt, H.R. (1979). Late Pleistocene and Holocene distributional history of the deciduous forest in the southeastern United States. *Veröff. Geobot. Inst. ETH, Stift. Rübel* 68: 79–107.

Delcourt, P.A., and Delcourt, H.R. (1981). Vegetation maps for eastern North America: 40,000 yr B. P. to the present. *In* R. Romans (ed.), *Proc. 1980 Geobotany Conference,* Plenum, New York.

Delcourt, P.A. and Delcourt, H.R. (1987). *Long-Term Forest Dynamics of the Temperate Zone*. New York: Springer-Verlag 439 pp.

Delcourt, H.R. and Delcourt, P.A. (1988). Quaternary landscape ecology: relevant scales in space and time. *Landsc. Ecol.* 2: 23–44.

Delcourt, H.R. and Delcourt, P.A. (1991). *Quaternary Ecology, a Paleoecological Perspective*. New York: Chapman and Hall 242 pp.

Delisle, Z.J. and Parshall, T. (2018). The effects of oriental bittersweet on native trees in a New England floodplain. *Northeast. Nat.* 25: 188–196.

Denevan, W.M. (1992). The pristine myth: the landscape of the Americas in 1492. *Ann. Assoc. Am. Geogr.* 82: 369–385.

Denton, S.R. and Barnes, B.V. (1988). An ecological climatic classification of Michigan: a quantitative approach. *For. Sci.* 34: 119–138.

D'Eon, R.G., Glenn, S.M., Parfitt, I., and Fortin, M.-J. (2002). Landscape connectivity as a function of scale and organism vagility in a real forested landscape. *Conserv. Ecol.* 6: 10.

Despain, D.G. (1990). *Yellowstone Vegetation: Consequences of Environment and History in a Natural Setting*. Boulder, Colorado, USA: Rinehart 239 pp.

Desprez-Loustau, M.L., Marcais, B., Nageleisen, L.M. et al. (2006). Interactive effects of drought and pathogens in forest trees. *Ann. For. Sci.* 63: 597–612.

Dewan, S., Mijnsbrugge, K.V., De Frenne, P. et al. (2018). Maternal temperature during seed maturation affects seed germination and timing of bud set in seedlings of European black poplar. *For. Ecol. Manag.* 410: 126–135.

DeWoody, J., Rowe, C.A., Hipkins, V.D., and Mock, K.E. (2008). "Pando" lives: molecular genetic evidence of a giant aspen clone in Central Utah. *West. N. Am. Nat.* 68: Article 8.

Diamond, J.S., Epstein, J.M., Cohen, M.J. et al. (2021). A little relief: ecological functions and autogenesis of wetland microtopography. *Wiley Interdiscip. Rev. Water* 8: e1493.

Diaz, J.A., Carbonell, R., Virgos, E. et al. (2000). Effects of forest fragmentation on the distribution of the lizard *Psammodromus algirus*. *Anim. Conserv.* 3: 235–240.

Dieterich, H. (1970). Die Bedeutung der Vegetationskunde für die forstliche Standortskunde. *Der Biologieunterricht.* 6: 48–60.

Dieterich, J.H. and Swetnam, T.W. (1984). Dendrochronology of a fire-scarred ponderosa pine. *For. Sci.* 30: 238–247.

Dieterich, H., Müller, S., and Schlenker, G. (1970). *Urwald von morgen*. Stuttgart, Germany: Verlag Eugen Ulmer 174 pp.

DiFazio, S.P., Slavov, G.T., Burczyk, J. et al. (2004). Gene flow from tree plantations and implications for transgenic risk assessment. In: *Forest Biotechnology for the 21st Century* (ed. C. Walter and M. Carson). Kerala, India: Research Signpost 446 pp.

Divíšek, J., Chytrý, M., Beckage, B. et al. (2018). Similarity of introduced plant species to native ones facilitates naturalization, but differences enhance invasion success. *Nat. Commun.* 9: 1–10.

Dobbs, R.C. (1976). White spruce seed dispersal in central British Columbia. *For. Chron.* 52: 225–228.

Dobzhansky, T. (1951). *Genetics and the Origin of Species*, 3e. New York: Columbia Univ. Press 364 pp.

Dobzhansky, T. (1968). Adaptedness and fitness. In: *Population Biology and Evolution* (ed. R.C. Lewontin), 109–121. Syracuse, New York: Syracuse Univ. Press.

Dodd, J.C., Burton, C.C., Burns, R.G., and Jefferies, P. (1987). Phosphatase activity associated with the roots and rhizosphere of plants infected with vesicular-arbuscular mycorrhizal fungi. *New Phytol.* 107: 163–172.

Dodds, K.J., Aoki, C.F., Arango-Velez, A. et al. (2018). Expansion of southern pine beetle into northeastern forests: management and impact of a primary bark beetle in a new region. *J. For.* 116: 178–191.

Dolan, B.J. and Parker, G.R. (2005). Ecosystem classification in a flat, highly fragmented region of Indiana, USA. *For. Ecol. Manag.* 219: 109–131.

Domec, J.C., Rivera, L.N., King, J.S. et al. (2013). Hemlock woolly adelgid (*Adelges tsugae*) infestation affects water and carbon relations of eastern hemlock (*Tsuga canadensis*) and Carolina hemlock (*Tsuga caroliniana*). *New Phytol.* 199: 452–463.

Donato, D.C., Harvey, B.J., and Turner, M.G. (2016). Regeneration of montane forests 24 years after the 1988 Yellowstone fires: a fire-catalyzed shift in lower treelines? *Ecosphere* 7: e01410.

Donnegan, J.A. and Rebertus, A.J. (1999). Rates and mechanisms of subalpine forest succession along an environmental gradient. *Ecology* 80: 1370–1384.

Dooley, S.R. and Treseder, K.K. (2012). The effect of fire on microbial biomass: a meta-analysis of field studies. *Biogeochemistry* 109: 49–61.

Doolittle, W.T. (1957). Site index of scarlet and black oak in relation to southern Appalachian soil and topography. *For. Sci.* 3: 114–124.

Downs, A.A. (1938). Glaze damage in the birch-beech-maple-hemlock type of Pennsylvania and New York. *J. For.* 36: 63–70.

Downs, R.J. (1962). Photocontrol of growth and dormancy in woody plants. In: *Tree Growth* (ed. T.T. Kozlowski), 131–149. New York: Ronald Press.

Downs, A.A. and McQuilken, W.E. (1944). Seed production of southern Appalachian oaks. *J. For.* 53: 439–441.

Drew, T.J. and Flewelling, J.W. (1977). Some recent Japanese theories of yield-density relationships and their application to Monterey pine plantations. *For. Sci.* 23: 517–534.

Drew, T.J. and Flewelling, J.W. (1979). Stand density management: an alternative approach and its application to Douglas-fir plantations. *For. Sci.* 25: 518–532.

Drobyshev, I., Goebel, P.C., Hix, D.M. et al. (2008). Pre-and post-European settlement fire history of red pine dominated forest ecosystems of Seney National Wildlife Refuge, Upper Michigan. *Can. J. For. Res.* 38: 2497–2514.

Drury, W.H. and Nisbet, I.C.T. (1973). Succession. *J. Arnold Arbor.* 54: 331–368.

Du Pisani, J.A. (2006). Sustainable development—historical roots of the concept. *Environ. Sci.* 3: 83–96.

Duffield, J.W. and Snyder, E.B. (1958). Benefits from hybridizing American forest trees. *J. For.* 56: 809–815.

Duffy, D.C. and Meier, A.J. (1992). Do Appalachian herbaceous understories ever recover from clearcutting? *Conserv. Biol.* 6: 196–201.

DuMerle, P. (1988). Phenological resistance of oaks to the green oak leafroller, *Tortrix viridana* (Lepidoptera: Tortricidae). In: *Mechanisms of Woody Plant Defenses Against Insects: Search for Pattern* (ed. W.J. Mattson, J. Levieux and C. Bernard-Dagan), 215–226. New York: Springer-Verlag.

Dumont, E., Bahrman, N., Goulas, E. et al. (2011). A proteomic approach to decipher chilling response from cold acclimation in pea (*Pisum sativum* L.). *Plant Sci.* 180: 86–98.

Dumroese, R.K., William, M.I., Stanturf, J.A., and Clair, J.B.S. (2015). Considerations for restoring temperate forests of tomorrow: Forest restoration, assisted migration, and bioengineering. *New For.* 46: 947–964.

Dunn, P.H. and DeBano, L.F. (1977). Fire's effect on biological and chemical properties of chaparral soils. *In* H.A. Mooney, C.E. Conrad. (tech. coords.). *Proceedings of a Symposium on Environmental Conservation: Fire and Fuel Management in Mediterranean Ecosystems*. August 4–5, 1988, Palo Alto, CA. Washington, D.C. USDA For. Serv. WO-3.

Dusenge, M.E., Duarte, A.G., and Way, D.A. (2019). Plant carbon metabolism and climate change: elevated CO_2 and temperature impacts on photosynthesis, photorespiration and respiration. *New Phytol.* 221: 32–49.

Duvigneaud, P. (1946). La variabilité des associations végétales. *Bull. Soc. R. Bot. Belg.* 78: 107–134.

Dwyer, L.M. and Merriam, G. (1981). Influence of topographic heterogeneity on deciduous litter decomposition. *Oikos* 37: 228–237.

Dyksterhuis, E.J. (1949). Condition and management of range land based on quantitative ecology. *J. Range Manag.* 2: 104–115.

Dyksterhuis, E.J. (1958). Ecological principles in range evaluation. *Bot. Rev.* 24: 253–272.

Dyrness, C.T. (1976). Effect of wildfire on soil wettability in the high Cascades of Oregon. USDA For. Serv. Res. Paper PNW-202. Pacific Northwest For. and Rge. Exp. Sta., Portland, OR. 18 pp.

Easley, A.T., Passineau, J.F., and Driver, B.L. (comps.). (1990). The use of wilderness for personal growth, therapy, and education. USDA. For. Serv., Gen. Tech. Rep. RM-193. Rocky Mountain Res. Sta., Ft. Collins, CO. 197 pp.

Easterling, D.R., Kunkel, K.E., Arnold, J.R. et al. (2017). Precipitation change in the United States. In: *Climate Science Special Report: Fourth National Climate Assessment, Volume I* (ed. D.J. Wuebbles, D.W. Fahey, K.A. Hibbard, et al.), 207–230. Washington, DC, USA: U.S. Global Change Research Program.

Eckenwalder, J.E. (1977). North American cottonwoods (*Populus,* Salicaceae) of sections *Abaso* and *Aigeiros*. *J. Arnold Arbor.* 58: 193–208.

Eckenwalder, J.E. (1996). Systematics and evolution of *Populus*. In: *Biology of Populus and its Implications for Management and Conservation* (ed. R.F. Stettler, H.D. Bradshaw Jr., P.E. Heilman and T.M. Hinckley). Ottawa: NRC Res. Press 539 pp.

Eckert, A.J., Bower, A.D., Gonzalez-Martinez, S.C. et al. (2010). Back to nature: ecological genomics of loblolly pine (*Pinus taeda*, Pinaceae). *Mol. Ecol.* 19: 3789–3805.

Eckert, A.J., Maloney, P.E., Vogler, D.R. et al. (2015). Local adaptation at fine spatial scales: an example from sugar pine (*Pinus lambertiana*, Pinaceae). *Tree Genet. Genomes* 11: 42–58.

Ecoregions Working Group (1989). Ecoclimatic regions of Canada, first approximation. Ecoregions Working Group of the Canada Committee on Ecological Land Classification. Ecol. Land Classif. Series, No. 23, Can. Wildlife Serv., Env. Canada, Ottawa, Ontario. 119 pp. and map.

Edwards, C.A. (1974). Macroarthropods. In: *Biology of Plant Litter Decomposition* (ed. C.H. Dickinson and G.J.F. Pugh), 533–554. New York: Academic Press.

Edwards, P.J. and Abivardi, C. (1998). The value of biodiversity: where ecology and economy blend. *Biol. Conserv.* 83: 239–246.

Edwards, C.A. and Bohlen, P.J. (1996). *Biology and Ecology of Earthworms*. London: Chapman and Hall 426 pp.

Egler, F.E. (1954). Vegetation science concepts I. initial floristic composition, a factor in old-field vegetation development. *Vegetatio* 4: 412–417.

Egler, F. E. (1968). The contumacious continuum. *In* P. Dansereau (ed.), The continuum concept of vegetation: responses. *Bot. Rev.* 34:253–332.

Egler, F.E. (1977). *The Nature of Vegetation, its Management and Mismanagement*. Bridgewater, CO: Conn. Cons. Assoc. 527 pp.

Eis, S. (1976). Association of western white pine cone crops with weather variables. *Can. J. For. Res.* 6: 6–12.

Ekberg, I., Eriksson, G., and Gormling, I. (1979). Photoperiodic reactions in conifer species. *Holarct. Ecol.* 2: 255–263.

Ekins, P., Simon, S., Deutsch, L. et al. (2003). A framework for the practical application of the concepts of critical natural capital and strong sustainability. *Ecol. Econ.* 44: 165–185.

El-Kassaby, Y.A., Dunsworth, B.G., and Krakowski, J. (2003). Genetic evaluation of alternative silvicultural systems in coastal montane forests: western hemlock and amabilis fir. *Theor. Appl. Genet.* 107: 598–610.

Elkington, J. (2004). Enter the triple bottom line. In: *The Triple Bottom Line: Does it all Add Up?* (ed. A. Henriques and J. Richardson). London: Routledge Publishng 209 pp.

Elkins, J.W. and Rosen, R. (1989). *Summary Report 1988: Geophysical Monitoring for Climate Change*. ERL, Boulder, CO: NOAA 142 pp.

Ellenberg, H. (1968). Wege der Geobotanik zum Verstandnis der Pflanzendecke. *Naturwissenschaften* 55: 462–470.

Ellenberg, H. (1974). Zeigerwerte der Gefässpfanzen Mitteleuropas. *Scripta Geobot.* 9: 1–97.

Ellenberg, H. (1988). *Vegetation Ecology of Central Europe*, 4e. Cambridge: Cambridge Univ. Press 731 pp.

Ellis, M., von Dohlen, C. D., Anderson, J. E., and Romme, W. H. (1994). Some important factors affecting density of lodgepole pine seedlings following the 1988 Yellowstone fires. *In* E. G. Despain (ed.), *Conf. Proc. Plants and Their Environments*. Tech. Report NPS/NRYELL/NRTR-93/XX. USDI, Nat. Park Service.

Ellison, A.M., Bank, M.S., Clinton, B.D. et al. (2005). Loss of foundation species: consequences for the structure and dynamics of forested ecosystems. *Front. Ecol. Environ.* 3: 479–486.

Ellison, A.M., Barker Plotkin, A.A., Khalid, S., and (2016). Foundation species loss and biodiversity of the herbaceous layer in New England forests. *Forests* 7: 9.

Ellstrand, N.C. (1992). Gene flow among seed plant populations. In: *Population Genetics of Forest Trees* (ed. W.T. Adams, S.H. Strauss, D.L. Copes and A.R. Griffin), 241–256. Boston, MA: Kluwer.

Ellsworth, D.S. and Reich, P.B. (1993). Canopy structure and vertical patterns of photosynthesis and related leaf traits in a deciduous forest. *Oecologia* 96: 69–178.

Elston, J.J. and Hewitt, D.G. (2010). Intake of mast by wildlife in Texas and the potential for competition with wild boars. *Southwest. Nat.* 55: 57–66.

Endara, M.-J. and Coley, P.D. (2011). The resource availability hypothesis revisited: a meta-analysis. *Funct. Ecol.* 25: 89–398.

Engeman, R.M., Stevens, A., Allen, J. et al. (2007). Feral swine management for conservation of an imperiled wetland habitat: Florida's vanishing seepage slopes. *Biol. Conserv.* 134: 440–446.

Engler, A. (1905). Einfluss der Provenienz des Samens auf die Eigenschaften der forstlichen Holzgewächse. *Mitt, schweiz. Centralanst. forstl. Versuchsw.* 8:81–236.

Engler, A. (1908). Tatsachen, Hypothsen und Irrtümer auf dem Gebiete der Samenprovenienz-Frage. *Forstwiss. Centralbl.* 30: 295–314.

Epanchin-Niell, R.S. and Liebhold, A.M. (2015). Benefits of invasion prevention: effect of time lags, spread rates, and damage persistence. *Ecol. Econ.* 116: 146–153.

Erickson, A., Nitschke, C., Coops, N. et al. (2015). Past-century decline in forest regeneration potential across a latitudinal and elevational gradient in Canada. *Ecol. Model.* 313: 94–102.

Esau, K. (1977). *Anatomy of Seed Plants*. New York: Wiley 550 pp.

Eschner, A.R. and Patric, J.H. (1982). Debris avalanches in eastern upland forests. *J. For.* 80: 343–347.

Eschrich, W., Burchardt, R., and Essiamah, S. (1989). The induction of sun and shade leaves of the European beech (*Fagus sylvatica* L.): anatomical studies. *Trees* 3: 1–10.

Eshelman, K.R., Wagner, R.E., and Secrist, F.M. (1989). Vegetation classification—problems, principles, and proposals. *In* D. E. Ferguson, P. Morgan, and F. D. Johnson (comps.), *Symp. Proc. Land Classifications Based on Vegetation: Applications for Resource Management*. USDA For. Serv. Gen. Tech. Report INT-257. Intermountain Res. Sta., Ogden, UT. 315 pp.

Eshleman, K.N., Morgan, R.P., Webb, J.R. et al. (1998). Temporal patterns of nitrogen leakage from mid-Appalachian forested watersheds: role of insect defoliation. *Water Resour. Res.* 34: 2005–2016.

Evans, K.L., Greenwood, J.J., and Gaston, K.J. (2005). Dissecting the species–energy relationship. *Proc. R. Soc. B Biol. Sci.* 272: 2155–2163.

Ewel, K.C. (1990). Swamps. In: *Ecosystems of Florida* (ed. R.L. Myers and J.J. Ewel). Orlando: Univ. Central Florida Press.

Eyre, F.H. (1980). *Forest Cover Types of the United States and Canada*. Bethesda, MD: Soc. Am. For 148 pp.

Faegri, K. and Iverson, J. (1989). *Textbook of Pollen Analysis*, 4e. New York: Wiley 338 pp.

Fahrig, L. (2003). Effects of habitat fragmentation on biodiversity. *Annu. Rev. Ecol. Evol. Syst.* 34: 487–515.

Fairbridge, R.W. (ed.) (1972). *The Encyclopedia of Geochemistry and Environmental Sciences*. New York: Van Nostrand Reinhold 1321 pp.

Fall, M.W., Avery, M.L., Campbell, T.A. et al. (2011). Rodents and other vertebrate invaders in the United States. In: *Biological Invasions: Economic and Environmental Costs of Alien Plant, Animal, and Microbe Species*, 2e (ed. D. Pimentel), 381–410. Boca Raton, FL: CRC Press.

FAO (1980). *Poplars and Willows in Wood Production and Land Use*. Rome: Food and Agr. Organization of the United Nations 328 pp.

Farhat, P., Hidalgo, O., Robert, T. et al. (2019). Polyploidy in the conifer genus Juniperus: an unexpectedly high rate. *Front. Plant Sci.* 10: 676.

Farley, J. (2012). Ecosystem services: the economics debate. *Ecosyst. Serv.* 1: 40–49.

Fassnacht, K.S. and Gower, S.T. (1997). Interrelationships between the edaphic and stand characteristics, leaf area index and above-ground net primary productivity of upland forest ecosystems in north Central Wisconsin. *Can. J. For. Res.* 27: 1058–1067.

Fastie, C. (1995). Causes and ecosystem consequences of multiple pathways of primary succession at Glacier Bay, Alaska. *Ecology* 76: 1899–1916.

Fearer, T.M., Norman, G.W., Pack, J.C. Sr. et al. (2008). Influence of physiographic and climatic factors on spatial patterns of acorn production in Maryland and Virginia, USA. *J. Biogeogr.* 35: 2012–2025.

Fearnside, P.M. (2021). The intrinsic value of Amazon biodiversity. *Biodivers. Conserv.* 30: 1199–1202.

Fedrowitz, K., Koricheva, J., Baker, S.C. et al. (2014). Can retention forestry help conserve biodiversity? A meta-analysis. *J. Appl. Ecol.* 51: 1669–1679.

Feeny, P. (1976). Plant apparency and chemical defense. In: *Biochemical Interaction between Plants and Insects* (ed. J.W. Wallace and R.L. Mansell), 1–40. Boston, MA: Springer.

Fei, S., Desprez, J.M., Potter, K.M. et al. (2017). Divergence of species responses to climate change. *Sci. Adv.* 3: e1603055.

Fei, S., Jo, I., Guo, Q. et al. (2018). Impacts of climate on the biodiversity-productivity relationship in natural forests. *Nat. Commun.* 9: 1–7.

Fenneman, N.M. (1931). *Physiography of Western United States*. New York: McGraw-Hill 534 pp.

Fenneman, N.M. (1938). *Physiography of Eastern United States*. New York: McGraw-Hill 691 pp.

Ferguson, D.E., Morgan, P., and Johnson, F.D. (comps.) (1989). *Proceedings—Land Classifications Based on Vegetation: Applications for Resource Management*. USDA For. Serv. Gen. Tech. Report INT-257. Intermountain Res. Sta., Ogden, UT. 315 pp.

Field, C. and Mooney, H.A. (1986). The photosynthesis-nitrogen relationship in wild plants. In: *On the Economy of Plant Form and Function* (ed. T.J. Givnish), 25–55. Cambridge: Cambridge Univ. Press.

Findlay, B.F. (1976). Recent developments in eco-climatic classifications. In: *Ecological (Biophysical) Land Classification in Canada* (ed. J. Thie and G. Ironside), 121–127. Environment Canada, Ottawa: Lands Directorate.

Finegan, B. (1984). Forest succession. *Nature* 312: 109–114.

Finlay, R.D., Ek, H., Odham, G., and Söderström, B. (1989). Uptake, translocation and assimilation of nitrogen from 15N-labeled ammonium and nitrate sources by intact ectomycorrhizal systems of *Fagus sylvatica* and *Paxillus involutus*. *New Phytol.* 113: 47–55.

Finney, M.A. and Martin, R.E. (1989). Fire history in a *Sequoia sempervirens* forest at Salt Point State Park, California. *Can. J. For. Res.* 19: 1451–1457.

Finnigan, J.J. and Brunet, Y. (1995). Turbulent airflow in forests on flat and hilly terrain. In: *Wind and Trees* (ed. M.P. Coutts and J. Grace), 3–40. Cambridge: Cambridge Univ. Press.

Finzi, A.C., Raymer, P.C., Giasson, M.A., and Orwig, D.A. (2014). Net primary production and soil respiration in New England hemlock forests affected by the hemlock woolly adelgid. *Ecosphere* 5 (8): 1–16.

Fiorucci, A.S. and Fankhauser, C. (2017). Plant strategies for enhancing access to sunlight. *Curr. Biol.* 27: R931–R940.

Firestone, M.K. and Davidson, E.A. (1989). Microbial basis of NO and N_2O production and consumption in soil. In: *Exchange of Trace Gases between Terrestrial Ecosystems and the Atmosphere* (ed. M.O. Andreae and D.S. Schimel). New York: Wiley.

Fischer, W.C. and Bradley, A. F. (1987). Fire ecology of western Montana forest habitat types. USDA For. Serv. Gen. Tech. Report INT-223. Int. Res. Sta., Odgen, UT. 95 pp.

Fischer, J., Brosi, B., Daily, G. et al. (2008). Should agricultural policies encourage land sparing or wildlife-friendly farming? *Ecol. Environ.* 6: 380–385.

Fisher, W.C. (1989). The fire effects information system: a comprehensive vegetation knowledge base. *In* D.E. Ferguson, P. Morgan, and F.D. Johnson (comps.), *Proceedings—Land Classifications Based on Vegetation: Applications for Resource Management*. USDA For. Serv. Gen. Tech. Report INT-257. Intermountain Res. Sta., Ogden, UT.

Fisher, F.M., Parker, L.W., Anderson, J.P., and Whitford, W.G. (1987). Nitrogen mineralizaiton in a desert soil: interacting effects of soil moisture and nitrogen fertilizer. *Soil Sci. Soc. Am. J.* 1033–1041.

Fisher, J.I., Mustard, J.F., and Vadeboncoeur, M.A. (2006). Green leaf phenology at Landsat resolution: scaling from the field to the satellite. *Remote Sens. Environ.* 100: 265–279.

Fisichelli, N., Wright, A., Rice, K. et al. (2014a). First-year seedlings and climate change: species-specific responses of 15 North American tree species. *Oikos* 123: 1331–1340.

Fisichelli, N.A., Frelich, L.E., and Reich, P.B. (2014b). Temperate tree expansion into adjacent boreal forest patches facilitated by warmer temperatures. *Ecography* 37: 152–161.

Fitter, A.H. (1987). An architectural approach to the comparative ecology of plant root systems. *New Phytol.* 106 (Supp): 61–77.

Fitter, A.H. and Hay, R.K.M. (1987). *Environmental Physiology of Plants*, 2e. New York: Academic Press 423 pp.

Fitter, A.H. and Stickland, T.R. (1991). Architectural analysis of plant root systems 2. Influence of nutrient supply on architecture in contrasting plant species. *New Phytol.* 118: 383–389.

Flanagan, P.W. and Van Cleve, K. (1983). Nutrient cycling in relation to decomposition and organic matter quality in taiga ecosystems. *Can. J. For. Res.* 13: 795–817.

Flannigan, M.D. (1993). Fire regime and the abundance of red pine. *Int. J. Wildland Fire* 3: 241–247.

Fleischner, T.L. (1994). Ecological costs of livestock grazing in western North America. *Conserv. Biol.* 8: 629–644.

Flint, H.R. (1930). Fire as a factor in the management of north Idaho national forests. *Northwest Sci.* 4: 12–15.

Flint, H.L. (1972). Cold hardiness of twigs of *Quercus rubra* L. as a function of geographic origin. *Ecology* 53: 1163–1170.

Floate, K.D. and Whitham, T.G. (1993). The "hybrid bridge" hypothesis: host shifting via plant hybrid swarms. *Am. Nat.* 141: 651–662.

Floate, K.D., Kearsley, M.J.C., and Whitham, T.G. (1993). Elevated herbivory in plant hybrid zones: *Chrysomela confluens, Populus* and phenological sinks. *Ecology* 74: 2056–2065.

Flower, C.E., Knight, K.S., and Gonzalez-Meler, M.A. (2013). Impacts of the emerald ash borer (*Agrilus planipennis* Fairmaire) induced ash (*Fraxinus* spp.) mortality on forest carbon cycling and successional dynamics in the eastern United States. *Biol. Invasions* 15: 931–944.

Foggin, G.T. III and DeBano, L.F. (1971). Some geographic implications of water-repellent soils. *Prof. Geogr.* 23: 347–350.

Ford-Robertson, F.C. (ed.) (1983). *Terminology of Forest Science Technology Practice and Products*, 2nd Printing. Washington, D.C: Soc. Am. Foresters 370 pp.

Forman, R.T.T. (1995). *Land Mosaics, the Ecology of Landscapes and Regions*. Cambridge: Cambridge Univ. Press 632 pp.

Forman, R.T.T. and Godron, M. (1986). *Landscape Ecology*. New York: Wiley New York. 619 pp.

Forrester, D.I., Ammer, C., Annighöfer, P.J. et al. (2018). Effects of crown architecture and stand structure on light absorption in mixed and monospecific *Fagus sylvatica* and *Pinus sylvestris* forests along a productivity and climate gradient through Europe. *J. Ecol.* 106: 746–760.

Foster, D.R. (1988a). Disturbance history, community organization and vegetation dynamics of the old-growth Pisgah forest, southwestern New Hampshire, U.S.A. *J. Ecol.* 76: 105–134.

Foster, D.R. (1988b). Species and stand response to catastrophic wind in central New England, U.S.A. *J. Ecol.* 76: 135–151.

Foster, D.R. and Boose, E.R. (1992). Patterns of forest damage resulting from catastrophic wind in central New England, USA. *J. Ecol.* 80: 70–98.

Foster, D.R. and Boose, E.R. (1995). Hurricane disturbance regimes in temperate and tropical forest ecosystems. In: *Wind and Trees* (ed. M.P. Coutts and J. Grace), 305–339. Cambridge: Cambridge Univ. Press.

Foster, D.R., Knight, D.H., and Franklin, J.F. (1998). Landscape patterns and legacies resulting from large, infrequent forest disturbances. *Ecosystems* 1: 497–510.

Fotis, A.T., Morin, T.H., Fahey, R.T. et al. (2018). Forest structure in space and time: biotic and abiotic determinants of canopy complexity and their effects on net pri mary productivity. *Agric. For. Meterol.* 250: 181–191.

Fowler, D.P. (1965a). Effects of inbreeding in red pine, *Pinus resinosa* Ait. II. Pollination studies. *Silvae Genet.* 14: 12–23.

Fowler, D.P. (1965b). Effects of inbreeding in red pine, *Pinus resinosa* Ait. IV. Comparison with other Northeastern *Pinus* species. *Silvae Genet.* 14: 76–81.

Fowler, D. (1980). Removal of sulphur and nitrogen compounds from the atmosphere in rain and by dry deposition. In: *Ecological Effects of Acid Precipitation* (ed. D. Drablos and A. Tollan), 22–32. Oslo, Norway: SNSF (Acid Precipitation-Effects on Forests and Fish) Project.

Fowler, D.P. and Mullin, R.E. (1977). Upland-lowland ecotypes not well developed in black spruce in northern Ontario. *Can. J. For. Res.* 7: 35–10.

Fox, J.T. (1977). Alternation and coexistence of tree species. *Am. Nat.* 111: 69–89.

Foy, C.D., Chaney, O., and White, M.C. (1978). The physiology of metal toxicity of plants. *Ann. Rev. Plant Physiol.* 29: 511–566.

Fralish, J.S., Crooks, F.B., Chambers, J.L., and Harty, F.M. (1991). Comparison of presettlement, second-growth and old-growth forest on six site types in the Illinois Shawnee Hills. *Am. Midl. Nat.* 125: 294–309.

Fralish, J.S., Mclntosh, R.P., and Loucks, O.L. (eds.). (1993). *John T. Curtis, Fifty Years of Wisconsin Plant Ecology*. Wisc. Acad. Sci., Arts & Letters, Madison. 339 pp.

Franklin, J.F. (1968). Cone production by upper-slope conifers. USDA For. Serv. Res. Paper PNW-60. Pacific Northwest For. and Rge. Exp. Sta., Portland, OR. 21 pp.

Franklin, E.C. (1970). *Survey of Mutant Forms and Inbreeding Depression in Species of the Family Pinaceae*. Vol. 61, USDA For. Serv, Southeastern For. Exp. Sta., Asheville, NC. 21 pp.

Franklin, J.F. (1990). Biological legacies: a critical management concept from Mount St. Helens. *In Trans. Fifty-fifth North American Wildlife and Natural Resources Conference*. Wildlife Manage. Inst., Washington, D.C.

Franklin, J.F. (1993). Preserving biodiversity: species, ecosystems, or landscapes? *Ecol. Appl.* 3: 202–205.

Franklin, J.F. (1995). Sustainability of managed temperate forest ecosystems. In: *Defining and Measuring Sustainability: The Biogeophysical Foundations* (ed. M. Munasinghe and W. Shearer), 355–385. Washington, D.C: World Bank.

Franklin, J.F. (1997). Ecosystem management: an overview. *In* M. Boyce (ed.), *Proc. Symp. Ecosystem Management: Applications for Sustainable Forest and Wildlife Resources*. Yale Univ. Press, New Haven, CT.

Franklin, J.F. (2003). Challenges to temperate forest stewardship – focusing on the future. In: *Towards Forest Sustainability* (ed. D.B. Lindenmayer and J.F. Franklin), 1–13. Washington, DC: Island Press.

Franklin, J.F. and Dyrness, C.T. (1973). Natural vegetation of Oregon and Washington. USDA For. Serv. Gen. Tech. Rep. PNW-8, Pacific Northwest For. Rge. Exp. Sta., Portland OR. 417 pp.

Franklin, J.F. and Dyrness, C.T. (1980). *Natural Vegetation of Oregon and Washington*. Corvallis: Oregon Univ. Press 452 pp.

Franklin, J.F. and Johnson, K.N. (2012). A restoration framework for federal forests in the Pacific northwest. *J. For.* 110: 429–439.

Franklin, J.F. and MacMahon, J.A. (2000). Messages from a mountain. *Science* 288: 1183–1184.

Franklin, J.F., Cromack, K. Jr., Denison, W. et al. (1981). Ecological characteristics of old-growth Douglas-fir forests. USDA For. Serv. Gen. Tech. Report PNW-118. Pacific Northwest For. and Rge. Exp. Sta., Corvallis OR. 48 pp.

Franklin, J.F., MacMahon, J.A., Swanson, F.J., and Sedell, J.R. (1985). Ecosystem responses to the eruption of Mount St. Helens. *Natl. Geogr. Res.* 1: 196–215.

Franklin, J.F., Shugart, H.H., and Harmon, M.E. (1987). Tree death as an ecological process. *Bioscience* 37: 550–556.

Franklin, J.F., Frenzen, P.M., and Swanson, F.J. (1988). Re-creation of ecosystems at Mount St. Helens: contrasts in artificial and natural approaches. In: *Rehabilitating Damaged Ecosystems*, vol. *2* (ed. J. Cairns Jr.), 287–334. Boca Raton, FL: CRC Press.

Franklin, J.F., Berg, D.R., Thornburgh, D.A., and Tappeiner, J.C. (1997). Alternative silvicultural approaches to timber harvesting: variable retention harvest systems. In: *Creating a Forestry for the 21st Century* (ed. K.A. Kohm and J.F. Franklin), 111–139. Washington, D.C: Island Press.

Franklin, J.F., Lindenmayer, D., MacMahon, J.A. et al. (2000). Threads of continuity. *Conserv. Biol. Pract.* 1: 8–16.

Franklin, J.F., Johnson, K.N., and Johnson, D.L. (2018). *Ecological Forest Management*. Long Grove, IL: Waveland Press 646 pp.

Fraser, E.C., Lieffers, V.J., and Landhäusser, S.M. (2006). Carbohydrate transfer through root grafts to support shaded trees. *Tree Physiol.* 26: 1019–1023.

Frederickson, M.E. (2017). Mutualisms are not on the verge of breakdown. *Trends Ecol. Evol.* 32: 727–734.

Fredriksen, R.L. (1972). Nutrient budget of a Douglas-fir forest on an experimental watershed in western Oregon. *In* J.F. Franklin. L.J. Dempster, and R.H. Waring (eds.), *Symp. Proc. Research on Coniferous Forest Ecosystems*. USDA For. Serv., Pacific Northwest For. and Rge. Exp. Sta., Portland, OR.

Frelich, L.E. (1992). The relationship of natural disturbances to white pine stand development. *In* R. A. Stine, M.J. Baughman (eds.). *White Pine Symposium Proceedings: History, Ecology, Policy and Management*. Department of Forest Resources, College of Natural Resources, University of Minnesota, and Minnesota Extension Service. St. Paul, Minnesota.

Frelich, L.E. (2002). *Forest Dynamics and Disturbance Regimes*. Cambridge, New York. 266 pp.

Frelich, L.E. and Lorimer, C.G. (1985). Current and predicted long-term effects of deer browsing in hemlock forests in Michigan, USA. *Biol. Conserv.* 34: 99–120.

Frelich, L.E. and Lorimer, C.G. (1991). Natural disturbance regimes in hemlock-hardwood forests of the upper Great Lakes region. *Ecol. Monogr.* 61: 145–164.

Frelich, L.E., Hale, C.M., Scheu, S. et al. (2006). Earthworm invasion into previously earthworm-free temperate and boreal forests. *Biol. Invasions* 8: 1235–1245.

Freud, S. (1917). *Introductory Lectures on Psychoanalysis*. New York: Liveright 656 pp.

Frey, S.J., Hadley, A.S., and Betts, M.G. (2016). Microclimate predicts within-season distribution dynamics of montane forest birds. *Divers. Distrib.* 22: 944–959.

Frissell, S.S. Jr. (1973). The importance of fire as a natural ecological factor in Itasca State Park, Minnesota. *Quat. Res.* 3: 397–407.

Fritts, H. (1976). *Tree Rings and Climate*. New York: Academic Press 567 pp.

Froelich, N.J. and Schmid, H.P. (2006). Flow divergence and density flows above and below a deciduous forest: part II. Below-canopy thermotopographic flows. *Agric. For. Meteorol.* 138: 29–43.

Frye, R.J. and Quinn, J.A. (1979). Forest development in relation to topography and soils on a floodplain of the Raritan River, New Jersey. *Bull. Torrey Bot. Club* 106: 334–345.

Fuller, W.H., Shannon, S., and Burgess, P.S. (1955). Effect of burning on certain forest soils of northern Arizona. *For. Sci.* 1: 44–50.

Funk, J.L. and Vitousek, P.M. (2007). Resource-use efficiency and plant invasion in low-resource systems. *Nature* 446: 1079–1081.

Furnier, G.R., Knowles, P., Clyde, M.A., and Dancik, B.P. (1987). Effects of avian seed dispersal on the genetic structure of whitebark pine populations. *Evolution* 4: 607–612.

Furniss, M. (1972). Observations of resistance and susceptibility to Douglas-fir beetles. *In* Program Abstracts, Second North American For. Biol. Workshop, Oregon State Univ., Corvallis. p. 24.

Gabriel, D., Sait, S.M., Hodgson, J.A. et al. (2010). Scale matters: the impact of organic farming on biodiversity at different spatial scales. *Ecol. Lett.* 13: 858–869.

Gadgil, M. and Berkes, F. (1991). Traditional resource management systems. *Resour. Manage. Optimization* 8: 127–141.

Galbraith-Kent, S.L. and Handel, S.N. (2008). Invasive *Acer platanoides* inhibits native sapling growth in forest understorey communities. *J. Ecol.* 96: 293–302.

Gallagher, R.V., Randall, R.P., and Leishman, M.R. (2015). Trait differences between naturalized and invasive plant species independent of residence time and phylogeny. *Conserv. Biol.* 29: 360–369.

Galston, A.W. (1994). *Life Processes of Plants*. New York: W. H. Freeman 245 pp.

Gandhi, K.J. and Herms, D.A. (2010). Direct and indirect effects of alien insect herbivores on ecological processes and interactions in forests of eastern North America. *Biol. Invasions* 12: 389–405.

Garland, J.A. (1977). The dry deposition of sulphur dioxide to land and water surfaces. *Proc. R. Soc. Lond.* 12: 245–268.

Garner, W.W. (1923). Further studies in photoperiodism in relation to hydrogen-ion concentration of the cell-sap and the carbohydrate content of the plant. *J. Agric. Res.* 23: 871–920.

Garner, W.W. and Allard, H.A. (1920). Effect of the relative length of day and night and other factors of the environment on growth and reproduction in plants. *J. Agric. Res.* 18: 553–606.

Gashwiler, J.S. (1967). Conifer seed survival in a western Oregon clearcut. *Ecology* 48: 431–438.

Gašparíková, O., Čiamporová, M., Mistrík, I., and Baluška, F. (2001). *Recent Advances of Plant Root Structure and Function*. New York: Kluwer Academic Publishers 192 pp.

Gaston, K.J. and Spicer, J.I. (2014). *Biodiversity: An Introduction*. Wiley 208 pp.

Gaston, K.J., Warren, P.H., Devine-Wright, P. et al. (2007). Psychological benefits of greenspace increase with biodiversity. *Biol. Lett.* 3: 390–394.

Gates, D.M. (1968). Energy exchange between organism and environment. In: *Biometeorology* (ed. W.P. Lowry). Corvallis: Oregon State Univ. Press.

Gates, D.M. (1980). *Biophysical Ecology*. New York: Springer-Verlag 611 pp.

Gaul, D., Hertel, D., and Leuschner, C. (2008). Effects of experimental soil frost on the fine-root system of mature Norway spruce. *J. Plant Nutr. Soil Sci.* 171: 690–698.

Gaylord, M.L., Kolb, T.E., Pockman, W.T. et al. (2013). Drought predisposes piñon–juniper woodlands to insect attacks and mortality. *New Phytol.* 198: 567–578.

Gehring, C.A. and Whitham, T.G. (1991). Herbivore-driven mycorrhizal mutualism in insect-susceptible pinyon pine. *Nature* 353: 556–557.

Geiger, R., Aron, R.H., and Todhunter, P. (2009). *The Climate near the Ground*, 7e. New York: Rowman & Littlefield Publishers 623 pp.

Genet, H., Bréda, N., and Dufrêne, E. (2010). Age-related variation in carbon allocation at tree and stand scales in beech (*Fagus sylvatica* L.) and sessile oak (*Quercus petraea* (Matt.) Liebl.) using a chronosequence approach. *Tree Physiol.* 30: 177–192.

Genre, A., Lanfranco, L., Perotto, S., and Bonfante, P. (2020). Unique and common traits in mycorrhizal symbioses. *Nat. Rev. Microbiol.* 18: 649–660.

Gentry, A.H. (1986). Endemism in tropical versus temperate plant communities. In: *Conservation Biology: The Science of Scarcity and Diversity* (ed. M. Soulé). Sunderland, MA: Sinauer Associates.

Gentry, A.H. and Dodson, C.H. (1987). Diversity and biogeography of neotropical vascular epiphytes. *Ann. Missouri Bot. Gard.* 74: 205–233.

George, M.F. and Burke, M.J. (1976). The occurrence of deep supercooling in cold hardy plants. *Curr. Adv. Plant Sci.* 22: 349–360.

George, M.F. and Burke, M.J. (1977). Supercooling in overwintering azalea flower buds: Additional freezing parameters. *Plant Physiol.* 59: 319–325.

George, M.F., Burke, M.J., Pellet, H.M., and Johnson, A.G. (1974). Low temperature exotherms and woody plant distribution. *Hortic. Sci.* 6: 519–522.

George, J.P., Grabner, M., Karanitsch-Ackerl, S. et al. (2017). Genetic variation, phenotypic stability, and repeatability of drought response in European larch throughout 50 years in a common garden experiment. *Tree Physiol.* 37: 33–46.

Gerhardt, F. and Foster, D.R. (2002). Physiographical and historical effects on forest vegetation in central New England, USA. *J. Biogeogr.* 29: 1421–1437.

Gersper, P.L. and Holowaychuk, N. (1971). Some effects of stem flow from forest canopy trees on chemical properties of soils. *Ecology* 52: 691–702.

Ghazoul, J. (2007). Recognizing the complexities of ecosystem management and the ecosystem service concept. *Gaia* 16: 215–221.

Ghilarov, A.M. (2000). Ecosystem functioning and intrinsic value of biodiversity. *Oikos* 90: 408–412.

Gholz, H.L. (1982). Environmental limits on above-ground net primary production, leaf area, and biomass in vegetation zones of the Pacific Northwest. *Ecology* 63: 469–481.

Gholz, H.L. and Fisher, R.L. (1982). Organic matter production and distribution in slash pine (*Pinus elliottii)* plantations. *Ecology* 63: 1827–1839.

Gholz, H.L., Wedin, D.A., Smitherman, S.M. et al. (2000). Long-term dynamics of pine and hardwood litter in contrasting environments: toward a global model of decomposition. *Glob. Chang. Biol.* 6: 751–765.

Gill, C.J. (1970). The flooding tolerance of woody species—a review. *For. Abstr.* 31: 671–688.

Gill, A. M. (1977). Plant traits adapted to fires in Mediterranean land ecosystems. *In* H. Mooney and C. E. Conrad (eds.), *Proc. Symp. The Environmental Consequences of Fire and Fuel Management in Mediterranean Ecosystems*. USDA, For. Serv. Gen. Tech. Report WO-3, Washington, D.C.

Gill, D.S. and Marks, P.L. (1991). Tree and shrub seedling colonization of old fields in Central New York. *Ecol. Monogr.* 61: 183–205.

Gill, A.M., Groves, R.H., and Noble, I.R. (1981). *Fire and the Australian Biota*. Canberra: Australian Acad. Sci 582 pp.

Gill, D.S., Amthor, J.S., and Bormann, F.H. (1998). Leaf phenology, photosynthesis, and the persistence of saplings and shrubs in a mature northern hardwood forest. *Tree Physiol.* 18: 281–289.

Gill, A.L., Gallinat, A.S., Sanders-DeMott, R. et al. (2015). Changes in autumn senescence in northern hemisphere deciduous trees: a meta-analysis of autumn phenology studies. *Ann. Bot.* 116: 875–888.

Gillespie, J.H. (2004). *Population Genetics: A Concise Guide*, 2e. Baltimore, MD: Johns Hopkins University Press 232 pp.

Gillett, N.P., Weaver, A.J., Zwiers, F.W., and Flannigan, M.D. (2004). Detecting the effect of climate change on Canadian forest fires. *Geophys. Res. Lett.* 31: L18211.

Gillman, L.N. and Wright, S.D. (2006). The influence of productivity on the species richness of plants: a critical assessment. *Ecology* 87: 1234–1243.

Girard, F., Payette, S., and Gagnon, R. (2009). Origin of the lichen–spruce woodland in the closed-crown forest zone of eastern Canada. *Glob. Ecol. Biogeogr.* 18: 291–303.

Givnish, T.J. (1981). Serotiny, geography, and fire in the pine barrens of New Jersey. *Evolution* 35: 101–123.

Givnish, T.J. (1988). Adaptation to sun and shade: a whole-plant perspective. *Aust. J. Plant Physiol.* 15: 63–92.

Givnish, T.J. (2002). Adaptive significance of evergreen vs. deciduous leaves: solving the triple paradox. *Silva Fenn.* 36: 703–743.

Givnish, T.J., Montgomery, R.A., and Goldstein, G. (2004). Adaptive radiation of photosynthetic physiology in the Hawaiian lobeliads: light regimes, static light responses, and whole-plant compensation points. *Am. J. Bot.* 91: 228–246.

Gleason, H.A. (1917). The structure and development of the plant association. *Bull. Torrey Bot. Club* 44: 463–481.

Gleason, H.A. (1926). The individualistic concept of the plant association. *Bull. Torrey Bot. Club* 53: 7–26.

Gleason, H.A. (1936). Is the synusia an association? *Ecology* 17: 444–451.

Gleason, H.A. (1939). The individualistic concept of the plant association. *Am. Midl. Nat.* 21: 92–110.

Glenn-Lewin, D.C. and van der Maarel, E. (1992). Patterns and processes of vegetation dynamics. In: *Plant Succession: Theory and Prediction* (ed. D.C. Glenn-Lewin, R.K. Peet and T.T. Veblen), 13–59. London: Chapman and Hall.

Glenn-Lewin, D.C., Peet, R.K., and Veblen, T.T. (ed.) (1992). *Plant Succession: Theory and Prediction*. London: Chapman and Hall 352 pp.

Glinski, R. L. (1977). Regeneration and distribution of sycamores and cottonwood trees along Sonoita Creek, Santa Cruz County, Arizona. *In* R. R. Johnson, D. A. Jones (tech. coords). *Importance, Preservation, and Management of Riparian Habitat: A Symposium*. U.S. For. Serv. Gen. Tech. Rep. RM-43. Rocky Mountain For. Range Exp. Sta., Fort Collins, CO.

Glynn, C., Herms, D.A., Orians, C.M. et al. (2007). Testing the growth differentiation balance hypothesis: dynamic responses of willows to nutrient availability. *New Phytol.* 176: 623–634.

Godwin, H. (1956). *The History of the British Flora*. Cambridge: Cambridge Univ. Press 384 pp.

Goebel, P.C., Palik, B.J., Kirkman, L.K. et al. (2001). Forest ecosystems of a Lower Gulf Coastal Plain landscape: multifactor classification and analysis. *J. Torr. Bot. Soc.* 128: 47–75.

Goldblatt, P. (1980). Polyploidy in angiosperms: monocotyledons. In: *Polyploidy* (ed. W.H. Lewis), 219–239. New York: Plenum Press.

Golley, F.B. (1977). *Ecological Succession*. Stroudsburg, PA: Dowden, Hutchinson, and Ross 373 pp.

Golley, F.B. (ed.) (1983). *Ecosystems of the World, Tropical Rain Forest Ecosystems*, vol. 14A. New York: Elsevier 382 pp.

Golley, F.B. (1993). *A History of the Ecosystem Concept in Ecology*. New Haven, CT: Yale Univ. Press 254 pp.

Good, N.F. (1968). A study of natural replacement of chestnut in six stands in the highlands of New Jersey. *Bull. Torrey Bot. Club* 95: 240–253.

Goodall, D., Lugo, A., Brinson, M., and Brown, S. (1990). *Ecosystems of the World, Forested Wetlands*, vol. 15. New York: Elsevier 527 pp.

Gordon, D., Rosati, A., Damiano, C., and Dejong, T.M. (2006). Seasonal effects of light exposure, temperature, trunk growth and plant carbohydrate status on the initiation and growth of epicormic shoots in *Prunus persica*. *J. Hortic. Sci. Biotechnol.* 81: 421–428.

Gosz, J.R., Holmes, R.T., Likens, G.E., and Bormann, F.H. (1978). The flow of energy in a forest ecosystem. *Sci. Am.* 328: 92–102.

Gough, C.M., Atkins, J.W., Fahey, R.T., and Hardiman, B.S. (2019). High rates of primary production in structurally complex forests. *Ecology* 100: 1–6.

Gough, C.M., Atkins, J.W., Fahey, R.T. et al. (2020). Community and structural constraints on the complexity of eastern North American forests. *Glob. Ecol. Biogeogr.* 29: 2107–2118.

Gowdy, J.M. (1997). The value of biodiversity: markets, society, and ecosystems. *Land Econ.* 73: 25–41.

Graber, R.E. (1970). Natural seed fall in white pine (*Pinus strobus* L.) stands of varying density. USDA For. Serv. Res. Note NE-119. Northeastern For. Exp. Sta., Upper Darby, PA. 6 pp.

Grace, J. (1989). Tree lines. *Phil. Trans. R. Soc. Lond. B* 324: 233–245.

Grace, J.B., Anderson, T.M., Seabloom, E.W. et al. (2016). Integrative modelling reveals mechanisms linking productivity and plant species richness. *Nature* 529: 390–393.

Gradmann, R. (1898). *Das Pflanzenleben der Schwäbischen Alb*. Verlag Schwäbischen Albvereins, Tübingen. Vol 1:1–376; Vol 2:1–376.

Gradowski, T., Lieffers, V.J., Landhäusser, S.M. et al. (2010). Regeneration of *Populus* nine years after variable retention harvest in boreal mixedwood forests. *For. Ecol. Manag.* 259: 383–389.

Graham, A. (1999). *Late Cretaceous and Cenozoic history of North American vegetation*. New York: Oxford University Press 350 pp.

Graham, B.F. Jr. and Bormann, F.H. (1966). Natural root grafts. *Bot. Rev.* 32: 255–292.

Granhall, U. (1981). Biological nitrogen fixation in relation to environmental factors and functioning of natural ecosystems. In: *Terrestrial Nitrogen Cycles*. Ecol. Bull. 33 (ed. F.E. Clark and T. Rosswell), 131–144. Stockholm: Swedish Natural Sci. Res. Council.

Grant, V. (1963). *The Origin of Adaptations*. New York: Columbia Univ. Press 606 pp.

Grant, V. (1971). *Plant Speciation*. New York: Columbia Univ. Press 435 pp.

Grant, V. (1977). *Organismic Evolution*. San Francisco: W. H. Freeman 418 pp.

Grant, M.C. (1993). The trembling giant. *Discover* 14: 83–89.

Grant, M.C., Mitton, J.B., and Linhart, Y.B. (1992). Even larger organisms. *Nature* 360: 216.

Gratkowski, H.J. (1956). Windthrow around staggered settings in old-growth Douglas-fir. *For. Sci.* 2: 60–74.

Graumlich, L.J., Brubaker, L.B., and Grier, C.C. (1989). Long-term trends in forest net primary productivity: Cascade Mountains, Washington. *Ecology* 70: 405–410.

Gray, S.B. and Brady, S.M. (2016). Plant developmental responses to climate change. *Dev. Biol.* 419: 64–77.

Gray, L.K., Gylander, T., Mbogga, M.S. et al. (2011). Assisted migration to address climate change: recommendations for aspen reforestation in western Canada. *Ecol. Appl.* 21: 1591–1603.

Green, D.G. (1981). Time series and postglacial forest ecology. *Quat. Res.* 15: 265–277.

Green, D.C. and Grigal, D.F. (1979). Jack pine biomass accretion on shallow and deep soils in Minnesota. *Soil Sci. Soc. Amer. J.* 43: 1233–1237.

Green, D.C. and Grigal, D.F. (1980). Nutrient accumulations in jack pine stands on deep and shallow soils over bedrock. *For. Sci.* 26: 325–333.

Green, S.J. and Grosholz, E.D. (2021). Functional eradication as a framework for invasive species control. *Front. Ecol. Environ.* 19: 98–107.

Green, R.E., Connell, S.J., Scharlemann, J.P., and Balmford, A. (2005). Farming and the fate of wild nature. *Science* 307: 550–555.

Greenwood, M.S. and Hutchison, K.W. (1993). Maturation as a developmental process. In: *Clonal Forestry I, Genetics and Biotechnology* (ed. M.R. Ahuja and W.J. Libby), 14–33. Berlin: Springer-Verlag.

Greer, B.T., Still, C., Howe, G.T. et al. (2016). Populations of aspen (*Populus tremuloides* Michx.) with different evolutionary histories differ in their climate occupancy. *Ecol. Evol.* 6: 3032–3039.

Greer, B.T., Still, C., Cullinan, G.L. et al. (2018). Polyploidy influences plant–environment interactions in quaking aspen (*Populus tremuloides* Michx.). *Tree Physiol.* 38: 630–640.

Gregory, S.V., Swanson, F.J., McKee, W.A., and Cummins, K.W. (1991). An ecosystem perspective of riparian zones. *Bioscience* 41: 540–551.

Gregow, H., Laaksonen, A., and Alper, M. (2017). Increasing large scale windstorm damage in western, central and northern European forests, 1951–2010. *Sci. Rep.* 7: 46397.

Grier, C.C. (1975). Wildfire effects on nutrient distribution and leaching in a coniferous ecosystem. *Can. J. For. Res.* 5: 599–607.

Grier, C. C, Lee, K. M., Nadkarni, N. M. et al. (1989). Productivity of forests of the United States and its relation to soil and site factors and management: a review. USDA For. Serv. Gen. Tech. Report PNW-GTR-222. Pacific Northwest Res. Sta., Corvallis, OR. 51 pp.

Griffin, G.J. (1992). American chestnut survival in understory mesic sites following the chestnut blight pandemic. *Can. J. Bot.* 70: 1950–1956.

Grigal, D. F. (1984). Shortcomings of soil surveys for forest management. *In* J. G. Bockheim (ed.), *Symp. Proc. Forest Land Classification: Experience, Problems, Perspectives*. NCR-102 North Central For. Soils Com., Soc. Am. For., USDA For. Serv., and USDA Soil Cons. Serv. Madison, WI.

Grime, J.P. (1966). Shade avoidance and shade tolerance in flowering plants. In: *Light as an Ecological Factor* (ed. R. Bainbridge, G.C. Evans and O. Rackham), 187–207. England: Blackwell, Oxford.

Grime, J.P. (1977). Evidence for the existence of three primary strategies in plants and its relevance to ecological and evolutionary theory. *Am. Nat.* 111: 1169–1194.

Grime, J.P. (1979). *Plant Strategies and Vegetation Processes*. New York: Wiley 222 pp.

Grime, J.P. (1988). The C-S-R model of primary plant strategies—origins, implications and tests. In: *Plant Evolutionary Biology* (ed. L.D. Gottlieb and S.K. Jain), 371–393. New York: Chapman and Hall.

Grimm, R.C. (1984). Fire and other factors controlling the Big Woods vegetation of Minnesota in the mid-nineteenth century. *Ecol. Monogr.* 54: 291–311.

Grimm, V. and Wissel, C. (1997). Babel, or the ecological stability discussions: an inventory and analysis of terminology and a guide for avoiding confusion. *Oecologia* 109: 323–334.

Groffman, P.M. and Tiedje, J.M. (1989). Denitrification in north temperate forests soils: spatial and temporal patterns at the landscape and seasonal scales. *Soil Biol. Biochem.* 21: 613–620.

Groffman, P.M., Tiedje, J.M., Mokma, D.L., and Simkins, S. (1992). Regional scale analysis of denitrification in north temperate forest soils. *Landsc. Ecol.* 7: 45–53.

Groner, E. and Novoplansky, A. (2003). Reconsidering diversity–productivity relationships: directness of productivity estimates matters. *Ecol. Lett.* 6: 695–699.

Groom, M.J., Meffe, G.K., Carroll, C.R., and Andelman, S.J. (2006). *Principles of Conservation Biology*, 3e. Sunderland, MA: Sinauer Associates 816 pp.

Gross, H.L. (1972). Crown deterioration and reduced growth associated with excessive seed production by birch. *Can. J. Bot.* 50: 2431–2437.

Gross, H.L. and Harnden, A. A. (1968). Dieback and abnormal growth of yellow birch induced by heavy fruiting. Canada Dept. For. and Rural Develop., For. Res Lab. Info. Report O-X-79, Sault Ste. Marie, Ontario. 12 pp.

Grotkopp, E. and Rejmánek, M. (2007). High seedling relative growth rate and specific leaf area are traits of invasive species: phylogenetically independent contrasts of woody angiosperms. *Am. J. Bot.* 94: 526–532.

Grubb, P.J. (1977). The maintenance of species-richness in plant communities: the importance of the regeneration niche. *Biol. Rev.* 52: 107–145.

Grubb, P.J. (1985). Plant populations and vegetation in relation to habitat, disturbance and competition: problems of generalization. In: *The Population Structure of Vegetation* (Handbook of Vegetation Science, Part III) (ed. J. White), 23–36. Dordrecht: W. Junk.

Grubb, P.J. (1988). The uncoupling of disturbance and recruitment, two kinds of seed bank, and persistence of plant populations at the regional and local scales. *Ann. Zool. Fenn.* 25: 23–36.

Grumbine, R.E. (1994). What is ecosystem management? *Conserv. Biol.* 8: 27–38.

Grytnes, J.A. and Vetaas, O.R. (2002). Species richness and altitude: a comparison between null models and interpolated plant species richness along the Himalayan altitudinal gradient, Nepal. *Am. Nat.* 159: 294–304.

Gu, L., Hanson, P.J., Post, W.M. et al. (2008). The 2007 eastern US spring freeze: increased cold damage in a warming world? *Bioscience* 58: 253–262.

Guarino, R., Willner, W., Pignatti, S. et al. (2018). Spatio-temporal variations in the application of the Braun-Blanquet approach in Europe. *Phytocoenologia* 48: 239–250.

Gunderson, C.A., Edwards, N.T., Walker, A.V. et al. (2012). Forest phenology and a warmer climate–growing season extension in relation to climatic provenance. *Glob. Chang. Biol.* 18: 2008–2025.

Guo, J.P. and Wang, J.R. (2006). Comparison of height growth and growth intercept models of jack pine plantations and natural stands in northern Ontario. *Can. J. For. Res.* 36: 2179–2188.

Gurevitch, J. and Padilla, D. (2004). Are invasive species a major cause of extinctions? *Trends Ecol. Evol.* 19: 470–474.

Gurnell, A.M. and Grabowski, R.C. (2016). Vegetation–hydrogeomorphology interactions in a low-energy, human-impacted river. *River Res. Appl.* 32: 202–215.

Gurnell, A.M., Bertoldi, W., and Corenblit, D. (2012). Changing river channels: the roles of hydrological processes, plants and pioneer landforms in humid temperate, mixed load, gravel bed rivers. *Earth-Sci. Rev.* 111: 129–141.

Gurnell, A.M., Corenblit, D., García de Jalón, D. et al. (2016). A conceptual model of vegetation–hydrogeomorphology interactions within river corridors. *River Res. Appl.* 32: 142–163.

Gustafson, F.G. (1943). Influence of light upon tree growth. *J. For.* 41: 212–213.

Gustafson, E.J. and Crow, T.R. (1994). Modeling the effects of forest harvesting on landscape structure and the spatial distribution of cowbird brood parasitism. *Landsc. Ecol.* 9: 237–248.

Haack, R.A. and Byler, J.W. (1993). Insects and pathogens, regulators of forest ecosystems. *J. For.* 91: 32–37.

Haase, E.F. (1970). Environmental fluctuations on south-facing slopes in the Santa Catalina Mountains of Arizona. *Ecology* 51: 959–974.

Habeck, J.R. (1958). White cedar ecotypes in Wisconsin. *Ecology* 39: 457–463.

Habeck, J.R. and Mutch, R.W. (1973). Fire-dependent forests in the northern Rocky Mountains. *Quat. Res.* 3: 408–424.

Hack, J.T. and Goodlett, J.C. (1960). Geomorphology and forest ecology of a mountain region in the central Appalachians. U. S. Geol. Surv. Prof. Paper 347. 66 pp + map.

Hacke, U.G., Sperry, J.S., Pockman, W.T. et al. (2001). Trends in wood density and structure are linked to prevention of xylem implosion by negative pressure. *Oecologia* 126: 457–461.

Hacket-Pain, A.J., Lageard, J.G.A., and Thomas, P.A. (2017). Drought and reproductive effort interact to control growth of a temperate broadleaved tree species (*Fagus sylvatica*). *Tree Physiol.* 37: 744–754.

Hagemeier, M. and Leuschner, C. (2019). Leaf and crown optical properties of five early-, mid-and late-successional temperate tree species and their relation to sapling light demand. *Forests* 10: 925.

Hagen, J.B. (1992). *An Entangled Bank*. New Brunswick, NJ: Rutgers Univ. Press 245 pp.

Hagglund, B. (1981). Evaluation of forest site productivity. *For. Abstr.* 42: 515–527.

Hagmann, R.K., Hessburg, P.F., Prichard, S.J. et al. (2021). Evidence for widespread changes in the structure, composition, and fire regimes of western North American forests. *Ecol. Appl.* 31: e02431.

Halbwachs, H., Brandl, R., and Bässler, C. (2015). Spore wall traits of ectomycorrhizal and saprotrophic agarics may mirror their distinct lifestyles. *Fungal Ecol.* 17: 197–204.

Hale, C.M., Frelich, L.E., Reich, P.B., and Pastor, J. (2005a). Effects of European earthworm invasion on soil characteristics in northern hardwood forests of Minnesota, USA. *Ecosystems* 8: 911–927.

Hale, C.M., Frelich, L.E., and Reich, P.B. (2005b). Exotic European earthworm invasion dynamics in northern hardwood forests of Minnesota, USA. *Ecol. Appl.* 15: 848–860.

Hale, C.M., Frelich, L.E., and Reich, P.B. (2006). Changes in hardwood forest understory plant communities in response to European earthworm invasions. *Ecology* 87: 1637–1649.

Hall, F.C. (1989). Plant community classification: from concept to application. *In* D.E. Ferguson,

E.E., P. Morgan, and F.D. Johnson (comps.), *Proceedings—Land Classifications Based on Vegetation: Applications for Resource Management*. USDA For. Serv. Gen. Tech. Report INT-257. Intermountain Res. Sta., Ogden, UT. 315 pp.

Hall, A.R., Miller, A.D., Leggett, H.C. et al. (2012). Diversity-disturbance relationships: frequency and intensity interact. *Biol. Lett.* 8: 768–771.

Hallé, F., Oldeman, R.A.A., and Tomlinson, P.B. (1978). *Tropical Trees and Forests*. Berlin: Springer-Verlag 441 pp.

Haller, J.R. (1965). The role of 2-needle fascicles in the adaptation and evolution of ponderosa pine. *Brittonia* 17: 354–382.

Halliday, W.E.D. (1937). A forest classification for Canada. *For. Serv. Bull.* 89; Can. Dept. Mines and Resources. 50 pp.

Halofsky, J.E., Andrews-Key, S.A., Edwards, J.E. et al. (2018). Adapting forest management to climate change: the state of science and applications in Canada and the United States. *For. Ecol. Manag.* 421: 84–97.

Halpern, C.B. and Harmon, M.E. (1983). Early plant succession on the Muddy River mudflow, Mount St. Helens. *Am. Midl. Nat.* 110: 97–106.

Halpern, C.B. and Spies, T.A. (1995). Plant species diversity in natural and managed forest of the Pacific Northwest. *Ecol. Appl.* 5: 913–934.

Hamanishi, E.T. and Campbell, M.M. (2011). Genome-wide responses to drought in forest trees. *Forestry* 84: 273–283.

Hamilton, M.B. (2009). *Population Genetics*. Hoboken, NJ: Wiley Blackwell 424 pp.

Hammerson, G.A. (1994). Beaver (*Caston canadensis*): ecosystem alterations, management, and monitoring. *Nat. Areas J.* 14: 44–57.

Hammitt, W. E. and Barnes, B.V. (1989). Composition and structure of an old-growth oak-hickory forest in southern Michigan over 20 years. *In* G. Rink and C. A. Budelsky (eds.), *Proc. Seventh Central Hardwoods Conference*. USDA For. Serv. Gen. Tech. Report NC-132. North Central For. Exp. Sta., St. Paul, MN.

Hampf, F.E. (ed.). (1965). Site index curves for some forest species in the eastern United States. USDA For. Serv., Eastern Region. Upper Darby, PA. 43 pp.

Hamrick, J.L. (1989). Isozymes and the analysis of genetic structure in plant populations. In: *Isozymes in Plant Biology* (ed. D.E. Soltis and P.S. Soltis), 87–105. Portland, OR: Dioscorides Press.

Hamrick, J.L. (2004). Response of forest trees to global environmental changes. *For. Ecol. Manag.* 197: 323–335.

Hamrick, J.L. and Godt, M.J.W. (1989). Allozyme diversity in plant species. In: *Plant Populations Genetics, Breeding, and Genetic Resources* (ed. A.H.D. Brown, M.T. Clegg, A.L. Kahler and B.S. Weir), 43–63. Sunderland, MA: Sinauer.

Hamrick, J.L. and Godt, M.J.W. (1996). Effects of life history traits on genetic diversity in plant species. *Philos. Trans. R. Soc. B: Biol. Sci.* 351: 1291–1298.

Hamrick, J.L., Godt, M.J.W., and Sherman-Broyles, S.L. (1992). Factors influencing levels of genetic diversity in woody plant species. In: *Population Genetics of Forest Trees* (ed. W.T. Adams, S.H. Strauss, D.L. Copes and A.R. Griffin), 95–124. Boston, MA: Kluwer.

Han, Q. and Kabeya, D. (2017). Recent developments in understanding mast seeding in relation to dynamics of carbon and nitrogen resources in temperate trees. *Ecol. Res.* 32: 771–778.

Handler, S., Pike, C., and St. Clair, B. (2018). Assisted migration. USDA Forest Service Climate Change Resource Center. http://www.fs.usda.gov/ccrc/topics/assisted-migration. Accessed 01-13-2022.

Hanes, T. (1988). California chaparral. *In* M.G. Barbour and J. Major (eds.), *Terrestrial Vegetation of California*. California Native Plant Society, Special Publ. No. 9. Berkeley, CA.

Hanks, J.P., Fitzhugh, E., and Hanks, S. R. (1983). A habitat type classification system for ponderosa pine forests of northern Arizona. USDA For. Serv. Gen. Tech. Report RM-97. Rocky Mountain For. and Rge. Exp. Sta., Fort Collins, CO. 22 pp.

Hanley, N. and Perrings, C. (2019). The economic value of biodiversity. *Ann. Rev. Resour. Econ.* 11: 355–375.

Hanley, M.E., Lamont, B.B., Fairbanks, M.M., and Rafferty, C.M. (2007). Plant structural traits and their role in anti-herbivore defence. *Perspect. Plant Ecol. Evol. Syst.* 8: 157–178.

Hannah, L. (2021). *Climate Change Biology*, 3e. Cambridge, MA: Academic Press 528 pp.

Hanover, J.W. (1975). Physiology of tree resistance to insects. *Annu. Rev. Entomol.* 20: 75–95.

Hansen, H.P. (1947). Postglacial forest succession, climate, and chronology in the Pacific Northwest. *Trans. Am. Philos. Soc.* 37 (1): 130 pp.

Hansen, H.P. (1955). Postglacial forests in south central and central British Columbia. *Am. J. Sci.* 253: 640–658.

Hansen, E.M. (2014). Forest development and carbon dynamics after mountain pine beetle outbreaks. *For. Sci.* 60: 476–488.

Hansen, A.J., Spies, T.A., Swanson, F.J. et al. (1991). Conserving biodiversity in managed forests. *Bioscience* 41: 382–392.

Hansen, J.W., Challinor, A., Ines, A. et al. (2006). Translating climate forecasts into agricultural terms: advances and challenges. *Clim. Res.* 33: 27–41.

Hanson, A.D. and Hitz, W.D. (1982). Metabolic responses of mesophytes to plant water deficits. *Annu. Rev. Plant Physiol.* 33: 163–203.

Hardig, T.M., Brunsfeld, S.J., Fritz, R.S. et al. (2000). Morphological and molecular evidence for hybridization and introgression in a willow (*Salix*) hybrid zone. *Mol. Ecol.* 9: 9–24.

Hardin, J.W. (1979). Patterns of variation in foliar trichomes of eastern North American *Quercus*. *Am. J. Bot.* 66: 576–585.

Hardy, J.P., Groffman, P.M., Fitzhugh, R.D. et al. (2001). Snow depth manipulation and its influence on soil frost and water dynamics in a northern hardwood forest. *Biogeochemistry* 56: 151–174.

Hare, F.K. and Ritchie, J.C. (1972). The boreal bioclimates. *Geogr. Rev.* 62: 333–365.

Hari, P. and Häkkinen, R. (1991). The utilization of old phenological time series of budburst to compare models describing annual cycles of plants. *Tree Physiol.* 8: 281–287.

Harley, J.L. and Smith, S.E. (1983). *Mycorrhizal Symbiosis*. New York: Academic Press 483 pp.

Harley, P., Guenther, A., and Zimmerman, P. (1996). Effects of light, temperature and canopy position on net photosynthesis and isoprene emission from sweetgum (*Liquidambar styraciflua*) leaves. *Tree Physiol.* 16: 25–32.

Harlow, W.M., Harrar, E.S., Hardin, J.W., and White, F.M. (1996). *Textbook of Dendrology*, 8e. New York: McGraw-Hill 501 pp.

Harmon, M.E. (1984). Survival of trees after low-intensity surface fires in Great Smoky Mountains National Park. *Ecology* 65: 796–802.

Harmon, M.E. (1987). The influence of litter and humus accumulations and canopy openness on *Picea sitchensis* (Bong.) Carr. and *Tsuga heterophylla* (Raf.) Sarg. seedlings growing on logs. *Can. J. For. Res.* 17: 1475–1479.

Harmon, M.E. and Franklin, J.F. (1989). Tree seedlings on logs in *Picea-Tsuga* forests of Oregon and Washington. *Ecology* 70: 48–59.

Harmon, M.E. and Hua, C. (1991). Coarse woody debris dynamics in two old-growth ecosystems. *Bioscience* 41: 604–610.

Harmon, M.E., Franklin, J.F., Swanson, F.J. et al. (1986). Ecology of coarse woody debris in temperate ecosystems. *Adv. Ecol. Res.* 15: 133–302.

Harmon, M.E., Ferrell, W.K., and Franklin, J.F. (1990). Effects on carbon storage of conversion of old-growth forests to young forests. *Science* 247: 699–702.

Harmon, M.E., Bond-Lamberty, B., Tang, J., and Vargas, R. (2011). Heterotrophic respiration in disturbed forests: a review with examples from North America. *J. Geophys. Res.* 116: G00K04.

Harper, J.L. (1961). Approaches to the study of plant competition. *In* F.L. Milthorpe (ed.), *Mechanisms in Biological Competition*. Symp. Soc. Exp. Biology 15:1–39.

Harper, R. M. (1962). Historical notes on the relation of fires to forests. *In Proc. Annual Tall Timbers Fire Ecology Conf.* 1:11–29. Tall Timbers Res. Sta. Talahassee, FL.

Harper, J.L. (1977). *Population Biology of Plants*. London: Academic Press 892 pp.

Harper, J.L. and White, J. (1974). The demography of plants. *Annu. Rev. Ecol. Syst.* 5: 419–463.

Harrington, T.C. (1986). Growth decline of wind-exposed red spruce and balsam fir in the White Mountains. *Can. J. For. Res.* 16: 232–238.

Harrington, W. (1991). Wildlife: severe decline and partial recovery. In: *America's Renewable Resources: Historical Trends and Current Challenges* (ed. K.D. Frederick and R.A. Sedjo), 205–248. Washington, D.C: Resources for the Future.

Harrington, C.A. and Gould, P.J. (2015). Tradeoffs between chilling and forcing in satisfying dormancy requirements for Pacific Northwest tree species. *Front. Plant Sci.* 6: 120.

Harrington, C.A., Gould, P.J., and Clair, J.B.S. (2010). Modeling the effects of winter environment on dormancy release of Douglas-fir. *For. Ecol. Manag.* 259: 798–808.

Harris, T.M. (1958). Forest fire in the mesozoic. *J. Ecol.* 46: 447–453.

Harris, L.D. (1984). *The Fragmented Forest*. Chicago: Univ. Chicago Press 211 pp.

Harris, J.R. and Fanelli, J. (1999). Root and shoot growth of pot-in-pot red and sugar maple. *J. Environ. Hortic.* 17: 80–83.

Harris, L.D. and Silva-Lopez, G. (1992). Forest fragmentation and the conservation of biological diversity. In: *Conservation Biology* (ed. P.L. Fiedler and S.K. Jain), 197–237. New York: Chapman and Hall.

Harris, J.R., Bassuk, N.L., Zobel, R.W., and Whitlow, T.H. (1995). Root and shoot growth periodicity of green ash, scarlet oak, Turkish hazelnut, and tree lilac. *J. Am. Soc. Hortic. Sci.* 120: 211–216.

Harris, N.L., Gibbs, D.A., Baccini, A. et al. (2021). Global maps of twenty-first century forest carbon fluxes. *Nat. Clim. Chang.* 11: 234–240.

Harshberger, J.W. (1899). Thermotropic movement of leaves of *Rhododendron maximum*. *Proc. Natl. Acad. Sci. U. S. A.* 214–222.

Hart, J.W. (1988). *Light and Plant Growth*. Boston, MA: Unwin Hyman 204 pp.

Hart, S.C., Stark, J. M., Davidson, E. A., and Firestone, M. K. (1994). Nitrogen mineralization, immobilization, and nitrification. *In* R. W. Weaver, S. Angle, P. Bottomley, D. Bezdicek, S. Smith, A. Tabatabai, and A. Wollum (eds.), *Methods of Soil Analysis: Part 2. Microbiological and Biochemical Properties*. Soil Sci. Soc. Amer. Book Series, No. 5., Soil Sci. Soc. Amer., Segoe, WS.

Hart, S.J., Henkelman, J., McLoughlin, P.D. et al. (2019). Examining forest resilience to changing fire frequency in a fire-prone region of boreal forest. *Glob. Chang. Biol.* 25: 869–884.

Hartesveldt, R.J., and Harvey, H.T. (1967). The fire ecology of sequoia regeneration. *In Proc. Annual Tall Timbers Fire Ecology Conf.* 7:65–77. Tall Timbers Res. Sta., Tallahassee, FL.

Hartl, D.L. (2020). *A Primer of Population Genetics and Genomics*, 4e. Oxford, UK: Oxford University Press 320 pp.

Hartl, D.L. and Clark, A.G. (2006). *Principles of Population Genetics*, 4e. Sunderland, MA: Sinauer Associates 672 pp.

Harvey, C.A. (2000). Colonization of agricultural windbreaks by forest trees: effects of connectivity and remnant trees. *Ecol. Appl.* 10: 1762–1773.

Harvey, B.J., Donato, D.C., and Turner, M.G. (2016a). Burn me twice, shame on who? Interactions between successive forest fires across a temperate mountain region. *Ecology* 97: 2272–2282.

Harvey, B.J., Donato, D.C., and Turner, M.G. (2016b). High and dry: Post-fire tree seedling establishment in subalpine forests decreases with post-fire drought and large stand-replacing burn patches. *Glob. Ecol. Biogeogr.* 25: 655–669.

Hase, H. and Foelster, H. (1983). Impact of plantation forestry with teak (*Tectonia grandis*) on the nutrient status of young alluvial soils in west Venezuela. *For. Ecol. Manag.* 6: 33–57.

Hattas, D., Scogings, P.F., and Julkunen-Tiitto, R. (2017). Does the growth differentiation balance hypothesis explain allocation to secondary metabolites in *Combretum apiculatum*, an African savanna woody species? *J. Chem. Ecol.* 43: 153–163.

Haugo, R.D., Kellogg, B.S., Cansler, C.A. et al. (2019). The missing fire: quantifying human exclusion of wildfire in Pacific Northwest forests, USA. *Ecosphere* 10: e02702.

Hawk, G.M. and Zobel, D.B. (1974). Forest succession on alluvial landforms of the McKenzie River valley, Oregon. *Northwest Sci.* 48: 245–265.

Hawkins, W.A. and Graham, R.C. (2017). Soil mineralogy of a vernal pool catena in southern California. *Soil Sci. Soc. Am. J.* 81: 214–223.

Hayhoe, K., Wuebbles, D.J., Easterling, D.R. et al. (2018). Our changing climate. In: *Impacts, Risks, and Adaptation in the United States: Fourth National Climate Assessment, Volume II* (ed. D.R. Reidmiller, C.W. Avery, D.R. Easterling, et al.), 72–144. Washington, DC, USA: U.S. Global Change Research Program.

He, H.S., Gustafson, E.J., and Lischke, H. (2017). Modeling forest landscapes in a changing climate: theory and application. *Landsc. Ecol.* 32: 1299–1305.

Heberling, J.M. and Fridley, J.D. (2013). Resource-use strategies of native and invasive plants in eastern North American forests. *New Phytol.* 200: 523–533.

Heberling, J.M., Cassidy, S.T., Fridley, J.D., and Kalisz, S. (2019). Carbon gain phenologies of spring-flowering perennials in a deciduous forest indicate a novel niche for a widespread invader. *New Phytol.* 221: 778–788.

Hector, A., Hautier, Y., Saner, P. et al. (2010). General stabilizing effects of plant diversity on grassland productivity through population asynchrony and overyielding. *Ecology* 91: 2213–2220.

Hedrick, P.W. (2011). *Genetics of Populations*, 4e. Sudbury, MA: Jones and Bartlett 675 pp.

Heikkenen, H.J., Scheckler, S.E., Egan, P.J.J. Jr., and Williams, C.B. Jr. (1986). Incomplete abscission of needle clusters and resin release from artificially water-stressed loblolly pine (*Pinus taeda*): a component for plant-animal interactions. *Am. J. Bot.* 73: 1384–1392.

Heinselman, M.L. (1973). Fire in the virgin forests of the Boundary Waters Canoe Area, Minnesota. *Quat. Res.* 3: 329–382.

Heinselman, M.L. (1981). Fire and succession in the conifer forests of northern North America. In: *Forest Succession: Concepts and Application* (ed. D.C. West, H.H. Shugart and D.B. Botkin), 374–405. New York: Springer-Verlag.

Heinze, J., Gensch, S., Weber, E., and Joshi, J. (2017). Soil temperature modifies effects of soil biota on plant growth. *J. Plant Biol.* 10: 808–821.

Helgath, S. F. (1975). Trial deterioration in the Selway-Bitterroot Wilderness. USDA For. Serv. Res. Note INT-193. Intermountain For. and Rge. Exp. Sta., Ogden, UT. 15 pp.

Helgerson, O.T. (1989). Heat damage in tree seedlings and its prevention. *New For.* 3: 333–358.

Hellmers, H. (1962). Temperature effect upon optimum tree growth. In: *Tree Growth* (ed. T.T. Kozlowski), 275–287. New York: Ronald Press.

Hellmers, H. (1966a). Temperature action and interaction of temperature regimes in the growth of red fir seedlings. *For. Sci.* 12: 90–96.

Hellmers, H. (1966b). Growth response of redwood seedlings to thermoperiodism. *For. Sci.* 12: 276–283.

Hellmers, H. and Sundahl, W.P. (1959). Response of *Sequoia sempervirens* (D. Don) Endl. and *Pseudotsuga menziesii* (Mirb. Franco) seedlings to temperature. *Nature* 184: 1247–1248.

Hellmers, H., Genthe, M.K., and Ronco, F. (1970). Temperature affects growth and development of Engelmann spruce. *For. Sci.* 16: 447–452.

Hember, R.A., Kurz, M.A., and Coops, N.C. (2017). Relationships between individual-tree mortality and water-balance variables indicate positive trends in water stress-induced tree mortality across North America. *Glob. Chang. Biol.* 23: 1691–1710.

Henderson, G.S., Swank, W.T., Waide, J.B., and Grier, C.C. (1978). Nutrient budgets of Appalachian and cascade region watersheds: a comparison. *For. Sci.* 24: 385–397.

Hendrick, R.L. and Pregitzer, K.S. (1992). The demography of fine roots in a northern hardwood forest. *Ecology* 73: 1094–1104.

Hendricks, J.J., Nadelhoffer, K.J., and Aber, J.D. (1993). Assessing the role of fine roots in carbon and nitrogen cycling. *Trends Ecol. Evol.* 8: 174–178.

Henkel, T.K., Chambers, J.Q., and Baker, D.A. (2016). Delayed tree mortality and Chinese tallow (*Triadica sebifera*) population explosion in a Louisiana bottomland hardwood forest following Hurricane Katrina. *For. Ecol. Manag.* 378: 222–232.

Henry, H.A.L. and Aarssen, L.W. (2001). Inter- and intraspecific relationships between shade tolerance and shade avoidance in temperate trees. *Oikos* 93: 477–487.

Henry, J.A., Portier, J.M., and Coyne, J. (1994). *The Climate and Weather of Florida*. Sarasota, FL: Pineapple Press 279 pp.

Hermann, R.K. and Lavender, D.P. (1968). Early growth of Douglas-fir from various altitudes and aspects in southern Oregon. *Silvae Genet.* 17: 143–151.

Hermes, K. (1955). Die Lage der oberen Waldgrenze in den Gebirgen der Erde und ihr Abstand zur Schneegrenze (KoÈlner geographische Arbeiten 5). Geographisches Institut, UniversitaÈt KoÈln.

Herms, D.A. and Mattson, W.J. (1992). The dilemma of plants: to grow or defend. *Q. Rev. Biol.* 67: 283–335.

Herms, D.A. and McCullough, D.G. (2014). Emerald ash borer invasion of North America: history, biology, ecology, impacts, and management. *Annu. Rev. Entomol.* 59: 13–30.

Herrel, A., Joly, D., and Danchin, E. (2020). Epigenetics in ecology and evolution. *Funct. Ecol.* 34: 381–384.

Heslop-Harrison, J. (1964). Forty years of genecology. *Adv. Ecol. Res.* 2: 159–247.

Heslop-Harrison, J. (1967). *New Concepts in Flowering-Plant Taxonomy*. Cambridge, MA: Harvard Univ. Press 134 pp.

Hessburg, P.F., Miller, C.I., Parks, S.A. et al. (2019). Climate, environment, and disturbance history govern resilience of western North American forests. *Front. Ecol. Evol.* 7: 239.

Hett, J.M. (1971). A dynamic analysis of age in sugar maple seedlings. *Ecology* 52: 1071–1074.

Heusser, C. J. (1960). *Late-Pleistocene environments of North Pacific North America*. Amer. Geog. Soc. Spec. Publ. 35. 308 pp.

Heusser, C.J. (1965). A Pleistocene phytogeographical sketch of the Pacific Northwest and Alaska. In: *The Quaternary of the United States* (ed. H.E. Wright Jr. and D.G. Frey), 469–483. Princeton, NJ: Princeton Univ. Press.

Heusser, C.J. (1983). Vegetational history of the northwestern United States including Alaska. In: *The Late Pleistocene*, Vol. 1 of H. E. Wright, Jr., (ed.), *Late-Quaternary Environments of the United States* (ed. S.C. Porter). Minneapolis: Univ. Minnesota Press.

Hewitt, N. and Kellman, M. (2002). Tree seed dispersal among forest fragments: II. Dispersal abilities and biogeographical controls. *J. Biogeogr.* 29: 351–363.

Hewlett, J.D. and Hibbert, A.R. (1963). Moisture and energy conditions within a sloping soil mass during drainage. *J. Geophys. Res.* 68: 1081–1087.

Heyerdahl, E.K., Brubaker, L.B., and Agee, J.K. (2001). Spatial controls of historical fire regimes: a multiscale example from the Interior West, USA. *Ecology* 82: 660–678.

Hibbs, D.E. (1981). Leader growth and the architecture of three North American hemlocks. *Can. J. Bot.* 59: 476–480.

Hicke, J.A., Allen, C.D., Desai, A.R. et al. (2012). Effects of biotic disturbances on forest carbon cycling in the United States and Canada. *Glob. Chang. Biol.* 18: 7–34.

Hicke, J.A., Meddens, A.J., and Kolden, C.A. (2016). Recent tree mortality in the western United States from bark beetles and forest fires. *For. Sci.* 62: 141–153.

Higuera, P.E., Shuman, B.N., and Wolf, K.D. (2021). Rocky Mountain subalpine forests now burning more than any time in recent millennia. *Proc. Natl. Acad. Sci.* 118: e2103135118.

Hilker, M. and Schmülling, T. (2019). Stress priming, memory, and signalling in plants. *Plant Cell Environ.* 42: 753–761.

Hills, G.A. (1952). The classification and evaluation of site for forestry. Ontario Dept. Lands and For., Res. Rept. 24. 41 pp.

Hills, G.A. (1960). Regional site research. *For. Chron.* 36: 401–423.

Hills, G.A. (1977). An integrated iterative holistic approach to ecosystem classification. *In* J. Thie and G. Ironside (eds.), *Ecological (Biophysical) Land Classification in Canada*. Ecological land classification series, No. 1, Lands Directorate, Env. Canada, Ottawa.

Hills, G. A. and Pierpoint, G. (1960). Forest site evaluation in Ontario. Ontario Dept. Lands and For., Res. Rept. 42. 64 pp.

Hilmers, T., Friess, N., Bässler, C. et al. (2018). Biodiversity along temperate forest succession. *J. Appl. Ecol.* 55: 2756–2766.

Himelick, E.B. and Neely, D. (1962). Root-grafting of city-planted American elms. *Plant Dis. Rep.* 46: 86–87.

Hinckley, T.M., Teskey, R.O., Duhme, F., and Richter, H. (1981). Temperate hardwood forests. In: *Water Deficits and Plant Growth*, vol. 6 (ed. T.T. Kozlowski), 153–208. New York: Academic Press.

Hinds, T.E., Hawksworth, F.G., and Davidson, R.W. (1965). Beetle-killed Engelmann spruce: its deterioration in Colorado. *J. For.* 63: 536–542.

Hinesley, L.E. (1981). Initial growth of Fraser fir seedlings at different day/night temperatures. *For. Sci.* 27: 545–550.

Hix, D.M. (1988). Multifactor classification and analysis of upland hardwood forest ecosystems of the Kickapoo River watershed, southwestern Wisconsin. *Can. J. For. Res.* 18: 1405–1415.

Hix, D.M. and Barnes, B.V. (1984). Effects of clear-cutting on the vegetation and soil of an eastern hemlock dominated ecosystem, western Upper Michigan. *Can. J. For. Res.* 14: 914–923.

Hobbie, E.A., Tingey, D.T., Rygiewicz, P.T. et al. (2002). Contributions of current year photosynthate to fine roots estimated using a 13 C-depleted CO_2 source. *Plant Soil* 247: 233–242.

Hobbs, R.J. (2007). Setting effective and realistic restoration goals: key directions for research. *Restoration Ecology* 15 (2): 354–357.

Hobbs, R.J. and Humphries, S.E. (1995). An integrated approach to the ecology and management of plant invasions. *Conserv. Biol.* 9: 761–770.

Hobbs, R.J., Arico, S., Aronson, J. et al. (2006). Novel ecosystems: theoretical and management aspects of the new ecological world order. *Glob. Ecol. Biogeogr.* 15: 1–7.

Hocker, H.W. Jr. (1956). Certain aspects of climate as related to the distribution of loblolly pine. *Ecology* 37: 824–834.

Hoegh-Guldberg, O., Hughes, L., McIntyre, S. et al. (2008). Assisted colonization and rapid climate change. *Science* 321: 345–346.

Holderidge, L.R. (1947). Determination of world plant formations from simple climatic data. *Science* 105: 367–368.

Holderidge, L.R. (1967). *Life Zone Ecology*. Tropical Sci. Center, San Jose, Costa Rica. 206 pp.

Hollender, C.A. and Dardick, C. (2015). Molecular basis of angiosperm tree architecture. *New Phytol.* 206: 541–556.

Holliday, J.A., Zhou, L., Bawa, R. et al. (2016). Evidence for extensive parallelism but divergent genomic architecture of adaptation along altitudinal and latitudinal gradients in *Populus trichocarpa*. *New Phytol.* 209: 1240–1251.

Holling, C.S. and Meffe, G.K. (1996). Command and control and the pathology of natural resource management. *Conserv. Biol.* 10: 328–337.

Holly, D.C., Ervin, G.N., Jackson, C.R. et al. (2009). Effect of an invasive grass on ambient rates of decomposition and microbial community structure: a search for causality. *Biol. Invasions* 11: 1855–1868.

Holmsgaard, E. and Bang, C. (1989). Loss of volume increment due to cone production in Norway spruce. *Forstl. Forsøgsvawsen i Danmark* 42: 217–231.

Holsinger, L., Parks, S.A., and Miller, C. (2016). Weather, fuels, and topography impede wildland fire spread in western US landscapes. *For. Ecol. Manag.* 380: 59–69.

Holthuijzen, A.M.A., Sharik, T.L., and Fraser, J.D. (1987). Dispersal of eastern red cedar (*Juniperus virginiana*) into pastures: an overview. *Can. J. Bot.* 65: 1092–1095.

Holzinger, B., Hulber, K., Camenisch, M., and Grabherr, G. (2008). Changes in plant species richness over the last century in the eastern Swiss Alps: elevational gradient, bedrock effects and migration rates. *Plant Ecol.* 195: 179–196.

Hook, D.D. (1984). Waterlogging tolerance of lowland tree species of the South. *South. J. Appl. For.* 8: 136–149.

Hooper, D.U., Chapin, F.S. III, Ewel, J.J. et al. (2005). Effects of biodiversity on ecosystem functioning: a consensus of current knowledge. *Ecol. Monogr.* 75: 3–35.

Hooper, D.U., Adair, E.C., Cardinale, B.J. et al. (2012). A global synthesis reveals biodiversity loss as a major driver of ecosystem change. *Nature* 486: 105–108.

Horn, M.H.S.B., Correa, P., Parolin, B.J.A. et al. (2011). Seed dispersal by fishes in tropical and temperate fresh waters: the growing evidence. *Acta Oecol.* 37: 561–577.

Horton, K.W. (1956). The ecology of lodgepole pine in Alberta and its role in forest succession. *For. Br. Can. Tech.* Note 45. 29 pp.

Hosie, R.C. (1969). *Native Trees of Canada*, 7e. Ottawa, Canada: Can. For. Serv. 380 pp.

Host, G.E., Pregitzer, K.S., Ramm, C.W. et al. (1987). Landform-mediated differences in successional pathways among upland forest ecosystems in northwestern Lower Michigan. *For. Sci.* 33: 445–457.

Host, G.E., Pregitzer, K.S., Ramm, C.W. et al. (1988). Variation in over-story biomass among glacial landforms and ecological land units in northwestern Lower Michigan. *Can. J. For. Res.* 18: 659–668.

Hough, A.F. and Forbes, R.D. (1943). The ecology and silvics of forests in the high plateaus of Pennsylvania. *Ecol. Monogr.* 13: 299–320.

Houlton, B.Z., Morford, S.L., and Dahlgren, R.A. (2008). Convergent evidence for nitrogen sources in the Earth's surface environment. *Science* 360: 58–62.

Howe, H.F. (1993). Specialized and generalized dispersal systems: where does "the paradigm" stand? *Vegetatio* 107/108: 3–13.

Hsu, C.Y., Liu, Y., Luthe, D.S., and Yuceer, C. (2006). Poplar FT2 shortens the juvenile phase and promotes seasonal flowering. *Plant Cell* 18: 1846–1861.

Hu, Z., Michaletz, S.T., Johnson, D.J. et al. (2018). Traits drive global wood decomposition rates more than climate. *Glob. Chang. Biol.* 24: 5259–5269.

Hubbes, M. (1999). The American elm and Dutch elm disease. *For. Chron.* 75: 265–273.

Huber, O. (1978). Light compensation point of vascular plants of a tropical cloud forest and an ecological interpretation. *Photosynthetica* 12: 382–390.

Huberty, A.F. and Denno, R.F. (2004). Plant water stress and its consequences for herbivorous insects: a new synthesis. *Ecology* 85: 1383–1398.

Huenneke, L.F. and Sharitz, R.R. (1986). Microsite abundance and distribution of woody seedlings in a South Carolina cypress-tupelo swamp. *Am. Midi. Nat.* 115: 328–335.

Hufkens, K., Friedl, M.A., Keenan, T.F. et al. (2012). Ecological impacts of a widespread frost event following early spring leaf-out. *Glob. Chang. Biol.* 18: 2365–2377.

Huggett, R.J. (1995). *Geoecology, an Evolutionary Approach*. New York: Routledge 320 pp.

Hulme, P.E. (2009). Trade, transport and trouble: managing invasive species pathways in an era of globalization. *J. Appl. Ecol.* 46: 10–18.

Hunter, M.L. Jr. and Gibbs, J.P. (2021). *Fundamentals of Conservation Biology*, 4e. New York: Wiley Blackwell 672 pp.

Hunter, A.F. and Lechowicz, M.J. (1992). Predicting the timing of budburst in temperate trees. *J. Appl. Ecol.* 29: 597–604.

Hunter, M.L. and Schmiegelow, F.K. (2011). *Wildlife, Forests, and Forestry Principles of Managing Forests for Biological Diversity*, 2e. Upper Saddle River, New Jersey, USA: Prentice Hall 259 pp.

Huntly, N. (1991). Herbivores and the dynamics of communities and ecosystems. *Annu. Rev. Ecol. Syst.* 22: 477–503.

Hupp, C.R. and Osterkamp, W.R. (1985). Bottomland vegetation distribution along Passage Creek, Virginia, in relation to fluvial landforms. *Ecology* 66: 670–681.

Huston, M. (1979). A general hypothesis of species diversity. *American Naturalist* 113: 81–101.

Huston, M.A. (1993). Biological diversity, soils, and economics. *Science* 262: 1676–1680.

Huston, M.A. (1994). *Biological Diversity*. Cambridge: Cambridge Univ. Press 681 pp.

Huston, M.A. (2014). Disturbance, productivity, and species diversity: empiricism vs. logic in ecological theory. *Ecology* 95: 2382–2396.

Huston, M. and Smith, T. (1987). Plant succession: life history and competition. *Am. Nat.* 130: 168–198.

Huston, M., DeAngelis, D., and Post, W. (1988). New computer models unify ecological theory. *Bioscience* 38: 682–691.

Hutchings, M.J., John, E.A., and Stewart, A.J.A. (ed.) (2000). *The Ecological Consequences of Environmental Heterogeneity*. Oxford, UK: Blackwell Science 434 pp.

Hutchins, H.E. and Lanner, R.M. (1982). The central role of Clark's nutcracker in the dispersal and establishment of whitebark pine. *Oecologia* 55: 192–201.

Hutchinson, B.A. and Matt, D.R. (1976). Beam enrichment of diffuse radiation in a deciduous forest. *Agric. Meteorol.* 18: 255–265.

Huxley, J.S. (1938). Clines, an auxiliary taxonomic principle. *Nature* 142: 219–220.

Huxley, J.S. (1939). Clines, an auxiliary method in taxonomy. *Bijd. Dierkunde* 27: 491–520.

Huxley, C.R. and Cutler, D.F. (1991). *Ant–Plant Interactions*. Oxford: Oxford Univ. Press 601 pp.

Ibáñez, I., Katz, D.S., and Lee, B.R. (2017). The contrasting effects of short-term climate change on the early recruitment of tree species. *Oecologia* 184: 701–713.

Illick, J. S. (1914). *Pennsylvania Trees*. Pa. Dept. For. Bull. 11, 231 pp.

Illick, J.S. (1921). Replacement of the chestnut. *J. For.* 19: 105–114.

Intergovernmental Panel on Climate Change (2013). IPCC, 2013: Climate Change 2013: The Physical Science Basis. In: *Fifth Assessment Report of the Intergovernmental Panel on Climate Change* (ed. T.F. Stocker, D. Qin, G.-K. Plattner, et al.). Cambridge, UK: Cambridge University Press.

Intergovernmental Panel on Climate Change (2014). Climate change 2014: Impacts, adaptation, and vulnerability. In: *Fifth Assessment Report of the Intergovernmental Panel on Climate Change* (ed. V.R. Barros, C.B. Field, D.J. Dokken, et al.). Cambridge, UK: Cambridge University Press.

Intergovernmental Panel on Climate Change (2021). Climate change 2021: The physical science basis. In: *Sixth Assessment Report of the Intergovernmental Panel on Climate Change* (ed. V.P. Masson-Delmotte, P. Zhai, A. Pirani, et al.). Cambridge, UK: Cambridge University Press.

Isaac, L. A. (1938). Factors affecting establishment of Douglas-fir seedlings. U.S. Dept. Agr. Circ. 486. 45 pp.

Ishii, H.T., Ford, E.D., and Kennedy, M.C. (2007). Physiological and ecological implications of adaptive reiteration as a mechanism for crown maintenance and longevity. *Tree Physiol.* 27: 455–462.

Iversen, C.M., Murphy, M.T., Allen, M.F. et al. (2012). Advancing the use of minirhizotrons in wetlands. *Plant Soil* 352: 23–39.

Iverson, L.R., Dale, M.F., Scott, C.T., and Prasad, A. (1997). A GIS-derived integrated moisture index to predict forest composition and productivity of Ohio forests (USA). *Landsc. Ecol.* 12: 331–348.

Iverson, L.R., Schwartz, M.W., and Prasad, A.M. (2004). How fast and far might tree species migrate in the eastern United States due to climate change? *Glob. Ecol. Biogeogr.* 13: 209–219.

Iverson, L.R., Prasad, A.M., Matthews, S.N., and Peters, M. (2008). Estimating potential habitat for 134 eastern US tree species under six climate scenarios. *For. Ecol. Manag.* 254: 390–406.

Iverson, L.R., Peters, M.P., Prasad, A.M., and Matthews, S.N. (2019a). Analysis of climate change impacts on tree species of the eastern US: Results of DISTRIB-II modeling. *Forests* 10: 302.

Iverson, L.R., Prasad, A.M., Peters, M.P., and Matthews, S.N. (2019b). Facilitating adaptive forest management under climate change: A spatially specific synthesis of 125 species for habitat changes and assisted migration over the eastern United States. *Forests* 10: 989.

Ives, A.R. and Carpenter, S.R. (2007). Stability and diversity of ecosystems. *Science* 317: 58–62.

Jackson, L.W.R. (1952). Radial growth of forest trees in the Georgia Piedmont. *Ecology* 33: 336–341.

Jackson, W.R. (1967). Effect of shade tolerance on leaf structure of deciduous tree species. *Ecology* 48: 498–499.

Jackson, S.T. (2004). Late Quaternary biogeography: linking biotic responses to environmental variability across timescales. In: *Frontiers of Biogeography* (ed. M. Lomolino and L. Heaney), 47–65. Sunderland, MA: Sinauer Associates.

Jackson, R.B., Mooney, H.A., and Schulze, E.D. (1997). A global budget for fine root biomass, surface area, and nutrient contents. *Proc. Natl. Acad. Sci.* 94: 7362–7366.

Jackson, S.T., Webb, R.S., Anderson, K.H. et al. (2000). Vegetation and environment in eastern North America during the last glacial maximum. *Quart. Sci. Rev.* 19: 489–508.

Jacobs, D.F., Cole, D.W., and McBride, J.R. (1985). Fire history and perpetuation of natural coast redwood ecosystems. *J. For.* 83: 494–497.

Jacobson, R.B., Miller, A.J., and Smith, J.A. (1989). The role of catastrophic geomorphic events in central Appalachian landscape evolution. *Geomorphology* 2: 257–284.

Jacoby, G.C., Williams, P.L., and Buckley, B.M. (1992). Tree ring correlation between prehistoric landslides and abrupt tectonic events in Seattle, Washington. *Science* 258: 1621–1623.

Jakubos, B. and Romme, W.H. (1993). Invasion of subalpine meadows by lodgepole pine in Yellowstone National Park, Wyoming, USA. *Arct. Alp. Res.* 25: 382–390.

James, S. (1984). Lignotubers and burls: their structure, function, and ecological significance in Mediterranean ecosystems. *Bot. Rev.* 50: 225–245.

Janos, D.P. (1987). VA mycorrhizas in humid tropical systems. In: *Ecophysiology of VA Mycorrhizal Plants* (ed. G.R. Safer). Boca Raton, FL: CRC Press 224 pp.

Janowiak, M.K., Swanston, C.W., Nagel, L.M. et al. (2014). A practical approach for translating climate change adaptation principles into forest management actions. *J. For.* 112: 424–433.

Janzen, D.H. (1969). Seed-eaters versus seed size, number, toxicity and dispersal. *Evolution* 23: 1–27.

Janzen, D.H. (1983). Dispersal of seeds by vertebrate guts. In: *Coevolution* (ed. D.J. Futuyma and M. Slatkin), 232–262. Sunderland, MA: Sinauer Assoc.

Janzen, D.H. (1985). The natural history of mutualisms. In: *The Biology of Mutualism* (ed. D.H. Boucher), 40–99. London: Croom Helm.

Janzen, D.H. and Martin, P.S. (1982). Neotropical anacronisms: the fruits the gomphotheres ate. *Science* 215: 19–27.

Jaramillo-Correa, J.P., Beaulieu, J., Khasa, D.P., and Bousquet, J. (2009). Inferring the past from the present phylogeographic structure of North American forest trees: seeing the forest for the genes. *Can. J. For. Res.* 39: 286–307.

Jarvis, P.G. and Laverenz, J.W. (1983). Productivity of temperate, deciduous, and evergreen forests. In: *Encyclopedia of Plant Physiology*, New Series, *Vol. 12D* (ed. O.L. Lange, P.S. Nobel, C.B. Osmond and H. Ziegler), 233–280. New York: Springer-Verlag.

Jenkins, C.N., Van Houtan, K.S., Pimm, S.L., and Sexton, J.O. (2015). US protected lands mismatch biodiversity priorities. *Proc. Natl. Acad. Sci.* 112: 5081–5086.

Jenny, H. (1941). *Factors of Soil Formation*. New York: McGraw-Hill 281 pp.

Jenny, H. (1980). *The Soil Resource: Origin and Behavior*. New York: Springer-Verlag 377 pp.

Jenny, H., Arkley, R.J., and Schultz, A.M. (1969). The pygmy forest-podsol ecosystem and its dune associates of the Mendocino coast. *Madroño* 20: 60–74.

Jiang, L. and Pu, Z.C. (2009). Different effects of species diversity on temporal stability in single-trophic and multitrophic communities. *Am. Nat.* 174: 651–659.

Jiao, Y., Wickett, N.J., Ayyampalayam, S. et al. (2011). Ancestral polyploidy in seed plants and angiosperms. *Nature* 473: 97–100.

Jiggins, C.D. (2019). Can genomics shed light on the origin of species? *PLoS Biol.* 17: e3000394.

Jo, I., Fridley, J.D., and Frank, D.A. (2015). Linking above- and belowground resource use strategies for native and invasive species of temperate deciduous forests. *Biol. Invasions* 17: 1545–1554.

Jo, I., Fridley, J.D., and Frank, D.A. (2017). Invasive plants accelerate nitrogen cycling: evidence from experimental woody monocultures. *J. Ecol.* 105: 1105–1110.

Jochner, S. and Menzel, A. (2015). Urban phenological studies–Past, present, future. *Environ. Pollut.* 203: 250–261.

Johnsen, O., Daehlen, O.G., Østreng, G., and Skrøppa, T. (2005). Daylength and temperature during seed production interactively affect adaptive performance of *Picea abies* progenies. *New Phytol.* 168: 589–596.

Johnson, P.W. (1972). Factors affecting buttressing in *Triplochiton scleroxylon* K. Schum. *Ghana J. Agric. Sci.* 5: 13–21.

Johnson, E.A. (1992). *Fire and Vegetation Dynamics: Studies from the North American Boreal Forest*. Cambridge: Cambridge Univ. Press 129 pp.

Johnson, W.C. and Adkisson, C.S. (1985). Airlifting the oaks. *Nat. Hist.* 10: 41–46.

Johnson, E.A. and Miyanishi, K. (ed.) (2001). *Forest Fires: Behavior and Ecological Effects*. New York: Academic Press 594 pp.

Johnson, W.C. and Webb, T. III (1989). The role of blue jays (*Cyanocitta cristata* L.) in the postglacial dispersal of fagaceous trees in eastern North America. *J. Biogeogr.* 16: 561–571.

Johnson, W.C., Burgess, R.L., and Keammerer, W.R. (1976). Forest overstory vegetation and environment on the Missouri River floodplain in North Dakota. *Ecol. Monogr.* 46: 59–84.

Johnson, M.G., Phillips, D.L., Tingey, D.T., and Storm, M.J. (2000). Effects of elevated CO_2, N-fertilization, and season on survival of ponderosa pine fine roots. *Can. J. For. Res.* 30: 220–228.

Johnson, E.A., Miyanishi, K., and K. (2008). Testing the assumptions of chronosequences in succession. *Ecol. Lett.* 11: 419–431.

Johnson, P.S., Shifley, S.R., Rogers, R. et al. (2019). *The Ecology and Silviculture of Oaks*, 3e. Boston, MA: CAB International 648 pp.

Johnstone, J.F., Hollingsworth, T.N., Chapin, F.S. III, and Mack, M.C. (2010). Changes in fire regime break the legacy lock on successional trajectories in Alaskan boreal forest. *Glob. Chang. Biol.* 16: 1281–1295.

Johnstone, J.F., Allen, C.D., Franklin, J.F. et al. (2016). Changing disturbance regimes, ecological memory, and forest resilience. *Front. Ecol. Environ.* 14: 369–378.

Jolley, D.B., Ditchkoff, S.S., Sparklin, B.D. et al. (2010). Estimate of herpetofauna depredation by a population of wild pigs. *J. Mammal.* 91: 519–524.

Jones, J.R. (1969). Review and comparison of site evaluation methods. USDA For. Serv. Res. Paper RM-51. Rocky Mountain For. and Rge. Exp. Sta., Fort Collins, CO. 27 pp.

Jones, S.M. (1991). Landscape ecosystem classification for South Carolina. *In* D.L. Mengel and D. T. Tew (eds.), *Symp. Proc., Ecological Land Classification: Applications to Identify the Productive Potential of Southern Forests*. USDA For. Serv. Gen. Tech. Report SE-68. Southeastern For. Exp. Sta., Asheville, NC.

Jones, S.M. and Lloyd, F.T. (1993). Landscape ecosystem classification: the first step toward ecosystem management in the southeastern United States. In: *Defining Sustainable Forestry* (ed. G.H. Aplet, N. Johnson, J.T. Olson and V.A. Sample). Washington D.C.: Island Press 231 pp.

Jones, R.K., Pierpoint, G., Wickware, G.M. et al. (1983). Field guide to forest ecosystem classification for the Clay Belt. Site Region 3E. Ont. Min. Nat. Res. Toronto, ON. 123 pp.

Jordan, C., Caskey, W., Escalante, G. et al. (1982). The nitrogen cycle in a "Terra Firme" rainforest on oxisol in Amazon territory of Venezuela. *Plant Soil* 67: 325–332.

Jorgensen, J.R. and Hodges, C.S. Jr. (1970). Microbial characteristics of a forest soil after twenty years of prescribed burning. *Mycologia* 62: 721–726.

Jouzel, J., Masson-Delmotte, V., Cattani, O. et al. (2007). Orbital and millennial Antarctic climate variability over the past 800,000 years. *Science* 317: 793–796.

Jucker, T., Bouriaud, O., and Coomes, D.A. (2015). Crown plasticity enables trees to optimize canopy packing in mixed-species forests. *Funct. Ecol.* 29: 1078–1086.

Jull, L.G., Frank, A., Blazich, L., and Hinesley, E. (1999). Seedling growth of Atlantic white-cedar as influenced by photoperiod and day/night temperature. *J. Environ. Hortic.* 17: 107–113.

Jung, J.-H., Domijan, M., Klose, C. et al. (2016). Phytochromes function as thermosensors in Arabidopsis. *Science* 354: 886–889.

Jurasinski, G., Retzer, V., and Beierkuhnlein, C. (2009). Inventory, differentiation, and proportional diversity: a consistent terminology for quantifying species diversity. *Oecologia* 159: 15–26.

Jurik, T.W., Webber, J.A., and Gates, D.M. (1988). Effects of temperature and light on photosynthesis of dominant species of a northern hardwood forest. *Bot. Gaz.* 149: 203–208.

Justus, J., Colyvan, M., Regan, H., and Maguire, L. (2009). Buying into conservation: intrinsic versus instrumental value. *Trends Ecol. Evol.* 24: 187–191.

Kalberer, S.R., Wisniewski, M., and Arora, R. (2006). Deacclimation and reacclimation of cold-hardy plants: current understanding and emerging concepts. *Plant Sci.* 171: 3–16.

Kalela, E.K. (1957). Über Veränderungen in den Wurzelverhältnissen der Kiefernbestände im Laufe der Vegetationsperiode. *Acta For. Fenn.* 65: 1–41.

Kamminga-Van Wijk, C. and Prins, H. (1993). The kinetics of NH_4^+ and NO_3^- uptake by Douglas-fir from single N-solutions and from solutions containing both NH_4^+ and NO_3^-. *Plant Soil* 151: 91–96.

Kaplan, S. (1992). Environmental preference in a knowledge-seeking, knowledge-using organism. In: *The Adapted Mind: Evolutionary Psychology and the Generation of Culture* (ed. J.H. Barkow, L. Cosmides and J. Tooby), 581–598. New York: Oxford Univ. Press.

Kashian, D.M. (2016). Sprouting and seed production may promote persistence of green ash in the presence of the emerald ash borer. *Ecosphere* 7: e01332.

Kashian, D.M. and Barnes, B.V. (2021). Tree growth and survival over 61 years at the Second International Larch Provenance Test in southeastern Michigan, USA. *Silvae Genet.* 70: 9–21.

Kashian, D.M. and Witter, J.A. (2011). Assessing the potential for ash canopy tree replacement via current regeneration following emerald ash borer-caused mortality on southeastern Michigan landscapes. *For. Eco. Manag.* 261: 480–488.

Kashian, D.M., Barnes, B.V. (2000). Landscape influence on the spatial and temporal distribution of the Kirtland's warbler at the Bald Hill burn, northern Lower Michigan, USA. *Can. J. For. Res.* 30: 1895–1904.

Kashian, D.M., Barnes, B.V., and Walker, W.S. (2003). Landscape ecosystems of northern Lower Michigan and the occurrence and management of the Kirtland's warbler. *For. Sci.* 49: 140–159.

Kashian, D.M., Tinker, D.B., Turner, M.G., and Scarpace, F.L. (2004). Spatial heterogeneity of lodgepole pine sapling densities following the 1988 fires in Yellowstone National Park, Wyoming, U.S.A. *Can. J. For. Res.* 34: 2263–2276.

Kashian, D.M., Turner, M.G., and Romme, W.H. (2005a). Variability in leaf area and stemwood increment along a 300-year lodgepole pine chronosequence. *Ecosystems* 8: 48–61.

Kashian, D.M., Turner, M.G., Romme, W.H., and Lorimer, C.G. (2005b). Variability and convergence in stand structural development on a fire-dominated subalpine landscape. *Ecology* 86: 643–654.

Kashian, D.M., Romme, W.H., Tinker, D.B. et al. (2006). Carbon storage on landscapes with stand-replacing fires. *Bioscience* 56: 598–606.

Kashian, D.M., Romme, W.H., and Regan, C.M. (2007). Reconciling divergent interpretations of quaking aspen decline on the northern Colorado Front Range. *Ecol. Appl.* 17: 1296–1311.

Kashian, D.M., Corace, R.G. III, Shartell, L.M. et al. (2012). Variability and persistence of post-fire biological legacies in jack pine-dominated ecosystems of northern Lower Michigan. *For. Ecol. Manag.* 263: 148–158.

Kashian, D.M., Romme, W.H., Tinker, D.B. et al. (2013). Post-fire changes in forest carbon storage over a 300-year chronosequence of *Pinus contorta*-dominated forests. *Ecol. Monogr.* 83: 49–66.

Kashian, D.M., Sosin, J.R., Huber, P.W. et al. (2017). A neutral modeling approach for designing spatially heterogeneous jack pine plantations in northern Lower Michigan, USA. *Landsc. Ecol.* 32: 1117–1131.

Kauffman, J.B. and Krueger, W.C. (1984). Livestock impacts on riparian ecosystems and streamside management implications: A review. *J. Range Manag.* 37: 430–437.

Kauffman, J.B. and Uhl, C. (1990). Interactions of anthropogenic activities, fire, and rain forests in the Amazon Basin. In: *Fire in the Tropical Biota*. (Ecological Studies 84) (ed. J.G. Goldammer), 117–134. New York: Springer-Verlag.

Kauffman, J.B., Krueger, W.C., and Vavra, M. (1983). Effects of late season cattle grazing on riparian plant communities. *Rangel. Ecol. Manag./J. Range Manag. Archives* 36: 685–691.

Kaufmann, J., Bork, E.W., Alexander, M.J., and Blenis, P.V. (2013). Habitat selection by cattle in foothill landscapes following variable harvest of aspen forest. *For. Ecol. Manag.* 306: 15–22.

Kay, C.E. (1994). Aboriginal overkill, the role of Native Americans in structuring western ecosystems. *Hum. Nat.* 5: 359–398.

Kay, C.E. (1997). Is aspen doomed? *J. For.* 95: 4–11.

Kaye, M.W., Binkley, D., and Stohlgren, T.J. (2005). Effects of conifers and elk browsing on quaking aspen forests in the central Rocky Mountains, USA. *Ecol. Appl.* 15: 1284–1295.

Kayll, A.J. (1974). Use of fire in land management. In: *Fire and Ecosystems* (ed. T.T. Kozlowski and C.E. Ahlgren), 483–511. New York: Academic Press.

Keane, R.E., Morgan, P., and Running, S.W. (1996). FIRE-BGC—a mechanistic ecological process model for simulating fire succession on coniferous forest landscapes of the northern Rocky Mountains. USDA For. Serv. Res. Paper INT-RP-484. Int. Res. Sta., Odgen, UT. 122 pp.

Keane, R.E., Ryan, K.C., Veblen, T.T. et al. (2002). *Cascading Effects of Fire Exclusion in Rocky Mountain Ecosystems: A Literature Eeview*. USDA For. Serv. Gen. Tech. Rep. RMRS-GTR-91. Rocky Mountain Res. Sta. Fort Collins, CO. 24 p.

Kearsley, M.J.C. and Whitham, T.G. (1989). Developmental changes in resistance to herbivory: implications for individuals and populations. *Ecology* 70: 422–434.

Keeley, J.E. and Keeley, S.C. (1988). Chaparral. In: *North American Terrestrial Vegetation* (ed. M.G. Barbour and W.D. Billings), 203–254. Cambridge: Cambridge Univ. Press.

Keeley, J.E. and Pausas, J.G. (2019). Distinguishing disturbance from perturbations in fire-prone ecosystems. *Int. J. Wildland Fire* 28: 282–287.

Keeling, E.G., Sala, A., and DeLuca, T.H. (2006). Effects of fire exclusion on forest structure and composition in unlogged ponderosa pine/Douglas-fir forests. *For. Ecol. Manag.* 237: 418–428.

Keenan, T.F., Hollinger, D.Y., Bohrer, G. et al. (2013). Increase in forest water-use efficiency as atmospheric carbon dioxide concentrations rise. *Nature* 499: 324–327.

Keenan, T.F., Gray, J., Friedl, M.A. et al. (2014). Net carbon uptake has increased through warming-induced changes in temperate forest phenology. *Nat. Clim. Chang.* 4: 598–604.

Keesing, F. and Ostfeld, R.S. (2015). Is biodiversity good for your health? *Science* 349: 235–236.

Keesing, F. and Ostfeld, R.S. (2021). Impacts of biodiversity and biodiversity loss on zoonotic diseases. *Proc. Natl. Acad. Sci.* 118: e2023540118.

Keesing, F., Belden, L.K., Daszak, P. et al. (2010). Impacts of biodiversity on the emergence and transmission of infectious diseases. *Nature* 468 (7324): 647–652.

Keever, C. (1953). Present composition of some stands of the former oak, chestnut forest in the southern Blue Ridge Mountains. *Ecology* 34: 44–54.

Kellogg, L.F. (1939). Site index curves for plantation black walnut in the Central States region. *Central States For. Exp. Sta., Res. Note* 35: 3 pp.

Kelly, D. and Sullivan, J. J. (1997). Quantifying the benefits of mast seeding on predator satiation and wind pollination in *Chionochloa pallens* (Poaceae). 78:143–150.

Kelly, D., Hart, D.E., and Allen, R.B. (2001). Evaluating the wind pollination benefits of mast seeding. *Ecology* 82: 117–126.

Kelsall, J.P., Telfer, E.S., and Wright, T.D. (1977). The effects of fire on the ecology of the Boreal Forest, with particular reference to the Canadian north: a review and selected bibliography. Can. Wildlife Serv., Ottawa. Occ. Paper 32. 58 pp.

Kelsey, E.P., Cann, M.D., Lupo, K.M., and Haddad, L.J. (2019). Synoptic to microscale processes affecting the evolution of a cold-air pool in a northern New England forested mountain valley. *J. Appl. Meteorol. Climatol.* 58: 1309–1324.

Kemp, K.B., Higuera, P.E., Morgan, P., and Abatzoglou, J.T. (2019). Climate will increasingly determine post-fire tree regeneration success in low-elevation forests, Northern Rockies, USA. *Ecosphere* 10: e02568.

Kemperman, J.A. and Barnes, B.V. (1976). Clone size in American aspens. *Can. J. Bot.* 54: 2603–2607.

Kenoyer, L.A. (1933). Forest distribution in southwestern Michigan as interpreted from the original land survey (1826–32). *Pap. Mich. Acad. Sci. Arts Lett.* 19: 107–111.

Keyes, M.R. and Grier, C.C. (1981). Above-and below-ground net production in 40-year-old Douglas-fir stands on low and high productivity sites. *Can. J. For. Res.* 11: 599–605.

Khoshoo, T.N. (1959). Polyploidy in gymnosperms. *Evolution* 13: 24–39.

Kikuzawa, K. (2003). Phenological and morphological adaptations to the light environment in two woody and two herbaceous plant species. *Funct. Ecol.* 17: 29–38.

Kilburn, P.D. (1960). Effects of logging and fire on xerophytic forests in northern Michigan. *Bull. Torrey Bot. Club* 6: 402–405.

Kilgore, B.M. (1972). Fire's role in a sequoia forest. *Naturalist* 23: 26–37.

Kilgore, B.M. (1973). The ecological role of fire in Sierran conifer forests: its application to national park management. *Quat. Res.* 3: 496–513.

Kilgore, B.M. (1975). Restoring fire to National Park wilderness. *Am. For.* 81: 16–19.

Kilgore, B.M. (1976a). Fire management in the National Parks: an overview. *In Proc. Annual Tall Timbers Fire Ecology Conference and Fire and Land Management Symposium,* 14:45–57. Tall Timbers Res. Sta., Tallahassee, FL.

Kilgore, B.M. (1976b). America's renewable resource potential—1975: the turning point. *Proc. Soc. Am. For.* 1975: 178–188.

Kilgore, B.M. (1976c). From fire control to fire management: an ecological basis for policies. *In Trans. 41st North American Wildlife and Natural Resources Conference.* Wildlife Management Institute, Washington, D.C.

Kilgore, B.M. and Sando, R.W. (1975). Crown-fire potential in a sequoia forest after prescribed burning. *For. Sci.* 21: 83–87.

Killingbeck, K.T. (1996). Nutrients in senesced leaves: keys to the search for potential resorption and resorption proficiency. *Ecology* 77: 1716–1727.

Kilvitis, H.J., Alvarez, M., Foust, C.M. et al. (2014). Ecological epigenetics. In: *Ecological Genomics. Advances in Experimental Medicine and Biology*, vol. 781 (ed. C. Landry and N. Aubin-Horth), 191–210. New York: Springer.

King, D.I. and Schlossberg, S. (2014). Synthesis of the conservation value of the early-successional stage in forests of eastern North America. *For. Ecol. Manag.* 324: 186–195.

Kinzig, A.P. (2009). Ecosystem services. In: *The Princeton Guide to Ecology* (ed. S.A. Levin, S.R. Carpenter, H.C.J. Godfray, et al.), 573–578. Princeton, NJ: Princeton University Press.

Kitajima, K. (1994). Relative importance of photosynthetic traits and allocation patterns as correlates of seedling shade tolerance of 13 tropical trees. *Oecologia* 98: 419–428.

Klappa, C.F. (1980). Rhizoliths in terrestrial carbonates: classification. recognition, genesis and significance. *Sedimentology* 27: 613–629.

Klein Goldewijk, K. (2001). Estmating global land use over the past 300 years: the HYDE database. *Glob. Biogeochem. Cycles* 15: 417–433.

Kleinschmit, J. and Bastien, J.C. (1992). IUFRO's role in Douglas-fir (*Pseudotsuga menziesii* (Mirb.) Franco) tree improvement. *Silvae Genet.* 41: 161–173.

Klijn, F. and Udo de Haes, H.A. (1994). A hierarchical approach to ecosystems and its implications for ecological land classification. *Landsc. Ecol.* 9: 89–104.

Kling, M.M. and Ackerly, D.D. (2021). Global wind patterns shape genetic differentiation, asymmetric gene flow, and genetic diversity in trees. *Proc. Natl. Acad. Sci.* 118: e2017317118.

Klinge, H. and Rodrigues, W.A. (1968a). Litter production in an area of Amazonian terra firme forest. Part I. Litter-fall, organic carbon and total nitrogen contents of litter. *Amazoniana* 1: 287–302.

Klinge, H. and Rodrigues, W.A. (1968b). Litter production in an area of Amazonian terra firme forest. Part II. Mineral nutrient content of the litter. *Amazoniana* 1: 303–310.

Klinge, H., Rodrigues, W.A., Brunig, E., and Fittkau, E.J. (1975). Biomass and structure in a central Amazon rain forest. In: *Tropical Ecological Systems: Trends in Terrestrial and Aquatic Research* (ed. F.B. Golley and E. Medina), 115–122. New York: Springer-Verlag.

Klinka, K., Krajina, V.J., Ceska, A., and Scagel, A.M. (1989). *Indicator Plants of Coastal British Columbia*. Vancouver: Univ, British Columbia Press. 288 pp.

Knapp, A.K. and Carter, G.A. (1998). Variability in leaf optical properties among 26 species from a broad range of habitats. *Am. J. Bot.* 85: 940–946.

Knight, H. (1966). Loss of nitrogen from the forest floor by burning. *For. Chron.* 42: 149–152.

Knight, D.H., Jones, G.P., Reiners, W.A., and Romme, W.H. (2014). *Mountains and Plains, The Ecology of Wyoming Landscapes*, 2e. New Haven, CT: Yale Univ. Press 404 pp.

Knoepp, J.D., Turner, D.P., and Tingey, D.T. (1993). Effects of ammonium and nitrate on nutrient uptake and activity of nitrogen assimilation enzymes in western hemlock. *For. Ecol. Manag.* 59: 179–191.

Knops, J.M., Koenig, W.D., and Carmen, W.J. (2007). Negative correlation does not imply a tradeoff between growth and reproduction in California oaks. *Proc. Natl. Acad. Sci.* 104: 16982–16985.

Knowles, R. (1981). Denitrification. In: *Soil Biochemistry Vol. 5* (ed. E.A. Paul and J.N. Ladd), 315–329. New York: Dekker.

Knowles, P. and Grant, M.C. (1985). Genetic variation of lodgepole pine over time and microgeographical space. *Can. J. Bot.* 63: 722–727.

Kobe, R.K. (1997). Carbohydrate allocation to storage as a basis of interspecific variation in sapling survivorship and growth. *Oikos* 80: 226–233.

Kobe, R.K., Pacala, S.W., Silander, J.A., and Canham, C.D. (1995). Juvenile tree survivorship as a component of shade tolerance. *Ecol. Appl.* 5: 517–532.

Koch, J., Pearson, D.E., Huebner, C.D. et al. (2021). Restoration of landscapes and habitats affected by established invasive species. In: *Invasive Species in Forests and Rangelands of the United States* (ed. T.M. Poland, T. Patel-Weynand, D.M. Finch, et al.), 185–202. New York: Springer.

Koenig, W.D. and Knops, J.M.H. (1998). Scale of mast-seeding and tree-ring growth. *Nature* 396: 225–226.

Kohm, K.A. and Franklin, J.F. (1997). Introduction. In: *Creating a Forestry for the 21st Century* (ed. K.A. Kohm and J.F. Franklin), 1–5. Washington, D.C.: Island Press.

Kolb, T.E. and Teulon, D.A.J. (1991). Relationship between sugar maple budburst phenology and pear thrips damage. *Can. J. For. Res.* 21: 1043–1048.

Kolb, T.E., Fettig, C.J., Ayres, M.P. et al. (2016). Observed and anticipated impacts of drought on forest insects and diseases in the United States. *For. Ecol. Manag.* 380: 321–334.

Kolka, R.K., D'Amato, A.W., Wagenbrenner, J.W. et al. (2018). Review of ecosystem level impacts of emerald ash borer on black ash wetlands: what does the future hold? *Forests* 9: 179.

Kollas, C., Körner, C., and Randin, C.F. (2014). Spring frost and growing season length co-control the cold range limits of broad-leaved trees. *J. Biogeogr.* 41: 773–783.

Komarek, E.V. (1971). Principles of fires ecology and fire management in relation to the Alaskan environment. *In* C.W. Slaughter, R.A. Barney, and G.M. Hansen (eds.), *Symp. Proc. Fire in the Northern Environment*. USDA For. Serv. Pacific Northwest For. and Range Exp. Sta., Portland, OR.

Komarek, E.V. (1973). Ancient fires. *In Proc. Annual Tall Timbers Fire Ecology Conf.* 12:219–240. Tall Timbers Res. Sta., Tallahassee, FL.

Köppen, W. (1931). *Grundriss der Klimakunde*. Berlin: Walter de Gruyter 388 pp.

Körner, C. (1998). A re-assessment of high elevation treeline positions and their explanation. *Oecologia* 115: 445–459.

Körner, C. (2017). A matter of tree longevity. *Science* 355: 130–131.

Körner, C. and Paulsen, J. (2004). A world-wide study of high-altitude treeline temperatures. *J. Biogeogr.* 31: 713–732.

Korstian, C.F. and Coile, T.S. (1938). Plant competition in forest stands. *Duke Univ. School For. Bull.* 3. 125 pp.

Korstian, C.F. and Stickel, P.W. (1927). The natural replacement of blight-killed chestnut in the hardwood forests of the Northeast. *J. Agric. Res.* 34: 631–648.

Koski, V. (1970). A study of pollen dispersal as a mechanism of gene flow in conifers. *Comm. Inst. For. Fenn.* 70: 1–78.

Koski, V. and Tallqvist, J.R. (1978). Results of long-time measurements of the quantity of flowering and seed crop of forest trees. *Folia Forestalia* No. 364:1–60

Kosola, K.R., Dickmann, D.I., Paul, E.A., and Parry, D. (2001). Repeated insect defoliation effects on growth, nitrogen acquisition, carbohydrates, and root demography of poplars. *Oecologia* 129: 65–74.

Kovač, M., Kutnar, L., and Hladnik, D. (2016). Assessing biodiversity and conservation status of the Natura 2000 forest habitat types: Tools for designated forestlands stewardship. *For. Ecol. Manag.* 359: 256–267.

Kovač, M., Gasparini, P., Notarangelo, M. et al. (2020). Towards a set of national forest inventory indicators to be used for assessing the conservation status of the habitats directive forest habitat types. *J. Nat. Conserv.* 53: 125747.

Kozlowski, T.T. (1949). Light and water in relation to growth and competition of Piedmont forest tree species. *Ecol. Monogr.* 19: 207–231.

Kozlowski, T.T. (1971a). *Growth and Development of Trees, I. Seed Germination, Ontogeny, and Shoot Growth*. New York: Academic Press 443 pp.

Kozlowski, T.T. (1971b). *Growth and Development of Trees, II. Cambial Growth, Root Growth, and Reproductive Growth*. New York: Academic Press 520 pp.

Kozlowski, T.T. and Ahlgren, C.E. (ed.) (1974). *Fire and Ecosystems*. New York: Academic Press 542 pp.

Kozlowski, T.T. and Keller, T. (1966). Food relations of woody plants. *Bot. Rev.* 32: 293–382.

Kozlowski, T.T. and Pallardy, S.G. (1997a). *Physiology of Woody Plants*, 2e. San Diego, CA: Academic Press 411 pp.

Kozlowski, T.T. and Pallardy, S.G. (1997b). *Growth Control in Woody Plants*. San Diego, CA: Academic Press 641 pp.

Kozlowski, T.T. and Scholtes, W.H. (1948). Growth of roots and root hairs of pine and hardwood seedlings in the Piedmont. *J. For.* 46: 750–754.

Kozlowski, T.T., Kramer, P.J., and Pallardy, S.G. (1991). *The Physiological Ecology of Woody Plants*. New York: Academic Press 657 pp.

Kraft, N.J., Adler, P.B., Godoy, O. et al. (2015). Community assembly, coexistence and the environmental filtering metaphor. *Funct. Ecol.* 29: 592–599.

Kramer, P.J. (1943). Amount and duration of growth of various species of tree seedlings. *Plant Physiol.* 18: 239–251.

Kramer, P.J. (1957). Some effects of various combinations of day and night temperatures and photoperiod on the growth of loblolly pine seedlings. *For. Sci.* 3: 45–55.

Kramer, P.J. (1969). *Plant and Soil Water Relationships: A Modern Synthesis*. New York: McGraw-Hill 482 pp.

Kramer, P.J. (1983). *Water Relations of Plants*. New York: Academic Press 489 pp.

Kramer, P.J. and Boyer, J.S. (1995). *Water Relations of Plants and Soils*. New York: Academic Press 495.

Kramer, P.J. and Clark, W.S. (1947). A comparison of photosynthesis in individual pine needles and entire seedlings at various light intensities. *Plant Physiol.* 22: 51–57.

Kramer, P.J. and Decker, J.P. (1944). Relation between light intensity and rate of photosynthesis of loblolly pine and certain hardwoods. *Plant Physiol.* 19: 350–358.

Krawchuk, M.A., Haire, S.L., Coop, J. et al. (2016). Topographic and fire weather controls of fire refugia in forested ecosystems of northwestern North America. *Ecosphere* 7: e01632.

Krawchuk, M.A., Meigs, G.W., Cartwright, J.M. et al. (2020). Disturbance refugia within mosaics of forest fire, drought, and insect outbreaks. *Front. Ecol. Environ.* 18: 235–244.

Krebs, C.J. (1989). *Ecological Methodology*. New York: Harper and Row 654 pp.

Krebs, C.J. (1998). *Ecological Methodology*, 2e. San Francisco: Benjamin Cummings 624 pp.

Kreyling, J., Buhk, C., Backhaus, S. et al. (2014). Local adaptations to frost in marginal and central populations of the dominant forest tree *Fagus sylvatica* L. as affected by temperature and extreme drought in common garden experiments. *Ecol. Evol.* 4: 594–605.

Kriebel. H.B. (1957). Patterns of genetic variation in sugar maple. *Ohio Agr. Exp. Sta. Res. Bull.* 791. 56 pp.

Kroehler, C.J. and Linkins, A.E. (1988). The root surface phosphatases of *Eriophorum vaginatum:* effects of temperature, pH, substrate concentration, and inorganic phosphorus. *Plant Soil* 105: 3–10.

Krzic, M., Newman, R.F., Broersma, K., and Bomke, A.A. (1999). Soil compaction of forest plantations of interior British Columbia. *Rangel. Ecol. Manag./J. Range Manag. Archives* 52 (6): 671–677.

Kubiszewski, I., Costanza, R., Anderson, S., and Sutton, P. (2020). The future value of ecosystem services: Global scenarios and national implications. *Ecosyst. Serv.* 26: 289–301.

Küchler, A. W. (1964). The potential natural vegetation of the conterminous United States. *Am. Geogr. Soc. Spec. Publ.* No. 36. 154 pp.

Kulakowski, D. and Jarvis, D. (2013). Low-severity fires increase susceptibility of lodgepole pine to mountain pine beetle outbreaks in Colorado. *For. Ecol. Manag.* 289: 544–550.

Kulakowski, D. and Veblen, T.T. (2007). Effect of prior disturbances on the extent and severity of wildfire in Colorado subalpine forests. *Ecology* 88: 759–769.

Kulakowski, D., Jarvis, D., Veblen, T.T., and Smith, J. (2012). Stand-replacing fires reduce susceptibility to mountain pine beetle outbreaks in Colorado. *J. Biogeogr.* 39: 2052–2060.

Kulakowski, D., Matthews, C., Jarvis, D., and Veblen, T.T. (2013). Compounded disturbances in subalpine forests in western Colorado favour future dominance by quaking aspen *Populus tremuloides*. *J. Veg. Sci.* 24: 168–176.

Kulakowski, D., Kaye, M.W., and Kashian, D.M. (2013). Long-term aspen cover change in the western US. *For. Ecol. Manag.* 299: 52–59.

Kulakowski, D., Veblen, T.T., and Bebi, P. (2016). Fire severity controlled susceptibility to a 1940s spruce beetle outbreak in Colorado, USA. *PLoS One* 11: e0158138.

Kumschick, S., Gaertner, M., Vila, M. et al. (2015). Ecological impacts of alien species: Quantification, scope, caveats, and recommendations. *Bioscience* 65: 55–63.

Kunkel, K.E., Palecki, M., Ensor, L. et al. (2009). Trends in twentieth-century U.S. snowfall using a quality-controlled dataset. *J. Atmos. Ocean. Technol.* 26: 33–44.

Kupfer, J.A., Myers, A.T., McLane, S.E., and Melton, G.N. (2008). Patterns of forest damage in a southern Mississippi landscape caused by Hurricane Katrina. *Ecosystems* 11: 45–60.

Kurz, W.A., Dymond, C.C., Stinson, G. et al. (2008a). Mountain pine beetle and forest carbon feedback to climate change. *Nature* 452: 987–990.

Kurz, W.A., Stinson, G., Rampley, G.J. et al. (2008b). Risk of natural disturbances makes future contribution of Canada's forests to the global carbon cycle highly uncertain. *Proc. Natl. Acad. Sci.* 105: 1551–1555.

Kusbach, J.V.A. and Mikeska, M. (2003). Czech forest ecosystem classification. *J. For. Sci.* 49: 85–93.

Kvaalen, H. and Johnsen, Ø. (2008). Timing of bud set in *Picea abies* is regulated by a memory of temperature during zygotic and somatic embryogenesis. *New Phytol.* 177: 49–59.

Ladefoged, K. (1952). The periodicity of wood formation. *Copenh. Biol. Skr.* 7 (3): 1–98.

Lahti, T. (1995). *Understorey vegetation as an indicator of forest site potential in southern Finland*. Finnish Soc: For. Sci., Finnish For. Res. Inst 68 pp.

LaMarche, V.C. and Hirschboeck, K.K. (1984). Frost rings in trees as records of major volcanic eruptions. *Nature* 307: 121–126.

Lamarque, L.J., Delzon, S., and Lortie, C.J. (2011). Tree invasions: a comparative test of the dominant hypotheses and functional traits. *Biol. Invasions* 13: 1969–1989.

Lambers, H. and Oliveira, R.S. (2019). *Plant Physiological Ecology*, 3e. New York: Springer-Verlag 763 pp.

Lambert, J.M. and Dale, M.B. (1964). The use of statistics in phytosociology. *Adv. Ecol. Res.* 2: 55–99.

Lämke, J. and Bäurle, I. (2017). Epigenetic and chromatin-based mechanisms in environmental stress adaptation and stress memory in plants. *Genome Biol.* 18: 124.

Lamothe, K.A., Somers, K.M., and Jackson, D.A. (2019). Linking the ball-and-cup analogy and ordination trajectories to describe ecosystem stability, resistance, and resilience. *Ecosphere* 10: e02629.

Landres, P.B., Morgan, P., and Swanson, F.J. (1999). Overview of the use of natural variability concepts in managing ecological systems. *Ecol. Appl.* 9: 1179–1188.

Landsberg, J.J. (1995). Forest canopies. In: *Encyclopedia of Environmental Biology* (ed. W.A. Nierenberg). New York: Academic Press.

Lang, G.A., Early, J.D., Martin, G.C., and Darnell, R.L. (1987). Endo-, Para-, and eco dormancy: physiological terminology and classification for dormancy research. *Hortic. Sci.* 22: 371–378.

Langford, A.N. and Buell, M.F. (1969). Integration, identity and stability in the plant association. *Adv. Ecol. Res.* 6: 83–135.

Langlet, O. (1959). A cline or not a cline—a question of scots pine. *Silvae Genet.* 8: 13–22.

Langlet, O. (1971). Two hundred years' genecology. *Taxon* 20: 653–721.

Langston, N. (1995). *Forest Dreams, Forest Nightmares, the Paradox of Old Growth in the Inland West.* Seattle: Univ. Washington Press 368 pp.

Lanner, R.M. (1966). Needed: a new approach to the study of pollen dispersion. *Silvae Genet.* 15: 50–52.

Lanner, R.M. (1980). Avian seed dispersal as a factor in the ecology and evolution of limber and whitebark pines. In B.P. Dancik and K.O. Higginbotham (eds.), *Proc. Sixth North American For. Biol. Workshop*. Univ. Alberta, Edmonton, Alberta, Canada.

Lanner, R.M. (1981). *The Piñon Pine: A Natural and Cultural History*. Reno: Univ, Nevada Press 208 pp.

Lanner, R.M. (1982). Adaptations of whitebark pine for seed dispersal by Clark's nutcracker. *Can. J. For. Res.* 12: 391–402.

Lanner, R.M. (1990). Biology, taxonomy, evolution, and geography of stone pines of the world. *In* W. Schmidt and K. McDonald (comps.), *Proc. Symp. Whitebark Pine Ecosystems: Ecology and Management of a High-Mountain Resource,* USDA For. Serv. Gen. Tech. Report INT-270, Intermountain Exp. Sta., Odgen, UT.

Lapin, M. (1990). *The Landscape Ecosystem Groups of the University of Michigan Biological Station: Classification, Mapping, and Analysis of Ecological Diversity*. Master's Thesis, University of Michigan, School of Natural Resources. 140 pp.

Lapin, M. and Barnes, B.V. (1995). Using the landscape ecosystem approach to assess species and ecosystem diversity. *Conserv. Biol.* 9: 1148–1158.

Lapointe, M. and Brisson, J. (2011). Tar spot disease on Norway maple in North America: quantifying the impacts of a Reunion between an invasive tree species and its adventive natural enemy in an urban forest. *Ecoscience* 18: 63–69.

Larcher, W. (1969). The effect of environmental and physiological variables on the carbon dioxide gas exchange of trees. *Photosynthetica* 3: 167–198.

Larcher, W. (1995). *Physiological Plant Ecology*, 3e. New York: Springer-Verlag 506 pp.

Larcher, W. (2003). *Physiological Plant Ecology*, 4e. New York: Springer-Verlag 534 pp.

Larcher, W. and Bauer, H. (1981). Ecological significance of resistance to low temperature. In: *Physiological Plant Ecology I, Responses to the Physical Environment*, New Series, vol. 12A (ed. O.L. Lange, P.S. Nobel, C.B. Osmond and H. Ziegler). New York: Springer-Verlag.

Larsen, B.B., Miller, E.C., Rhodes, M.K., and Wiens, J.J. (2017). Inordinate fondness multiplied and redistributed: the number of species on earth and the new pie of life. *Q. Rev. Biol.* 92: 229–265.

Larson, P.R. (1963b). Stem form development of forest trees. *For. Sci. Monogr.* 5. 42 pp.

Larson, M.M. (1967). Effect of temperature on initial development of ponderosa pine seedlings from three sources. *For. Sci.* 13: 286–294.

Larson, M.M. (1970). Root regeneration and early growth of red oak seedlings: influence of soil temperature. *For. Sci.* 16: 442–446.

Larsson, S. (1989). Stressful times for the plant stress—insect performance hypothesis. *Oikos* 56: 277–283.

Lassoie, J.P. (1982). Physiological processes in Douglas-fir. In: *Analysis of Coniferous Forest Ecosystems in the Western United States*. US/IBP Synthesis Series (ed. R.L. Edmonds), 126–185. Stroudsburg. PA: Dowden, Hutchinson and Ross.

Latham, R.E. and Ricklefs, R.E. (1993a). Global patterns of tree species richness in moist forests: energy-diversity theory does not account for variation in species richness. *Oikos* 67: 325–333.

Latham, R.E. and Ricklefs, R.E. (1993b). Continental comparisons of temperate-zone tree species diversity. In: *Species Diversity in Ecological Communities: Historical and Geographical Perspectives* (ed. R. Ricklefs and D. Schluter), 294–313. Chicago: Univ. Chicago Press.

Lavender, D.P., and Overton, W.S. (1972). Thermoperiods and soil temperatures as they affect growth and dormancy of Douglas-fir seedlings of different geographic origin. Oregon State Univ., For. Res. Lab. Res. Paper 13. 26 pp.

Lavoie, N., Venzina, L.-P., and Margolis, H. (1992). Absorption and assimilation of nitrate and ammonium ions by jack pine seedlings. *Tree Physiol.* 11: 171–183.

Lawrence, W.H. and Rediske, J.H. (1962). Fate of sown Douglas-fir seed. *For. Sci.* 8: 210–218.

Lawton, J.H. (1993). Range, population abundance and conservation. *Trends Ecol. Evol.* 8: 409–413.

Layser, E.F. (1974). Vegetative classification: its application to forestry in the northern Rocky Mountains. *J. For.* 73: 354–357.

Leak, W.B. (1975). Age distribution in virgin red spruce and northern hardwoods. *Ecology* 56: 1451–1454.

Leak, W. B. (1978). Relationship of species and site index to habitat in the White Mountains of New Hampshire. USDA For. Serv. Res. Paper NE-397. Northeastern For. Exp. Sta., Broomall, PA. 9 pp.

Leak, W. B. (1982). Habitat mapping and interpretation in New England. USDA For. Serv. Res. Paper NE-496. Northeastern For. Exp. Sta. Upper Darby, PA. 10 pp.

Ledig, F.T. (1992). Human impacts on genetic diversity in forest ecosystems. *Oikos* 63: 87–108.

Ledig, F.T. and Fryer, J.H. (1972). A pocket of variability in *Pinus rigida*. *Evolution* 26: 259–266.

Ledig, F.T. and Perry, T.O. (1969). Net assimilation rate and growth in loblolly pine seedlings. *For. Sci.* 15: 431–438.

Lee, B.R. and Ibáñez, I. (2021a). Spring phenological escape is critical for the survival of temperate tree seedlings. *Funct. Ecol.* 35: 1848–1861.

Lee, B.R. and Ibáñez, I. (2021b). Improved phenological escape can help temperate tree seedlings maintain demographic performance under climate change conditions. *Glob. Chang. Biol.* 27: 3883–3897.

Lefcheck, J.S., Byrnes, J.E., Isbell, F. et al. (2015). Biodiversity enhances ecosystem multifunctionality across trophic levels and habitats. *Nat. Commun.* 6: 1–7.

Legris, M., Klose, C., Burgie, E.S. et al. (2016). Phytochrome B integrates light and temperature signals in Arabidopsis. *Science* 354: 897–900.

Lemieux, G.J. (1965). Soil-vegetation relationships in the northern hardwoods of Quebec. In: *Forest–Soil Relationships in North America* (ed. C.T. Youngberg). Corvallis: Oregon State Univ. Press.

Lennartz, M.R. (1988). The red-cockaded woodpecker: old-growth species in a second-growth landscape. *Nat. Areas J.* 8: 160–165.

Lenoir, J., Gégout, J.C., Marquet, P.A. et al. (2008). A significant upward shift in plant species optimum elevation during the 20th century. *Science* 320: 1768–1771.

Leopold, A. (1949). *A Sand County Almanac*. Reprinted 1966. Oxford Univ. Press, New York. 189 pp.

Lesser, M.R. and Jackson, S.T. (2013). Contributions of long-distance dispersal to population growth in colonising *Pinus ponderosa* populations. *Ecol. Lett.* 16: 380–389.

Letourneau, D.K., Jedlicka, J.A., Bothwell, S.G., and Moreno, C.R. (2009). Effects of natural enemy biodiversity on the suppression of arthropod herbivores in terrestrial ecosystems. *Annu. Rev. Ecol. Evol. Syst.* 40: 573–592.

Levin, S.A. (2012). The challenge of sustainability: lessons from an evolutionary perspective. In: *Sustainability Science: The Emerging Paradigm and the Urban Environment* (ed. M.P. Weinstein and R.E. Turner). New York: Springer.

Levine, J.M., Adler, P.B., and Yelenik, S.G. (2004). A meta-analysis of biotic resistance to exotic plant invasions. *Ecol. Lett.* 7: 975–989.

Levins, R. (1969). Some demographic and genetic consequences of environmental heterogeneity for biological control. *Bull. Entomol. Soc. Am.* 15: 237–240.

Levins, R. (1970). Extinction. In: *Some Mathematical Problems in Biology* (ed. M. Gerstenhaber). RI: American Mathematical Society, Providence.

Levitt, J. (1980a). *Responses of Plants to Environmental Stresses. Vol. 1. Chilling, Freezing, and High Temperature Stress*. New York: Academic Press 497 pp.

Levitt, J. (1980b). *Responses of Plants to Environmental Stresses. Vol. 2. Water, Radiation, Salt, and Other Stresses*. New York: Academic Press 607 pp.

Levsen, N.D. and Mort, M.E. (2009). Inter-simple sequence repeat (ISSR) and morphological variation in the western North American range of *Chrysosplenium tetrandrum* (Saxifragaceae). *Botany* 87 (8): 780–790.

Lev-Yadun, S. and Sprugel, D. (2011). Why should trees have natural root grafts? *Tree Physiol.* 31: 575–578.

Lexer, C., Fay, M.F., Joseph, J.A. et al. (2005). Barrier to gene flow between two ecologically divergent *Populus* species, *P. alba* (white poplar) and *P. tremula* (European aspen): the role of ecology and life history in gene introgression. *Mol. Ecol.* 14: 1045–1057.

Li, W.-L., Berlyn, G.P., and Ashton, P.M.S. (1996). Polyploids and their structural and physiological characteristics relative to water deficit in *Betula papyrifera* (Betulaceae). *Am. J. Bot.* 83: 15–20.

Li, J., Li, G., Wang, H., and Wang, D.X. (2011). Phytochrome signaling mechanisms. *Arabidopsis Book* 9: e0148.

Liang, J., Watson, J.V., Zhou, M., and Lei, X. (2016a). Effects of productivity on biodiversity in forest ecosystems across the United States and China. *Conserv. Biol.* 30: 308–317.

Liang, J., Crowther, T.W., Picard, N. et al. (2016b). Positive biodiversity-productivity relationship predominant in global forests. *Science* 354: 6309.

Liao, C., Peng, R., Luo, Y. et al. (2008). Altered ecosystem carbon and nitrogen cycles by plant invasion: a meta-analysis. *New Phytol.* 177: 706–714.

Lichti, N.I., Steele, M.A., and Swihart, R.K. (2017). Seed fate and decision-making processes in scatter-hoarding rodents. *Biol. Rev.* 92: 474–504.

Lieberman, M., Lieberman, D., and Peralta, R. (1989). Forests are not just Swiss cheese: canopy stereogeometry of non-gaps in tropical forests. *Ecology* 70: 550–552.

Liebhold, A.M., MacDonald, W.L., Bergdahl, D., and Mastro, V.C. (1995). Invasion by exotic forest pests: a threat to forest ecosystems. *For. Sci. Monogr.* 30: 49 pp.

Liebhold, A.M., McCullough, D.G., Blackburn, L.M. et al. (2013). A highly aggregated geographical distribution of forest pest invasions in the USA. *Divers. Distrib.* 19: 1208–1216.

Liebhold, A.M., Brockerhoff, E.G., Kalisz, S. et al. (2017). Biological invasions in forest ecosystems. *Biol. Invasions* 19: 3437–3458.

Lieth, H. (1973). Primary production in terrestrial ecosystems. *Hum. Ecol.* 1: 303–332.

Ligon, J.D. (1978). Reproductive interdependence of Pinon jays and pinon pines. *Ecol. Monogr.* 48: 111–126.

Ligon, J.D. and Stacey, P.B. (1995). Land use, lag times and the detection of demographic change: the case of the acorn woodpecker. *Conserv. Biol.* 10: 840–846.

Likens, G.E., Bormann, F.H., Pierce, R.S. et al. (1977). *Biogeochemistry of a Forested Ecosystem*. New York: Springer-Verlag 146 pp.

Likens, G.E., Bormann, F.H., Pierce, R.S., and Reiners, W.A. (1978). Recovery of a deforested ecosystem. *Science* 199: 492–496.

Lind, B.M., Friedline, C.J., Wegrzyn, J.L. et al. (2017). Water availability drives signatures of local adaptation in whitebark pine (*Pinus albicaulis* Engelm.) across fine spatial scales of the Lake Tahoe Basin, USA. *Mol. Ecol.* 26: 3168–3185.

Lindberg, S.E. and Garten, C.T. (1988). Sources of sulphur in forest canopy throughfall. *Nature* 336: 148–151.

Lindenmayer, D.B. and Cunningham, S.A. (2013). Six principles for managing forests as ecologically sustainable ecosystems. *Landsc. Ecol.* 28: 1099–1110.

Lindenmayer, D.B. and Fischer, J. (2013). *Habitat Fragmentation and Landscape Change: An Ecological and Conservation Synthesis*. Washington, DC: Island Press 352 pp.

Lindenmayer, D.B. and Franklin, J.F. (2002). *Conserving Forest Biodiversity: A Comprehensive Multiscaled Approach*. Washington, DC: Island Press 352 pp.

Lindenmayer, D.B. and Franklin, J.F. (ed.) (2003). *Towards Forest Sustainability*. Washington, DC: Island Press 212 pp.

Lindenmayer, D., Franklin, J., Lõhmus, A. et al. (2012). A major shift to the retention approach for forestry can help resolve some global forest sustainability issues. *Conserv. Lett.* 5: 421–431.

Linder, S. (1971). Photosynthetic action spectra of scots pine needles of different ages from seedlings grown under different nursery conditions. *Physiol. Plant.* 25: 58–63.

Linder, S. and Axelsson, B. (1982). Changes in carbon uptake and allocation patterns as a result of irrigation and fertilization in a young *Pinus sylvestris* stand. *In* R. H. Waring (ed.), *Carbon Uptake and Allocation: Key to Management of Subalpine Forest Ecosystems*. IUFRO Workshop, Forest Resources Laboratory, Oregon State Univ., Corvallis.

Linder, S., McDonald, J., and Lohammar, T. (1981). Effects of nitrogen status and irradiance during cultivation on photosynthesis and respiration of birch seedlings. Energy Forest Project Tech. Rep. 12. Swedish Univ. Agricultural Science, Uppsala, Sweden. 19 pp.

Lindsey, A. A., and Escobar, L. K. (1976). *Eastern Deciduous Forest, II, Beech-Maple Region*. U.S. Dep. Interior, Natl. Park Serv., Nat. Hist. Theme Stud. No. 3. NPS Publ. No. 148. 238 pp.

Lindsey, A.A. and Sawyer, J.O. Jr. (1970). Vegetation-climate relationships in the eastern United States. *Proc. Indiana Acad. Sci.* 80: 210–214.

Lindsey, A.A., Petty, R.O., Sterling, D.K., and Van Asdall, W. (1961). Vegetation and environment along the Wabash and Tippecanoe Rivers. *Ecol. Monogr.* 31: 105–156.

Linhart, Y.B. (1989). Interactions between genetic and ecological patchiness in forest trees and their dependent species. In: *Evolutionary Ecology of Plants* (ed. J.H. Bock and Y.B. Linhart), 393–430. Boulder, CO: Westview Press.

Linhart, Y.B., Mitton, J.B., Sturgeon, K.B., and Davis, M.L. (1981). Genetic variation in space and time in a population of ponderosa pine. *Heredity* 48: 407–426.

Lippincott, C.L. (2000). Effects of *Imperata cylindrica* (L.) Beauv. (Cogongrass) invasion on fire regime in Florida sandhill (USA). *Nat. Areas J.* 20: 140–149.

Litschert, S.E., Brown, T.C., and Theobald, D.M. (2012). Historic and future extent of wildfires in the southern Rockies Ecoregion, USA. *For. Ecol. Manag.* 269: 124–133.

Little, S. (1974). Effects of fire on temperate forests: northeastern United States. In: *Fire and Ecosystems* (ed. T.T. Kozlowski and C.E. Ahlgren), 225–250. New York: Academic Press.

Little, E.L., Jr. (1979). *Checklist of United States Trees*. USDA For. Serv. Handbook No. 541, Washington D.C. 374 pp.

Little, E.L. Jr., Brinkman, A., and McComb, A.L. (1957). Two natural Iowa hybrid poplars. *For. Sci.* 3: 253–262.

Little, S. and Moore, E.B. (1949). The ecological role of prescribed burns in the pine-oak forests of southern New Jersey. *Ecology* 30: 223–233.

Litton, C.M., Ryan, M.G., Knight, D.H., and Stahl, P.D. (2003). Soil-surface carbon dioxide efflux and microbial biomass in relation to tree density 13 years after a stand replacing fire in a lodgepole pine ecosystem. *Glob. Chang. Biol.* 9: 680–696.

Liu, H., Bauer, L.S., Gao, R. et al. (2003). Exploratory survey for the emerald ash borer, *Agrilus planipennis* (Coleoptera: Buprestidae), and its natural enemies in China. *Gt. Lakes Entomol.* 36: 11.

Liu, H., Platt, S.G., and Borg, C.K. (2004). Seed dispersal by the Florida box turtle (*Terrapene carolina* bauri) in pine Rockland forests of the lower Florida keys, United States. *Oecologia* 138: 539–546.

Liu, J.G., Han, X., Yang, T. et al. (2019). Genome-wide transcriptional adaptation to salt stress in Populus. *BMC Plant Biol.* 19: 367.

Lloyd, A.H. and Fastie, C.L. (2002). Spatial and temporal variability in the growth and climate response of treeline trees in Alaska. *Clim. Chang.* 52: 481–509.

Lloyd, W.J. and Lemmon, P.E. (1970). Rectifying azimuth (of aspect) in studies of soil-site index relationships. In: *Tree Growth and Forest Soils* (ed. C.T. Youngberg and C.B. Davey), 435–448. Corvallis: Oregon State Univ. Press.

Loach, K. (1967). Shade tolerance on tree seedling. I. Leaf photosynthesis and respiration in plants raised under artificial shade. *New Phytol.* 66: 607–621.

Loehle, C. (1986). Phototropism of whole trees: effects of habitat and growth form. *Am. Midi. Nat.* 116: 190–196.

Loehle, C. and Jones, R.H. (1990). Adaptive significance of root grafting in trees. *Funct. Ecol.* 4: 268–271.

Loehle, C. and Wein, G. (1994). Landscape habitat diversity: a multiscale information theory approach. *Ecol. Model.* 73: 311–329.

Logan, K.T. (1965). Growth of tree seedlings as affected by light intensity. I. White birch, yellow birch, sugar maple, and silver maple. Dept. For. Canada, Publ. 1121. 16 pp.

Logan, K.T. (1966a). Growth of tree seedlings as affected by light intensity. II. Red pine, white pine, jack pine and eastern larch. Dept. For. Canada, Publ. 1160. 19 pp.

Logan, K.T. (1966b). Growth of tree seedlings as affected by light intensity. III. Basswood and white elm. Dept. For. Canada, Publ. 1176. 15 pp.

Logan, K.T. (1970). Adaptations of the photosynthetic apparatus of sun- and shade-grown yellow birch (*Betula alleghaniensis* Britt.). *Can. J. Bot.* 48: 1681–1688.

Logan, K.T. and Krotkov, G. (1969). Adaptations of the photosynthetic mechanism of sugar maple (*Acer saccharum*) seedlings grown in various light intensities. *Physiol. Plant.* 22: 104–116.

Logan, J.A. and Powell, J.A. (2001). Ghost forests, global warming and the mountain pine beetle (Coleoptera: Scolytidae). *Am. Entomol.* 47: 160–173.

Logan, J.A., Macfarlane, W.W., and Willcox, L. (2010). Whitebark pine vulnerability to climate-driven mountain pine beetle disturbance in the greater Yellowstone ecosystem. *Ecol. Appl.* 20: 895–902.

Long, J.N. (1980). Productivity of western coniferous forests. In: *Analysis of Coniferous Forest Ecosystems in the Western United States*. US/IBP Synthesis Series (ed. R.L. Edmonds), 89–125. Stroudsburg, PA: Dowden, Hutchinson and Ross.

Lonsdale, W.M. (1990). The self-thinning rule: dead or alive? *Ecology* 71: 1373–1388.

Loope, L., Duever, M., Herndon, A. et al. (1994). Hurricane impact on uplands and freshwater swamp forest. *BioScience* 44: 238–246.

Loreau, M. and De Mazancourt, C. (2013). Biodiversity and ecosystem stability: a synthesis of underlying mechanisms. *Ecol. Lett.* 16: 106–115.

Loreau, M., Naeem, S., Inchausti, P. et al. (2001). Biodiversity and ecosystem functioning: current knowledge and future challenges. *Science* 294: 804–808.

Lorimer, C.G. (1977). The presettlement forest and natural disturbance cycle of northeastern Maine. *Ecology* 58: 139–148.

Lorimer, C.G. (1980). Age structure and disturbance history of a southern Appalachian virgin forest. *Ecology* 61: 1169–1184.

Lorimer, C.G. (1984). Development of the red maple understory in northeastern oak forests. *For. Sci.* 30: 3–22.

Lorimer, C.G. (1993). Causes of the oak regeneration problem. *In* D. Loftis, D. and C.E. McGee (eds.), *Symp. Proc., Oak Regeneration: Serious Problems, Practical Recommendations*. USDA For. Serv. Gen. Tech. Report SE-84. Southeastern For. Exp. Sta., Asheville, NC.

Lorimer, C.G., Frelich, L.E., and Nordheim, E.V. (1988). Estimating gap origin probabilities for canopy trees. *Ecology* 69: 778–785.

Lotspeich, F.B. (1980). Watersheds as the basic ecosystem: this conceptual framework provides a basis for a natural classification system. *Water Resour. Bull.* 16: 581–586.

Loucks, O.L. (1962). *A Forest Classification for the Maritime Provinces*. For. Res. Br., Can. Dept. For. 167 pp.

Loucks, O.L. (1970). Evolution of diversity, efficiency and community stability. *Am. Zool.* 10: 17–25.

Louma, D.L., Stockdale, C.A., Molina, R., and Eberhart, J.L. (2006). The spatial influence of *Pseudotsuga menziesii* retention trees on ectomycorrhizal diversity. *Can. J. For. Res.* 36: 2561–2573.

Lovett, G.M. (1994). Atmospheric deposition of nutrients and pollutants in North America: an ecological perspective. *Ecol. Appl.* 4: 629–650.

Lovett, G.M. and Kinsmann, D. (1990). Atmospheric pollutant deposition to high-elevation ecosystems. *Atmos. Environ.* 24A: 2767–2786.

Lovett, G.M., Jones, C.G., Turner, M.G., and Weathers, K.C. (ed.) (2005). *Ecosystem Function in Heterogeneous Landscapes*. New York: Springer 489 pp.

Lovett, G.M., Canham, C.D., Arthur, M.A. et al. (2006). Forest ecosystem responses to exotic pests and pathogens in eastern North America. *Bioscience* 56: 395–405.

Lovett, G.M., Arthur, M.A., Weathers, K.C., and Griffin, J.M. (2010). Long-term changes in forest carbon and nitrogen cycling caused by an introduced pest/pathogen complex. *Ecosystems* 13: 1188–1200.

Lowry, W.P. (1966). Apparent meteorological requirements for abundant cone crop in Douglas-fir. *For. Sci.* 12: 185–192.

Luckman, B.H. (1986). Reconstruction of Little ice age events in the Canadian Rocky Mountains. *Géog. Phys. Quatern.* 40: 17–28.

Lüdi, W. (1945). Besiedlung und Vegetationsentwicklung auf den jungen Seitenmoränen des Grossen Aletschgletschers mit einem Vergleich der Besiedlung im Vorfeld des Rhonegletschers und des Oberen Grindelwaldgletschers. *Bericht über das Geobotanische Forschungsinstitut Rübel*. Zürich, 1944:35–112.

Lüthi, D., Le Floch, M., Bereiter, B. et al. (2008). High-resolution carbon dioxide concentration record 650,000–800,000 years before present. *Nature* 453: 379–382.

Lutz, H.J. (1930). The vegetation of Heart's Content, a virgin forest in northwestern Pennsylvania. *Ecology* 11: 1–29.

Lutz, H.J. (1934). Ecological relations in the pitch pine plains of southern New Jersey. *Yale School of For. Bull.* 38. 80 pp.

Lutz, H.J. (1940). Disturbance of forest soil resulting from the uprooting of trees. *Yale Univ. School For. Bull.* 45. 37 pp.

Lutz, H.J. (1956). Ecological effects of forest fires in the interior of Alaska. *U.S. Dept. Agr. Tech. Bull.* 1133. 121 pp.

Lyford, W.H. (1980). Development of the root system of northern red oak (*Quercus rubra* L.). *Harvard Forest Paper* No. 21. 30 pp.

Lyford, W.H. and MacLean, D.W. (1966). Mound and pit microrelief in relation to soil disturbance and tree distribution in New Brunswick, Canada. *Harvard Forest Paper* No. 15. 18 pp.

Lyford, W.R. and Wilson, B.F. (1966). Controlled growth of forest tree roots: technique and application. *Harvard Forest Paper* No. 16. Harvard Univ. Petersham, Mass. 12 pp.

Lynch, H.J., Renkin, R.A., Crabtree, R.L., and Moorcroft, P.R. (2006). The influence of previous mountain pine beetle (*Dendroctonus ponderosae*) activity on the 1988 Yellowstone fires. *Ecosystems* 9: 1318–1327.

Lyr, H. and Hoffmann, G. (1967). Growth rates and growth periodicity of tree roots. *Int. Rev. For. Res.* 2: 181–236.

Lytle, D.A. and Poff, N.L. (2004). Adaptation to natural flow regimes. *Trends Ecol. Evol.* 19: 94–100.

MacArthur, R.H. and MacArthur, J. (1961). On bird species diversity. *Ecology* 42: 594–598.

MacArthur, R.H. and Wilson, E.O. (1967). *The Theory of Island Biogeography*. Princeton, NJ: Princeton University Press 224 pp.

MacDonald, G.M. and Cwynar, L.C. (1985). A fossil pollen based reconstruction of the late quaternary history of lodgepole pine (*Pinus contorta* ssp. *latifolia*) in the western interior of Canada. *Can. J. For. Res.* 15: 1039–1044.

Mace, G.M., Norris, K., and Fitter, A.H. (2012). Biodiversity and ecosystem services: a multilayered relationship. *Trends Ecol. Evol.* 27: 19–26.

Mack, R.N., Simberloff, D., Lonsdale, W.M. et al. (2000). Biotic invasions: causes, epidemiology, global consequences, and control. *Ecol. Appl.* 10: 689–710.

Mackey, R.L. and Currie, D.J. (2001). The diversity-disturbance relationship: is it generally strong and peaked? *Ecology* 82: 3479–3492.

Mackey, H.E. Jr. and Sivec, N. (1973). The present composition of a former oak-chestnut forest in the Allegheny Mountains of western Pennsylvania. *Ecology* 54: 915–919.

MacKinnon, A., Meidinger, D., and Klinka, K. (1992). Use of biogeoclimatic ecosystem classification system in British Columbia. *For. Chron.* 68: 100–120.

Maguire, W.P. (1955). Radiation, surface temperatures, and seedling survival. *For. Sci.* 1: 277–285.

Magurran, A.E. (1988). *Ecological Diversity and Its Measurement*. Princeton, NJ: Princeton Univ. Press 179 pp.

Magurran, A.E. (2013). *Measuring Biological Diversity*. New York: Wiley 264 pp.

Magurran, A.E. and McGill, B.J. (ed.) (2011). *Biological Diversity: Frontiers in Measurement and Assessment*. Oxford, UK: Oxford University Press 368 pp.

Maherali, H., Walden, A.E., and Husband, B.C. (2009). Genome duplication and the evolution of physiological responses to water stress. *New Phytol.* 184: 721–731.

Mahrt, L., Lee, X., Black, A. et al. (2000). Nocturnal mixing in a forest subcanopy. *Agric. For. Meteorol.* 101: 67–78.

Mailly, D. and Gaudreault, M. (2005). Growth intercept models for black spruce, jack pine and balsam fir in Quebec. *For. Chron.* 81: 104–113.

Major, J. (1951). A functional, factorial approach to plant ecology. *Ecology* 32: 392–412.

Major, E.J. (1990). Water stress in Sitka spruce and its effect on the green spruce aphid Elatobium. *Popul. Dyn. Insects* 1: 85–93.

Malanson, G.P. (1993). *Riparian Landscapes*. Cambridge: Cambridge Univ. Press 296 pp.

Mallet, J. (2006). What does Drosophila genetics tell us about speciation? *Trends Ecol. Evol.* 21: 386–393.

Mallet, J., Besansky, N., and Hahn, M.W. (2016). How reticulated are species? *Bioessays* 38: 140–149.

Marchand, D.E. (1971). Rates and modes of denudation, White Mountains, eastern California. *Am. J. Sci.* 270: 109–135.

Marcot, B.C. (1997). Biodiversity of old forests of the west: lessons from our elders. In: *Creating a Forestry for the 21st Century* (ed. K.A. Kohm and J.F. Franklin), 87–105. Washington, DC: Island Press.

Margalef, R. (1972). Homage to Evelyn Hutchinson, or why is there an upper limit to diversity. *Trans. Connecticut Acad. Arts Sci.* 44: 211–235.

Marion, G.M. and Black, C.H. (1988). Potentially available nitrogen and phosphorus along a chaparral fire cycle chronosequence. *Soil Sci. Soc. Am. J.* 52: 1155–1162.

Marks, P.L. (1974). The role of pin cherry (*Prunus pensylvanica* L.) in the maintenance of stability in northern hardwood ecosystems. *Ecol. Monogr.* 44: 73–88.

Marks, P.L. (1975). On the relation between extension growth and successional status of deciduous trees of the northeastern United States. *Bull. Torrey Bot. Club* 102: 172–177.

Marks, P.L. and Harcombe, P.A. (1981). Community diversity of coastal plain forests in southern East Texas. *Ecology* 56: 1004–1008.

Marlon, J.R., Bartlein, P.J., Daniau, A.-L. et al. (2013). Global biomass burning: a synthesis and review of Holocene paleofire records and their controls. *Quat. Sci. Rev.* 65: 5–25.

Marquis, D.A. (1974). The impact of deer browsing on Allegheny hardwood regeneration. USDA For. Serv. Res. Paper NE-308. Northeastern For. Exp. Sta., Upper Darby, PA. 8 pp.

Marquis, D.A. 1975. The Allegheny hardwood forests of Pennsylvania. USDA For. Serv. Gen. Tech. Report NE-15. Northeastern For. Exp. Sta., Upper Darby, PA. 32 pp.

Marquis, D.A. (1981). Effect of deer browsing on timber in Allegheny hardwood forests of northwestern Pennsylvania. USDA For. Serv. Res. Paper NE-475. Northeastern For. Exp. Sta., Upper Darby, PA. 10 pp.

Marr, J.W. (1977). The development and movement of tree islands near the upper limit of tree growth in the southern Rocky Mountains. *Ecology* 58: 1159–1164.

Marshall, P.E. and Kozlowski, T.T. (1976). Importance of photosynthetic cotyledons for early growth of woody angiosperms. *Physiol. Plant.* 37: 336–340.

Marshall, J.D. and Toffel, M.W. (2005). Framing the elusive concept of sustainability: a sustainability hierarchy. *Environ. Sci. Technol.* 39: 673–682.

Marshall, I.B., Schut, P.H., and Ballard, M. (1999). A national ecological framework for Canada: Attribute data. Agriculture and Agri-Food Canada, Research Branch, Centre for Land and Biological Resources Research and Environment Canada, State of the Environment Directorate, Ecozone Analysis Branch. Ottawa/Hull. http://sis.agr.gc.ca/cansis/nsdb/ecostrat/1999report/index.html (accessed July 13, 2021).

Martin, P.H. (1999). Norway maple (*Acer platanoides*) invasion of a natural forest stand: understory consequence and regeneration pattern. *Biol. Invasions* 1: 215–222.

Martin, K.L. and Goebel, P.C. (2013). The foundation species influence of eastern hemlock (*Tsuga canadensis*) on biodiversity and ecosystem function on the unglaciated Allegheny Plateau. *For. Ecol. Manag.* 289: 143–152.

Martin, R.E., Cooper, R.W., Crow, A.B. et al. (1977). Report of task force on prescribed burning. *J. For.* 75: 297–301.

Martin, J.L., Gower, S.T., Plaut, J., and Holmes, B. (2005). Carbon pools in a boreal mixedwood logging chronosequence. *Glob. Chang. Biol.* 11: 1883–1894.

Martín, D., Vázquez-Piqué, J., Carevic, F.S. et al. (2015). Trade-off between stem growth and acorn production in holm oak. *Trees* 29: 825–834.

Marx, D.H. and Beattie, D.J. (1977). Mycorrhizae—promising aid to timber growers. *For. Farmer* 36: 6–9.

Maser, C. and Maser, Z. (1988). Interactions among squirrels, mycorrhizal fungi, and coniferous forests in Oregon. *Great Basin Nat.* 48: 358–369.

Maser, C., Trappe, J.M., and Nussbaum, R.A. (1978). Fungal—small mammal interrelationships with emphasis on Oregon conifer forests. *Ecology* 59: 799–809.

Maser, C.R., Tarrant, F., Trappe, J.M., and Franklin, J.F. (eds.). (1988). From the forest to the sea: A story of fallen trees. USDA For. Serv. Gen. Tech. Report PNW-229. Washington, D.C. 153 pp.

Masterson, J. (1994). Stomatal size in fossil plants: evidence for polyploidy in majority of angiosperms. *Science* 264: 421–424.

Matthews, J.D. (1963). Factors affecting the production of seed by forest trees. *For. Abstr.* 24 (1): i–xiii.

Matthews, J.A. (1992). *The Ecology of Recently Deglaciated Terrain*. Cambridge: Cambridge Univ. Press 386 pp.

Mattson, W.J. and Addy, N.D. (1975). Phytophagous insects as regulators of forest primary production. *Science* 190: 515–522.

Mattson, W.J. and Haack, R.A. (1987). The role of drought in outbreaks of plant-eating insects. *Bioscience* 37: 110–118.

Mattson, W.J. and Shriner, D.S. (eds.). (2001). *Northern Minnesota Independence Day Storm: A Research Needs Assessment*. USDA For. Serv. Gen. Tech. Report NC-GTR-216. North Central Res. Sta., St. Paul, MN. 65 pp.

Mattson, W.J., Lawrence, R.K., Haack, R.A. et al. (1988). Defensive strategies of woody plants against different insect-feeding guilds in relation to plant ecological strategies and intimacy of association with insects. In: *Mechanism of Woody Plant Defenses against Insects* (ed. W.J. Mattson, J. Levieux and C. Bernard-Dagan), 3–38. New York: Springer Verlag.

Mattson, W. J., Herms, D. A., Witter, J. A., and Allen, D. C. (1991). Woody plant grazing systems: North American outbreak folivores and their host plants. *In* Y. N. Baranchikov, W. J. Mattson, F. P. Hain, and T. L. Payne (eds.), *Forest Insect Guilds: Patterns of Interaction with Host Trees.* USDA For. Serv. Gen. Tech. Report NE-153. Northeastern For. Exp. Sta., Radnor, PA.

Mattson, W., Vanhanen, H., Veteli, T. et al. (2007). Few immigrant phytophagous insects on woody plants in Europe: legacy of the European crucible? *Biol. Invasions* 9: 957–974.

Mátyás, C. and Yeatman, C.W. (1992). Effect of geographical transfer on growth and survival of jack pine (*Pinus banksiana* Lamb.) populations. *Silvae Genet.* 41: 370–376.

May, R.M. (1974). *Stability and Complexity in Model Ecosystems*. Princeton, NJ: Princeton University Press 304 pp.

May, R.M. (2011). Why worry about how many species and their loss? *PLoS Biol.* 9: e1001130.

May, R.M. and McLean, A.R. (2007). *Theoretical Ecology: Principles and Applications*, 3e. Oxford, UK: Oxford University Press.

Mayer, P.M., Reynolds, S.K. Jr., McCutchen, M.D., and Canfield, T.J. (2007). Meta-analysis of nitrogen removal in riparian buffers. *J. Environ. Qual.* 36: 1172–1180.

Mayerfeld, D., Rickenbach, M., and Rissman, A. (2016). Overcoming history: attitudes of resource professionals and farmers toward silvopasture in southwest Wisconsin. *Agrofor. Syst.* 90: 723–736.

Mayfield, A.E. III, Seybold, S.J., Haag, W.R. et al. (2021). Impacts of invasive species in terrestrial and aquatic systems in the United States. In: *Invasive Species in Forests and Rangelands of the United States* (ed. T.M. Poland, T. Patel-Weynand, D.M. Finch, et al.), 5–40. New York: Springer.

McAndrews, J.H. (1966). Postglacial history of prairie, savanna and forest in northwestern Minnesota. *Torreya Bot. Club Mem.* 22 (2): 72 pp.

McBride, J. (1973). Natural replacement of disease-killed elms. *Am. Midi. Nat.* 90: 300–306.

McCann, K.S. (2000). The diversity-stability debate. *Nature* 405: 228–233.

McClanahan, T.R. (1986). The effect of a seed source on primary succession in a forest ecosystem. *Vegetatio* 65: 175–178.

McCree, K.J. (1972). The action spectrum, absorptance and quantum yield of photosynthesis in crop plants. *Agric. Meteorol.* 9: 191–216.

McCullough, D.G., Work, T.T., Cavey, J.F. et al. (2006). Interceptions of nonindigenous plant pests at US ports of entry and border crossings over a 17-year period. *Biol. Invasions* 8: 611–630.

McCune, B. and Allen, T.F.H. (1985). Forest dynamics in the bitterroot canyons, Montana. *Can. J. Bot.* 63: 377–383.

McCune, B. and Cottam, G. (1985). The successional status of a southern Wisconsin oak woods. *Ecology* 66: 1270–1278.

McDermott, R. E. (1954). Seedling tolerance as a factor in bottomland timber succession. *Mo. Agr. Exp. Sta. Res. Bull.* 557. 11 pp.

McDowell, N.G., Allen, C.D., Anderson-Teixeira, K. et al. (2020). Pervasive shifts in forest dynamics in a changing world. *Science* 368: eaaz9463.

McElhinny, C., Gibbons, P., and Brack, C. (2006). An objective and quantitative methodology for constructing an index of stand structural complexity. *For. Ecol. Manag.* 235: 54–71.

McEvoy, P.M., McAdam, J.H., Mosquera-Losada, M., and Rigueiro-Rodriguez, A. (2006). Tree regeneration and sapling damage of pedunculate oak *Quercus robur* in a grazed forest in Galicia, NW Spain: A comparison of continuous and rotational grazing systems. *Agrofor. Syst.* 66: 85–92.

McFarlane, R.W. (1992). *A Stillness in the Pines: The Ecology of the Red-Cocaded Woodpecker*. New York: W. W. Norton 270 pp.

McIntire, E.J. and Fajardo, A. (2014). Facilitation as a ubiquitous driver of biodiversity. *New Phytol.* 201: 403–416.

McIntosh, R.P. (1967). The continuum concept of vegetation. *Bot. Rev.* 33: 130–187.

McIntosh, R.P. (1968). Reply. *In* P. Dansereau (ed.), The continuum concept of vegetation: responses. *Bot. Rev*. 34:253–332.

McIntosh, R.P. (1993). The continuum continued: John T. Curtis' influence on ecology. In: *Fifty Years of Wisconsin Plant Ecology* (ed. J.S. Fralish, R.P. McIntosh and O.L. Loucks), 122–195. Madison: Univ. Wisconsin. Press.

McIntyre, L. (1985). Humboldt's way. *Natl. Geogr.* 168 (3): 318–351.

McKenney, D.W., Pedlar, J.H., Lawrence, K. et al. (2007a). Potential impacts of climate change on the distribution of North American trees. *Bioscience* 57: 939–948.

McKenney, D.W., Pedlar, J.H., Lawrence, K. et al. (2007b). Beyond traditional hardiness zones: using climate envelopes to map plant range limits. *Bioscience* 57: 929–937.

McKenney, D.W., Pedlar, J.H., Rood, R.B., and Price, D. (2011). Revisiting projected shifts in the climate envelopes of North American trees using updated general circulation models. *Glob. Chang. Biol.* 17: 2720–2730.

McKenzie, D., Miller, C., and Falk, D.A. (ed.) (2011). *The Landscape Ecology of Fire*. New York: Springer 332 pp.

McKevlin, M.R., Hook, D.D., and McKee, W.H. Jr. (1995). Growth and nutrient use efficiency of water tupelo seedlings in flooded and well-drained soil. *Tree Physiol.* 15: 753–758.

McKnight, J.S., Hook, D.D., Langdon, O.G., and Johnson, R.L. (1981). Flood tolerance and related characteristics of trees of the bottomland forests of the southern United States. In: *Wetlands of Bottomland Hardwood Forests*, vol. 11 (ed. J.R. Clark and J. Benforado), 29–69. Amsterdam: Elsevier.

McLachlan, J.S. and Clark, J.S. (2004). Reconstructing historical ranges with fossil data at continental scales. *For. Ecol. Manag.* 197: 139–147.

McLachlan, J.S., Clark, J.S., and Manos, P.S. (2005). Molecular indicators of tree migration capacity under rapid climate change. *Ecology* 86: 2088–2098.

McLane, S.C. and Aitken, S.N. (2012). Whitebark pine (*Pinus albicaulis*) assisted migration potential: testing establishment north of the species range. *Ecol. Appl.* 22: 142–153.

McLaughlin, B.C., Ackerly, D.D., Klos, P.Z. et al. (2017). Hydrologic refugia, plants, and climate change. *Glob. Chang. Biol.* 23: 2941–2961.

McNab, W.H. (1987). Yellow-poplar site quality related to slope type in mountainous terrain. *Northern J. Appl. For.* 4: 189–192.

McNab, W.H. (1989). Terrain shape index: quantifying effect of minor landforms on tree height. *For. Sci.* 35: 91–104.

McNab, W.H. (1991). Land classification in the Blue Ridge Province: state-of-the-science report. *In* D.L. Mengel and D.T. Tew (eds.), *Symp. Proc. Ecological Land Classification: Applications to Identify the Productive Potential of Southern Forests.* USDA For. Serv. Gen. Tech. Report SE-68. Southeastern For. Exp. Sta., Asheville, NC.

McNab, W.H. (1993). A topographic index to quantify the effect of mesoscale landform on site productivity. *Can. J. For. Res.* 23: 1100–1107.

McNab, W.H., Browning, S.A., Simon, S.A., and Fouts, P.E. (1999). An unconventional approach to ecosystem unit classification in western North Carolina, USA. *For. Ecol. Manag.* 114: 405–420.

Meades, W.J. and Roberts, B.A. (1992). A review of forest site classification activities in Newfoundland and Labrador. *For. Chron.* 68: 25–33.

Meehl, G.A., Washington, W.M., Ammann, C.M. et al. (2004). Combinations of natural and anthropogenic forcings in twentieth-century climate. *J. Clim.* 17: 3721–3727.

Meeker, D.O. Jr. and Merkel, D.L. (1984). Climax theories and a recommendation for vegetation classification—a viewpoint. *J. Range Manag.* 37: 427–430.

Meeuwig, R. O. (1971). Infiltration and water repellency in granitic soils. USDA For. Serv. Res. Paper INT-111. Intermountain For. and Rge. Exp. Sta., Ogden, UT. 20 pp.

Meffe, G.K. and Carroll, C.R. (1997). *Principles of Conservation Biology*, 2e, 673. Sunderland, MA: Sinauer Associates.

Meier, A.R., Saunders, M.R., and Michler, C.H. (2012). Epicormic buds in trees: a review of bud establishment, development, and dormancy release. *Tree Physiol.* 32: 565–584.

Meigs, G.W., Dunn, C.J., Parks, S.A., and Krawchuk, M.A. (2020). Influence of topography and fuels on fire refugia probability under varying fire weather conditions in forests of the Pacific northwest, USA. *Can. J. For. Res.* 50: 636–647.

Meiners, T.M., Smith, D.W., Sharik, T.L., and Beck, D.E. (1984). Soil and plant water stress in an Appalachian oak forest in relation to topography and stand age. *Plant Soil* 80: 171–179.

Melgoza, G., Nowak, R.S., and Tausch, R.J. (1990). Soil water exploitation after fire: competition between *Bromus tectorum* (cheatgrass) and two native species. *Oecologia* 83: 7–13.

Melillo, J.M. (1981). Nitrogen cycling in deciduous forests. In: *Terrestrial Nitrogen Cycles*. Ecol. Bull. 33 (ed. F.E. Clark and T. Rosswell), 427–442. Stockholm: Swedish Natural Sci. Res. Council.

Melillo, J.M., Aber, J.D., and Muratore, J.F. (1982). Nitrogen and lignin control of hardwood leaf litter decomposition dynamics. *Ecology* 63: 621–626.

Melillo, J.M., Aber, J.D., Steudler, P.A., and Schimel, J.P. (1983). Denitrification potentials in a successional sequence of northern hardwood forest stands. In: *Environmental Biogeochemistry*. Ecol. Bull. 35. (ed. R. Hallberg), 217–228. Stockholm: Swedish Natural Sci. Res. Council.

Melillo, J.M., McGuire, A.D., Kicklighter, D.W. et al. (1993). Climate change and terrestrial net primary productivity. *Nature* 363: 234–363.

Melles, S.J., Fortin, M.-J., Lindsay, K., and Badzinski, D. (2011). Expanding northward: influence of climate change, forest connectivity, and population processes on a threatened species' range shift. *Glob. Chang. Biol.* 17: 17–31.

Mendelsohn, R. and Balick, M.J. (1995). The value of undiscovered pharmaceuticals in tropical forests. *Econ. Bot.* 49: 223–228.

Méndez-Toribio, M., Meave, J.A., Zermeño-Hernández, I., and Ibarra-Manríquez, G. (2016). Effects of slope aspect and topographic position on environmental variables, disturbance regime and tree community attributes in a seasonal tropical dry forest. *J. Veg. Sci.* 27: 1094–1103.

Messier, C. and Bellefleur, P. (1988). Light quantity and quality on the forest floor of pioneer and climax stages in a birch-beech-sugar maple stand. *Can. J. For. Res.* 18: 615–622.

Metz, L.J., Lotti, T., and Klawitter, R.A. (1961). Some effects of prescribed burning on coastal plain forest soil. USDA For. Serv. Sta. Paper 133. Southeastern For. Exp. Sta., Asheville, NC. 10 pp.

Meyers, R. K., Zahner, R., and Jones, S. M. (1986). Forest habitat regions of South Carolina. Clemson Univ., Dept. Forestry Res. Series No. 42. 31 pp. + map.

Michalak, J.L., Lawler, J.J., Roberts, D.R., and Carroll, C. (2018). Distribution and protection of climatic refugia in North America. *Conserv. Biol.* 32: 1414–1425.

Miles, J. (1979). *Vegetation Dynamics*. London: Chapman and Hall 80 pp.

Miles, J. (1985). The pedogenic effects of different species and vegetation types and the implications of succession. *J. Soil Sci.* 36: 571–584.

Miles, J. (1987). Vegetation succession: past and present perceptions. In: *Colonization, Succession and Stability* (ed. A.J. Gray, M.J. Crawley and P.J. Edwards), 1–30. Oxford: Blackwell.

Millar, C.I., Swanston, C.W., and Peterson, D.L. (2014). Adapting to climate change. In: *Climate Change and United States Forests* (ed. D.L. Peterson, J.M. Vose and T. Patel-Weynand), 183–222. New York: Springer.

Millennium Ecosystem Assessment (2005). *Ecosystems and Human Well Being: Current State and Trends*. Washington, DC: Island Press 815 pp.

Miller, J.R. and Bestelmeyer, B.T. (2016). What's wrong with novel ecosystems, really? *Restor. Ecol.* 24: 577–582.

Miller, G.T. Jr. (1992). *Living in the Environment: An Introduction to Environmental Science*, 7e. Belmont, CA: Wadsworth 705 pp.

Miller, H.A. and Lamb, S.H. (1985). *Oaks of North America*. Happy Camp, CA: Naturegraph Publ 327 pp.

Miller, C. and Urban, D.L. (2000). Connectivity of forest fuels and surface fire regimes. *Landsc. Ecol.* 15: 145–154.

Miller, B., Reading, R., Trittholt, J.S. et al. (1999). Using focal species in the design of nature reserve networks. *Wild Earth* 8: 81–92.

Miller, T.E., Burns, J.H., Munguia, P. et al. (2005). A critical review of twenty years' use of the resource-ratio theory. *Am. Nat.* 165: 439–448.

Miller, T.E., Burns, J.H., Munguia, P. et al. (2007). Evaluating support for the resource-ratio hypothesis: a reply to Wilson et al. *Am. Nat.* 169: 707–708.

Miller, A., Reilly, D., Bauman, S., and Shea, K. (2012). Interactions between frequency and size of disturbance affect competitive outcomes. *Ecol. Res.* 27: 783–791.

Mills, L.S., Soulé, M.E., and Doak, D.F. (1993). The keystone-species concept in ecology and conservation. *Bioscience* 43: 219–224.

Miniat, C.F., Fraterrigo, J.M., Brantley, S.T. et al. (2021). Impacts of invasive species on forest and grassland ecosystem processes in the United States. In: *Invasive Species in Forests and Rangelands of the United States* (ed. T.M. Poland, T. Patel-Weynand, D.M. Finch, et al.), 41–55. New York: Springer.

Minteer, B.A. and Collins, J.P. (2010). Move it or lose it? The ecological ethics of relocating species under climate change. *Ecol. Appl.* 20: 1801–1804.

Mirkin, B.M., Naumova, L.G., Martynenko, V.B., and Shirokikh, P.S. (2015). Contribution of the Braun-Blanquet syntaxonomy to research on successions of plant communities. *Russ. J. Ecol.* 46: 303–308.

Mirov, N.T. (1967). *The Genus "Pinus."*. New York: Ronald Press 602 pp.

Mittelbach, G.G., Steiner, C.F., Scheiner, S.M. et al. (2001). What is the observed relationship between species richness and productivity? *Ecology* 82: 2381–2396.

Mitton, J.B. and Ferrenberg, S.M. (2012). Mountain pine beetle develops an unprecedented summer generation in response to climate warming. *Am. Nat.* 179: E163–E171.

Mitton, J.B. and Grant, M.C. (1980). Observations on the ecology and evolution of quaking aspen, *Populus tremuloides,* in the Colorado front range. *Am. J. Bot.* 67: 202–209.

Miyazaki, Y. (2013). Dynamics of internal carbon resources during masting behavior in trees. *Ecol. Res.* 28: 143–150.

Mladenoff, D.J. and Steams, F. (1993). Eastern hemlock regeneration and deer browsing in the northern Great Lakes region: a re-examination and model simulation. *Conserv. Biol.* 7: 889–900.

Mock, K.E., Rowe, C.A., Hooten, M.B. et al. (2008). Clonal dynamics in western north American aspen (*Populus tremuloides*). *Mol. Ecol.* 17: 4827–4844.

Mock, K.E., Callahan, C.M., Islam-Faridi, M.N. et al. (2012). Widespread triploidy in western north American aspen (*Populus tremuloides*). *PLoS One* 7: e48406.

Mody, K., Eichenberger, D., and Dorn, S. (2009). Stress magnitude matters: different intensities of pulsed water stress produce non-monotonic resistance responses of host plants to insect herbivores. *Ecological Entomology* 34 (1): 133–143.

Moehring, D.H., Grano, C.X., and Bassett, J.R. (1966). Properties of forested loess soils after repeated prescribed burns. USDA For. Serv. Res. Note SO-40. Southern For. Exp. Sta., New Orleans, LA. 4 pp.

Mohn, C.A. and Pauley, S. (1969). Early performance of cottonwood seed sources in Minnesota. *Minn. For. Notes* 207: 4 pp.

Monsen, S.B. and Shaw, N.L. (2000). Big sagebrush (*Artemisia tridentata*) communities-ecology, importance and restoration. *In* L. Wagner, D. Neuman (eds.). *Striving for Restoration, Fostering Technology and Policy for Reestablishing Ecological Function*. Proceedings of the Billings Land Reclamation Symposium, March 20–24, 2000, Billings, MT. Reclamation Research Unit Publication No. 00-01. Bozeman, MT: Montana State University.

Monserud, R.A. (1984). Problems with site index: an opinionated review. *In* J.G. Bockheim (ed.), *Symp Proc. Forest Land Classification: Experience, Problems, Perspectives,* NCR-102 North Central For. Soil Com., Soc. Am. For., USDA For. Serv., USDA Cons. Serv., Madison, WI.

Monserud, R.A. and Rehfeldt, G.E. (1990). Genetic and environmental components of variation of site index in inland Douglas-fir. *For. Sci.* 36: 1–9.

Montgomery, D.R., Grant, G.E., and Sullivan, K. (1995). Watershed analysis as a framework for implementing ecosystem management. *Water Resour. Bull.* 31: 369–386.

Montgomery, R.A., Rice, K.E., Stefanski, A. et al. (2020). Phenological responses of temperate and boreal trees to warming depend on ambient spring temperatures, leaf habit, and geographic range. *Proc. Natl. Acad. Sci.* 117: 10397–10405.

Moola, F.M. and Vasseur, L. (2008). The importance of clonal growth to the recovery of *Gaultheria procumbens* L. (Ericaceae) after forest disturbance. *Plant Ecol.* 201: 319–337.

Mooney, H.A. (1972). The carbon balance of plants. *Annu. Rev. Ecol. Syst.* 3: 315–346.

Moore, J.R. and Watt, M.S. (2015). Modelling the influence of predicted future climate change on the risk of wind damage within New Zealand's planted forests. *Glob. Chang. Biol.* 21: 3021–3035.

Moos, C., Khelidj, N., Guisan, A. et al. (2021). A quantitative assessment of rockfall influence on forest structure in the Swiss Alps. *Eur. J. For. Res.* 140: 91–104.

Mopper, S. and Whitham, T.G. (1992). The plant stress paradox: effects on pinyon sawfly sex ratios and fecundity. *Ecology* 73: 515–525.

Mopper, S., Maschinski, J., Cobb, N., and Whitham, T.G. (1991a). A new look at habitat structure: consequences of herbivore-modified plant architecture. In: *Habitat Complexity: The Physical Arrangement of Objects in Space* (ed. S. Bell, E. McCoy and H. Mushinsky), 260–280. New York: Chapman and Hall.

Mora, C., Tittensor, D.P., Adl, S. et al. (2011). How many species are there on earth and in the ocean? *PLoS Biol.* 9: e1001127.

Moran, E., Lauder, J., Musser, C. et al. (2017). The genetics of drought tolerance in conifers. *New Phytol.* 216: 1034–1048.

Morecroft, M.D., Taylor, M.E., and Oliver, H.R. (1998). Air and soil microclimates of deciduous woodland compared to an open site. *Agric. For. Meteorol.* 90: 141–156.

Morelli, T.L., Daly, C., Dobrowski, S.Z. et al. (2016). Managing climate change refugia for climate adaptation. *PLoS One* 11: e0159909.

Morelli, T.L., Barrows, C.W., Ramirez, A.R. et al. (2020). Climate-change refugia: biodiversity in the slow lane. *Front. Ecol. Environ.* 18: 228–234.

Morgan, P. and Bunting, S. C. (1990). Fire effects in whitebark pine forests. *In* W.C. Schmidt and K. J. McDonald (comps.), *Symp. Proc. Whitebark Pine Ecosystems: Ecology and Management of a High-Mountain Resource.* USDA For. Serv. Gen Tech. Rep. INT-270, Intermountain Res. Sta., Odgen, UT.

Morgenstern, E.K. (1996). *Geographic Variation in Forest Trees.* Vancouver, B.C.: Univ. British Columbia Press 209 pp.

Morgenstern, E.K. and Mullin, T.J. (1990). Growth and survival of black spruce in the range wide provenance study. *Can. J. For. Res.* 20: 130–143.

Mori, A.S., Lertzman, K.P., and Gustafsson, L. (2017). Biodiversity and ecosystem services in forest ecosystems: a research agenda for applied forest ecology. *J. Appl. Ecol.* 54: 12–27.

Morin, R.S. and Liebhold, A.M. (2016). Invasive forest defoliator contributes to the impending downward trend of oak dominance in eastern North America. *Forestry* 89: 284–289.

Morrison, P.H. and Swanson, F.J. (1990). Fire history and pattern in a Cascade Range landscape. USDA For. Serv. Gen. Tech. Report PNW-GTR-254, Pacific Northwest Res. Sta., Corvallis, OR. 77 pp.

Mosley, J.C., Bunting, S.C., and Manoukian, M.E. (1999). Cheatgrass. In: *Biology and Management of Noxious Rangeland Weeds* (ed. R.L. Sheley and J.K. Petroff), 175–188. Corvallis, OR: Oregon State University Press.

Mosquin, T. (1966). Reproductive specialization as a factor in the evolution of the Canadian flora. In: *The Evolution of Canada's Flora* (ed. R.L. Taylor and R.A. Ludwig), 43–65. Toronto: Univ. Toronto Press.

Moss, C.E. (1913). *Vegetation of the Peak District.* Cambridge: Cambridge Univ. Press 235 pp.

Mott, K.A. and Parkhurst, D.F. (1991). Stomatal responses to humidity in air and helox. *Plant Cell Environ.* 14: 509–515.

Mount, A.B. (1964). The interdependence of the eucalyptus and forest fires in southern Australia. *Aust. For.* 28: 166–172.

Mueggler, W.F. (1985). Vegetation associations. *In* N.V. DeByle and R. P. Winokur (eds.), *Aspen: Ecology and Management in the Western United States.* USDA For. Serv. Gen. Tech. Report RM-119. Rocky Mountain For. and Rge. Exp. Sta., Ft. Collins, CO.

Mueggler, W.R. and Campbell, R.B. (1986). Aspen community types of Utah. USDA For. Serv. Res. Paper INT-362. Intermountain Res. Sta., Ogden, UT. 69 pp.

Mueller-Dombois, D. (1987). Natural dieback in forests. *Bioscience* 37: 575–583.

Mueller-Dombois, D. and Ellenberg, H. (1974). *Aims and Methods of Vegetation Ecology.* New York: Wiley 547 pp.

Muffler, L., Beierkuhnlein, C., Aas, G. et al. (2016). Distribution ranges and spring phenology explain late frost sensitivity in 170 woody plants from the northern hemisphere. *Glob. Ecol. Biogeogr.* 25: 1061–1071.

Mühlhäusser, G. and Müller, S. (1995). Das südwestdeutsche Verfahren der Forstlichen Standortskartierung. *In* Arbeitsgruppe Biozönosenkunde. 1995. *Ansätze für eine Regionale Biotop und Biozönosenkunde von Baden-Württemberg*. Mitt. Forst. Versuchs u. Forschungsanst. Baden-Württemberg. Freiburg, Germany. 166 pp.

Mühlhäusser, G., Hubner, W., and Sturmmer, G. (1983). Die Forstliche Standortskarte 1:10,000 nach dem Baden-Württembergischen Verfahren. *Mitt. Verein forstl. Standortsk. u. Forstpflz.* 30:3–13.

Muir, P.S. and Lotan, J.E. (1985). Disturbance history and serotiny in *Pinus contorta* in western Montana. *Ecology* 66: 1658–1668.

Munoz, S.E., Gajewski, K., and Peros, M.C. (2010). Synchronous environmental and cultural change in the prehistory of the northeastern United States. *Proc. Natl. Acad. Sci.* 107: 22008–22013.

Musselman, R.C., Lester, D.T., and Adams, M.S. (1975). Localized ecotypes of *Thuja occidentalis* L. in Wisconsin. *Ecology* 56: 647–655.

Mutch, R.W. (1970). Wildland fires and ecosystems—a hypothesis. *Ecology* 51: 1046–1051.

Mutch, R.W. (1976). Fire management and land use planning today: tradition and change in the Forest Service. *Western Wildlands* 3: 13–19.

Myers, R.L. (1990). Scrub and high pine. In: *Ecosystems of Florida* (ed. R.L. Myers and J.J. Ewel). Orlando: Univ. Central Florida Press.

Myers, R.L. and Ewel, J.J. (ed.) (1990). *Ecosystems of Florida*. Orlando: Univ. Central Florida Press 765 pp.

Myers, J.A. and Kitajima, K. (2007). Carbohydrate storage enhances seedling shade and stress tolerance in a neotropical forest. *J. Ecol.* 95: 383–395.

Myneni, R., Keeling, C., Tucker, C. et al. (1997). Increased plant growth in the northern high latitudes from 1981 to 1991. *Nature* 386: 698–702.

Nadelhoffer, K.J., Aber, J.D., and Melillo, J.M. (1985). Fine roots, net primary production, and soil nitrogen availability: a new hypothesis. *Ecology* 66: 1377–1390.

Naeem, S. (2009). Biodiversity, ecosystem function, and ecosystem services. In: *The Princeton Guide to Ecology* (ed. S.A. Levin, S.R. Carpenter, H.C.J. Godfray, et al.), 584–590. Princeton, NJ: Princeton University Press.

Naeem, S., Duffy, J.E., and Zavaleta, E. (2012). The functions of biological diversity in an age of extinction. *Science* 336: 1401–1406.

Naess, A. (1986). Intrinsic value: will the defenders of nature please rise? In: *Conservation Biology: The Science of Scarcity and Diversity* (ed. M.E. Soulé), 504–515. Sunderland, MA: Sinauer Associates.

Nanninga, C., Buyarski, C.R., Pretorius, A.M., and Montgomery, R.A. (2017). Increased exposure to chilling advances the time to budburst in North American tree species. *Tree Physiol.* 37: 1727–1738.

Nanson, G.C. and Croke, J.C. (1992). A genetic classification of floodplains. *Geomorphology* 4: 459–486.

National Drought Mitigation Center (2021). Maps and data. https://droughtmonitor.unl.edu. Accessed 01-21-2021.

National Oceanic and Atmospheric Administration (2021a). National Centers for Environmental Information. https://www.ncei.noaa.gov. Accessed 09-19-2021.

National Oceanic and Atmospheric Administration (2021b). U.S. Climate Extremes Index. http://www.ncdc.noaa.gov/extremes/cei. Accessed 09-19-2021.

Naveh, Z. and Lieberman, A.S. (1994). *Landscape Ecology: Theory and Application*, 2e. New York: Springer 360 pp.

Neary, D.G., Klopatek, C.C., DeBano, L.F., and Ffolliott, P.F. (1999). Fire effects on belowground sustainability: a review and synthesis. *For. Ecol. Manag.* 122: 51–71.

Neary, D.G., Ryan, K.C., and DeBano, L.F. (2005). Wildland fire in ecosystems: effects of fire on soils and water. USDA For. Serv. Gen. Tech. Rep. RMRS-GTR-42-vol. 4. Rocky Mountain Research Station, Ogden, UT. 250 pp.

Neilson, R.P. and Wullstein, L.H. (1980). Catkin freezing and acorn production in gambel oak in Utah, 1978. *Am. J. Bot.* 67: 426–428.

Neilson, R.P., Pitelka, L.F., Solomon, A.M. et al. (2005). Forecasting regional to global plant migration in response to climate change. *Bioscience* 55: 749–759.

Neiman, K.E. (1988). Soil characteristics as an aid to identifying forest habitat types in northern Idaho. USDA For. Serv. Res. Pap. INT-390. Intermountain Res. Sta., Ogden, UT. 16 pp.

Nesbitt, L., Hotte, N., Barron, S. et al. (2017). The social and economic value of cultural ecosystem services provided by urban forests in North America: A review and suggestions for future research. *Urban For. Urban Green.* 25: 103–111.

Neumayer, E. (2013). *Weak Versus Strong Sustainability: Exploring the Limits of Two Opposing Paradigms*, 4e. Cheltenham, UK: Edward Elgar Publishing 296 pp.

Newman, J.A., Anand, M., Henry, H.A.L. et al. (2011). *Climate Change Biology*. Boston, MA: CAB International 288 pp.

Nichols, G.E. (1923). A working basis for the ecological classification of plant communities. *Ecology* 4: 154–179.

Nicolai, V. (1986). The bark of trees: thermal properties, microclimate and fauna. *Oecologia* 69: 148–160.

Nicolini, E., Chanson, B., and Bonne, F. (2001). Stem growth and epicormic branch formation in understorey beech trees (*Fagus sylvatica* L.). *Ann. Bot.* 87: 737–750.

Nicoll, B.C. and Ray, D. (1996). Adaptive growth of tree root systems in response to wind action and site conditions. *Tree Physiol.* 16: 891–898.

Nicolson, M. (1990). Henry Allan Gleason and the individualistic hypothesis: the structure of a botanist's career. *Bot. Rev.* 56: 91–161.

Nielsen, U.N., Wall, D.H., and Six, J. (2015). Soil biodiversity and the environment. *Annu. Rev. Env. Resour.* 40: 63–90.

Nielsen, U.N., Wall, D.H., and Six, J. (2015). Soil biodiversity and the environment. *Annu. Rev. Environ. Resour.* 40: 63–90.

Nienstaedt, H. (1974). Genetic variation in some phenological characteristics of forest trees. In: *Phenology and Seasonality Modeling* (ed. H. Lieth), 589–400. New York: Springer-Verlag.

Niering, W.A. and Egler, F.E. (1955). A shrub community of *Viburnum lentago*, stable for twenty-five years. *Ecology* 36: 356–360.

Niering, W.A. and Goodwin, R.H. (1974). Creation of relatively stable shrublands with herbicides: arresting succession on rights-of-way and pasture-land. *Ecology* 55: 784–795.

Niinemets, Ü. and Valladares, F. (2006). Tolerance to shade, drought, and waterlogging of temperate northern hemisphere trees and shrubs. *Ecol. Monogr.* 76: 521–547.

Nikles, D.G. (1970). Breeding for growth and yield. *Unasylva* 24 (2–3): 9–22.

Nilsen, E.T. (1985). Seasonal and diurnal leaf movements of *Rhododendron maximum* L. in contrasting irradiance environments. *Oecologia* 65: 296–302.

Nilsen, E.T. (1987). Influence of water relations and temperature on leaf movements of Rhododendron species. *Plant Physiology* 83: 607–612.

Nilsen, E.T. (1992). Thermonastic leaf movements: a synthesis of research with Rhododendron. *Bot. J. Linn. Soc.* 110: 205–233.

Nilsson, S.G. and Wästljung, U. (1987). Seed predation and cross-pollination in mast-seeding beech (Fagus sylvatica) patches. *Ecology* 68: 260–265.

Nobel, P.S. (2009). *Physicochemical and Biophysical Plant Physiology*, 4e. New York: Academic Press 604 pp.

Noble, I.R. and Slatyer, R.O. (1977). Post-fire succession of plants in Mediterranean ecosystems. *In* H.A. Mooney and C. E. Conrad (eds.), Symp. Proc. *Environmental Consequences of Fire and Fire Management in Mediterranean Ecosystems.* USDA For. Serv. Gen. Tech. Report WO-3. Washington, D.C.

Noble, I.R. and Slatyer, R.O. (1980). The use of vital attributes to predict successional changes in plant communities subject to recurrent disturbances. *Vegetatio* 43: 5–21.

Norby, R.J. and Zak, D.R. (2011). Ecological lessons from free-air CO_2 enrichment (FACE) experiments. *Annu. Rev. Ecol. Evol. Syst.* 42: 181–203.

Norby, R.J., Warren, J.M., Iversen, C.M. et al. (2010). CO_2 enhancement of forest productivity constrained by limited nitrogen availability. *Proc. Natl. Acad. Sci.* 107: 19368–19373.

Norman, S.A., Schaetzl, R.J., and Small, T.W. (1995). Effects of slope angle on mass movement by tree uprooting. *Geomorphology* 14: 19–27.

North, M.P., Stephens, S.L., Collins, B.M. et al. (2015). Reform forest fire management. *Science* 349: 1280–1281.

Norton, B.G. (1982). Environmental ethics and nonhuman rights. *Environ. Ethics* 4: 17–36.

Norton, D.A. and Kelly, D. (1988). Mast seeding over 33 years by *Dacrydium cupressinum* lamb, (rimu) (Podocarpaceae) in New Zealand: the importance of economies of scale. *Funct. Ecol.* 2: 399–408.

Noss, R. F., LaRoe, E. T., III, and Scott, J. M. (1995). Endangered ecosystems of the United States: a preliminary assessment of loss and degradation. USDI, Nat. Biol. Serv., Biol. Report 28. Washington D.C. 58 pp.

Nowacki, G.J. and Abrams, M.D. (1991). Community and edaphic analysis of mixed oak forests in the ridge and valley province of central Pennsylvania. *In* L.H. McCormick and K.W. Gottschalk (eds.), *Proc. Eighth Central Hardwood Conference.* USDA, For. Serv. Gen. Tech. Report NE-148. Northeastern For. Exp. Sta., Broomall, PA.

Nowacki, G.J. and Abrams, M.D. (2008). The demise of fire and "mesophication" of forests in the eastern United States. *Bioscience* 58: 123–138.

Nowacki, G.J., Abrams, M.D., and Lorimer, C.G. (1990). Composition, structure, and historical development of northern red oak stands along an edaphic gradient in north-Central Wisconsin. *For. Sci.* 36: 276–292.

Noy-Meir, I. and van der Maarel, E. (1987). Relations between community theory and community analysis in vegetation science: some historical perspectives. *Vegetatio* 69: 5–15.

Nybom, H. (2004). Comparison of different nuclear DNA markers for estimating intraspecific genetic diversity in plants. *Mol. Ecol.* 13: 1143–1155.

Nye, P.H. (1977). The rate-limiting step in plant nutrient absorption from soil. *Soil Sci.* 123: 292–297.

Nye, P.H. and Tinker, P.B. (1977). *Solute Movement in the Soil–Root System.* Oxford: Blackwell 342 pp.

Nyland, R.D., Kenefic, L.S., Bohn, K.K., and Stout, S.L. (2016). *Silviculture: Concepts and Applications*, 3e. Long Grove, IL: Waveland Press 680 pp.

Oberdorfer, E. (1990). *Pflanzensoziologische Exkursionsflora.* Stuttgart: Verlag Eugen Ulmer 1050 pp.

Obeso, J.R. (2002). The costs of reproduction in plants. *New Phytol.* 155: 321–348.

O'Dea, M.E., Zasada, J.C., and Tappeiner, J.C. III (1995). Vine maple clone growth and reproduction in managed and unmanaged coastal Oregon Douglas--fir forests. *Ecol. Appl.* 5: 63–73.

Oehri, J., Schmid, B., Schaepman-Strub, G., and Niklaus, P.A. (2017). Biodiversity promotes primary productivity and growing season lengthening at the landscape scale. *Proc. Natl. Acad. Sci.* 114: 10160–10165.

Ohmart, R.D. and Anderson, B.W. (1982). North American desert riparian ecosystems. In: *Reference Handbook on the Deserts of North America* (ed. G.L. Bender), 433–479. Westport, CT: Greenwood Press.

Oke, T.R. (1987). *Boundary Layer Climates*, 2e. Oxfordshire, England, UK: Routledge Publishing 464 pp.

Oldeman, R.A.A. (1990). *Forests: Elements of Silvology*. New York: Springer-Verlag 624 pp.

Oldemeyer, J.L., Franzmann, A.W., Brundage, A.L. et al. (1977). Browse quality and the Kenai moose population. *J. Wildl. Manag.* 41: 533–542.

Oliver, C.D. (1981). Forest development in North America following major disturbances. *For. Ecol. Manag.* 3: 153–168.

Oliver, C.D. and Larson, B.C. (1996). *Forest Stand Dynamics: Updated Edition*. New York: Wiley 519 pp.

Öllerer, K., Varga, A., Kirby, K. et al. (2019). Beyond the obvious impact of domestic livestock grazing on temperate forest vegetation–A global review. *Biol. Conserv.* 237: 209–219.

Ollinger, S.V., Aber, J.D., Lovett, G.M. et al. (1993). A spatial model of atmospheric deposition for the northeastern U.S. *Ecol. Appl.* 3: 459–472.

Olson, J.S. (1958). Rates of succession and soil changes on southern Lake Michigan sand dunes. *Bot. Gaz.* 119: 125–170.

Olson, J.S. (1963). Energy storage and the balance of producers and decomposers in ecological systems. *Ecology* 44: 322–331.

Olson, S.R., Reiners, W.A., Cronan, C.S., and Lang, G.E. (1981). The chemistry and flux of through-fall and stemflow in subalpine balsam fir forests. *Holarct. Ecol.* 4: 291–300.

Omernik, J.M. (1987). Ecoregions of the conterminous United States. *Ann. Assoc. Am. Geogr.* 77: 118–125.

Omernik, J.M. (1995). Ecoregions: A spatial framework for environmental management. In: *Biological Assessment and Criteria: Tools for Water Resource Planning and Decision Making* (ed. W.S. Davis and T.P. Simon). Boca Raton, FL: Lewis Publishers 432 pp.

Omernik, J.M. (2004). Perspectives on the nature and definition of ecological regions. *Environ. Manag.* 34 (Supplement 1): S27–S38.

Omernik, J.M. and Bailey, R.G. (1997). Distinguishing between watersheds and ecoregions. *J. Am. Water Res. Assoc.* 33: 935–949.

Omernik, J.M. and Griffith, G.E. (2014). Ecoregions of the conterminous United States: evolution of a hierarchical spatial framework. *Environ. Manag.* 54: 1249–1266.

O'Neill, R.V. and DeAngelis, D.L. (1981). Comparative productivity and biomass relations of forest ecosystems. *In* D.E. Reichle (ed.), *Dynamic Properties of Forest Ecosystems*, Int. Biol. Programme 23. Cambridge Univ. Press, Cambridge.

Ontl, T.A., Janowiak, M.K., Swanston, C.W. et al. (2020). Forest management for carbon sequestration and climate adaptation. *J. For.* 118: 86–101.

Oosting, H.J. (1956). *The Study of Plant Communities*, 2e. San Francisco CA: Freeman 440 pp.

Opdam, P. and Wascher, D. (2004). Climate change meets habitat fragmentation: linking landscape and biogeographical scale levels in research and conservation. *Biol. Conserv.* 117: 285–297.

Ordonez, A. (2014). Functional and phylogenetic similarity of alien plants to cooccurring natives. *Ecology* 95: 1191–1202.

Orodho, A.B., Trlica, M.J., and Bonham, C.D. (1990). Long-term heavy-grazing effects on soil and vegetation in the four corners region. *Southwest. Nat.* 35: 9–14.

Orwig, D.A., Cobb, R.C., D'Amato, A.W. et al. (2008). Multi-year ecosystem response to hemlock woolly adelgid infestation in southern New England forests. *Can. J. For. Res.* 38: 834–843.

Osawa, A. and Allen, R.B. (1993). Allometric theory explains self-thinning relationships of mountain beech and red pine. *Ecology* 74: 1020–1032.

Osterkamp, W.R. and Hupp, E.R. (1984). Geomorphic and vegetative characteristics along three northern Virginia streams. *Bull. Geol. Soc. Am.* 95: 1093–1101.

Ostfeld, R.S. and Keesing, F. (2000). Biodiversity and disease risk: the case of Lyme disease. *Conserv. Biol.* 14: 722–728.

Oswalt, S., Oswalt, C., Crall, A. et al. (2021). Inventory and monitoring of invasive species. In: *Invasive Species in Forests and Rangelands of the United States* (ed. T.M. Poland, T. Patel-Weynand, D.M. Finch, et al.), 231–242. New York: Springer.

Otto, S.P. and Whitton, J. (2000). Polyploid incidence and evolution. *Annu. Rev. Genet.* 34: 401–437.

Ouarmim, S., Asselin, H., Hely, C. et al. (2014). Long-term dynamics of fire refuges in boreal mixedwood forests. *J. Quat. Sci.* 29: 123–129.

Ovington, J.D. (ed.) (1983). *Ecosystems of the World, Temperate Broad-Leaved Evergreen Forests, Vol. 10*. New York: Elsevier 242 pp.

Owens, J.N. (1991). Flowering and seed set. In: *Physiology of Trees* (ed. A.S. Raghavendra), 247–271. New York: Wiley.

Owens, J.N., Molder, M., and Langer, H. (1977). Bud development in *Picea glauca*. I. Annual growth cycle of vegetative buds and shoot elongation as they relate to date and temperature sums. *Can. J. Bot.* 55: 2728–2745.

Pabst, R.J. and Spies, T.A. (2001). Ten years of vegetation succession on a debris-flow deposit in Oregon. *J. Am. Water Res. Assoc.* 37: 1693–1708.

Paillet, F.L. (2002). Chestnut: history and ecology of a transformed species. *J. Biogeogr.* 29: 1517–1530.

Paine, R.T. (1969). A note on trophic complexity and community stability. *Am. Nat.* 103: 91–93.

Paine, R.T., Tegner, M.J., and Johnson, E.A. (1998). Compounded perturbations yield ecological surprises. *Ecosystems* 1: 535–545.

Palik, B.J. and Pregitzer, K.S. (1994). White pine seed-tree legacies in an aspen landscape: influences on post-disturbance white pine population structure. *For. Ecol. Manag.* 67: 191–201.

Palik, B.J., Goebel, P.C., Kirkman, L.K., and West, L. (2000). Using landscape hierarchies to guide restoration of disturbed ecosystems. *Ecol. Appl.* 10: 89–202.

Palik, B.J., D'Amato, A.W., Franklin, J.F., and Johnson, K.N. (2021). *Ecological Silviculture.* Long Grove, IL: Waveland Press 343 pp.

Palik, B.J., D'Amato, A.W., and Slesak, R.A. (2021). Wide-spread vulnerability of black ash (*Fraxinus nigra* marsh.) wetlands in Minnesota USA to loss of tree dominance from invasive emerald ash borer. *Forestry* 94: 455–463.

Pallardy, S.G. (2008). *Physiology of Woody Plants*, 3e. Burlington, MA: Academic Press 454 pp.

Pallardy, S.G., Nigh, T.A., and Garrett, H.E. (1988). Changes in forest composition in central Missouri: 1968–1982. *Am. Midi Nat.* 120: 380–389.

Pan, Y., Birdsey, R.A., Fang, J. et al. (2011). A large and persistent carbon sink in the world's forests. *Science* 33: 988–993.

Paquette, A. and Messier, C. (2010). The role of plantations in managing the world's forests in the Anthropocene. *Front. Ecol. Environ.* 8: 27–34.

Paquette, A. and Messier, C. (2011). The effect of biodiversity on tree productivity: from temperate to boreal forests. *Glob. Ecol. Biogeogr.* 20: 170–180.

Paradis, A., Elkinton, J., Hayhoe, K., and Buonaccorsi, J. (2008). Role of winter temperature and climate change on the survival and future range expansion of the hemlock woolly adelgid (*Adelges tsugae*) in eastern North America. *Mitig. Adapt. Strateg. Glob. Chang.* 13: 541–554.

Park, Y.S. and Fowler, D.P. (1988). Geographic variation of black spruce tested in the maritimes. *Can. J. For. Res.* 18: 106–114.

Park, A. and Talbot, C. (2018). Information underload: ecological complexity, incomplete knowledge, and data deficits create challenges for the assisted migration of forest trees. *Bioscience* 68: 251–263.

Park, A., Puettmann, K., Wilson, E. et al. (2014). Can boreal and temperate forest management be adapted to the uncertain ties of 21st century climate change? *Crit. Rev. Plant Sci.* 33: 251–285.

Parker, A.J. and Parker, K.C. (1994). Structural variability of mature lodgepole pine stands on gently sloping terrain in Taylor Park basin, Colorado. *Can. J. For. Res.* 24: 2020–2029.

Parker, I.M., Simberloff, D., Lonsdale, W.M. et al. (1999). Impact: toward a framework for understanding the ecological effects of invaders. *Biol. Invasions* 1: 3–19.

Parks, S.A., Miller, C., Nelson, C.R., and Holden, Z.A. (2014). Previous fires moderate burn severity of subsequent wildland fires in two large western US wilderness areas. *Ecosystems* 17: 29–42.

Parsons, W.F., Knight, D.H., and Miller, S.L. (1994). Root gap dynamics in lodgepole pine forest: nitrogen transformations in gaps of different size. *Ecol. Appl.* 4: 354–362.

Pascual, J., Cañal, M.J., Correia, B. et al. (2014). Can epigenetics help forest plants to adapt to climate change? In: *Epigenetics in Plants of Agronomic Importance: Fundamentals and Applications* (ed. R. Alvarez-Venegas, C. De la Peña and J.A. Casas-Mollano), 125–146. New York: Springer.

Pastor, J. and Post, W.M. (1986). Influence of climate, soil moisture, and succession on forest carbon and nitrogen cycles. *Biogeochemistry* 2: 3–27.

Pastor, J., Aber, J.D., McClaugherty, C.A., and Melillo, J.M. (1984). Aboveground production and N and P cycling along a nitrogen mineralization gradient on Blackhawk Island, Wisconsin. *Ecology* 65: 256–268.

Patric, J. H. and Black, P.E. (1968). Potential evapotranspiration and climate in Alaska by Thornthwaite's classification. USDA For. Serv. Res. Paper PNW-71. Pacific Northwest For. and Rge. Exp. Sta., Portland, OR. 28 pp.

Patric, J.H. and Helvey, J.D. (1986). Some effects of grazing on soil and water in the eastern forest. USDA For. Serv. Gen. Tech. Report NE-115, Northeastern For. Exp. Sta., Broomall, PA. 24 pp.

Patterson, W.A. III and Backman, A.E. (1988). Fire and disease history of forests. In: *Vegetation History* (ed. B. Huntley and T. Webb III), 603–632. Dordrecht: Kluwer.

Paul, E.A. and Clark, F.E. (1996). *Soil Microbiology and Biochemistry*, 2e. New York: Academic Press 340 pp.

Paul, C., Hanley, N., Meyer, S.T. et al. (2020). On the functional relationship between biodiversity and economic value. *Sci. Adv.* 6: eaax7712.

Pauley, S.S. (1958). Photoperiodism in relation to tree improvement. In: *The Physiology of Forest Trees* (ed. K.V. Thimann), 557–571. New York: Ronald Press.

Pauley, S.S. and Perry, T.O. (1954). Ecotypic variation of the photoperiodic response in *Populus*. *J. Arnold Arbor.* 35: 167–188.

Pausas, J.G. and Ribeiro, E. (2013). The global fire-productivity relationship. *Glob. Ecol. Biogeogr.* 22: 728–736.

Pavlik, B.M., Muick, P.C., Johnson, S.G., and Popper, M. (1991). *Oaks of California*. Los Olivos, CA: Cachuna Press 184 pp.

Pearce, F. (1997). Lightning sparks pollution rethink. *New Scientist* 153(2066): 15.

Pearcy, R.W., Chazdon, R.L., Gross, L.J. et al. (1994). Photosynthetic utilization of sunflecks: a temporally patchy resource on a time scale of seconds to minutes. In: *Exploitation of Environmental Heterogeneity by Plants*. (ed. M.M.Caldwell and R.W. Pearcy), 175–208, San Diego: Academic Press.

Pearsall, D.R. (1995). *Landscape Ecosystems of the University of Michigan Biological Station: Ecosystem Diversity and Ground-Cover Diversity*. Ph.D. Thesis. University of Michigan, Ann Arbor. 396 pp.

Pearse, I.S., Koenig, W.D., and Kelly, D. (2016). Mechanisms of mast seeding: resources, weather, cues, and selection. *New Phytol.* 212: 546–562.

Pearson, G.A. (1936). Some observations on the reaction of pine seedlings to shade. *Ecology* 17: 270–276.

Pearson, G.A. (1940). Shade effects in ponderosa pine. *J. For.* 38: 778–780.

Pec, G.J., Karst, J., Sywenky, A.N. et al. (2015). Rapid increases in forest understory diversity and productivity following a mountain pine beetle (*Dendroctonus ponderosae*) outbreak in pine forests. *PLoS One* 10: e0124691.

Pedlar, J.H., McKenney, D.W., Beaulieu, J. et al. (2011). The implementation of assisted migration in Canadian forests. *For. Chron.* 87: 766–777.

Pedlar, J.H., McKenney, D.W., Aubin, I. et al. (2012). Placing forestry in the assisted migration debate. *Bioscience* 62: 835–842.

Peet, R.K. (1974). The measurement of species diversity. *Annu. Rev. Ecol. Syst.* 5: 285–307.

Peet, R.K. (1978). Forest vegetation of the Colorado front range: patterns of species diversity. *Vegetatio* 37: 65–78.

Peet, R.K. (1981). Forest vegetation of the Colorado front range: composition and dynamics. *Vegetatio* 45: 3–75.

Peet, R.K. (1992). Community structure and ecosystem function. In: *Plant Succession: Theory and Prediction* (ed. D.C. Glenn-Lewin, R.K. Peet and T.T. Veblen), 103–151. London: Chapman and Hall.

Peet, R.K. and Christensen, N.L. (1980a). Succession: a population process. *Vegetatio* 43: 131–140.

Peet, R.K., and Christensen, N.L. (1980b). Hardwood forest vegetation of the North Carolina piedmont. *Veröff. Geobot. Inst. Eddg. Tech. Hochsch.*, Stift. Rübel, Zürich 69:14–39.

Peet, R.K. and Christensen, N.L. (1987). Competition and tree death. *Bioscience* 37: 586–595.

Pellitier, P.T. and Zak, D.R. (2018). Ectomycorrhizal fungi and the liberation of nitrogen from soil organic matter. *New Phytol.* 217: 68–73.

Pellitier, P.T., Zak, D.R., Argiroff, W.A., and Upchurch, R.A. (2021). Coupled shifts in ectomycorrhizal communities and plant uptake of organic nitrogen along a soil gradient: an isotopic perspective. *Ecosystems* 24: 1976–1990.

Penning de Vries, F.W.T. (1975). The cost of maintenance processes in plant cells. *Ann. Bot.* 39: 77–92.

Peñuelas, J., Prieto, P., Beier, C. et al. (2007). Response of plant species richness and primary productivity in shrublands along a north–south gradient in Europe to seven years of experimental warming and drought: reductions in primary productivity in the heat and drought year of 2003. *Glob. Chang. Biol.* 13: 2563–2581.

Perlin, J. (1989). *A Forest Journey, the Role of Wood in the Development of Civilization*. Cambridge, MA: Harvard Univ. Press 445 pp.

Perrings, C. (2007). Future challenges. *Proc. Natl. Acad. Sci.* 104: 15179–15180.

Perry, P.O., Sellers, H.E., and Blanchard, C.O. (1969). Estimation of photosynthetically active radiation under a forest canopy with chlorophyll extracts and from basal area measurements. *Ecology* 50: 39–44.

Perry, T.O. (1962). Racial variation in the day and night temperature requirements or red maple and loblolly pine. *For. Sci.* 8: 336–344.

Perry, T.O. (1971). Dormancy of trees in winter. *Science* 171: 29–36.

Perry, T.O. and Wang, C.W. (1960). Genetic variation in the winter chilling requirement for date of dormancy break for *Acer rubrum*. *Ecology* 41: 790–794.

Perry, D.A., Hessburg, P.F., Skinner, C.N. et al. (2011). The ecology of mixed severity fire regimes in Washington, Oregon, and northern California. *For. Ecol. Manag.* 262: 703–717.

Pesendorfer, M.B., Sillett, T.S., Koenig, W.D., and Morrison, S.A. (2016). Scatter-hoarding corvids as seed dispersers for oaks and pines: a review of a widely distributed mutualism and its utility to habitat restoration. *Condor: Ornithol. Appl.* 118: 215–237.

Peters, M.P., Prasad, A.M., Matthews, S.N., and Iverson, L.R. (2020). Climate change tree atlas, Version 4. U.S. Forest Service, Northern Research Station and Northern Institute of Applied Climate Science, Delaware, OH. https://www.nrs.fs.fed.us/atlas. Accessed 01-17-2022.

Peterson, C.J. (2000). Catastrophic wind damage to North American forests and the potential impact of climate change. *Sci. Total Environ.* 262: 287–311.

Peterson, C.J. and Pickett, S.T.A. (1995). Forest reorganization: a case study in an old-growth forest catastrophic blowdown. *Ecology* 76: 763–774.

Peterson, C.J., Carson, W.P., McCarthy, B.C., and Pickett, S.T.A. (1990). Microsite variation and soil dynamics within newly created treefall pits and mounds. *Oikos* 58: 39–46.

Peterson, B.J., Wollheim, W.M., Mulholland, P.J. et al. (2001). Control of nitrogen export from watersheds by headwater streams. *Science* 292: 86–90.

Petit, R. and Hampe, A. (2006). Some evolutionary consequences of being a tree. *Annu. Rev. Ecol. Evol. Syst.* 37: 187–214.

Petit, J.R., Jouzel, J., Raynaud, D. et al. (1999). Climate and atmospheric history of the past 420,000 years from the Vostok ice core, Antarctica. *Nature* 399: 429–436.

Petit, R.J., Bialozyt, R., Garnier-Géré, P., and P, and A. Hampe. (2004). Ecology and genetics of tree invasions: from recent introductions to quaternary migrations. *For. Ecol. Manag.* 197: 117–137.

Petrie, M.D., Bradford, J.B., Hubbard, R.M. et al. (2017). Climate change may restrict dryland forest regeneration in the 21st century. *Ecology* 98: 1548–1559.

Petty, R.O., and Jackson, M.T. (1966). Plant communities. In: *Natural Features of Indiana*. (ed. A.A. Lindsey), 264–296, Indianapolis: Ind. Acad. Sci.

Pfister, R.D. and Arno, S.F. (1980). Classifying forest habitats based on potential climax vegetation. *For. Sci.* 26: 52–70.

Pfister, R., Kovalchik, B. L., Arno, S. F., and Presby, R. C. (1977). Forest habitat types of Montana. USDA For. Serv. Gen. Tech. Report INT-34. Intermountain For. and Rge. Exp. Sta., Ogden, UT. 174 pp.

Phalan, B., Balmford, A., Green, R.E., and Scharlemann, J.P.W. (2011). Minimising the harm to biodiversity of producing more food globally. *Food Policy* 36: S62–S71.

Phillips, J. (1931). The biotic community. *J. Ecol.* 19: 1–24.

Phillips, J. (1934-35). Succession, development, the climax, and the complex organism; an analysis of concepts. *J. Ecol.* 22/23: 554–571. /210–246.

Phillips, W.S. (1963). Depth of roots in soil. *Ecology* 44: 424.

Piao, S., Liu, Q., Chen, A. et al. (2019). Plant phenology and global climate change: current progresses and challenges. *Glob. Chang. Biol.* 25: 1922–1940.

Pickett, S.T.A. (1988). Space-for-time substitution as an alternative to long-term studies. In: *Long-Term Studies in Ecology: Approaches and Alternatives* (ed. G.E. Likens), 110–135. New York: Springer-Verlag.

Pickett, S.T.A. and White, P.S. (ed.) (1985). *The Ecology of Natural Disturbance and Patch Dynamics*. New York: Academic Press 472 pp.

Pickett, S.T.A., Collins, S.L., and Armesto, J.J. (1987). Models, mechanisms and pathways of succession. *Bot. Rev.* 53: 335–371.

Pigott, C.D. and Huntley, J.P. (1981). Factors controlling the distribution of *Tilia cordata* at the northern limits of its geographical range. III. Nature and causes of seed sterility. *New Phytol.* 87: 817–839.

Pimentel, D., Zuniga, R., and Morrison, D. (2005). Update on the environmental and economic costs associated with alien-invasive species in the United States. *Ecol. Econ.* 52: 273–288.

Pimm, S.L., Davis, G.E., Loope, L. et al. (1994). Hurricane Andrew. *Bioscience* 44: 224–229.

Pinter, N., Fiedel, S., and Keeley, J.E. (2011). Fire and vegetation shifts in the Americas at the vanguard of Paleoindian migration. *Quat. Sci. Rev.* 30: 269–272.

Piotto, D. (2008). A meta-analysis comparing tree growth in monocultures and mixed plantations. *For. Ecol. Manag.* 255: 781–786.

Piovesan, G. and Adams, J.M. (2001). Masting behaviour in beech: linking reproduction and climatic variation. *Can. J. Bot.* 79: 1039–1047.

Platnik, N.I. (1991). Patterns of biodiversity: tropical vs. temperate. *J. Nat. Hist.* 25: 1083–1088.

Platnik, N.I. (1992). Patterns of biodiversity. In: *Systematics, Ecology, and the Biodiversity Crisis* (ed. N. Eldregde), 15–24. New York: Columbia University Press.

Platt, W.J. (1975). The colonization and formation of equilibrium plant species associations on badger disturbances in a tall-grass prairie. *Ecol. Monogr.* 45: 285–305.

Platt, W.J. and Gottschalk, R.M. (2001). Effects of exotic grasses on potential fine fuel loads in the groundcover of south Florida slash pine savannas. *Int. J. Wildland Fire* 10: 155–159.

Plotkin, A.B., Schoonmaker, P., Leon, B., and Foster, D. (2017). Microtopography and ecology of pit-mound structures in second-growth versus old-growth forests. *For. Ecol. Manag.* 404: 14–23.

Podani, J. (2006). Braun-Blanquet's legacy and data analysis in vegetation science. *J. Veg. Sci.* 17: 113–117.

Poland, T.M., Patel-Weynand, T., Finch, D.M. et al. (ed.) (2021). *Invasive Species in Forests and Rangelands of the United States: A Comprehensive Science Synthesis for the United States Forest Sector*. New York: Springer 500 pp.

Poore, M.E.D. (1955). The use of phytosociological methods in ecological investigations. *J. Ecol.* 43: 226–269.

Porte, A., Huard, F., and Dreyfus, P. (2004). Microclimate beneath pine plantation, semi-mature pine plantation and mixed broadleaved-pine forest. *Agric. For. Meteorol.* 126: 175–182.

Post, E. (2013). *Ecology of Climate Change*. Princeton, NJ: Princeton University Press 408 pp.

Pott, R. (2011). Phytosociology: A modern geobotanical method. *Plant Biosyst.* 145: 9–18.

Potts, A.S. and Hunter, M.D. (2021). Unraveling the roles of genotype and environment in the expression of plant defense phenotypes. *Ecol. Evol.* 11: 8542–8561.

Poulson, T.L. and Platt, W.J. (1996). Replacement patterns of beech and sugar maple in Warren Woods, Michigan. *Ecology* 77: 1234–1253.

Poulson, T.L. and White, W.B. (1969). The cave environment. *Science* 165: 971–981.

Povak, N.A., Hessburg, P.F., and Salter, R.B. (2018). Evidence for scale-dependent topographic controls on wildfire spread. *Ecosphere* 9: e02443.

Power, M.E., Tilman, D., Estes, J.A. et al. (1996). Challenges in the quest for keystones. *Bioscience* 46: 609–620.

Prach, K. and Walker, L.R. (2020). *Comparative Plant Succession among Terrestrial Biomes of the World*. Cambridge, UK: Cambridge University Press 412 pp.

Prasad, A., Pedlar, J., Peters, M. et al. (2020). Combining US and Canadian forest inventories to assess habitat suitability and migration potential of 25 tree species under climate change. *Divers. Distrib.* 26: 1142–1159.

Pratt, R.B., Jacobsen, A.L., Ewers, F.W., and Davis, S.D. (2007). Relationships among xylem transport, biomechanics and storage in stems and roots of nine Rhamnaceae species of the California chaparral. *New Phytol.* 174: 787–798.

Pregitzer, K.S. and Barnes, B.V. (1984). Classification and comparison of the upland hardwood and conifer ecosystems of the Cyrus H. McCormick experimental Forest. Upper peninsula, Michigan. *Can. J. For. Res.* 14: 362–375.

Pregitzer, K.S. and Friend, A.L. (1996). The structure and function of *Populus* root systems. In: *Biology of Populus and its Implications for Management and Conservation*. Nat. Res (ed. R.F. Stettler, H.D. Bradshaw Jr., P.E. Heilman and T.M. Hinckley), 331–354. Ottawa: Council Canada Press.

Pregitzer, K.S., Barnes, B.V., and Lemme, G.E. (1983). Relationship of topography to soils and vegetation in an upper Michigan ecosystem. *Soil Sci. Soc. Am. J.* 47: 117–123.

Pregitzer, K.S., Hendrick, R.L., and Fogel, R. (1992). The demography of fine roots in responses to patches of water and nitrogen. *New Phytol.* 125: 575–580.

Pregitzer, K.S., Zak, D.R., Curtis, P.S. et al. (1995). Atmospheric CO_2, soil nitrogen, and turnover of fine roots. *New Phytol.* 129: 579–585.

Pregitzer, K.S., DeForest, J.L., Burton, A.J. et al. (2002). Fine root architecture of nine North American trees. *Ecol. Monogr.* 72: 293–309.

Pregitzer, C.C., Bailey, J.K., Schweitzer, J.A., and J. A. (2013). Genetic by environment interactions affect plant-soil linkages. *Ecol. Evol.* 3: 2322–2333.

Pretzsch, H. (2014). Canopy space filling and tree crown morphology in mixed-species stands compared with monocultures. *For. Ecol. Manag.* 327: 251–264.

Prevéy, J.S., Harrington, C.A., and Clair, J.B.S. (2018). The timing of flowering in Douglas-fir is determined by cool-season temperatures and genetic variation. *For. Ecol. Manag.* 409: 729–739.

Prieto, I., Almagro, M., Bastida, F., and Querejeta, J.I. (2019). Altered leaf litter quality exacerbates the negative impact of climate change on decomposition. *J. Ecol.* 107: 2364–2382.

Primack, R.B. (2010). *Essentials of Conservation Biology*, 5e. Sunderland, MA: Sinauer Associates 601 pp.

Pritchett, W.L. and Smith, W.H. (1970). Fertilizing slash pine on sandy soils of the lower coastal plain. In: *Tree Growth and Forest Soils* (ed. C.T. Youngberg and C.B. Davey). Corvallis: Oregon State Univ. Press.

Puettmann, K.J., Coates, K.D., and Messier, C. (2008). *A Critique of Silviculture: Managing for Complexity*. Washington, DC: Island Press 206 pp.

Pyne, S.J. (1982). *Fire in America*. Princeton, NJ: Princeton Univ. Press 654 pp.

Pyne, S.J. (2019). *Fire: A Brief History*, 2e. Seattle: University of Washington Press 240 pp.

Pyšek, P., Jarošík, V., Hulme, P.E. et al. (2012). A global assessment of invasive plant impacts on resident species, communities and ecosystems: the interaction of impact measures, invading species' traits and environment. *Glob. Chang. Biol.* 18: 1725–1737.

Quamme, H.A. (1985). Avoidance of freezing injury in woody plants by deep supercooling. *Acta Hortic.* 168: 11–30.

Quick, B.E. (1923). A comparative study of the distribution of the climax association in southern Michigan. *Pap. Mich. Acad. Sci. Arts Lett.* 3: 211–244.

Quijas, S., Schmid, B., and Balvanera, P. (2010). Plant diversity enhances provision of ecosystem services: A new synthesis. *Basic Appl. Ecol.* 11: 582–593.

Quine, C., Coutts, M., Gardiner, B., and Pyatt, G. (1995). Forests and wind: management to minimize damage. *For. Comm. Bull.* 114. HMSO, London. 27 pp.

Radeloff, V.C., Mladenoff, D.J., Guries, R.P., and Boyce, M.S. (2004). Spatial patterns of cone serotiny in *Pinus banksiana* in relation to fire disturbance. *For. Ecol. Manag.* 189: 133–141.

Radeloff, V.C., Hammer, R.B., and Stewart, S.I. (2005). Rural and suburban sprawl in the US Midwest from 1940 to 2000 and its relation to forest fragmentation. *Conserv. Biol.* 19: 793–805.

Raffa, K.F., Aukema, B.H., Bentz, B.J. et al. (2008). Cross-scale drivers of natural disturbances prone to anthropogenic amplification: the dynamics of bark beetle eruptions. *Bioscience* 58: 501–517.

Raghavendra, A.S. (ed.) (1991). *Physiology of Trees*. New York: Wiley 509 pp.

Raich, J.W. and Nadelhoffer, K.J. (1989). Belowground carbon allocation in forest ecosystems: global trends. *Ecology* 70: 1346–1354.

Ralph, C.J. (1985). Habitat association patterns of forest and steppe birds of northern Patagonia, Argentina. *Condor* 87: 471–483.

Ralston, C.W. (1964). Evaluation of forest site productivity. *Int. Rev. For. Res.* 1: 171–201.

Rammer, W., Braziunas, K.H., Hansen, W.D. et al. (2021). Widespread regeneration failure in forests of greater Yellowstone under scenarios of future climate and fire. *Glob. Chang. Biol.* 27: 4339–4351.

Ratnam, W., Rajora, O.P., Finkeldey, R. et al. (2014). Genetic effects of forest management practices: global synthesis and perspectives. *For. Ecol. Manag.* 333: 52–65.

Raup, D.M. (1986). Biological extinction in earth history. *Science* 231: 1528–1533.

Read, R.A. (1952). Tree species occurrences as influenced by geology and soil on an Ozark north slope. *Ecology* 33: 239–246.

Read, R.A. (1980). Genetic variation in seedling progeny of ponderosa pine provenances. *For. Sci. Monogr.* 23: 59 pp.

Regan, T. (1981). The nature and possibility of an environmental ethic. *Environ. Ethics* 3: 19–34.

Rehfeldt, G.E. (1986a). Adaptive variation in *Pinus ponderosa* from intermountain regions, I. Snake and Salmon River basins. *For. Sci.* 32: 79–92.

Rehfeldt, G.E. (1986b). Adaptive variation in *Pinus ponderosa* from intermountain regions. II. Middle Columbia River system. USDA For. Serv. Res. Paper INT-373. Intermountain For. Res., Sta., Odgen, UT. 9 pp.

Rehfeldt, G.E. (1988). Ecological genetics *of Pinus contorta* from the Rocky Mountains (USA): a synthesis. *Silvae Genet.* 37: 131–135.

Rehfeldt, G.E. (1989). Ecological adaptation in Douglas-fir (*Pseudotsuga menziesii* var. *glauca*): a synthesis. *For. Ecol. Manag.* 28: 203–215.

Rehfeldt, G.E. (1990). Genetic differentiation among populations of *Pinus ponderosa* from the upper Colorado River basin. *Bot. Gaz.* 151: 125–137.

Rehfeldt, G.E., Stage, A.R., and Bingham, R.T. (1971). Strobili development in western white pine: periodicity, prediction, and association with weather. *For. Sci.* 17: 454–461.

Reich, P.B., Uhl, C., Walters, M.B., and Ellsworth, D.S. (1991). Leaf life-span as a determinant of leaf structure and function among 23 species in Amazonian forest communities. *Oecologia* 86: 16–24.

Reich, P.B., Walters, M.B., and Ellsworth, D.S. (1992). Leaf life-span in relation to leaf, plant and stand characteristics among diverse ecosystems. *Ecol. Monogr.* 62: 365–392.

Reich, P.B., Walters, M.B., Kloeppel, B.D., and Ellsworth, D.S. (1995). Different photosynthesis-nitrogen relations in deciduous hardwood and evergreen coniferous tree species. *Oecologia* 104: 24–30.

Reich, P.B., Grigal, D.F., Aber, J.D., and Gower, S.T. (1997). Nitrogen mineralization and productivity in 50 hardwood and conifer stands on diverse soils. *Ecology* 78: 335–347.

Reich, P.B., Ellsworth, D.S., Walters, M.B. et al. (1999). Generality of leaf trait relationships: A test across six biomes. *Ecology* 80: 1955–1969.

Reich, P.B., Wright, I.J., Cavender-Bares, J. et al. (2003). The evolution of plant functional variation: traits, spectra, and strategies. *Int. J. Plant Sci.* 164: S143–S164.

Reich, P.B., Hobbie, S.E., Lee, T. et al. (2006). Nitrogen limitation constrains sustainability of ecosystem response to CO_2. *Nature* 440: 922–925.

Reichle, D.E. (ed.) (1981). *Dynamic Properties of Forest Ecosystems*. International Biosphere Programme 23. Cambridge: Cambridge Univ. Press 683 pp.

Reinartz, J.A. and Popp, J.W. (1987). Structure of clones of northern prickly-ash (*Xanthoxylum americanum*). *Am. J. Bot.* 74: 415–428.

Reineke, L.H. (1933). Perfecting a stand-density index for even-aged forests. *J. Agric. Res.* 46: 627–638.

Reiners, W.A. and Lang, G.E. (1979). Vegetational patterns and processes in the balsam fir zone, White Mountains, New Hampshire. *Ecology* 60: 403–417.

Rejmánek, M. and Richardson, D.M. (1996). What attributes make some plant species more invasive? *Ecology* 77: 1655–1661.

Remington, C.L. (1968). Suture-zones of hybrid interaction between recently joined biotas. *Evol. Biol.* 2: 321–428.

Renkin, R., Despain, D., and Clark, D. (1994). Aspen seedlings following the 1988 Yellowstone fires. *In* E. G. Despain (ed.), *Conf. Proc., Plants and Their Environments*. Tech. Report NPS/NRYELL/NRTR-93/XX. USDI, Nat. Park Service.

Rennie, P.J. (1962). Methods of assessing forest site capacity. *Trans. 7th Inter. Soc. Soil Sci., Comm.* IV and V, pp. 3–18.

Rey, O., Eizaguirre, C., Angers, B. et al. (2020). Linking epigenetics and biological conservation: towards a conservation epigenetics perspective. *Funct. Ecol.* 34: 414–427.

Reyer, C.P., Brouwers, N., Rammig, A. et al. (2015). Forest resilience and tipping points at different spatio-temporal scales: approaches and challenges. *J. Ecol.* 103: 5–15.

Reyers, B., Polasky, S., Tallis, H. et al. (2012). Finding common ground for biodiversity and ecosystem services. *Bioscience* 62: 503–507.

Rhoades, D.F. (1976). The anti-herbivore defenses of Larrea. In: *The Biology and Chemistry of the Creosote Bush, A Desert Shrub* (ed. T.J. Mabry, J.H. Hunziker and D.R. DiFeo), 3–54. Stroudsburg, PA: Dowden, Hutchinson, and Ross.

Rhoades, D.F. and Cates, R.G. (1976). Toward a general theory of plant antiherbivore chemistry. In: *Biochemical Interaction between Plants and Insects* (ed. J.W. Wallace and R.L. Mansell). Boston, MA: Springer.

Ricart, R.D., Pearsall, D.R., and Curtis, P.S. (2020). Multidecadal shifts in forest plant diversity and community composition across glacial landforms in northern lower Michigan, USA. *Can. J. For. Res.* 50: 126–135.

Ricciardi, A. and Simberloff, D. (2009). Assisted colonization is not a viable conservation strategy. *Trends Ecol. Evol.* 24: 248–253.

Rice, S.K., Westerman, B., and Federici, R. (2004). Impacts of the exotic, nitrogen-fixing black locust (*Robinia pseudoacacia*) on nitrogen-cycling in a pine–oak ecosystem. *Plant Ecol.* 174: 97–107.

Rice, A., Šmarda, P., Novosolov, M. et al. (2019). The global biogeography of polyploid plants. *Nat. Ecol. Evol.* 3: 265–273.

Richards, E.J. (2006). Inherited epigenetic variation – revisiting soft inheritance. *Nat. Rev. Genet.* 7: 395–401.

Richards, N.A. and Stone, E.L. (1964). The application of soil survey to planting site selection: an example from the Allegheny uplands of New York. *J. For.* 62: 475–480.

Richards, C.L., Bossdorf, O., and Pigliucci, M. (2010). What role does heritable epigenetic variation play in phenotypic evolution? *Bioscience* 60: 232–237.

Richardson, S.D. (1956). Studies of root growth of *Acer saccharinum* L. IV: the effect of differential shoot and root temperature on root growth. *Proc. K. Ned. Akad. Wet.* 59: 428–438.

Richardson, A.D., Hufkens, K., Milliman, T. et al. (2018). Ecosystem warming extends vegetation activity but heightens vulnerability to cold temperatures. *Nature* 560: 368–371.

Ricketts, T.H. (2001). The matrix matters: effective isolation in fragmented landscapes. *Am. Nat.* 158: 87–99.

Ricklefs, R. and Schluter, D. (1993). *Species Diversity in Ecological Communities: Historical and Geographical Perspectives*. Chicago: Univ. Chicago Press 414 pp.

Ripple, W.J. and Beschta, R.L. (2003). Wolf reintroduction, predation risk, and cottonwood recovery in Yellowstone National Park. *For. Ecol. Manag.* 184: 299–313.

Ripple, W.J. and Beschta, R.L. (2012). Trophic cascades in Yellowstone: the first 15 years after wolf reintroduction. *Biol. Conserv.* 145: 205–213.

Ripple, W.J., Beschta, R.L., and Painter, L.E. (2015). Trophic cascades from wolves to alders in Yellowstone. *For. Ecol. Manag.* 354: 254–260.

Risser, P.G., Karr, J.R., and Forman, R.T.T. (1984). Landscape ecology: directions and approaches. *Illinois Natural History Survey*, Number 2. Champaign, IL.

Ritchie, M.E., Tilman, D., and Knops, J.M.H. (1998). Herbivore effects and plant and nitrogen dynamics in oak savanna. *Ecology* 79: 165–177.

Robbins, C.S., Dawson, D.K., and Dowell, B.A. (1989). Habitat area requirements of breeding forest birds in the middle Atlantic states. *Wildl. Monogr.* 103: 1–34.

Robertson, G.P. (1982). Factors regulating nitrification in primary and secondary succession. *Ecology* 63: 1561–1573.

Robertson, G.P. and Tiedje, J.M. (1984). Denitrification and nitrous oxide production in successional and old-growth Michigan forests. *Soil Sci. Soc. Am. J.* 48: 383–389.

Robertson, G.P. and Tiedje, J.M. (1988). Denitrification in a humid tropical rainforest. *Nature* 336: 756–759.

Robertson, G.P. and Vitousek, P.M. (1981). Nitrification potentials in primary and secondary succession. *Ecology* 62: 376–386.

Robinson, S.K., Thompson, R.R. III, Donovan, T.M. et al. (1995). Regional forest fragmentation and the nesting success of migratory birds. *Science* 267: 1987–1990.

Robledo-Arnuncio, J.J. and Gil, L. (2005). Patterns of pollen dispersal in a small population of *Pinus sylvestris* L. revealed by total-exclusion paternity analysis. *Heredity* 94: 13–22.

Rodman, K.C., Veblen, T.T., Battaglia, M.A. et al. (2020). A changing climate is snuffing out post-fire recovery in montane forests. *Glob. Ecol. Biogeogr.* 29: 2039–2051.

Roe, A.L. (1967). Seed dispersal in a bumper spruce seed year. USDA For. Serv. Res. Paper INT-39. Intermountain For. and Rge. Exp. Sta., Ogden, UT. 10 pp.

Roesch, L.F., Fulthorpe, R.R., Riva, A. et al. (2007). Pyrosequencing enumerates and contrasts soil microbial diversity. *ISME J.* 1: 283–290.

Rogers, W.S. and Booth, G.A. (1960). The roots of fruit trees. *Sci. Hortic.* 14: 27–34.

Rohde, K. (1992). Latitudinal gradients in species diversity: the search for the primary cause. *Oikos* 65: 514–527.

Rohde, A. and Bhalerao, R.P. (2007). Plant dormancy in the perennial context. *Trends Plant Sci.* 12: 217–223.

Röhrig, E. and Ulrich, B. (1991). *Ecosystems of the World, Temperate Deciduous Forests. Vol. 7.* New York: Elsevier 635 pp.

Roland, J. (1993). Large-scale forest fragmentation increases the duration of tent caterpillar outbreak. *Oecologia* 93: 25–30.

Román-Palacios, C. and Wiens, J.J. (2020). Recent responses to climate change reveal the drivers of species extinction and survival. *Proc. Natl. Acad. Sci.* 117: 4211–4217.

Romme, W.H. (1982). Fire and landscape diversity in subalpine forests of Yellowstone National Park. *Ecol. Monogr.* 52: 199–221.

Romme, W.H. and Despain, D.G. (1989). Historical perspective on the Yellowstone fires of 1988. *Bioscience* 39: 695–699.

Romme, W.H. and Knight, D.H. (1982). Landscape diversity, the concept applied to Yellowstone Park. *Bioscience* 32: 664–670.

Romme, W.H., Knight, D.H., and Yavitt, J.B. (1986). Mountain pine beetle outbreaks in the Rocky Mountains: regulators of primary productivity? *Am. Nat.* 127: 484–494.

Romme, W.H., Turner, M.G., Tuskan, G.A., and Reed, R.A. (2005). Establishment, persistence, and growth of aspen (*Populus tremuloides*) seedlings in Yellowstone National Park. *Ecology* 86: 404–418.

Ronov, A.B. and Yaroshevsky, A.A. (1972). Earth's crust geochemistry. In: *The Encyclopedia of Geochemistry and Environmental Sciences* (ed. R.W. Fairbridge), 243–254. New York: Van Nostrand Reinhold.

Roschanski, A.M., Csilléry, K., Liepelt, S. et al. (2016). Evidence of divergent selection at landscape and local scales in *Abies alba* mill. in the French Mediterranean Alps. *Mol. Ecol.* 25: 776–794.

Rosenvald, R. and Lõhmus, A. (2008). For what, when, and where is green-tree retention better than clear-cutting? A review of the biodiversity aspects. *For. Ecol. Manag.* 255: 1–15.

Rosenzweig, M.L. (1992). Species diversity gradients: we know more and less than we thought. *J. Mammal.* 73: 715–730.

Rosenzweig, M.L. (1995). *Species Diversity in Space and Time*. Cambridge, UK: Cambridge University Press 460 pp.

Rosenzwieg, M. (1968). Net primary productivity of terrestrial communities: prediction from climatological data. *Am. Nat.* 102: 67–74.

Roskoski, J.P. (1980). Nitrogen fixation hardwood forests of the northeastern United States. *Plant Soil* 54: 33–44.

Ross, M.S., Sharik, T.L., and Smith, D.Wm. (1982). Age structural relationships of tree populations in an Appalachian oak forest. *Bull. Torrey Bot. Club* 109: 287–298.

Ross, S.D., Pharis, R.P., and Binder, W.D. (1983). Growth regulators and conifers: their physiology and potential uses in forestry. In: *Plant Growth Regulating Chemicals, Vol. 2* (ed. L.G. Nickell). Boca Raton, FL: CRC Press.

Ross, M.S., O'Brien, J.J., and Flynn, L.J. (1992). Ecological site classification of Florida keys terrestrial habitats. *Biotropica* 24: 486–502.

Roth, I. (1990). Leaf structure of a Venezuelan cloud forest in relation to microclimate. *Encyclopedia of Plant Anatomy*, Vol. 14, Part 1. Gebrüder Borntraeger, Berlin. 244 pp.

Rother, M.T., Veblen, T.T., and Furman, L.G. (2015). A field experiment informs expected patterns of conifer regeneration after disturbance under changing climate conditions. *Can. J. For. Res.* 45: 1607–1616.

Rothstein, D.E., Zak, D.R., and Pregitzer, K.S. (1996). Nitrate deposition in northern hardwood forests and the nitrogen metabolism of *Acer saccharum* marsh. *Oecologia* 108: 338–344.

Rothstein, D.E., Yermakov, Z., and Buell, A.L. (2004). Loss and recovery of ecosystem carbon pools following stand-replacing wildfire in Michigan jack pine forests. *Can. J. For. Res.* 34: 1908–1918.

Rousi, M., Tahvanainen, J., and Uotila, I. (1989). Inter- and intraspecific variation in the resistance of winter-dormant birch (*Betula* spp.) against browsing by the mountain hare. *Holarct. Ecol.* 12: 187–192.

Rousi, M., Tahvanainen, J., and Uotila, I. (1991). A mechanism of resistance to hare browsing in winter-dormant European white birch (*Betula pendula*). *Am. Nat.* 137: 64–82.

Rout, T.M., Moore, J.L., and McCarthy, M.A. (2014). Prevent, search or destroy? A partially observable model for invasive species management. *J. Appl. Ecol.* 51: 804–813.

Rowe, J.S. (1956). Uses of undergrowth species in forestry. *Ecology* 37: 461–473.

Rowe, J.S. (1961a). The level-of-integration concept and ecology. *Ecology* 42: 420–427.

Rowe, J.S. (1961b). Critique of some vegetational concepts as applied to forests of northwestern Alberta. *Can. J. Bot.* 39: 1007–1017.

Rowe, J.S. (1962). Soil, site and land classification. *For. Chron.* 38: 420–432.

Rowe, J.S. (1964). Environmental preconditioning, with special reference to forestry. *Ecology* 45: 399–403.

Rowe, J.S. (1966). Phytogeographic zonation: an ecological appreciation. In: *The Evolution of Canada's Flora* (ed. R.L. Taylor and R.A. Ludwig), 12–27. Univ. Toronto Press.

Rowe, J.S. (1969). Plant community as a landscape feature. *In* K.N.H. Greenidge (ed.), *Symp. Proc. Terrestrial Plant Ecology*, Nova Scotia Museum, Halifax.

Rowe, J.S. (1972). *Forest Regions of Canada*. Can. For. Serv., Dept. Env. Publ. No. 1300, Ottawa. 172 pp + map.

Rowe, J.S. (1983). Concepts of fire effects on plant individuals and species. In: *The Role of Fire in Northern Circumpolar Ecosystems* (ed. R.W. Wein and D.A. MacLean), 135–154. New York: Wiley.

Rowe, J.S. (1984a). Understanding forest landscapes: what you conceive is what you get. The Leslie L. Schaffer Lectureship in Forest Science. Vancouver, B. C., Canada.

Rowe, J.S. (1984b). Forestland classification: limitations of the use of vegetation. *In* J.G. Bockheim (ed.), *Symp. Proc. Forest Land Classification: Experience, Problems, Perspectives*. NCR-102 North Central For. Soils Comm., Soc. Am. For., USDA For. Serv., and USDA Soil Cons. Serv., Madison, WI.

Rowe, J.S. (1989). The importance of conserving systems. In: *Endangered Spaces: The Future for Canada's Wilderness* (ed. M. Hummel), 228–235. Toronto: Key Porter Books.

Rowe, J. S. (1990). *Home Place*. NeWest Publishers, Edmonton, Alberta, Canada. Canadian Parks and Wilderness Society, Henderson Book Series No. 12. 253 pp.

Rowe, J.S. (1992a). Prologue. *For. Chron.* 68: 22–24.

Rowe, J.S. (1992b). The ecosystem approach to forestland management. *For. Chron.* 68: 222–224.

Rowe, J.S. (1992c). The integration of ecological studies. *Funct. Ecol.* 6: 115–119.

Rowe, J.S. (1994). A new paradigm for forestry. *For. Chron.* 70: 565–568.

Rowe, J.S. (1997). The necessity of protecting ecoscapes. *Global Biodivers.* 7: 9–12.

Rowe, J.S. (1998). "Earth" as the metaphor for "life". *Bioscience* 48: 428–429.

Rowe, J.S. (2000). An earth-based ethic for humanity. Natur und Kultur: Transdisciplinare Zeitschrift fur okologische. *Nachhaltigkeit* 1: 106–120.

Rowe, J.S. and Barnes, B.V. (1994). Geo-ecosystems and bio-ecosystems. *Ecol. Soc. Am. Bull.* 75: 40–41.

Rowe, J.S. and Scotter, G.W. (1973). Fire in the boreal forest. *Quat. Res.* 3: 444–464.

Rowe, J.S. and Sheard, J.W. (1981). Ecological land classification: a survey approach. *Environ. Manag.* 5: 451–464.

Royall, P.D., Delcourt, P.A., and Delcourt, H.R. (1991). Late quaternary paleoecology and paleoenvironments of the Central Mississippi alluvial valley. *Bull. Geol. Soc. Am.* 103: 157–170.

Rubec, C.D.A. (1992). *Thirty Years of Ecological Land Surveys in Canada, from 1960 to 1990 Symp. Proc. Landscape Approaches to Wildlife and Ecosystem Management* (ed. G.B. Ingram and M.R. Moss). Morin Heights, Canada: Polysci. Publ. Inc.

Rudis, V.A. (1995). Regional forest fragmentation effects on bottomland hardwood community types and resource values. *Landsc. Ecol.* 10: 291–307.

Rudolph, T.D. (1964). Lammas growth and prolepsis in jack pine in the Lake States. *For. Sci. Monogr.* 6: 70 pp.

Rundel, P.W. (1972). Habitat restriction in giant sequoia: the environmental control of grove boundaries. *Am. Midi. Nat.* 87: 81–99.

Rundel, P.W. (1981). Fire as an ecological factor. In: *Physiological Plant Ecology I. Responses to the Physical Environment*, New Series Vol. 12A (ed. O.L. Lange, P.S. Nobel, C.B. Osmond and H. Ziegler). New York: Springer-Verlag.

Rundel, P.W. (1983). Impact of fire on nutrient cycles in Mediterranean-type ecosystems with reference to chaparral. In: *Mediterranean-Type Ecosystems: The Role of Nutrients* (ed. F.J. Kruger, D.T. Mitchell and J.U.M. Jarvis). New York: Springer-Verlag 570 pp.

Runge, M. and Rode, M.W. (1991). Effects of soil acidity on plant associations. In: *Soil Acidity* (ed. B. Ulrich and M.E. Sumner), 183–202. Berlin: Springer-Verlag.

Runkle, J.R. (1981). Gap regeneration in some old-growth forests of the eastern United States. *Ecology* 62: 1041–1051.

Runkle, J.R. (1982). Patterns of disturbance in some old-growth mesic forests of eastern North America. *Ecology* 63: 1533–1546.

Runkle, J.R. (1985). Disturbance regimes in temperate forests. In: *The Ecology of Natural Disturbance and Patch Dynamics* (ed. S.T.A. Pickett and P.S. White), 17–33. New York: Academic Press.

Runkle, J.R. (1990). Gap dynamics in an Ohio *Acer-Fagus* forest and speculations on the geography of disturbance. *Can. J. For. Res.* 20: 632–641.

Running, S.W., Nemani, R.R., Peterson, D.I. et al. (1989). Mapping regional forest evapotranspiration and photosynthesis by coupling satellite data with ecosystem simulation. *Ecology* 70: 1090–1101.

Rupp, D.E., Shafer, S.L., Daly, C. et al. (2020). Temperature gradients and inversions in a forested Cascade Range basin: synoptic to local-scale controls. *J. Geophys. Res. Atmos.* 125: e2020JD032686.

Rushton, B.S. (1993). Natural hybridization within the genus *Quercus* L. *Ann. Sci. For.* 50: 73s–90s.

Rustad, L.E.J.L., Campbell, J., Marion, G. et al. (2001). A meta-analysis of the response of soil respiration, net nitrogen mineralization, and aboveground plant growth to experimental ecosystem warming. *Oecologia* 126: 543–562.

Ryan, M.G. (1988). *The importance of maintenance respiration by the living cells in sapwood of subalpine conifers.* Ph.D. Dissertation. Oregon State Univ., Corvallis. 104 pp.

Ryan, M.G. (1991). Effects of climate change on plant respiration. *Ecol. Appl.* 1: 157–167.

Ryan, M.G., Linder, S., Vose, J.M., and Hubbard, R.M. (1994). Dark respiration in pines. *Ecol. Bull.* 43: 50–63.

Ryan, M.G., Harmon, M.E., Birdsey, R.A. et al. (2010). A synthesis of the science on forests and carbon for US forests. *Issues Ecol.* 13: 1–16.

Ryan, K.C., Knapp, E.E., and Varner, J.M. (2013). Prescribed fire in North American forests and woodlands: history, current practice, and challenges. *Front. Ecol. Environ.* 11: 15–24.

Rychert, R., Skujins, J., Sorensen, D., and Prochella, D. (1978). Nitrogen fixation by lichens and free-living microorganisms in deserts. In: *Nitrogen in Desert Ecosystems* (ed. N.E. West and J. Skujins), 20–30. Stroudsburg, PA: Dowden, Hutchinson, and Ross.

Sackett, S.S. and Haase, S.M. (1992). Measuring soil and tree temperatures during prescribed fires with thermocouple probes. USDA For. Serv. Gen. Tech. Rep. PSW-131, Pacific Southwest Res. Sta., Albany, CA. 15 pp.

Saeki, I., Dick, C.W., Barnes, B.V., and Murakami, N. (2011). Comparative phylogeography of red maple (*Acer rubrum* L.) and silver maple (*Acer saccharinum* L.): impacts of habitat specialization, hybridization and glacial history. *J. Biogeogr.* 38: 992–1005.

Sáenz-Romero, C., O'Neill, G., Aitken, S.N., and Lindig-Cisneros, R. (2021). Assisted migration field tests in Canada and Mexico: lessons, limitations, and challenges. *Forests* 12: 9.

Sagnard, F., Oddou-Muratorio, S., Pichot, C. et al. (2011). Effects of seed dispersal, adult tree and seedling density on the spatial genetic structure of regeneration at fine temporal and spatial scales. *Tree Genet. Genomes* 7: 37–48.

Sakai, A. and Larcher, W. (1987). *Frost Survival of Plants.* New York: Springer-Verlag 321 pp.

Sakai, A. and Weiser, C.J. (1973). Freezing resistance of trees in North America with reference to tree regions. *Ecology* 54: 118–126.

Salisbury, F.B. and Ross, C.W. (1992). *Plant Physiology*, 4e. Belmont, CA: Wadsworth 682 pp.

Salles, J.M. (2011). Valuing biodiversity and ecosystem services: why put economic values on nature? *C. R. Biol.* 334: 469–482.

Samman, S. and Logan, J.A. (2000). *Assessment and Response to Bark Beetle Outbreaks in the Rocky Mountain Area*. USDA For. Serv. Gen. Tech. Rep. RMRS-GTR-62. Rocky Mountain Res. Sta., Ogden, Utah. 46 pp.

Šamonil, P., Schaetzl, R.J., Valtera, M. et al. (2013). Crossdating of disturbances by tree uprooting: can treethrow microtopography persist for 6000 years? *For. Ecol. Manag.* 307: 123–135.

Šamonil, P., Daněk, P., Schaetzl, R.J. et al. (2015). Soil mixing and genesis as affected by tree uprooting in three temperate forests. *Eur. J. Soil Sci.* 66: 589–603.

Sampson, H.C. (1930). Succession in the swamp forest formation in northern Ohio. *Ohio J. Sci.* 30: 340–356.

Sanchez, P.A. (1976). *Properties and Management of Tropical Soils*. New York: Wiley 619 pp.

Sánchez-Pinillos, M., Leduc, A., Ameztegui, A. et al. (2019). Resistance, resilience or change: post-disturbance dynamics of boreal forests after insect outbreaks. *Ecosystems* 22: 1886–1901.

Sankey, T.T., Montagne, C., Graumlich, L. et al. (2006). Twentieth century forest-grassland ecotone shift in Montana under differing livestock grazing pressure. *For. Ecol. Manag.* 234: 282–292.

Sarvas, R. (1962). Investigations on the flowering and seed crop of *Pinus silvestris*. *Comm. Inst. For. Fenn.* 53: 1–198.

Sarvas, R. (1969). Genetical adaptation of forest trees to the heat factor of the climate. Second world consultation on forest tree breeding, Washington, DC. FO-FTB-69-2/15, 11 pp.

Sato, K. and Iwasa, Y. (1993). Modelling of wave regeneration in subalpine *Abies* forests: population dynamics with spatial structure. *Ecology* 74: 1538–1550.

Saunders, D.A., Hobbs, R.H., and Margules, C.R. (1991). Biological consequences of ecosystem fragmentation: a review. *Conserv. Biol.* 5: 18–32.

Savage, M. and Swetnam, T.W. (1990). Early 19th-century fire decline following sheep pasturing in a Navajo ponderosa pine forest. *Ecology* 71: 2374–2378.

Savolainen, O., Pyhäjärvi, T., and Knürr, T. (2007). Gene flow and local adaptation in trees. *Annu. Rev. Ecol. Evol. Syst.* 38: 595–619.

Sax, D.F. (2002). Native and naturalized plant diversity are positively correlated in scrub communities of California and Chile. *Divers. Distrib.* 8: 193–210.

Schaetzl, R.J. and Follmer, L.R. (1990). Longevity of treethrow microtopography: implications for mass wasting. *Geomorphology* 3: 113–123.

Schaetzl, R.J., Johnson, D.L., Burns, S.F., and Small, T.W. (1989a). Tree uprooting: review of terminology, process, and environmental implications. *Can. J. For. Res.* 19: 1–11.

Schaetzl, R.J., Burns, S.F., Johnson, D.L., and Small, T.W. (1989b). Tree uprooting: review of impact on forest ecology. *Vegetatio* 79: 165–176.

Schaetzl, R.J., Burns, S.F., Small, T.W., and Johnson, D.L. (1990). Tree uprooting: review of types and patterns of soil disturbance. *Phys. Geogr.* 11: 277–291.

Schaffalitzky De Muckadell, M. (1959). Investigations on aging of apical meristems in woody plants and its importance in silviculture. *Det Forstl. Forggsv. i Danmark* 25:309–455.

Schaffalitzky De Muckadell, M. (1962). Environmental factors in development stages of trees. In: *Tree Growth* (ed. T.T. Kozlowski), 289–298. New York: Ronald Press.

Scheller, R.M. (2018). The challenges of forest modeling given climate change. *Landsc. Ecol.* 33: 1481–1488.

Scheller, R.M., Domingo, J.B., Sturtevant, B.R. et al. (2007). Design, development, and application of LANDIS-II, a spatial landscape simulation model with flexible temporal and spatial resolution. *Ecol. Model.* 201: 409–419.

Schier, G.A. (1975). Deterioration of aspen clones in the Middle Rocky Mountains. USDA For. Serv. Res. Paper INT-170. Intermountain For. and Rge. Exp. Sta., Ogden, Utah. 14 pp.

Schimel, J.P., Firestone, M.K., and Killham, K.S. (1984). Identification of heterotrophic nitrification in a Sierran forest soil. *Appl. Environ. Microbiol.* 48: 802–806.

Schimel, D.S., Coleman, D.C., and Horton, K.A. (1985). Organic matter dynamics in paired range-land and cropland toposequences in North Dakota. *Geoderma* 36: 201–214.

Schimper, A.F.W. (1898). *Pflanzen-Geographie auf Physiologischer Grundlage*. Jena: Gustav Fischer 876 pp.

Schlenker, G. (1960). Zum Problem der Einordnung klimatischer Unterschiede in das System der Waldstandorte Baden-Württembergs. *Mitt. Vereins forstl. Standortsk. Forstpflz.* 9:3–15.

Schlenker, G. (1964). Entwicklung des in Südwestdeutschland angewandten Verfahrens der forstlichen Standortskunde. *In Standort, Wald und Waldwirschaft in Oberschwabèn, "Oberschwäbische Fichtenreviere."* Stuttgart.

Schlesinger, W.H. (1978). On the relative dominance of shrubs in Okefenokee Swamp. *Am. Nat.* 112: 949–954.

Schlesinger, W.H. and Marks, P.L. (1977). Mineral cycling and the niche of Spanish moss, *Tillandsia usneoides* L. *Am. J. Bot.* 64: 1254–1262.

Schlichting, C.D. (1986). The evolution of phenotypic plasticity in plants. *Annu. Rev. Ecol. Syst.* 17: 667–693.

Schmidt, H. (1970). *Versuche über die Pollenverteilung in einem Kiefernbestand*. Diss. Forstl. Fak. Univ. Göttingen, Germany.

Schmidt, W.C. and Dufour, W.P. (1975). Building a natural area system for Montana. *Western Wildlands* 2: 20–29.

Schmidt, W.C. and Shearer, R.C. (1971). Ponderosa pine seed—for animals or trees? USDA For. Serv. Res.

Paper INT-112. Intermountain For. and Rge. Exp. Sta., Ogden, Utah. 14 pp.

Schmitz, O.J., Kalies, E.L., and Booth, M.G. (2006). Alternative dynamic regimes and trophic control of plant succession. *Ecosystems* 9: 659–672.

Schoenike, R.R. (1976). Geographical variations in jack pine (*Pinus banksiana*). *Univ. Minnesota. Agr. Exp. Sta. Tech. Bull.* 304. 47 pp.

Schoennagel, T., Turner, M.G., and Romme, W.H. (2003). The influence of fire interval and serotiny on postfire lodgepole pine density in Yellowstone National Park. *Ecology* 84: 2967–2978.

Schoennagel, T., Veblen, T.T., and Romme, W.H. (2004). The interaction of fire, fuels, and climate across Rocky Mountain forests. *Bioscience* 54: 661–676.

Schowalter, T.D. (1989). Canopy arthropod community structure and herbivory in old-growth and regenerating forests in western Oregon. *Can. J. For. Res.* 19: 318–322.

Schrodt, F., Bailey, J.J., Kissling, W.D. et al. (2019). To advance sustainable stewardship, we must document not only biodiversity but geodiversity. *Proc. Natl. Acad. Sci.* 116: 16155–161658.

Schüle, W. (1990). Landscape and climate in prehistory: interactions of wildlife, man, and fire. In: *Fire in the Tropical Biota* (Ecological Studies 84) (ed. J.G. Goldammer), 273–318. New York: Springer-Verlag.

Schuler, T.M. and Smith, F.W. (1988). Effect of species mix on size/density and leaf-area relations in southwest pinyon/juniper woodlands. *For. Ecol. Manag.* 25: 211–220.

Schullery, P. (1989). The fires and fire policy. *Bioscience* 39: 686–694.

Schulze, E.D., Fuchs, M.I., and Fuchs, M. (1977). Spacial distribution of photosynthetic capacity and performance in a montane spruce forest of northern Germany. I. Biomass distribution and daily CO_2 uptake in different crown layers. *Oecologia* 29: 43–61.

Schupp, E.W., Jordano, P., and Gómez, J.M. (2010). Seed dispersal effectiveness revisited: a conceptual review. *New Phytol.* 188: 333–353.

Schupp, E.W., Jordano, P., and Gómez, J.M. (2017). A general framework for effectiveness concepts in mutualisms. *Ecol. Lett.* 20: 577–590.

Schuster, R.L., Logan, R.L., and Pringle, P.T. (1992). Prehistoric rock avalanches in the Olympic Mountains, Washington. *Science* 258: 1620–1621.

Schuur, E.A., McGuire, A.D., Schädel, C. et al. (2015). Climate change and the permafrost carbon feedback. *Nature* 520: 171–179.

Schwartz, M.D., Ahas, R., and Aasa, A. (2006). Onset of spring starting earlier across the northern hemisphere. *Glob. Chang. Biol.* 12: 343–351.

Schweingruber, F.H. (1989). *Tree Rings: Basics and Applications of Dendrochronology*. New York: Springer 290 pp.

Schwertmann, U. and Taylor, R.M. (1989). Iron oxides. In: *Minerals in Soil Environments*, 2e (ed. J.B. Dixon and S.B. Weed). Madison, WI: Soil Sci. Soc. Amer.

Sconiers, W.B. and Eubanks, M.D. (2017). Not all droughts are created equal? The effects of stress severity on insect herbivore abundance. *Arthropod Plant Interact.* 11: 45–60.

Scott, D.A., Proctor, J., and Thompson, J. (1992). Ecological studies on a lowland evergreen rain forest on Maraca Island, Brazil. II. Litter and nutrient cycling. *J. Ecol.* 80: 705–717.

Searle, E.B. and Chen, H.Y. (2018). Temporal declines in tree longevity associated with faster lifetime growth rates in boreal forests. *Environ. Res. Lett.* 13: e125003.

Sebald, O. (1964). Ökologische Artengruppen für den Wuchsbezirk "Oberer Neckar." *Mitt. Vereins forstl. Standortsk. Forstpflz.* 14:60–63.

Segraves, K.A. and Anneberg, T.J. (2016). Species interactions and plant polyploidy. *Am. J. Bot.* 103: 1326–1335.

Seidl, R., Schelhaas, M.J., Rammer, W., and Verkerk, P.J. (2014). Increasing forest disturbances in Europe and their impact on carbon storage. *Nat. Clim. Chang.* 4: 806–810.

Seidl, R., Thom, D., Kautz, M. et al. (2017). Forest disturbances under climate change. *Nat. Clim. Chang.* 7: 395–402.

Sercu, B.K., Baeten, L., van Coillie, F. et al. (2017). How tree species identity and diversity affect light transmittance to the understory in mature temperate forests. *Ecol. Evol.* 7: 10861–10870.

Seymour, R.S. and Hunter, M.L. (1999). Principles of ecological forestry. In: *Maintaining Biodiversity in Forest Ecosystems* (ed. M.L. Hunter), 22–61. Cambridge: Cambridge University Press.

Shaffer, M. (2015). Changing filters. *Conserv. Biol.* 29: 611–612.

Sharik, T.L. and Barnes, B.V. (1976). Phenology of shoot growth among diverse populations of yellow birch (*Betula alleghaniensis*) and sweet birch (*B. lenta*). *Can. J. Bot.* 54: 2122–2129.

Sharik, T.L., Feret, P.P., and Dyer, R.W. (1990). Recovery of the endangered Virginia roundleaf birch (*Betula uber*): a decade of effort. *In* R.S. Mitchell, C.J. Sheviak, and D.J. Leopold (eds.), *Ecosystem Management: Rare Species and Significant Habitats*. Proc. 15th Ann. Natural Areas Conference. New York State Museum Bull. 471.

Sharov, A.A., Leonard, D., Liebhold, A.M. et al. (2002). "Slow the spread": a national program to contain the gypsy moth. *J. For.* 100: 30–36.

Sharrow, S.H. (2007). Soil compaction by grazing livestock in silvopastures as evidenced by changes in soil physical properties. *Agrofor. Syst.* 71: 215–223.

Shaw, K.L. and Mullen, S.P. (2011). Genes versus phenotypes in the study of speciation. *Genetica* 139: 649–661.

Shea, K.L. (1990). Genetic variation between and within populations of Engelmann spruce and sub-alpine fir. *Genome* 33: 1–8.

Sheil, D. and Bongers, F. (2020). Interpreting forest diversity-productivity relationships: volume values, disturbance histories and alternative inferences. *For. Ecosyst.* 7: 6.

Shimwell, D.W. (1971). *The Description and Classification of Vegetation*. Seattle: Univ. Washington Press 322 pp.

Shirley, H.L. (1945b). Reproduction of upland conifers in the Lake states as affected by root competition and light. *Am. Midi. Nat.* 33: 537–612.

Shiva, V. (1990). Biodiversity, biotechnology, and profit: the need for a peoples' plan to protect biological diversity. *Ecologist* 20: 44–17.

Short, H.L. (1976). Composition and squirrel use of acorns of black and white oak groups. *J. Wildl. Manag.* 40: 479–483.

Shotola, S.J., Weaver, G.T., Robertson, P.A., and Ashby, W.C. (1992). Sugar maple invasion of an old-growth oak-hickory forest in southwestern Illinois. *Am. Midl. Nat.* 127: 125–138.

Show, S.B., and Kotok, E.I. (1929). Cover type and fire control in the National Forests of northern California. USDA, Dept. Bull. 1495. Washington, D.C. 35 pp.

Shrader-Frechette, K.S. and McCoy, E.D. (1993). *Method in Ecology: Strategies for Conservation*. Cambridge: Cambridge Univ. Press 328 pp.

Shuman, B., Webb III, T., Bartlein, P. et al. (2002). The anatomy of a climatic oscillation: vegetation change in eastern North America during the Younger Dryas chronozone. *Quaternary Science Reviews* 21: 1777–1791.

Siemann, E., Carrillo, J.A., Gabler, C.A. et al. (2009). Experimental test of the impacts of feral hogs on forest dynamics and processes in the southeastern US. *For. Ecol. Manag.* 258: 546–553.

Silen, R.R. (1978). Genetics of Douglas-fir. USDA For. Serv. Res. Paper W0–35, Washington D.C. 34 pp.

Silkworth, D.R. and Grigal, D.F. (1982). Determining and evaluating nutrient losses from whole-tree harvesting of aspen. *Soil Sci. Soc. Am. J.* 46: 626–631.

Sillett, S.C., McCune, B., Peck, J.E. et al. (2000). Dispersal limitations of epiphytic lichens result in species dependent on old-growth forests. *Ecol. Appl.* 10: 789–799.

Silvertown, J.W. (1980). The evolutionary ecology of mast seeding in trees. *Biol. J. Linn. Soc.* 14: 235–250.

Silvertown, J.W. and Lovett-Doust, J. (1993). *Introduction to Plant Population Biology*. Oxford: Blackwell 210pp.

Silvester, W.B., Sollins, P., Verhoeven, T., and Cline, S.P. (1982). Nitrogen fixation and acetylene reduction in decaying conifer boles: effects of incubation time, aeration, and moisture content. *Can. J. For. Res.* 12: 646–652.

Simard, A.J., Haines, D.A., Blank, R.W., and Frost, J.S. (1983). The Mack Lake fire. USDA For. Serv. Gen. Tech. Report NC-83. North Central For. Exp. Sta. St. Paul, MN. 36 pp.

Simard, M., Romme, W.H., Griffin, J.M., and Turner, M.G. (2011). Do mountain pine beetle outbreaks change the probability of active crown fire in lodgepole pine forests? *Ecol. Monogr.* 81: 3–24.

Simberloff, D. (2013). *Invasive Species: What Everyone Needs to Know*. Oxford University Press 352 pp.

Simberloff, D. (2015). Non-native invasive species and novel ecosystems. *F1000Prime Reports* 7:47.

Simberloff, D., Murcia, C., and Aronson, C. (2015). "Novel ecosystems" are a Trojan horse for conservation. http://ensia.com/voices/novel-ecosystems-are-a-trojanhorse-for-conservation (accessed 20 Nov 2021).

Simpson, E.H. (1949). Measurement of diversity. *Nature* 163: 688.

Simpson, G.G. (1952). How many species? *Evolution* 6: 342.

Simpson, T.A. (1990). *Landscape Ecosystems and Cover Types of the Reserve Area and Adjoining Land of the Huron Mountain Club, Marquette Co., Michigan*. Ph.D. Thesis, University of Michigan, Ann Arbor. 384 pp.

Simpson, T.A., Stuart, P.E., and Barnes, B.V. (1990). Landscape ecosystems and cover types of the Reserve Area and adjoining lands of the Huron Mountain Club, Marquette Co., MI. *Huron Mountain Wildlife Foundation Occasional Paper* No. 4. 128 pp.

Sims, R.A. and Uhlig, P. (1992). The current status of forest site classification in Ontario. *For. Chron.* 68: 64–77.

Sims, R.A., Towill, W.D., Baldwin, K.A., and Wickware, G.M. (1989). *Field Guide to the Forest Ecosystem Classification for Northwestern Ontario*. Ont. Min. Nat. Res. Toronto, ON. 191 pp.

Singer, J.A., Turnbull, R., Foster, M. et al. (2019). Sudden aspen decline: a review of pattern and process in a changing climate. *Forests* 10: 671.

Sirén, G. (1955). The development of spruce forest on raw humus sites in northern Finland and its ecology. *Acta For. Fenn.* 62 (4): 1–408.

Sittaro, F., Paquette, A., Messier, C., and Nock, C.A. (2017). Tree range expansion in eastern North America fails to keep pace with climate warming at northern range limits. *Glob. Chang. Biol.* 23: 3292–3301.

Skroch, M. and López-Hoffman, L. (2009). Saving nature under the big tent of ecosystem services: a response to Adams and Redford. *Conserv. Biol.* 24: 325–327.

Slaughter, C.W., Barnes, R.J., and Hansen, G.M. (eds.). (1971). *Fire in the Northern Environment—A*

Symposium. USDA For. Serv. Pacific Northwest For. and Rge. Exp. Sta., Portland, OR. 275 pp.

Smalley, G.W. (1984). Classification and evaluation of forest sites in the Cumberland Mountains. USDA For. Serv. Gen. Tech. Report SO-50. Southern For. Exp. Sta., New Orleans, LA. 84 pp.

Smalley, G.W. (1991). No more plots; go with what you know: developing a forest land classification system for the Interior Uplands. *In* D.L. Mengel and D.T. Tew (eds.), *Symp. Proc. Ecological Land Classification: Applications to Identify and Productive Potential of Southern Forests*. USDA For. Serv. Gen. Tech. Report SE-68. Southeastern For. Exp. Sta., Asheville, NC.

Smilanich, A.M., Fincher, R.M., and Dyer, L.A. (2016). Does plant apparency matter? Thirty years of data provide limited support but reveal clear patterns of the effects of plant chemistry on herbivores. *New Phytol.* 210: 1044–1057.

Smith, D.M. (1951). The influence of seedbed conditions on the regeneration of eastern white pine. *Conn. Agr. Exp. Sta. Bull.* 545: 61 pp.

Smith, R.H. (1966). Resin quality as a factor in the resistance of pines to bark beetles. In: *Breeding Pest-Resistant Trees* (ed. H.D. Gerhold, R.E. McDermott, E.J. Schreiner and J.A. Winieski), 189–196. New York: Pergamon Press.

Smith, R.H. (1977). Monoterpenes of ponderosa pine xylem resin in western United States. USDA For. Serv., Tech Bull. 1532. 48 pp.

Smith, H. (1982). Light quality, photoperception and plant strategy. *Annu. Rev. Plant Physiol.* 33: 481–518.

Smith, C.C. (1990). The advantage of mast years for wind pollination. *Am. Nat.* 136: 154–166.

Smith, B.R. and Blumstein, D.T. (2008). Fitness consequences of personality: a meta-analysis. *Behav. Ecol.* 19: 448–455.

Smith, W.B., and Brand, G.J. (1983). Allometric biomass equations for 98 species of herbs, shrubs and small trees. USDA For. Serv. Note NC-299. North Central For. Exp. Sta., St. Paul, MN. 8 pp.

Smith, D.D. and Follmer, D. (1972). Food preferences of squirrels. *Ecology* 53: 82–91.

Smith, T. and Huston, M. (1989). A theory of the spatial and temporal dynamics of plant communities. *Plant Ecol.* 83: 49–69.

Smith, T.J. III, Robblee, M.B., Wanless, H.R., and Doyle, T.W. (1994). Mangroves, hurricanes, and lightning strikes. *Bioscience* 44: 256–262.

Smith, D.M., Larson, B.C., Kelty, M.J., and Ashton, P.M.S. (1997). *The Practice of Silviculture*, 9e. New York: Wiley 537 pp.

Smith, N.R., Kishchuk, B.E., and Mohn, W.W. (2008). Effects of wildfire and harvest disturbances on forest soil bacterial communities. *Appl. Environ. Microbiol.* 74: 216–224.

Smith, R.S., Blaze, J.A., and Byers, J.F. (2020). Negative indirect effects of hurricanes on recruitment of range-expanding mangroves. *Mar. Ecol. Prog. Ser.* 644: 65–74.

Snaydon, R.W. and Davies, M.S. (1972). Rapid population differentiation in a mosaic environment. II. Morphological variation in *Anthoxanthum odoratum*. *Evolution* 26: 390–405.

Snaydon, R.W. and Davies, M.S. (1976). Rapid population differentiation in a mosaic environment. IV. Populations *of Anthoxanthum odoratum* at sharp boundaries. *Heredity* 37: 9–25.

Sobhani, V.M., Barrett, M., and Peterson, C.J. (2014). Robust prediction of treefall pit and mound sizes from tree size across 10 forest blowdowns in eastern North America. *Ecosystems* 17: 837–850.

Soil Survey Staff (1975). *Soil Taxonomy*. USDA Agr. Handbook 436. Washington, D.C. 754 pp.

Solbrig, O.T. (1970). *Principles and Methods of Plant Biosystematics*. Toronto: Collier-Macmillan 226 pp.

Solbrig, O.T. (1991). The origin and function of biodiversity. *Environment* 33 (16–20): 34–38.

Sollars, E.S.A. and Buggs, R.J.A. (2018). Genome-wide epigenetic variation among ash trees differing in susceptibility to a fungal disease. *BMC Genom.* 19: 502.

Sollins, P., Robertson, G.P., and Uehara, G. (1988). Nutrient mobility in variable-and permanent-charge soils. *Biogeochemistry* 6: 181–199.

Solórzano, L.A. and Páez-Acosta, G.I. (2009). Forests. In: *The Princeton Guide to Ecology* (ed. S.A. Levin, S.R. Carpenter, H.C.J. Godfray, et al.), 606–613. Princeton, NJ: Princeton University Press.

Soltis, P.S. (2005). Ancient and recent polyploidy in angiosperms. *New Phytol.* 166: 5–8.

Soltis, D.E., Morris, A.B., McLachlan, J.S. et al. (2006). Comparative phylogeography of unglaciated eastern North America. *Mol. Ecol.* 15: 4261–4293.

Soltis, D.E., Albert, V.A., Leebens-Mack, J. et al. (2009). Polyploidy and angiosperm diversification. *Am. J. Bot.* 96: 336–348.

Soltis, D.E., Visger, C.J., and Soltis, P.S. (2014). The polyploidy revolution then. . . and now: Stebbins revisited. *Am. J. Bot.* 101: 1057–1078.

Soltis, P.S., Marchant, D.B., Van de Peer, Y., and Soltis, D.E. (2015). Polyploidy and genome evolution in plants. *Curr. Opin. Genet. Dev.* 35: 119–125.

Soolanayakanahally, R.Y., Guy, R.D., Silim, S.N., and Song, M. (2013). Timing of photoperiodic competency causes phenological mismatch in balsam poplar (*Populus balsamifera* L.). *Plant Cell Environ.* 36: 116–127.

Sork, V.L., Bramble, J., and Owen, S. (1993). Ecology of mast-fruiting in three species of North American deciduous oaks. *Ecology* 74: 528–541.

Sork, V.L., Aitken, S.N., Dyer, R.J. et al. (2013). Putting the landscape into the genomics of trees: approaches for understanding local adaptation and population responses to changing climate. *Tree Genet. Genomes* 9: 901–911.

Sow, M.D., Allona, I., Ambroise, C. et al. (2018). Epigenetics in forest trees: state of the art and potential implications for breeding and management in a context of climate change. *Adv. Bot. Res.* 88: 387–453.

Sow, M.D., Le Gac, A.-L., Fichot, R. et al. (2021). RNAi suppression of DNA methylation affects the drought stress response and genome integrity in transgenic poplar. *New Phytol.* 232: 80–97.

Speer, J.H. (2010). *Fundamentals of Tree-Ring Research*. Tucson, AZ: University of Arizona Press 360 pp.

Sperry, J.S., Hacke, U.G., and Pittermann, J. (2006). Size and function in conifer tracheids and angiosperm vessels. *Am. J. Bot.* 93: 1490–1500.

Sperry, J.S., Venturas, M.D., Todd, H.N. et al. (2019). The impact of rising CO2 and acclimation on the response of US forests to global warming. *Proc. Natl. Acad. Sci.* 116: 25734–25744.

Spies, T.A. (1983). *Classification and Analysis of Forest Ecosystems of the Sylvania Recreation Area, Upper Michigan*. Ph.D. Thesis, University of Michigan, Ann Arbor. 321 pp.

Spies, T.A. (1997). Forest stand structure, composition, and function. In: *Creating a Forestry for the 21st Century* (ed. K.A. Kohm and J.F. Franklin), 11–30. Washington, D.C: Island Press.

Spies, T.A. and Barnes, B.V. (1981). A morphological analysis of *Populus alba, P. grandidentata* and their natural hybrids in southeastern Michigan. *Silvae Genet.* 30: 102–106.

Spies, T.A. and Barnes, B.V. (1985a). A multi-factor ecological classification of the northern hardwood and conifer ecosystems of the Sylvania Recreation Area, Upper Peninsula, Michigan. *Can. J. For. Res.* 15: 949–960.

Spies, T.A. and Barnes, B.V. (1985b). Ecological species groups of upland northern hardwood-hemlock forest ecosystems of the Sylvania Recreation Area, Upper Peninsula, Michigan. *Can. J. For. Res.* 15: 961–972.

Spies, T.A. and Franklin, J.F. (1989). Gap characteristics and vegetation response in tall coniferous forests. *Ecology* 70: 543–545.

Spies, T.A. and Franklin, J.F. (1991). The structure of natural young, mature, and old-growth forests in Washington and Oregon. *In* L.F. Ruggiero, K.B. Aubry, A.B. Carey, and M. H. Huff (tech. coords.), *Wildlife and Vegetation of Unmanaged Douglas-fir forests*. USDA For. Serv. PNW-GTR-385. Pacific Northwest Exp. Sta., Portland, OR.

Spies, T.A., Franklin, J.F., and Klopsch, M. (1990). Canopy gaps in Douglas-fir forests of the Cascade Mountains. *Can. J. For. Res.* 20: 649–658.

Sposito, G. (1989). *The Chemistry of Soils*. New York: Oxford Univ. Press 277 pp.

Sprugel, D.G. (1976). Dynamic structure of wave-regenerated *Abies balsamea* forests in the northeastern United States. *J. Ecol.* 64: 889–911.

Sprugel, D.G. and Bormann, F.H. (1981). Natural disturbance and the steady state in high-altitude balsam fir forests. *Science* 211: 390–393.

Spurr, S.H. (1945). A new definition of silviculture. *J. For.* 43: 44.

Spurr, S.H. (1952a). Origin of the concept of forest succession. *Ecology* 33: 426–427.

Spurr, S.H. (1952b). *Forest Inventory*. New York: Ronald Press 476 pp.

Spurr, S.H. (1954). The forests of Itasca in the nineteenth century as related to fire. *Ecology* 35: 21–25.

Spurr, S.H. (1956a). Natural restocking of forests following the 1938 hurricane in central New England. *Ecology* 37: 433–451.

Spurr, S.H. (1956b). Forest associations in the Harvard Forest. *Ecol. Monogr.* 26: 245–262.

Spurr, S.H. (1957). Local climate in the Harvard Forest. *Ecology* 38: 37–46.

Spurr, S.H. (1963). Growth of Douglas-fir in New Zealand. New Zealand For. Serv. For. Res. Inst. Tech. Paper 43. 54 pp.

Spurr, S.H. (1964). *Forest Ecology*. New York: Ronald Press 352 pp.

Spurr, S.H. and Barnes, B.V. (1973). *Forest Ecology*, 2e. New York: Ronald Press 571 pp.

Spurr, S.H. and Barnes, B.V. (1980). *Forest Ecology*, 3e. New York: Wiley 687 pp.

Squillace, A.E. (1966). Geographic variation in slash pine. *For. Sci. Monogr.* 10. 56 pp.

St. Clair, B.J., Mandel, N.L., and Vance-Borland, K.W. (2005). Genecology of Douglas-fir in western Oregon and Washington. *Ann. Bot.* 96: 1199–1214.

Staaf, H. and Berg, B. (1982). Accumulation and release of plant nutrients in decomposing Scots pine needle litter. Long-term decomposition in a Scots pine forest II. *Can. J. Bot.* 60: 1561–1568.

Stacey, P.B. and Koenig, W.D. (1984). Cooperative breeding in the acorn woodpecker. *Sci. Am.* 252: 114–121.

Stage, A.R. (1989). Utility of vegetation-based land classes for predicting forest regeneration and growth. *In* D.E. Ferguson, P. Morgan, F.D. Johnson (comps.), Proceedings—*Land Classifications Based on Vegetation: Applications for Resource Management*. USDA For. Serv. Gen. Tech. Report INT-257. Intermountain Res. Sta., Ogden, UT.

Staley, D.M., Kean, J.W., and Rengers, F.K. (2020). The recurrence interval of post-fire debris-flow generating rainfall in the southwestern United States. *Geomorphology* 370: 107392.

Stambaugh, M.C., Guyette, R.P., Grabner, K., and Kolaks, J. (2006). Understanding Ozark forest litter variability through a synthesis of accumulation rates and fire events. *In* B.W. Butler, P.I. Andrews (comps.). *Fuels Management- How To Measure Success*. Conference Proceedings. 28–30 March, 2006. Portland, OR. USDA For. Serv. Gen. Tech.

Rep. RMRS-P-41. Rocky Mountain Res. Sta., Fort Collins, CO.

Stamp, N. (2003). Out of the quagmire of plant defense hypotheses. *Quart. Rev. Bot.* 78: 23–55.

Stamp, N. (2004). Can the growth-differentiation balance hypothesis be tested rigorously? *Oikos* 107: 439–448.

Stanturf, J.A., Wade, D.D., Waldrop, T.A. et al. 2002. Fire in southern forest landscapes. *In* D.M. Wear, J. Greis (eds.). *Southern Forest Resource Assessment.* USDA For. Serv. Gen. Tech. Rep. SRS-53. Southern Res. Sta., Asheville, NC.

Stark, N. (1968). Seed ecology of *Sequoiadendron giganteum. Madroño* 19: 267–277.

Stark, N.M. (1977). Fire and nutrient cycling in a Douglas-fir/larch forest. *Ecology* 58: 16–30.

Stebbins, G.L. (1950). *Variation and Evolution in Plants.* New York: Columbia Univ. Press 643 pp.

Stebbins, G.L. (1969). The significance of hybridization for plant taxonomy and evolution. *Taxon* 18: 26–35.

Stebbins, G.L. (1970). Variation and evolution in plants: progress during the past twenty years. In: *Essays in Evolution and Genetics* (ed. M.K. Hecht and W.C. Steere), 173–208. New York: Appleton-Century-Crofts.

Stebbins, G.L. (1971). *Processes of Organic Evolution*, 2e. Englewood Cliffs, NJ: Prentice-Hall 193 pp.

Stebbins, G.L. (1985). Polyploidy, hybridization, and the invasion of new habitats. *Ann. Missouri Bot. Gard.* 72: 824–832.

Steele, B.M. and Cooper, S.V. (1986). Predicting site index and height for selected tree species of northern Idaho. USDA For. Serv. Res. Paper INT-126. Int. For. and Rge. Exp. Sta., Ogden, UT. 16 p.

Steele, M.A., Knowles, T., Bridle, K., and Simms, E.L. (1993). Tannins and partial consumption of acorns: implications for dispersal of oaks by seed predators. *Am. Midl. Nat.* 130: 229–238.

Steinbrenner, E.C. (1951). Effect of grazing on floristic composition and soil properties of farm woodlands in southern Wisconsin. *J. For.* 49: 906–910.

Stephenson, S.L. (1974). Ecological composition of some former oak-chestnut communities in western Virginia. *Castanea* 39: 278–286.

Stephenson, N. L. and Brigham, C. (2021). Preliminary estimates of sequoia mortality in the 2020 Castle Fire. Sequoia and Kings Canyon National Parks: National Park Service. https://www.nps.gov/articles/000/preliminary-estimates-of-sequoia-mortality-in-the-2020-castle-fire.htm. Accessed 01-18-2022.

Sterba, H. and Monserud, R.A. (1993). The maximum density concept applied to uneven-aged mixed-species stands. *For. Sci.* 39: 432–452.

Sterck, F. (2005). Woody tree architecture. In: *Plant Architecture and its Manipulation* (ed. C. Turnbull), 209–237. Oxford, UK: Blackwell.

Stevens-Rumann, C.S. and Morgan, P. (2019). Tree regeneration following wildfires in the western US: a review. *Fire Ecol.* 15: 1–17.

Stevens-Rumann, C.S., Kemp, K.B., Higuera, P.E. et al. (2018). Evidence for declining forest resilience to wildfires under climate change. *Ecol. Lett.* 21: 243–252.

Stewart, G.H., Rose, A.B., and Veblen, T.T. (1991). Forest development in canopy gaps in old-growth beech (*Nothofagus*) forests, *New Zealand. J. Veg. Sci.* 2: 679–690.

Stewart, J.R., Lister, A.M., Barnes, I., and Dalén, L. (2010). Refugia revisited: individualistic responses of species in space and time. *Proc. R. Soc. B Biol. Sci.* 277: 661–671.

Stoeckeler, J.H. (1948). The growth of quaking aspen as affected by soil properties and fire. *J. For.* 46: 727–737.

Stoeckeler, J.H. (1960). Soil factors affecting the growth of quaking aspen forests in the Lake states. *Univ. Minnesota Agr. Exp. Sta. Tech.* Bull. 233. 48 pp.

Stohlgren, T.J. (2007). *Measuring Plant Diversity: Lessons from the Field.* New York: Oxford University Press 408 pp.

Stone, E.L. (1974). The communal root system of red pine: growth of girdled trees. *For. Sci.* 20: 294–305.

Stone, E.L. (1975). Windthrow influences on spatial heterogeneity in a forest soil. *Mitt. Eid. Anst. forstl. Versuchsw.* 51:77–87.

Stone, D.M. (1977). Leaf dispersal in a pole-sized maple stand. *Can. J. For. Res.* 7: 189–192.

Stone, E.L. and Kalisz, P.J. (1991). On the maximum extent of tree roots. *For. Ecol. Manag.* 46: 59–102.

Stone, E.L. and Stone, M.H. (1954). Root collar sprouts in pine. *J. For.* 52: 487–491.

Storch, F., Dormann, C.F., and Bauhus, J. (2018). Quantifying forest structural diversity based on large-scale inventory data: a new approach to support biodiversity monitoring. *For. Ecosyst.* 5: 34.

Stottlemyer, A.D., Shelburne, V.B., Waldrop, T.A. et al. (2009). Fuel characterization in the southern Appalachian Mountains: an application of landscape ecosystem classification. *Int. J. Wildland Fire* 18: 423–429.

Stowe, K., Marquis, R., Hochwender, C., and Simms, E. (2000). The evolutionary ecology of tolerance to consumer damage. *Annu. Rev. Ecol. Syst.* 31: 565–595.

Stralberg, D., Carroll, C., Pedlar, J.H. et al. (2018). Macrorefugia for North American trees and songbirds: climatic limiting factors and multi-scale topographic influences. *Glob. Ecol. Biogeogr.* 27: 690–703.

Strauss, S.Y., Webb, C.O., and Salamin, N. (2006). Exotic taxa less related to native species are more invasive. *Proc. Natl. Acad. Sci.* 103: 5841–5845.

Strimbeck, R.G., Schaberg, P.G., Fossdal, C.G. et al. (2015). Extreme low temperature tolerance in woody plants. *Front. Plant Sci.* 6: 1–15.

Stromberg, J.C., Chew, M.K., Nagler, P.I., and Glenn, E.P. (2009). Changing perceptions of change: the role of scientists in *Tamarix* and river management. *Restor. Ecol.* 17: 177–186.

Strong, W.L., Oswald, E.T., and Downing, D.J. (eds.). (1990). The Canadian vegetation classification system. First Approx. Ecological Land Classif. Series, No. 25, Env. Canada, Ottawa. 22 pp.

Stuckey, R.L. (1981). Origin and development of the concept of the Prairie Peninsula. *In* R.L. Stuckey and K.J. Reese (eds.), *The Prairie Peninsula—In the "Shadow" of Transeau*. Ohio Biol. Surv. Biol. Notes 15.

Sturrock, R.N., Frankel, S.J., Brown, A.V. et al. (2011). Climate change and forest diseases. *Plant Pathol.* 60: 133–149.

Sturtevant, B.R. and Seagle, S.W. (2004). Comparing estimates of forest site quality in old second-growth oak forests. *For. Ecol. Manag.* 191: 311–328.

Sturtevant, B.R., Bissonette, J.A., Long, J.N., and Roberts, D.W. (1997). Coarse woody debris as a function of age, stand structure, and disturbance in boreal Newfoundland. *Ecol. Appl.* 7: 702–712.

Suarez-Gonzalez, A., Lexer, C., and Cronk, Q.C. (2018). Adaptive introgression: a plant perspective. *Biol. Lett.* 14: 20170688.

Sukachev, V.N. and Dylis, N.V. (1964). *Fundamentals of Forest Biogeocoenology*. (trans. J. M. MacLennan). Edinburgh: Oliver and Boyd, Ltd. 672 pp.

Svensson, J.R., Lindegarth, M., Jonsso, P.R., and Pavia, H. (2012). Disturbance-diversity models: what do they really predict and how are they tested? *Proc. R. Soc. B Biol. Sci.* 279: 2163–2170.

Swain, A.M. (1973). A history of fire and vegetation in northeast Minnesota as recorded in lake sediments. *Quat. Res.* 3: 383–396.

Swain, A.M. (1978). Environmental changes during the past 2000 years in north-central Wisconsin: analysis of pollen, charcoal, and seeds from varved lake sediments. *Quat. Res.* 10: 55–68.

Swank, W.T. and Crossley, D.A. (ed.) (1988). *Forest Hydrology and Ecology at Coweeta*. New York: Springer-Verlag 469 pp.

Swanson, F.J. (1981). Fire and geomorphic processes. *In Proc. Fire Regimes and Ecosystems*. USDA For Serv. Gen. Tech;. Report WO-26. Washington, D.C.

Swanson, F.J., Fredriksen, R.L., and McCorison, F.M. (1982a). Material transfer in a western Oregon forested watershed. In: *Analysis of Coniferous Forest Ecosystems in the Western United States*. US/IBP Synthesis Series 14 (ed. R.L. Edmonds), 233–266. Stroudsburg, PA: Hutchinson Ross.

Swanson, F.J., Gregory, S.V., Sedell, J.R., and Campbell, A.G. (1982b). Land-water interactions: the riparian zone. In: *Analysis of Coniferous Forest Ecosystems in the Western United States*. US/IBP Synthesis Series 14 (ed. R.L. Edmonds), 267–291. Stroudsburg, PA: Hutchinson Ross.

Swanson, F.J., Kratz, T.K., Caine, N., and Woodmansee, R.G. (1988). Landform effects on ecological processes and features. *Bioscience* 38: 92–98.

Swanson, F.J., Franklin, F.J., and Sedell, J.R. (1990). Landscape patterns, disturbance, and management in the Pacific Northwest, USA. In: *Changing Landscapes: An Ecological Perspective* (ed. I.S. Zonneveld and R.T.T. Forman), 191–213. New York: Springer-Verlag.

Swanson, M.E., Franklin, J.F., Beschta, R.L. et al. (2011). The forgotten stage of forest succession: early-successional ecosystems on forest sites. *Front. Ecol. Environ.* 9: 117–125.

Swanston, C.W., Janowiak, M.K., Brandt, L.A. et al. (2016). *Forest Adaptation Resources: Climate Change Tools and Approaches for Land Managers*. USDA For. Serv. Gen. Tech. Rep. NRS-GTR-87-2. Northern Res. Sta., Newtown Square, PA. 161 pp.

Swenson, J.J. and Waring, R.H. (2006). Modelled photosynthesis predicts woody plant richness at three geographic scales across the northwestern United States. *Glob. Ecol. Biogeogr.* 15: 470–485.

Swetnam, T.W. (1993). Fire history and climate change in giant sequoia groves. *Science* 262: 885–889.

Swetnam, T.W., Baisan, C.H., Caprio, A.C. et al. (2009). Multi-millennial fire history of the Giant Forest, Sequoia National Park, California, USA. *Fire Ecol.* 5: 120–150.

Swift, M.J., Heal, O.W., and Anderson, J.M. (1979). *Decomposition in Terrestrial Ecosystems*. Oxford, UK: Blackwell Scientific 372 pp.

Swihart, R.K. and Bryant, J.P. (2001). Importance of biogeography and ontogeny of woody plants in winter herbivory by mammals. *J. Mammal.* 82: 1–21.

Taiz, L. and Zeiger, E. (1991). *Plant Physiology*. Redwood City, CA: Benjamin/Cummings 559 pp.

Tallis, J.H. (1991). *Plant Community History*. New York: Chapman and Hall 398 pp.

Tansley, A.G. (1929). Succession: the concept and its values. *In* B.M. Duggar (ed.), *Proc. International Congress of Plant Sciences,* Ithaca, NY, 1926. George Banta Publ. Co., Menasha, WS.

Tansley, A.G. (1935). The use and abuse of vegetational concepts and terms. *Ecology* 16: 284–307.

Tansley, A.G. (1939). *The British Islands and their Vegetation*. Cambridge: Cambridge Univ. Press 930 pp.

Tappeiner, J., Zasada, J., Ryan, P., and Newton, M. (1991). Salmonberry clonal and populations structure: the basis for a persistent cover. *Ecology* 72: 609–618.

Tardif, J., Flannigan, M., and Bergeron, Y. (2001). An analysis of the daily radial activity of 7 boreal tree species, northwestern Quebec. *Environ. Monit. Assess.* 67: 141–160.

Tarrant, R.F. (1956b). Effect of slash burning on some physical soil properties. *For. Sci.* 2: 18–22.

Tate, K.W., Dudley, D.M., McDougald, N.K., and George, M.R. (2004). Effect of canopy and grazing on soil bulk density. *J. Range Manag.* 57: 411–417.

Taub, D.R. and Goldberg, D. (1996). Root system topology of plants from habitats of differing soil resource availability. *Funct. Ecol.* 10: 258–264.

Taulavuori, K.M.J., Laine, K., and Taulavuori, E.B. (2002). Artificial deacclimation response of *Vaccinium myrtillus* in mid-winter. *Ann. Bot. Fenn.* 39: 143–147.

Taylor, A.R. (1974). Forest fire. In: *Yearbook of Science and Technology* (ed. D. N. Lapedes and). London: McGraw-Hill.

Taylor, T.M.C. (1959). The taxonomic relationship between *Picea glauca* (Moench) Voss and *P. engelmannii* parry. *Madroño* 15: 111–115.

Taylor, S.W. and Carroll, A.L. (2004). Disturbance, forest age, and mountain pine beetle outbreak dynamics in BC: a historical perspective. *In* T.L. Shore, J.E. Brooks, J.E. Stone (eds.). *Mountain Pine Beetle Symposium: Challenges and Solutions*. Information Report BC-X-399. Natural Resources Canada, Canadian Forest Service, Pacific Forestry Center, Victoria, BC.

Taylor, A.H. and Skinner, C.N. (1998). Fire history and landscape dynamics in a late-successional reserve, Klamath Mountains, California, USA. *For. Ecol. Manag.* 111: 285–301.

Taylor, A.H. and Skinner, C.N. (2003). Spatial patterns and controls on historical fire regimes and forest structure in the Klamath Mountains. *Ecological Applications* 13 (3): 704–719.

Taylor, S., Carroll, A., Alfaro, R., and Safranyik, L. (2006). Forest, climate and mountain pine beetle outbreak dynamics in western Canada. In: *The Mountain Pine Beetle: A Synthesis of Biology, Management, and Impacts on Lodgepole Pine* (ed. L. Safranyik and B. Wilson). Pacific Forestry Centre, Victoria, BC: Natural Resources Canada, Canadian Forest Service.

Teipner, C.L., Garton, E.O., and Nelson, L. Jr. (1983). Pocket gophers in forest ecosystems. USDA For. Serv. Gen. Tech. Report INT-154. Intermountain For. and Rge. Exp. Sta., Ogden, UT. 53 pp.

Tepley, A.J., Thompson, J.R., Epstein, H.E., and Anderson-Teixeira, K.J. (2017). Vulnerability to forest loss through altered postfire recovery dynamics in a warming climate in the Klamath Mountains. *Glob. Chang. Biol.* 23: 4117–4132.

Terborgh, J. (1989). *Where Have all the Birds Gone?* Princeton, NJ: Princeton Univ. Press 207 pp.

The Nature Conservancy (1996). *Conservation by design: A framework for mission success. Arlington, VA* (Updated 2011). Spatial data available at http://maps.tnc.org. Accessed 05-09-2021.

Thom, D. and Seidl, R. (2016). Natural disturbance impacts on ecosystem services and biodiversity in temperate and boreal forests. *Biol. Rev.* 91: 760–781.

Thom, D., Rammer, W., and Seidl, R. (2017). The impact of future forest dynamics on climate: interactive effects of changing vegetation and disturbance regimes. *Ecol. Monogr.* 87: 665–684.

Thomas, S.C. and Winner, W.E. (2000). Leaf area index of an old-growth Douglas-fir forest estimated from direct structural measurements in the canopy. *Can. J. For. Res.* 30: 1922–1930.

Thompson, S.C.G. and Barton, M.A. (1994). Ecocentric and anthropocentric attitudes toward the environment. *J. Environ. Psychol.* 14: 149–158.

Thompson, M.P., MacGregor, D.G., Dunn, C.J. et al. (2018). Rethinking the wildland fire management system. *J. For.* 116: 382–390.

Thomsen, M.S., Altieri, A.H., Angelini, C. et al. (2018). Secondary foundation species enhance biodiversity. *Nat. Ecol. Evol.* 2: 634–639.

Thomson, A.M., Dick, C.W., and Dayanandan, S. (2015). A similar phylogeographical structure among sympatric North American birches (*Betula*) is better explained by introgression than by shared biogeographical history. *J. Biogeogr.* 42: 339–350.

Thoreau, H.D. (1993). *Faith in a Seed*. Washington D.C: Island Press 283 pp.

Tiedemann, A.R., Clary, W.P., and Barbour, R.J. (1987). Underground systems of Gambel oak (*Quercus gambelii*) in Central Utah. *Am. J. Bot.* 74: 1065–1071.

Tiedje, J.M. (1988). Ecology of denitrification and dissimilatory nitrate reduction to ammonium. In: *Biology of Anaerobic Microorganisms* (ed. A.J.B. Zehnder), 179–244. New York: Wiley.

Tiedje, J.M., Simkins, S., and Groffman, P.M. (1989). Perspectives on measurement of denitrification in field including recommended protocols for acetylene based methods. *Plant Soil* 115: 261–284.

Tilghman, N.G. (1989). Impacts of white-tailed deer on forest regeneration in northwestern Pennsylvania. *J. Wildl. Manag.* 53: 524–532.

Tilman, D. (1982). *Resource Competition and Community Structure*. Princeton, NJ: Princeton University Press 296 pp.

Tilman, D. (1985). The resource-ratio hypothesis of plant succession. *Am. Nat.* 125: 827–852.

Tilman, D. (1988). *Plant Strategies and the Dynamics and Structure of Plant Communities*. Princeton, NJ: Princeton Univ. Press 360 pp.

Tilman, D. (1996). Biodiversity: population versus ecosystem stability. *Ecology* 77: 350–363.

Tilman, D., Reich, P.B., and Knops, J.M.H. (2006). Biodiversity and ecosystem stability in a decade-long grassland experiment. *Nature* 441: 629–632.

Tilman, D., Reich, P.B., and Isbell, F. (2012). Biodiversity impacts ecosystem productivity as much as resources, disturbance, or herbivory. *Proc. Natl. Acad. Sci.* 109: 10394–10397.

Tinker, D.B., Romme, W.H., Hargrove, W.W. et al. (1994). Landscape-scale heterogeneity in lodgepole pine serotiny. *Can. J. For. Res.* 24: 897–903.

Tinker, D.B., Romme, W.H., and Despain, D.G. (2003). Historic range of variability in landscape structure in subalpine forests of the Greater Yellowstone Area, USA. *Landsc. Ecol.* 18: 427–439.

Titus, J.H. (1990). Microtopography and woody plant regeneration in a hardwood floodplain swamp in Florida. *Bull. Torrey Bot. Club* 117: 429–127.

Tomlinson, P.B. (1983). Tree architecture. *Am. Scientist* 71: 141–149.

Toumey, J.W. and Kienholz, R. (1931). Trenched plots under forest canopies. *Yale Univ. School For. Bull.* 30. 31 pp.

Tovar-Sanchez, E. and Oyama, K. (2006). Effect of hybridization of the *Quercus crassifolia*× *Quercus crassipes* complex on the community structure of endophagous insects. *Oecologia* 147: 702–713.

Townsend, A.M., Bentz, S.E., and Johnson, G.R. (1995). Variation in response of selected American elm clones to *Ophlostoma ulmi*. *J. Environ. Hortic.* 13: 126–128.

Transeau, E.N. (1935). The prairie peninsula. *Ecology* 16: 423–137.

Trappe, J.M. and Fogel, R.D. (1977). Ecosystematic function of mycorrhizae. *In* J.K. Marshall (ed.), *The Belowground Ecosystem: A Synthesis of Plant-Associated Processes*. Range Sci. Dep. Sci. Ser. No. 26. Colorado State Univ., Ft. Collins.

Trense, D. and Tietze, D.T. (2018). Studying speciation: genomic essentials and approaches. In: *Bird Species* (ed. D. Tietze), 39–61. New York: Springer.

Treseder, K.K., Mack, M.C., and Cross, A. (2004). Relationships among fires, fungi, and soil dynamics in Alaskan boreal forests. *Ecol. Appl.* 14: 1826–1838.

Treshow, M. (1970). *Environment and Plant Response*. New York: McGraw-Hill 422 pp.

Trewartha, G.T. (1968). *An Introduction to Climate*, 4e. New York: McGraw-Hill 408 pp.

Trimble, G.R. Jr. and Weitzman, S. (1956). Site index studies of upland oaks in the northern Appalachians. *For. Sci.* 2: 162–173.

Tritton, L.M. and Hornbeck, J.W. (1982). Biomass equations for major tree species of the Northeast. USDA For. Serv. Gen. Tech. Report NE-69. Northeastern For. Exp. Sta., Upper Darby, PA. 52 pp.

Troeger, R. (1960). Kiefernprovenienzversuche. I. Teil. Der grosse Kiefernprovenienzversuch im südwürttembergischen Forstbezirk Schussenried. *AFJZ* 131 (3–4): 49–59.

Trofymow, J.A., Addison, J., Blackwell, B.A. et al. (2003). Attributes and indicators of old growth and successional Douglas-fir forests on Vancouver Island. *Environ. Rev.* 11: S187–S204.

Troll, C. (1939). *Luftbildplan and ökologische Bodenforschung*, 241–298. Berlin: Zeitschrift der Gesellschaft für Erdkunde.

Troll, C. (1963a). Landscape ecology and land development with special reference to the tropics. *J. Trop. Geogr.* 17: 1–11.

Troll, C. (1963b). Über Landschafts-Sukession. *Arbeiten zur Rheinschen Landeskunde Bonn*. 19:5–12.

Troll, C. (1968). Landschaftsökologie. *In* R. Tüxen (ed.), *Pflanzensoziologie und Landschaftsökologie. Berichte das Internalen Symposiums der Internationalen Vereinigung für Vegetationskunde, Stolzenau/Weser 1963*. W. Junk, The Hague.

Troll, C. (1971). Landscape ecology (geo-ecology) and bioceonology—a terminology study. *Geoforum* 8: 43–46.

Trouet, V. (2020). *Tree Story: The History of the World Written in Rings*. Baltimore, MD: Johns Hopkins University Press 256 pp.

Tubbs, C.H. (1965). Influence of temperature and early spring conditions on sugar maple and yellow birch germination in upper Michigan. USDA For. Serv. Res. Note LS-72. North Central For. Exp. Sta., St. Paul, MN. 2 pp.

Tucker, G.F. and Emmington, W.H. (1977). Morphological changes in leaves of residual western hemlock after clear and shelterwood cutting. *For. Sci.* 23: 195–203.

Tucker, M.M., Corace, R.G., Cleland, D.T., and Kashian, D.M. (2016). Long-term effects of managing for an endangered songbird on the heterogeneity of a fire-prone landscape. *Landsc. Ecol.* 31: 2445–2458.

Tukey, H.B. (1970). The leaching of substances from plants. *Ann. Rev. Plant Physiology* 21: 305–324.

Turetsky, M.R., Abbott, B.W., Jones, M.C. et al. (2019). Permafrost collapse is accelerating carbon release. *Nature* 569: 32–34.

Turnbull, C.G. (ed.) (2005). *Plant Architecture and its Manipulation*. Oxford, UK: Blackwell 336 pp.

Turner, M.G. (2010). Disturbance and landscape dynamics in a changing world. *Ecology* 91: 2833–2849.

Turner, M.G. and Chapin, F.S. III (2005). Causes and consequences of spatial heterogeneity in ecosystem function. In: *Ecosytem Function in Heterogeneous Landscapes* (ed. G.M. Lovett, C.G. Jones, M.G. Turner and K.C. Weathers). New York: Springer.

Turner, M.G. and Gardner, R.H. (ed.) (1991). *Quantitative Methods in Landscape Ecology: The Analysis and Interpretation of Landscape Heterogeneity*. New York: Springer-Verlag 536 pp.

Turner, M.G. and Gardner, R.H. (2015). *Landscape Ecology in Theory and Practice*, 2e. New York: Springer-Verlag 482 pp.

Turner, M.G. and Romme, W.H. (1994). Landscape dynamics in crown fire ecosystems. *Landsc. Ecol.* 9: 59–77.

Turner, M.G., Gardner, R.H., Dale, V.H., and O'Neill, R.V. (1989). Predicting the spread of disturbance across heterogeneous landscapes. *Oikos* 55: 121–129.

Turner, M.G., Hargrove, W.W., Gardner, R.H., and Romme, W.H. (1994). Effects of fire on landscape heterogeneity in Yellowstone National Park, Wyoming. *J. Veg. Sci.* 5: 731–742.

Turner, J.R.G., Lennon, J.J., and Greenwood, J.J.D. (1996). Does climate cause the global biodiversity gradient? In: *Aspects of the Genesis and Maintenance of Biological Diversity* (ed. M.E. Hochberg, J. Clobert and R. Barbault), 199–220. Oxford, UK: Oxford University Press.

Turner, M.G., Romme, W.H., Gardner, R.H., and Hargrove, W.W. (1997). Effects of fire size and pattern on early succession in Yellowstone National Park. *Ecol. Monogr.* 67: 411–433.

Turner, M.G., Romme, W.H., and Gardner, R.H. (1999). Prefire heterogeneity, fire severity and plant reestablishment in subalpine forests of Yellowstone National Park, Wyoming. *Int. J. Wildland Fire* 9: 21–36.

Turner, M.G., Romme, W.H., and Tinker, D.B. (2003a). Surprises and lessons from the 1988 Yellowstone fires. *Front. Ecol. Environ.* 1: 351–358.

Turner, M.G., Romme, W.H., and Reed, R.A. (2003b). Postfire aspen seedling recruitment across the Yellowstone (USA) landscape. *Landsc. Ecol.* 18: 127–140.

Turner, M.G., Tinker, D.B., Romme, W.H. et al. (2004). Landscape patterns of sapling density, leaf area, and aboveground net primary production in postfire lodgepole pine forests, Yellowstone National Park (USA). *Ecosystems* 7: 751–775.

Turner, M.G., Donato, D.C., and Romme, W.H. (2013). Consequences of spatial heterogeneity for ecosystem services in changing forest landscapes: priorities for future research. *Landsc. Ecol.* 28: 1081–1097.

Turner, M.G., Whitby, T.G., Tinker, D.B., and Romme, W.H. (2016). Twenty-four years after the Yellowstone Fires: are postfire lodgepole pine stands converging in structure and function? *Ecology* 97: 1260–1273.

Turner, M.G., Braziunas, K.H., Hansen, W.D., and Harvey, B.J. (2019). Short-interval severe fire erodes the resilience of subalpine lodgepole pine forests. *Proc. Natl. Acad. Sci.* 116: 11319–11328.

Turreson, G. (1922a). The species and the variety as ecological units. *Hereditas* 3: 100–113.

Turreson, G. (1922b). The genotypical response of the plant species to the habitat. *Hereditas* 3: 211–350.

Turreson, G. (1923). The scope and import of genecology. *Hereditas* 4: 171–176.

U. S. Department of Agriculture (1974). *Seeds of Woody Plants in the United States*. USDA For. Serv. Agr. Handbook No. 450. Washington, D.C. 883 pp.

U. S. Department of Agriculture (1990). *Silvics of North America*. USDA For. Serv. Agr. Handbook 654. Washington, D.C. Vol. 1, Conifers, 675 pp; Vol. 2, Hardwoods, 877 pp.

U. S. Department of Agriculture Natural Resources Conservation Service (2020). Snow telemetry (SNOTEL) and snow course data and products. http://www.wcc.nrcs.usda.gov/snow/index.html. Accessed 09-21-2021.

U. S. Global Change Research Program (2017). Climate science special report: Fourth National Climate Assessment, Volume I. https://science2017.globalchange.gov. Accessed 9/19/2021.

U. S. National Vegetation Classification (2019). United States National Vegetation Classification Database, V2.03. Federal Geographic Data Committee, Vegetation Subcommittee, Washington DC. [http://usnvc.org] (accessed 05-09-2021).

Uehara, G. and Gillman, G. (1981). *The Mineralogy, Chemistry and Physics of Tropical Soils with Variable Charge Clays*. Boulder, CO: Westview Press 170 pp.

Uemura, M., Tominaga, Y., Nakagawara, C. et al. (2006). Responses of the plasma membrane to low temperatures. *Physiol. Plant.* 126: 81–89.

United Nations Environment Programme-World Conservation Monitoring Centre, Internationla Union for Conservation of Nature, National Geographic Society. 2020. Protected Planet Live Report 2020. UNEP-WCMC, IUCN and NGS: Cambridge UK: Gland, Switzerland and Washington, DC, USA. https://livereport.protectedplanet.net. Accessed 07-28-2021.

Urban, M.C. (2015). Accelerating extinction risk from climate change. *Science* 348: 571–573.

Urgenson, L.S., Halpern, C.B., and Anderson, P.D. (2013). Twelve-year responses of planted and naturally regenerating conifers to variable-retention harvest in the Pacific Northwest, USA. *Can. J. For. Res.* 43: 46–55.

Usbeck, T., Wohlgemuth, T., Dobbertin, M. et al. (2010). Increasing storm damage to forests in Switzerland from 1858 to 2007. *Agric. For. Meteorol.* 150: 47–55.

Ustin, S.L., Woodward, R.A., and Barbour, M.G. (1984). Relationships between sunfleck dynamics and red fir seedling distribution. *Ecology* 65: 1420–1428.

Vale, T.R. (1982). *Plants and People: Vegetation Change in North America*. Washington, DC: Associatio of American Geographers 88 pp.

Valladares, F. and Niinemets, Ü. (2008). Shade tolerance, a key plant feature of complex nature and consequences. *Annu. Rev. Ecol. Evol. Syst.* 39: 237–257.

Van Auken, O.W. (2009). Causes and consequences of woody plant encroachment into western North American grasslands. *Journal of environmental management* 90 (10): 2931–2942.

Van Buskirk, E.K., Reddy, A.K., Nagatani, A., and Chen, M. (2014). Photobody localization of phytochrome B is tightly correlated with prolonged and light-dependent inhibition of hypocotyl elongation in the dark. *Plant Physiol.* 165: 595–607.

Van Cleve, K., Oliver, L., Schlentner, R. et al. (1983). Productivity and nutrient cycling in taiga forests. *Can. J. For. Res.* 13: 747–766.

Van Cleve, K., Chapin, F.S. III, Flanagan, P.W. et al. (ed.) (1986). *Forest Ecosystems in the Alaskan Taiga*. New York: Springer-Verlag 230 pp.

Van de Peer, Y., Mizrachi, E., and Marchal, K. (2017). The evolutionary significance of polyploidy. *Nat. Rev. Genet.* 18: 411–424.

Van den Berg, L.J.L., Bullock, J.M., Clarke, R.T. et al. (2001). Territory selection by the Dartford warbler (*Sylvia undata*) in Dorset, England: the role of vegetation type, habitat fragmentation and population size. *Biol. Conserv.* 101: 217–228.

Van der Heijden, M.G., Martin, F.M., Selosse, M.A., and Sanders, I.R. (2015). Mycorrhizal ecology and evolution: the past, the present, and the future. *New Phytol.* 205: 1406–1423.

Van der Maarel, E. (1975). The Braun-Blanquet approach in perspective. *Vegetatio* 30: 213–219.

Van der Pijl, L. (1957). The dispersal of plants by bats. *Acta Bot. Neerl.* 6: 291–315.

Van der Pijl, L. (1972). *Principles of Dispersal in Higher Plants*. New York: Springer-Verlag 162 pp.

Van Dyke, F. and Lamb, R.L. (2020). *Conservation Biology: Foundations, Concepts, Applications*, 3e. New York: Springer 644 pp.

Van Kleunen, M., Weber, E., and Fischer, M. (2010). A meta-analysis of trait differences between invasive and non-invasive plant species. *Ecol. Lett.* 13: 235–245.

Van Mantgem, P.J., Stephenson, N.L., Byrne, J.C. et al. (2009). Widespread increase of tree mortality rates in the western United States. *Science* 323: 521–524.

Van Ommen Kloeke, A.E.E., Douma, J.C., Ordoñez, J.C. et al. (2012). Global quantification of contrasting leaf life span strategies for deciduous and evergreen species in response to environmental conditions. *Global Ecol. Biogeogr.* 21: 224–235.

Van Veen, J.A., Ladd, J.N., and Frissel, M.J. (1984). Modelling C & N turnover through the microbial biomass in soil. *Plant Soil* 76: 257–274.

Van Wagner, C.E. (1970). Fire and red pine. *In Proc. Annual Tall Timbers Fire Ecology Conf.* 10:211–219. Tall Timbers Res. Sta., Tallahassee, FL.

Van Wagner, C.E. (1983). Fire behaviour in northern conifer forests and shrublands. In: *The Role of Fire in Northern Circumpolar Ecosystems* (ed. R.W. Wein and D.A. MacLean), 65–80. New York: Wiley.

Van Zandt, P.A. (2007). Plant defense, growth, and habitat: a comparative assessment of constitutive and induced resistance. *Ecology* 88: 1984–1993.

Vander Wall, S.B. (1990). *Food Hoarding in Animals*. Chicago, IL, USA: University of Chicago Press 453 pp.

Vander Wall, S.B. (2010). How plants manipulate the scatter-hoarding behaviour of seed-dispersing animals. *Philos. Trans. R. Soc. B* 365: 989–997.

Vander Wall, S.B. and Balda, R.P. (1977). Co-adaptations of the Clark's nutcracker and the piñon pine for efficient seed harvest and dispersal. *Ecol. Monogr.* 47: 89–111.

Vanderwel, M.C., Coomes, D.A., and Purves, D.W. (2013). Quantifying variation in forest disturbance, and its effects on aboveground biomass dynamics, across the eastern United States. *Glob. Chang. Biol.* 19: 1504–1517.

Vankat, J.L. (1979). *The Natural Vegetation of North America*. New York: Wiley 486 pp.

Varela, E., Górriz-Mifsud, E., Ruiz-Mirazo, J., and López-i-Gelats, F. (2018). Payment for targeted grazing: integrating local shepherds into wildfire prevention. *Forests* 9: 464.

Varnell, L.M. (1998). The relationship between inundation history and baldcypress stem form in a Virginia floodplain swamp. *Wetlands* 18: 176–183.

Vašutová, M., Mleczko, P., López-García, A. et al. (2019). Taxi drivers: the role of animals in transporting mycorrhizal fungi. *Mycorrhiza* 29: 413–434.

Veatch, J.O. (1953). *Soils and Land of Michigan*. East Lansing: Michigan State College Press 241 pp.

Veatch, J.O. (1959). Presettlement forest in Michigan. Michigan State Univ., Dept. Res Development. East Lansing. Map.

Veblen, T.T. (1987). Trees of the trembling earth. *Nat. Hist.* 96: 43–46.

Veblen, T.T. (1992). Regeneration dynamics. In: *Plant Succession, Theory and Prediction* (ed. D.C. Glenn-Lewin, R.K. Peet and T.T. Veblen), 152–187. London: Chapman & Hall.

Veblen, T.T. (2000). Disturbance patterns in southern Rocky Mountain forests. Forest fragmentation in the southern Rocky Mountains. University Press of Colorado, Boulder, Colorado, USA, pp.31-54.

Veblen, T.T. and Lorenz, D.C. (1986). Anthropogenic disturbance and recovery patterns in montane forests, Colorado Front Range. *Phys. Geogr.* 7: 1–24.

Veblen, T.T. and Lorenz, D.C. (1988). Recent vegetation changes along the forest/steppe ecotone of northern Patagonia. *Annals Assoc. Amer. Geographers* 78: 93–111.

Veblen, T.T. and Lorenz, D.C. (1991). *The Colorado Front Range, A Century of Ecological Change*. Salt Lake City: Univ. Utah Press 210 pp.

Veblen, T.T., Hadley, K.S., and Reid, M.S. (1991a). Disturbance and stand development of a Colorado subalpine forest. *J. Biogeogr.* 18: 707–716.

Veblen, T.T., Hadley, K.S., Reid, M.S., and Rebertus, A.J. (1991b). The response of subalpine forests to spruce beetle outbreak in Colorado. *Ecology* 72: 213–231.

Veblen, T.T., Hadley, K.S., Nel, E.M. et al. (1994). Disturbance regime and disturbance interactions in a Rocky Mountain subalpine forest. *J. Ecol.* 82: 125–135.

Veblen, T.T., Kitzberger, T., and Donnegan, J. (2000). Climatic and human influences on fire regimes in ponderosa pine forests in the Colorado Front Range. *Ecol. Appl.* 10: 1178–1195.

Vegis, A. (1964). Dormancy in higher plants. *Annu. Rev. Plant Physiol.* 15: 185–224.

Venette, R.C., Gordon, D.R., Juzwik, J. et al. (2021). Early intervention strategies for invasive species management: connections between assessment, prevention efforts, eradication, and other rapid responses. In: *Invasive Species in Forests and Rangelands of the United States* (ed. T.M. Poland, T. Patel-Weynand, D.M. Finch, et al.), 111–131. New York: Springer.

Venterea, R.T., Groffman, P.M., Verchot, L.V. et al. (2003). Nitrogen oxide gas emissions from temperate forest soils receiving long-term nitrogen inputs. *Glob. Chang. Biol.* 9: 346–357.

Verrall, A.F. and Graham, T.W. (1935). The transmission of *Ceratostomella ulmi* through root grafts. *Phytopathology* 25: 1039–1040.

Vézina, P.E. and Boulter, D.W.K. (1966). The spectral composition of near ultraviolet and visible radiation beneath forest canopies. *Can. J. Bot.* 44: 1267–1284.

Viereck, L.A. (1970). Forest succession and soil development adjacent to the Chena River in interior Alaska. *Arct. Alp. Res.* 2: 1–26.

Viereck, L.A. (1973). Wildfire in the taiga of Alaska. *Quat. Res.* 3: 465–495.

Viereck, L.A. (1982). Effects of fire and firelines on active layer thickness and soil temperatures in interior Alaska. *In* H. M. French (ed.), *Proc. Fourth Can. Permafrost Conference,* Calgary. Natl. Res. Council Can. Ottawa.

Viereck, L.A. (1983). The effects of fire in black spruce ecosystems of Alaska and northern Canada. In: *The Role of Fire in Northern Circumpolar Ecosystems* (ed. R.W. Wein and D.A. MacLean), 210–220. New York: Wiley.

Viereck, L.A. and Foote, J.M. (1970). The status of *Populus balsamifera* and *P. trichocarpa* in Alaska. *Canad. Field-Nat.* 84: 169–174.

Viereck, L. A. Dyrness, C. T., and Van Cleve, K. (1984). Potential use of the Alaska vegetation system as an indicator of forest site productivity in interior Alaska. *In* M. Murray (ed.), *Proc. Workshop, Forest Classification at High Latitudes as an Aid to Regeneration*. USDA For. Serv. Gen. Tech. Report PNW-177. Pacific Northwest For. and Rge. Exp. Sta., Portland, OR.

Vierling, L.A. and Wessman, C.A. (2000). Photosynthetically active radiation heterogeneity within a monodominant Congolese rain forest canopy. *Agric. For. Meteorol.* 103: 265–278.

Viers, S.D. (1980). The influence of fire in coast redwood forests. *In* M.A. Stokes and J.H. Dieterich (eds.), *Proc. Fire History Workshop*. USDA For Serv. Gen Tech. Report. RM-81, Rocky Mountain For. and Rge. Exp. Sta., Fort Collins, CO.

Vince-Prue, D. (1975). *Photoperiodism in Plants.* New York: McGraw-Hill 444 pp.

Viro, P.J. (1974). Effects of forest fire on soil. In: *Fire and Ecosystems* (ed. T.T. Kozlowski and C.E. Ahlgren), 7–45. New York: Academic Press.

Vitasse, Y., Lenz, A., and Körner, C. (2014). The interaction between freezing tolerance and phenology in temperate deciduous trees. *Front. Plant Sci.* 5: 541.

Vitousek, P.M. (1982). Nutrient cycling and nutrient use efficiency. *Am. Nat.* 119: 553–572.

Vitousek, P.M. (1984). Litterfall, nutrient cycling, and nutrient limitation in tropical forests. *Ecology* 65: 285–298.

Vitousek, P.M. and Matson, P.A. (1984). Mechanisms of nitrogen retention in forest ecosystems: a field experiment. *Science* 225: 51–52.

Vitousek, P.M. and Melillo, J.M. (1979). Nitrate loss from disturbed forests: patterns and mechanisms. *For. Sci.* 25: 605–619.

Vitousek, P.M. and Reiners, W.A. (1975). Ecosystem succession and nutrient retention: a hypothesis. *Bioscience* 25: 376–381.

Vitousek, P.M. and Sanford, R.L. Jr. (1986). Nutrient cycling in moist tropical forests. *Annu. Rev. Ecol. Syst.* 17: 137–167.

Vitousek, P.M., Gosz, J.R., Grier, C.C. et al. (1982). A comparative analysis of potential nitrification and nitrate mobility in forest ecosystems. *Ecol. Monogr.* 52: 155–177.

Vitousek, P.M., Walker, L.R., Whiteaker, L.D., and Mueller-Dombois, D. (1987). Biological invasion by *Myrica faya* alters ecosystem development in Hawaii. *Science* 238: 802–804.

Vitousek, P.M., Mooney, H.A., Lubchenco, J., and Melillo, J.M. (1997). Human domination of Earth's ecosystems. *Science* 277: 494–499.

Vitra, A., Lenz, A., and Vitasse, Y. (2017). Frost hardening and dehardening potential in temperate trees from winter to budburst. *New Phytol.* 216: 113–123.

Vitt, P., Havens, K., Kramer, A. et al. (2010). Assisted migration of plants: changes in latitudes, changes in attitudes. *Biol. Conserv.* 143: 18–27.

Vizcaíno-Palomar, N., Revuelta-Eugercios, B., Zavala, M.A. et al. (2014). The role of population origin and microenvironment in seedling emergence and early survival in Mediterranean maritime pine (*Pinus pinaster* Aiton). *PLoS One* 9: e109132.

Vogelmann, J.E. (1995). Assessment of forest fragmentation in southern New England using remote sensing and geographic information systems technology. *Conserv. Biol.* 9: 439–449.

Vogl, R.J. (1964). Vegetational history of the Crex Meadows, a prairie savanna in northwestern Wisconsin. *Am. Midi. Nat.* 72: 157–175.

Vogl, R. J. (1970). Fire and the northern Wisconsin pine barrens. *In Proc. Annual Tall Timbers Fire Ecology Conf.* 6:47–96. Tall Timbers Res. Sta., Tallahassee, FL.

Vogl, R.J. (1973). Ecology of knobcone pine in the Santa Ana Mountains, California. *Ecol. Monogr.* 43: 125–143.

Vogl, R.J. and Ryder, C. (1969). Effects of slash burning on conifer reproduction in Montana's Mission Range. *Northwest Sci.* 43: 135–147.

Vogl, R.J., Armstrong, W.P., White, K.L., and Cole, K.L. (1988). The closed-cone pines and cypresses. *In* M.G. Barbour and J. Major (eds), *Terrestrial Vegetation of California*. CA. Native Plant Soc, Special Publ. No. 9. Berkeley.

Vogt, K.A., Grier, C.C., Meier, C.E., and Keyes, M.R. (1983). Organic matter and nutrient dynamics in forest floors in young and mature *Abies amabilis* stands in western Washington. *Ecol. Monogr.* 53: 139–157.

Vogt, K.A., Grier, C.C., and Vogt, D.J. (1986). Production, turnover, and nutrient dynamics of above- and belowground detritus in world forests. *Adv. Ecol. Res.* 15: 303–377.

Voigt, G.K. (1968). Variation in nutrient uptake by trees. *In Forest Fertilization*. Tennessee Valley Authority, Muscle Shoals, Ala.

Von Arx, G., Dobbertin, M., and Rebetez, M. (2012). Spatio-temporal effects of forest canopy on understory microclimate in a long-term experiment in Switzerland. *Agric. For. Meteorol.* 166-7: 144–155.

Von Arx, G., Pannatier, E.G., Thimonier, A., and Rebetez, M. (2013). Microclimate in forests with varying leaf area index and soil moisture: potential implications for seeding establishment in a changing climate. *J. Ecol.* 101: 1201–1213.

Vose, J. M., Clinton, B. D., and Swank, W. T. (1994). Fire, drought, and forest management influences on pine/hardwood ecosystems in the southern Appalachians. *In Proc. 12th Int. Conf. on Fire and Forest Meteorology, Jekyll Island, GA*. Soc. Am. For., Bethesda, MD.

Vose, J.M., Clark, J.S., Luce, C.H., and Patel-Weynand, T. (ed.) (2016). *Effects of Drought on Forests and Rangelands in the United States: A Comprehensive Science Synthesis*. USDA For. Serv. Gen Tech. Rep. WO 93b. Washington, DC: Washington Office 289 pp.

Voss, E.G. (1972). *Michigan Flora. I. Gymnosperms and Monocots*. Cranbrook Inst. Sci. and Univ. Mich. Herbarium. Bloomfield Hills, MI. 488 pp.

Voss, E.G. and Crow, G.E. (1976). Across Michigan by covered wagon: a botanical expedition in 1888. *Mich. Bot.* 15: 3–70.

Wagner, F.H. (1969). Ecosystem concepts in fish and game management. In: *The Ecosystem Concept in Natural Resource Management* (ed. G.M. Van Dyne), 259–307. New York: Academic Press.

Wagner, W.H. Jr. (1983). Reticulistics: the recognition of hybrids and their role in cladistics and classification. In: *Advances in Cladistics, Vol. 2* (ed. N.I. Platnick and V.A. Funk), 63–79. New York: Columbia Univ. Press.

Wagner, W.H. Jr. and Schoen, D.J. (1976). Shingle oak (*Quercus imbricaria)* and its hybrids in Michigan. *Mich. Bot.* 15: 141–155.

Wagner, W.H. Jr., Taylor, S.R., Grieve, G. et al. (1988). Simple-leaved ashes (*Fraxinus:* Oleaceae) in Michigan. *Mich. Bot.* 27: 119–134.

Wahlenberg, W.G. (1949). Forest succession in the southern Piedmont region. *J. For.* 47: 713–715.

Waide, J.B., Caskey, W.H., Todd, R.L., and Boring, L.R. (1988). Changes in soil nitrogen pools and transformations following forest clearcutting. In: *Forest Hydrology and Ecology at Coweeta*. (Ecological Studies 66) (ed. W.T. Swank and D.A. Crossley Jr.), 221–232. New York: Springer-Verlag.

Waide, R.B., Willig, M.R., Steiner, C.F. et al. (1999). The relationship between productivity and species richness. *Annu. Rev. Ecol. Syst.* 30: 257–300.

Wakeley, P.C. (1954). Planting the southern pines. *USDA For. Serv. Agr. Monogr.* No. 18. Washington, D. C. 233 pp.

Wakeley, P.D. and Marrero, J. (1958). Five-year intercept as site index in southern pine plantations. *J. For.* 56: 332–336.

Waldrop, M.P. and Harden, J.W. (2008). Interactive effects of wildfire and permafrost on microbial communities and soil processes in an Alaskan black spruce forest. *Glob. Chang. Biol.* 14: 2591–2602.

Walker, L.R. and Chapin, F.S. III (1986). Physiological controls over seedling growth in primary succession on an Alaskan floodplain. *Ecology* 67: 1508–1523.

Walker, L.R. and Chapin, F.S. III (1987). Interactions among processes controlling successional change. *Oikos* 50: 131–135.

Walker, L.R. and Wardle, D.A. (2014). Plant succession as an integrator of contrasting ecological time scales. *Trends Ecol. Evol.* 29: 504–510.

Walker, L.R., Zasada, J.C., and Chapin, F.S. III (1986). The role of life history processes in primary succession on an Alaskan floodplain. *Ecology* 67: 1243–1253.

Walker, W.S., Barnes, B.V., and Kashian, D.M. (2003). Landscape ecosystems of the Mack Lake burn, northern Lower Michigan, and the occurrence of the Kirtland's warbler. *For. Sci.* 49: 119–139.

Walker, L.R., Wardle, D.A., Bardgett, R.D., and Clarkson, B.D. (2010). The use of chronosequences in studies of ecological succession and soil development. *J. Ecol.* 98: 725–736.

Wall, D.H., Bradford, M.A., St. John, M.G. et al. (2008). Global decomposition experiment shows soil animal impacts on decomposition are climate-dependent. *Glob. Chang. Biol.* 14: 2661–2677.

Wallace, L.L. and Dunn, E.L. (1980). Comparative photosynthesis of three gap phase successional tree species. *Oecologia* 45: 331–340.

Walters, M.B. and Reich, P.B. (1996). Are shade tolerance, survival, and growth linked? Low light and nitrogen effects on hardwood seedlings. *Ecology* 77: 841–853.

Walters, M.B. and Reich, P.B. (1997). Growth *of Acer saccharum* seedlings in deeply shaded understories of northern Wisconsin: effects of nitrogen and water availability. *Can. J. For. Res.* 27: 237–247.

Walters, M.B. and Reich, P.B. (2000a). Seed size, nitrogen supply, and growth rate affect tree seedling survival in deep shade. *Ecology* 81: 1887–1901.

Walters, M.B. and Reich, P.B. (2000b). Trade-offs in low-light CO2 exchange: a component of variation in shade tolerance among cold temperate tree seedlings. *Funct. Ecol.* 14: 155–165.

Walters, M.B., Kruger, E.L., and Reich, P.B. (1993). Growth, biomass distribution and CO_2 exchange of northern hardwood seedlings in high and low light: relationships with successional status and shade tolerance. *Oecologia* 94: 7–16.

Wan, S., Hui, D., and Luo, Y. (2001). Fire effects on nitrogen pools and dynamics in terrestrial ecosystems: a meta-analysis. *Ecol. Appl.* 11: 1349–1365.

Wang, F. and Xu, Y.J. (2009). Hurricane Katrina-induced forest damage in relation to ecological factors at landscape scale. *Environ. Monit. Assess.* 156: 491–507.

Wang, W.J., He, H.S., Thompson, F.R. et al. (2017). Changes in forest biomass and tree species distribution under climate change in the northeastern United States. *Landsc. Ecol.* 32: 1399–1413.

Wang, S., Zhang, Y., Ju, W. et al. (2020). Recent global decline of CO_2 fertilization effects on vegetation photosynthesis. *Science* 370: 1295–1300.

Wani, M.C., Taylor, H.L., Wall, M.E. et al. (1971). Plant antitumor agents. VI. The isolation and structure of taxol, a novel antileukemic and antitumor agent from *Taxus brevifolia*. *J. Am. Chem. Soc.* 93: 2325–2327.

Ward, J.V., Tockner, K., Arscott, D.B., and Claret, C. (2002). Riverine landscape diversity. *Freshw. Biol.* 47: 517–539.

Wardle, P. (1968). Engelmann spruce (*Picea engelmannii* Engel.) at its upper limits in the Front Range, Colorado. *Ecology* 49: 483–495.

Wardle, P. (1985). New Zealand timberlines. 3. A synthesis. *New Zealand J. Bot.* 23: 263–271.

Wardle, D.A. (1992). A comparative assessment of the factors which influence microbial biomass carbon and nitrogen levels in soil. *Biol. Rev.* 67: 321–358.

Wardle, D.A. and Peltzer, D.A. (2017). Impacts of invasive biota in forest ecosystems in an aboveground–belowground context. *Biol. Invasions* 19: 3301–3316.

Wareing, P.F. (1959). Problems of juvenility and flowering in trees. *J. Linn. Soc. Lond. Bot.* 56: 282–289.

Wareing, P.F. (1987). Phase change and vegetative propagation. In: *Improving Vegetatively Propagated Crops* (ed. A.J. Abbott and R.K. Atkin), 262–270. London: Academic Press.

Wareing, P.F. and Robinson, L.W. (1963). Juvenility problems in woody plants. *Rep. For. Res.* 125–127.

Waring, R.H. (1989). Ecosystems: fluxes of matter and energy. In: *Ecological Concepts* (ed. J.M. Cherrett), 17–41. Oxford: Blackwell.

Waring, R.H. (1991). Responses of evergreen trees to multiple stresses. In: *Response of Plants to Multiple Stresses* (ed. H.A. Mooney, W.E. Winner and E.J. Pell), 371–390. New York: Academic Press.

Waring, R.H. and Pitman, G.B. (1985). Modifying lodgepole pine stands to change susceptibility to mountain pine beetle attack. *Ecology* 66: 889–897.

Waring, R.H. and Schlesinger, W.H. (1985). *Forest Ecosystems: Concepts and Management*. New York: Academic Press 340 pp.

Waring, R.H., McDonald, A.J.S., Larsson, S. et al. (1985). Differences in chemical composition of plants grown at constant relative growth rates with stable mineral nutrition. *Oecologia* 66: 157–160.

Warming, E. (1909). *Oecology of Plants, an Introduction to the Study of Plant Communities*. Oxford: Clarendon Press 422 pp.

Watson, D.M. and Rawsthorne, J. (2013). Mistletoe specialist frugivores: latterday 'Johnny Appleseeds' or self-serving market gardeners? *Oecologia* 172: 925–932.

Watt, A.S. (1925). On the ecology of British beech woods with special reference to their regeneration. Part II: the development and structure of beech communities on the Sussex Downs. *J. Ecol.* 13: 27–73.

Watt, A.S. (1947). Pattern and process in the plant community. *J. Ecol.* 35: 1–22.

Watts, W.A. (1970). The full-glacial vegetation of northwestern Georgia. *Ecology* 51: 17–33.

Watts, W.A. (1979). Late quaternary vegetation of central Appalachia and the New Jersey Coastal Plain. *Ecol. Monogr.* 49: 427–469.

Watts, W.A. and Stuiver, M. (1980). Late Wisconsin climate of northern Florida and the origin of species-rich deciduous forest. *Science* 210: 325–327.

Waughman, G.J., French, R.J., and Jones, K. (1981). Nitrogen fixation in some terrestrial environments. In: *Nitrogen Fixation, Vol. I: Ecology* (ed. W.J. Broughton), 135–192. Oxford: Clarendon Press.

Way, D.A. and Oren, R. (2010). Differential responses to changes in growth temperature between trees from different functional groups and biomes: a review and synthesis of data. *Tree Physiol.* 30: 669–688.

Way, D.A. and Pearcy, R.W. (2012). Sunflecks in trees and forests: from photosynthetic physiology to global change biology. *Tree Physiol.* 32: 1066–1081.

Weaver, H. (1951). Fire as an ecological factor in southwestern ponderosa pine forests. *J. For.* 49: 93–98.

Weaver, H. (1974). Effects of fire on temperate forest: western United States. In: *Fire and Ecosystems* (ed. T.T. Kozlowski and C.E. Ahlgren), 279–319. New York: Academic Press.

Weaver, J.E. and Clements, F.E. (1929). *Plant Ecology*. New York: McGraw-Hill 520 pp.

Webb, S.L. (1986). Potential role of passenger pigeons and other vertebrates in the rapid Holocene migrations of nut trees. *Quat. Res.* 26: 367–375.

Webb, T. III, Cushing, E.J., and Wright, H.E. Jr. (1983). Holocene changes in the vegetation of the Midwest. In: *Late-Quaternary Environments of the United States, Vol. 2, The Holocene* (ed. H.E. Wright Jr.), 142–165. Minneapolis: Univ. Minnesota Press.

Webb, L.J., Tracey, J.G., and Williams, W.T. (1972). Regeneration and pattern in the subtropical rain forest. *J. Ecol.* 6: 675–695.

Webb, T.I.I.I., Bartlein, P.J., Harrison, S.P., and Anderson, K.H. (1993). Vegetation, lake levels, and climate in eastern North America for the past 18,000 years. In: *Global Climates since the Last Glacial Maximum* (ed. H.E. Wright Jr., J.E. Kutzbach, T. Webb III, et al.), 415–467. Minneapolis: Univ. Minnesota Press.

Webster, C.R. and Lorimer, C.G. (2005). Minimum opening sizes for canopy recruitment of midtolerant tree species: a retrospective approach. *Ecol. Appl.* 15: 1245–1262.

Webster, C.R., Nelson, K., and Wangen, S.R. (2005). Stand dynamics of an insular population of an invasive tree, *Acer platanoides*. *For. Ecol. Manag.* 208: 85–99.

Webster, J.R., Morkeski, K., Wojculewski, C.A. et al. (2012). Effects of hemlock mortality on streams in the southern Appalachian Mountains. *Am. Midl. Nat.* 168: 112–131.

Wehner, M.F., Arnold, J.R., Knutson, T. et al. (2017). Droughts, floods, and wildfires. In: *Climate Science Special Report: Fourth National Climate Assessment, Volume I* (ed. D.J. Wuebbles, D.W. Fahey, K.A. Hibbard, et al.), 231–256. Washington, DC: U.S. Global Change Research Program.

Weidman, R.H. (1939). Evidences of racial influence in a 25-year test of ponderosa pine. *J. Agric. Res.* 59: 855–887.

Weigelt, A., Mommer, L., Andraczek, K. et al. (2021). An integrated framework for plant form and function: a belowground perspective. *New Phytol.* 232: 42–59.

Weil, R.R. and Brady, N.C. (2017). *The Nature and Properties of Soils*, 15e. Essex, England, UK: Pearson 1104 pp.

Wein, R.W. and MacLean, D.A. (1983). *The Role of Fire in Northern Circumpolar Ecosystems*. New York: Wiley 322 pp.

Weiser, C.J. (1970). Cold resistance and injury in woody plants. *Science* 169: 1269–1278.

Welbourn, M.L., Stone, E.L., and Lassoie, J.P. (1981). Distribution of net litter inputs with respect to slope position and wind direction. *For. Sci.* 27: 651–659.

Wellner, C.A. (1970). Fire history in the northern Rocky Mountains. *In Symp. Proc. The Role of Fire in the Intermountain West*. Intermountain Fire Research Council, Univ. Montana, School of Forestry, Missoula.

Wellner, C.A. (1989). Classification of habitat types in the western United States. *In* D.E. Ferguson, P. Morgan, and F.D. Johnson (comps.), *Proceedings—Land Classifications Based on Vegetation: Applications for Resource Management*. USDA For. Serv. Gen. Tech. Report INT-257. Intermountain Res. Sta., Ogden, UT.

Wells, P.V. (1965). Scarp woodlands, transported grassland soils, and concept of grassland climate in the Great Plains region. *Science* 148: 246–249.

Wells, P.V. (1970). Postglacial vegetational history of the Great Plains. *Science* 167: 1574–1582.

Wells, G.L. (1992). *The Aeolian Landscapes of North America for the Late Pleistocene*. PhD Thesis, University of Oxford, U.K. 256 pp.

Wells, O.O. and Wakeley, P.C. (1966). Geographic variation in survival, growth and fusiform rust infection of planted loblolly pine. *For. Sci. Monogr.* 11. 40 pp.

Wells, O.O., Switzer, G.L., and Schmidtling, R.C. (1991). Geographic variation in Mississippi loblolly pine and sweetgum. *Silvae Genet.* 40: 105–119.

Wendel, G.W. (1987). Abundance and distribution of vegetation under four hardwood stands in north-central West Virginia. USDA For. Serv. Res. Paper NE-607. Northeastern For. Exp. Sta., Broomall, PA. 6 pp.

Wertz, W.A. and Arnold, J.F. (1972). Land systems inventory. USDA For. Serv., Intermountain Region, Ogden, UT. 12 pp.

Wertz, W.A. and Arnold, J.F. (1975). Land stratification for land-use planning. In: *Forest Soils and Forest Land Management* (ed. B. Bernier and C.H. Winget), 617–629. Quebec: Laval Univ. Press.

West, D.C., Shugart, H.H., and Botkin, D.B. (1981). Introduction. In: *Forest Succession: Concepts and Application* (ed. D.C. West, H.H. Shugart and D.B. Botkin). New York: Springer-Verlag.

Westerling, A.L., Hidalgo, H.G., Cayan, D.R., and Swetnam, T.W. (2006). Warming and earlier spring increase western US forest wildfire activity. *Science* 313: 940–943.

Westerling, A.L., Turner, M.G., Smithwick, E.A.H. et al. (2011). Continued warming could transform Greater Yellowstone fire regimes by mid-21st century. *Proc. Natl. Acad. Sci.* 108: 13165–13170.

Westhoff, V. and van der Maarel, E. (1973). The Braun-Blanquet approach. In: *Handbook of Vegetation Science, Part V, Ordination and Classification of Communities* (ed. R. Tüxen), 287–399. The Hague: Junk.

Westman, W.E. (1990). Managing for biodiversity. *Bioscience* 40: 26–33.

Westoby, J. (1989). *Introduction to World Forestry: People and their Trees*. Hoboken, NJ: Blackwell 240 pp.

Westphal, M.I., Browne, M., MacKinnon, K., and Noble, I. (2008). The link between international trade and the global distribution of invasive alien species. *Biol. Invasions* 10: 391–398.

Wharton, C.H., Kitchens, W.M., Pendleton, E.C., and Sipe, T.W. (1982). The ecology of bottomland hardwood swamps of the Southeast: a community profile. U. S. Fish and Wildlife Service, FWS/OBS-81/37. 133 pp.

Wheeler, N.C. and Critchfield, W.B. (1985). The distribution and botanical characteristics of lodgepole pine: biogeographical and management implications. *In* D.M. Baumgartner, R.G. Krebill, J.T. Arnott, and G.F. Weetman (eds.), *Symp. Proc. Lodgepole Pine, The Species and Its Management.* Washington State Univ., Pullman.

Whelan, R.J. (1995). *The Ecology of Fire*. Cambridge: Cambridge Univ. Press 346 pp.

Whitcomb, R.F., Robbins, C.S., Lynch, J.F. et al. (1981). Effects of forest fragmentation on avifauna of the eastern deciduous forest. In: *Forest Island Dynamics in Man-Dominated Landscapes* (ed. R.L. Burgess and D.M. Sharpe), 125–205. New York: Springer-Verlag.

White, T.C.R. (1978). The importance of a relative shortage of food in animal ecology. *Oecologia* 33: 71–86.

White, P.S. (1979). Pattern, process and natural disturbance in vegetation. *Bot. Rev.* 45: 229–299.

White, A.S. (1983). The effects of thirteen years of annual prescribed burning on a *Quercus ellipsoidalis* community in Minnesota. *Ecology* 64: 1081–1085.

White, J. (1985). The thinning rule and its application to mixtures of plant populations. In: *Studies on Plant Demography* (ed. J. White), 291–309. New York: Academic Press.

White, A.S. (1986). Prescribed burning for oak savanna restoration in central Minnesota. USDA For. Serv. Res. Paper NC-266. North Central For. Res. Sta., St. Paul, MN. 12 pp.

White, C.S. (1986). Volatile and water-soluble inhibitors of nitrogen mineralization and nitrification in a ponderosa pine ecosystem. *Biol. Fertil. Soils* 2: 97–104.

White, C.S. (1988). Nitrification inhibition by monoterpeniods: theoretical mode of action based on molecular structures. *Ecology* 69: 1631–1633.

White, P.S. and Jentsch, A. (2001). The search for generality in studies of disturbance and ecosystem dynamics. *Prog. Bot.* 62: 399–450.

White, T.L., Adams, W.T., and Neale, D.B. (2007). *Forest Genetics*. Cambridge, MA: CABI Publishing 704 pp.

Whitehead, D.R. (1981). Late-Pleistocene vegetational changes in northeastern North Carolina. *Ecol. Monogr.* 51: 451–471.

Whitham, T.G. (1989). Plant hybrid zones as sinks for pests. *Science* 244: 1490–1493.

Whitham, T.G. and Mopper, S. (1985). Chronic herbivory: impacts on architecture and sex expression of pinyon pine. *Science* 228: 1089–1091.

Whitham, T.G. and Slobodchikoff, C.N. (1981). Evolution by individuals, plant-herbivore interactions, and mosaics of genetic variability: the adaptive significance of somatic mutations in plants. *Oecologia* 49: 287–292.

Whitham, T.G., Morrow, P.A., and Potts, B.M. (1991). Conservation of hybrid plants. *Science* 254: 779–780.

Whitham, T.G., Morrow, P.A., and Potts, B.M. (1994). Plant hybrid zones as centers of biodiversity: the herbivore community of two endemic Tasmanian eucalypts. *Oecologia* 97: 481–190.

Whitlock, C. and Bartlein, P.J. (1997). Vegetation and climate change in northwest America during the past 125 kyr. *Nature* 388: 57–61.

Whitney, G.G. (1994). *From Coastal Wilderness to Fruited Plain: A History of Environmental Change in Temperate North America from 1500 to the Present.* Cambridge: Cambridge Univ. Press 451 pp.

Whitney, H.E. and Johnson, W.C. (1984). Ice storms and forest succession in southwestern Virginia. *Bull. Torrey Bot. Club* 111: 429–437.

Whitney, K., Randell, R.A., and Rieseberg, L. (2006). Adaptive introgression of herbivore resistance traits in the weedy sunflower *Helianthus annuus*. *Am. Nat.* 167: 794–807.

Whittaker, R.H. (1953). A consideration of climax theory: the climax as a population and pattern. *Ecol. Monogr.* 23: 41–78.

Whittaker, R.H. (1956). Vegetation of the Great Smoky Mountains. *Ecol. Monogr.* 26: 1–80.

Whittaker, R.H. (1960). Vegetation of the Siskiyou Mountains, Oregon and California. *Ecol. Monogr.* 30: 279–338.

Whittaker, R.H. (1962). Classification of natural communities. *Bot. Rev.* 28: 1–239.

Whittaker, R.H. (1966). Forest dimensions and production in the Great Smoky Mountains. *Ecology* 47: 103–121.

Whittaker, R.H. (1967). Gradient analysis of vegetation. *Biol. Rev.* 42: 207–264.

Whittaker, R.H. (1972). Evolution and measurement of species diversity. *Taxon* 21: 213–251.

Whittaker, R.H. (1975). *Communities and Ecosystems*, 2e. New York: MacMillan 385 pp.

Whittaker, R.H. (1977). Evolution of species diversity in land communities. In: *Evolutionary Biology*, vol. 10 (ed. M.K. Hecht, W.C. Steere and B. Wallace), 1–67. New York: Plenum.

Whittet, R., Cavers, S., Cottrell, J., and Ennos, R. (2016). Seed sourcing for woodland creation in an era of uncertainty: an analysis of the options for Great Britain. *Forestry* 90: 163–173.

Wicken, E. B. (1986). Terrestrial ecozones of Canada. Environment Canada. Ecological Land Classification Series No. 19. Lands Directorate, Ottawa. 26 pp.

Wicklow, D. (1975). Fire as an environmental cue initiating ascomycete development in a tallgrass prairie. *Mycologia* 67: 852–862.

Wickware, G., and Rubec, C.D.A. (1989). Ecoregions of Ontario. Ecological Land Classification Series No. 26, Env. Canada, Ottawa. 37 pp.

Wiens, J.A. (2005). Toward a unified landscape ecology. In: *Issues and Perspectives in Landscape Ecology* (ed. J.A. Wiens and M.R. Moss), 365–373. Cambridge, UK: Cambridge University Press.

Wiens, J.A. and Moss, M.R. (2005). *Issues and Perspectives in Landscape Ecology*. Cambridge, UK: Cambridge University Press 412 pp.

Wiersma, J.H. (1962). Enkele quantitative aspecten van het exotenvraagstuk. *Ned. Bosb. Tijdschr.* 34: 175–184.

Wiersma, J.H. (1963). A new method of dealing with results of provenance tests. *Silvae Genet.* 12: 200–205.

Wiersum, K.F. (1995). 200 years of sustainability in forestry: lessons from history. *Environ. Manag.* 19: 321–329.

Wignall, T.A., Browning, G., and Mackenzie, K.A.D. (1987). The physiology of epicormic bud emergence in pedunculate oak (*Quercus robur* L.). Responses to partial notch girdling in thinned and unthinned stands. *Forestry* 60: 45–56.

Wilcove, D.S. (1989). Protecting biodiversity in multiple-use lands: lessons from the US Forest Service. *Trends Ecol. Evol.* 4: 385–388.

Wilcove, D.S., McLellan, C.H., and Dobson, A.P. (1986). Habitat fragmentation in the temperate zone. In: *Conservation Biology, the Science of Scarcity and Diversity* (ed. M.E. Soulé), 237–256. Sunderland, MA: Sinauer Associates.

Wilcove, D.S., Rothstein, D., Dubow, J. et al. (1998). Quantifying threats to imperiled species in the United States. *Bioscience* 48: 607–615.

Wilde, S.A. (1958). *Forest Soils: Their Properties and Relation to Silviculture*. New York: Ronald Press 537 pp.

Wilkin, K., Ackerly, D., and Stephens, S. (2016). Climate change refugia, fire ecology and management. *Forests* 7: 77.

Will, R.E., Wilson, S.M., Zou, C.B., and Hennessey, T.C. (2013). Increased vapor pressure deficit due to higher temperature leads to greater transpiration and faster mortality during drought for tree seedlings common to the forest–grassland ecotone. *New Phytol.* 200: 366–374.

Williams, A.P. and Abatzoglou, J.T. (2016). Recent advances and remaining uncertainties in resolving past and future climate effects on global fire activity. *Curr. Clim. Change Rep.* 2: 1–14.

Williams, M.A. and Baker, W.L. (2012). Spatially extensive reconstructions show variable-severity fire and heterogeneous structure in historical western United States dry forests. *Glob. Ecol. Biogeogr.* 21: 1042–1052.

Williams, M.I. and Dumroese, R.K. (2013). Preparing for climate change: forestry and assisted migration. *J. For.* 111: 287–229.

Williams, G.J. III and McMillan, C. (1971). Phenology of six United States provenances of *Liquidamhar stryraciflua* under controlled conditions. *Am. J. Bot.* 58: 24–31.

Williams, J.W. and Jackson, S.T. (2007). Novel climates, no-analog communities, and ecological surprises. *Front. Ecol. Environ.* 5: 475–482.

Williams, J.W., Shuman, B.N., Webb, T. III et al. (2004). Late-Quaternary vegetation dynamics in North America: scaling from taxa to biomes. *Ecol. Monogr.* 74: 309–334.

Williams, C.A., Collatz, G.J., Masek, J., and Goward, S.N. (2012). Carbon consequences of forest disturbance and recovery across the conterminous United States. *Glob. Biogeochem. Cycles* 26: GB1005.

Williams, C.A., Gu, H., MacLean, R. et al. (2016). Disturbance and the carbon balance of US forests: a quantitative review of impacts from harvests, fires, insects, and droughts. *Glob. Planet. Chang.* 143: 66–80.

Williamson, M. (1996). *Biological Invasions*. New York: Springer 256 pp.

Williamson, G.B. and Black, E.M. (1981). High temperature of forest fires under pines as a selective advantage over oaks. *Nature* 293: 643–644.

Willmer, P. (2011). *Pollination and Floral Ecology*. Princeton, NJ: Princeton University Press 832 pp.

Wilmers, C.C., Post, E., Peterson, R.O., and Vucetich, J.A. (2006). Predator disease out-break modulates top-down, bottom-up and climatic effects on herbivore population dynamics. *Ecol. Lett.* 9: 383–389.

Wilson, B.F. (1984). *The Growing Tree*. Amherst: Univ. Massachusetts Press 138 pp.

Wilson, E.O. (1992). *The Diversity of Life*. Harvard University, Cambridge, MA: Belknap Press 432 pp.

Wilson, B.F. (2000). Apical control of branch growth and angle in woody plants. *Am. J. Bot.* 87: 601–607.

Wilson, E.O. (2010). *The Diversity of Life*, 2e. Cambridge, MA: Belknap Press, Harvard University 440 pp.

Wilson, M.V. and Shmida, A. (1984). Measuring beta diversity with presence-absence data. *J. Ecol.* 72: 1055–1064.

Wilson, C.M., Schaeffer, R.N., Hickin, M.L. et al. (2018). Chronic impacts of invasive herbivores on a foundational forest species: a whole-tree perspective. *Ecology* 99: 1783–1791.

Wingfield, M.J., Garnas, J.R., Hajek, A. et al. (2016). Novel and co-evolved associations between insects and microorganisms as drivers of forest pestilence. *Biol. Invasions* 18: 1045–1056.

Wistendahl, W.A. (1958). The flood plain of the Raritan River, New Jersey. *Ecol. Monogr.* 28: 129–153.

With, K.A. (2019). *Essenitals of Landscape Ecology*. Oxford, UK: Oxford University Press 656 pp.

With, K.A., Cadaret, S.J., and Davis, C. (1999). Movement responses to patch structure in experimental fractal landscapes. *Ecology* 80: 1340–1353.

Witter, J.A. and Waisanen, L.A. (1978). The effect of differential flushing times among trembling aspen clones on tortricid caterpillar populations. *Environ. Entomol.* 7: 139–143.

Wofsy, S.P., Goulden, M.L., Munger, J.W. et al. (1993). Net exchange of CO_2 in a mid-latitude forest. *Science* 260: 1314–1317.

Wohl, E. (2013). Fluvial geomorphology. In: *Treatise on Geomorphology*, vol. 9 (ed. J. Schroder). San Diego, CA: Academic Press 860 pp.

Wolda, H. (1981). Similarity indices, sample size and diversity. *Oecologia* 50: 296–302.

Wolf, M. and Weissing, F.J. (2012). Animal personalities: consequences for ecology and evolution. *Trends Ecol. Evol.* 27: 452–461.

Wolfe, J., Bryant, G., and Koster, K.L. (2002). What is 'unfreezable water', how unfreezable is it and how much is there? *CryoLetters* 23: 157–166.

Wood, D.M. and del Moral, R. (1987). Mechanisms of early primary succession in subalpine habitats on Mount St. Helens. *Ecology* 68: 780–790.

Wood, T., Bormann, F.H., and Voigt, G.T. (1984). Phosphorus cycling in a northern hardwood forest: biological and chemical control. *Science* 223: 391–393.

Wood, T.E., Takebayashi, N., Barker, M.S. et al. (2009). The frequency of polyploid speciation in vascular plants. *Proc. Natl. Acad. Sci.* 106: 13875–13879.

Woodall, C.W., Oswalt, C.M., Westfall, J.A. et al. (2009). An indicator of tree migration in forests of the eastern United States. *For. Ecol. Manag.* 257: 1434–1444.

Woodall, C.W., Zhu, K., Westfall, J.A. et al. (2013). Assessing the stability of tree ranges and influence of disturbance in eastern US forests. *For. Ecol. Manag.* 291: 172–180.

Woodall, C.W., Westfall, J.A., D'Amato, A.W. et al. (2018). Decadal changes in tree range stability across forests of the eastern US. *For. Ecol. Manag.* 429: 503–510.

Woodcock, D. and Shier, A. (2002). Wood specific gravity and its radial variations: the many ways to make a tree. *Trees* 16: 437–443.

Woods, F.W. (1953). Disease as a factor in the evolution of forest composition. *J. For.* 51: 871–873.

Woods, F.W. (1957). Factors limiting root penetration in deep sands of the southeastern coastal plain. *Ecology* 38: 357–359.

Woods, K.D. (1979). Reciprocal replacement and the maintenance of codominance in a beech-maple forest. *Oikos* 33: 31–39.

Woods, K.D. (1984). Patterns of tree replacement: canopy effects on understory pattern in hemlock-northern hardwood forests. *Vegetatio* 56: 87–107.

Woods, F.W. and Shanks, R.E. (1959). Natural replacement of chestnut by other species in he Great Smoky Mountains National Park. *Ecology* 40: 349–361.

Woodward, F.I. (1987). *Climate and Plant Distribution*. Cambridge: Cambridge Univ. Press 174 pp.

Woodwell, G. M., and D. B. Botkin. (1970). Metabolism of terrestrial ecosystems by gas exchange techniques: The Brookhaven approach. *In* D. Reichle (ed.), *Analysis of Temperate Forest Ecosystems*. (Ecological Studies, Vol. 1) Springer-Verlag, New York.

Worrall, J. (1983). Temperature–bud-burst relationships in amabilis and subalpine fir provenance tests replicated at different elevations. *Silvae Genet.* 32: 203–209.

Worrall, J.J., Egeland, L., Eager, T. et al. (2008). Rapid mortality of *Populus tremuloides* in southwestern Colorado, USA. *For. Ecol. Manag.* 255: 686–696.

Worrall, J.J., Rehfeldt, G.E., Hamann, A. et al. (2013). Recent declines of Populus tremuloides in North America linked to climate. *For. Ecol. Manag.* 299: 35–51.

Worster, D. (1994). *Nature's Economy*, 2e. Cambridge: Cambridge Univ. Press 505 pp.

Wright, J.W (1970). Genetics of eastern white pine (*Pinus strobus* L.). USDA For. Serv. Res. Paper WO-9. Washington, D.C. 16 pp.

Wright, J.W. (1976). *Introduction to Forest Genetics*. New York: Academic Press 463 pp.

Wright, H.E. Jr. (1971). Late quaternary vegetational history of North America. In: *The Late Cenozoic Glacial Ages* (ed. K.K. Turekian), 425–464. New Haven, CT: Yale Univ. Press.

Wright, H.E. Jr. and Heinselman, M.L. (ed.) (1973). The ecological role of fire in natural conifer forests of western and northern North America. *Quat. Res.* 3: 317–513.

Wright, J.W., Pauley, S.S., Polk, R.B. et al. (1966). Performance of Scotch pine varieties in the North Central Region. *Silvae Genet.* 15: 101–110.

Wright, J.W., Wilson, L.F., and Randall, W. (1967). Differences among Scotch pine varieties in susceptibility to European pine sawfly. *For. Sci.* 13: 175–181.

Wright, R.A., Wein, R.W., and Dancik, B.P. (1992). Population differentiation in seedling root size between adjacent stands of jack pine. *For. Sci.* 38: 777–785.

Wright, D.H., Curie, D.J., and Maurer, B.A. (1993). Energy supply and patterns of species richness on local and regional scales. In: *Species Diversity in Ecological Communities: Historical and Geographical Perspectives* (ed. R.E. Ricklefs and D. Schluter), 66–74. Chicago, IL: University of Chicago Press.

Wright, I.J., Recih, P.B., Westoby, M. et al. (2004). The worldwide leaf economics spectrum. *Nature* 428: 821.

Wu, J. (2013). Landscape sustainability science: ecosystem services and human well-being in changing landscapes. *Landsc. Ecol.* 28: 999–1023.

Wu, J. and Hobbs, R.J. (ed.) (2007). *Key Topics in Landscape Ecology*. Cambridge, UK: Cambridge University Press 314 pp.

Wuerthner, G. (1988). *Yellowstone and the Fires of Change*. Salt Lake City, UT: Haggis House Publ.

Yaffee, S.L. (1994). *The Wisdom of the Spotted Owl*. Washington, D.C.: Island Press 430 pp.

Yahner, R.H. (1988). Changes in wildlife communities near edges. *Conserv. Biol.* 2: 333–339.

Yahner, R.H. (1995). *Eastern Deciduous Forest, Ecology and Wildlife Conservation*. Minneapolis: Univ. Minnesota Press 220 pp.

Yakovlev, I.A., Fossdal, C.G., and Johnsen, O. (2010). MicroRNAs, the epigenetic memory and climatic adaptation in Norway spruce. *New Phytol.* 187: 1154–1169.

Yakovlev, I., Fossdal, C., Skrøppa, T. et al. (2012). An adaptive epigenetic memory in conifers with important implications for seed production. *Seed Sci. Res.* 22: 63–76.

Yeboah, D. and Chen, H.Y.H. (2016). Diversity–disturbance relationship in forest landscapes. *Landsc. Ecol.* 31: 981–987.

Yocom, H.A. (1968). Shortleaf pine seed dispersal. *J. For.* 66: 422.

Youngberg, C.T. and Wollum, A.G. (1976). Nitrogen accretion in developing *Ceonothus velutinus* stands. *Soil Sci. Soc. Am. Proc.* 40: 109–112.

Youngblood, A.P. and Mauk, R.L. (1985). Coniferous forest habitat types of central and southern Utah. USDA For. Serv. Gen. Tech. Report INT-187. Intermountain For. Exp. Sta., Ogden, UT. 89 pp.

Zachariassen, K.E. and Kristiansen, E. (2000). Ice nucleation and antinuclea tion in nature. *Cryobiology* 41: 257–279.

Zahner, R. (1958). Site-quality relationships of pine forests in southern Arkansas and northern Louisiana. *For. Sci.* 4: 162–176.

Zahner, R. (1968). Water deficits and growth of trees. In: *Water Deficits and Plant Growth. II* (ed. T.T. Kozlowski), 191–254. New York: Academic Press.

Zahner, R. and Crawford, N.A. (1965). The clonal concept in aspen site relations. In: *Tree Growth and Forest Soils* (ed. C.T. Youngberg and C.B. Davey), 230–243. Corvallis: Oregon State Univ. Press.

Zahner, R. and Stage, A.R. (1966). A procedure for calculating daily moisture stress and its utility in regressions of tree growth on weather. *Ecology* 47: 64–74.

Zak, D.R. and Pregitzer, K.S. (1990). Spatial and temporal variability of nitrogen cycling in northern Lower Michigan. *For. Sci.* 36: 367–380.

Zak, D.R., Pregitzer, K.S., and Host, G.E. (1986). Landscape variation in nitrogen mineralization and nitrification. *Can. J. For. Res.* 16: 1258–1263.

Zak, D.R., Host, G.E., and Pregitzer, K.S. (1989). Regional variability in nitrogen mineralization, nitrification, and overstory biomass in northern Lower Michigan. *Can. J. For. Res.* 19: 1521–1526.

Zak, D.R., Groffman, P.M., Pregitzer, K.S. et al. (1990). The vernal dam: plant-microbe competition for nitrogen in northern hardwood forests. *Ecology* 71: 651–656.

Zak, D.R., Tilman, D., Parmenter, R.R. et al. (1994). Plant production and soil microorgansims in late-successional ecosystems: a continental-scale study. *Ecology* 75: 2333–2347.

Zasada, J. (1985). Production, dispersal, and germination, and first year seedling survival of white spruce and birch in the Rosie Creek burn. *In* G.P. Juday and C.T. Dyrness (eds.), *Early Results of the Rosie Creek Research Project—1984.* Agr. and For. Exp. Sta., Univ. Alaska, Fairbanks, AK. Misc. Publ. 85-2.

Zasada, J.C., Sharik, T.L., and Nygren, M. (1992). The reproductive process in boreal forest trees. In: *A Systems Analysis of the Global Boreal Forest* (ed. H.H. Shugart, R. Leemans and G.B. Bonan), 211–233. New York: Cambridge Univ. Press.

Zhang, D., Hui, D., Luo, Y., and Zhou, G. (2008). Rates of litter decomposition in terrestrial ecosystems: global patterns and controlling factors. *J. Plant Ecol.* 1: 85–93.

Zhang, Y., Chen, H.Y.H., and Reich, P.B. (2012). Forest productivity increases with evenness, species richness and trait variation: a global meta-analysis. *J. Ecol.* 100: 742–749.

Zheng, Z., Li, Y., Li, M. et al. (2021). Whole-genome diversification analysis of the hornbeam species reveals speciation and adaptation among closely related species. *Front. Plant Sci.* 12: 1–10.

Zhu, K., Woodall, C.W., and Clark, J.S. (2012). Failure to migrate: lack of tree range expansion in response to climate change. *Glob. Chang. Biol.* 18: 1042–1052.

Zhu, K., Woodall, C.W., Ghosh, S. et al. (2014). Dual impacts of climate change: forest migration and turnover through life history. *Glob. Chang. Biol.* 20: 251–264.

Zimmermann, M.H. (1971). Transport in the xylem. In: *Trees: Structure and Function* (ed. M.H. Zimmermann and C.L. Brown), 169–220. New York: Springer-Verlag.

Zimmermann, M.H. and Brown, C.L. (1971). *Trees: Structure and Function*. New York: Springer-Verlag 336 pp.

Zobel, D.B. (1969). Factors affecting the distribution of *Pinus pungens,* an Appalachian endemic. *Ecol. Monogr.* 39: 303–333.

Zobel, D.B. and Antos, J.A. (1992). Survival of plants buried for eight growing seasons by volcanic tephra. *Ecology* 73: 698–701.

Zobel, B. and Talbert, J. (1984). *Applied Forest Tree Improvement*. New York: Wiley 505 pp.

Zogg, G.P. and Barnes, B.V. (1995). Ecological classification and analysis of wetland ecosystems, northern lower Michigan. *Can. J. For. Res.* 25: 1865–1875.

Zohner, C.M., Benito, B.M., Svenning, J.-C., et al. (2016). Day length unlikely to constrain climate-driven shifts in leaf-out times of northern woody plants. *Nat. Clim. Chang.* 6: 1120–1123.

Zohner, C.M., Mo, L., and Renner, S.S. (2018). Global warming reduces leaf-out and flowering synchrony among individuals. *elife* 7: e40214.

Zou, X., Theiss, C., and Barnes, B.V. (1992). Pattern of Kirtland's warbler occurrence in relation to the landscape structure of its summer habitat in northern Lower Michigan. *Landsc. Ecol.* 6: 221–231.

Zverev, V., Zvereva, E.L., and Kozlov, M.V. (2017). Ontogenetic changes in insect herbivory in birch (*Betula pubesecens*): the importance of plant apparency. *Funct. Ecol.* 31: 2224–2232.

Zweifel, R., Bohm, J.P., and Hasler, R. (2002). Midday stomatal closure in Norway spruce-reactions in the upper and lower crown. *Tree Physiol.* 22: 1125–1136.

Zwolak, R. and Sih, A. (2020). Animal personalities and seed dispersal: a conceptual review. *Funct. Ecol.* 34: 1294–1310.

Scientific Names of Trees and Shrubs

Acacia	*Acacia* spp.
Ailanthus; tree-of-heaven	*Ailanthus altissima* (Mill.) Swingle
Alaska-cedar	*Chamaecyparis nootkatensis* (D. Don) Spach
Alder, green or Sitka	*Alnus crispa* (Ait.) Pursh.
Alder, red or Oregon	*Alnus rubra* (Bong.)
Alder, Sitka	*Alnus sinuata* (Reg.) Rydb.
Alder, speckled	*Alnus rugosa* (Du Roi) Spreng.
Apple, common	*Malus pumila* Miller
Arborvitae	*Thuja occidentalis* L.
Ash, black	*Fraxinus nigra* Marsh.
Ash, blue	*Fraxinus quadrangulata* Michx.
Ash, European	*Fraxinus excelsior* L.
Ash, red or green	*Fraxinus pennsylvanica* Marsh.
Ash, white	*Fraxinus americana* L.
Ash, prickly-	*Zanthoxylum americanum* Mill.
Aspen, bigtooth	*Populus grandidentata* Michx.
Aspen, European	*Populus tremula* L.
Aspen, trembling or quaking	*Populus tremuloides* Michx.
Baldcypress	*Taxodium distichum* (L.) Rich.
Basswood	*Tilia americana* L.
Beech, American	*Fagus grandifolia* Ehrh.
Beech, Antarctic	*Nothofagus antarctica* (Forst.) Oerst.
Beech, blue-	*Carpinus caroliniana* Walt.
Beech, European	*Fagus sylvatica* L.
Bigtree; giant sequoia	*Sequoiadendron giganteum* (Lindl.) Buchholz
Birch, black or sweet	*Betula lenta* L.
Birch, bog, swamp, or low	*Betula pumila* L.
Birch, European or silver	*Betula pendula* Roth.
Birch, gray	*Betula populifolia* Marsh.

Forest Ecology, Fifth Edition. Daniel M. Kashian, Donald R. Zak, Burton V. Barnes, and Stephen H. Spurr.

Birch, paper or white	*Betula papyrifera* Marsh.
Birch, river or red	*Betula nigra* L.
Birch, Virginia round-leaf	*Betula uber* (Ashe) Fern.
Birch, yellow	*Betula alleghaniensis* Britton
Bitterbrush	*Purshia tridentata* (Pursh) DC.
Bittersweet, Asian or Oriental	*Celastrus orbiculatus* Thunb.
Blackgum	*Nyssa sylvatica* Marsh.
Blueberry, lowbush	*Vaccinium angustifolium* Aiton
Boxelder	*Acer negundo* L.
Buckeye, Ohio	*Aesculus glabra* Willd.
Buckeye, painted	*Aesculus sylvatica* Bartr.
Buckeye, red	*Aesculus pavia* L.
Buckeye, yellow	*Aesculus octandra* Marsh.
Buckthorn, alder	*Rhamnus alnifolia* L'Héritier
Buckthorn, common	*Rhamnus cathartica* L.
Buckthorn, glossy	*Frangula* alnus Mill
Butternut	*Juglans cinerea* L.
Catalpa, northern	*Catalpa speciosa* Warder
Catalpa, southern	*Catalpa bignonioides* Walter
Ceanothus, redstem	*Ceanothus sanguineus* Pursh.
Cedar, Alaska-	*Chamaecyparis nootkatensis* (D. Don) Spach
Cedar, Atlantic white-	*Chamaecyparis thyoides* (L.) B. S. P.
Cedar, eastern red-	*Juniperus virginiana* L.
Cedar, incense	*Libocedrus decurrens* Torr.
Cedar, northern white-	*Thuja occidentalis* L.
Cedar, Port Orford-	*Chamaecyparis lawsoniana* (A. Murr.) Parl.
Cedar, western red-	*Thuja plicata* Donn
Cherry, black	*Prunus serotina* Ehrh.
Cherry, choke	*Prunus virginiana* L.
Cherry, pin or fire	*Prunus pensylvanica* L. f.
Chestnut, American	*Castanea dentata* (Marsh.) Borkh.
Chinquapin, golden	*Chrysolepis chrysophylla* Douglas ex Hook.
Chinquapin, Ozark	*Castanea ozarkensis* Ashe
Cottonwood, black	*Populus trichocarpa* Torr. & Gray
Cottonwood, eastern	*Populus deltoides* Bartr.
Cottonwood, European; black poplar	*Populus nigra* L.
Cottonwood, Fremont	*Populus fremontii* Wats.
Cottonwood, narrowleaf	*Populus angustifolia* James
Cottonwood, plains	*Populus deltoides var. occidentalis* Rydb.
Creeping strawberry-bush	*Euonymus obovata* Nutt.
Creosote bush	*Larrea* spp.
Cypress, Arizona	*Cupressus arizonica* Greens

Cypress, Mexican	*Cupressus lusitanica* Miller
Cypress, pond	*Taxodium distichum* var. *imbricarium* (Nutt.) Croon (syn = *T. ascendens* Brong.)
Dawn redwood	*Metasequoia glyptostroboides* H. H. Hu & Cheng
Dogwood, alternate-leaf	*Cornus alternifolia* L.
Dogwood, flowering	*Cornus florida* L.
Dogwood, gray	*Cornus foemina* Miller
Dogwood, red-osier	*Cornus stolonifera* Michx.
Dogwood, roundleaf	*Cornus rugosa* Lamarck
Dogwood, silky	*Cornus amomum* Miller
Douglas-fir	*Pseudotsuga menziesii* (Mirb.) Franco
Elm, American	*Ulmus americana* L.
Elm, rock	*Ulmus thomasii* Sarg.
Elm, Siberian	*Ulmus pumila* L.
Elm, slippery	*Ulmus rubra* Muhl.
Elm, winged	*Ulmus alata* Michx.
Eucalyptus	*Eucalyptus* spp.
Fetterbush	*Leucothoe* spp.
Fir, alpine or subalpine	*Abies lasiocarpa* (Hook.) Nutt.
Fir, balsam	*Abies balsamea* (L.) Mill.
Fir, European silver	*Abies alba* Mill.
Fir, Fraser	*Abies fraseri* (Pursh.) Poir.
Fir, grand or lowland white	*Abies grandis* (Dougl.) Lindl.
Fir, noble	*Abies procera* Rehd.
Fir, Pacific silver	*Abies amabilis* (Dougl.) Forbes
Fir, red	*Abies magnifica* A. Murr.
Fir, white	*Abies concolor* (Gord. & Glend.) Lindl.
Giant sequoia; bigtree	*Sequoiadendron giganteum* (Lindl.) Buchholz
Ginkgo; maidenhair tree	*Ginkgo biloba* L.
Gum, black; black tupelo	*Nyssa sylvatica* Marshall
Hackberry	*Celtis occidentalis* L.
Hackberry, dwarf	*Celtis tenuifolia* Nuttall
Hackberry, iguana	*Celtis iguanaea* (Jacq.) Sarg.
Haw, black	*Viburnum prunifolium* L.
Haw, dotted	*Crataegus punctata* Jacquin
Hawthorn	*Crataegus* spp.
Hazel, beaked	*Corylus cornuta* Marsh.
Hemlock, eastern	*Tsuga canadensis* (L.) Carr.
Hemlock, mountain	*Tsuga mertensiana* (Bong.) Carr.
Hemlock, western	*Tsuga heterophylla* (Raf.) Sarg.
Hickory, bitternut	*Carya cordiformis* (Wangenh.) K. Koch
Hickory, mockernut	*Carya tomentosa* Nutt.

Hickory, pignut	*Carya glabra* (Mill.) Sweet
Hickory, shagbark	*Carya ovata* (Mill.) K. Koch
Hickory, shellbark	*Carya laciniosa* (Michaux f.) G. Don
Hickory, water	*Carya aquatica* (Michx. f.) Nutt.
Hobblebush	*Viburnum alnifolium* Marsh.
Holly, American	*Ilex opaca* Ait.
Honeysuckle, Amur	*Lonicera maackii* (Rupr.) Maxim.
Hop-hornbeam; ironwood	*Ostrya virginiana* (Mill.) K. Koch
Hornbeam	*Carpinus betulus* L.
Horse-chestnut	*Aesculus hippocastanum* L.
Ironwood; hop-hornbeam	*Ostrya virginiana* (Mill.) K. Koch
Juniper, alligator	*Juniperus deppeana* Steud.
Juniper, common	*Juniperus communis* L.
Juniper, ground	*Juniperus communis* L. var. depressa Pursh.
Juniper, one-seed	*Juniperus monosperma* (Engelm.) Sarg.
Juniper, Rocky Mountain	*Juniperus scopulorum* Sarg.
Juniper, Utah	*Juniperus osteosperma* (Torr.) Little
Juniper, western	*Juniperus occidentalis* Hook.
Kentucky coffeetree	*Gymnocladus dioicus* (L.) K. Koch
Kudzu	*Pueraria lobata* (Willd.) Ohwi
Larch, eastern; tamarack	*Larix laricina* (Du Roi) K. Koch
Larch, European	*Larix decidua* Mill.
Larch, subalpine	*Larix lyallii* Parl.
Larch, western	*Larix occidentalis* Nutt.
Laurel, mountain	*Kalmia latifolia* L.
Leatherleaf	*Chamaedaphne calyculata* (L.) Moench
Lilac	*Syringa vulgaris* L.
Locust, black	*Robinia pseudoacacia* L.
Locust, honey	*Gleditsia triacanthos* L.
Madrone, Pacific	*Arbutus menziesii* Pursh.
Magnolia, Fraser	*Magnolia fraseri* Walt.
Magnolia, southern or evergreen	*Magnolia grandiflora* L.
Magnolia, sweetbay	*Magnolia virginiana* L.
Magnolia, umbrella	*Magnolia tripetala* L.
Maidenhair tree; gingko	*Ginkgo biloba* L.
Mangrove, black	*Avicennia nitida* Jacq.
Mangrove, red	*Rhizophora mangle* L.
Maple, ash-leaf	*Acer negundo* L.
Maple, bigleaf	*Acer macrophyllum* Pursh.
Maple, bigtooth	*Acer grandidentatum* Nutt.
Maple, black	*Acer nigrum* Michaux f.
Maple, mountain	*Acer spicatum* Lam.

Maple, Norway	*Acer platanoides* L.
Maple, red	*Acer rubrum* L.
Maple, silver	*Acer saccharinum* L.
Maple, striped	*Acer pensylvanicum* L.
Maple, sugar	*Acer saccharum* Marsh.
Maple, vine	*Acer circinatum* Pursh.
Mesquite	*Prosopis juliflora* (Sw.) DC.
Mexican cypress	*Cupressus lusitanica* Miller
Mimosa	*Albizia julibrissin* Durazz.
Mountain-ash, American	*Sorbus americana* Marsh.
Mountain-ash, showy	*Sorbus decora* (Sarg.) Schneid.
Mountain laurel	*Kalmia latifolia* L.
Mulberry, red	*Morus rubra* L.
Mulberry, white	*Morus alba* L.
Myrtle, California wax	*Myrica californica* Cham. & Schltdl.
Nannyberry	*Viburnum lentago* L.
Oak, bear	*Quercus ilicifolia* Wangenh.
Oak, black	*Quercus velutina* Lam.
Oak, blackjack	*Quercus marilandica* Muenchh.
Oak, bur	*Quercus macrocarpa* Michx.
Oak, California black	*Quercus kelloggii* Newb.
Oak, California live	*Quercus agrifolia* Née
Oak, canyon live	*Quercus chrysolepis* Liebm.
Oak, cherrybark	*Quercus pagoda Raf.* (syn. *Q. falcata* var. *pagodifolia* 111.)
Oak, chestnut or rock chestnut	*Quercus prinus* L.
Oak, chinquapin or yellow	*Quercus muehlenbergii* Engelm.
Oak, diamond leaf	*Quercus laurifolia* Michx.
Oak, dwarf chestnut	*Quercus prinoides* Willd.
Oak, Emory	*Quercus emoryi* Torr.
Oak, English, European, or common	*Quercus robur* L.
Oak, Gambel	*Quercus gambelii* Nutt.
Oak, holm	*Quercus ilex* L.
Oak, interior live	*Quercus wislizeni* A. DC.
Oak, live	*Quercus virginiana* Mill.
Oak, northern pin	*Quercus ellipsoidalis* E. J. Hill
Oak, Oregon	*Quercus garryana* Dougl.
Oak, overcup	*Quercus lyrata* Walt.
Oak, pin	*Quercus palustris* Muenchh.
Oak, post	*Quercus stellata* Wangenh.
Oak, northern red	*Quercus rubra* L.
Oak, sand live	*Quercus geminata* Small, [syn. *Q. virginiana* var. *maritima* (Michx.) Sarg.]

Oak, scarlet	*Quercus coccinea* Muenchh.
Oak, shingle	*Quercus imbricaria* Michx.
Oak, southern red	*Quercus falcata* Michx.
Oak, swamp chestnut	*Quercus michauxii* Nutt.
Oak, swamp laurel	*Quercus laurifolia* Michx.
Oak, swamp white	*Quercus bicolor* Willd.
Oak, turkey	*Quercus laevis* Walt.
Oak, water	*Quercus nigra* L.
Oak, white	*Quercus alba* L.
Oak, willow	*Quercus phellos* L.
Olive, Russian	*Elaeagnus angustifolia* L.
Osage-orange	*Madura pomifera* (Raf.) C. K. Schneid.
Pacific silver fir	*Abies amabilis* (Dougl.) Forbes
Palm, cabbage	*Sabal palmetto* (Walt.) Lodd.
Palmetto, cabbage	*Sabal palmetto* (Walt.) Lodd.
Palmetto, saw	*Serenoa repens* (Bartr.) Small
Pawpaw	*Asimina triloba* (L.) Dunal
Pine, bishop	*Pinus muricata* D. Don
Pine, black or Austrian	*Pinus nigra* Arnold
Pine, bristlecone	*Pinus aristata* Engelm.
Pine, Caribbean	*Pinus caribaea* Morelet
Pine, digger	*Pinus sabiniana* Dougl.
Pine, eastern white	*Pinus strobus* L.
Pine, erectcone or Calabrian	*Pinus brutia* Ten.
Pine, Jack	*Pinus banksiana* Lamb.
Pine, Jeffrey	*Pinus jeffreyi* Grev. & Balf.
Pine, knobcone	*Pinus attenuata* Lemmon
Pine, limber	*Pinus flexilis* James
Pine, loblolly	*Pinus taeda* L.
Pine, lodgepole	*Pinus contorta* Dougl.
Coastal lodgepole pine	*Pinus contorta* ssp. *contorta*
Mendocino White Plains lodgepole pine	*Pinus contorta* ssp. *bolanderi*
Rocky Mountain-Intermountain lodgepole pine	*Pinus contorta* ssp. *latifolia*
Sierra–Cascade lodgepole pine	*Pinus contorta* ssp. *murrayana*
Pine, longleaf	*Pinus palustris* Mill.
Pine, Mexican pinyon (piñon)	*Pinus cembroides* Zucc.
Pine, Mexican white	*Pinus ayacahuite* Ehrenb.
Pine, Monterey or radiata	*Pinus radiata* D. Don
Pine, mugo or mountain	*Pinus mugo* Turra. [P. *montana* Mill.]
Pine, patula	*Pinus patula* Schl. & Cham.

Pine, pinyon (piñon)	*Pinus edulis* Engelm. or *P. monophylla* Torr. & Frem. or *P. cembroides* Zucc. or *P. quadrifolia* Parl. ex Sudw.
Pine, pitch	*Pinus rigida* Mill.
Pine, pond	*Pinus serotina* Michx.
Pine, ponderosa	*Pinus ponderosa* Laws.
Pine, red	*Pinus resinosa* Ait.
Pine, sand	*Pinus clausa* (Chapm.) Vasey
Pine, Scots or Scotch	*Pinus sylvestris* L.
Pine, shortleaf	*Pinus echinata* Mill.
Pine, slash	*Pinus elliottii* Engelm. var. *elliottii*
Pine, sugar	*Pinus lambertiana* Dougl.
Pine, stone or Swiss stone	*Pinus cembra* L.
Pine, table mountain	*Pinus pungens* Lamb.
Pine, Torrey	*Pinus torreyana* Parry ex Carr.
Pine, Virginia	*Pinus virginiana* Mill.
Pine, western white	*Pinus monticola* Dougl.
Pine, whitebark	*Pinus albicaulis* Engelm.
Planetree, oriental	*Platanus orientalis* L.
Plum, Allegheny	*Prunus alleghaniensis* Porter
Poplar, balsam	*Populus balsamifera* L.
Poplar, black hybrid	*Populus* x *euramericana*
Poplar, European white	*Populus alba* L.
Poplar, Lombardy	*Populus nigra* 'italica' Muenchh.
Prickly-ash	*Zanthoxylum americanum* Mill.
Privet, swamp	*Forestiera acuminata* (Michx.) Poir.
Redbud, eastern	*Cercis canadensis* L.
Redwood	*Sequoia sempervirens* (D. Don) Endl.
Redwood, dawn	*Metasequoia glyptostroboides* H. H. Hu & Cheng
Rhododendron, rosebay	*Rhododendron maximum* L.
Rose	*Rosa spp.*
Sagebrush, big	*Artemisia tridentata* Nutt.
Saguaro; giant cactus	*Cereus giganteus* Engelm.
Salmonberry	*Rubus spectabilis* Pursh.
Sassafras	*Sassafras albidum* (Nutt.) Nees
Sequoia, giant; bigtree	*Sequoiadendron giganteum* (Lindl.) Buchholz
Serviceberry, downy	*Amelanchier arborea* (Michaux f.) Fernald
Serviceberry, smooth	*Amelanchier laevis* Wieg.
Sheepberry	*Viburnum lentago* L.
Silver bell; Carolina silverbell	*Halesia carolina* L.
Snowbrush	*Ceanothus velutinus* Doug. ex Hook

Sourwood	*Oxydendrum arboreum* (L.) DC.
Spruce, black	*Picea mariana* (Mill.) B. S. P.
Spruce, Colorado blue	*Picea pungens* Engelm.
Spruce, Engelmann	*Picea engelmannii* Parry
Spruce, Norway or European	*Picea abies* (L.) Karst.
Spruce, red	*Picea rubens* Sarg.
Spruce, Sitka	*Picea sitchensis* (Bong.) Carr.
Spruce, white	*Picea glauca* (Moench) Voss
Sugarberry	*Celtis laevigata* Willd.
Sumac, smooth	*Rhus glabra* L.
Sweetgum	*Liquidambar styraciflua* L.
Sycamore, American	*Platanus occidentalis* L.
Tallow, Chinese	*Triadica sebifera* (L.) Small
Tamarack; eastern larch	*Larix laricina* (Du Roi) K. Koch
Tamarix, five-stamen	*Tamarix pentandra* Pall.
Torreya, Florida	*Torreya taxifolia Arn.*
Tree-of-heaven; ailanthus	*Ailanthus altissima* (Mill.) Swingle
Tuliptree; yellow-poplar	*Liriodendron tulipifera* L.
Tupelo, water or swamp	*Nyssa aquatica* L.
Virginia creeper	*Parthenocissus quinquefolia* (L.) Planchon
Walnut, black	*Juglans nigra* L.
Willow, black	*Salix nigra* Marsh.
Willow, crack	*Salix fragilis* L.
Willow, peachleaf	*Salix amygdaloides* Andersson
Willow, sandbar	*Salix interior* Rowlee
Willow, weeping	*Salix babylonica* L.
Witch-hazel	*Hamamelis virginiana* L.
Yellow-poplar; tuliptree	*Liriodendron tulipifera* L.
Yew, Canada	*Taxus canadensis* Marshall
Yew, Florida	*Taxus floridana* Nutt.
Yew, Pacific	Taxus brevifolia Nutt.

Index

Note: Page numbers followed by "*f*" and "*t*" figures and tables, respectively.

Forest Ecology, Fifth Edition. Daniel M. Kashian, Donald R. Zak, Burton V. Barnes, and Stephen H. Spurr.

D

G

M

N

O

P

T

X

Y